CEAC
(SCA)

GORDIAN®

Electrical Costs with RSMeans data

Joseph Kelble, Senior Editor

2018
41st annual edition

Chief Data Officer
Noam Reininger

Engineering Director
Bob Mewis, CCP

Contributing Editors
Christopher Babbitt
Sam Babbitt
Michelle Curran
Matthew Doheny *(8)*
Cheryl Elsmore
Linval Gentles
John Gomes *(13, 41)*
Derrick Hale, PE *(2, 31, 32, 33, 34, 35, 44, 46)*
Wafaa Hamitou *(11, 12)*

Joseph Kelble *(14, 21, 22, 23, 25, 26, 27, 28, 48)*
Charles Kibbee *(1, 4)*
Gerard Lafond, PE
Thomas Lane *(6, 7)*
Genevieve Medeiros
Elisa Mello
Ken Monty
Marilyn Phelan, AIA *(9, 10)*
Stephen C. Plotner *(3, 5)*
Callum Riley
Stephen Rosenberg
Jeff Sessions
Gabe Sirota

Matthew Sorrentino
Kevin Souza
Keegan Spraker
Tim Tonello
Jen Walsh
David Yazbek

Product Manager
Andrea Sillah

Production Manager
Debbie Panarelli

Production
Jonathan Forgit
Mary Lou Geary

Sharon Larsen
Sheryl Rose

Technical Support
Judy Abbruzzese
Gary L. Hoitt

Cover Design
Blaire Collins

Data Analytics
Tim Duggan
Todd Glowac
Matthew Kelliher-Gibson

Numbers in italics are the divisional responsibilities for each editor. Please contact the designated editor directly with any questions.

Gordian RSMeans data
Construction Publishers & Consultants
1099 Hingham Street, Suite 201
Rockland, MA 02370
United States of America
1-800-448-8182
www.RSMeans.com

Copyright 2017 by The Gordian Group Inc.
All rights reserved.
Cover photo © iStock.com/roman023

Printed in the United States of America
ISSN 1540-6318
ISBN 978-1-946872-04-3

D1088138

$266.99 per copy (in United States)
Price is subject to change without prior notice.
0038

Related Data and Services

Our engineers recommend the following products and services to complement *Electrical Costs with RSMeans data:*

Annual Cost Data Books

2018 Electrical Change Order Costs with RSMeans data
2018 Building Construction Costs with RSMeans data
2018 Mechanical Costs with RSMeans data

Reference Books

Electrical Estimating Methods
Estimating Building Costs
RSMeans Estimating Handbook
Green Building: Project Planning & Estimating
How to Estimate with RSMeans data
Plan Reading & Material Takeoff
Project Scheduling & Management for Construction

Seminars and In-House Training

Unit Price Estimating
Training for our online estimation solution
Practical Project Management for Construction Professionals
Scheduling with MSProject for Construction Professionals
Mechanical & Electrical Estimating

RSMeans data Online

For access to the latest cost data, an intuitive search, and an easy-to-use estimate builder, take advantage of the time savings available from our online application. To learn more visit: www.RSMeans.com/2018online.

Enterprise Solutions

Building owners, facility managers, building product manufacturers, and attorneys across the public and private sectors engage with RSMeans data Enterprise to solve unique challenges where trusted construction cost data is critical. To learn more visit: www.RSMeans.com/Enterprise.

Custom Built Data Sets

Building and Space Models: Quickly plan construction costs across multiple locations based on geography, project size, building system component, product options, and other variables for precise budgeting and cost control.

Predictive Analytics: Accurately plan future builds with custom graphical interactive dashboards, negotiate future costs of tenant build-outs, and identify and compare national account pricing.

Consulting

Building Product Manufacturing Analytics: Validate your claims and assist with new product launches.

Third-Party Legal Resources: Used in cases of construction cost or estimate disputes, construction product failure vs. installation failure, eminent domain, class action construction product liability, and more.

API

For resellers or internal application integration, RSMeans data is offered via API. Deliver Unit, Assembly, and Square Foot Model data within your interface. To learn more about how you can provide your customers with the latest in localized construction cost data visit: www.RSMeans.com/API.

Table of Contents

Foreword

The Value of RSMeans data from Gordian

Since 1942, RSMeans data has been the industry-standard materials, labor, and equipment cost information database for contractors, facility owners and managers, architects, engineers, and anyone else that requires the latest localized construction cost information. Over 75 years later, the objective remains the same: to provide facility and construction professionals with the most current and comprehensive construction cost database possible.

With the constant influx of new construction methods and materials, in addition to ever-changing labor and material costs, last year's cost data is not reliable for today's designs, estimates, or budgets. The RSMeans data engineers invest over 22,000 hours in cost research annually and apply real-world construction experience to identify and quantify new building products and methodologies, adjust productivity rates, and adjust costs to local market conditions across the nation. This unparalleled construction cost expertise is why so many facility and construction professionals rely on RSMeans data year over year.

About Gordian

Gordian originated in the spirit of innovation and a strong commitment to helping clients reach and exceed their construction goals. In 1982, Gordian's Chairman and Founder, Harry H. Mellon, created Job Order Contracting while serving as Chief Engineer at the Supreme Headquarters Allied Powers Europe. Job Order Contracting is a unique indefinite delivery/indefinite quantity (IDIQ) process, which enables facility owners to complete a substantial number of repair, maintenance, and construction projects with a single, competitively awarded contract. Realizing facility and infrastructure owners across various industries could greatly benefit from the time and cost saving advantages of this innovative construction procurement solution, he established Gordian in 1990.

Continuing the commitment to providing the most relevant and accurate facility and construction data, software, and expertise in the industry, Gordian enhanced the fortitude of its data with the acquisition of RSMeans in 2014. And in an effort to expand its facility management capabilities, Gordian acquired Sightlines, the leading provider of facilities benchmarking data and analysis, in 2015.

Our Offerings

Gordian is the leader in facility and construction cost data, software, and expertise for all phases of the building life cycle. From planning to design, procurement, construction, and operations, Gordian's solutions help clients maximize efficiency, optimize cost savings, and increase building quality with its highly specialized data engineers, software, and unique proprietary data sets.

Our Commitment

At Gordian, we do more than talk about the quality of our data and the usefulness of its application. We stand behind all of our RSMeans data—from historical cost indexes to construction materials and techniques—to craft current costs and predict future trends. If you have any questions about our products or services, please call us toll-free at 800-448-8182 or visit our website at www.gordian.com.

MasterFormat® 2014/ MasterFormat® 2016 Comparison Table

This table compares the 2014 edition of the Construction Specifications Institute's MasterFormat® to the expanded 2016 edition. For your convenience, all revised 2014 numbers and titles are listed along with the corresponding 2016 numbers and titles. In some cases, a designation of RSMeans is used to identify sections of data numbered exclusively in RSMeans products.

CSI 2014 MF ID	CSI 2014 MF Description	2014 Designation	CSI 2016 MF ID	CSI 2016 MF Description	2016 Designation
35 20 23.23	Hydraulic Dredging	CSI	35 24 13.13	Cutter Suction Dredging	RSMeans
35 20 23.13	Mechanical Dredging	CSI	35 24 23.13	Mechanical Dredging	CSI
35 20 23	Dredging	CSI	35 24 23	Clamshell Dredging	RSMeans
35 20 16.73	Slide Gates	RSMeans	35 22 73.16	Slide Gates	CSI
35 20 16.69	Knife Gates	RSMeans	35 22 69.16	Knife Gates	RSMeans
35 20 16.66	Flap Gates	RSMeans	35 22 66.16	Flap Gates	RSMeans
35 20 16.63	Canal Gates	RSMeans	35 22 63.16	Canal Gates	RSMeans
35 20 16.26	Hydraulic Sluice Gates	CSI	35 22 26.16	Hydraulic Sluice Gates	RSMeans
35 20 16	Hydraulic Gates	CSI	35 22 26	Sluice Gates	RSMeans
35 20	Waterway and Marine Construction and Equipment	CSI	35 22	Hydraulic Gates	RSMeans
33 72 33.46	Substation Converter Stations	CSI	33 78 33.46	Substation Converter Stations	RSMeans
33 72 33.36	Cable Trays For Utility Substations	CSI	26 05 36.36	Cable Trays For Utility Substations	CSI
33 72 33.33	Raceway/Boxes For Utility Substations	CSI	26 05 33.33	Raceway/Boxes For Utility Substations	CSI
33 52 43.13	Aviation Fuel Piping	CSI	33 52 13.43	Aviation Fuel Piping	RSMeans
33 52 16.13	Gasoline Piping	CSI	33 52 13.16	Gasoline Piping	RSMeans
33 52 13.14	Petroleum Products	RSMeans	33 52 13.19	Petroleum Products	RSMeans
33 51 33.10	Piping, Valves & Meters, Gas Distribution	RSMeans	33 59 33.10	Piping, Valves & Meters, Gas Distribution	CSI
33 51 33	Natural-Gas Metering	CSI	33 59 33	Natural-Gas Metering	CSI
33 51 13.30	Piping, Gas Service & Distribution, Steel	RSMeans	33 52 16.16	Piping, Gas Service & Distribution, Steel	RSMeans
33 51 13.20	Piping, Gas Service & Distribution, Steel	RSMeans	33 52 16.13	Steel Natural Gas Piping	RSMeans
33 51 13.10	Piping, Gas Service and Distribution, Polyethylene	RSMeans	33 52 16.20	Piping, Gas Service And Distribution, Polyethylene	CSI
33 49 23	Storm Drainage Water Retention Structures	CSI	33 46 23	Modular Buried Stormwater Storage Units	CSI
33 49 13	Storm Drainage Manholes, Frames, and Covers	CSI	33 05 61	Concrete Manholes	RSMeans
33 47 19	Water Ponds and Reservoirs	CSI	33 46 11	Stormwater Ponds	CSI
33 47 13.54	Garden Ponds	RSMeans	13 12 13.54	Garden Ponds	CSI
33 47 13.53	Reservoir Liners HDPE	CSI	31 05 19.53	Reservoir Liners HDPE	RSMeans
33 47 13	Pond and Reservoir Liners	CSI	31 05 19	Geosynthetics for Earthwork	RSMeans
33 46 26.10	Geotextiles For Subsurface Drainage	RSMeans	33 41 23.19	Geosynthetic Drainage Layers	RSMeans
33 46 26	Geotextile Subsurface Drainage Filtration	CSI	33 41 23	Drainage Layers	CSI
33 46 16	Subdrainage Piping	CSI	33 41 16	Subdrainage Piping	CSI
33 46	Subdrainage	CSI	33 41	Subdrainage	CSI
33 44 16	Utility Trench Drains	CSI	33 42 36	Stormwater Trench Drains	CSI
33 44 13	Utility Area Drains	CSI	33 42 33	Stormwater Curbside Drains and Inlets	RSMeans
33 42 16.15	Oval Arch Culverts	RSMeans	33 42 13.15	Oval Arch Culverts	RSMeans
33 42 16.13	Culverts & Box Trench Sections	CSI	33 42 13.14	Culverts & Box Trench Sections	RSMeans
33 41 13	Public Storm Utility Drainage Piping	CSI	33 42 11	Stormwater Gravity Piping	CSI
33 41	Storm Utility Drainage Piping	CSI	33 42	Stormwater Conveyance	CSI
33 36 50	Drainage Field Systems	RSMeans	33 34 51	Drainage Field System	CSI
33 36 33.13	Utility Septic Tank Tile Drainage Field	CSI	33 34 51.13	Utility Septic Tank Tile Drainage Field	RSMeans
33 36 19	Utility Septic Tank Effluent Filter	CSI	33 34 16	Septic Tank Effluent Filters	CSI
33 36 13.19	Polyethylene Utility Septic Tank	CSI	33 34 13.33	Polyethylene Septic Tanks	CSI
33 36 13.13	Concrete Utility Septic Tank	CSI	33 34 13.13	Concrete Septic Tanks	RSMeans
33 36 13	Utility Septic Tank and Effluent Wet Wells	CSI	33 34 13	Septic Tanks	CSI
33 36	Utility Septic Tanks	CSI	33 34	Onsite Wastewater Disposal	CSI
33 31 13	Public Sanitary Utility Sewerage Piping	CSI	33 31 11	Public Sanitary Sewerage Gravity Piping	CSI
33 21 13	Public Water Supply Wells	CSI	33 11 13	Potable Water Supply Wells	RSMeans
33 21	Water Supply Wells	CSI	33 11	Groundwater Sources	CSI
33 16 19.50	Elevated Water Storage Tanks	RSMeans	33 16 11.50	Elevated Water Storage Tanks	CSI
33 16 19	Elevated Water Utility Storage Tanks	CSI	33 16 11	Elevated Composite Water Storage Tanks	CSI
33 16 13.29	Wood Water Storage Tanks	RSMeans	33 16 59.29	Wood Water Storage Tanks	CSI
33 16 13.23	Plastic-Coated Fabric Pillow Water Tanks	RSMeans	33 16 56.23	Plastic-Coated Fabric Pillow Water Tanks	CSI
33 16 13.19	Horizontal Plastic Water Tanks	CSI	33 16 56.19	Horizontal Plastic Water Tanks	CSI
33 16 13.16	Prestressed Conc. Water Storage Tanks	CSI	33 16 36.16	Prestressed Conc. Water Storage Tanks	RSMeans
33 16 13.13	Steel Water Storage Tanks	CSI	33 16 23.13	Steel Water Storage Tanks	RSMeans
33 16 13	Aboveground Water Utility Storage Tanks	CSI	33 16 23	Ground-Level Steel Water Storage Tanks	CSI
33 12 19.10	Fire Hydrants	RSMeans	33 14 19.30	Fire Hydrants	CSI
33 12 16.20	Valves	RSMeans	33 14 19.20	Valves	CSI

Number	Description	Source	Number	Description	Source
33 12 16.10	Valves	RSMeans	33 14 19.10	Valves	CSI
33 12 16	Water Utility Distribution Valves	CSI	33 14 19	Valves and Hydrants for Water Utility Service	CSI
33 12 13.15	Tapping, Crosses and Sleeves	RSMeans	33 14 17.15	Tapping, Crosses and Sleeves	CSI
33 12 13	Water Service Connections	CSI	33 14 17	Site Water Utility Service Laterals	RSMeans
33 11 13	Public Water Utility Distribution Piping	CSI	33 14 13	Public Water Utility Distribution Piping	RSMeans
33 11	Water Utility Distribution Piping	CSI	33 14	Water Utility Transmission and Distribution	CSI
33 05 26	Utility Identification	CSI	33 05 97	Identification and Signage for Utilities	RSMeans
33 05 23.22	Directional Drilling	RSMeans	33 05 07.13	Utility Directional Drilling	RSMeans
33 05 23.20	Horizontal Boring	RSMeans	33 05 07.23	Utility Boring and Jacking	CSI
33 05 23.19	Microtunneling	CSI	33 05 07.36	Microtunneling	CSI
33 05 23	Trenchless Utility Installation	CSI	33 05 07	Trenchless Installation of Utility Piping	RSMeans
33 05 16	Utility Structures	CSI	33 05 63	Concrete Vaults and Chambers	RSMeans
33 01 30.71	Rehabilitation of Sewer Utilities	CSI	33 01 30.23	Pipe Bursting	RSMeans
33 01 30.16	TV Inspection of Sewer Pipelines	CSI	33 01 30.11	Television Inspection of Sewers	CSI
32 31 13.30	Fence, Chain Link, Gates & Posts	RSMeans	32 31 11.10	Gate Operators	CSI
31 71 21.10	Cut and Cover Tunnels	RSMeans	31 71 23.10	Cut and Cover Tunnels	CSI
31 71 21	Tunnel Excavation by Cut and Cover	RSMeans	31 71 23	Tunneling by Cut and Cover	CSI
28 46 13	Hard-Wired Detention Monitoring & Control Systems	CSI	28 52 11	Detention Monitoring and Control Systems	CSI
28 46	Electronic Detention Monitoring & Control Systems	CSI	28 52	Detention Security Systems	CSI
28 41 13	Building Systems	RSMeans	28 33 11	Electronic Structural Monitoring Systems	RSMeans
28 39 10	Notification Systems	RSMeans	28 47 12	Notification Systems	CSI
28 39	Mass Notification Systems	CSI	28 47	Mass Notification	RSMeans
28 33 33	Gas Detection Sensors	CSI	28 42 15	Gas Detection Sensors	CSI
28 33	Gas Detection and Alarm	CSI	28 42	Gas Detection and Alarm	RSMeans
28 32 33	Radiation Detection Sensors	CSI	28 41 15	Radiation Detection Sensors	CSI
28 31 49.50	Carbon-Monoxide Detectors	RSMeans	28 46 11.21	Carbon-Monoxide Detection Sensors	CSI
28 31 46.50	Smoke Detectors	RSMeans	28 46 11.27	Other Sensors	CSI
28 31 43	Fire Detection Sensors	CSI	28 46 11	Fire Sensors and Detectors	CSI
28 31 23	Fire Det. & Alarm Annunciation Panels & Fire Station	CSI	28 46 21	Fire Alarm	CSI
28 23 19.10	Digital Video Recorder (DVR)	RSMeans	28 05 19.11	Digital Video Recorders	RSMeans
28 23 19	Digital Video Recorders & Analog Recording Devices	CSI	28 05 19	Storage Appliances for Electronic Safety & Security	RSMeans
28 16 16	Intrusion Detection Systems Infrastructure	CSI	28 31 16	Intrusion Detection Systems Infrastructure	RSMeans
28 13 53.39	Security Access Full Body Imaging Machine	RSMeans	28 18 15.39	Security Access Full Body Imaging Machine	CSI
28 13 53.36	Security Access Debugging Kit	RSMeans	28 18 53.36	Security Access Debugging Kit	CSI
28 13 53.33	Security Access Counterfeit Money Detector	RSMeans	28 18 53.33	Security Access Counterfeit Money Detector	CSI
28 13 53.23	Security Access Explosive Detection Equipment	CSI	28 18 15.23	Security Access Explosive Detection Equipment	CSI
28 13 53.16	Security Access X-Ray Equipment	CSI	28 18 13.16	Security Access X-Ray Equipment	CSI
28 13 53.13	Security Access Metal Detectors	CSI	28 18 11.13	Security Access Metal Detectors	RSMeans
28 13 53	Security Access Detection	CSI	28 18 11	Security Access Metal Detectors	CSI
28 13 23.50	Vehicle Barriers	RSMeans	28 19 15.50	Vehicle Barriers	CSI
28 13 23	Access Control Remote Devices	RSMeans	28 19 15	Perimeter Vehicle Access Management Systems	RSMeans
28 13	Access Control	CSI	28 19	Access Control Vehicle Identification Systems	CSI
28 05 13.23	Fire Alarm Communications Conductors and Cables	CSI	27 15 01.19	Fire Alarm Communications Conductors and Cables	RSMeans
28 05 13.10	Alarm & Communications Cable	RSMeans	27 15 01.11	Conductors & Cables For Electronic Safety & Security	CSI
28 01 30.51	Maint. and Admin. of Elec. Detection and Alarm	CSI	28 01 80.51	Maint. & Administration of Fire Detection & Alarm	RSMeans
28 01 30	Operation and Maint. of Elec. Detection and Alarm	CSI	28 01 80	Operation and Maint. of Fire Detection and Alarm	RSMeans
26 56 19.20	Roadway Luminaire	RSMeans	26 56 21.20	Roadway Luminaire	RSMeans
26 56 19.10	Roadway Lighting Fixtures	RSMeans	26 56 21.10	LED Exterior Lighting	RSMeans
26 56 16.55	Parking LED Lighting	RSMeans	26 56 19.60	Parking LED Lighting	RSMeans
26 53 13.10	Exit Lighting Fixtures	RSMeans	26 52 13.16	Exit Signs	CSI
26 26 13.10	Power Distribution Unit	RSMeans	26 27 33.20	Power Distribution Unit	RSMeans
13 34 23.35	Geodesic Domes	RSMeans	13 33 13.35	Geodesic Domes	CSI
13 34 23.15	Domes	RSMeans	13 34 56.15	Domes	CSI
11 14 13.13	Portable Posts and Railings	CSI	11 14 19.13	Portable Posts and Railings	CSI
10 21 13.20	Plastic Toilet Compartment Components	RSMeans	10 21 14.19	Plastic Toilet Compartment Components	CSI
10 21 13.17	Plastic-Laminate Clad Toilet Compartment Components	RSMeans	10 21 14.16	Plastic-Laminate Clad Toilet Compartment Components	CSI
10 21 13.14	Metal Toilet Compartment Components	RSMeans	10 21 14.13	Metal Toilet Compartment Components	CSI
08 74 23.50	Security Access Control Accessories	RSMeans	28 15 11.19	Security Access Control Accessories	CSI
08 74 19.50	Biometric Identity Access	RSMeans	28 15 11.15	Biometric Identity Devices	RSMeans
08 74 16.50	Keypad Access	RSMeans	28 15 11.13	Keypads	RSMeans
08 74 13	Card Key Access Control Hardware	CSI	28 15 11	Integrated Credential Readers & Field Entry Mgmt	RSMeans
08 74	Access Control Hardware	CSI	28 15	Access Control Hardware Devices	CSI
08 56 63	Detention Windows	CSI	11 98 21	Detention Windows	RSMeans
08 34 63	Detention Doors and Frames	CSI	11 98 12	Detention Doors and Frames	RSMeans
07 72 73.10	Pitch Pockets, Variable Sizes	RSMeans	07 71 16.20	Pitch Pockets, Variable Sizes	CSI
05 53 13.70	Expanded Steel Grating, at Ground	RSMeans	05 53 19.20	Expanded Grating, Steel	CSI
02 85 33	Removal and Disposal of Materials with Mold	CSI	02 87 13.33	Removal and Disposal of Materials with Mold	RSMeans
02 85 16	Mold Remediation Preparation and Containment	CSI	02 87 13	Mold Remediation	RSMeans
02 85	Mold Remediation	CSI	02 87	Biohazard Remediation	CSI

For additional tools that help with the utilization of the Construction Specifications Institute's 2016 edition of MasterFormat® please visit the following website: http://www.masterformat.com/revisions/

How the Cost Data Is Built: An Overview

Unit Prices*

All cost data have been divided into 50 divisions according to the MasterFormat® system of classification and numbering.

Assemblies*

The cost data in this section have been organized in an "Assemblies" format. These assemblies are the functional elements of a building and are arranged according to the 7 elements of the UNIFORMAT II classification system. For a complete explanation of a typical "Assembly", see "RSMeans data: Assemblies—How They Work."

*Residential Models**

Model buildings for four classes of construction—economy, average, custom, and luxury—are developed and shown with complete costs per square foot.

*Commercial/Industrial/ Institutional Models**

This section contains complete costs for 77 typical model buildings expressed as costs per square foot.

*Green Commercial/Industrial/ Institutional Models**

This section contains complete costs for 25 green model buildings expressed as costs per square foot.

*References**

This section includes information on Equipment Rental Costs, Crew Listings, Historical Cost Indexes, City Cost Indexes, Location Factors, Reference Tables, and Change Orders, as well as a listing of abbreviations.

- **Equipment Rental Costs:** Included are the average costs to rent and operate hundreds of pieces of construction equipment.
- **Crew Listings:** This section lists all the crews referenced in the cost data. A crew is composed of more than one trade classification and/or the addition of power equipment to any trade classification. Power equipment is included in the cost of the crew. Costs are shown both with bare labor rates and with the installing contractor's overhead and profit added. For each, the total crew cost per eight-hour day and the composite cost per labor-hour are listed.

Unit Cost data

Assembly Cost data

Square Foot Models

- **Historical Cost Indexes:** These indexes provide you with data to adjust construction costs over time.
- **City Cost Indexes:** All costs in this data set are U.S. national averages. Costs vary by region. You can adjust for this by CSI Division to over 730 cities in 900+ 3-digit zip codes throughout the U.S. and Canada by using this data.
- **Location Factors:** You can adjust total project costs to over 730 cities in 900+ 3-digit zip codes throughout the U.S. and Canada by using the weighted number, which applies across all divisions.
- **Reference Tables:** At the beginning of selected major classifications in the Unit Prices are reference numbers indicators. These numbers refer you to related information in the Reference Section. In this section, you'll find reference tables, explanations, and estimating information that support how we develop the unit price data, technical data, and estimating procedures.
- **Change Orders:** This section includes information on the factors that influence the pricing of change orders.

- **Abbreviations:** A listing of abbreviations used throughout this information, along with the terms they represent, is included.

Index (printed versions only)

A comprehensive listing of all terms and subjects will help you quickly find what you need when you are not sure where it occurs in MasterFormat®.

Conclusion

This information is designed to be as comprehensive and easy to use as possible.

> The Construction Specifications Institute (CSI) and Construction Specifications Canada (CSC) have produced the 2016 edition of MasterFormat®, a system of titles and numbers used extensively to organize construction information.
>
> All unit prices in the RSMeans cost data are now arranged in the 50-division MasterFormat® 2016 system.

---

* Not all information is available in all data sets

Note: The material prices in RSMeans cost data are "contractor's prices." They are the prices that contractors can expect to pay at the lumberyards, suppliers'/distributors' warehouses, etc. Small orders of specialty items would be higher than the costs shown, while very large orders, such as truckload lots, would be less. The variation would depend on the size, timing, and negotiating power of the contractor. The labor costs are primarily for new construction or major renovation rather than repairs or minor alterations. With reasonable exercise of judgment, the figures can be used for any building work.

Estimating with RSMeans data: Unit Prices

Following these steps will allow you to complete an accurate estimate using RSMeans data Unit Prices.

1. Scope Out the Project

- Think through the project and identify the CSI divisions needed in your estimate.
- Identify the individual work tasks that will need to be covered in your estimate.
- The Unit Price data have been divided into 50 divisions according to CSI MasterFormat® 2016.
- In printed versions, the Unit Price Section Table of Contents on page 1 may also be helpful when scoping out your project.
- Experienced estimators find it helpful to begin with Division 2 and continue through completion. Division 1 can be estimated after the full project scope is known.

2. Quantify

- Determine the number of units required for each work task that you identified.
- Experienced estimators include an allowance for waste in their quantities. (Waste is not included in our Unit Price line items unless otherwise stated.)

3. Price the Quantities

- Use the search tools available to locate individual Unit Price line items for your estimate.
- Reference Numbers indicated within a Unit Price section refer to additional information that you may find useful.
- The crew indicates who is performing the work for that task. Crew codes are expanded in the Crew Listings in the Reference Section to include all trades and equipment that comprise the crew.
- The Daily Output is the amount of work the crew is expected to complete in one day.
- The Labor-Hours value is the amount of time it will take for the crew to install one unit of work.
- The abbreviated Unit designation indicates the unit of measure upon which the crew, productivity, and prices are based.
- Bare Costs are shown for materials, labor, and equipment needed to complete the Unit Price line item. Bare costs do not include waste, project overhead, payroll insurance, payroll taxes, main office overhead, or profit.
- The Total Incl O&P cost is the billing rate or invoice amount of the installing contractor or subcontractor who performs the work for the Unit Price line item.

4. Multiply

- Multiply the total number of units needed for your project by the Total Incl O&P cost for each Unit Price line item.
- Be careful that your take off unit of measure matches the unit of measure in the Unit column.
- The price you calculate is an estimate for a completed item of work.
- Keep scoping individual tasks, determining the number of units required for those tasks, matching each task with individual Unit Price line items, and multiplying quantities by Total Incl O&P costs.
- An estimate completed in this manner is priced as if a subcontractor, or set of subcontractors, is performing the work. The estimate does not yet include Project Overhead or Estimate Summary components such as general contractor markups on subcontracted work, general contractor office overhead and profit, contingency, and location factors.

5. Project Overhead

- Include project overhead items from Division 1–General Requirements.
- These items are needed to make the job run. They are typically, but not always, provided by the general contractor. Items include, but are not limited to, field personnel, insurance, performance bond, permits, testing, temporary utilities, field office and storage facilities, temporary scaffolding and platforms, equipment mobilization and demobilization, temporary roads and sidewalks, winter protection, temporary barricades and fencing, temporary security, temporary signs, field engineering and layout, final cleaning, and commissioning.
- Each item should be quantified and matched to individual Unit Price line items in Division 1, then priced and added to your estimate.
- An alternate method of estimating project overhead costs is to apply a percentage of the total project cost, usually 5% to 15% with an average of 10% (see General Conditions).
- Include other project related expenses in your estimate such as:
  - Rented equipment not itemized in the Crew Listings
  - Rubbish handling throughout the project (see section 02 41 19.19)

6. Estimate Summary

- Include sales tax as required by laws of your state or county.
- Include the general contractor's markup on self-performed work, usually 5% to 15% with an average of 10%.
- Include the general contractor's markup on subcontracted work, usually 5% to 15% with an average of 10%.
- Include the general contractor's main office overhead and profit:
  - RSMeans data provides general guidelines on the general contractor's main office overhead (see section 01 31 13.60 and Reference Number R013113-50).
  - Markups will depend on the size of the general contractor's operations, projected annual revenue, the level of risk, and the level of competition in the local area and for this project in particular.
- Include a contingency, usually 3% to 5%, if appropriate.
- Adjust your estimate to the project's location by using the City Cost Indexes or the Location Factors in the Reference Section:
  - Look at the rules in "How to Use the City Cost Indexes" to see how to apply the Indexes for your location.
  - When the proper Index or Factor has been identified for the project's location, convert it to a multiplier by dividing it by 100, then multiply that multiplier by your estimated total cost. The original estimated total cost will now be adjusted up or down from the national average to a total that is appropriate for your location.

How to Use the Cost Data: The Details

What's Behind the Numbers? The Development of Cost data

RSMeans data engineers continually monitor developments in the construction industry in order to ensure reliable, thorough, and up-to-date cost information. While overall construction costs may vary relative to general economic conditions, price fluctuations within the industry are dependent upon many factors. Individual price variations may, in fact, be opposite to overall economic trends. Therefore, costs are constantly tracked, and complete updates are performed yearly. Also, new items are frequently added in response to changes in materials and methods.

Costs in U.S. Dollars

All costs represent U.S. national averages and are given in U.S. dollars. The City Cost Index (CCI) with RSMeans data can be used to adjust costs to a particular location. The CCI for Canada can be used to adjust U.S. national averages to local costs in Canadian dollars. No exchange rate conversion is necessary because it has already been factored in.

G The processes or products identified by the green symbol in our publications have been determined to be environmentally responsible and/or resource-efficient solely by RSMeans data engineering staff. The inclusion of the green symbol does not represent compliance with any specific industry association or standard.

Material Costs

RSMeans data engineers contact manufacturers, dealers, distributors, and contractors all across the U.S. and Canada to determine national average material costs. If you have access to current material costs for your specific location, you may wish to make adjustments to reflect differences from the national average. Included within material costs are fasteners for a normal installation. RSMeans data engineers use manufacturers' recommendations, written specifications, and/or standard construction practices for the sizing and spacing of fasteners. Adjustments to material costs may be required for your specific application or location. The manufacturer's warranty is assumed. Extended warranties are not included in the material costs. **Material costs do not include sales tax.**

Labor Costs

Labor costs are based upon a mathematical average of trade-specific wages in 30 major U.S. cities. The type of wage (union, open shop, or residential) is identified on the inside back cover of printed publications or is selected by the estimator when using the electronic products. Markups for the wages can also be found on the inside back cover of printed publications and/or under the labor references found in the electronic products.

- If wage rates in your area vary from those used, or if rate increases are expected within a given year, labor costs should be adjusted accordingly.

Labor costs reflect productivity based on actual working conditions. In addition to actual installation, these figures include time spent during a normal weekday on tasks, such as material receiving and handling, mobilization at site, site movement, breaks, and cleanup.

Productivity data is developed over an extended period so as not to be influenced by abnormal variations and reflects a typical average.

Equipment Costs

Equipment costs include not only rental but also operating costs for equipment under normal use. The operating costs include parts and labor for routine servicing, such as the repair and replacement of pumps, filters, and worn lines. Normal operating expendables, such as fuel, lubricants, tires, and electricity (where applicable), are also included. Extraordinary operating expendables with highly variable wear patterns, such as diamond bits and blades, are excluded. These costs are included under materials. Equipment rental rates are obtained from industry sources throughout North America—contractors, suppliers, dealers, manufacturers, and distributors.

Rental rates can also be treated as reimbursement costs for contractor-owned equipment. Owned equipment costs include depreciation, loan payments, interest, taxes, insurance, storage, and major repairs.

Equipment costs do not include operators' wages.

Equipment Cost/Day—The cost of equipment required for each crew is included in the Crew Listings in the Reference Section (small tools that are considered essential everyday tools are not listed out separately). The Crew Listings itemize specialized tools and heavy equipment along with labor trades. The daily cost of itemized equipment included in a crew is based on dividing the weekly bare rental rate by 5 (number of working days per week), then adding the hourly operating cost times 8 (the number of hours per day). This Equipment Cost/Day is shown in the last column of the Equipment Rental Costs in the Reference Section.

Mobilization, Demobilization—The cost to move construction equipment from an equipment yard or rental company to the job site and back again is not included in equipment costs. Mobilization (to the site) and demobilization (from the site) costs can be found in the Unit Price Section. If a piece of equipment is already at the job site, it is not appropriate to utilize mobilization or demobilization costs again in an estimate.

Overhead and Profit

Total Cost including O&P for the installing contractor is shown in the last column of the Unit Price and/or Assemblies. This figure is the sum of the bare material cost plus 10% for profit, the bare labor cost plus total overhead and profit, and the bare equipment cost plus 10% for profit. Details for the calculation of overhead and profit on labor are shown on the inside back cover of the printed product and in the Reference Section of the electronic product.

General Conditions

Cost data in this data set are presented in two ways: Bare Costs and Total Cost including O&P (Overhead and Profit). General Conditions, or General Requirements, of the contract should also be added to the Total Cost including O&P when applicable. Costs for General Conditions are listed in Division 1 of the Unit Price Section and in the Reference Section.

General Conditions for the installing contractor may range from 0% to 10% of the Total Cost including O&P. For the general or prime contractor, costs for General Conditions may range from 5% to 15% of the Total Cost including O&P, with a figure of 10% as the most typical allowance. If applicable, the Assemblies and Models sections use costs that include the installing contractor's overhead and profit (O&P).

Factors Affecting Costs

Costs can vary depending upon a number of variables. Here's a listing of some factors that affect costs and points to consider.

Quality—The prices for materials and the workmanship upon which productivity is based represent sound construction work. They are also in line with industry standard and manufacturer specifications and are frequently used by federal, state, and local governments.

Overtime—We have made no allowance for overtime. If you anticipate premium time or work beyond normal working hours, be sure to make an appropriate adjustment to your labor costs.

Productivity—The productivity, daily output, and labor-hour figures for each line item are based on an eight-hour work day in daylight hours in moderate temperatures, and up to a 14' working height unless otherwise indicated. For work that extends beyond normal work hours or is performed under adverse conditions, productivity may decrease.

Size of Project—The size, scope of work, and type of construction project will have a significant impact on cost. Economies of scale can reduce costs for large projects. Unit costs can often run higher for small projects.

Location—Material prices are for metropolitan areas. However, in dense urban areas, traffic and site storage limitations may increase costs. Beyond a 20-mile radius of metropolitan areas, extra trucking or transportation charges may also increase the material costs slightly. On the other hand, lower wage rates may be in effect. Be sure to consider both of these factors when preparing an estimate, particularly if the job site is located in a central city or remote rural location. In addition, highly specialized subcontract items may require travel and per-diem expenses for mechanics.

Other Factors—

- season of year
- contractor management
- weather conditions
- local union restrictions
- building code requirements
- availability of:
  - adequate energy
  - skilled labor
  - building materials
- owner's special requirements/restrictions
- safety requirements
- environmental considerations
- access

Unpredictable Factors—General business conditions influence "in-place" costs of all items. Substitute materials and construction methods may have to be employed. These may affect the installed cost and/or life cycle costs. Such factors may be difficult to evaluate and cannot necessarily be predicted on the basis of the job's location in a particular section of the country. Thus, where these factors apply, you may find significant but unavoidable cost variations for which you will have to apply a measure of judgment to your estimate.

Rounding of Costs

In printed publications only, all unit prices in excess of $5.00 have been rounded to make them easier to use and still maintain adequate precision of the results.

How Subcontracted Items Affect Costs

A considerable portion of all large construction jobs is usually subcontracted. In fact, the percentage done by subcontractors is constantly increasing and may run over 90%. Since the workers employed by these companies do nothing else but install their particular product, they soon become experts in that line. The result is, installation by these firms is accomplished so efficiently that the total in-place cost, even with the general contractor's overhead and profit, is no more, and often less, than if the principal contractor had handled the installation. Companies that deal with construction specialties are anxious to have their product perform well and, consequently, the installation will be the best possible.

Contingencies

The allowance for contingencies generally provides for unforeseen construction difficulties. On alterations or repair jobs, 20% is not too much. If drawings are final and only field contingencies are being considered, 2% or 3% is probably sufficient, and often nothing needs to be added. Contractually, changes in plans will be covered by extras. The contractor should consider inflationary price trends and possible material shortages during the course of the job. These escalation factors are dependent upon both economic conditions and the anticipated time between the estimate and actual construction. If drawings are not complete or approved, or a budget cost is wanted, it is wise to add 5% to 10%. Contingencies, then, are a matter of judgment.

Important Estimating Considerations

The productivity, or daily output, of each craftsman or crew assumes a well-managed job where tradesmen with the proper tools and equipment, along with the appropriate construction materials, are present. Included are daily set-up and cleanup time, break time, and plan layout time. Unless otherwise indicated, time for material movement on site (for items that can be transported by hand) of up to 200' into the building and to the first or second floor is also included. If material has to be transported by other means, over greater distances, or to higher floors, an additional allowance should be considered by the estimator.

While horizontal movement is typically a sole function of distances, vertical transport introduces other variables that can significantly impact productivity. In an occupied building, the use of elevators (assuming access, size, and required protective measures are acceptable) must be understood at the time of the estimate. For new construction, hoist wait and cycle times can easily be 15 minutes and may result in scheduled access extending beyond the normal work day. Finally, all vertical transport will impose strict weight limits likely to preclude the use of any motorized material handling.

The productivity, or daily output, also assumes installation that meets manufacturer/designer/ standard specifications. A time allowance for quality control checks, minor adjustments, and any task required to ensure the proper function or operation is also included. For items that require connections to services, time is included for positioning, leveling, securing the unit, and for making all the necessary connections (and start up where applicable), ensuring a complete installation. Estimating of the services themselves (electrical, plumbing, water, steam, hydraulics, dust collection, etc.) is separate.

In some cases, the estimator must consider the use of a crane and an appropriate crew for the installation of large or heavy items. For those situations where a crane is not included in the assigned crew and as part of the line item cost,

then equipment rental costs, mobilization and demobilization costs, and operator and support personnel costs must be considered.

Labor-Hours

The labor-hours expressed in this publication are derived by dividing the total daily labor-hours for the crew by the daily output. Based on average installation time and the assumptions listed above, the labor-hours include: direct labor, indirect labor, and nonproductive time. A typical day for a craftsman might include, but is not limited to:

- Direct Work
  - ☐ Measuring and layout
  - ☐ Preparing materials
  - ☐ Actual installation
  - ☐ Quality assurance/quality control
- Indirect Work
  - ☐ Reading plans or specifications
  - ☐ Preparing space
  - ☐ Receiving materials
  - ☐ Material movement
  - ☐ Giving or receiving instruction
  - ☐ Miscellaneous
- Non-Work
  - ☐ Chatting
  - ☐ Personal issues
  - ☐ Breaks
  - ☐ Interruptions (i.e., sickness, weather, material or equipment shortages, etc.)

If any of the items for a typical day do not apply to the particular work or project situation, the estimator should make any necessary adjustments.

Final Checklist

Estimating can be a straightforward process provided you remember the basics. Here's a checklist of some of the steps you should remember to complete before finalizing your estimate.

Did you remember to:

- factor in the City Cost Index for your locale?
- take into consideration which items have been marked up and by how much?
- mark up the entire estimate sufficiently for your purposes?
- read the background information on techniques and technical matters that could impact your project time span and cost?
- include all components of your project in the final estimate?
- double check your figures for accuracy?
- call RSMeans data engineers if you have any questions about your estimate or the data you've used? Remember, Gordian stands behind all of our products, including our extensive RSMeans data solutions. If you have any questions about your estimate, about the costs you've used from our data, or even about the technical aspects of the job that may affect your estimate, feel free to call the Gordian RSMeans editors at 1-800-448-8182.

Unit Price Section

Table of Contents

1

RSMeans data: Unit Prices— How They Work

All RSMeans data: Unit Prices are organized in the same way.

03 30 Cast-In-Place Concrete

03 30 53 – Miscellaneous Cast-In-Place Concrete

03 30 53.40 Concrete In Place	Crew	Daily Output	Labor-Hours	Unit	Material	2018 Bare Costs Labor	Equipment	Total	Total Incl O&P
0010 **CONCRETE IN PLACE**									
0020 Including forms (4 uses), Grade 60 rebar, concrete (Portland cement									
0050 Type I), placement and finishing unless otherwise indicated									
3540 Equipment pad (3000 psi), 3' x 3' x 6" thick	C-14H	45	1.067	Ea.	44.50	52.50	.57	97.57	130
3550 4' x 4' x 6" thick		30	1.600		69	79	.85	148.85	197
3560 5' x 5' x 8" thick		18	2.667		126	132	1.41	259.41	340
3570 6' x 6' x 8" thick		14	3.429		173	169	1.82	343.82	450
3580 8' x 8' x 10" thick		8	6		370	296	3.18	669.18	860
3590 10' x 10' x 12" thick		5	9.600		645	475	5.10	1,125.10	1,425
3800 Footings (3000 psi), spread under 1 C.Y.	C-14C	28	4	C.Y.	185	192	.92	377.92	495
3825 1 C.Y. to 5 C.Y.		43	2.605		219	125	.60	344.60	430
3850 Over 5 C.Y. R033053-60		75	1.493		203	72	.34	275.34	330
3900 Footings, strip (3000 psi), 18" x 9", unreinforced	C-14L	40	2.400		146	113	.65	259.65	335
3920 18" x 9", reinforced	C-14C	35	3.200		169	154	.74	323.74	420

It is important to understand the structure of RSMeans data: Unit Prices, so that you can find information easily and use it correctly.

1 Line Numbers

Line Numbers consist of 12 characters, which identify a unique location in the database for each task. The first 6 or 8 digits conform to the Construction Specifications Institute MasterFormat® 2016. The remainder of the digits are a further breakdown in order to arrange items in understandable groups of similar tasks. Line numbers are consistent across all all of our publications, so a line number in any of our products will always refer to the same item of work.

2 Descriptions

Descriptions are shown in a hierarchical structure to make them readable. In order to read a complete description, read up through the indents to the top of the section. Include everything that is above and to the left that is not contradicted by information below. For instance, the complete description for line 03 30 53.40 3550 is "Concrete in place, including forms (4 uses), Grade 60 rebar, concrete (Portland cement Type 1), placement and finishing unless otherwise indicated; Equipment pad (3000 psi), 4' × 4' × 6" thick."

3 RSMeans data

When using **RSMeans data**, it is important to read through an entire section to ensure that you use the data that most closely matches your work. Note that sometimes there is additional information shown in the section that may improve your price. There are frequently lines that further describe, add to, or adjust data for specific situations.

4 Reference Information

Gordian's RSMeans engineers have created **reference** information to assist you in your estimate. **If** there is information that applies to a section, it will be indicated at the start of the section.

5 Crews

Crews include labor and/or equipment necessary to accomplish each task. In this case, Crew C-14H is used. Gordian's RSMeans staff selects a crew to represent the workers and equipment that are typically used for that task. In this case, Crew C-14H consists of one carpenter foreman (outside), two carpenters, one rodman, one laborer, one cement finisher, and one gas engine vibrator. Details of all crews can be found in the Reference Section.

Crews - Standard

Crew C-14H	Bare Costs		Incl. Subs O&P		Cost Per Labor-Hour	
Crew No.	Hr.	Daily	Hr.	Daily	Bare Costs	Incl. O&P
1 Carpenter Foreman (outside)	$52.70	$421.60	$80.25	$642.00	$49.36	$74.88
2 Carpenters	50.70	811.20	77.20	1235.20		
1 Rodman (reinf.)	54.65	437.20	83.45	667.60		
1 Laborer	39.85	318.80	60.70	485.60		
1 Cement Finisher	47.55	380.40	70.45	563.60		
1 Gas Engine Vibrator		25.60		28.16	.53	.59
48 L.H., Daily Totals		$2394.80		$3622.16	$49.89	$75.46

6 ## Daily Output

The **Daily Output** is the amount of work that the crew can do in a normal 8-hour workday, including mobilization, layout, movement of materials, and cleanup. In this case, crew C-14H can install thirty 4' × 4' × 6" thick concrete pads in a day. Daily output is variable and based on many factors, including the size of the job, location, and environmental conditions. RSMeans data represents work done in daylight (or adequate lighting) and temperate conditions.

7 ## Labor-Hours

The figure in the **Labor-Hours** column is the amount of labor required to perform one unit of work—in this case the amount of labor required to construct one 4' × 4' equipment pad. This figure is calculated by dividing the number of hours of labor in the crew by the daily output (48 labor-hours divided by 30 pads = 1.6 hours of labor per pad). Multiply 1.6 times 60 to see the value in minutes: 60 × 1.6 = 96

minutes. Note: the labor-hour figure is not dependent on the crew size. A change in crew size will result in a corresponding change in daily output, but the labor-hours per unit of work will not change.

8 ## Unit of Measure

All RSMeans data: Unit Prices include the typical **Unit of Measure** used for estimating that item. For concrete-in-place the typical unit is cubic yards (C.Y.) or each (Ea.). For installing broadloom carpet it is square yard, and for gypsum board it is square foot. The estimator needs to take special care that the unit in the data matches the unit in the take-off. Unit conversions may be found in the Reference Section.

9 ## Bare Costs

Bare Costs are the costs of materials, labor, and equipment that the installing contractor pays. They represent the cost, in U.S. dollars, for one unit of work. They do not include any markups for profit or labor burden.

10 ## Bare Total

The **Total column** represents the total bare cost for the installing contractor in U.S. dollars. In this case, the sum of $69 for material + $79 for labor + $.85 for equipment is $148.85.

11 ## Total Incl O&P

The **Total Incl O&P column** is the total cost, including overhead and profit, that the installing contractor will charge the customer. This represents the cost of materials plus 10% profit, the cost of labor plus labor burden and 10% profit, and the cost of equipment plus 10% profit. It does not include the general contractor's overhead and profit. Note: See the inside back cover of the printed product or the Reference Section of the electronic product for details of how the labor burden is calculated.

National Average

*The RSMeans data in our print publications represents a "national average" cost. This data should be modified to the project location using the **City Cost Indexes** or **Location Factors** tables found in the Reference Section. Use the Location Factors to adjust estimate totals if the project covers multiple trades. Use the City Cost Indexes (CCI) for single trade*

projects or projects where a more detailed analysis is required. All figures in the two tables are derived from the same research. The last row of data in the CCI—the weighted average—is the same as the numbers reported for each location in the location factor table.

RSMeans data: Unit Prices— How They Work (Continued)

Project Name: Pre-Engineered Steel Building			Architect: As Shown					
Location:	Anywhere, USA						01/01/18	STD
Line Number	Description	Qty	Unit	Material	Labor	Equipment	SubContract	Estimate Total
03 30 53.40 3940	Strip footing, 12" x 24", reinforced	34	C.Y.	$5,406.00	$3,808.00	$18.36	$0.00	
03 30 53.40 3950	Strip footing, 12" x 36", reinforced	15	C.Y.	$2,295.00	$1,350.00	$6.45	$0.00	
03 11 13.65 3000	Concrete slab edge forms	500	L.F.	$145.00	$1,280.00	$0.00	$0.00	
03 22 11.10 0200	Welded wire fabric reinforcing	150	C.S.F.	$2,940.00	$4,200.00	$0.00	$0.00	
03 31 13.35 0300	Ready mix concrete, 4000 psi for slab on grade	278	C.Y.	$35,584.00	$0.00	$0.00	$0.00	
03 31 13.70 4300	Place, strike off & consolidate concrete slab	278	C.Y.	$0.00	$5,031.80	$130.66	$0.00	
03 35 13.30 0250	Machine float & trowel concrete slab	15,000	S.F.	$0.00	$9,450.00	$300.00	$0.00	
03 15 16.20 0140	Cut control joints in concrete slab	950	L.F.	$47.50	$399.00	$57.00	$0.00	
03 39 23.13 0300	Sprayed concrete curing membrane	150	C.S.F.	$1,815.00	$1,005.00	$0.00	$0.00	
Division 03	**Subtotal**			**$48,232.50**	**$26,523.80**	**$512.47**	**$0.00**	**$75,268.77**
08 36 13.10 2650	Manual 10' x 10' steel sectional overhead door	8	Ea.	$10,400.00	$3,600.00	$0.00	$0.00	
08 36 13.10 2860	Insulation and steel back panel for OH door	800	S.F.	$4,000.00	$0.00	$0.00	$0.00	
Division 08	**Subtotal**			**$14,400.00**	**$3,600.00**	**$0.00**	**$0.00**	**$18,000.00**
13 34 19.50 1100	Pre-Engineered Steel Building, 100' x 150' x 24'	15,000	SF Flr.	$0.00	$0.00	$0.00	$367,500.00	
13 34 19.50 6050	Framing for PESB door opening, 3' x 7'	4	Opng.	$0.00	$0.00	$0.00	$2,240.00	
13 34 19.50 6100	Framing for PESB door opening, 10' x 10'	8	Opng.	$0.00	$0.00	$0.00	$9,200.00	
13 34 19.50 6200	Framing for PESB window opening, 4' x 3'	6	Opng.	$0.00	$0.00	$0.00	$3,330.00	
13 34 19.50 5750	PESB door, 3' x 7', single leaf	4	Opng.	$2,620.00	$700.00	$0.00	$0.00	
13 34 19.50 7750	PESB sliding window, 4' x 3' with screen	6	Opng.	$2,550.00	$600.00	$45.30	$0.00	
13 34 19.50 6550	PESB gutter, eave type, 26 ga., painted	300	L.F.	$2,220.00	$819.00	$0.00	$0.00	
13 34 19.50 8650	PESB roof vent, 12" wide x 10' long	15	Ea.	$555.00	$3,285.00	$0.00	$0.00	
13 34 19.50 6900	PESB insulation, vinyl faced, 4" thick	27,400	S.F.	$13,152.00	$9,590.00	$0.00	$0.00	
Division 13	**Subtotal**			**$21,097.00**	**$14,994.00**	**$45.30**	**$382,270.00**	**$418,406.30**
			Subtotal	$83,729.50	$45,117.80	$557.77	$382,270.00	$511,675.07
Division 01	**General Requirements @ 7%**			5,861.07	3,158.25	39.04	26,758.90	
			Estimate Subtotal	$89,590.57	$48,276.05	$596.81	$409,028.90	$511,675.07
			Sales Tax @ 5%	4,479.53		29.84	10,225.72	
			Subtotal A	94,070.09	48,276.05	626.65	419,254.62	
			GC O & P	9,407.01	25,634.58	62.67	41,925.46	
			Subtotal B	103,477.10	73,910.63	689.32	461,180.08	$639,257.13
			Contingency @ 5%					31,962.86
			Subtotal C					$671,219.99
			Bond @ $12/1000 +10% O&P					8,860.10
			Subtotal D					$680,080.09
			Location Adjustment Factor		102.30			15,641.84
			Grand Total					**$695,721.94**

This estimate is based on an interactive spreadsheet. You are free to download it and adjust it to your methodology.
A copy of this spreadsheet is available at **www.RSMeans.com/2018books.**

Sample Estimate

This sample demonstrates the elements of an estimate, including a tally of the RSMeans data lines and a summary of the markups on a contractor's work to arrive at a total cost to the owner. The Location Factor with RSMeans data is added at the bottom of the estimate to adjust the cost of the work to a specific location.

---

1 Work Performed

The body of the estimate shows the RSMeans data selected, including the line number, a brief description of each item, its take-off unit and quantity, and the bare costs of materials, labor, and equipment. This estimate also includes a column titled "SubContract." This data is taken from the column "Total Incl O&P" and represents the total that a subcontractor would charge a general contractor for the work, including the sub's markup for overhead and profit.

2 Division 1, General Requirements

This is the first division numerically but the last division estimated. Division 1 includes project-wide needs provided by the general contractor. These requirements vary by project but may include temporary facilities and utilities, security, testing, project cleanup, etc. For small projects a percentage can be used—typically between 5% and 15% of project cost. For large projects the costs may be itemized and priced individually.

3 Sales Tax

If the work is subject to state or local sales taxes, the amount must be added to the estimate. Sales tax may be added to material costs, equipment costs, and subcontracted work. In this case, sales tax was added in all three categories. It was assumed that approximately half the subcontracted work would be material cost, so the tax was applied to 50% of the subcontract total.

4 GC O&P

This entry represents the general contractor's markup on material, labor, equipment, and subcontractor costs. Our standard markup on materials, equipment, and subcontracted work is 10%. In this estimate, the markup on the labor performed by the GC's workers uses "Skilled Workers Average" shown in Column F on the table "Installing Contractor's Overhead & Profit," which can be found on the inside back cover of the printed product or in the Reference Section of the electronic product.

5 Contingency

A factor for contingency may be added to any estimate to represent the cost of unknowns that may occur between the time that the estimate is performed and the time the project is constructed. The amount of the allowance will depend on the stage of design at which the estimate is done and the contractor's assessment of the risk involved. Refer to section 01 21 16.50 for contingency allowances.

6 Bonds

Bond costs should be added to the estimate. The figures here represent a typical performance bond, ensuring that if the general contractor does not complete the obligations in the construction contract the bonding company will pay the cost for completion of the work.

7 Location Adjustment

Published prices are based on national average costs. If necessary, adjust the total cost of the project using a location factor from the "Location Factor" table or the "City Cost Index" table. Use location factors if the work is general, covering multiple trades. If the work is by a single trade (e.g., masonry) use the more specific data found in the "City Cost Indexes."

Estimating Tips
01 20 00 Price and Payment Procedures

- Allowances that should be added to estimates to cover contingencies and job conditions that are not included in the national average material and labor costs are shown in Section 01 21.

- When estimating historic preservation projects (depending on the condition of the existing structure and the owner's requirements), a 15%–20% contingency or allowance is recommended, regardless of the stage of the drawings.

01 30 00 Administrative Requirements

- Before determining a final cost estimate, it is a good practice to review all the items listed in Subdivisions 01 31 and 01 32 to make final adjustments for items that may need customizing to specific job conditions.

- Requirements for initial and periodic submittals can represent a significant cost to the General Requirements of a job. Thoroughly check the submittal specifications when estimating a project to determine any costs that should be included.

01 40 00 Quality Requirements

- All projects will require some degree of quality control. This cost is not included in the unit cost of construction listed in each division. Depending upon the terms of the contract, the various costs of inspection and testing can be the responsibility of either the owner or the contractor. Be sure to include the required costs in your estimate.

01 50 00 Temporary Facilities and Controls

- Barricades, access roads, safety nets, scaffolding, security, and many more requirements for the execution of a safe project are elements of direct cost. These costs can easily be overlooked when preparing an estimate. When looking through the major classifications of this subdivision, determine which items apply to each division in your estimate.

- Construction equipment rental costs can be found in the Reference Section in Section 01 54 33. Operators' wages are not included in equipment rental costs.

- Equipment mobilization and demobilization costs are not included in equipment rental costs and must be considered separately.

- The cost of small tools provided by the installing contractor for his workers is covered in the "Overhead" column on the "Installing Contractor's Overhead and Profit" table that lists labor trades, base rates, and markups and, therefore, is included in the "Total Incl. O&P" cost of any unit price line item.

01 70 00 Execution and Closeout Requirements

- When preparing an estimate, thoroughly read the specifications to determine the requirements for Contract Closeout. Final cleaning, record documentation, operation and maintenance data, warranties and bonds, and spare parts and maintenance materials can all be elements of cost for the completion of a contract. Do not overlook these in your estimate.

Reference Numbers
Reference numbers are shown at the beginning of some major classifications. These numbers refer to related items in the Reference Section. The reference information may be an estimating procedure, an alternate pricing method, or technical information.

Note: Not all subdivisions listed here necessarily appear. ∎

Did you know?

RSMeans data is available through our online application with 24/7 access:

- Search for unit prices by keyword
- Leverage the most up-to-date data
- Build and export estimates

Try it free for 30 days!
www.rsmeans.com/2018freetrial

01 11 Summary of Work

01 11 31 – Professional Consultants

01 11 31.20 Construction Management Fees	Crew	Daily Output	Labor-Hours	Unit	Material	2018 Bare Costs Labor	Equipment	Total	Total Incl O&P
0010 **CONSTRUCTION MANAGEMENT FEES**									
0020 $1,000,000 job, minimum				Project				4.50%	4.50%
0050 Maximum								7.50%	7.50%
0060 For work to $100,000								10%	10%
0300 $50,000,000 job, minimum								2.50%	2.50%
0350 Maximum								4%	4%

01 11 31.30 Engineering Fees	Crew	Daily Output	Labor-Hours	Unit	Material	2018 Bare Costs Labor	Equipment	Total	Total Incl O&P
0010 **ENGINEERING FEES** R011110-30									
0020 Educational planning consultant, minimum				Project				.50%	.50%
0100 Maximum				"				2.50%	2.50%
0200 Electrical, minimum				Contrct				4.10%	4.10%
0300 Maximum								10.10%	10.10%
0400 Elevator & conveying systems, minimum								2.50%	2.50%
0500 Maximum								5%	5%
0600 Food service & kitchen equipment, minimum								8%	8%
0700 Maximum								12%	12%
1000 Mechanical (plumbing & HVAC), minimum								4.10%	4.10%
1100 Maximum								10.10%	10.10%

01 21 Allowances

01 21 16 – Contingency Allowances

01 21 16.50 Contingencies	Crew	Daily Output	Labor-Hours	Unit	Material	2018 Bare Costs Labor	Equipment	Total	Total Incl O&P
0010 **CONTINGENCIES**, Add to estimate									
0020 Conceptual stage				Project				20%	20%
0050 Schematic stage								15%	15%
0100 Preliminary working drawing stage (Design Dev.)								10%	10%
0150 Final working drawing stage								3%	3%

01 21 53 – Factors Allowance

01 21 53.50 Factors	Crew	Daily Output	Labor-Hours	Unit	Material	2018 Bare Costs Labor	Equipment	Total	Total Incl O&P
0010 **FACTORS** Cost adjustments									
0100 Add to construction costs for particular job requirements									
0500 Cut & patch to match existing construction, add, minimum				Costs	2%	3%			
0550 Maximum					5%	9%			
0800 Dust protection, add, minimum					1%	2%			
0850 Maximum					4%	11%			
1100 Equipment usage curtailment, add, minimum					1%	1%			
1150 Maximum					3%	10%			
1400 Material handling & storage limitation, add, minimum					1%	1%			
1450 Maximum					6%	7%			
1700 Protection of existing work, add, minimum					2%	2%			
1750 Maximum					5%	7%			
2000 Shift work requirements, add, minimum						5%			
2050 Maximum						30%			
2300 Temporary shoring and bracing, add, minimum					2%	5%			
2350 Maximum					5%	12%			

01 21 53 – Factors Allowance

01 21 53.60 Security Factors	Crew	Daily Output	Labor-Hours	Unit	Material	2018 Bare Costs Labor	Equipment	Total	Total Incl O&P
0010 **SECURITY FACTORS** R012153-60									
0100 Additional costs due to security requirements									
0110 Daily search of personnel, supplies, equipment and vehicles									
0120 Physical search, inventory and doc of assets, at entry				Costs		30%			
0130 At entry and exit						50%			
0140 Physical search, at entry						6.25%			
0150 At entry and exit						12.50%			
0160 Electronic scan search, at entry						2%			
0170 At entry and exit						4%			
0180 Visual inspection only, at entry						.25%			
0190 At entry and exit						.50%			
0200 ID card or display sticker only, at entry						.12%			
0210 At entry and exit						.25%			
0220 Day 1 as described below, then visual only for up to 5 day job duration									
0230 Physical search, inventory and doc of assets, at entry				Costs		5%			
0240 At entry and exit						10%			
0250 Physical search, at entry						1.25%			
0260 At entry and exit						2.50%			
0270 Electronic scan search, at entry						.42%			
0280 At entry and exit						.83%			
0290 Day 1 as described below, then visual only for 6-10 day job duration									
0300 Physical search, inventory and doc of assets, at entry				Costs		2.50%			
0310 At entry and exit						5%			
0320 Physical search, at entry						.63%			
0330 At entry and exit						1.25%			
0340 Electronic scan search, at entry						.21%			
0350 At entry and exit						.42%			
0360 Day 1 as described below, then visual only for 11-20 day job duration									
0370 Physical search, inventory and doc of assets, at entry				Costs		1.25%			
0380 At entry and exit						2.50%			
0390 Physical search, at entry						.31%			
0400 At entry and exit						.63%			
0410 Electronic scan search, at entry						.10%			
0420 At entry and exit						.21%			
0430 Beyond 20 days, costs are negligible									
0440 Escort required to be with tradesperson during work effort				Costs		6.25%			

01 21 55 – Job Conditions Allowance

01 21 55.50 Job Conditions	Crew	Daily Output	Labor-Hours	Unit	Material	2018 Bare Costs Labor	Equipment	Total	Total Incl O&P
0010 **JOB CONDITIONS** Modifications to applicable									
0020 cost summaries									
0100 Economic conditions, favorable, deduct				Project				2%	2%
0200 Unfavorable, add								5%	5%
0300 Hoisting conditions, favorable, deduct								2%	2%
0400 Unfavorable, add								5%	5%
0700 Labor availability, surplus, deduct								1%	1%
0800 Shortage, add								10%	10%
0900 Material storage area, available, deduct								1%	1%
1000 Not available, add								2%	2%
1100 Subcontractor availability, surplus, deduct								5%	5%
1200 Shortage, add								12%	12%
1300 Work space, available, deduct								2%	2%
1400 Not available, add								5%	5%

01 21 Allowances

01 21 57 – Overtime Allowance

01 21 57.50 Overtime		Crew	Daily Output	Labor-Hours	Unit	Material	2018 Bare Costs Labor	Equipment	Total	Total Incl O&P
0010	**OVERTIME** for early completion of projects or where	R012909-90								
0020	labor shortages exist, add to usual labor, up to					Costs	100%			

01 21 63 – Taxes

01 21 63.10 Taxes

0010	**TAXES**	R012909-80								
0020	Sales tax, State, average					%	5.08%			
0050	Maximum	R012909-85					7.50%			
0200	Social Security, on first $118,500 of wages							7.65%		
0300	Unemployment, combined Federal and State, minimum	R012909-86						.60%		
0350	Average							9.60%		
0400	Maximum							12%		

01 31 Project Management and Coordination

01 31 13 – Project Coordination

01 31 13.20 Field Personnel

		Crew	Daily Output	Labor-Hours	Unit	Material	Labor	Equipment	Total	Total Incl O&P
0010	**FIELD PERSONNEL**									
0020	Clerk, average				Week		485		485	740
0100	Field engineer, junior engineer						1,150		1,150	1,750
0120	Engineer						1,500		1,500	2,300
0140	Senior engineer						1,700		1,700	2,600
0160	General purpose laborer, average						1,600		1,600	2,425
0180	Project manager, minimum						2,125		2,125	3,275
0200	Average						2,450		2,450	3,750
0220	Maximum						2,800		2,800	4,275
0240	Superintendent, minimum						2,075		2,075	3,175
0260	Average						2,275		2,275	3,500
0280	Maximum						2,600		2,600	3,975
0290	Timekeeper, average						1,325		1,325	2,025

01 31 13.30 Insurance

0010	**INSURANCE**	R013113-40								
0020	Builders risk, standard, minimum				Job				.24%	.24%
0050	Maximum	R013113-50							.64%	.64%
0200	All-risk type, minimum								.25%	.25%
0250	Maximum	R013113-60							.62%	.62%
0400	Contractor's equipment floater, minimum				Value				.50%	.50%
0450	Maximum				"				1.50%	1.50%
0600	Public liability, average				Job				2.02%	2.02%
0800	Workers' compensation & employer's liability, average									
0850	by trade, carpentry, general				Payroll		13.05%			
1000	Electrical						5.22%			
1150	Insulation						10.90%			
1450	Plumbing						6.94%			
1550	Sheet metal work (HVAC)						8.27%			

01 31 13.50 General Contractor's Mark-Up

0010	**GENERAL CONTRACTOR'S MARK-UP** on Change Orders									
0200	Extra work, by subcontractors, add				%				10%	10%
0250	By General Contractor, add								15%	15%
0400	Omitted work, by subcontractors, deduct all but								5%	5%
0450	By General Contractor, deduct all but								7.50%	7.50%

For customer support on your Electrical Costs with RSMeans data, call 800.448.8182.

01 31 Project Management and Coordination

01 31 13 – Project Coordination

01 31 13.50 General Contractor's Mark-Up		Crew	Daily Output	Labor-Hours	Unit	Material	2018 Bare Costs Labor	Equipment	Total	Total Incl O&P
0600	Overtime work, by subcontractors, add				%				15%	15%
0650	By General Contractor, add								10%	10%

01 31 13.80 Overhead and Profit										
0010	**OVERHEAD & PROFIT** Allowance to add to items in this	R013113-50								
0020	book that do not include Subs O&P, average				%				25%	
0100	Allowance to add to items in this book that	R013113-55								
0110	do include Subs O&P, minimum				%				5%	5%
0150	Average								10%	10%
0200	Maximum								15%	15%
0300	Typical, by size of project, under $100,000								30%	
0350	$500,000 project								25%	
0400	$2,000,000 project								20%	
0450	Over $10,000,000 project								15%	

01 31 13.90 Performance Bond										
0010	**PERFORMANCE BOND**	R013113-80								
0020	For buildings, minimum				Job				.60%	.60%
0100	Maximum				"				2.50%	2.50%

01 32 Construction Progress Documentation

01 32 33 – Photographic Documentation

01 32 33.50 Photographs		Crew	Daily Output	Labor-Hours	Unit	Material	2018 Bare Costs Labor	Equipment	Total	Total Incl O&P
0010	**PHOTOGRAPHS**									
0020	8" x 10", 4 shots, 2 prints ea., std. mounting				Set	540			540	595
0100	Hinged linen mounts					545			545	600
0200	8" x 10", 4 shots, 2 prints each, in color					500			500	550
0300	For I.D. slugs, add to all above					5.05			5.05	5.55
1500	Time lapse equipment, camera and projector, buy				Ea.	2,700			2,700	2,975
1550	Rent per month				"	1,275			1,275	1,400
1700	Cameraman and processing, black & white				Day	1,275			1,275	1,400
1720	Color				"	1,475			1,475	1,625

01 41 Regulatory Requirements

01 41 26 – Permit Requirements

01 41 26.50 Permits		Crew	Daily Output	Labor-Hours	Unit	Material	2018 Bare Costs Labor	Equipment	Total	Total Incl O&P
0010	**PERMITS**									
0020	Rule of thumb, most cities, minimum				Job				.50%	.50%
0100	Maximum				"				2%	2%

For customer support on your Electrical Costs with RSMeans data, call 800.448.8182.

11

01 51 13 – Temporary Electricity

01 51 13.50 Temporary Power Equip (Pro-Rated Per Job)	Crew	Daily Output	Labor-Hours	Unit	Material	2018 Bare Costs Labor	Equipment	Total	Total Incl O&P
0010 **TEMPORARY POWER EQUIP (PRO-RATED PER JOB)** R015113-65									
0020 Service, overhead feed, 3 use									
0030 100 Amp	1 Elec	1.25	6.400	Ea.	395	370		765	990
0040 200 Amp		1	8		470	465		935	1,225
0050 400 Amp		.75	10.667		830	620		1,450	1,850
0060 600 Amp		.50	16		1,200	930		2,130	2,700
0100 Underground feed, 3 use									
0110 100 Amp	1 Elec	2	4	Ea.	375	233		608	760
0120 200 Amp		1.15	6.957		480	405		885	1,125
0130 400 Amp		1	8		840	465		1,305	1,625
0140 600 Amp		.75	10.667		1,025	620		1,645	2,075
0150 800 Amp		.50	16		1,500	930		2,430	3,050
0160 1000 Amp		.35	22.857		1,675	1,325		3,000	3,825
0170 1200 Amp		.25	32		1,850	1,850		3,700	4,825
0180 2000 Amp		.20	40		2,250	2,325		4,575	5,950
0200 Transformers, 3 use									
0210 30 kVA	1 Elec	1	8	Ea.	1,775	465		2,240	2,675
0220 45 kVA		.75	10.667		2,125	620		2,745	3,275
0230 75 kVA		.50	16		3,600	930		4,530	5,375
0240 112.5 kVA		.40	20		3,875	1,175		5,050	6,000
0250 Feeder, PVC, CU wire in trench									
0260 60 Amp	1 Elec	96	.083	L.F.	2.62	4.85		7.47	10.15
0270 100 Amp		85	.094		5.05	5.50		10.55	13.75
0280 200 Amp		59	.136		12.60	7.90		20.50	25.50
0290 400 Amp		42	.190		32.50	11.10		43.60	52
0300 Feeder, PVC, aluminum wire in trench									
0310 60 Amp	1 Elec	96	.083	L.F.	2.70	4.85		7.55	10.20
0320 100 Amp		85	.094		3.05	5.50		8.55	11.55
0330 200 Amp		59	.136		7.05	7.90		14.95	19.60
0340 400 Amp		42	.190		15	11.10		26.10	33
0350 Feeder, EMT, CU wire									
0360 60 Amp	1 Elec	90	.089	L.F.	3.22	5.15		8.37	11.30
0370 100 Amp		80	.100		5.95	5.80		11.75	15.25
0380 200 Amp		60	.133		13.20	7.75		20.95	26
0390 400 Amp		35	.229		34	13.30		47.30	57
0400 Feeder, EMT, aluminum wire									
0410 60 Amp	1 Elec	90	.089	L.F.	4.39	5.15		9.54	12.60
0420 100 Amp		80	.100		4.57	5.80		10.37	13.70
0430 200 Amp		60	.133		9.10	7.75		16.85	21.50
0440 400 Amp		35	.229		17.10	13.30		30.40	38.50
0500 Equipment, 3 use									
0510 Spider box, 50 Amp	1 Elec	8	1	Ea.	725	58		783	885
0520 Lighting cord, 100'		8	1		121	58		179	220
0530 Light stanchion		8	1		77	58		135	172
0540 Temporary cords, 100', 3 use									
0550 Feeder cord, 50 Amp	1 Elec	16	.500	Ea.	263	29		292	335
0560 Feeder cord, 100 Amp		12	.667		1,075	39		1,114	1,225
0570 Tap cord, 50 Amp		12	.667		605	39		644	725
0580 Tap cord, 100 Amp		6	1.333		1,075	77.50		1,152.50	1,300
0590 Temporary cords, 50', 3 use									
0600 Feeder cord, 50 Amp	1 Elec	16	.500	Ea.	293	29		322	365
0610 Feeder cord, 100 Amp		12	.667		600	39		639	720
0620 Tap cord, 50 Amp		12	.667		293	39		332	380

For customer support on your Electrical Costs with RSMeans data, call 800.448.8182.

01 51 Temporary Utilities

01 51 13 – Temporary Electricity

01 51 13.50 Temporary Power Equip (Pro-Rated Per Job)	Crew	Daily Output	Labor-Hours	Unit	Material	2018 Bare Costs Labor	Equipment	Total	Total Incl O&P
0630	Tap cord, 100 Amp	1 Elec	6	1.333	Ea.	600	77.50		677.50	775
0700	Connections									
0710	Compressor or pump									
0720	30 Amp	1 Elec	7	1.143	Ea.	15.65	66.50		82.15	117
0730	60 Amp		5.30	1.509		28.50	88		116.50	162
0740	100 Amp		4	2		64	116		180	245
0750	Tower crane									
0760	60 Amp	1 Elec	4.50	1.778	Ea.	28.50	103		131.50	186
0770	100 Amp	"	3	2.667	"	64	155		219	305
0780	Manlift									
0790	Single	1 Elec	3	2.667	Ea.	33	155		188	268
0800	Double	"	2	4	"	68	233		301	425
0810	Welder with disconnect									
0820	50 Amp	1 Elec	5	1.600	Ea.	180	93		273	335
0830	100 Amp		3.80	2.105		315	123		438	530
0840	200 Amp		2.50	3.200		570	186		756	910
0850	400 Amp		1	8		1,325	465		1,790	2,150
0860	Office trailer									
0870	60 Amp	1 Elec	4.50	1.778	Ea.	50	103		153	210
0880	100 Amp		3	2.667		91	155		246	330
0890	200 Amp		2	4		295	233		528	675
0900	Lamping, add per floor				Total				625	690
0910	Maintenance, total temp. power cost				Job				5%	5%

01 51 13.80 Temporary Utilities

	01 51 13.80 Temporary Utilities	Crew	Daily Output	Labor-Hours	Unit	Material	2018 Bare Costs Labor	Equipment	Total	Total Incl O&P
0010	**TEMPORARY UTILITIES**									
0350	Lighting, lamps, wiring, outlets, 40,000 S.F. building, 8 strings	1 Elec	34	.235	CSF Flr	5.60	13.70		19.30	26.50
0360	16 strings	"	17	.471		11.20	27.50		38.70	53.50
0400	Power for temp lighting only, 6.6 KWH, per month								.92	1.01
0430	11.8 KWH, per month								1.65	1.82
0450	23.6 KWH, per month								3.30	3.63
0600	Power for job duration incl. elevator, etc., minimum								47	51.50
0650	Maximum								110	121
1000	Toilet, portable, see Equip. Rental 01 54 33 in Reference Section									

01 52 Construction Facilities

01 52 13 – Field Offices and Sheds

01 52 13.20 Office and Storage Space

	01 52 13.20 Office and Storage Space	Crew	Daily Output	Labor-Hours	Unit	Material	2018 Bare Costs Labor	Equipment	Total	Total Incl O&P
0010	**OFFICE AND STORAGE SPACE**									
0020	Office trailer, furnished, no hookups, 20' x 8', buy	2 Skwk	1	16	Ea.	8,900	840		9,740	11,100
0250	Rent per month					198			198	218
0300	32' x 8', buy	2 Skwk	.70	22.857		14,200	1,200		15,400	17,400
0350	Rent per month					247			247	272
0400	50' x 10', buy	2 Skwk	.60	26.667		29,300	1,400		30,700	34,400
0450	Rent per month					355			355	395
0500	50' x 12', buy	2 Skwk	.50	32		25,900	1,675		27,575	31,100
0550	Rent per month					450			450	495
0700	For air conditioning, rent per month, add					50			50	55
0800	For delivery, add per mile				Mile	12			12	13.20
0890	Delivery each way				Ea.	2,725			2,725	3,000
0900	Bunk house trailer, 8' x 40' duplex dorm with kitchen, no hookups, buy	2 Carp	1	16		87,000	810		87,810	96,500
0910	9 man with kitchen and bath, no hookups, buy		1	16		89,000	810		89,810	99,000

For customer support on your Electrical Costs with RSMeans data, call 800.448.8182.

13

01 52 Construction Facilities

01 52 13 – Field Offices and Sheds

01 52 13.20 Office and Storage Space	Crew	Daily Output	Labor-Hours	Unit	Material	2018 Bare Costs Labor	Equipment	Total	Total Incl O&P
0920	18 man sleeper with bath, no hookups, buy	2 Carp	1	16	Ea.	96,000	810		96,810	106,500
1000	Portable buildings, prefab, on skids, economy, 8' x 8'		265	.060	S.F.	25	3.06		28.06	32
1100	Deluxe, 8' x 12'		150	.107	"	28	5.40		33.40	39.50
1200	Storage boxes, 20' x 8', buy	2 Skwk	1.80	8.889	Ea.	3,325	465		3,790	4,375
1250	Rent per month						84.50		84.50	93
1300	40' x 8', buy	2 Skwk	1.40	11.429		3,875	600		4,475	5,200
1350	Rent per month					110			110	121

01 52 13.40 Field Office Expense

		Crew	Daily Output	Labor-Hours	Unit	Material	Labor	Equipment	Total	Total Incl O&P
0010	**FIELD OFFICE EXPENSE**									
0100	Office equipment rental average				Month	205			205	226
0120	Office supplies, average				"	82			82	90
0125	Office trailer rental, see Section 01 52 13.20									
0140	Telephone bill; avg. bill/month incl. long dist.				Month	86			86	94.50
0160	Lights & HVAC				"	161			161	177

01 54 Construction Aids

01 54 16 – Temporary Hoists

01 54 16.50 Weekly Forklift Crew

		Crew	Daily Output	Labor-Hours	Unit	Material	Labor	Equipment	Total	Total Incl O&P
0010	**WEEKLY FORKLIFT CREW**									
0100	All-terrain forklift, 45' lift, 35' reach, 9000 lb. capacity	A-3P	.20	40	Week		2,050	2,425	4,475	5,775

01 54 19 – Temporary Cranes

01 54 19.50 Daily Crane Crews

		Crew	Daily Output	Labor-Hours	Unit	Material	Labor	Equipment	Total	Total Incl O&P
0010	**DAILY CRANE CREWS** for small jobs, portal to portal R015433-15									
0100	12-ton truck-mounted hydraulic crane	A-3H	1	8	Day		450	630	1,080	1,375
0200	25-ton	A-3I	1	8			450	760	1,210	1,500
0300	40-ton	A-3J	1	8			450	1,325	1,775	2,125
0400	55-ton	A-3K	1	16			840	1,500	2,340	2,925
0500	80-ton	A-3L	1	16			840	2,250	3,090	3,750
0900	If crane is needed on a Saturday, Sunday or Holiday									
0910	At time-and-a-half, add				Day		50%			
0920	At double time, add				"		100%			

01 54 19.60 Monthly Tower Crane Crew

		Crew	Daily Output	Labor-Hours	Unit	Material	Labor	Equipment	Total	Total Incl O&P
0010	**MONTHLY TOWER CRANE CREW**, excludes concrete footing									
0100	Static tower crane, 130' high, 106' jib, 6200 lb. capacity	A-3N	.05	176	Month		9,875	26,000	35,875	43,500

01 54 23 – Temporary Scaffolding and Platforms

01 54 23.70 Scaffolding

		Crew	Daily Output	Labor-Hours	Unit	Material	Labor	Equipment	Total	Total Incl O&P
0010	**SCAFFOLDING** R015423-10									
0015	Steel tube, regular, no plank, labor only to erect & dismantle									
0090	Building exterior, wall face, 1 to 5 stories, 6'-4" x 5' frames	3 Carp	8	3	C.S.F.		152		152	232
0200	6 to 12 stories	4 Carp	8	4			203		203	310
0301	13 to 20 stories	5 Clab	8	5			199		199	305
0460	Building interior, wall face area, up to 16' high	3 Carp	12	2			101		101	154
0560	16' to 40' high		10	2.400			122		122	185
0800	Building interior floor area, up to 30' high		150	.160	C.C.F.		8.10		8.10	12.35
0900	Over 30' high	4 Carp	160	.200	"		10.15		10.15	15.45
0906	Complete system for face of walls, no plank, material only rent/mo				C.S.F.	33			33	36
0908	Interior spaces, no plank, material only rent/mo				C.C.F.	3.77			3.77	4.14
0910	Steel tubular, heavy duty shoring, buy									
0920	Frames 5' high 2' wide				Ea.	99.50			99.50	109

14

01 54 23.70 Scaffolding	Crew	Daily Output	Labor-Hours	Unit	Material	2018 Bare Costs Labor	Equipment	Total	Total Incl O&P
0925 5' high 4' wide				Ea.	114			114	125
0930 6' high 2' wide					115			115	127
0935 6' high 4' wide					126			126	139
0940 Accessories									
0945 Cross braces				Ea.	20			20	22
0950 U-head, 8" x 8"					22			22	24
0955 J-head, 4" x 8"					16			16	17.60
0960 Base plate, 8" x 8"					17.70			17.70	19.45
0965 Leveling jack					39			39	42.50
1000 Steel tubular, regular, buy									
1100 Frames 3' high 5' wide				Ea.	91			91	100
1150 5' high 5' wide					107			107	117
1200 6'-4" high 5' wide					99.50			99.50	109
1350 7'-6" high 6' wide					170			170	187
1500 Accessories, cross braces					16			16	17.60
1550 Guardrail post					20			20	22
1600 Guardrail 7' section					8.05			8.05	8.85
1650 Screw jacks & plates					25.50			25.50	28
1700 Sidearm brackets					22.50			22.50	24.50
1750 8" casters					37			37	40.50
1800 Plank 2" x 10" x 16'-0"					66			66	72.50
1900 Stairway section					283			283	310
1910 Stairway starter bar					32			32	35
1920 Stairway inside handrail					53			53	58
1930 Stairway outside handrail					84			84	92.50
1940 Walk-thru frame guardrail					41.50			41.50	46
2000 Steel tubular, regular, rent/mo.									
2100 Frames 3' high 5' wide				Ea.	4.38			4.38	4.82
2150 5' high 5' wide					4.38			4.38	4.82
2200 6'-4" high 5' wide					5.25			5.25	5.80
2250 7'-6" high 6' wide					9.75			9.75	10.75
2500 Accessories, cross braces					.88			.88	.97
2550 Guardrail post					.88			.88	.97
2600 Guardrail 7' section					.88			.88	.97
2650 Screw jacks & plates					1.75			1.75	1.93
2700 Sidearm brackets					1.75			1.75	1.93
2750 8" casters					7			7	7.70
2800 Outrigger for rolling tower					2.63			2.63	2.89
2850 Plank 2" x 10" x 16'-0"					9.75			9.75	10.75
2900 Stairway section					32.50			32.50	36
2940 Walk-thru frame guardrail					2.19			2.19	2.41
3000 Steel tubular, heavy duty shoring, rent/mo.									
3250 5' high 2' & 4' wide				Ea.	8.30			8.30	9.15
3300 6' high 2' & 4' wide					8.30			8.30	9.15
3500 Accessories, cross braces					.88			.88	.97
3600 U-head, 8" x 8"					2.44			2.44	2.68
3650 J-head, 4" x 8"					2.44			2.44	2.68
3700 Base plate, 8" x 8"					.88			.88	.97
3750 Leveling jack					2.44			2.44	2.68
5700 Planks, 2" x 10" x 16'-0", labor only to erect & remove to 50' H	3 Carp	72	.333			16.90		16.90	25.50
5800 Over 50' high	4 Carp	80	.400			20.50		20.50	31

For customer support on your Electrical Costs with RSMeans data, call 800.448.8182.

15

01 54 23 – Temporary Scaffolding and Platforms

01 54 23.75 Scaffolding Specialties	Crew	Daily Output	Labor-Hours	Unit	Material	2018 Bare Costs Labor	Equipment	Total	Total Incl O&P
0010 **SCAFFOLDING SPECIALTIES**									
1500 Sidewalk bridge using tubular steel scaffold frames including									
1510 planking (material cost is rent/month)	3 Carp	45	.533	L.F.	8.25	27		35.25	50
1600 For 2 uses per month, deduct from all above					50%				
1700 For 1 use every 2 months, add to all above					100%				
1900 Catwalks, 20" wide, no guardrails, 7' span, buy				Ea.	150			150	165
2000 10' span, buy					211			211	232
3720 Putlog, standard, 8' span, with hangers, buy					75.50			75.50	83
3750 12' span, buy					99.50			99.50	109
3760 Trussed type, 16' span, buy					255			255	281
3790 22' span, buy					275			275	300
3795 Rent per month					32			32	35
3800 Rolling ladders with handrails, 30" wide, buy, 2 step					291			291	320
4000 7 step					815			815	900
4050 10 step					1,175			1,175	1,275
4100 Rolling towers, buy, 5' wide, 7' long, 10' high					1,300			1,300	1,425
4200 For additional 5' high sections, to buy					245			245	270
4300 Complete incl. wheels, railings, outriggers,									
4350 21' high, to buy				Ea.	2,225			2,225	2,450
4400 Rent/month = 5% of purchase cost				"	196			196	216

01 54 26 – Temporary Swing Staging

01 54 26.50 Swing Staging	Crew	Daily Output	Labor-Hours	Unit	Material	2018 Bare Costs Labor	Equipment	Total	Total Incl O&P
0010 **SWING STAGING**, 500 lb. cap., 2' wide to 24' long, hand operated									
0020 steel cable type, with 60' cables, buy				Ea.	5,725			5,725	6,300
0030 Rent per month				"	570			570	630
2200 Move swing staging (setup and remove)	E-4	2	16	Move		880	49.50	929.50	1,500

01 54 36 – Equipment Mobilization

01 54 36.50 Mobilization	Crew	Daily Output	Labor-Hours	Unit	Material	2018 Bare Costs Labor	Equipment	Total	Total Incl O&P
0010 **MOBILIZATION** (Use line item again for demobilization) R015436-50									
0015 Up to 25 mi. haul dist. (50 mi. RT for mob/demob crew)									
1200 Small equipment, placed in rear of, or towed by pickup truck	A-3A	4	2	Ea.		103	31.50	134.50	190
1300 Equipment hauled on 3-ton capacity towed trailer	A-3Q	2.67	3			154	57	211	295
1400 20-ton capacity	B-34U	2	8			390	212	602	820
1500 40-ton capacity	B-34N	2	8			400	330	730	960
1600 50-ton capacity	B-34V	1	24			1,225	910	2,135	2,850
1700 Crane, truck-mounted, up to 75 ton (driver only)	1 Eqhv	4	2			112		112	169
1800 Over 75 ton (with chase vehicle)	A-3E	2.50	6.400			325	50.50	375.50	545
2400 Crane, large lattice boom, requiring assembly	B-34W	.50	144			7,025	6,725	13,750	18,000
2500 For each additional 5 miles haul distance, add						10%	10%		
3000 For large pieces of equipment, allow for assembly/knockdown									
3100 For mob/demob of micro-tunneling equip, see Section 33 05 23.19									

01 55 Vehicular Access and Parking

01 55 23 – Temporary Roads

01 55 23.50 Roads and Sidewalks	Crew	Daily Output	Labor-Hours	Unit	Material	2018 Bare Costs Labor	Equipment	Total	Total Incl O&P
0010 **ROADS AND SIDEWALKS** Temporary									
0050 Roads, gravel fill, no surfacing, 4" gravel depth	B-14	715	.067	S.Y.	3.36	2.83	.44	6.63	8.50
0100 8" gravel depth	"	615	.078	"	6.70	3.29	.51	10.50	12.95
1000 Ramp, 3/4" plywood on 2" x 6" joists, 16" OC	2 Carp	300	.053	S.F.	1.56	2.70		4.26	5.85
1100 On 2" x 10" joists, 16" OC	"	275	.058	"	2.31	2.95		5.26	7.05

01 56 Temporary Barriers and Enclosures

01 56 13 – Temporary Air Barriers

01 56 13.90 Winter Protection

01 56 13.90 Winter Protection	Crew	Daily Output	Labor-Hours	Unit	Material	2018 Bare Costs Labor	Equipment	Total	Total Incl O&P
0010 **WINTER PROTECTION**									
0100 Framing to close openings	2 Clab	500	.032	S.F.	.45	1.28		1.73	2.43
0200 Tarpaulins hung over scaffolding, 8 uses, not incl. scaffolding		1500	.011		.25	.43		.68	.93
0300 Prefab fiberglass panels, steel frame, 8 uses		1200	.013		2.52	.53		3.05	3.58

01 56 16 – Temporary Dust Barriers

01 56 16.10 Dust Barriers, Temporary

01 56 16.10 Dust Barriers, Temporary	Crew	Daily Output	Labor-Hours	Unit	Material	2018 Bare Costs Labor	Equipment	Total	Total Incl O&P
0010 **DUST BARRIERS, TEMPORARY**									
0020 Spring loaded telescoping pole & head, to 12', erect and dismantle	1 Clab	240	.033	Ea.		1.33		1.33	2.02
0025 Cost per day (based upon 250 days)				Day	.29			.29	.31
0030 To 21', erect and dismantle	1 Clab	240	.033	Ea.		1.33		1.33	2.02
0035 Cost per day (based upon 250 days)				Day	.58			.58	.63
0040 Accessories, caution tape reel, erect and dismantle	1 Clab	480	.017	Ea.		.66		.66	1.01
0045 Cost per day (based upon 250 days)				Day	.34			.34	.37
0060 Foam rail and connector, erect and dismantle	1 Clab	240	.033	Ea.		1.33		1.33	2.02
0065 Cost per day (based upon 250 days)				Day	.12			.12	.13
0070 Caution tape	1 Clab	384	.021	C.L.F.	3.13	.83		3.96	4.71
0080 Zipper, standard duty		60	.133	Ea.	7.30	5.30		12.60	16.10
0090 Heavy duty		48	.167	"	9.75	6.65		16.40	21
0100 Polyethylene sheet, 4 mil		37	.216	Sq.	2.58	8.60		11.18	15.95
0110 6 mil		37	.216	"	3.73	8.60		12.33	17.20
1000 Dust partition, 6 mil polyethylene, 1" x 3" frame	2 Carp	2000	.008	S.F.	.32	.41		.73	.97
1080 2" x 4" frame	"	2000	.008	"	.35	.41		.76	1
1085 Negative air machine, 1800 CFM				Ea.	860			860	950
1090 Adhesive strip application, 2" width	1 Clab	192	.042	C.L.F.	6.55	1.66		8.21	9.75

01 56 23 – Temporary Barricades

01 56 23.10 Barricades

01 56 23.10 Barricades	Crew	Daily Output	Labor-Hours	Unit	Material	2018 Bare Costs Labor	Equipment	Total	Total Incl O&P
0010 **BARRICADES**									
0020 5' high, 3 rail @ 2" x 8", fixed	2 Carp	20	.800	L.F.	6.05	40.50		46.55	68.50
0150 Movable		30	.533		5	27		32	46.50
1000 Guardrail, wooden, 3' high, 1" x 6" on 2" x 4" posts		200	.080		1.32	4.06		5.38	7.65
1100 2" x 6" on 4" x 4" posts		165	.097		2.54	4.92		7.46	10.30
1200 Portable metal with base pads, buy					14.25			14.25	15.65
1250 Typical installation, assume 10 reuses	2 Carp	600	.027		2.30	1.35		3.65	4.59
1300 Barricade tape, polyethylene, 7 mil, 3" wide x 500' long roll				Ea.	25			25	27.50
3000 Detour signs, set up and remove									
3010 Reflective aluminum, MUTCD, 24" x 24", post mounted	1 Clab	20	.400	Ea.	2.54	15.95		18.49	27.50
4000 Roof edge portable barrier stands and warning flags, 50 uses	1 Rohe	9100	.001	L.F.	.06	.03		.09	.12
4010 100 uses	"	9100	.001	"	.03	.03		.06	.08

For customer support on your Electrical Costs with RSMeans data, call 800.448.8182.

17

01 56 Temporary Barriers and Enclosures

01 56 26 – Temporary Fencing

01 56 26.50 Temporary Fencing	Crew	Daily Output	Labor-Hours	Unit	Material	2018 Bare Costs Labor	Equipment	Total	Total Incl O&P
0010 **TEMPORARY FENCING**									
0020 Chain link, 11 ga., 4' high	2 Clab	400	.040	L.F.	1.63	1.59		3.22	4.22
0100 6' high		300	.053		4.55	2.13		6.68	8.25
0200 Rented chain link, 6' high, to 1000' (up to 12 mo.)		400	.040		2.99	1.59		4.58	5.70
0250 Over 1000' (up to 12 mo.)		300	.053		3.19	2.13		5.32	6.75
0350 Plywood, painted, 2" x 4" frame, 4' high	A-4	135	.178		6.35	8.55		14.90	19.90
0400 4" x 4" frame, 8' high	"	110	.218		12.10	10.45		22.55	29
0500 Wire mesh on 4" x 4" posts, 4' high	2 Carp	100	.160		10.35	8.10		18.45	24
0550 8' high	"	80	.200		15.60	10.15		25.75	32.50

01 56 29 – Temporary Protective Walkways

01 56 29.50 Protection	Crew	Daily Output	Labor-Hours	Unit	Material	2018 Bare Costs Labor	Equipment	Total	Total Incl O&P
0010 **PROTECTION**									
0020 Stair tread, 2" x 12" planks, 1 use	1 Carp	75	.107	Tread	5.35	5.40		10.75	14.10
0100 Exterior plywood, 1/2" thick, 1 use		65	.123		1.85	6.25		8.10	11.55
0200 3/4" thick, 1 use		60	.133		2.77	6.75		9.52	13.35
2200 Sidewalks, 2" x 12" planks, 2 uses		350	.023	S.F.	.89	1.16		2.05	2.74
2300 Exterior plywood, 2 uses, 1/2" thick		750	.011		.31	.54		.85	1.16
2400 5/8" thick		650	.012		.38	.62		1	1.37
2500 3/4" thick		600	.013		.46	.68		1.14	1.54

01 58 Project Identification

01 58 13 – Temporary Project Signage

01 58 13.50 Signs	Crew	Daily Output	Labor-Hours	Unit	Material	2018 Bare Costs Labor	Equipment	Total	Total Incl O&P
0010 **SIGNS**									
0020 High intensity reflectorized, no posts, buy				Ea.	25			25	27.50

01 66 Product Storage and Handling Requirements

01 66 19 – Material Handling

01 66 19.10 Material Handling	Crew	Daily Output	Labor-Hours	Unit	Material	2018 Bare Costs Labor	Equipment	Total	Total Incl O&P
0010 **MATERIAL HANDLING**									
0020 Above 2nd story, via stairs, per C.Y. of material per floor	2 Clab	145	.110	C.Y.		4.40		4.40	6.70
0030 Via elevator, per C.Y. of material		240	.067			2.66		2.66	4.05
0050 Distances greater than 200', per C.Y. of material per each addl 200'		300	.053			2.13		2.13	3.24

01 74 Cleaning and Waste Management

01 74 13 – Progress Cleaning

01 74 13.20 Cleaning Up	Crew	Daily Output	Labor-Hours	Unit	Material	2018 Bare Costs Labor	Equipment	Total	Total Incl O&P
0010 **CLEANING UP**									
0020 After job completion, allow, minimum				Job				.30%	.30%
0040 Maximum				"				1%	1%
0050 Cleanup of floor area, continuous, per day, during const.	A-5	24	.750	M.S.F.	2.41	30.50	1.97	34.88	51
0100 Final by GC at end of job	"	11.50	1.565	"	2.54	63	4.10	69.64	103

01 91 Commissioning

01 91 13 – General Commissioning Requirements

01 91 13.50 Building Commissioning	Crew	Daily Output	Labor-Hours	Unit	Material	2018 Bare Costs Labor	Equipment	Total	Total Incl O&P
0010 **BUILDING COMMISSIONING**									
0100 Systems operation and verification during turnover				%				.25%	.25%
0150 Including all systems subcontractors								.50%	.50%
0200 Systems design assistance, operation, verification and training								.50%	.50%
0250 Including all systems subcontractors								1%	1%

01 93 Facility Maintenance

01 93 13 – Facility Maintenance Procedures

01 93 13.16 Electrical Facilities Maintenance

		Crew	Daily Output	Labor-Hours	Unit	Material	2018 Bare Costs Labor	Equipment	Total	Total Incl O&P
0010	**ELECTRICAL FACILITIES MAINTENANCE**									
0700	Cathodic protection systems									
0720	Check and adjust reading on rectifier	1 Elec	20	.400	Ea.		23.50		23.50	35
0730	Check pipe to soil potential		20	.400			23.50		23.50	35
0740	Replace lead connection		4	2			116		116	174
0800	Control device, install		5.70	1.404			81.50		81.50	122
0810	Disassemble, clean and reinstall		7	1.143			66.50		66.50	99.50
0820	Replace		10.70	.748			43.50		43.50	65
0830	Trouble shoot		10	.800			46.50		46.50	69.50
0900	Demolition, for electrical demolition see Section 26 05 05.10									
1000	Distribution systems and equipment install or repair a breaker									
1010	In power panels up to 200 amps	1 Elec	7	1.143	Ea.		66.50		66.50	99.50
1020	Over 200 amps		2	4			233		233	350
1030	Reset breaker or replace fuse		20	.400			23.50		23.50	35
1100	Megger test MCC (each stack)		4	2			116		116	174
1110	MCC vacuum and clean (each stack)		5.30	1.509			88		88	131
2500	Remove and replace or maint., road fixture & lamp		3	2.667		1,225	155		1,380	1,575
2510	Fluorescent fixture		7	1.143		79	66.50		145.50	187
2515	Relamp (fluor.) facility area each tube, spot	1 Clab	24	.333		2.23	13.30		15.53	22.50
2516	Group		100	.080		2.23	3.19		5.42	7.30
2518	Fluorescent fixture, clean (area)		44	.182		.04	7.25		7.29	11.10
2520	Incandescent fixture	1 Elec	11	.727		56	42.50		98.50	125
2530	Lamp (incandescent or fluorescent)	1 Clab	60	.133		2.23	5.30		7.53	10.55
2535	Replace cord in socket lamp	1 Elec	13	.615		1.39	36		37.39	55
2540	Ballast electronic type for two tubes		8	1		43.50	58		101.50	135
2541	Starter		30	.267		.86	15.50		16.36	24
2545	Replace other lighting parts		11	.727		20	42.50		62.50	85.50
2550	Switch		11	.727		2.91	42.50		45.41	66.50
2555	Receptacle		11	.727		9.05	42.50		51.55	73.50
2560	Floodlight		4	2		231	116		347	430
2570	Christmas lighting, indoor, per string	1 Clab	16	.500			19.95		19.95	30.50
2580	Outdoor		13	.615			24.50		24.50	37.50
2590	Test battery operated emergency lights		40	.200			7.95		7.95	12.15
2600	Repair and replace component in communication system	1 Elec	6	1.333		53.50	77.50		131	175
2700	Repair misc. appliances (incl. clocks, vent fan, blower, etc.)	"	6	1.333			77.50		77.50	116
2710	Reset clocks & timers	1 Clab	50	.160			6.40		6.40	9.70
2720	Adjust time delay relays	1 Elec	16	.500			29		29	43.50
2730	Test specific gravity of lead-acid batteries	1 Clab	80	.100			3.99		3.99	6.05
3000	Motors and generators									
3020	Disassemble, clean and reinstall motor, up to 1/4 HP	1 Elec	4	2	Ea.		116		116	174
3030	Up to 3/4 HP		3	2.667			155		155	232
3040	Up to 10 HP		2	4			233		233	350

For customer support on your Electrical Costs with RSMeans data, call 800.448.8182.

19

01 93 13 – Facility Maintenance Procedures

01 93 13.16 Electrical Facilities Maintenance		Crew	Daily Output	Labor-Hours	Unit	Material	2018 Bare Costs		Total	Total Incl O&P
							Labor	Equipment		
3050	Replace part, up to 1/4 HP	1 Elec	6	1.333	Ea.		77.50		77.50	116
3060	Up to 3/4 HP		4	2			116		116	174
3070	Up to 10 HP		3	2.667			155		155	232
3080	Megger test motor windings		5.33	1.501			87.50		87.50	131
3082	Motor vibration check		16	.500			29		29	43.50
3084	Oil motor bearings		25	.320			18.60		18.60	28
3086	Run test emergency generator for 30 minutes		11	.727			42.50		42.50	63.50
3090	Rewind motor, up to 1/4 HP		3	2.667			155		155	232
3100	Up to 3/4 HP		2	4			233		233	350
3110	Up to 10 HP		1.50	5.333			310		310	465
3150	Generator, repair or replace part		4	2			116		116	174
3160	Repair DC generator		2	4			233		233	350
4000	Stub pole, install or remove		3	2.667			155		155	232
4500	Transformer maintenance up to 15 kVA		2.70	2.963			172		172	258

For customer support on your Electrical Costs with RSMeans data, call 800.448.8182.

Estimating Tips
02 30 00 Subsurface Investigation

In preparing estimates on structures involving earthwork or foundations, all information concerning soil characteristics should be obtained. Look particularly for hazardous waste, evidence of prior dumping of debris, and previous stream beds.

02 40 00 Demolition and Structure Moving

The costs shown for selective demolition do not include rubbish handling or disposal. These items should be estimated separately using RSMeans data or other sources.

- Historic preservation often requires that the contractor remove materials from the existing structure, rehab them, and replace them. The estimator must be aware of any related measures and precautions that must be taken when doing selective demolition and cutting and patching. Requirements may include special handling and storage, as well as security.

- In addition to Subdivision 02 41 00, you can find selective demolition items in each division. Example: Roofing demolition is in Division 7.
- Absent of any other specific reference, an approximate demolish-in-place cost can be obtained by halving the new-install labor cost. To remove for reuse, allow the entire new-install labor figure.

02 40 00 Building Deconstruction

This section provides costs for the careful dismantling and recycling of most low-rise building materials.

02 50 00 Containment of Hazardous Waste

This section addresses on-site hazardous waste disposal costs.

02 80 00 Hazardous Material Disposal/Remediation

This subdivision includes information on hazardous waste handling, asbestos remediation, lead remediation, and mold remediation. See reference numbers

R028213-20 and R028319-60 for further guidance in using these unit price lines.

02 90 00 Monitoring Chemical Sampling, Testing Analysis

This section provides costs for on-site sampling and testing hazardous waste.

Reference Numbers

Reference numbers are shown at the beginning of some major classifications. These numbers refer to related items in the Reference Section. The reference information may be an estimating procedure, an alternate pricing method, or technical information.

Note: Not all subdivisions listed here necessarily appear. ■

02 41 Demolition

02 41 13 – Selective Site Demolition

		Daily Output	Labor-Hours	Unit	Material	2018 Bare Costs			Total Incl O&P
						Labor	Equipment	Total	
02 41 13.17 Demolish, Remove Pavement and Curb	Crew								
0010 **DEMOLISH, REMOVE PAVEMENT AND CURB**	R024119-10								
5010 Pavement removal, bituminous roads, up to 3" thick	B-38	690	.058	S.Y.		2.63	1.63	4.26	5.75
5050 4"-6" thick		420	.095			4.32	2.68	7	9.50
5100 Bituminous driveways		640	.063			2.83	1.76	4.59	6.20
5200 Concrete to 6" thick, hydraulic hammer, mesh reinforced		255	.157			7.10	4.41	11.51	15.65
5300 Rod reinforced		200	.200			9.05	5.65	14.70	19.95
5400 Concrete, 7"-24" thick, plain		33	1.212	C.Y.		55	34	89	121
5500 Reinforced		24	1.667	"		75.50	47	122.50	167
5600 With hand held air equipment, bituminous, to 6" thick	B-39	1900	.025	S.F.		1.06	.12	1.18	1.76
5700 Concrete to 6" thick, no reinforcing		1600	.030			1.26	.15	1.41	2.08
5800 Mesh reinforced		1400	.034			1.44	.17	1.61	2.37
5900 Rod reinforced		765	.063			2.64	.31	2.95	4.35
02 41 13.23 Utility Line Removal									
0010 **UTILITY LINE REMOVAL**									
0015 No hauling, abandon catch basin or manhole	B-6	7	3.429	Ea.		150	44.50	194.50	276
0020 Remove existing catch basin or manhole, masonry		4	6			262	78	340	485
0030 Catch basin or manhole frames and covers, stored		13	1.846			80.50	24	104.50	149
0040 Remove and reset		7	3.429			150	44.50	194.50	276
02 41 13.30 Minor Site Demolition									
0010 **MINOR SITE DEMOLITION**	R024119-10								
4000 Sidewalk removal, bituminous, 2" thick	B-6	350	.069	S.Y.		2.99	.89	3.88	5.50
4010 2-1/2" thick		325	.074			3.23	.96	4.19	5.95
4050 Brick, set in mortar		185	.130			5.65	1.69	7.34	10.45
4100 Concrete, plain, 4"		160	.150			6.55	1.95	8.50	12.10
4110 Plain, 5"		140	.171			7.50	2.23	9.73	13.80
4120 Plain, 6"		120	.200			8.75	2.60	11.35	16.10
4200 Mesh reinforced, concrete, 4"		150	.160			7	2.08	9.08	12.90
4210 5" thick		131	.183			8	2.39	10.39	14.75
4220 6" thick		112	.214			9.35	2.79	12.14	17.25

02 41 19 – Selective Demolition

		Daily Output	Labor-Hours	Unit	Material	2018 Bare Costs			Total Incl O&P
						Labor	Equipment	Total	
02 41 19.13 Selective Building Demolition	Crew								
0010 **SELECTIVE BUILDING DEMOLITION**									
0020 Costs related to selective demolition of specific building components									
0025 are included under Common Work Results (XX 05)									
0030 in the component's appropriate division.									
02 41 19.16 Selective Demolition, Cutout									
0010 **SELECTIVE DEMOLITION, CUTOUT**	R024119-10								
0020 Concrete, elev. slab, light reinforcement, under 6 C.F.	B-9	65	.615	C.F.		25	3.60	28.60	41.50
0050 Light reinforcing, over 6 C.F.	"	75	.533	"		21.50	3.12	24.62	36
6000 Walls, interior, not including re-framing,									
6010 openings to 5 S.F.									
6100 Drywall to 5/8" thick	1 Clab	24	.333	Ea.		13.30		13.30	20
6200 Paneling to 3/4" thick		20	.400			15.95		15.95	24.50
6300 Plaster, on gypsum lath		20	.400			15.95		15.95	24.50
6340 On wire lath		14	.571			23		23	34.50
02 41 19.19 Selective Demolition									
0010 **SELECTIVE DEMOLITION,** Rubbish Handling	R024119-10								
0020 The following are to be added to the demolition prices									
0600 Dumpster, weekly rental, 1 dump/week, 6 C.Y. capacity (2 tons)				Week	415			415	455
0700 10 C.Y. capacity (3 tons)					480			480	530
0725 20 C.Y. capacity (5 tons)	R024119-20				565			565	625

02 41 Demolition

02 41 19 – Selective Demolition

02 41 19.19 Selective Demolition

		Crew	Daily Output	Labor-Hours	Unit	Material	2018 Bare Costs Labor	2018 Bare Costs Equipment	Total	Total Incl O&P
0800	30 C.Y. capacity (7 tons)				Week	730			730	800
0840	40 C.Y. capacity (10 tons)					775			775	850
2000	Load, haul, dump and return, 0'-50' haul, hand carried	2 Clab	24	.667	C.Y.		26.50		26.50	40.50
2005	Wheeled		37	.432			17.25		17.25	26.50
2040	0'-100' haul, hand carried		16.50	.970			38.50		38.50	59
2045	Wheeled		25	.640			25.50		25.50	39
2050	Forklift	A-3R	25	.320			16.40	5.95	22.35	31.50
2080	Haul and return, add per each extra 100' haul, hand carried	2 Clab	35.50	.451			17.95		17.95	27.50
2085	Wheeled		54	.296			11.80		11.80	18
2120	For travel in elevators, up to 10 floors, add		140	.114			4.55		4.55	6.95
2130	0'-50' haul, incl. up to 5 riser stairs, hand carried		23	.696			27.50		27.50	42
2135	Wheeled		35	.457			18.20		18.20	28
2140	6-10 riser stairs, hand carried		22	.727			29		29	44
2145	Wheeled		34	.471			18.75		18.75	28.50
2150	11-20 riser stairs, hand carried		20	.800			32		32	48.50
2155	Wheeled		31	.516			20.50		20.50	31.50
2160	21-40 riser stairs, hand carried		16	1			40		40	60.50
2165	Wheeled		24	.667			26.50		26.50	40.50
2170	0-100' haul, incl. 5 riser stairs, hand carried		15	1.067			42.50		42.50	65
2175	Wheeled		23	.696			27.50		27.50	42
2180	6-10 riser stairs, hand carried		14	1.143			45.50		45.50	69.50
2185	Wheeled		21	.762			30.50		30.50	46.50
2190	11-20 riser stairs, hand carried		12	1.333			53		53	81
2195	Wheeled		18	.889			35.50		35.50	54
2200	21-40 riser stairs, hand carried		8	2			79.50		79.50	121
2205	Wheeled		12	1.333			53		53	81
2210	Haul and return, add per each extra 100' haul, hand carried		35.50	.451			17.95		17.95	27.50
2215	Wheeled		54	.296			11.80		11.80	18
2220	For each additional flight of stairs, up to 5 risers, add		550	.029	Flight		1.16		1.16	1.77
2225	6-10 risers, add		275	.058			2.32		2.32	3.53
2230	11-20 risers, add		138	.116			4.62		4.62	7.05
2235	21-40 risers, add		69	.232			9.25		9.25	14.10
3000	Loading & trucking, including 2 mile haul, chute loaded	B-16	45	.711	C.Y.		30	12.05	42.05	58.50
3040	Hand loading truck, 50' haul	"	48	.667			28	11.30	39.30	55
3080	Machine loading truck	B-17	120	.267			11.80	5.40	17.20	24
5000	Haul, per mile, up to 8 C.Y. truck	B-34B	1165	.007			.32	.47	.79	.99
5100	Over 8 C.Y. truck	"	1550	.005			.24	.35	.59	.75

02 41 19.20 Selective Demolition, Dump Charges

		Crew	Daily Output	Labor-Hours	Unit	Material	Labor	Equipment	Total	Total Incl O&P
0010	**SELECTIVE DEMOLITION, DUMP CHARGES** R024119-10									
0020	Dump charges, typical urban city, tipping fees only									
0100	Building construction materials				Ton	74			74	81
0200	Trees, brush, lumber					63			63	69.50
0300	Rubbish only					63			63	69.50
0500	Reclamation station, usual charge					74			74	81

02 41 19.27 Selective Demolition, Torch Cutting

		Crew	Daily Output	Labor-Hours	Unit	Material	Labor	Equipment	Total	Total Incl O&P
0010	**SELECTIVE DEMOLITION, TORCH CUTTING** R024119-10									
0020	Steel, 1" thick plate	E-25	333	.024	L.F.	.87	1.36	.04	2.27	3.22
0040	1" diameter bar	"	600	.013	Ea.	.15	.76	.02	.93	1.41
1000	Oxygen lance cutting, reinforced concrete walls									
1040	12"-16" thick walls	1 Clab	10	.800	L.F.		32		32	48.50
1080	24" thick walls	"	6	1.333	"		53		53	81

For customer support on your Electrical Costs with RSMeans data, call 800.448.8182.

23

02 82 13 – Asbestos Abatement

02 82 13.47 Asbestos Waste Pkg., Handling, and Disp.	Crew	Daily Output	Labor-Hours	Unit	Material	2018 Bare Costs Labor	Equipment	Total	Total Incl O&P
0010 **ASBESTOS WASTE PACKAGING, HANDLING, AND DISPOSAL**									
0100 Collect and bag bulk material, 3 C.F. bags, by hand	A-9	400	.160	Ea.	.86	9		9.86	14.95
0200 Large production vacuum loader	A-12	880	.073		1	4.10	.82	5.92	8.35
1000 Double bag and decontaminate	A-9	960	.067		.86	3.76		4.62	6.80
2000 Containerize bagged material in drums, per 3 C.F. drum	"	800	.080		18.70	4.51		23.21	27.50
3000 Cart bags 50' to dumpster	2 Asbe	400	.040			2.25		2.25	3.50
5000 Disposal charges, not including haul, minimum				C.Y.				61	67
5020 Maximum				"				355	395
9000 For type B (supplied air) respirator equipment, add				%				10%	10%

For customer support on your Electrical Costs with RSMeans data, call 800.448.8182.

Estimating Tips
General

- Carefully check all the plans and specifications. Concrete often appears on drawings other than structural drawings, including mechanical and electrical drawings for equipment pads. The cost of cutting and patching is often difficult to estimate. See Subdivision 03 81 for Concrete Cutting, Subdivision 02 41 19.16 for Cutout Demolition, Subdivision 03 05 05.10 for Concrete Demolition, and Subdivision 02 41 19.19 for Rubbish Handling (handling, loading, and hauling of debris).

- Always obtain concrete prices from suppliers near the job site. A volume discount can often be negotiated, depending upon competition in the area. Remember to add for waste, particularly for slabs and footings on grade.

03 10 00 Concrete Forming and Accessories

- A primary cost for concrete construction is forming. Most jobs today are constructed with prefabricated forms. The selection of the forms best suited for the job and the total square feet of forms required for efficient concrete forming and placing are key elements in estimating concrete construction. Enough forms must be available for erection to make efficient use of the concrete placing equipment and crew.

- Concrete accessories for forming and placing depend upon the systems used. Study the plans and specifications to ensure that all special accessory requirements have been included in the cost estimate, such as anchor bolts, inserts, and hangers.

- Included within costs for forms-in-place are all necessary bracing and shoring.

03 20 00 Concrete Reinforcing

- Ascertain that the reinforcing steel supplier has included all accessories, cutting, bending, and an allowance for lapping, splicing, and waste. A good rule of thumb is 10% for lapping, splicing, and waste. Also, 10% waste should be allowed for welded wire fabric.

- The unit price items in the subdivisions for Reinforcing In Place, Glass Fiber Reinforcing, and Welded Wire Fabric include the labor to install accessories such as beam and slab bolsters, high chairs, and bar ties and tie wire. The material cost for these accessories is not included; they may be obtained from the Accessories Subdivisions.

03 30 00 Cast-In-Place Concrete

- When estimating structural concrete, pay particular attention to requirements for concrete additives, curing methods, and surface treatments. Special consideration for climate, hot or cold, must be included in your estimate. Be sure to include requirements for concrete placing equipment and concrete finishing.

- For accurate concrete estimating, the estimator must consider each of the following major components individually: forms, reinforcing steel, ready-mix concrete, placement of the concrete, and finishing of the top surface. For faster estimating, Subdivision 03 30 53.40 for Concrete-In-Place can be used; here, various items of concrete work are presented that include the costs of all five major components (unless specifically stated otherwise).

03 40 00 Precast Concrete
03 50 00 Cast Decks and Underlayment

- The cost of hauling precast concrete structural members is often an important factor. For this reason, it is important to get a quote from the nearest supplier. It may become economically feasible to set up precasting beds on the site if the hauling costs are prohibitive.

Reference Numbers

Reference numbers are shown at the beginning of some major classifications. These numbers refer to related items in the Reference Section. The reference information may be an estimating procedure, an alternate pricing method, or technical information.

Note: Not all subdivisions listed here necessarily appear. ∎

03 01 Maintenance of Concrete

03 01 30 – Maintenance of Cast-In-Place Concrete

03 01 30.62 Concrete Patching	Crew	Daily Output	Labor-Hours	Unit	Material	2018 Bare Costs Labor	Equipment	Total	Total Incl O&P
0010 **CONCRETE PATCHING**									
0100　Floors, 1/4" thick, small areas, regular grout	1 Cefi	170	.047	S.F.	1.54	2.24		3.78	5
0150　　Epoxy grout	"	100	.080	"	8.65	3.80		12.45	15.20
2000　Walls, including chipping, cleaning and epoxy grout									
2100　　1/4" deep	1 Cefi	65	.123	S.F.	7.45	5.85		13.30	16.80
2150　　1/2" deep		50	.160		14.85	7.60		22.45	27.50
2200　　3/4" deep	↓	40	.200	↓	22.50	9.50		32	38.50

03 11 Concrete Forming

03 11 13 – Structural Cast-In-Place Concrete Forming

03 11 13.40 Forms In Place, Equipment Foundations

	Crew	Daily Output	Labor-Hours	Unit	Material	2018 Bare Costs Labor	Equipment	Total	Total Incl O&P
0010 **FORMS IN PLACE, EQUIPMENT FOUNDATIONS**									
0020　1 use	C-2	160	.300	SFCA	2.74	14.75		17.49	25.50
0050　2 use		190	.253		1.51	12.45		13.96	20.50
0100　3 use		200	.240		1.10	11.80		12.90	19.20
0150　4 use	↓	205	.234	↓	.90	11.55		12.45	18.55

03 11 13.45 Forms In Place, Footings

	Crew	Daily Output	Labor-Hours	Unit	Material	2018 Bare Costs Labor	Equipment	Total	Total Incl O&P
0010 **FORMS IN PLACE, FOOTINGS**									
0020　Continuous wall, plywood, 1 use	C-1	375	.085	SFCA	6.60	4.10		10.70	13.50
0050　　2 use		440	.073		3.62	3.49		7.11	9.30
0100　　3 use		470	.068		2.63	3.27		5.90	7.90
0150　　4 use		485	.066		2.15	3.17		5.32	7.20
5000　Spread footings, job-built lumber, 1 use		305	.105		2.10	5.05		7.15	9.95
5050　　2 use		371	.086		1.17	4.14		5.31	7.60
5100　　3 use		401	.080		.84	3.83		4.67	6.75
5150　　4 use	↓	414	.077	↓	.68	3.71		4.39	6.40

03 11 13.65 Forms In Place, Slab On Grade

	Crew	Daily Output	Labor-Hours	Unit	Material	2018 Bare Costs Labor	Equipment	Total	Total Incl O&P
0010 **FORMS IN PLACE, SLAB ON GRADE**									
3000　Edge forms, wood, 4 use, on grade, to 6" high	C-1	600	.053	L.F.	.29	2.56		2.85	4.22
6000　Trench forms in floor, wood, 1 use		160	.200	SFCA	1.89	9.60		11.49	16.70
6050　　2 use		175	.183		1.04	8.80		9.84	14.50
6100　　3 use		180	.178		.75	8.55		9.30	13.85
6150　　4 use	↓	185	.173	↓	.61	8.30		8.91	13.30

03 15 Concrete Accessories

03 15 05 – Concrete Forming Accessories

03 15 05.75 Sleeves and Chases

	Crew	Daily Output	Labor-Hours	Unit	Material	2018 Bare Costs Labor	Equipment	Total	Total Incl O&P
0010 **SLEEVES AND CHASES**									
0100　Plastic, 1 use, 12" long, 2" diameter	1 Carp	100	.080	Ea.	1.89	4.06		5.95	8.30
0150　　4" diameter		90	.089		5.30	4.51		9.81	12.70
0200　　6" diameter		75	.107		9.30	5.40		14.70	18.50
0250　　12" diameter		60	.133		27.50	6.75		34.25	41
5000　Sheet metal, 2" diameter [G]		100	.080		1.49	4.06		5.55	7.85
5100　　4" diameter [G]		90	.089		1.86	4.51		6.37	8.90
5150　　6" diameter [G]		75	.107		1.86	5.40		7.26	10.30
5200　　12" diameter [G]		60	.133		3.72	6.75		10.47	14.40
6000　Steel pipe, 2" diameter [G]		100	.080		2.98	4.06		7.04	9.50
6100　　4" diameter [G]		90	.089		10.85	4.51		15.36	18.80
6150　　6" diameter [G]		75	.107		35	5.40		40.40	47

03 15 Concrete Accessories

03 15 05 – Concrete Forming Accessories

03 15 05.75 Sleeves and Chases		Crew	Daily Output	Labor-Hours	Unit	Material	2018 Bare Costs Labor	Equipment	Total	Total Incl O&P
6200	12" diameter	G	1 Carp	60	.133	Ea.	83.50	6.75		90.25	102

03 15 16 – Concrete Construction Joints

03 15 16.20 Control Joints, Saw Cut

		Crew	Daily Output	Labor-Hours	Unit	Material	Labor	Equipment	Total	Total Incl O&P
0010	**CONTROL JOINTS, SAW CUT**									
0100	Sawcut control joints in green concrete									
0120	1" depth	C-27	2000	.008	L.F.	.03	.38	.05	.46	.66
0140	1-1/2" depth		1800	.009		.05	.42	.06	.53	.75
0160	2" depth	↓	1600	.010	↓	.07	.48	.06	.61	.85
0180	Sawcut joint reservoir in cured concrete									
0182	3/8" wide x 3/4" deep, with single saw blade	C-27	1000	.016	L.F.	.05	.76	.10	.91	1.30
0184	1/2" wide x 1" deep, with double saw blades		900	.018		.10	.85	.11	1.06	1.49
0186	3/4" wide x 1-1/2" deep, with double saw blades	↓	800	.020		.21	.95	.13	1.29	1.78
0190	Water blast joint to wash away laitance, 2 passes	C-29	2500	.003			.13	.03	.16	.22
0200	Air blast joint to blow out debris and air dry, 2 passes	C-28	2000	.004	↓		.19	.01	.20	.29
0300	For backer rod, see Section 07 91 23.10									
0342	For joint sealant, see Section 07 92 13.20									

03 15 19 – Cast-In Concrete Anchors

03 15 19.10 Anchor Bolts

			Crew	Daily Output	Labor-Hours	Unit	Material	Labor	Equipment	Total	Total Incl O&P
0010	**ANCHOR BOLTS**										
0015	Made from recycled materials										
0025	Single bolts installed in fresh concrete, no templates										
0030	Hooked w/nut and washer, 1/2" diameter, 8" long	G	1 Carp	132	.061	Ea.	1.42	3.07		4.49	6.25
0040	12" long	G		131	.061		1.58	3.10		4.68	6.45
0050	5/8" diameter, 8" long	G		129	.062		3.80	3.14		6.94	8.95
0060	12" long	G		127	.063		4.67	3.19		7.86	10
0070	3/4" diameter, 8" long	G		127	.063		4.67	3.19		7.86	10
0080	12" long	G		125	.064	↓	5.85	3.24		9.09	11.35

03 15 19.45 Machinery Anchors

			Crew	Daily Output	Labor-Hours	Unit	Material	Labor	Equipment	Total	Total Incl O&P
0010	**MACHINERY ANCHORS**, heavy duty, incl. sleeve, floating base nut,										
0020	lower stud & coupling nut, fiber plug, connecting stud, washer & nut.										
0030	For flush mounted embedment in poured concrete heavy equip. pads.										
0200	Stud & bolt, 1/2" diameter	G	E-16	40	.400	Ea.	52.50	22.50	2.46	77.46	97
0300	5/8" diameter	G		35	.457		61	25.50	2.81	89.31	112
0500	3/4" diameter	G		30	.533		72.50	29.50	3.28	105.28	132
0600	7/8" diameter	G		25	.640		82.50	35.50	3.94	121.94	153
0800	1" diameter	G		20	.800		88.50	44.50	4.92	137.92	176
0900	1-1/4" diameter	G	↓	15	1.067	↓	118	59.50	6.55	184.05	234

03 21 Reinforcement Bars

03 21 11 – Plain Steel Reinforcement Bars

03 21 11.60 Reinforcing In Place

			Crew	Daily Output	Labor-Hours	Unit	Material	Labor	Equipment	Total	Total Incl O&P
0010	**REINFORCING IN PLACE**, 50-60 ton lots, A615 Grade 60										
0020	Includes labor, but not material cost, to install accessories										
0030	Made from recycled materials										
0502	Footings, #4 to #7	G	4 Rodm	4200	.008	Lb.	.48	.42		.90	1.17
0552	#8 to #18	G		7200	.004		.48	.24		.72	.90
0602	Slab on grade, #3 to #7	G	↓	4200	.008	↓	.48	.42		.90	1.17
0900	For other than 50-60 ton lots										
1000	Under 10 ton job, #3 to #7, add						25%	10%			
1010	#8 to #18, add						20%	10%			

03 21 Reinforcement Bars

03 21 11 – Plain Steel Reinforcement Bars

03 21 11.60 Reinforcing In Place	Crew	Daily Output	Labor-Hours	Unit	Material	2018 Bare Costs Labor	Equipment	Total	Total Incl O&P
1050	10-50 ton job, #3 to #7, add					10%			
1060	#8 to #18, add					5%			
1100	60-100 ton job, #3 to #7, deduct					5%			
1110	#8 to #18, deduct					10%			
1150	Over 100 ton job, #3 to #7, deduct					10%			
1160	#8 to #18, deduct					15%			

03 22 Fabric and Grid Reinforcing

03 22 11 – Plain Welded Wire Fabric Reinforcing

03 22 11.10 Plain Welded Wire Fabric

0010	PLAIN WELDED WIRE FABRIC ASTM A185									
0020	Includes labor, but not material cost, to install accessories									
0050	Sheets									
0100	6 x 6 - W1.4 x W1.4 (10 x 10) 21 lb./C.S.F.	G	2 Rodm	35	.457	C.S.F.	15.10	25	40.10	54.50

03 22 13 – Galvanized Welded Wire Fabric Reinforcing

03 22 13.10 Galvanized Welded Wire Fabric

0010	GALVANIZED WELDED WIRE FABRIC									
0100	Add to plain welded wire pricing for galvanized welded wire				Lb.	.23		.23	.25

03 22 16 – Epoxy-Coated Welded Wire Fabric Reinforcing

03 22 16.10 Epoxy-Coated Welded Wire Fabric

0010	EPOXY-COATED WELDED WIRE FABRIC									
0100	Add to plain welded wire pricing for epoxy-coated welded wire				Lb.	.39		.39	.43

03 30 Cast-In-Place Concrete

03 30 53 – Miscellaneous Cast-In-Place Concrete

03 30 53.40 Concrete In Place

		Crew	Daily Output	Labor-Hours	Unit	Material	Labor	Equipment	Total	Total Incl O&P
0010	CONCRETE IN PLACE									
0020	Including forms (4 uses), Grade 60 rebar, concrete (Portland cement									
0050	Type I), placement and finishing unless otherwise indicated									
3540	Equipment pad (3000 psi), 3' x 3' x 6" thick	C-14H	45	1.067	Ea.	44.50	52.50	.57	97.57	130
3550	4' x 4' x 6" thick		30	1.600		69	79	.85	148.85	197
3560	5' x 5' x 8" thick		18	2.667		126	132	1.41	259.41	340
3570	6' x 6' x 8" thick		14	3.429		173	169	1.82	343.82	450
3580	8' x 8' x 10" thick		8	6		370	296	3.18	669.18	860
3590	10' x 10' x 12" thick		5	9.600		645	475	5.10	1,125.10	1,425
3800	Footings (3000 psi), spread under 1 C.Y.	C-14C	28	4	C.Y.	185	192	.92	377.92	495
3825	1 C.Y. to 5 C.Y.		43	2.605		219	125	.60	344.60	430
3850	Over 5 C.Y.		75	1.493		203	72	.34	275.34	330
3900	Footings, strip (3000 psi), 18" x 9", unreinforced	C-14L	40	2.400		146	113	.65	259.65	335
3920	18" x 9", reinforced	C-14C	35	3.200		169	154	.74	323.74	420
3925	20" x 10", unreinforced	C-14L	45	2.133		142	100	.58	242.58	310
3930	20" x 10", reinforced	C-14C	40	2.800		161	135	.64	296.64	385
3935	24" x 12", unreinforced	C-14L	55	1.745		140	82	.47	222.47	280
3940	24" x 12", reinforced	C-14C	48	2.333		159	112	.54	271.54	345
3945	36" x 12", unreinforced	C-14L	70	1.371		136	64.50	.37	200.87	248
3950	36" x 12", reinforced	C-14C	60	1.867		153	90	.43	243.43	305
4000	Foundation mat (3000 psi), under 10 C.Y.		38.67	2.896		221	139	.67	360.67	455
4050	Over 20 C.Y.		56.40	1.986		196	95.50	.46	291.96	360

03 30 Cast-In-Place Concrete

03 30 53 – Miscellaneous Cast-In-Place Concrete

03 30 53.40 Concrete In Place

		Crew	Daily Output	Labor-Hours	Unit	Material	2018 Bare Costs Labor	2018 Bare Costs Equipment	Total	Total Incl O&P
4650	Slab on grade (3500 psi), not including finish, 4" thick	C-14E	60.75	1.449	C.Y.	145	71	.42	216.42	268
4700	6" thick	"	92	.957	↓	140	47	.28	187.28	226
4701	Thickened slab edge (3500 psi), for slab on grade poured									
4702	monolithically with slab; depth is in addition to slab thickness;									
4703	formed vertical outside edge, earthen bottom and inside slope									
4705	8" deep x 8" wide bottom, unreinforced	C-14L	2190	.044	L.F.	3.97	2.06	.01	6.04	7.50
4710	8" x 8", reinforced	C-14C	1670	.067		6.20	3.22	.02	9.44	11.75
4715	12" deep x 12" wide bottom, unreinforced	C-14L	1800	.053		8.15	2.51	.01	10.67	12.80
4720	12" x 12", reinforced	C-14C	1310	.086		12.25	4.11	.02	16.38	19.75
4725	16" deep x 16" wide bottom, unreinforced	C-14L	1440	.067		13.80	3.13	.02	16.95	20
4730	16" x 16", reinforced	C-14C	1120	.100		18.70	4.81	.02	23.53	28
4735	20" deep x 20" wide bottom, unreinforced	C-14L	1150	.083		21	3.92	.02	24.94	29
4740	20" x 20", reinforced	C-14C	920	.122		27	5.85	.03	32.88	39
4745	24" deep x 24" wide bottom, unreinforced	C-14L	930	.103		29.50	4.85	.03	34.38	40
4750	24" x 24", reinforced	C-14C	740	.151	↓	38	7.30	.03	45.33	52.50

03 31 Structural Concrete

03 31 13 – Heavyweight Structural Concrete

03 31 13.35 Heavyweight Concrete, Ready Mix

		Crew	Daily Output	Labor-Hours	Unit	Material	2018 Bare Costs Labor	2018 Bare Costs Equipment	Total	Total Incl O&P
0010	**HEAVYWEIGHT CONCRETE, READY MIX**, delivered									
0012	Includes local aggregate, sand, Portland cement (Type I) and water									
0015	Excludes all additives and treatments									
0020	2000 psi				C.Y.	115			115	127
0100	2500 psi					119			119	130
0150	3000 psi					121			121	133
0200	3500 psi					124			124	137
0300	4000 psi					128			128	140
1000	For high early strength (Portland cement Type III), add					10%				
1300	For winter concrete (hot water), add					5.35			5.35	5.90
1410	For mid-range water reducer, add					3.59			3.59	3.95
1420	For high-range water reducer/superplasticizer, add					6.30			6.30	6.90
1430	For retarder, add					3.23			3.23	3.55
1440	For non-Chloride accelerator, add					6.40			6.40	7.05
1450	For Chloride accelerator, per 1%, add					3.86			3.86	4.25
1460	For fiber reinforcing, synthetic (1 lb./C.Y.), add					8			8	8.80
1500	For Saturday delivery, add				↓	8.50			8.50	9.35
1510	For truck holding/waiting time past 1st hour per load, add				Hr.	105			105	116
1520	For short load (less than 4 C.Y.), add per load				Ea.	86			86	94.50
2000	For all lightweight aggregate, add				C.Y.	45%				

03 31 13.70 Placing Concrete

		Crew	Daily Output	Labor-Hours	Unit	Material	2018 Bare Costs Labor	2018 Bare Costs Equipment	Total	Total Incl O&P
0010	**PLACING CONCRETE**									
0020	Includes labor and equipment to place, level (strike off) and consolidate									
1900	Footings, continuous, shallow, direct chute	C-6	120	.400	C.Y.		16.60	.43	17.03	25.50
1950	Pumped	C-20	150	.427			18.25	6.20	24.45	34.50
2000	With crane and bucket	C-7	90	.800			35	11.90	46.90	65.50
2100	Footings, continuous, deep, direct chute	C-6	140	.343			14.20	.37	14.57	22
2150	Pumped	C-20	160	.400			17.10	5.85	22.95	32.50
2200	With crane and bucket	C-7	110	.655			28.50	9.75	38.25	53.50
2400	Footings, spread, under 1 C.Y., direct chute	C-6	55	.873			36	.93	36.93	56
2450	Pumped	C-20	65	.985			42	14.35	56.35	80
2500	With crane and bucket	C-7	45	1.600	↓		69.50	24	93.50	131

For customer support on your Electrical Costs with RSMeans data, call 800.448.8182.

29

03 31 Structural Concrete

03 31 13 – Heavyweight Structural Concrete

03 31 13.70 Placing Concrete

		Crew	Daily Output	Labor-Hours	Unit	Material	2018 Bare Costs Labor	Equipment	Total	Total Incl O&P
2600	Over 5 C.Y., direct chute	C-6	120	.400	C.Y.		16.60	.43	17.03	25.50
2650	Pumped	C-20	150	.427			18.25	6.20	24.45	34.50
2700	With crane and bucket	C-7	100	.720			31.50	10.70	42.20	59.50
2900	Foundation mats, over 20 C.Y., direct chute	C-6	350	.137			5.70	.15	5.85	8.75
2950	Pumped	C-20	400	.160			6.85	2.33	9.18	12.90
3000	With crane and bucket	C-7	300	.240			10.45	3.57	14.02	19.70

03 35 Concrete Finishing

03 35 13 – High-Tolerance Concrete Floor Finishing

03 35 13.30 Finishing Floors, High Tolerance

		Crew	Daily Output	Labor-Hours	Unit	Material	2018 Bare Costs Labor	Equipment	Total	Total Incl O&P
0010	**FINISHING FLOORS, HIGH TOLERANCE**									
0012	Finishing of fresh concrete flatwork requires that concrete									
0013	first be placed, struck off & consolidated									
0015	Basic finishing for various unspecified flatwork									
0100	Bull float only	C-10	4000	.006	S.F.		.27		.27	.40
0125	Bull float & manual float		2000	.012			.54		.54	.81
0150	Bull float, manual float & broom finish, w/edging & joints		1850	.013			.58		.58	.87
0200	Bull float, manual float & manual steel trowel		1265	.019			.85		.85	1.27
0210	For specified Random Access Floors in ACI Classes 1, 2, 3 and 4 to achieve									
0215	Composite Overall Floor Flatness and Levelness values up to FF35/FL25									
0250	Bull float, machine float & machine trowel (walk-behind)	C-10C	1715	.014	S.F.		.63	.02	.65	.97
0300	Power screed, bull float, machine float & trowel (walk-behind)	C-10D	2400	.010			.45	.05	.50	.72
0350	Power screed, bull float, machine float & trowel (ride-on)	C-10E	4000	.006			.27	.06	.33	.47
0352	For specified Random Access Floors in ACI Classes 5, 6, 7 and 8 to achieve									
0354	Composite Overall Floor Flatness and Levelness values up to FF50/FL50									
0356	Add for two-dimensional restraightening after power float	C-10	6000	.004	S.F.		.18		.18	.27
0358	For specified Random or Defined Access Floors in ACI Class 9 to achieve									
0360	Composite Overall Floor Flatness and Levelness values up to FF100/FL100									
0362	Add for two-dimensional restraightening after bull float & power float	C-10	3000	.008	S.F.		.36		.36	.54
0364	For specified Superflat Defined Access Floors in ACI Class 9 to achieve									
0366	Minimum Floor Flatness and Levelness values of FF100/FL100									
0368	Add for 2-dim'l restraightening after bull float, power float, power trowel	C-10	2000	.012	S.F.		.54		.54	.81

03 54 Cast Underlayment

03 54 16 – Hydraulic Cement Underlayment

03 54 16.50 Cement Underlayment

		Crew	Daily Output	Labor-Hours	Unit	Material	2018 Bare Costs Labor	Equipment	Total	Total Incl O&P
0010	**CEMENT UNDERLAYMENT**									
2510	Underlayment, P.C. based, self-leveling, 4100 psi, pumped, 1/4" thick	C-8	20000	.003	S.F.	1.68	.12	.04	1.84	2.09
2520	1/2" thick		19000	.003		3.36	.13	.05	3.54	3.94
2530	3/4" thick		18000	.003		5.05	.14	.05	5.24	5.80
2540	1" thick		17000	.003		6.70	.15	.05	6.90	7.70
2550	1-1/2" thick		15000	.004		10.10	.17	.06	10.33	11.40
2560	Hand placed, 1/2" thick	C-18	450	.020		3.36	.80	.13	4.29	5.05
2610	Topping, P.C. based, self-leveling, 6100 psi, pumped, 1/4" thick	C-8	20000	.003		2.47	.12	.04	2.63	2.95
2620	1/2" thick		19000	.003		4.93	.13	.05	5.11	5.70
2630	3/4" thick		18000	.003		7.40	.14	.05	7.59	8.40
2660	1" thick		17000	.003		9.85	.15	.05	10.05	11.15
2670	1-1/2" thick		15000	.004		14.80	.17	.06	15.03	16.60
2680	Hand placed, 1/2" thick	C-18	450	.020		4.93	.80	.13	5.86	6.80

03 63 Epoxy Grouting

03 63 05 – Grouting of Dowels and Fasteners

03 63 05.10 Epoxy Only

	Crew	Daily Output	Labor-Hours	Unit	Material	2018 Bare Costs Labor	Equipment	Total	Total Incl O&P
0010 **EPOXY ONLY**									
1500 Chemical anchoring, epoxy cartridge, excludes layout, drilling, fastener									
1530 For fastener 3/4" diam. x 6" embedment	2 Skwk	72	.222	Ea.	5.20	11.65		16.85	23.50
1535 1" diam. x 8" embedment		66	.242		7.80	12.70		20.50	28
1540 1-1/4" diam. x 10" embedment		60	.267		15.65	13.95		29.60	38.50
1545 1-3/4" diam. x 12" embedment		54	.296		26	15.50		41.50	52.50
1550 14" embedment		48	.333		31.50	17.45		48.95	61
1555 2" diam. x 12" embedment		42	.381		41.50	19.95		61.45	76.50
1560 18" embedment		32	.500		52	26		78	97.50

03 81 Concrete Cutting

03 81 13 – Flat Concrete Sawing

03 81 13.50 Concrete Floor/Slab Cutting

	Crew	Daily Output	Labor-Hours	Unit	Material	2018 Bare Costs Labor	Equipment	Total	Total Incl O&P
0010 **CONCRETE FLOOR/SLAB CUTTING**									
0050 Includes blade cost, layout and set-up time									
0300 Saw cut concrete slabs, plain, up to 3" deep	B-89	1060	.015	L.F.	.12	.72	.40	1.24	1.67
0320 Each additional inch of depth		3180	.005		.04	.24	.13	.41	.56
0400 Mesh reinforced, up to 3" deep		980	.016		.14	.78	.43	1.35	1.80
0420 Each additional inch of depth		2940	.005		.05	.26	.14	.45	.60
0500 Rod reinforced, up to 3" deep		800	.020		.17	.96	.53	1.66	2.21
0520 Each additional inch of depth		2400	.007		.06	.32	.18	.56	.73

03 81 13.75 Concrete Saw Blades

	Crew	Daily Output	Labor-Hours	Unit	Material	2018 Bare Costs Labor	Equipment	Total	Total Incl O&P
0010 **CONCRETE SAW BLADES**									
3000 Blades for saw cutting, included in cutting line items									
3020 Diamond, 12" diameter				Ea.	220			220	242
3040 18" diameter					415			415	460
3080 24" diameter					695			695	765
3120 30" diameter					980			980	1,075
3160 36" diameter					1,300			1,300	1,425
3200 42" diameter					2,550			2,550	2,800

03 82 Concrete Boring

03 82 13 – Concrete Core Drilling

03 82 13.10 Core Drilling

	Crew	Daily Output	Labor-Hours	Unit	Material	2018 Bare Costs Labor	Equipment	Total	Total Incl O&P
0010 **CORE DRILLING**									
0015 Includes bit cost, layout and set-up time									
0020 Reinforced concrete slab, up to 6" thick									
0100 1" diameter core	B-89A	17	.941	Ea.	.18	43.50	6.60	50.28	74
0150 For each additional inch of slab thickness in same hole, add		1440	.011		.03	.51	.08	.62	.90
0200 2" diameter core		16.50	.970		.29	44.50	6.80	51.59	76.50
0250 For each additional inch of slab thickness in same hole, add		1080	.015		.05	.68	.10	.83	1.20
0300 3" diameter core		16	1		.39	46	7	53.39	78.50
0350 For each additional inch of slab thickness in same hole, add		720	.022		.07	1.02	.16	1.25	1.81
0500 4" diameter core		15	1.067		.50	49	7.50	57	84
0550 For each additional inch of slab thickness in same hole, add		480	.033		.08	1.54	.23	1.85	2.70
0700 6" diameter core		14	1.143		.77	52.50	8	61.27	90
0750 For each additional inch of slab thickness in same hole, add		360	.044		.13	2.05	.31	2.49	3.61
0900 8" diameter core		13	1.231		1.18	56.50	8.65	66.33	97.50
0950 For each additional inch of slab thickness in same hole, add		288	.056		.20	2.56	.39	3.15	4.56

For customer support on your Electrical Costs with RSMeans data, call 800.448.8182.

31

03 82 Concrete Boring

03 82 13 – Concrete Core Drilling

03 82 13.10 Core Drilling

		Crew	Daily Output	Labor-Hours	Unit	Material	2018 Bare Costs Labor	2018 Bare Costs Equipment	Total	Total Incl O&P
1100	10" diameter core	B-89A	12	1.333	Ea.	1.49	61.50	9.35	72.34	106
1150	For each additional inch of slab thickness in same hole, add		240	.067		.25	3.07	.47	3.79	5.50
1300	12" diameter core		11	1.455		1.83	67	10.20	79.03	115
1350	For each additional inch of slab thickness in same hole, add		206	.078		.30	3.58	.54	4.42	6.40
1500	14" diameter core		10	1.600		2.12	74	11.20	87.32	128
1550	For each additional inch of slab thickness in same hole, add		180	.089		.35	4.10	.62	5.07	7.35
1700	18" diameter core		9	1.778		2.92	82	12.45	97.37	142
1750	For each additional inch of slab thickness in same hole, add		144	.111		.49	5.10	.78	6.37	9.25
1754	24" diameter core		8	2		4.14	92	14	110.14	161
1756	For each additional inch of slab thickness in same hole, add		120	.133		.69	6.15	.93	7.77	11.20
1760	For horizontal holes, add to above						20%	20%		
1770	Prestressed hollow core plank, 8" thick									
1780	1" diameter core	B-89A	17.50	.914	Ea.	.24	42	6.40	48.64	72
1790	For each additional inch of plank thickness in same hole, add		3840	.004		.03	.19	.03	.25	.35
1794	2" diameter core		17.25	.928		.39	43	6.50	49.89	73
1796	For each additional inch of plank thickness in same hole, add		2880	.006		.05	.26	.04	.35	.48
1800	3" diameter core		17	.941		.52	43.50	6.60	50.62	74.50
1810	For each additional inch of plank thickness in same hole, add		1920	.008		.07	.38	.06	.51	.72
1820	4" diameter core		16.50	.970		.67	44.50	6.80	51.97	76.50
1830	For each additional inch of plank thickness in same hole, add		1280	.013		.08	.58	.09	.75	1.07
1840	6" diameter core		15.50	1.032		1.03	47.50	7.25	55.78	81.50
1850	For each additional inch of plank thickness in same hole, add		960	.017		.13	.77	.12	1.02	1.44
1860	8" diameter core		15	1.067		1.57	49	7.50	58.07	85
1870	For each additional inch of plank thickness in same hole, add		768	.021		.20	.96	.15	1.31	1.85
1880	10" diameter core		14	1.143		1.99	52.50	8	62.49	91.50
1890	For each additional inch of plank thickness in same hole, add		640	.025		.25	1.15	.18	1.58	2.22
1900	12" diameter core		13.50	1.185		2.44	54.50	8.30	65.24	95.50
1910	For each additional inch of plank thickness in same hole, add		548	.029		.30	1.35	.20	1.85	2.63

03 82 16 – Concrete Drilling

03 82 16.10 Concrete Impact Drilling

		Crew	Daily Output	Labor-Hours	Unit	Material	2018 Bare Costs Labor	2018 Bare Costs Equipment	Total	Total Incl O&P
0010	**CONCRETE IMPACT DRILLING**									
0020	Includes bit cost, layout and set-up time, no anchors									
0050	Up to 4" deep in concrete/brick floors/walls									
0100	Holes, 1/4" diameter	1 Carp	75	.107	Ea.	.07	5.40		5.47	8.35
0150	For each additional inch of depth in same hole, add		430	.019		.02	.94		.96	1.46
0200	3/8" diameter		63	.127		.06	6.45		6.51	9.85
0250	For each additional inch of depth in same hole, add		340	.024		.01	1.19		1.20	1.84
0300	1/2" diameter		50	.160		.06	8.10		8.16	12.40
0350	For each additional inch of depth in same hole, add		250	.032		.01	1.62		1.63	2.49
0400	5/8" diameter		48	.167		.10	8.45		8.55	12.95
0450	For each additional inch of depth in same hole, add		240	.033		.03	1.69		1.72	2.60
0500	3/4" diameter		45	.178		.13	9		9.13	13.85
0550	For each additional inch of depth in same hole, add		220	.036		.03	1.84		1.87	2.85
0600	7/8" diameter		43	.186		.18	9.45		9.63	14.55
0650	For each additional inch of depth in same hole, add		210	.038		.05	1.93		1.98	2.99
0700	1" diameter		40	.200		.16	10.15		10.31	15.60
0750	For each additional inch of depth in same hole, add		190	.042		.04	2.14		2.18	3.29
0800	1-1/4" diameter		38	.211		.29	10.65		10.94	16.55
0850	For each additional inch of depth in same hole, add		180	.044		.07	2.25		2.32	3.51
0900	1-1/2" diameter		35	.229		.50	11.60		12.10	18.20
0950	For each additional inch of depth in same hole, add		165	.048		.12	2.46		2.58	3.88
1000	For ceiling installations, add						40%			

Estimating Tips

05 05 00 Common Work Results for Metals

- Nuts, bolts, washers, connection angles, and plates can add a significant amount to both the tonnage of a structural steel job and the estimated cost. As a rule of thumb, add 10% to the total weight to account for these accessories.

- Type 2 steel construction, commonly referred to as "simple construction," consists generally of field-bolted connections with lateral bracing supplied by other elements of the building, such as masonry walls or x-bracing. The estimator should be aware, however, that shop connections may be accomplished by welding or bolting. The method may be particular to the fabrication shop and may have an impact on the estimated cost.

05 10 00 Structural Steel

- Steel items can be obtained from two sources: a fabrication shop or a metals service center. Fabrication shops can fabricate items under more controlled conditions than crews in the field can. They are also more efficient and can produce items more economically. Metal service centers serve as a source of long mill shapes to both fabrication shops and contractors.

- Most line items in this structural steel subdivision, and most items in 05 50 00 Metal Fabrications, are indicated as being shop fabricated. The bare material cost for these shop fabricated items is the "Invoice Cost" from the shop and includes the mill base price of steel plus mill extras, transportation to the shop, shop drawings and detailing where warranted, shop fabrication and handling, sandblasting and a shop coat of primer paint, all necessary structural bolts, and delivery to the job site. The bare labor cost and bare equipment

cost for these shop fabricated items are for field installation or erection.

- Line items in Subdivision 05 12 23.40 Lightweight Framing, and other items scattered in Division 5, are indicated as being field fabricated. The bare material cost for these field fabricated items is the "Invoice Cost" from the metals service center and includes the mill base price of steel plus mill extras, transportation to the metals service center, material handling, and delivery of long lengths of mill shapes to the job site. Material costs for structural bolts and welding rods should be added to the estimate. The bare labor cost and bare equipment cost for these items are for both field fabrication and field installation or erection, and include time for cutting, welding, and drilling in the fabricated metal items. Drilling into concrete and fasteners to fasten field fabricated items to other work are not included and should be added to the estimate.

05 20 00 Steel Joist Framing

- In any given project the total weight of open web steel joists is determined by the loads to be supported and the design. However, economies can be realized in minimizing the amount of labor used to place the joists. This is done by maximizing the joist spacing, and therefore minimizing the number of joists required to be installed on the job. Certain spacings and locations may be required by the design, but in other cases maximizing the spacing and keeping it as uniform as possible will keep the costs down.

05 30 00 Steel Decking

- The takeoff and estimating of a metal deck involves more than the area of the floor or roof and the type of deck specified or shown on the drawings. Many different sizes and types of openings may exist. Small openings

for individual pipes or conduits may be drilled after the floor/roof is installed, but larger openings may require special deck lengths as well as reinforcing or structural support. The estimator should determine who will be supplying this reinforcing. Additionally, some deck terminations are part of the deck package, such as screed angles and pour stops, and others will be part of the steel contract, such as angles attached to structural members and cast-in-place angles and plates. The estimator must ensure that all pieces are accounted for in the complete estimate.

05 50 00 Metal Fabrications

- The most economical steel stairs are those that use common materials, standard details, and most importantly, a uniform and relatively simple method of field assembly. Commonly available A36/A992 channels and plates are very good choices for the main stringers of the stairs, as are angles and tees for the carrier members. Risers and treads are usually made by specialty shops, and it is most economical to use a typical detail in as many places as possible. The stairs should be pre-assembled and shipped directly to the site. The field connections should be simple and straightforward enough to be accomplished efficiently, and with minimum equipment and labor.

Reference Numbers

Reference numbers are shown at the beginning of some major classifications. These numbers refer to related items in the Reference Section. The reference information may be an estimating procedure, an alternate pricing method, or technical information.

Note: Not all subdivisions listed here necessarily appear. ∎

05 05 19 – Post-Installed Concrete Anchors

05 05 19.10 Chemical Anchors

		Crew	Daily Output	Labor-Hours	Unit	Material	2018 Bare Costs Labor	Equipment	Total	Total Incl O&P
0010	**CHEMICAL ANCHORS**									
0020	Includes layout & drilling									
1430	Chemical anchor, w/rod & epoxy cartridge, 3/4" diameter x 9-1/2" long	B-89A	27	.593	Ea.	10.85	27.50	4.15	42.50	58
1435	1" diameter x 11-3/4" long		24	.667		20	30.50	4.67	55.17	74.50
1440	1-1/4" diameter x 14" long		21	.762		38	35	5.35	78.35	101
1445	1-3/4" diameter x 15" long		20	.800		69.50	37	5.60	112.10	139
1450	18" long		17	.941		83	43.50	6.60	133.10	165
1455	2" diameter x 18" long		16	1		113	46	7	166	202
1460	24" long	↓	15	1.067	↓	147	49	7.50	203.50	244

05 05 19.20 Expansion Anchors

		Crew	Daily Output	Labor-Hours	Unit	Material	2018 Bare Costs Labor	Equipment	Total	Total Incl O&P
0010	**EXPANSION ANCHORS**									
0100	Anchors for concrete, brick or stone, no layout and drilling									
0200	Expansion shields, zinc, 1/4" diameter, 1-5/16" long, single [G]	1 Carp	90	.089	Ea.	.45	4.51		4.96	7.35
0300	1-3/8" long, double [G]		85	.094		.58	4.77		5.35	7.90
0400	3/8" diameter, 1-1/2" long, single [G]		85	.094		.69	4.77		5.46	8
0500	2" long, double [G]		80	.100		1.15	5.05		6.20	8.95
0600	1/2" diameter, 2-1/16" long, single [G]		80	.100		1.26	5.05		6.31	9.10
0700	2-1/2" long, double [G]		75	.107		2.13	5.40		7.53	10.60
0800	5/8" diameter, 2-5/8" long, single [G]		75	.107		2.18	5.40		7.58	10.65
0900	2-3/4" long, double [G]		70	.114		2.92	5.80		8.72	12
1000	3/4" diameter, 2-3/4" long, single [G]		70	.114		3.29	5.80		9.09	12.40
1100	3-15/16" long, double [G]	↓	65	.123	↓	5.45	6.25		11.70	15.50
2100	Hollow wall anchors for gypsum wall board, plaster or tile									
2300	1/8" diameter, short [G]	1 Carp	160	.050	Ea.	.33	2.54		2.87	4.22
2400	Long [G]		150	.053		.32	2.70		3.02	4.47
2500	3/16" diameter, short [G]		150	.053		.51	2.70		3.21	4.68
2600	Long [G]		140	.057		.67	2.90		3.57	5.15
2700	1/4" diameter, short [G]		140	.057		.73	2.90		3.63	5.20
2800	Long [G]		130	.062		.80	3.12		3.92	5.65
3000	Toggle bolts, bright steel, 1/8" diameter, 2" long [G]		85	.094		.25	4.77		5.02	7.55
3100	4" long [G]		80	.100		.29	5.05		5.34	8
3400	1/4" diameter, 3" long [G]		75	.107		.41	5.40		5.81	8.70
3500	6" long [G]		70	.114		.59	5.80		6.39	9.45
3600	3/8" diameter, 3" long [G]		70	.114		.86	5.80		6.66	9.75
3700	6" long [G]		60	.133		1.50	6.75		8.25	11.95
3800	1/2" diameter, 4" long [G]		60	.133		1.65	6.75		8.40	12.10
3900	6" long [G]	↓	50	.160	↓	2.08	8.10		10.18	14.65
4000	Nailing anchors									
4100	Nylon nailing anchor, 1/4" diameter, 1" long	1 Carp	3.20	2.500	C	23	127		150	218
4200	1-1/2" long		2.80	2.857		25.50	145		170.50	250
4300	2" long		2.40	3.333		27.50	169		196.50	287
4400	Metal nailing anchor, 1/4" diameter, 1" long [G]		3.20	2.500		22.50	127		149.50	218
4500	1-1/2" long [G]		2.80	2.857		29	145		174	253
4600	2" long [G]	↓	2.40	3.333	↓	35.50	169		204.50	296
5000	Screw anchors for concrete, masonry,									
5100	stone & tile, no layout or drilling included									
5700	Lag screw shields, 1/4" diameter, short [G]	1 Carp	90	.089	Ea.	.47	4.51		4.98	7.35
5900	3/8" diameter, short [G]		85	.094		.72	4.77		5.49	8.05
6100	1/2" diameter, short [G]		80	.100		1.03	5.05		6.08	8.85
6300	5/8" diameter, short [G]		70	.114		1.42	5.80		7.22	10.35
6600	Lead, #6 & #8, 3/4" long [G]		260	.031		.19	1.56		1.75	2.59
6700	#10 - #14, 1-1/2" long [G]		200	.040		.38	2.03		2.41	3.51

34

05 05 Common Work Results for Metals

05 05 19 – Post-Installed Concrete Anchors

05 05 19.20 Expansion Anchors

		Crew	Daily Output	Labor-Hours	Unit	Material	2018 Bare Costs Labor	Equipment	Total	Total Incl O&P
6800	#16 & #18, 1-1/2" long	1 Carp	160	.050	Ea.	.41	2.54		2.95	4.31
6900	Plastic, #6 & #8, 3/4" long		260	.031		.05	1.56		1.61	2.44
7100	#10 & #12, 1" long		220	.036		.07	1.84		1.91	2.89
8000	Wedge anchors, not including layout or drilling									
8050	Carbon steel, 1/4" diameter, 1-3/4" long G	1 Carp	150	.053	Ea.	.51	2.70		3.21	4.68
8100	3-1/4" long G		140	.057		.67	2.90		3.57	5.15
8150	3/8" diameter, 2-1/4" long G		145	.055		.54	2.80		3.34	4.86
8200	5" long G		140	.057		.95	2.90		3.85	5.45
8250	1/2" diameter, 2-3/4" long G		140	.057		1.07	2.90		3.97	5.60
8300	7" long G		125	.064		1.82	3.24		5.06	6.95
8350	5/8" diameter, 3-1/2" long G		130	.062		1.81	3.12		4.93	6.75
8400	8-1/2" long G		115	.070		3.86	3.53		7.39	9.60
8450	3/4" diameter, 4-1/4" long G		115	.070		2.75	3.53		6.28	8.35
8500	10" long G		95	.084		6.25	4.27		10.52	13.35
8550	1" diameter, 6" long G		100	.080		8.70	4.06		12.76	15.80
8575	9" long G		85	.094		11.35	4.77		16.12	19.70
8600	12" long G		75	.107		12.25	5.40		17.65	21.50
8650	1-1/4" diameter, 9" long G		70	.114		22	5.80		27.80	33
8700	12" long G		60	.133		28	6.75		34.75	41
8750	For type 303 stainless steel, add					350%				
8800	For type 316 stainless steel, add					450%				
8950	Self-drilling concrete screw, hex washer head, 3/16" diam. x 1-3/4" long G	1 Carp	300	.027	Ea.	.19	1.35		1.54	2.27
8960	2-1/4" long G		250	.032		.21	1.62		1.83	2.70
8970	Phillips flat head, 3/16" diam. x 1-3/4" long G		300	.027		.19	1.35		1.54	2.27
8980	2-1/4" long G		250	.032		.21	1.62		1.83	2.70

05 05 21 – Fastening Methods for Metal

05 05 21.15 Drilling Steel

		Crew	Daily Output	Labor-Hours	Unit	Material	2018 Bare Costs Labor	Equipment	Total	Total Incl O&P
0010	**DRILLING STEEL**									
1910	Drilling & layout for steel, up to 1/4" deep, no anchor									
1920	Holes, 1/4" diameter	1 Sswk	112	.071	Ea.	.10	3.90		4	6.50
1925	For each additional 1/4" depth, add		336	.024		.10	1.30		1.40	2.24
1930	3/8" diameter		104	.077		.09	4.20		4.29	6.95
1935	For each additional 1/4" depth, add		312	.026		.09	1.40		1.49	2.39
1940	1/2" diameter		96	.083		.11	4.55		4.66	7.55
1945	For each additional 1/4" depth, add		288	.028		.11	1.52		1.63	2.60
1950	5/8" diameter		88	.091		.16	4.97		5.13	8.30
1955	For each additional 1/4" depth, add		264	.030		.16	1.66		1.82	2.89
1960	3/4" diameter		80	.100		.20	5.45		5.65	9.15
1965	For each additional 1/4" depth, add		240	.033		.20	1.82		2.02	3.20
1970	7/8" diameter		72	.111		.27	6.05		6.32	10.25
1975	For each additional 1/4" depth, add		216	.037		.27	2.02		2.29	3.60
1980	1" diameter		64	.125		.23	6.85		7.08	11.40
1985	For each additional 1/4" depth, add		192	.042		.23	2.28		2.51	3.97
1990	For drilling up, add					40%				

05 05 23 – Metal Fastenings

05 05 23.10 Bolts and Hex Nuts

		Crew	Daily Output	Labor-Hours	Unit	Material	2018 Bare Costs Labor	Equipment	Total	Total Incl O&P
0010	**BOLTS & HEX NUTS**, Steel, A307									
0100	1/4" diameter, 1/2" long G	1 Sswk	140	.057	Ea.	.06	3.12		3.18	5.15
0200	1" long G		140	.057		.07	3.12		3.19	5.20
0300	2" long G		130	.062		.10	3.36		3.46	5.60
0400	3" long G		130	.062		.15	3.36		3.51	5.65
0500	4" long G		120	.067		.17	3.64		3.81	6.15

05 05 23 – Metal Fastenings

	05 05 23.10 Bolts and Hex Nuts		Crew	Daily Output	Labor-Hours	Unit	Material	2018 Bare Costs Labor	2018 Bare Costs Equipment	Total	Total Incl O&P
0600	3/8" diameter, 1" long	G	1 Sswk	130	.062	Ea.	.14	3.36		3.50	5.65
0700	2" long	G		130	.062		.18	3.36		3.54	5.70
0800	3" long	G		120	.067		.24	3.64		3.88	6.20
0900	4" long	G		120	.067		.30	3.64		3.94	6.30
1000	5" long	G		115	.070		.38	3.80		4.18	6.60
1100	1/2" diameter, 1-1/2" long	G		120	.067		.40	3.64		4.04	6.40
1200	2" long	G		120	.067		.46	3.64		4.10	6.45
1300	4" long	G		115	.070		.75	3.80		4.55	7.05
1400	6" long	G		110	.073		1.05	3.97		5.02	7.65
1500	8" long	G		105	.076		1.38	4.16		5.54	8.30
1600	5/8" diameter, 1-1/2" long	G		120	.067		.85	3.64		4.49	6.90
1700	2" long	G		120	.067		.94	3.64		4.58	7
1800	4" long	G		115	.070		1.34	3.80		5.14	7.70
1900	6" long	G		110	.073		1.72	3.97		5.69	8.40
2000	8" long	G		105	.076		2.55	4.16		6.71	9.60
2100	10" long	G		100	.080		3.20	4.37		7.57	10.65
2200	3/4" diameter, 2" long	G		120	.067		1.15	3.64		4.79	7.20
2300	4" long	G		110	.073		1.65	3.97		5.62	8.30
2400	6" long	G		105	.076		2.12	4.16		6.28	9.15
2500	8" long	G		95	.084		3.20	4.60		7.80	11
2600	10" long	G		85	.094		4.20	5.15		9.35	13
2700	12" long	G		80	.100		4.92	5.45		10.37	14.35
2800	1" diameter, 3" long	G		105	.076		2.89	4.16		7.05	9.95
2900	6" long	G		90	.089		4.20	4.86		9.06	12.55
3000	12" long	G		75	.107		7.50	5.85		13.35	17.80
3100	For galvanized, add						75%				
3200	For stainless, add						350%				

05 05 23.30 Lag Screws

			Crew	Daily Output	Labor-Hours	Unit	Material	Labor	Equipment	Total	Total Incl O&P
0010	**LAG SCREWS**										
0020	Steel, 1/4" diameter, 2" long	G	1 Carp	200	.040	Ea.	.10	2.03		2.13	3.20
0100	3/8" diameter, 3" long	G		150	.053		.28	2.70		2.98	4.43
0200	1/2" diameter, 3" long	G		130	.062		.66	3.12		3.78	5.50
0300	5/8" diameter, 3" long	G		120	.067		1.29	3.38		4.67	6.55

05 05 23.35 Machine Screws

			Crew	Daily Output	Labor-Hours	Unit	Material	Labor	Equipment	Total	Total Incl O&P
0010	**MACHINE SCREWS**										
0020	Steel, round head, #8 x 1" long	G	1 Carp	4.80	1.667	C	4.06	84.50		88.56	133
0110	#8 x 2" long	G		2.40	3.333		8.10	169		177.10	266
0200	#10 x 1" long	G		4	2		5.05	101		106.05	160
0300	#10 x 2" long	G		2	4		8.75	203		211.75	320

05 05 23.50 Powder Actuated Tools and Fasteners

			Crew	Daily Output	Labor-Hours	Unit	Material	Labor	Equipment	Total	Total Incl O&P
0010	**POWDER ACTUATED TOOLS & FASTENERS**										
0020	Stud driver, .22 caliber, single shot					Ea.	152			152	167
0100	.27 caliber, semi automatic, strip					"	460			460	505
0300	Powder load, single shot, .22 cal, power level 2, brown					C	5.65			5.65	6.25
0400	Strip, .27 cal, power level 4, red						8.15			8.15	9
0600	Drive pin, .300 x 3/4" long	G	1 Carp	4.80	1.667		4.39	84.50		88.89	134
0700	.300 x 3" long with washer	G	"	4	2		13.40	101		114.40	169

05 05 23.70 Structural Blind Bolts

			Crew	Daily Output	Labor-Hours	Unit	Material	Labor	Equipment	Total	Total Incl O&P
0010	**STRUCTURAL BLIND BOLTS**										
0100	1/4" diameter x 1/4" grip	G	1 Sswk	240	.033	Ea.	1.73	1.82		3.55	4.88
0150	1/2" grip	G		216	.037		1.86	2.02		3.88	5.35
0200	3/8" diameter x 1/2" grip	G		232	.034		3.12	1.88		5	6.50

05 05 Common Work Results for Metals

05 05 23 – Metal Fastenings

05 05 23.70 Structural Blind Bolts		Crew	Daily Output	Labor-Hours	Unit	Material	2018 Bare Costs Labor	Equipment	Total	Total Incl O&P
0250	3/4" grip	G 1 Sswk	208	.038	Ea.	3.27	2.10		5.37	7.05
0300	1/2" diameter x 1/2" grip	G	224	.036		6.05	1.95		8	9.85
0350	3/4" grip	G	200	.040		8.20	2.19		10.39	12.55
0400	5/8" diameter x 3/4" grip	G	216	.037		10.90	2.02		12.92	15.30
0450	1" grip	G	192	.042		14.80	2.28		17.08	19.95

05 12 Structural Steel Framing

05 12 23 – Structural Steel for Buildings

05 12 23.60 Pipe Support Framing

		Crew	Daily Output	Labor-Hours	Unit	Material	2018 Bare Costs Labor	Equipment	Total	Total Incl O&P
0010	**PIPE SUPPORT FRAMING**									
0020	Under 10#/L.F., shop fabricated	G E-4	3900	.008	Lb.	1.80	.45	.03	2.28	2.75
0200	10.1 to 15#/L.F.	G	4300	.007		1.78	.41	.02	2.21	2.65
0400	15.1 to 20#/L.F.	G	4800	.007		1.75	.37	.02	2.14	2.55
0600	Over 20#/L.F.	G	5400	.006		1.72	.33	.02	2.07	2.45

05 35 Raceway Decking Assemblies

05 35 13 – Steel Cellular Decking

05 35 13.50 Cellular Decking

		Crew	Daily Output	Labor-Hours	Unit	Material	2018 Bare Costs Labor	Equipment	Total	Total Incl O&P
0010	**CELLULAR DECKING**									
0015	Made from recycled materials									
0200	Cellular units, galv, 1-1/2" deep, Type BC, 20-20 ga., over 15 squares	G E-4	1460	.022	S.F.	10.10	1.21	.07	11.38	13.15
0250	18-20 ga.	G	1420	.023		11.45	1.24	.07	12.76	14.70
0300	18-18 ga.	G	1390	.023		11.75	1.27	.07	13.09	15.05
0320	16-18 ga.	G	1360	.024		14	1.30	.07	15.37	17.60
0340	16-16 ga.	G	1330	.024		15.60	1.33	.07	17	19.40
0400	3" deep, Type NC, galvanized, 20-20 ga.	G	1375	.023		11.10	1.28	.07	12.45	14.40
0500	18-20 ga.	G	1350	.024		13.40	1.31	.07	14.78	16.95
0600	18-18 ga.	G	1290	.025		13.35	1.37	.08	14.80	17
0700	16-18 ga.	G	1230	.026		15.05	1.44	.08	16.57	19
0800	16-16 ga.	G	1150	.028		16.40	1.53	.09	18.02	20.50
1000	4-1/2" deep, Type JC, galvanized, 18-20 ga.	G	1100	.029		15.45	1.60	.09	17.14	19.70
1100	18-18 ga.	G	1040	.031		15.35	1.70	.09	17.14	19.75
1200	16-18 ga.	G	980	.033		17.30	1.80	.10	19.20	22
1300	16-16 ga.	G	935	.034		18.85	1.89	.11	20.85	24
1900	For multi-story or congested site, add						50%			

For customer support on your Electrical Costs with RSMeans data, call 800.448.8182.

37

Division Notes

	CREW	DAILY OUTPUT	LABOR-HOURS	UNIT	BARE COSTS				TOTAL INCL O&P
					MAT.	LABOR	EQUIP.	TOTAL	

Estimating Tips
06 05 00 Common Work Results for Wood, Plastics, and Composites

- Common to any wood-framed structure are the accessory connector items such as screws, nails, adhesives, hangers, connector plates, straps, angles, and hold-downs. For typical wood-framed buildings, such as residential projects, the aggregate total for these items can be significant, especially in areas where seismic loading is a concern. For floor and wall framing, the material cost is based on 10 to 25 lbs. of accessory connectors per MBF. Hold-downs, hangers, and other connectors should be taken off by the piece.

  Included with material costs are fasteners for a normal installation. Gordian's RSMeans engineers use manufacturers' recommendations, written specifications, and/or standard construction practice for the sizing and spacing of fasteners. Prices for various fasteners are shown for informational purposes only. Adjustments should be made if unusual fastening conditions exist.

06 10 00 Carpentry

- Lumber is a traded commodity and therefore sensitive to supply and demand in the marketplace. Even with "budgetary" estimating of wood-framed projects, it is advisable to call local suppliers for the latest market pricing.

- The common quantity unit for wood-framed projects is "thousand board feet" (MBF). A board foot is a volume of wood—1" x 1' x 1' or 144 cubic inches. Board-foot quantities are generally calculated using nominal material dimensions— dressed sizes are ignored. Board foot per lineal foot of any stick of lumber can be calculated by dividing the nominal cross-sectional area by 12. As an example, 2,000 lineal feet of 2 x 12 equates to 4 MBF by dividing the nominal area, 2 x 12, by 12, which equals 2, and multiplying by 2,000 to give 4,000 board feet. This simple rule applies to all nominal dimensioned lumber.

- Waste is an issue of concern at the quantity takeoff for any area of construction. Framing lumber is sold in even foot lengths, i.e., 8', 10', 12', 14', 16' and depending on spans, wall heights, and the grade of lumber, waste is inevitable. A rule of thumb for lumber waste is 5%–10% depending on material quality and the complexity of the framing.

- Wood in various forms and shapes is used in many projects, even where the main structural framing is steel, concrete, or masonry. Plywood as a back-up partition material and 2x boards used as blocking and cant strips around roof edges are two common examples. The estimator should ensure that the costs of all wood materials are included in the final estimate.

06 20 00 Finish Carpentry

- It is necessary to consider the grade of workmanship when estimating labor costs for erecting millwork and an interior finish. In practice, there are three grades: premium, custom, and economy. The RSMeans daily output for base and case moldings is in the range of 200 to 250 L.F. per carpenter per day. This is appropriate for most average custom-grade projects. For premium projects, an adjustment to productivity of 25%–50% should be made, depending on the complexity of the job.

Reference Numbers

Reference numbers are shown at the beginning of some major classifications. These numbers refer to related items in the Reference Section. The reference information may be an estimating procedure, an alternate pricing method, or technical information.

Note: Not all subdivisions listed here necessarily appear. ∎

06 16 36 – Wood Panel Product Sheathing

06 16 36.10 Sheathing	Crew	Daily Output	Labor-Hours	Unit	Material	2018 Bare Costs Labor	Equipment	Total	Total Incl O&P
0010 **SHEATHING**									
0012 Plywood on roofs, CDX									
0030 5/16" thick	2 Carp	1600	.010	S.F.	.59	.51		1.10	1.41
0035 Pneumatic nailed		1952	.008		.59	.42		1.01	1.27
0050 3/8" thick		1525	.010		.60	.53		1.13	1.47
0055 Pneumatic nailed		1860	.009		.60	.44		1.04	1.32
0100 1/2" thick		1400	.011		.62	.58		1.20	1.56
0105 Pneumatic nailed		1708	.009		.62	.48		1.10	1.40
0200 5/8" thick		1300	.012		.77	.62		1.39	1.79
0205 Pneumatic nailed		1586	.010		.77	.51		1.28	1.62
0300 3/4" thick		1200	.013		.92	.68		1.60	2.05
0305 Pneumatic nailed		1464	.011		.92	.55		1.47	1.86
0500 Plywood on walls, with exterior CDX, 3/8" thick		1200	.013		.60	.68		1.28	1.69
0505 Pneumatic nailed		1488	.011		.60	.55		1.15	1.49
0600 1/2" thick		1125	.014		.62	.72		1.34	1.78
0605 Pneumatic nailed		1395	.011		.62	.58		1.20	1.57
0700 5/8" thick		1050	.015		.77	.77		1.54	2.02
0705 Pneumatic nailed		1302	.012		.77	.62		1.39	1.79
0800 3/4" thick		975	.016		.92	.83		1.75	2.29
0805 Pneumatic nailed		1209	.013		.92	.67		1.59	2.04

Estimating Tips
07 10 00 Dampproofing and Waterproofing
- Be sure of the job specifications before pricing this subdivision. The difference in cost between waterproofing and dampproofing can be great. Waterproofing will hold back standing water. Dampproofing prevents the transmission of water vapor. Also included in this section are vapor retarding membranes.

07 20 00 Thermal Protection
- Insulation and fireproofing products are measured by area, thickness, volume or R-value. Specifications may give only what the specific R-value should be in a certain situation. The estimator may need to choose the type of insulation to meet that R-value.

07 30 00 Steep Slope Roofing
07 40 00 Roofing and Siding Panels
- Many roofing and siding products are bought and sold by the square. One square is equal to an area that measures 100 square feet.

  This simple change in unit of measure could create a large error if the estimator is not observant. Accessories necessary for a complete installation must be figured into any calculations for both material and labor.

07 50 00 Membrane Roofing
07 60 00 Flashing and Sheet Metal
07 70 00 Roofing and Wall Specialties and Accessories
- The items in these subdivisions compose a roofing system. No one component completes the installation, and all must be estimated. Built-up or single-ply membrane roofing systems are made up of many products and installation trades. Wood blocking at roof perimeters or penetrations, parapet coverings, reglets, roof drains, gutters, downspouts, sheet metal flashing, skylights, smoke vents, and roof hatches all need to be considered along with the roofing material. Several different installation trades will need to work together on the roofing system. Inherent difficulties in the scheduling and coordination of various trades must be accounted for when estimating labor costs.

07 90 00 Joint Protection
- To complete the weather-tight shell, the sealants and caulkings must be estimated. Where different materials meet—at expansion joints, at flashing penetrations, and at hundreds of other locations throughout a construction project—caulking and sealants provide another line of defense against water penetration. Often, an entire system is based on the proper location and placement of caulking or sealants. The detailed drawings that are included as part of a set of architectural plans show typical locations for these materials. When caulking or sealants are shown at typical locations, this means the estimator must include them for all the locations where this detail is applicable. Be careful to keep different types of sealants separate, and remember to consider backer rods and primers if necessary.

Reference Numbers
Reference numbers are shown at the beginning of some major classifications. These numbers refer to related items in the Reference Section. The reference information may be an estimating procedure, an alternate pricing method, or technical information.

Note: Not all subdivisions listed here necessarily appear. ■

07 84 13 – Penetration Firestopping

07 84 13.10 Firestopping		Crew	Daily Output	Labor-Hours	Unit	Material	2018 Bare Costs Labor	Equipment	Total	Total Incl O&P
0010	**FIRESTOPPING** R078413-30									
0100	Metallic piping, non insulated									
0110	Through walls, 2" diameter	1 Carp	16	.500	Ea.	3.63	25.50		29.13	42.50
0120	4" diameter		14	.571		6.35	29		35.35	51
0130	6" diameter		12	.667		9.30	34		43.30	62
0140	12" diameter		10	.800		18.60	40.50		59.10	82.50
0150	Through floors, 2" diameter		32	.250		1.85	12.70		14.55	21.50
0160	4" diameter		28	.286		3.25	14.50		17.75	25.50
0170	6" diameter		24	.333		4.74	16.90		21.64	30.50
0180	12" diameter		20	.400		9.55	20.50		30.05	41.50
0190	Metallic piping, insulated									
0200	Through walls, 2" diameter	1 Carp	16	.500	Ea.	6.35	25.50		31.85	45.50
0210	4" diameter		14	.571		9.30	29		38.30	54.50
0220	6" diameter		12	.667		12.50	34		46.50	65.50
0230	12" diameter		10	.800		21.50	40.50		62	85.50
0240	Through floors, 2" diameter		32	.250		3.25	12.70		15.95	23
0250	4" diameter		28	.286		4.74	14.50		19.24	27
0260	6" diameter		24	.333		6.40	16.90		23.30	32.50
0270	12" diameter		20	.400		9.75	20.50		30.25	42
0280	Non metallic piping, non insulated									
0290	Through walls, 2" diameter	1 Carp	12	.667	Ea.	46	34		80	102
0300	4" diameter		10	.800		89.50	40.50		130	160
0310	6" diameter		8	1		172	50.50		222.50	266
0330	Through floors, 2" diameter		16	.500		30	25.50		55.50	71.50
0340	4" diameter		6	1.333		57.50	67.50		125	166
0350	6" diameter		6	1.333		111	67.50		178.50	225
0370	Ductwork, insulated & non insulated, round									
0380	Through walls, 6" diameter	1 Carp	12	.667	Ea.	12.50	34		46.50	65.50
0390	12" diameter		10	.800		21.50	40.50		62	85.50
0400	18" diameter		8	1		26	50.50		76.50	106
0410	Through floors, 6" diameter		16	.500		6.40	25.50		31.90	45.50
0420	12" diameter		14	.571		10.90	29		39.90	56
0430	18" diameter		12	.667		13.25	34		47.25	66
0440	Ductwork, insulated & non insulated, rectangular									
0450	With stiffener/closure angle, through walls, 6" x 12"	1 Carp	8	1	Ea.	16.70	50.50		67.20	95.50
0460	12" x 24"		6	1.333		33.50	67.50		101	140
0470	24" x 48"		4	2		67	101		168	228
0480	With stiffener/closure angle, through floors, 6" x 12"		10	.800		8.35	40.50		48.85	71
0490	12" x 24"		8	1		16.70	50.50		67.20	95.50
0500	24" x 48"		6	1.333		33.50	67.50		101	140
0510	Multi trade openings									
0520	Through walls, 6" x 12"	1 Carp	2	4	Ea.	46.50	203		249.50	360
0530	12" x 24"	"	1	8		152	405		557	785
0540	24" x 48"	2 Carp	1	16		540	810		1,350	1,825
0550	48" x 96"	"	.75	21.333		2,025	1,075		3,100	3,875
0560	Through floors, 6" x 12"	1 Carp	2	4		42	203		245	355
0570	12" x 24"	"	1	8		58	405		463	685
0580	24" x 48"	2 Carp	.75	21.333		139	1,075		1,214	1,800
0590	48" x 96"	"	.50	32		330	1,625		1,955	2,850
0600	Structural penetrations, through walls									
0610	Steel beams, W8 x 10	1 Carp	8	1	Ea.	70.50	50.50		121	155
0620	W12 x 14		6	1.333		102	67.50		169.50	215
0630	W21 x 44		5	1.600		146	81		227	284

07 84 Firestopping

07 84 13 – Penetration Firestopping

07 84 13.10 Firestopping		Crew	Daily Output	Labor-Hours	Unit	Material	2018 Bare Costs Labor	Equipment	Total	Total Incl O&P
0640	W36 x 135	1 Carp	3	2.667	Ea.	226	135		361	455
0650	Bar joists, 18" deep		6	1.333		41.50	67.50		109	149
0660	24" deep		6	1.333		55.50	67.50		123	164
0670	36" deep		5	1.600		83	81		164	216
0680	48" deep		4	2		111	101		212	276
0690	Construction joints, floor slab at exterior wall									
0700	Precast, brick, block or drywall exterior									
0710	2" wide joint	1 Carp	125	.064	L.F.	9.80	3.24		13.04	15.70
0720	4" wide joint	"	75	.107	"	15.30	5.40		20.70	25
0730	Metal panel, glass or curtain wall exterior									
0740	2" wide joint	1 Carp	40	.200	L.F.	9.80	10.15		19.95	26
0750	4" wide joint	"	25	.320	"	15.30	16.20		31.50	41.50
0760	Floor slab to drywall partition									
0770	Flat joint	1 Carp	100	.080	L.F.	13.60	4.06		17.66	21
0780	Fluted joint		50	.160		15.15	8.10		23.25	29
0790	Etched fluted joint		75	.107		19.05	5.40		24.45	29.50
0800	Floor slab to concrete/masonry partition									
0810	Flat joint	1 Carp	75	.107	L.F.	13.60	5.40		19	23
0820	Fluted joint	"	50	.160	"	15.50	8.10		23.60	29.50
0830	Concrete/CMU wall joints									
0840	1" wide	1 Carp	100	.080	L.F.	19.40	4.06		23.46	27.50
0850	2" wide		75	.107		24	5.40		29.40	35
0860	4" wide		50	.160		33.50	8.10		41.60	49.50
0870	Concrete/CMU floor joints									
0880	1" wide	1 Carp	200	.040	L.F.	19.40	2.03		21.43	24.50
0890	2" wide		150	.053		24	2.70		26.70	30.50
0900	4" wide		100	.080		34.50	4.06		38.56	43.50

07 91 Preformed Joint Seals

07 91 13 – Compression Seals

07 91 13.10 Compression Seals		Crew	Daily Output	Labor-Hours	Unit	Material	2018 Bare Costs Labor	Equipment	Total	Total Incl O&P
0010	**COMPRESSION SEALS**									
4900	O-ring type cord, 1/4"	1 Bric	472	.017	L.F.	.44	.85		1.29	1.79
4910	1/2"		440	.018		1.20	.91		2.11	2.72
4920	3/4"		424	.019		2.35	.95		3.30	4.04
4930	1"		408	.020		4.07	.99		5.06	6
4940	1-1/4"		384	.021		7.75	1.05		8.80	10.10
4950	1-1/2"		368	.022		9.50	1.09		10.59	12.10
4960	1-3/4"		352	.023		16.20	1.14		17.34	19.60
4970	2"		344	.023		22	1.17		23.17	26

07 91 23 – Backer Rods

07 91 23.10 Backer Rods		Crew	Daily Output	Labor-Hours	Unit	Material	2018 Bare Costs Labor	Equipment	Total	Total Incl O&P
0010	**BACKER RODS**									
0030	Backer rod, polyethylene, 1/4" diameter	1 Bric	4.60	1.739	C.L.F.	2.37	87.50		89.87	137
0050	1/2" diameter		4.60	1.739		4.28	87.50		91.78	139
0070	3/4" diameter		4.60	1.739		6.75	87.50		94.25	141
0090	1" diameter		4.60	1.739		12.10	87.50		99.60	147

For customer support on your Electrical Costs with RSMeans data, call 800.448.8182.

43

07 91 Preformed Joint Seals

07 91 26 – Joint Fillers

07 91 26.10 Joint Fillers	Crew	Daily Output	Labor-Hours	Unit	Material	2018 Bare Costs Labor	Equipment	Total	Total Incl O&P
0010	**JOINT FILLERS**								
4360	Butyl rubber filler, 1/4" x 1/4"	1 Bric	290	.028	L.F.	.22	1.39		1.61	2.37
4365	1/2" x 1/2"		250	.032		.89	1.61		2.50	3.44
4370	1/2" x 3/4"		210	.038		1.34	1.92		3.26	4.40
4375	3/4" x 3/4"		230	.035		2.01	1.75		3.76	4.89
4380	1" x 1"		180	.044		2.68	2.24		4.92	6.35
4390	For coloring, add					12%				
4980	Polyethylene joint backing, 1/4" x 2"	1 Bric	2.08	3.846	C.L.F.	13.50	193		206.50	310
4990	1/4" x 6"		1.28	6.250	"	29	315		344	510
5600	Silicone, room temp vulcanizing foam seal, 1/4" x 1/2"		1312	.006	L.F.	.45	.31		.76	.97
5610	1/2" x 1/2"		656	.012		.90	.61		1.51	1.93
5620	1/2" x 3/4"		442	.018		1.35	.91		2.26	2.88
5630	3/4" x 3/4"		328	.024		2.03	1.23		3.26	4.12
5640	1/8" x 1"		1312	.006		.45	.31		.76	.97
5650	1/8" x 3"		442	.018		1.35	.91		2.26	2.88
5670	1/4" x 3"		295	.027		2.71	1.36		4.07	5.05
5680	1/4" x 6"		148	.054		5.40	2.72		8.12	10.10
5690	1/2" x 6"		82	.098		10.85	4.91		15.76	19.40
5700	1/2" x 9"		52.50	.152		16.25	7.65		23.90	29.50
5710	1/2" x 12"		33	.242		21.50	12.20		33.70	42.50

07 92 Joint Sealants

07 92 13 – Elastomeric Joint Sealants

07 92 13.20 Caulking and Sealant Options

		Crew	Daily Output	Labor-Hours	Unit	Material	Labor	Equipment	Total	Total Incl O&P
0010	**CAULKING AND SEALANT OPTIONS**									
0050	Latex acrylic based, bulk				Gal.	29.50			29.50	32.50
0055	Bulk in place 1/4" x 1/4" bead	1 Bric	300	.027	L.F.	.09	1.34		1.43	2.15
0060	1/4" x 3/8"		294	.027		.16	1.37		1.53	2.27
0065	1/4" x 1/2"		288	.028		.21	1.40		1.61	2.37
0075	3/8" x 3/8"		284	.028		.23	1.42		1.65	2.43
0080	3/8" x 1/2"		280	.029		.31	1.44		1.75	2.54
0085	3/8" x 5/8"		276	.029		.39	1.46		1.85	2.66
0095	3/8" x 3/4"		272	.029		.47	1.48		1.95	2.77
0100	1/2" x 1/2"		275	.029		.42	1.46		1.88	2.70
0105	1/2" x 5/8"		269	.030		.52	1.50		2.02	2.86
0110	1/2" x 3/4"		263	.030		.62	1.53		2.15	3.03
0115	1/2" x 7/8"		256	.031		.73	1.57		2.30	3.21
0120	1/2" x 1"		250	.032		.83	1.61		2.44	3.37
0125	3/4" x 3/4"		244	.033		.94	1.65		2.59	3.55
0130	3/4" x 1"		225	.036		1.25	1.79		3.04	4.11
0135	1" x 1"		200	.040		1.66	2.01		3.67	4.91
0190	Cartridges				Gal.	32.50			32.50	36
0200	11 fl. oz. cartridge				Ea.	2.81			2.81	3.09
0600	1/2" x 1/2"	1 Bric	275	.029	L.F.	.46	1.46		1.92	2.74
3200	Polyurethane, 1 or 2 component				Gal.	53.50			53.50	58.50
3300	Cartridges				"	67			67	73.50
3500	Bulk, in place, 1/4" x 1/4"	1 Bric	300	.027	L.F.	.17	1.34		1.51	2.24
3655	1/2" x 1/4"		288	.028		.35	1.40		1.75	2.52
3800	3/4" x 3/8"		272	.029		.79	1.48		2.27	3.12
3900	1" x 1/2"		250	.032		1.39	1.61		3	3.99

07 92 Joint Sealants

07 92 19 – Acoustical Joint Sealants

07 92 19.10 Acoustical Sealant	Crew	Daily Output	Labor-Hours	Unit	Material	2018 Bare Costs Labor	Equipment	Total	Total Incl O&P
0010 **ACOUSTICAL SEALANT**									
0020 Acoustical sealant, elastomeric, cartridges				Ea.	8.45			8.45	9.25
0025 In place, 1/4" x 1/4"	1 Bric	300	.027	L.F.	.34	1.34		1.68	2.43
0030 1/4" x 1/2"		288	.028		.69	1.40		2.09	2.90
0035 1/2" x 1/2"		275	.029		1.38	1.46		2.84	3.75
0040 1/2" x 3/4"		263	.030		2.07	1.53		3.60	4.61
0045 3/4" x 3/4"		244	.033		3.10	1.65		4.75	5.95
0050 1" x 1"		200	.040		5.50	2.01		7.51	9.15

For customer support on your Electrical Costs with RSMeans data, call 800.448.8182.

45

Division Notes

	CREW	DAILY OUTPUT	LABOR-HOURS	UNIT	BARE COSTS				TOTAL INCL O&P
					MAT.	LABOR	EQUIP.	TOTAL	

Estimating Tips
08 10 00 Doors and Frames
All exterior doors should be addressed for their energy conservation (insulation and seals).

- Most metal doors and frames look alike, but there may be significant differences among them. When estimating these items, be sure to choose the line item that most closely compares to the specification or door schedule requirements regarding:
  - type of metal
  - metal gauge
  - door core material
  - fire rating
  - finish
- Wood and plastic doors vary considerably in price. The primary determinant is the veneer material. Lauan, birch, and oak are the most common veneers. Other variables include the following:
  - hollow or solid core
  - fire rating
  - flush or raised panel
  - finish
- Door pricing includes bore for cylindrical locksets and mortise for hinges.

08 30 00 Specialty Doors and Frames
- There are many varieties of special doors, and they are usually priced per each. Add frames, hardware, or operators required for a complete installation.

08 40 00 Entrances, Storefronts, and Curtain Walls
- Glazed curtain walls consist of the metal tube framing and the glazing material. The cost data in this subdivision is presented for the metal tube framing alone or the composite wall. If your estimate requires a detailed takeoff of the framing, be sure to add the glazing cost and any tints.

08 50 00 Windows
- Steel windows are unglazed and aluminum can be glazed or unglazed. Some metal windows are priced without glass. Refer to 08 80 00 Glazing for glass pricing. The grade C indicates commercial grade windows, usually ASTM C-35.
- All wood windows and vinyl are priced preglazed. The glazing is insulating glass. Add the cost of screens and grills if required and not already included.

08 70 00 Hardware
- Hardware costs add considerably to the cost of a door. The most efficient method to determine the hardware requirements for a project is to review the door and hardware schedule together. One type of door may have different hardware, depending on the door usage.
- Door hinges are priced by the pair, with most doors requiring 1-1/2 pairs per door. The hinge prices do not include installation labor, because it is included in door installation.

Hinges are classified according to the frequency of use, base material, and finish.

08 80 00 Glazing
- Different openings require different types of glass. The most common types are:
  - float
  - tempered
  - insulating
  - impact-resistant
  - ballistic-resistant
- Most exterior windows are glazed with insulating glass. Entrance doors and window walls, where the glass is less than 18" from the floor, are generally glazed with tempered glass. Interior windows and some residential windows are glazed with float glass.
- Coastal communities require the use of impact-resistant glass, dependent on wind speed.
- The insulation or 'u' value is a strong consideration, along with solar heat gain, to determine total energy efficiency.

Reference Numbers
Reference numbers are shown at the beginning of some major classifications. These numbers refer to related items in the Reference Section. The reference information may be an estimating procedure, an alternate pricing method, or technical information.

Note: Not all subdivisions listed here necessarily appear. ∎

08 31 Access Doors and Panels

08 31 13 – Access Doors and Frames

08 31 13.10 Types of Framed Access Doors	Crew	Daily Output	Labor-Hours	Unit	Material	2018 Bare Costs Labor	Equipment	Total	Total Incl O&P
0010 **TYPES OF FRAMED ACCESS DOORS**									
4000 Recessed door for drywall									
4100 Metal 12" x 12"	1 Carp	6	1.333	Ea.	74.50	67.50		142	185
4150 12" x 24"		5.50	1.455		119	74		193	243
4200 24" x 36"	↓	5	1.600	↓	168	81		249	310

Estimating Tips
General
- Room Finish Schedule: A complete set of plans should contain a room finish schedule. If one is not available, it would be well worth the time and effort to obtain one.

09 20 00 Plaster and Gypsum Board
- Lath is estimated by the square yard plus a 5% allowance for waste. Furring, channels, and accessories are measured by the linear foot. An extra foot should be allowed for each accessory miter or stop.
- Plaster is also estimated by the square yard. Deductions for openings vary by preference, from zero deduction to 50% of all openings over 2 feet in width. The estimator should allow one extra square foot for each linear foot of horizontal interior or exterior angle located below the ceiling level. Also, double the areas of small radius work.
- Drywall accessories, studs, track, and acoustical caulking are all measured by the linear foot. Drywall taping is figured by the square foot. Gypsum wallboard is estimated by the square foot. No material deductions should be made for door or window openings under 32 S.F.

09 60 00 Flooring
- Tile and terrazzo areas are taken off on a square foot basis. Trim and base materials are measured by the linear foot. Accent tiles are listed per each. Two basic methods of installation are used. Mud set is approximately 30% more expensive than thin set. The cost of grout is included with tile unit price lines unless otherwise noted. In terrazzo work, be sure to include the linear footage of embedded decorative strips, grounds, machine rubbing, and power cleanup.
- Wood flooring is available in strip, parquet, or block configuration. The latter two types are set in adhesives with quantities estimated by the square foot. The laying pattern will influence labor costs and material waste. In addition to the material and labor for laying wood floors, the estimator must make allowances for sanding and finishing these areas, unless the flooring is prefinished.
- Sheet flooring is measured by the square yard. Roll widths vary, so consideration should be given to use the most economical width, as waste must be figured into the total quantity. Consider also the installation methods available—direct glue down or stretched. Direct glue-down installation is assumed with sheet carpet unit price lines unless otherwise noted.

09 70 00 Wall Finishes
- Wall coverings are estimated by the square foot. The area to be covered is measured—length by height of the wall above the baseboards—to calculate the square footage of each wall. This figure is divided by the number of square feet in the single roll which is being used. Deduct, in full, the areas of openings such as doors and windows. Where a pattern match is required allow 25%–30% waste.

09 80 00 Acoustic Treatment
- Acoustical systems fall into several categories. The takeoff of these materials should be by the square foot of area with a 5% allowance for waste. Do not forget about scaffolding, if applicable, when estimating these systems.

09 90 00 Painting and Coating
- A major portion of the work in painting involves surface preparation. Be sure to include cleaning, sanding, filling, and masking costs in the estimate.
- Protection of adjacent surfaces is not included in painting costs. When considering the method of paint application, an important factor is the amount of protection and masking required. These must be estimated separately and may be the determining factor in choosing the method of application.

Reference Numbers
Reference numbers are shown at the beginning of some major classifications. These numbers refer to related items in the Reference Section. The reference information may be an estimating procedure, an alternate pricing method, or technical information.

Note: Not all subdivisions listed here necessarily appear. ■

Did you know?

RSMeans data is available through our online application with 24/7 access:
- Search for unit prices by keyword
- Leverage the most up-to-date data
- Build and export estimates

Try it free for 30 days!
www.rsmeans.com/2018freetrial

09 22 Supports for Plaster and Gypsum Board

09 22 03 – Fastening Methods for Finishes

09 22 03.20 Drilling Plaster/Drywall	Crew	Daily Output	Labor-Hours	Unit	Material	2018 Bare Costs Labor	Equipment	Total	Total Incl O&P
0010 **DRILLING PLASTER/DRYWALL**									
1100 Drilling & layout for drywall/plaster walls, up to 1" deep, no anchor									
1200 Holes, 1/4" diameter	1 Carp	150	.053	Ea.	.01	2.70		2.71	4.13
1300 3/8" diameter		140	.057		.01	2.90		2.91	4.42
1400 1/2" diameter		130	.062		.01	3.12		3.13	4.76
1500 3/4" diameter		120	.067		.02	3.38		3.40	5.15
1600 1" diameter		110	.073		.02	3.69		3.71	5.60
1700 1-1/4" diameter		100	.080		.04	4.06		4.10	6.25
1800 1-1/2" diameter		90	.089		.06	4.51		4.57	6.90
1900 For ceiling installations, add						40%			

09 57 Special Function Ceilings

09 57 53 – Security Ceiling Assemblies

09 57 53.10 Ceiling Assem., Security, Radio Freq. Shielding	Crew	Daily Output	Labor-Hours	Unit	Material	2018 Bare Costs Labor	Equipment	Total	Total Incl O&P
0010 **CEILING ASSEMBLY, SECURITY, RADIO FREQUENCY SHIELDING**									
0020 Prefabricated, galvanized steel	2 Carp	375	.043	SF Surf	5.30	2.16	·	7.46	9.10
0050 5 oz., copper ceiling panel		155	.103		4.48	5.25		9.73	12.90
0110 12 oz., copper ceiling panel		140	.114		9.05	5.80		14.85	18.75
0250 Ceiling hangers	E-1	45	.533	Ea.	33.50	29	2.19	64.69	85
0300 Shielding transition, 11 ga. preformed angles	"	1365	.018	L.F.	33.50	.95	.07	34.52	38

09 69 Access Flooring

09 69 13 – Rigid-Grid Access Flooring

09 69 13.10 Access Floors	Crew	Daily Output	Labor-Hours	Unit	Material	2018 Bare Costs Labor	Equipment	Total	Total Incl O&P
0010 **ACCESS FLOORS**									
0015 Access floor pkg. including panel, pedestal, & stringers 1500 lbs. load									
0100 Package pricing, conc. fill panels, no fin., 6" ht.	4 Carp	750	.043	S.F.	16.70	2.16		18.86	21.50
0105 12" ht.		750	.043		17.10	2.16		19.26	22
0110 18" ht.		750	.043		17.15	2.16		19.31	22
0115 24" ht.		750	.043		17.55	2.16		19.71	22.50
0120 Package pricing, steel panels, no fin., 6" ht.		750	.043		13.35	2.16		15.51	17.95
0125 12" ht.		750	.043		16.75	2.16		18.91	21.50
0130 18" ht.		750	.043		16.85	2.16		19.01	22
0135 24" ht.		750	.043		16.95	2.16		19.11	22
0140 Package pricing, wood core panels, no fin., 6" ht.		750	.043		13.50	2.16		15.66	18.15
0145 12" ht.		750	.043		13.75	2.16		15.91	18.40
0150 18" ht.		750	.043		13.75	2.16		15.91	18.40
0155 24" ht.		750	.043		14	2.16		16.16	18.70
0160 Pkg. pricing, conc. fill pnls, no stringers, no fin., 6" ht, 1250 lbs. load		700	.046		10.80	2.32		13.12	15.40
0165 12" ht.		700	.046		10	2.32		12.32	14.55
0170 18" ht.		700	.046		10.95	2.32		13.27	15.55
0175 24" ht.		700	.046		11.90	2.32		14.22	16.60
0250 Panels, 2' x 2' conc. fill, no fin.	2 Carp	500	.032		23	1.62		24.62	27.50
0255 With 1/8" high pressure laminate		500	.032		56	1.62		57.62	64.50
0260 Metal panels with 1/8" high pressure laminate		500	.032		66.50	1.62		68.12	75.50
0265 Wood core panels with 1/8" high pressure laminate		500	.032		43.50	1.62		45.12	50.50
0400 Aluminum panels, no fin.		500	.032		33.50	1.62		35.12	39
0600 For carpet covering, add					9.05			9.05	10
0700 For vinyl floor covering, add					9.35			9.35	10.30

09 69 13.10 Access Floors		Crew	Daily Output	Labor-Hours	Unit	Material	2018 Bare Costs Labor	Equipment	Total	Total Incl O&P
0900	For high pressure laminate covering, add				S.F.	7.70			7.70	8.45
0910	For snap on stringer system, add	2 Carp	1000	.016	▼	1.59	.81		2.40	2.99
1000	Machine cutouts after initial installation	1 Carp	50	.160	Ea.	20	8.10		28.10	34.50
1050	Pedestals, 6" to 12"	2 Carp	85	.188		8.65	9.55		18.20	24
1100	Air conditioning grilles, 4" x 12"	1 Carp	17	.471		72.50	24		96.50	117
1150	4" x 18"	"	14	.571	▼	99.50	29		128.50	154
1200	Approach ramps, steel	2 Carp	60	.267	S.F.	26.50	13.50		40	49.50
1300	Aluminum	"	40	.400	"	34	20.50		54.50	68.50
1500	Handrail, 2 rail, aluminum	1 Carp	15	.533	L.F.	117	27		144	170

For customer support on your Electrical Costs with RSMeans data, call 800.448.8182.

51

Division Notes

	CREW	DAILY OUTPUT	LABOR-HOURS	UNIT	BARE COSTS				TOTAL INCL O&P
					MAT.	LABOR	EQUIP.	TOTAL	

Estimating Tips
General
- The items in this division are usually priced per square foot or each.
- Many items in Division 10 require some type of support system or special anchors that are not usually furnished with the item. The required anchors must be added to the estimate in the appropriate division.
- Some items in Division 10, such as lockers, may require assembly before installation. Verify the amount of assembly required. Assembly can often exceed installation time.

10 20 00 Interior Specialties
- Support angles and blocking are not included in the installation of toilet compartments, shower/dressing compartments, or cubicles. Appropriate line items from Division 5 or 6 may need to be added to support the installations.
- Toilet partitions are priced by the stall. A stall consists of a side wall, pilaster, and door with hardware. Toilet tissue holders and grab bars are extra.
- The required acoustical rating of a folding partition can have a significant impact on costs. Verify the sound transmission coefficient rating of the panel priced against the specification requirements.

- Grab bar installation does not include supplemental blocking or backing to support the required load. When grab bars are installed at an existing facility, provisions must be made to attach the grab bars to a solid structure.

Reference Numbers
Reference numbers are shown at the beginning of some major classifications. These numbers refer to related items in the Reference Section. The reference information may be an estimating procedure, an alternate pricing method, or technical information.

Note: Not all subdivisions listed here necessarily appear. ■

Did you know?

RSMeans data is available through our online application with 24/7 access:

- Search for unit prices by keyword
- Leverage the most up-to-date data
- Build and export estimates

Try it free for 30 days!
www.rsmeans.com/2018freetrial

10 28 16 – Bath Accessories

10 28 16.20 Medicine Cabinets	Crew	Daily Output	Labor-Hours	Unit	Material	2018 Bare Costs Labor	Equipment	Total	Total Incl O&P
0010 **MEDICINE CABINETS**									
0020 With mirror, sst frame, 16" x 22", unlighted	1 Carp	14	.571	Ea.	99	29		128	153
0100 Wood frame		14	.571		136	29		165	194
0300 Sliding mirror doors, 20" x 16" x 4-3/4", unlighted		7	1.143		124	58		182	225
0400 24" x 19" x 8-1/2", lighted		5	1.600		206	81		287	350
0600 Triple door, 30" x 32", unlighted, plywood body		7	1.143		350	58		408	475
0700 Steel body		7	1.143		375	58		433	500
0900 Oak door, wood body, beveled mirror, single door		7	1.143		175	58		233	280
1000 Double door		6	1.333		370	67.50		437.50	510
1200 Hotel cabinets, stainless, with lower shelf, unlighted		10	.800		218	40.50		258.50	300
1300 Lighted		5	1.600		340	81		421	500

For customer support on your Electrical Costs with RSMeans data, call 800.448.8182.

Estimating Tips
General

- The items in this division are usually priced per square foot or each. Many of these items are purchased by the owner for installation by the contractor. Check the specifications for responsibilities and include time for receiving, storage, installation, and mechanical and electrical hookups in the appropriate divisions.

- Many items in Division 11 require some type of support system that is not usually furnished with the item. Examples of these systems include blocking for the attachment of casework and support angles for ceiling-hung projection screens. The required blocking or supports must be added to the estimate in the appropriate division.

- Some items in Division 11 may require assembly or electrical hookups. Verify the amount of assembly required or the need for a hard electrical connection and add the appropriate costs.

Reference Numbers

Reference numbers are shown at the beginning of some major classifications. These numbers refer to related items in the Reference Section. The reference information may be an estimating procedure, an alternate pricing method, or technical information.

Note: Not all subdivisions listed here necessarily appear. ■

11 12 Parking Control Equipment

11 12 13 – Parking Key and Card Control Units

11 12 13.10 Parking Control Units	Crew	Daily Output	Labor-Hours	Unit	Material	2018 Bare Costs Labor	Equipment	Total	Total Incl O&P
0010 **PARKING CONTROL UNITS**									
5100 Card reader	1 Elec	2	4	Ea.	1,400	233		1,633	1,900
5120 Proximity with customer display	2 Elec	1	16		6,000	930		6,930	8,000
6000 Parking control software, basic functionality	1 Elec	.50	16		24,000	930		24,930	27,800
6020 Multi-function	"	.20	40	↓	98,000	2,325		100,325	111,500

11 12 16 – Parking Ticket Dispensers

11 12 16.10 Ticket Dispensers

	Crew	Daily Output	Labor-Hours	Unit	Material	Labor	Equipment	Total	Total Incl O&P
0010 **TICKET DISPENSERS**									
5900 Ticket spitter with time/date stamp, standard	2 Elec	2	8	Ea.	6,675	465		7,140	8,025
5920 Mag stripe encoding	"	2	8	"	19,000	465		19,465	21,500

11 12 26 – Parking Fee Collection Equipment

11 12 26.13 Parking Fee Coin Collection Equipment

	Crew	Daily Output	Labor-Hours	Unit	Material	Labor	Equipment	Total	Total Incl O&P
0010 **PARKING FEE COIN COLLECTION EQUIPMENT**									
5200 Cashier booth, average	B-22	1	30	Ea.	10,600	1,400	194	12,194	14,100
5300 Collector station, pay on foot	2 Elec	.20	80		112,000	4,650		116,650	130,000
5320 Credit card only	"	.50	32	↓	20,700	1,850		22,550	25,500

11 12 26.23 Fee Equipment

	Crew	Daily Output	Labor-Hours	Unit	Material	Labor	Equipment	Total	Total Incl O&P
0010 **FEE EQUIPMENT**									
5600 Fee computer	1 Elec	1.50	5.333	Ea.	12,200	310		12,510	13,900

11 12 33 – Parking Gates

11 12 33.13 Lift Arm Parking Gates

	Crew	Daily Output	Labor-Hours	Unit	Material	Labor	Equipment	Total	Total Incl O&P
0010 **LIFT ARM PARKING GATES**									
5000 Barrier gate with programmable controller	2 Elec	3	5.333	Ea.	3,050	310		3,360	3,825
5020 Industrial		3	5.333		5,250	310		5,560	6,250
5050 Non-programmable, with reader and 12' arm		3	5.333		1,925	310		2,235	2,600
5500 Exit verifier	↓	1	16		19,500	930		20,430	22,800
5700 Full sign, 4" letters	1 Elec	2	4		1,250	233		1,483	1,725
5800 Inductive loop	2 Elec	4	4		210	233		443	580
5950 Vehicle detector, microprocessor based	1 Elec	3	2.667	↓	440	155		595	715

11 21 Retail and Service Equipment

11 21 73 – Commercial Laundry and Dry Cleaning Equipment

11 21 73.13 Dry Cleaning Equipment

	Crew	Daily Output	Labor-Hours	Unit	Material	Labor	Equipment	Total	Total Incl O&P
0010 **DRY CLEANING EQUIPMENT**									
2000 Dry cleaners, electric, 20 lb. capacity, not incl. rough-in	L-1	.20	80	Ea.	33,700	4,825		38,525	44,300
2050 25 lb. capacity		.17	94.118		51,500	5,675		57,175	65,000
2100 30 lb. capacity		.15	107		54,000	6,425		60,425	69,000
2150 60 lb. capacity	↓	.09	178	↓	78,500	10,700		89,200	102,000

11 21 73.19 Finishing Equipment

	Crew	Daily Output	Labor-Hours	Unit	Material	Labor	Equipment	Total	Total Incl O&P
0010 **FINISHING EQUIPMENT**									
3500 Folders, blankets & sheets, minimum	1 Elec	.17	47.059	Ea.	35,100	2,750		37,850	42,700
3700 King size with automatic stacker		.10	80		63,500	4,650		68,150	77,000
3800 For conveyor delivery, add	↓	.45	17.778	↓	16,100	1,025		17,125	19,400

11 21 73.23 Commercial Ironing Equipment

	Crew	Daily Output	Labor-Hours	Unit	Material	Labor	Equipment	Total	Total Incl O&P
0010 **COMMERCIAL IRONING EQUIPMENT**									
4500 Ironers, institutional, 110", single roll	1 Elec	.20	40	Ea.	34,000	2,325		36,325	40,900

11 21 Retail and Service Equipment

11 21 73 – Commercial Laundry and Dry Cleaning Equipment

11 21 73.26 Commercial Washers and Extractors

		Crew	Daily Output	Labor-Hours	Unit	Material	2018 Bare Costs Labor	Equipment	Total	Total Incl O&P
0010	**COMMERCIAL WASHERS AND EXTRACTORS**, not including rough-in									
6000	Combination washer/extractor, 20 lb. capacity	L-6	1.50	8	Ea.	6,575	485		7,060	7,950
6100	30 lb. capacity		.80	15		10,000	910		10,910	12,400
6200	50 lb. capacity		.68	17.647		12,700	1,075		13,775	15,600
6300	75 lb. capacity		.30	40		18,700	2,425		21,125	24,300
6350	125 lb. capacity		.16	75		29,100	4,550		33,650	39,000

11 21 73.33 Coin-Operated Laundry Equipment

		Crew	Daily Output	Labor-Hours	Unit	Material	2018 Bare Costs Labor	Equipment	Total	Total Incl O&P
0010	**COIN-OPERATED LAUNDRY EQUIPMENT**									
5290	Clothes washer									
5300	Commercial, coin operated, average	1 Plum	3	2.667	Ea.	1,325	166		1,491	1,700

11 30 Residential Equipment

11 30 13 – Residential Appliances

11 30 13.15 Cooking Equipment

		Crew	Daily Output	Labor-Hours	Unit	Material	2018 Bare Costs Labor	Equipment	Total	Total Incl O&P
0010	**COOKING EQUIPMENT**									
0020	Cooking range, 30" free standing, 1 oven, minimum	2 Clab	10	1.600	Ea.	470	64		534	610
0050	Maximum		4	4		2,300	159		2,459	2,775
0700	Free standing, 1 oven, 21" wide range, minimum		10	1.600		475	64		539	615
0750	21" wide, maximum		4	4		660	159		819	970
0900	Countertop cooktops, 4 burner, standard, minimum	1 Elec	6	1.333		325	77.50		402.50	475
0950	Maximum		3	2.667		1,750	155		1,905	2,150
1050	As above, but with grill and griddle attachment, minimum		6	1.333		1,450	77.50		1,527.50	1,700
1100	Maximum		3	2.667		3,625	155		3,780	4,200
1200	Induction cooktop, 30" wide		3	2.667		1,475	155		1,630	1,850
1250	Microwave oven, minimum		4	2		97	116		213	281
1300	Maximum		2	4		460	233		693	855

11 30 13.17 Kitchen Cleaning Equipment

		Crew	Daily Output	Labor-Hours	Unit	Material	2018 Bare Costs Labor	Equipment	Total	Total Incl O&P
0010	**KITCHEN CLEANING EQUIPMENT**									
2750	Dishwasher, built-in, 2 cycles, minimum	L-1	4	4	Ea.	296	241		537	685
2800	Maximum		2	8		455	480		935	1,225
2950	4 or more cycles, minimum		4	4		400	241		641	800
2960	Average		4	4		535	241		776	945
3000	Maximum		2	8		1,175	480		1,655	2,025

11 30 13.18 Waste Disposal Equipment

		Crew	Daily Output	Labor-Hours	Unit	Material	2018 Bare Costs Labor	Equipment	Total	Total Incl O&P
0010	**WASTE DISPOSAL EQUIPMENT**									
3300	Garbage disposal, sink type, minimum	L-1	10	1.600	Ea.	106	96.50		202.50	260
3350	Maximum	"	10	1.600	"	204	96.50		300.50	370

11 30 13.19 Kitchen Ventilation Equipment

		Crew	Daily Output	Labor-Hours	Unit	Material	2018 Bare Costs Labor	Equipment	Total	Total Incl O&P
0010	**KITCHEN VENTILATION EQUIPMENT**									
4150	Hood for range, 2 speed, vented, 30" wide, minimum	L-3	5	3.200	Ea.	93	176		269	370
4200	Maximum		3	5.333		910	293		1,203	1,450
4300	42" wide, minimum		5	3.200		170	176		346	450
4330	Custom		5	3.200		1,650	176		1,826	2,075
4350	Maximum		3	5.333		2,000	293		2,293	2,650
4500	For ventless hood, 2 speed, add					17.45			17.45	19.20
4650	For vented 1 speed, deduct from maximum					63			63	69

11 30 13.24 Washers

		Crew	Daily Output	Labor-Hours	Unit	Material	2018 Bare Costs Labor	Equipment	Total	Total Incl O&P
0010	**WASHERS**									
5000	Residential, 4 cycle, average	1 Plum	3	2.667	Ea.	975	166		1,141	1,325

For customer support on your Electrical Costs with RSMeans data, call 800.448.8182.

57

11 30 Residential Equipment

11 30 15 – Miscellaneous Residential Appliances

11 30 15.23 Water Heaters

		Crew	Daily Output	Labor-Hours	Unit	Material	2018 Bare Costs Labor	2018 Bare Costs Equipment	Total	Total Incl O&P
0010	**WATER HEATERS**									
6900	Electric, glass lined, 30 gallon, minimum	L-1	5	3.200	Ea.	900	193		1,093	1,275
6950	Maximum		3	5.333		1,250	320		1,570	1,850
7100	80 gallon, minimum		2	8		1,675	480		2,155	2,550
7150	Maximum	↓	1	16	↓	2,325	965		3,290	4,000

11 30 15.43 Air Quality

		Crew	Daily Output	Labor-Hours	Unit	Material	2018 Bare Costs Labor	2018 Bare Costs Equipment	Total	Total Incl O&P
0010	**AIR QUALITY**									
2450	Dehumidifier, portable, automatic, 15 pint	1 Elec	4	2	Ea.	205	116		321	400
2550	40 pint		3.75	2.133		250	124		374	460
3550	Heater, electric, built-in, 1250 watt, ceiling type, minimum		4	2		117	116		233	300
3600	Maximum		3	2.667		213	155		368	465
3700	Wall type, minimum		4	2		197	116		313	390
3750	Maximum		3	2.667		194	155		349	445
3900	1500 watt wall type, with blower		4	2		188	116		304	380
3950	3000 watt	↓	3	2.667		480	155		635	760
4850	Humidifier, portable, 8 gallons/day					148			148	162
5000	15 gallons/day				↓	191			191	210

11 32 Unit Kitchens

11 32 13 – Metal Unit Kitchens

11 32 13.10 Commercial Unit Kitchens

		Crew	Daily Output	Labor-Hours	Unit	Material	2018 Bare Costs Labor	2018 Bare Costs Equipment	Total	Total Incl O&P
0010	**COMMERCIAL UNIT KITCHENS**									
1500	Combination range, refrigerator and sink, 30" wide, minimum	L-1	2	8	Ea.	1,225	480		1,705	2,075
1550	Maximum		1	16		1,075	965		2,040	2,625
1570	60" wide, average		1.40	11.429		1,275	690		1,965	2,425
1590	72" wide, average		1.20	13.333		1,700	800		2,500	3,075
1600	Office model, 48" wide		2	8		1,550	480		2,030	2,450
1620	Refrigerator and sink only	↓	2.40	6.667	↓	2,625	400		3,025	3,475
1640	Combination range, refrigerator, sink, microwave									
1660	Oven and ice maker	L-1	.80	20	Ea.	4,825	1,200		6,025	7,100

11 41 Foodservice Storage Equipment

11 41 13 – Refrigerated Food Storage Cases

11 41 13.20 Refrigerated Food Storage Equipment

		Crew	Daily Output	Labor-Hours	Unit	Material	2018 Bare Costs Labor	2018 Bare Costs Equipment	Total	Total Incl O&P
0010	**REFRIGERATED FOOD STORAGE EQUIPMENT**									
2350	Cooler, reach-in, beverage, 6' long	Q-1	6	2.667	Ea.	3,200	149		3,349	3,750

11 42 Food Preparation Equipment

11 42 10 – Commercial Food Preparation Equipment

11 42 10.10 Choppers, Mixers and Misc. Equipment	Crew	Daily Output	Labor-Hours	Unit	Material	2018 Bare Costs Labor	Equipment	Total	Total Incl O&P
0010 **CHOPPERS, MIXERS AND MISC. EQUIPMENT**									
1850 Coffee urn, twin 6 gallon urns	1 Plum	2	4	Ea.	2,300	249		2,549	2,900
3800 Food mixers, bench type, 20 quarts	L-7	7	4		3,175	195		3,370	3,800
3900 60 quarts	"	5	5.600		12,000	273		12,273	13,600

11 44 Food Cooking Equipment

11 44 13 – Commercial Ranges

11 44 13.10 Cooking Equipment

11 44 13.10 Cooking Equipment	Crew	Daily Output	Labor-Hours	Unit	Material	2018 Bare Costs Labor	Equipment	Total	Total Incl O&P
0010 **COOKING EQUIPMENT**									
0020 Bake oven, gas, one section	Q-1	8	2	Ea.	5,725	112		5,837	6,475
1300 Broiler, without oven, standard	"	8	2		3,875	112		3,987	4,425
1550 Infrared	L-7	4	7		7,425	340		7,765	8,700
6350 Kettle, w/steam jacket, tilting, w/positive lock, SS, 20 gallons		7	4		8,700	195		8,895	9,875
6600 60 gallons		6	4.667		20,000	227		20,227	22,300
8850 Steamer, electric 27 KW		7	4		11,800	195		11,995	13,200
9100 Electric, 10 KW or gas 100,000 BTU		5	5.600		7,200	273		7,473	8,325
9150 Toaster, conveyor type, 16-22 slices/minute					1,250			1,250	1,375
9200 For deluxe models of above equipment, add					75%				
9400 Rule of thumb: Equipment cost based									
9410 on kitchen work area									
9420 Office buildings, minimum	L-7	77	.364	S.F.	94	17.70		111.70	131
9450 Maximum		58	.483		159	23.50		182.50	211
9550 Public eating facilities, minimum		77	.364		123	17.70		140.70	163
9600 Maximum		46	.609		201	29.50		230.50	266
9750 Hospitals, minimum		58	.483		127	23.50		150.50	176
9800 Maximum		39	.718		234	35		269	310

11 46 Food Dispensing Equipment

11 46 16 – Service Line Equipment

11 46 16.10 Commercial Food Dispensing Equipment

11 46 16.10 Commercial Food Dispensing Equipment	Crew	Daily Output	Labor-Hours	Unit	Material	2018 Bare Costs Labor	Equipment	Total	Total Incl O&P
0010 **COMMERCIAL FOOD DISPENSING EQUIPMENT**									
3300 Food warmer, counter, 1.2 KW				Ea.	720			720	790
3550 1.6 KW					2,200			2,200	2,425
5750 Ice dispenser 567 pound	Q-1	6	2.667		5,450	149		5,599	6,225

11 48 Foodservice Cleaning and Disposal Equipment

11 48 13 – Commercial Dishwashers

11 48 13.10 Dishwashers

11 48 13.10 Dishwashers		Crew	Daily Output	Labor-Hours	Unit	Material	2018 Bare Costs Labor	Equipment	Total	Total Incl O&P
0010 **DISHWASHERS**										
2700 Dishwasher, commercial, rack type										
2720 10 to 12 racks/hour		Q-1	3.20	5	Ea.	3,400	280		3,680	4,150
2730 Energy star rated, 35 to 40 racks/hour	G		1.30	12.308		4,250	690		4,940	5,700
2740 50 to 60 racks/hour	G		1.30	12.308		11,100	690		11,790	13,200

For customer support on your Electrical Costs with RSMeans data, call 800.448.8182.

59

11 52 Audio-Visual Equipment

11 52 16 – Projectors

11 52 16.10 Movie Equipment

		Crew	Daily Output	Labor-Hours	Unit	Material	2018 Bare Costs Labor	2018 Bare Costs Equipment	Total	Total Incl O&P
0010	**MOVIE EQUIPMENT**									
0020	Changeover, minimum				Ea.	515			515	565
0100	Maximum					975			975	1,075
0800	Lamphouses, incl. rectifiers, xenon, 1,000 watt	1 Elec	2	4		7,200	233		7,433	8,250
0900	1,600 watt		2	4		7,675	233		7,908	8,800
1000	2,000 watt		1.50	5.333		8,350	310		8,660	9,650
1100	4,000 watt		1.50	5.333		11,500	310		11,810	13,100
3700	Sound systems, incl. amplifier, mono, minimum		.90	8.889		3,725	515		4,240	4,875
3800	Dolby/Super Sound, maximum		.40	20		20,100	1,175		21,275	23,900
4100	Dual system, 2 channel, front surround, minimum		.70	11.429		5,150	665		5,815	6,675
4200	Dolby/Super Sound, 4 channel, maximum		.40	20		18,800	1,175		19,975	22,400
5300	Speakers, recessed behind screen, minimum		2	4		1,175	233		1,408	1,650
5400	Maximum		1	8		3,450	465		3,915	4,475
7000	For automation, varying sophistication, minimum		1	8	System	2,650	465		3,115	3,625
7100	Maximum	2 Elec	.30	53.333	"	6,200	3,100		9,300	11,500

11 52 16.20 Movie Equipment- Digital

		Crew	Daily Output	Labor-Hours	Unit	Material	2018 Bare Costs Labor	2018 Bare Costs Equipment	Total	Total Incl O&P
0010	**MOVIE EQUIPMENT- DIGITAL**									
1000	Digital 2K projection system, 98" DMD	1 Elec	2	4	Ea.	49,000	233		49,233	54,500
1100	OEM lens		2	4		5,975	233		6,208	6,925
2000	Pedestal with power distribution		2	4		2,300	233		2,533	2,875
3000	Software		2	4		1,950	233		2,183	2,500

11 53 Laboratory Equipment

11 53 19 – Laboratory Sterilizers

11 53 19.13 Sterilizers

		Crew	Daily Output	Labor-Hours	Unit	Material	2018 Bare Costs Labor	2018 Bare Costs Equipment	Total	Total Incl O&P
0010	**STERILIZERS**									
0700	Glassware washer, undercounter, minimum	L-1	1.80	8.889	Ea.	6,675	535		7,210	8,150
0710	Maximum	"	1	16	"	14,600	965		15,565	17,500

11 53 33 – Emergency Safety Appliances

11 53 33.13 Emergency Equipment

		Crew	Daily Output	Labor-Hours	Unit	Material	2018 Bare Costs Labor	2018 Bare Costs Equipment	Total	Total Incl O&P
0010	**EMERGENCY EQUIPMENT**									
1400	Safety equipment, eye wash, hand held				Ea.	415			415	455
1450	Deluge shower				"	820			820	900

11 53 43 – Service Fittings and Accessories

11 53 43.13 Fittings

		Crew	Daily Output	Labor-Hours	Unit	Material	2018 Bare Costs Labor	2018 Bare Costs Equipment	Total	Total Incl O&P
0010	**FITTINGS**									
8000	Alternate pricing method: as percent of lab furniture									
8050	Installation, not incl. plumbing & duct work				% Furn.				22%	22%
8100	Plumbing, final connections, simple system								10%	10%
8110	Moderately complex system								15%	15%
8120	Complex system								20%	20%
8150	Electrical, simple system								10%	10%
8160	Moderately complex system								20%	20%
8170	Complex system								35%	35%

11 61 Broadcast, Theater, and Stage Equipment

11 61 33 – Rigging Systems and Controls

11 61 33.10 Controls	Crew	Daily Output	Labor-Hours	Unit	Material	2018 Bare Costs Labor	Equipment	Total	Total Incl O&P
0010 **CONTROLS**									
0050 Control boards with dimmers and breakers, minimum	1 Elec	1	8	Ea.	14,600	465		15,065	16,800
0100 Average		.50	16		49,300	930		50,230	55,500
0150 Maximum		.20	40		140,000	2,325		142,325	157,500

11 66 Athletic Equipment

11 66 43 – Interior Scoreboards

11 66 43.10 Scoreboards

	Crew	Daily Output	Labor-Hours	Unit	Material	2018 Bare Costs Labor	Equipment	Total	Total Incl O&P
0010 **SCOREBOARDS**									
7000 Baseball, minimum	R-3	1.30	15.385	Ea.	3,700	890	99.50	4,689.50	5,500
7200 Maximum		.05	400		20,700	23,200	2,600	46,500	60,500
7300 Football, minimum		.86	23.256		5,125	1,350	151	6,626	7,850
7400 Maximum		.20	100		19,000	5,800	650	25,450	30,300
7500 Basketball (one side), minimum		2.07	9.662		2,475	560	62.50	3,097.50	3,625
7600 Maximum		.30	66.667		7,550	3,875	430	11,855	14,600
7700 Hockey-basketball (four sides), minimum		.25	80		9,300	4,650	520	14,470	17,700
7800 Maximum		.15	133		16,200	7,725	865	24,790	30,400

11 71 Medical Sterilizing Equipment

11 71 10 – Medical Sterilizers & Distillers

11 71 10.10 Sterilizers and Distillers

	Crew	Daily Output	Labor-Hours	Unit	Material	2018 Bare Costs Labor	Equipment	Total	Total Incl O&P
0010 **STERILIZERS AND DISTILLERS**									
6200 Steam generators, electric 10 kW to 180 kW, freestanding									
6250 Minimum	1 Elec	3	2.667	Ea.	11,100	155		11,255	12,500
6300 Maximum	"	.70	11.429	"	23,400	665		24,065	26,700

11 72 Examination and Treatment Equipment

11 72 53 – Treatment Equipment

11 72 53.13 Medical Treatment Equipment

	Crew	Daily Output	Labor-Hours	Unit	Material	2018 Bare Costs Labor	Equipment	Total	Total Incl O&P
0010 **MEDICAL TREATMENT EQUIPMENT**									
6700 Surgical lights, doctor's office, single arm	2 Elec	2	8	Ea.	3,025	465		3,490	4,025
6750 Dual arm	"	1	16	"	5,225	930		6,155	7,150

11 76 Operating Room Equipment

11 76 10 – Operating Room Equipment

11 76 10.10 Surgical Equipment

	Crew	Daily Output	Labor-Hours	Unit	Material	2018 Bare Costs Labor	Equipment	Total	Total Incl O&P
0010 **SURGICAL EQUIPMENT**									
6800 Surgical lights, major operating room, dual head, minimum	2 Elec	1	16	Ea.	5,050	930		5,980	6,975
6850 Maximum	"	1	16	"	37,500	930		38,430	42,600

For customer support on your Electrical Costs with RSMeans data, call 800.448.8182.

61

11 78 Mortuary Equipment

11 78 13 – Mortuary Refrigerators

11 78 13.10 Mortuary and Autopsy Equipment	Crew	Daily Output	Labor-Hours	Unit	Material	2018 Bare Costs Labor	Equipment	Total	Total Incl O&P
0010 **MORTUARY AND AUTOPSY EQUIPMENT**									
0015 Autopsy table, standard	1 Plum	1	8	Ea.	9,825	495		10,320	11,600

11 78 16 – Crematorium Equipment

11 78 16.10 Crematory

	Crew	Daily Output	Labor-Hours	Unit	Material	Labor	Equipment	Total	Total Incl O&P
0010 **CREMATORY**									
1500 Crematory, not including building, 1 place	Q-3	.20	160	Ea.	77,000	9,475		86,475	99,000
1750 2 place	"	.10	320	"	110,000	18,900		128,900	149,000

11 82 Facility Solid Waste Handling Equipment

11 82 26 – Facility Waste Compactors

11 82 26.10 Compactors

	Crew	Daily Output	Labor-Hours	Unit	Material	Labor	Equipment	Total	Total Incl O&P
0010 **COMPACTORS**									
0020 Compactors, 115 volt, 250 lbs./hr., chute fed	L-4	1	24	Ea.	12,500	1,150		13,650	15,500
1000 Heavy duty industrial compactor, 0.5 C.Y. capacity		1	24		10,700	1,150		11,850	13,600
1050 1.0 C.Y. capacity		1	24		16,400	1,150		17,550	19,800
1100 3.0 C.Y. capacity		.50	48		27,400	2,275		29,675	33,600
1150 5.0 C.Y. capacity		.50	48		34,500	2,275		36,775	41,500
1200 Combination shredder/compactor (5,000 lbs./hr.)		.50	48		67,000	2,275		69,275	77,500
1400 For handling hazardous waste materials, 55 gallon drum packer, std.					20,900			20,900	23,000
1410 55 gallon drum packer w/HEPA filter					26,100			26,100	28,700
1420 55 gallon drum packer w/charcoal & HEPA filter					34,800			34,800	38,200
1430 All of the above made explosion proof, add					1,525			1,525	1,675
5800 Shredder, industrial, minimum					25,200			25,200	27,700
5850 Maximum					134,000			134,000	147,500
5900 Baler, industrial, minimum					10,000			10,000	11,000
5950 Maximum				"	585,500			585,500	644,500

Estimating Tips
General
- The items in this division are usually priced per square foot or each. Most of these items are purchased by the owner and installed by the contractor. Do not assume the items in Division 12 will be purchased and installed by the contractor. Check the specifications for responsibilities and include receiving, storage, installation, and mechanical and electrical hookups in the appropriate divisions.
- Some items in this division require some type of support system that is not usually furnished with the item. Examples of these systems include blocking for the attachment of casework and heavy drapery rods. The required blocking must be added to the estimate in the appropriate division.

Reference Numbers
Reference numbers are shown at the beginning of some major classifications. These numbers refer to related items in the Reference Section. The reference information may be an estimating procedure, an alternate pricing method, or technical information.

Note: Not all subdivisions listed here necessarily appear. ■

12 46 Furnishing Accessories

12 46 19 – Clocks

12 46 19.50 Wall Clocks	Crew	Daily Output	Labor-Hours	Unit	Material	2018 Bare Costs Labor	Equipment	Total	Total Incl O&P
0010 **WALL CLOCKS**									
0080 12" diameter, single face	1 Elec	8	1	Ea.	102	58		160	200
0100 Double face	"	6.20	1.290	"	159	75		234	287

Estimating Tips
General

- The items and systems in this division are usually estimated, purchased, supplied, and installed as a unit by one or more subcontractors. The estimator must ensure that all parties are operating from the same set of specifications and assumptions, and that all necessary items are estimated and will be provided. Many times the complex items and systems are covered, but the more common ones, such as excavation or a crane, are overlooked for the very reason that everyone assumes nobody could miss them. The estimator should be the central focus and be able to ensure that all systems are complete.

- Another area where problems can develop in this division is at the interface between systems. The estimator must ensure, for instance, that anchor bolts, nuts, and washers are estimated and included for the air-supported structures and pre-engineered buildings to be bolted to their foundations. Utility supply is a common area where essential items or pieces of equipment can be missed or overlooked, because each subcontractor may feel it is another's responsibility. The estimator should also be aware of certain items which may be supplied as part of a package but installed by others, and ensure that the installing contractor's estimate includes the cost of installation. Conversely, the estimator must also ensure that items are not costed by two different subcontractors, resulting in an inflated overall estimate.

13 30 00 Special Structures

- The foundations and floor slab, as well as rough mechanical and electrical, should be estimated, as this work is required for the assembly and erection of the structure. Generally, as noted in the data set, the pre-engineered building comes as a shell. Pricing is based on the size and structural design parameters stated in the reference section. Additional features, such as windows and doors with their related structural framing, must also be included by the estimator. Here again, the estimator must have a clear understanding of the scope of each portion of the work and all the necessary interfaces.

Reference Numbers

Reference numbers are shown at the beginning of some major classifications. These numbers refer to related items in the Reference Section. The reference information may be an estimating procedure, an alternate pricing method, or technical information.

Note: Not all subdivisions listed here necessarily appear. ■

13 11 13 – Below-Grade Swimming Pools

13 11 13.50 Swimming Pools	Crew	Daily Output	Labor-Hours	Unit	Material	2018 Bare Costs Labor	Equipment	Total	Total Incl O&P
0010 **SWIMMING POOLS** Residential in-ground, vinyl lined									
0020 Concrete sides, w/equip, sand bottom	B-52	300	.187	SF Surf	26	8.65	2.01	36.66	44.50
0100 Metal or polystyrene sides	B-14	410	.117		22	4.93	.76	27.69	32.50
0200 Add for vermiculite bottom				↓	1.67			1.67	1.84
0500 Gunite bottom and sides, white plaster finish									
0600 12' x 30' pool	B-52	145	.386	SF Surf	49	17.90	4.15	71.05	85
0720 16' x 32' pool		155	.361		44	16.75	3.88	64.63	78.50
0750 20' x 40' pool	↓	250	.224	↓	39.50	10.40	2.41	52.31	61.50
0810 Concrete bottom and sides, tile finish									
0820 12' x 30' pool	B-52	80	.700	SF Surf	49.50	32.50	7.55	89.55	111
0830 16' x 32' pool		95	.589		40.50	27.50	6.35	74.35	93.50
0840 20' x 40' pool	↓	130	.431	↓	32.50	20	4.63	57.13	71
1100 Motel, gunite with plaster finish, incl. medium									
1150 capacity filtration & chlorination	B-52	115	.487	SF Surf	60	22.50	5.25	87.75	106
1200 Municipal, gunite with plaster finish, incl. high									
1250 capacity filtration & chlorination	B-52	100	.560	SF Surf	77.50	26	6	109.50	132
1350 Add for formed gutters				L.F.	114			114	126
1360 Add for stainless steel gutters				"	340			340	370
1700 Filtration and deck equipment only, as % of total				Total				20%	20%
1800 Deck equipment, rule of thumb, 20' x 40' pool				SF Pool				1.18	1.30
1900 5,000 S.F. pool				"				1.73	1.90

13 11 23 – On-Grade Swimming Pools

13 11 23.50 Swimming Pools	Crew	Daily Output	Labor-Hours	Unit	Material	2018 Bare Costs Labor	Equipment	Total	Total Incl O&P
0010 **SWIMMING POOLS** Residential above ground, steel construction									
0100 Round, 15' diam.	B-80A	3	8	Ea.	785	320	79.50	1,184.50	1,450
0120 18' diam.		2.50	9.600		930	385	95	1,410	1,725
0140 21' diam.		2	12		1,075	480	119	1,674	2,025
0160 24' diam.		1.80	13.333		1,175	530	132	1,837	2,225
0180 27' diam.		1.50	16		1,500	640	159	2,299	2,825
0200 30' diam.		1	24		1,650	955	238	2,843	3,500
0220 Oval, 12' x 24'		2.30	10.435		1,600	415	104	2,119	2,525
0240 15' x 30'		1.80	13.333		2,525	530	132	3,187	3,725
0260 18' x 33'	↓	1	24	↓	2,775	955	238	3,968	4,750

13 11 46 – Swimming Pool Accessories

13 11 46.50 Swimming Pool Equipment	Crew	Daily Output	Labor-Hours	Unit	Material	2018 Bare Costs Labor	Equipment	Total	Total Incl O&P
0010 **SWIMMING POOL EQUIPMENT**									
0020 Diving stand, stainless steel, 3 meter	2 Carp	.40	40	Ea.	16,700	2,025		18,725	21,500
2100 Lights, underwater, 12 volt, with transformer, 300 watt	1 Elec	1	8		360	465		825	1,100
2200 110 volt, 500 watt, standard		1	8		295	465		760	1,025
2400 Low water cutoff type	↓	1	8	↓	282	465		747	1,000
2800 Heaters, see Section 23 52 28.10									

13 12 Fountains

13 12 13 – Exterior Fountains

13 12 13.10 Outdoor Fountains		Crew	Daily Output	Labor-Hours	Unit	Material	2018 Bare Costs Labor	Equipment	Total	Total Incl O&P
0010	**OUTDOOR FOUNTAINS**									
0100	Outdoor fountain, 48" high with bowl and figures	2 Clab	2	8	Ea.	570	320		890	1,100
0200	Commercial, concrete or cast stone, 40-60" H, simple		2	8		775	320		1,095	1,350
0220	Average		2	8		1,350	320		1,670	1,950
0240	Ornate		2	8		2,625	320		2,945	3,350
0260	Metal, 72" high		2	8		1,200	320		1,520	1,800
0280	90" high		2	8		1,875	320		2,195	2,550
0300	120" high		2	8		5,000	320		5,320	5,975
0320	Resin or fiberglass, 40-60" H, wall type		2	8		480	320		800	1,025
0340	Waterfall type		2	8		1,000	320		1,320	1,575

13 12 23 – Interior Fountains

13 12 23.10 Indoor Fountains

	13 12 23.10 Indoor Fountains	Crew	Daily Output	Labor-Hours	Unit	Material	2018 Bare Costs Labor	Equipment	Total	Total Incl O&P
0010	**INDOOR FOUNTAINS**									
0100	Commercial, floor type, resin or fiberglass, lighted, cascade type	2 Clab	2	8	Ea.	286	320		606	800
0120	Tiered type		2	8		285	320		605	800
0140	Waterfall type		2	8		275	320		595	785

13 21 Controlled Environment Rooms

13 21 13 – Clean Rooms

13 21 13.50 Clean Room Components

	13 21 13.50 Clean Room Components	Crew	Daily Output	Labor-Hours	Unit	Material	2018 Bare Costs Labor	Equipment	Total	Total Incl O&P
0010	**CLEAN ROOM COMPONENTS**									
1100	Clean room, soft wall, 12' x 12', Class 100	1 Carp	.18	44.444	Ea.	16,600	2,250		18,850	21,700
1110	Class 1,000		.18	44.444		15,700	2,250		17,950	20,600
1120	Class 10,000		.21	38.095		13,600	1,925		15,525	17,900
1130	Class 100,000		.21	38.095		13,000	1,925		14,925	17,300
2800	Ceiling grid support, slotted channel struts 4'-0" OC, ea. way				S.F.				5.90	6.50
3000	Ceiling panel, vinyl coated foil on mineral substrate									
3020	Sealed, non-perforated				S.F.				1.27	1.40
4000	Ceiling panel seal, silicone sealant, 150 L.F./gal.	1 Carp	150	.053	L.F.	.29	2.70		2.99	4.44
4100	Two sided adhesive tape	"	240	.033	"	.13	1.69		1.82	2.71
4200	Clips, one per panel				Ea.	1.02			1.02	1.12
6000	HEPA filter, 2' x 4', 99.97% eff., 3" dp beveled frame (silicone seal)					390			390	425
6040	6" deep skirted frame (channel seal)					475			475	525
6100	99.99% efficient, 3" deep beveled frame (silicone seal)					495			495	545
6140	6" deep skirted frame (channel seal)					480			480	525
6200	99.999% efficient, 3" deep beveled frame (silicone seal)					535			535	590
6240	6" deep skirted frame (channel seal)					515			515	565
7000	Wall panel systems, including channel strut framing									
7020	Polyester coated aluminum, particle board				S.F.				18.20	18.20
7100	Porcelain coated aluminum, particle board								32	32
7400	Wall panel support, slotted channel struts, to 12' high								16.35	16.35

For customer support on your Electrical Costs with RSMeans data, call 800.448.8182.

67

13 24 16 – Saunas

13 24 16.50 Saunas and Heaters	Crew	Daily Output	Labor-Hours	Unit	Material	2018 Bare Costs Labor	2018 Bare Costs Equipment	Total	Total Incl O&P
0010 **SAUNAS AND HEATERS**									
0020 Prefabricated, incl. heater & controls, 7' high, 6' x 4', C/C	L-7	2.20	12.727	Ea.	5,125	620		5,745	6,600
0050 6' x 4', C/P		2	14		4,325	680		5,005	5,775
0400 6' x 5', C/C		2	14		5,950	680		6,630	7,575
0450 6' x 5', C/P		2	14		5,175	680		5,855	6,725
0600 6' x 6', C/C		1.80	15.556		7,000	755		7,755	8,850
0650 6' x 6', C/P		1.80	15.556		5,425	755		6,180	7,100
0800 6' x 9', C/C		1.60	17.500		7,900	850		8,750	10,000
0850 6' x 9', C/P		1.60	17.500		6,900	850		7,750	8,900
1000 8' x 12', C/C		1.10	25.455		10,500	1,250		11,750	13,500
1050 8' x 12', C/P		1.10	25.455		9,300	1,250		10,550	12,100
1200 8' x 8', C/C		1.40	20		8,275	975		9,250	10,600
1250 8' x 8', C/P		1.40	20		7,375	975		8,350	9,575
1400 8' x 10', C/C		1.20	23.333		9,050	1,125		10,175	11,700
1450 8' x 10', C/P		1.20	23.333		7,975	1,125		9,100	10,500
1600 10' x 12', C/C		1	28		12,200	1,375		13,575	15,500
1650 10' x 12', C/P		1	28		10,800	1,375		12,175	14,000
2500 Heaters only (incl. above), wall mounted, to 200 C.F.					775			775	855
2750 To 300 C.F.					1,050			1,050	1,150
3000 Floor standing, to 720 C.F., 10,000 watts, w/controls	1 Elec	3	2.667		2,500	155		2,655	2,975
3250 To 1,000 C.F., 16,000 watts	"	3	2.667		3,625	155		3,780	4,200

13 24 26 – Steam Baths

13 24 26.50 Steam Baths and Components	Crew	Daily Output	Labor-Hours	Unit	Material	2018 Bare Costs Labor	2018 Bare Costs Equipment	Total	Total Incl O&P
0010 **STEAM BATHS AND COMPONENTS**									
0020 Heater, timer & head, single, to 140 C.F.	1 Plum	1.20	6.667	Ea.	2,325	415		2,740	3,175
0500 To 300 C.F.		1.10	7.273		2,600	450		3,050	3,525
1000 Commercial size, with blow-down assembly, to 800 C.F.		.90	8.889		5,725	550		6,275	7,125
1500 To 2,500 C.F.		.80	10		8,000	620		8,620	9,725
2000 Multiple, motels, apts., 2 baths, w/blow-down assm., 500 C.F.	Q-1	1.30	12.308		6,675	690		7,365	8,375
2500 4 baths	"	.70	22.857		10,700	1,275		11,975	13,600

13 47 Facility Protection

13 47 13 – Cathodic Protection

13 47 13.16 Cathodic Prot. for Underground Storage Tanks

13 47 13.16 Cathodic Prot. for Underground Storage Tanks	Crew	Daily Output	Labor-Hours	Unit	Material	2018 Bare Costs Labor	2018 Bare Costs Equipment	Total	Total Incl O&P
0010 **CATHODIC PROTECTION FOR UNDERGROUND STORAGE TANKS**									
1000 Anodes, magnesium type, 9 #	R-15	18.50	2.595	Ea.	35	148	15.30	198.30	277
1010 17 #		13	3.692		74.50	211	22	307.50	420
1020 32 #		10	4.800		117	274	28.50	419.50	570
1030 48 #		7.20	6.667		161	380	39.50	580.50	790
1100 Graphite type w/epoxy cap, 3" x 60" (32 #)	R-22	8.40	4.438		120	236		356	485
1110 4" x 80" (68 #)		6	6.213		232	330		562	750
1120 6" x 72" (80 #)		5.20	7.169		1,500	380		1,880	2,225
1130 6" x 36" (45 #)		9.60	3.883		760	207		967	1,150
2000 Rectifiers, silicon type, air cooled, 28 V/10 A	R-19	3.50	5.714		2,050	335		2,385	2,750
2010 20 V/20 A		3.50	5.714		2,100	335		2,435	2,825
2100 Oil immersed, 28 V/10 A		3	6.667		2,825	390		3,215	3,675
2110 20 V/20 A		3	6.667		3,000	390		3,390	3,875
3000 Anode backfill, coke breeze	R-22	3850	.010	Lb.	.28	.52		.80	1.08
4000 Cable, HMWPE, No. 8		2.40	15.533	M.L.F.	570	825		1,395	1,850
4010 No. 6		2.40	15.533		845	825		1,670	2,150

13 47 Facility Protection

13 47 13 – Cathodic Protection

13 47 13.16 Cathodic Prot. for Underground Storage Tanks	Crew	Daily Output	Labor-Hours	Unit	Material	2018 Bare Costs Labor	Equipment	Total	Total Incl O&P
4020	No. 4	R-22	2.40	15.533	M.L.F.	1,275	825		2,100	2,625
4030	No. 2		2.40	15.533		1,975	825		2,800	3,400
4040	No. 1		2.20	16.945		2,700	905		3,605	4,325
4050	No. 1/0		2.20	16.945		3,500	905		4,405	5,200
4060	No. 2/0		2.20	16.945		4,975	905		5,880	6,825
4070	No. 4/0	↓	2	18.640	↓	6,275	995		7,270	8,400
5000	Test station, 7 terminal box, flush curb type w/lockable cover	R-19	12	1.667	Ea.	76.50	97		173.50	230
5010	Reference cell, 2" diam. PVC conduit, cplg., plug, set flush	"	4.80	4.167	"	151	243		394	530

13 49 Radiation Protection

13 49 13 – Integrated X-Ray Shielding Assemblies

13 49 13.50 Lead Sheets

		Crew	Daily Output	Labor-Hours	Unit	Material	2018 Bare Costs Labor	Equipment	Total	Total Incl O&P
0010	**LEAD SHEETS**									
0300	Lead sheets, 1/16" thick	2 Lath	135	.119	S.F.	10.90	5.85		16.75	20.50
0400	1/8" thick		120	.133		22	6.55		28.55	34
0500	Lead shielding, 1/4" thick		135	.119		42.50	5.85		48.35	55
0550	1/2" thick	↓	120	.133	↓	79.50	6.55		86.05	97
0950	Lead headed nails (average 1 lb. per sheet)				Lb.	8.40			8.40	9.25
1000	Butt joints in 1/8" lead or thicker, 2" batten strip x 7' long	2 Lath	240	.067	Ea.	31	3.28		34.28	39
1200	X-ray protection, average radiography or fluoroscopy									
1210	room, up to 300 S.F. floor, 1/16" lead, economy	2 Lath	.25	64	Total	11,100	3,150		14,250	17,000
1500	7'-0" walls, deluxe	"	.15	107	"	13,400	5,250		18,650	22,500
1600	Deep therapy X-ray room, 250 kV capacity,									
1800	up to 300 S.F. floor, 1/4" lead, economy	2 Lath	.08	200	Total	31,100	9,825		40,925	48,700
1900	7'-0" walls, deluxe	"	.06	267	"	38,400	13,100		51,500	61,500

13 49 19 – Lead-Lined Materials

13 49 19.50 Shielding Lead

		Crew	Daily Output	Labor-Hours	Unit	Material	2018 Bare Costs Labor	Equipment	Total	Total Incl O&P
0010	**SHIELDING LEAD**									
0100	Laminated lead in wood doors, 1/16" thick, no hardware				S.F.	53.50			53.50	59
0200	Lead lined door frame, not incl. hardware,									
0210	1/16" thick lead, butt prepared for hardware	1 Lath	2.40	3.333	Ea.	860	164		1,024	1,175
0850	Window frame with 1/16" lead and voice passage, 36" x 60"	2 Glaz	2	8		4,525	390		4,915	5,575
0870	24" x 36" frame		4	4	↓	2,350	194		2,544	2,875
0900	Lead gypsum board, 5/8" thick with 1/16" lead		160	.100	S.F.	12.10	4.85		16.95	20.50
0910	1/8" lead	↓	140	.114		24.50	5.55		30.05	35.50
0930	1/32" lead	2 Lath	200	.080	↓	8.50	3.93		12.43	15.15

13 49 21 – Lead Glazing

13 49 21.50 Lead Glazing

		Crew	Daily Output	Labor-Hours	Unit	Material	2018 Bare Costs Labor	Equipment	Total	Total Incl O&P
0010	**LEAD GLAZING**									
0600	Lead glass, 1/4" thick, 2.0 mm LE, 12" x 16"	2 Glaz	13	1.231	Ea.	390	59.50		449.50	520
0700	24" x 36"		8	2		1,375	97		1,472	1,650
0800	36" x 60"	↓	2	8	↓	3,775	390		4,165	4,775
2000	X-ray viewing panels, clear lead plastic									
2010	7 mm thick, 0.3 mm LE, 2.3 lb./S.F.	H-3	139	.115	S.F.	271	4.97		275.97	305
2020	12 mm thick, 0.5 mm LE, 3.9 lb./S.F.		82	.195		385	8.40		393.40	435
2030	18 mm thick, 0.8 mm LE, 5.9 lb./S.F.		54	.296		440	12.80		452.80	505
2040	22 mm thick, 1.0 mm LE, 7.2 lb./S.F.		44	.364		575	15.70		590.70	655
2050	35 mm thick, 1.5 mm LE, 11.5 lb./S.F.		28	.571		880	24.50		904.50	1,000
2060	46 mm thick, 2.0 mm LE, 15.0 lb./S.F.	↓	21	.762	↓	1,150	33		1,183	1,300
2090	For panels 12 S.F. to 48 S.F., add crating charge				Ea.				50	50

For customer support on your Electrical Costs with RSMeans data, call 800.448.8182.

69

13 49 23 – Integrated RFI/EMI Shielding Assemblies

13 49 23.50 Modular Shielding Partitions	Crew	Daily Output	Labor-Hours	Unit	Material	2018 Bare Costs Labor	2018 Bare Costs Equipment	Total	Total Incl O&P
0010 **MODULAR SHIELDING PARTITIONS**									
4000 X-ray barriers, modular, panels mounted within framework for									
4002 attaching to floor, wall or ceiling, upper portion is clear lead									
4005 plastic window panels 48"H, lower portion is opaque leaded									
4008 steel panels 36"H, structural supports not incl.									
4010 1-section barrier, 36"W x 84"H overall									
4020 0.5 mm LE panels	H-3	6.40	2.500	Ea.	8,800	108		8,908	9,850
4030 0.8 mm LE panels		6.40	2.500		9,500	108		9,608	10,600
4040 1.0 mm LE panels		5.33	3.002		11,100	130		11,230	12,400
4050 1.5 mm LE panels		5.33	3.002		14,800	130		14,930	16,500
4060 2-section barrier, 72"W x 84"H overall									
4070 0.5 mm LE panels	H-3	4	4	Ea.	12,800	173		12,973	14,300
4080 0.8 mm LE panels		4	4		14,100	173		14,273	15,800
4090 1.0 mm LE panels		3.56	4.494		17,300	194		17,494	19,400
5000 1.5 mm LE panels		3.20	5		24,700	216		24,916	27,400
5010 3-section barrier, 108"W x 84"H overall									
5020 0.5 mm LE panels	H-3	3.20	5	Ea.	19,100	216		19,316	21,300
5030 0.8 mm LE panels		3.20	5		21,200	216		21,416	23,600
5040 1.0 mm LE panels		2.67	5.993		26,000	259		26,259	29,000
5050 1.5 mm LE panels		2.46	6.504		37,000	281		37,281	41,100
7000 X-ray barriers, mobile, mounted within framework w/casters on									
7005 bottom, clear lead plastic window panels on upper portion,									
7010 opaque on lower, 30"W x 75"H overall, incl. framework									
7020 24"H upper w/0.5 mm LE, 48"H lower w/0.8 mm LE	1 Carp	16	.500	Ea.	4,225	25.50		4,250.50	4,700
7030 48"W x 75"H overall, incl. framework									
7040 36"H upper w/0.5 mm LE, 36"H lower w/0.8 mm LE	1 Carp	16	.500	Ea.	6,675	25.50		6,700.50	7,400
7050 36"H upper w/1.0 mm LE, 36"H lower w/1.5 mm LE	"	16	.500	"	7,900	25.50		7,925.50	8,750
7060 72"W x 75"H overall, incl. framework									
7070 36"H upper w/0.5 mm LE, 36"H lower w/0.8 mm LE	1 Carp	16	.500	Ea.	7,900	25.50		7,925.50	8,750
7080 36"H upper w/1.0 mm LE, 36"H lower w/1.5 mm LE	"	16	.500	"	9,900	25.50		9,925.50	10,900

13 49 33 – Radio Frequency Shielding

13 49 33.50 Shielding, Radio Frequency	Crew	Daily Output	Labor-Hours	Unit	Material	2018 Bare Costs Labor	2018 Bare Costs Equipment	Total	Total Incl O&P
0010 **SHIELDING, RADIO FREQUENCY**									
0020 Prefabricated, galvanized steel	2 Carp	375	.043	SF Surf	5.30	2.16		7.46	9.10
0040 5 oz., copper floor panel		480	.033		4.48	1.69		6.17	7.50
0050 5 oz., copper wall/ceiling panel		155	.103		4.48	5.25		9.73	12.90
0100 12 oz., copper floor panel		470	.034		9.05	1.73		10.78	12.60
0110 12 oz., copper wall/ceiling panel		140	.114		9.05	5.80		14.85	18.75
0150 Door, copper/wood laminate, 4' x 7'		1.50	10.667	Ea.	8,400	540		8,940	10,100

Estimating Tips

Pipe for fire protection and all uses is located in Subdivisions 21 11 13 and 22 11 13.

The labor adjustment factors listed in Subdivision 22 01 02.20 also apply to Division 21.

Many, but not all, areas in the U.S. require backflow protection in the fire system. Insurance underwriters may have specific requirements for the type of materials to be installed or design requirements based on the hazard to be protected. Local jurisdictions may have requirements not covered by code. It is advisable to be aware of any special conditions.

For your reference, the following is a list of the most applicable Fire Codes and Standards which may be purchased from the NFPA, 1 Batterymarch Park, Quincy, MA 02169-7471.

- NFPA 1: Uniform Fire Code
- NFPA 10: Portable Fire Extinguishers
- NFPA 11: Low-, Medium-, and High-Expansion Foam
- NFPA 12: Carbon Dioxide Extinguishing Systems (Also companion 12A)
- NFPA 13: Installation of Sprinkler Systems (Also companion 13D, 13E, and 13R)
- NFPA 14: Installation of Standpipe and Hose Systems
- NFPA 15: Water Spray Fixed Systems for Fire Protection
- NFPA 16: Installation of Foam-Water Sprinkler and Foam-Water Spray Systems
- NFPA 17: Dry Chemical Extinguishing Systems (Also companion 17A)
- NFPA 18: Wetting Agents
- NFPA 20: Installation of Stationary Pumps for Fire Protection

- NFPA 22: Water Tanks for Private Fire Protection
- NFPA 24: Installation of Private Fire Service Mains and their Appurtenances
- NFPA 25: Inspection, Testing and Maintenance of Water-Based Fire Protection

Reference Numbers

Reference numbers are shown at the beginning of some major classifications. These numbers refer to related items in the Reference Section. The reference information may be an estimating procedure, an alternate pricing method, or technical information.

Note: Not all subdivisions listed here necessarily appear. ■

21 21 Carbon-Dioxide Fire-Extinguishing Systems

21 21 16 – Carbon-Dioxide Fire-Extinguishing Equipment

21 21 16.50 CO2 Fire Extinguishing System	Crew	Daily Output	Labor-Hours	Unit	Material	2018 Bare Costs Labor	Equipment	Total	Total Incl O&P
0010 **CO₂ FIRE EXTINGUISHING SYSTEM**									
0100 Control panel, single zone with batteries (2 zones det., 1 suppr.)	1 Elec	1	8	Ea.	920	465		1,385	1,725
0150 Multizone (4) with batteries (8 zones det., 4 suppr.)	"	.50	16		2,775	930		3,705	4,450
1000 Dispersion nozzle, CO₂, 3" x 5"	1 Plum	18	.444		144	27.50		171.50	200
2000 Extinguisher, CO₂ system, high pressure, 75 lb. cylinder	Q-1	6	2.667		1,300	149		1,449	1,675
2100 100 lb. cylinder	"	5	3.200		1,675	179		1,854	2,125
3000 Electro/mechanical release	L-1	4	4		935	241		1,176	1,375
3400 Manual pull station	1 Plum	6	1.333		79	83		162	212
4000 Pneumatic damper release	"	8	1		155	62		217	264

21 22 Clean-Agent Fire-Extinguishing Systems

21 22 16 – Clean-Agent Fire-Extinguishing Equipment

21 22 16.50 Clean-Agent Extinguishing Systems

	Crew	Daily Output	Labor-Hours	Unit	Material	2018 Bare Costs Labor	Equipment	Total	Total Incl O&P
0010 **CLEAN-AGENT EXTINGUISHING SYSTEMS**									
0020 FM200 fire extinguishing system									
1100 Dispersion nozzle FM200, 1-1/2"	1 Plum	14	.571	Ea.	182	35.50		217.50	254
2400 Extinguisher, FM200 system, filled, with mounting bracket									
2460 26 lb. container	Q-1	8	2	Ea.	1,975	112		2,087	2,325
2480 44 lb. container		7	2.286		2,700	128		2,828	3,175
2500 63 lb. container		6	2.667		3,100	149		3,249	3,650
2520 101 lb. container		5	3.200		4,900	179		5,079	5,650
2540 196 lb. container		4	4		6,625	224		6,849	7,600
6000 FM200 system, simple nozzle layout, with broad dispersion				C.F.	1.76			1.76	1.94
6010 Extinguisher, FM200 system, filled, with mounting bracket									
6020 Complex nozzle layout and/or including underfloor dispersion				C.F.	3.50			3.50	3.85
6100 20,000 C.F. 2 exits, 8' clng					2			2	2.20
6200 100,000 C.F. 4 exits, 8' clng					1.80			1.80	1.98
6300 250,000 C.F. 6 exits, 8' clng					1.52			1.52	1.67
7010 HFC-227ea fire extinguishing system									
7100 Cylinders with clean-agent									
7110 Does not include pallete jack/fork lift rental fees									
7120 70 lb. cyl, w/35 lb. agent, no solenoid	Q-12	14	1.143	Ea.	2,400	61.50		2,461.50	2,750
7130 70 lb. cyl w/70 lb. agent, no solenoid		10	1.600		3,075	86		3,161	3,525
7140 70 lb. cyl w/35 lb. agent, w/solenoid		14	1.143		3,275	61.50		3,336.50	3,700
7150 70 lb. cyl w/70 lb. agent, w/solenoid		10	1.600		4,150	86		4,236	4,675
7220 250 lb. cyl, w/125 lb. agent, no solenoid		8	2		5,425	108		5,533	6,125
7230 250 lb. cyl, w/250 lb. agent, no solenoid		5	3.200		5,425	172		5,597	6,200
7240 250 lb. cyl, w/125 lb. agent, w/solenoid		8	2		7,100	108		7,208	7,975
7250 250 lb. cyl, w/250 lb. agent, w/solenoid		5	3.200		10,200	172		10,372	11,500
7320 560 lb. cyl, w/300 lb. agent, no solenoid		4	4		10,200	216		10,416	11,500
7330 560 lb. cyl, w/560 lb. agent, no solenoid		2.50	6.400		15,200	345		15,545	17,300
7340 560 lb. cyl, w/300 lb. agent, w/solenoid		4	4		13,100	216		13,316	14,700
7350 560 lb. cyl, w/560 lb. agent, w/solenoid		2.50	6.400		19,500	345		19,845	22,000
7420 1,200 lb. cyl, w/600 lb. agent, no solenoid	Q-13	4	8		19,200	455		19,655	21,800
7430 1,200 lb. cyl, w/1,200 lb. agent, no solenoid		3	10.667		30,900	610		31,510	34,900
7440 1,200 lb. cyl, w/600 lb. agent, w/solenoid		4	8		45,700	455		46,155	50,500
7450 1,200 lb. cyl, w/1,200 lb. agent, w/solenoid		3	10.667		71,500	610		72,110	79,500
7500 Accessories									
7510 Dispersion nozzle	1 Spri	16	.500	Ea.	95	30		125	150
7520 Agent release panel	1 Elec	4	2		60	116		176	240
7530 Maintenance switch	"	6	1.333		37.50	77.50		115	157

21 22 Clean-Agent Fire-Extinguishing Systems

21 22 16 – Clean-Agent Fire-Extinguishing Equipment

21 22 16.50 Clean-Agent Extinguishing Systems	Crew	Daily Output	Labor-Hours	Unit	Material	2018 Bare Costs Labor	Equipment	Total	Total Incl O&P
7540	Solenoid valve, 12v dc	1 Spri	8	1	Ea.	232	60		292	345
7550	12v ac		8	1		470	60		530	610
7560	12v dc, explosion proof	↓	8	1	↓	390	60		450	520

For customer support on your Electrical Costs with RSMeans data, call 800.448.8182.

73

Division Notes

	CREW	DAILY OUTPUT	LABOR-HOURS	UNIT	BARE COSTS				TOTAL INCL O&P
					MAT.	LABOR	EQUIP.	TOTAL	

Estimating Tips
22 10 00 Plumbing Piping and Pumps

This subdivision is primarily basic pipe and related materials. The pipe may be used by any of the mechanical disciplines, i.e., plumbing, fire protection, heating, and air conditioning.

Note: CPVC plastic piping approved for fire protection is located in 21 11 13.

- The labor adjustment factors listed in Subdivision 22 01 02.20 apply throughout Divisions 21, 22, and 23. CAUTION: the correct percentage may vary for the same items. For example, the percentage add for the basic pipe installation should be based on the maximum height that the installer must install for that particular section. If the pipe is to be located 14' above the floor but it is suspended on threaded rod from beams, the bottom flange of which is 18' high (4' rods), then the height is actually 18' and the add is 20%. The pipe cover, however, does not have to go above the 14', and so the add should be 10%.

- Most pipe is priced first as straight pipe with a joint (coupling, weld, etc.) every 10' and a hanger usually every 10'. There are exceptions with hanger spacing such as for cast iron pipe (5')

and plastic pipe (3 per 10'). Following each type of pipe there are several lines listing sizes and the amount to be subtracted to delete couplings and hangers. This is for pipe that is to be buried or supported together on trapeze hangers. The reason that the couplings are deleted is that these runs are usually long, and frequently longer lengths of pipe are used. By deleting the couplings, the estimator is expected to look up and add back the correct reduced number of couplings.

- When preparing an estimate, it may be necessary to approximate the fittings. Fittings usually run between 25% and 50% of the cost of the pipe. The lower percentage is for simpler runs, and the higher number is for complex areas, such as mechanical rooms.

- For historic restoration projects, the systems must be as invisible as possible, and pathways must be sought for pipes, conduit, and ductwork. While installations in accessible spaces (such as basements and attics) are relatively straightforward to estimate, labor costs may be more difficult to determine when delivery systems must be concealed.

22 40 00 Plumbing Fixtures

- Plumbing fixture costs usually require two lines: the fixture itself and its "rough-in, supply, and waste."

- In the Assemblies Section (Plumbing D2010) for the desired fixture, the System Components Group at the center of the page shows the fixture on the first line. The rest of the list (fittings, pipe, tubing, etc.) will total up to what we refer to in the Unit Price section as "Rough-in, supply, waste, and vent." Note that for most fixtures we allow a nominal 5' of tubing to reach from the fixture to a main or riser.

- Remember that gas- and oil-fired units need venting.

Reference Numbers

Reference numbers are shown at the beginning of some major classifications. These numbers refer to related items in the Reference Section. The reference information may be an estimating procedure, an alternate pricing method, or technical information.

Note: Not all subdivisions listed here necessarily appear. ■

22 01 02.10 Boilers, General	Crew	Daily Output	Labor-Hours	Unit	Material	2018 Bare Costs Labor	Equipment	Total	Total Incl O&P
0010 **BOILERS, GENERAL**, Prices do not include flue piping, elec. wiring,									
0020 gas or oil piping, boiler base, pad, or tankless unless noted									
0100 Boiler H.P.: 10 KW = 34 lb./steam/hr. = 33,475 BTU/hr.									
0150 To convert SFR to BTU rating: Hot water, 150 x SFR;									
0160 Forced hot water, 180 x SFR; steam, 240 x SFR									

22 01 02.20 Labor Adjustment Factors	Crew	Daily Output	Labor-Hours	Unit	Material	2018 Bare Costs Labor	Equipment	Total	Total Incl O&P
0010 **LABOR ADJUSTMENT FACTORS** (For Div. 21, 22 and 23) R220102-20									
0100 Labor factors, The below are reasonable suggestions, however									
0110 each project must be evaluated for its own peculiarities, and									
0120 the adjustments be increased or decreased depending on the									
0130 severity of the special conditions.									
1000 Add to labor for elevated installation (Above floor level)									
1080 10' to 14.5' high						10%			
1100 15' to 19.5' high						20%			
1120 20' to 24.5' high						25%			
1140 25' to 29.5' high						35%			
1160 30' to 34.5' high						40%			
1180 35' to 39.5' high						50%			
1200 40' and higher						55%			
2000 Add to labor for crawl space									
2100 3' high						40%			
2140 4' high						30%			
3000 Add to labor for multi-story building									
3010 For new construction (No elevator available)									
3100 Add for floors 3 thru 10						5%			
3110 Add for floors 11 thru 15						10%			
3120 Add for floors 16 thru 20						15%			
3130 Add for floors 21 thru 30						20%			
3140 Add for floors 31 and up						30%			
3170 For existing structure (Elevator available)									
3180 Add for work on floor 3 and above						2%			
4000 Add to labor for working in existing occupied buildings									
4100 Hospital						35%			
4140 Office building						25%			
4180 School						20%			
4220 Factory or warehouse						15%			
4260 Multi dwelling						15%			
5000 Add to labor, miscellaneous									
5100 Cramped shaft						35%			
5140 Congested area						15%			
5180 Excessive heat or cold						30%			
9000 Labor factors, The above are reasonable suggestions, however									
9010 each project should be evaluated for its own peculiarities.									
9100 Other factors to be considered are:									
9140 Movement of material and equipment through finished areas									
9180 Equipment room									
9220 Attic space									
9260 No service road									
9300 Poor unloading/storage area									
9340 Congested site area/heavy traffic									

22 05 23 – General-Duty Valves for Plumbing Piping

22 05 23.20 Valves, Bronze	Crew	Daily Output	Labor-Hours	Unit	Material	2018 Bare Costs Labor	2018 Bare Costs Equipment	Total	Total Incl O&P
0010 **VALVES, BRONZE**									
8800 Water heater water & gas safety shut off									
8810 Protection against a leaking water heater									
8814 Shut off valve	1 Plum	16	.500	Ea.	185	31		216	251
8818 Water heater dam		32	.250		31.50	15.55		47.05	58
8822 Gas control wiring harness	↓	32	.250	↓	19.75	15.55		35.30	45
8830 Whole house flood safety shut off									
8834 Connections									
8838 3/4" NPT	1 Plum	12	.667	Ea.	935	41.50		976.50	1,100
8842 1" NPT		11	.727		960	45		1,005	1,125
8846 1-1/4" NPT	↓	10	.800		995	49.50		1,044.50	1,175

22 05 29 – Hangers and Supports for Plumbing Piping and Equipment

22 05 29.10 Hangers & Supp. for Plumb'g/HVAC Pipe/Equip.	Crew	Daily Output	Labor-Hours	Unit	Material	2018 Bare Costs Labor	2018 Bare Costs Equipment	Total	Total Incl O&P
0010 **HANGERS AND SUPPORTS FOR PLUMB'G/HVAC PIPE/EQUIP.**									
0011 TYPE numbers per MSS-SP58									
0050 Brackets									
0060 Beam side or wall, malleable iron, TYPE 34									
0070 3/8" threaded rod size	1 Plum	48	.167	Ea.	4.07	10.35		14.42	20
0080 1/2" threaded rod size		48	.167		3.78	10.35		14.13	19.75
0090 5/8" threaded rod size		48	.167		11	10.35		21.35	27.50
0100 3/4" threaded rod size		48	.167		18.90	10.35		29.25	36.50
0110 7/8" threaded rod size	↓	48	.167		12.50	10.35		22.85	29.50
0120 For concrete installation, add				↓		30%			
0150 Wall, welded steel, medium, TYPE 32									
0160 0 size, 12" wide, 18" deep	1 Plum	34	.235	Ea.	178	14.60		192.60	217
0170 1 size, 18" wide, 24" deep		34	.235		211	14.60		225.60	254
0180 2 size, 24" wide, 30" deep	↓	34	.235	↓	279	14.60		293.60	325
0300 Clamps									
0310 C-clamp, for mounting on steel beam flange, w/locknut, TYPE 23									
0320 3/8" threaded rod size	1 Plum	160	.050	Ea.	3.93	3.11		7.04	9
0330 1/2" threaded rod size		160	.050		4.88	3.11		7.99	10.05
0340 5/8" threaded rod size		160	.050		5.30	3.11		8.41	10.55
0350 3/4" threaded rod size	↓	160	.050	↓	6.80	3.11		9.91	12.20
0750 Riser or extension pipe, carbon steel, TYPE 8									
0760 3/4" pipe size	1 Plum	48	.167	Ea.	3.29	10.35		13.64	19.20
0770 1" pipe size		47	.170		3.42	10.60		14.02	19.70
0780 1-1/4" pipe size		46	.174		5.50	10.80		16.30	22.50
0790 1-1/2" pipe size		45	.178		9.85	11.05		20.90	27.50
0800 2" pipe size		43	.186		5.55	11.55		17.10	23.50
0810 2-1/2" pipe size		41	.195		5.40	12.15		17.55	24
0820 3" pipe size		40	.200		5.10	12.45		17.55	24.50
0830 3-1/2" pipe size		39	.205		5.45	12.75		18.20	25
0840 4" pipe size		38	.211		8.95	13.10		22.05	29.50
0850 5" pipe size		37	.216		11.35	13.45		24.80	32.50
0860 6" pipe size	↓	36	.222	↓	15.10	13.80		28.90	37.50
1150 Insert, concrete									
1160 Wedge type, carbon steel body, malleable iron nut, galvanized									
1170 1/4" threaded rod size	1 Plum	96	.083	Ea.	9.55	5.20		14.75	18.30
1180 3/8" threaded rod size		96	.083		19.40	5.20		24.60	29.50
1190 1/2" threaded rod size		96	.083		10.65	5.20		15.85	19.55
1200 5/8" threaded rod size		96	.083		60	5.20		65.20	74
1210 3/4" threaded rod size	↓	96	.083		11.95	5.20		17.15	21

For customer support on your Electrical Costs with RSMeans data, call 800.448.8182.

77

22 05 29.10 Hangers & Supp. for Plumb'g/HVAC Pipe/Equip.	Crew	Daily Output	Labor-Hours	Unit	Material	2018 Bare Costs Labor	Equipment	Total	Total Incl O&P
1220	7/8" threaded rod size	1 Plum	96	.083	Ea.	12.35	5.20		17.55	21.50
2650	Rods, carbon steel									
2660	Continuous thread									
2670	1/4" thread size	1 Plum	144	.056	L.F.	2.18	3.45		5.63	7.60
2680	3/8" thread size		144	.056		2.33	3.45		5.78	7.75
2690	1/2" thread size		144	.056		3.67	3.45		7.12	9.25
2700	5/8" thread size		144	.056		5.20	3.45		8.65	10.90
2710	3/4" thread size		144	.056		9.15	3.45		12.60	15.30
2720	7/8" thread size		144	.056		11.50	3.45		14.95	17.85
2725	1/4" thread size, bright finish		144	.056		1.24	3.45		4.69	6.55
2726	1/2" thread size, bright finish		144	.056		4.23	3.45		7.68	9.85
2730	For galvanized, add					40%				
4400	U-bolt, carbon steel									
4410	Standard, with nuts, TYPE 42									
4420	1/2" pipe size	1 Plum	160	.050	Ea.	1.12	3.11		4.23	5.90
4430	3/4" pipe size		158	.051		1.12	3.15		4.27	5.95
4450	1" pipe size		152	.053		1.16	3.27		4.43	6.20
4460	1-1/4" pipe size		148	.054		1.47	3.36		4.83	6.65
4470	1-1/2" pipe size		143	.056		1.56	3.48		5.04	6.95
4480	2" pipe size		139	.058		1.69	3.58		5.27	7.25
4490	2-1/2" pipe size		134	.060		2.75	3.71		6.46	8.65
4500	3" pipe size		128	.063		3.10	3.88		6.98	9.25
4510	3-1/2" pipe size		122	.066		3.34	4.08		7.42	9.80
4520	4" pipe size		117	.068		3.40	4.25		7.65	10.15
4530	5" pipe size		114	.070		4	4.36		8.36	10.95
4540	6" pipe size		111	.072		7.60	4.48		12.08	15.10
4580	For plastic coating on 1/2" thru 6" size, add					150%				
8800	Wire cable support system									
8810	Cable with hook terminal and locking device									
8830	2 mm (.079") diam. cable (100 lb. cap.)									
8840	1 m (3.3') length, with hook	1 Shee	96	.083	Ea.	3.91	4.98		8.89	11.90
8850	2 m (6.6') length, with hook		84	.095		4.45	5.70		10.15	13.60
8860	3 m (9.9') length, with hook		72	.111		3.75	6.65		10.40	14.30
8870	5 m (16.4') length, with hook	Q-9	60	.267		4.81	14.35		19.16	27.50
8880	10 m (32.8') length, with hook	"	30	.533		7	28.50		35.50	51.50
8900	3 mm (.118") diam. cable (200 lb. cap.)									
8910	1 m (3.3') length, with hook	1 Shee	96	.083	Ea.	7.85	4.98		12.83	16.25
8920	2 m (6.6') length, with hook		84	.095		8.70	5.70		14.40	18.30
8930	3 m (9.9') length, with hook		72	.111		9.80	6.65		16.45	21
8940	5 m (16.4') length, with hook	Q-9	60	.267		11.90	14.35		26.25	35
8950	10 m (32.8') length, with hook	"	30	.533		15.20	28.50		43.70	61
9000	Cable system accessories									
9010	Anchor bolt, 3/8", with nut	1 Shee	140	.057	Ea.	1.07	3.42		4.49	6.40
9020	Air duct corner protector		160	.050		1.18	2.99		4.17	5.85
9030	Air duct support attachment		140	.057		1.95	3.42		5.37	7.35
9040	Flange clip, hammer-on style									
9044	For flange thickness 3/32"-9/64", 160 lb. cap.	1 Shee	180	.044	Ea.	.37	2.66		3.03	4.47
9048	For flange thickness 1/8"-1/4", 200 lb. cap.		160	.050		.30	2.99		3.29	4.89
9052	For flange thickness 5/16"-1/2", 200 lb. cap.		150	.053		.61	3.19		3.80	5.55
9056	For flange thickness 9/16"-3/4", 200 lb. cap.		140	.057		.82	3.42		4.24	6.10
9060	Wire insulation protection tube		180	.044	L.F.	.45	2.66		3.11	4.56
9070	Wire cutter				Ea.	42.50			42.50	46.50

22 33 13 – Instantaneous Electric Domestic Water Heaters

22 33 13.10 Hot Water Dispensers

	Crew	Daily Output	Labor-Hours	Unit	Material	2018 Bare Costs Labor	Equipment	Total	Total Incl O&P
0010 HOT WATER DISPENSERS									
0160 Commercial, 100 cup, 11.3 amp	1 Plum	14	.571	Ea.	455	35.50		490.50	555
3180 Household, 60 cup	"	14	.571	"	239	35.50		274.50	315

22 33 30 – Residential, Electric Domestic Water Heaters

22 33 30.13 Residential, Small-Capacity Elec. Water Heaters

	Crew	Daily Output	Labor-Hours	Unit	Material	2018 Bare Costs Labor	Equipment	Total	Total Incl O&P
0010 RESIDENTIAL, SMALL-CAPACITY ELECTRIC DOMESTIC WATER HEATERS									
1000 Residential, electric, glass lined tank, 5 yr., 10 gal., single element R224000-10	1 Plum	2.30	3.478	Ea.	430	216		646	795
1040 20 gallon, single element R224000-20		2.20	3.636		570	226		796	965
1060 30 gallon, double element		2.20	3.636		1,000	226		1,226	1,450
1080 40 gallon, double element		2	4		1,075	249		1,324	1,550
1100 52 gallon, double element		2	4		1,200	249		1,449	1,700
1120 66 gallon, double element		1.80	4.444		1,650	276		1,926	2,225
1140 80 gallon, double element		1.60	5		1,850	310		2,160	2,525
1180 120 gallon, double element		1.40	5.714		2,575	355		2,930	3,350

22 33 33 – Light-Commercial Electric Domestic Water Heaters

22 33 33.10 Commercial Electric Water Heaters

	Crew	Daily Output	Labor-Hours	Unit	Material	2018 Bare Costs Labor	Equipment	Total	Total Incl O&P
0010 COMMERCIAL ELECTRIC WATER HEATERS									
4000 Commercial, 100° rise. NOTE: for each size tank, a range of									
4010 heaters between the ones shown are available									
4020 Electric									
4100 5 gal., 3 kW, 12 GPH, 208 volt	1 Plum	2	4	Ea.	4,775	249		5,024	5,625
4120 10 gal., 6 kW, 25 GPH, 208 volt		2	4		5,275	249		5,524	6,175
4130 30 gal., 24 kW, 98 GPH, 208 volt		1.92	4.167		8,700	259		8,959	9,975
4136 40 gal., 36 kW, 148 GPH, 208 volt		1.88	4.255		10,400	264		10,664	11,900
4140 50 gal., 9 kW, 37 GPH, 208 volt		1.80	4.444		7,275	276		7,551	8,425
4160 50 gal., 36 kW, 148 GPH, 208 volt		1.80	4.444		11,200	276		11,476	12,700
4180 80 gal., 12 kW, 49 GPH, 208 volt		1.50	5.333		8,925	330		9,255	10,300
4200 80 gal., 36 kW, 148 GPH, 208 volt		1.50	5.333		12,400	330		12,730	14,100
4220 100 gal., 36 kW, 148 GPH, 208 volt		1.20	6.667		12,900	415		13,315	14,800
4240 120 gal., 36 kW, 148 GPH, 208 volt		1.20	6.667		13,500	415		13,915	15,400
4260 150 gal., 15 kW, 61 GPH, 480 volt		1	8		31,600	495		32,095	35,500
4280 150 gal., 120 kW, 490 GPH, 480 volt		1	8		44,400	495		44,895	49,700
4300 200 gal., 15 kW, 61 GPH, 480 volt	Q-1	1.70	9.412		34,200	525		34,725	38,400
4320 200 gal., 120 kW, 490 GPH, 480 volt		1.70	9.412		46,800	525		47,325	52,500
4340 250 gal., 15 kW, 61 GPH, 480 volt		1.50	10.667		35,300	595		35,895	39,700
4360 250 gal., 150 kW, 615 GPH, 480 volt		1.50	10.667		51,500	595		52,095	57,500
4380 300 gal., 30 kW, 123 GPH, 480 volt		1.30	12.308		38,900	690		39,590	43,700
4400 300 gal., 180 kW, 738 GPH, 480 volt		1.30	12.308		56,000	690		56,690	62,500
4420 350 gal., 30 kW, 123 GPH, 480 volt		1.10	14.545		41,100	815		41,915	46,400
4440 350 gal., 180 kW, 738 GPH, 480 volt		1.10	14.545		58,000	815		58,815	64,500
4460 400 gal., 30 kW, 123 GPH, 480 volt		1	16		46,600	895		47,495	52,500
4480 400 gal., 210 kW, 860 GPH, 480 volt		1	16		67,000	895		67,895	75,000
4500 500 gal., 30 kW, 123 GPH, 480 volt		.80	20		54,500	1,125		55,625	61,000
4520 500 gal., 240 kW, 984 GPH, 480 volt		.80	20		89,000	1,125		90,125	99,000
4540 600 gal., 30 kW, 123 GPH, 480 volt	Q-2	1.20	20		62,000	1,150		63,150	70,500
4560 600 gal., 300 kW, 1,230 GPH, 480 volt		1.20	20		94,500	1,150		95,650	106,000
4580 700 gal., 30 kW, 123 GPH, 480 volt		1	24		65,500	1,400		66,900	74,000
4600 700 gal., 300 kW, 1,230 GPH, 480 volt		1	24		98,500	1,400		99,900	110,000
4620 800 gal., 60 kW, 245 GPH, 480 volt		.90	26.667		70,500	1,550		72,050	80,000
4640 800 gal., 300 kW, 1,230 GPH, 480 volt		.90	26.667		100,500	1,550		102,050	113,000
4660 1,000 gal., 60 kW, 245 GPH, 480 volt		.70	34.286		77,500	2,000		79,500	88,000
4680 1,000 gal., 480 kW, 1,970 GPH, 480 volt		.70	34.286		128,500	2,000		130,500	144,500

For customer support on your Electrical Costs with RSMeans data, call 800.448.8182.

79

22 33 Electric Domestic Water Heaters

22 33 33 – Light-Commercial Electric Domestic Water Heaters

22 33 33.10 Commercial Electric Water Heaters	Crew	Daily Output	Labor-Hours	Unit	Material	2018 Bare Costs Labor	Equipment	Total	Total Incl O&P
4700	1,200 gal., 60 kW, 245 GPH, 480 volt	Q-2	.60	40	Ea.	87,500	2,325		89,825	100,000
4720	1,200 gal., 480 kW, 1,970 GPH, 480 volt		.60	40		133,000	2,325		135,325	149,500
4740	1,500 gal., 60 kW, 245 GPH, 480 volt		.50	48		115,000	2,775		117,775	130,500
4760	1,500 gal., 480 kW, 1,970 GPH, 480 volt		.50	48		160,000	2,775		162,775	180,000
5400	Modulating step control for under 90 kW, 2-5 steps	1 Elec	5.30	1.509		790	88		878	1,000
5440	For above 90 kW, 1 through 5 steps beyond standard, add		3.20	2.500		230	146		376	470
5460	For above 90 kW, 6 through 10 steps beyond standard, add		2.70	2.963		450	172		622	750
5480	For above 90 kW, 11 through 18 steps beyond standard, add		1.60	5		670	291		961	1,175

22 34 Fuel-Fired Domestic Water Heaters

22 34 30 – Residential Gas Domestic Water Heaters

22 34 30.13 Residential, Atmos, Gas Domestic Wtr Heaters

		Crew	Daily Output	Labor-Hours	Unit	Material	2018 Bare Costs Labor	Equipment	Total	Total Incl O&P
0010	**RESIDENTIAL, ATMOSPHERIC, GAS DOMESTIC WATER HEATERS**									
2000	Gas fired, foam lined tank, 10 yr., vent not incl.									
2040	30 gallon	1 Plum	2	4	Ea.	1,775	249		2,024	2,325
2060	40 gallon		1.90	4.211		1,775	262		2,037	2,350
2080	50 gallon		1.80	4.444		1,925	276		2,201	2,550
2090	60 gallon		1.70	4.706		1,725	292		2,017	2,325
2100	75 gallon		1.50	5.333		2,700	330		3,030	3,475
2120	100 gallon		1.30	6.154		2,850	380		3,230	3,700
2900	Water heater, safety-drain pan, 26" round		20	.400		19.90	25		44.90	59.50
3000	Tank leak safety, water & gas shut off see 22 05 23.20 8800									

22 34 46 – Oil-Fired Domestic Water Heaters

22 34 46.10 Residential Oil-Fired Water Heaters

		Crew	Daily Output	Labor-Hours	Unit	Material	2018 Bare Costs Labor	Equipment	Total	Total Incl O&P
0010	**RESIDENTIAL OIL-FIRED WATER HEATERS**									
3000	Oil fired, glass lined tank, 5 yr., vent not included, 30 gallon	1 Plum	2	4	Ea.	1,275	249		1,524	1,775
3040	50 gallon		1.80	4.444		1,575	276		1,851	2,150
3060	70 gallon		1.50	5.333		2,200	330		2,530	2,925

Estimating Tips

The labor adjustment factors listed in Subdivision 22 01 02.20 also apply to Division 23.

23 10 00 Facility Fuel Systems

- The prices in this subdivision for above- and below-ground storage tanks do not include foundations or hold-down slabs, unless noted. The estimator should refer to Divisions 3 and 31 for foundation system pricing. In addition to the foundations, required tank accessories, such as tank gauges, leak detection devices, and additional manholes and piping, must be added to the tank prices.

23 50 00 Central Heating Equipment

- When estimating the cost of an HVAC system, check to see who is responsible for providing and installing the temperature control system. It is possible to overlook controls, assuming that they would be included in the electrical estimate.

- When looking up a boiler, be careful on specified capacity. Some manufacturers rate their products on output while others use input.

- Include HVAC insulation for pipe, boiler, and duct (wrap and liner).

- Be careful when looking up mechanical items to get the correct pressure rating and connection type (thread, weld, flange).

23 70 00 Central HVAC Equipment

- Combination heating and cooling units are sized by the air conditioning requirements. (See Reference No. R236000-20 for the preliminary sizing guide.)

- A ton of air conditioning is nominally 400 CFM.

- Rectangular duct is taken off by the linear foot for each size, but its cost is usually estimated by the pound. Remember that SMACNA standards now base duct on internal pressure.

- Prefabricated duct is estimated and purchased like pipe: straight sections and fittings.

- Note that cranes or other lifting equipment are not included on any lines in Division 23. For example, if a crane is required to lift a heavy piece of pipe into place high above a gym floor, or to put a rooftop unit on the roof of a four-story building, etc., it must be added. Due to the potential for extreme variation—from nothing additional required to a major crane or helicopter—we feel that including a nominal amount for "lifting contingency" would be useless and detract from the accuracy of the estimate. When using equipment rental cost data from RSMeans, do not forget to include the cost of the operator(s).

Reference Numbers

Reference numbers are shown at the beginning of some major classifications. These numbers refer to related items in the Reference Section. The reference information may be an estimating procedure, an alternate pricing method, or technical information.

Note: Not all subdivisions listed here necessarily appear. ∎

23 05 Common Work Results for HVAC

23 05 02 – HVAC General

23 05 02.10 Air Conditioning, General	Crew	Daily Output	Labor- Hours	Unit	Material	2018 Bare Costs Labor	Equipment	Total	Total Incl O&P
0010 **AIR CONDITIONING, GENERAL** Prices are for standard efficiencies (SEER 13)									
0020 for upgrade to SEER 14 add					10%				

23 09 Instrumentation and Control for HVAC

23 09 23 – Direct-Digital Control System for HVAC

23 09 23.10 Control Components/DDC Systems

	Crew	Daily Output	Labor- Hours	Unit	Material	2018 Bare Costs Labor	Equipment	Total	Total Incl O&P
0010 **CONTROL COMPONENTS/DDC SYSTEMS** (Sub's quote incl. M & L)									
0100 Analog inputs									
0110 Sensors (avg. 50' run in 1/2" EMT)									
0120 Duct temperature				Ea.				415	415
0130 Space temperature								665	665
0140 Duct humidity, +/- 3%								695	695
0150 Space humidity, +/- 2%								1,075	1,075
0160 Duct static pressure								565	565
0170 CFM/transducer								765	765
0180 KW/transducer								1,350	1,350
0182 KWH totalization (not incl. elec. meter pulse xmtr.)								625	625
0190 Space static pressure				▼				1,075	1,075
1000 Analog outputs (avg. 50' run in 1/2" EMT)									
1010 P/I transducer				Ea.				635	635
1020 Analog output, matl. in MUX								305	305
1030 Pneumatic (not incl. control device)								645	645
1040 Electric (not incl. control device)				▼				380	380
2000 Status (alarms)									
2100 Digital inputs (avg. 50' run in 1/2" EMT)									
2110 Freeze				Ea.				435	435
2120 Fire								395	395
2130 Differential pressure (air)								595	595
2140 Differential pressure (water)								975	975
2150 Current sensor								435	435
2160 Duct high temperature thermostat								570	570
2170 Duct smoke detector				▼				705	705
2200 Digital output (avg. 50' run in 1/2" EMT)									
2210 Start/stop				Ea.				340	340
2220 On/off (maintained contact)				"				585	585
3000 Controller MUX panel, incl. function boards									
3100 48 point				Ea.				5,275	5,275
3110 128 point				"				7,225	7,225
3200 DDC controller (avg. 50' run in conduit)									
3210 Mechanical room									
3214 16 point controller (incl. 120 volt/1 phase power supply)				Ea.				3,275	3,275
3229 32 point controller (incl. 120 volt/1 phase power supply)				"				5,425	5,425
3230 Includes software programming and checkout									
3260 Space									
3266 VAV terminal box (incl. space temp. sensor)				Ea.				840	840
3280 Host computer (avg. 50' run in conduit)									
3281 Package complete with PC, keyboard,									
3282 printer, monitor, basic software				Ea.				3,150	3,150
4000 Front end costs									
4100 Computer (P.C.) with software program				Ea.				6,350	6,350
4200 Color graphics software				▼				3,925	3,925

23 09 23 – Direct-Digital Control System for HVAC

23 09 23.10 Control Components/DDC Systems		Crew	Daily Output	Labor-Hours	Unit	Material	2018 Bare Costs Labor	Equipment	Total	Total Incl O&P
4300	Color graphics slides				Ea.				490	490
4350	Additional printer				↓				980	980
4400	Communications trunk cable				L.F.				3.80	3.80
4500	Engineering labor (not incl. dftg.)				Point				94	94
4600	Calibration labor								120	120
4700	Start-up, checkout labor				↓				120	120
4800	Programming labor, as req'd									
5000	Communications bus (data transmission cable)									
5010	#18 twisted shielded pair in 1/2" EMT conduit				C.L.F.				380	380
8000	Applications software									
8050	Basic maintenance manager software (not incl. data base entry)				Ea.				1,950	1,950
8100	Time program				Point				6.85	6.85
8120	Duty cycle								13.65	13.65
8140	Optimum start/stop								41.50	41.50
8160	Demand limiting								20.50	20.50
8180	Enthalpy program				↓				41.50	41.50
8200	Boiler optimization				Ea.				1,225	1,225
8220	Chiller optimization				"				1,625	1,625
8240	Custom applications									
8260	Cost varies with complexity									

23 09 43 – Pneumatic Control System for HVAC

23 09 43.10 Pneumatic Control Systems

			Crew	Daily Output	Labor-Hours	Unit	Material	2018 Bare Costs Labor	Equipment	Total	Total Incl O&P
0010	**PNEUMATIC CONTROL SYSTEMS**										
0011	Including a nominal 50' of tubing. Add control panelboard if req'd.										
0100	Heating and ventilating, split system										
0200	Mixed air control, economizer cycle, panel readout, tubing										
0220	Up to 10 tons	G	Q-19	.68	35.294	Ea.	4,450	2,025		6,475	7,925
0240	For 10 to 20 tons	G		.63	37.915		4,750	2,175		6,925	8,450
0260	For over 20 tons	G		.58	41.096		5,125	2,350		7,475	9,175
0270	Enthalpy cycle, up to 10 tons			.50	48.387		4,900	2,775		7,675	9,525
0280	For 10 to 20 tons			.46	52.174		5,275	2,975		8,250	10,300
0290	For over 20 tons		↓	.42	56.604	↓	5,750	3,250		9,000	11,200
0300	Heating coil, hot water, 3 way valve,										
0320	Freezestat, limit control on discharge, readout		Q-5	.69	23.088	Ea.	3,300	1,300		4,600	5,625
0500	Cooling coil, chilled water, room										
0520	Thermostat, 3 way valve		Q-5	2	8	Ea.	1,475	455		1,930	2,300
0600	Cooling tower, fan cycle, damper control,										
0620	Control system including water readout in/out at panel		Q-19	.67	35.821	Ea.	5,850	2,050		7,900	9,525
1000	Unit ventilator, day/night operation,										
1100	freezestat, ASHRAE, cycle 2		Q-19	.91	26.374	Ea.	3,250	1,500		4,750	5,850
2000	Compensated hot water from boiler, valve control,										
2100	readout and reset at panel, up to 60 GPM		Q-19	.55	43.956	Ea.	6,025	2,525		8,550	10,400
2120	For 120 GPM			.51	47.059		6,500	2,700		9,200	11,200
2140	For 240 GPM			.49	49.180		6,750	2,825		9,575	11,700
3000	Boiler room combustion air, damper to 5 S.F., controls			1.37	17.582		2,925	1,000		3,925	4,725
3500	Fan coil, heating and cooling valves, 4 pipe control system			3	8		1,300	460		1,760	2,125
3600	Heat exchanger system controls		↓	.86	27.907	↓	2,850	1,600		4,450	5,525
3900	Multizone control (one per zone), includes thermostat, damper										
3910	motor and reset of discharge temperature		Q-5	.51	31.373	Ea.	2,925	1,775		4,700	5,900
4000	Pneumatic thermostat, including controlling room radiator valve		"	2.43	6.593		880	375		1,255	1,525
4040	Program energy saving optimizer	G	Q-19	1.21	19.786		7,325	1,125		8,450	9,750
4060	Pump control system		"	3	8	↓	1,350	460		1,810	2,175

For customer support on your Electrical Costs with RSMeans data, call 800.448.8182.

83

23 09 43 – Pneumatic Control System for HVAC

23 09 43.10 Pneumatic Control Systems		Crew	Daily Output	Labor-Hours	Unit	Material	2018 Bare Costs Labor	2018 Bare Costs Equipment	Total	Total Incl O&P
4080	Reheat coil control system, not incl. coil	Q-5	2.43	6.593	Ea.	1,150	375		1,525	1,825
4500	Air supply for pneumatic control system									
4600	Tank mounted duplex compressor, starter, alternator,									
4620	piping, dryer, PRV station and filter									
4630	1/2 HP	Q-19	.68	35.139	Ea.	10,800	2,000		12,800	14,900
4640	3/4 HP		.64	37.383		11,400	2,150		13,550	15,800
4650	1 HP		.61	39.539		12,500	2,250		14,750	17,100
4660	1-1/2 HP		.58	41.739		13,300	2,375		15,675	18,200
4680	3 HP		.55	43.956		18,100	2,525		20,625	23,700
4690	5 HP		.42	57.143		31,400	3,275		34,675	39,400
4800	Main air supply, includes 3/8" copper main and labor	Q-5	1.82	8.791	C.L.F.	365	500		865	1,150
4810	If poly tubing used, deduct								30%	30%
7000	Static pressure control for air handling unit, includes pressure									
7010	sensor, receiver controller, readout and damper motors	Q-19	.64	37.383	Ea.	8,600	2,150		10,750	12,700
7020	If return air fan requires control, add								70%	70%
8600	VAV boxes, incl. thermostat, damper motor, reheat coil & tubing	Q-5	1.46	10.989		1,350	625		1,975	2,450
8610	If no reheat coil, deduct								204	204

23 09 53 – Pneumatic and Electric Control System for HVAC

23 09 53.10 Control Components

		Crew	Daily Output	Labor-Hours	Unit	Material	2018 Bare Costs Labor	2018 Bare Costs Equipment	Total	Total Incl O&P
0010	**CONTROL COMPONENTS**									
0700	Controller, receiver									
0850	Electric, single snap switch	1 Elec	4	2	Ea.	490	116		606	710
0860	Dual snap switches	"	3	2.667	"	655	155		810	950
3590	Sensor, electric operated									
3620	Humidity	1 Elec	8	1	Ea.	262	58		320	375
3650	Pressure		8	1		271	58		329	385
3680	Temperature		10	.800		119	46.50		165.50	201
5000	Thermostats									
5200	24 hour, automatic, clock [G]	1 Shee	8	1	Ea.	235	60		295	350
5220	Electric, low voltage, 2 wire	1 Elec	13	.615		62	36		98	122
5230	3 wire	"	10	.800		54.50	46.50		101	130
7090	Valves, motor controlled, including actuator									
7100	Electric motor actuated									
7200	Brass, two way, screwed									
7210	1/2" pipe size	L-6	36	.333	Ea.	355	20.50		375.50	420
7220	3/4" pipe size		30	.400		600	24.50		624.50	695
7230	1" pipe size		28	.429		715	26		741	825
7240	1-1/2" pipe size		19	.632		820	38.50		858.50	960
7250	2" pipe size		16	.750		1,150	45.50		1,195.50	1,350
7350	Brass, three way, screwed									
7360	1/2" pipe size	L-6	33	.364	Ea.	325	22		347	390
7370	3/4" pipe size		27	.444		355	27		382	430
7380	1" pipe size		25.50	.471		405	28.50		433.50	490
7384	1-1/4" pipe size		21	.571		460	35		495	560
7390	1-1/2" pipe size		17	.706		505	43		548	620
7400	2" pipe size		14	.857		765	52		817	920
7550	Iron body, two way, flanged									
7560	2-1/2" pipe size	L-6	4	3	Ea.	1,150	182		1,332	1,525
7570	3" pipe size		3	4		1,175	243		1,418	1,650
7580	4" pipe size		2	6		2,250	365		2,615	3,000
7850	Iron body, three way, flanged									
7860	2-1/2" pipe size	L-6	3	4	Ea.	1,150	243		1,393	1,650

23 09 Instrumentation and Control for HVAC

23 09 53 – Pneumatic and Electric Control System for HVAC

23 09 53.10 Control Components	Crew	Daily Output	Labor-Hours	Unit	Material	2018 Bare Costs Labor	Equipment	Total	Total Incl O&P
7870	3" pipe size	L-6	2.50	4.800	Ea.	1,300	292		1,592	1,900
7880	4" pipe size	↓	2	6	↓	1,675	365		2,040	2,400

23 34 HVAC Fans

23 34 13 – Axial HVAC Fans

23 34 13.10 Axial Flow HVAC Fans

		Crew	Daily Output	Labor-Hours	Unit	Material	2018 Bare Costs Labor	Equipment	Total	Total Incl O&P
0010	**AXIAL FLOW HVAC FANS**									
0020	Air conditioning and process air handling									
1500	Vaneaxial, low pressure, 2,000 CFM, 1/2 HP	Q-20	3.60	5.556	Ea.	2,375	305		2,680	3,050
1520	4,000 CFM, 1 HP		3.20	6.250		2,425	340		2,765	3,200
1540	8,000 CFM, 2 HP		2.80	7.143		3,500	390		3,890	4,450
1560	16,000 CFM, 5 HP	↓	2.40	8.333	↓	4,900	455		5,355	6,100

23 34 14 – Blower HVAC Fans

23 34 14.10 Blower Type HVAC Fans

		Crew	Daily Output	Labor-Hours	Unit	Material	2018 Bare Costs Labor	Equipment	Total	Total Incl O&P
0010	**BLOWER TYPE HVAC FANS**									
2000	Blowers, direct drive with motor, complete									
2020	1,045 CFM @ 0.5" S.P., 1/5 HP	Q-20	18	1.111	Ea.	365	61		426	495
2040	1,385 CFM @ 0.5" S.P., 1/4 HP		18	1.111		360	61		421	490
2060	1,640 CFM @ 0.5" S.P., 1/3 HP		18	1.111		325	61		386	450
2080	1,760 CFM @ 0.5" S.P., 1/2 HP	↓	18	1.111	↓	335	61		396	465
2090	4 speed									
2100	1,164 to 1,739 CFM @ 0.5" S.P., 1/3 HP	Q-20	16	1.250	Ea.	340	68.50		408.50	480
2120	1,467 to 2,218 CFM @ 1.0" S.P., 3/4 HP	"	14	1.429	"	390	78		468	550
2500	Ceiling fan, right angle, extra quiet, 0.10" S.P.									
2520	95 CFM	Q-20	20	1	Ea.	291	54.50		345.50	405
2540	210 CFM		19	1.053		345	57.50		402.50	470
2560	385 CFM		18	1.111		435	61		496	575
2580	885 CFM		16	1.250		860	68.50		928.50	1,050
2600	1,650 CFM		13	1.538		1,200	84		1,284	1,425
2620	2,960 CFM	↓	11	1.818		1,575	99.50		1,674.50	1,900
2680	For speed control switch, add	1 Elec	16	.500	↓	159	29		188	218
7500	Utility set, steel construction, pedestal, 1/4" S.P.									
7520	Direct drive, 150 CFM, 1/8 HP	Q-20	6.40	3.125	Ea.	925	171		1,096	1,275
7540	485 CFM, 1/6 HP		5.80	3.448		1,050	189		1,239	1,425
7560	1,950 CFM, 1/2 HP		4.80	4.167		1,350	228		1,578	1,850
7580	2,410 CFM, 3/4 HP		4.40	4.545		2,525	249		2,774	3,150
7600	3,328 CFM, 1-1/2 HP	↓	3	6.667	↓	2,500	365		2,865	3,300
7680	V-belt drive, drive cover, 3 phase									
7700	800 CFM, 1/4 HP	Q-20	6	3.333	Ea.	940	182		1,122	1,300
7720	1,300 CFM, 1/3 HP		5	4		985	219		1,204	1,400
7740	2,000 CFM, 1 HP		4.60	4.348		1,175	238		1,413	1,625
7760	2,900 CFM, 3/4 HP		4.20	4.762		1,575	260		1,835	2,125
7780	3,600 CFM, 3/4 HP		4	5		1,925	274		2,199	2,550
7800	4,800 CFM, 1 HP		3.50	5.714		2,275	315		2,590	3,000
7820	6,700 CFM, 1-1/2 HP		3	6.667		2,825	365		3,190	3,650
7830	7,500 CFM, 2 HP		2.50	8		3,825	440		4,265	4,900
7840	11,000 CFM, 3 HP		2	10		5,125	545		5,670	6,450
7860	13,000 CFM, 3 HP		1.60	12.500		5,200	685		5,885	6,775
7880	15,000 CFM, 5 HP		1	20		5,375	1,100		6,475	7,575
7900	17,000 CFM, 7-1/2 HP	↓	.80	25	↓	5,750	1,375		7,125	8,400

For customer support on your Electrical Costs with RSMeans data, call 800.448.8182.

85

23 34 14 – Blower HVAC Fans

23 34 14.10 Blower Type HVAC Fans	Crew	Daily Output	Labor-Hours	Unit	Material	2018 Bare Costs Labor	2018 Bare Costs Equipment	Total	Total Incl O&P
7920 20,000 CFM, 7-1/2 HP	Q-20	.80	25	Ea.	6,850	1,375		8,225	9,625

23 34 16 – Centrifugal HVAC Fans

23 34 16.10 Centrifugal Type HVAC Fans

	Crew	Daily Output	Labor-Hours	Unit	Material	2018 Bare Costs Labor	2018 Bare Costs Equipment	Total	Total Incl O&P
0010 **CENTRIFUGAL TYPE HVAC FANS**									
0200 In-line centrifugal, supply/exhaust booster									
0220 aluminum wheel/hub, disconnect switch, 1/4" S.P.									
0240 500 CFM, 10" diameter connection	Q-20	3	6.667	Ea.	1,575	365		1,940	2,275
0260 1,380 CFM, 12" diameter connection		2	10		1,600	545		2,145	2,575
0280 1,520 CFM, 16" diameter connection		2	10		1,650	545		2,195	2,625
0300 2,560 CFM, 18" diameter connection		1	20		1,800	1,100		2,900	3,625
0320 3,480 CFM, 20" diameter connection		.80	25		2,075	1,375		3,450	4,375
0326 5,080 CFM, 20" diameter connection	↓	.75	26.667	↓	2,175	1,450		3,625	4,625
3500 Centrifugal, airfoil, motor and drive, complete									
3520 1,000 CFM, 1/2 HP	Q-20	2.50	8	Ea.	2,025	440		2,465	2,900
3540 2,000 CFM, 1 HP		2	10		2,300	545		2,845	3,350
3560 4,000 CFM, 3 HP		1.80	11.111		2,900	610		3,510	4,125
3580 8,000 CFM, 7-1/2 HP		1.40	14.286		4,275	780		5,055	5,875
3600 12,000 CFM, 10 HP	↓	1	20	↓	5,800	1,100		6,900	8,050
5000 Utility set, centrifugal, V belt drive, motor									
5020 1/4" S.P., 1,200 CFM, 1/4 HP	Q-20	6	3.333	Ea.	1,875	182		2,057	2,325
5040 1,520 CFM, 1/3 HP		5	4		2,450	219		2,669	3,025
5060 1,850 CFM, 1/2 HP		4	5		2,425	274		2,699	3,075
5080 2,180 CFM, 3/4 HP		3	6.667		2,875	365		3,240	3,725
5100 1/2" S.P., 3,600 CFM, 1 HP		2	10		2,975	545		3,520	4,100
5120 4,250 CFM, 1-1/2 HP		1.60	12.500		3,600	685		4,285	5,025
5140 4,800 CFM, 2 HP		1.40	14.286		4,375	780		5,155	5,975
5160 6,920 CFM, 5 HP		1.30	15.385		5,300	840		6,140	7,100
5180 7,700 CFM, 7-1/2 HP	↓	1.20	16.667	↓	6,350	910		7,260	8,350
5200 For explosion proof motor, add					15%				
5500 Fans, industrial exhauster, for air which may contain granular matl.									
5520 1,000 CFM, 1-1/2 HP	Q-20	2.50	8	Ea.	3,075	440		3,515	4,075
5540 2,000 CFM, 3 HP		2	10		3,750	545		4,295	4,950
5560 4,000 CFM, 7-1/2 HP		1.80	11.111		5,375	610		5,985	6,850
5580 8,000 CFM, 15 HP		1.40	14.286		6,950	780		7,730	8,825
5600 12,000 CFM, 30 HP	↓	1	20	↓	10,700	1,100		11,800	13,500
7000 Roof exhauster, centrifugal, aluminum housing, 12" galvanized									
7020 curb, bird screen, back draft damper, 1/4" S.P.									
7100 Direct drive, 320 CFM, 11" sq. damper	Q-20	7	2.857	Ea.	770	156		926	1,075
7120 600 CFM, 11" sq. damper		6	3.333		985	182		1,167	1,350
7140 815 CFM, 13" sq. damper		5	4		985	219		1,204	1,400
7160 1,450 CFM, 13" sq. damper		4.20	4.762		1,575	260		1,835	2,125
7180 2,050 CFM, 16" sq. damper		4	5		1,925	274		2,199	2,550
7200 V-belt drive, 1,650 CFM, 12" sq. damper		6	3.333		1,425	182		1,607	1,825
7220 2,750 CFM, 21" sq. damper		5	4		1,675	219		1,894	2,175
7230 3,500 CFM, 21" sq. damper		4.50	4.444		1,900	243		2,143	2,475
7240 4,910 CFM, 23" sq. damper		4	5		2,300	274		2,574	2,950
7260 8,525 CFM, 28" sq. damper		3	6.667		3,075	365		3,440	3,925
7280 13,760 CFM, 35" sq. damper		2	10		4,250	545		4,795	5,500
7300 20,558 CFM, 43" sq. damper	↓	1	20		8,425	1,100		9,525	10,900
7320 For 2 speed winding, add					15%				
7340 For explosion proof motor, add					755			755	835
7360 For belt driven, top discharge, add				↓	15%				

23 34 HVAC Fans

23 34 16 – Centrifugal HVAC Fans

23 34 16.10 Centrifugal Type HVAC Fans		Crew	Daily Output	Labor-Hours	Unit	Material	2018 Bare Costs Labor	Equipment	Total	Total Incl O&P
8500	Wall exhausters, centrifugal, auto damper, 1/8" S.P.									
8520	Direct drive, 610 CFM, 1/20 HP	Q-20	14	1.429	Ea.	410	78		488	570
8540	796 CFM, 1/12 HP		13	1.538		970	84		1,054	1,200
8560	822 CFM, 1/6 HP		12	1.667		1,025	91		1,116	1,300
8580	1,320 CFM, 1/4 HP		12	1.667		1,375	91		1,466	1,675
8600	1,756 CFM, 1/4 HP		11	1.818		1,225	99.50		1,324.50	1,500
8620	1,983 CFM, 1/4 HP		10	2		1,250	109		1,359	1,550
8640	2,900 CFM, 1/2 HP		9	2.222		1,325	122		1,447	1,625
8660	3,307 CFM, 3/4 HP		8	2.500		1,425	137		1,562	1,775
9500	V-belt drive, 3 phase									
9520	2,800 CFM, 1/4 HP	Q-20	9	2.222	Ea.	1,850	122		1,972	2,225
9540	3,740 CFM, 1/2 HP		8	2.500		1,925	137		2,062	2,325
9560	4,400 CFM, 3/4 HP		7	2.857		1,950	156		2,106	2,375
9580	5,700 CFM, 1-1/2 HP		6	3.333		2,025	182		2,207	2,500

23 34 23 – HVAC Power Ventilators

23 34 23.10 HVAC Power Circulators and Ventilators		Crew	Daily Output	Labor-Hours	Unit	Material	2018 Bare Costs Labor	Equipment	Total	Total Incl O&P
0010	**HVAC POWER CIRCULATORS AND VENTILATORS**									
3000	Paddle blade air circulator, 3 speed switch									
3020	42", 5,000 CFM high, 3,000 CFM low G	1 Elec	2.40	3.333	Ea.	123	194		317	425
3040	52", 6,500 CFM high, 4,000 CFM low G	"	2.20	3.636	"	156	212		368	485
3100	For antique white motor, same cost									
3200	For brass plated motor, same cost									
3300	For light adaptor kit, add G				Ea.	39.50			39.50	43.50
6000	Propeller exhaust, wall shutter									
6020	Direct drive, one speed, 0.075" S.P.									
6100	653 CFM, 1/30 HP	Q-20	10	2	Ea.	202	109		311	390
6120	1,033 CFM, 1/20 HP		9	2.222		300	122		422	515
6140	1,323 CFM, 1/15 HP		8	2.500		335	137		472	580
6160	2,444 CFM, 1/4 HP		7	2.857		420	156		576	700
6300	V-belt drive, 3 phase									
6320	6,175 CFM, 3/4 HP	Q-20	5	4	Ea.	1,275	219		1,494	1,725
6340	7,500 CFM, 3/4 HP		5	4		1,350	219		1,569	1,800
6360	10,100 CFM, 1 HP		4.50	4.444		1,550	243		1,793	2,075
6380	14,300 CFM, 1-1/2 HP		4	5		1,825	274		2,099	2,425
6400	19,800 CFM, 2 HP		3	6.667		2,150	365		2,515	2,900
6420	26,250 CFM, 3 HP		2.60	7.692		2,275	420		2,695	3,150
6440	38,500 CFM, 5 HP		2.20	9.091		2,675	495		3,170	3,700
6460	46,000 CFM, 7-1/2 HP		2	10		2,875	545		3,420	3,975
6480	51,500 CFM, 10 HP		1.80	11.111		2,975	610		3,585	4,200
6650	Residential, bath exhaust, grille, back draft damper									
6660	50 CFM	Q-20	24	.833	Ea.	52.50	45.50		98	127
6670	110 CFM		22	.909		104	49.50		153.50	190
6680	Light combination, squirrel cage, 100 watt, 70 CFM		24	.833		114	45.50		159.50	195
6700	Light/heater combination, ceiling mounted									
6710	70 CFM, 1,450 watt	Q-20	24	.833	Ea.	140	45.50		185.50	224
6800	Heater combination, recessed, 70 CFM		24	.833		69.50	45.50		115	146
6820	With 2 infrared bulbs		23	.870		97	47.50		144.50	180
6900	Kitchen exhaust, grille, complete, 160 CFM		22	.909		105	49.50		154.50	192
6910	180 CFM		20	1		106	54.50		160.50	200
6920	270 CFM		18	1.111		230	61		291	345
6930	350 CFM		16	1.250		145	68.50		213.50	264
6940	Residential roof jacks and wall caps									

23 34 23.10 HVAC Power Circulators and Ventilators	Crew	Daily Output	Labor-Hours	Unit	Material	2018 Bare Costs Labor	Equipment	Total	Total Incl O&P
6944	Wall cap with back draft damper								
6946	3" & 4" diam. round duct	1 Shee	11	.727	Ea.	25	43.50		68.50	94
6948	6" diam. round duct	"	11	.727	"	72.50	43.50		116	146
6958	Roof jack with bird screen and back draft damper								
6960	3" & 4" diam. round duct	1 Shee	11	.727	Ea.	24.50	43.50		68	93.50
6962	3-1/4" x 10" rectangular duct	"	10	.800	"	44	48		92	122
6980	Transition								
6982	3-1/4" x 10" to 6" diam. round	1 Shee	20	.400	Ea.	32	24		56	71.50
8020	Attic, roof type								
8030	Aluminum dome, damper & curb								
8080	12" diameter, 1,000 CFM (gravity)	1 Elec	10	.800	Ea.	600	46.50		646.50	730
8090	16" diameter, 1,500 CFM (gravity)		9	.889		725	51.50		776.50	880
8100	20" diameter, 2,500 CFM (gravity)		8	1		890	58		948	1,075
8110	26" diameter, 4,000 CFM (gravity)		7	1.143		1,075	66.50		1,141.50	1,275
8120	32" diameter, 6,500 CFM (gravity)		6	1.333		1,475	77.50		1,552.50	1,750
8130	38" diameter, 8,000 CFM (gravity)		5	1.600		2,200	93		2,293	2,575
8140	50" diameter, 13,000 CFM (gravity)		4	2		3,175	116		3,291	3,675
8160	Plastic, ABS dome								
8180	1,050 CFM	1 Elec	14	.571	Ea.	177	33.50		210.50	245
8200	1,600 CFM	"	12	.667	"	265	39		304	350
8240	Attic, wall type, with shutter, one speed								
8250	12" diameter, 1,000 CFM	1 Elec	14	.571	Ea.	380	33.50		413.50	470
8260	14" diameter, 1,500 CFM		12	.667		415	39		454	515
8270	16" diameter, 2,000 CFM		9	.889		465	51.50		516.50	595
8290	Whole house, wall type, with shutter, one speed								
8300	30" diameter, 4,800 CFM	1 Elec	7	1.143	Ea.	1,000	66.50		1,066.50	1,200
8310	36" diameter, 7,000 CFM		6	1.333		1,075	77.50		1,152.50	1,325
8320	42" diameter, 10,000 CFM		5	1.600		1,225	93		1,318	1,500
8330	48" diameter, 16,000 CFM		4	2		1,525	116		1,641	1,850
8340	For two speed, add					91			91	100
8350	Whole house, lay-down type, with shutter, one speed								
8360	30" diameter, 4,500 CFM	1 Elec	8	1	Ea.	1,075	58		1,133	1,250
8370	36" diameter, 6,500 CFM		7	1.143		1,150	66.50		1,216.50	1,350
8380	42" diameter, 9,000 CFM		6	1.333		1,250	77.50		1,327.50	1,500
8390	48" diameter, 12,000 CFM		5	1.600		1,425	93		1,518	1,725
8440	For two speed, add					68.50			68.50	75.50
8450	For 12 hour timer switch, add	1 Elec	32	.250		68.50	14.55		83.05	97.50

23 52 Heating Boilers
23 52 13 – Electric Boilers

23 52 13.10 Electric Boilers, ASME

		Crew	Daily Output	Labor-Hours	Unit	Material	2018 Bare Costs Labor	Equipment	Total	Total Incl O&P
0010	**ELECTRIC BOILERS, ASME**, Standard controls and trim									
1000	Steam, 6 KW, 20.5 MBH	Q-19	1.20	20	Ea.	4,250	1,150		5,400	6,400
1040	9 KW, 30.7 MBH		1.20	20		4,200	1,150		5,350	6,325
1060	18 KW, 61.4 MBH		1.20	20		4,325	1,150		5,475	6,475
1080	24 KW, 81.8 MBH		1.10	21.818		5,000	1,250		6,250	7,375
1120	36 KW, 123 MBH		1.10	21.818		5,525	1,250		6,775	7,950
1160	60 KW, 205 MBH		1	24		6,925	1,375		8,300	9,650
1220	112 KW, 382 MBH		.75	32		9,475	1,825		11,300	13,200
1240	148 KW, 505 MBH		.65	36.923		9,975	2,100		12,075	14,200
1260	168 KW, 573 MBH		.60	40		20,400	2,300		22,700	25,900

For customer support on your Electrical Costs with RSMeans data, call 800.448.8182.

23 52 13.10 Electric Boilers, ASME		Crew	Daily Output	Labor-Hours	Unit	Material	2018 Bare Costs Labor	Equipment	Total	Total Incl O&P
1280	222 KW, 758 MBH	Q-19	.55	43.636	Ea.	23,800	2,500		26,300	30,000
1300	296 KW, 1,010 MBH		.45	53.333		25,000	3,050		28,050	32,200
1320	300 KW, 1,023 MBH		.40	60		26,400	3,425		29,825	34,200
1340	370 KW, 1,263 MBH		.35	68.571		29,200	3,925		33,125	38,000
1360	444 KW, 1,515 MBH		.30	80		31,400	4,575		35,975	41,400
1380	518 KW, 1,768 MBH	Q-21	.36	88.889		32,600	5,225		37,825	43,800
1400	592 KW, 2,020 MBH		.34	94.118		35,400	5,525		40,925	47,200
1420	666 KW, 2,273 MBH		.32	100		36,800	5,875		42,675	49,300
1460	740 KW, 2,526 MBH		.28	114		38,300	6,700		45,000	52,000
1480	814 KW, 2,778 MBH		.25	128		40,600	7,500		48,100	56,000
1500	962 KW, 3,283 MBH		.22	145		44,900	8,525		53,425	62,000
1520	1,036 KW, 3,536 MBH		.20	160		47,600	9,375		56,975	66,500
1540	1,110 KW, 3,788 MBH		.19	168		49,600	9,875		59,475	69,500
1560	2,070 KW, 7,063 MBH		.18	178		68,000	10,400		78,400	90,500
1580	2,250 KW, 7,677 MBH		.17	188		80,000	11,000		91,000	104,500
1600	2,340 KW, 7,984 MBH		.16	200		86,500	11,700		98,200	112,500
2000	Hot water, 7.5 KW, 25.6 MBH	Q-19	1.30	18.462		5,125	1,050		6,175	7,225
2020	15 KW, 51.2 MBH		1.30	18.462		5,175	1,050		6,225	7,275
2040	30 KW, 102 MBH		1.20	20		5,525	1,150		6,675	7,800
2060	45 KW, 164 MBH		1.20	20		5,625	1,150		6,775	7,900
2070	60 KW, 205 MBH		1.20	20		5,750	1,150		6,900	8,025
2080	75 KW, 256 MBH		1.10	21.818		6,125	1,250		7,375	8,625
2100	90 KW, 307 MBH		1.10	21.818		6,100	1,250		7,350	8,600
2120	105 KW, 358 MBH		1	24		6,500	1,375		7,875	9,200
2140	120 KW, 410 MBH		.90	26.667		6,625	1,525		8,150	9,575
2160	135 KW, 461 MBH		.75	32		7,525	1,825		9,350	11,100
2180	150 KW, 512 MBH		.65	36.923		7,775	2,100		9,875	11,700
2200	165 KW, 563 MBH		.60	40		7,900	2,300		10,200	12,100
2220	296 KW, 1,010 MBH		.55	43.636		15,700	2,500		18,200	21,000
2280	370 KW, 1,263 MBH		.40	60		17,300	3,425		20,725	24,300
2300	444 KW, 1,515 MBH		.35	68.571		20,200	3,925		24,125	28,100
2340	518 KW, 1,768 MBH	Q-21	.44	72.727		22,400	4,275		26,675	31,000
2360	592 KW, 2,020 MBH		.43	74.419		25,400	4,375		29,775	34,500
2400	666 KW, 2,273 MBH		.40	80		26,500	4,700		31,200	36,300
2420	740 KW, 2,526 MBH		.39	82.051		27,700	4,800		32,500	37,700
2440	814 KW, 2,778 MBH		.38	84.211		29,600	4,950		34,550	39,900
2460	888 KW, 3,031 MBH		.37	86.486		31,600	5,075		36,675	42,300
2480	962 KW, 3,283 MBH		.36	88.889		33,000	5,225		38,225	44,200
2500	1,036 KW, 3,536 MBH		.34	94.118		35,200	5,525		40,725	47,100
2520	1,110 KW, 3,788 MBH		.33	96.970		38,400	5,675		44,075	51,000
2540	1,440 KW, 4,915 MBH		.32	100		42,900	5,875		48,775	56,000
2560	1,560 KW, 5,323 MBH		.31	103		46,800	6,050		52,850	60,500
2580	1,680 KW, 5,733 MBH		.30	107		50,000	6,250		56,250	64,500
2600	1,800 KW, 6,143 MBH		.29	110		53,500	6,475		59,975	68,000
2620	1,980 KW, 6,757 MBH		.28	114		58,000	6,700		64,700	74,000
2640	2,100 KW, 7,167 MBH		.27	119		63,500	6,950		70,450	80,500
2660	2,220 KW, 7,576 MBH		.26	123		65,000	7,225		72,225	82,500
2680	2,400 KW, 8,191 MBH		.25	128		68,000	7,500		75,500	86,000
2700	2,610 KW, 8,905 MBH		.24	133		72,500	7,825		80,325	91,500
2720	2,790 KW, 9,519 MBH		.23	139		75,500	8,150		83,650	95,500
2740	2,970 KW, 10,133 MBH		.21	152		77,500	8,925		86,425	98,500
2760	3,150 KW, 10,748 MBH		.19	168		82,500	9,875		92,375	106,000
2780	3,240 KW, 11,055 MBH		.18	178		84,500	10,400		94,900	108,500

For customer support on your Electrical Costs with RSMeans data, call 800.448.8182.

89

23 52 Heating Boilers

23 52 13 – Electric Boilers

23 52 13.10 Electric Boilers, ASME

		Crew	Daily Output	Labor-Hours	Unit	Material	2018 Bare Costs Labor	Equipment	Total	Total Incl O&P
2800	3,420 KW, 11,669 MBH	Q-21	.17	188	Ea.	90,000	11,000		101,000	115,500
2820	3,600 KW, 12,283 MBH		.16	200		94,000	11,700		105,700	121,000

23 52 28 – Swimming Pool Boilers

23 52 28.10 Swimming Pool Heaters

		Crew	Daily Output	Labor-Hours	Unit	Material	Labor	Equipment	Total	Total Incl O&P
0010	**SWIMMING POOL HEATERS**, Not including wiring, external									
0020	piping, base or pad									
2000	Electric, 12 KW, 4,800 gallon pool	Q-19	3	8	Ea.	2,100	460		2,560	3,000
2020	15 KW, 7,200 gallon pool		2.80	8.571		2,150	490		2,640	3,075
2040	24 KW, 9,600 gallon pool		2.40	10		2,475	570		3,045	3,575
2060	30 KW, 12,000 gallon pool		2	12		2,525	685		3,210	3,800
2080	36 KW, 14,400 gallon pool		1.60	15		2,900	860		3,760	4,500
2100	57 KW, 24,000 gallon pool		1.20	20		3,650	1,150		4,800	5,725
9000	To select pool heater: 12 BTUH x S.F. pool area									
9010	X temperature differential = required output									
9050	For electric, KW = gallons x 2.5 divided by 1,000									
9100	For family home type pool, double the									
9110	Rated gallon capacity = 1/2°F rise per hour									

23 54 Furnaces

23 54 13 – Electric-Resistance Furnaces

23 54 13.10 Electric Furnaces

		Crew	Daily Output	Labor-Hours	Unit	Material	Labor	Equipment	Total	Total Incl O&P
0010	**ELECTRIC FURNACES**, Hot air, blowers, std. controls									
1000	Electric, UL listed									
1070	10.2 MBH	Q-20	4.40	4.545	Ea.	335	249		584	750
1080	17.1 MBH		4.60	4.348		415	238		653	815
1100	34.1 MBH		4.40	4.545		520	249		769	950
1120	51.6 MBH		4.20	4.762		565	260		825	1,025
1140	68.3 MBH		4	5		590	274		864	1,075
1160	85.3 MBH		3.80	5.263		610	288		898	1,100

23 82 Convection Heating and Cooling Units

23 82 16 – Air Coils

23 82 16.20 Duct Heaters

		Crew	Daily Output	Labor-Hours	Unit	Material	Labor	Equipment	Total	Total Incl O&P
0010	**DUCT HEATERS**, Electric, 480 V, 3 Ph.									
0020	Finned tubular insert, 500°F									
0100	8" wide x 6" high, 4.0 kW	Q-20	16	1.250	Ea.	795	68.50		863.50	980
0120	12" high, 8.0 kW		15	1.333		1,325	73		1,398	1,550
0140	18" high, 12.0 kW		14	1.429		1,850	78		1,928	2,150
0160	24" high, 16.0 kW		13	1.538		2,375	84		2,459	2,750
0180	30" high, 20.0 kW		12	1.667		2,900	91		2,991	3,350
0300	12" wide x 6" high, 6.7 kW		15	1.333		845	73		918	1,050
0320	12" high, 13.3 kW		14	1.429		1,375	78		1,453	1,625
0340	18" high, 20.0 kW		13	1.538		1,925	84		2,009	2,225
0360	24" high, 26.7 kW		12	1.667		2,450	91		2,541	2,850
0380	30" high, 33.3 kW		11	1.818		3,000	99.50		3,099.50	3,450
0500	18" wide x 6" high, 13.3 kW		14	1.429		910	78		988	1,125
0520	12" high, 26.7 kW		13	1.538		1,575	84		1,659	1,850
0540	18" high, 40.0 kW		12	1.667		2,075	91		2,166	2,425
0560	24" high, 53.3 kW		11	1.818		2,775	99.50		2,874.50	3,200

For customer support on your Electrical Costs with RSMeans data, call 800.448.8182.

23 82 16 – Air Coils

23 82 16.20 Duct Heaters		Crew	Daily Output	Labor-Hours	Unit	Material	2018 Bare Costs Labor	Equipment	Total	Total Incl O&P
0580	30" high, 66.7 kW	Q-20	10	2	Ea.	3,450	109		3,559	3,975
0700	24" wide x 6" high, 17.8 kW		13	1.538		1,000	84		1,084	1,225
0720	12" high, 35.6 kW		12	1.667		1,700	91		1,791	2,000
0740	18" high, 53.3 kW		11	1.818		2,375	99.50		2,474.50	2,775
0760	24" high, 71.1 kW		10	2		3,075	109		3,184	3,550
0780	30" high, 88.9 kW		9	2.222		3,775	122		3,897	4,325
0900	30" wide x 6" high, 22.2 kW		12	1.667		1,050	91		1,141	1,325
0920	12" high, 44.4 kW		11	1.818		1,800	99.50		1,899.50	2,125
0940	18" high, 66.7 kW		10	2		2,525	109		2,634	2,950
0960	24" high, 88.9 kW		9	2.222		3,250	122		3,372	3,750
0980	30" high, 111.0 kW		8	2.500		3,975	137		4,112	4,575
1400	Note decreased kW available for									
1410	each duct size at same cost									
1420	See line 5000 for modifications and accessories									
2000	Finned tubular flange with insulated									
2020	terminal box, 500°F									
2100	12" wide x 36" high, 54 kW	Q-20	10	2	Ea.	3,925	109		4,034	4,500
2120	40" high, 60 kW		9	2.222		4,500	122		4,622	5,125
2200	24" wide x 36" high, 118.8 kW		9	2.222		5,075	122		5,197	5,750
2220	40" high, 132 kW		8	2.500		5,000	137		5,137	5,700
2400	36" wide x 8" high, 40 kW		11	1.818		1,975	99.50		2,074.50	2,325
2420	16" high, 80 kW		10	2		2,650	109		2,759	3,100
2440	24" high, 120 kW		9	2.222		3,475	122		3,597	4,000
2460	32" high, 160 kW		8	2.500		4,475	137		4,612	5,100
2480	36" high, 180 kW		7	2.857		5,500	156		5,656	6,275
2500	40" high, 200 kW		6	3.333		6,100	182		6,282	6,975
2600	40" wide x 8" high, 45 kW		11	1.818		2,075	99.50		2,174.50	2,450
2620	16" high, 90 kW		10	2		2,875	109		2,984	3,325
2640	24" high, 135 kW		9	2.222		3,650	122		3,772	4,175
2660	32" high, 180 kW		8	2.500		4,850	137		4,987	5,525
2680	36" high, 202.5 kW		7	2.857		5,575	156		5,731	6,375
2700	40" high, 225 kW		6	3.333		6,350	182		6,532	7,275
2800	48" wide x 8" high, 54.8 kW		10	2		2,175	109		2,284	2,575
2820	16" high, 109.8 kW		9	2.222		3,025	122		3,147	3,500
2840	24" high, 164.4 kW		8	2.500		3,850	137		3,987	4,450
2860	32" high, 219.2 kW		7	2.857		5,150	156		5,306	5,875
2880	36" high, 246.6 kW		6	3.333		5,900	182		6,082	6,775
2900	40" high, 274 kW		5	4		6,675	219		6,894	7,675
3000	56" wide x 8" high, 64 kW		9	2.222		2,475	122		2,597	2,900
3020	16" high, 128 kW		8	2.500		3,450	137		3,587	4,000
3040	24" high, 192 kW		7	2.857		4,175	156		4,331	4,825
3060	32" high, 256 kW		6	3.333		5,825	182		6,007	6,700
3080	36" high, 288 kW		5	4		6,700	219		6,919	7,675
3100	40" high, 320 kW		4	5		7,400	274		7,674	8,575
3200	64" wide x 8" high, 74 kW		8	2.500		2,550	137		2,687	3,025
3220	16" high, 148 kW		7	2.857		3,550	156		3,706	4,125
3240	24" high, 222 kW		6	3.333		4,500	182		4,682	5,225
3260	32" high, 296 kW		5	4		6,050	219		6,269	6,975
3280	36" high, 333 kW		4	5		7,200	274		7,474	8,350
3300	40" high, 370 kW		3	6.667		7,950	365		8,315	9,300
3800	Note decreased kW available for									
3820	each duct size at same cost									
5000	Duct heater modifications and accessories									

For customer support on your Electrical Costs with RSMeans data, call 800.448.8182.

91

23 82 16 – Air Coils

23 82 16.20 Duct Heaters		Crew	Daily Output	Labor-Hours	Unit	Material	2018 Bare Costs Labor	2018 Bare Costs Equipment	Total	Total Incl O&P
5120	T.C.O. limit auto or manual reset	Q-20	42	.476	Ea.	130	26		156	183
5140	Thermostat		28	.714		555	39		594	675
5160	Overheat thermocouple (removable)		7	2.857		790	156		946	1,100
5180	Fan interlock relay		18	1.111		190	61		251	300
5200	Air flow switch		20	1		164	54.50		218.50	264
5220	Split terminal box cover		100	.200		54.50	10.95		65.45	76.50
8000	To obtain BTU multiply kW by 3413									

23 83 Radiant Heating Units

23 83 33 – Electric Radiant Heaters

23 83 33.10 Electric Heating

		Crew	Daily Output	Labor-Hours	Unit	Material	2018 Bare Costs Labor	2018 Bare Costs Equipment	Total	Total Incl O&P
0010	**ELECTRIC HEATING**, not incl. conduit or feed wiring									
1100	Rule of thumb: Baseboard units, including control	1 Elec	4.40	1.818	kW	109	106		215	277
1300	Baseboard heaters, 2' long, 350 watt		8	1	Ea.	27.50	58		85.50	118
1400	3' long, 750 watt		8	1		31.50	58		89.50	122
1600	4' long, 1,000 watt		6.70	1.194		37	69.50		106.50	145
1800	5' long, 935 watt		5.70	1.404		46.50	81.50		128	173
2000	6' long, 1,500 watt		5	1.600		51.50	93		144.50	196
2200	7' long, 1,310 watt		4.40	1.818		54.50	106		160.50	218
2400	8' long, 2,000 watt		4	2		62.50	116		178.50	243
2600	9' long, 1,680 watt		3.60	2.222		85	129		214	287
2800	10' long, 1,875 watt		3.30	2.424		154	141		295	380
2950	Wall heaters with fan, 120 to 277 volt									
3160	Recessed, residential, 750 watt	1 Elec	6	1.333	Ea.	93	77.50		170.50	218
3170	1,000 watt		6	1.333		93	77.50		170.50	218
3180	1,250 watt		5	1.600		100	93		193	249
3190	1,500 watt		4	2		100	116		216	284
3210	2,000 watt		4	2		93	116		209	276
3230	2,500 watt		3.50	2.286		232	133		365	455
3240	3,000 watt		3	2.667		360	155		515	625
3250	4,000 watt		2.70	2.963		365	172		537	660
3260	Commercial, 750 watt		6	1.333		198	77.50		275.50	335
3270	1,000 watt		6	1.333		198	77.50		275.50	335
3280	1,250 watt		5	1.600		198	93		291	355
3290	1,500 watt		4	2		198	116		314	390
3300	2,000 watt		4	2		198	116		314	390
3310	2,500 watt		3.50	2.286		198	133		331	415
3320	3,000 watt		3	2.667		420	155		575	690
3330	4,000 watt		2.70	2.963		490	172		662	795
3600	Thermostats, integral		16	.500		27.50	29		56.50	73.50
3800	Line voltage, 1 pole		8	1		15.70	58		73.70	104
3810	2 pole		8	1		21	58		79	110
4000	Heat trace system, 400 degree R238313-10									
4020	115 V, 2.5 watts/L.F.	1 Elec	530	.015	L.F.	10.90	.88		11.78	13.30
4030	5 watts/L.F. R238313-20		530	.015		10.90	.88		11.78	13.30
4050	10 watts/L.F.		530	.015		10.90	.88		11.78	13.30
4060	208 V, 5 watts/L.F.		530	.015		10.90	.88		11.78	13.30
4080	480 V, 8 watts/L.F.		530	.015		10.90	.88		11.78	13.30
4200	Heater raceway									
4260	Heat transfer cement									
4280	1 gallon				Ea.	62			62	68.50

23 83 33.10 Electric Heating	Crew	Daily Output	Labor-Hours	Unit	Material	2018 Bare Costs Labor	Equipment	Total	Total Incl O&P
4300	5 gallon				Ea.	242			242	266
4320	Cable tie									
4340	3/4" pipe size	1 Elec	470	.017	Ea.	.02	.99		1.01	1.50
4360	1" pipe size		444	.018		.02	1.05		1.07	1.59
4380	1-1/4" pipe size		400	.020		.02	1.16		1.18	1.76
4400	1-1/2" pipe size		355	.023		.02	1.31		1.33	1.98
4420	2" pipe size		320	.025		.05	1.46		1.51	2.24
4440	3" pipe size		160	.050		.05	2.91		2.96	4.41
4460	4" pipe size		100	.080		.06	4.66		4.72	7
4480	Thermostat NEMA 3R, 22 amp, 0-150 degree, 10' cap.		8	1		201	58		259	310
4500	Thermostat NEMA 4X, 25 amp, 40 degree, 5-1/2' cap.		7	1.143		246	66.50		312.50	370
4520	Thermostat NEMA 4X, 22 amp, 25-325 degree, 10' cap.		7	1.143		625	66.50		691.50	785
4540	Thermostat NEMA 4X, 22 amp, 15-140 degree		6	1.333		545	77.50		622.50	715
4580	Thermostat NEMA 4, 7, 9, 22 amp, 25-325 degree, 10' cap.		3.60	2.222		815	129		944	1,100
4600	Thermostat NEMA 4, 7, 9, 22 amp, 15-140 degree		3	2.667		800	155		955	1,100
4720	Fiberglass application tape, 36 yard roll		11	.727		77.50	42.50		120	149
5000	Radiant heating ceiling panels, 2' x 4', 500 watt		16	.500		380	29		409	465
5050	750 watt		16	.500		480	29		509	575
5200	For recessed plaster frame, add		32	.250		133	14.55		147.55	168
5300	Infrared quartz heaters, 120 volts, 1,000 watt		6.70	1.194		335	69.50		404.50	475
5350	1,500 watt		5	1.600		335	93		428	510
5400	240 volts, 1,500 watt		5	1.600		335	93		428	510
5450	2,000 watt		4	2		335	116		451	545
5500	3,000 watt		3	2.667		385	155		540	650
5550	4,000 watt		2.60	3.077		385	179		564	690
5570	Modulating control	▼	.80	10	▼	134	580		714	1,025
5600	Unit heaters, heavy duty, with fan & mounting bracket									
5650	Single phase, 208-240-277 volt, 3 kW	1 Elec	6	1.333	Ea.	695	77.50		772.50	880
5750	5 kW		5.50	1.455		760	84.50		844.50	960
5800	7 kW		5	1.600		1,250	93		1,343	1,525
5850	10 kW		4	2		1,025	116		1,141	1,300
5950	15 kW		3.80	2.105		1,675	123		1,798	2,025
6000	480 volt, 3 kW		6	1.333		495	77.50		572.50	660
6020	4 kW		5.80	1.379		755	80.50		835.50	950
6040	5 kW		5.50	1.455		750	84.50		834.50	950
6060	7 kW		5	1.600		1,125	93		1,218	1,400
6080	10 kW		4	2		1,175	116		1,291	1,475
6100	13 kW		3.80	2.105		1,775	123		1,898	2,125
6120	15 kW		3.70	2.162		2,450	126		2,576	2,875
6140	20 kW		3.50	2.286		2,850	133		2,983	3,350
6300	3 phase, 208-240 volt, 5 kW		5.50	1.455		580	84.50		664.50	765
6320	7 kW		5	1.600		905	93		998	1,125
6340	10 kW		4	2		890	116		1,006	1,150
6360	15 kW		3.70	2.162		1,500	126		1,626	1,875
6380	20 kW		3.50	2.286		2,725	133		2,858	3,200
6400	25 kW		3.30	2.424		3,425	141		3,566	3,975
6500	480 volt, 5 kW		5.50	1.455		740	84.50		824.50	940
6520	7 kW		5	1.600		1,075	93		1,168	1,325
6540	10 kW		4	2		885	116		1,001	1,150
6560	13 kW		3.80	2.105		1,575	123		1,698	1,900
6580	15 kW		3.70	2.162		1,475	126		1,601	1,825
6600	20 kW		3.50	2.286		2,275	133		2,408	2,700
6620	25 kW	▼	3.30	2.424	▼	3,225	141		3,366	3,750

23 83 33 – Electric Radiant Heaters

23 83 33.10 Electric Heating		Crew	Daily Output	Labor-Hours	Unit	Material	2018 Bare Costs Labor	Equipment	Total	Total Incl O&P
6630	30 kW	1 Elec	3	2.667	Ea.	3,400	155		3,555	3,975
6640	40 kW		2	4		4,275	233		4,508	5,050
6650	50 kW		1.60	5		6,125	291		6,416	7,175
6800	Vertical discharge heaters, with fan									
6820	Single phase, 208-240-277 volt, 10 kW	1 Elec	4	2	Ea.	1,000	116		1,116	1,275
6840	15 kW		3.70	2.162		1,650	126		1,776	2,000
6900	3 phase, 208-240 volt, 10 kW		4	2		945	116		1,061	1,225
6920	15 kW		3.70	2.162		1,600	126		1,726	1,975
6940	20 kW		3.50	2.286		2,925	133		3,058	3,400
6960	25 kW		3.30	2.424		3,425	141		3,566	3,975
6980	30 kW		3	2.667		3,975	155		4,130	4,600
7000	40 kW		2	4		4,475	233		4,708	5,275
7020	50 kW		1.60	5		6,375	291		6,666	7,425
7100	480 volt, 10 kW		4	2		1,100	116		1,216	1,375
7120	15 kW		3.70	2.162		1,775	126		1,901	2,150
7140	20 kW		3.50	2.286		2,850	133		2,983	3,350
7160	25 kW		3.30	2.424		3,425	141		3,566	3,975
7180	30 kW		3	2.667		3,975	155		4,130	4,600
7200	40 kW		2	4		4,275	233		4,508	5,050
7220	50 kW		1.60	5		6,125	291		6,416	7,175
7410	Sill height convector heaters, 5" high x 2' long, 500 watt		6.70	1.194		330	69.50		399.50	470
7420	3' long, 750 watt		6.50	1.231		390	71.50		461.50	535
7430	4' long, 1,000 watt		6.20	1.290		455	75		530	610
7440	5' long, 1,250 watt		5.50	1.455		515	84.50		599.50	690
7450	6' long, 1,500 watt		4.80	1.667		585	97		682	785
7460	8' long, 2,000 watt		3.60	2.222		790	129		919	1,075
7470	10' long, 2,500 watt		3	2.667		985	155		1,140	1,300
7900	Cabinet convector heaters, 240 volt, three phase,									
7920	3' long, 2,000 watt	1 Elec	5.30	1.509	Ea.	2,225	88		2,313	2,550
7940	3,000 watt		5.30	1.509		2,300	88		2,388	2,675
7960	4,000 watt		5.30	1.509		2,375	88		2,463	2,725
7980	6,000 watt		4.60	1.739		2,700	101		2,801	3,125
8000	8,000 watt		4.60	1.739		2,750	101		2,851	3,175
8020	4' long, 4,000 watt		4.60	1.739		2,400	101		2,501	2,800
8040	6,000 watt		4	2		2,500	116		2,616	2,925
8060	8,000 watt		4	2		2,575	116		2,691	3,000
8080	10,000 watt		4	2		2,600	116		2,716	3,050
8100	Available also in 208 or 277 volt									
8200	Cabinet unit heaters, 120 to 277 volt, 1 pole,									
8220	wall mounted, 2 kW	1 Elec	4.60	1.739	Ea.	2,150	101		2,251	2,525
8230	3 kW		4.60	1.739		2,250	101		2,351	2,625
8240	4 kW		4.40	1.818		2,325	106		2,431	2,700
8250	5 kW		4.40	1.818		2,400	106		2,506	2,800
8260	6 kW		4.20	1.905		2,450	111		2,561	2,875
8270	8 kW		4	2		2,500	116		2,616	2,925
8280	10 kW		3.80	2.105		2,550	123		2,673	3,000
8290	12 kW		3.50	2.286		2,600	133		2,733	3,050
8300	13.5 kW		2.90	2.759		3,125	161		3,286	3,700
8310	16 kW		2.70	2.963		3,150	172		3,322	3,725
8320	20 kW		2.30	3.478		4,000	202		4,202	4,700
8330	24 kW		1.90	4.211		4,075	245		4,320	4,850
8350	Recessed, 2 kW		4.40	1.818		2,150	106		2,256	2,525
8370	3 kW		4.40	1.818		2,250	106		2,356	2,625

For customer support on your Electrical Costs with RSMeans data, call 800.448.8182.

23 83 33.10 Electric Heating		Crew	Daily Output	Labor-Hours	Unit	Material	2018 Bare Costs Labor	Equipment	Total	Total Incl O&P
8380	4 kW	1 Elec	4.20	1.905	Ea.	2,325	111		2,436	2,725
8390	5 kW		4.20	1.905		2,400	111		2,511	2,800
8400	6 kW		4	2		2,450	116		2,566	2,875
8410	8 kW		3.80	2.105		2,550	123		2,673	2,975
8420	10 kW		3.50	2.286		3,100	133		3,233	3,600
8430	12 kW		2.90	2.759		3,100	161		3,261	3,675
8440	13.5 kW		2.70	2.963		3,125	172		3,297	3,700
8450	16 kW		2.30	3.478		3,325	202		3,527	3,950
8460	20 kW		1.90	4.211		3,800	245		4,045	4,550
8470	24 kW		1.60	5		3,850	291		4,141	4,650
8490	Ceiling mounted, 2 kW		3.20	2.500		2,150	146		2,296	2,600
8510	3 kW		3.20	2.500		2,250	146		2,396	2,700
8520	4 kW		3	2.667		2,325	155		2,480	2,775
8530	5 kW		3	2.667		2,400	155		2,555	2,850
8540	6 kW		2.80	2.857		2,400	166		2,566	2,900
8550	8 kW		2.40	3.333		2,500	194		2,694	3,050
8560	10 kW		2.20	3.636		2,550	212		2,762	3,125
8570	12 kW		2	4		2,600	233		2,833	3,200
8580	13.5 kW		1.50	5.333		2,650	310		2,960	3,375
8590	16 kW		1.30	6.154		2,725	360		3,085	3,500
8600	20 kW		.90	8.889		3,800	515		4,315	4,950
8610	24 kW		.60	13.333		3,850	775		4,625	5,375
8630	208 to 480 V, 3 pole									
8650	Wall mounted, 2 kW	1 Elec	4.60	1.739	Ea.	2,300	101		2,401	2,675
8670	3 kW		4.60	1.739		2,400	101		2,501	2,775
8680	4 kW		4.40	1.818		2,450	106		2,556	2,850
8690	5 kW		4.40	1.818		2,525	106		2,631	2,925
8700	6 kW		4.20	1.905		2,550	111		2,661	2,975
8710	8 kW		4	2		2,625	116		2,741	3,075
8720	10 kW		3.80	2.105		2,675	123		2,798	3,125
8730	12 kW		3.50	2.286		2,725	133		2,858	3,200
8740	13.5 kW		2.90	2.759		2,775	161		2,936	3,300
8750	16 kW		2.70	2.963		2,825	172		2,997	3,375
8760	20 kW		2.30	3.478		3,925	202		4,127	4,625
8770	24 kW		1.90	4.211		4,175	245		4,420	4,950
8790	Recessed, 2 kW		4.40	1.818		2,300	106		2,406	2,675
8810	3 kW		4.40	1.818		2,325	106		2,431	2,700
8820	4 kW		4.20	1.905		2,425	111		2,536	2,825
8830	5 kW		4.20	1.905		2,450	111		2,561	2,850
8840	6 kW		4	2		2,475	116		2,591	2,875
8850	8 kW		3.80	2.105		2,550	123		2,673	2,975
8860	10 kW		3.50	2.286		2,725	133		2,858	3,175
8870	12 kW		2.90	2.759		2,825	161		2,986	3,375
8880	13.5 kW		2.70	2.963		2,575	172		2,747	3,075
8890	16 kW		2.30	3.478		4,200	202		4,402	4,900
8900	20 kW		1.90	4.211		4,250	245		4,495	5,050
8920	24 kW		1.60	5		4,325	291		4,616	5,175
8940	Ceiling mount, 2 kW		3.20	2.500		2,400	146		2,546	2,850
8950	3 kW		3.20	2.500		2,275	146		2,421	2,725
8960	4 kW		3	2.667		2,550	155		2,705	3,025
8970	5 kW		3	2.667		2,625	155		2,780	3,125
8980	6 kW		2.80	2.857		2,650	166		2,816	3,150
8990	8 kW		2.40	3.333		2,750	194		2,944	3,325

For customer support on your Electrical Costs with RSMeans data, call 800.448.8182.

95

23 83 33 – Electric Radiant Heaters

23 83 33.10 Electric Heating		Crew	Daily Output	Labor-Hours	Unit	Material	2018 Bare Costs Labor	Equipment	Total	Total Incl O&P
9000	10 kW	1 Elec	2.20	3.636	Ea.	2,500	212		2,712	3,075
9020	13.5 kW		1.50	5.333		2,875	310		3,185	3,650
9030	16 kW		1.30	6.154		4,200	360		4,560	5,125
9040	20 kW		.90	8.889		4,250	515		4,765	5,450
9060	24 kW		.60	13.333		4,725	775		5,500	6,350
9230	13.5 kW, 40,956 BTU		2.20	3.636		1,550	212		1,762	2,025
9250	24 kW, 81,912 BTU	↓	2	4	↓	1,600	233		1,833	2,125

Estimating Tips

- When estimating material costs for special systems, it is always prudent to obtain manufacturers' quotations for equipment prices and special installation requirements that may affect the total cost.

- For cost modifications for elevated tray installation, add the percentages to labor according to the height of the installation and only to the quantities exceeding the different height levels, not to the total tray quantities. Refer to 26 01 02.20 for labor adjustment factors.

- Do not overlook the costs for equipment used in the installation.

Reference Numbers

Reference numbers are shown at the beginning of some major classifications. These numbers refer to related items in the Reference Section. The reference information may be an estimating procedure, an alternate pricing method, or technical information.

Note: Not all subdivisions listed here necessarily appear. ■

25 05 Common Work Results for Integrated Automation

25 05 28 – Pathways for Integrated Automation

25 05 28.39 Surface Raceways for Integrated Automation	Crew	Daily Output	Labor-Hours	Unit	Material	2018 Bare Costs Labor	Equipment	Total	Total Incl O&P
0010 **SURFACE RACEWAYS FOR INTEGRATED AUTOMATION**									
1000 Din rail, G type, 15 mm, steel, galvanized, 3', for termination blocks	1 Elec	48	.167	Ea.	10.65	9.70		20.35	26
1010 32 mm, steel, galvanized, 3', for termination blocks		48	.167		16.35	9.70		26.05	32.50
1020 35 mm, steel, galvanized, 3', for termination blocks		48	.167		14.55	9.70		24.25	30.50
1030 Din rail, rail mounted, 35 mm, aluminum, 3', for termination blocks		48	.167		20.50	9.70		30.20	37
1040 35 mm, aluminum, 6', for termination blocks		48	.167		25.50	9.70		35.20	43
2000 Terminal block, Din mounted, 35 mm, medium screw, connector end bracket		48	.167		2.19	9.70		11.89	16.90
2010 35 mm, large screw, connector end bracket		48	.167		3.43	9.70		13.13	18.25
2020 Large screw, connector end bracket		48	.167		4.21	9.70		13.91	19.15
3000 Terminal block, feed through, 35 mm, Din mounted, 600V, 20A		48	.167		1.04	9.70		10.74	15.65
3010 35 mm, Din mounted, 600V, 30A		48	.167		1.29	9.70		10.99	15.90
3020 Terminal block, fuse holder, LED indicator, Din mounted, 110V, 6A		48	.167		7.55	9.70		17.25	23
3030 Din mounted, 220V, 6A		48	.167		7.70	9.70		17.40	23
3040 Din mounted, 300V, 6A		48	.167		7.10	9.70		16.80	22.50

25 36 Integrated Automation Instrumentation & Terminal Devices

25 36 19 – Integrated Automation Current Sensors

25 36 19.10 Integrated Automation Current Sensors

	Crew	Daily Output	Labor-Hours	Unit	Material	2018 Bare Costs Labor	Equipment	Total	Total Incl O&P
0010 **INTEGRATED AUTOMATION CURRENT SENSORS**									
1000 Current limiter inrush, 100 to 480 VAC, 10A, power supply	1 Elec	48	.167	Ea.	80	9.70		89.70	103

Estimating Tips
26 05 00 Common Work Results for Electrical

- Conduit should be taken off in three main categories—power distribution, branch power, and branch lighting—so the estimator can concentrate on systems and components, therefore making it easier to ensure all items have been accounted for.

- For cost modifications for elevated conduit installation, add the percentages to labor according to the height of installation and only to the quantities exceeding the different height levels, not to the total conduit quantities. Refer to 26 01 02.20 for labor adjustment factors.

- Remember that aluminum wiring of equal ampacity is larger in diameter than copper and may require larger conduit.

- If more than three wires at a time are being pulled, deduct percentages from the labor hours of that grouping of wires.

- When taking off grounding systems, identify separately the type and size of wire, and list each unique type of ground connection.

- The estimator should take the weights of materials into consideration when completing a takeoff. Topics to consider include: How will the materials be supported? What methods of support are available? How high will the support structure have to reach? Will the final support structure be able to withstand the total burden? Is the support material included or separate from the fixture, equipment, and material specified?

- Do not overlook the costs for equipment used in the installation. If scaffolding or highlifts are available in the field, contractors may use them in lieu of the proposed ladders and rolling staging.

26 20 00 Low-Voltage Electrical Transmission

- Supports and concrete pads may be shown on drawings for the larger equipment, or the support system may be only a piece of plywood for the back of a panelboard. In either case, they must be included in the costs.

26 40 00 Electrical and Cathodic Protection

- When taking off cathodic protection systems, identify the type and size of cable, and list each unique type of anode connection.

26 50 00 Lighting

- Fixtures should be taken off room by room using the fixture schedule, specifications, and the ceiling plan. For large concentrations of lighting fixtures in the same area, deduct the percentages from labor hours.

Reference Numbers
Reference numbers are shown at the beginning of some major classifications. These numbers refer to related items in the Reference Section. The reference information may be an estimating procedure, an alternate pricing method, or technical information.

Note: Not all subdivisions listed here necessarily appear. ∎

26 01 02 – Labor Adjustment

26 01 02.20 Labor Adjustment Factors	Crew	Daily Output	Labor-Hours	Unit	Material	2018 Bare Costs Labor	Equipment	Total	Total Incl O&P
0010 **LABOR ADJUSTMENT FACTORS** (For Div. 26, 27 and 28) R260519-90									
0100 Subtract from labor for Economy of Scale for Wire									
0110 4-5 wires						25%			
0120 6-10 wires						30%			
0130 11-15 wires						35%			
0140 over 15 wires						40%			
0150 Labor adjustment factors (For Div. 26, 27, 28 and 48)									
0200 Labor factors: The below are reasonable suggestions, but									
0210 each project must be evaluated for its own peculiarities, and									
0220 the adjustments be increased or decreased depending on the									
0230 severity of the special conditions.									
1000 Add to labor for elevated installation (above floor level)									
1010 10' to 14.5' high						10%			
1020 15' to 19.5' high						20%			
1030 20' to 24.5' high						25%			
1040 25' to 29.5' high						35%			
1050 30' to 34.5' high						40%			
1060 35' to 39.5' high						50%			
1070 40' and higher						55%			
2000 Add to labor for crawl space									
2010 3' high						40%			
2020 4' high						30%			
3000 Add to labor for multi-story building									
3100 For new construction (No elevator available)									
3110 Add for floors 3 thru 10						5%			
3120 Add for floors 11 thru 15						10%			
3130 Add for floors 16 thru 20						15%			
3140 Add for floors 21 thru 30						20%			
3150 Add for floors 31 and up						30%			
3200 For existing structure (Elevator available)									
3210 Add for work on floor 3 and above						2%			
4000 Add to labor for working in existing occupied buildings									
4010 Hospital						35%			
4020 Office building						25%			
4030 School						20%			
4040 Factory or warehouse						15%			
4050 Multi-dwelling						15%			
5000 Add to labor, miscellaneous									
5010 Cramped shaft						35%			
5020 Congested area						15%			
5030 Excessive heat or cold						30%			
6000 Labor factors: the above are reasonable suggestions, but									
6100 each project should be evaluated for its own peculiarities									
6200 Other factors to be considered are:									
6210 Movement of material and equipment through finished areas						10%			
6220 Equipment room min security direct access w/authorization						15%			
6230 Attic space						25%			
6240 No service road						25%			
6250 Poor unloading/storage area, no hydraulic lifts or jacks						20%			
6260 Congested site area/heavy traffic						20%			
7000 Correctional facilities (no compounding division 1 adjustment factors)									
7010 Minimum security w/facilities escort						30%			
7020 Medium security w/facilities and correctional officer escort						40%			

26 01 Operation and Maintenance of Electrical Systems

26 01 02 – Labor Adjustment

26 01 02.20 Labor Adjustment Factors	Crew	Daily Output	Labor-Hours	Unit	Material	2018 Bare Costs		Total	Total Incl O&P
						Labor	Equipment		
7030	Max security w/facilities & correctional officer escort (no inmate contact)						50%			

26 01 40 – Operation and Maintenance of Electrical Protection Systems

26 01 40.51 Electrical Systems Repair and Replacement

		Crew	Daily Output	Labor-Hours	Unit	Material	Labor	Equipment	Total	Total Incl O&P
0010	**ELECTRICAL SYSTEMS REPAIR AND REPLACEMENT**									
3000	Remove and replace (reinstall), switch cover	1 Elec	60	.133	Ea.		7.75		7.75	11.60
3020	Outlet cover	"	60	.133	"		7.75		7.75	11.60

26 01 50 – Operation and Maintenance of Lighting

26 01 50.81 Luminaire Replacement

		Crew	Daily Output	Labor-Hours	Unit	Material	Labor	Equipment	Total	Total Incl O&P
0010	**LUMINAIRE REPLACEMENT**									
0100	Fluorescent fixtures, elevate from work area	1 Elec	96	.083	Ea.	88.50	4.85		93.35	105
0962	Remove and install new replacement lens, 2' x 2' troffer		16	.500		15.90	29		44.90	61
0964	2' x 4' troffer		16	.500		14.90	29		43.90	60
3200	Remove and replace (reinstall), lighting fixture	↓	4	2	↓		116		116	174

26 05 Common Work Results for Electrical

26 05 05 – Selective Demolition for Electrical

26 05 05.10 Electrical Demolition

		Crew	Daily Output	Labor-Hours	Unit	Material	Labor	Equipment	Total	Total Incl O&P
0010	**ELECTRICAL DEMOLITION** R260105-30									
0020	Electrical demolition, conduit to 10' high, incl. fittings & hangers									
0100	Rigid galvanized steel, 1/2" to 1" diameter	1 Elec	242	.033	L.F.		1.92		1.92	2.88
0120	1-1/4" to 2"	"	200	.040			2.33		2.33	3.48
0140	2-1/2" to 3-1/2"	2 Elec	302	.053			3.08		3.08	4.61
0160	4" to 6"	"	160	.100			5.80		5.80	8.70
0170	PVC #40, 1/2" to 1"	1 Elec	410	.020			1.14		1.14	1.70
0172	1-1/4" to 2"		350	.023			1.33		1.33	1.99
0174	2-1/2"		250	.032			1.86		1.86	2.78
0176	3" to 3-1/2"	2 Elec	340	.047			2.74		2.74	4.09
0178	4" to 6"	"	230	.070			4.05		4.05	6.05
0200	Electric metallic tubing (EMT), 1/2" to 1"	1 Elec	394	.020			1.18		1.18	1.77
0220	1-1/4" to 1-1/2"		326	.025			1.43		1.43	2.14
0240	2" to 3"	↓	236	.034			1.97		1.97	2.95
0260	3-1/2" to 4"	2 Elec	310	.052	↓		3		3	4.49
0270	Armored cable (BX) avg. 50' runs									
0280	#14, 2 wire	1 Elec	690	.012	L.F.		.67		.67	1.01
0290	#14, 3 wire		571	.014			.82		.82	1.22
0300	#12, 2 wire		605	.013			.77		.77	1.15
0310	#12, 3 wire		514	.016			.91		.91	1.35
0320	#10, 2 wire		514	.016			.91		.91	1.35
0330	#10, 3 wire		425	.019			1.10		1.10	1.64
0340	#8, 3 wire	↓	342	.023	↓		1.36		1.36	2.03
0350	Non metallic sheathed cable (Romex)									
0360	#14, 2 wire	1 Elec	720	.011	L.F.		.65		.65	.97
0370	#14, 3 wire		657	.012			.71		.71	1.06
0380	#12, 2 wire		629	.013			.74		.74	1.11
0390	#10, 3 wire	↓	450	.018			1.03		1.03	1.55
0400	Wiremold raceway, including fittings & hangers									
0420	No. 3000	1 Elec	250	.032	L.F.		1.86		1.86	2.78
0440	No. 4000		217	.037			2.15		2.15	3.21
0460	No. 6000		166	.048			2.80		2.80	4.19
0462	Plugmold with receptacle	↓	114	.070			4.08		4.08	6.10

For customer support on your Electrical Costs with RSMeans data, call 800.448.8182.

101

26 05 05 – Selective Demolition for Electrical

26 05 05.10 Electrical Demolition	Crew	Daily Output	Labor-Hours	Unit	Material	2018 Bare Costs Labor	2018 Bare Costs Equipment	Total	Total Incl O&P
0465	Telephone/power pole	1 Elec	12	.667	Ea.		39		39	58
0470	Non-metallic, straight section		480	.017	L.F.		.97		.97	1.45
0500	Channels, steel, including fittings & hangers									
0520	3/4" x 1-1/2"	1 Elec	308	.026	L.F.		1.51		1.51	2.26
0540	1-1/2" x 1-1/2"		269	.030			1.73		1.73	2.59
0560	1-1/2" x 1-7/8"		229	.035			2.03		2.03	3.04
0600	Copper bus duct, indoor, 3 phase									
0610	Including hangers & supports									
0620	225 amp	2 Elec	135	.119	L.F.		6.90		6.90	10.30
0640	400 amp		106	.151			8.80		8.80	13.15
0660	600 amp		86	.186			10.85		10.85	16.20
0680	1,000 amp		60	.267			15.50		15.50	23
0700	1,600 amp		40	.400			23.50		23.50	35
0720	3,000 amp		10	1.600			93		93	139
0800	Plug-in switches, 600 V 3 ph., incl. disconnecting									
0820	wire, conduit terminations, 30 amp	1 Elec	15.50	.516	Ea.		30		30	45
0840	60 amp		13.90	.576			33.50		33.50	50
0850	100 amp		10.40	.769			45		45	67
0860	200 amp		6.20	1.290			75		75	112
0880	400 amp	2 Elec	5.40	2.963			172		172	258
0900	600 amp		3.40	4.706			274		274	410
0920	800 amp		2.60	6.154			360		360	535
0940	1,200 amp		2	8			465		465	695
0960	1,600 amp		1.70	9.412			550		550	820
1010	Safety switches, 250 or 600 V, incl. disconnection									
1050	of wire & conduit terminations									
1100	30 amp	1 Elec	12.30	.650	Ea.		38		38	56.50
1120	60 amp		8.80	.909			53		53	79
1140	100 amp		7.30	1.096			64		64	95.50
1160	200 amp		5	1.600			93		93	139
1180	400 amp	2 Elec	6.80	2.353			137		137	205
1200	600 amp	"	4.60	3.478			202		202	305
1202	Explosion proof, 60 amp	1 Elec	8.80	.909			53		53	79
1204	100 amp		5	1.600			93		93	139
1206	200 amp		3.33	2.402			140		140	209
1210	Panel boards, incl. removal of all breakers,									
1220	conduit terminations & wire connections									
1230	3 wire, 120/240 V, 100A, to 20 circuits	1 Elec	2.60	3.077	Ea.		179		179	268
1240	200 amps, to 42 circuits	2 Elec	2.60	6.154			360		360	535
1241	225 amps, to 42 circuits		5.60	2.857			166		166	249
1250	400 amps, to 42 circuits		2.20	7.273			425		425	635
1260	4 wire, 120/208 V, 125A, to 20 circuits	1 Elec	2.40	3.333			194		194	290
1270	200 amps, to 42 circuits	2 Elec	2.40	6.667			390		390	580
1280	400 amps, to 42 circuits		1.92	8.333			485		485	725
1285	600 amps, to 42 circuits		1.60	10			580		580	870
1300	Transformer, dry type, 1 phase, incl. removal of									
1320	supports, wire & conduit terminations									
1340	1 kVA	1 Elec	7.70	1.039	Ea.		60.50		60.50	90.50
1360	5 kVA		4.70	1.702			99		99	148
1380	10 kVA		3.60	2.222			129		129	193
1400	37.5 kVA	2 Elec	3	5.333			310		310	465
1420	75 kVA	"	2.50	6.400			370		370	555
1440	3 phase to 600 V, primary									

26 05 05.10 Electrical Demolition	Crew	Daily Output	Labor-Hours	Unit	Material	2018 Bare Costs Labor	Equipment	Total	Total Incl O&P
1460	3 kVA	1 Elec	3.87	2.067	Ea.		120		120	180
1480	15 kVA	2 Elec	3.67	4.360			254		254	380
1490	25 kVA		3.51	4.558			265		265	395
1500	30 kVA		3.42	4.678			272		272	405
1510	45 kVA		3.18	5.031			293		293	440
1520	75 kVA		2.69	5.948			345		345	515
1530	112.5 kVA	R-3	2.90	6.897			400	44.50	444.50	650
1540	150 kVA		2.70	7.407			430	48	478	700
1550	300 kVA		1.80	11.111			645	72	717	1,050
1560	500 kVA		1.40	14.286			830	92.50	922.50	1,350
1570	750 kVA		1.10	18.182			1,050	118	1,168	1,700
1600	Pull boxes & cabinets, sheet metal, incl. removal									
1620	of supports and conduit terminations									
1640	6" x 6" x 4"	1 Elec	31.10	.257	Ea.		14.95		14.95	22.50
1660	12" x 12" x 4"		23.30	.343			20		20	30
1680	24" x 24" x 6"		12.30	.650			38		38	56.50
1700	36" x 36" x 8"		7.70	1.039			60.50		60.50	90.50
1720	Junction boxes, 4" sq. & oct.		80	.100			5.80		5.80	8.70
1740	Handy box		107	.075			4.35		4.35	6.50
1760	Switch box		107	.075			4.35		4.35	6.50
1780	Receptacle & switch plates		257	.031			1.81		1.81	2.71
1790	Receptacles & switches, 15 to 30 amp		135	.059			3.45		3.45	5.15
1800	Wire, THW-THWN-THHN, removed from									
1810	in place conduit, to 10' high									
1830	#14	1 Elec	65	.123	C.L.F.		7.15		7.15	10.70
1840	#12		55	.145			8.45		8.45	12.65
1850	#10		45.50	.176			10.25		10.25	15.30
1860	#8		40.40	.198			11.50		11.50	17.25
1870	#6		32.60	.245			14.30		14.30	21.50
1880	#4	2 Elec	53	.302			17.55		17.55	26.50
1890	#3		50	.320			18.60		18.60	28
1900	#2		44.60	.359			21		21	31
1910	1/0		33.20	.482			28		28	42
1920	2/0		29.20	.548			32		32	47.50
1930	3/0		25	.640			37.50		37.50	55.50
1940	4/0		22	.727			42.50		42.50	63.50
1950	250 kcmil		20	.800			46.50		46.50	69.50
1960	300 kcmil		19	.842			49		49	73.50
1970	350 kcmil		18	.889			51.50		51.50	77.50
1980	400 kcmil		17	.941			55		55	82
1990	500 kcmil		16.20	.988			57.50		57.50	86
2000	Interior fluorescent fixtures, incl. supports									
2010	& whips, to 10' high									
2100	Recessed drop-in 2' x 2', 2 lamp	2 Elec	35	.457	Ea.		26.50		26.50	40
2110	2' x 2', 4 lamp		30	.533			31		31	46.50
2120	2' x 4', 2 lamp		33	.485			28		28	42
2140	2' x 4', 4 lamp		30	.533			31		31	46.50
2160	4' x 4', 4 lamp		20	.800			46.50		46.50	69.50
2180	Surface mount, acrylic lens & hinged frame									
2200	1' x 4', 2 lamp	2 Elec	44	.364	Ea.		21		21	31.50
2210	6" x 4', 2 lamp		44	.364			21		21	31.50
2220	2' x 2', 2 lamp		44	.364			21		21	31.50
2260	2' x 4', 4 lamp		33	.485			28		28	42

For customer support on your Electrical Costs with RSMeans data, call 800.448.8182.

103

26 05 05 – Selective Demolition for Electrical

26 05 05.10 Electrical Demolition	Crew	Daily Output	Labor-Hours	Unit	Material	2018 Bare Costs Labor	Equipment	Total	Total Incl O&P
2280	4' x 4', 4 lamp	2 Elec	23	.696	Ea.		40.50		40.50	60.50
2281	4' x 4', 6 lamp		23	.696			40.50		40.50	60.50
2300	Strip fixtures, surface mount									
2320	4' long, 1 lamp	2 Elec	53	.302	Ea.		17.55		17.55	26.50
2340	4' long, 2 lamp		50	.320			18.60		18.60	28
2360	8' long, 1 lamp		42	.381			22		22	33
2380	8' long, 2 lamp		40	.400			23.50		23.50	35
2400	Pendant mount, industrial, incl. removal									
2410	of chain or rod hangers, to 10' high									
2420	4' long, 2 lamp	2 Elec	35	.457	Ea.		26.50		26.50	40
2421	4' long, 4 lamp		35	.457			26.50		26.50	40
2440	8' long, 2 lamp		27	.593			34.50		34.50	51.50
2460	Interior incandescent, surface, ceiling									
2470	or wall mount, to 12' high									
2480	Metal cylinder type, 75 Watt	2 Elec	62	.258	Ea.		15		15	22.50
2500	150 Watt		62	.258			15		15	22.50
2502	300 Watt		53	.302			17.55		17.55	26.50
2520	Metal halide, high bay									
2540	400 Watt	2 Elec	15	1.067	Ea.		62		62	93
2560	1,000 Watt		12	1.333			77.50		77.50	116
2580	150 Watt, low bay		20	.800			46.50		46.50	69.50
2585	Globe type, ceiling mount		62	.258			15		15	22.50
2586	Can type, recessed mount		62	.258			15		15	22.50
2600	Exterior fixtures, incandescent, wall mount									
2620	100 Watt	2 Elec	50	.320	Ea.		18.60		18.60	28
2640	Quartz, 500 Watt		33	.485			28		28	42
2660	1,500 Watt		27	.593			34.50		34.50	51.50
2680	Wall pack, mercury vapor									
2700	175 Watt	2 Elec	25	.640	Ea.		37.50		37.50	55.50
2720	250 Watt	"	25	.640			37.50		37.50	55.50
7000	Weatherhead/mast, 2"	1 Elec	16	.500			29		29	43.50
7002	3"		10	.800			46.50		46.50	69.50
7004	3-1/2"		8.50	.941			55		55	82
7006	4"		8	1			58		58	87
7100	Service entry cable, #6, +#6 neutral		420	.019	L.F.		1.11		1.11	1.66
7102	#4, +#4 neutral		360	.022			1.29		1.29	1.93
7104	#2, +#4 neutral		345	.023			1.35		1.35	2.02
7106	#2, +#2 neutral		330	.024			1.41		1.41	2.11
9900	Add to labor for higher elevated installation									
9905	10' to 14.5' high, add						10%			
9910	15' to 20' high, add						20%			
9920	20' to 25' high, add						25%			
9930	25' to 30' high, add						35%			
9940	30' to 35' high, add						40%			
9950	35' to 40' high, add						50%			
9960	Over 40' high, add						55%			

26 05 05.15 Electrical Demolition, Grounding

		Crew	Daily Output	Labor-Hours	Unit	Material	Labor	Equipment	Total	Total Incl O&P
0010	**ELECTRICAL DEMOLITION, GROUNDING** Addition									
0100	Ground clamp, bronze	1 Elec	64	.125	Ea.		7.30		7.30	10.90
0140	Water pipe ground clamp, bronze, heavy duty		24	.333			19.40		19.40	29
0150	Ground rod, 8' to 10'		13	.615			36		36	53.50
0200	Ground wire, bare armored	2 Elec	900	.018	L.F.		1.03		1.03	1.55

26 05 05.15 Electrical Demolition, Grounding

		Crew	Daily Output	Labor-Hours	Unit	Material	2018 Bare Costs Labor	Equipment	Total	Total Incl O&P
0240	Bare copper or aluminum	2 Elec	2800	.006	L.F.		.33		.33	.50

26 05 05.20 Electrical Demolition, Wiring Methods

			Crew	Daily Output	Labor-Hours	Unit	Material	2018 Bare Costs Labor	Equipment	Total	Total Incl O&P
0010	**ELECTRICAL DEMOLITION, WIRING METHODS** Addition										
0100	Armored cable, w/PVC jacket, in cable tray										
0110	#6	R024119-10	1 Elec	9.30	.860	C.L.F.		50		50	75
0120	#4		2 Elec	16.20	.988			57.50		57.50	86
0130	#2			13.80	1.159			67.50		67.50	101
0140	#1			12	1.333			77.50		77.50	116
0150	1/0			10.80	1.481			86		86	129
0160	2/0			10.20	1.569			91.50		91.50	136
0170	3/0			9.60	1.667			97		97	145
0180	4/0			9	1.778			103		103	155
0190	250 kcmil		3 Elec	10.80	2.222			129		129	193
0210	350 kcmil			9.90	2.424			141		141	211
0230	500 kcmil			9	2.667			155		155	232
0240	750 kcmil			8.10	2.963			172		172	258
1100	Control cable, 600 V or less										
1110	3 wires		1 Elec	24	.333	C.L.F.		19.40		19.40	29
1120	5 wires			20	.400			23.50		23.50	35
1130	7 wires			16.50	.485			28		28	42
1140	9 wires			15	.533			31		31	46.50
1150	12 wires		2 Elec	26	.615			36		36	53.50
1160	15 wires			22	.727			42.50		42.50	63.50
1170	19 wires			19	.842			49		49	73.50
1180	25 wires			16	1			58		58	87
1500	Mineral insulated (MI) cable, 600 V										
1510	#10		1 Elec	4.80	1.667	C.L.F.		97		97	145
1520	#8			4.50	1.778			103		103	155
1530	#6			4.20	1.905			111		111	166
1540	#4		2 Elec	7.20	2.222			129		129	193
1550	#2			6.60	2.424			141		141	211
1560	#1			6.30	2.540			148		148	221
1570	1/0			6	2.667			155		155	232
1580	2/0			5.70	2.807			163		163	244
1590	3/0			5.40	2.963			172		172	258
1600	4/0			4.80	3.333			194		194	290
1610	250 kcmil		3 Elec	7.20	3.333			194		194	290
1620	500 kcmil		"	5.90	4.068			237		237	355
2000	Shielded cable, XLP shielding, to 35 kV										
2010	#4		2 Elec	13.20	1.212	C.L.F.		70.50		70.50	105
2020	#1			12	1.333			77.50		77.50	116
2030	1/0			11.40	1.404			81.50		81.50	122
2040	2/0			10.80	1.481			86		86	129
2050	4/0			9.60	1.667			97		97	145
2060	250 kcmil		3 Elec	13.50	1.778			103		103	155
2070	350 kcmil			11.70	2.051			119		119	178
2080	500 kcmil			11.20	2.143			125		125	186
2090	750 kcmil			10.80	2.222			129		129	193
3000	Modular flexible wiring										
3010	Cable set		1 Elec	120	.067	Ea.		3.88		3.88	5.80
3020	Conversion module			48	.167			9.70		9.70	14.50
3030	Switching assembly			96	.083			4.85		4.85	7.25

For customer support on your Electrical Costs with RSMeans data, call 800.448.8182.

105

26 05 05.20 Electrical Demolition, Wiring Methods	Crew	Daily Output	Labor-Hours	Unit	Material	2018 Bare Costs Labor	Equipment	Total	Total Incl O&P
3200	Undercarpet									
3210	Power or telephone, flat cable	1 Elec	1320	.006	L.F.		.35		.35	.53
3220	Transition block assemblies w/fitting		75	.107	Ea.		6.20		6.20	9.30
3230	Floor box with fitting		60	.133			7.75		7.75	11.60
3300	Data system, cable with connection		50	.160			9.30		9.30	13.90
4000	Cable tray, including fitting & support									
4010	Galvanized steel, 6" wide	2 Elec	310	.052	L.F.		3		3	4.49
4020	9" wide		295	.054			3.16		3.16	4.72
4030	12" wide		285	.056			3.27		3.27	4.88
4040	18" wide		270	.059			3.45		3.45	5.15
4050	24" wide		260	.062			3.58		3.58	5.35
4060	30" wide		240	.067			3.88		3.88	5.80
4070	36" wide		220	.073			4.23		4.23	6.35
4110	Aluminum, 6" wide		420	.038			2.22		2.22	3.31
4120	9" wide		415	.039			2.24		2.24	3.35
4130	12" wide		390	.041			2.39		2.39	3.57
4140	18" wide		370	.043			2.52		2.52	3.76
4150	24" wide		350	.046			2.66		2.66	3.98
4160	30" wide		325	.049			2.87		2.87	4.28
4170	36" wide		300	.053			3.10		3.10	4.64
4310	Cable channel, aluminum, 4" wide straight	1 Elec	240	.033			1.94		1.94	2.90
4320	Cable tray fittings, 6" to 9" wide	2 Elec	27	.593	Ea.		34.50		34.50	51.50
4330	12" to 18" wide	"	19	.842	"		49		49	73.50
5000	Conduit nipples, with locknuts and bushings									
5020	1/2"	1 Elec	108	.074	Ea.		4.31		4.31	6.45
5040	3/4"		96	.083			4.85		4.85	7.25
5060	1"		81	.099			5.75		5.75	8.60
5080	1-1/4"		69	.116			6.75		6.75	10.10
5100	1-1/2"		60	.133			7.75		7.75	11.60
5120	2"		54	.148			8.60		8.60	12.90
5140	2-1/2"		45	.178			10.35		10.35	15.45
5160	3"		36	.222			12.95		12.95	19.35
5180	3-1/2"		33	.242			14.10		14.10	21
5200	4"		27	.296			17.25		17.25	26
5220	5"		21	.381			22		22	33
5240	6"		18	.444			26		26	38.50
5500	Electric nonmetallic tubing (ENT), flexible, 1/2" to 1" diameter		690	.012	L.F.		.67		.67	1.01
5510	1-1/4" to 2" diameter		300	.027			1.55		1.55	2.32
5600	Flexible metallic tubing, steel, 3/8" to 3/4" diameter		600	.013			.78		.78	1.16
5610	1" to 1-1/4" diameter		260	.031			1.79		1.79	2.68
5620	1-1/2" to 2" diameter		140	.057			3.33		3.33	4.97
5630	2-1/2" diameter		100	.080			4.66		4.66	6.95
5640	3" to 3-1/2" diameter	2 Elec	150	.107			6.20		6.20	9.30
5650	4" diameter	"	100	.160			9.30		9.30	13.90
5700	Sealtite flexible conduit, 3/8" to 3/4" diameter	1 Elec	420	.019			1.11		1.11	1.66
5710	1" to 1-1/4" diameter		180	.044			2.59		2.59	3.87
5720	1-1/2" to 2" diameter		120	.067			3.88		3.88	5.80
5730	2-1/2" diameter		80	.100			5.80		5.80	8.70
5740	3" diameter	2 Elec	150	.107			6.20		6.20	9.30
5750	4" diameter	"	100	.160			9.30		9.30	13.90
5800	Wiring duct, plastic, 1-1/2" to 2-1/2" wide	1 Elec	360	.022			1.29		1.29	1.93
5810	3" wide		330	.024			1.41		1.41	2.11
5820	4" wide		300	.027			1.55		1.55	2.32

26 05 05.20 Electrical Demolition, Wiring Methods

		Crew	Daily Output	Labor-Hours	Unit	Material	2018 Bare Costs Labor	Equipment	Total	Total Incl O&P
6000	Floor box and carpet flange	1 Elec	9	.889	Ea.		51.50		51.50	77.50
6300	Wireway, with fittings and supports, to 10' high									
6310	2-1/2" x 2-1/2"	1 Elec	180	.044	L.F.		2.59		2.59	3.87
6320	4" x 4"	"	160	.050			2.91		2.91	4.35
6330	6" x 6"	2 Elec	240	.067			3.88		3.88	5.80
6340	8" x 8"		160	.100			5.80		5.80	8.70
6350	10" x 10"		120	.133			7.75		7.75	11.60
6360	12" x 12"		80	.200			11.65		11.65	17.40
6400	Cable reel with receptacle, 120 V/208 V		13	1.231	Ea.		71.50		71.50	107
6500	Equipment connection, to 10 HP	1 Elec	23	.348			20		20	30.50
6510	To 15 to 30 HP		18	.444			26		26	38.50
6520	To 40 to 60 HP		15	.533			31		31	46.50
6530	To 75 to 100 HP		10	.800			46.50		46.50	69.50
7008	Receptacle, explosion proof, 120 V, to 30 A		16	.500			29		29	43.50
7020	Dimmer switch, 2,000 W or less		62	.129			7.50		7.50	11.25

26 05 05.25 Electrical Demolition, Electrical Power

		Crew	Daily Output	Labor-Hours	Unit	Material	2018 Bare Costs Labor	Equipment	Total	Total Incl O&P
0010	**ELECTRICAL DEMOLITION, ELECTRICAL POWER** R024119-10									
0100	Meter centers and sockets									
0120	Meter socket, 4 terminal R260105-30	1 Elec	8.40	.952	Ea.		55.50		55.50	83
0140	Trans-socket, 13 terminal, 400 A	"	3.60	2.222			129		129	193
0160	800 A	2 Elec	4.40	3.636			212		212	315
0200	Meter center, 400 A		5.80	2.759			161		161	240
0210	600 A		4	4			233		233	350
0220	800 A		3.30	4.848			282		282	420
0230	1,200 A		2.80	5.714			335		335	495
0240	1,600 A		2.50	6.400			370		370	555
0300	Base meter devices, 3 meter		3.60	4.444			259		259	385
0310	4 meter		3.30	4.848			282		282	420
0320	5 meter		2.90	5.517			320		320	480
0330	6 meter		2.20	7.273			425		425	635
0340	7 meter		2	8			465		465	695
0350	8 meter		1.90	8.421			490		490	735
0400	Branch meter devices									
0410	Socket w/circuit breaker 200 A, 2 meter	2 Elec	3.30	4.848	Ea.		282		282	420
0420	3 meter		2.90	5.517			320		320	480
0430	4 meter		2.60	6.154			360		360	535
0450	Main circuit breaker, 400 A		5.80	2.759			161		161	240
0460	600 A		4	4			233		233	350
0470	800 A		3.30	4.848			282		282	420
0480	1,200 A		2.80	5.714			335		335	495
0490	1,600 A		2.50	6.400			370		370	555
0500	Main lug terminal box, 800 A		3.40	4.706			274		274	410
0510	1,200 A		2.60	6.154			360		360	535
1000	Motors, 230/460 V, 60 Hz, 3/4 HP	1 Elec	10.70	.748			43.50		43.50	65
1010	5 HP		9	.889			51.50		51.50	77.50
1020	10 HP		8	1			58		58	87
1030	15 HP		6.40	1.250			73		73	109
1040	20 HP	2 Elec	10.40	1.538			89.50		89.50	134
1050	50 HP		9.60	1.667			97		97	145
1060	75 HP		5.60	2.857			166		166	249
1070	100 HP	3 Elec	5.40	4.444			259		259	385
1080	150 HP		3.60	6.667			390		390	580

For customer support on your Electrical Costs with RSMeans data, call 800.448.8182.

107

26 05 05.25 Electrical Demolition, Electrical Power	Crew	Daily Output	Labor-Hours	Unit	Material	2018 Bare Costs Labor	Equipment	Total	Total Incl O&P
1090	200 HP	3 Elec	3	8	Ea.		465		465	695
1200	Variable frequency drive, 460 V, for 5 HP motor size	1 Elec	3.20	2.500			146		146	218
1210	10 HP motor size	"	2.70	2.963			172		172	258
1220	20 HP motor size	2 Elec	3.60	4.444			259		259	385
1230	50 HP motor size	"	2.10	7.619			445		445	665
1240	75 HP motor size	R-3	2.20	9.091			525	59	584	855
1250	100 HP motor size		2	10			580	65	645	940
1260	150 HP motor size		2	10			580	65	645	940
1270	200 HP motor size	↓	1.70	11.765	↓		680	76	756	1,100
2000	Generator set w/accessories, 3 phase 4 wire, 277/480 V									
2010	7.5 kW	3 Elec	2	12	Ea.		700		700	1,050
2020	20 kW		1.70	14.118			820		820	1,225
2040	30 kW		1.33	18.045			1,050		1,050	1,575
2060	50 kW		1	24			1,400		1,400	2,100
2080	100 kW		.75	32			1,850		1,850	2,775
2100	150 kW		.63	38.095			2,225		2,225	3,325
2120	250 kW		.59	40.678			2,375		2,375	3,550
2140	400 kW		.50	48			2,800		2,800	4,175
2160	500 kW	↓	.44	54.545			3,175		3,175	4,750
2180	750 kW	4 Elec	.56	57.143			3,325		3,325	4,975
2200	1,000 kW	"	.46	69.565	↓		4,050		4,050	6,050
3000	Uninterruptible power supply system (UPS)									
3010	Single phase, 120 V, 1 kVA	1 Elec	3.20	2.500	Ea.		146		146	218
3020	2 kVA	2 Elec	3.60	4.444			259		259	385
3030	5 kVA	3 Elec	2.50	9.600			560		560	835
3040	10 kVA		2.30	10.435			605		605	910
3050	15 kVA	↓	1.80	13.333	↓		775		775	1,150
4000	Transformer, incl. support, wire & conduit termination									
4010	Buck-boost, single phase, 120/240 V, 0.1 kVA	1 Elec	25	.320	Ea.		18.60		18.60	28
4020	0.5 kVA		12.50	.640			37.50		37.50	55.50
4030	1 kVA		6.30	1.270			74		74	110
4040	5 kVA	↓	3.80	2.105			123		123	183
4800	5 kV or 15 kV primary, 277/480 V second, 112.5 kVA	R-3	3.50	5.714			330	37	367	535
4810	150 kVA		2.70	7.407			430	48	478	700
4820	225 kVA		2.30	8.696			505	56.50	561.50	815
4830	500 kVA		1.45	13.793			800	89.50	889.50	1,300
4840	750 kVA		1.33	15.038			870	97.50	967.50	1,400
4850	1,000 kVA		1.25	16			930	104	1,034	1,525
4860	2,000 kVA		1.05	19.048			1,100	123	1,223	1,775
4870	3,000 kVA	↓	.75	26.667			1,550	173	1,723	2,525
6010	Isolation panel, 3 kVA	1 Elec	2.30	3.478			202		202	305
6020	5 kVA		2.20	3.636			212		212	315
6030	7.5 kVA		2.10	3.810			222		222	330
6040	10 kVA		1.80	4.444			259		259	385
6050	15 kVA	↓	1.40	5.714	↓		335		335	495
7000	Power filters & conditioners									
7100	Automatic voltage regulator	2 Elec	6	2.667	Ea.		155		155	232
7200	Capacitor, 1 kVAR	1 Elec	8.60	.930			54		54	81
7210	5 kVAR		5.80	1.379			80.50		80.50	120
7220	10 kVAR		4.80	1.667			97		97	145
7230	15 kVAR		4.20	1.905			111		111	166
7240	20 kVAR		3.50	2.286			133		133	199
7250	30 kVAR		3.40	2.353			137		137	205

26 05 05.25 Electrical Demolition, Electrical Power

		Crew	Daily Output	Labor-Hours	Unit	Material	2018 Bare Costs Labor	2018 Bare Costs Equipment	Total	Total Incl O&P
7260	50 kVAR	1 Elec	3.20	2.500	Ea.		146		146	218
7400	Computer isolator transformer									
7410	Single phase, 120/240 V, 0.5 kVAR	1 Elec	12.80	.625	Ea.		36.50		36.50	54.50
7420	1 kVAR		8.50	.941			55		55	82
7430	2.5 kVAR		6.40	1.250			73		73	109
7440	5 kVAR		3.70	2.162			126		126	188
7500	Computer regulator transformer									
7510	Single phase, 240 V, 0.5 kVAR	1 Elec	8.50	.941	Ea.		55		55	82
7520	1 kVAR		6.40	1.250			73		73	109
7530	2 kVAR		3.20	2.500			146		146	218
7540	Single phase, plug-in unit 120 V, 0.5 kVAR		26	.308			17.90		17.90	27
7550	1 kVAR		17	.471			27.50		27.50	41
7600	Power conditioner transformer									
7610	Single phase 115 V - 240 V, 3 kVA	2 Elec	5.10	3.137	Ea.		183		183	273
7620	5 kVA	"	3.70	4.324			252		252	375
7630	7.5 kVA	3 Elec	4.80	5			291		291	435
7640	10 kVA	"	4.30	5.581			325		325	485
7700	Transient suppressor/voltage regulator									
7710	Single phase 115 or 220 V, 1 kVA	1 Elec	8.50	.941	Ea.		55		55	82
7720	2 kVA		7.30	1.096			64		64	95.50
7730	4 kVA		6.80	1.176			68.50		68.50	102
7800	Transient voltage suppressor transformer									
7810	Single phase, 115 or 220 V, 3.6 kVA	1 Elec	12.80	.625	Ea.		36.50		36.50	54.50
7820	7.2 kVA		11.50	.696			40.50		40.50	60.50
7830	14.4 kVA		10.20	.784			45.50		45.50	68
7840	Single phase, plug-in, 120 V, 1.8 kVA		26	.308			17.90		17.90	27
8000	Power measurement & control									
8010	Switchboard instruments, 3 phase 4 wire, indicating unit	1 Elec	25	.320	Ea.		18.60		18.60	28
8020	Recording unit		12	.667			39		39	58
8100	3 current transformers, 3 phase 4 wire, 5 to 800 A		6.40	1.250			73		73	109
8110	1,000 to 1,500 A		4.20	1.905			111		111	166
8120	2,000 to 4,000 A		3.20	2.500			146		146	218

26 05 05.30 Electrical Demolition, Transmission and Distribution

		Crew	Daily Output	Labor-Hours	Unit	Material	2018 Bare Costs Labor	2018 Bare Costs Equipment	Total	Total Incl O&P
0010	**ELECTRICAL DEMOLITION, TRANSMISSION & DISTRIBUTION**									
0100	Load interrupter switch, 600 A, NEMA 1, 4.8 kV	R-3	1.33	15.038	Ea.		870	97.50	967.50	1,400
0120	13.8 kV R024119-10	"	1.27	15.748			915	102	1,017	1,475
0200	Lightning arrester, 4.8 kV	1 Elec	9	.889			51.50		51.50	77.50
0220	13.8 kV R260105-30		6.67	1.199			70		70	104
0300	Alarm or option items		3.33	2.402			140		140	209

26 05 05.35 Electrical Demolition, L.V. Distribution

		Crew	Daily Output	Labor-Hours	Unit	Material	2018 Bare Costs Labor	2018 Bare Costs Equipment	Total	Total Incl O&P
0010	**ELECTRICAL DEMOLITION, L.V. DISTRIBUTION** Addition R024119-10									
0100	Circuit breakers in enclosure									
0120	Enclosed (NEMA 1), 600 V, 3 pole, 30 A R260105-30	1 Elec	12.30	.650	Ea.		38		38	56.50
0140	60 A		8.80	.909			53		53	79
0160	100 A		7.30	1.096			64		64	95.50
0180	225 A		5	1.600			93		93	139
0200	400 A	2 Elec	6.80	2.353			137		137	205
0220	600 A		4.60	3.478			202		202	305
0240	800 A		3.60	4.444			259		259	385
0260	1,200 A		3.10	5.161			300		300	450
0280	1,600 A		2.80	5.714			335		335	495
0300	2,000 A		2.50	6.400			370		370	555

For customer support on your Electrical Costs with RSMeans data, call 800.448.8182.

109

26 05 05 – Selective Demolition for Electrical

26 05 05.35 Electrical Demolition, L.V. Distribution	Crew	Daily Output	Labor-Hours	Unit	Material	2018 Bare Costs Labor	Equipment	Total	Total Incl O&P
0400	Enclosed (NEMA 7), 600 V, 3 pole, 50 A	1 Elec	8.80	.909	Ea.		53		53	79
0410	100 A		5.80	1.379			80.50		80.50	120
0420	150 A		3.80	2.105			123		123	183
0430	250 A	2 Elec	6.20	2.581			150		150	225
0440	400 A	"	4.60	3.478			202		202	305
0460	Manual motor starter, NEMA 1	1 Elec	24	.333			19.40		19.40	29
0480	NEMA 4 or NEMA 7		15	.533			31		31	46.50
0500	Time switches, single pole single throw		15.40	.519			30		30	45
0520	Photo cell		30	.267			15.50		15.50	23
0540	Load management device, 4 loads		7.70	1.039			60.50		60.50	90.50
0550	8 loads		3.80	2.105			123		123	183
0560	Master light control panel	2 Elec	2	8			465		465	695
0600	Transfer switches, enclosed, 30 A	1 Elec	12	.667			39		39	58
0610	60 A		9.50	.842			49		49	73.50
0620	100 A		6.50	1.231			71.50		71.50	107
0630	150 A	2 Elec	12	1.333			77.50		77.50	116
0640	260 A		10	1.600			93		93	139
0660	400 A		8	2			116		116	174
0670	600 A		5	3.200			186		186	278
0680	800 A		4	4			233		233	350
0690	1,200 A		3.50	4.571			266		266	400
0700	1,600 A		3	5.333			310		310	465
0710	2,000 A		2.50	6.400			370		370	555
0800	Air terminal with base and connector	1 Elec	24	.333			19.40		19.40	29
1000	Enclosed controller, NEMA 1, 30 A		13.80	.580			33.50		33.50	50.50
1020	60 A		11.50	.696			40.50		40.50	60.50
1040	100 A		9.60	.833			48.50		48.50	72.50
1060	150 A		7.70	1.039			60.50		60.50	90.50
1080	200 A		5.40	1.481			86		86	129
1100	400 A	2 Elec	6.90	2.319			135		135	202
1120	600 A		4.60	3.478			202		202	305
1140	800 A		3.80	4.211			245		245	365
1160	1,200 A		3.10	5.161			300		300	450
1200	Control station, NEMA 1	1 Elec	30	.267			15.50		15.50	23
1220	NEMA 7		23	.348			20		20	30.50
1230	Control switches, push button		69	.116			6.75		6.75	10.10
1240	Indicating light unit		120	.067			3.88		3.88	5.80
1250	Relay		15	.533			31		31	46.50
2000	Motor control center components									
2020	Starter, NEMA 1, size 1	1 Elec	9	.889	Ea.		51.50		51.50	77.50
2040	Size 2	2 Elec	13.30	1.203			70		70	105
2060	Size 3		6.70	2.388			139		139	208
2080	Size 4		5.30	3.019			176		176	263
2100	Size 5		3.30	4.848			282		282	420
2120	NEMA 7, size 1	1 Elec	8.70	.920			53.50		53.50	80
2140	Size 2	2 Elec	12.70	1.260			73.50		73.50	110
2160	Size 3		6.30	2.540			148		148	221
2180	Size 4		5	3.200			186		186	278
2200	Size 5		3.20	5			291		291	435
2300	Fuse, light contactor, NEMA 1, 30 A	1 Elec	9	.889			51.50		51.50	77.50
2310	60 A		6.70	1.194			69.50		69.50	104
2320	100 A		3.30	2.424			141		141	211
2330	200 A		2.70	2.963			172		172	258

For customer support on your Electrical Costs with RSMeans data, call 800.448.8182.

26 05 05.35 Electrical Demolition, L.V. Distribution	Crew	Daily Output	Labor-Hours	Unit	Material	2018 Bare Costs Labor	Equipment	Total	Total Incl O&P
2350	Motor control center, incoming section	2 Elec	4	4	Ea.		233		233	350
2352	Structure per section	"	4.20	3.810			222		222	330
2400	Starter & structure, 10 HP	1 Elec	9	.889			51.50		51.50	77.50
2410	25 HP	2 Elec	13.30	1.203			70		70	105
2420	50 HP		6.70	2.388			139		139	208
2430	75 HP		5.30	3.019			176		176	263
2440	100 HP		4.70	3.404			198		198	296
2450	200 HP		3.30	4.848			282		282	420
2460	400 HP		2.70	5.926			345		345	515
2500	Motor starter & control									
2510	Motor starter, NEMA 1, 5 HP	1 Elec	8.90	.899	Ea.		52.50		52.50	78
2520	10 HP	"	6.20	1.290			75		75	112
2530	25 HP	2 Elec	8.40	1.905			111		111	166
2540	50 HP		6.90	2.319			135		135	202
2550	100 HP		4.60	3.478			202		202	305
2560	200 HP		3.50	4.571			266		266	400
2570	400 HP		3.10	5.161			300		300	450
2610	Motor starter, NEMA 7, 5 HP	1 Elec	6.20	1.290			75		75	112
2620	10 HP	"	4.20	1.905			111		111	166
2630	25 HP	2 Elec	6.90	2.319			135		135	202
2640	50 HP		4.60	3.478			202		202	305
2650	100 HP		3.50	4.571			266		266	400
2660	200 HP		1.90	8.421			490		490	735
2710	Combination control unit, NEMA 1, 5 HP	1 Elec	6.90	1.159			67.50		67.50	101
2720	10 HP	"	5	1.600			93		93	139
2730	25 HP	2 Elec	7.70	2.078			121		121	181
2740	50 HP		5.10	3.137			183		183	273
2750	100 HP		3.10	5.161			300		300	450
2810	NEMA 7, 5 HP	1 Elec	5	1.600			93		93	139
2820	10 HP	"	3.90	2.051			119		119	178
2830	25 HP	2 Elec	5.10	3.137			183		183	273
2840	50 HP		3.10	5.161			300		300	450
2850	100 HP		2.30	6.957			405		405	605
2860	200 HP		1.50	10.667			620		620	930
3000	Panelboard or load center circuit breaker									
3010	Bolt-on or plug in, 15 A to 50 A	1 Elec	20	.400	Ea.		23.50		23.50	35
3020	60 A to 70 A		16	.500			29		29	43.50
3030	Bolt-on, 80 A to 100 A		14	.571			33.50		33.50	49.50
3040	Up to 250 A		7.30	1.096			64		64	95.50
3050	Motor operated, 30 A		13	.615			36		36	53.50
3060	60 A		10	.800			46.50		46.50	69.50
3070	100 A		8	1			58		58	87
3200	Switchboard circuit breaker									
3210	15 A to 60 A	1 Elec	19	.421	Ea.		24.50		24.50	36.50
3220	70 A to 100 A		14	.571			33.50		33.50	49.50
3230	125 A to 400 A		10	.800			46.50		46.50	69.50
3240	450 A to 600 A		5.30	1.509			88		88	131
3250	700 A to 800 A		4.30	1.860			108		108	162
3260	1,000 A		3.30	2.424			141		141	211
3270	1,200 A		2.70	2.963			172		172	258
3500	Switchboard, incoming section, 400 A	2 Elec	3.70	4.324			252		252	375
3510	600 A		3.30	4.848			282		282	420
3520	800 A		2.90	5.517			320		320	480

26 05 05.35 Electrical Demolition, L.V. Distribution

		Crew	Daily Output	Labor-Hours	Unit	Material	2018 Bare Costs Labor	Equipment	Total	Total Incl O&P
3530	1,200 A	2 Elec	2.40	6.667	Ea.		390		390	580
3540	1,600 A		2.20	7.273			425		425	635
3550	2,000 A		2.10	7.619			445		445	665
3560	3,000 A		1.90	8.421			490		490	735
3570	4,000 A		1.70	9.412			550		550	820
3610	Distribution section, 600 A		4	4			233		233	350
3620	800 A		3.60	4.444			259		259	385
3630	1,200 A		3.10	5.161			300		300	450
3640	1,600 A		2.90	5.517			320		320	480
3650	2,000 A		2.70	5.926			345		345	515
3710	Transition section, 600 A		3.80	4.211			245		245	365
3720	800 A		3.30	4.848			282		282	420
3730	1,200 A		2.70	5.926			345		345	515
3740	1,600 A		2.40	6.667			390		390	580
3750	2,000 A		2.20	7.273			425		425	635
3760	2,500 A		2.10	7.619			445		445	665
3770	3,000 A		1.90	8.421			490		490	735
4000	Bus duct, aluminum or copper, 30 A	1 Elec	200	.040	L.F.		2.33		2.33	3.48
4020	60 A		160	.050			2.91		2.91	4.35
4040	100 A		140	.057			3.33		3.33	4.97
5000	Feedrail, trolley busway, up to 60 A		160	.050			2.91		2.91	4.35
5020	100 A		120	.067			3.88		3.88	5.80
5040	Busway, 50 A		200	.040			2.33		2.33	3.48
6000	Fuse, 30 A		133	.060	Ea.		3.50		3.50	5.25
6010	60 A		133	.060			3.50		3.50	5.25
6020	100 A		105	.076			4.43		4.43	6.65
6030	200 A		95	.084			4.90		4.90	7.35
6040	400 A		80	.100			5.80		5.80	8.70
6050	600 A		53	.151			8.80		8.80	13.15
6060	601 A to 1,200 A		42	.190			11.10		11.10	16.55
6070	1,500 A to 1,600 A		34	.235			13.70		13.70	20.50
6080	1,800 A to 2,500 A		26	.308			17.90		17.90	27
6090	4,000 A		21	.381			22		22	33
6100	4,500 A to 5,000 A		18	.444			26		26	38.50
6120	6,000 A		15	.533			31		31	46.50
6200	Fuse plug or fustat		100	.080			4.66		4.66	6.95

26 05 05.50 Electrical Demolition, Lighting

		Crew	Daily Output	Labor-Hours	Unit	Material	2018 Bare Costs Labor	Equipment	Total	Total Incl O&P
0010	**ELECTRICAL DEMOLITION, LIGHTING** Addition R024119-10									
0100	Fixture hanger, flexible, 1/2" diameter	1 Elec	36	.222	Ea.		12.95		12.95	19.35
0120	3/4" diameter R260105-30	"	30	.267	"		15.50		15.50	23
3000	Light pole, anchor base, excl. concrete bases									
3010	Metal light pole, 10'	2 Elec	24	.667	Ea.		39		39	58
3020	16'	"	18	.889			51.50		51.50	77.50
3030	20'	R-3	8.70	2.299			133	14.90	147.90	216
3035	25'		8.60	2.326			135	15.05	150.05	219
3038	30'		6	3.333			193	21.50	214.50	315
3039	35'		6	3.333			193	21.50	214.50	315
3040	40'		6	3.333			193	21.50	214.50	315
3100	Wood light pole, 10'	2 Elec	36	.444			26		26	38.50
3120	20'	"	24	.667			39		39	58
3140	Bollard light, 42"	1 Elec	9	.889			51.50		51.50	77.50
3160	Walkway luminaire	"	8.10	.988			57.50		57.50	86

26 05 05 – Selective Demolition for Electrical

26 05 05.50 Electrical Demolition, Lighting

		Crew	Daily Output	Labor-Hours	Unit	Material	2018 Bare Costs Labor	Equipment	Total	Total Incl O&P
4000	Explosion proof									
4010	Metal halide, 175 W	1 Elec	8.70	.920	Ea.		53.50		53.50	80
4020	250 W	"	8.10	.988			57.50		57.50	86
4030	400 W	2 Elec	14.40	1.111			64.50		64.50	96.50
4050	High pressure sodium, 70 W	1 Elec	9	.889			51.50		51.50	77.50
4060	100 W		9	.889			51.50		51.50	77.50
4070	150 W		8.10	.988			57.50		57.50	86
4100	Incandescent		8.70	.920			53.50		53.50	80
4200	Fluorescent		8.10	.988			57.50		57.50	86
5000	Ballast, fluorescent fixture		24	.333			19.40		19.40	29
5040	High intensity discharge fixture		24	.333			19.40		19.40	29
5300	Exit and emergency lighting									
5310	Exit light	1 Elec	24	.333	Ea.		19.40		19.40	29
5320	Emergency battery pack lighting unit		12	.667			39		39	58
5330	Remote lamp only		80	.100			5.80		5.80	8.70
5340	Self-contained fluorescent lamp pack		30	.267			15.50		15.50	23
5500	Track lighting, 8' section	2 Elec	40	.400			23.50		23.50	35
5510	Track lighting fixture	1 Elec	64	.125			7.30		7.30	10.90
5800	Energy saving devices									
5810	Occupancy sensor	1 Elec	21	.381	Ea.		22		22	33
5820	Automatic wall switch		72	.111			6.45		6.45	9.65
5830	Remote power pack		30	.267			15.50		15.50	23
5840	Photoelectric control		24	.333			19.40		19.40	29
5850	Fixture whip		100	.080			4.66		4.66	6.95
6000	Lamps									
6010	Fluorescent	1 Elec	200	.040	Ea.		2.33		2.33	3.48
6030	High intensity discharge lamp, up to 400 W		68	.118			6.85		6.85	10.25
6040	Up to 1,000 W		45	.178			10.35		10.35	15.45
6050	Quartz		90	.089			5.15		5.15	7.75
6070	Incandescent		360	.022			1.29		1.29	1.93
6080	Exterior, PAR		290	.028			1.61		1.61	2.40
6090	Guards for fluorescent lamp		24	.333			19.40		19.40	29

26 05 13 – Medium-Voltage Cables

26 05 13.10 Cable Terminations

		Crew	Daily Output	Labor-Hours	Unit	Material	2018 Bare Costs Labor	Equipment	Total	Total Incl O&P
0010	**CABLE TERMINATIONS**, 5 kV to 35 kV									
0100	Indoor, insulation diameter range 0.64" to 1.08"									
0300	Padmount, 5 kV	1 Elec	8	1	Ea.	77	58		135	172
0400	10 kV		6.40	1.250		100	73		173	219
0500	15 kV		6	1.333		124	77.50		201.50	252
0600	25 kV		5.60	1.429		184	83		267	325
0700	Insulation diameter range 1.05" to 1.8"									
0800	Padmount, 5 kV	1 Elec	8	1	Ea.	43.50	58		101.50	135
0900	10 kV		6	1.333		141	77.50		218.50	271
1000	15 kV		5.60	1.429		145	83		228	283
1100	25 kV		5.30	1.509		213	88		301	365
1200	Insulation diameter range 1.53" to 2.32"									
1300	Padmount, 5 kV	1 Elec	7.40	1.081	Ea.	142	63		205	250
1400	10 kV		5.60	1.429		171	83		254	310
1500	15 kV		5.30	1.509		177	88		265	325
1600	25 kV		5	1.600		305	93		398	475
1700	Outdoor systems, #4 stranded to 1/0 stranded									
1800	5 kV	1 Elec	7.40	1.081	Ea.	101	63		164	206

For customer support on your Electrical Costs with RSMeans data, call 800.448.8182.

113

26 05 13 – Medium-Voltage Cables

26 05 13.10 Cable Terminations

		Crew	Daily Output	Labor-Hours	Unit	Material	2018 Bare Costs Labor	2018 Bare Costs Equipment	Total	Total Incl O&P
1900	15 kV	1 Elec	5.30	1.509	Ea.	121	88		209	264
2000	25 kV		5	1.600		180	93		273	335
2100	35 kV		4.80	1.667		330	97		427	510
2200	#1 solid to 4/0 stranded, 5 kV		6.90	1.159		144	67.50		211.50	260
2300	15 kV		5	1.600		172	93		265	330
2400	25 kV		4.80	1.667		234	97		331	405
2500	35 kV		4.60	1.739		335	101		436	515
2600	2/0 solid to 350 kcmil stranded, 5 kV		6.40	1.250		180	73		253	305
2700	15 kV		4.80	1.667		194	97		291	360
2800	25 kV		4.60	1.739		300	101		401	480
2900	35 kV		4.40	1.818		395	106		501	590
3000	400 kcmil compact to 750 kcmil stranded, 5 kV		6	1.333		171	77.50		248.50	305
3100	15 kV		4.60	1.739		234	101		335	410
3200	25 kV		4.40	1.818		340	106		446	535
3300	35 kV		4.20	1.905		350	111		461	550
3400	1,000 kcmil, 5 kV		5.60	1.429		240	83		323	390
3500	15 kV		4.40	1.818		315	106		421	510
3600	25 kV		4.20	1.905		350	111		461	550
3700	35 kV		4	2		485	116		601	705

26 05 13.16 Medium-Voltage, Single Cable

		Crew	Daily Output	Labor-Hours	Unit	Material	2018 Bare Costs Labor	2018 Bare Costs Equipment	Total	Total Incl O&P
0010	**MEDIUM-VOLTAGE, SINGLE CABLE** Splicing & terminations not included									
0040	Copper, XLP shielding, 5 kV, #6	2 Elec	4.40	3.636	C.L.F.	165	212		377	495
0050	#4		4.40	3.636		214	212		426	550
0100	#2		4	4		262	233		495	640
0200	#1		4	4		315	233		548	695
0400	1/0		3.80	4.211		345	245		590	745
0600	2/0		3.60	4.444		420	259		679	845
0800	4/0		3.20	5		565	291		856	1,050
1000	250 kcmil	3 Elec	4.50	5.333		675	310		985	1,200
1200	350 kcmil		3.90	6.154		875	360		1,235	1,500
1400	500 kcmil		3.60	6.667		1,125	390		1,515	1,825
1600	15 kV, ungrounded neutral, #1	2 Elec	4	4		355	233		588	740
1800	1/0		3.80	4.211		425	245		670	830
2000	2/0		3.60	4.444		480	259		739	915
2200	4/0		3.20	5		640	291		931	1,150
2400	250 kcmil	3 Elec	4.50	5.333		710	310		1,020	1,250
2600	350 kcmil		3.90	6.154		895	360		1,255	1,525
2800	500 kcmil		3.60	6.667		1,100	390		1,490	1,775
3000	25 kV, grounded neutral, 1/0	2 Elec	3.60	4.444		575	259		834	1,025
3200	2/0		3.40	4.706		635	274		909	1,100
3400	4/0		3	5.333		795	310		1,105	1,350
3600	250 kcmil	3 Elec	4.20	5.714		985	335		1,320	1,575
3800	350 kcmil		3.60	6.667		1,150	390		1,540	1,850
3900	500 kcmil		3.30	7.273		1,350	425		1,775	2,125
4000	35 kV, grounded neutral, 1/0	2 Elec	3.40	4.706		620	274		894	1,100
4200	2/0		3.20	5		715	291		1,006	1,225
4400	4/0		2.80	5.714		900	335		1,235	1,475
4600	250 kcmil	3 Elec	3.90	6.154		1,050	360		1,410	1,675
4800	350 kcmil		3.30	7.273		1,275	425		1,700	2,025
5000	500 kcmil		3	8		1,500	465		1,965	2,350
5050	Aluminum, XLP shielding, 5 kV, #2	2 Elec	5	3.200		200	186		386	500
5070	#1		4.40	3.636		206	212		418	540

26 05 13 – Medium-Voltage Cables

26 05 13.16 Medium-Voltage, Single Cable

		Crew	Daily Output	Labor-Hours	Unit	Material	2018 Bare Costs Labor	Equipment	Total	Total Incl O&P
5090	1/0	2 Elec	4	4	C.L.F.	239	233		472	615
5100	2/0		3.80	4.211		268	245		513	660
5150	4/0		3.60	4.444		315	259		574	735
5200	250 kcmil	3 Elec	4.80	5		385	291		676	855
5220	350 kcmil		4.50	5.333		450	310		760	960
5240	500 kcmil		3.90	6.154		580	360		940	1,175
5260	750 kcmil		3.60	6.667		765	390		1,155	1,425
5300	15 kV aluminum, XLP, #1	2 Elec	4.40	3.636		256	212		468	595
5320	1/0		4	4		264	233		497	640
5340	2/0		3.80	4.211		315	245		560	710
5360	4/0		3.60	4.444		350	259		609	770
5380	250 kcmil	3 Elec	4.80	5		420	291		711	895
5400	350 kcmil		4.50	5.333		470	310		780	980
5420	500 kcmil		3.90	6.154		650	360		1,010	1,250
5440	750 kcmil		3.60	6.667		860	390		1,250	1,525

26 05 19 – Low-Voltage Electrical Power Conductors and Cables

26 05 19.13 Undercarpet Electrical Power Cables

		Crew	Daily Output	Labor-Hours	Unit	Material	2018 Bare Costs Labor	Equipment	Total	Total Incl O&P
0010	**UNDERCARPET ELECTRICAL POWER CABLES** R260519-80									
0020	Power System									
0100	Cable flat, 3 conductor, #12, w/attached bottom shield	1 Elec	982	.008	L.F.	5.35	.47		5.82	6.60
0200	Shield, top, steel		1768	.005	"	4.96	.26		5.22	5.85
0250	Splice, 3 conductor		48	.167	Ea.	15.10	9.70		24.80	31
0300	Top shield		96	.083		1.34	4.85		6.19	8.70
0350	Tap		40	.200		19.40	11.65		31.05	39
0400	Insulating patch, splice, tap & end		48	.167		47.50	9.70		57.20	67
0450	Fold		230	.035			2.02		2.02	3.03
0500	Top shield, tap & fold		96	.083		1.34	4.85		6.19	8.70
0700	Transition, block assembly		77	.104		69.50	6.05		75.55	85.50
0750	Receptacle frame & base		32	.250		38	14.55		52.55	64
0800	Cover receptacle		120	.067		3.25	3.88		7.13	9.40
0850	Cover blank		160	.050		3.83	2.91		6.74	8.55
0860	Receptacle, direct connected, single		25	.320		81.50	18.60		100.10	118
0870	Dual		16	.500		134	29		163	191
0880	Combination high & low, tension		21	.381		98.50	22		120.50	142
0900	Box, floor with cover		20	.400		82.50	23.50		106	126
0920	Floor service w/barrier		4	2		234	116		350	430
1000	Wall, surface, with cover		20	.400		54.50	23.50		78	95
1100	Wall, flush, with cover		20	.400		38	23.50		61.50	77
1450	Cable flat, 5 conductor #12, w/attached bottom shield		800	.010	L.F.	8.80	.58		9.38	10.50
1550	Shield, top, steel		1768	.005	"	8.70	.26		8.96	10
1600	Splice, 5 conductor		48	.167	Ea.	24.50	9.70		34.20	41.50
1650	Top shield		96	.083		1.34	4.85		6.19	8.70
1700	Tap		48	.167		32	9.70		41.70	49.50
1750	Insulating patch, splice tap & end		83	.096		55	5.60		60.60	69
1800	Transition, block assembly		77	.104		50	6.05		56.05	64
1850	Box, wall, flush with cover		20	.400		51	23.50		74.50	91
1900	Cable flat, 4 conductor #12		933	.009	L.F.	6.20	.50		6.70	7.55
1950	3 conductor #10		982	.008		6	.47		6.47	7.30
1960	4 conductor #10		933	.009		7.85	.50		8.35	9.40
1970	5 conductor #10		884	.009		9.60	.53		10.13	11.35
2500	Telephone System									
2510	Transition fitting wall box, surface	1 Elec	24	.333	Ea.	62	19.40		81.40	97

For customer support on your Electrical Costs with RSMeans data, call 800.448.8182.

115

26 05 19.13 Undercarpet Electrical Power Cables		Crew	Daily Output	Labor-Hours	Unit	Material	2018 Bare Costs Labor	Equipment	Total	Total Incl O&P
2520	Flush	1 Elec	24	.333	Ea.	62	19.40		81.40	97
2530	Flush, for PC board		24	.333		62	19.40		81.40	97
2540	Floor service box	↓	4	2		240	116		356	440
2550	Cover, surface					16.05			16.05	17.70
2560	Flush					16.05			16.05	17.70
2570	Flush for PC board					16.05			16.05	17.70
2700	Floor fitting w/duplex jack & cover	1 Elec	21	.381		47.50	22		69.50	85
2720	Low profile		53	.151		17.35	8.80		26.15	32.50
2740	Miniature w/duplex jack		53	.151		26.50	8.80		35.30	42
2760	25 pair kit		21	.381		50	22	·	72	88
2780	Low profile		53	.151		17.75	8.80		26.55	32.50
2800	Call director kit for 5 cable		19	.421		79.50	24.50		104	124
2820	4 pair kit		19	.421		95	24.50		119.50	141
2840	3 pair kit		19	.421		100	24.50		124.50	147
2860	Comb. 25 pair & 3 conductor power		21	.381		79.50	22		101.50	121
2880	5 conductor power		21	.381		91	22		113	133
2900	PC board, 8 per 3 pair		161	.050		69	2.89		71.89	80.50
2920	6 per 4 pair		161	.050		69	2.89		71.89	80.50
2940	3 pair adapter		161	.050		62.50	2.89		65.39	73.50
2950	Plug		77	.104		3.07	6.05		9.12	12.45
2960	Couplers		321	.025	↓	8.65	1.45		10.10	11.65
3000	Bottom shield for 25 pair cable		4420	.002	L.F.	.87	.11		.98	1.12
3020	4 pair		4420	.002		.40	.11		.51	.60
3040	Top shield for 25 pair cable		4420	.002	↓	.87	.11		.98	1.12
3100	Cable assembly, double-end, 50', 25 pair		11.80	.678	Ea.	243	39.50		282.50	325
3110	3 pair		23.60	.339		75	19.75		94.75	112
3120	4 pair		23.60	.339	↓	79.50	19.75		99.25	117
3140	Bulk 3 pair		1473	.005	L.F.	1.29	.32		1.61	1.89
3160	4 pair	↓	1473	.005	"	1.59	.32		1.91	2.22
3500	Data System									
3520	Cable 25 conductor w/connection 40', 75 ohm	1 Elec	14.50	.552	Ea.	72	32		104	127
3530	Single lead		22	.364		196	21		217	248
3540	Dual lead	↓	22	.364	↓	244	21		265	300
3560	Shields same for 25 conductor as 25 pair telephone									
3570	Single & dual, none required									
3590	BNC coax connectors, plug	1 Elec	40	.200	Ea.	11.20	11.65		22.85	30
3600	TNC coax connectors, plug	"	40	.200	"	15.70	11.65		27.35	34.50
3700	Cable-bulk									
3710	Single lead	1 Elec	1473	.005	L.F.	2.72	.32		3.04	3.46
3720	Dual lead	"	1473	.005	"	3.94	.32		4.26	4.80
3730	Hand tool crimp				Ea.	485			485	535
3740	Hand tool notch				"	19.75			19.75	21.50
3750	Boxes & floor fitting same as telephone									
3790	Data cable notching, 90°	1 Elec	97	.082	Ea.		4.80		4.80	7.15
3800	180°		60	.133	↓		7.75		7.75	11.60
8100	Drill floor		160	.050	↓	3.92	2.91		6.83	8.65
8200	Marking floor		1600	.005	L.F.		.29		.29	.44
8300	Tape, hold down		6400	.001	"	.17	.07		.24	.30
8350	Tape primer, 500' per can	↓	96	.083	Ea.	54.50	4.85		59.35	67.50
8400	Tool, splicing				"	171			171	188

26 05 19.20 Armored Cable	Crew	Daily Output	Labor-Hours	Unit	Material	2018 Bare Costs Labor	2018 Bare Costs Equipment	Total	Total Incl O&P
0010 **ARMORED CABLE** R260519-20									
0050 600 volt, copper (BX), #14, 2 conductor, solid	1 Elec	2.40	3.333	C.L.F.	41.50	194		235.50	335
0100 3 conductor, solid		2.20	3.636		64.50	212		276.50	385
0120 4 conductor, solid		2	4		90.50	233		323.50	450
0150 #12, 2 conductor, solid		2.30	3.478		41.50	202		243.50	350
0200 3 conductor, solid		2	4		71	233		304	430
0220 4 conductor, solid		1.80	4.444		97.50	259		356.50	490
0250 #10, 2 conductor, solid		2	4		80.50	233		313.50	440
0300 3 conductor, solid		1.60	5		110	291		401	555
0320 4 conductor, solid		1.40	5.714		180	335		515	695
0340 #8, 2 conductor, stranded		1.50	5.333		244	310		554	735
0350 3 conductor, stranded		1.30	6.154		244	360		604	805
0370 4 conductor, stranded		1.10	7.273		350	425		775	1,025
0380 #6, 2 conductor, stranded		1.30	6.154		255	360		615	815
0390 #4, 3 conductor, stranded		1.40	5.714		470	335		805	1,000
0400 3 conductor with PVC jacket, in cable tray, #6	↓	3.10	2.581		545	150		695	825
0450 #4	2 Elec	5.40	2.963		665	172		837	990
0500 #2		4.60	3.478		800	202		1,002	1,175
0550 #1		4	4		1,025	233		1,258	1,475
0600 1/0		3.60	4.444		1,100	259		1,359	1,575
0650 2/0		3.40	4.706		1,325	274		1,599	1,850
0700 3/0		3.20	5		1,775	291		2,066	2,375
0750 4/0	↓	3	5.333		2,125	310		2,435	2,800
0800 250 kcmil	3 Elec	3.60	6.667		2,625	390		3,015	3,475
0850 350 kcmil		3.30	7.273		3,300	425		3,725	4,275
0900 500 kcmil	↓	3	8		4,575	465		5,040	5,725
0910 4 conductor with PVC jacket, in cable tray, #6	1 Elec	2.70	2.963		885	172		1,057	1,225
0920 #4	2 Elec	4.60	3.478		820	202		1,022	1,200
0930 #2		4	4		1,275	233		1,508	1,750
0940 #1		3.60	4.444		1,450	259		1,709	1,975
0950 1/0		3.40	4.706		1,425	274		1,699	1,950
0960 2/0		3.20	5		1,750	291		2,041	2,350
0970 3/0		3	5.333		2,300	310		2,610	3,025
0980 4/0	↓	2.40	6.667		2,675	390		3,065	3,525
0990 250 kcmil	3 Elec	3.30	7.273		3,225	425		3,650	4,175
1000 350 kcmil		3	8		4,325	465		4,790	5,475
1010 500 kcmil	↓	2.70	8.889	↓	6,000	515		6,515	7,375
1050 5 kV, copper, 3 conductor with PVC jacket,									
1060 non-shielded, in cable tray, #4	2 Elec	380	.042	L.F.	7.60	2.45		10.05	12
1100 #2		360	.044		9.85	2.59		12.44	14.70
1200 #1		300	.053		12.55	3.10		15.65	18.45
1400 1/0		290	.055		14.50	3.21		17.71	21
1600 2/0		260	.062		16.75	3.58		20.33	24
2000 4/0	↓	240	.067		22.50	3.88		26.38	30.50
2100 250 kcmil	3 Elec	330	.073		30.50	4.23		34.73	40
2150 350 kcmil		315	.076		37.50	4.43		41.93	48
2200 500 kcmil	↓	270	.089	↓	51.50	5.15		56.65	64.50
2400 15 kV, copper, 3 conductor with PVC jacket galv., steel armored									
2500 grounded neutral, in cable tray, #2	2 Elec	300	.053	L.F.	16.25	3.10		19.35	22.50
2600 #1		280	.057		17	3.33		20.33	23.50
2800 1/0		260	.062		19.75	3.58		23.33	27.50
2900 2/0		220	.073	↓	26	4.23		30.23	35

For customer support on your Electrical Costs with RSMeans data, call 800.448.8182.

117

26 05 19 – Low-Voltage Electrical Power Conductors and Cables

26 05 19.20 Armored Cable		Crew	Daily Output	Labor-Hours	Unit	Material	2018 Bare Costs Labor	Equipment	Total	Total Incl O&P
3000	4/0	2 Elec	190	.084	L.F.	29.50	4.90		34.40	39.50
3100	250 kcmil	3 Elec	270	.089		32.50	5.15		37.65	44
3150	350 kcmil		240	.100		38.50	5.80		44.30	50.50
3200	500 kcmil		210	.114		51.50	6.65		58.15	67
3400	15 kV, copper, 3 conductor with PVC jacket,									
3450	ungrounded neutral, in cable tray, #2	2 Elec	260	.062	L.F.	17.20	3.58		20.78	24.50
3500	#1		230	.070		19.05	4.05		23.10	27
3600	1/0		200	.080		22	4.66		26.66	31
3700	2/0		190	.084		27	4.90		31.90	37
3800	4/0		160	.100		32.50	5.80		38.30	44
4000	250 kcmil	3 Elec	210	.114		37.50	6.65		44.15	51.50
4050	350 kcmil		195	.123		49.50	7.15		56.65	65
4100	500 kcmil		180	.133		60.50	7.75		68.25	78
4200	600 volt, aluminum, 3 conductor in cable tray with PVC jacket									
4300	#2	2 Elec	540	.030	L.F.	3.93	1.72		5.65	6.90
4400	#1		460	.035		4.80	2.02		6.82	8.35
4500	1/0		400	.040		6	2.33		8.33	10.10
4600	2/0		360	.044		6.05	2.59		8.64	10.50
4700	3/0		340	.047		7.10	2.74		9.84	11.90
4800	4/0		320	.050		8.60	2.91		11.51	13.80
4900	250 kcmil	3 Elec	450	.053		10.30	3.10		13.40	16
5000	350 kcmil		360	.067		12.30	3.88		16.18	19.35
5200	500 kcmil		330	.073		15.30	4.23		19.53	23
5300	750 kcmil		285	.084		19.70	4.90		24.60	29
5400	600 volt, aluminum, 4 conductor in cable tray with PVC jacket									
5410	#2	2 Elec	520	.031	L.F.	4.74	1.79		6.53	7.90
5430	#1		440	.036		5.75	2.12		7.87	9.50
5450	1/0		380	.042		6.75	2.45		9.20	11.05
5470	2/0		340	.047		6.90	2.74		9.64	11.70
5480	3/0		320	.050		8.10	2.91		11.01	13.25
5500	4/0		300	.053		9.50	3.10		12.60	15.10
5520	250 kcmil	3 Elec	420	.057		10.20	3.33		13.53	16.15
5540	350 kcmil		330	.073		13.20	4.23		17.43	21
5560	500 kcmil		300	.080		16.50	4.66		21.16	25
5580	750 kcmil		270	.089		22.50	5.15		27.65	33
5600	5 kV, aluminum, unshielded in cable tray, #2 with PVC jacket	2 Elec	380	.042		6.05	2.45		8.50	10.30
5700	#1 with PVC jacket		360	.044		6.80	2.59		9.39	11.30
5800	1/0 with PVC jacket		300	.053		6.95	3.10		10.05	12.30
6000	2/0 with PVC jacket		290	.055		7.10	3.21		10.31	12.60
6200	3/0 with PVC jacket		260	.062		8.50	3.58		12.08	14.70
6300	4/0 with PVC jacket		240	.067		10	3.88		13.88	16.80
6400	250 kcmil with PVC jacket	3 Elec	330	.073		10.95	4.23		15.18	18.40
6500	350 kcmil with PVC jacket		315	.076		12.85	4.43		17.28	21
6600	500 kcmil with PVC jacket		300	.080		15.25	4.66		19.91	23.50
6800	750 kcmil with PVC jacket		270	.089		18.80	5.15		23.95	28.50
6900	15 kV, aluminum, shielded-grounded, #2 with PVC jacket	2 Elec	320	.050		13.80	2.91		16.71	19.55
7000	#1 with PVC jacket		300	.053		14.20	3.10		17.30	20
7200	1/0 with PVC jacket		280	.057		15.30	3.33		18.63	22
7300	2/0 with PVC jacket		260	.062		15.55	3.58		19.13	22.50
7400	3/0 with PVC jacket		240	.067		17.45	3.88		21.33	25
7500	4/0 with PVC jacket		220	.073		17.90	4.23		22.13	26
7600	250 kcmil with PVC jacket	3 Elec	300	.080		19.65	4.66		24.31	28.50
7700	350 kcmil with PVC jacket		270	.089		23	5.15		28.15	33.50

26 05 19.20 Armored Cable	Crew	Daily Output	Labor-Hours	Unit	Material	2018 Bare Costs Labor	Equipment	Total	Total Incl O&P
7800	500 kcmil with PVC jacket	3 Elec	240	.100	L.F.	27.50	5.80		33.30	39
8000	750 kcmil with PVC jacket		204	.118		33	6.85		39.85	46.50
8200	15 kV, aluminum, shielded-ungrounded, #1 with PVC jacket	2 Elec	250	.064		17.30	3.72		21.02	24.50
8300	1/0 with PVC jacket		230	.070		17.90	4.05		21.95	26
8400	2/0 with PVC jacket		210	.076		19.65	4.43		24.08	28
8500	3/0 with PVC jacket		200	.080		19.95	4.66		24.61	29
8600	4/0 with PVC jacket		190	.084		21.50	4.90		26.40	31.50
8700	250 kcmil with PVC jacket	3 Elec	270	.089		23.50	5.15		28.65	33.50
8800	350 kcmil with PVC jacket		240	.100		26.50	5.80		32.30	37.50
8900	500 kcmil with PVC jacket		210	.114		32	6.65		38.65	45.50
8950	750 kcmil with PVC jacket		174	.138		40	8.05		48.05	56
9010	600 volt, copper (MC) steel clad, #14, 2 wire	R-1A	8.16	1.961	C.L.F.	42.50	103		145.50	201
9020	3 wire		7.21	2.219		75.50	116		191.50	257
9030	4 wire		6.69	2.392		184	125		309	390
9040	#12, 2 wire		7.21	2.219		54.50	116		170.50	234
9050	3 wire		6.69	2.392		70.50	125		195.50	265
9060	4 wire		6.04	2.649		116	139		255	335
9070	#10, 2 wire		6.04	2.649		103	139		242	320
9080	3 wire		5.65	2.832		133	148		281	370
9090	4 wire		5.23	3.059		216	160		376	475
9100	#8, 2 wire, stranded		3.94	4.061		158	213		371	495
9110	3 wire, stranded		3.40	4.706		193	246		439	580
9120	4 wire, stranded		2.80	5.714		266	299		565	735
9130	#6, 2 wire, stranded		3.45	4.638		208	243		451	595
9200	600 volt, copper (MC) aluminum clad, #14, 2 wire		8.16	1.961		41.50	103		144.50	200
9210	3 wire		7.21	2.219		73	116		189	255
9220	4 wire		6.69	2.392		88.50	125		213.50	285
9230	#12, 2 wire		7.21	2.219		46.50	116		162.50	225
9240	3 wire		6.69	2.392		7.35	125		132.35	195
9250	4 wire		6.04	2.649		131	139		270	350
9260	#10, 2 wire		6.04	2.649		85	139		224	300
9270	3 wire		5.65	2.832		116	148		264	350
9280	4 wire		5.23	3.059		208	160		368	470
9600	Alum (MC) aluminum clad, #6, 3 conductor w/#6 grnd		2.75	5.818		297	305		602	780
9610	4 conductor w/#6 grnd		2.29	6.987		223	365		588	790
9620	#4, 3 conductor w/#6 grnd	R-1B	3.13	7.668		213	385		598	815
9630	4 conductor w/#6 grnd		2.67	8.989		231	455		686	935
9640	#2, 3 conductor w/#4 grnd		2.59	9.266		425	465		890	1,175
9650	4 conductor w/#4 grnd		2.21	10.860		475	550		1,025	1,350
9660	#1, 3 conductor w/#4 grnd		2.15	11.163		315	565		880	1,200
9670	4 conductor w/#4 grnd		1.85	12.973		645	655		1,300	1,700
9680	1/0, 3 conductor w/#4 grnd		2.16	11.111		360	560		920	1,225
9690	4 conductor w/#4 grnd		1.73	13.873		440	700		1,140	1,525
9700	2/0, 3 conductor w/#4 grnd		1.91	12.565		655	635		1,290	1,675
9710	4 conductor w/#4 grnd		1.53	15.686		490	790		1,280	1,725
9720	3/0, 3 conductor w/#4 grnd		1.75	13.714		840	690		1,530	1,950
9730	4 conductor w/#4 grnd		1.40	17.143		610	865		1,475	1,975
9740	4/0, 3 conductor w/#2 grnd		1.54	15.584		945	785		1,730	2,225
9750	4 conductor w/#2 grnd		1.23	19.512		395	985		1,380	1,900
9760	250 kcmil, 3 conductor w/#1 grnd	R-1C	1.90	16.842		1,200	880	68.50	2,148.50	2,725
9770	4 conductor w/#1 grnd		1.52	21.053		1,625	1,100	85.50	2,810.50	3,550
9780	350 kcmil, 3 conductor w/1/0 grnd		1.76	18.182		1,375	950	74	2,399	3,000
9790	4 conductor w/1/0 grnd		1.41	22.695		1,150	1,200	92.50	2,442.50	3,150

For customer support on your Electrical Costs with RSMeans data, call 800.448.8182.

119

26 05 19 – Low-Voltage Electrical Power Conductors and Cables

26 05 19.20 Armored Cable		Crew	Daily Output	Labor-Hours	Unit	Material	2018 Bare Costs Labor	2018 Bare Costs Equipment	Total	Total Incl O&P
9800	500 kcmil, 3 conductor w/#1 grnd	R-1C	1.64	19.512	C.L.F.	1,900	1,025	79.50	3,004.50	3,700
9810	4 conductor w/2/0 grnd		1.31	24.427		1,600	1,275	99.50	2,974.50	3,775
9840	750 kcmil, 3 conductor w/1/0 grnd		1.31	24.427		2,875	1,275	99.50	4,249.50	5,175
9850	4 conductor w/3/0 grnd		1.04	30.769		3,425	1,600	125	5,150	6,325

26 05 19.25 Cable Connectors

		Crew	Daily Output	Labor-Hours	Unit	Material	2018 Bare Costs Labor	2018 Bare Costs Equipment	Total	Total Incl O&P
0010	**CABLE CONNECTORS**									
0100	600 volt, nonmetallic, #14-2 wire	1 Elec	160	.050	Ea.	1.54	2.91		4.45	6.05
0200	#14-3 wire to #12-2 wire		133	.060		1.54	3.50		5.04	6.95
0300	#12-3 wire to #10-2 wire		114	.070		1.54	4.08		5.62	7.80
0400	#10-3 wire to #14-4 and #12-4 wire		100	.080		1.54	4.66		6.20	8.65
0500	#8-3 wire to #10-4 wire		80	.100		2.90	5.80		8.70	11.90
0600	#6-3 wire		40	.200		4.03	11.65		15.68	22
0800	SER, 3 #8 insulated + 1 #8 ground		32	.250		3.39	14.55		17.94	25.50
0900	3 #6 + 1 #6 ground		24	.333		4.24	19.40		23.64	33.50
1000	3 #4 + 1 #6 ground		22	.364		4.47	21		25.47	36.50
1100	3 #2 + 1 #4 ground		20	.400		8.35	23.50		31.85	44
1200	3 1/0 + 1 #2 ground		18	.444		17.95	26		43.95	58
1400	3 2/0 + 1 #1 ground		16	.500		24	29		53	70
1600	3 4/0 + 1 #2/0 ground		14	.571		28	33.50		61.50	80.50
1800	600 volt, armored, #14-2 wire		80	.100		1.23	5.80		7.03	10.05
2200	#14-4, #12-3 and #10-2 wire		40	.200		.92	11.65		12.57	18.40
2400	#12-4, #10-3 and #8-2 wire		32	.250		3.24	14.55		17.79	25.50
2600	#8-3 and #10-4 wire		26	.308		4.02	17.90		21.92	31.50
2650	#8-4 wire		22	.364		5.65	21		26.65	37.50
2652	Non-PVC jacket connector, #6-3 wire, #6-4 wire		22	.364		53	21		74	90
2660	1/0-3 wire		11	.727		70.50	42.50		113	142
2670	300 kcmil-3 wire		6	1.333		107	77.50		184.50	233
2680	700 kcmil-3 wire		4	2		165	116		281	355
2700	PVC jacket connector, #6-3 wire, #6-4 wire		16	.500		5.55	29		34.55	49.50
2800	#4-3 wire, #4-4 wire		16	.500		5.55	29		34.55	49.50
2900	#2-3 wire		12	.667		5.55	39		44.55	64
3000	#1-3 wire, #2-4 wire		12	.667		14.45	39		53.45	74
3200	1/0-3 wire		11	.727		14.45	42.50		56.95	79.50
3400	2/0-3 wire, 1/0-4 wire		10	.800		14.45	46.50		60.95	85.50
3500	3/0-3 wire, 2/0-4 wire		9	.889		16.80	51.50		68.30	96
3600	4/0-3 wire, 3/0-4 wire		7	1.143		16.80	66.50		83.30	118
3800	250 kcmil-3 wire, 4/0-4 wire		6	1.333		29.50	77.50		107	149
4000	350 kcmil-3 wire, 250 kcmil-4 wire		5	1.600		29.50	93		122.50	172
4100	350 kcmil-4 wire		4	2		216	116		332	410
4200	500 kcmil-3 wire		4	2		216	116		332	410
4250	500 kcmil-4 wire, 750 kcmil-3 wire		3.50	2.286		252	133		385	475
4300	750 kcmil-4 wire		3	2.667		252	155		407	510
4400	5 kV, armored, #4		8	1		68	58		126	162
4600	#2		8	1		68	58		126	162
4800	#1		8	1		87	58		145	183
5000	1/0		6.40	1.250		108	73		181	228
5200	2/0		5.30	1.509		108	88		196	250
5500	4/0		4	2		144	116		260	335
5600	250 kcmil		3.60	2.222		144	129		273	350
5650	350 kcmil		3.20	2.500		173	146		319	410
5700	500 kcmil		2.50	3.200		218	186		404	520
5720	750 kcmil		2.20	3.636		265	212		477	605

26 05 19 – Low-Voltage Electrical Power Conductors and Cables

26 05 19.25 Cable Connectors		Crew	Daily Output	Labor-Hours	Unit	Material	2018 Bare Costs Labor	Equipment	Total	Total Incl O&P
5750	1,000 kcmil	1 Elec	2	4	Ea.	325	233		558	710
5800	15 kV, armored, #1		4	2		130	116		246	315
5900	1/0		4	2		163	116		279	355
6000	3/0		3.60	2.222		251	129		380	470
6100	4/0		3.40	2.353		285	137		422	520
6200	250 kcmil		3.20	2.500		310	146		456	560
6300	350 kcmil		2.70	2.963		290	172		462	580
6400	500 kcmil		2	4		385	233		618	775

26 05 19.30 Cable Splicing

		Crew	Daily Output	Labor-Hours	Unit	Material	2018 Bare Costs Labor	Equipment	Total	Total Incl O&P
0010	**CABLE SPLICING** URD or similar, ideal conditions									
0100	#6 stranded to #1 stranded, 5 kV	1 Elec	4	2	Ea.	149	116		265	340
0120	15 kV		3.60	2.222		211	129		340	425
0140	25 kV		3.20	2.500		258	146		404	500
0200	#1 stranded to 4/0 stranded, 5 kV		3.60	2.222		157	129		286	365
0210	15 kV		3.20	2.500		228	146		374	470
0220	25 kV		2.80	2.857		258	166		424	530
0300	4/0 stranded to 500 kcmil stranded, 5 kV		3.30	2.424		157	141		298	385
0310	15 kV		2.90	2.759		370	161		531	645
0320	25 kV		2.50	3.200		435	186		621	760
0400	500 kcmil, 5 kV		3.20	2.500		200	146		346	435
0410	15 kV		2.80	2.857		370	166		536	660
0420	25 kV		2.30	3.478		435	202		637	785
0500	600 kcmil, 5 kV		2.90	2.759		200	161		361	460
0510	15 kV		2.40	3.333		455	194		649	790
0520	25 kV		2	4		495	233		728	895
0600	750 kcmil, 5 kV		2.60	3.077		200	179		379	485
0610	15 kV		2.20	3.636		455	212		667	815
0620	25 kV		1.90	4.211		495	245		740	910
0700	1,000 kcmil, 5 kV		2.30	3.478		200	202		402	525
0710	15 kV		1.90	4.211		495	245		740	910
0720	25 kV		1.60	5		530	291		821	1,025

26 05 19.35 Cable Terminations

		Crew	Daily Output	Labor-Hours	Unit	Material	2018 Bare Costs Labor	Equipment	Total	Total Incl O&P
0010	**CABLE TERMINATIONS**									
0015	Wire connectors, screw type, #22 to #14	1 Elec	260	.031	Ea.	.06	1.79		1.85	2.75
0020	#18 to #12		240	.033		.07	1.94		2.01	2.98
0035	#16 to #10		230	.035		.27	2.02		2.29	3.33
0040	#14 to #8		210	.038		.30	2.22		2.52	3.64
0045	#12 to #6		180	.044		.45	2.59		3.04	4.37
0050	Terminal lugs, solderless, #16 to #10		50	.160		.34	9.30		9.64	14.25
0100	#8 to #4		30	.267		.68	15.50		16.18	24
0150	#2 to #1		22	.364		.89	21		21.89	32.50
0200	1/0 to 2/0		16	.500		1.56	29		30.56	45
0250	3/0		12	.667		4.16	39		43.16	62.50
0300	4/0		11	.727		4.37	42.50		46.87	68.50
0350	250 kcmil		9	.889		3.29	51.50		54.79	81
0400	350 kcmil		7	1.143		4.24	66.50		70.74	104
0450	500 kcmil		6	1.333		7.35	77.50		84.85	124
0500	600 kcmil		5.80	1.379		8.30	80.50		88.80	129
0550	750 kcmil		5.20	1.538		10.90	89.50		100.40	146
0600	Split bolt connectors, tapped, #6		16	.500		3.15	29		32.15	47
0650	#4		14	.571		3.63	33.50		37.13	53.50
0700	#2		12	.667		5.50	39		44.50	64

For customer support on your Electrical Costs with RSMeans data, call 800.448.8182.

121

26 05 19 – Low-Voltage Electrical Power Conductors and Cables

26 05 19.35 Cable Terminations		Crew	Daily Output	Labor-Hours	Unit	Material	2018 Bare Costs Labor	Equipment	Total	Total Incl O&P
0750	#1	1 Elec	11	.727	Ea.	7.45	42.50		49.95	71.50
0800	1/0		10	.800		7.45	46.50		53.95	77.50
0850	2/0		9	.889		11.25	51.50		62.75	90
0900	3/0		7.20	1.111		25	64.50		89.50	124
1000	4/0		6.40	1.250		30.50	73		103.50	143
1100	250 kcmil		5.70	1.404		30.50	81.50		112	156
1200	300 kcmil		5.30	1.509		31.50	88		119.50	166
1400	350 kcmil		4.60	1.739		31.50	101		132.50	186
1500	500 kcmil	↓	4	2	↓	69.50	116		185.50	251
1600	Crimp 1 hole lugs, copper or aluminum, 600 volt									
1620	#14	1 Elec	60	.133	Ea.	.55	7.75		8.30	12.20
1630	#12		50	.160		1.09	9.30		10.39	15.10
1640	#10		45	.178		1.09	10.35		11.44	16.65
1780	#8		36	.222		1.43	12.95		14.38	21
1800	#6		30	.267		1.82	15.50		17.32	25
2000	#4		27	.296		3.14	17.25		20.39	29.50
2200	#2		24	.333		2.90	19.40		22.30	32
2400	#1		20	.400		5.25	23.50		28.75	41
2500	1/0		17.50	.457		5.95	26.50		32.45	46.50
2600	2/0		15	.533		7.65	31		38.65	55
2800	3/0		12	.667		6.85	39		45.85	65.50
3000	4/0		11	.727		6	42.50		48.50	70
3200	250 kcmil		9	.889		7.40	51.50		58.90	85.50
3400	300 kcmil		8	1		11.35	58		69.35	99.50
3500	350 kcmil		7	1.143		11.85	66.50		78.35	113
3600	400 kcmil		6.50	1.231		16.10	71.50		87.60	125
3800	500 kcmil		6	1.333		17.75	77.50		95.25	136
4000	600 kcmil		5.80	1.379		26.50	80.50		107	149
4200	700 kcmil		5.50	1.455		29	84.50		113.50	159
4400	750 kcmil	↓	5.20	1.538	↓	29	89.50		118.50	166
4500	Crimp 2-way connectors, copper or alum., 600 volt									
4510	#14	1 Elec	60	.133	Ea.	2.45	7.75		10.20	14.30
4520	#12		50	.160		2.60	9.30		11.90	16.75
4530	#10		45	.178		2.74	10.35		13.09	18.45
4540	#8		27	.296		2.77	17.25		20.02	29
4600	#6		25	.320		3.76	18.60		22.36	32
4800	#4		23	.348		4.06	20		24.06	35
5000	#2		20	.400		4.54	23.50		28.04	40
5200	#1		16	.500		5.90	29		34.90	50
5400	1/0		13	.615		5.60	36		41.60	59.50
5420	2/0		12	.667		5.95	39		44.95	64.50
5440	3/0		11	.727		7.45	42.50		49.95	71.50
5460	4/0		10	.800		7.80	46.50		54.30	78
5480	250 kcmil		9	.889		13.70	51.50		65.20	92.50
5500	300 kcmil		8.50	.941		16.30	55		71.30	100
5520	350 kcmil		8	1		16.45	58		74.45	105
5540	400 kcmil		7.30	1.096		21.50	64		85.50	119
5560	500 kcmil		6.20	1.290		26	75		101	141
5580	600 kcmil		5.50	1.455		46	84.50		130.50	178
5600	700 kcmil		4.50	1.778		52	103		155	212
5620	750 kcmil		4	2		54	116		170	234
7000	Compression equipment adapter, aluminum wire, #6		30	.267		7.75	15.50		23.25	31.50
7020	#4		27	.296		7.75	17.25		25	34.50

122

For customer support on your Electrical Costs with RSMeans data, call 800.448.8182.

26 05 19.35 Cable Terminations

		Crew	Daily Output	Labor-Hours	Unit	Material	2018 Bare Costs Labor	Equipment	Total	Total Incl O&P
7040	#2	1 Elec	24	.333	Ea.	6.45	19.40		25.85	36
7060	#1		20	.400		7.90	23.50		31.40	43.50
7080	1/0		18	.444		10.40	26		36.40	50
7100	2/0		15	.533		14.35	31		45.35	62.50
7140	4/0		11	.727		17.45	42.50		59.95	82.50
7160	250 kcmil		9	.889		15.85	51.50		67.35	95
7180	300 kcmil		8	1		18.90	58		76.90	108
7200	350 kcmil		7	1.143		24.50	66.50		91	127
7220	400 kcmil		6.50	1.231		31	71.50		102.50	141
7240	500 kcmil		6	1.333		30	77.50		107.50	149
7260	600 kcmil		5.80	1.379		36	80.50		116.50	160
7280	750 kcmil		5.20	1.538		42	89.50		131.50	181
8000	Compression tool, hand					1,350			1,350	1,475
8100	Hydraulic					1,775			1,775	1,950
8500	Hydraulic dies					330			330	365

26 05 19.50 Mineral Insulated Cable

		Crew	Daily Output	Labor-Hours	Unit	Material	2018 Bare Costs Labor	Equipment	Total	Total Incl O&P
0010	**MINERAL INSULATED CABLE** 600 volt									
0100	1 conductor, #12	1 Elec	1.60	5	C.L.F.	350	291		641	820
0200	#10		1.60	5		575	291		866	1,075
0400	#8		1.50	5.333		640	310		950	1,175
0500	#6		1.40	5.714		760	335		1,095	1,325
0600	#4	2 Elec	2.40	6.667		1,025	390		1,415	1,700
0800	#2		2.20	7.273		1,450	425		1,875	2,225
0900	#1		2.10	7.619		1,675	445		2,120	2,500
1000	1/0		2	8		1,950	465		2,415	2,850
1100	2/0		1.90	8.421		2,325	490		2,815	3,300
1200	3/0		1.80	8.889		2,800	515		3,315	3,850
1400	4/0		1.60	10		3,225	580		3,805	4,425
1410	250 kcmil	3 Elec	2.40	10		3,625	580		4,205	4,875
1420	350 kcmil		1.95	12.308		4,150	715		4,865	5,650
1430	500 kcmil		1.95	12.308		5,350	715		6,065	6,950
1500	2 conductor, #12	1 Elec	1.40	5.714		930	335		1,265	1,525
1600	#10		1.20	6.667		1,150	390		1,540	1,825
1800	#8		1.10	7.273		1,225	425		1,650	1,975
2000	#6		1.05	7.619		1,800	445		2,245	2,650
2100	#4	2 Elec	2	8		2,450	465		2,915	3,375
2200	3 conductor, #12	1 Elec	1.20	6.667		1,175	390		1,565	1,850
2400	#10		1.10	7.273		1,325	425		1,750	2,075
2600	#8		1.05	7.619		1,650	445		2,095	2,500
2800	#6		1	8		2,175	465		2,640	3,100
3000	#4	2 Elec	1.80	8.889		2,725	515		3,240	3,750
3100	4 conductor, #12	1 Elec	1.20	6.667		1,225	390		1,615	1,925
3200	#10		1.10	7.273		1,475	425		1,900	2,250
3400	#8		1	8		1,925	465		2,390	2,825
3600	#6		.90	8.889		2,500	515		3,015	3,525
3620	7 conductor, #12		1.10	7.273		1,575	425		2,000	2,350
3640	#10		1	8		2,050	465		2,515	2,950
3800	Terminations, 600 volt, 1 conductor, #12		8	1	Ea.	19.65	58		77.65	109
4000	#10		7.60	1.053		19.65	61.50		81.15	113
4100	#8		7.30	1.096		19.65	64		83.65	117
4200	#6		6.70	1.194		19.65	69.50		89.15	126
4400	#4		6.20	1.290		19.65	75		94.65	134

For customer support on your Electrical Costs with RSMeans data, call 800.448.8182.

123

26 05 19.50 Mineral Insulated Cable

		Crew	Daily Output	Labor-Hours	Unit	Material	2018 Bare Costs Labor	Equipment	Total	Total Incl O&P
4600	#2	1 Elec	5.70	1.404	Ea.	29.50	81.50		111	155
4800	#1		5.30	1.509		29.50	88		117.50	164
5000	1/0		5	1.600		29.50	93		122.50	172
5100	2/0		4.70	1.702		29.50	99		128.50	181
5200	3/0		4.30	1.860		29.50	108		137.50	195
5400	4/0		4	2		65.50	116		181.50	246
5410	250 kcmil		4	2		65.50	116		181.50	246
5420	350 kcmil		4	2		103	116		219	287
5430	500 kcmil		4	2		103	116		219	287
5500	2 conductor, #12		6.70	1.194		19.65	69.50		89.15	126
5600	#10		6.40	1.250		29.50	73		102.50	142
5800	#8		6.20	1.290		29.50	75		104.50	145
6000	#6		5.70	1.404		29.50	81.50		111	155
6200	#4		5.30	1.509		65.50	88		153.50	203
6400	3 conductor, #12		5.70	1.404		29.50	81.50		111	155
6500	#10		5.50	1.455		29.50	84.50		114	160
6600	#8		5.20	1.538		29.50	89.50		119	167
6800	#6		4.80	1.667		29.50	97		126.50	178
7200	#4		4.60	1.739		65.50	101		166.50	223
7400	4 conductor, #12		4.60	1.739		33	101		134	187
7500	#10		4.40	1.818		33	106		139	194
7600	#8		4.20	1.905		33	111		144	202
8400	#6		4	2		69	116		185	250
8500	7 conductor, #12		3.50	2.286		33	133		166	235
8600	#10		3	2.667		71	155		226	310
8800	Crimping tool, plier type					73.50			73.50	81
9000	Stripping tool					278			278	305
9200	Hand vise					83			83	91

26 05 19.55 Non-Metallic Sheathed Cable

		Crew	Daily Output	Labor-Hours	Unit	Material	2018 Bare Costs Labor	Equipment	Total	Total Incl O&P
0010	**NON-METALLIC SHEATHED CABLE** 600 volt									
0100	Copper with ground wire (Romex)									
0150	#14, 2 conductor	1 Elec	2.70	2.963	C.L.F.	18.35	172		190.35	278
0200	3 conductor		2.40	3.333		25.50	194		219.50	320
0220	4 conductor		2.20	3.636		39.50	212		251.50	360
0250	#12, 2 conductor		2.50	3.200		24	186		210	305
0300	3 conductor		2.20	3.636		40	212		252	360
0320	4 conductor		2	4		60	233		293	415
0350	#10, 2 conductor		2.20	3.636		41	212		253	360
0400	3 conductor		1.80	4.444		62.50	259		321.50	455
0420	4 conductor		1.60	5		90	291		381	535
0430	#8, 2 conductor		1.60	5		75.50	291		366.50	520
0450	3 conductor		1.50	5.333		117	310		427	595
0500	#6, 3 conductor		1.40	5.714		197	335		532	710
0520	#4, 3 conductor	2 Elec	2.40	6.667		405	390		795	1,025
0540	#2, 3 conductor	"	2.20	7.273		785	425		1,210	1,500
0550	SE type SER aluminum cable, 3 RHW and									
0600	1 bare neutral, 3 #8 & 1 #8	1 Elec	1.60	5	C.L.F.	56.50	291		347.50	495
0650	3 #6 & 1 #6	"	1.40	5.714		67	335		402	570
0700	3 #4 & 1 #6	2 Elec	2.40	6.667		71.50	390		461.50	660
0750	3 #2 & 1 #4		2.20	7.273		124	425		549	770
0800	3 #1/0 & 1 #2		2	8		182	465		647	895
0850	3 #2/0 & 1 #1		1.80	8.889		212	515		727	1,000

26 05 19.55 Non-Metallic Sheathed Cable	Crew	Daily Output	Labor-Hours	Unit	Material	2018 Bare Costs Labor	Equipment	Total	Total Incl O&P
0900	3 #4/0 & 1 #2/0	2 Elec	1.60	10	C.L.F.	295	580		875	1,200
1000	URD - triplex underground distribution cable, alum. 2 #4 + #4 neutral		2.80	5.714		77	335		412	580
1010	2 #2 + #4 neutral		2.65	6.038		106	350		456	640
1020	2 #2 + #2 neutral		2.55	6.275		96.50	365		461.50	650
1030	2 1/0 + #2 neutral		2.40	6.667		116	390		506	705
1040	2 1/0 + 1/0 neutral		2.30	6.957		113	405		518	730
1050	2 2/0 + #1 neutral		2.20	7.273		132	425		557	780
1060	2 2/0 + 2/0 neutral		2.10	7.619		148	445		593	830
1070	2 3/0 + 1/0 neutral		1.95	8.205		164	480		644	895
1080	2 3/0 + 3/0 neutral		1.95	8.205		169	480		649	900
1090	2 4/0 + 2/0 neutral		1.85	8.649		186	505		691	955
1100	2 4/0 + 4/0 neutral		1.85	8.649		206	505		711	975
1450	UF underground feeder cable, copper with ground, #14, 2 conductor	1 Elec	4	2		41	116		157	219
1500	#12, 2 conductor		3.50	2.286		34	133		167	237
1550	#10, 2 conductor		3	2.667		53	155		208	290
1600	#14, 3 conductor		3.50	2.286		37.50	133		170.50	241
1650	#12, 3 conductor		3	2.667		53.50	155		208.50	291
1700	#10, 3 conductor		2.50	3.200		84.50	186		270.50	370
1710	#8, 3 conductor		2	4		136	233		369	500
1720	#6, 3 conductor		1.80	4.444		225	259		484	635
2400	SEU service entrance cable, copper 2 conductors, #8 + #8 neutral		1.50	5.333		101	310		411	575
2600	#6 + #8 neutral		1.30	6.154		127	360		487	675
2800	#6 + #6 neutral		1.30	6.154		149	360		509	700
3000	#4 + #6 neutral	2 Elec	2.20	7.273		195	425		620	850
3200	#4 + #4 neutral		2.20	7.273		213	425		638	870
3400	#3 + #5 neutral		2.10	7.619		335	445		780	1,025
3600	#3 + #3 neutral		2.10	7.619		264	445		709	955
3800	#2 + #4 neutral		2	8		294	465		759	1,025
4000	#1 + #1 neutral		1.90	8.421		340	490		830	1,100
4200	1/0 + 1/0 neutral		1.80	8.889		955	515		1,470	1,825
4400	2/0 + 2/0 neutral		1.70	9.412		655	550		1,205	1,550
4600	3/0 + 3/0 neutral		1.60	10		805	580		1,385	1,750
4620	4/0 + 4/0 neutral		1.45	11.034		2,150	640		2,790	3,300
4800	Aluminum 2 conductors, #8 + #8 neutral	1 Elec	1.60	5		50	291		341	490
5000	#6 + #6 neutral	"	1.40	5.714		57.50	335		392.50	560
5100	#4 + #6 neutral	2 Elec	2.50	6.400		66.50	370		436.50	630
5200	#4 + #4 neutral		2.40	6.667		65	390		455	650
5300	#2 + #4 neutral		2.30	6.957		80.50	405		485.50	695
5400	#2 + #2 neutral		2.20	7.273		93.50	425		518.50	740
5450	1/0 + #2 neutral		2.10	7.619		124	445		569	800
5500	1/0 + 1/0 neutral		2	8		130	465		595	840
5550	2/0 + #1 neutral		1.90	8.421		137	490		627	885
5600	2/0 + 2/0 neutral		1.80	8.889		146	515		661	935
5800	3/0 + 1/0 neutral		1.70	9.412		216	550		766	1,050
6000	3/0 + 3/0 neutral		1.70	9.412		186	550		736	1,025
6200	4/0 + 2/0 neutral		1.60	10		191	580		771	1,075
6400	4/0 + 4/0 neutral		1.60	10		207	580		787	1,100
6500	Service entrance cap for copper SEU									
6600	100 amp	1 Elec	12	.667	Ea.	8.10	39		47.10	67
6700	150 amp		10	.800		10.90	46.50		57.40	81.50
6800	200 amp		8	1		17.80	58		75.80	107

For customer support on your Electrical Costs with RSMeans data, call 800.448.8182.

125

26 05 19.70 Portable Cord

		Crew	Daily Output	Labor-Hours	Unit	Material	2018 Bare Costs Labor	2018 Bare Costs Equipment	Total	Total Incl O&P
0010	**PORTABLE CORD** 600 volt									
0100	Type SO, #18, 2 conductor	1 Elec	980	.008	L.F.	.29	.47		.76	1.03
0110	3 conductor		980	.008		.33	.47		.80	1.07
0120	#16, 2 conductor		840	.010		.38	.55		.93	1.25
0130	3 conductor		840	.010		4.07	.55		4.62	5.30
0140	4 conductor		840	.010		.58	.55		1.13	1.47
0240	#14, 2 conductor		840	.010		.53	.55		1.08	1.41
0250	3 conductor		840	.010		.72	.55		1.27	1.62
0260	4 conductor		840	.010		.91	.55		1.46	1.83
0280	#12, 2 conductor		840	.010		.78	.55		1.33	1.69
0290	3 conductor		840	.010		.90	.55		1.45	1.82
0300	4 conductor		840	.010		1.48	.55		2.03	2.46
0320	#10, 2 conductor		765	.010		.84	.61		1.45	1.83
0330	3 conductor		765	.010		1.26	.61		1.87	2.30
0340	4 conductor		765	.010		1.29	.61		1.90	2.33
0360	#8, 2 conductor		555	.014		1.84	.84		2.68	3.27
0370	3 conductor		540	.015		2.29	.86		3.15	3.81
0380	4 conductor		525	.015		2.30	.89		3.19	3.86
0400	#6, 2 conductor		525	.015		2.64	.89		3.53	4.23
0410	3 conductor		490	.016		3.20	.95		4.15	4.94
0420	4 conductor		415	.019		3.50	1.12		4.62	5.55
0440	#4, 2 conductor	2 Elec	830	.019		7.95	1.12		9.07	10.45
0450	3 conductor		700	.023		5.80	1.33		7.13	8.40
0460	4 conductor		660	.024		5.90	1.41		7.31	8.55
0480	#2, 2 conductor		450	.036		5.05	2.07		7.12	8.65
0490	3 conductor		350	.046		6.85	2.66		9.51	11.55
0500	4 conductor		280	.057		8.15	3.33		11.48	13.95
2000	See 26 27 26.20 for Wiring Devices Elements									

26 05 19.75 Modular Flexible Wiring System

		Crew	Daily Output	Labor-Hours	Unit	Material	2018 Bare Costs Labor	2018 Bare Costs Equipment	Total	Total Incl O&P
0010	**MODULAR FLEXIBLE WIRING SYSTEM**									
0020	Commercial system grid ceiling									
0100	Conversion Module	1 Elec	32	.250	Ea.	7.90	14.55		22.45	30.50
0120	Fixture cable, for fixture to fixture, 3 conductor, 15' long		40	.200		34	11.65		45.65	55
0150	Extender cable, 3 conductor, 15' long		48	.167		32	9.70		41.70	49.50
0200	Switch drop, 1 level, 9' long		32	.250		22.50	14.55		37.05	46.50
0220	2 level, 9' long		32	.250		30	14.55		44.55	55.50
0250	Power tee, 9' long		32	.250		28.50	14.55		43.05	53
1020	Industrial system open ceiling									
1100	Converter, interface between hardwiring and modular wiring	1 Elec	32	.250	Ea.	17.20	14.55		31.75	41
1120	Fixture cable, for fixture to fixture, 3 conductor, 21' long		32	.250		110	14.55		124.55	143
1125	3 conductor, 25' long		24	.333		127	19.40		146.40	169
1130	3 conductor, 31' long		19	.421		113	24.50		137.50	161
1150	Fixture cord drop, 10' long		40	.200		30.50	11.65		42.15	51

26 05 19.90 Wire

		Crew	Daily Output	Labor-Hours	Unit	Material	2018 Bare Costs Labor	2018 Bare Costs Equipment	Total	Total Incl O&P
0010	**WIRE**, normal installation conditions in wireway, conduit, cable tray									
0020	600 volt, copper type THW, solid, #14 R260519-90	1 Elec	13	.615	C.L.F.	5.90	36		41.90	60
0030	#12		11	.727		9.65	42.50		52.15	74
0040	#10		10	.800		15.75	46.50		62.25	87
0050	Stranded, #14		13	.615		8.60	36		44.60	63
0100	#12		11	.727		13.20	42.50		55.70	78
0120	#10 R260519-92		10	.800		21	46.50		67.50	92.50
0140	#8		8	1		34.50	58		92.50	125

26 05 19.90 Wire		Crew	Daily Output	Labor-Hours	Unit	Material	2018 Bare Costs Labor	Equipment	Total	Total Incl O&P
0160	#6	R260519-93	1 Elec	6.50	1.231	C.L.F.	43.50	71.50		115	155
0180	#4		2 Elec	10.60	1.509		69.50	88		157.50	207
0200	#3	R260519-94		10	1.600		105	93		198	254
0220	#2			9	1.778		102	103		205	267
0240	#1	R260533-22		8	2		136	116		252	325
0260	1/0			6.60	2.424		216	141		357	450
0280	2/0			5.80	2.759		205	161		366	465
0300	3/0			5	3.200		235	186		421	535
0350	4/0			4.40	3.636		435	212		647	790
0400	250 kcmil		3 Elec	6	4		350	233		583	735
0420	300 kcmil			5.70	4.211		380	245		625	780
0450	350 kcmil			5.40	4.444		465	259		724	895
0480	400 kcmil			5.10	4.706		540	274		814	1,000
0490	500 kcmil			4.80	5		690	291		981	1,200
0500	600 kcmil			3.90	6.154		1,150	360		1,510	1,800
0510	750 kcmil			3.30	7.273		1,325	425		1,750	2,075
0520	1,000 kcmil			2.70	8.889		1,950	515		2,465	2,900
0540	600 volt, aluminum type THHN, stranded, #6		1 Elec	8	1		24	58		82	114
0560	#4		2 Elec	13	1.231		30	71.50		101.50	140
0580	#2			10.60	1.509		41	88		129	176
0600	#1			9	1.778		60	103		163	221
0620	1/0			8	2		72	116		188	254
0640	2/0			7.20	2.222		86	129		215	288
0680	3/0			6.60	2.424		107	141		248	330
0700	4/0			6.20	2.581		119	150		269	355
0720	250 kcmil		3 Elec	8.70	2.759		144	161		305	400
0740	300 kcmil			8.10	2.963		199	172		371	475
0760	350 kcmil			7.50	3.200		201	186		387	500
0780	400 kcmil			6.90	3.478		229	202		431	555
0800	500 kcmil			6	4		258	233		491	635
0850	600 kcmil			5.70	4.211		325	245		570	725
0880	700 kcmil			5.10	4.706		400	274		674	850
0900	750 kcmil			4.80	5		400	291		691	875
0910	1,000 kcmil			3.78	6.349		595	370		965	1,200
0920	600 volt, copper type THWN-THHN, solid, #14		1 Elec	13	.615		6.55	36		42.55	60.50
0940	#12			11	.727		9.60	42.50		52.10	74
0960	#10			10	.800		14.25	46.50		60.75	85
1000	Stranded, #14			13	.615		7.35	36		43.35	61.50
1200	#12			11	.727		10.50	42.50		53	75
1250	#10			10	.800		16.10	46.50		62.60	87
1300	#8			8	1		28.50	58		86.50	118
1350	#6			6.50	1.231		48.50	71.50		120	160
1400	#4		2 Elec	10.60	1.509		69	88		157	207
1450	#3			10	1.600		93.50	93		186.50	242
1500	#2			9	1.778		117	103		220	283
1550	#1			8	2		151	116		267	340
1600	1/0			6.60	2.424		195	141		336	425
1650	2/0			5.80	2.759		221	161		382	485
1700	3/0			5	3.200		283	186		469	590
2000	4/0			4.40	3.636		350	212		562	700
2200	250 kcmil		3 Elec	6	4		430	233		663	825
2400	300 kcmil			5.70	4.211		505	245		750	920
2600	350 kcmil			5.40	4.444		590	259		849	1,025

26 05 19.90 Wire		Crew	Daily Output	Labor-Hours	Unit	Material	2018 Bare Costs Labor	Equipment	Total	Total Incl O&P
2700	400 kcmil	3 Elec	5.10	4.706	C.L.F.	675	274		949	1,150
2800	500 kcmil		4.80	5		840	291		1,131	1,350
2802	600 kcmil		3.90	6.154		755	360		1,115	1,375
2804	750 kcmil		3.30	7.273		1,400	425		1,825	2,150
2805	1,000 kcmil		1.94	12.371		2,075	720		2,795	3,350
2900	600 volt, copper type XHHW, solid, #14	1 Elec	13	.615		11.75	36		47.75	66.50
2920	#12		11	.727		18.40	42.50		60.90	83.50
2940	#10		10	.800		28.50	46.50		75	101
3000	Stranded, #14		13	.615		9.15	36		45.15	63.50
3020	#12		11	.727		13.80	42.50		56.30	78.50
3040	#10		10	.800		20.50	46.50		67	92
3060	#8		8	1		31	58		89	121
3080	#6		6.50	1.231		51.50	71.50		123	164
3100	#4	2 Elec	10.60	1.509		79	88		167	218
3110	#3		10	1.600		98	93		191	247
3120	#2		9	1.778		123	103		226	290
3140	#1		8	2		168	116		284	360
3160	1/0		6.60	2.424		208	141		349	440
3180	2/0		5.80	2.759		261	161		422	525
3200	3/0		5	3.200		325	186		511	640
3220	4/0		4.40	3.636		410	212		622	765
3240	250 kcmil	3 Elec	6	4		455	233		688	850
3260	300 kcmil		5.70	4.211		510	245		755	930
3280	350 kcmil		5.40	4.444		600	259		859	1,050
3300	400 kcmil		5.10	4.706		680	274		954	1,150
3320	500 kcmil		4.80	5		850	291		1,141	1,375
3340	600 kcmil		3.90	6.154		925	360		1,285	1,550
3360	750 kcmil		3.30	7.273		1,550	425		1,975	2,350
3380	1,000 kcmil		2.40	10		1,975	580		2,555	3,050
5020	600 volt, aluminum type XHHW, stranded, #6	1 Elec	8	1		24	58		82	114
5040	#4	2 Elec	13	1.231		30	71.50		101.50	140
5060	#2		10.60	1.509		41	88		129	176
5080	#1		9	1.778		59.50	103		162.50	221
5100	1/0		8	2		70.50	116		186.50	252
5120	2/0		7.20	2.222		88.50	129		217.50	291
5140	3/0		6.60	2.424		104	141		245	325
5160	4/0		6.20	2.581		115	150		265	350
5180	250 kcmil	3 Elec	8.70	2.759		148	161		309	405
5200	300 kcmil		8.10	2.963		193	172		365	470
5220	350 kcmil		7.50	3.200		203	186		389	500
5240	400 kcmil		6.90	3.478		228	202		430	555
5260	500 kcmil		6	4		249	233		482	625
5280	600 kcmil		5.70	4.211		325	245		570	720
5300	700 kcmil		5.40	4.444		365	259		624	785
5320	750 kcmil		5.10	4.706		370	274		644	815
5340	1,000 kcmil		3.60	6.667		545	390		935	1,175
5390	600 volt, copper type XLPE-USE (RHW), solid, #14	1 Elec	12	.667		13.45	39		52.45	73
5400	#12		11	.727		17.35	42.50		59.85	82.50
5420	#10		10	.800		23	46.50		69.50	94.50
5440	Stranded, #14		13	.615		15.15	36		51.15	70
5460	#12		11	.727		18.15	42.50		60.65	83.50
5480	#10		10	.800		29.50	46.50		76	102
5500	#8		8	1		34	58		92	125

26 05 19 – Low-Voltage Electrical Power Conductors and Cables

26 05 19.90 Wire		Crew	Daily Output	Labor-Hours	Unit	Material	2018 Bare Costs Labor	Equipment	Total	Total Incl O&P
5520	#6	1 Elec	6.50	1.231	C.L.F.	57	71.50		128.50	170
5540	#4	2 Elec	10.60	1.509		86.50	88		174.50	226
5560	#2		9	1.778		141	103		244	310
5580	#1		8	2		190	116		306	385
5600	1/0		6.60	2.424		237	141		378	470
5620	2/0		5.80	2.759		315	161		476	585
5640	3/0		5	3.200		390	186		576	710
5660	4/0		4.40	3.636		425	212		637	780
5680	250 kcmil	3 Elec	6	4		495	233		728	895
5700	300 kcmil		5.70	4.211		615	245		860	1,050
5720	350 kcmil		5.40	4.444		695	259		954	1,150
5740	400 kcmil		5.10	4.706		795	274		1,069	1,275
5760	500 kcmil		4.80	5		835	291		1,126	1,350
5780	600 kcmil		3.90	6.154		1,450	360		1,810	2,125
5800	750 kcmil		3.30	7.273		1,675	425		2,100	2,475
5820	1,000 kcmil		2.70	8.889		2,225	515		2,740	3,225
5840	600 volt, aluminum type XLPE-USE (RHW), stranded, #6	1 Elec	8	1		30	58		88	120
5860	#4	2 Elec	13	1.231		35.50	71.50		107	146
5880	#2		10.60	1.509		50.50	88		138.50	187
5900	#1		9	1.778		72	103		175	234
5920	1/0		8	2		83	116		199	265
5940	2/0		7.20	2.222		97	129		226	300
5960	3/0		6.60	2.424		114	141		255	335
5980	4/0		6.20	2.581		127	150		277	365
6000	250 kcmil	3 Elec	8.70	2.759		172	161		333	430
6020	300 kcmil		8.10	2.963		223	172		395	505
6040	350 kcmil		7.50	3.200		225	186		411	525
6060	400 kcmil		6.90	3.478		279	202		481	610
6080	500 kcmil		6	4		305	233		538	685
6100	600 kcmil		5.70	4.211		390	245		635	795
6110	700 kcmil		5.40	4.444		415	259		674	840
6120	750 kcmil		5.10	4.706		455	274		729	915

26 05 23 – Control-Voltage Electrical Power Cables

26 05 23.10 Control Cable

		Crew	Daily Output	Labor-Hours	Unit	Material	2018 Bare Costs Labor	Equipment	Total	Total Incl O&P
0010	**CONTROL CABLE**									
0020	600 volt, copper, #14 THWN wire with PVC jacket, 2 wires	1 Elec	9	.889	C.L.F.	37.50	51.50		89	119
0030	3 wires		8	1		51	58		109	143
0100	4 wires		7	1.143		54.50	66.50		121	160
0150	5 wires		6.50	1.231		75	71.50		146.50	189
0200	6 wires		6	1.333		103	77.50		180.50	229
0300	8 wires		5.30	1.509		133	88		221	277
0400	10 wires		4.80	1.667		154	97		251	315
0500	12 wires		4.30	1.860		199	108		307	380
0600	14 wires		3.80	2.105		214	123		337	420
0700	16 wires		3.50	2.286		242	133		375	465
0800	18 wires		3.30	2.424		272	141		413	510
0810	19 wires		3.10	2.581		287	150		437	540
0900	20 wires		3	2.667		300	155		455	565
1000	22 wires		2.80	2.857		335	166		501	615

For customer support on your Electrical Costs with RSMeans data, call 800.448.8182.

129

26 05 23.20 Special Wires and Fittings	Crew	Daily Output	Labor-Hours	Unit	Material	2018 Bare Costs Labor	Equipment	Total	Total Incl O&P
0010 **SPECIAL WIRES & FITTINGS**									
0100 Fixture TFFN 600 volt 90°C stranded, #18	1 Elec	13	.615	C.L.F.	9.50	36		45.50	64
0150 #16		13	.615		12.85	36		48.85	67.50
0500 Thermostat, jacket non-plenum, twisted, #18-2 conductor		8	1		10	58		68	98
0550 #18-3 conductor		7	1.143		12.30	66.50		78.80	113
0600 #18-4 conductor		6.50	1.231		18.15	71.50		89.65	127
0650 #18-5 conductor		6	1.333		21.50	77.50		99	140
0700 #18-6 conductor		5.50	1.455		25.50	84.50		110	156
0750 #18-7 conductor		5	1.600		31.50	93		124.50	174
0800 #18-8 conductor		4.80	1.667		32.50	97		129.50	181
2460 Tray cable, type TC, copper, #16-2 conductor		9.40	.851		28	49.50		77.50	105
2464 #16-3 conductor		8.40	.952		45	55.50		100.50	133
2468 #16-4 conductor		7.30	1.096		44.50	64		108.50	145
2472 #16-5 conductor		6.70	1.194		43	69.50		112.50	151
2476 #16-7 conductor		5.50	1.455		58.50	84.50		143	192
2480 #16-9 conductor		5	1.600		73	93		166	220
2484 #16-12 conductor	2 Elec	8.80	1.818		118	106		224	288
2488 #16-15 conductor		7.20	2.222		128	129		257	335
2492 #16-19 conductor		6.60	2.424		158	141		299	385
2496 #16-25 conductor		6.40	2.500		222	146		368	465
2500 #14-2 conductor	1 Elec	9	.889		28	51.50		79.50	109
2520 #14-3 conductor		8	1		39.50	58		97.50	131
2540 #14-4 conductor		7	1.143		53	66.50		119.50	158
2560 #14-5 conductor		6.50	1.231		55.50	71.50		127	168
2564 #14-7 conductor		5.30	1.509		91.50	88		179.50	232
2568 #14-9 conductor	2 Elec	9.60	1.667		109	97		206	265
2572 #14-12 conductor		8.60	1.860		141	108		249	315
2576 #14-15 conductor		7.20	2.222		175	129		304	385
2578 #14-19 conductor		6.20	2.581		220	150		370	465
2582 #14-25 conductor		5.30	3.019		288	176		464	580
2590 #12-2 conductor	1 Elec	8.40	.952		48	55.50		103.50	136
2592 #12-3 conductor		7.60	1.053		63	61.50		124.50	161
2594 #12-4 conductor		6.60	1.212		91.50	70.50		162	205
2596 #12-5 conductor		6.20	1.290		106	75		181	228
2598 #12-7 conductor	2 Elec	10.40	1.538		114	89.50		203.50	260
2602 #12-9 conductor		8.80	1.818		208	106		314	385
2604 #12-12 conductor		7.80	2.051		226	119		345	425
2606 #12-15 conductor		6.80	2.353		263	137		400	495
2608 #12-19 conductor		6	2.667		320	155		475	585
2610 #12-25 conductor	3 Elec	7.70	3.117		435	181		616	750
2618 #10-2 conductor	1 Elec	8	1		74	58		132	168
2622 #10-3 conductor	"	7.30	1.096		84.50	64		148.50	189
2624 #10-4 conductor	2 Elec	12.80	1.250		141	73		214	264
2626 #10-5 conductor		11.80	1.356		136	79		215	268
2628 #10-7 conductor		9.40	1.702		197	99		296	365
2630 #10-9 conductor		8.40	1.905		229	111		340	420
2632 #10-12 conductor		7.20	2.222		245	129		374	460
2640 300 V, copper braided shield, PVC jacket									
2650 2 conductor #18 stranded	1 Elec	7	1.143	C.L.F.	26	66.50		92.50	128
2660 3 conductor #18 stranded		6	1.333		39.50	77.50		117	159
2670 4 conductor #18 stranded		6	1.333		161	77.50		238.50	293
3000 Strain relief grip for cable									

26 05 23 – Control-Voltage Electrical Power Cables

26 05 23.20 Special Wires and Fittings	Crew	Daily Output	Labor-Hours	Unit	Material	2018 Bare Costs Labor	Equipment	Total	Total Incl O&P
3050	Cord, top, #12-3	1 Elec	40	.200	Ea.	9.50	11.65		21.15	28
3060	#12-4		40	.200		9.50	11.65		21.15	28
3070	#12-5		39	.205		11.70	11.95		23.65	30.50
3100	#10-3		39	.205		11.70	11.95		23.65	30.50
3110	#10-4		38	.211		11.70	12.25		23.95	31
3120	#10-5		38	.211		13.95	12.25		26.20	33.50
3200	Bottom, #12-3		40	.200		23.50	11.65		35.15	43.50
3210	#12-4		40	.200		23.50	11.65		35.15	43.50
3220	#12-5		39	.205		24.50	11.95		36.45	45
3230	#10-3		39	.205		24.50	11.95		36.45	45
3300	#10-4		38	.211		24.50	12.25		36.75	45.50
3310	#10-5		38	.211		30.50	12.25		42.75	52
3400	Cable ties, standard, 4" length		190	.042		.15	2.45		2.60	3.83
3410	7" length		160	.050		.18	2.91		3.09	4.55
3420	14.5" length		90	.089		.41	5.15		5.56	8.20
3430	Heavy, 14.5" length		80	.100		.61	5.80		6.41	9.35
3500	Cable gland, nylon, 1/4 NPT thread		60	.133		1.99	7.75		9.74	13.80
3510	Cable gland, nylon, 3/8 NPT thread		60	.133		2.29	7.75		10.04	14.10
3520	Cable gland, nylon, 1/2 NPT thread		60	.133		2.87	7.75		10.62	14.75
3530	Cable gland, nylon, 3/4 NPT thread		60	.133		4.29	7.75		12.04	16.30
3540	Cable gland, nylon, 1" NPT thread		60	.133		7.20	7.75		14.95	19.50
3550	Cable gland, nylon, 1-1/4 NPT thread		60	.133		8	7.75		15.75	20.50
3970	U-cable guards									
3980	.75" x 5' U-cable guard,14 gauge	1 Elec	24	.333	Ea.	11.05	19.40		30.45	41
3990	.75" x 8' U-cable guard,14 gauge		24	.333		17.55	19.40		36.95	48.50
4000	1 1/8" x 5' U-cable guard		24	.333		13.45	19.40		32.85	44
4010	1 1/8" x 8' U-cable guard		24	.333		18.90	19.40		38.30	50
4020	2 3/16" x 5' U-cable guard		24	.333		26	19.40		45.40	57.50
4030	2 3/16" x 8' U-cable guard		24	.333		27	19.40		46.40	58.50
4040	3 3/16" x 5' U-cable guard		24	.333		32	19.40		51.40	64.50
4050	3 3/16" x 8' U-cable guard		24	.333		35.50	19.40		54.90	68
4060	4" x 8' U-cable guard		24	.333		43	19.40		62.40	76.50
4070	2" x 9' flanged-cable guard		20	.400		78.50	23.50		102	122
4080	3" x 9' flanged-cable guard		20	.400		89	23.50		112.50	133
4090	3.5" x 9' flanged-cable guard		20	.400		113	23.50		136.50	159
5000	2" x 5' flanged-cable guard, extension		20	.400		65.50	23.50		89	107
5010	3" x 5' flanged-cable guard, extension		20	.400		96.50	23.50		120	141
5020	.05" x 8' U-cable guard, black plastic		26	.308		4.01	17.90		21.91	31.50
5030	.75" x 8' U-cable guard, black plastic		26	.308		7.10	17.90		25	35
5040	1" x 8' U-cable guard, black plastic		26	.308		10.50	17.90		28.40	38.50

26 05 26 – Grounding and Bonding for Electrical Systems

26 05 26.80 Grounding

		Crew	Daily Output	Labor-Hours	Unit	Material	2018 Bare Costs Labor	Equipment	Total	Total Incl O&P
0010	**GROUNDING** R260526-80									
0030	Rod, copper clad, 8' long, 1/2" diameter	1 Elec	5.50	1.455	Ea.	19.90	84.50		104.40	149
0040	5/8" diameter		5.50	1.455		21.50	84.50		106	151
0050	3/4" diameter		5.30	1.509		35	88		123	170
0080	10' long, 1/2" diameter		4.80	1.667		22.50	97		119.50	170
0090	5/8" diameter		4.60	1.739		24	101		125	177
0100	3/4" diameter		4.40	1.818		42	106		148	205
0130	15' long, 3/4" diameter		4	2		49	116		165	228
0150	Coupling, bronze, 1/2" diameter					4.56			4.56	5
0160	5/8" diameter					5.30			5.30	5.85

For customer support on your Electrical Costs with RSMeans data, call 800.448.8182.

131

26 05 26.80 Grounding		Crew	Daily Output	Labor-Hours	Unit	Material	2018 Bare Costs Labor	Equipment	Total	Total Incl O&P
0170	3/4" diameter				Ea.	14.30			14.30	15.70
0190	Drive studs, 1/2" diameter					11.30			11.30	12.45
0210	5/8" diameter					19.10			19.10	21
0220	3/4" diameter					20.50			20.50	22.50
0230	Clamp, bronze, 1/2" diameter	1 Elec	32	.250		4.68	14.55		19.23	27
0240	5/8" diameter		32	.250		4.81	14.55		19.36	27.50
0250	3/4" diameter		32	.250		5.60	14.55		20.15	28
0260	Wire ground bare armored, #8-1 conductor		2	4	C.L.F.	69.50	233		302.50	425
0270	#6-1 conductor		1.80	4.444		84.50	259		343.50	480
0280	#4-1 conductor		1.60	5		134	291		425	580
0320	Bare copper wire, #14 solid		14	.571		5.95	33.50		39.45	56
0330	#12		13	.615		14.10	36		50.10	69
0340	#10		12	.667		19.50	39		58.50	79.50
0350	#8		11	.727		24.50	42.50		67	90
0360	#6		10	.800		38.50	46.50		85	112
0370	#4		8	1		99	58		157	196
0380	#2		5	1.600		127	93		220	279
0390	Bare copper wire, stranded, #8		11	.727		43	42.50		85.50	111
0400	#6		10	.800		41	46.50		87.50	115
0450	#4	2 Elec	16	1		84.50	58		142.50	180
0600	#2		10	1.600		89	93		182	237
0650	#1		9	1.778		128	103		231	295
0700	1/0		8	2		145	116		261	335
0750	2/0		7.20	2.222		259	129		388	480
0800	3/0		6.60	2.424		315	141		456	555
1000	4/0		5.70	2.807		440	163		603	730
1200	250 kcmil	3 Elec	7.20	3.333		445	194		639	780
1210	300 kcmil		6.60	3.636		505	212		717	870
1220	350 kcmil		6	4		465	233		698	860
1230	400 kcmil		5.70	4.211		540	245		785	960
1240	500 kcmil		5.10	4.706		715	274		989	1,200
1260	750 kcmil		3.60	6.667		1,150	390		1,540	1,850
1270	1,000 kcmil		3	8		1,500	465		1,965	2,350
1360	Bare aluminum, stranded, #6	1 Elec	9	.889		14.60	51.50		66.10	93.50
1370	#4	2 Elec	16	1		29.50	58		87.50	120
1380	#2		13	1.231		40.50	71.50		112	152
1390	#1		10.60	1.509		42	88		130	177
1400	1/0		9	1.778		53.50	103		156.50	214
1410	2/0		8	2		67.50	116		183.50	249
1420	3/0		7.20	2.222		91	129		220	293
1430	4/0		6.60	2.424		96.50	141		237.50	315
1440	250 kcmil	3 Elec	9.30	2.581		120	150		270	355
1450	300 kcmil		8.70	2.759		141	161		302	395
1460	400 kcmil		7.50	3.200		181	186		367	480
1470	500 kcmil		6.90	3.478		216	202		418	545
1480	600 kcmil		6	4		246	233		479	620
1490	700 kcmil		5.70	4.211		270	245		515	660
1500	750 kcmil		5.10	4.706		280	274		554	720
1510	1,000 kcmil		4.80	5		350	291		641	825
1800	Water pipe ground clamps, heavy duty									
2000	Bronze, 1/2" to 1" diameter	1 Elec	8	1	Ea.	29.50	58		87.50	120
2100	1-1/4" to 2" diameter		8	1		28.50	58		86.50	119
2200	2-1/2" to 3" diameter		6	1.333		68	77.50		145.50	191

26 05 26.80 Grounding		Crew	Daily Output	Labor-Hours	Unit	Material	2018 Bare Costs Labor	Equipment	Total	Total Incl O&P
2400	Grounding, exothermic welding									
2500	Exothermic welding reusable mold, cable to cable, parallel, vertical	1 Elec	8	1	Ea.	345	58		403	465
2505	Splice single		8	1		294	58		352	410
2510	Splice single to double		8	1		345	58		403	465
2520	Cable to cable, termination, Tee		8	1		340	58		398	460
2530	Cable to rod, termination, Tee		8	1		268	58		326	380
2535	Cable to rebar over, termination, Tee		8	1		86	58		144	182
2540	Cable to rod, termination, 90 Deg		8	1		335	58		393	455
2550	Cable to verticle flat steel		8	1		345	58		403	460
2555	Cable to 45 Deg		8	1		350	58		408	470
2730	Exothermic weld, 4/0 wire to 1" ground rod		7	1.143		9.55	66.50		76.05	110
2740	4/0 wire to building steel		7	1.143		9.55	66.50		76.05	110
2750	4/0 wire to motor frame		7	1.143		9.55	66.50		76.05	110
2760	4/0 wire to 4/0 wire		7	1.143		9.55	66.50		76.05	110
2770	4/0 wire to #4 wire		7	1.143		9.55	66.50		76.05	110
2780	4/0 wire to #8 wire		7	1.143		9.55	66.50		76.05	110
2790	Mold, reusable, for above					142			142	157
2800	Brazed connections, #6 wire	1 Elec	12	.667		16.15	39		55.15	76
3000	#2 wire		10	.800		21.50	46.50		68	93.50
3100	3/0 wire		8	1		32	58		90	123
3200	4/0 wire		7	1.143		36.50	66.50		103	140
3400	250 kcmil wire		5	1.600		43	93		136	186
3600	500 kcmil wire		4	2		53	116		169	232
3700	Insulated ground wire, copper #14		13	.615	C.L.F.	7.35	36		43.35	61.50
3710	#12		11	.727		10.50	42.50		53	75
3720	#10		10	.800		16.10	46.50		62.60	87
3730	#8		8	1		28.50	58		86.50	118
3740	#6		6.50	1.231		48.50	71.50		120	160
3750	#4	2 Elec	10.60	1.509		69	88		157	207
3770	#2		9	1.778		117	103		220	283
3780	#1		8	2		151	116		267	340
3790	1/0		6.60	2.424		195	141		336	425
3800	2/0		5.80	2.759		221	161		382	485
3810	3/0		5	3.200		283	186		469	590
3820	4/0		4.40	3.636		350	212		562	700
3830	250 kcmil	3 Elec	6	4		430	233		663	825
3840	300 kcmil		5.70	4.211		505	245		750	920
3850	350 kcmil		5.40	4.444		590	259		849	1,025
3860	400 kcmil		5.10	4.706		675	274		949	1,150
3870	500 kcmil		4.80	5		840	291		1,131	1,350
3880	600 kcmil		3.90	6.154		1,150	360		1,510	1,800
3890	750 kcmil		3.30	7.273		1,325	425		1,750	2,075
3900	1,000 kcmil		2.70	8.889		1,950	515		2,465	2,900
3960	Insulated ground wire, aluminum, #6	1 Elec	8	1		24	58		82	114
3970	#4	2 Elec	13	1.231		30	71.50		101.50	140
3980	#2		10.60	1.509		41	88		129	176
3990	#1		9	1.778		60	103		163	221
4000	1/0		8	2		72	116		188	254
4010	2/0		7.20	2.222		86	129		215	288
4020	3/0		6.60	2.424		107	141		248	330
4030	4/0		6.20	2.581		119	150		269	355
4040	250 kcmil	3 Elec	8.70	2.759		144	161		305	400
4050	300 kcmil		8.10	2.963		199	172		371	475

For customer support on your Electrical Costs with RSMeans data, call 800.448.8182.

133

26 05 Common Work Results for Electrical

26 05 26 – Grounding and Bonding for Electrical Systems

26 05 26.80 Grounding		Crew	Daily Output	Labor-Hours	Unit	Material	2018 Bare Costs Labor	Equipment	Total	Total Incl O&P
4060	350 kcmil	3 Elec	7.50	3.200	C.L.F.	201	186		387	500
4070	400 kcmil		6.90	3.478		229	202		431	555
4080	500 kcmil		6	4		258	233		491	635
4090	600 kcmil		5.70	4.211		325	245		570	725
4100	700 kcmil		5.10	4.706		400	274		674	850
4110	750 kcmil		4.80	5		400	291		691	875
5000	Copper electrolytic ground rod system									
5010	Includes augering hole, mixing bentonite clay,									
5020	Installing rod, and terminating ground wire									
5100	Straight vertical type, 2" diam.									
5120	8.5' long, clamp connection	1 Elec	2.67	2.996	Ea.	730	174		904	1,075
5130	With exothermic weld connection		1.95	4.103		740	239		979	1,175
5140	10' long		2.35	3.404		860	198		1,058	1,250
5150	With exothermic weld connection		1.78	4.494		865	262		1,127	1,350
5160	12' long		2.16	3.704		1,075	216		1,291	1,525
5170	With exothermic weld connection		1.67	4.790		1,100	279		1,379	1,625
5180	20' long		1.74	4.598		1,400	268		1,668	1,950
5190	With exothermic weld connection		1.40	5.714		1,425	335		1,760	2,050
5195	40' long with exothermic weld connection	2 Elec	2	8		3,075	465		3,540	4,100
5200	L-shaped, 2" diam.									
5220	4' vert. x 10' horiz., clamp connection	1 Elec	5.33	1.501	Ea.	1,325	87.50		1,412.50	1,575
5230	With exothermic weld connection	"	3.08	2.597	"	1,350	151		1,501	1,700
5300	Protective box at grade level, with breather slots									
5320	Round 12" long, fiberlyte	1 Elec	32	.250	Ea.	70.50	14.55		85.05	100
5330	Concrete	"	16	.500		111	29		140	166
5400	Bentonite clay, 50# bag, 1 per 10' of rod					49			49	54
5500	Equipotential earthing bar	1 Elec	2	4		142	233		375	505
7000	Exothermic welding kit, multi vertical	"	8	1		600	58		658	745

26 05 29 – Hangers and Supports for Electrical Systems

26 05 29.20 Hangers

		Crew	Daily Output	Labor-Hours	Unit	Material	2018 Bare Costs Labor	Equipment	Total	Total Incl O&P
0010	**HANGERS** R260533-21									
0015	See section 22 05 29.10 for additional items									
0030	Conduit supports									
0050	Strap w/2 holes, rigid steel conduit									
0100	1/2" diameter	1 Elec	470	.017	Ea.	.09	.99		1.08	1.58
0150	3/4" diameter		440	.018		.10	1.06		1.16	1.69
0200	1" diameter		400	.020		.14	1.16		1.30	1.89
0300	1-1/4" diameter		355	.023		.23	1.31		1.54	2.21
0350	1-1/2" diameter		320	.025		.30	1.46		1.76	2.51
0400	2" diameter		266	.030		.40	1.75		2.15	3.06
0500	2-1/2" diameter		160	.050		.60	2.91		3.51	5
0550	3" diameter		133	.060		.88	3.50		4.38	6.20
0600	3-1/2" diameter		100	.080		1.79	4.66		6.45	8.90
0650	4" diameter		80	.100		1.03	5.80		6.83	9.85
0700	EMT, 1/2" diameter		470	.017		.16	.99		1.15	1.66
0800	3/4" diameter		440	.018		.15	1.06		1.21	1.75
0850	1" diameter		400	.020		.26	1.16		1.42	2.03
0900	1-1/4" diameter		355	.023		.35	1.31		1.66	2.35
0950	1-1/2" diameter		320	.025		.38	1.46		1.84	2.60
1000	2" diameter		266	.030		.48	1.75		2.23	3.15
1100	2-1/2" diameter		160	.050		1.15	2.91		4.06	5.60
1150	3" diameter		133	.060		1.48	3.50		4.98	6.90

26 05 29.20 Hangers	Crew	Daily Output	Labor-Hours	Unit	Material	2018 Bare Costs Labor	Equipment	Total	Total Incl O&P
1200	3-1/2" diameter	1 Elec	100	.080	Ea.	2.05	4.66		6.71	9.20
1250	4" diameter		80	.100		2.38	5.80		8.18	11.30
1400	Hanger, with bolt, 1/2" diameter		200	.040		.34	2.33		2.67	3.85
1450	3/4" diameter		190	.042		.36	2.45		2.81	4.06
1500	1" diameter		176	.045		.46	2.65		3.11	4.46
1550	1-1/4" diameter		160	.050		.57	2.91		3.48	4.98
1600	1-1/2" diameter		140	.057		.76	3.33		4.09	5.80
1650	2" diameter		130	.062		1.25	3.58		4.83	6.75
1700	2-1/2" diameter		100	.080		1.37	4.66		6.03	8.45
1750	3" diameter		64	.125		1.92	7.30		9.22	13
1800	3-1/2" diameter		50	.160		1.68	9.30		10.98	15.75
1850	4" diameter		40	.200		2.60	11.65		14.25	20.50
1900	Riser clamps, conduit, 1/2" diameter		40	.200		7.25	11.65		18.90	25.50
1950	3/4" diameter		36	.222		7.50	12.95		20.45	27.50
2000	1" diameter		30	.267		7.90	15.50		23.40	31.50
2100	1-1/4" diameter		27	.296		9.70	17.25		26.95	36.50
2150	1-1/2" diameter		27	.296		10.05	17.25		27.30	37
2200	2" diameter		20	.400		10.80	23.50		34.30	47
2250	2-1/2" diameter		20	.400		12.10	23.50		35.60	48.50
2300	3" diameter		18	.444		12.85	26		38.85	52.50
2350	3-1/2" diameter		18	.444		13.15	26		39.15	53
2400	4" diameter		14	.571		14.65	33.50		48.15	65.50
2500	Threaded rod, painted, 1/4" diameter		260	.031	L.F.	1.40	1.79		3.19	4.22
2600	3/8" diameter		200	.040		5.05	2.33		7.38	9.05
2700	1/2" diameter		140	.057		5.45	3.33		8.78	10.95
2800	5/8" diameter		100	.080		6.65	4.66		11.31	14.30
2900	3/4" diameter		60	.133		6.40	7.75		14.15	18.60
2940	Couplings painted, 1/4" diameter				C	277			277	305
2960	3/8" diameter					435			435	480
2970	1/2" diameter					565			565	625
2980	5/8" diameter					730			730	805
2990	3/4" diameter					930			930	1,025
3000	Nuts, galvanized, 1/4" diameter					9.95			9.95	10.95
3050	3/8" diameter					17.15			17.15	18.85
3100	1/2" diameter					43.50			43.50	48
3150	5/8" diameter					154			154	170
3200	3/4" diameter					129			129	142
3250	Washers, galvanized, 1/4" diameter					14.05			14.05	15.45
3300	3/8" diameter					19.05			19.05	21
3350	1/2" diameter					37			37	40.50
3400	5/8" diameter					83.50			83.50	92
3450	3/4" diameter					144			144	158
3500	Lock washers, galvanized, 1/4" diameter					9.25			9.25	10.20
3550	3/8" diameter					15.45			15.45	17
3600	1/2" diameter					21.50			21.50	23.50
3650	5/8" diameter					38.50			38.50	42.50
3700	3/4" diameter					72.50			72.50	79.50
3705	Metal channel slotted, 304 stainless steel, aluminum									
3710	304 stainless steel, 1' long, 12GA, 1-5/8" x 1-5/8"	1 Elec	24	.333	L.F.	19.80	19.40		39.20	51
3715	304 stainless steel, 1'6" long, 12GA, 1-5/8" x 1-5/8"		24	.333		26.50	19.40		45.90	58
3720	304 stainless steel, 2' long, 12GA, 1-5/8" x 1-5/8"		24	.333		39.50	19.40		58.90	72.50
3725	304 stainless steel, 3' long, 12GA, 1-5/8" x 1-5/8"		24	.333		53	19.40		72.40	87
3730	304 stainless steel, 4' long, 12GA, 1-5/8" x 1-5/8"		24	.333		62.50	19.40		81.90	98

For customer support on your Electrical Costs with RSMeans data, call 800.448.8182.

135

26 05 29 – Hangers and Supports for Electrical Systems

26 05 29.20 Hangers		Crew	Daily Output	Labor-Hours	Unit	Material	2018 Bare Costs Labor	Equipment	Total	Total Incl O&P
3735	304 stainless steel, 5' long, 12GA, 1-5/8" x 1-5/8"	1 Elec	24	.333	L.F.	95	19.40		114.40	134
3740	304 stainless steel, 10' long, 12GA, 1-5/8" x 1-5/8"		24	.333		112	19.40		131.40	152
3745	Aluminum, 1' length, 12GA, 1-5/8" x 1-5/8"		24	.333		5.70	19.40		25.10	35.50
3750	Aluminum, 1'6" length, 12GA, 1-5/8" x 1-5/8"		24	.333		8.55	19.40		27.95	38.50
3755	Aluminum, 2' length, 12GA, 1-5/8" x 1-5/8"		24	.333		11.45	19.40		30.85	41.50
3760	Aluminum, 3' length, 12GA, 1-5/8" x 1-5/8"		24	.333		17.15	19.40		36.55	48
3765	Aluminum, 4' length, 12GA, 1-5/8" x 1-5/8"		24	.333		23	19.40		42.40	54
3770	Aluminum, 5' length, 12GA, 1-5/8" x 1-5/8"		24	.333		27	19.40		46.40	59
3775	Aluminum, 10' length, 12GA, 1-5/8" x 1-5/8"		24	.333		46.50	19.40		65.90	80.50
3800	Channels, steel, 3/4" x 1-1/2", 14 ga.		80	.100		2.61	5.80		8.41	11.55
3900	1-1/2" x 1-1/2", 12 ga.		70	.114		3.41	6.65		10.06	13.70
4000	1-7/8" x 1-1/2"		60	.133		16.30	7.75		24.05	29.50
4100	3" x 1-1/2"		50	.160		18.25	9.30		27.55	34
4110	1-5/8" x 13/16", 16 ga.		80	.100		5.25	5.80		11.05	14.50
4114	Trapeze channel support, 12" wide, steel, 12 ga.		8.80	.909	Ea.	23.50	53		76.50	105
4116	18" wide, steel		8.30	.964	"	25.50	56		81.50	112
4120	1-5/8" x 1-5/8", 14 ga.		70	.114	L.F.	6.15	6.65		12.80	16.75
4130	1-5/8" x 7/8", 12 ga.		70	.114		4.74	6.65		11.39	15.15
4140	1-5/8" x 1-3/8", 12 ga.		60	.133		7.20	7.75		14.95	19.55
4150	1-5/8" x 1-5/8", 12 ga.		50	.160		5.90	9.30		15.20	20.50
4160	Flat plate fitting, 2 hole, 3-1/2", material only				Ea.	3.08			3.08	3.39
4170	3 hole, 1-7/16" x 4-1/8"					3.84			3.84	4.22
4200	Spring nuts, long, 1/4"	1 Elec	120	.067		.64	3.88		4.52	6.50
4250	3/8"		100	.080		.70	4.66		5.36	7.70
4300	1/2"		80	.100		.78	5.80		6.58	9.55
4350	Spring nuts, short, 1/4"		120	.067		1.26	3.88		5.14	7.20
4400	3/8"		100	.080		1.59	4.66		6.25	8.70
4450	1/2"		80	.100		1.67	5.80		7.47	10.55
4500	Closure strip		200	.040	L.F.	3	2.33		5.33	6.80
4550	End cap		60	.133	Ea.	1.07	7.75		8.82	12.80
4600	End connector 3/4" conduit		40	.200		7.20	11.65		18.85	25.50
4650	Junction box, 1 channel		16	.500		31	29		60	77.50
4700	2 channel		14	.571		53	33.50		86.50	108
4750	3 channel		12	.667		68	39		107	133
4800	4 channel		10	.800		77	46.50		123.50	155
4850	Splice plate		40	.200		9.45	11.65		21.10	28
4900	Continuous concrete insert, 1-1/2" deep, 1' long		16	.500		19.85	29		48.85	65.50
4950	2' long		14	.571		24.50	33.50		58	76.50
5000	3' long		12	.667		23.50	39		62.50	84
5050	4' long		10	.800		34	46.50		80.50	107
5100	6' long		8	1		51	58		109	143
5150	3/4" deep, 1' long		16	.500		17.90	29		46.90	63
5200	2' long		14	.571		22.50	33.50		56	74
5250	3' long		12	.667		23	39		62	83.50
5300	4' long		10	.800		30.50	46.50		77	103
5350	6' long		8	1		46	58		104	138
5400	90° angle fitting 2-1/8" x 2-1/8"		60	.133		3.67	7.75		11.42	15.65
5450	Supports, suspension rod type, small		60	.133		42.50	7.75		50.25	58.50
5500	Large		40	.200		50.50	11.65		62.15	73
5550	Beam clamp, small		60	.133		23	7.75		30.75	37
5600	Large		40	.200		25	11.65		36.65	45
5650	U-support, small		60	.133		4.55	7.75		12.30	16.60
5700	Large		40	.200		9.85	11.65		21.50	28

26 05 29.20 Hangers	Crew	Daily Output	Labor-Hours	Unit	Material	2018 Bare Costs Labor	Equipment	Total	Total Incl O&P
5750 Concrete insert, cast, for up to 1/2" threaded rod	1 Elec	16	.500	Ea.	9.45	29		38.45	54
5800 Beam clamp, 1/4" clamp, for 1/4" threaded drop rod		32	.250		2.36	14.55		16.91	24.50
5900 3/8" clamp, for 3/8" threaded drop rod		32	.250		3.17	14.55		17.72	25.50
6000 Strap, rigid conduit, 1/2" diameter		540	.015		1.13	.86		1.99	2.53
6050 3/4" diameter		440	.018		1.17	1.06		2.23	2.87
6100 1" diameter		420	.019		1.42	1.11		2.53	3.22
6150 1-1/4" diameter		400	.020		1.64	1.16		2.80	3.54
6200 1-1/2" diameter		400	.020		1.85	1.16		3.01	3.78
6250 2" diameter		267	.030		2.07	1.74		3.81	4.89
6300 2-1/2" diameter		267	.030		2.25	1.74		3.99	5.10
6350 3" diameter		160	.050		2.60	2.91		5.51	7.20
6400 3-1/2" diameter		133	.060		3.14	3.50		6.64	8.70
6450 4" diameter		100	.080		3.31	4.66		7.97	10.60
6500 5" diameter		80	.100		8.45	5.80		14.25	17.95
6550 6" diameter		60	.133		13.45	7.75		21.20	26.50
6600 EMT, 1/2" diameter		540	.015		.87	.86		1.73	2.25
6650 3/4" diameter		440	.018		.91	1.06		1.97	2.58
6700 1" diameter		420	.019		1.01	1.11		2.12	2.77
6750 1-1/4" diameter		400	.020		1.21	1.16		2.37	3.07
6800 1-1/2" diameter		400	.020		1.50	1.16		2.66	3.39
6850 2" diameter		267	.030		1.55	1.74		3.29	4.32
6900 2-1/2" diameter		267	.030		1.86	1.74		3.60	4.66
6950 3" diameter		160	.050		2.09	2.91		5	6.65
6970 3-1/2" diameter		133	.060		2.89	3.50		6.39	8.45
6990 4" diameter		100	.080		2.86	4.66		7.52	10.10
7000 Clip, 1 hole for rigid conduit, 1/2" diameter		500	.016		.06	.93		.99	1.46
7050 3/4" diameter		470	.017		.69	.99		1.68	2.24
7100 1" diameter		440	.018		1.69	1.06		2.75	3.44
7150 1-1/4" diameter		400	.020		1.63	1.16		2.79	3.53
7200 1-1/2" diameter		355	.023		1.89	1.31		3.20	4.04
7250 2" diameter		320	.025		1.68	1.46		3.14	4.03
7300 2-1/2" diameter		266	.030		4.65	1.75		6.40	7.70
7350 3" diameter		160	.050		4.38	2.91		7.29	9.15
7400 3-1/2" diameter		133	.060		6.40	3.50		9.90	12.30
7450 4" diameter		100	.080		13.30	4.66		17.96	21.50
7500 5" diameter		80	.100		99	5.80		104.80	118
7550 6" diameter		60	.133		119	7.75		126.75	143
7600 Hammer on purlin clip, 1/4" hole, 1/16" flange		60	.133		.52	7.75		8.27	12.15
7610 1/4" hole threaded,1/4" flange, rod hanger		60	.133		1.16	7.75		8.91	12.90
7620 3/8" hole threaded,1/4" flange, rod hanger		60	.133		1.15	7.75		8.90	12.85
7630 Z purlin clip, 1/4" & 3/8" flange threaded rod hanger		60	.133		1.06	7.75		8.81	12.75
7640 1/2" rigid, 3/4" conduit, 1" flange		60	.133		1.18	7.75		8.93	12.90
7650 1" conduit, 1" flange		60	.133		1.39	7.75		9.14	13.15
7660 1-1/4" conduit, 1" flange		60	.133		1.64	7.75		9.39	13.40
7670 Push-in type conduit clip 3/4" EMT, with 9/32" hole		60	.133		.39	7.75		8.14	12.05
7680 3/4" EMT, 1/2" rigid		60	.133		1.06	7.75		8.81	12.75
7690 Push-in type tandem conduit clip 3/4" - 1" EMT, 1/2" - 3/4" rigid		60	.133		1.08	7.75		8.83	12.80
7820 Conduit hangers, with bolt & 12" rod, 1/2" diameter		150	.053		5.40	3.10		8.50	10.60
7830 3/4" diameter		145	.055		5.40	3.21		8.61	10.75
7840 1" diameter		135	.059		5.50	3.45		8.95	11.20
7850 1-1/4" diameter		120	.067		5.60	3.88		9.48	12
7860 1-1/2" diameter		110	.073		5.80	4.23		10.03	12.75
7870 2" diameter		100	.080		6.70	4.66		11.36	14.30

For customer support on your Electrical Costs with RSMeans data, call 800.448.8182.

137

26 05 29 – Hangers and Supports for Electrical Systems

26 05 29.20 Hangers		Crew	Daily Output	Labor-Hours	Unit	Material	2018 Bare Costs Labor	Equipment	Total	Total Incl O&P
7880	2-1/2" diameter	1 Elec	80	.100	Ea.	6.80	5.80		12.60	16.20
7890	3" diameter		60	.133		7.35	7.75		15.10	19.70
7900	3-1/2" diameter		45	.178		7.15	10.35		17.50	23.50
7910	4" diameter		35	.229		8.05	13.30		21.35	29
7920	5" diameter		30	.267		9.25	15.50		24.75	33
7930	6" diameter		25	.320		12.10	18.60		30.70	41.50
7950	Jay clamp, 1/2" diameter		32	.250		5.95	14.55		20.50	28.50
7960	3/4" diameter		32	.250		5.85	14.55		20.40	28.50
7970	1" diameter		32	.250		4.70	14.55		19.25	27
7980	1-1/4" diameter		30	.267		7.95	15.50		23.45	32
7990	1-1/2" diameter		30	.267		8.45	15.50		23.95	32.50
8000	2" diameter		30	.267		11.20	15.50		26.70	35.50
8010	2-1/2" diameter		28	.286		19.40	16.65		36.05	46.50
8020	3" diameter		28	.286		19.40	16.65		36.05	46.50
8030	3-1/2" diameter		25	.320		36.50	18.60		55.10	68
8040	4" diameter		25	.320		36.50	18.60		55.10	68
8050	5" diameter		20	.400		97.50	23.50		121	142
8060	6" diameter		16	.500		186	29		215	248
8070	Channels, 3/4" x 1-1/2" w/12" rods for 1/2" to 1" conduit		30	.267		8.95	15.50		24.45	33
8080	1-1/2" x 1-1/2" w/12" rods for 1-1/4" to 2" conduit		28	.286		9.30	16.65		25.95	35.50
8090	1-1/2" x 1-1/2" w/12" rods for 2-1/2" to 4" conduit		26	.308		10.90	17.90		28.80	39
8100	1-1/2" x 1-7/8" w/12" rods for 5" to 6" conduit		24	.333		39	19.40		58.40	72
8110	Beam clamp, conduit, plastic coated steel, 1/2" diam.		30	.267		24.50	15.50		40	50
8120	3/4" diameter		30	.267		27.50	15.50		43	53
8130	1" diameter		30	.267		28.50	15.50		44	54
8140	1-1/4" diameter		28	.286		35.50	16.65		52.15	64
8150	1-1/2" diameter		28	.286		43	16.65		59.65	72.50
8160	2" diameter		28	.286		57	16.65		73.65	88
8170	2-1/2" diameter		26	.308		53.50	17.90		71.40	86
8180	3" diameter		26	.308		70.50	17.90		88.40	105
8190	3-1/2" diameter		23	.348		72.50	20		92.50	110
8200	4" diameter		23	.348		81	20		101	120
8210	5" diameter		18	.444		203	26		229	263
8220	Channels, plastic coated									
8250	3/4" x 1-1/2", w/12" rods for 1/2" to 1" conduit	1 Elec	28	.286	Ea.	35	16.65		51.65	63.50
8260	1-1/2" x 1-1/2", w/12" rods for 1-1/4" to 2" conduit		26	.308		39.50	17.90		57.40	70.50
8270	1-1/2" x 1-1/2", w/12" rods for 2-1/2" to 3-1/2" conduit		24	.333		43	19.40		62.40	76.50
8280	1-1/2" x 1-7/8", w/12" rods for 4" to 5" conduit		22	.364		71.50	21		92.50	111
8290	1-1/2" x 1-7/8", w/12" rods for 6" conduit		20	.400		78.50	23.50		102	122
8320	Conduit hangers, plastic coated steel, with bolt & 12" rod, 1/2" diam.		140	.057		20.50	3.33		23.83	27.50
8330	3/4" diameter		135	.059		21	3.45		24.45	28
8340	1" diameter		125	.064		21.50	3.72		25.22	29
8350	1-1/4" diameter		110	.073		22.50	4.23		26.73	31.50
8360	1-1/2" diameter		100	.080		25.50	4.66		30.16	35
8370	2" diameter		90	.089		28	5.15		33.15	39
8380	2-1/2" diameter		70	.114		33	6.65		39.65	46.50
8390	3" diameter		50	.160		40	9.30		49.30	58
8400	3-1/2" diameter		35	.229		40.50	13.30		53.80	64.50
8410	4" diameter		25	.320		59.50	18.60		78.10	93.50
8420	5" diameter		20	.400		65	23.50		88.50	107
9000	Parallel type, conduit beam clamp, 1/2"		32	.250		6.60	14.55		21.15	29.50
9010	3/4"		32	.250		3.88	14.55		18.43	26.50
9020	1"		32	.250		4.38	14.55		18.93	27

26 05 29.20 Hangers

		Crew	Daily Output	Labor-Hours	Unit	Material	2018 Bare Costs Labor	Equipment	Total	Total Incl O&P
9030	1-1/4"	1 Elec	30	.267	Ea.	9.35	15.50		24.85	33.50
9040	1-1/2"		30	.267		10.65	15.50		26.15	34.50
9050	2"		30	.267		9.25	15.50		24.75	33
9060	2-1/2"		28	.286		15.25	16.65		31.90	42
9070	3"		28	.286		17.50	16.65		34.15	44.50
9090	4"		25	.320		14.70	18.60		33.30	44
9110	Right angle, conduit beam clamp, 1/2"		32	.250		2.47	14.55		17.02	24.50
9120	3/4"		32	.250		2.66	14.55		17.21	25
9130	1"		32	.250		2.36	14.55		16.91	24.50
9140	1-1/4"		30	.267		3.30	15.50		18.80	26.50
9150	1-1/2"		30	.267		3.70	15.50		19.20	27
9160	2"		30	.267		5.45	15.50		20.95	29
9170	2-1/2"		28	.286		6.65	16.65		23.30	32.50
9180	3"		28	.286		7.50	16.65		24.15	33.50
9190	3-1/2"		25	.320		9.15	18.60		27.75	38
9200	4"		25	.320		9.80	18.60		28.40	39
9230	Adjustable, conduit hanger, 1/2"		32	.250		7.50	14.55		22.05	30.50
9240	3/4"		32	.250		5.70	14.55		20.25	28.50
9250	1"		32	.250		8	14.55		22.55	31
9260	1-1/4"		30	.267		8.10	15.50		23.60	32
9270	1-1/2"		30	.267		9.65	15.50		25.15	33.50
9280	2"		30	.267		4.17	15.50		19.67	27.50
9290	2-1/2"		28	.286		17.30	16.65		33.95	44
9300	3"		28	.286		18.95	16.65		35.60	46
9310	3-1/2"		25	.320		25	18.60		43.60	55
9320	4"		25	.320		30	18.60		48.60	61
9330	5"		20	.400		38.50	23.50		62	77.50
9340	6"		16	.500		49	29		78	97.50
9350	Combination conduit hanger, 3/8"		32	.250		14.20	14.55		28.75	37.50
9360	Adjustable flange, 3/8"		32	.250		16.65	14.55		31.20	40.50

26 05 29.30 Fittings and Channel Support

		Crew	Daily Output	Labor-Hours	Unit	Material	2018 Bare Costs Labor	Equipment	Total	Total Incl O&P
0010	**FITTINGS & CHANNEL SUPPORT**									
0020	Rooftop channel support									
0200	2-7/8" L x 1-5/8" W, 12 ga. pre galv., dbl. base	1 Elec	46	.174	Ea.	4.07	10.10		14.17	19.65
0210	2-7/8" L x 1-5/8" W, 12 ga. hot dip galv., dbl. base		46	.174		2.37	10.10		12.47	17.75
0220	2-7/16" L x 1-5/8" W, 12 ga. pre galv., sngl base		48	.167		1.95	9.70		11.65	16.65
0230	2-7/16" L x 1-5/8" W, 12 ga. hot dip galv., sngl base		48	.167		6.95	9.70		16.65	22
0240	13/16" L x 1-5/8" W, 14 ga. pre galv., sngl base		56	.143		1.66	8.30		9.96	14.30
0250	13/16" L x 1-5/8" W, 14 ga. hot dip galv., sngl base		56	.143		6.80	8.30		15.10	19.90
0260	1-5/8" L x 1-5/8" W, 12 ga. pre galv., sngl base		54	.148		1.80	8.60		10.40	14.90
0270	1-5/8" L x 1-5/8" W, 12 ga. hot dip galv., sngl base		54	.148		1.90	8.60		10.50	15
0280	1-5/8" x 28", 12 ga. pre galv., dbl. H block base		42	.190		4.12	11.10		15.22	21
0290	1-5/8" x 36", 12 ga. pre galv., dbl. H block base		40	.200		4.36	11.65		16.01	22
0300	1-5/8" x 42", 12 ga. pre galv., dbl. H block base		34	.235		4.90	13.70		18.60	26
0310	1-5/8" x 50", 12 ga. pre galv., dbl. H block base		38	.211		5.20	12.25		17.45	24
0320	1-5/8" x 60", 12 ga. pre galv., dbl. H block base		36	.222		5.75	12.95		18.70	25.50
0330	13/16" x 8" H, threaded rod pre galv., H block base		35	.229		2.77	13.30		16.07	23
0340	13/16" x 12" H, threaded rod pre galv., H block base		34.60	.231		3.46	13.45		16.91	24
0350	1-5/8" x 16" H, threaded rod dbl. H block base		34	.235		5.15	13.70		18.85	26
0360	1-5/8" x 12" H, pre galv., dbl. H block base		34.60	.231		9.90	13.45		23.35	31
0370	1-5/8" x 24"H, pre galv., dbl. H block base		34	.235		10.45	13.70		24.15	32
0380	3-1/4" x 24"H, pre galv., dbl. H block base		32	.250		23.50	14.55		38.05	48

For customer support on your Electrical Costs with RSMeans data, call 800.448.8182.

139

26 05 29 – Hangers and Supports for Electrical Systems

26 05 29.30 Fittings and Channel Support		Crew	Daily Output	Labor-Hours	Unit	Material	2018 Bare Costs Labor	Equipment	Total	Total Incl O&P
0390	3-1/4" x 36"H, pre galv., dbl. H block base	1 Elec	30	.267	Ea.	26	15.50		41.50	51.50

26 05 33 – Raceway and Boxes for Electrical Systems

26 05 33.13 Conduit

		Crew	Daily Output	Labor-Hours	Unit	Material	2018 Bare Costs Labor	Equipment	Total	Total Incl O&P
0010	**CONDUIT** To 10' high, includes 2 terminations, 2 elbows,	R260533-20								
0020	11 beam clamps, and 11 couplings per 100 L.F.									
0300	Aluminum, 1/2" diameter	1 Elec	100	.080	L.F.	1.72	4.66		6.38	8.85
0500	3/4" diameter	R260533-25	90	.089		2.28	5.15		7.43	10.25
0700	1" diameter		80	.100		2.82	5.80		8.62	11.80
1000	1-1/4" diameter		70	.114		3.91	6.65		10.56	14.25
1030	1-1/2" diameter		65	.123		4.70	7.15		11.85	15.85
1050	2" diameter		60	.133		6.40	7.75		14.15	18.60
1070	2-1/2" diameter		50	.160		9.25	9.30		18.55	24
1100	3" diameter	2 Elec	90	.178		12.15	10.35		22.50	29
1130	3-1/2" diameter		80	.200		16	11.65		27.65	35
1140	4" diameter		70	.229		18.05	13.30		31.35	40
1150	5" diameter		50	.320		38	18.60		56.60	70
1160	6" diameter		40	.400		54.50	23.50		78	95
1161	Field bends, 45° to 90°, 1/2" diameter	1 Elec	53	.151	Ea.		8.80		8.80	13.15
1162	3/4" diameter		47	.170			9.90		9.90	14.80
1163	1" diameter		44	.182			10.60		10.60	15.80
1164	1-1/4" diameter		23	.348			20		20	30.50
1165	1-1/2" diameter		21	.381			22		22	33
1166	2" diameter		16	.500			29		29	43.50
1170	Elbows, 1/2" diameter		40	.200		5.40	11.65		17.05	23.50
1200	3/4" diameter		32	.250		7.15	14.55		21.70	30
1230	1" diameter		28	.286		10.10	16.65		26.75	36
1250	1-1/4" diameter		24	.333		16.15	19.40		35.55	47
1270	1-1/2" diameter		20	.400		21.50	23.50		45	58.50
1300	2" diameter		16	.500		31	29		60	78
1330	2-1/2" diameter		12	.667		56.50	39		95.50	120
1350	3" diameter		8	1		71.50	58		129.50	166
1370	3-1/2" diameter		6	1.333		115	77.50		192.50	243
1400	4" diameter		5	1.600		148	93		241	300
1410	5" diameter		4	2		430	116		546	650
1420	6" diameter		2.50	3.200		565	186		751	900
1430	Couplings, 1/2" diameter		320	.025		1.81	1.46		3.27	4.17
1450	3/4" diameter		192	.042		2.75	2.43		5.18	6.65
1470	1" diameter		160	.050		3.58	2.91		6.49	8.30
1500	1-1/4" diameter		120	.067		4.22	3.88		8.10	10.45
1530	1-1/2" diameter		96	.083		4.96	4.85		9.81	12.70
1550	2" diameter		80	.100		7.15	5.80		12.95	16.60
1570	2-1/2" diameter		69	.116		15.25	6.75		22	27
1600	3" diameter		64	.125		20.50	7.30		27.80	33.50
1630	3-1/2" diameter		56	.143		27.50	8.30		35.80	43
1650	4" diameter		53	.151		29	8.80		37.80	45
1670	5" diameter		48	.167		101	9.70		110.70	127
1690	6" diameter		46	.174		164	10.10		174.10	195
1691	See note on line 26 05 33.13 9995	R260533-30								
1750	Rigid galvanized steel, 1/2" diameter	1 Elec	90	.089	L.F.	2.31	5.15		7.46	10.30
1770	3/4" diameter		80	.100		4.57	5.80		10.37	13.75
1800	1" diameter		65	.123		6.75	7.15		13.90	18.15
1830	1-1/4" diameter		60	.133		4.74	7.75		12.49	16.80

26 05 33.13 Conduit		Crew	Daily Output	Labor-Hours	Unit	Material	2018 Bare Costs Labor	Equipment	Total	Total Incl O&P
1850	1-1/2" diameter	1 Elec	55	.145	L.F.	7.90	8.45		16.35	21.50
1870	2" diameter		45	.178		9.95	10.35		20.30	26.50
1900	2-1/2" diameter		35	.229		11.80	13.30		25.10	33
1930	3" diameter	2 Elec	50	.320		13.70	18.60		32.30	43
1950	3-1/2" diameter		44	.364		17.40	21		38.40	50.50
1970	4" diameter		40	.400		20	23.50		43.50	57
1980	5" diameter		30	.533		37.50	31		68.50	88
1990	6" diameter		20	.800		54	46.50		100.50	129
1991	Field bends, 45° to 90°, 1/2" diameter	1 Elec	44	.182	Ea.		10.60		10.60	15.80
1992	3/4" diameter		40	.200			11.65		11.65	17.40
1993	1" diameter		36	.222			12.95		12.95	19.35
1994	1-1/4" diameter		19	.421			24.50		24.50	36.50
1995	1-1/2" diameter		18	.444			26		26	38.50
1996	2" diameter		13	.615			36		36	53.50
2000	Elbows, 1/2" diameter		32	.250		4.16	14.55		18.71	26.50
2030	3/4" diameter		28	.286		4.02	16.65		20.67	29.50
2050	1" diameter		24	.333		6.70	19.40		26.10	36.50
2070	1-1/4" diameter		18	.444		9.60	26		35.60	49
2100	1-1/2" diameter		16	.500		17.25	29		46.25	62.50
2130	2" diameter		12	.667		25	39		64	85.50
2150	2-1/2" diameter		8	1		30.50	58		88.50	121
2170	3" diameter		6	1.333		42	77.50		119.50	162
2200	3-1/2" diameter		4.20	1.905		66.50	111		177.50	239
2220	4" diameter		4	2		76	116		192	258
2230	5" diameter		3.50	2.286		204	133		337	425
2240	6" diameter		2	4		365	233		598	750
2250	Couplings, 1/2" diameter		267	.030		1.02	1.74		2.76	3.73
2270	3/4" diameter		160	.050		1.47	2.91		4.38	5.95
2300	1" diameter		133	.060		2.09	3.50		5.59	7.55
2330	1-1/4" diameter		100	.080		2.37	4.66		7.03	9.55
2350	1-1/2" diameter		80	.100		4.69	5.80		10.49	13.85
2370	2" diameter		67	.119		6.15	6.95		13.10	17.20
2400	2-1/2" diameter		57	.140		9.85	8.15		18	23
2430	3" diameter		53	.151		12.65	8.80		21.45	27
2450	3-1/2" diameter		47	.170		15.80	9.90		25.70	32
2470	4" diameter		44	.182		26	10.60		36.60	44.50
2480	5" diameter		40	.200		38	11.65		49.65	59.50
2490	6" diameter		38	.211		57	12.25		69.25	81
2491	See note on line 26 05 33.13 9995 R260533-30									
2500	Steel, intermediate conduit (IMC), 1/2" diameter	1 Elec	100	.080	L.F.	1.69	4.66		6.35	8.80
2530	3/4" diameter		90	.089		2.05	5.15		7.20	10
2550	1" diameter		70	.114		2.58	6.65		9.23	12.80
2570	1-1/4" diameter		65	.123		3.58	7.15		10.73	14.65
2600	1-1/2" diameter		60	.133		4.76	7.75		12.51	16.85
2630	2" diameter		50	.160		6.15	9.30		15.45	20.50
2650	2-1/2" diameter		40	.200		9.15	11.65		20.80	27.50
2670	3" diameter	2 Elec	60	.267		12.10	15.50		27.60	36.50
2700	3-1/2" diameter		54	.296		16.90	17.25		34.15	44.50
2730	4" diameter		50	.320		11.70	18.60		30.30	41
2731	Field bends, 45° to 90°, 1/2" diameter	1 Elec	44	.182	Ea.		10.60		10.60	15.80
2732	3/4" diameter		40	.200			11.65		11.65	17.40
2733	1" diameter		36	.222			12.95		12.95	19.35
2734	1-1/4" diameter		19	.421			24.50		24.50	36.50

For customer support on your Electrical Costs with RSMeans data, call 800.448.8182.

141

26 05 33.13 Conduit		Crew	Daily Output	Labor-Hours	Unit	Material	2018 Bare Costs Labor	2018 Bare Costs Equipment	Total	Total Incl O&P
2735	1-1/2" diameter	1 Elec	18	.444	Ea.		26		26	38.50
2736	2" diameter		13	.615			36		36	53.50
2750	Elbows, 1/2" diameter		32	.250		6.45	14.55		21	29
2770	3/4" diameter		28	.286		11.80	16.65		28.45	38
2800	1" diameter		24	.333		13.05	19.40		32.45	43.50
2830	1-1/4" diameter		18	.444		16.50	26		42.50	56.50
2850	1-1/2" diameter		16	.500		32.50	29		61.50	79
2870	2" diameter		12	.667		57.50	39		96.50	122
2900	2-1/2" diameter		8	1		40	58		98	131
2930	3" diameter		6	1.333		56	77.50		133.50	178
2950	3-1/2" diameter		4.20	1.905		190	111		301	375
2970	4" diameter		4	2		79	116		195	261
3000	Couplings, 1/2" diameter		293	.027		1.02	1.59		2.61	3.50
3030	3/4" diameter		176	.045		1.47	2.65		4.12	5.55
3050	1" diameter		147	.054		2.09	3.17		5.26	7.05
3070	1-1/4" diameter		110	.073		2.37	4.23		6.60	8.95
3100	1-1/2" diameter		88	.091		4.69	5.30		9.99	13.05
3130	2" diameter		73	.110		6.15	6.40		12.55	16.35
3150	2-1/2" diameter		63	.127		9.85	7.40		17.25	22
3170	3" diameter		59	.136		12.65	7.90		20.55	25.50
3200	3-1/2" diameter		52	.154		15.80	8.95		24.75	31
3230	4" diameter		49	.163		26	9.50		35.50	42.50
3231	See note on line 26 05 33.13 9995 R260533-30									
4100	Rigid steel, plastic coated, 40 mil thick									
4130	1/2" diameter	1 Elec	80	.100	L.F.	8	5.80		13.80	17.50
4150	3/4" diameter		70	.114		9.20	6.65		15.85	20
4170	1" diameter		55	.145		11.15	8.45		19.60	25
4200	1-1/4" diameter		50	.160		14.60	9.30		23.90	30
4230	1-1/2" diameter		45	.178		17.20	10.35		27.55	34.50
4250	2" diameter		35	.229		22.50	13.30		35.80	44.50
4270	2-1/2" diameter		25	.320		32	18.60		50.60	63
4300	3" diameter	2 Elec	44	.364		39	21		60	74.50
4330	3-1/2" diameter		40	.400		47.50	23.50		71	87.50
4350	4" diameter		36	.444		56	26		82	100
4370	5" diameter		30	.533		120	31		151	179
4400	Elbows, 1/2" diameter	1 Elec	28	.286	Ea.	17.45	16.65		34.10	44
4430	3/4" diameter		24	.333		18.20	19.40		37.60	49
4450	1" diameter		18	.444		21.50	26		47.50	62
4470	1-1/4" diameter		16	.500		26.50	29		55.50	72.50
4500	1-1/2" diameter		12	.667		32.50	39		71.50	94
4530	2" diameter		8	1		42	58		100	133
4550	2-1/2" diameter		6	1.333		86	77.50		163.50	211
4570	3" diameter		4.20	1.905		134	111		245	315
4600	3-1/2" diameter		4	2		189	116		305	380
4630	4" diameter		3.80	2.105		193	123		316	395
4650	5" diameter		3.50	2.286		490	133		623	740
4680	Couplings, 1/2" diameter		213	.038		4.99	2.19		7.18	8.75
4700	3/4" diameter		128	.063		5.80	3.64		9.44	11.85
4730	1" diameter		107	.075		6.95	4.35		11.30	14.15
4750	1-1/4" diameter		80	.100		8.65	5.80		14.45	18.25
4770	1-1/2" diameter		64	.125		10.35	7.30		17.65	22.50
4800	2" diameter		53	.151		14.60	8.80		23.40	29
4830	2-1/2" diameter		46	.174		34.50	10.10		44.60	53

26 05 33.13 Conduit

		Crew	Daily Output	Labor-Hours	Unit	Material	2018 Bare Costs Labor	Equipment	Total	Total Incl O&P
4850	3" diameter	1 Elec	43	.186	Ea.	39	10.85		49.85	58.50
4870	3-1/2" diameter		38	.211		51	12.25		63.25	75
4900	4" diameter		36	.222		56.50	12.95		69.45	82
4950	5" diameter		32	.250		216	14.55		230.55	259
4951	See note on line 26 05 33.13 9995	R260533-30								
5000	Electric metallic tubing (EMT), 1/2" diameter	1 Elec	170	.047	L.F.	.80	2.74		3.54	4.96
5020	3/4" diameter		130	.062		1.13	3.58		4.71	6.60
5040	1" diameter		115	.070		1.84	4.05		5.89	8.10
5060	1-1/4" diameter		100	.080		2.99	4.66		7.65	10.25
5080	1-1/2" diameter		90	.089		2.82	5.15		7.97	10.85
5100	2" diameter		80	.100		3.85	5.80		9.65	12.95
5120	2-1/2" diameter		60	.133		4.43	7.75		12.18	16.45
5140	3" diameter	2 Elec	100	.160		5.55	9.30		14.85	20
5160	3-1/2" diameter		90	.178		6.95	10.35		17.30	23
5180	4" diameter		80	.200		12.10	11.65		23.75	30.50
5200	Field bends, 45° to 90°, 1/2" diameter	1 Elec	89	.090	Ea.		5.25		5.25	7.80
5220	3/4" diameter		80	.100			5.80		5.80	8.70
5240	1" diameter		73	.110			6.40		6.40	9.55
5260	1-1/4" diameter		38	.211			12.25		12.25	18.30
5280	1-1/2" diameter		36	.222			12.95		12.95	19.35
5300	2" diameter		26	.308			17.90		17.90	27
5320	Offsets, 1/2" diameter		65	.123			7.15		7.15	10.70
5340	3/4" diameter		62	.129			7.50		7.50	11.25
5360	1" diameter		53	.151			8.80		8.80	13.15
5380	1-1/4" diameter		30	.267			15.50		15.50	23
5400	1-1/2" diameter		28	.286			16.65		16.65	25
5420	2" diameter		20	.400			23.50		23.50	35
5700	Elbows, 1" diameter		40	.200		3.58	11.65		15.23	21.50
5720	1-1/4" diameter		32	.250		4.17	14.55		18.72	26.50
5740	1-1/2" diameter		24	.333		4.91	19.40		24.31	34.50
5760	2" diameter		20	.400		7.05	23.50		30.55	43
5780	2-1/2" diameter		12	.667		19.50	39		58.50	79.50
5800	3" diameter		9	.889		26.50	51.50		78	107
5820	3-1/2" diameter		7	1.143		35	66.50		101.50	138
5840	4" diameter		6	1.333		41.50	77.50		119	162
6200	Couplings, set screw, steel, 1/2" diameter		235	.034		1.58	1.98		3.56	4.70
6220	3/4" diameter		188	.043		2.28	2.48		4.76	6.20
6240	1" diameter		157	.051		3.69	2.97		6.66	8.50
6260	1-1/4" diameter		118	.068		7.50	3.95		11.45	14.15
6280	1-1/2" diameter		94	.085		1.74	4.95		6.69	9.30
6300	2" diameter		78	.103		12.95	5.95		18.90	23
6320	2-1/2" diameter		67	.119		5.45	6.95		12.40	16.40
6340	3" diameter		59	.136		5.25	7.90		13.15	17.55
6360	3-1/2" diameter		47	.170		7.20	9.90		17.10	23
6380	4" diameter		43	.186		50	10.85		60.85	71
6381	See note on line 26 05 33.13 9995	R260533-30								
6500	Box connectors, set screw, steel, 1/2" diameter	1 Elec	120	.067	Ea.	.38	3.88		4.26	6.20
6520	3/4" diameter		110	.073		.62	4.23		4.85	7.05
6540	1" diameter		90	.089		1.17	5.15		6.32	9.05
6560	1-1/4" diameter		70	.114		2.40	6.65		9.05	12.60
6580	1-1/2" diameter		60	.133		2.82	7.75		10.57	14.70
6600	2" diameter		50	.160		3.85	9.30		13.15	18.15
6620	2-1/2" diameter		36	.222		7.75	12.95		20.70	28

26 05 33 – Raceway and Boxes for Electrical Systems

26 05 33.13 Conduit		Crew	Daily Output	Labor-Hours	Unit	Material	2018 Bare Costs Labor	Equipment	Total	Total Incl O&P
6640	3" diameter	1 Elec	27	.296	Ea.	13.85	17.25		31.10	41.50
6680	3-1/2" diameter		21	.381		10.85	22		32.85	45
6700	4" diameter		16	.500		12.55	29		41.55	57.50
6740	Insulated box connectors, set screw, steel, 1/2" diameter		120	.067		.68	3.88		4.56	6.55
6760	3/4" diameter		110	.073		1.60	4.23		5.83	8.10
6780	1" diameter		90	.089		2.76	5.15		7.91	10.80
6800	1-1/4" diameter		70	.114		2.57	6.65		9.22	12.80
6820	1-1/2" diameter		60	.133		4.42	7.75		12.17	16.45
6840	2" diameter		50	.160		6.50	9.30		15.80	21
6860	2-1/2" diameter		36	.222		19.80	12.95		32.75	41.50
6880	3" diameter		27	.296		31.50	17.25		48.75	60.50
6900	3-1/2" diameter		21	.381		55	22		77	93.50
6920	4" diameter		16	.500		43.50	29		72.50	91.50
7000	EMT to conduit adapters, 1/2" diameter (compression)		70	.114		3.56	6.65		10.21	13.85
7020	3/4" diameter		60	.133		4.89	7.75		12.64	17
7040	1" diameter		50	.160		7.50	9.30		16.80	22
7060	1-1/4" diameter		40	.200		12	11.65		23.65	30.50
7080	1-1/2" diameter		30	.267		14.35	15.50		29.85	39
7100	2" diameter		25	.320		24.50	18.60		43.10	55
7200	EMT to Greenfield adapters, 1/2" to 3/8" diameter (compression)		90	.089		2.13	5.15		7.28	10.10
7220	1/2" diameter		90	.089		4.45	5.15		9.60	12.65
7240	3/4" diameter		80	.100		5.80	5.80		11.60	15.10
7260	1" diameter		70	.114		15.30	6.65		21.95	27
7270	1-1/4" diameter		60	.133		18	7.75		25.75	31.50
7280	1-1/2" diameter		50	.160		21	9.30		30.30	37
7290	2" diameter		40	.200		30	11.65		41.65	51
7400	EMT, LB, LR or LL fittings with covers, 1/2" diameter, set screw		24	.333		5	19.40		24.40	34.50
7420	3/4" diameter		20	.400		6.40	23.50		29.90	42
7440	1" diameter		16	.500		9.15	29		38.15	53.50
7450	1-1/4" diameter		13	.615		13.90	36		49.90	69
7460	1-1/2" diameter		11	.727		17.95	42.50		60.45	83.50
7470	2" diameter		9	.889		30	51.50		81.50	111
7600	EMT, "T" fittings with covers, 1/2" diameter, set screw		16	.500		5.25	29		34.25	49.50
7620	3/4" diameter		15	.533		6.85	31		37.85	54
7640	1" diameter		12	.667		8.30	39		47.30	67
7650	1-1/4" diameter		11	.727		13.30	42.50		55.80	78
7660	1-1/2" diameter		10	.800		16.45	46.50		62.95	87.50
7670	2" diameter		8	1		22.50	58		80.50	112
8000	EMT, expansion fittings, no jumper, 1/2" diameter		24	.333		63	19.40		82.40	98
8020	3/4" diameter		20	.400		82	23.50		105.50	126
8040	1" diameter		16	.500		103	29		132	157
8060	1-1/4" diameter		13	.615		135	36		171	203
8080	1-1/2" diameter		11	.727		185	42.50		227.50	267
8100	2" diameter		9	.889		271	51.50		322.50	375
8110	2-1/2" diameter		7	1.143		375	66.50		441.50	510
8120	3" diameter		6	1.333		560	77.50		637.50	730
8140	4" diameter		5	1.600		805	93		898	1,025
8200	Split adapter, 1/2" diameter		110	.073		2.90	4.23		7.13	9.55
8210	3/4" diameter		90	.089		2.79	5.15		7.94	10.80
8220	1" diameter		70	.114		4.86	6.65		11.51	15.30
8230	1-1/4" diameter		60	.133		7.60	7.75		15.35	19.95
8240	1-1/2" diameter		50	.160		11.35	9.30		20.65	26.50
8250	2" diameter		36	.222		32	12.95		44.95	54.50

26 05 33.13 Conduit		Crew	Daily Output	Labor-Hours	Unit	Material	2018 Bare Costs Labor	Equipment	Total	Total Incl O&P
8300	1 hole clips, 1/2" diameter	1 Elec	500	.016	Ea.	.41	.93		1.34	1.84
8320	3/4" diameter		470	.017		.23	.99		1.22	1.73
8340	1" diameter		444	.018		.34	1.05		1.39	1.94
8360	1-1/4" diameter		400	.020		.48	1.16		1.64	2.27
8380	1-1/2" diameter		355	.023		.97	1.31		2.28	3.03
8400	2" diameter		320	.025		1.33	1.46		2.79	3.64
8420	2-1/2" diameter		266	.030		3.66	1.75		5.41	6.65
8440	3" diameter		160	.050		4.19	2.91		7.10	8.95
8460	3-1/2" diameter		133	.060		8	3.50		11.50	14.05
8480	4" diameter		100	.080		7.40	4.66		12.06	15.10
8500	Clamp back spacers, 1/2" diameter		500	.016		.85	.93		1.78	2.33
8510	3/4" diameter		470	.017		1	.99		1.99	2.58
8520	1" diameter		444	.018		1.70	1.05		2.75	3.44
8530	1-1/4" diameter		400	.020		3.99	1.16		5.15	6.15
8540	1-1/2" diameter		355	.023		6.80	1.31		8.11	9.45
8550	2" diameter		320	.025		5.40	1.46		6.86	8.15
8560	2-1/2" diameter		266	.030		12.25	1.75		14	16.10
8570	3" diameter		160	.050		31	2.91		33.91	39
8580	3-1/2" diameter		133	.060		40	3.50		43.50	49.50
8590	4" diameter		100	.080		91	4.66		95.66	107
8600	Offset connectors, 1/2" diameter		40	.200		3.22	11.65		14.87	21
8610	3/4" diameter		32	.250		3.69	14.55		18.24	26
8620	1" diameter		24	.333		4.36	19.40		23.76	34
8650	90° pulling elbows, female, 1/2" diameter, with gasket		24	.333		7.25	19.40		26.65	37
8660	3/4" diameter		20	.400		9.70	23.50		33.20	45.50
8700	Couplings, compression, 1/2" diameter, steel		78	.103		1.51	5.95		7.46	10.55
8710	3/4" diameter		67	.119		1.72	6.95		8.67	12.30
8720	1" diameter		59	.136		2.99	7.90		10.89	15.10
8730	1-1/4" diameter		47	.170		6.05	9.90		15.95	21.50
8740	1-1/2" diameter		38	.211		8.90	12.25		21.15	28
8750	2" diameter		31	.258		12.05	15		27.05	36
8760	2-1/2" diameter		24	.333		47.50	19.40		66.90	81
8770	3" diameter		21	.381		27	22		49	62.50
8780	3-1/2" diameter		19	.421		102	24.50		126.50	149
8790	4" diameter		16	.500		86	29		115	138
8791	See note on line 26 05 33.13 9995 R260533-30									
8800	Box connectors, compression, 1/2" diam., steel	1 Elec	120	.067	Ea.	2.22	3.88		6.10	8.25
8810	3/4" diameter		110	.073		3.06	4.23		7.29	9.70
8820	1" diameter		90	.089		4.43	5.15		9.58	12.60
8830	1-1/4" diameter		70	.114		11.80	6.65		18.45	23
8840	1-1/2" diameter		60	.133		13.55	7.75		21.30	26.50
8850	2" diameter		50	.160		18.05	9.30		27.35	34
8860	2-1/2" diameter		36	.222		48.50	12.95		61.45	72.50
8870	3" diameter		27	.296		65.50	17.25		82.75	98
8880	3-1/2" diameter		21	.381		102	22		124	145
8890	4" diameter		16	.500		103	29		132	157
8900	Box connectors, insulated compression, 1/2" diam., steel		120	.067		1.85	3.88		5.73	7.85
8910	3/4" diameter		110	.073		2.10	4.23		6.33	8.65
8920	1" diameter		90	.089		3.93	5.15		9.08	12.05
8930	1-1/4" diameter		70	.114		8.05	6.65		14.70	18.85
8940	1-1/2" diameter		60	.133		11.95	7.75		19.70	25
8950	2" diameter		50	.160		17.60	9.30		26.90	33.50
8960	2-1/2" diameter		36	.222		64	12.95		76.95	90

For customer support on your Electrical Costs with RSMeans data, call 800.448.8182.

145

26 05 33.13 Conduit		Crew	Daily Output	Labor-Hours	Unit	Material	2018 Bare Costs Labor	Equipment	Total	Total Incl O&P
8970	3" diameter	1 Elec	27	.296	Ea.	59	17.25		76.25	90.50
8980	3-1/2" diameter		21	.381		95.50	22		117.50	138
8990	4" diameter		16	.500	▼	88	29		117	141
9100	PVC, schedule 40, 1/2" diameter		190	.042	L.F.	.89	2.45		3.34	4.64
9110	3/4" diameter		145	.055		1	3.21		4.21	5.90
9120	1" diameter		125	.064		1.52	3.72		5.24	7.20
9130	1-1/4" diameter		110	.073		1.80	4.23		6.03	8.35
9140	1-1/2" diameter		100	.080		2.09	4.66		6.75	9.25
9150	2" diameter		90	.089		2.72	5.15		7.87	10.75
9160	2-1/2" diameter	▼	65	.123		3.28	7.15		10.43	14.30
9170	3" diameter	2 Elec	110	.145		4.58	8.45		13.03	17.70
9180	3-1/2" diameter		100	.160		6.25	9.30		15.55	21
9190	4" diameter		90	.178		5.45	10.35		15.80	21.50
9200	5" diameter		70	.229		7.80	13.30		21.10	28.50
9210	6" diameter	▼	60	.267	▼	11.35	15.50		26.85	35.50
9220	Elbows, 1/2" diameter	1 Elec	50	.160	Ea.	.62	9.30		9.92	14.60
9225	3/4" diameter		42	.190		.78	11.10		11.88	17.40
9230	1" diameter		35	.229		1.30	13.30		14.60	21.50
9235	1-1/4" diameter		28	.286		2.17	16.65		18.82	27.50
9240	1-1/2" diameter	▼	20	.400		2.22	23.50		25.72	37.50
9245	2" diameter	R-1A	36.40	.440		2.91	23		25.91	37.50
9250	2-1/2" diameter		26.70	.599		6.60	31.50		38.10	54.50
9255	3" diameter		22.90	.699		10.50	36.50		47	66
9260	3-1/2" diameter		22.20	.721		13.65	38		51.65	71.50
9265	4" diameter		18.20	.879		17.05	46		63.05	88
9270	5" diameter		12.10	1.322		26	69.50		95.50	133
9275	6" diameter	▼	11.10	1.441		36	75.50		111.50	153
9312	Couplings, 1/2" diameter	1 Elec	50	.160		.11	9.30		9.41	14
9314	3/4" diameter		42	.190		.21	11.10		11.31	16.80
9316	1" diameter		35	.229		.24	13.30		13.54	20
9318	1-1/4" diameter		28	.286		.35	16.65		17	25.50
9320	1-1/2" diameter	▼	20	.400		.40	23.50		23.90	35.50
9322	2" diameter	R-1A	36.40	.440		1.15	23		24.15	36
9324	2-1/2" diameter		26.70	.599		.90	31.50		32.40	48
9326	3" diameter		22.90	.699		1.46	36.50		37.96	56
9328	3-1/2" diameter		22.20	.721		1.78	38		39.78	58.50
9330	4" diameter		18.20	.879		2.32	46		48.32	71.50
9332	5" diameter		12.10	1.322		6	69.50		75.50	111
9334	6" diameter	▼	11.10	1.441	▼	7.65	75.50		83.15	121
9335	See note on line 26 05 33.13 9995	R260533-30								
9340	Field bends, 45° & 90°, 1/2" diameter	1 Elec	45	.178	Ea.		10.35		10.35	15.45
9350	3/4" diameter		40	.200			11.65		11.65	17.40
9360	1" diameter		35	.229			13.30		13.30	19.90
9370	1-1/4" diameter		32	.250			14.55		14.55	22
9380	1-1/2" diameter		27	.296			17.25		17.25	26
9390	2" diameter		20	.400			23.50		23.50	35
9400	2-1/2" diameter		16	.500			29		29	43.50
9410	3" diameter		13	.615			36		36	53.50
9420	3-1/2" diameter		12	.667			39		39	58
9430	4" diameter		10	.800			46.50		46.50	69.50
9440	5" diameter		9	.889			51.50		51.50	77.50
9450	6" diameter		8	1			58		58	87
9460	PVC adapters, 1/2" diameter	▼	50	.160	▼	.21	9.30		9.51	14.15

26 05 33.13 Conduit		Crew	Daily Output	Labor-Hours	Unit	Material	2018 Bare Costs Labor	Equipment	Total	Total Incl O&P
9470	3/4" diameter	1 Elec	42	.190	Ea.	.34	11.10		11.44	16.90
9480	1" diameter		38	.211		.48	12.25		12.73	18.85
9490	1-1/4" diameter		35	.229		.65	13.30		13.95	20.50
9500	1-1/2" diameter		32	.250		.79	14.55		15.34	23
9510	2" diameter		27	.296		1.11	17.25		18.36	27
9520	2-1/2" diameter		23	.348		1.88	20		21.88	32.50
9530	3" diameter		18	.444		2.63	26		28.63	41.50
9540	3-1/2" diameter		13	.615		5.35	36		41.35	59.50
9550	4" diameter		11	.727		4.40	42.50		46.90	68.50
9560	5" diameter		8	1		8.30	58		66.30	96
9570	6" diameter		6	1.333		19	77.50		96.50	137
9580	PVC-LB, LR or LL fittings & covers									
9590	1/2" diameter	1 Elec	20	.400	Ea.	2.74	23.50		26.24	38
9600	3/4" diameter		16	.500		4	29		33	48
9610	1" diameter		12	.667		4.15	39		43.15	62.50
9620	1-1/4" diameter		9	.889		7.20	51.50		58.70	85.50
9630	1-1/2" diameter		7	1.143		8.50	66.50		75	109
9640	2" diameter		6	1.333		12.30	77.50		89.80	130
9650	2-1/2" diameter		6	1.333		39.50	77.50		117	160
9660	3" diameter		5	1.600		39	93		132	182
9670	3-1/2" diameter		4	2		40	116		156	218
9680	4" diameter		3	2.667		61	155		216	300
9690	PVC-tee fitting & cover									
9700	1/2"	1 Elec	14	.571	Ea.	4.77	33.50		38.27	55
9710	3/4"		13	.615		4.83	36		40.83	59
9720	1"		10	.800		5.10	46.50		51.60	75
9730	1-1/4"		9	.889		9.80	51.50		61.30	88.50
9740	1-1/2"		8	1		10.35	58		68.35	98.50
9750	2"		7	1.143		16.25	66.50		82.75	117
9760	PVC-reducers, 3/4" x 1/2" diameter					1.08			1.08	1.19
9770	1" x 1/2" diameter					2.40			2.40	2.64
9780	1" x 3/4" diameter					2.67			2.67	2.94
9790	1-1/4" x 3/4" diameter					2.83			2.83	3.11
9800	1-1/4" x 1" diameter					3.66			3.66	4.03
9810	1-1/2" x 1-1/4" diameter					3.59			3.59	3.95
9820	2" x 1-1/4" diameter					4.48			4.48	4.93
9830	2-1/2" x 2" diameter					14.90			14.90	16.35
9840	3" x 2" diameter					12.80			12.80	14.10
9850	4" x 3" diameter					15.75			15.75	17.35
9860	Cement, quart					14.70			14.70	16.15
9870	Gallon					69.50			69.50	76.50
9880	Heat bender, to 6" diameter					1,400			1,400	1,550
9900	Add to labor for higher elevated installation									
9905	10' to 14.5' high, add						10%			
9910	15' to 20' high, add						20%			
9920	20' to 25' high, add						25%			
9930	25' to 30' high, add						35%			
9940	30' to 35' high, add						40%			
9950	35' to 40' high, add						50%			
9960	Over 40' high, add						55%			
9995	Do not include labor when adding couplings to a fitting installation	R260533-30								

For customer support on your Electrical Costs with RSMeans data, call 800.448.8182.

147

26 05 33.14 Conduit	Crew	Daily Output	Labor-Hours	Unit	Material	2018 Bare Costs Labor	Equipment	Total	Total Incl O&P
0010 **CONDUIT** To 10' high, includes 11 couplings per 100'									
0200 Electric metallic tubing, 1/2" diameter	1 Elec	435	.018	L.F.	.60	1.07		1.67	2.26
0220 3/4" diameter		253	.032		.93	1.84		2.77	3.77
0240 1" diameter		207	.039		1.54	2.25		3.79	5.05
0260 1-1/4" diameter		173	.046		2.63	2.69		5.32	6.90
0280 1-1/2" diameter		153	.052		2.43	3.04		5.47	7.20
0300 2" diameter		130	.062		3.38	3.58		6.96	9.05
0320 2-1/2" diameter		92	.087		3.55	5.05		8.60	11.45
0340 3" diameter	2 Elec	148	.108		4.37	6.30		10.67	14.20
0360 3-1/2" diameter		134	.119		5.70	6.95		12.65	16.65
0380 4" diameter		114	.140		10.60	8.15		18.75	24
0500 Steel rigid galvanized, 1/2" diameter	1 Elec	146	.055		1.56	3.19		4.75	6.50
0520 3/4" diameter		125	.064		3.83	3.72		7.55	9.75
0540 1" diameter		93	.086		6.10	5		11.10	14.20
0560 1-1/4" diameter		88	.091		3.64	5.30		8.94	11.90
0580 1-1/2" diameter		80	.100		6.60	5.80		12.40	15.95
0600 2" diameter		65	.123		8.15	7.15		15.30	19.65
0620 2-1/2" diameter		48	.167		9.90	9.70		19.60	25.50
0640 3" diameter	2 Elec	64	.250		11.45	14.55		26	34.50
0660 3-1/2" diameter		60	.267		14.80	15.50		30.30	39.50
0680 4" diameter		52	.308		16.90	17.90		34.80	45.50
0700 5" diameter		50	.320		31	18.60		49.60	62
0720 6" diameter		48	.333		41.50	19.40		60.90	74.50
1000 Steel intermediate conduit (IMC), 1/2" diameter R260533-25	1 Elec	155	.052		1.02	3		4.02	5.60
1010 3/4" diameter		130	.062		1.19	3.58		4.77	6.65
1020 1" diameter		100	.080		1.80	4.66		6.46	8.95
1030 1-1/4" diameter		93	.086		2.40	5		7.40	10.15
1040 1-1/2" diameter		85	.094		3.21	5.50		8.71	11.75
1050 2" diameter		70	.114		4.06	6.65		10.71	14.40
1060 2-1/2" diameter		53	.151		7.75	8.80		16.55	21.50
1070 3" diameter	2 Elec	80	.200		10.15	11.65		21.80	28.50
1080 3-1/2" diameter		70	.229		12.15	13.30		25.45	33.50
1090 4" diameter		60	.267		9	15.50		24.50	33

26 05 33.15 Conduit Nipples

	Crew	Daily Output	Labor-Hours	Unit	Material	2018 Bare Costs Labor	Equipment	Total	Total Incl O&P
0010 **CONDUIT NIPPLES** With locknuts and bushings									
0100 Aluminum, 1/2" diameter, close	1 Elec	36	.222	Ea.	5.95	12.95		18.90	26
0120 1-1/2" long		36	.222		6	12.95		18.95	26
0140 2" long		36	.222		6.05	12.95		19	26
0160 2-1/2" long		36	.222		9.05	12.95		22	29.50
0180 3" long		36	.222		6.75	12.95		19.70	27
0200 3-1/2" long		36	.222		6.50	12.95		19.45	26.50
0220 4" long		36	.222		6.60	12.95		19.55	26.50
0240 5" long		36	.222		7.30	12.95		20.25	27.50
0260 6" long		36	.222		7.45	12.95		20.40	27.50
0280 8" long		36	.222		8.85	12.95		21.80	29
0300 10" long		36	.222		9.80	12.95		22.75	30
0320 12" long		36	.222		16.55	12.95		29.50	37.50
0340 3/4" diameter, close		32	.250		8.20	14.55		22.75	31
0360 1-1/2" long		32	.250		8.25	14.55		22.80	31
0380 2" long		32	.250		8.30	14.55		22.85	31
0400 2-1/2" long		32	.250		8.60	14.55		23.15	31.50
0420 3" long		32	.250		8.80	14.55		23.35	31.50

26 05 33.15 Conduit Nipples		Crew	Daily Output	Labor-Hours	Unit	Material	2018 Bare Costs Labor	Equipment	Total	Total Incl O&P
0440	3-1/2" long	1 Elec	32	.250	Ea.	11.60	14.55		26.15	35
0460	4" long		32	.250		9.75	14.55		24.30	32.50
0480	5" long		32	.250		9.50	14.55		24.05	32.50
0500	6" long		32	.250		10.75	14.55		25.30	34
0520	8" long		32	.250		11.80	14.55		26.35	35
0540	10" long		32	.250		13.10	14.55		27.65	36.50
0560	12" long		32	.250		14.75	14.55		29.30	38.50
0580	1" diameter, close		27	.296		12	17.25		29.25	39
0600	2" long		27	.296		17.95	17.25		35.20	46
0620	2-1/2" long		27	.296		12.70	17.25		29.95	40
0640	3" long		27	.296		13.40	17.25		30.65	41
0660	3-1/2" long		27	.296		18.05	17.25		35.30	46
0680	4" long		27	.296		13.95	17.25		31.20	41.50
0700	5" long		27	.296		15.95	17.25		33.20	43.50
0720	6" long		27	.296		16.40	17.25		33.65	44
0740	8" long		27	.296		24.50	17.25		41.75	53
0760	10" long		27	.296		33	17.25		50.25	62.50
0780	12" long		27	.296		33	17.25		50.25	62.50
0800	1-1/4" diameter, close		23	.348		17.30	20		37.30	49.50
0820	2" long		23	.348		22	20		42	54.50
0840	2-1/2" long		23	.348		17.50	20		37.50	50
0860	3" long		23	.348		18.65	20		38.65	51
0880	3-1/2" long		23	.348		18.85	20		38.85	51.50
0900	4" long		23	.348		19.20	20		39.20	51.50
0920	5" long		23	.348		21	20		41	53.50
0940	6" long		23	.348		23.50	20		43.50	56
0960	8" long		23	.348		33.50	20		53.50	67.50
0980	10" long		23	.348		27	20		47	60
1000	12" long		23	.348		30	20		50	63.50
1020	1-1/2" diameter, close		20	.400		24.50	23.50		48	62
1040	2" long		20	.400		24	23.50		47.50	61.50
1060	2-1/2" long		20	.400		24	23.50		47.50	61.50
1080	3" long		20	.400		24	23.50		47.50	61.50
1100	3-1/2" long		20	.400		26.50	23.50		50	64
1120	4" long		20	.400		27.50	23.50		51	65
1140	5" long		20	.400		28	23.50		51.50	66
1160	6" long		20	.400		29	23.50		52.50	67
1180	8" long		20	.400		33	23.50		56.50	71.50
1200	10" long		20	.400		38	23.50		61.50	77
1220	12" long		20	.400		40.50	23.50		64	79.50
1240	2" diameter, close		18	.444		36	26		62	78
1260	2-1/2" long		18	.444		35	26		61	77
1280	3" long		18	.444		36.50	26		62.50	78.50
1300	3-1/2" long		18	.444		37.50	26		63.50	79.50
1320	4" long		18	.444		38	26		64	80
1340	5" long		18	.444		40	26		66	82.50
1360	6" long		18	.444		42.50	26		68.50	85
1380	8" long		18	.444		47.50	26		73.50	91
1400	10" long		18	.444		50	26		76	93.50
1420	12" long		18	.444		58	26		84	103
1440	2-1/2" diameter, close		15	.533		71.50	31		102.50	126
1460	3" long		15	.533		71.50	31		102.50	125
1480	3-1/2" long		15	.533		73.50	31		104.50	127

For customer support on your Electrical Costs with RSMeans data, call 800.448.8182.

149

26 05 33.15 Conduit Nipples	Crew	Daily Output	Labor-Hours	Unit	Material	2018 Bare Costs Labor	Equipment	Total	Total Incl O&P
1500	4" long	1 Elec	15	.533	Ea.	80	31		111	135
1520	5" long		15	.533		90.50	31		121.50	146
1540	6" long		15	.533		78.50	31		109.50	133
1560	8" long		15	.533		82.50	31		113.50	137
1580	10" long		15	.533		103	31		134	160
1600	12" long		15	.533		93	31		124	149
1620	3" diameter, close		12	.667		87.50	39		126.50	155
1640	3" long		12	.667		95.50	39		134.50	163
1660	3-1/2" long		12	.667		117	39		156	186
1680	4" long		12	.667		118	39		157	187
1700	5" long		12	.667		94	39		133	161
1720	6" long		12	.667		101	39		140	169
1740	8" long		12	.667		111	39		150	180
1760	10" long		12	.667		147	39		186	219
1780	12" long		12	.667		123	39		162	193
1800	3-1/2" diameter, close		11	.727		183	42.50		225.50	265
1820	4" long		11	.727		191	42.50		233.50	274
1840	5" long		11	.727		191	42.50		233.50	274
1860	6" long		11	.727		202	42.50		244.50	286
1880	8" long		11	.727		183	42.50		225.50	265
1900	10" long		11	.727		212	42.50		254.50	297
1920	12" long		11	.727		285	42.50		327.50	380
1940	4" diameter, close		9	.889		202	51.50		253.50	300
1960	4" long		9	.889		209	51.50		260.50	310
1980	5" long		9	.889		223	51.50		274.50	325
2000	6" long		9	.889		267	51.50		318.50	370
2020	8" long		9	.889		231	51.50		282.50	330
2040	10" long		9	.889		246	51.50		297.50	350
2060	12" long		9	.889		249	51.50		300.50	350
2080	5" diameter, close		7	1.143		435	66.50		501.50	575
2100	5" long		7	1.143		545	66.50		611.50	700
2120	6" long		7	1.143		530	66.50		596.50	685
2140	8" long		7	1.143		560	66.50		626.50	715
2160	10" long		7	1.143		485	66.50		551.50	630
2180	12" long		7	1.143		500	66.50		566.50	650
2200	6" diameter, close		6	1.333		700	77.50		777.50	885
2220	5" long		6	1.333		760	77.50		837.50	950
2240	6" long		6	1.333		690	77.50		767.50	875
2260	8" long		6	1.333		820	77.50		897.50	1,025
2280	10" long		6	1.333		870	77.50		947.50	1,075
2300	12" long		6	1.333		875	77.50		952.50	1,075
2320	Rigid galvanized steel, 1/2" diameter, close		32	.250		2.58	14.55		17.13	25
2340	1-1/2" long		32	.250		2.77	14.55		17.32	25
2360	2" long		32	.250		2.89	14.55		17.44	25
2380	2-1/2" long		32	.250		2.99	14.55		17.54	25.50
2400	3" long		32	.250		3.11	14.55		17.66	25.50
2420	3-1/2" long		32	.250		3.24	14.55		17.79	25.50
2440	4" long		32	.250		3.37	14.55		17.92	25.50
2460	5" long		32	.250		3.55	14.55		18.10	26
2480	6" long		32	.250		3.94	14.55		18.49	26.50
2500	8" long		32	.250		5.60	14.55		20.15	28
2520	10" long		32	.250		6.50	14.55		21.05	29
2540	12" long		32	.250		6.85	14.55		21.40	29.50

26 05 33.15 Conduit Nipples	Crew	Daily Output	Labor-Hours	Unit	Material	2018 Bare Costs Labor	Equipment	Total	Total Incl O&P
2560	3/4" diameter, close	1 Elec	27	.296	Ea.	3.33	17.25		20.58	29.50
2580	2" long		27	.296		3.55	17.25		20.80	30
2600	2-1/2" long		27	.296		3.70	17.25		20.95	30
2620	3" long		27	.296		3.83	17.25		21.08	30
2640	3-1/2" long		27	.296		3.90	17.25		21.15	30.50
2660	4" long		27	.296		4.13	17.25		21.38	30.50
2680	5" long		27	.296		4.41	17.25		21.66	31
2700	6" long		27	.296		4.78	17.25		22.03	31.50
2720	8" long		27	.296		6.50	17.25		23.75	33
2740	10" long		27	.296		7.85	17.25		25.10	34.50
2760	12" long		27	.296		8	17.25		25.25	35
2780	1" diameter, close		23	.348		5.55	20		25.55	36.50
2800	2" long		23	.348		5.70	20		25.70	37
2820	2-1/2" long		23	.348		5.85	20		25.85	37
2840	3" long		23	.348		6.10	20		26.10	37
2860	3-1/2" long		23	.348		6.45	20		26.45	37.50
2880	4" long		23	.348		6.60	20		26.60	38
2900	5" long		23	.348		6.95	20		26.95	38
2920	6" long		23	.348		7.30	20		27.30	38.50
2940	8" long		23	.348		9.25	20		29.25	40.50
2960	10" long		23	.348		10.85	20		30.85	42.50
2980	12" long		23	.348		11.80	20		31.80	43.50
3000	1-1/4" diameter, close		20	.400		7.70	23.50		31.20	43.50
3020	2" long		20	.400		7.90	23.50		31.40	43.50
3040	3" long		20	.400		8.35	23.50		31.85	44
3060	3-1/2" long		20	.400		8.65	23.50		32.15	44.50
3080	4" long		20	.400		8.90	23.50		32.40	45
3100	5" long		20	.400		9.45	23.50		32.95	45.50
3120	6" long		20	.400		9.95	23.50		33.45	46
3140	8" long		20	.400		12.75	23.50		36.25	49
3160	10" long		20	.400		14.70	23.50		38.20	51
3180	12" long		20	.400		16.10	23.50		39.60	52.50
3200	1-1/2" diameter, close		18	.444		11	26		37	50.50
3220	2" long		18	.444		11.15	26		37.15	51
3240	2-1/2" long		18	.444		11.50	26		37.50	51
3260	3" long		18	.444		11.75	26		37.75	51.50
3280	3-1/2" long		18	.444		12.25	26		38.25	52
3300	4" long		18	.444		12.60	26		38.60	52.50
3320	5" long		18	.444		13.20	26		39.20	53
3340	6" long		18	.444		14.25	26		40.25	54
3360	8" long		18	.444		17.35	26		43.35	57.50
3380	10" long		18	.444		19.25	26		45.25	59.50
3400	12" long		18	.444		20	26		46	60.50
3420	2" diameter, close		16	.500		14.10	29		43.10	59
3440	2-1/2" long		16	.500		14.65	29		43.65	59.50
3460	3" long		16	.500		15.30	29		44.30	60.50
3480	3-1/2" long		16	.500		15.90	29		44.90	61
3500	4" long		16	.500		16.45	29		45.45	61.50
3520	5" long		16	.500		17.40	29		46.40	62.50
3540	6" long		16	.500		18.25	29		47.25	63.50
3560	8" long		16	.500		21.50	29		50.50	67.50
3580	10" long		16	.500		23.50	29		52.50	69.50
3600	12" long		16	.500		25.50	29		54.50	71.50

For customer support on your Electrical Costs with RSMeans data, call 800.448.8182.

151

26 05 33.15 Conduit Nipples	Crew	Daily Output	Labor-Hours	Unit	Material	2018 Bare Costs Labor	Equipment	Total	Total Incl O&P
3620	2-1/2" diameter, close	1 Elec	13	.615	Ea.	39.50	36		75.50	97
3640	3" long		13	.615		43	36		79	101
3660	3-1/2" long		13	.615		45	36		81	103
3680	4" long		13	.615		45.50	36		81.50	104
3700	5" long		13	.615		48	36		84	107
3720	6" long		13	.615		49.50	36		85.50	108
3740	8" long		13	.615		54.50	36		90.50	114
3760	10" long		13	.615		57.50	36		93.50	117
3780	12" long		13	.615		62	36		98	122
3800	3" diameter, close		12	.667		46	39		85	109
3820	3" long		12	.667		47	39		86	110
3900	3-1/2" long		12	.667		47.50	39		86.50	111
3920	4" long		12	.667		49	39		88	112
3940	5" long		12	.667		51	39		90	115
3960	6" long		12	.667		53.50	39		92.50	117
3980	8" long		12	.667		59	39		98	123
4000	10" long		12	.667		63.50	39		102.50	128
4020	12" long		12	.667		70	39		109	135
4040	3-1/2" diameter, close		10	.800		73.50	46.50		120	151
4060	4" long		10	.800		77	46.50		123.50	154
4080	5" long		10	.800		79.50	46.50		126	157
4100	6" long		10	.800		82	46.50		128.50	160
4120	8" long		10	.800		87	46.50		133.50	166
4140	10" long		10	.800		93	46.50		139.50	172
4160	12" long		10	.800		99	46.50		145.50	179
4180	4" diameter, close		8	1		114	58		172	213
4200	4" long		8	1		116	58		174	215
4220	5" long		8	1		118	58		176	217
4240	6" long		8	1		120	58		178	219
4260	8" long		8	1		123	58		181	223
4280	10" long		8	1		128	58		186	227
4300	12" long		8	1		132	58		190	232
4320	5" diameter, close		6	1.333		182	77.50		259.50	315
4340	5" long		6	1.333		194	77.50		271.50	330
4360	6" long		6	1.333		199	77.50		276.50	335
4380	8" long		6	1.333		203	77.50		280.50	340
4400	10" long		6	1.333		217	77.50		294.50	355
4420	12" long		6	1.333		233	77.50		310.50	370
4440	6" diameter, close		5	1.600		305	93		398	475
4460	5" long		5	1.600		325	93		418	495
4480	6" long		5	1.600		330	93		423	505
4500	8" long		5	1.600		340	93		433	515
4520	10" long		5	1.600		355	93		448	530
4540	12" long		5	1.600		365	93		458	545
4560	Plastic coated, 40 mil thick, 1/2" diameter, 2" long		32	.250		17.40	14.55		31.95	41
4580	2-1/2" long		32	.250		18.75	14.55		33.30	42.50
4600	3" long		32	.250		19	14.55		33.55	43
4680	3-1/2" long		32	.250		20.50	14.55		35.05	44.50
4700	4" long		32	.250		20.50	14.55		35.05	44.50
4720	5" long		32	.250		20.50	14.55		35.05	44.50
4740	6" long		32	.250		21	14.55		35.55	45.50
4760	8" long		32	.250		22.50	14.55		37.05	46.50
4780	10" long		32	.250		25	14.55		39.55	49.50

26 05 33.15 Conduit Nipples		Crew	Daily Output	Labor-Hours	Unit	Material	2018 Bare Costs Labor	Equipment	Total	Total Incl O&P
4800	12" long	1 Elec	32	.250	Ea.	31.50	14.55		46.05	56.50
4820	3/4" diameter, 2" long		26	.308		18.50	17.90		36.40	47.50
4840	2-1/2" long		26	.308		19.10	17.90		37	48
4860	3" long		26	.308		20.50	17.90		38.40	49.50
4880	3-1/2" long		26	.308		21.50	17.90		39.40	50.50
4900	4" long		26	.308		22.50	17.90		40.40	51.50
4920	5" long		26	.308		22.50	17.90		40.40	52
4940	6" long		26	.308		30	17.90		47.90	60
4960	8" long		26	.308		23.50	17.90		41.40	53
4980	10" long		26	.308		24	17.90		41.90	53.50
5000	12" long		26	.308		25.50	17.90		43.40	55
5020	1" diameter, 2" long		22	.364		21	21		42	54.50
5040	2-1/2" long		22	.364		22	21		43	55.50
5060	3" long		22	.364		22.50	21		43.50	56.50
5080	3-1/2" long		22	.364		25	21		46	59
5100	4" long		22	.364		25	21		46	59
5120	5" long		22	.364		24.50	21		45.50	58.50
5140	6" long		22	.364		25.50	21		46.50	59.50
5160	8" long		22	.364		27	21		48	61
5180	10" long		22	.364		29	21		50	63.50
5200	12" long		22	.364		31.50	21		52.50	66
5220	1-1/4" diameter, 2" long		18	.444		25.50	26		51.50	66.50
5240	2-1/2" long		18	.444		28	26		54	69
5260	3" long		18	.444		28	26		54	69.50
5280	3-1/2" long		18	.444		29.50	26		55.50	70.50
5300	4" long		18	.444		33.50	26		59.50	75
5320	5" long		18	.444		31	26		57	72.50
5340	6" long		18	.444		31	26		57	72.50
5360	8" long		18	.444		32.50	26		58.50	74
5380	10" long		18	.444		36.50	26		62.50	78.50
5400	12" long		18	.444		43.50	26		69.50	86.50
5420	1-1/2" diameter, 2" long		16	.500		30	29		59	76.50
5440	2-1/2" long		16	.500		32	29		61	78.50
5460	3" long		16	.500		31	29		60	77.50
5480	3-1/2" long		16	.500		33.50	29		62.50	80.50
5500	4" long		16	.500		33.50	29		62.50	80
5520	5" long		16	.500		34	29		63	81
5540	6" long		16	.500		37	29		66	84.50
5560	8" long		16	.500		40	29		69	87
5580	10" long		16	.500		48.50	29		77.50	97
5600	12" long		16	.500		57	29		86	106
5620	2" diameter, 2-1/2" long		14	.571		36	33.50		69.50	89
5640	3" long		14	.571		37.50	33.50		71	90.50
5660	3-1/2" long		14	.571		39	33.50		72.50	92.50
5680	4" long		14	.571		44	33.50		77.50	98
5700	5" long		14	.571		43.50	33.50		77	97.50
5720	6" long		14	.571		45.50	33.50		79	99.50
5740	8" long		14	.571		50	33.50		83.50	105
5760	10" long		14	.571		62.50	33.50		96	119
5780	12" long		14	.571		70	33.50		103.50	127
5800	2-1/2" diameter, 3-1/2" long		12	.667		79.50	39		118.50	145
5820	4" long		12	.667		82	39		121	148
5840	5" long		12	.667		92.50	39		131.50	160

For customer support on your Electrical Costs with RSMeans data, call 800.448.8182.

153

26 05 33.15 Conduit Nipples	Crew	Daily Output	Labor-Hours	Unit	Material	2018 Bare Costs Labor	Equipment	Total	Total Incl O&P
5860	6" long	1 Elec	12	.667	Ea.	93	39		132	161
5880	8" long		12	.667		106	39		145	175
5900	10" long		12	.667		119	39		158	188
5920	12" long		12	.667		128	39		167	199
5940	3" diameter, 3-1/2" long		11	.727		95	42.50		137.50	168
5960	4" long		11	.727		92	42.50		134.50	165
5980	5" long		11	.727		99.50	42.50		142	174
6000	6" long		11	.727		106	42.50		148.50	181
6020	8" long		11	.727		128	42.50		170.50	205
6040	10" long		11	.727		135	42.50		177.50	212
6060	12" long		11	.727		160	42.50		202.50	240
6080	3-1/2" diameter, 4" long		9	.889		132	51.50		183.50	223
6100	5" long		9	.889		134	51.50		185.50	225
6120	6" long		9	.889		139	51.50		190.50	231
6140	8" long		9	.889		161	51.50		212.50	255
6160	10" long		9	.889		181	51.50		232.50	277
6180	12" long		9	.889		201	51.50		252.50	299
6200	4" diameter, 4" long		7.50	1.067		182	62		244	293
6220	5" long		7.50	1.067		184	62		246	295
6240	6" long		7.50	1.067		200	62		262	315
6260	8" long		7.50	1.067		215	62		277	330
6280	10" long		7.50	1.067		267	62		329	385
6300	12" long		7.50	1.067		275	62		337	395
6320	5" diameter, 5" long		5.50	1.455		152	84.50		236.50	295
6340	6" long		5.50	1.455		234	84.50		318.50	385
6360	8" long		5.50	1.455		242	84.50		326.50	395
6380	10" long		5.50	1.455		254	84.50		338.50	405
6400	12" long		5.50	1.455		277	84.50		361.50	430
6420	6" diameter, 5" long		4.50	1.778		360	103		463	550
6440	6" long		4.50	1.778		370	103		473	565
6460	8" long		4.50	1.778		390	103		493	580
6480	10" long		4.50	1.778		405	103		508	600
6500	12" long		4.50	1.778		425	103		528	620

26 05 33.16 Boxes for Electrical Systems

		Crew	Daily Output	Labor-Hours	Unit	Material	2018 Bare Costs Labor	Equipment	Total	Total Incl O&P
0010	**BOXES FOR ELECTRICAL SYSTEMS** R260533-65									
0020	Pressed steel, octagon, 4"	1 Elec	20	.400	Ea.	2.86	23.50		26.36	38
0040	For Romex or BX		20	.400		4.59	23.50		28.09	40
0050	For Romex or BX, with bracket		20	.400		6.85	23.50		30.35	42.50
0060	Covers, blank		64	.125		.88	7.30		8.18	11.85
0100	Extension rings		40	.200		3.68	11.65		15.33	21.50
0150	Square, 4"		20	.400		5.25	23.50		28.75	41
0160	For Romex or BX		20	.400		9.85	23.50		33.35	46
0170	For Romex or BX, with bracket		20	.400		8.80	23.50		32.30	44.50
0200	Extension rings		40	.200		3.79	11.65		15.44	21.50
0220	2-1/8" deep, 1" KO		20	.400		2.94	23.50		26.44	38
0250	Covers, blank		64	.125		.67	7.30		7.97	11.65
0260	Raised device		64	.125		1.19	7.30		8.49	12.20
0300	Plaster rings		64	.125		1.54	7.30		8.84	12.60
0350	Square, 4-11/16"		20	.400		3.77	23.50		27.27	39
0370	2-1/8" deep, 3/4" to 1-1/4" KO		20	.400		4.63	23.50		28.13	40
0400	Extension rings		40	.200		8.50	11.65		20.15	27
0450	Covers, blank		53	.151		1.18	8.80		9.98	14.45

154

For customer support on your Electrical Costs with RSMeans data, call 800.448.8182.

26 05 33.16 Boxes for Electrical Systems	Crew	Daily Output	Labor-Hours	Unit	Material	2018 Bare Costs Labor	Equipment	Total	Total Incl O&P
0460	Raised device	1 Elec	53	.151	Ea.	5.65	8.80		14.45	19.35
0500	Plaster rings		53	.151		5.75	8.80		14.55	19.50
0550	Handy box		27	.296		1.99	17.25		19.24	28
0560	Covers, device		64	.125		1	7.30		8.30	12
0600	Extension rings		54	.148		2.61	8.60		11.21	15.75
0650	Switchbox		27	.296		4.61	17.25		21.86	31
0660	Romex or BX		27	.296		6.25	17.25		23.50	33
0670	with bracket		27	.296		5.55	17.25		22.80	32
0680	Partition, metal		27	.296		3.12	17.25		20.37	29.50
0700	Masonry, 1 gang, 2-1/2" deep		27	.296		6.70	17.25		23.95	33.50
0710	3-1/2" deep		27	.296		6.60	17.25		23.85	33.50
0750	2 gang, 2-1/2" deep		20	.400		14.20	23.50		37.70	50.50
0760	3-1/2" deep		20	.400		9.45	23.50		32.95	45.50
0800	3 gang, 2-1/2" deep		13	.615		14.40	36		50.40	69.50
0850	4 gang, 2-1/2" deep		10	.800		19.95	46.50		66.45	91.50
0860	5 gang, 2-1/2" deep		9	.889		30	51.50		81.50	111
0870	6 gang, 2-1/2" deep		8	1		51	58		109	143
0880	Masonry thru-the-wall, 1 gang, 4" block		16	.500		29	29		58	75.50
0890	6" block		16	.500		40.50	29		69.50	88
0900	8" block		16	.500		48	29		77	96.50
0920	2 gang, 6" block		16	.500		60.50	29		89.50	110
0940	Bar hanger with 3/8" stud, for wood and masonry boxes		53	.151		4.90	8.80		13.70	18.55
0950	Concrete, set flush, 4" deep		20	.400		8.85	23.50		32.35	44.50
1000	Plate with 3/8" stud		80	.100		8.95	5.80		14.75	18.55
1100	Concrete, floor, 1 gang		5.30	1.509		94.50	88		182.50	235
1150	2 gang		4	2		143	116		259	330
1200	3 gang		2.70	2.963		209	172		381	490
1250	For duplex receptacle, pedestal mounted, add		24	.333		96.50	19.40		115.90	135
1270	Flush mounted, add		27	.296		30	17.25		47.25	59
1300	For telephone, pedestal mounted, add		30	.267		102	15.50		117.50	135
1350	Carpet flange, 1 gang		53	.151		52.50	8.80		61.30	71
1400	Cast, 1 gang, FS (2" deep), 1/2" hub		12	.667		21.50	39		60.50	81.50
1410	3/4" hub		12	.667		23	39		62	83.50
1420	FD (2-11/16" deep), 1/2" hub		12	.667		18.95	39		57.95	79
1430	3/4" hub		12	.667		21.50	39		60.50	81.50
1450	2 gang, FS, 1/2" hub		10	.800		39.50	46.50		86	113
1460	3/4" hub		10	.800		41.50	46.50		88	115
1470	FD, 1/2" hub		10	.800		49.50	46.50		96	124
1480	3/4" hub		10	.800		47.50	46.50		94	122
1500	3 gang, FS, 3/4" hub		9	.889		64	51.50		115.50	148
1510	Switch cover, 1 gang, FS		64	.125		5.70	7.30		13	17.15
1520	2 gang		53	.151		8.95	8.80		17.75	23
1530	Duplex receptacle cover, 1 gang, FS		64	.125		5.75	7.30		13.05	17.20
1540	2 gang, FS		53	.151		8.90	8.80		17.70	23
1542	Weatherproof blank cover, 1 gang		64	.125		1.22	7.30		8.52	12.25
1544	2 gang		53	.151		2.52	8.80		11.32	15.90
1550	Weatherproof switch cover, 1 gang		64	.125		5.50	7.30		12.80	16.95
1554	2 gang		53	.151		9.90	8.80		18.70	24
1600	Weatherproof receptacle cover, 1 gang		64	.125		7.95	7.30		15.25	19.60
1604	2 gang		53	.151		7.90	8.80		16.70	22
1620	Weatherproof receptacle cover, tamper resistant, 1 gang		58	.138		13.75	8.05		21.80	27
1624	2 gang		48	.167		27	9.70		36.70	44
1750	FSC, 1 gang, 1/2" hub		11	.727		24.50	42.50		67	90.50

26 05 33.16 Boxes for Electrical Systems	Crew	Daily Output	Labor-Hours	Unit	Material	2018 Bare Costs Labor	2018 Bare Costs Equipment	Total	Total Incl O&P
1760	3/4" hub	1 Elec	11	.727	Ea.	26.50	42.50		69	92.50
1770	2 gang, 1/2" hub		9	.889		45	51.50		96.50	127
1780	3/4" hub		9	.889		47.50	51.50		99	130
1790	FDC, 1 gang, 1/2" hub		11	.727		28	42.50		70.50	94.50
1800	3/4" hub		11	.727		29.50	42.50		72	96
1810	2 gang, 1/2" hub		9	.889		57.50	51.50		109	141
1820	3/4" hub		9	.889		50.50	51.50		102	133
1850	Weatherproof in-use cover, 1 gang		64	.125		24.50	7.30		31.80	38
1870	2 gang		53	.151		24.50	8.80		33.30	39.50
2000	Poke-thru fitting, fire rated, for 3-3/4" floor		6.80	1.176		142	68.50		210.50	258
2040	For 7" floor		6.80	1.176		194	68.50		262.50	315
2100	Pedestal, 15 amp, duplex receptacle & blank plate		5.25	1.524		140	88.50		228.50	287
2120	Duplex receptacle and telephone plate		5.25	1.524		140	88.50		228.50	287
2140	Pedestal, 20 amp, duplex recept. & phone plate		5	1.600		141	93		234	295
2160	Telephone plate, both sides		5.25	1.524		134	88.50		222.50	280
2200	Abandonment plate		32	.250		42	14.55		56.55	68.50

26 05 33.17 Outlet Boxes, Plastic

26 05 33.17 Outlet Boxes, Plastic		Crew	Daily Output	Labor-Hours	Unit	Material	2018 Bare Costs Labor	2018 Bare Costs Equipment	Total	Total Incl O&P
0010	**OUTLET BOXES, PLASTIC**									
0050	4" diameter, round with 2 mounting nails	1 Elec	25	.320	Ea.	2.73	18.60		21.33	31
0100	Bar hanger mounted		25	.320		5.30	18.60		23.90	34
0200	4", square with 2 mounting nails		25	.320		5.30	18.60		23.90	34
0300	Plaster ring		64	.125		2.02	7.30		9.32	13.10
0400	Switch box with 2 mounting nails, 1 gang		30	.267		3.03	15.50		18.53	26.50
0500	2 gang		25	.320		3.70	18.60		22.30	32
0600	3 gang		20	.400		4.92	23.50		28.42	40.50
0700	Old work box		30	.267		3.91	15.50		19.41	27.50
1400	PVC, FSS, 1 gang, 1/2" hub		14	.571		12.75	33.50		46.25	63.50
1410	3/4" hub		14	.571		9.65	33.50		43.15	60
1420	FD, 1 gang for variable terminations		14	.571		8.05	33.50		41.55	58.50
1450	FS, 2 gang for variable terminations		12	.667		10.20	39		49.20	69.50
1480	Weatherproof blank cover, FS, 1 gang		64	.125		3.75	7.30		11.05	15.05
1500	2 gang		53	.151		4.22	8.80		13.02	17.80
1510	Weatherproof switch cover, FS, 1 gang		64	.125		8.10	7.30		15.40	19.80
1520	2 gang		53	.151		15.20	8.80		24	30
1530	Weatherproof duplex receptacle cover, FS, 1 gang		64	.125		12.25	7.30		19.55	24.50
1540	2 gang		53	.151		13.65	8.80		22.45	28
1750	FSC, 1 gang, 1/2" hub		13	.615		8.05	36		44.05	62.50
1760	3/4" hub		13	.615		8.75	36		44.75	63
1770	FSC, 2 gang, 1/2" hub		11	.727		14.95	42.50		57.45	80
1780	3/4" hub		11	.727		13.15	42.50		55.65	78
1790	FDC, 1 gang, 1/2" hub		13	.615		9.25	36		45.25	63.50
1800	3/4" hub		13	.615		10.55	36		46.55	65
1810	Weatherproof, T box w/3 holes		14	.571		10.60	33.50		44.10	61
1820	4" diameter round w/5 holes		14	.571		9.95	33.50		43.45	60.50
1850	In-use cover, 1 gang		64	.125		9.35	7.30		16.65	21
1870	2 gang		53	.151		12	8.80		20.80	26.50

26 05 33.18 Pull Boxes

26 05 33.18 Pull Boxes		Crew	Daily Output	Labor-Hours	Unit	Material	2018 Bare Costs Labor	2018 Bare Costs Equipment	Total	Total Incl O&P
0010	**PULL BOXES**	R260533-70								
0100	Steel, pull box, NEMA 1, type SC, 6" W x 6" H x 4" D	1 Elec	8	1	Ea.	9.75	58		67.75	97.50
0180	8" W x 6" H x 4" D		8	1		14.65	58		72.65	103
0200	8" W x 8" H x 4" D		8	1		12.55	58		70.55	101
0210	10" W x 10" H x 4" D		7	1.143		17.85	66.50		84.35	119

26 05 33.18 Pull Boxes		Crew	Daily Output	Labor-Hours	Unit	Material	2018 Bare Costs Labor	2018 Bare Costs Equipment	Total	Total Incl O&P
0220	12" W x 12" H x 4" D	1 Elec	6.50	1.231	Ea.	21.50	71.50		93	131
0230	15" W x 15" H x 4" D		5.20	1.538		48.50	89.50		138	187
0240	18" W x 18" H x 4" D		4.40	1.818		38.50	106		144.50	201
0250	6" W x 6" H x 6" D		8	1		15.05	58		73.05	104
0260	8" W x 8" H x 6" D		7.50	1.067		15.15	62		77.15	110
0270	10" W x 10" H x 6" D		5.50	1.455		19.60	84.50		104.10	149
0300	10" W x 12" H x 6" D		5.30	1.509		28.50	88		116.50	162
0310	12" W x 12" H x 6" D		5.20	1.538		25	89.50		114.50	162
0320	15" W x 15" H x 6" D		4.60	1.739		35.50	101		136.50	190
0330	18" W x 18" H x 6" D		4.20	1.905		46.50	111		157.50	217
0340	24" W x 24" H x 6" D		3.20	2.500		92	146		238	320
0350	12" W x 12" H x 8" D		5	1.600		32	93		125	174
0360	15" W x 15" H x 8" D		4.50	1.778		45.50	103		148.50	205
0370	18" W x 18" H x 8" D		4	2		59.50	116		175.50	240
0380	24" W x 18" H x 6" D		3.70	2.162		88	126		214	285
0400	16" W x 20" H x 8" D		4	2		82	116		198	264
0500	20" W x 24" H x 8" D		3.20	2.500		96	146		242	325
0510	24" W x 24" H x 8" D		3	2.667		104	155		259	345
0600	24" W x 36" H x 8" D		2.70	2.963		151	172		323	425
0610	30" W x 30" H x 8" D		2.70	2.963		198	172		370	475
0620	36" W x 36" H x 8" D		2	4		222	233		455	595
0630	24" W x 24" H x 10" D		2.50	3.200		215	186		401	515
0650	Pull box, hinged, NEMA 1, 6" W x 6" H x 4" D		8	1		13	58		71	101
0660	8" W x 8" H x 4" D		8	1		17.70	58		75.70	106
0670	10" W x 10" H x 4" D		7	1.143		22	66.50		88.50	124
0680	12" W x 12" H x 4" D		6	1.333		29	77.50		106.50	148
0690	15" W x 15" H x 4" D		5.20	1.538		37.50	89.50		127	176
0700	18" W x 18" H x 4" D		4.40	1.818		35	106		141	197
0710	6" W x 6" H x 6" D		8	1		15.85	58		73.85	104
0720	8" W x 8" H x 6" D		7.50	1.067		21.50	62		83.50	117
0730	10" W x 10" H x 6" D		5.50	1.455		27	84.50		111.50	157
0740	12" W x 12" H x 6" D		5.20	1.538		34.50	89.50		124	172
0800	12" W x 16" H x 6" D		4.70	1.702		51.50	99		150.50	205
0810	15" W x 15" H x 6" D		4.60	1.739		42	101		143	197
0820	18" W x 18" H x 6" D		4.20	1.905		59.50	111		170.50	232
1000	20" W x 20" H x 6" D		3.60	2.222		121	129		250	325
1010	24" W x 24" H x 6" D		3.20	2.500		125	146		271	355
1020	12" W x 12" H x 8" D		5	1.600		66.50	93		159.50	213
1030	15" W x 15" H x 8" D		4.50	1.778		86	103		189	250
1040	18" W x 18" H x 8" D		4	2		121	116		237	305
1200	20" W x 20" H x 8" D		3.20	2.500		151	146		297	385
1210	24" W x 24" H x 8" D		3	2.667		169	155		324	420
1220	30" W x 30" H x 8" D		2.70	2.963		244	172		416	525
1400	24" W x 36" H x 8" D		2.70	2.963		237	172		409	520
1600	24" W x 42" H x 8" D		2	4		355	233		588	740
1610	36" W x 36" H x 8" D		2	4		320	233		553	700
2100	Pull box, NEMA 3R, type SC, raintight & weatherproof									
2150	6" L x 6" W x 6" D	1 Elec	10	.800	Ea.	16.65	46.50		63.15	88
2200	8" L x 6" W x 6" D		8	1		24	58		82	114
2250	10" L x 6" W x 6" D		7	1.143		32	66.50		98.50	135
2300	12" L x 12" W x 6" D		5	1.600		48	93		141	192
2350	16" L x 16" W x 6" D		4.50	1.778		89	103		192	253
2400	20" L x 20" W x 6" D		4	2		89	116		205	272

For customer support on your Electrical Costs with RSMeans data, call 800.448.8182.

157

26 05 33 – Raceway and Boxes for Electrical Systems

26 05 33.18 Pull Boxes		Crew	Daily Output	Labor-Hours	Unit	Material	2018 Bare Costs Labor	Equipment	Total	Total Incl O&P
2450	24" L x 18" W x 8" D	1 Elec	3	2.667	Ea.	146	155		301	395
2500	24" L x 24" W x 10" D		2.50	3.200		310	186		496	620
2550	30" L x 24" W x 12" D		2	4		385	233		618	775
2600	36" L x 36" W x 12" D		1.50	5.333		450	310		760	960
2800	Cast iron, pull boxes for surface mounting									
3000	NEMA 4, watertight & dust tight									
3050	6" L x 6" W x 6" D	1 Elec	4	2	Ea.	273	116		389	475
3100	8" L x 6" W x 6" D		3.20	2.500		380	146		526	635
3150	10" L x 6" W x 6" D		2.50	3.200		425	186		611	750
3200	12" L x 12" W x 6" D		2.30	3.478		775	202		977	1,150
3250	16" L x 16" W x 6" D		1.30	6.154		710	360		1,070	1,325
3300	20" L x 20" W x 6" D		.80	10		865	580		1,445	1,825
3350	24" L x 18" W x 8" D		.70	11.429		2,600	665		3,265	3,850
3400	24" L x 24" W x 10" D		.50	16		4,900	930		5,830	6,800
3450	30" L x 24" W x 12" D		.40	20		6,275	1,175		7,450	8,650
3500	36" L x 36" W x 12" D		.20	40		3,575	2,325		5,900	7,400
3510	NEMA 4 clamp cover, 6" L x 6" W x 4" D		4	2		184	116		300	375
3520	8" L x 6" W x 4" D		4	2		235	116		351	435
4000	NEMA 7, explosion proof									
4050	6" L x 6" W x 6" D	1 Elec	2	4	Ea.	695	233		928	1,125
4100	8" L x 6" W x 6" D		1.80	4.444		1,000	259		1,259	1,475
4150	10" L x 6" W x 6" D		1.60	5		1,300	291		1,591	1,850
4200	12" L x 12" W x 6" D		1	8		2,350	465		2,815	3,300
4250	16" L x 14" W x 6" D		.60	13.333		3,400	775		4,175	4,900
4300	18" L x 18" W x 8" D		.50	16		6,175	930		7,105	8,175
4350	24" L x 18" W x 8" D		.40	20		7,825	1,175		9,000	10,400
4400	24" L x 24" W x 10" D		.30	26.667		10,700	1,550		12,250	14,000
4450	30" L x 24" W x 12" D		.20	40		14,100	2,325		16,425	19,000
5000	NEMA 9, dust tight 6" L x 6" W x 6" D		3.20	2.500		400	146		546	660
5050	8" L x 6" W x 6" D		2.70	2.963		480	172		652	790
5100	10" L x 6" W x 6" D		2	4		640	233		873	1,050
5150	12" L x 12" W x 6" D		1.60	5		1,225	291		1,516	1,775
5200	16" L x 16" W x 6" D		1	8		2,025	465		2,490	2,925
5250	18" L x 18" W x 8" D		.70	11.429		3,150	665		3,815	4,475
5300	24" L x 18" W x 8" D		.60	13.333		4,450	775		5,225	6,050
5350	24" L x 24" W x 10" D		.40	20		6,075	1,175		7,250	8,425
5400	30" L x 24" W x 12" D		.30	26.667		9,050	1,550		10,600	12,300
6000	J.I.C. wiring boxes, NEMA 12, dust tight & drip tight									
6050	6" L x 8" W x 4" D	1 Elec	10	.800	Ea.	40.50	46.50		87	115
6100	8" L x 10" W x 4" D		8	1		50.50	58		108.50	143
6150	12" L x 14" W x 6" D		5.30	1.509		86	88		174	226
6200	14" L x 16" W x 6" D		4.70	1.702		100	99		199	258
6250	16" L x 20" W x 6" D		4.40	1.818		219	106		325	400
6300	24" L x 30" W x 6" D		3.20	2.500		320	146		466	575
6350	24" L x 30" W x 8" D		2.90	2.759		470	161		631	755
6400	24" L x 36" W x 8" D		2.70	2.963		365	172		537	665
6450	24" L x 42" W x 8" D		2.30	3.478		415	202		617	760
6500	24" L x 48" W x 8" D		2	4		450	233		683	845

26 05 33.23 Wireway		Crew	Daily Output	Labor-Hours	Unit	Material	2018 Bare Costs Labor	Equipment	Total	Total Incl O&P
0010	**WIREWAY** to 10' high [R260533-60]									
0020	For higher elevations, see Section 26 05 36.40									
0100	NEMA 1, screw cover w/fittings and supports, 2-1/2" x 2-1/2"	1 Elec	45	.178	L.F.	12.20	10.35		22.55	29
0200	4" x 4"	"	40	.200		13.15	11.65		24.80	32
0400	6" x 6"	2 Elec	60	.267		21	15.50		36.50	46
0600	8" x 8"		40	.400		28	23.50		51.50	65.50
0620	10" x 10"		30	.533		47.50	31		78.50	98.50
0640	12" x 12"		20	.800		54.50	46.50		101	130
0800	Elbows, 90°, 2-1/2"	1 Elec	24	.333	Ea.	32.50	19.40		51.90	64.50
1000	4"		20	.400		36	23.50		59.50	74.50
1200	6"		18	.444		41	26		67	83.50
1400	8"		16	.500		66.50	29		95.50	117
1420	10"		12	.667		89	39		128	156
1440	12"		10	.800		117	46.50		163.50	199
1500	Elbows, 45°, 2-1/2"		24	.333		29.50	19.40		48.90	61.50
1510	4"		20	.400		39.50	23.50		63	78.50
1520	6"		18	.444		43.50	26		69.50	86.50
1530	8"		16	.500		66.50	29		95.50	117
1540	10"		12	.667		84.50	39		123.50	151
1550	12"		10	.800		169	46.50		215.50	255
1600	"T" box, 2-1/2"		18	.444		37	26		63	79.50
1800	4"		16	.500		47	29		76	95
2000	6"		14	.571		52	33.50		85.50	107
2200	8"		12	.667		93	39		132	160
2220	10"		10	.800		115	46.50		161.50	197
2240	12"		8	1		144	58		202	245
2300	Cross, 2-1/2"		16	.500		41.50	29		70.50	89
2310	4"		14	.571		50.50	33.50		84	105
2320	6"		12	.667		62.50	39		101.50	127
2400	Panel adapter, 2-1/2"		24	.333		11.35	19.40		30.75	41.50
2600	4"		20	.400		10.85	23.50		34.35	47
2800	6"		18	.444		11.60	26		37.60	51.50
3000	8"		16	.500		16	29		45	61
3020	10"		14	.571		29.50	33.50		63	82
3040	12"		12	.667		37.50	39		76.50	99
3200	Reducer, 4" to 2-1/2"		24	.333		12.10	19.40		31.50	42.50
3400	6" to 4"		20	.400		25.50	23.50		49	63
3600	8" to 6"		18	.444		31	26		57	72.50
3620	10" to 8"		16	.500		44.50	29		73.50	92
3640	12" to 10"		14	.571		49	33.50		82.50	103
3780	End cap, 2-1/2"		24	.333		4.89	19.40		24.29	34.50
3800	4"		20	.400		5.80	23.50		29.30	41.50
4000	6"		18	.444		6.85	26		32.85	46
4200	8"		16	.500		9.10	29		38.10	53.50
4220	10"		14	.571		15	33.50		48.50	66
4240	12"		12	.667		20	39		59	80
4300	U-connector, 2-1/2"		200	.040		4.39	2.33		6.72	8.30
4320	4"		200	.040		5.30	2.33		7.63	9.30
4340	6"		180	.044		6.65	2.59		9.24	11.15
4360	8"		170	.047		10.90	2.74		13.64	16.10
4380	10"		150	.053		14.95	3.10		18.05	21
4400	12"		130	.062		21	3.58		24.58	29

For customer support on your Electrical Costs with RSMeans data, call 800.448.8182.

159

26 05 33 – Raceway and Boxes for Electrical Systems

26 05 33.23 Wireway		Crew	Daily Output	Labor-Hours	Unit	Material	2018 Bare Costs Labor	Equipment	Total	Total Incl O&P
4420	Hanger, 2-1/2"	1 Elec	100	.080	Ea.	12.60	4.66		17.26	21
4430	4"		100	.080		12.60	4.66		17.26	21
4440	6"		80	.100		13.85	5.80		19.65	24
4450	8"		65	.123		24.50	7.15		31.65	37.50
4460	10"		50	.160		44.50	9.30		53.80	63
4470	12"		40	.200		65	11.65		76.65	89
4475	NEMA 3R, screw cover w/fittings and supports, 4" x 4"		36	.222	L.F.	16.30	12.95		29.25	37.50
4480	6" x 6"	2 Elec	55	.291		18.10	16.95		35.05	45.50
4485	8" x 8"		36	.444		32	26		58	73.50
4490	12" x 12"		18	.889		57.50	51.50		109	141
4500	Hinged cover, with fittings and supports, 2-1/2" x 2-1/2"	1 Elec	60	.133		18.05	7.75		25.80	31.50
4520	4" x 4"	"	45	.178		14.20	10.35		24.55	31
4540	6" x 6"	2 Elec	80	.200		21.50	11.65		33.15	41.50
4560	8" x 8"		60	.267		38	15.50		53.50	65
4580	10" x 10"		50	.320		49	18.60		67.60	82
4600	12" x 12"		24	.667		73.50	39		112.50	139
4700	Elbows 90°, 2-1/2" x 2-1/2"	1 Elec	32	.250	Ea.	46.50	14.55		61.05	73.50
4720	4"		27	.296		75.50	17.25		92.75	109
4730	6"		23	.348		54	20		74	89.50
4740	8"		18	.444		68	26		94	114
4750	10"		14	.571		111	33.50		144.50	172
4760	12"		12	.667		179	39		218	254
4800	Tee box, hinged cover, 2-1/2" x 2-1/2"		23	.348		62.50	20		82.50	99.50
4810	4"		20	.400		87.50	23.50		111	132
4820	6"		18	.444		62	26		88	107
4830	8"		16	.500		167	29		196	228
4840	10"		12	.667		163	39		202	238
4860	12"		10	.800		210	46.50		256.50	300
4880	Cross box, hinged cover, 2-1/2" x 2-1/2"		18	.444		61.50	26		87.50	106
4900	4"		16	.500		91	29		120	144
4920	6"		13	.615		106	36		142	171
4940	8"		11	.727		153	42.50		195.50	232
4960	10"		10	.800		241	46.50		287.50	335
4980	12"		9	.889		241	51.50		292.50	345
5000	NEMA 12, hinged cover, 2-1/2" x 2-1/2"		40	.200	L.F.	22	11.65		33.65	41.50
5020	4" x 4"		35	.229		25	13.30		38.30	47.50
5040	6" x 6"	2 Elec	60	.267		38.50	15.50		54	65.50
5060	8" x 8"	"	50	.320		51	18.60		69.60	84
5120	Elbows 90°, flanged, 2-1/2" x 2-1/2"	1 Elec	23	.348	Ea.	54.50	20		74.50	90.50
5140	4"		20	.400		68.50	23.50		92	110
5160	6"		18	.444		87	26		113	134
5180	8"		15	.533		127	31		158	186
5240	Tee box, flanged, 2-1/2" x 2-1/2"		18	.444		74.50	26		100.50	121
5260	4"		16	.500		87	29		116	140
5280	6"		15	.533		118	31		149	176
5300	8"		13	.615		171	36		207	242
5360	Cross box, flanged, 2-1/2" x 2-1/2"		15	.533		115	31		146	174
5380	4"		13	.615		128	36		164	194
5400	6"		12	.667		166	39		205	240
5420	8"		10	.800		210	46.50		256.50	300
5480	Flange gasket, 2-1/2"		160	.050		2.70	2.91		5.61	7.30
5500	4"		80	.100		3.74	5.80		9.54	12.80
5520	6"		53	.151		5.15	8.80		13.95	18.85

26 05 33.23 Wireway	Crew	Daily Output	Labor-Hours	Unit	Material	2018 Bare Costs Labor	Equipment	Total	Total Incl O&P
5530	8"	1 Elec	40	.200	Ea.	6.70	11.65		18.35	25

26 05 33.25 Conduit Fittings for Rigid Galvanized Steel

		Crew	Daily Output	Labor-Hours	Unit	Material	Labor	Equipment	Total	Total Incl O&P
0010	**CONDUIT FITTINGS FOR RIGID GALVANIZED STEEL**									
0050	Standard, locknuts, 1/2" diameter				Ea.	.21			.21	.23
0100	3/4" diameter					.38			.38	.42
0300	1" diameter					.61			.61	.67
0500	1-1/4" diameter					.75			.75	.83
0700	1-1/2" diameter					1.30			1.30	1.43
1000	2" diameter					1.75			1.75	1.93
1030	2-1/2" diameter					4.67			4.67	5.15
1050	3" diameter					6.05			6.05	6.65
1070	3-1/2" diameter					9.05			9.05	10
1100	4" diameter					11.95			11.95	13.15
1110	5" diameter					26			26	28.50
1120	6" diameter					48			48	52.50
1130	Bushings, plastic, 1/2" diameter	1 Elec	40	.200		.08	11.65		11.73	17.50
1150	3/4" diameter		32	.250		.09	14.55		14.64	22
1170	1" diameter		28	.286		.19	16.65		16.84	25
1200	1-1/4" diameter		24	.333		.23	19.40		19.63	29.50
1230	1-1/2" diameter		18	.444		.33	26		26.33	39
1250	2" diameter		15	.533		.55	31		31.55	47
1270	2-1/2" diameter		13	.615		.98	36		36.98	54.50
1300	3" diameter		12	.667		1.26	39		40.26	59.50
1330	3-1/2" diameter		11	.727		1.35	42.50		43.85	65
1350	4" diameter		9	.889		1.64	51.50		53.14	79.50
1360	5" diameter		7	1.143		10.65	66.50		77.15	111
1370	6" diameter		5	1.600		28.50	93		121.50	171
1390	Steel, 1/2" diameter		40	.200		.64	11.65		12.29	18.10
1400	3/4" diameter		32	.250		.71	14.55		15.26	23
1430	1" diameter		28	.286		1.29	16.65		17.94	26.50
1450	Steel insulated, 1-1/4" diameter		24	.333		3.79	19.40		23.19	33
1470	1-1/2" diameter		18	.444		4.87	26		30.87	44
1500	2" diameter		15	.533		7.20	31		38.20	54.50
1530	2-1/2" diameter		13	.615		14.10	36		50.10	69
1550	3" diameter		12	.667		18.05	39		57.05	78
1570	3-1/2" diameter		11	.727		26	42.50		68.50	92.50
1600	4" diameter		9	.889		28.50	51.50		80	109
1610	5" diameter		7	1.143		60.50	66.50		127	166
1620	6" diameter		5	1.600		168	93		261	325
1630	Sealing locknuts, 1/2" diameter		40	.200		1.27	11.65		12.92	18.80
1650	3/4" diameter		32	.250		1.48	14.55		16.03	23.50
1670	1" diameter		28	.286		2.25	16.65		18.90	27.50
1700	1-1/4" diameter		24	.333		3.46	19.40		22.86	33
1730	1-1/2" diameter		18	.444		4.40	26		30.40	43.50
1750	2" diameter		15	.533		5.65	31		36.65	52.50
1760	Grounding bushing, insulated, 1/2" diameter		32	.250		5.30	14.55		19.85	28
1770	3/4" diameter		28	.286		5.10	16.65		21.75	30.50
1780	1" diameter		20	.400		6.05	23.50		29.55	41.50
1800	1-1/4" diameter		18	.444		8.20	26		34.20	47.50
1830	1-1/2" diameter		16	.500		9.80	29		38.80	54.50
1850	2" diameter		13	.615		12.35	36		48.35	67
1870	2-1/2" diameter		12	.667		20	39		59	80

For customer support on your Electrical Costs with RSMeans data, call 800.448.8182.

161

26 05 33.25 Conduit Fittings for Rigid Galvanized Steel	Crew	Daily Output	Labor-Hours	Unit	Material	2018 Bare Costs Labor	Equipment	Total	Total Incl O&P
1900	3" diameter	1 Elec	11	.727	Ea.	23	42.50		65.50	89
1930	3-1/2" diameter		9	.889		25.50	51.50		77	106
1950	4" diameter		8	1		38.50	58		96.50	130
1960	5" diameter		6	1.333		77	77.50		154.50	201
1970	6" diameter		4	2		110	116		226	295
1990	Coupling, with set screw, 1/2" diameter		50	.160		3.34	9.30		12.64	17.55
2000	3/4" diameter		40	.200		4.21	11.65		15.86	22
2030	1" diameter		35	.229		7.05	13.30		20.35	27.50
2050	1-1/4" diameter		28	.286		10.40	16.65		27.05	36.50
2070	1-1/2" diameter		23	.348		13.70	20		33.70	45.50
2090	2" diameter		20	.400		30.50	23.50		54	68.50
2100	2-1/2" diameter		18	.444		64	26		90	109
2110	3" diameter		15	.533		84	31		115	139
2120	3-1/2" diameter		12	.667		119	39		158	189
2130	4" diameter		10	.800		166	46.50		212.50	253
2140	5" diameter		9	.889		234	51.50		285.50	335
2150	6" diameter		8	1		345	58		403	465
2160	Box connector with set screw, plain, 1/2" diameter		70	.114		2.13	6.65		8.78	12.30
2170	3/4" diameter		60	.133		3.09	7.75		10.84	15
2180	1" diameter		50	.160		4.77	9.30		14.07	19.15
2190	Insulated, 1-1/4" diameter		40	.200		8.70	11.65		20.35	27
2200	1-1/2" diameter		30	.267		13.90	15.50		29.40	38.50
2210	2" diameter		20	.400		28.50	23.50		52	66
2220	2-1/2" diameter		18	.444		76.50	26		102.50	123
2230	3" diameter		15	.533		112	31		143	170
2240	3-1/2" diameter		12	.667		166	39		205	241
2250	4" diameter		10	.800		162	46.50		208.50	248
2260	5" diameter		9	.889		385	51.50		436.50	500
2270	6" diameter		8	1		340	58		398	455
2280	LB, LR or LL fittings & covers, 1/2" diameter		16	.500		7.10	29		36.10	51.50
2290	3/4" diameter		13	.615		8.35	36		44.35	62.50
2300	1" diameter		11	.727		13.05	42.50		55.55	78
2330	1-1/4" diameter		8	1		19.45	58		77.45	109
2350	1-1/2" diameter		6	1.333		25	77.50		102.50	144
2370	2" diameter		5	1.600		40.50	93		133.50	184
2380	2-1/2" diameter		4	2		87.50	116		203.50	270
2390	3" diameter		3.50	2.286		116	133		249	325
2400	3-1/2" diameter		3	2.667		188	155		343	440
2410	4" diameter		2.50	3.200		205	186		391	505
2420	T fittings, with cover, 1/2" diameter		12	.667		9.25	39		48.25	68
2430	3/4" diameter		11	.727		11.35	42.50		53.85	76
2440	1" diameter		9	.889		18.10	51.50		69.60	97.50
2450	1-1/4" diameter		6	1.333		22	77.50		99.50	140
2470	1-1/2" diameter		5	1.600		31	93		124	173
2500	2" diameter		4	2		41.50	116		157.50	220
2510	2-1/2" diameter		3.50	2.286		94.50	133		227.50	305
2520	3" diameter		3	2.667		122	155		277	365
2530	3-1/2" diameter		2.50	3.200		209	186		395	505
2540	4" diameter		2	4		193	233		426	560
2550	Nipples chase, plain, 1/2" diameter		40	.200		.39	11.65		12.04	17.85
2560	3/4" diameter		32	.250		1.25	14.55		15.80	23.50
2570	1" diameter		28	.286		2.06	16.65		18.71	27.50
2600	Insulated, 1-1/4" diameter		24	.333		7.35	19.40		26.75	37

26 05 33.25 Conduit Fittings for Rigid Galvanized Steel		Crew	Daily Output	Labor-Hours	Unit	Material	2018 Bare Costs Labor	Equipment	Total	Total Incl O&P
2630	1-1/2" diameter	1 Elec	18	.444	Ea.	9.35	26		35.35	49
2650	2" diameter		15	.533		13.05	31		44.05	61
2660	2-1/2" diameter		12	.667		39	39		78	101
2670	3" diameter		10	.800		38.50	46.50		85	112
2680	3-1/2" diameter		9	.889		52	51.50		103.50	135
2690	4" diameter		8	1		81	58		139	176
2700	5" diameter		7	1.143		284	66.50		350.50	410
2710	6" diameter		6	1.333		435	77.50		512.50	590
2720	Nipples offset, plain, 1/2" diameter		40	.200		4.21	11.65		15.86	22
2730	3/4" diameter		32	.250		4.46	14.55		19.01	27
2740	1" diameter		24	.333		5.30	19.40		24.70	35
2750	Insulated, 1-1/4" diameter		20	.400		6.85	23.50		30.35	42.50
2760	1-1/2" diameter		18	.444		8.30	26		34.30	47.50
2770	2" diameter		16	.500		11.65	29		40.65	56.50
2780	3" diameter		14	.571		79.50	33.50		113	137
2850	Coupling, expansion, 1/2" diameter		12	.667		47	39		86	110
2880	3/4" diameter		10	.800		52.50	46.50		99	128
2900	1" diameter		8	1		62.50	58		120.50	156
2920	1-1/4" diameter		6.40	1.250		80	73		153	197
2940	1-1/2" diameter		5.30	1.509		115	88		203	258
2960	2" diameter		4.60	1.739		166	101		267	335
2980	2-1/2" diameter		3.60	2.222		262	129		391	480
3000	3" diameter		3	2.667		320	155		475	580
3020	3-1/2" diameter		2.80	2.857		480	166		646	780
3040	4" diameter		2.40	3.333		550	194		744	895
3060	5" diameter		2	4		1,000	233		1,233	1,450
3080	6" diameter		1.80	4.444		1,675	259		1,934	2,200
3100	Expansion deflection, 1/2" diameter		12	.667		259	39		298	345
3120	3/4" diameter		12	.667		292	39		331	380
3140	1" diameter		10	.800		315	46.50		361.50	415
3160	1-1/4" diameter		6.40	1.250		440	73		513	595
3180	1-1/2" diameter		5.30	1.509		425	88		513	595
3200	2" diameter		4.60	1.739		600	101		701	810
3220	2-1/2" diameter		3.60	2.222		810	129		939	1,075
3240	3" diameter		3	2.667		1,075	155		1,230	1,400
3260	3-1/2" diameter		2.80	2.857		1,150	166		1,316	1,500
3280	4" diameter		2.40	3.333		1,300	194		1,494	1,725
3300	5" diameter		2	4		2,475	233		2,708	3,075
3320	6" diameter		1.80	4.444		3,775	259		4,034	4,525
3340	Ericson, 1/2" diameter		16	.500		3.24	29		32.24	47
3360	3/4" diameter		14	.571		3.97	33.50		37.47	54
3380	1" diameter		11	.727		6.55	42.50		49.05	70.50
3400	1-1/4" diameter		8	1		14.10	58		72.10	103
3420	1-1/2" diameter		7	1.143		16	66.50		82.50	117
3440	2" diameter		5	1.600		31	93		124	173
3460	2-1/2" diameter		4	2		61.50	116		177.50	242
3480	3" diameter		3.50	2.286		106	133		239	315
3500	3-1/2" diameter		3	2.667		122	155		277	365
3520	4" diameter		2.70	2.963		148	172		320	420
3540	5" diameter		2.50	3.200		293	186		479	605
3560	6" diameter		2.30	3.478		405	202		607	750
3580	Split, 1/2" diameter		32	.250		3.65	14.55		18.20	26
3600	3/4" diameter		27	.296		4.60	17.25		21.85	31

26 05 33.25 Conduit Fittings for Rigid Galvanized Steel		Crew	Daily Output	Labor-Hours	Unit	Material	2018 Bare Costs Labor	Equipment	Total	Total Incl O&P
3620	1" diameter	1 Elec	20	.400	Ea.	8.60	23.50		32.10	44.50
3640	1-1/4" diameter		16	.500		12.20	29		41.20	57
3660	1-1/2" diameter		14	.571		14.20	33.50		47.70	65
3680	2" diameter		12	.667		27	39		66	88
3700	2-1/2" diameter		10	.800		68	46.50		114.50	145
3720	3" diameter		9	.889		91	51.50		142.50	178
3740	3-1/2" diameter		8	1		146	58		204	248
3760	4" diameter		7	1.143		168	66.50		234.50	285
3780	5" diameter		6	1.333		287	77.50		364.50	430
3800	6" diameter		5	1.600		375	93		468	550
4600	Reducing bushings, 3/4" to 1/2" diameter		54	.148		2.25	8.60		10.85	15.40
4620	1" to 3/4" diameter		46	.174		2.16	10.10		12.26	17.55
4640	1-1/4" to 1" diameter		40	.200		4.43	11.65		16.08	22.50
4660	1-1/2" to 1-1/4" diameter		36	.222		5.95	12.95		18.90	26
4680	2" to 1-1/2" diameter		32	.250		11.70	14.55		26.25	35
4740	2-1/2" to 2" diameter		30	.267		14.15	15.50		29.65	38.50
4760	3" to 2-1/2" diameter		28	.286		17.60	16.65		34.25	44.50
4800	Through-wall seal, 1/2" diameter		8	1		67	58		125	161
4820	3/4" diameter		7.50	1.067		257	62		319	375
4840	1" diameter		6.50	1.231		258	71.50		329.50	390
4860	1-1/4" diameter		5.50	1.455		295	84.50		379.50	450
4880	1-1/2" diameter		5	1.600		395	93		488	575
4900	2" diameter		4.20	1.905		286	111		397	480
4920	2-1/2" diameter		3.50	2.286		485	133		618	730
4940	3" diameter		3	2.667		515	155		670	795
4960	3-1/2" diameter		2.50	3.200		780	186		966	1,150
4980	4" diameter		2	4		795	233		1,028	1,225
5000	5" diameter		1.50	5.333		815	310		1,125	1,350
5020	6" diameter	▼	1	8	▼	815	465		1,280	1,600
5100	Cable supports, 2 or more wires									
5120	1-1/2" diameter	1 Elec	8	1	Ea.	150	58		208	252
5140	2" diameter		6	1.333		204	77.50		281.50	340
5160	2-1/2" diameter		4	2		220	116		336	415
5180	3" diameter		3.50	2.286		281	133		414	510
5200	3-1/2" diameter		2.60	3.077		365	179		544	670
5220	4" diameter		2	4		460	233		693	855
5240	5" diameter		1.50	5.333		630	310		940	1,150
5260	6" diameter		1	8		1,025	465		1,490	1,825
5280	Service entrance cap, 1/2" diameter		16	.500		4.99	29		33.99	49
5300	3/4" diameter		13	.615		5.95	36		41.95	60
5320	1" diameter		10	.800		4.88	46.50		51.38	75
5340	1-1/4" diameter		8	1		4.87	58		62.87	92.50
5360	1-1/2" diameter		6.50	1.231		9.85	71.50		81.35	118
5380	2" diameter		5.50	1.455		20	84.50		104.50	149
5400	2-1/2" diameter		4	2		64.50	116		180.50	245
5420	3" diameter		3.40	2.353		71.50	137		208.50	284
5440	3-1/2" diameter		3	2.667		122	155		277	365
5460	4" diameter		2.70	2.963		151	172		323	425
5750	90° pull elbows steel, female, 1/2" diameter		16	.500		7.65	29		36.65	52
5760	3/4" diameter		13	.615		8.80	36		44.80	63
5780	1" diameter		11	.727		13.60	42.50		56.10	78.50
5800	1-1/4" diameter		8	1		19.65	58		77.65	109
5820	1-1/2" diameter	▼	6	1.333	▼	26	77.50		103.50	145

26 05 33.25 Conduit Fittings for Rigid Galvanized Steel	Crew	Daily Output	Labor-Hours	Unit	Material	2018 Bare Costs Labor	Equipment	Total	Total Incl O&P
5840	2" diameter	1 Elec	5	1.600	Ea.	42.50	93		135.50	186
6000	Explosion proof, flexible coupling									
6010	1/2" diameter, 4" long	1 Elec	12	.667	Ea.	108	39		147	177
6020	6" long		12	.667		116	39		155	186
6050	12" long		12	.667		163	39		202	237
6070	18" long		12	.667		203	39		242	281
6090	24" long		12	.667		229	39		268	310
6110	30" long		12	.667		270	39		309	355
6130	36" long		12	.667		300	39		339	390
6140	3/4" diameter, 4" long		10	.800		122	46.50		168.50	205
6150	6" long		10	.800		138	46.50		184.50	222
6180	12" long		10	.800		193	46.50		239.50	282
6200	18" long		10	.800		259	46.50		305.50	355
6220	24" long		10	.800		305	46.50		351.50	405
6240	30" long		10	.800		305	46.50		351.50	410
6260	36" long		10	.800		380	46.50		426.50	485
6270	1" diameter, 6" long		8	1		276	58		334	390
6300	12" long		8	1		330	58		388	450
6320	18" long		8	1		420	58		478	550
6340	24" long		8	1		480	58		538	610
6360	30" long		8	1		565	58		623	710
6380	36" long		8	1		580	58		638	725
6390	1-1/4" diameter, 12" long		6.40	1.250		475	73		548	630
6410	18" long		6.40	1.250		605	73		678	775
6430	24" long		6.40	1.250		720	73		793	900
6450	30" long		6.40	1.250		840	73		913	1,025
6470	36" long		6.40	1.250		1,175	73		1,248	1,400
6480	1-1/2" diameter, 12" long		5.30	1.509		635	88		723	825
6500	18" long		5.30	1.509		740	88		828	945
6520	24" long		5.30	1.509		940	88		1,028	1,150
6540	30" long		5.30	1.509		1,125	88		1,213	1,375
6560	36" long		5.30	1.509		1,150	88		1,238	1,375
6570	2" diameter, 12" long		4.60	1.739		805	101		906	1,025
6590	18" long		4.60	1.739		1,050	101		1,151	1,325
6610	24" long		4.60	1.739		1,150	101		1,251	1,400
6630	30" long		4.60	1.739		1,650	101		1,751	1,975
6650	36" long		4.60	1.739		1,525	101		1,626	1,825
7000	Close up plug, 1/2" diameter, explosion proof		40	.200		2.42	11.65		14.07	20
7010	3/4" diameter		32	.250		2.88	14.55		17.43	25
7020	1" diameter		28	.286		3.38	16.65		20.03	28.50
7030	1-1/4" diameter		24	.333		3.58	19.40		22.98	33
7040	1-1/2" diameter		18	.444		4.98	26		30.98	44
7050	2" diameter		15	.533		8.65	31		39.65	56
7060	2-1/2" diameter		13	.615		13.15	36		49.15	68
7070	3" diameter		12	.667		20	39		59	80
7080	3-1/2" diameter		11	.727		23	42.50		65.50	89
7090	4" diameter		9	.889		32.50	51.50		84	113
7091	Elbow female, 45°, 1/2"		16	.500		14.15	29		43.15	59
7092	3/4"		13	.615		17.35	36		53.35	72.50
7093	1"		11	.727		20.50	42.50		63	86
7094	1-1/4"		8	1		28.50	58		86.50	119
7095	1-1/2"		6	1.333		32	77.50		109.50	152
7096	2"		5	1.600		36	93		129	179

For customer support on your Electrical Costs with RSMeans data, call 800.448.8182.

165

26 05 33.25 Conduit Fittings for Rigid Galvanized Steel	Crew	Daily Output	Labor-Hours	Unit	Material	2018 Bare Costs Labor	Equipment	Total	Total Incl O&P
7097	2-1/2"	1 Elec	4.50	1.778	Ea.	107	103		210	272
7098	3"		4.20	1.905		120	111		231	298
7099	3-1/2"		4	2		162	116		278	350
7100	4"		3.80	2.105		195	123		318	395
7101	90°, 1/2"		16	.500		13.60	29		42.60	58.50
7102	3/4"		13	.615		14.80	36		50.80	70
7103	1"		11	.727		19.90	42.50		62.40	85.50
7104	1-1/4"		8	1		32	58		90	122
7105	1-1/2"		6	1.333		46.50	77.50		124	168
7106	2"		5	1.600		74	93		167	221
7107	2-1/2"		4.50	1.778		157	103		260	330
7110	Elbows 90°, long male & female, 1/2" diameter, explosion proof		16	.500		19.95	29		48.95	65.50
7120	3/4" diameter		13	.615		22	36		58	77.50
7130	1" diameter		11	.727		31	42.50		73.50	97.50
7140	1-1/4" diameter		8	1		36.50	58		94.50	127
7150	1-1/2" diameter		6	1.333		38	77.50		115.50	158
7160	2" diameter		5	1.600		56	93		149	201
7170	Capped elbow, 1/2" diameter, explosion proof		11	.727		21.50	42.50		64	87.50
7180	3/4" diameter		8	1		26.50	58		84.50	116
7190	1" diameter		6	1.333		29	77.50		106.50	148
7200	1-1/4" diameter		5	1.600		59.50	93		152.50	205
7210	Pulling elbow, 1/2" diameter, explosion proof		11	.727		88	42.50		130.50	160
7220	3/4" diameter		8	1		163	58		221	266
7230	1" diameter		6	1.333		238	77.50		315.50	380
7240	1-1/4" diameter		5	1.600		285	93		378	455
7250	1-1/2" diameter		5	1.600		330	93		423	505
7260	2" diameter		4	2		380	116		496	595
7270	2-1/2" diameter		3.50	2.286		855	133		988	1,150
7280	3" diameter		3	2.667		1,175	155		1,330	1,525
7290	3-1/2" diameter		2.50	3.200		1,500	186		1,686	1,925
7300	4" diameter		2.20	3.636		1,600	212		1,812	2,075
7310	LB conduit body, 1/2" diameter		11	.727		53.50	42.50		96	123
7320	3/4" diameter		8	1		61	58		119	154
7330	T conduit body, 1/2" diameter		9	.889		56	51.50		107.50	139
7340	3/4" diameter		6	1.333		66.50	77.50		144	189
7350	Explosion proof, round box w/cover, 3 threaded hubs, 1/2" diameter		8	1		69.50	58		127.50	164
7351	3/4" diameter		8	1		68.50	58		126.50	163
7352	1" diameter		7.50	1.067		71	62		133	171
7353	1-1/4" diameter		7	1.143		110	66.50		176.50	221
7354	1-1/2" diameter		7	1.143		262	66.50		328.50	390
7355	2" diameter		6	1.333		293	77.50		370.50	435
7356	Round box w/cover & mtng flange, 3 threaded hubs, 1/2" diameter		8	1		53.50	58		111.50	146
7357	3/4" diameter		8	1		54	58		112	147
7358	4 threaded hubs, 1" diameter		7	1.143		65	66.50		131.50	171
7400	Unions, 1/2" diameter		20	.400		13.35	23.50		36.85	49.50
7410	3/4" - 1/2" diameter		16	.500		18.95	29		47.95	64.50
7420	3/4" diameter		16	.500		18.50	29		47.50	64
7430	1" diameter		14	.571		33.50	33.50		67	86.50
7440	1-1/4" diameter		12	.667		51.50	39		90.50	115
7450	1-1/2" diameter		10	.800		63.50	46.50		110	140
7460	2" diameter		8.50	.941		84.50	55		139.50	175
7480	2-1/2" diameter		8	1		119	58		177	217
7490	3" diameter		7	1.143		187	66.50		253.50	305

26 05 33.25 Conduit Fittings for Rigid Galvanized Steel	Crew	Daily Output	Labor-Hours	Unit	Material	2018 Bare Costs Labor	Equipment	Total	Total Incl O&P
7500	3-1/2" diameter	1 Elec	6	1.333	Ea.	340	77.50		417.50	490
7510	4" diameter		5	1.600		325	93		418	495
7680	Reducer, 3/4" to 1/2"		54	.148		3.16	8.60		11.76	16.40
7690	1" to 1/2"		46	.174		4.25	10.10		14.35	19.85
7700	1" to 3/4"		46	.174		4.18	10.10		14.28	19.75
7710	1-1/4" to 3/4"		40	.200		5.80	11.65		17.45	24
7720	1-1/4" to 1"		40	.200		6.60	11.65		18.25	24.50
7730	1-1/2" to 1"		36	.222		10.80	12.95		23.75	31.50
7740	1-1/2" to 1-1/4"		36	.222		10.25	12.95		23.20	30.50
7750	2" to 3/4"		32	.250		19.20	14.55		33.75	43
7760	2" to 1-1/4"		32	.250		12.55	14.55		27.10	36
7770	2" to 1-1/2"		32	.250		16.45	14.55		31	40
7780	2-1/2" to 1-1/2"		30	.267		18.90	15.50		34.40	44
7790	3" to 2"		30	.267		22.50	15.50		38	48
7800	3-1/2" to 2-1/2"		28	.286		45	16.65		61.65	74.50
7810	4" to 3"		28	.286		50.50	16.65		67.15	80.50
7820	Sealing fitting, vertical/horizontal, 1/2" diameter		14.50	.552		15.50	32		47.50	65
7830	3/4" diameter		13.30	.602		18.35	35		53.35	72.50
7840	1" diameter		11.40	.702		22.50	41		63.50	86
7850	1-1/4" diameter		10	.800		27	46.50		73.50	99.50
7860	1-1/2" diameter		8.80	.909		42.50	53		95.50	126
7870	2" diameter		8	1		55	58		113	148
7880	2-1/2" diameter		6.70	1.194		73	69.50		142.50	184
7890	3" diameter		5.70	1.404		96	81.50		177.50	228
7900	3-1/2" diameter		4.70	1.702		264	99		363	440
7910	4" diameter		4	2		380	116		496	595
7920	Sealing hubs, 1" by 1-1/2"		12	.667		28.50	39		67.50	89.50
7930	1-1/4" by 2"		10	.800		51	46.50		97.50	126
7940	1-1/2" by 2"		9	.889		71.50	51.50		123	156
7950	2" by 2-1/2"		8	1		93.50	58		151.50	190
7960	3" by 4"		7	1.143		164	66.50		230.50	281
7970	4" by 5"		6	1.333		330	77.50		407.50	480
7980	Drain, 1/2"		32	.250		100	14.55		114.55	132
7990	Breather, 1/2"		32	.250		100	14.55		114.55	132
7991	Explosion proof sealant compound, hub, fittings, 60 min set time, 1lb. pail		48	.167		17.90	9.70		27.60	34
7992	5 lb. pail		48	.167		46	9.70		55.70	65
7993	2 oz. tube		50	.160		56	9.30		65.30	75.50
7994	6 oz. tube		50	.160		112	9.30		121.30	137
7995	2.0 oz. cartridge		50	.160		34	9.30		43.30	51.50
7996	6.0 oz. cartridge		50	.160		70	9.30		79.30	91
7997	2.0 oz. box		50	.160		23	9.30		32.30	39
7998	8.0 oz. box		50	.160		77	9.30		86.30	98.50
7999	1.0 lb. box		50	.160		126	9.30		135.30	153
8000	Plastic coated 40 mil thick									
8010	LB, LR or LL conduit body w/cover, 1/2" diameter	1 Elec	13	.615	Ea.	46.50	36		82.50	105
8020	3/4" diameter		11	.727		64	42.50		106.50	134
8030	1" diameter		8	1		72.50	58		130.50	167
8040	1-1/4" diameter		6	1.333		104	77.50		181.50	231
8050	1-1/2" diameter		5	1.600		129	93		222	281
8060	2" diameter		4.50	1.778		189	103		292	365
8070	2-1/2" diameter		4	2		325	116		441	535
8080	3" diameter		3.50	2.286		425	133		558	665
8090	3-1/2" diameter		3	2.667		595	155		750	885

For customer support on your Electrical Costs with RSMeans data, call 800.448.8182.

167

26 05 33.25 Conduit Fittings for Rigid Galvanized Steel	Crew	Daily Output	Labor-Hours	Unit	Material	2018 Bare Costs Labor	Equipment	Total	Total Incl O&P
8100	4" diameter	1 Elec	2.50	3.200	Ea.	675	186		861	1,025
8150	T conduit body with cover, 1/2" diameter		11	.727		55.50	42.50		98	125
8160	3/4" diameter		9	.889		71	51.50		122.50	156
8170	1" diameter		6	1.333		84.50	77.50		162	209
8180	1-1/4" diameter		5	1.600		118	93		211	269
8190	1-1/2" diameter		4.50	1.778		144	103		247	315
8200	2" diameter		4	2		211	116		327	405
8210	2-1/2" diameter		3.50	2.286		330	133		463	565
8220	3" diameter		3	2.667		445	155		600	720
8230	3-1/2" diameter		2.50	3.200		645	186		831	990
8240	4" diameter		2	4		715	233		948	1,150
8300	FS conduit body, 1 gang, 3/4" diameter		11	.727		63.50	42.50		106	133
8310	1" diameter		10	.800		57.50	46.50		104	133
8350	2 gang, 3/4" diameter		9	.889		96.50	51.50		148	184
8360	1" diameter		8	1		118	58		176	217
8400	Duplex receptacle cover		64	.125		41	7.30		48.30	56
8410	Switch cover		64	.125		58	7.30		65.30	74.50
8420	Switch, vaportight cover		53	.151		179	8.80		187.80	210
8430	Blank, cover		64	.125		35.50	7.30		42.80	50
8520	FSC conduit body, 1 gang, 3/4" diameter		10	.800		69	46.50		115.50	146
8530	1" diameter		9	.889		69	51.50		120.50	154
8550	2 gang, 3/4" diameter		8	1		114	58		172	212
8560	1" diameter		7	1.143		122	66.50		188.50	235
8590	Conduit hubs, 1/2" diameter		18	.444		36	26		62	78
8600	3/4" diameter		16	.500		42	29		71	89.50
8610	1" diameter		14	.571		52.50	33.50		86	108
8620	1-1/4" diameter		12	.667		60	39		99	124
8630	1-1/2" diameter		10	.800		72.50	46.50		119	149
8640	2" diameter		8.80	.909		99	53		152	188
8650	2-1/2" diameter		8.50	.941		154	55		209	251
8660	3" diameter		8	1		209	58		267	315
8670	3-1/2" diameter		7.50	1.067		266	62		328	385
8680	4" diameter		7	1.143		330	66.50		396.50	465
8690	5" diameter		6	1.333		365	77.50		442.50	515
8700	Plastic coated 40 mil thick									
8710	Pipe strap, stamped 1 hole, 1/2" diameter	1 Elec	470	.017	Ea.	12.50	.99		13.49	15.25
8720	3/4" diameter		440	.018		11.50	1.06		12.56	14.25
8730	1" diameter		400	.020		12.95	1.16		14.11	15.95
8740	1-1/4" diameter		355	.023		19	1.31		20.31	23
8750	1-1/2" diameter		320	.025		21	1.46		22.46	25.50
8760	2" diameter		266	.030		28.50	1.75		30.25	34
8770	2-1/2" diameter		200	.040		32.50	2.33		34.83	39.50
8780	3" diameter		133	.060		44	3.50		47.50	54
8790	3-1/2" diameter		110	.073		79	4.23		83.23	93.50
8800	4" diameter		90	.089		85.50	5.15		90.65	102
8810	5" diameter		70	.114		136	6.65		142.65	160
8840	Clamp back spacers, 3/4" diameter		440	.018		17.15	1.06		18.21	20.50
8850	1" diameter		400	.020		23.50	1.16		24.66	27.50
8860	1-1/4" diameter		355	.023		26	1.31		27.31	30.50
8870	1-1/2" diameter		320	.025		39.50	1.46		40.96	45.50
8880	2" diameter		266	.030		54.50	1.75		56.25	62.50
8900	3" diameter		133	.060		81.50	3.50		85	95
8920	4" diameter		90	.089		61	5.15		66.15	75

For customer support on your Electrical Costs with RSMeans data, call 800.448.8182.

26 05 33 – Raceway and Boxes for Electrical Systems

26 05 33.25 Conduit Fittings for Rigid Galvanized Steel		Crew	Daily Output	Labor-Hours	Unit	Material	2018 Bare Costs Labor	Equipment	Total	Total Incl O&P
8950	Touch-up plastic coating, spray, 12 oz.				Ea.	54.50			54.50	60
8960	Sealing fittings, 1/2" diameter	1 Elec	11	.727		58	42.50		100.50	128
8970	3/4" diameter		9	.889		63.50	51.50		115	148
8980	1" diameter		7.50	1.067		76	62		138	177
8990	1-1/4" diameter		6.50	1.231		89.50	71.50		161	206
9000	1-1/2" diameter		5.50	1.455		124	84.50		208.50	263
9010	2" diameter		4.80	1.667		165	97		262	325
9020	2-1/2" diameter		4	2		205	116		321	400
9030	3" diameter		3.50	2.286		254	133		387	480
9040	3-1/2" diameter		3	2.667		640	155		795	935
9050	4" diameter		2.50	3.200		1,125	186		1,311	1,500
9060	5" diameter		1.70	4.706		2,125	274		2,399	2,750
9070	Unions, 1/2" diameter		18	.444		52	26		78	95.50
9080	3/4" diameter		15	.533		59	31		90	112
9090	1" diameter		13	.615		72.50	36		108.50	134
9100	1-1/4" diameter		11	.727		118	42.50		160.50	193
9110	1-1/2" diameter		9.50	.842		159	49		208	249
9120	2" diameter		8	1		194	58		252	300
9130	2-1/2" diameter		7.50	1.067		258	62		320	375
9140	3" diameter		6.80	1.176		320	68.50		388.50	450
9150	3-1/2" diameter		5.80	1.379		435	80.50		515.50	600
9160	4" diameter		4.80	1.667		580	97		677	785
9170	5" diameter		4	2		955	116		1,071	1,225

26 05 33.30 Electrical Nonmetallic Tubing (ENT)

26 05 33.30 Electrical Nonmetallic Tubing (ENT)		Crew	Daily Output	Labor-Hours	Unit	Material	2018 Bare Costs Labor	Equipment	Total	Total Incl O&P
0010	**ELECTRICAL NONMETALLIC TUBING (ENT)**									
0050	Flexible, 1/2" diameter	1 Elec	270	.030	L.F.	.54	1.72		2.26	3.17
0100	3/4" diameter		230	.035		.75	2.02		2.77	3.86
0200	1" diameter		145	.055		1.40	3.21		4.61	6.35
0210	1-1/4" diameter		125	.064		1.28	3.72		5	6.95
0220	1-1/2" diameter		100	.080		2.09	4.66		6.75	9.25
0230	2" diameter		75	.107		2.56	6.20		8.76	12.10
0300	Connectors, to outlet box, 1/2" diameter		230	.035	Ea.	.91	2.02		2.93	4.03
0310	3/4" diameter		210	.038		1.61	2.22		3.83	5.10
0320	1" diameter		200	.040		2.39	2.33		4.72	6.10
0400	Couplings, to conduit, 1/2" diameter		145	.055		.86	3.21		4.07	5.75
0410	3/4" diameter		130	.062		1.13	3.58		4.71	6.60
0420	1" diameter		125	.064		2.14	3.72		5.86	7.90

26 05 33.33 Raceway/Boxes for Utility Substations

26 05 33.33 Raceway/Boxes for Utility Substations		Crew	Daily Output	Labor-Hours	Unit	Material	2018 Bare Costs Labor	Equipment	Total	Total Incl O&P
0010	**RACEWAY/BOXES FOR UTILITY SUBSTATIONS**									
7000	Conduit, conductors, and insulators									
7100	Conduit, metallic	R-11	560	.100	Lb.	3.02	5.55	1.18	9.75	12.90
7110	Non-metallic	"	800	.070	"	9.30	3.87	.83	14	16.95
7190	See Section 26 05 33									
7200	Wire and cable	R-11	700	.080	Lb.	9.85	4.43	.95	15.23	18.50
7290	See Section 26 05 19									

26 05 33.35 Flexible Metallic Conduit

26 05 33.35 Flexible Metallic Conduit		Crew	Daily Output	Labor-Hours	Unit	Material	2018 Bare Costs Labor	Equipment	Total	Total Incl O&P
0010	**FLEXIBLE METALLIC CONDUIT**									
0050	Steel, 3/8" diameter	1 Elec	200	.040	L.F.	.38	2.33		2.71	3.90
0100	1/2" diameter		200	.040		.42	2.33		2.75	3.94
0200	3/4" diameter		160	.050		.58	2.91		3.49	4.99
0250	1" diameter		100	.080		1	4.66		5.66	8.05
0300	1-1/4" diameter		70	.114		1.39	6.65		8.04	11.50

For customer support on your Electrical Costs with RSMeans data, call 800.448.8182.

169

26 05 33.35 Flexible Metallic Conduit	Crew	Daily Output	Labor-Hours	Unit	Material	2018 Bare Costs Labor	Equipment	Total	Total Incl O&P
0350	1-1/2" diameter	1 Elec	50	.160	L.F.	2.26	9.30		11.56	16.40
0370	2" diameter		40	.200		2.79	11.65		14.44	20.50
0380	2-1/2" diameter		30	.267		3.35	15.50		18.85	26.50
0390	3" diameter	2 Elec	50	.320		5.75	18.60		24.35	34.50
0400	3-1/2" diameter		40	.400		6.60	23.50		30.10	42.50
0410	4" diameter		30	.533		7.45	31		38.45	54.50
0420	Connectors, plain, 3/8" diameter	1 Elec	100	.080	Ea.	1.40	4.66		6.06	8.50
0430	1/2" diameter		80	.100		1.48	5.80		7.28	10.35
0440	3/4" diameter		70	.114		1.71	6.65		8.36	11.85
0450	1" diameter		50	.160		3.90	9.30		13.20	18.20
0452	1-1/4" diameter		45	.178		4.80	10.35		15.15	21
0454	1-1/2" diameter		40	.200		6.90	11.65		18.55	25
0456	2" diameter		28	.286		9.70	16.65		26.35	35.50
0458	2-1/2" diameter		25	.320		17.15	18.60		35.75	47
0460	3" diameter		20	.400		24	23.50		47.50	61.50
0462	3-1/2" diameter		16	.500		84.50	29		113.50	137
0464	4" diameter		13	.615		93.50	36		129.50	157
0490	Insulated, 1" diameter		40	.200		7.75	11.65		19.40	26
0500	1-1/4" diameter		40	.200		10.45	11.65		22.10	29
0550	1-1/2" diameter		32	.250		19.45	14.55		34	43.50
0600	2" diameter		23	.348		25	20		45	58
0610	2-1/2" diameter		20	.400		49	23.50		72.50	89
0620	3" diameter		17	.471		100	27.50		127.50	151
0630	3-1/2" diameter		13	.615		269	36		305	350
0640	4" diameter		10	.800		265	46.50		311.50	360
0650	Connectors 90°, plain, 3/8" diameter		80	.100		2.23	5.80		8.03	11.15
0660	1/2" diameter		60	.133		4.33	7.75		12.08	16.35
0700	3/4" diameter		50	.160		7.05	9.30		16.35	21.50
0750	1" diameter		40	.200		11.05	11.65		22.70	29.50
0790	Insulated, 1" diameter		40	.200		12.65	11.65		24.30	31.50
0800	1-1/4" diameter		30	.267		20.50	15.50		36	46
0850	1-1/2" diameter		23	.348		45	20		65	80
0900	2" diameter		18	.444		61	26		87	106
0910	2-1/2" diameter		16	.500		127	29		156	184
0920	3" diameter		14	.571		160	33.50		193.50	226
0930	3-1/2" diameter		11	.727		450	42.50		492.50	560
0940	4" diameter		8	1		670	58		728	825
0960	Couplings, to flexible conduit, 1/2" diameter		50	.160		.79	9.30		10.09	14.75
0970	3/4" diameter		40	.200		1.35	11.65		13	18.90
0980	1" diameter		35	.229		2.19	13.30		15.49	22.50
0990	1-1/4" diameter		28	.286		5.10	16.65		21.75	30.50
1000	1-1/2" diameter		23	.348		6.85	20		26.85	38
1010	2" diameter		20	.400		13.65	23.50		37.15	50
1020	2-1/2" diameter		18	.444		36.50	26		62.50	79
1030	3" diameter		15	.533		92.50	31		123.50	149
1032	Aluminum, 3/8" diameter		210	.038	L.F.	.40	2.22		2.62	3.75
1034	1/2" diameter		210	.038		.44	2.22		2.66	3.79
1036	3/4" diameter		165	.048		.59	2.82		3.41	4.87
1038	1" diameter		105	.076		1.12	4.43		5.55	7.90
1040	1-1/4" diameter		75	.107		1.59	6.20		7.79	11.05
1042	1-1/2" diameter		53	.151		2.30	8.80		11.10	15.70
1044	2" diameter		42	.190		2.94	11.10		14.04	19.80
1046	2-1/2" diameter		32	.250		3.51	14.55		18.06	26

26 05 33.35 Flexible Metallic Conduit		Crew	Daily Output	Labor-Hours	Unit	Material	2018 Bare Costs Labor	Equipment	Total	Total Incl O&P
1048	3" diameter	2 Elec	53	.302	L.F.	5.70	17.55		23.25	33
1050	3-1/2" diameter		42	.381		6.55	22		28.55	40
1052	4" diameter		32	.500		7.35	29		36.35	51.50
1070	Sealtite, 3/8" diameter	1 Elec	140	.057		.67	3.33		4	5.70
1080	1/2" diameter		140	.057		.91	3.33		4.24	5.95
1090	3/4" diameter		100	.080		1.23	4.66		5.89	8.30
1100	1" diameter		70	.114		2.35	6.65		9	12.55
1200	1-1/4" diameter		50	.160		2.87	9.30		12.17	17.05
1300	1-1/2" diameter		40	.200		3.06	11.65		14.71	21
1400	2" diameter		30	.267		3.90	15.50		19.40	27.50
1410	2-1/2" diameter		27	.296		6.80	17.25		24.05	33.50
1420	3" diameter	2 Elec	50	.320		8.05	18.60		26.65	37
1440	4" diameter	"	30	.533		19.80	31		50.80	68.50
1490	Connectors, plain, 3/8" diameter	1 Elec	70	.114	Ea.	2.31	6.65		8.96	12.50
1500	1/2" diameter		70	.114		2.82	6.65		9.47	13.05
1700	3/4" diameter		50	.160		3.98	9.30		13.28	18.30
1900	1" diameter		40	.200		6.95	11.65		18.60	25
1910	Insulated, 1" diameter		40	.200		8.65	11.65		20.30	27
2000	1-1/4" diameter		32	.250		10.25	14.55		24.80	33.50
2100	1-1/2" diameter		27	.296		16.45	17.25		33.70	44
2200	2" diameter		20	.400		32	23.50		55.50	70.50
2210	2-1/2" diameter		15	.533		78	31		109	133
2220	3" diameter		12	.667		109	39		148	178
2240	4" diameter		8	1		202	58		260	310
2290	Connectors, 90°, 3/8" diameter		70	.114		4.60	6.65		11.25	15
2300	1/2" diameter		70	.114		4.53	6.65		11.18	14.95
2400	3/4" diameter		50	.160		6.90	9.30		16.20	21.50
2600	1" diameter		40	.200		13.55	11.65		25.20	32.50
2790	Insulated, 1" diameter		40	.200		9.80	11.65		21.45	28
2800	1-1/4" diameter		32	.250		15.45	14.55		30	39
3000	1-1/2" diameter		27	.296		23.50	17.25		40.75	52
3100	2" diameter		20	.400		27	23.50		50.50	64.50
3110	2-1/2" diameter		14	.571		128	33.50		161.50	191
3120	3" diameter		11	.727		340	42.50		382.50	440
3140	4" diameter		7	1.143		173	66.50		239.50	290
4300	Coupling sealtite to rigid, 1/2" diameter		20	.400		3.64	23.50		27.14	39
4500	3/4" diameter		18	.444		5	26		31	44
4800	1" diameter		14	.571		6.55	33.50		40.05	57
4900	1-1/4" diameter		12	.667		11.65	39		50.65	71
5000	1-1/2" diameter		11	.727		21.50	42.50		64	87
5100	2" diameter		10	.800		31	46.50		77.50	104
5110	2-1/2" diameter		9.50	.842		128	49		177	215
5120	3" diameter		9	.889		133	51.50		184.50	225
5130	3-1/2" diameter		9	.889		159	51.50		210.50	253
5140	4" diameter		8.50	.941		184	55		239	284

26 05 33.95 Cutting and Drilling

		Crew	Daily Output	Labor-Hours	Unit	Material	2018 Bare Costs Labor	Equipment	Total	Total Incl O&P
0010	**CUTTING AND DRILLING**									
0100	Hole drilling to 10' high, concrete wall									
0110	8" thick, 1/2" pipe size	R-31	12	.667	Ea.	.24	39	3.96	43.20	62.50
0120	3/4" pipe size		12	.667		.24	39	3.96	43.20	62.50
0130	1" pipe size		9.50	.842		.39	49	5	54.39	79.50
0140	1-1/4" pipe size		9.50	.842		.39	49	5	54.39	79.50

For customer support on your Electrical Costs with RSMeans data, call 800.448.8182.

171

26 05 33.95 Cutting and Drilling	Crew	Daily Output	Labor-Hours	Unit	Material	2018 Bare Costs Labor	Equipment	Total	Total Incl O&P
0150	1-1/2" pipe size	R-31	9.50	.842	Ea.	.39	49	5	54.39	79.50
0160	2" pipe size		4.40	1.818		.52	106	10.80	117.32	170
0170	2-1/2" pipe size		4.40	1.818		.52	106	10.80	117.32	170
0180	3" pipe size		4.40	1.818		.52	106	10.80	117.32	170
0190	3-1/2" pipe size		3.30	2.424		.67	141	14.40	156.07	228
0200	4" pipe size		3.30	2.424		.67	141	14.40	156.07	228
0500	12" thick, 1/2" pipe size		9.40	.851		.37	49.50	5.05	54.92	80
0520	3/4" pipe size		9.40	.851		.37	49.50	5.05	54.92	80
0540	1" pipe size		7.30	1.096		.58	64	6.50	71.08	103
0560	1-1/4" pipe size		7.30	1.096		.58	64	6.50	71.08	103
0570	1-1/2" pipe size		7.30	1.096		.58	64	6.50	71.08	103
0580	2" pipe size		3.60	2.222		.78	129	13.20	142.98	208
0590	2-1/2" pipe size		3.60	2.222		.78	129	13.20	142.98	208
0600	3" pipe size		3.60	2.222		.78	129	13.20	142.98	208
0610	3-1/2" pipe size		2.80	2.857		1.01	166	16.95	183.96	269
0630	4" pipe size		2.50	3.200		1.01	186	19	206.01	300
0650	16" thick, 1/2" pipe size		7.60	1.053		.49	61.50	6.25	68.24	99
0670	3/4" pipe size		7	1.143		.49	66.50	6.80	73.79	107
0690	1" pipe size		6	1.333		.77	77.50	7.90	86.17	126
0710	1-1/4" pipe size		5.50	1.455		.77	84.50	8.65	93.92	137
0730	1-1/2" pipe size		5.50	1.455		.77	84.50	8.65	93.92	137
0750	2" pipe size		3	2.667		1.04	155	15.85	171.89	251
0770	2-1/2" pipe size		2.70	2.963		1.04	172	17.60	190.64	279
0790	3" pipe size		2.50	3.200		1.04	186	19	206.04	300
0810	3-1/2" pipe size		2.30	3.478		1.34	202	20.50	223.84	330
0830	4" pipe size		2	4		1.34	233	24	258.34	375
0850	20" thick, 1/2" pipe size		6.40	1.250		.61	73	7.45	81.06	118
0870	3/4" pipe size		6	1.333		.61	77.50	7.90	86.01	125
0890	1" pipe size		5	1.600		.96	93	9.50	103.46	151
0910	1-1/4" pipe size		4.80	1.667		.96	97	9.90	107.86	157
0930	1-1/2" pipe size		4.60	1.739		.96	101	10.35	112.31	163
0950	2" pipe size		2.70	2.963		1.30	172	17.60	190.90	279
0970	2-1/2" pipe size		2.40	3.333		1.30	194	19.80	215.10	315
0990	3" pipe size		2.20	3.636		1.30	212	21.50	234.80	340
1010	3-1/2" pipe size		2	4		1.68	233	24	258.68	380
1030	4" pipe size		1.70	4.706		1.68	274	28	303.68	440
1050	24" thick, 1/2" pipe size		5.50	1.455		.73	84.50	8.65	93.88	137
1070	3/4" pipe size		5.10	1.569		.73	91.50	9.30	101.53	147
1090	1" pipe size		4.30	1.860		1.16	108	11.05	120.21	175
1110	1-1/4" pipe size		4	2		1.16	116	11.90	129.06	188
1130	1-1/2" pipe size		4	2		1.16	116	11.90	129.06	188
1150	2" pipe size		2.40	3.333		1.56	194	19.80	215.36	315
1170	2-1/2" pipe size		2.20	3.636		1.56	212	21.50	235.06	340
1190	3" pipe size		2	4		1.56	233	24	258.56	380
1210	3-1/2" pipe size		1.80	4.444		2.01	259	26.50	287.51	415
1230	4" pipe size		1.50	5.333		2.01	310	31.50	343.51	500
1500	Brick wall, 8" thick, 1/2" pipe size		18	.444		.24	26	2.64	28.88	41.50
1520	3/4" pipe size		18	.444		.24	26	2.64	28.88	41.50
1540	1" pipe size		13.30	.602		.39	35	3.57	38.96	57
1560	1-1/4" pipe size		13.30	.602		.39	35	3.57	38.96	57
1580	1-1/2" pipe size		13.30	.602		.39	35	3.57	38.96	57
1600	2" pipe size		5.70	1.404		.52	81.50	8.35	90.37	132
1620	2-1/2" pipe size		5.70	1.404		.52	81.50	8.35	90.37	132

26 05 33.95 Cutting and Drilling		Crew	Daily Output	Labor-Hours	Unit	Material	2018 Bare Costs Labor	Equipment	Total	Total Incl O&P
1640	3" pipe size	R-31	5.70	1.404	Ea.	.52	81.50	8.35	90.37	132
1660	3-1/2" pipe size		4.40	1.818		.67	106	10.80	117.47	171
1680	4" pipe size		4	2		.67	116	11.90	128.57	188
1700	12" thick, 1/2" pipe size		14.50	.552		.37	32	3.28	35.65	52
1720	3/4" pipe size		14.50	.552		.37	32	3.28	35.65	52
1740	1" pipe size		11	.727		.58	42.50	4.32	47.40	69
1760	1-1/4" pipe size		11	.727		.58	42.50	4.32	47.40	69
1780	1-1/2" pipe size		11	.727		.58	42.50	4.32	47.40	69
1800	2" pipe size		5	1.600		.78	93	9.50	103.28	150
1820	2-1/2" pipe size		5	1.600		.78	93	9.50	103.28	150
1840	3" pipe size		5	1.600		.78	93	9.50	103.28	150
1860	3-1/2" pipe size		3.80	2.105		1.01	123	12.50	136.51	198
1880	4" pipe size		3.30	2.424		1.01	141	14.40	156.41	228
1900	16" thick, 1/2" pipe size		12.30	.650		.49	38	3.86	42.35	61.50
1920	3/4" pipe size		12.30	.650		.49	38	3.86	42.35	61.50
1940	1" pipe size		9.30	.860		.77	50	5.10	55.87	81.50
1960	1-1/4" pipe size		9.30	.860		.77	50	5.10	55.87	81.50
1980	1-1/2" pipe size		9.30	.860		.77	50	5.10	55.87	81.50
2000	2" pipe size		4.40	1.818		1.04	106	10.80	117.84	171
2010	2-1/2" pipe size		4.40	1.818		1.04	106	10.80	117.84	171
2030	3" pipe size		4.40	1.818		1.04	106	10.80	117.84	171
2050	3-1/2" pipe size		3.30	2.424		1.34	141	14.40	156.74	228
2070	4" pipe size		3	2.667		1.34	155	15.85	172.19	251
2090	20" thick, 1/2" pipe size		10.70	.748		.61	43.50	4.44	48.55	70.50
2110	3/4" pipe size		10.70	.748		.61	43.50	4.44	48.55	70.50
2130	1" pipe size		8	1		.96	58	5.95	64.91	94.50
2150	1-1/4" pipe size		8	1		.96	58	5.95	64.91	94.50
2170	1-1/2" pipe size		8	1		.96	58	5.95	64.91	94.50
2190	2" pipe size		4	2		1.30	116	11.90	129.20	188
2210	2-1/2" pipe size		4	2		1.30	116	11.90	129.20	188
2230	3" pipe size		4	2		1.30	116	11.90	129.20	188
2250	3-1/2" pipe size		3	2.667		1.68	155	15.85	172.53	251
2270	4" pipe size		2.70	2.963		1.68	172	17.60	191.28	279
2290	24" thick, 1/2" pipe size		9.40	.851		.73	49.50	5.05	55.28	80.50
2310	3/4" pipe size		9.40	.851		.73	49.50	5.05	55.28	80.50
2330	1" pipe size		7.10	1.127		1.16	65.50	6.70	73.36	107
2350	1-1/4" pipe size		7.10	1.127		1.16	65.50	6.70	73.36	107
2370	1-1/2" pipe size		7.10	1.127		1.16	65.50	6.70	73.36	107
2390	2" pipe size		3.60	2.222		1.56	129	13.20	143.76	209
2410	2-1/2" pipe size		3.60	2.222		1.56	129	13.20	143.76	209
2430	3" pipe size		3.60	2.222		1.56	129	13.20	143.76	209
2450	3-1/2" pipe size		2.80	2.857		2.01	166	16.95	184.96	270
2470	4" pipe size		2.50	3.200		2.01	186	19	207.01	300
3000	Knockouts to 8' high, metal boxes & enclosures									
3020	With hole saw, 1/2" pipe size	1 Elec	53	.151	Ea.		8.80		8.80	13.15
3040	3/4" pipe size		47	.170			9.90		9.90	14.80
3050	1" pipe size		40	.200			11.65		11.65	17.40
3060	1-1/4" pipe size		36	.222			12.95		12.95	19.35
3070	1-1/2" pipe size		32	.250			14.55		14.55	22
3080	2" pipe size		27	.296			17.25		17.25	26
3090	2-1/2" pipe size		20	.400			23.50		23.50	35
4010	3" pipe size		16	.500			29		29	43.50
4030	3-1/2" pipe size		13	.615			36		36	53.50

For customer support on your Electrical Costs with RSMeans data, call 800.448.8182.

173

26 05 33 – Raceway and Boxes for Electrical Systems

26 05 33.95 Cutting and Drilling	Crew	Daily Output	Labor-Hours	Unit	Material	2018 Bare Costs Labor	Equipment	Total	Total Incl O&P
4050	4" pipe size	1 Elec	11	.727	Ea.		42.50		42.50	63.50

26 05 36 – Cable Trays for Electrical Systems

26 05 36.10 Cable Tray Ladder Type

		Crew	Daily Output	Labor-Hours	Unit	Material	2018 Bare Costs Labor	Equipment	Total	Total Incl O&P
0010	**CABLE TRAY LADDER TYPE** w/ftngs. & supports, 4" dp., to 15' elev.									
0100	For higher elevations, see Section 26 05 36.40									
0160	Galvanized steel tray R260536-10									
0170	4" rung spacing, 6" wide	2 Elec	98	.163	L.F.	13.60	9.50		23.10	29
0180	9" wide R260536-11		92	.174		14.85	10.10		24.95	31.50
0200	12" wide		86	.186		16.40	10.85		27.25	34
0400	18" wide		82	.195		19	11.35		30.35	38
0600	24" wide		78	.205		22	11.95		33.95	42
0650	30" wide		68	.235		27.50	13.70		41.20	51
0800	6" rung spacing, 6" wide		100	.160		12.40	9.30		21.70	27.50
0850	9" wide		94	.170		13.60	9.90		23.50	30
0860	12" wide		88	.182		14.30	10.60		24.90	31.50
0870	18" wide		84	.190		16.30	11.10		27.40	34.50
0880	24" wide		80	.200		18.25	11.65		29.90	37.50
0890	30" wide		70	.229		21.50	13.30		34.80	43.50
0910	9" rung spacing, 6" wide		102	.157		11.55	9.15		20.70	26.50
0920	9" wide		98	.163		11.85	9.50		21.35	27
0930	12" wide		94	.170		13.60	9.90		23.50	30
0940	18" wide		90	.178		15.35	10.35		25.70	32.50
0950	24" wide		86	.186		17.50	10.85		28.35	35.50
0960	30" wide		80	.200		19.15	11.65		30.80	38.50
0980	12" rung spacing, 6" wide		106	.151		11.35	8.80		20.15	25.50
0990	9" wide		104	.154		11.65	8.95		20.60	26
1000	12" wide		100	.160		12.30	9.30		21.60	27.50
1010	18" wide		96	.167		13.20	9.70		22.90	29
1020	24" wide		94	.170		14.15	9.90		24.05	30.50
1030	30" wide		88	.182		16.10	10.60		26.70	33.50
1041	18" rung spacing, 6" wide		108	.148		11.30	8.60		19.90	25.50
1042	9" wide		106	.151		11.45	8.80		20.25	26
1043	12" wide		102	.157		12.20	9.15		21.35	27
1044	18" wide		98	.163		12.80	9.50		22.30	28.50
1045	24" wide		96	.167		13.20	9.70		22.90	29
1046	30" wide		90	.178	▼	13.95	10.35		24.30	31
1050	Elbows horiz. 9" rung spacing, 90°, 12" radius, 6" wide		9.60	1.667	Ea.	52.50	97		149.50	203
1060	9" wide		8.40	1.905		55.50	111		166.50	228
1070	12" wide		7.60	2.105		60.50	123		183.50	250
1080	18" wide		6.20	2.581		79.50	150		229.50	315
1090	24" wide		5.40	2.963		89	172		261	355
1100	30" wide		4.80	3.333		118	194		312	420
1120	90°, 24" radius, 6" wide		9.20	1.739		108	101		209	270
1130	9" wide		8	2		111	116		227	297
1140	12" wide		7.20	2.222		115	129		244	320
1150	18" wide		5.80	2.759		124	161		285	375
1160	24" wide		5	3.200		134	186		320	425
1170	30" wide		4.40	3.636		145	212		357	475
1190	90°, 36" radius, 6" wide		8.80	1.818		119	106		225	289
1200	9" wide		7.60	2.105		126	123		249	320
1210	12" wide		6.80	2.353		132	137		269	350
1220	18" wide		5.40	2.963	▼	151	172		323	425

26 05 36.10 Cable Tray Ladder Type	Crew	Daily Output	Labor-Hours	Unit	Material	2018 Bare Costs Labor	Equipment	Total	Total Incl O&P
1230	24" wide	2 Elec	4.60	3.478	Ea.	166	202		368	485
1240	30" wide		4	4		199	233		432	570
1260	45°, 12" radius, 6" wide		13.20	1.212		40.50	70.50		111	150
1270	9" wide		11	1.455		42	84.50		126.50	174
1280	12" wide		9.60	1.667		44	97		141	194
1290	18" wide		7.60	2.105		46.50	123		169.50	234
1300	24" wide		6.20	2.581		59.50	150		209.50	291
1310	30" wide		5.40	2.963		76	172		248	340
1330	45°, 24" radius, 6" wide		12.80	1.250		51.50	73		124.50	166
1340	9" wide		10.60	1.509		54.50	88		142.50	191
1350	12" wide		9.20	1.739		61.50	101		162.50	219
1360	18" wide		7.20	2.222		63	129		192	262
1370	24" wide		5.80	2.759		76	161		237	325
1380	30" wide		5	3.200		90.50	186		276.50	380
1400	45°, 36" radius, 6" wide		12.40	1.290		73.50	75		148.50	193
1410	9" wide		10.20	1.569		78	91.50		169.50	222
1420	12" wide		8.80	1.818		79.50	106		185.50	246
1430	18" wide		6.80	2.353		86	137		223	300
1440	24" wide		5.40	2.963		91	172		263	360
1450	30" wide	▼	4.60	3.478	▼	118	202		320	435
1470	Elbows horizontal, 4" rung spacing, use 9" rung x 1.50									
1480	6" rung spacing, use 9" rung x 1.20									
1490	12" rung spacing, use 9" rung x 0.93									
1500	Elbows vertical, 9" rung spacing, 90°, 12" radius, 6" wide	2 Elec	9.60	1.667	Ea.	72	97		169	224
1510	9" wide		8.40	1.905		73.50	111		184.50	247
1520	12" wide		7.60	2.105		76	123		199	267
1530	18" wide		6.20	2.581		79.50	150		229.50	315
1540	24" wide		5.40	2.963		83	172		255	350
1550	30" wide		4.80	3.333		89	194		283	390
1570	24" radius, 6" wide		9.20	1.739		102	101		203	263
1580	9" wide		8	2		105	116		221	290
1590	12" wide		7.20	2.222		108	129		237	310
1600	18" wide		5.80	2.759		115	161		276	365
1610	24" wide		5	3.200		121	186		307	410
1620	30" wide		4.40	3.636		126	212		338	455
1640	36" radius, 6" wide		8.80	1.818		137	106		243	310
1650	9" wide		7.60	2.105		142	123		265	340
1660	12" wide		6.80	2.353		143	137		280	365
1670	18" wide		5.40	2.963		156	172		328	430
1680	24" wide		4.60	3.478		158	202		360	480
1690	30" wide	▼	4	4	▼	178	233		411	545
1710	Elbows vertical, 4" rung spacing, use 9" rung x 1.25									
1720	6" rung spacing, use 9" rung x 1.15									
1730	12" rung spacing, use 9" rung x 0.90									
1740	Tee horizontal, 9" rung spacing, 12" radius, 6" wide	2 Elec	5	3.200	Ea.	111	186		297	400
1750	9" wide		4.60	3.478		119	202		321	435
1760	12" wide		4.40	3.636		126	212		338	455
1770	18" wide		4	4		143	233		376	510
1780	24" wide		3.60	4.444		169	259		428	570
1790	30" wide		3.40	4.706		202	274		476	630
1810	24" radius, 6" wide		4.60	3.478		173	202		375	495
1820	9" wide		4.20	3.810		181	222		403	530
1830	12" wide	▼	4	4	▼	188	233		421	555

For customer support on your Electrical Costs with RSMeans data, call 800.448.8182.

175

26 05 36.10 Cable Tray Ladder Type		Crew	Daily Output	Labor-Hours	Unit	Material	2018 Bare Costs Labor	Equipment	Total	Total Incl O&P
1840	18" wide	2 Elec	3.60	4.444	Ea.	204	259		463	610
1850	24" wide		3.20	5		221	291		512	680
1860	30" wide		3	5.333		240	310		550	730
1880	36" radius, 6" wide		4.20	3.810		259	222		481	615
1890	9" wide		3.80	4.211		271	245		516	665
1900	12" wide		3.60	4.444		288	259		547	700
1910	18" wide		3.20	5		310	291		601	775
1920	24" wide		2.80	5.714		350	335		685	880
1930	30" wide		2.60	6.154		380	360		740	955
1980	Tee vertical, 9" rung spacing, 12" radius, 6" wide		5.40	2.963		189	172		361	465
1990	9" wide		5.20	3.077		191	179		370	480
2000	12" wide		5	3.200		193	186		379	490
2010	18" wide		4.60	3.478		194	202		396	520
2020	24" wide		4.40	3.636		199	212		411	535
2030	30" wide		4	4		212	233		445	585
2050	24" radius, 6" wide		5	3.200		340	186		526	655
2060	9" wide		4.80	3.333		345	194		539	670
2070	12" wide		4.60	3.478		350	202		552	690
2080	18" wide		4.20	3.810		360	222		582	725
2090	24" wide		4	4		370	233		603	760
2100	30" wide		3.60	4.444		380	259		639	805
2120	36" radius, 6" wide		4.60	3.478		635	202		837	1,000
2130	9" wide		4.40	3.636		645	212		857	1,025
2140	12" wide		4.20	3.810		655	222		877	1,050
2150	18" wide		3.80	4.211		660	245		905	1,100
2160	24" wide		3.60	4.444		670	259		929	1,125
2170	30" wide	▼	3.20	5	▼	695	291		986	1,200
2190	Tee, 4" rung spacing, use 9" rung x 1.30									
2200	6" rung spacing, use 9" rung x 1.20									
2210	12" rung spacing, use 9" rung x 0.90									
2220	Cross horizontal, 9" rung spacing, 12" radius, 6" wide	2 Elec	4	4	Ea.	137	233		370	500
2230	9" wide		3.80	4.211		144	245		389	525
2240	12" wide		3.60	4.444		158	259		417	560
2250	18" wide		3.40	4.706		172	274		446	600
2260	24" wide		3	5.333		191	310		501	675
2270	30" wide		2.80	5.714		231	335		566	750
2290	24" radius, 6" wide		3.60	4.444		267	259		526	680
2300	9" wide		3.40	4.706		275	274		549	715
2310	12" wide		3.20	5		283	291		574	745
2320	18" wide		3	5.333		300	310		610	800
2330	24" wide		2.60	6.154		320	360		680	890
2340	30" wide		2.40	6.667		340	390		730	955
2360	36" radius, 6" wide		3.20	5		370	291		661	845
2370	9" wide		3	5.333		390	310		700	890
2380	12" wide		2.80	5.714		405	335		740	940
2390	18" wide		2.60	6.154		445	360		805	1,025
2400	24" wide		2.20	7.273		495	425		920	1,175
2410	30" wide	▼	2	8	▼	540	465		1,005	1,275
2430	Cross horizontal, 4" rung spacing, use 9" rung x 1.30									
2440	6" rung spacing, use 9" rung x 1.20									
2450	12" rung spacing, use 9" rung x 0.90									
2460	Reducer, 9" to 6" wide tray	2 Elec	13	1.231	Ea.	76.50	71.50		148	191
2470	12" to 9" wide tray	▼	12	1.333	▼	78	77.50		155.50	202

26 05 36.10 Cable Tray Ladder Type		Crew	Daily Output	Labor-Hours	Unit	Material	2018 Bare Costs Labor	Equipment	Total	Total Incl O&P
2480	18" to 12" wide tray	2 Elec	10.40	1.538	Ea.	78	89.50		167.50	220
2490	24" to 18" wide tray		9	1.778		79.50	103		182.50	243
2500	30" to 24" wide tray		8	2		81	116		197	264
2510	36" to 30" wide tray		7	2.286		90.50	133		223.50	299
2511	Reducer, 18" to 6" wide tray		10.40	1.538		81	89.50		170.50	224
2512	24" to 12" wide tray		9	1.778		83	103		186	246
2513	30" to 18" wide tray		8	2		86	116		202	269
2514	30" to 12" wide tray		8	2		89	116		205	272
2515	36" to 24" wide tray		7	2.286		92.50	133		225.50	300
2516	36" to 18" wide tray		7	2.286		92.50	133		225.50	300
2517	36" to 12" wide tray		7	2.286		92.50	133		225.50	300
2520	Dropout or end plate, 6" wide		32	.500		8.50	29		37.50	53
2530	9" wide		28	.571		10.05	33.50		43.55	60.50
2540	12" wide		26	.615		11	36		47	65.50
2550	18" wide		22	.727		12.90	42.50		55.40	77.50
2560	24" wide		20	.800		15.60	46.50		62.10	86.50
2570	30" wide		18	.889		17.05	51.50		68.55	96.50
2590	Tray connector		48	.333	▼	15.10	19.40		34.50	45.50
3200	Aluminum tray, 4" deep, 6" rung spacing, 6" wide		134	.119	L.F.	13.40	6.95		20.35	25
3210	9" wide		128	.125		14.20	7.30		21.50	26.50
3220	12" wide		124	.129		15	7.50		22.50	28
3230	18" wide		114	.140		16.70	8.15		24.85	30.50
3240	24" wide		106	.151		19.40	8.80		28.20	34.50
3250	30" wide		100	.160		21.50	9.30		30.80	37.50
3270	9" rung spacing, 6" wide		140	.114		10.85	6.65		17.50	22
3280	9" wide		134	.119		11.35	6.95		18.30	23
3290	12" wide		130	.123		11.60	7.15		18.75	23.50
3300	18" wide		122	.131		13.55	7.65		21.20	26.50
3310	24" wide		116	.138		15.75	8.05		23.80	29.50
3320	30" wide		108	.148		17.40	8.60		26	32
3340	12" rung spacing, 6" wide		146	.110		11.55	6.40		17.95	22.50
3350	9" wide		140	.114		11.70	6.65		18.35	23
3360	12" wide		134	.119		12.05	6.95		19	23.50
3370	18" wide		128	.125		12.75	7.30		20.05	25
3380	24" wide		124	.129		14.20	7.50		21.70	27
3390	30" wide		114	.140		15.20	8.15		23.35	29
3401	18" rung spacing, 6" wide		150	.107		11.45	6.20		17.65	22
3402	9" wide tray		144	.111		11.60	6.45		18.05	22.50
3403	12" wide tray		140	.114		12.05	6.65		18.70	23
3404	18" wide tray		134	.119		12.70	6.95		19.65	24.50
3405	24" wide tray		130	.123		13.65	7.15		20.80	26
3406	30" wide tray		120	.133	▼	14.20	7.75		21.95	27
3410	Elbows horiz., 9" rung spacing, 90°, 12" radius, 6" wide		9.60	1.667	Ea.	51.50	97		148.50	202
3420	9" wide		8.40	1.905		53	111		164	224
3430	12" wide		7.60	2.105		62.50	123		185.50	252
3440	18" wide		6.20	2.581		70.50	150		220.50	305
3450	24" wide		5.40	2.963		87.50	172		259.50	355
3460	30" wide		4.80	3.333		92.50	194		286.50	390
3480	24" radius, 6" wide		9.20	1.739		91.50	101		192.50	252
3490	9" wide		8	2		95.50	116		211.50	279
3500	12" wide		7.20	2.222		100	129		229	305
3510	18" wide		5.80	2.759		109	161		270	360
3520	24" wide		5	3.200		119	186		305	410

26 05 36 – Cable Trays for Electrical Systems

26 05 36.10 Cable Tray Ladder Type		Crew	Daily Output	Labor-Hours	Unit	Material	2018 Bare Costs Labor	Equipment	Total	Total Incl O&P
3530	30" wide	2 Elec	4.40	3.636	Ea.	132	212		344	460
3550	90°, 36" radius, 6" wide		8.80	1.818		101	106		207	269
3560	9" wide		7.60	2.105		103	123		226	296
3570	12" wide		6.80	2.353		116	137		253	330
3580	18" wide		5.40	2.963		129	172		301	400
3590	24" wide		4.60	3.478		146	202		348	465
3600	30" wide		4	4		161	233		394	525
3620	45°, 12" radius, 6" wide		13.20	1.212		34	70.50		104.50	143
3630	9" wide		11	1.455		35	84.50		119.50	166
3640	12" wide		9.60	1.667		36.50	97		133.50	185
3650	18" wide		7.60	2.105		43	123		166	231
3660	24" wide		6.20	2.581		53.50	150		203.50	284
3670	30" wide		5.40	2.963		61.50	172		233.50	325
3690	45°, 24" radius, 6" wide		12.80	1.250		45	73		118	159
3700	9" wide		10.60	1.509		51.50	88		139.50	188
3710	12" wide		9.20	1.739		53	101		154	209
3720	18" wide		7.20	2.222		56.50	129		185.50	255
3730	24" wide		5.80	2.759		65.50	161		226.50	310
3740	30" wide		5	3.200		78	186		264	365
3760	45°, 36" radius, 6" wide		12.40	1.290		61.50	75		136.50	180
3770	9" wide		10.20	1.569		63	91.50		154.50	205
3780	12" wide		8.80	1.818		64	106		170	229
3790	18" wide		6.80	2.353		70.50	137		207.50	283
3800	24" wide		5.40	2.963		86.50	172		258.50	355
3810	30" wide		4.60	3.478		91.50	202		293.50	405
3830	Elbows horizontal, 4" rung spacing, use 9" rung x 1.50									
3840	6" rung spacing, use 9" rung x 1.20									
3850	12" rung spacing, use 9" rung x 0.93									
3860	Elbows vertical, 9" rung spacing, 90°, 12" radius, 6" wide	2 Elec	9.60	1.667	Ea.	71.50	97		168.50	224
3870	9" wide		8.40	1.905		75.50	111		186.50	249
3880	12" wide		7.60	2.105		76.50	123		199.50	268
3890	18" wide		6.20	2.581		79	150		229	310
3900	24" wide		5.40	2.963		84	172		256	350
3910	30" wide		4.80	3.333		88	194		282	385
3930	24" radius, 6" wide		9.20	1.739		78	101		179	237
3940	9" wide		8	2		80.50	116		196.50	263
3950	12" wide		7.20	2.222		83	129		212	284
3960	18" wide		5.80	2.759		88	161		249	335
3970	24" wide		5	3.200		93	186		279	380
3980	30" wide		4.40	3.636		98	212		310	425
4000	36" radius, 6" wide		8.80	1.818		121	106		227	291
4010	9" wide		7.60	2.105		123	123		246	320
4020	12" wide		6.80	2.353		131	137		268	350
4030	18" wide		5.40	2.963		136	172		308	405
4040	24" wide		4.60	3.478		141	202		343	460
4050	30" wide		4	4		152	233		385	515
4070	Elbows vertical, 4" rung spacing, use 9" rung x 1.25									
4080	6" rung spacing, use 9" rung x 1.15									
4090	12" rung spacing, use 9" rung x 0.90									
4100	Tee horizontal, 9" rung spacing, 12" radius, 6" wide	2 Elec	5	3.200	Ea.	79.50	186		265.50	365
4110	9" wide		4.60	3.478		85.50	202		287.50	400
4120	12" wide		4.40	3.636		94.50	212		306.50	420
4130	18" wide		4.20	3.810		103	222		325	445

26 05 36.10 Cable Tray Ladder Type		Crew	Daily Output	Labor-Hours	Unit	Material	2018 Bare Costs Labor	Equipment	Total	Total Incl O&P
4140	24" wide	2 Elec	4	4	Ea.	113	233		346	475
4150	30" wide		3.60	4.444		132	259		391	530
4170	24" radius, 6" wide		4.60	3.478		136	202		338	455
4180	9" wide		4.20	3.810		143	222		365	490
4190	12" wide		4	4		147	233		380	510
4200	18" wide		3.80	4.211		161	245		406	540
4210	24" wide		3.60	4.444		176	259		435	580
4220	30" wide		3.20	5		191	291		482	645
4240	36" radius, 6" wide		4.20	3.810		192	222		414	540
4250	9" wide		3.80	4.211		197	245		442	580
4260	12" wide		3.60	4.444		216	259		475	625
4270	18" wide		3.40	4.706		227	274		501	660
4280	24" wide		3.20	5		251	291		542	710
4290	30" wide		2.80	5.714		320	335		655	845
4310	Tee vertical, 9" rung spacing, 12" radius, 6" wide		5.40	2.963		178	172		350	455
4320	9" wide		5.20	3.077		185	179		364	470
4330	12" wide		5	3.200		187	186		373	485
4340	18" wide		4.60	3.478		190	202		392	515
4350	24" wide		4.40	3.636		205	212		417	540
4360	30" wide		4.20	3.810		209	222		431	560
4380	24" radius, 6" wide		5	3.200		305	186		491	615
4390	9" wide		4.80	3.333		305	194		499	625
4400	12" wide		4.60	3.478		310	202		512	650
4410	18" wide		4.20	3.810		320	222		542	680
4420	24" wide		4	4		330	233		563	710
4430	30" wide		3.80	4.211		335	245		580	735
4450	36" radius, 6" wide		4.60	3.478		665	202		867	1,025
4460	9" wide		4.40	3.636		680	212		892	1,075
4470	12" wide		4.20	3.810		705	222		927	1,100
4480	18" wide		3.80	4.211		710	245		955	1,150
4490	24" wide		3.60	4.444		730	259		989	1,200
4500	30" wide	▼	3.40	4.706	▼	740	274		1,014	1,225
4520	Tees, 4" rung spacing, use 9" rung x 1.30									
4530	6" rung spacing, use 9" rung x 1.20									
4540	12" rung spacing, use 9" rung x 0.90									
4550	Cross horizontal, 9" rung spacing, 12" radius, 6" wide	2 Elec	4.40	3.636	Ea.	111	212		323	435
4560	9" wide		4.20	3.810		123	222		345	465
4570	12" wide		4	4		137	233		370	500
4580	18" wide		3.60	4.444		151	259		410	550
4590	24" wide		3.40	4.706		156	274		430	580
4600	30" wide		3	5.333		183	310		493	665
4620	24" radius, 6" wide		4	4		212	233		445	585
4630	9" wide		3.80	4.211		220	245		465	605
4640	12" wide		3.60	4.444		226	259		485	635
4650	18" wide		3.20	5		237	291		528	695
4660	24" wide		3	5.333		259	310		569	750
4670	30" wide		2.60	6.154		276	360		636	840
4690	36" radius, 6" wide		3.60	4.444		251	259		510	660
4700	9" wide		3.40	4.706		261	274		535	695
4710	12" wide		3.20	5		276	291		567	740
4720	18" wide		2.80	5.714		295	335		630	820
4730	24" wide		2.60	6.154		365	360		725	935
4740	30" wide	▼	2.20	7.273	▼	405	425		830	1,075

For customer support on your Electrical Costs with RSMeans data, call 800.448.8182.

179

26 05 36.10 Cable Tray Ladder Type		Crew	Daily Output	Labor-Hours	Unit	Material	2018 Bare Costs Labor	2018 Bare Costs Equipment	Total	Total Incl O&P
4760	Cross horizontal, 4" rung spacing, use 9" rung x 1.30									
4770	6" rung spacing, use 9" rung x 1.20									
4780	12" rung spacing, use 9" rung x 0.90									
4790	Reducer, 9" to 6" wide tray	2 Elec	16	1	Ea.	59	58		117	152
4800	12" to 9" wide tray		14	1.143		61.50	66.50		128	167
4810	18" to 12" wide tray		12.40	1.290		61.50	75		136.50	180
4820	24" to 18" wide tray		10.60	1.509		63	88		151	200
4830	30" to 24" wide tray		9.20	1.739		64	101		165	222
4840	36" to 30" wide tray		8	2		68	116		184	249
4841	Reducer, 18" to 6" wide tray		12.40	1.290		61.50	75		136.50	180
4842	24" to 12" wide tray		10.60	1.509		63	88		151	200
4843	30" to 18" wide tray		9.20	1.739		64	101		165	222
4844	30" to 12" wide tray		9.20	1.739		64	101		165	222
4845	36" to 24" wide tray		8	2		68	116		184	249
4846	36" to 18" wide tray		8	2		68	116		184	249
4847	36" to 12" wide tray		8	2		68	116		184	249
4850	Dropout or end plate, 6" wide		32	.500		7.80	29		36.80	52
4860	9" wide tray		28	.571		8.35	33.50		41.85	58.50
4870	12" wide tray		26	.615		9.05	36		45.05	63.50
4880	18" wide tray		22	.727		12.05	42.50		54.55	77
4890	24" wide tray		20	.800		14.20	46.50		60.70	85
4900	30" wide tray		18	.889		16.80	51.50		68.30	96
4920	Tray connector		48	.333		11.30	19.40		30.70	41.50
8000	Elbow 36" radius horiz., 60°, 6" wide tray		10.60	1.509		85.50	88		173.50	225
8010	9" wide tray		9	1.778		93	103		196	257
8020	12" wide tray		7.80	2.051		95.50	119		214.50	283
8030	18" wide tray		6.20	2.581		102	150		252	335
8040	24" wide tray		5	3.200		119	186		305	410
8050	30" wide tray		4.40	3.636		122	212		334	450
8060	30°, 6" wide tray		14	1.143		65.50	66.50		132	172
8070	9" wide tray		11.40	1.404		69	81.50		150.50	198
8080	12" wide tray		9.80	1.633		70.50	95		165.50	220
8090	18" wide tray		7.40	2.162		74	126		200	270
8100	24" wide tray		5.80	2.759		90.50	161		251.50	340
8110	30" wide tray		4.80	3.333		93	194		287	390
8120	Adjustable, 6" wide tray		12.40	1.290		86.50	75		161.50	208
8130	9" wide tray		10.20	1.569		91.50	91.50		183	237
8140	12" wide tray		8.80	1.818		93	106		199	260
8150	18" wide tray		6.80	2.353		102	137		239	315
8160	24" wide tray		5.40	2.963		108	172		280	375
8170	30" wide tray		4.60	3.478		121	202		323	440
8180	Wye 36" radius horiz., 45°, 6" wide tray		4.60	3.478		94	202		296	410
8190	9" wide tray		4.40	3.636		99.50	212		311.50	425
8200	12" wide tray		4.20	3.810		103	222		325	445
8210	18" wide tray		3.80	4.211		117	245		362	495
8220	24" wide tray		3.60	4.444		131	259		390	530
8230	30" wide tray		3.40	4.706		146	274		420	570
8240	Elbow 36" radius vert. in/outside, 60°, 6" wide tray		10.60	1.509		94	88		182	235
8250	9" wide tray		9	1.778		98	103		201	263
8260	12" wide tray		7.80	2.051		99.50	119		218.50	287
8270	18" wide tray		6.20	2.581		103	150		253	340
8280	24" wide tray		5	3.200		109	186		295	400
8290	30" wide tray		4.40	3.636		117	212		329	445

For customer support on your Electrical Costs with RSMeans data, call 800.448.8182.

26 05 Common Work Results for Electrical

26 05 36 – Cable Trays for Electrical Systems

26 05 36.10 Cable Tray Ladder Type		Crew	Daily Output	Labor-Hours	Unit	Material	2018 Bare Costs Labor	Equipment	Total	Total Incl O&P
8300	45°, 6" wide tray	2 Elec	12.40	1.290	Ea.	85	75		160	206
8310	9" wide tray		10.20	1.569		90	91.50		181.50	235
8320	12" wide tray		8.80	1.818		91.50	106		197.50	259
8330	18" wide tray		6.80	2.353		94	137		231	310
8340	24" wide tray		5.40	2.963		99.50	172		271.50	365
8350	30" wide tray		4.60	3.478		102	202		304	415
8360	30°, 6" wide tray		14	1.143		74.50	66.50		141	182
8370	9" wide tray		11.40	1.404		76	81.50		157.50	206
8380	12" wide tray		9.80	1.633		81	95		176	231
8390	18" wide tray		7.40	2.162		82.50	126		208.50	279
8400	24" wide tray		5.80	2.759		84	161		245	335
8410	30" wide tray		4.80	3.333		89	194		283	390
8660	Adjustable, 6" wide tray		12.40	1.290		85	75		160	206
8670	9" wide tray		10.20	1.569		90	91.50		181.50	235
8680	12" wide tray		8.80	1.818		91.50	106		197.50	259
8690	18" wide tray		6.80	2.353		102	137		239	315
8700	24" wide tray		5.40	2.963		108	172		280	375
8710	30" wide tray		4.60	3.478		119	202		321	435
8720	Cross, vertical, 24" radius, 6" wide tray		3.60	4.444		645	259		904	1,100
8730	9" wide tray		3.40	4.706		655	274		929	1,125
8740	12" wide tray		3.20	5		675	291		966	1,175
8750	18" wide tray		2.80	5.714		705	335		1,040	1,275
8760	24" wide tray		2.60	6.154		730	360		1,090	1,325
8770	30" wide tray		2.20	7.273		770	425		1,195	1,475
9200	Splice plate	1 Elec	48	.167	Pr.	11.80	9.70		21.50	27.50
9210	Expansion joint		48	.167		14.20	9.70		23.90	30
9220	Horizontal hinged		48	.167		11.80	9.70		21.50	27.50
9230	Vertical hinged		48	.167		14.20	9.70		23.90	30
9240	Ladder hanger, vertical		28	.286	Ea.	3.51	16.65		20.16	29
9250	Ladder to channel connector		24	.333		41	19.40		60.40	74.50
9260	Ladder to box connector, 30" wide		19	.421		41	24.50		65.50	82
9270	24" wide		20	.400		41	23.50		64.50	80.50
9280	18" wide		21	.381		40	22		62	77
9290	12" wide		22	.364		37	21		58	72.50
9300	9" wide		23	.348		33.50	20		53.50	67
9310	6" wide		24	.333		29.50	19.40		48.90	61.50
9320	Ladder floor flange		24	.333		26.50	19.40		45.90	58.50
9330	Cable roller for tray, 30" wide		10	.800		243	46.50		289.50	340
9340	24" wide		11	.727		207	42.50		249.50	291
9350	18" wide		12	.667		197	39		236	275
9360	12" wide		13	.615		155	36		191	224
9370	9" wide		14	.571		137	33.50		170.50	201
9380	6" wide		15	.533		111	31		142	170
9390	Pulley, single wheel		12	.667		254	39		293	340
9400	Triple wheel		10	.800		505	46.50		551.50	630
9440	Nylon cable tie, 14" long		80	.100		.58	5.80		6.38	9.35
9450	Ladder, hold down clamp		60	.133		7.55	7.75		15.30	19.90
9460	Cable clamp		60	.133		7.65	7.75		15.40	20
9470	Wall bracket, 30" wide tray		19	.421		45	24.50		69.50	86
9480	24" wide tray		20	.400		35	23.50		58.50	73.50
9490	18" wide tray		21	.381		31.50	22		53.50	67.50
9500	12" wide tray		22	.364		29	21		50	63.50
9510	9" wide tray		23	.348		25.50	20		45.50	59

For customer support on your Electrical Costs with RSMeans data, call 800.448.8182.

181

26 05 36 – Cable Trays for Electrical Systems

26 05 36.10 Cable Tray Ladder Type		Crew	Daily Output	Labor-Hours	Unit	Material	2018 Bare Costs Labor	Equipment	Total	Total Incl O&P
9520	6" wide tray	1 Elec	24	.333	Ea.	25	19.40		44.40	56.50

26 05 36.20 Cable Tray Solid Bottom

26 05 36.20 Cable Tray Solid Bottom		Crew	Daily Output	Labor-Hours	Unit	Material	2018 Bare Costs Labor	Equipment	Total	Total Incl O&P
0010	**CABLE TRAY SOLID BOTTOM** w/ftngs. & supports, 3" deep, to 15' high									
0200	For higher elevations, see Section 26 05 36.40									
0220	Galvanized steel, tray, 6" wide R260536-10	2 Elec	120	.133	L.F.	10.80	7.75		18.55	23.50
0240	12" wide		100	.160		13.85	9.30		23.15	29
0260	18" wide		70	.229		16.85	13.30		30.15	38.50
0280	24" wide R260536-11		.60	.267		19.70	15.50		35.20	44.50
0300	30" wide		50	.320		23.50	18.60		42.10	54
0340	Elbow horizontal, 90°, 12" radius, 6" wide		9.60	1.667	Ea.	82	97		179	235
0360	12" wide		6.80	2.353		91.50	137		228.50	305
0370	18" wide		5.40	2.963		105	172		277	375
0380	24" wide		4.40	3.636		128	212		340	455
0390	30" wide		3.80	4.211		153	245		398	535
0420	24" radius, 6" wide		9.20	1.739		117	101		218	279
0440	12" wide		6.40	2.500		135	146		281	365
0450	18" wide		5	3.200		158	186		344	450
0460	24" wide		4	4		183	233		416	550
0470	30" wide		3.40	4.706		214	274		488	645
0500	36" radius, 6" wide		8.80	1.818		170	106		276	345
0520	12" wide		6	2.667		192	155		347	445
0530	18" wide		4.60	3.478		235	202		437	565
0540	24" wide		3.60	4.444		254	259		513	665
0550	30" wide		3	5.333		305	310		615	805
0580	Elbow vertical, 90°, 12" radius, 6" wide		9.60	1.667		99	97		196	254
0600	12" wide		6.80	2.353		108	137		245	325
0610	18" wide		5.40	2.963		116	172		288	385
0620	24" wide		4.40	3.636		124	212		336	450
0630	30" wide		3.80	4.211		132	245		377	510
0670	24" radius, 6" wide		9.20	1.739		140	101		241	305
0690	12" wide		6.40	2.500		154	146		300	390
0700	18" wide		5	3.200		167	186		353	460
0710	24" wide		4	4		181	233		414	550
0720	30" wide		3.40	4.706		192	274		466	620
0750	36" radius, 6" wide		8.80	1.818		192	106		298	370
0770	12" wide		6.60	2.424		214	141		355	445
0780	18" wide		4.60	3.478		235	202		437	565
0790	24" wide		3.60	4.444		257	259		516	665
0800	30" wide		3	5.333		280	310		590	775
0840	Tee horizontal, 12" radius, 6" wide		5	3.200		115	186		301	405
0860	12" wide		4	4		124	233		357	485
0870	18" wide		3.40	4.706		148	274		422	575
0880	24" wide		2.80	5.714		165	335		500	675
0890	30" wide		2.60	6.154		189	360		549	745
0940	24" radius, 6" wide		4.60	3.478		178	202		380	500
0960	12" wide		3.60	4.444		208	259		467	615
0970	18" wide		3	5.333		232	310		542	720
0980	24" wide		2.40	6.667		315	390		705	925
0990	30" wide		2.20	7.273		340	425		765	1,000
1020	36" radius, 6" wide		4.20	3.810		280	222		502	640
1040	12" wide		3.20	5		315	291		606	780
1050	18" wide		2.60	6.154		340	360		700	910

26 05 Common Work Results for Electrical

26 05 36 – Cable Trays for Electrical Systems

26 05 36.20 Cable Tray Solid Bottom		Crew	Daily Output	Labor-Hours	Unit	Material	2018 Bare Costs Labor	Equipment	Total	Total Incl O&P
1060	24" wide	2 Elec	2.20	7.273	Ea.	440	425		865	1,125
1070	30" wide		2	8		475	465		940	1,225
1100	Tee vertical, 12" radius, 6" wide		5	3.200		189	186		375	485
1120	12" wide		4	4		197	233		430	565
1130	18" wide		3.60	4.444		200	259		459	605
1140	24" wide		3.40	4.706		219	274		493	650
1150	30" wide		3	5.333		235	310		545	725
1180	24" radius, 6" wide		4.60	3.478		279	202		481	610
1200	12" wide		3.60	4.444		291	259		550	705
1210	18" wide		3.20	5		305	291		596	770
1220	24" wide		3	5.333		320	310		630	820
1230	30" wide		2.60	6.154		355	360		715	925
1260	36" radius, 6" wide		4.20	3.810		435	222		657	810
1280	12" wide		3.20	5		445	291		736	920
1290	18" wide		2.80	5.714		475	335		810	1,025
1300	24" wide		2.60	6.154		490	360		850	1,075
1310	30" wide		2.20	7.273		575	425		1,000	1,275
1340	Cross horizontal, 12" radius, 6" wide		4	4		137	233		370	500
1360	12" wide		3.40	4.706		150	274		424	575
1370	18" wide		2.80	5.714		175	335		510	685
1380	24" wide		2.40	6.667		197	390		587	795
1390	30" wide		2	8		219	465		684	935
1420	24" radius, 6" wide		3.60	4.444		250	259		509	660
1440	12" wide		3	5.333		280	310		590	775
1450	18" wide		2.40	6.667		315	390		705	925
1460	24" wide		2	8		400	465		865	1,125
1470	30" wide		1.80	8.889		435	515		950	1,250
1500	36" radius, 6" wide		3.20	5		410	291		701	885
1520	12" wide		2.60	6.154		450	360		810	1,025
1530	18" wide		2	8		490	465		955	1,225
1540	24" wide		1.80	8.889		580	515		1,095	1,425
1550	30" wide		1.60	10		630	580		1,210	1,575
1580	Drop out or end plate, 6" wide		32	.500		16.70	29		45.70	62
1600	12" wide		26	.615		21	36		57	76.50
1610	18" wide		22	.727		22.50	42.50		65	88
1620	24" wide		20	.800		26.50	46.50		73	98.50
1630	30" wide		18	.889		29.50	51.50		81	110
1660	Reducer, 12" to 6" wide		12	1.333		80.50	77.50		158	205
1680	18" to 12" wide		10.60	1.509		82	88		170	221
1700	18" to 6" wide		10.60	1.509		82	88		170	221
1720	24" to 18" wide		9.20	1.739		85	101		186	245
1740	24" to 12" wide		9.20	1.739		85	101		186	245
1760	30" to 24" wide		8	2		91.50	116		207.50	275
1780	30" to 18" wide		8	2		91.50	116		207.50	275
1800	30" to 12" wide		8	2		91.50	116		207.50	275
1820	36" to 30" wide		7.20	2.222		94.50	129		223.50	297
1840	36" to 24" wide		7.20	2.222		94.50	129		223.50	297
1860	36" to 18" wide		7.20	2.222		94.50	129		223.50	297
1880	36" to 12" wide		7.20	2.222		94.50	129		223.50	297
2000	Aluminum tray, 6" wide		150	.107	L.F.	10.30	6.20		16.50	20.50
2020	12" wide		130	.123		13.50	7.15		20.65	25.50
2030	18" wide		100	.160		16.95	9.30		26.25	32.50
2040	24" wide		90	.178		21.50	10.35		31.85	39

For customer support on your Electrical Costs with RSMeans data, call 800.448.8182.

183

26 05 36.20 Cable Tray Solid Bottom		Crew	Daily Output	Labor-Hours	Unit	Material	2018 Bare Costs Labor	Equipment	Total	Total Incl O&P
2050	30" wide	2 Elec	70	.229	L.F.	25	13.30		38.30	47.50
2080	Elbow horizontal, 90°, 12" radius, 6" wide		9.60	1.667	Ea.	83	97		180	236
2100	12" wide		7.60	2.105		93	123		216	286
2110	18" wide		6.80	2.353		114	137		251	330
2120	24" wide		5.80	2.759		131	161		292	385
2130	30" wide		5	3.200		159	186		345	455
2160	24" radius, 6" wide		9.20	1.739		118	101		219	281
2180	12" wide		7.20	2.222		139	129		268	345
2190	18" wide		6.40	2.500		159	146		305	395
2200	24" wide		5.40	2.963		197	172		369	475
2210	30" wide		4.60	3.478		225	202		427	555
2240	36" radius, 6" wide		8.80	1.818		174	106		280	350
2260	12" wide		6.80	2.353		209	137		346	435
2270	18" wide		6	2.667		249	155		404	505
2280	24" wide		5	3.200		266	186		452	570
2290	30" wide		4.20	3.810		300	222		522	660
2320	Elbow vertical, 90°, 12" radius, 6" wide		9.60	1.667		98.50	97		195.50	253
2340	12" wide		7.60	2.105		102	123		225	295
2350	18" wide		6.80	2.353		116	137		253	330
2360	24" wide		5.80	2.759		122	161		283	375
2370	30" wide		5	3.200		129	186		315	420
2400	24" radius, 6" wide		9.20	1.739		139	101		240	305
2420	12" wide		7.20	2.222		152	129		281	360
2430	18" wide		6.40	2.500		163	146		309	395
2440	24" wide		5.40	2.963		177	172		349	450
2450	30" wide		4.60	3.478		188	202		390	510
2480	36" radius, 6" wide		8.80	1.818		184	106		290	360
2500	12" wide		6.80	2.353		206	137		343	430
2510	18" wide		6	2.667		225	155		380	480
2520	24" wide		5	3.200		234	186		420	535
2530	30" wide		4.20	3.810		259	222		481	615
2560	Tee horizontal, 12" radius, 6" wide		5	3.200		128	186		314	420
2580	12" wide		4.40	3.636		155	212		367	485
2590	18" wide		4	4		174	233		407	540
2600	24" wide		3.60	4.444		195	259		454	600
2610	30" wide		3	5.333		225	310		535	715
2640	24" radius, 6" wide		4.60	3.478		211	202		413	540
2660	12" wide		4	4		238	233		471	610
2670	18" wide		3.60	4.444		269	259		528	680
2680	24" wide		3	5.333		330	310		640	830
2690	30" wide		2.40	6.667		375	390		765	995
2720	36" radius, 6" wide		4.20	3.810		345	222		567	710
2740	12" wide		3.60	4.444		375	259		634	800
2750	18" wide		3.20	5		435	291		726	910
2760	24" wide		2.60	6.154		505	360		865	1,100
2770	30" wide		2	8		555	465		1,020	1,300
2800	Tee vertical, 12" radius, 6" wide		5	3.200		183	186		369	480
2820	12" wide		4.40	3.636		187	212		399	520
2830	18" wide		4.20	3.810		194	222		416	545
2840	24" wide		4	4		206	233		439	575
2850	30" wide		3.60	4.444		225	259		484	635
2880	24" radius, 6" wide		4.60	3.478		269	202		471	600
2900	12" wide		4	4		285	233		518	665

26 05 36.20 Cable Tray Solid Bottom

		Crew	Daily Output	Labor-Hours	Unit	Material	2018 Bare Costs Labor	Equipment	Total	Total Incl O&P
2910	18" wide	2 Elec	3.80	4.211	Ea.	305	245		550	700
2920	24" wide		3.60	4.444		330	259		589	750
2930	30" wide		3.20	5		360	291		651	835
2960	36" radius, 6" wide		4.20	3.810		420	222		642	790
2980	12" wide		3.40	4.706		440	274		714	895
2990	18" wide		3.40	4.706		465	274		739	920
3000	24" wide		3.20	5		480	291		771	960
3010	30" wide		2.80	5.714		515	335		850	1,075
3040	Cross horizontal, 12" radius, 6" wide		4.40	3.636		163	212		375	495
3060	12" wide		4	4		184	233		417	550
3070	18" wide		3.40	4.706		209	274		483	640
3080	24" wide		2.80	5.714		238	335		573	755
3090	30" wide		2.60	6.154		269	360		629	830
3120	24" radius, 6" wide		4	4		285	233		518	665
3140	12" wide		3.60	4.444		330	259		589	750
3150	18" wide		3	5.333		360	310		670	860
3160	24" wide		2.40	6.667		435	390		825	1,050
3170	30" wide		2.20	7.273		480	425		905	1,150
3200	36" radius, 6" wide		3.60	4.444		500	259		759	930
3220	12" wide		3.20	5		525	291		816	1,000
3230	18" wide		2.60	6.154		580	360		940	1,175
3240	24" wide		2	8		635	465		1,100	1,400
3250	30" wide		1.80	8.889		685	515		1,200	1,525
3280	Dropout, or end plate, 6" wide		32	.500		16.40	29		45.40	61.50
3300	12" wide		26	.615		18.55	36		54.55	74
3310	18" wide		22	.727		22	42.50		64.50	88
3320	24" wide		20	.800		24.50	46.50		71	96.50
3330	30" wide		18	.889		28.50	51.50		80	109
3380	Reducer, 12" to 6" wide		14	1.143		80	66.50		146.50	188
3400	18" to 12" wide		12	1.333		83	77.50		160.50	207
3420	18" to 6" wide		12	1.333		83	77.50		160.50	207
3440	24" to 18" wide		10.60	1.509		88	88		176	228
3460	24" to 12" wide		10.60	1.509		88	88		176	228
3480	30" to 24" wide		9.20	1.739		96	101		197	256
3500	30" to 18" wide		9.20	1.739		96	101		197	256
3520	30" to 12" wide		9.20	1.739		96	101		197	256
3540	36" to 30" wide		8	2		101	116		217	285
3560	36" to 24" wide		8	2		102	116		218	286
3580	36" to 18" wide		8	2		102	116		218	286
3600	36" to 12" wide		8	2		104	116		220	289

26 05 36.30 Cable Tray Trough

			Crew	Daily Output	Labor-Hours	Unit	Material	2018 Bare Costs Labor	Equipment	Total	Total Incl O&P
0010	**CABLE TRAY TROUGH** vented, w/ftngs. & supports, 6" deep, to 10' high										
0020	For higher elevations, see Section 26 05 36.40										
0200	Galvanized steel, tray, 6" wide	R260536-10	2 Elec	90	.178	L.F.	13.50	10.35		23.85	30.50
0240	12" wide			80	.200		14.75	11.65		26.40	33.50
0260	18" wide			70	.229		22	13.30		35.30	44
0280	24" wide	R260536-11		60	.267		48	15.50		63.50	76
0300	30" wide			50	.320		48.50	18.60		67.10	81
0340	Elbow horizontal, 90°, 12" radius, 6" wide			7.60	2.105	Ea.	107	123		230	300
0360	12" wide			5.60	2.857		127	166		293	390
0370	18" wide			4.40	3.636		146	212		358	475
0380	24" wide			3.60	4.444		181	259		440	585

For customer support on your Electrical Costs with RSMeans data, call 800.448.8182.

185

26 05 36.30 Cable Tray Trough		Crew	Daily Output	Labor-Hours	Unit	Material	2018 Bare Costs Labor	Equipment	Total	Total Incl O&P
0390	30" wide	2 Elec	3.20	5	Ea.	204	291		495	660
0420	24" radius, 6" wide		7.20	2.222		158	129		287	365
0440	12" wide		5.20	3.077		183	179		362	470
0450	18" wide		4	4		219	233		452	590
0460	24" wide		3.20	5		248	291		539	710
0470	30" wide		2.80	5.714		286	335		621	810
0500	36" radius, 6" wide		6.80	2.353		217	137		354	445
0520	12" wide		4.80	3.333		247	194		441	560
0530	18" wide		3.60	4.444		296	259		555	710
0540	24" wide		2.80	5.714		305	335		640	830
0550	30" wide		2.40	6.667		345	390		735	960
0580	Elbow vertical, 90°, 12" radius, 6" wide		7.60	2.105		124	123		247	320
0600	12" wide		5.60	2.857		137	166		303	400
0610	18" wide		4.40	3.636		139	212		351	470
0620	24" wide		3.60	4.444		160	259		419	560
0630	30" wide		3.20	5		163	291		454	615
0660	24" radius, 6" wide		7.20	2.222		177	129		306	390
0680	12" wide		5.20	3.077		187	179		366	475
0690	18" wide		4	4		208	233		441	580
0700	24" wide		3.20	5		215	291		506	670
0710	30" wide		2.80	5.714		240	335		575	760
0740	36" radius, 6" wide		6.80	2.353		239	137		376	470
0760	12" wide		4.80	3.333		254	194		448	570
0770	18" wide		3.60	4.444		282	259		541	695
0780	24" wide		2.80	5.714		300	335		635	825
0790	30" wide		2.40	6.667		310	390		700	920
0820	Tee horizontal, 12" radius, 6" wide		4	4		144	233		377	510
0840	12" wide		3.20	5		161	291		452	615
0850	18" wide		2.80	5.714		182	335		517	695
0860	24" wide		2.40	6.667		211	390		601	810
0870	30" wide		2.20	7.273		240	425		665	900
0900	24" radius, 6" wide		3.60	4.444		239	259		498	650
0920	12" wide		2.80	5.714		261	335		596	780
0930	18" wide		2.40	6.667		292	390		682	900
0940	24" wide		2	8		380	465		845	1,100
0950	30" wide		1.80	8.889		410	515		925	1,225
0980	36" radius, 6" wide		3.20	5		345	291		636	815
1000	12" wide		2.40	6.667		390	390		780	1,000
1010	18" wide		2	8		430	465		895	1,175
1020	24" wide		1.60	10		515	580		1,095	1,425
1030	30" wide		1.40	11.429		545	665		1,210	1,600
1060	Tee vertical, 12" radius, 6" wide		4	4		227	233		460	600
1080	12" wide		3.20	5		230	291		521	690
1090	18" wide		3	5.333		237	310		547	725
1100	24" wide		2.80	5.714		250	335		585	770
1110	30" wide		2.60	6.154		265	360		625	825
1140	24" radius, 6" wide		3.60	4.444		305	259		564	720
1160	12" wide		2.80	5.714		320	335		655	850
1170	18" wide		2.60	6.154		335	360		695	905
1180	24" wide		2.40	6.667		360	390		750	980
1190	30" wide		2.20	7.273		380	425		805	1,050
1220	36" radius, 6" wide		3.20	5		495	291		786	980
1240	12" wide		2.40	6.667		510	390		900	1,150

26 05 36.30 Cable Tray Trough		Crew	Daily Output	Labor-Hours	Unit	Material	2018 Bare Costs Labor	Equipment	Total	Total Incl O&P
1250	18" wide	2 Elec	2.20	7.273	Ea.	525	425		950	1,225
1260	24" wide		2	8		555	465		1,020	1,300
1270	30" wide		1.80	8.889		645	515		1,160	1,475
1300	Cross horizontal, 12" radius, 6" wide		3.20	5		179	291		470	630
1320	12" wide		2.80	5.714		180	335		515	695
1330	18" wide		2.40	6.667		202	390		592	800
1340	24" wide		2	8		212	465		677	930
1350	30" wide		1.80	8.889		238	515		753	1,025
1380	24" radius, 6" wide		2.80	5.714		281	335		616	805
1400	12" wide		2.40	6.667		294	390		684	905
1410	18" wide		2	8		325	465		790	1,050
1420	24" wide		1.60	10		410	580		990	1,325
1430	30" wide		1.40	11.429		455	665		1,120	1,500
1460	36" radius, 6" wide		2.40	6.667		480	390		870	1,100
1480	12" wide		2	8		500	465		965	1,250
1490	18" wide		1.60	10		510	580		1,090	1,425
1500	24" wide		1.20	13.333		610	775		1,385	1,825
1510	30" wide		1	16		670	930		1,600	2,150
1540	Dropout or end plate, 6" wide		26	.615		21.50	36		57.50	77
1560	12" wide		22	.727		26.50	42.50		69	92.50
1580	18" wide		20	.800		29	46.50		75.50	101
1600	24" wide		18	.889		33	51.50		84.50	114
1620	30" wide		16	1		35.50	58		93.50	126
1660	Reducer, 12" to 6" wide		9.40	1.702		87	99		186	244
1680	18" to 12" wide		8.40	1.905		90	111		201	265
1700	18" to 6" wide		8.40	1.905		90.50	111		201.50	266
1720	24" to 18" wide		7.20	2.222		96.50	129		225.50	299
1740	24" to 12" wide		7.20	2.222		96.50	129		225.50	299
1760	30" to 24" wide		6.40	2.500		99.50	146		245.50	330
1780	30" to 18" wide		6.40	2.500		99.50	146		245.50	330
1800	30" to 12" wide		6.40	2.500		99.50	146		245.50	330
1820	36" to 30" wide		5.80	2.759		107	161		268	355
1840	36" to 24" wide		5.80	2.759		107	161		268	355
1860	36" to 18" wide		5.80	2.759		108	161		269	360
1880	36" to 12" wide		5.80	2.759		108	161		269	360
2000	Aluminum, tray, vented, 6" wide		120	.133	L.F.	14.70	7.75		22.45	28
2010	9" wide		110	.145		16.65	8.45		25.10	31
2020	12" wide		100	.160		19.65	9.30		28.95	35.50
2030	18" wide		90	.178		22	10.35		32.35	40
2040	24" wide		80	.200		26.50	11.65		38.15	46.50
2050	30" wide		70	.229		35	13.30		48.30	58.50
2080	Elbow horiz., 90°, 12" radius, 6" wide		7.60	2.105	Ea.	92	123		215	284
2090	9" wide		7	2.286		100	133		233	310
2100	12" wide		6.20	2.581		106	150		256	340
2110	18" wide		5.60	2.857		124	166		290	385
2120	24" wide		4.60	3.478		144	202		346	465
2130	30" wide		4	4		177	233		410	545
2160	24" radius, 6" wide		7.20	2.222		133	129		262	340
2180	12" wide		5.80	2.759		156	161		317	410
2190	18" wide		5.20	3.077		179	179		358	465
2200	24" wide		4.20	3.810		205	222		427	555
2210	30" wide		3.60	4.444		234	259		493	645
2240	36" radius, 6" wide		6.80	2.353		192	137		329	415

For customer support on your Electrical Costs with RSMeans data, call 800.448.8182.

187

26 05 36.30 Cable Tray Trough		Crew	Daily Output	Labor-Hours	Unit	Material	2018 Bare Costs Labor	Equipment	Total	Total Incl O&P
2260	12" wide	2 Elec	5.40	2.963	Ea.	218	172		390	500
2270	18" wide		4.80	3.333		248	194		442	565
2280	24" wide		3.80	4.211		272	245		517	665
2290	30" wide		3.40	4.706		320	274		594	760
2320	Elbow vertical, 90°, 12" radius, 6" wide		7.60	2.105		111	123		234	305
2330	9" wide		7	2.286		119	133		252	330
2340	12" wide		6.20	2.581		120	150		270	355
2350	18" wide		5.60	2.857		124	166		290	385
2360	24" wide		4.60	3.478		137	202		339	455
2370	30" wide		4	4		144	233		377	510
2400	24" radius, 6" wide		7.20	2.222		151	129		280	360
2420	12" wide		5.80	2.759		164	161		325	420
2430	18" wide		5.20	3.077		178	179		357	465
2440	24" wide		4.20	3.810		181	222		403	530
2450	30" wide		3.60	4.444		191	259		450	595
2480	36" radius, 6" wide		6.80	2.353		191	137		328	415
2500	12" wide		5.40	2.963		204	172		376	480
2510	18" wide		4.80	3.333		229	194		423	540
2520	24" wide		3.80	4.211		247	245		492	635
2530	30" wide		3.40	4.706		272	274		546	710
2560	Tee horizontal, 12" radius, 6" wide		4	4		140	233		373	505
2570	9" wide		3.80	4.211		144	245		389	525
2580	12" wide		3.60	4.444		156	259		415	555
2590	18" wide		3.20	5		181	291		472	635
2600	24" wide		2.80	5.714		208	335		543	725
2610	30" wide		2.40	6.667		223	390		613	825
2640	24" radius, 6" wide		3.60	4.444		213	259		472	620
2660	12" wide		3.20	5		242	291		533	700
2670	18" wide		2.80	5.714		267	335		602	790
2680	24" wide		2.40	6.667		335	390		725	945
2690	30" wide		2	8		360	465		825	1,100
2720	36" radius, 6" wide		3.20	5		340	291		631	810
2740	12" wide		2.80	5.714		395	335		730	930
2750	18" wide		2.40	6.667		445	390		835	1,075
2760	24" wide		2	8		520	465		985	1,275
2770	30" wide		1.60	10		555	580		1,135	1,475
2800	Tee vertical, 12" radius, 6" wide		4	4		196	233		429	565
2810	9" wide		3.80	4.211		197	245		442	580
2820	12" wide		3.60	4.444		212	259		471	620
2830	18" wide		3.40	4.706		215	274		489	645
2840	24" wide		3.20	5		218	291		509	675
2850	30" wide		3	5.333		229	310		539	715
2880	24" radius, 6" wide		3.60	4.444		277	259		536	690
2900	12" wide		3.20	5		299	291		590	765
2910	18" wide		3	5.333		320	310		630	815
2920	24" wide		2.80	5.714		340	335		675	870
2930	30" wide		2.60	6.154		365	360		725	940
2960	36" radius, 6" wide		3.20	5		420	291		711	895
2980	12" wide		2.80	5.714		445	335		780	985
2990	18" wide		2.60	6.154		460	360		820	1,050
3000	24" wide		2.40	6.667		495	390		885	1,125
3010	30" wide		2.20	7.273		535	425		960	1,225
3040	Cross horizontal, 12" radius, 6" wide		3.60	4.444		174	259		433	575

26 05 Common Work Results for Electrical

26 05 36 – Cable Trays for Electrical Systems

26 05 36.30 Cable Tray Trough	Crew	Daily Output	Labor-Hours	Unit	Material	Labor	Equipment	Total	Total Incl O&P
3050	9" wide	2 Elec	3.40	4.706	Ea.	187	274		461	615
3060	12" wide		3.20	5		192	291		483	645
3070	18" wide		2.80	5.714		201	335		536	715
3080	24" wide		2.40	6.667		224	390		614	825
3090	30" wide		2.20	7.273		287	425		712	950
3120	24" radius, 6" wide		3.20	5		315	291		606	780
3140	12" wide		2.80	5.714		340	335		675	870
3150	18" wide		2.40	6.667		370	390		760	985
3160	24" wide		2	8		425	465		890	1,175
3170	30" wide		1.80	8.889		460	515		975	1,275
3200	36" radius, 6" wide		2.80	5.714		510	335		845	1,050
3220	12" wide		2.40	6.667		535	390		925	1,175
3230	18" wide		2	8		580	465		1,045	1,325
3240	24" wide		1.60	10		675	580		1,255	1,600
3250	30" wide		1.40	11.429		730	665		1,395	1,800
3280	Dropout, or end plate, 6" wide		26	.615		17.75	36		53.75	73
3300	12" wide		22	.727		21	42.50		63.50	86.50
3310	18" wide		20	.800		25.50	46.50		72	97.50
3320	24" wide		18	.889		30.50	51.50		82	111
3330	30" wide		16	1		32.50	58		90.50	123
3370	Reducer, 9" to 6" wide		12	1.333		87	77.50		164.50	212
3380	12" to 6" wide		11.40	1.404		91.50	81.50		173	223
3390	12" to 9" wide		11.40	1.404		91.50	81.50		173	223
3400	18" to 12" wide		9.60	1.667		98	97		195	253
3420	18" to 6" wide		9.60	1.667		98	97		195	253
3430	18" to 9" wide		9.60	1.667		98	97		195	253
3440	24" to 18" wide		8.40	1.905		105	111		216	281
3460	24" to 12" wide		8.40	1.905		105	111		216	281
3470	24" to 9" wide		8.40	1.905		106	111		217	282
3475	24" to 6" wide		8.40	1.905		106	111		217	282
3480	30" to 24" wide		7.20	2.222		107	129		236	310
3500	30" to 18" wide		7.20	2.222		107	129		236	310
3520	30" to 12" wide		7.20	2.222		110	129		239	315
3540	36" to 30" wide		6.40	2.500		111	146		257	340
3560	36" to 24" wide		6.40	2.500		111	146		257	340
3580	36" to 18" wide		6.40	2.500		111	146		257	340
3600	36" to 12" wide		6.40	2.500		111	146		257	340
3610	Elbow horizontal, 60°, 12" radius, 6" wide		7.80	2.051		76	119		195	262
3620	9" wide		7.20	2.222		83	129		212	284
3630	12" wide		6.40	2.500		91.50	146		237.50	320
3640	18" wide		5.80	2.759		94.50	161		255.50	345
3650	24" wide		4.80	3.333		120	194		314	420
3680	Elbow horizontal, 45°, 12" radius, 6" wide		8	2		69	116		185	250
3690	9" wide		7.40	2.162		69	126		195	264
3700	12" wide		6.60	2.424		73	141		214	292
3710	18" wide		6	2.667		82.50	155		237.50	325
3720	24" wide		5	3.200		96	186		282	385
3750	Elbow horizontal, 30°, 12" radius, 6" wide		8.20	1.951		57	114		171	233
3760	9" wide		7.60	2.105		60	123		183	249
3770	12" wide		6.80	2.353		66	137		203	278
3780	18" wide		6.20	2.581		69	150		219	300
3790	24" wide		5.20	3.077		75.50	179		254.50	350
3820	Elbow vertical, 60° in/outside, 12" radius, 6" wide		7.80	2.051		94	119		213	281

For customer support on your Electrical Costs with RSMeans data, call 800.448.8182.

189

26 05 36.30 Cable Tray Trough	Crew	Daily Output	Labor-Hours	Unit	Material	2018 Bare Costs Labor	Equipment	Total	Total Incl O&P
3830	9" wide	2 Elec	7.20	2.222	Ea.	95.50	129		224.50	298
3840	12" wide		6.40	2.500		96.50	146		242.50	325
3850	18" wide		5.80	2.759		101	161		262	350
3860	24" wide		4.80	3.333		105	194		299	405
3890	Elbow vertical, 45° in/outside, 12" radius, 6" wide		8	2		81	116		197	263
3900	9" wide		7.40	2.162		78	126		204	274
3910	12" wide		6.60	2.424		83	141		224	300
3920	18" wide		6	2.667		85.50	155		240.50	325
3930	24" wide		5	3.200		94	186		280	380
3960	Elbow vertical, 30° in/outside, 12" radius, 6" wide		8.20	1.951		65	114		179	242
3970	9" wide		7.60	2.105		69	123		192	259
3980	12" wide		6.80	2.353		70.50	137		207.50	283
3990	18" wide		6.20	2.581		73.50	150		223.50	305
4000	24" wide		5.20	3.077		74.50	179		253.50	350
4250	Reducer, left or right hand, 24" to 18" wide		8.40	1.905		99.50	111		210.50	275
4260	24" to 12" wide		8.40	1.905		99.50	111		210.50	275
4270	24" to 9" wide		8.40	1.905		99.50	111		210.50	275
4280	24" to 6" wide		8.40	1.905		101	111		212	277
4290	18" to 12" wide		9.60	1.667		93	97		190	247
4300	18" to 9" wide		9.60	1.667		93	97		190	247
4310	18" to 6" wide		9.60	1.667		93	97		190	247
4320	12" to 9" wide		11.40	1.404		89	81.50		170.50	220
4330	12" to 6" wide		11.40	1.404		89	81.50		170.50	220
4340	9" to 6" wide	▼	12	1.333		86	77.50		163.50	211
4350	Splice plate	1 Elec	48	.167		8.25	9.70		17.95	23.50
4360	Splice plate, expansion joint		48	.167		8.45	9.70		18.15	24
4370	Splice plate, hinged, horizontal		48	.167		6.80	9.70		16.50	22
4380	Vertical		48	.167		9.65	9.70		19.35	25
4390	Trough, hanger, vertical		28	.286		27	16.65		43.65	55
4400	Box connector, 24" wide		20	.400		34	23.50		57.50	72.50
4410	18" wide		21	.381		30.50	22		52.50	66.50
4420	12" wide		22	.364		29.50	21		50.50	63.50
4430	9" wide		23	.348		28	20		48	61
4440	6" wide		24	.333		26.50	19.40		45.90	58
4450	Floor flange		24	.333		28	19.40		47.40	59.50
4460	Hold down clamp		60	.133		3.19	7.75		10.94	15.10
4520	Wall bracket, 24" wide tray		20	.400		31	23.50		54.50	69.50
4530	18" wide tray		21	.381		30.50	22		52.50	66.50
4540	12" wide tray		22	.364		16.20	21		37.20	49.50
4550	9" wide tray		23	.348		14.45	20		34.45	46.50
4560	6" wide tray	▼	24	.333		13.15	19.40		32.55	43.50
5000	Cable channel aluminum, vented, 1-1/4" deep, 4" wide, straight		80	.100	L.F.	7.60	5.80		13.40	17.05
5010	Elbow horizontal, 36" radius, 90°		5	1.600	Ea.	131	93		224	283
5020	60°		5.50	1.455		101	84.50		185.50	238
5030	45°		6	1.333		82	77.50		159.50	206
5040	30°		6.50	1.231		72	71.50		143.50	186
5050	Adjustable		6	1.333		68.50	77.50		146	192
5060	Elbow vertical, 36" radius, 90°		5	1.600		141	93		234	294
5070	60°		5.50	1.455		112	84.50		196.50	251
5080	45°		6	1.333		92	77.50		169.50	217
5090	30°		6.50	1.231		82	71.50		153.50	197
5100	Adjustable		6	1.333		68.50	77.50		146	192
5110	Splice plate, hinged, horizontal	▼	48	.167		9.65	9.70		19.35	25

26 05 36.30 Cable Tray Trough

		Crew	Daily Output	Labor-Hours	Unit	Material	2018 Bare Costs Labor	Equipment	Total	Total Incl O&P
5120	Splice plate, hinged, vertical	1 Elec	48	.167	Ea.	13.75	9.70		23.45	29.50
5130	Hanger, vertical		28	.286		15.15	16.65		31.80	41.50
5140	Single		28	.286		23	16.65		39.65	50.50
5150	Double		20	.400		23.50	23.50		47	61
5160	Channel to box connector		24	.333		31	19.40		50.40	63.50
5170	Hold down clip		80	.100		4.37	5.80		10.17	13.50
5180	Wall bracket, single		28	.286		10.25	16.65		26.90	36.50
5190	Double		20	.400		13.15	23.50		36.65	49.50
5200	Cable roller		16	.500		161	29		190	221
5210	Splice plate		48	.167		4.95	9.70		14.65	19.95

26 05 36.36 Cable Trays for Utility Substations

		Crew	Daily Output	Labor-Hours	Unit	Material	2018 Bare Costs Labor	Equipment	Total	Total Incl O&P
0010	**CABLE TRAYS FOR UTILITY SUBSTATIONS**									
7700	Cable tray	R-11	40	1.400	L.F.	17.50	77.50	16.55	111.55	153
7790	See Section 26 05 36									

26 05 36.40 Cable Tray, Covers and Dividers

		Crew	Daily Output	Labor-Hours	Unit	Material	2018 Bare Costs Labor	Equipment	Total	Total Incl O&P
0010	**CABLE TRAY, COVERS AND DIVIDERS** To 10' high R260536-10									
0011	For higher elevations, see lines 9900 – 9960									
0100	Covers, ventilated galv. steel, straight, 6" wide tray size	2 Elec	520	.031	L.F.	8.30	1.79		10.09	11.80
0200	9" wide tray size		460	.035		8.85	2.02		10.87	12.80
0300	12" wide tray size R260536-11		400	.040		15.15	2.33		17.48	20
0400	18" wide tray size		300	.053		32.50	3.10		35.60	40
0500	24" wide tray size		220	.073		47	4.23		51.23	58.50
0600	30" wide tray size		180	.089		23	5.15		28.15	33
1000	Elbow horizontal, 90°, 12" radius, 6" wide tray size		150	.107	Ea.	39.50	6.20		45.70	53
1020	9" wide tray size		128	.125		44	7.30		51.30	59.50
1040	12" wide tray size		108	.148		46.50	8.60		55.10	64
1060	18" wide tray size		84	.190		64	11.10		75.10	87
1080	24" wide tray size		66	.242		77.50	14.10		91.60	107
1100	30" wide tray size		60	.267		98.50	15.50		114	131
1160	24" radius, 6" wide tray size		136	.118		66	6.85		72.85	83
1180	9" wide tray size		116	.138		67.50	8.05		75.55	86.50
1200	12" wide tray size		96	.167		75	9.70		84.70	97
1220	18" wide tray size		76	.211		93	12.25		105.25	120
1240	24" wide tray size		60	.267		113	15.50		128.50	147
1260	30" wide tray size		52	.308		148	17.90		165.90	190
1320	36" radius, 6" wide tray size		120	.133		98.50	7.75		106.25	120
1340	9" wide tray size		104	.154		108	8.95		116.95	132
1360	12" wide tray size		84	.190		116	11.10		127.10	144
1380	18" wide tray size		72	.222		147	12.95		159.95	181
1400	24" wide tray size		52	.308		175	17.90		192.90	220
1420	30" wide tray size		46	.348		208	20		228	260
1480	Elbow horizontal, 45°, 12" radius, 6" wide tray size		150	.107		30	6.20		36.20	42.50
1500	9" wide tray size		128	.125		35.50	7.30		42.80	50.50
1520	12" wide tray size		108	.148		39	8.60		47.60	55.50
1540	18" wide tray size		88	.182		47.50	10.60		58.10	68
1560	24" wide tray size		76	.211		55.50	12.25		67.75	79.50
1580	30" wide tray size		66	.242		65.50	14.10		79.60	93
1640	24" radius, 6" wide tray size		136	.118		41.50	6.85		48.35	56
1660	9" wide tray size		116	.138		46.50	8.05		54.55	63
1680	12" wide tray size		96	.167		52	9.70		61.70	72
1700	18" wide tray size		80	.200		60.50	11.65		72.15	84
1720	24" wide tray size		70	.229		69	13.30		82.30	96

For customer support on your Electrical Costs with RSMeans data, call 800.448.8182.

191

26 05 36.40 Cable Tray, Covers and Dividers	Crew	Daily Output	Labor-Hours	Unit	Material	2018 Bare Costs Labor	Equipment	Total	Total Incl O&P
1740	30" wide tray size	2 Elec	60	.267	Ea.	88	15.50		103.50	120
1800	36" radius, 6" wide tray size		120	.133		61	7.75		68.75	78.50
1820	9" wide tray size		104	.154		69	8.95		77.95	89
1840	12" wide tray size		84	.190		76	11.10		87.10	100
1860	18" wide tray size		76	.211		87	12.25		99.25	114
1880	24" wide tray size		62	.258		106	15		121	139
1900	30" wide tray size		52	.308		115	17.90		132.90	154
1960	Elbow vertical, 90°, 12" radius, 6" wide tray size		150	.107		35	6.20		41.20	48
1980	9" wide tray size		128	.125		35.50	7.30		42.80	50.50
2000	12" wide tray size		108	.148		38.50	8.60		47.10	55.50
2020	18" wide tray size		88	.182		44	10.60		54.60	64
2040	24" wide tray size		68	.235		44.50	13.70		58.20	69.50
2060	30" wide tray size		60	.267		49	15.50		64.50	77
2120	24" radius, 6" wide tray size		136	.118		43	6.85		49.85	57.50
2140	9" wide tray size		116	.138		47.50	8.05		55.55	64
2160	12" wide tray size		96	.167		49	9.70		58.70	68.50
2180	18" wide tray size		80	.200		65.50	11.65		77.15	89.50
2200	24" wide tray size		62	.258		73	15		88	103
2220	30" wide tray size		52	.308		80	17.90		97.90	115
2280	36" radius, 6" wide tray size		120	.133		48.50	7.75		56.25	65
2300	9" wide tray size		104	.154		62	8.95		70.95	81.50
2320	12" wide tray size		84	.190		69	11.10		80.10	92
2340	18" wide tray size		76	.211		87	12.25		99.25	114
2350	24" wide tray size		54	.296		96.50	17.25		113.75	132
2360	30" wide tray size		46	.348		115	20		135	158
2400	Tee horizontal, 12" radius, 6" wide tray size		92	.174		63.50	10.10		73.60	85
2410	9" wide tray size		80	.200		65.50	11.65		77.15	89.50
2420	12" wide tray size		68	.235		74	13.70		87.70	102
2430	18" wide tray size		60	.267		89.50	15.50		105	121
2440	24" wide tray size		52	.308		112	17.90		129.90	151
2460	30" wide tray size		36	.444		132	26		158	184
2500	24" radius, 6" wide tray size		88	.182		102	10.60		112.60	128
2510	9" wide tray size		76	.211		108	12.25		120.25	137
2520	12" wide tray size		64	.250		111	14.55		125.55	144
2530	18" wide tray size		56	.286		143	16.65		159.65	182
2540	24" wide tray size		48	.333		219	19.40		238.40	270
2560	30" wide tray size		32	.500		250	29		279	320
2600	36" radius, 6" wide tray size		84	.190		174	11.10		185.10	208
2610	9" wide tray size		72	.222		177	12.95		189.95	214
2620	12" wide tray size		60	.267		197	15.50		212.50	240
2630	18" wide tray size		52	.308		230	17.90		247.90	280
2640	24" wide tray size		44	.364		305	21		326	365
2660	30" wide tray size		28	.571		330	33.50		363.50	410
2700	Cross horizontal, 12" radius, 6" wide tray size		68	.235		94.50	13.70		108.20	125
2710	9" wide tray size		64	.250		100	14.55		114.55	132
2720	12" wide tray size		60	.267		112	15.50		127.50	146
2730	18" wide tray size		52	.308		132	17.90		149.90	172
2740	24" wide tray size		36	.444		156	26		182	210
2760	30" wide tray size		30	.533		177	31		208	242
2800	24" radius, 6" wide tray size		64	.250		170	14.55		184.55	209
2810	9" wide tray size		60	.267		185	15.50		200.50	227
2820	12" wide tray size		56	.286		200	16.65		216.65	245
2830	18" wide tray size		48	.333		239	19.40		258.40	292

26 05 36.40 Cable Tray, Covers and Dividers	Crew	Daily Output	Labor-Hours	Unit	Material	2018 Bare Costs Labor	Equipment	Total	Total Incl O&P
2840	24" wide tray size	2 Elec	32	.500	Ea.	290	29		319	365
2860	30" wide tray size		26	.615		320	36		356	410
2900	36" radius, 6" wide tray size		60	.267		291	15.50		306.50	345
2910	9" wide tray size		56	.286		290	16.65		306.65	345
2920	12" wide tray size		52	.308		315	17.90		332.90	370
2930	18" wide tray size		44	.364		355	21		376	425
2940	24" wide tray size		28	.571		445	33.50		478.50	540
2960	30" wide tray size		22	.727		480	42.50		522.50	595
3000	Reducer, 9" to 6" wide tray size		128	.125		40.50	7.30		47.80	55.50
3010	12" to 6" wide tray size		108	.148		42.50	8.60		51.10	59.50
3020	12" to 9" wide tray size		108	.148		42.50	8.60		51.10	59.50
3030	18" to 12" wide tray size		88	.182		45	10.60		55.60	65.50
3050	18" to 6" wide tray size		88	.182		45	10.60		55.60	65.50
3060	24" to 18" wide tray size		80	.200		63	11.65		74.65	87
3070	24" to 12" wide tray size		80	.200		57.50	11.65		69.15	81
3090	30" to 24" wide tray size		70	.229		66.50	13.30		79.80	93.50
3100	30" to 18" wide tray size		70	.229		66.50	13.30		79.80	93.50
3110	30" to 12" wide tray size		70	.229		57.50	13.30		70.80	83.50
3140	36" to 30" wide tray size		64	.250		73	14.55		87.55	102
3150	36" to 24" wide tray size		64	.250		73	14.55		87.55	102
3160	36" to 18" wide tray size		64	.250		73	14.55		87.55	102
3170	36" to 12" wide tray size		64	.250		73	14.55		87.55	102
3250	Covers, aluminum, straight, 6" wide tray size		520	.031	L.F.	4.38	1.79		6.17	7.50
3270	9" wide tray size		460	.035		5.30	2.02		7.32	8.85
3290	12" wide tray size		400	.040		6.30	2.33		8.63	10.45
3310	18" wide tray size		320	.050		8.35	2.91		11.26	13.50
3330	24" wide tray size		260	.062		10.40	3.58		13.98	16.80
3350	30" wide tray size		200	.080		11.50	4.66		16.16	19.60
3400	Elbow horizontal, 90°, 12" radius, 6" wide tray size		150	.107	Ea.	32.50	6.20		38.70	45.50
3410	9" wide tray size		128	.125		34	7.30		41.30	48.50
3420	12" wide tray size		108	.148		37	8.60		45.60	53.50
3430	18" wide tray size		88	.182		48	10.60		58.60	69
3440	24" wide tray size		70	.229		60.50	13.30		73.80	86.50
3460	30" wide tray size		64	.250		72.50	14.55		87.05	102
3500	24" radius, 6" wide tray size		136	.118		42	6.85		48.85	56.50
3510	9" wide tray size		116	.138		52	8.05		60.05	69
3520	12" wide tray size		96	.167		57.50	9.70		67.20	77.50
3530	18" wide tray size		80	.200		68	11.65		79.65	92
3540	24" wide tray size		64	.250		84	14.55		98.55	115
3560	30" wide tray size		56	.286		103	16.65		119.65	138
3600	36" radius, 6" wide tray size		120	.133		73.50	7.75		81.25	92.50
3610	9" wide tray size		104	.154		80	8.95		88.95	101
3620	12" wide tray size		84	.190		91	11.10		102.10	117
3630	18" wide tray size		76	.211		108	12.25		120.25	137
3640	24" wide tray size		56	.286		132	16.65		148.65	170
3660	30" wide tray size		50	.320		152	18.60		170.60	195
3700	Elbow horizontal, 45°, 12" radius, 6" wide tray size		150	.107		24	6.20		30.20	36
3710	9" wide tray size		128	.125		24.50	7.30		31.80	38
3720	12" wide tray size		108	.148		27.50	8.60		36.10	43.50
3730	18" wide tray size		88	.182		31.50	10.60		42.10	51
3740	24" wide tray size		80	.200		35.50	11.65		47.15	56.50
3760	30" wide tray size		70	.229		44.50	13.30		57.80	69
3800	24" radius, 6" wide tray size		136	.118		26.50	6.85		33.35	39.50

For customer support on your Electrical Costs with RSMeans data, call 800.448.8182.

193

26 05 36.40 Cable Tray, Covers and Dividers	Crew	Daily Output	Labor-Hours	Unit	Material	2018 Bare Costs Labor	Equipment	Total	Total Incl O&P
3810	9" wide tray size	2 Elec	116	.138	Ea.	33.50	8.05		41.55	48.50
3820	12" wide tray size		96	.167		34.50	9.70		44.20	52.50
3830	18" wide tray size		80	.200		42	11.65		53.65	63.50
3840	24" wide tray size		72	.222		52	12.95		64.95	76.50
3860	30" wide tray size		64	.250		60	14.55		74.55	88
3900	36" radius, 6" wide tray size		120	.133		46	7.75		53.75	62.50
3910	9" wide tray size		104	.154		50.50	8.95		59.45	69
3920	12" wide tray size		84	.190		53	11.10		64.10	75
3930	18" wide tray size		76	.211		64	12.25		76.25	88.50
3940	24" wide tray size		64	.250		77.50	14.55		92.05	107
3960	30" wide tray size		56	.286		88.50	16.65		105.15	122
3970	36" wide tray size		50	.320		101	18.60		119.60	139
4000	Elbow vertical, 90°, 12" radius, 6" wide tray size		150	.107		27.50	6.20		33.70	40
4010	9" wide tray size		128	.125		28	7.30		35.30	41.50
4020	12" wide tray size		108	.148		30.50	8.60		39.10	46.50
4030	18" wide tray size		88	.182		35.50	10.60		46.10	55
4040	24" wide tray size		70	.229		36.50	13.30		49.80	60.50
4060	30" wide tray size		64	.250		37.50	14.55		52.05	63
4070	36" wide tray size		54	.296		42	17.25		59.25	72
4100	24" radius, 6" wide tray size		136	.118		31.50	6.85		38.35	45.50
4110	9" wide tray size		116	.138		34.50	8.05		42.55	50
4120	12" wide tray size		96	.167		37.50	9.70		47.20	55.50
4130	18" wide tray size		80	.200		45	11.65		56.65	67
4140	24" wide tray size		64	.250		49.50	14.55		64.05	76.50
4160	30" wide tray size		56	.286		60	16.65		76.65	91
4170	36" wide tray size		48	.333		65.50	19.40		84.90	101
4200	36" radius, 6" wide tray size		120	.133		35.50	7.75		43.25	50.50
4210	9" wide tray size		104	.154		43.50	8.95		52.45	61.50
4220	12" wide tray size		84	.190		50.50	11.10		61.60	72
4230	18" wide tray size		76	.211		61	12.25		73.25	85.50
4240	24" wide tray size		56	.286		72	16.65		88.65	104
4260	30" wide tray size		50	.320		91	18.60		109.60	128
4270	36" wide tray size		44	.364		98	21		119	140
4300	Tee horizontal, 12" radius, 6" wide tray size		108	.148		45	8.60		53.60	62.50
4310	9" wide tray size		88	.182		48	10.60		58.60	69
4320	12" wide tray size		80	.200		53.50	11.65		65.15	76.50
4330	18" wide tray size		68	.235		63.50	13.70		77.20	90.50
4340	24" wide tray size		56	.286		80.50	16.65		97.15	114
4360	30" wide tray size		44	.364		94.50	21		115.50	136
4370	36" wide tray size		36	.444		114	26		140	165
4400	24" radius, 6" wide tray size		96	.167		73.50	9.70		83.20	95
4410	9" wide tray size		80	.200		77.50	11.65		89.15	102
4420	12" wide tray size		72	.222		86.50	12.95		99.45	115
4430	18" wide tray size		60	.267		101	15.50		116.50	134
4440	24" wide tray size		48	.333		161	19.40		180.40	207
4460	30" wide tray size		40	.400		178	23.50		201.50	231
4470	36" wide tray size		32	.500		199	29		228	262
4500	36" radius, 6" wide tray size		88	.182		129	10.60		139.60	158
4510	9" wide tray size		72	.222		132	12.95		144.95	164
4520	12" wide tray size		64	.250		145	14.55		159.55	181
4530	18" wide tray size		56	.286		165	16.65		181.65	206
4540	24" wide tray size		44	.364		209	21		230	262
4560	30" wide tray size		36	.444		235	26		261	298

194

For customer support on your Electrical Costs with RSMeans data, call 800.448.8182.

26 05 Common Work Results for Electrical

26 05 36 – Cable Trays for Electrical Systems

26 05 36.40 Cable Tray, Covers and Dividers	Crew	Daily Output	Labor-Hours	Unit	Material	2018 Bare Costs Labor	Equipment	Total	Total Incl O&P
4570 36" wide tray size	2 Elec	28	.571	Ea.	265	33.50		298.50	340
4600 Cross horizontal, 12" radius, 6" wide tray size		80	.200		69	11.65		80.65	93.50
4610 9" wide tray size		72	.222		73	12.95		85.95	100
4620 12" wide tray size		64	.250		81.50	14.55		96.05	112
4630 18" wide tray size		56	.286		99	16.65		115.65	134
4640 24" wide tray size		48	.333		114	19.40		133.40	154
4660 30" wide tray size		40	.400		136	23.50		159.50	184
4670 36" wide tray size		32	.500		156	29		185	216
4700 24" radius, 6" wide tray size		72	.222		128	12.95		140.95	160
4710 9" wide tray size		64	.250		137	14.55		151.55	173
4720 12" wide tray size		56	.286		147	16.65		163.65	187
4730 18" wide tray size		48	.333		172	19.40		191.40	218
4740 24" wide tray size		40	.400		205	23.50		228.50	260
4760 30" wide tray size		32	.500		240	29		269	310
4770 36" wide tray size		24	.667		260	39		299	345
4800 36" radius, 6" wide tray size		64	.250		209	14.55		223.55	252
4810 9" wide tray size		56	.286		218	16.65		234.65	265
4820 12" wide tray size		50	.320		226	18.60		244.60	277
4830 18" wide tray size		44	.364		249	21		270	305
4840 24" wide tray size		36	.444		305	26		331	375
4860 30" wide tray size		28	.571		355	33.50		388.50	440
4870 36" wide tray size		22	.727		385	42.50		427.50	490
4900 Reducer, 9" to 6" wide tray size		128	.125		35	7.30		42.30	49.50
4910 12" to 6" wide tray size		108	.148		36.50	8.60		45.10	53
4920 12" to 9" wide tray size		108	.148		36.50	8.60		45.10	53
4930 18" to 12" wide tray size		88	.182		40	10.60		50.60	60
4950 18" to 6" wide tray size		88	.182		40	10.60		50.60	60
4960 24" to 18" wide tray size		80	.200		51	11.65		62.65	74
4970 24" to 12" wide tray size		80	.200		44	11.65		55.65	66
4990 30" to 24" wide tray size		70	.229		53.50	13.30		66.80	78.50
5000 30" to 18" wide tray size		70	.229		53.50	13.30		66.80	78.50
5010 30" to 12" wide tray size		70	.229		53.50	13.30		66.80	78.50
5040 36" to 30" wide tray size		64	.250		59	14.55		73.55	87
5050 36" to 24" wide tray size		64	.250		59	14.55		73.55	87
5060 36" to 18" wide tray size		64	.250		59	14.55		73.55	87
5070 36" to 12" wide tray size		64	.250		59	14.55		73.55	87
5710 Tray cover hold down clamp	1 Elec	60	.133		9.45	7.75		17.20	22
8000 Divider strip, straight, galvanized, 3" deep		200	.040	L.F.	5.05	2.33		7.38	9.05
8020 4" deep		180	.044		6.15	2.59		8.74	10.65
8040 6" deep		160	.050		8.05	2.91		10.96	13.20
8060 Aluminum, straight, 3" deep		210	.038		5.05	2.22		7.27	8.85
8080 4" deep		190	.042		6.25	2.45		8.70	10.55
8100 6" deep		170	.047		7.95	2.74		10.69	12.85
8110 Divider strip, vertical fitting, 3" deep									
8120 12" radius, galvanized, 30°	1 Elec	28	.286	Ea.	24.50	16.65		41.15	52
8140 45°		27	.296		30.50	17.25		47.75	59.50
8160 60°		26	.308		32.50	17.90		50.40	62.50
8180 90°		25	.320		40.50	18.60		59.10	72.50
8200 Aluminum, 30°		29	.276		16.30	16.05		32.35	42
8220 45°		28	.286		19.20	16.65		35.85	46
8240 60°		27	.296		22.50	17.25		39.75	50.50
8260 90°		26	.308		28	17.90		45.90	58
8280 24" radius, galvanized, 30°		25	.320		36.50	18.60		55.10	68

For customer support on your Electrical Costs with RSMeans data, call 800.448.8182.

26 05 36.40 Cable Tray, Covers and Dividers		Crew	Daily Output	Labor-Hours	Unit	Material	2018 Bare Costs Labor	Equipment	Total	Total Incl O&P
8300	45°	1 Elec	24	.333	Ea.	39	19.40		58.40	71.50
8320	60°		23	.348		49	20		69	84.50
8340	90°		22	.364		66.50	21		87.50	105
8360	Aluminum, 30°		26	.308		25.50	17.90		43.40	55
8380	45°		25	.320		29.50	18.60		48.10	60.50
8400	60°		24	.333		36.50	19.40		55.90	69.50
8420	90°		23	.348		50	20		70	85.50
8440	36" radius, galvanized, 30°		22	.364		46.50	21		67.50	82.50
8460	45°		21	.381		54	22		76	92.50
8480	60°		20	.400		63.50	23.50		87	105
8500	90°		19	.421		88	24.50		112.50	133
8520	Aluminum, 30°		23	.348		38.50	20		58.50	73
8540	45°		22	.364		50	21		71	86.50
8560	60°		21	.381		64.50	22		86.50	104
8570	90°		20	.400		84.50	23.50		108	128
8590	Divider strip, vertical fitting, 4" deep									
8600	12" radius, galvanized, 30°	1 Elec	27	.296	Ea.	29	17.25		46.25	58
8610	45°		26	.308		34	17.90		51.90	64
8620	60°		25	.320		38	18.60		56.60	70
8630	90°		24	.333		46.50	19.40		65.90	80
8640	Aluminum, 30°		28	.286		21.50	16.65		38.15	49
8650	45°		27	.296		25	17.25		42.25	53.50
8660	60°		26	.308		28.50	17.90		46.40	58.50
8670	90°		25	.320		34	18.60		52.60	65.50
8680	24" radius, galvanized, 30°		24	.333		46.50	19.40		65.90	80
8690	45°		23	.348		58.50	20		78.50	94.50
8700	60°		22	.364		66	21		87	104
8710	90°		21	.381		88	22		110	130
8720	Aluminum, 30°		25	.320		34	18.60		52.60	65.50
8730	45°		24	.333		41.50	19.40		60.90	74.50
8740	60°		23	.348		48.50	20		68.50	84
8750	90°		22	.364		66.50	21		87.50	105
8760	36" radius, galvanized, 30°		23	.348		55	20		75	91
8770	45°		22	.364		63	21		84	101
8780	60°		21	.381		79	22		101	120
8790	90°		20	.400		105	23.50		128.50	151
8800	Aluminum, 30°		24	.333		53	19.40		72.40	87.50
8810	45°		23	.348		68	20		88	105
8820	60°		22	.364		83	21		104	123
8830	90°		21	.381		105	22		127	148
8840	Divider strip, vertical fitting, 6" deep									
8850	12" radius, galvanized, 30°	1 Elec	24	.333	Ea.	33.50	19.40		52.90	66
8860	45°		23	.348		37.50	20		57.50	71.50
8870	60°		22	.364		43	21		64	78.50
8880	90°		21	.381		53.50	22		75.50	92
8890	Aluminum, 30°		25	.320		25	18.60		43.60	55.50
8900	45°		24	.333		29.50	19.40		48.90	61.50
8910	60°		23	.348		31	20		51	64.50
8920	90°		22	.364		36.50	21		57.50	72
8930	24" radius, galvanized, 30°		23	.348		46.50	20		66.50	81.50
8940	45°		22	.364		58.50	21		79.50	95.50
8950	60°		21	.381		66.50	22		88.50	107
8960	90°		20	.400		88	23.50		111.50	132

26 05 Common Work Results for Electrical

26 05 36 – Cable Trays for Electrical Systems

26 05 36.40 Cable Tray, Covers and Dividers

		Crew	Daily Output	Labor-Hours	Unit	Material	2018 Bare Costs Labor	Equipment	Total	Total Incl O&P
8970	Aluminum, 30°	1 Elec	24	.333	Ea.	34.50	19.40		53.90	67
8980	45°		23	.348		47.50	20		67.50	82.50
8990	60°		22	.364		52	21		73	88.50
9000	90°		21	.381		70.50	22		92.50	111
9010	36" radius, galvanized, 30°		22	.364		54	21		75	91
9020	45°		21	.381		66.50	22		88.50	107
9030	60°		20	.400		88	23.50		111.50	132
9040	90°		19	.421		115	24.50		139.50	164
9050	Aluminum, 30°		23	.348		53.50	20		73.50	89.50
9060	45°		22	.364		72	21		93	111
9070	60°		21	.381		85	22		107	127
9080	90°		20	.400		102	23.50		125.50	147
9120	Divider strip, horizontal fitting, galvanized, 3" deep		33	.242		33	14.10		47.10	57.50
9130	4" deep		30	.267		37	15.50		52.50	64
9140	6" deep		27	.296		47.50	17.25		64.75	78
9150	Aluminum, 3" deep		35	.229		24.50	13.30		37.80	47
9160	4" deep		32	.250		27	14.55		41.55	52
9170	6" deep		29	.276		36	16.05		52.05	63.50
9300	Divider strip protector		300	.027	L.F.	2.95	1.55		4.50	5.55
9310	Fastener, ladder tray				Ea.	.52			.52	.57
9320	Trough or solid bottom tray				"	.36			.36	.40
9900	Add to labor for higher elevated installation									
9905	10' to 14.5' high, add						10%			
9910	15' to 20' high, add						20%			
9920	20' to 25' high, add						25%			
9930	25' to 30' high, add						35%			
9940	30' to 35' high, add						40%			
9950	35' to 40' high, add						50%			
9960	Over 40' high, add						55%			

26 05 39 – Underfloor Raceways for Electrical Systems

26 05 39.30 Conduit In Concrete Slab

		Crew	Daily Output	Labor-Hours	Unit	Material	2018 Bare Costs Labor	Equipment	Total	Total Incl O&P
0010	**CONDUIT IN CONCRETE SLAB** Including terminations, R260539-30									
0020	fittings and supports									
3230	PVC, schedule 40, 1/2" diameter	1 Elec	270	.030	L.F.	.49	1.72		2.21	3.11
3250	3/4" diameter		230	.035		.56	2.02		2.58	3.65
3270	1" diameter		200	.040		.77	2.33		3.10	4.32
3300	1-1/4" diameter		170	.047		1.01	2.74		3.75	5.20
3330	1-1/2" diameter		140	.057		1.21	3.33		4.54	6.30
3350	2" diameter		120	.067		1.52	3.88		5.40	7.45
3370	2-1/2" diameter		90	.089		2.49	5.15		7.64	10.50
3400	3" diameter	2 Elec	160	.100		3.22	5.80		9.02	12.25
3430	3-1/2" diameter		120	.133		4.09	7.75		11.84	16.10
3440	4" diameter		100	.160		4.41	9.30		13.71	18.75
3450	5" diameter		80	.200		6.95	11.65		18.60	25
3460	6" diameter		60	.267		9.25	15.50		24.75	33
3530	Sweeps, 1" diameter, 30" radius	1 Elec	32	.250	Ea.	9.20	14.55		23.75	32
3550	1-1/4" diameter		24	.333		9.45	19.40		28.85	39.50
3570	1-1/2" diameter		21	.381		10.30	22		32.30	44.50
3600	2" diameter		18	.444		11.30	26		37.30	51
3630	2-1/2" diameter		14	.571		17.90	33.50		51.40	69
3650	3" diameter		10	.800		61.50	46.50		108	137
3670	3-1/2" diameter		8	1		64.50	58		122.50	158

For customer support on your Electrical Costs with RSMeans data, call 800.448.8182.

197

26 05 39 – Underfloor Raceways for Electrical Systems

26 05 39.30 Conduit In Concrete Slab

		Crew	Daily Output	Labor-Hours	Unit	Material	2018 Bare Costs Labor	2018 Bare Costs Equipment	Total	Total Incl O&P
3700	4" diameter	1 Elec	7	1.143	Ea.	25.50	66.50		92	128
3710	5" diameter	↓	6	1.333		45.50	77.50		123	166
3730	Couplings, 1/2" diameter					.13			.13	.14
3750	3/4" diameter					.16			.16	.18
3770	1" diameter					.25			.25	.28
3800	1-1/4" diameter					.38			.38	.42
3830	1-1/2" diameter					.46			.46	.51
3850	2" diameter					.65			.65	.72
3870	2-1/2" diameter					1.20			1.20	1.32
3900	3" diameter					1.81			1.81	1.99
3930	3-1/2" diameter					2.29			2.29	2.52
3950	4" diameter					2.75			2.75	3.03
3960	5" diameter					5.80			5.80	6.40
3970	6" diameter					9.05			9.05	9.95
4030	End bells, 1" diameter, PVC	1 Elec	60	.133		2.78	7.75		10.53	14.65
4050	1-1/4" diameter		53	.151		3.59	8.80		12.39	17.10
4100	1-1/2" diameter		48	.167		2.89	9.70		12.59	17.70
4150	2" diameter		34	.235		4.48	13.70		18.18	25.50
4170	2-1/2" diameter		27	.296		3.64	17.25		20.89	30
4200	3" diameter		20	.400		5.75	23.50		29.25	41.50
4250	3-1/2" diameter		16	.500		5.70	29		34.70	50
4300	4" diameter		14	.571		7.10	33.50		40.60	57.50
4310	5" diameter		12	.667		10.60	39		49.60	69.50
4320	6" diameter		9	.889		12.40	51.50		63.90	91
4350	Rigid galvanized steel, 1/2" diameter		200	.040	L.F.	1.99	2.33		4.32	5.65
4400	3/4" diameter		170	.047		4.27	2.74		7.01	8.80
4450	1" diameter		130	.062		6.60	3.58		10.18	12.60
4500	1-1/4" diameter		110	.073		4.21	4.23		8.44	11
4600	1-1/2" diameter		100	.080		7.30	4.66		11.96	15
4800	2" diameter		90	.089		9.05	5.15		14.20	17.70

26 05 39.40 Conduit In Trench

		Crew	Daily Output	Labor-Hours	Unit	Material	2018 Bare Costs Labor	2018 Bare Costs Equipment	Total	Total Incl O&P
0010	**CONDUIT IN TRENCH** Includes terminations and fittings R260539-40									
0020	Does not include excavation or backfill, see Section 31 23 16									
0200	Rigid galvanized steel, 2" diameter	1 Elec	150	.053	L.F.	8.70	3.10		11.80	14.25
0400	2-1/2" diameter	"	100	.080		10.70	4.66		15.36	18.70
0600	3" diameter	2 Elec	160	.100		12.55	5.80		18.35	22.50
0800	3-1/2" diameter		140	.114		16.55	6.65		23.20	28
1000	4" diameter		100	.160		18.90	9.30		28.20	35
1200	5" diameter		80	.200		36.50	11.65		48.15	57.50
1400	6" diameter		60	.267		51.50	15.50		67	79.50

26 05 43 – Underground Ducts and Raceways for Electrical Systems

26 05 43.10 Trench Duct

		Crew	Daily Output	Labor-Hours	Unit	Material	2018 Bare Costs Labor	2018 Bare Costs Equipment	Total	Total Incl O&P
0010	**TRENCH DUCT** Steel with cover									
0020	Standard adjustable, depths to 4"									
0100	Straight, single compartment, 9" wide	2 Elec	40	.400	L.F.	254	23.50		277.50	315
0200	12" wide		32	.500		330	29		359	405
0400	18" wide		26	.615		365	36		401	455
0600	24" wide		22	.727		430	42.50		472.50	540
0700	27" wide		21	.762		215	44.50		259.50	305
0800	30" wide		20	.800		230	46.50		276.50	325
1000	36" wide		16	1		264	58		322	375
1020	Two compartment, 9" wide		38	.421		123	24.50		147.50	172

26 05 Common Work Results for Electrical

26 05 43 – Underground Ducts and Raceways for Electrical Systems

26 05 43.10 Trench Duct		Crew	Daily Output	Labor-Hours	Unit	Material	2018 Bare Costs Labor	Equipment	Total	Total Incl O&P
1030	12" wide	2 Elec	30	.533	L.F.	144	31		175	205
1040	18" wide		24	.667		193	39		232	270
1050	24" wide		20	.800		248	46.50		294.50	345
1060	30" wide		18	.889		305	51.50		356.50	415
1070	36" wide		14	1.143		350	66.50		416.50	485
1090	Three compartment, 9" wide		36	.444		153	26		179	207
1100	12" wide		28	.571		171	33.50		204.50	238
1110	18" wide		22	.727		214	42.50		256.50	300
1120	24" wide		18	.889		269	51.50		320.50	375
1130	30" wide		16	1		325	58		383	440
1140	36" wide		12	1.333		380	77.50		457.50	535
1200	Horizontal elbow, 9" wide		5.40	2.963	Ea.	415	172		587	720
1400	12" wide		4.60	3.478		445	202		647	795
1600	18" wide		4	4		590	233		823	995
1800	24" wide		3.20	5		840	291		1,131	1,350
1900	27" wide		3	5.333		995	310		1,305	1,575
2000	30" wide		2.60	6.154		1,125	360		1,485	1,750
2200	36" wide		2.40	6.667		1,475	390		1,865	2,200
2220	Two compartment, 9" wide		3.80	4.211		675	245		920	1,100
2230	12" wide		3	5.333		745	310		1,055	1,275
2240	18" wide		2.40	6.667		895	390		1,285	1,575
2250	24" wide		2	8		1,100	465		1,565	1,900
2260	30" wide		1.80	8.889		1,475	515		1,990	2,375
2270	36" wide		1.60	10		1,775	580		2,355	2,825
2290	Three compartment, 9" wide		3.60	4.444		680	259		939	1,125
2300	12" wide		2.80	5.714		760	335		1,095	1,325
2310	18" wide		2.20	7.273		925	425		1,350	1,650
2320	24" wide		1.80	8.889		1,175	515		1,690	2,050
2330	30" wide		1.60	10		1,525	580		2,105	2,550
2350	36" wide		1.40	11.429		1,850	665		2,515	3,050
2400	Vertical elbow, 9" wide		5.40	2.963		168	172		340	440
2600	12" wide		4.60	3.478		158	202		360	480
2800	18" wide		4	4		181	233		414	550
3000	24" wide		3.20	5		225	291		516	685
3100	27" wide		3	5.333		235	310		545	725
3200	30" wide		2.60	6.154		248	360		608	810
3400	36" wide		2.40	6.667		273	390		663	880
3600	Cross, 9" wide		4	4		680	233		913	1,100
3800	12" wide		3.20	5		720	291		1,011	1,225
4000	18" wide		2.60	6.154		865	360		1,225	1,475
4200	24" wide		2.20	7.273		1,075	425		1,500	1,800
4300	27" wide		2.20	7.273		1,250	425		1,675	2,000
4400	30" wide		2	8		1,375	465		1,840	2,200
4600	36" wide		1.80	8.889		1,700	515		2,215	2,650
4620	Two compartment, 9" wide		3.80	4.211		685	245		930	1,125
4630	12" wide		3	5.333		720	310		1,030	1,250
4640	18" wide		2.40	6.667		870	390		1,260	1,550
4650	24" wide		2	8		1,100	465		1,565	1,925
4660	30" wide		1.80	8.889		1,475	515		1,990	2,375
4670	36" wide		1.60	10		1,775	580		2,355	2,825
4690	Three compartment, 9" wide		3.60	4.444		695	259		954	1,150
4700	12" wide		2.80	5.714		795	335		1,130	1,375
4710	18" wide		2.20	7.273		945	425		1,370	1,675

For customer support on your Electrical Costs with RSMeans data, call 800.448.8182.

199

26 05 43.10 Trench Duct		Crew	Daily Output	Labor-Hours	Unit	Material	2018 Bare Costs Labor	Equipment	Total	Total Incl O&P
4720	24" wide	2 Elec	1.80	8.889	Ea.	1,175	515		1,690	2,075
4730	30" wide		1.60	10		1,525	580		2,105	2,550
4740	36" wide		1.40	11.429		1,900	665		2,565	3,075
4800	End closure, 9" wide		14.40	1.111		41	64.50		105.50	142
5000	12" wide		12	1.333		47	77.50		124.50	168
5200	18" wide		10	1.600		72	93		165	219
5400	24" wide		8	2		95	116		211	278
5500	27" wide		7	2.286		110	133		243	320
5600	30" wide		6.60	2.424		145	141		286	370
5800	36" wide		5.80	2.759		141	161		302	395
6000	Tees, 9" wide		4	4		400	233		633	790
6200	12" wide		3.60	4.444		465	259		724	895
6400	18" wide		3.20	5		595	291		886	1,100
6600	24" wide		3	5.333		855	310		1,165	1,400
6700	27" wide		2.80	5.714		945	335		1,280	1,550
6800	30" wide		2.60	6.154		1,100	360		1,460	1,750
7000	36" wide		2	8		1,450	465		1,915	2,300
7020	Two compartment, 9" wide		3.80	4.211		465	245		710	875
7030	12" wide		3.40	4.706		500	274		774	960
7040	18" wide		3	5.333		665	310		975	1,200
7050	24" wide		2.80	5.714		900	335		1,235	1,475
7060	30" wide		2.40	6.667		1,225	390		1,615	1,925
7070	36" wide		1.90	8.421		1,525	490		2,015	2,400
7090	Three compartment, 9" wide		3.60	4.444		530	259		789	970
7100	12" wide		3.20	5		555	291		846	1,050
7110	18" wide		2.80	5.714		705	335		1,040	1,275
7120	24" wide		2.60	6.154		970	360		1,330	1,600
7130	30" wide		2.20	7.273		1,275	425		1,700	2,025
7140	36" wide		1.80	8.889		1,625	515		2,140	2,550
7200	Riser, and cabinet connector, 9" wide		5.40	2.963		175	172		347	450
7400	12" wide		4.60	3.478		204	202		406	530
7600	18" wide		4	4		205	233		438	575
7800	24" wide		3.20	5		305	291		596	770
7900	27" wide		3	5.333		305	310		615	800
8000	30" wide		2.60	6.154		350	360		710	920
8200	36" wide		2	8		405	465		870	1,150
8400	Insert assembly, cell to conduit adapter, 1-1/4"	1 Elec	16	.500		69.50	29		98.50	120
8500	Adjustable partition	"	320	.025	L.F.	24.50	1.46		25.96	29
8600	Depth of duct over 4", per 1", add					11.10			11.10	12.20
8700	Support post	1 Elec	240	.033		25.50	1.94		27.44	31
8800	Cover double tile trim, 2 sides					39.50			39.50	43.50
8900	4 sides					113			113	124
9160	Trench duct 3-1/2" x 4-1/2", add					10.55			10.55	11.60
9170	Trench duct 4" x 5", add					10.55			10.55	11.60
9200	For carpet trim, add					36			36	39.50
9210	For double carpet trim, add					109			109	120

26 05 43.20 Underfloor Duct

			Crew	Daily Output	Labor-Hours	Unit	Material	Labor	Equipment	Total	Total Incl O&P
0010	**UNDERFLOOR DUCT**	R260543-50									
0100	Duct, 1-3/8" x 3-1/8" blank, standard		2 Elec	160	.100	L.F.	12.85	5.80		18.65	23
0200	1-3/8" x 7-1/4" blank, super duct			120	.133		30	7.75		37.75	44.50
0400	7/8" or 1-3/8" insert type, 24" OC, 1-3/8" x 3-1/8", std.			140	.114		20	6.65		26.65	32
0600	1-3/8" x 7-1/4", super duct			100	.160		35	9.30		44.30	52.50

200

For customer support on your Electrical Costs with RSMeans data, call 800.448.8182.

26 05 43.20 Underfloor Duct	Crew	Daily Output	Labor-Hours	Unit	Material	2018 Bare Costs Labor	Equipment	Total	Total Incl O&P
0800 Junction box, single duct, 1 level, 3-1/8"	1 Elec	4	2	Ea.	450	116		566	670
0820 3-1/8" x 7-1/4"		4	2		560	116		676	790
0840 2 level, 3-1/8" upper & lower		3.20	2.500		520	146		666	790
0860 3-1/8" upper, 7-1/4" lower		2.70	2.963		520	172		692	830
0880 Carpet pan for above		80	.100		375	5.80		380.80	425
0900 Terrazzo pan for above		67	.119		850	6.95		856.95	940
1000 Junction box, single duct, 1 level, 7-1/4"		2.70	2.963		525	172		697	835
1020 2 level, 7-1/4" upper & lower		2.70	2.963		600	172		772	920
1040 2 duct, two 3-1/8" upper & lower		3.20	2.500		770	146		916	1,075
1200 1 level, 2 duct, 3-1/8"		3.20	2.500		590	146		736	870
1220 Carpet pan for above boxes		80	.100		370	5.80		375.80	415
1240 Terrazzo pan for above boxes		67	.119		860	6.95		866.95	955
1260 Junction box, 1 level, two 3-1/8" x one 3-1/8" + one 7-1/4"		2.30	3.478		1,000	202		1,202	1,400
1280 2 level, two 3-1/8" upper, one 3-1/8" + one 7-1/4" lower		2	4		1,100	233		1,333	1,575
1300 Carpet pan for above boxes		80	.100		370	5.80		375.80	415
1320 Terrazzo pan for above boxes		67	.119		860	6.95		866.95	955
1400 Junction box, 1 level, 2 duct, 7-1/4"		2.30	3.478		1,450	202		1,652	1,900
1420 Two 3-1/8" + one 7-1/4"		2	4		1,525	233		1,758	2,025
1440 Carpet pan for above		80	.100		370	5.80		375.80	415
1460 Terrazzo pan for above		67	.119		860	6.95		866.95	955
1580 Junction box, 1 level, one 3-1/8" + one 7-1/4" x same		2.30	3.478		1,000	202		1,202	1,400
1600 Triple duct, 3-1/8"		2.30	3.478		1,000	202		1,202	1,400
1700 Junction box, 1 level, one 3-1/8" + two 7-1/4"		2	4		1,700	233		1,933	2,225
1720 Carpet pan for above		80	.100		370	5.80		375.80	415
1740 Terrazzo pan for above		67	.119		860	6.95		866.95	955
1800 Insert to conduit adapter, 3/4" & 1"		32	.250		37	14.55		51.55	63
2000 Support, single cell		27	.296		55	17.25		72.25	86.50
2200 Super duct		16	.500		56	29		85	105
2400 Double cell		16	.500		55.50	29		84.50	105
2600 Triple cell		11	.727		53.50	42.50		96	123
2800 Vertical elbow, standard duct		10	.800		99.50	46.50		146	179
3000 Super duct		8	1		99.50	58		157.50	196
3200 Cabinet connector, standard duct		32	.250		79	14.55		93.55	109
3400 Super duct		27	.296		77	17.25		94.25	111
3600 Conduit adapter, 1" to 1-1/4"		32	.250		74.50	14.55		89.05	104
3800 2" to 1-1/4"		27	.296		89.50	17.25		106.75	125
4000 Outlet, low tension (tele, computer, etc.)		8	1		105	58		163	202
4200 High tension, receptacle (120 volt)		8	1		108	58		166	206
4300 End closure, standard duct		160	.050		4.34	2.91		7.25	9.10
4310 Super duct		160	.050		8.20	2.91		11.11	13.40
4350 Elbow, horiz., standard duct		26	.308		255	17.90		272.90	305
4360 Super duct		26	.308		245	17.90		262.90	297
4380 Elbow, offset, standard duct		26	.308		99.50	17.90		117.40	136
4390 Super duct		26	.308		101	17.90		118.90	138
4400 Marker screw assembly for inserts		50	.160		18.85	9.30		28.15	34.50
4410 Y take off, standard duct		26	.308		154	17.90		171.90	196
4420 Super duct		26	.308		154	17.90		171.90	196
4430 Box opening plug, standard duct		160	.050		14.20	2.91		17.11	20
4440 Super duct		160	.050		14.20	2.91		17.11	20
4450 Sleeve coupling, standard duct		160	.050		37.50	2.91		40.41	46
4460 Super duct		160	.050		50	2.91		52.91	59.50
4470 Conduit adapter, standard duct, 3/4"		32	.250		78	14.55		92.55	108
4480 1" or 1-1/4"		32	.250		76.50	14.55		91.05	106

26 05 43 – Underground Ducts and Raceways for Electrical Systems

26 05 43.20 Underfloor Duct	Crew	Daily Output	Labor-Hours	Unit	Material	2018 Bare Costs Labor	Equipment	Total	Total Incl O&P
4500	1-1/2"	1 Elec	32	.250	Ea.	75	14.55		89.55	105

26 05 80 – Wiring Connections

26 05 80.10 Motor Connections

		Crew	Daily Output	Labor-Hours	Unit	Material	2018 Bare Costs Labor	Equipment	Total	Total Incl O&P
0010	**MOTOR CONNECTIONS**									
0020	Flexible conduit and fittings, 115 volt, 1 phase, up to 1 HP motor	1 Elec	8	1	Ea.	4.87	58		62.87	92.50
0050	2 HP motor		6.50	1.231		9.30	71.50		80.80	117
0100	3 HP motor		5.50	1.455		8.20	84.50		92.70	136
0110	230 volt, 3 phase, 3 HP motor		6.78	1.180		5.90	68.50		74.40	110
0112	5 HP motor		5.47	1.463		5	85		90	133
0114	7-1/2 HP motor		4.61	1.735		8.55	101		109.55	160
0120	10 HP motor		4.20	1.905		15.90	111		126.90	184
0150	15 HP motor		3.30	2.424		15.90	141		156.90	229
0200	25 HP motor		2.70	2.963		24.50	172		196.50	285
0400	50 HP motor		2.20	3.636		49	212		261	370
0600	100 HP motor		1.50	5.333		103	310		413	580
1500	460 volt, 5 HP motor, 3 phase		8	1		5.25	58		63.25	93
1520	10 HP motor		8	1		5.25	58		63.25	93
1530	25 HP motor		6	1.333		10.10	77.50		87.60	127
1540	30 HP motor		6	1.333		10.10	77.50		87.60	127
1550	40 HP motor		5	1.600		15.15	93		108.15	156
1560	50 HP motor		5	1.600		19.05	93		112.05	160
1570	60 HP motor		3.80	2.105		21	123		144	206
1580	75 HP motor		3.50	2.286		29	133		162	231
1590	100 HP motor		2.50	3.200		43	186		229	325
1600	125 HP motor		2	4		49.50	233		282.50	405
1610	150 HP motor		1.80	4.444		53	259		312	445
1620	200 HP motor		1.50	5.333		85.50	310		395.50	560
2005	460 volt, 5 HP motor, 3 phase, w/sealtite		8	1		8.70	58		66.70	96.50
2010	10 HP motor		8	1		8.70	58		66.70	96.50
2015	25 HP motor		6	1.333		15.65	77.50		93.15	133
2020	30 HP motor		6	1.333		15.65	77.50		93.15	133
2025	40 HP motor		5	1.600		28.50	93		121.50	170
2030	50 HP motor		5	1.600		30.50	93		123.50	173
2035	60 HP motor		3.80	2.105		36	123		159	223
2040	75 HP motor		3.50	2.286		42	133		175	245
2045	100 HP motor		2.50	3.200		64	186		250	350
2055	150 HP motor		1.80	4.444		68	259		327	460
2060	200 HP motor		1.50	5.333		214	310		524	700

Note: Row 0050 references R260580-75

26 05 90 – Residential Applications

26 05 90.10 Residential Wiring

		Crew	Daily Output	Labor-Hours	Unit	Material	2018 Bare Costs Labor	Equipment	Total	Total Incl O&P
0010	**RESIDENTIAL WIRING**									
0020	20' avg. runs and #14/2 wiring incl. unless otherwise noted									
1000	Service & panel, includes 24' SE-AL cable, service eye, meter,									
1010	Socket, panel board, main bkr., ground rod, 15 or 20 amp									
1020	1-pole circuit breakers, and misc. hardware									
1100	100 amp, with 10 branch breakers	1 Elec	1.19	6.723	Ea.	325	390		715	945
1110	With PVC conduit and wire		.92	8.696		360	505		865	1,150
1120	With RGS conduit and wire		.73	10.959		560	640		1,200	1,575
1150	150 amp, with 14 branch breakers		1.03	7.767		850	450		1,300	1,600
1170	With PVC conduit and wire		.82	9.756		920	570		1,490	1,850
1180	With RGS conduit and wire		.67	11.940		1,175	695		1,870	2,350
1200	200 amp, with 18 branch breakers	2 Elec	1.80	8.889		1,000	515		1,515	1,875

For customer support on your Electrical Costs with RSMeans data, call 800.448.8182.

26 05 90.10 Residential Wiring	Crew	Daily Output	Labor-Hours	Unit	Material	2018 Bare Costs Labor	2018 Bare Costs Equipment	Total	Total Incl O&P
1220 With PVC conduit and wire	2 Elec	1.46	10.959	Ea.	1,075	640		1,715	2,125
1230 With RGS conduit and wire	↓	1.24	12.903		1,400	750		2,150	2,675
1800 Lightning surge suppressor	1 Elec	32	.250	↓	76	14.55		90.55	106
2000 Switch devices									
2100 Single pole, 15 amp, ivory, with a 1-gang box, cover plate,									
2110 Type NM (Romex) cable	1 Elec	17.10	.468	Ea.	15.50	27		42.50	57.50
2120 Type MC cable		14.30	.559		25	32.50		57.50	76
2130 EMT & wire		5.71	1.401		36	81.50		117.50	162
2150 3-way, #14/3, type NM cable		14.55	.550		9.80	32		41.80	59
2170 Type MC cable		12.31	.650		22.50	38		60.50	81.50
2180 EMT & wire		5	1.600		30	93		123	172
2200 4-way, #14/3, type NM cable		14.55	.550		18.45	32		50.45	68.50
2220 Type MC cable		12.31	.650		31.50	38		69.50	91
2230 EMT & wire		5	1.600		38.50	93		131.50	182
2250 S.P., 20 amp, #12/2, type NM cable		13.33	.600		11.10	35		46.10	64
2270 Type MC cable		11.43	.700		19.65	40.50		60.15	82.50
2280 EMT & wire		4.85	1.649		33.50	96		129.50	181
2290 S.P. rotary dimmer, 600 W, no wiring		17	.471		30	27.50		57.50	74
2300 S.P. rotary dimmer, 600 W, type NM cable		14.55	.550		33.50	32		65.50	85
2320 Type MC cable		12.31	.650		43.50	38		81.50	104
2330 EMT & wire		5	1.600		55	93		148	200
2350 3-way rotary dimmer, type NM cable		13.33	.600		21.50	35		56.50	75.50
2370 Type MC cable		11.43	.700		31	40.50		71.50	95
2380 EMT & wire	↓	4.85	1.649	↓	43	96		139	191
2400 Interval timer wall switch, 20 amp, 1-30 min., #12/2									
2410 Type NM cable	1 Elec	14.55	.550	Ea.	59.50	32		91.50	114
2420 Type MC cable		12.31	.650		65.50	38		103.50	129
2430 EMT & wire	↓	5	1.600	↓	82	93		175	229
2500 Decorator style									
2510 S.P., 15 amp, type NM cable	1 Elec	17.10	.468	Ea.	19.35	27		46.35	62
2520 Type MC cable		14.30	.559		29	32.50		61.50	80.50
2530 EMT & wire		5.71	1.401		39.50	81.50		121	166
2550 3-way, #14/3, type NM cable		14.55	.550		13.70	32		45.70	63
2570 Type MC cable		12.31	.650		26.50	38		64.50	85.50
2580 EMT & wire		5	1.600		33.50	93		126.50	176
2600 4-way, #14/3, type NM cable		14.55	.550		22.50	32		54.50	72.50
2620 Type MC cable		12.31	.650		35	38		73	95
2630 EMT & wire		5	1.600		42.50	93		135.50	186
2650 S.P., 20 amp, #12/2, type NM cable		13.33	.600		14.95	35		49.95	68.50
2670 Type MC cable		11.43	.700		23.50	40.50		64	87
2680 EMT & wire		4.85	1.649		37.50	96		133.50	186
2700 S.P., slide dimmer, type NM cable		17.10	.468		30.50	27		57.50	74
2720 Type MC cable		14.30	.559		40	32.50		72.50	92.50
2730 EMT & wire		5.71	1.401		52	81.50		133.50	179
2750 S.P., touch dimmer, type NM cable		17.10	.468		46.50	27		73.50	92
2770 Type MC cable		14.30	.559		56	32.50		88.50	111
2780 EMT & wire		5.71	1.401		68	81.50		149.50	197
2800 3-way touch dimmer, type NM cable		13.33	.600		49.50	35		84.50	106
2820 Type MC cable		11.43	.700		59	40.50		99.50	126
2830 EMT & wire	↓	4.85	1.649	↓	71	96		167	222
3000 Combination devices									
3100 S.P. switch/15 amp recpt., ivory, 1-gang box, plate									
3110 Type NM cable	1 Elec	11.43	.700	Ea.	20.50	40.50		61	83.50

26 05 90.10 Residential Wiring	Crew	Daily Output	Labor-Hours	Unit	Material	2018 Bare Costs Labor	Equipment	Total	Total Incl O&P
3120	Type MC cable	1 Elec	10	.800	Ea.	30	46.50		76.50	103
3130	EMT & wire		4.40	1.818		42	106		148	204
3150	S.P. switch/pilot light, type NM cable		11.43	.700		21	40.50		61.50	84
3170	Type MC cable		10	.800		30.50	46.50		77	103
3180	EMT & wire		4.43	1.806		42.50	105		147.50	204
3190	2-S.P. switches, 2-#14/2, no wiring		14	.571		12.85	33.50		46.35	63.50
3200	2-S.P. switches, 2-#14/2, type NM cables		10	.800		22	46.50		68.50	94
3220	Type MC cable		8.89	.900		36.50	52.50		89	119
3230	EMT & wire		4.10	1.951		43.50	114		157.50	218
3250	3-way switch/15 amp recpt., #14/3, type NM cable		10	.800		26.50	46.50		73	99
3270	Type MC cable		8.89	.900		39.50	52.50		92	122
3280	EMT & wire		4.10	1.951		46.50	114		160.50	222
3300	2-3 way switches, 2-#14/3, type NM cables		8.89	.900		34.50	52.50		87	117
3320	Type MC cable		8	1		55.50	58		113.50	148
3330	EMT & wire		4	2		53	116		169	233
3350	S.P. switch/20 amp recpt., #12/2, type NM cable		10	.800		28	46.50		74.50	101
3370	Type MC cable		8.89	.900		34	52.50		86.50	116
3380	EMT & wire		4.10	1.951		50.50	114		164.50	226
3400	Decorator style									
3410	S.P. switch/15 amp recpt., type NM cable	1 Elec	11.43	.700	Ea.	24	40.50		64.50	87.50
3420	Type MC cable		10	.800		34	46.50		80.50	107
3430	EMT & wire		4.40	1.818		45.50	106		151.50	209
3450	S.P. switch/pilot light, type NM cable		11.43	.700		25	40.50		65.50	88.50
3470	Type MC cable		10	.800		34.50	46.50		81	108
3480	EMT & wire		4.40	1.818		46.50	106		152.50	209
3500	2-S.P. switches, 2-#14/2, type NM cables		10	.800		26	46.50		72.50	98
3520	Type MC cable		8.89	.900		40	52.50		92.50	123
3530	EMT & wire		4.10	1.951		47.50	114		161.50	222
3550	3-way/15 amp recpt., #14/3, type NM cable		10	.800		30.50	46.50		77	103
3570	Type MC cable		8.89	.900		43.50	52.50		96	126
3580	EMT & wire		4.10	1.951		50.50	114		164.50	226
3650	2-3 way switches, 2-#14/3, type NM cables		8.89	.900		38.50	52.50		91	121
3670	Type MC cable		8	1		59	58		117	152
3680	EMT & wire		4	2		57	116		173	237
3700	S.P. switch/20 amp recpt., #12/2, type NM cable		10	.800		32	46.50		78.50	105
3720	Type MC cable		8.89	.900		38	52.50		90.50	120
3730	EMT & wire		4.10	1.951		54.50	114		168.50	230
4000	Receptacle devices									
4010	Duplex outlet, 15 amp recpt., ivory, 1-gang box, plate									
4015	Type NM cable	1 Elec	14.55	.550	Ea.	8.40	32		40.40	57
4020	Type MC cable		12.31	.650		17.95	38		55.95	76.50
4030	EMT & wire		5.33	1.501		28.50	87.50		116	163
4050	With #12/2, type NM cable		12.31	.650		9.50	38		47.50	67
4070	Type MC cable		10.67	.750		18.05	43.50		61.55	85
4080	EMT & wire		4.71	1.699		32	99		131	184
4100	20 amp recpt., #12/2, type NM cable		12.31	.650		17.25	38		55.25	75.50
4120	Type MC cable		10.67	.750		26	43.50		69.50	93.50
4130	EMT & wire		4.71	1.699		40	99		139	192
4140	For GFI see Section 26 05 90.10 line 4300 below									
4150	Decorator style, 15 amp recpt., type NM cable	1 Elec	14.55	.550	Ea.	12.25	32		44.25	61.50
4170	Type MC cable		12.31	.650		22	38		60	80.50
4180	EMT & wire		5.33	1.501		32.50	87.50		120	167
4200	With #12/2, type NM cable		12.31	.650		13.35	38		51.35	71

26 05 Common Work Results for Electrical

26 05 90 – Residential Applications

26 05 90.10 Residential Wiring		Crew	Daily Output	Labor-Hours	Unit	Material	2018 Bare Costs Labor	Equipment	Total	Total Incl O&P
4220	Type MC cable	1 Elec	10.67	.750	Ea.	22	43.50		65.50	89
4230	EMT & wire		4.71	1.699		36	99		135	188
4250	20 amp recpt., #12/2, type NM cable		12.31	.650		21	38		59	80
4270	Type MC cable		10.67	.750		29.50	43.50		73	97.50
4280	EMT & wire		4.71	1.699		44	99		143	196
4300	GFI, 15 amp recpt., type NM cable		12.31	.650		21	38		59	79.50
4320	Type MC cable		10.67	.750		30.50	43.50		74	98.50
4330	EMT & wire		4.71	1.699		41	99		140	193
4350	GFI with #12/2, type NM cable		10.67	.750		22	43.50		65.50	89
4370	Type MC cable		9.20	.870		30.50	50.50		81	109
4380	EMT & wire		4.21	1.900		44.50	111		155.50	214
4400	20 amp recpt., #12/2, type NM cable		10.67	.750		48.50	43.50		92	118
4420	Type MC cable		9.20	.870		57	50.50		107.50	138
4430	EMT & wire		4.21	1.900		71	111		182	243
4500	Weather-proof cover for above receptacles, add	↓	32	.250	↓	2.10	14.55		16.65	24.50
4550	Air conditioner outlet, 20 amp-240 volt recpt.									
4560	30' of #12/2, 2 pole circuit breaker									
4570	Type NM cable	1 Elec	10	.800	Ea.	58	46.50		104.50	134
4580	Type MC cable		9	.889		68.50	51.50		120	153
4590	EMT & wire		4	2		81	116		197	264
4600	Decorator style, type NM cable		10	.800		63	46.50		109.50	139
4620	Type MC cable		9	.889		73	51.50		124.50	158
4630	EMT & wire	↓	4	2	↓	86	116		202	269
4650	Dryer outlet, 30 amp-240 volt recpt., 20' of #10/3									
4660	2 pole circuit breaker									
4670	Type NM cable	1 Elec	6.41	1.248	Ea.	53.50	72.50		126	168
4680	Type MC cable		5.71	1.401		62.50	81.50		144	191
4690	EMT & wire	↓	3.48	2.299	↓	75	134		209	282
4700	Range outlet, 50 amp-240 volt recpt., 30' of #8/3									
4710	Type NM cable	1 Elec	4.21	1.900	Ea.	82	111		193	255
4720	Type MC cable		4	2		128	116		244	315
4730	EMT & wire		2.96	2.703		106	157		263	350
4750	Central vacuum outlet, type NM cable		6.40	1.250		50	73		123	164
4770	Type MC cable		5.71	1.401		64.50	81.50		146	193
4780	EMT & wire	↓	3.48	2.299	↓	83	134		217	291
4800	30 amp-110 volt locking recpt., #10/2 circ. bkr.									
4810	Type NM cable	1 Elec	6.20	1.290	Ea.	58	75		133	176
4820	Type MC cable		5.40	1.481		76	86		162	213
4830	EMT & wire	↓	3.20	2.500	↓	93.50	146		239.50	320
4900	Low voltage outlets									
4910	Telephone recpt., 20' of 4/C phone wire	1 Elec	26	.308	Ea.	8	17.90		25.90	36
4920	TV recpt., 20' of RG59U coax wire, F type connector	"	16	.500	"	15.55	29		44.55	60.50
4950	Door bell chime, transformer, 2 buttons, 60' of bellwire									
4970	Economy model	1 Elec	11.50	.696	Ea.	52	40.50		92.50	118
4980	Custom model		11.50	.696		94.50	40.50		135	165
4990	Luxury model, 3 buttons	↓	9.50	.842	↓	185	49		234	278
6000	Lighting outlets									
6050	Wire only (for fixture), type NM cable	1 Elec	32	.250	Ea.	5.75	14.55		20.30	28.50
6070	Type MC cable		24	.333		11.70	19.40		31.10	42
6080	EMT & wire		10	.800		21.50	46.50		68	93.50
6100	Box (4"), and wire (for fixture), type NM cable		25	.320		14.55	18.60		33.15	44
6120	Type MC cable		20	.400		20.50	23.50		44	57.50
6130	EMT & wire	↓	11	.727	↓	30.50	42.50		73	97

For customer support on your Electrical Costs with RSMeans data, call 800.448.8182.

205

26 05 90.10 Residential Wiring	Crew	Daily Output	Labor-Hours	Unit	Material	2018 Bare Costs Labor	Equipment	Total	Total Incl O&P
6200	Fixtures (use with line 6050 or 6100 above)									
6210	Canopy style, economy grade	1 Elec	40	.200	Ea.	23	11.65		34.65	43
6220	Custom grade		40	.200		53.50	11.65		65.15	76.50
6250	Dining room chandelier, economy grade		19	.421		72	24.50		96.50	116
6260	Custom grade		19	.421		305	24.50		329.50	370
6270	Luxury grade		15	.533		955	31		986	1,100
6310	Kitchen fixture (fluorescent), economy grade		30	.267		71.50	15.50		87	102
6320	Custom grade		25	.320		137	18.60		155.60	179
6350	Outdoor, wall mounted, economy grade		30	.267		28.50	15.50		44	54.50
6360	Custom grade		30	.267		121	15.50		136.50	156
6370	Luxury grade		25	.320		248	18.60		266.60	300
6410	Outdoor PAR floodlights, 1 lamp, 150 watt		20	.400		26	23.50		49.50	63.50
6420	2 lamp, 150 watt each		20	.400		45.50	23.50		69	85
6425	Motion sensing, 2 lamp, 150 watt each		20	.400		88.50	23.50		112	133
6430	For infrared security sensor, add		32	.250		99.50	14.55		114.05	132
6450	Outdoor, quartz-halogen, 300 watt flood		20	.400		33	23.50		56.50	71.50
6600	Recessed downlight, round, pre-wired, 50 or 75 watt trim		30	.267		62.50	15.50		78	91.50
6610	With shower light trim		30	.267		86	15.50		101.50	118
6620	With wall washer trim		28	.286		79.50	16.65		96.15	113
6630	With eye-ball trim		28	.286		66.50	16.65		83.15	98
6700	Porcelain lamp holder		40	.200		2.92	11.65		14.57	20.50
6710	With pull switch		40	.200		11.30	11.65		22.95	30
6750	Fluorescent strip, 2-20 watt tube, wrap around diffuser, 24"		24	.333		46	19.40		65.40	80
6760	1-34 watt tube, 48"		24	.333		105	19.40		124.40	145
6770	2-34 watt tubes, 48"		20	.400		122	23.50		145.50	169
6800	Bathroom heat lamp, 1-250 watt		28	.286		31	16.65		47.65	59.50
6810	2-250 watt lamps		28	.286		73	16.65		89.65	106
6820	For timer switch, see Section 26 05 90.10 line 2400									
6900	Outdoor post lamp, incl. post, fixture, 35' of #14/2									
6910	Type NM cable	1 Elec	3.50	2.286	Ea.	300	133		433	530
6920	Photo-eye, add		27	.296		27.50	17.25		44.75	56
6950	Clock dial time switch, 24 hr., w/enclosure, type NM cable		11.43	.700		68.50	40.50		109	136
6970	Type MC cable		11	.727		78	42.50		120.50	150
6980	EMT & wire		4.85	1.649		88.50	96		184.50	242
7000	Alarm systems									
7050	Smoke detectors, box, #14/3, type NM cable	1 Elec	14.55	.550	Ea.	33.50	32		65.50	85
7070	Type MC cable		12.31	.650		44	38		82	105
7080	EMT & wire		5	1.600		51	93		144	195
7090	For relay output to security system, add					10.75			10.75	11.85
8000	Residential equipment									
8050	Disposal hook-up, incl. switch, outlet box, 3' of flex									
8060	20 amp-1 pole circ. bkr., and 25' of #12/2									
8070	Type NM cable	1 Elec	10	.800	Ea.	28	46.50		74.50	101
8080	Type MC cable		8	1		37.50	58		95.50	128
8090	EMT & wire		5	1.600		54.50	93		147.50	199
8100	Trash compactor or dishwasher hook-up, incl. outlet box,									
8110	3' of flex, 15 amp-1 pole circ. bkr., and 25' of #14/2									
8120	Type NM cable	1 Elec	10	.800	Ea.	15.50	46.50		62	86.50
8130	Type MC cable		8	1		26.50	58		84.50	116
8140	EMT & wire		5	1.600		41	93		134	184
8150	Hot water sink dispenser hook-up, use line 8100									
8200	Vent/exhaust fan hook-up, type NM cable	1 Elec	32	.250	Ea.	5.75	14.55		20.30	28.50
8220	Type MC cable		24	.333		11.70	19.40		31.10	42

206

For customer support on your Electrical Costs with RSMeans data, call 800.448.8182.

26 05 Common Work Results for Electrical

26 05 90 – Residential Applications

26 05 90.10 Residential Wiring	Crew	Daily Output	Labor-Hours	Unit	Material	2018 Bare Costs Labor	Equipment	Total	Total Incl O&P
8230 EMT & wire	1 Elec	10	.800	Ea.	21.50	46.50		68	93.50
8250 Bathroom vent fan, 50 CFM (use with above hook-up)									
8260 Economy model	1 Elec	15	.533	Ea.	19.40	31		50.40	68
8270 Low noise model		15	.533		44	31		75	95
8280 Custom model	↓	12	.667	↓	121	39		160	191
8300 Bathroom or kitchen vent fan, 110 CFM									
8310 Economy model	1 Elec	15	.533	Ea.	68	31		99	122
8320 Low noise model	"	15	.533	"	92	31		123	148
8350 Paddle fan, variable speed (w/o lights)									
8360 Economy model (AC motor)	1 Elec	10	.800	Ea.	110	46.50		156.50	191
8362 With light kit		10	.800		149	46.50		195.50	234
8370 Custom model (AC motor)		10	.800		264	46.50		310.50	360
8372 With light kit		10	.800		305	46.50		351.50	405
8380 Luxury model (DC motor)		8	1		315	58		373	435
8382 With light kit		8	1		355	58		413	475
8390 Remote speed switch for above, add	↓	12	.667	↓	33.50	39		72.50	95
8500 Whole house exhaust fan, ceiling mount, 36", variable speed									
8510 Remote switch, incl. shutters, 20 amp-1 pole circ. bkr.									
8520 30' of #12/2, type NM cable	1 Elec	4	2	Ea.	1,225	116		1,341	1,525
8530 Type MC cable		3.50	2.286		1,250	133		1,383	1,575
8540 EMT & wire	↓	3	2.667	↓	1,250	155		1,405	1,600
8600 Whirlpool tub hook-up, incl. timer switch, outlet box									
8610 3' of flex, 20 amp-1 pole GFI circ. bkr.									
8620 30' of #12/2, type NM cable	1 Elec	5	1.600	Ea.	125	93		218	277
8630 Type MC cable		4.20	1.905		131	111		242	310
8640 EMT & wire	↓	3.40	2.353	↓	145	137		282	365
8650 Hot water heater hook-up, incl. 1-2 pole circ. bkr., box;									
8660 3' of flex, 20' of #10/2, type NM cable	1 Elec	5	1.600	Ea.	28.50	93		121.50	171
8670 Type MC cable		4.20	1.905		41	111		152	211
8680 EMT & wire	↓	3.40	2.353	↓	47.50	137		184.50	258
9000 Heating/air conditioning									
9050 Furnace/boiler hook-up, incl. firestat, local on-off switch									
9060 Emergency switch, and 40' of type NM cable	1 Elec	4	2	Ea.	52	116		168	231
9070 Type MC cable		3.50	2.286		66	133		199	272
9080 EMT & wire	↓	1.50	5.333	↓	89	310		399	565
9100 Air conditioner hook-up, incl. local 60 amp disc. switch									
9110 3' sealtite, 40 amp, 2 pole circuit breaker									
9130 40' of #8/2, type NM cable	1 Elec	3.50	2.286	Ea.	134	133		267	345
9140 Type MC cable		3	2.667		201	155		356	455
9150 EMT & wire	↓	1.30	6.154	↓	182	360		542	735
9200 Heat pump hook-up, 1-40 & 1-100 amp 2 pole circ. bkr.									
9210 Local disconnect switch, 3' sealtite									
9220 40' of #8/2 & 30' of #3/2									
9230 Type NM cable	1 Elec	1.30	6.154	Ea.	475	360		835	1,050
9240 Type MC cable		1.08	7.407		525	430		955	1,225
9250 EMT & wire	↓	.94	8.511	↓	490	495		985	1,275
9500 Thermostat hook-up, using low voltage wire									
9520 Heating only, 25' of #18-3	1 Elec	24	.333	Ea.	5.70	19.40		25.10	35.50
9530 Heating/cooling, 25' of #18-4	"	20	.400	"	7.15	23.50		30.65	43

For customer support on your Electrical Costs with RSMeans data, call 800.448.8182.

207

26 09 13 – Electrical Power Monitoring

26 09 13.10 Switchboard Instruments		Crew	Daily Output	Labor-Hours	Unit	Material	2018 Bare Costs Labor	Equipment	Total	Total Incl O&P
0010	**SWITCHBOARD INSTRUMENTS** 3 phase, 4 wire R260913-80									
0100	AC indicating, ammeter & switch	1 Elec	8	1	Ea.	2,875	58		2,933	3,250
0200	Voltmeter & switch		8	1		2,900	58		2,958	3,250
0300	Wattmeter		8	1		4,100	58		4,158	4,600
0400	AC recording, ammeter		4	2		7,300	116		7,416	8,225
0500	Voltmeter		4	2		7,300	116		7,416	8,225
0600	Ground fault protection, zero sequence		2.70	2.963		6,450	172		6,622	7,350
0700	Ground return path		2.70	2.963		6,450	172		6,622	7,350
0800	3 current transformers, 5 to 800 amp		2	4		3,000	233		3,233	3,650
0900	1,000 to 1,500 amp		1.30	6.154		4,325	360		4,685	5,275
1200	2,000 to 4,000 amp		1	8		5,100	465		5,565	6,300
1300	Fused potential transformer, maximum 600 volt		8	1		1,125	58		1,183	1,325

26 09 13.20 Voltage Monitor Systems

26 09 13.20 Voltage Monitor Systems		Crew	Daily Output	Labor-Hours	Unit	Material	2018 Bare Costs Labor	Equipment	Total	Total Incl O&P
0010	**VOLTAGE MONITOR SYSTEMS** (test equipment)									
0100	AC voltage monitor system, 120/240 V, one-channel				Ea.	2,675			2,675	2,925
0110	Modem adapter					335			335	365
0120	Add-on detector only					1,400			1,400	1,550
0150	AC voltage remote monitor sys., 3 channel, 120, 230, or 480 V					4,850			4,850	5,325
0160	With internal modem					5,100			5,100	5,625
0170	Combination temperature and humidity probe					750			750	825
0180	Add-on detector only					3,500			3,500	3,850
0190	With internal modem					3,825			3,825	4,200

26 09 13.30 Smart Metering

26 09 13.30 Smart Metering			Crew	Daily Output	Labor-Hours	Unit	Material	2018 Bare Costs Labor	Equipment	Total	Total Incl O&P
0010	**SMART METERING**, In panel										
0100	Single phase, 120/208 volt, 100 amp	G	1 Elec	8.78	.911	Ea.	325	53		378	435
0120	200 amp	G		8.78	.911		335	53		388	450
0200	277 volt, 100 amp	G		8.78	.911		360	53		413	475
0220	200 amp	G		8.78	.911		395	53		448	510
1100	Three phase, 120/208 volt, 100 amp	G		4.69	1.706		675	99.50		774.50	890
1120	200 amp	G		4.69	1.706		800	99.50		899.50	1,025
1130	400 amp	G		4.69	1.706		800	99.50		899.50	1,025
1140	800 amp	G		4.69	1.706		805	99.50		904.50	1,025
1150	1,600 amp	G		4.69	1.706		830	99.50		929.50	1,050
1200	277/480 volt, 100 amp	G		4.69	1.706		870	99.50		969.50	1,100
1220	200 amp	G		4.69	1.706		880	99.50		979.50	1,125
1230	400 amp	G		4.69	1.706		880	99.50		979.50	1,125
1240	800 amp	G		4.69	1.706		850	99.50		949.50	1,075
1250	1,600 amp	G		4.69	1.706		865	99.50		964.50	1,100
2000	Data recorder, 8 meters	G		10.97	.729		1,450	42.50		1,492.50	1,650
2100	16 meters	G		8.53	.938		3,600	54.50		3,654.50	4,025
3000	Software package, per meter, basic	G					254			254	279
3100	Premium	G					665			665	735

26 09 23 – Lighting Control Devices

26 09 23.10 Energy Saving Lighting Devices

26 09 23.10 Energy Saving Lighting Devices			Crew	Daily Output	Labor-Hours	Unit	Material	2018 Bare Costs Labor	Equipment	Total	Total Incl O&P
0010	**ENERGY SAVING LIGHTING DEVICES**										
0100	Occupancy sensors, passive infrared ceiling mounted	G	1 Elec	7	1.143	Ea.	75	66.50		141.50	182
0110	Ultrasonic ceiling mounted	G		7	1.143		85.50	66.50		152	194
0120	Dual technology ceiling mounted	G		6.50	1.231		125	71.50		196.50	244
0150	Automatic wall switches	G		24	.333		75	19.40		94.40	112
0160	Daylighting sensor, manual control, ceiling mounted	G		7	1.143		108	66.50		174.50	218
0170	Remote and dimming control with remote controller	G		6.50	1.231		168	71.50		239.50	292
0200	Passive infrared ceiling mounted	G		6.50	1.231		33	71.50		104.50	143

26 09 23 – Lighting Control Devices

26 09 23.10 Energy Saving Lighting Devices		Crew	Daily Output	Labor-Hours	Unit	Material	2018 Bare Costs Labor	Equipment	Total	Total Incl O&P
0400	Remote power pack	G	1 Elec	10	.800	Ea.	33	46.50		79.50	106
0450	Photoelectric control, S.P.S.T. 120 V	G		8	1		21	58		79	110
0500	S.P.S.T. 208 V/277 V	G		8	1		27	58		85	117
0550	D.P.S.T. 120 V	G		6	1.333		194	77.50		271.50	330
0600	D.P.S.T. 208 V/277 V	G		6	1.333		204	77.50		281.50	340
0650	S.P.D.T. 208 V/277 V	G		6	1.333		212	77.50		289.50	350
0660	Daylight level sensor, wall mounted, on/off or dimming	G		8	1		197	58		255	305

26 09 26 – Lighting Control Panelboards

26 09 26.10 Lighting Control Relay Panel

		Crew	Daily Output	Labor-Hours	Unit	Material	2018 Bare Costs Labor	Equipment	Total	Total Incl O&P
0010	**LIGHTING CONTROL RELAY PANEL** with timeclock										
0100	4 Relay	G	1 Elec	2.50	3.200	Ea.	1,100	186		1,286	1,500
0110	8 Relay	G		2.30	3.478		1,225	202		1,427	1,650
0120	16 Relay	G		1.80	4.444		1,375	259		1,634	1,900
0130	24 Relay	G		1.50	5.333		1,600	310		1,910	2,225
0140	48 Relay	G		1	8		1,450	465		1,915	2,275
0200	Room Controller, switching only										
0210	1 Relay	G	1 Elec	3	2.667	Ea.	1,425	155		1,580	1,775
0220	2 Relay	G		3	2.667		1,350	155		1,505	1,700
0230	3 Relay	G		3	2.667		1,450	155		1,605	1,800
0240	Dimming										
0250	1 Relay	G	1 Elec	3	2.667	Ea.	1,275	155		1,430	1,625
0260	2 Relay	G		3	2.667		1,325	155		1,480	1,700
0270	3 Relay	G		3	2.667		1,450	155		1,605	1,800

26 09 36 – Modular Dimming Controls

26 09 36.13 Manual Modular Dimming Controls

		Crew	Daily Output	Labor-Hours	Unit	Material	2018 Bare Costs Labor	Equipment	Total	Total Incl O&P
0010	**MANUAL MODULAR DIMMING CONTROLS**										
2000	Lighting control module	G	1 Elec	2	4	Ea.	315	233		548	695

26 12 Medium-Voltage Transformers

26 12 19 – Pad-Mounted, Liquid-Filled, Medium-Voltage Transformers

26 12 19.10 Transformer, Oil-Filled

		Crew	Daily Output	Labor-Hours	Unit	Material	2018 Bare Costs Labor	Equipment	Total	Total Incl O&P
0010	**TRANSFORMER, OIL-FILLED** primary delta or Y, R262213-60										
0050	Pad mounted 5 kV or 15 kV, with taps, 277/480 V secondary, 3 phase										
0100	150 kVA		R-3	.65	30.769	Ea.	9,750	1,775	199	11,724	13,600
0110	225 kVA			.55	36.364		11,100	2,100	236	13,436	15,600
0200	300 kVA			.45	44.444		13,900	2,575	288	16,763	19,500
0300	500 kVA			.40	50		19,700	2,900	325	22,925	26,400
0400	750 kVA			.38	52.632		25,000	3,050	340	28,390	32,500
0500	1,000 kVA			.26	76.923		29,600	4,450	500	34,550	39,800
0600	1,500 kVA			.23	86.957		35,200	5,050	565	40,815	46,900
0700	2,000 kVA			.20	100		44,400	5,800	650	50,850	58,500
0710	2,500 kVA			.19	105		53,500	6,100	680	60,280	69,000
0720	3,000 kVA			.17	118		65,000	6,825	760	72,585	82,500
0800	3,750 kVA			.16	125		83,500	7,250	810	91,560	103,500
1990	Pole mounted distribution type, single phase										
2000	13.8 kV primary, 120/240 V secondary, 10 kVA		R-15	7.45	6.443	Ea.	1,075	370	38	1,483	1,775
2010	50 kVA			3.70	12.973		2,125	740	76.50	2,941.50	3,500
2020	100 kVA			2.75	17.455		3,725	995	103	4,823	5,700
2030	167 kVA			2.15	22.326		6,175	1,275	132	7,582	8,850
2900	2,400 V primary, 120/240 V secondary, 10 kVA			7.45	6.443		1,025	370	38	1,433	1,725

For customer support on your Electrical Costs with RSMeans data, call 800.448.8182.

209

26 12 Medium-Voltage Transformers

26 12 19 – Pad-Mounted, Liquid-Filled, Medium-Voltage Transformers

26 12 19.10 Transformer, Oil-Filled	Crew	Daily Output	Labor-Hours	Unit	Material	2018 Bare Costs Labor	Equipment	Total	Total Incl O&P
2910	15 kVA	R-15	6.70	7.164	Ea.	1,275	410	42	1,727	2,050
2920	25 kVA		6	8		1,575	455	47	2,077	2,475
2930	37.5 kVA		4.30	11.163		1,950	640	66	2,656	3,150
2940	50 kVA		3.70	12.973		2,300	740	76.50	3,116.50	3,700
2950	75 kVA		3	16		3,350	915	94	4,359	5,150
2960	100 kVA	▼	2.75	17.455	▼	3,450	995	103	4,548	5,400

26 12 19.20 Transformer, Liquid-Filled

		Crew	Daily Output	Labor-Hours	Unit	Material	2018 Bare Costs Labor	Equipment	Total	Total Incl O&P
0010	**TRANSFORMER, LIQUID-FILLED** Pad mounted									
0020	5 kV or 15 kV primary, 277/480 volt secondary, 3 phase									
0050	225 kVA	R-3	.55	36.364	Ea.	14,300	2,100	236	16,636	19,100
0100	300 kVA		.45	44.444		16,900	2,575	288	19,763	22,800
0200	500 kVA		.40	50		21,400	2,900	325	24,625	28,200
0250	750 kVA		.38	52.632		27,600	3,050	340	30,990	35,400
0300	1,000 kVA		.26	76.923		32,100	4,450	500	37,050	42,500
0350	1,500 kVA		.23	86.957		37,500	5,050	565	43,115	49,400
0400	2,000 kVA		.20	100		46,400	5,800	650	52,850	60,500
0450	2,500 kVA	▼	.19	105	▼	55,500	6,100	680	62,280	71,000

26 13 Medium-Voltage Switchgear

26 13 16 – Medium-Voltage Fusible Interrupter Switchgear

26 13 16.10 Switchgear

		Crew	Daily Output	Labor-Hours	Unit	Material	2018 Bare Costs Labor	Equipment	Total	Total Incl O&P
0010	**SWITCHGEAR**, Incorporate switch with cable connections, transformer,									
0100	& low voltage section									
0200	Load interrupter switch, 600 amp, 2 position									
0300	NEMA 1, 4.8 kV, 300 kVA & below w/CLF fuses	R-3	.40	50	Ea.	22,900	2,900	325	26,125	29,800
0400	400 kVA & above w/CLF fuses		.38	52.632		25,500	3,050	340	28,890	33,000
0500	Non fusible		.41	48.780		18,500	2,825	315	21,640	25,000
0600	13.8 kV, 300 kVA & below w/CLF fuses		.38	52.632		28,700	3,050	340	32,090	36,600
0700	400 kVA & above w/CLF fuses		.36	55.556		28,700	3,225	360	32,285	36,800
0800	Non fusible	▼	.40	50		21,800	2,900	325	25,025	28,600
0900	Cable lugs for 2 feeders 4.8 kV or 13.8 kV	1 Elec	8	1		695	58		753	850
1000	Pothead, one 3 conductor or three 1 conductor		4	2		3,350	116		3,466	3,850
1100	Two 3 conductor or six 1 conductor		2	4		6,600	233		6,833	7,600
1200	Key interlocks	▼	8	1	▼	770	58		828	935
1300	Lightning arresters, distribution class (no charge)									
1400	Intermediate class or line type 4.8 kV	1 Elec	2.70	2.963	Ea.	3,750	172		3,922	4,375
1500	13.8 kV		2	4		4,975	233		5,208	5,825
1600	Station class, 4.8 kV		2.70	2.963		6,425	172		6,597	7,325
1700	13.8 kV	▼	2	4		11,100	233		11,333	12,600
1800	Transformers, 4,800 volts to 480/277 volts, 75 kVA	R-3	.68	29.412		19,600	1,700	191	21,491	24,300
1900	112.5 kVA		.65	30.769		23,900	1,775	199	25,874	29,200
2000	150 kVA		.57	35.088		27,200	2,025	227	29,452	33,200
2100	225 kVA		.48	41.667		31,200	2,425	270	33,895	38,200
2200	300 kVA		.41	48.780		34,900	2,825	315	38,040	43,000
2300	500 kVA		.36	55.556		45,900	3,225	360	49,485	55,500
2400	750 kVA		.29	68.966		52,000	4,000	445	56,445	64,000
2500	13,800 volts to 480/277 volts, 75 kVA		.61	32.787		27,500	1,900	212	29,612	33,400
2600	112.5 kVA		.55	36.364		36,500	2,100	236	38,836	43,600
2700	150 kVA		.49	40.816		36,900	2,375	264	39,539	44,400
2800	225 kVA		.41	48.780		42,600	2,825	315	45,740	51,500
2900	300 kVA	▼	.37	54.054	▼	43,500	3,125	350	46,975	53,000

26 13 Medium-Voltage Switchgear

26 13 16 — Medium-Voltage Fusible Interrupter Switchgear

26 13 16.10 Switchgear		Crew	Daily Output	Labor-Hours	Unit	Material	2018 Bare Costs Labor	Equipment	Total	Total Incl O&P
3000	500 kVA	R-3	.31	64.516	Ea.	48,100	3,750	420	52,270	59,000
3100	750 kVA	↓	.26	76.923		53,000	4,450	500	57,950	65,000
3200	Forced air cooling & temperature alarm	1 Elec	1	8	↓	4,250	465		4,715	5,375
3300	Low voltage components									
3400	Maximum panel height 49-1/2", single or twin row									
3500	Breaker heights, type FA or FH, 6"									
3600	type KA or KH, 8"									
3700	type LA, 11"									
3800	type MA, 14"									
3900	Breakers, 2 pole, 15 to 60 amp, type FA	1 Elec	5.60	1.429	Ea.	565	83		648	745
4000	70 to 100 amp, type FA		4.20	1.905		710	111		821	950
4100	15 to 60 amp, type FH		5.60	1.429		860	83		943	1,075
4200	70 to 100 amp, type FH		4.20	1.905		1,025	111		1,136	1,300
4300	125 to 225 amp, type KA		3.40	2.353		1,025	137		1,162	1,350
4400	125 to 225 amp, type KH		3.40	2.353		2,525	137		2,662	3,000
4500	125 to 400 amp, type LA		2.50	3.200		2,725	186		2,911	3,250
4600	125 to 600 amp, type MA		1.80	4.444		5,775	259		6,034	6,725
4700	700 & 800 amp, type MA		1.50	5.333		7,300	310		7,610	8,500
4800	3 pole, 15 to 60 amp, type FA		5.30	1.509		685	88		773	885
4900	70 to 100 amp, type FA		4	2		845	116		961	1,100
5000	15 to 60 amp, type FH		5.30	1.509		1,125	88		1,213	1,375
5100	70 to 100 amp, type FH		4	2		1,275	116		1,391	1,575
5200	125 to 225 amp, type KA		3.20	2.500		1,600	146		1,746	2,000
5300	125 to 225 amp, type KH		3.20	2.500		2,925	146		3,071	3,450
5400	125 to 400 amp, type LA		2.30	3.478		3,325	202		3,527	3,950
5500	125 to 600 amp, type MA		1.60	5		5,575	291		5,866	6,550
5600	700 & 800 amp, type MA	↓	1.30	6.154	↓	6,825	360		7,185	8,025

26 22 Low-Voltage Transformers

26 22 13 — Low-Voltage Distribution Transformers

26 22 13.10 Transformer, Dry-Type

			Crew	Daily Output	Labor-Hours	Unit	Material	Labor	Equipment	Total	Total Incl O&P
0010	**TRANSFORMER, DRY-TYPE**	R262213-10									
0050	Single phase, 240/480 volt primary, 120/240 volt secondary										
0100	1 kVA	R262213-60	1 Elec	2	4	Ea.	380	233		613	770
0300	2 kVA			1.60	5		590	291		881	1,075
0500	3 kVA			1.40	5.714		735	335		1,070	1,300
0700	5 kVA			1.20	6.667		1,000	390		1,390	1,675
0900	7.5 kVA		2 Elec	2.20	7.273		1,375	425		1,800	2,150
1100	10 kVA			1.60	10		1,575	580		2,155	2,600
1300	15 kVA			1.20	13.333		2,125	775		2,900	3,475
1500	25 kVA			1	16		2,375	930		3,305	4,025
1700	37.5 kVA			.80	20		2,875	1,175		4,050	4,925
1900	50 kVA			.70	22.857		3,500	1,325		4,825	5,850
2100	75 kVA		↓	.65	24.615		4,500	1,425		5,925	7,100
2110	100 kVA		R-3	.90	22.222		5,900	1,300	144	7,344	8,550
2120	167 kVA		"	.80	25		9,750	1,450	162	11,362	13,100
2190	480 V primary, 120/240 V secondary, nonvent., 15 kVA		2 Elec	1.20	13.333		1,675	775		2,450	2,975
2200	25 kVA			.90	17.778		2,475	1,025		3,500	4,275
2210	37 kVA			.75	21.333		2,925	1,250		4,175	5,075
2220	50 kVA			.65	24.615		3,450	1,425		4,875	5,950
2230	75 kVA		↓	.60	26.667	↓	4,600	1,550		6,150	7,375

For customer support on your Electrical Costs with RSMeans data, call 800.448.8182.

211

26 22 13.10 Transformer, Dry-Type	Crew	Daily Output	Labor-Hours	Unit	Material	2018 Bare Costs Labor	2018 Bare Costs Equipment	Total	Total Incl O&P
2240	100 kVA	2 Elec	.50	32	Ea.	5,975	1,850		7,825	9,350
2250	Low operating temperature (80°C), 25 kVA		1	16		4,475	930		5,405	6,325
2260	37 kVA		.80	20		4,800	1,175		5,975	7,050
2270	50 kVA		.70	22.857		6,275	1,325		7,600	8,925
2280	75 kVA		.65	24.615		9,950	1,425		11,375	13,100
2290	100 kVA		.55	29.091		10,400	1,700		12,100	13,900
2300	3 phase, 480 volt primary, 120/208 volt secondary									
2310	Ventilated, 3 kVA	1 Elec	1	8	Ea.	1,125	465		1,590	1,950
2700	6 kVA		.80	10		1,275	580		1,855	2,275
2900	9 kVA		.70	11.429		1,275	665		1,940	2,400
3100	15 kVA	2 Elec	1.10	14.545		1,625	845		2,470	3,050
3300	30 kVA		.90	17.778		1,600	1,025		2,625	3,325
3500	45 kVA		.80	20		1,900	1,175		3,075	3,850
3700	75 kVA		.70	22.857		2,550	1,325		3,875	4,800
3900	112.5 kVA	R-3	.90	22.222		4,225	1,300	144	5,669	6,725
4100	150 kVA		.85	23.529		5,125	1,375	152	6,652	7,875
4300	225 kVA		.65	30.769		7,475	1,775	199	9,449	11,100
4500	300 kVA		.55	36.364		9,375	2,100	236	11,711	13,700
4700	500 kVA		.45	44.444		15,600	2,575	288	18,463	21,400
4800	750 kVA		.35	57.143		22,700	3,325	370	26,395	30,400
4820	1,000 kVA		.32	62.500		26,700	3,625	405	30,730	35,300
4850	K-4 rated, 15 kVA	2 Elec	1.10	14.545		3,600	845		4,445	5,225
4855	30 kVA		.90	17.778		5,375	1,025		6,400	7,450
4860	45 kVA		.80	20		6,450	1,175		7,625	8,825
4865	75 kVA		.70	22.857		9,850	1,325		11,175	12,800
4870	112.5 kVA	R-3	.90	22.222		12,900	1,300	144	14,344	16,300
4875	150 kVA		.85	23.529		16,900	1,375	152	18,427	20,800
4880	225 kVA		.65	30.769		23,500	1,775	199	25,474	28,700
4885	300 kVA		.55	36.364		30,800	2,100	236	33,136	37,300
4890	500 kVA		.45	44.444		43,000	2,575	288	45,863	51,500
4900	K-13 rated, 15 kVA	2 Elec	1.10	14.545		4,075	845		4,920	5,750
4905	30 kVA		.90	17.778		6,125	1,025		7,150	8,275
4910	45 kVA		.80	20		7,375	1,175		8,550	9,850
4915	75 kVA		.70	22.857		11,100	1,325		12,425	14,200
4920	112.5 kVA	R-3	.90	22.222		14,700	1,300	144	16,144	18,300
4925	150 kVA		.85	23.529		19,200	1,375	152	20,727	23,300
4930	225 kVA		.65	30.769		26,200	1,775	199	28,174	31,700
4935	300 kVA		.55	36.364		34,900	2,100	236	37,236	41,800
4940	500 kVA		.45	44.444		58,000	2,575	288	60,863	67,500
5020	480 volt primary, 120/208 volt secondary									
5030	Nonventilated, 15 kVA	2 Elec	1.10	14.545	Ea.	2,975	845		3,820	4,550
5040	30 kVA		.80	20		4,525	1,175		5,700	6,725
5050	45 kVA		.70	22.857		6,075	1,325		7,400	8,675
5060	75 kVA		.65	24.615		9,625	1,425		11,050	12,800
5070	112.5 kVA	R-3	.85	23.529		15,800	1,375	152	17,327	19,600
5081	150 kVA		.85	23.529		19,800	1,375	152	21,327	23,900
5090	225 kVA		.60	33.333		22,100	1,925	216	24,241	27,400
5100	300 kVA		.50	40		21,800	2,325	259	24,384	27,800
5200	Low operating temperature (80°C), 30 kVA	2 Elec	.90	17.778		4,875	1,025		5,900	6,925
5210	45 kVA		.80	20		5,875	1,175		7,050	8,200
5220	75 kVA		.70	22.857		8,825	1,325		10,150	11,700
5230	112.5 kVA	R-3	.90	22.222		11,800	1,300	144	13,244	15,100
5240	150 kVA		.85	23.529		15,100	1,375	152	16,627	18,800

26 22 13 – Low-Voltage Distribution Transformers

26 22 13.10 Transformer, Dry-Type

		Crew	Daily Output	Labor-Hours	Unit	Material	2018 Bare Costs Labor	Equipment	Total	Total Incl O&P
5250	225 kVA	R-3	.65	30.769	Ea.	20,500	1,775	199	22,474	25,500
5260	300 kVA		.55	36.364		27,600	2,100	236	29,936	33,800
5270	500 kVA		.45	44.444		37,000	2,575	288	39,863	44,900
5380	3 phase, 5 kV primary, 277/480 volt secondary									
5400	High voltage, 112.5 kVA	R-3	.85	23.529	Ea.	18,500	1,375	152	20,027	22,600
5410	150 kVA		.65	30.769		20,200	1,775	199	22,174	25,100
5420	225 kVA		.55	36.364		23,800	2,100	236	26,136	29,500
5430	300 kVA		.45	44.444		30,400	2,575	288	33,263	37,600
5440	500 kVA		.35	57.143		39,200	3,325	370	42,895	48,500
5450	750 kVA		.32	62.500		61,000	3,625	405	65,030	73,500
5460	1,000 kVA		.30	66.667		71,500	3,875	430	75,805	85,000
5470	1,500 kVA		.27	74.074		83,500	4,300	480	88,280	98,500
5480	2,000 kVA		.25	80		98,000	4,650	520	103,170	115,000
5490	2,500 kVA		.20	100		110,000	5,800	650	116,450	130,500
5500	3,000 kVA		.18	111		144,500	6,450	720	151,670	169,000
5590	15 kV primary, 277/480 volt secondary									
5600	High voltage, 112.5 kVA	R-3	.85	23.529	Ea.	28,200	1,375	152	29,727	33,300
5610	150 kVA		.65	30.769		33,000	1,775	199	34,974	39,200
5620	225 kVA		.55	36.364		36,400	2,100	236	38,736	43,500
5630	300 kVA		.45	44.444		42,900	2,575	288	45,763	51,500
5640	500 kVA		.35	57.143		54,500	3,325	370	58,195	65,000
5650	750 kVA		.32	62.500		71,000	3,625	405	75,030	84,000
5660	1,000 kVA		.30	66.667		81,000	3,875	430	85,305	95,500
5670	1,500 kVA		.27	74.074		93,000	4,300	480	97,780	109,500
5680	2,000 kVA		.25	80		103,000	4,650	520	108,170	121,000
5690	2,500 kVA		.20	100		119,500	5,800	650	125,950	141,000
5700	3,000 kVA		.18	111		142,500	6,450	720	149,670	167,500
6000	2400 volt primary, 480 volt secondary, 300 kVA		.45	44.444		30,400	2,575	288	33,263	37,600
6010	500 kVA		.35	57.143		39,200	3,325	370	42,895	48,500
6020	750 kVA		.32	62.500		56,500	3,625	405	60,530	68,000
9300	Energy efficient transformer 3 ph									
9400	15 kVA, 480VAC delta, 208Y/120VAC	R-3	1.50	13.333	Ea.	1,975	775	86.50	2,836.50	3,425
9405	30 kVA, 480VAC delta, 208Y/120VAC		1.50	13.333		2,525	775	86.50	3,386.50	4,025
9410	45 kVA, 480VAC delta, 208Y/120VAC		1.50	13.333		3,050	775	86.50	3,911.50	4,600
9415	75 kVA, 480VAC delta, 208Y/120VAC		1.25	16		4,500	930	104	5,534	6,475
9420	112 kVA, 480VAC delta, 208Y/120VAC		1.25	16		6,425	930	104	7,459	8,600
9425	150 kVA, 480VAC delta, 208Y/120VAC		1.25	16		10,700	930	104	11,734	13,300
9430	225 kVA, 480VAC delta, 208Y/120VAC		1.25	16		10,700	930	104	11,734	13,300
9435	300 kVA, 480VAC delta, 208Y/120VAC		1.25	16		15,900	930	104	16,934	19,000

26 22 13.20 Isolating Panels

		Crew	Daily Output	Labor-Hours	Unit	Material	2018 Bare Costs Labor	Equipment	Total	Total Incl O&P
0010	**ISOLATING PANELS** used with isolating transformers									
0020	For hospital applications									
0100	Critical care area, 8 circuit, 3 kVA	1 Elec	.58	13.793	Ea.	6,825	805		7,630	8,700
0200	5 kVA		.54	14.815		6,850	860		7,710	8,825
0400	7.5 kVA		.52	15.385		6,600	895		7,495	8,600
0600	10 kVA		.44	18.182		8,100	1,050		9,150	10,500
0800	Operating room power & lighting, 8 circuit, 3 kVA		.58	13.793		4,750	805		5,555	6,425
1000	5 kVA		.54	14.815		5,300	860		6,160	7,125
1200	7.5 kVA		.52	15.385		2,875	895		3,770	4,525
1400	10 kVA		.44	18.182		3,800	1,050		4,850	5,750
1600	X-ray systems, 15 kVA, 90 amp		.44	18.182		10,400	1,050		11,450	13,000
1800	25 kVA, 125 amp		.36	22.222		16,700	1,300		18,000	20,200

For customer support on your Electrical Costs with RSMeans data, call 800.448.8182.

213

26 22 13 – Low-Voltage Distribution Transformers

26 22 13.30 Isolating Transformer		Crew	Daily Output	Labor-Hours	Unit	Material	2018 Bare Costs Labor	Equipment	Total	Total Incl O&P
0010	**ISOLATING TRANSFORMER**									
0100	Single phase, 120/240 volt primary, 120/240 volt secondary									
0200	0.50 kVA R262213-60	1 Elec	4	2	Ea.	400	116		516	615
0400	1 kVA		2	4		935	233		1,168	1,375
0600	2 kVA		1.60	5		1,500	291		1,791	2,075
0800	3 kVA		1.40	5.714		835	335		1,170	1,425
1000	5 kVA		1.20	6.667		1,100	390		1,490	1,800
1200	7.5 kVA		1.10	7.273		1,400	425		1,825	2,150
1400	10 kVA		.80	10		1,775	580		2,355	2,825
1600	15 kVA		.60	13.333		2,275	775		3,050	3,650
1800	25 kVA		.50	16		3,275	930		4,205	5,000
1810	37.5 kVA	2 Elec	.80	20		5,425	1,175		6,600	7,725
1820	75 kVA	"	.65	24.615		8,150	1,425		9,575	11,100
1830	3 phase, 120/240 V primary, 120/240 V secondary, 112.5 kVA	R-3	.90	22.222		10,300	1,300	144	11,744	13,400
1840	150 kVA		.85	23.529		13,100	1,375	152	14,627	16,600
1850	225 kVA		.65	30.769		18,200	1,775	199	20,174	22,900
1860	300 kVA		.55	36.364		24,200	2,100	236	26,536	30,000
1870	500 kVA		.45	44.444		40,300	2,575	288	43,163	48,600
1880	750 kVA		.35	57.143		40,700	3,325	370	44,395	50,000

26 22 13.90 Transformer Handling

		Crew	Daily Output	Labor-Hours	Unit	Material	2018 Bare Costs Labor	Equipment	Total	Total Incl O&P
0010	**TRANSFORMER HANDLING** Add to normal labor cost in restricted areas									
5000	Transformers									
5150	15 kVA, approximately 200 pounds	2 Elec	2.70	5.926	Ea.		345		345	515
5160	25 kVA, approximately 300 pounds		2.50	6.400			370		370	555
5170	37.5 kVA, approximately 400 pounds		2.30	6.957			405		405	605
5180	50 kVA, approximately 500 pounds		2	8			465		465	695
5190	75 kVA, approximately 600 pounds		1.80	8.889			515		515	775
5200	100 kVA, approximately 700 pounds		1.60	10			580		580	870
5210	112.5 kVA, approximately 800 pounds	3 Elec	2.20	10.909			635		635	950
5220	125 kVA, approximately 900 pounds		2	12			700		700	1,050
5230	150 kVA, approximately 1,000 pounds		1.80	13.333			775		775	1,150
5240	167 kVA, approximately 1,200 pounds		1.60	15			875		875	1,300
5250	200 kVA, approximately 1,400 pounds		1.40	17.143			1,000		1,000	1,500
5260	225 kVA, approximately 1,600 pounds		1.30	18.462			1,075		1,075	1,600
5270	250 kVA, approximately 1,800 pounds		1.10	21.818			1,275		1,275	1,900
5280	300 kVA, approximately 2,000 pounds		1	24			1,400		1,400	2,100
5290	500 kVA, approximately 3,000 pounds		.75	32			1,850		1,850	2,775
5300	600 kVA, approximately 3,500 pounds		.67	35.821			2,075		2,075	3,125
5310	750 kVA, approximately 4,000 pounds		.60	40			2,325		2,325	3,475
5320	1,000 kVA, approximately 5,000 pounds		.50	48			2,800		2,800	4,175

26 22 16 – Low-Voltage Buck-Boost Transformers

26 22 16.10 Buck-Boost Transformer		Crew	Daily Output	Labor-Hours	Unit	Material	2018 Bare Costs Labor	Equipment	Total	Total Incl O&P
0010	**BUCK-BOOST TRANSFORMER** R262213-60									
0100	Single phase, 120/240 V primary, 12/24 V secondary									
0200	0.10 kVA	1 Elec	8	1	Ea.	105	58		163	203
0400	0.25 kVA		5.70	1.404		148	81.50		229.50	285
0600	0.50 kVA		4	2		201	116		317	395
0800	0.75 kVA		3.10	2.581		264	150		414	515
1000	1.0 kVA		2	4		278	233		511	655
1200	1.5 kVA		1.80	4.444		370	259		629	795
1400	2.0 kVA		1.60	5		515	291		806	1,000

26 22 Low-Voltage Transformers

26 22 16 – Low-Voltage Buck-Boost Transformers

26 22 16.10 Buck-Boost Transformer		Crew	Daily Output	Labor-Hours	Unit	Material	2018 Bare Costs Labor	Equipment	Total	Total Incl O&P
1600	3.0 kVA	1 Elec	1.40	5.714	Ea.	595	335		930	1,150
1800	5.0 kVA	↓	1.20	6.667		1,100	390		1,490	1,800
2000	3 phase, 240 V primary, 208/120 V secondary, 15 kVA	2 Elec	2.40	6.667		2,300	390		2,690	3,125
2200	30 kVA		1.60	10		2,575	580		3,155	3,700
2400	45 kVA		1.40	11.429		3,100	665		3,765	4,400
2600	75 kVA	↓	1.20	13.333		3,750	775		4,525	5,275
2800	112.5 kVA	R-3	1.40	14.286		4,650	830	92.50	5,572.50	6,475
3000	150 kVA		1.10	18.182		6,200	1,050	118	7,368	8,525
3200	225 kVA		1	20		8,075	1,150	130	9,355	10,800
3400	300 kVA	↓	.90	22.222	↓	11,000	1,300	144	12,444	14,200

26 24 Switchboards and Panelboards

26 24 13 – Switchboards

26 24 13.10 Incoming Switchboards

		Crew	Daily Output	Labor-Hours	Unit	Material	2018 Bare Costs Labor	Equipment	Total	Total Incl O&P
0010	**INCOMING SWITCHBOARDS** main service section R262419-84									
0100	Aluminum bus bars, not including CT's or PT's									
0200	No main disconnect, includes CT compartment									
0300	120/208 volt, 4 wire, 600 amp	2 Elec	1	16	Ea.	4,000	930		4,930	5,800
0400	800 amp		.88	18.182		4,000	1,050		5,050	5,975
0500	1,000 amp		.80	20		4,800	1,175		5,975	7,025
0600	1,200 amp		.72	22.222		4,800	1,300		6,100	7,200
0700	1,600 amp		.66	24.242		4,800	1,400		6,200	7,375
0800	2,000 amp		.62	25.806		5,175	1,500		6,675	7,925
1000	3,000 amp		.56	28.571		6,825	1,675		8,500	9,975
1200	277/480 volt, 4 wire, 600 amp		1	16		4,000	930		4,930	5,800
1300	800 amp		.88	18.182		4,000	1,050		5,050	5,975
1400	1,000 amp		.80	20		5,075	1,175		6,250	7,325
1500	1,200 amp		.72	22.222		5,075	1,300		6,375	7,500
1600	1,600 amp		.66	24.242		5,075	1,400		6,475	7,675
1700	2,000 amp		.62	25.806		5,175	1,500		6,675	7,925
1800	3,000 amp		.56	28.571		6,825	1,675		8,500	9,975
1900	4,000 amp	↓	.52	30.769	↓	8,450	1,800		10,250	12,000
2000	Fused switch & CT compartment									
2100	120/208 volt, 4 wire, 400 amp	2 Elec	1.12	14.286	Ea.	2,675	830		3,505	4,200
2200	600 amp		.94	17.021		3,175	990		4,165	4,975
2300	800 amp		.84	19.048		10,800	1,100		11,900	13,600
2400	1,200 amp		.68	23.529		14,100	1,375		15,475	17,600
2500	277/480 volt, 4 wire, 400 amp		1.14	14.035		3,600	815		4,415	5,175
2600	600 amp		.94	17.021		5,600	990		6,590	7,650
2700	800 amp		.84	19.048		10,800	1,100		11,900	13,600
2800	1,200 amp	↓	.68	23.529	↓	14,100	1,375		15,475	17,600
2900	Pressure switch & CT compartment									
3000	120/208 volt, 4 wire, 800 amp	2 Elec	.80	20	Ea.	9,700	1,175		10,875	12,500
3100	1,200 amp		.66	24.242		18,800	1,400		20,200	22,800
3200	1,600 amp		.62	25.806		20,000	1,500		21,500	24,300
3300	2,000 amp		.56	28.571		21,200	1,675		22,875	25,800
3310	2,500 amp		.50	32		26,100	1,850		27,950	31,500
3320	3,000 amp		.44	36.364		35,200	2,125		37,325	41,900
3330	4,000 amp		.40	40		45,200	2,325		47,525	53,000
3340	120/208 volt, 4 wire, 800 amp, with ground fault		.80	20		16,000	1,175		17,175	19,400
3350	1,200 amp, with ground fault	↓	.66	24.242	↓	20,700	1,400		22,100	24,900

For customer support on your Electrical Costs with RSMeans data, call 800.448.8182.

215

26 24 13.10 Incoming Switchboards		Crew	Daily Output	Labor-Hours	Unit	Material	2018 Bare Costs Labor	Equipment	Total	Total Incl O&P
3360	1,600 amp, with ground fault	2 Elec	.62	25.806	Ea.	22,500	1,500		24,000	27,000
3370	2,000 amp, with ground fault		.56	28.571		24,300	1,675		25,975	29,200
3400	277/480 volt, 4 wire, 800 amp, with ground fault		.80	20		16,000	1,175		17,175	19,400
3600	1,200 amp, with ground fault		.66	24.242		20,700	1,400		22,100	24,900
4000	1,600 amp, with ground fault		.62	25.806		22,500	1,500		24,000	27,000
4200	2,000 amp, with ground fault		.56	28.571		24,300	1,675		25,975	29,200
4400	Circuit breaker, molded case & CT compartment									
4600	3 pole, 4 wire, 600 amp	2 Elec	.94	17.021	Ea.	8,350	990		9,340	10,700
4800	800 amp		.84	19.048		10,000	1,100		11,100	12,700
5000	1,200 amp		.68	23.529		13,600	1,375		14,975	17,100
5100	Copper bus bars, not incl. CT's or PT's, add, minimum					15%				

26 24 13.20 In Plant Distribution Switchboards

		Crew	Daily Output	Labor-Hours	Unit	Material	2018 Bare Costs Labor	Equipment	Total	Total Incl O&P
0010	**IN PLANT DISTRIBUTION SWITCHBOARDS**									
0100	Main lugs only, to 600 volt, 3 pole, 3 wire, 200 amp	2 Elec	1.20	13.333	Ea.	1,250	775		2,025	2,525
0110	400 amp		1.20	13.333		1,250	775		2,025	2,525
0120	600 amp		1.20	13.333		1,300	775		2,075	2,575
0130	800 amp		1.08	14.815		1,400	860		2,260	2,850
0140	1,200 amp		.92	17.391		1,700	1,000		2,700	3,375
0150	1,600 amp		.86	18.605		2,225	1,075		3,300	4,075
0160	2,000 amp		.82	19.512		2,475	1,125		3,600	4,400
0250	To 480 volt, 3 pole, 4 wire, 200 amp		1.20	13.333		1,075	775		1,850	2,350
0260	400 amp		1.20	13.333		1,250	775		2,025	2,525
0270	600 amp		1.20	13.333		1,350	775		2,125	2,650
0280	800 amp		1.08	14.815		1,500	860		2,360	2,950
0290	1,200 amp		.92	17.391		1,850	1,000		2,850	3,550
0300	1,600 amp		.86	18.605		2,100	1,075		3,175	3,925
0310	2,000 amp		.82	19.512		2,450	1,125		3,575	4,400
0400	Main circuit breaker, to 600 volt, 3 pole, 3 wire, 200 amp		1.20	13.333		2,975	775		3,750	4,425
0410	400 amp		1.14	14.035		2,975	815		3,790	4,500
0420	600 amp		1.10	14.545		3,775	845		4,620	5,425
0430	800 amp		1.04	15.385		6,325	895		7,220	8,300
0440	1,200 amp		.88	18.182		8,275	1,050		9,325	10,700
0450	1,600 amp		.84	19.048		13,300	1,100		14,400	16,300
0460	2,000 amp		.80	20		14,200	1,175		15,375	17,400
0550	277/480 volt, 3 pole, 4 wire, 200 amp		1.20	13.333		3,150	775		3,925	4,600
0560	400 amp		1.14	14.035		3,150	815		3,965	4,675
0570	600 amp		1.10	14.545		3,950	845		4,795	5,600
0580	800 amp		1.04	15.385		6,625	895		7,520	8,650
0590	1,200 amp		.88	18.182		8,550	1,050		9,600	11,000
0600	1,600 amp		.84	19.048		13,300	1,100		14,400	16,300
0610	2,000 amp		.80	20		14,200	1,175		15,375	17,400
0700	Main fusible switch w/fuse, 208/240 volt, 3 pole, 3 wire, 200 amp		1.20	13.333		3,300	775		4,075	4,775
0710	400 amp		1.14	14.035		3,300	815		4,115	4,850
0720	600 amp		1.10	14.545		4,000	845		4,845	5,675
0730	800 amp		1.04	15.385		8,200	895		9,095	10,400
0740	1,200 amp		.88	18.182		9,525	1,050		10,575	12,100
0800	120/208, 120/240 volt, 3 pole, 4 wire, 200 amp		1.20	13.333		2,925	775		3,700	4,375
0810	400 amp		1.14	14.035		2,925	815		3,740	4,450
0820	600 amp		1.10	14.545		3,825	845		4,670	5,475
0830	800 amp		1.04	15.385		6,075	895		6,970	8,025
0840	1,200 amp		.88	18.182		7,050	1,050		8,100	9,325
0900	480 or 600 volt, 3 pole, 3 wire, 200 amp		1.20	13.333		3,175	775		3,950	4,625

26 24 13 – Switchboards

26 24 13.20 In Plant Distribution Switchboards	Crew	Daily Output	Labor-Hours	Unit	Material	2018 Bare Costs Labor	Equipment	Total	Total Incl O&P
0910	400 amp	2 Elec	1.14	14.035	Ea.	3,175	815		3,990	4,700
0920	600 amp		1.10	14.545		3,800	845		4,645	5,450
0930	800 amp		1.04	15.385		5,850	895		6,745	7,775
0940	1,200 amp		.88	18.182		6,775	1,050		7,825	9,025
1000	277 or 480 volt, 3 pole, 4 wire, 200 amp		1.20	13.333		3,275	775		4,050	4,775
1010	400 amp		1.14	14.035		3,275	815		4,090	4,850
1020	600 amp		1.10	14.545		3,975	845		4,820	5,650
1030	800 amp		1.04	15.385		6,075	895		6,970	8,050
1040	1,200 amp		.88	18.182		7,050	1,050		8,100	9,325
1120	1,600 amp		.76	21.053		12,900	1,225		14,125	16,000
1130	2,000 amp		.68	23.529		17,000	1,375		18,375	20,800
1150	Pressure switch, bolted, 3 pole, 208/240 volt, 3 wire, 800 amp		.96	16.667		9,550	970		10,520	12,000
1160	1,200 amp		.80	20		12,200	1,175		13,375	15,200
1170	1,600 amp		.76	21.053		13,900	1,225		15,125	17,100
1180	2,000 amp		.68	23.529		15,900	1,375		17,275	19,500
1200	120/208 or 120/240 volt, 3 pole, 4 wire, 800 amp		.96	16.667		7,550	970		8,520	9,750
1210	1,200 amp		.80	20		8,825	1,175		10,000	11,500
1220	1,600 amp		.76	21.053		13,900	1,225		15,125	17,100
1230	2,000 amp		.68	23.529		15,900	1,375		17,275	19,500
1300	480 or 600 volt, 3 wire, 800 amp		.96	16.667		9,550	970		10,520	12,000
1310	1,200 amp		.80	20		13,400	1,175		14,575	16,500
1320	1,600 amp		.76	21.053		15,100	1,225		16,325	18,400
1330	2,000 amp		.68	23.529		17,000	1,375		18,375	20,800
1400	277/480 volt, 4 wire, 800 amp		.96	16.667		9,550	970		10,520	12,000
1410	1,200 amp		.80	20		13,400	1,175		14,575	16,500
1420	1,600 amp		.76	21.053		15,100	1,225		16,325	18,400
1430	2,000 amp		.68	23.529		17,000	1,375		18,375	20,800
1500	Main ground fault protector, 1,200-2,000 amp		5.40	2.963		3,175	172		3,347	3,750
1600	Busway connection, 200 amp		5.40	2.963		460	172		632	765
1610	400 amp		4.60	3.478		460	202		662	810
1620	600 amp		4	4		460	233		693	855
1630	800 amp		3.20	5		460	291		751	940
1640	1,200 amp		2.60	6.154		460	360		820	1,050
1650	1,600 amp		2.40	6.667		950	390		1,340	1,625
1660	2,000 amp		2	8		950	465		1,415	1,750
1700	Shunt trip for remote operation, 200 amp		8	2		635	116		751	870
1710	400 amp		8	2		1,000	116		1,116	1,300
1720	600 amp		8	2		1,200	116		1,316	1,475
1730	800 amp		8	2		1,550	116		1,666	1,875
1740	1,200-2,000 amp		8	2		3,250	116		3,366	3,750
1800	Motor operated main breaker, 200 amp		8	2		3,150	116		3,266	3,625
1810	400 amp		8	2		3,150	116		3,266	3,625
1820	600 amp		8	2		3,225	116		3,341	3,725
1830	800 amp		8	2		3,225	116		3,341	3,725
1840	1,200-2,000 amp		8	2		3,300	116		3,416	3,800
1900	Current/potential transformer metering compartment, 200-800 amp		5.40	2.963		2,125	172		2,297	2,575
1940	1,200 amp		5.40	2.963		5,825	172		5,997	6,675
1950	1,600-2,000 amp		5.40	2.963		9,425	172		9,597	10,700
2000	With watt meter, 200-800 amp		4	4		8,050	233		8,283	9,225
2040	1,200 amp		4	4		9,950	233		10,183	11,400
2050	1,600-2,000 amp		4	4		9,950	233		10,183	11,400
2100	Split bus, 60-200 amp	1 Elec	5.30	1.509		191	88		279	340
2130	400 amp	2 Elec	4.60	3.478		330	202		532	670

26 24 13.20 In Plant Distribution Switchboards

		Crew	Daily Output	Labor-Hours	Unit	Material	2018 Bare Costs Labor	Equipment	Total	Total Incl O&P
2140	600 amp	2 Elec	3.60	4.444	Ea.	405	259		664	830
2150	800 amp		2.60	6.154		515	360		875	1,100
2170	1,200 amp		2	8		590	465		1,055	1,350
2250	Contactor control, 60 amp	1 Elec	2	4		1,250	233		1,483	1,725
2260	100 amp		1.50	5.333		1,400	310		1,710	2,025
2270	200 amp		1	8		2,125	465		2,590	3,025
2280	400 amp	2 Elec	1	16		6,450	930		7,380	8,500
2290	600 amp		.84	19.048		7,225	1,100		8,325	9,600
2300	800 amp		.72	22.222		8,550	1,300		9,850	11,400
2500	Modifier, two distribution sections, add		.80	20		2,950	1,175		4,125	5,000
2520	Three distribution sections, add		.40	40		5,650	2,325		7,975	9,700
2560	Auxiliary pull section, 20", add		2	8		1,075	465		1,540	1,875
2580	24", add		1.80	8.889		1,075	515		1,590	1,950
2600	30", add		1.60	10		1,075	580		1,655	2,050
2620	36", add		1.40	11.429		1,275	665		1,940	2,400
2640	Dog house, 12", add		2.40	6.667		221	390		611	825
2660	18", add		2	8		440	465		905	1,175
3000	Transition section between switchboard and transformer									
3050	or motor control center, 4 wire alum. bus, 600 amp	2 Elec	1.14	14.035	Ea.	2,175	815		2,990	3,625
3100	800 amp		1	16		2,425	930		3,355	4,075
3150	1,000 amp		.88	18.182		2,750	1,050		3,800	4,575
3200	1,200 amp		.80	20		3,000	1,175		4,175	5,075
3250	1,600 amp		.72	22.222		3,550	1,300		4,850	5,850
3300	2,000 amp		.66	24.242		4,100	1,400		5,500	6,600
3350	2,500 amp		.62	25.806		4,775	1,500		6,275	7,500
3400	3,000 amp		.56	28.571		5,500	1,675		7,175	8,525
4000	Weatherproof construction, per vertical section		1.76	9.091		2,500	530		3,030	3,550

26 24 13.30 Distribution Switchboards Section

		Crew	Daily Output	Labor-Hours	Unit	Material	2018 Bare Costs Labor	Equipment	Total	Total Incl O&P
0010	**DISTRIBUTION SWITCHBOARDS SECTION** R262419-80									
0100	Aluminum bus bars, not including breakers									
0160	Subfeed lug-rated at 60 amp	2 Elec	1.30	12.308	Ea.	1,150	715		1,865	2,325
0170	100 amp		1.26	12.698		1,450	740		2,190	2,675
0180	200 amp		1.20	13.333		1,300	775		2,075	2,600
0190	400 amp		1.10	14.545		1,475	845		2,320	2,900
0195	120/208 or 277/480 volt, 4 wire, 400 amp		1.10	14.545		1,450	845		2,295	2,875
0200	600 amp		1	16		1,525	930		2,455	3,075
0300	800 amp		.88	18.182		1,925	1,050		2,975	3,675
0400	1,000 amp		.80	20		2,550	1,175		3,725	4,550
0500	1,200 amp		.72	22.222		3,000	1,300		4,300	5,225
0600	1,600 amp		.66	24.242		4,000	1,400		5,400	6,475
0700	2,000 amp		.62	25.806		4,975	1,500		6,475	7,725
0800	2,500 amp		.60	26.667		5,975	1,550		7,525	8,900
0900	3,000 amp		.56	28.571		6,975	1,675		8,650	10,200
0950	4,000 amp		.52	30.769		7,975	1,800		9,775	11,500

26 24 13.40 Switchboards Feeder Section

		Crew	Daily Output	Labor-Hours	Unit	Material	2018 Bare Costs Labor	Equipment	Total	Total Incl O&P
0010	**SWITCHBOARDS FEEDER SECTION** group mounted devices R262419-82									
0030	Circuit breakers									
0160	FA frame, 15 to 60 amp, 240 volt, 1 pole	1 Elec	8	1	Ea.	114	58		172	212
0170	2 pole		7	1.143		275	66.50		341.50	400
0180	3 pole		5.30	1.509		425	88		513	595
0210	480 volt, 1 pole		8	1		166	58		224	269
0220	2 pole		7	1.143		475	66.50		541.50	620

26 24 13 - Switchboards

26 24 13.40 Switchboards Feeder Section	Crew	Daily Output	Labor-Hours	Unit	Material	2018 Bare Costs Labor	Equipment	Total	Total Incl O&P
0230	3 pole	1 Elec	5.30	1.509	Ea.	600	88		688	790
0260	600 volt, 2 pole		7	1.143		570	66.50		636.50	725
0270	3 pole		5.30	1.509		495	88		583	675
0280	FA frame, 70 to 100 amp, 240 volt, 1 pole		7	1.143		217	66.50		283.50	340
0310	2 pole		5	1.600		440	93		533	625
0320	3 pole		4	2		580	116		696	815
0330	480 volt, 1 pole		7	1.143		298	66.50		364.50	430
0360	2 pole		5	1.600		600	93		693	800
0370	3 pole		4	2		720	116		836	965
0380	600 volt, 2 pole		5	1.600		680	93		773	890
0410	3 pole		4	2		850	116		966	1,100
0420	KA frame, 70 to 225 amp		3.20	2.500		1,525	146		1,671	1,925
0430	LA frame, 125 to 400 amp		2.30	3.478		3,500	202		3,702	4,175
0460	MA frame, 450 to 600 amp		1.60	5		6,175	291		6,466	7,200
0470	700 to 800 amp		1.30	6.154		8,000	360		8,360	9,325
0480	MAL frame, 1,000 amp		1	8		8,300	465		8,765	9,850
0490	PA frame, 1,200 amp		.80	10		16,900	580		17,480	19,500
0500	Branch circuit, fusible switch, 600 volt, double 30/30 amp		4	2		845	116		961	1,100
0550	60/60 amp		3.20	2.500		870	146		1,016	1,175
0600	100/100 amp		2.70	2.963		1,100	172		1,272	1,450
0650	Single, 30 amp		5.30	1.509		695	88		783	895
0700	60 amp		4.70	1.702		825	99		924	1,050
0750	100 amp		4	2		1,050	116		1,166	1,325
0800	200 amp		2.70	2.963		1,350	172		1,522	1,750
0850	400 amp		2.30	3.478		2,475	202		2,677	3,025
0900	600 amp		1.80	4.444		3,025	259		3,284	3,700
0950	800 amp		1.30	6.154		5,100	360		5,460	6,125
1000	1,200 amp		.80	10		5,825	580		6,405	7,275
1080	Branch circuit, circuit breakers, high interrupting capacity									
1100	60 amp, 240, 480 or 600 volt, 1 pole	1 Elec	8	1	Ea.	320	58		378	435
1120	2 pole		7	1.143		820	66.50		886.50	1,000
1140	3 pole		5.30	1.509		650	88		738	845
1150	100 amp, 240, 480 or 600 volt, 1 pole		7	1.143		360	66.50		426.50	495
1160	2 pole		5	1.600		950	93		1,043	1,200
1180	3 pole		4	2		705	116		821	950
1200	225 amp, 240, 480 or 600 volt, 2 pole		3.50	2.286		2,250	133		2,383	2,675
1220	3 pole		3.20	2.500		2,200	146		2,346	2,625
1240	400 amp, 240, 480 or 600 volt, 2 pole		2.50	3.200		3,225	186		3,411	3,825
1260	3 pole		2.30	3.478		2,975	202		3,177	3,550
1280	600 amp, 240, 480 or 600 volt, 2 pole		1.80	4.444		4,100	259		4,359	4,875
1300	3 pole		1.60	5		3,625	291		3,916	4,425
1320	800 amp, 240, 480 or 600 volt, 2 pole		1.50	5.333		4,825	310		5,135	5,800
1340	3 pole		1.30	6.154		5,425	360		5,785	6,500
1360	1,000 amp, 240, 480 or 600 volt, 2 pole		1.10	7.273		6,700	425		7,125	8,000
1380	3 pole		1	8		5,875	465		6,340	7,175
1400	1,200 amp, 240, 480 or 600 volt, 2 pole		.90	8.889		6,550	515		7,065	8,000
1420	3 pole		.80	10		7,650	580		8,230	9,300
1700	Fusible switch, 240 V, 60 amp, 2 pole		3.20	2.500		430	146		576	690
1720	3 pole		3	2.667		590	155		745	880
1740	100 amp, 2 pole		2.70	2.963		435	172		607	740
1760	3 pole		2.50	3.200		660	186		846	1,000
1780	200 amp, 2 pole		2	4		910	233		1,143	1,350
1800	3 pole		1.90	4.211		1,075	245		1,320	1,550

For customer support on your Electrical Costs with RSMeans data, call 800.448.8182.

219

26 24 13 – Switchboards

26 24 13.40 Switchboards Feeder Section

		Crew	Daily Output	Labor-Hours	Unit	Material	2018 Bare Costs Labor	Equipment	Total	Total Incl O&P
1820	400 amp, 2 pole	1 Elec	1.50	5.333	Ea.	1,700	310		2,010	2,350
1840	3 pole		1.30	6.154		2,125	360		2,485	2,875
1860	600 amp, 2 pole		1	8		2,425	465		2,890	3,375
1880	3 pole		.90	8.889		2,975	515		3,490	4,050
1900	240-600 V, 800 amp, 2 pole		.70	11.429		5,400	665		6,065	6,950
1920	3 pole		.60	13.333		6,625	775		7,400	8,425
2000	600 V, 60 amp, 2 pole		3.20	2.500		650	146		796	935
2040	100 amp, 2 pole		2.70	2.963		670	172		842	995
2080	200 amp, 2 pole		2	4		1,150	233		1,383	1,625
2120	400 amp, 2 pole		1.50	5.333		2,300	310		2,610	3,000
2160	600 amp, 2 pole		1	8		2,800	465		3,265	3,775
2500	Branch circuit, circuit breakers, 60 amp, 600 volt, 3 pole		5.30	1.509		470	88		558	645
2520	240, 480 or 600 volt, 1 pole		8	1		100	58		158	197
2540	240 volt, 2 pole		7	1.143		172	66.50		238.50	289
2560	480 or 600 volt, 2 pole		7	1.143		365	66.50		431.50	500
2580	240 volt, 3 pole		5.30	1.509		415	88		503	590
2600	480 volt, 3 pole		5.30	1.509		485	88		573	665
2620	100 amp, 600 volt, 2 pole		5	1.600		465	93		558	655
2640	3 pole		4	2		590	116		706	825
2660	480 volt, 2 pole		5	1.600		400	93		493	580
2680	240 volt, 2 pole		5	1.600		229	93		322	390
2700	3 pole		4	2		370	116		486	580
2720	480 volt, 3 pole		4	2		540	116		656	770
2740	225 amp, 240, 480 or 600 volt, 2 pole		3.50	2.286		670	133		803	940
2760	3 pole		3.20	2.500		705	146		851	995
2780	400 amp, 240, 480 or 600 volt, 2 pole		2.50	3.200		1,400	186		1,586	1,825
2800	3 pole		2.30	3.478		1,625	202		1,827	2,100
2820	600 amp, 240 or 480 volt, 2 pole		1.80	4.444		2,300	259		2,559	2,900
2840	3 pole		1.60	5		2,825	291		3,116	3,525
2860	800 amp, 240, 480 or 600 volt, 2 pole		1.50	5.333		3,450	310		3,760	4,275
2880	3 pole		1.30	6.154		4,025	360		4,385	4,950
2900	1,000 amp, 240, 480 or 600 volt, 2 pole		1.10	7.273		4,225	425		4,650	5,275
2920	480 or 600 volt, 3 pole		1	8		4,850	465		5,315	6,050
2940	1,200 amp, 240, 480 or 600 volt, 2 pole		.90	8.889		5,975	515		6,490	7,350
2960	3 pole		.80	10		6,525	580		7,105	8,050
2980	600 volt, 3 pole		.80	10		6,525	580		7,105	8,050

26 24 16 – Panelboards

26 24 16.10 Load Centers

		Crew	Daily Output	Labor-Hours	Unit	Material	2018 Bare Costs Labor	Equipment	Total	Total Incl O&P
0010	**LOAD CENTERS** (residential type) R262416-50									
0100	3 wire, 120/240 V, 1 phase, including 1 pole plug-in breakers									
0200	100 amp main lugs, indoor, 8 circuits	1 Elec	1.40	5.714	Ea.	85.50	335		420.50	590
0300	12 circuits		1.20	6.667		145	390		535	740
0400	Rainproof, 8 circuits		1.40	5.714		108	335		443	615
0500	12 circuits		1.20	6.667		141	390		531	735
0600	200 amp main lugs, indoor, 16 circuits	R-1A	1.80	8.889		205	465		670	920
0700	20 circuits		1.50	10.667		189	560		749	1,050
0800	24 circuits		1.30	12.308		255	645		900	1,250
0900	30 circuits		1.20	13.333		291	700		991	1,375
1000	40 circuits		.80	20		405	1,050		1,455	2,025
1200	Rainproof, 16 circuits		1.80	8.889		207	465		672	920
1300	20 circuits		1.50	10.667		288	560		848	1,150
1400	24 circuits		1.30	12.308		282	645		927	1,275

For customer support on your Electrical Costs with RSMeans data, call 800.448.8182.

26 24 16.10 Load Centers		Crew	Daily Output	Labor-Hours	Unit	Material	2018 Bare Costs Labor	Equipment	Total	Total Incl O&P
1500	30 circuits	R-1A	1.20	13.333	Ea.	400	700		1,100	1,500
1600	40 circuits		.80	20		510	1,050		1,560	2,150
1800	400 amp main lugs, indoor, 42 circuits		.72	22.222		1,300	1,175		2,475	3,175
1900	Rainproof, 42 circuits		.72	22.222		1,650	1,175		2,825	3,550
2200	Plug in breakers, 20 amp, 1 pole, 4 wire, 120/208 volts									
2210	125 amp main lugs, indoor, 12 circuits	1 Elec	1.20	6.667	Ea.	182	390		572	780
2300	18 circuits		.80	10		216	580		796	1,100
2400	Rainproof, 12 circuits		1.20	6.667		228	390		618	830
2500	18 circuits		.80	10		256	580		836	1,150
2600	200 amp main lugs, indoor, 24 circuits	R-1A	1.30	12.308		335	645		980	1,325
2700	30 circuits		1.20	13.333		385	700		1,085	1,475
2800	36 circuits		1	16		345	840		1,185	1,625
2900	42 circuits		.80	20		500	1,050		1,550	2,125
3000	Rainproof, 24 circuits		1.30	12.308		270	645		915	1,250
3100	30 circuits		1.20	13.333		305	700		1,005	1,375
3200	36 circuits		1	16		390	840		1,230	1,675
3300	42 circuits		.80	20		515	1,050		1,565	2,150
3500	400 amp main lugs, indoor, 42 circuits		.72	22.222		1,250	1,175		2,425	3,125
3600	Rainproof, 42 circuits		.72	22.222		1,350	1,175		2,525	3,225
3700	Plug-in breakers, 20 amp, 1 pole, 3 wire, 120/240 volts									
3800	100 amp main breaker, indoor, 12 circuits	1 Elec	1.20	6.667	Ea.	156	390		546	750
3900	18 circuits	"	.80	10		202	580		782	1,100
4000	200 amp main breaker, indoor, 20 circuits	R-1A	1.50	10.667		281	560		841	1,150
4200	24 circuits		1.30	12.308		272	645		917	1,275
4300	30 circuits		1.20	13.333		340	700		1,040	1,425
4400	40 circuits		.90	17.778		420	930		1,350	1,850
4500	Rainproof, 20 circuits		1.50	10.667		360	560		920	1,225
4600	24 circuits		1.30	12.308		350	645		995	1,350
4700	30 circuits		1.20	13.333		475	700		1,175	1,575
4800	40 circuits		.90	17.778		600	930		1,530	2,050
5000	400 amp main breaker, indoor, 42 circuits		.72	22.222		3,000	1,175		4,175	5,050
5100	Rainproof, 42 circuits		.72	22.222		3,200	1,175		4,375	5,275
5300	Plug in breakers, 20 amp, 1 pole, 4 wire, 120/208 volts									
5400	200 amp main breaker, indoor, 30 circuits	R-1A	1.20	13.333	Ea.	395	700		1,095	1,475
5500	42 circuits		.80	20		475	1,050		1,525	2,100
5600	Rainproof, 30 circuits		1.20	13.333		815	700		1,515	1,950
5700	42 circuits		.80	20		540	1,050		1,590	2,175

26 24 16.20 Panelboard and Load Center Circuit Breakers

		Crew	Daily Output	Labor-Hours	Unit	Material	2018 Bare Costs Labor	Equipment	Total	Total Incl O&P
0010	**PANELBOARD AND LOAD CENTER CIRCUIT BREAKERS** R262416-50									
0050	Bolt-on, 10,000 amp I.C., 120 volt, 1 pole									
0100	15-50 amp	1 Elec	10	.800	Ea.	19.20	46.50		65.70	90.50
0200	60 amp		8	1		19.20	58		77.20	108
0300	70 amp		8	1		28	58		86	118
0350	240 volt, 2 pole									
0400	15-50 amp	1 Elec	8	1	Ea.	51	58		109	144
0500	60 amp		7.50	1.067		34.50	62		96.50	131
0600	80-100 amp		5	1.600		116	93		209	266
0700	3 pole, 15-60 amp		6.20	1.290		140	75		215	266
0800	70 amp		5	1.600		180	93		273	335
0900	80-100 amp		3.60	2.222		195	129		324	405
1000	22,000 amp I.C., 240 volt, 2 pole, 70-225 amp		2.70	2.963		555	172		727	875
1100	3 pole, 70-225 amp		2.30	3.478		275	202		477	605

For customer support on your Electrical Costs with RSMeans data, call 800.448.8182.

221

26 24 16.20 Panelboard and Load Center Circuit Breakers	Crew	Daily Output	Labor-Hours	Unit	Material	2018 Bare Costs Labor	Equipment	Total	Total Incl O&P
1200	14,000 amp I.C., 277 volts, 1 pole, 15-30 amp	1 Elec	8	1	Ea.	36	58		94	127
1300	22,000 amp I.C., 480 volts, 2 pole, 70-225 amp		2.70	2.963		505	172		677	815
1400	3 pole, 70-225 amp		2.30	3.478		770	202		972	1,150
2000	Plug-in panel or load center, 120/240 volt, to 60 amp, 1 pole		12	.667		6.15	39		45.15	65
2004	Circuit breaker, 120/240 volt, 20 A, 1 pole with NM cable		6.50	1.231		10.90	71.50		82.40	119
2006	30 A, 1 pole with NM cable		6.50	1.231		10.90	71.50		82.40	119
2010	2 pole		9	.889		26	.51.50		77.50	106
2014	50 A, 2 pole with NM cable		5.50	1.455		30.50	84.50		115	161
2020	3 pole		7.50	1.067		85.50	62		147.50	187
2030	100 amp, 2 pole		6	1.333		104	77.50		181.50	231
2040	3 pole		4.50	1.778		117	103		220	284
2050	150-200 amp, 2 pole		3	2.667		218	155		373	470
2060	Plug-in tandem, 120/240 V, 2-15 A, 1 pole		11	.727		21	42.50		63.50	86.50
2070	1-15 A & 1-20 A		11	.727		17.60	42.50		60.10	83
2080	2-20 A		11	.727		21	42.50		63.50	86.50
2082	Arc fault circuit interrupter, 120/240 V, 1-15 A & 1-20 A, 1 pole		11	.727		56	42.50		98.50	125
2100	High interrupting capacity, 120/240 volt, plug-in, 30 amp, 1 pole		12	.667		24	39		63	84.50
2110	60 amp, 2 pole		9	.889		24	51.50		75.50	104
2120	3 pole		7.50	1.067		283	62		345	405
2130	100 amp, 2 pole		6	1.333		141	77.50		218.50	271
2140	3 pole		4.50	1.778		405	103		508	600
2150	125 amp, 2 pole		3	2.667		785	155		940	1,100
2200	Bolt-on, 30 amp, 1 pole		10	.800		38	46.50		84.50	112
2210	60 amp, 2 pole		7.50	1.067		80.50	62		142.50	182
2220	3 pole		6.20	1.290		205	75		280	340
2230	100 amp, 2 pole		5	1.600		270	93		363	435
2240	3 pole		3.60	2.222		298	129		427	525
2300	Ground fault, 240 volt, 30 amp, 1 pole		7	1.143		87.50	66.50		154	196
2310	2 pole		6	1.333		156	77.50		233.50	288
2350	Key operated, 240 volt, 1 pole, 30 amp		7	1.143		123	66.50		189.50	236
2360	Switched neutral, 240 volt, 30 amp, 2 pole		6	1.333		40	77.50		117.50	160
2370	3 pole		5.50	1.455		66.50	84.50		151	200
2400	Shunt trip, for 240 volt breaker, 60 amp, 1 pole		4	2		71.50	116		187.50	253
2410	2 pole		3.50	2.286		71.50	133		204.50	278
2420	3 pole		3	2.667		71.50	155		226.50	310
2430	100 amp, 2 pole		3	2.667		71.50	155		226.50	310
2440	3 pole		2.50	3.200		71.50	186		257.50	355
2450	150 amp, 2 pole		2	4		170	233		403	535
2500	Auxiliary switch, for 240 volt breaker, 60 amp, 1 pole		4	2		65	116		181	246
2510	2 pole		3.50	2.286		65	133		198	271
2520	3 pole		3	2.667		65	155		220	305
2530	100 amp, 2 pole		3	2.667		65	155		220	305
2540	3 pole		2.50	3.200		65	186		251	350
2550	150 amp, 2 pole		2	4		108	233		341	470
2600	Panel or load center, 277/480 volt, plug-in, 30 amp, 1 pole		12	.667		52.50	39		91.50	116
2610	60 amp, 2 pole		9	.889		157	51.50		208.50	251
2620	3 pole		7.50	1.067		230	62		292	345
2650	Bolt-on, 60 amp, 2 pole		7.50	1.067		157	62		219	266
2660	3 pole		6.20	1.290		230	75		305	365
2700	I-line, 277/480 volt, 30 amp, 1 pole		8	1		60	58		118	153
2710	60 amp, 2 pole		7.50	1.067		211	62		273	325
2720	3 pole		6.20	1.290		274	75		349	410
2730	100 amp, 1 pole		7.50	1.067		116	62		178	220

26 24 16 – Panelboards

26 24 16.20 Panelboard and Load Center Circuit Breakers	Crew	Daily Output	Labor-Hours	Unit	Material	2018 Bare Costs Labor	Equipment	Total	Total Incl O&P
2740	2 pole	1 Elec	5	1.600	Ea.	274	93		367	440
2750	3 pole		3.50	2.286		325	133		458	555
2800	High interrupting capacity, 277/480 volt, plug-in, 30 amp, 1 pole		12	.667		305	39		344	395
2810	60 amp, 2 pole		9	.889		465	51.50		516.50	595
2820	3 pole		7	1.143		545	66.50		611.50	700
2830	Bolt-on, 30 amp, 1 pole		8	1		305	58		363	420
2840	60 amp, 2 pole		7.50	1.067		615	62		677	770
2850	3 pole		6.20	1.290		690	75		765	870
2900	H-line, 30 amp, 1 pole		8	1		305	58		363	420
2910	60 amp, 2 pole		7.50	1.067		430	62		492	570
2920	3 pole		6.20	1.290		480	75		555	635
2930	100 amp, 1 pole		7.50	1.067		490	62		552	635
2940	2 pole		5	1.600		490	93		583	680
2950	3 pole		3.60	2.222		540	129		669	790
2960	Shunt trip, 277/480 volt breaker, remote oper., 30 amp, 1 pole		4	2		325	116		441	530
2970	60 amp, 2 pole		3.50	2.286		475	133		608	725
2980	3 pole		3	2.667		540	155		695	820
2990	100 amp, 1 pole		3.50	2.286		380	133		513	620
3000	2 pole		3	2.667		540	155		695	820
3010	3 pole		2.50	3.200		590	186		776	925
3050	Under voltage trip, 277/480 volt breaker, 30 amp, 1 pole		4	2		325	116		441	530
3060	60 amp, 2 pole		3.50	2.286		475	133		608	725
3070	3 pole		3	2.667		540	155		695	820
3080	100 amp, 1 pole		3.50	2.286		380	133		513	620
3090	2 pole		3	2.667		540	155		695	820
3100	3 pole		2.50	3.200		590	186		776	925
3150	Motor operated, 277/480 volt breaker, 30 amp, 1 pole		4	2		575	116		691	810
3160	60 amp, 2 pole		3.50	2.286		730	133		863	1,000
3170	3 pole		3	2.667		790	155		945	1,100
3180	100 amp, 1 pole		3.50	2.286		630	133		763	895
3190	2 pole		3	2.667		790	155		945	1,100
3200	3 pole		2.50	3.200		840	186		1,026	1,200
3250	Panelboard spacers, per pole		40	.200		3.74	11.65		15.39	21.50
5110	NEMA 1 enclosure only, 600V, 3 p, 14k AIC, 100A	▼	2.20	3.636	▼	179	212		391	510

26 24 16.30 Panelboards Commercial Applications

		Crew	Daily Output	Labor-Hours	Unit	Material	2018 Bare Costs Labor	Equipment	Total	Total Incl O&P
0010	**PANELBOARDS COMMERCIAL APPLICATIONS** R262416-50									
0050	NQOD, w/20 amp 1 pole bolt-on circuit breakers									
0100	3 wire, 120/240 volts, 100 amp main lugs									
0150	10 circuits	1 Elec	1	8	Ea.	925	465		1,390	1,725
0200	14 circuits		.88	9.091		1,025	530		1,555	1,925
0250	18 circuits		.75	10.667		1,125	620		1,745	2,150
0300	20 circuits	▼	.65	12.308		1,250	715		1,965	2,450
0350	225 amp main lugs, 24 circuits	2 Elec	1.20	13.333		1,425	775		2,200	2,700
0400	30 circuits		.90	17.778		1,625	1,025		2,650	3,350
0450	36 circuits		.80	20		1,875	1,175		3,050	3,800
0500	38 circuits		.72	22.222		2,025	1,300		3,325	4,150
0550	42 circuits		.66	24.242		2,075	1,400		3,475	4,400
0600	4 wire, 120/208 volts, 100 amp main lugs, 12 circuits	1 Elec	1	8		995	465		1,460	1,800
0650	16 circuits		.75	10.667		1,150	620		1,770	2,175
0700	20 circuits		.65	12.308		1,325	715		2,040	2,525
0750	24 circuits		.60	13.333		1,275	775		2,050	2,550
0800	30 circuits	▼	.53	15.094		1,625	880		2,505	3,125

For customer support on your Electrical Costs with RSMeans data, call 800.448.8182.

223

26 24 16 – Panelboards

26 24 16.30 Panelboards Commercial Applications	Crew	Daily Output	Labor-Hours	Unit	Material	2018 Bare Costs Labor	Equipment	Total	Total Incl O&P
0850	225 amp main lugs, 32 circuits	2 Elec	.90	17.778	Ea.	1,825	1,025		2,850	3,575
0900	34 circuits		.84	19.048		1,875	1,100		2,975	3,700
0950	36 circuits		.80	20		1,925	1,175		3,100	3,850
1000	42 circuits		.68	23.529		2,150	1,375		3,525	4,400
1010	400 amp main lugs, 42 circs		.68	23.529		2,150	1,375		3,525	4,400
1040	225 amp main lugs, NEMA 7, 12 circuits		1	16		5,800	930		6,730	7,775
1100	24 circuits		.40	40		7,000	2,325		9,325	11,200
1200	NEHB, w/20 amp, 1 pole bolt-on circuit breakers									
1250	4 wire, 277/480 volts, 100 amp main lugs, 12 circuits	1 Elec	.88	9.091	Ea.	1,350	530		1,880	2,300
1300	20 circuits	"	.60	13.333		2,025	775		2,800	3,375
1350	225 amp main lugs, 24 circuits	2 Elec	.90	17.778		2,300	1,025		3,325	4,075
1400	30 circuits		.80	20		2,750	1,175		3,925	4,775
1448	32 circuits		4.90	3.265		3,200	190		3,390	3,775
1450	36 circuits		.72	22.222		3,200	1,300		4,500	5,425
1500	42 circuits		.60	26.667		3,650	1,550		5,200	6,325
1510	225 amp main lugs, NEMA 7, 12 circuits		.90	17.778		6,100	1,025		7,125	8,250
1590	24 circuits		.30	53.333		8,200	3,100		11,300	13,700
1600	NQOD panel, w/20 amp, 1 pole, circuit breakers									
1650	3 wire, 120/240 volt with main circuit breaker									
1700	100 amp main, 12 circuits	1 Elec	.80	10	Ea.	1,250	580		1,830	2,250
1750	20 circuits	"	.60	13.333		1,575	775		2,350	2,900
1800	225 amp main, 30 circuits	2 Elec	.68	23.529		2,975	1,375		4,350	5,325
1801	225 amp main, 32 circuits		5	3.200		2,975	186		3,161	3,550
1850	42 circuits		.52	30.769		3,450	1,800		5,250	6,450
1900	400 amp main, 30 circuits		.54	29.630		4,125	1,725		5,850	7,100
1950	42 circuits		.50	32		4,600	1,850		6,450	7,825
2000	4 wire, 120/208 volts with main circuit breaker									
2050	100 amp main, 24 circuits	1 Elec	.47	17.021	Ea.	1,925	990		2,915	3,575
2100	30 circuits	"	.40	20		2,075	1,175		3,250	4,050
2200	225 amp main, 32 circuits	2 Elec	.72	22.222		3,475	1,300		4,775	5,725
2250	42 circuits		.56	28.571		3,575	1,675		5,250	6,400
2300	400 amp main, 42 circuits		.48	33.333		5,075	1,950		7,025	8,500
2350	600 amp main, 42 circuits		.40	40		7,500	2,325		9,825	11,700
2400	NEHB, with 20 amp, 1 pole circuit breaker									
2450	4 wire, 277/480 volts with main circuit breaker									
2500	100 amp main, 24 circuits	1 Elec	.42	19.048	Ea.	2,650	1,100		3,750	4,550
2550	30 circuits	"	.38	21.053		3,100	1,225		4,325	5,225
2600	225 amp main, 30 circuits	2 Elec	.72	22.222		3,875	1,300		5,175	6,200
2650	42 circuits		.56	28.571		4,775	1,675		6,450	7,725
2700	400 amp main, 42 circuits		.46	34.783		5,750	2,025		7,775	9,350
2750	600 amp main, 42 circuits		.38	42.105		7,850	2,450		10,300	12,300
2900	Note: the following line items don't include branch circuit breakers									
2910	For branch circuit breakers information, see Section 26 24 16.20									
3010	Main lug, no main breaker, 240 volt, 1 pole, 3 wire, 100 amp	1 Elec	2.30	3.478	Ea.	595	202		797	960
3020	225 amp	2 Elec	2.40	6.667		710	390		1,100	1,375
3030	400 amp	"	1.80	8.889		1,050	515		1,565	1,925
3060	3 pole, 3 wire, 100 amp	1 Elec	2.30	3.478		640	202		842	1,000
3070	225 amp	2 Elec	2.40	6.667		755	390		1,145	1,400
3080	400 amp		1.80	8.889		1,125	515		1,640	2,025
3090	600 amp		1.60	10		1,300	580		1,880	2,325
3110	3 pole, 4 wire, 100 amp	1 Elec	2.30	3.478		705	202		907	1,075
3120	225 amp	2 Elec	2.40	6.667		885	390		1,275	1,550
3130	400 amp		1.80	8.889		1,125	515		1,640	2,025

26 24 16 – Panelboards

26 24 16.30 Panelboards Commercial Applications		Crew	Daily Output	Labor-Hours	Unit	Material	2018 Bare Costs Labor	Equipment	Total	Total Incl O&P
3140	600 amp	2 Elec	1.60	10	Ea.	1,300	580		1,880	2,325
3160	480 volt, 3 pole, 3 wire, 100 amp	1 Elec	2.30	3.478		845	202		1,047	1,225
3170	225 amp	2 Elec	2.40	6.667		1,025	390		1,415	1,700
3180	400 amp		1.80	8.889		1,425	515		1,940	2,350
3190	600 amp	↓	1.60	10		1,575	580		2,155	2,625
3210	277/480 volt, 3 pole, 4 wire, 100 amp	1 Elec	2.30	3.478		825	202		1,027	1,200
3220	225 amp	2 Elec	2.40	6.667		995	390		1,385	1,675
3230	400 amp		1.80	8.889		1,400	515		1,915	2,300
3240	600 amp	↓	1.60	10		1,550	580		2,130	2,575
3260	Main circuit breaker, 240 volt, 1 pole, 3 wire, 100 amp	1 Elec	2	4		795	233		1,028	1,225
3270	225 amp	2 Elec	2	8		1,625	465		2,090	2,475
3280	400 amp	"	1.60	10		2,550	580		3,130	3,675
3310	3 pole, 3 wire, 100 amp	1 Elec	2	4		890	233		1,123	1,325
3320	225 amp	2 Elec	2	8		1,850	465		2,315	2,750
3330	400 amp		1.60	10		2,925	580		3,505	4,075
3360	120/208 volt, 3 pole, 4 wire, 100 amp		4	4		910	233		1,143	1,350
3370	225 amp		2	8		1,850	465		2,315	2,750
3380	400 amp	↓	1.60	10		2,925	580		3,505	4,075
3410	480 volt, 3 pole, 3 wire, 100 amp	1 Elec	2	4		1,275	233		1,508	1,750
3420	225 amp	2 Elec	2	8		2,200	465		2,665	3,100
3430	400 amp		1.60	10		3,275	580		3,855	4,475
3460	277/480 volt, 3 pole, 4 wire, 100 amp		4	4		1,250	233		1,483	1,725
3470	225 amp		2	8		2,125	465		2,590	3,050
3480	400 amp	↓	1.60	10		3,300	580		3,880	4,500
3510	Main circuit breaker, HIC, 240 volt, 1 pole, 3 wire, 100 amp	1 Elec	2	4		1,225	233		1,458	1,675
3520	225 amp	2 Elec	2	8		3,175	465		3,640	4,200
3530	400 amp	"	1.60	10		4,250	580		4,830	5,550
3560	3 pole, 3 wire, 100 amp	1 Elec	2	4		1,350	233		1,583	1,825
3570	225 amp	2 Elec	2	8		3,600	465		4,065	4,650
3580	400 amp	"	1.60	10		4,750	580		5,330	6,100
3610	120/208 volt, 3 pole, 4 wire, 100 amp	1 Elec	2	4		1,350	233		1,583	1,825
3620	225 amp	2 Elec	2	8		3,600	465		4,065	4,650
3630	400 amp		1.60	10		4,750	580		5,330	6,100
3660	480 volt, 3 pole, 3 wire, 100 amp	1 Elec	2	4		2,025	233		2,258	2,575
3670	225 amp	2 Elec	2	8		3,975	465		4,440	5,075
3680	400 amp	"	1.60	10		5,075	580		5,655	6,450
3710	277/480 volt, 3 pole, 4 wire, 100 amp	1 Elec	2	4		1,925	233		2,158	2,475
3720	225 amp	2 Elec	2	8		3,800	465		4,265	4,875
3730	400 amp	"	1.60	10		5,025	580		5,605	6,425
3760	Main circuit breaker, shunt trip, 100 amp	1 Elec	1.20	6.667		1,000	390		1,390	1,675
3770	225 amp	2 Elec	1.60	10		2,300	580		2,880	3,400
3780	400 amp	"	1.40	11.429	↓	3,350	665		4,015	4,675

26 24 19 – Motor-Control Centers

26 24 19.20 Motor Control Center Components

		Crew	Daily Output	Labor-Hours	Unit	Material	2018 Bare Costs Labor	Equipment	Total	Total Incl O&P
0010	**MOTOR CONTROL CENTER COMPONENTS** R262419-60									
0100	Starter, size 1, FVNR, NEMA 1, type A, fusible	1 Elec	2.70	2.963	Ea.	1,450	172		1,622	1,825
0120	Circuit breaker		2.70	2.963		1,575	172		1,747	1,975
0140	Type B, fusible		2.70	2.963		1,575	172		1,747	2,000
0160	Circuit breaker		2.70	2.963		1,725	172		1,897	2,150
0180	NEMA 12, type A, fusible		2.60	3.077		1,475	179		1,654	1,875
0200	Circuit breaker		2.60	3.077		1,600	179		1,779	2,025
0220	Type B, fusible	↓	2.60	3.077	↓	1,600	179		1,779	2,050

For customer support on your Electrical Costs with RSMeans data, call 800.448.8182.

225

26 24 19.20 Motor Control Center Components	Crew	Daily Output	Labor-Hours	Unit	Material	2018 Bare Costs Labor	Equipment	Total	Total Incl O&P
0240 Circuit breaker	1 Elec	2.60	3.077	Ea.	1,750	179		1,929	2,200
0300 Starter, size 1, FVR, NEMA 1, type A, fusible		2	4		2,075	233		2,308	2,625
0320 Circuit breaker		2	4		2,075	233		2,308	2,625
0340 Type B, fusible		2	4		2,275	233		2,508	2,850
0360 Circuit breaker		2	4		2,275	233		2,508	2,850
0380 NEMA 12, type A, fusible		1.90	4.211		2,100	245		2,345	2,675
0400 Circuit breaker		1.90	4.211		2,100	245		2,345	2,675
0420 Type B, fusible		1.90	4.211		2,300	245		2,545	2,900
0440 Circuit breaker	▼	1.90	4.211	▼	2,300	245		2,545	2,900
0490 Starter size 1, 2 speed, separate winding									
0500 NEMA 1, type A, fusible	1 Elec	2.60	3.077	Ea.	2,725	179		2,904	3,275
0520 Circuit breaker		2.60	3.077		2,725	179		2,904	3,275
0540 Type B, fusible		2.60	3.077		3,000	179		3,179	3,575
0560 Circuit breaker		2.60	3.077		3,000	179		3,179	3,575
0580 NEMA 12, type A, fusible		2.50	3.200		2,775	186		2,961	3,325
0600 Circuit breaker		2.50	3.200		2,775	186		2,961	3,325
0620 Type B, fusible		2.50	3.200		3,050	186		3,236	3,625
0640 Circuit breaker	▼	2.50	3.200	▼	3,050	186		3,236	3,625
0650 Starter size 1, 2 speed, consequent pole									
0660 NEMA 1, type A, fusible	1 Elec	2.60	3.077	Ea.	2,725	179		2,904	3,275
0680 Circuit breaker		2.60	3.077		2,725	179		2,904	3,275
0700 Type B, fusible		2.60	3.077		3,000	179		3,179	3,575
0720 Circuit breaker		2.60	3.077		3,000	179		3,179	3,575
0740 NEMA 12, type A, fusible		2.50	3.200		2,775	186		2,961	3,325
0760 Circuit breaker		2.50	3.200		2,775	186		2,961	3,325
0780 Type B, fusible		2.50	3.200		3,025	186		3,211	3,600
0800 Circuit breaker	▼	2.50	3.200	▼	3,050	186		3,236	3,625
0810 Starter size 1, 2 speed, space only									
0820 NEMA 1, type A, fusible	1 Elec	16	.500	Ea.	610	29		639	715
0840 Circuit breaker		16	.500		610	29		639	715
0860 Type B, fusible		16	.500		610	29		639	715
0880 Circuit breaker		16	.500		610	29		639	715
0900 NEMA 12, type A, fusible		15	.533		635	31		666	745
0920 Circuit breaker		15	.533		635	31		666	745
0940 Type B, fusible		15	.533		635	31		666	745
0960 Circuit breaker	▼	15	.533		635	31		666	745
1100 Starter size 2, FVNR, NEMA 1, type A, fusible	2 Elec	4	4		1,625	233		1,858	2,150
1120 Circuit breaker		4	4		1,775	233		2,008	2,325
1140 Type B, fusible		4	4		1,800	233		2,033	2,325
1160 Circuit breaker		4	4		1,950	233		2,183	2,500
1180 NEMA 12, type A, fusible		3.80	4.211		1,650	245		1,895	2,200
1200 Circuit breaker		3.80	4.211		1,800	245		2,045	2,375
1220 Type B, fusible		3.80	4.211		1,800	245		2,045	2,375
1240 Circuit breaker		3.80	4.211		1,975	245		2,220	2,550
1300 FVR, NEMA 1, type A, fusible		3.20	5		2,750	291		3,041	3,450
1320 Circuit breaker		3.20	5		2,750	291		3,041	3,450
1340 Type B, fusible		3.20	5		3,025	291		3,316	3,750
1360 Circuit breaker		3.20	5		3,025	291		3,316	3,750
1380 NEMA type 12, type A, fusible		3	5.333		2,800	310		3,110	3,550
1400 Circuit breaker		3	5.333		2,800	310		3,110	3,550
1420 Type B, fusible		3	5.333		3,075	310		3,385	3,875
1440 Circuit breaker	▼	3	5.333	▼	3,075	310		3,385	3,875
1490 Starter size 2, 2 speed, separate winding									

26 24 19.20 Motor Control Center Components	Crew	Daily Output	Labor-Hours	Unit	Material	2018 Bare Costs Labor	2018 Bare Costs Equipment	Total	Total Incl O&P
1500	NEMA 1, type A, fusible	2 Elec	3.80	4.211	Ea.	3,075	245		3,320	3,750
1520	Circuit breaker		3.80	4.211		3,075	245		3,320	3,750
1540	Type B, fusible		3.80	4.211		3,375	245		3,620	4,100
1560	Circuit breaker		3.80	4.211		3,375	245		3,620	4,100
1570	NEMA 12, type A, fusible		3.60	4.444		3,125	259		3,384	3,825
1580	Circuit breaker		3.60	4.444		3,125	259		3,384	3,825
1600	Type B, fusible		3.60	4.444		3,425	259		3,684	4,150
1620	Circuit breaker		3.60	4.444		3,425	259		3,684	4,150
1630	Starter size 2, 2 speed, consequent pole									
1640	NEMA 1, type A, fusible	2 Elec	3.80	4.211	Ea.	3,500	245		3,745	4,225
1660	Circuit breaker		3.80	4.211		3,500	245		3,745	4,225
1680	Type B, fusible		3.80	4.211		3,700	245		3,945	4,450
1700	Circuit breaker		3.80	4.211		3,700	245		3,945	4,450
1720	NEMA 12, type A, fusible		3.80	4.211		3,500	245		3,745	4,225
1740	Circuit breaker		3.60	4.444		3,550	259		3,809	4,275
1760	Type B, fusible		3.60	4.444		3,750	259		4,009	4,500
1780	Circuit breaker		3.60	4.444		3,750	259		4,009	4,500
1830	Starter size 2, autotransformer									
1840	NEMA 1, type A, fusible	2 Elec	3.40	4.706	Ea.	5,450	274		5,724	6,400
1860	Circuit breaker		3.40	4.706		5,575	274		5,849	6,525
1880	Type B, fusible		3.40	4.706		5,950	274		6,224	6,950
1900	Circuit breaker		3.40	4.706		5,950	274		6,224	6,950
1920	NEMA 12, type A, fusible		3.20	5		5,550	291		5,841	6,525
1940	Circuit breaker		3.20	5		5,550	291		5,841	6,525
1960	Type B, fusible		3.20	5		6,050	291		6,341	7,075
1980	Circuit breaker		3.20	5		6,050	291		6,341	7,075
2030	Starter size 2, space only									
2040	NEMA 1, type A, fusible	1 Elec	16	.500	Ea.	610	29		639	715
2060	Circuit breaker		16	.500		610	29		639	715
2080	Type B, fusible		16	.500		610	29		639	715
2100	Circuit breaker		16	.500		610	29		639	715
2120	NEMA 12, type A, fusible		15	.533		635	31		666	745
2140	Circuit breaker		15	.533		635	31		666	745
2160	Type B, fusible		15	.533		635	31		666	745
2180	Circuit breaker		15	.533		635	31		666	745
2300	Starter size 3, FVNR, NEMA 1, type A, fusible	2 Elec	2	8		3,100	465		3,565	4,100
2320	Circuit breaker		2	8		2,750	465		3,215	3,725
2340	Type B, fusible		2	8		3,425	465		3,890	4,450
2360	Circuit breaker		2	8		3,025	465		3,490	4,050
2380	NEMA 12, type A, fusible		1.90	8.421		3,150	490		3,640	4,200
2400	Circuit breaker		1.90	8.421		2,800	490		3,290	3,800
2420	Type B, fusible		1.90	8.421		3,525	490		4,015	4,600
2440	Circuit breaker		1.90	8.421		3,075	490		3,565	4,125
2500	Starter size 3, FVR, NEMA 1, type A, fusible		1.60	10		4,225	580		4,805	5,525
2520	Circuit breaker		1.60	10		4,050	580		4,630	5,325
2540	Type B, fusible		1.60	10		4,600	580		5,180	5,950
2560	Circuit breaker		1.60	10		4,425	580		5,005	5,725
2580	NEMA 12, type A, fusible		1.50	10.667		4,325	620		4,945	5,675
2600	Circuit breaker		1.50	10.667		4,125	620		4,745	5,450
2620	Type B, fusible		1.50	10.667		4,700	620		5,320	6,075
2640	Circuit breaker		1.50	10.667		4,875	620		5,495	6,275
2690	Starter size 3, 2 speed, separate winding									
2700	NEMA 1, type A, fusible	2 Elec	2	8	Ea.	4,850	465		5,315	6,050

For customer support on your Electrical Costs with RSMeans data, call 800.448.8182.

227

26 24 19 – Motor-Control Centers

26 24 19.20 Motor Control Center Components	Crew	Daily Output	Labor-Hours	Unit	Material	2018 Bare Costs Labor	Equipment	Total	Total Incl O&P
2720	Circuit breaker	2 Elec	2	8	Ea.	4,300	465		4,765	5,425
2740	Type B, fusible		2	8		5,325	465		5,790	6,550
2760	Circuit breaker		2	8		4,700	465		5,165	5,875
2780	NEMA 12, type A, fusible		1.90	8.421		4,975	490		5,465	6,175
2800	Circuit breaker		1.90	8.421		4,375	490		4,865	5,550
2820	Type B, fusible		1.90	8.421		5,425	490		5,915	6,700
2840	Circuit breaker	↓	1.90	8.421	↓	4,775	490		5,265	6,000
2850	Starter size 3, 2 speed, consequent pole									
2860	NEMA 1, type A, fusible	2 Elec	2	8	Ea.	5,425	465		5,890	6,675
2880	Circuit breaker		2	8		4,850	465		5,315	6,050
2900	Type B, fusible		2	8		5,875	465		6,340	7,175
2920	Circuit breaker		2	8		5,250	465		5,715	6,475
2940	NEMA 12, type A, fusible		1.90	8.421		5,525	490		6,015	6,800
2960	Circuit breaker		1.90	8.421		4,950	490		5,440	6,175
2980	Type B, fusible		1.90	8.421		5,975	490		6,465	7,300
3000	Circuit breaker		1.90	8.421		5,350	490		5,840	6,600
3100	Starter size 3, autotransformer, NEMA 1, type A, fusible		1.60	10		6,825	580		7,405	8,375
3120	Circuit breaker		1.60	10		6,750	580		7,330	8,300
3140	Type B, fusible		1.60	10		7,375	580		7,955	9,000
3160	Circuit breaker		1.60	10		7,375	580		7,955	9,000
3180	NEMA 12, type A, fusible		1.50	10.667		6,875	620		7,495	8,500
3200	Circuit breaker		1.50	10.667		6,875	620		7,495	8,500
3220	Type B, fusible		1.50	10.667		7,500	620		8,120	9,175
3240	Circuit breaker		1.50	10.667		7,500	620		8,120	9,175
3260	Starter size 3, space only, NEMA 1, type A, fusible	1 Elec	15	.533		1,050	31		1,081	1,200
3280	Circuit breaker		15	.533		815	31		846	940
3300	Type B, fusible		15	.533		1,050	31		1,081	1,200
3320	Circuit breaker		15	.533		815	31		846	940
3340	NEMA 12, type A, fusible		14	.571		1,100	33.50		1,133.50	1,275
3360	Circuit breaker		14	.571		865	33.50		898.50	1,000
3380	Type B, fusible		14	.571		1,100	33.50		1,133.50	1,275
3400	Circuit breaker	↓	14	.571		865	33.50		898.50	1,000
3500	Starter size 4, FVNR, NEMA 1, type A, fusible	2 Elec	1.60	10		4,050	580		4,630	5,325
3520	Circuit breaker		1.60	10		3,675	580		4,255	4,900
3540	Type B, fusible		1.60	10		4,425	580		5,005	5,725
3560	Circuit breaker		1.60	10		4,050	580		4,630	5,325
3580	NEMA 12, type A, fusible		1.50	10.667		4,150	620		4,770	5,475
3600	Circuit breaker		1.50	10.667		3,725	620		4,345	5,025
3620	Type B, fusible		1.50	10.667		4,525	620		5,145	5,900
3640	Circuit breaker		1.50	10.667		4,100	620		4,720	5,425
3700	Starter size 4, FVR, NEMA 1, type A, fusible		1.20	13.333		5,525	775		6,300	7,225
3720	Circuit breaker		1.20	13.333		4,975	775		5,750	6,625
3740	Type B, fusible		1.20	13.333		6,025	775		6,800	7,775
3760	Circuit breaker		1.20	13.333		5,475	775		6,250	7,175
3780	NEMA 12, type A, fusible		1.16	13.793		5,650	805		6,455	7,425
3800	Circuit breaker		1.16	13.793		5,050	805		5,855	6,750
3820	Type B, fusible		1.16	13.793		5,100	805		5,905	6,825
3840	Circuit breaker	↓	1.16	13.793	↓	5,550	805		6,355	7,300
3890	Starter size 4, 2 speed, separate windings									
3900	NEMA 1, type A, fusible	2 Elec	1.60	10	Ea.	6,975	580		7,555	8,550
3920	Circuit breaker		1.60	10		5,225	580		5,805	6,625
3940	Type B, fusible		1.60	10		7,650	580		8,230	9,300
3960	Circuit breaker		1.60	10		5,725	580		6,305	7,175

For customer support on your Electrical Costs with RSMeans data, call 800.448.8182.

26 24 19.20 Motor Control Center Components	Crew	Daily Output	Labor-Hours	Unit	Material	2018 Bare Costs Labor	Equipment	Total	Total Incl O&P
3980	NEMA 12, type A, fusible	2 Elec	1.50	10.667	Ea.	7,100	620		7,720	8,725
4000	Circuit breaker		1.50	10.667		5,325	620		5,945	6,775
4020	Type B, fusible		1.50	10.667		7,775	620		8,395	9,475
4040	Circuit breaker		1.50	10.667		5,825	620		6,445	7,325
4050	Starter size 4, 2 speed, consequent pole									
4060	NEMA 1, type A, fusible	2 Elec	1.60	10	Ea.	8,150	580		8,730	9,825
4080	Circuit breaker		1.60	10		6,050	580		6,630	7,525
4100	Type B, fusible		1.60	10		8,950	580		9,530	10,700
4120	Circuit breaker		1.60	10		6,650	580		7,230	8,175
4140	NEMA 12, type A, fusible		1.50	10.667		8,275	620		8,895	10,000
4160	Circuit breaker		1.50	10.667		6,150	620		6,770	7,675
4180	Type B, fusible		1.50	10.667		9,075	620		9,695	10,900
4200	Circuit breaker		1.50	10.667		6,725	620		7,345	8,325
4300	Starter size 4, autotransformer, NEMA 1, type A, fusible		1.30	12.308		8,025	715		8,740	9,900
4320	Circuit breaker		1.30	12.308		8,075	715		8,790	9,975
4340	Type B, fusible		1.30	12.308		8,825	715		9,540	10,800
4360	Circuit breaker		1.30	12.308		8,825	715		9,540	10,800
4380	NEMA 12, type A, fusible		1.24	12.903		8,150	750		8,900	10,100
4400	Circuit breaker		1.24	12.903		8,200	750		8,950	10,200
4420	Type B, fusible		1.24	12.903		8,975	750		9,725	11,000
4440	Circuit breaker		1.24	12.903		8,950	750		9,700	11,000
4500	Starter size 4, space only, NEMA 1, type A, fusible	1 Elec	14	.571		1,450	33.50		1,483.50	1,650
4520	Circuit breaker		14	.571		1,050	33.50		1,083.50	1,200
4540	Type B, fusible		14	.571		1,450	33.50		1,483.50	1,650
4560	Circuit breaker		14	.571		1,050	33.50		1,083.50	1,200
4580	NEMA 12, type A, fusible		13	.615		1,550	36		1,586	1,750
4600	Circuit breaker		13	.615		1,100	36		1,136	1,275
4620	Type B, fusible		13	.615		1,550	36		1,586	1,750
4640	Circuit breaker		13	.615		1,100	36		1,136	1,275
4800	Starter size 5, FVNR, NEMA 1, type A, fusible	2 Elec	1	16		8,350	930		9,280	10,600
4820	Circuit breaker		1	16		5,925	930		6,855	7,900
4840	Type B, fusible		1	16		9,150	930		10,080	11,500
4860	Circuit breaker		1	16		6,500	930		7,430	8,575
4880	NEMA 12, type A, fusible		.96	16.667		8,450	970		9,420	10,800
4900	Circuit breaker		.96	16.667		6,025	970		6,995	8,075
4920	Type B, fusible		.96	16.667		9,275	970		10,245	11,700
4940	Circuit breaker		.96	16.667		6,625	970		7,595	8,750
5000	Starter size 5, FVR, NEMA 1, type A, fusible		.80	20		12,300	1,175		13,475	15,300
5020	Circuit breaker		.80	20		9,650	1,175		10,825	12,400
5040	Type B, fusible		.80	20		13,500	1,175		14,675	16,600
5060	Circuit breaker		.80	20		10,600	1,175		11,775	13,400
5080	NEMA 12, type A, fusible		.76	21.053		12,500	1,225		13,725	15,600
5100	Circuit breaker		.76	21.053		9,800	1,225		11,025	12,600
5120	Type B, fusible		.76	21.053		13,700	1,225		14,925	16,900
5140	Circuit breaker		.76	21.053		10,700	1,225		11,925	13,600
5190	Starter size 5, 2 speed, separate windings									
5200	NEMA 1, type A, fusible	2 Elec	1	16	Ea.	16,900	930		17,830	20,000
5220	Circuit breaker		1	16		12,600	930		13,530	15,300
5240	Type B, fusible		1	16		18,600	930		19,530	21,900
5260	Circuit breaker		1	16		13,900	930		14,830	16,700
5280	NEMA 12, type A, fusible		.96	16.667		17,200	970		18,170	20,400
5300	Circuit breaker		.96	16.667		12,800	970		13,770	15,600
5320	Type B, fusible		.96	16.667		18,800	970		19,770	22,200

26 24 19.20 Motor Control Center Components		Crew	Daily Output	Labor-Hours	Unit	Material	2018 Bare Costs Labor	Equipment	Total	Total Incl O&P
5340	Circuit breaker	2 Elec	.96	16.667	Ea.	14,000	970		14,970	16,900
5400	Starter size 5, autotransformer, NEMA 1, type A, fusible		.70	22.857		13,900	1,325		15,225	17,300
5420	Circuit breaker		.70	22.857		11,500	1,325		12,825	14,700
5440	Type B, fusible		.70	22.857		15,200	1,325		16,525	18,700
5460	Circuit breaker		.70	22.857		12,600	1,325		13,925	15,900
5480	NEMA 12, type A, fusible		.68	23.529		14,000	1,375		15,375	17,500
5500	Circuit breaker		.68	23.529		11,600	1,375		12,975	14,900
5520	Type B, fusible		.68	23.529		15,400	1,375		16,775	19,000
5540	Circuit breakers		.68	23.529		12,800	1,375		14,175	16,100
5600	Starter size 5, space only, NEMA 1, type A, fusible	1 Elec	12	.667		1,875	39		1,914	2,100
5620	Circuit breaker		12	.667		1,875	39		1,914	2,100
5640	Type B, fusible		12	.667		1,875	39		1,914	2,100
5660	Circuit breaker		12	.667		1,225	39		1,264	1,400
5680	NEMA 12, type A, fusible		11	.727		1,975	42.50		2,017.50	2,250
5700	Circuit breaker		11	.727		1,300	42.50		1,342.50	1,500
5720	Type B, fusible		11	.727		1,975	42.50		2,017.50	2,250
5740	Circuit breaker		11	.727		1,300	42.50		1,342.50	1,500
5800	Fuse, light contactor NEMA 1, type A, 30 amp		2.70	2.963		1,450	172		1,622	1,825
5820	60 amp		2	4		1,625	233		1,858	2,150
5840	100 amp		1	8		3,100	465		3,565	4,100
5860	200 amp		.80	10		7,200	580		7,780	8,800
5880	Type B, 30 amp		2.70	2.963		1,575	172		1,747	1,975
5900	60 amp		2	4		1,750	233		1,983	2,275
5920	100 amp		1	8		3,425	465		3,890	4,450
5940	200 amp	2 Elec	1.60	10		7,900	580		8,480	9,550
5960	NEMA 12, type A, 30 amp	1 Elec	2.60	3.077		1,475	179		1,654	1,875
5980	60 amp		1.90	4.211		1,650	245		1,895	2,200
6000	100 amp		.95	8.421		3,150	490		3,640	4,200
6020	200 amp	2 Elec	1.50	10.667		7,325	620		7,945	8,975
6040	Type B, 30 amp	1 Elec	2.60	3.077		1,600	179		1,779	2,025
6060	60 amp		1.90	4.211		1,775	245		2,020	2,325
6080	100 amp		.95	8.421		3,475	490		3,965	4,550
6100	200 amp	2 Elec	1.50	10.667		8,000	620		8,620	9,725
6200	Circuit breaker, light contactor NEMA 1, type A, 30 amp	1 Elec	2.70	2.963		1,575	172		1,747	1,975
6220	60 amp		2	4		1,775	233		2,008	2,325
6240	100 amp		1	8		2,750	465		3,215	3,725
6260	200 amp	2 Elec	1.60	10		5,975	580		6,555	7,450
6280	Type B, 30 amp	1 Elec	2.70	2.963		1,725	172		1,897	2,150
6300	60 amp		2	4		1,975	233		2,208	2,525
6320	100 amp		1	8		3,000	465		3,465	4,000
6340	200 amp	2 Elec	1.60	10		6,500	580		7,080	8,050
6360	NEMA 12, type A, 30 amp	1 Elec	2.60	3.077		1,600	179		1,779	2,025
6380	60 amp		1.90	4.211		1,800	245		2,045	2,375
6400	100 amp		.95	8.421		2,800	490		3,290	3,800
6420	200 amp	2 Elec	1.50	10.667		6,000	620		6,620	7,525
6440	Type B, 30 amp	1 Elec	2.60	3.077		1,875	179		2,054	2,325
6460	60 amp		1.90	4.211		2,000	245		2,245	2,575
6480	100 amp		.95	8.421		3,050	490		3,540	4,075
6500	200 amp	2 Elec	1.50	10.667		6,600	620		7,220	8,175
6600	Fusible switch, NEMA 1, type A, 30 amp	1 Elec	5.30	1.509		955	88		1,043	1,175
6620	60 amp		5	1.600		1,025	93		1,118	1,275
6640	100 amp		4	2		1,125	116		1,241	1,425
6660	200 amp		3.20	2.500		1,950	146		2,096	2,350

		Crew	Daily Output	Labor-Hours	Unit	Material	2018 Bare Costs Labor	Equipment	Total	Total Incl O&P
26 24 19.20	**Motor Control Center Components**									
6680	400 amp	2 Elec	4.60	3.478	Ea.	4,575	202		4,777	5,325
6700	600 amp		3.20	5		4,950	291		5,241	5,875
6720	800 amp	↓	2.60	6.154		13,600	360		13,960	15,400
6740	NEMA 12, type A, 30 amp	1 Elec	5.20	1.538		980	89.50		1,069.50	1,200
6760	60 amp		4.90	1.633		1,050	95		1,145	1,300
6780	100 amp		3.90	2.051		1,150	119		1,269	1,450
6800	200 amp	↓	3.10	2.581		2,000	150		2,150	2,425
6820	400 amp	2 Elec	4.40	3.636		4,675	212		4,887	5,475
6840	600 amp		3	5.333		5,100	310		5,410	6,075
6860	800 amp	↓	2.40	6.667		13,700	390		14,090	15,700
6900	Circuit breaker, NEMA 1, type A, 30 amp	1 Elec	5.30	1.509		875	88		963	1,100
6920	60 amp		5	1.600		875	93		968	1,100
6940	100 amp		4	2		875	116		991	1,150
6960	225 amp	↓	3.20	2.500		1,575	146		1,721	1,950
6980	400 amp	2 Elec	4.60	3.478		2,975	202		3,177	3,575
7000	600 amp		3.20	5		3,475	291		3,766	4,250
7020	800 amp	↓	2.60	6.154		7,275	360		7,635	8,525
7040	NEMA 12, type A, 30 amp	1 Elec	5.20	1.538		900	89.50		989.50	1,125
7060	60 amp		4.90	1.633		900	95		995	1,125
7080	100 amp		3.90	2.051		900	119		1,019	1,175
7100	225 amp	↓	3.10	2.581		1,600	150		1,750	1,975
7120	400 amp	2 Elec	4.40	3.636		3,025	212		3,237	3,650
7140	600 amp		3	5.333		3,525	310		3,835	4,350
7160	800 amp		2.40	6.667		7,400	390		7,790	8,725
7300	Incoming line, main lug only, 600 amp, alum., NEMA 1		1.60	10		1,100	580		1,680	2,100
7320	NEMA 12		1.50	10.667		1,125	620		1,745	2,175
7340	Copper, NEMA 1		1.60	10		1,175	580		1,755	2,150
7360	800 amp, alum., NEMA 1		1.50	10.667		2,775	620		3,395	3,975
7380	NEMA 12		1.40	11.429		2,825	665		3,490	4,100
7400	Copper, NEMA 1		1.50	10.667		2,900	620		3,520	4,100
7420	1,200 amp, copper, NEMA 1		1.40	11.429		3,000	665		3,665	4,300
7440	Incoming line, fusible switch, 400 amp, alum., NEMA 1		1.20	13.333		3,825	775		4,600	5,350
7460	NEMA 12		1.10	14.545		3,900	845		4,745	5,575
7480	Copper, NEMA 1		1.20	13.333		3,875	775		4,650	5,425
7500	600 amp, alum., NEMA 1		1.10	14.545		4,675	845		5,520	6,425
7520	NEMA 12		1	16		4,750	930		5,680	6,625
7540	Copper, NEMA 1		1.10	14.545		4,725	845		5,570	6,475
7560	Incoming line, circuit breaker, 225 amp, alum., NEMA 1		1.20	13.333		2,000	775		2,775	3,350
7580	NEMA 12		1.10	14.545		2,025	845		2,870	3,500
7600	Copper, NEMA 1		1.20	13.333		2,050	775		2,825	3,400
7620	400 amp, alum., NEMA 1		1.20	13.333		2,975	775		3,750	4,425
7640	NEMA 12		1.10	14.545		3,025	845		3,870	4,600
7660	Copper, NEMA 1		1.20	13.333		3,025	775		3,800	4,475
7680	600 amp, alum., NEMA 1		1.10	14.545		3,475	845		4,320	5,100
7700	NEMA 12		1	16		3,525	930		4,455	5,275
7720	Copper, NEMA 1		1.10	14.545		3,525	845		4,370	5,175
7740	800 amp, copper, NEMA 1	↓	.90	17.778		7,275	1,025		8,300	9,550
7760	Incoming line, for copper bus, add					110			110	121
7780	For 65,000 amp bus bracing, add					163			163	179
7800	For NEMA 3R enclosure, add					5,125			5,125	5,625
7820	For NEMA 12 enclosure, add					141			141	155
7840	For 1/4" x 1" ground bus, add	1 Elec	16	.500		91	29		120	144
7860	For 1/4" x 2" ground bus, add	"	12	.667		91	39		130	158

26 24 19.20 Motor Control Center Components

		Crew	Daily Output	Labor-Hours	Unit	Material	2018 Bare Costs Labor	2018 Bare Costs Equipment	Total	Total Incl O&P
7900	Main rating basic section, alum., NEMA 1, 800 amp	2 Elec	1.40	11.429	Ea.	297	665		962	1,325
7920	1,200 amp	"	1.20	13.333		585	775		1,360	1,800
7940	For copper bus, add					440			440	480
7960	For 65,000 amp bus bracing, add					294			294	325
7980	For NEMA 3R enclosure, add					5,125			5,125	5,625
8000	For NEMA 12, enclosure, add					141			141	155
8020	For 1/4" x 1" ground bus, add	1 Elec	16	.500		91	29		120	144
8040	For 1/4" x 2" ground bus, add		12	.667		91	39		130	158
8060	Unit devices, pilot light, standard		16	.500		78.50	29		107.50	130
8080	Pilot light, push to test		16	.500		110	29		139	165
8100	Pilot light, standard, and push button		12	.667		188	39		227	265
8120	Pilot light, push to test, and push button		12	.667		219	39		258	299
8140	Pilot light, standard, and select switch		12	.667		188	39		227	265
8160	Pilot light, push to test, and select switch		12	.667		219	39		258	299

26 24 19.30 Motor Control Center

		Crew	Daily Output	Labor-Hours	Unit	Material	2018 Bare Costs Labor	2018 Bare Costs Equipment	Total	Total Incl O&P
0010	**MOTOR CONTROL CENTER** Consists of starters & structures R262419-60									
0050	Starters, class 1, type B, comb. MCP, FVNR, with									
0100	control transformer, 10 HP, size 1, 12" high	1 Elec	2.70	2.963	Ea.	1,575	172		1,747	1,975
0200	25 HP, size 2, 18" high	2 Elec	4	4		1,775	233		2,008	2,325
0300	50 HP, size 3, 24" high		2	8		2,750	465		3,215	3,725
0350	75 HP, size 4, 24" high		1.60	10		3,675	580		4,255	4,900
0400	100 HP, size 4, 30" high		1.40	11.429		4,825	665		5,490	6,325
0500	200 HP, size 5, 48" high		1	16		7,225	930		8,155	9,325
0600	400 HP, size 6, 72" high		.80	20		14,800	1,175		15,975	18,100
0800	Structures, 600 amp, 22,000 rms, takes any									
0900	combination of starters up to 72" high	2 Elec	1.60	10	Ea.	1,950	580		2,530	3,000
1000	Back to back, 72" front & 66" back	"	1.20	13.333		2,700	775		3,475	4,125
1100	For copper bus, add per structure					240			240	264
1200	For NEMA 12, add per structure					139			139	152
1300	For 42,000 rms, add per structure					231			231	254
1400	For 100,000 rms, size 1 & 2, add					615			615	680
1500	Size 3, add					990			990	1,100
1600	Size 4, add					800			800	880
1700	For pilot lights, add per starter	1 Elec	16	.500		108	29		137	163
1800	For push button, add per starter		16	.500		108	29		137	163
1900	For auxiliary contacts, add per starter		16	.500		205	29		234	269

26 24 19.40 Motor Starters and Controls

		Crew	Daily Output	Labor-Hours	Unit	Material	2018 Bare Costs Labor	2018 Bare Costs Equipment	Total	Total Incl O&P
0010	**MOTOR STARTERS AND CONTROLS** R262419-65									
0050	Magnetic, FVNR, with enclosure and heaters, 480 volt									
0080	2 HP, size 00	1 Elec	3.50	2.286	Ea.	197	133		330	415
0100	5 HP, size 0		2.30	3.478		295	202		497	630
0200	10 HP, size 1		1.60	5		267	291		558	730
0300	25 HP, size 2 R263413-33	2 Elec	2.20	7.273		500	425		925	1,175
0400	50 HP, size 3		1.80	8.889		820	515		1,335	1,675
0500	100 HP, size 4		1.20	13.333		1,825	775		2,600	3,150
0600	200 HP, size 5		.90	17.778		4,250	1,025		5,275	6,225
0610	400 HP, size 6		.80	20		18,400	1,175		19,575	22,000
0620	NEMA 7, 5 HP, size 0	1 Elec	1.60	5		1,375	291		1,666	1,925
0630	10 HP, size 1	"	1.10	7.273		1,425	425		1,850	2,200
0640	25 HP, size 2	2 Elec	1.80	8.889		2,300	515		2,815	3,325
0650	50 HP, size 3		1.20	13.333		3,475	775		4,250	4,950
0660	100 HP, size 4		.90	17.778		5,600	1,025		6,625	7,700

26 24 19.40 Motor Starters and Controls		Crew	Daily Output	Labor-Hours	Unit	Material	2018 Bare Costs Labor	Equipment	Total	Total Incl O&P
0670	200 HP, size 5	2 Elec	.50	32	Ea.	13,400	1,850		15,250	17,500
0700	Combination, with motor circuit protectors, 5 HP, size 0	1 Elec	1.80	4.444		1,025	259		1,284	1,500
0800	10 HP, size 1	"	1.30	6.154		1,050	360		1,410	1,675
0900	25 HP, size 2	2 Elec	2	8		1,475	465		1,940	2,325
1000	50 HP, size 3		1.32	12.121		2,150	705		2,855	3,400
1200	100 HP, size 4	↓	.80	20		4,625	1,175		5,800	6,825
1220	NEMA 7, 5 HP, size 0	1 Elec	1.30	6.154		3,075	360		3,435	3,900
1230	10 HP, size 1	"	1	8		3,150	465		3,615	4,150
1240	25 HP, size 2	2 Elec	1.32	12.121		4,200	705		4,905	5,650
1250	50 HP, size 3		.80	20		6,925	1,175		8,100	9,350
1260	100 HP, size 4		.60	26.667		10,800	1,550		12,350	14,200
1270	200 HP, size 5	↓	.40	40		23,400	2,325		25,725	29,300
1400	Combination, with fused switch, 5 HP, size 0	1 Elec	1.80	4.444		590	259		849	1,025
1600	10 HP, size 1	"	1.30	6.154		630	360		990	1,225
1800	25 HP, size 2	2 Elec	2	8		1,025	465		1,490	1,825
2000	50 HP, size 3		1.32	12.121		1,725	705		2,430	2,950
2200	100 HP, size 4	↓	.80	20		3,025	1,175		4,200	5,075
2610	NEMA 4, with start-stop push button, size 1	1 Elec	1.30	6.154		1,575	360		1,935	2,275
2620	Size 2	2 Elec	2	8		2,125	465		2,590	3,025
2630	Size 3		1.32	12.121		3,350	705		4,055	4,750
2640	Size 4	↓	.80	20	↓	5,150	1,175		6,325	7,425
2650	NEMA 4, FVNR, including control transformer									
2660	Size 1	2 Elec	2.60	6.154	Ea.	1,625	360		1,985	2,300
2670	Size 2		2	8		2,300	465		2,765	3,225
2680	Size 3		1.32	12.121		3,675	705		4,380	5,075
2690	Size 4	↓	.80	20		5,650	1,175		6,825	7,975
2710	Magnetic, FVR, control circuit transformer, NEMA 1, size 1	1 Elec	1.30	6.154		950	360		1,310	1,575
2720	Size 2	2 Elec	2	8		1,525	465		1,990	2,375
2730	Size 3		1.32	12.121		2,300	705		3,005	3,600
2740	Size 4	↓	.80	20		5,025	1,175		6,200	7,275
2760	NEMA 4, size 1	1 Elec	1.10	7.273		1,375	425		1,800	2,125
2770	Size 2	2 Elec	1.60	10		2,200	580		2,780	3,300
2780	Size 3		1.20	13.333		3,300	775		4,075	4,775
2790	Size 4	↓	.70	22.857		6,800	1,325		8,125	9,475
2820	NEMA 12, size 1	1 Elec	1.10	7.273		1,125	425		1,550	1,850
2830	Size 2	2 Elec	1.60	10		1,775	580		2,355	2,825
2840	Size 3		1.20	13.333		2,775	775		3,550	4,225
2850	Size 4	↓	.70	22.857		5,750	1,325		7,075	8,325
2870	Combination FVR, fused, w/control XFMR & PB, NEMA 1, size 1	1 Elec	1	8		1,650	465		2,115	2,500
2880	Size 2	2 Elec	1.50	10.667		2,325	620		2,945	3,500
2890	Size 3		1.10	14.545		3,475	845		4,320	5,100
2900	Size 4	↓	.70	22.857		7,025	1,325		8,350	9,725
2910	NEMA 4, size 1	1 Elec	.90	8.889		2,425	515		2,940	3,450
2920	Size 2	2 Elec	1.40	11.429		3,550	665		4,215	4,900
2930	Size 3		1	16		5,575	930		6,505	7,550
2940	Size 4	↓	.60	26.667		9,275	1,550		10,825	12,500
2950	NEMA 12, size 1	1 Elec	1	8		1,850	465		2,315	2,750
2960	Size 2	2 Elec	1.40	11.429		2,625	665		3,290	3,875
2970	Size 3		1	16		3,850	930		4,780	5,625
2980	Size 4	↓	.60	26.667		7,625	1,550		9,175	10,700
3010	Manual, single phase, w/pilot, 1 pole, 120 V, NEMA 1	1 Elec	6.40	1.250		81	73		154	198
3020	NEMA 4		4	2		385	116		501	595
3030	2 pole, 120/240 V, NEMA 1		6.40	1.250		112	73		185	232

For customer support on your Electrical Costs with RSMeans data, call 800.448.8182.

233

26 24 19.40 Motor Starters and Controls		Crew	Daily Output	Labor-Hours	Unit	Material	2018 Bare Costs Labor	Equipment	Total	Total Incl O&P
3040	NEMA 4	1 Elec	4	2	Ea.	300	116		416	505
3041	3 phase, 3 pole, 600 V, NEMA 1		5.50	1.455		248	84.50		332.50	400
3042	NEMA 4		3.50	2.286		415	133		548	655
3043	NEMA 12		3.50	2.286		276	133		409	505
3070	Auxiliary contact, normally open					90.50			90.50	99.50
3500	Magnetic FVNR with NEMA 12, enclosure & heaters, 480 volt									
3600	5 HP, size 0	1 Elec	2.20	3.636	Ea.	250	212		462	590
3700	10 HP, size 1	"	1.50	5.333		375	310		685	880
3800	25 HP, size 2	2 Elec	2	8		705	465		1,170	1,475
3900	50 HP, size 3		1.60	10		1,075	580		1,655	2,075
4000	100 HP, size 4		1	16		2,600	930		3,530	4,250
4100	200 HP, size 5		.80	20		6,225	1,175		7,400	8,600
4200	Combination, with motor circuit protectors, 5 HP, size 0	1 Elec	1.70	4.706		830	274		1,104	1,325
4300	10 HP, size 1	"	1.20	6.667		860	390		1,250	1,525
4400	25 HP, size 2	2 Elec	1.80	8.889		1,275	515		1,790	2,200
4500	50 HP, size 3		1.20	13.333		2,100	775		2,875	3,450
4600	100 HP, size 4		.74	21.622		4,750	1,250		6,000	7,100
4700	Combination, with fused switch, 5 HP, size 0	1 Elec	1.70	4.706		720	274		994	1,200
4800	10 HP, size 1	"	1.20	6.667		750	390		1,140	1,400
4900	25 HP, size 2	2 Elec	1.80	8.889		1,150	515		1,665	2,025
5000	50 HP, size 3		1.20	13.333		1,825	775		2,600	3,150
5100	100 HP, size 4		.74	21.622		3,700	1,250		4,950	5,950
5200	Factory installed controls, adders to size 0 thru 5									
5300	Start-stop push button	1 Elec	32	.250	Ea.	52.50	14.55		67.05	79.50
5400	Hand-off-auto-selector switch		32	.250		52.50	14.55		67.05	79.50
5500	Pilot light		32	.250		98.50	14.55		113.05	130
5600	Start-stop-pilot		32	.250		151	14.55		165.55	188
5700	Auxiliary contact, NO or NC		32	.250		72	14.55		86.55	102
5800	NO-NC		32	.250		144	14.55		158.55	181
5810	Magnetic FVR, NEMA 7, w/heaters, size 1		.66	12.121		3,050	705		3,755	4,400
5830	Size 2	2 Elec	1.10	14.545		5,075	845		5,920	6,875
5840	Size 3		.70	22.857		8,175	1,325		9,500	11,000
5850	Size 4		.60	26.667		9,300	1,550		10,850	12,500
5860	Combination w/circuit breakers, heaters, control XFMR PB, size 1	1 Elec	.60	13.333		1,650	775		2,425	2,950
5870	Size 2	2 Elec	.80	20		2,100	1,175		3,275	4,075
5880	Size 3		.50	32		2,825	1,850		4,675	5,875
5890	Size 4		.40	40		5,575	2,325		7,900	9,600
5900	Manual, 240 volt, 0.75 HP motor	1 Elec	4	2		55.50	116		171.50	236
5910	2 HP motor		4	2		151	116		267	340
6000	Magnetic, 240 volt, 1 or 2 pole, 0.75 HP motor		4	2		197	116		313	390
6020	2 HP motor		4	2		217	116		333	415
6040	5 HP motor		3	2.667		310	155		465	570
6060	10 HP motor		2.30	3.478		770	202		972	1,150
6100	3 pole, 0.75 HP motor		3	2.667		197	155		352	450
6120	5 HP motor		2.30	3.478		267	202		469	600
6140	10 HP motor		1.60	5		500	291		791	985
6160	15 HP motor		1.60	5		500	291		791	985
6180	20 HP motor		1.10	7.273		820	425		1,245	1,525
6200	25 HP motor	2 Elec	2.20	7.273		820	425		1,245	1,525
6210	30 HP motor		1.80	8.889		820	515		1,335	1,675
6220	40 HP motor		1.80	8.889		1,825	515		2,340	2,775
6230	50 HP motor		1.80	8.889		1,825	515		2,340	2,775
6240	60 HP motor		1.20	13.333		4,250	775		5,025	5,825

26 24 19.40 Motor Starters and Controls		Crew	Daily Output	Labor-Hours	Unit	Material	2018 Bare Costs Labor	Equipment	Total	Total Incl O&P
6250	75 HP motor	2 Elec	1.20	13.333	Ea.	4,250	775		5,025	5,825
6260	100 HP motor		1.20	13.333		4,250	775		5,025	5,825
6270	125 HP motor		.90	17.778		11,900	1,025		12,925	14,700
6280	150 HP motor		.90	17.778		11,900	1,025		12,925	14,700
6290	200 HP motor		.90	17.778		11,900	1,025		12,925	14,700
6400	Starter & nonfused disconnect, 240 volt, 1-2 pole, 0 .75 HP motor	1 Elec	2	4		241	233		474	615
6410	2 HP motor		2	4		261	233		494	635
6420	5 HP motor		1.80	4.444		355	259		614	775
6430	10 HP motor		1.40	5.714		830	335		1,165	1,400
6440	3 pole, 0.75 HP motor		1.60	5		241	291		532	700
6450	5 HP motor		1.40	5.714		310	335		645	835
6460	10 HP motor		1.10	7.273		560	425		985	1,250
6470	15 HP motor		1	8		560	465		1,025	1,325
6480	20 HP motor	2 Elec	1.50	10.667		955	620		1,575	1,975
6490	25 HP motor		1.50	10.667		955	620		1,575	1,975
6500	30 HP motor		1.30	12.308		955	715		1,670	2,125
6510	40 HP motor		1.24	12.903		2,075	750		2,825	3,400
6520	50 HP motor		1.12	14.286		2,075	830		2,905	3,525
6530	60 HP motor		.90	17.778		4,500	1,025		5,525	6,500
6540	75 HP motor		.76	21.053		4,500	1,225		5,725	6,775
6550	100 HP motor		.70	22.857		4,500	1,325		5,825	6,950
6560	125 HP motor		.60	26.667		12,500	1,550		14,050	16,100
6570	150 HP motor		.52	30.769		12,500	1,800		14,300	16,500
6580	200 HP motor		.50	32		12,500	1,850		14,350	16,600
6600	Starter & fused disconnect, 240 volt, 1-2 pole, 0.75 HP motor	1 Elec	2	4		252	233		485	625
6610	2 HP motor		2	4		273	233		506	650
6620	5 HP motor		1.80	4.444		365	259		624	790
6630	10 HP motor		1.40	5.714		865	335		1,200	1,450
6640	3 pole, 0.75 HP motor		1.60	5		252	291		543	710
6650	5 HP motor		1.40	5.714		320	335		655	850
6660	10 HP motor		1.10	7.273		595	425		1,020	1,300
6690	15 HP motor		1	8		595	465		1,060	1,350
6700	20 HP motor	2 Elec	1.60	10		975	580		1,555	1,950
6710	25 HP motor		1.60	10		975	580		1,555	1,950
6720	30 HP motor		1.40	11.429		975	665		1,640	2,075
6730	40 HP motor		1.20	13.333		2,150	775		2,925	3,525
6740	50 HP motor		1.20	13.333		2,150	775		2,925	3,525
6750	60 HP motor		.90	17.778		4,600	1,025		5,625	6,600
6760	75 HP motor		.90	17.778		4,600	1,025		5,625	6,600
6770	100 HP motor		.70	22.857		4,600	1,325		5,925	7,050
6780	125 HP motor		.54	29.630		12,800	1,725		14,525	16,700
6790	Combination starter & nonfusible disconnect									
6800	240 volt, 1-2 pole, 0.75 HP motor	1 Elec	2	4	Ea.	675	233		908	1,100
6810	2 HP motor		2	4		515	233		748	920
6820	5 HP motor		1.50	5.333		660	310		970	1,200
6830	10 HP motor		1.20	6.667		1,025	390		1,415	1,700
6840	3 pole, 0.75 HP motor		1.80	4.444		625	259		884	1,075
6850	5 HP motor		1.30	6.154		660	360		1,020	1,250
6860	10 HP motor		1	8		1,025	465		1,490	1,825
6870	15 HP motor		1	8		1,025	465		1,490	1,825
6880	20 HP motor	2 Elec	1.32	12.121		1,675	705		2,380	2,900
6890	25 HP motor		1.32	12.121		1,675	705		2,380	2,900
6900	30 HP motor		1.32	12.121		1,675	705		2,380	2,900

For customer support on your Electrical Costs with RSMeans data, call 800.448.8182.

235

26 24 19.40 Motor Starters and Controls		Crew	Daily Output	Labor-Hours	Unit	Material	2018 Bare Costs Labor	Equipment	Total	Total Incl O&P
6910	40 HP motor	2 Elec	.80	20	Ea.	3,225	1,175		4,400	5,300
6920	50 HP motor		.80	20		3,225	1,175		4,400	5,300
6930	60 HP motor		.70	22.857		7,200	1,325		8,525	9,925
6940	75 HP motor		.70	22.857		7,200	1,325		8,525	9,925
6950	100 HP motor		.70	22.857		7,200	1,325		8,525	9,925
6960	125 HP motor		.60	26.667		19,000	1,550		20,550	23,200
6970	150 HP motor		.60	26.667		19,000	1,550		20,550	23,200
6980	200 HP motor		.60	26.667		19,000	1,550		20,550	23,200
6990	Combination starter and fused disconnect									
7000	240 volt, 1-2 pole, 0.75 HP motor	1 Elec	2	4	Ea.	645	233		878	1,050
7010	2 HP motor		2	4		645	233		878	1,050
7020	5 HP motor		1.50	5.333		680	310		990	1,200
7030	10 HP motor		1.20	6.667		1,050	390		1,440	1,725
7040	3 pole, 0.75 HP motor		1.80	4.444		645	259		904	1,100
7050	5 HP motor		1.30	6.154		680	360		1,040	1,275
7060	10 HP motor		1	8		1,050	465		1,515	1,850
7070	15 HP motor		1	8		1,050	465		1,515	1,850
7080	20 HP motor	2 Elec	1.32	12.121		1,750	705		2,455	2,975
7090	25 HP motor		1.32	12.121		1,750	705		2,455	2,975
7100	30 HP motor		1.32	12.121		1,750	705		2,455	2,975
7110	40 HP motor		.80	20		3,350	1,175		4,525	5,425
7120	50 HP motor		.80	20		3,350	1,175		4,525	5,425
7130	60 HP motor		.80	20		7,450	1,175		8,625	9,925
7140	75 HP motor		.70	22.857		7,450	1,325		8,775	10,200
7150	100 HP motor		.70	22.857		7,450	1,325		8,775	10,200
7160	125 HP motor		.70	22.857		19,700	1,325		21,025	23,600
7170	150 HP motor		.60	26.667		19,700	1,550		21,250	23,900
7180	200 HP motor		.60	26.667		19,700	1,550		21,250	23,900
7190	Combination starter & circuit breaker disconnect									
7200	240 volt, 1-2 pole, 0.75 HP motor	1 Elec	2	4	Ea.	650	233		883	1,075
7210	2 HP motor		2	4		650	233		883	1,075
7220	5 HP motor		1.50	5.333		685	310		995	1,225
7230	10 HP motor		1.20	6.667		1,050	390		1,440	1,725
7240	3 pole, 0.75 HP motor		1.80	4.444		670	259		929	1,125
7250	5 HP motor		1.30	6.154		705	360		1,065	1,300
7260	10 HP motor		1	8		1,075	465		1,540	1,875
7270	15 HP motor		1	8		1,075	465		1,540	1,875
7280	20 HP motor	2 Elec	1.32	12.121		1,800	705		2,505	3,050
7290	25 HP motor		1.32	12.121		1,800	705		2,505	3,050
7300	30 HP motor		1.32	12.121		1,800	705		2,505	3,050
7310	40 HP motor		.80	20		3,950	1,175		5,125	6,075
7320	50 HP motor		.80	20		3,950	1,175		5,125	6,075
7330	60 HP motor		.80	20		9,100	1,175		10,275	11,800
7340	75 HP motor		.70	22.857		9,100	1,325		10,425	12,000
7350	100 HP motor		.70	22.857		9,100	1,325		10,425	12,000
7360	125 HP motor		.70	22.857		19,700	1,325		21,025	23,700
7370	150 HP motor		.60	26.667		19,700	1,550		21,250	24,000
7380	200 HP motor		.60	26.667		19,700	1,550		21,250	24,000
7400	Magnetic FVNR with enclosure & heaters, 2 pole,									
7410	230 volt, 1 HP, size 00	1 Elec	4	2	Ea.	179	116		295	370
7420	2 HP, size 0		4	2		199	116		315	395
7430	3 HP, size 1		3	2.667		229	155		384	485
7440	5 HP, size 1P		3	2.667		293	155		448	555

For customer support on your Electrical Costs with RSMeans data, call 800.448.8182.

26 24 19.40 Motor Starters and Controls		Crew	Daily Output	Labor-Hours	Unit	Material	2018 Bare Costs Labor	Equipment	Total	Total Incl O&P
7450	115 volt, 1/3 HP, size 00	1 Elec	4	2	Ea.	179	116		295	370
7460	1 HP, size 0		4	2		199	116		315	395
7470	2 HP, size 1		3	2.667		229	155		384	485
7480	3 HP, size 1P		3	2.667		284	155		439	545
7500	3 pole, 480 volt, 600 HP, size 7	2 Elec	.70	22.857		15,800	1,325		17,125	19,400
7590	Magnetic FVNR with heater, NEMA 1									
7600	600 volt, 3 pole, 5 HP motor	1 Elec	2.30	3.478	Ea.	238	202		440	565
7610	10 HP motor	"	1.60	5		267	291		558	730
7620	25 HP motor	2 Elec	2.20	7.273		500	425		925	1,175
7630	30 HP motor		1.80	8.889		820	515		1,335	1,675
7640	40 HP motor		1.80	8.889		820	515		1,335	1,675
7650	50 HP motor		1.80	8.889		820	515		1,335	1,675
7660	60 HP motor		1.20	13.333		1,825	775		2,600	3,150
7670	75 HP motor		1.20	13.333		2,100	775		2,875	3,475
7680	100 HP motor		1.20	13.333		2,400	775		3,175	3,775
7690	125 HP motor		.90	17.778		4,150	1,025		5,175	6,125
7700	150 HP motor		.90	17.778		4,350	1,025		5,375	6,325
7710	200 HP motor		.90	17.778		4,550	1,025		5,575	6,550
7750	Starter & nonfused disconnect, 600 volt, 3 pole, 5 HP motor	1 Elec	1.40	5.714		320	335		655	845
7760	10 HP motor	"	1.10	7.273		350	425		775	1,025
7770	25 HP motor	2 Elec	1.50	10.667		585	620		1,205	1,575
7780	30 HP motor		1.30	12.308		955	715		1,670	2,125
7790	40 HP motor		1.30	12.308		955	715		1,670	2,125
7800	50 HP motor		1.30	12.308		955	715		1,670	2,125
7810	60 HP motor		.92	17.391		2,025	1,000		3,025	3,750
7820	75 HP motor		.92	17.391		2,325	1,000		3,325	4,075
7830	100 HP motor		.84	19.048		2,600	1,100		3,700	4,525
7840	125 HP motor		.70	22.857		4,475	1,325		5,800	6,925
7850	150 HP motor		.70	22.857		4,675	1,325		6,000	7,150
7860	200 HP motor		.60	26.667		4,875	1,550		6,425	7,675
7870	Starter & fused disconnect, 600 volt, 3 pole, 5 HP motor	1 Elec	1.40	5.714		385	335		720	915
7880	10 HP motor	"	1.10	7.273		410	425		835	1,100
7890	25 HP motor	2 Elec	1.50	10.667		645	620		1,265	1,650
7900	30 HP motor		1.30	12.308		995	715		1,710	2,175
7910	40 HP motor		1.30	12.308		995	715		1,710	2,175
7920	50 HP motor		1.30	12.308		995	715		1,710	2,175
7930	60 HP motor		.92	17.391		2,125	1,000		3,125	3,875
7940	75 HP motor		.92	17.391		2,425	1,000		3,425	4,200
7950	100 HP motor		.84	19.048		2,700	1,100		3,800	4,625
7960	125 HP motor		.70	22.857		4,600	1,325		5,925	7,075
7970	150 HP motor		.70	22.857		4,800	1,325		6,125	7,275
7980	200 HP motor		.60	26.667		5,000	1,550		6,550	7,825
7990	Combination starter and nonfusible disconnect									
8000	600 volt, 3 pole, 5 HP motor	1 Elec	1.80	4.444	Ea.	685	259		944	1,150
8010	10 HP motor	"	1.30	6.154		810	360		1,170	1,425
8020	25 HP motor	2 Elec	2	8		1,125	465		1,590	1,925
8030	30 HP motor		1.32	12.121		1,750	705		2,455	2,975
8040	40 HP motor		1.32	12.121		1,750	705		2,455	2,975
8050	50 HP motor		1.32	12.121		2,075	705		2,780	3,350
8060	60 HP motor		.80	20		3,350	1,175		4,525	5,425
8070	75 HP motor		.80	20		3,350	1,175		4,525	5,425
8080	100 HP motor		.80	20		4,125	1,175		5,300	6,275
8090	125 HP motor		.70	22.857		7,225	1,325		8,550	9,950

For customer support on your Electrical Costs with RSMeans data, call 800.448.8182.

237

26 24 19 – Motor-Control Centers

26 24 19.40 Motor Starters and Controls	Crew	Daily Output	Labor-Hours	Unit	Material	2018 Bare Costs Labor	Equipment	Total	Total Incl O&P
8100	150 HP motor	2 Elec	.70	22.857	Ea.	7,550	1,325		8,875	10,300
8110	200 HP motor	↓	.70	22.857	↓	7,925	1,325		9,250	10,700
8140	Combination starter and fused disconnect									
8150	600 volt, 3 pole, 5 HP motor	1 Elec	1.80	4.444	Ea.	645	259		904	1,100
8160	10 HP motor	"	1.30	6.154		705	360		1,065	1,300
8170	25 HP motor	2 Elec	2	8		1,150	465		1,615	1,950
8180	30 HP motor		1.32	12.121		1,775	705		2,480	3,025
8190	40 HP motor		1.32	12.121		1,925	705		2,630	3,175
8200	50 HP motor		1.32	12.121		1,925	705		2,630	3,175
8210	60 HP motor		.80	20		3,375	1,175		4,550	5,450
8220	75 HP motor		.80	20		3,375	1,175		4,550	5,450
8230	100 HP motor		.80	20		3,375	1,175		4,550	5,450
8240	125 HP motor		.70	22.857		7,450	1,325		8,775	10,200
8250	150 HP motor		.70	22.857		7,450	1,325		8,775	10,200
8260	200 HP motor	↓	.70	22.857	↓	7,450	1,325		8,775	10,200
8290	Combination starter & circuit breaker disconnect									
8300	600 volt, 3 pole, 5 HP motor	1 Elec	1.80	4.444	Ea.	940	259		1,199	1,400
8310	10 HP motor	"	1.30	6.154		915	360		1,275	1,525
8320	25 HP motor	2 Elec	2	8		1,375	465		1,840	2,200
8330	30 HP motor		1.32	12.121		1,800	705		2,505	3,050
8340	40 HP motor		1.32	12.121		1,800	705		2,505	3,050
8350	50 HP motor		1.32	12.121		1,825	705		2,530	3,050
8360	60 HP motor		.80	20		3,950	1,175		5,125	6,075
8370	75 HP motor		.80	20		3,950	1,175		5,125	6,075
8380	100 HP motor		.80	20		3,950	1,175		5,125	6,100
8390	125 HP motor		.70	22.857		9,100	1,325		10,425	12,000
8400	150 HP motor		.70	22.857		8,400	1,325		9,725	11,300
8410	200 HP motor	↓	.70	22.857	↓	9,150	1,325		10,475	12,100
8430	Starter & circuit breaker disconnect									
8440	600 volt, 3 pole, 5 HP motor	1 Elec	1.40	5.714	Ea.	755	335		1,090	1,325
8450	10 HP motor	"	1.10	7.273		785	425		1,210	1,500
8460	25 HP motor	2 Elec	1.50	10.667		1,025	620		1,645	2,050
8470	30 HP motor		1.30	12.308		1,450	715		2,165	2,675
8480	40 HP motor		1.30	12.308		1,450	715		2,165	2,675
8490	50 HP motor		1.30	12.308		1,450	715		2,165	2,675
8500	60 HP motor		.92	17.391		3,500	1,000		4,500	5,375
8510	75 HP motor		.92	17.391		3,800	1,000		4,800	5,700
8520	100 HP motor		.84	19.048		4,075	1,100		5,175	6,150
8530	125 HP motor		.70	22.857		5,850	1,325		7,175	8,425
8540	150 HP motor		.70	22.857		6,025	1,325		7,350	8,650
8550	200 HP motor	↓	.60	26.667		7,425	1,550		8,975	10,500
8900	240 volt, 1-2 pole, 0.75 HP motor	1 Elec	2	4		715	233		948	1,125
8910	2 HP motor		2	4		735	233		968	1,150
8920	5 HP motor		1.80	4.444		830	259		1,089	1,300
8930	10 HP motor		1.40	5.714		1,400	335		1,735	2,050
8950	3 pole, 0.75 HP motor		1.60	5		715	291		1,006	1,225
8970	5 HP motor		1.40	5.714		785	335		1,120	1,350
8980	10 HP motor		1.10	7.273		1,150	425		1,575	1,875
8990	15 HP motor	↓	1	8		1,150	465		1,615	1,950
9100	20 HP motor	2 Elec	1.50	10.667		1,550	620		2,170	2,625
9110	25 HP motor		1.50	10.667		1,550	620		2,170	2,625
9120	30 HP motor		1.30	12.308		1,550	715		2,265	2,775
9130	40 HP motor	↓	1.24	12.903		3,500	750		4,250	4,975

For customer support on your Electrical Costs with RSMeans data, call 800.448.8182.

26 24 Switchboards and Panelboards

26 24 19 – Motor-Control Centers

26 24 19.40 Motor Starters and Controls		Crew	Daily Output	Labor-Hours	Unit	Material	2018 Bare Costs Labor	Equipment	Total	Total Incl O&P
9140	50 HP motor	2 Elec	1.12	14.286	Ea.	3,500	830		4,330	5,100
9150	60 HP motor		.90	17.778		7,125	1,025		8,150	9,375
9160	75 HP motor		.76	21.053		7,125	1,225		8,350	9,650
9170	100 HP motor		.70	22.857		7,125	1,325		8,450	9,825
9180	125 HP motor		.60	26.667		16,100	1,550		17,650	20,000
9190	150 HP motor		.52	30.769		16,100	1,800		17,900	20,400
9200	200 HP motor		.50	32		16,100	1,850		17,950	20,500

26 25 Low-Voltage Enclosed Bus Assemblies

26 25 13 – Low-Voltage Busways

26 25 13.10 Aluminum Bus Duct

		Crew	Daily Output	Labor-Hours	Unit	Material	2018 Bare Costs Labor	Equipment	Total	Total Incl O&P
0010	**ALUMINUM BUS DUCT** 10 ft. long									
0050	Indoor 3 pole 4 wire, plug-in, straight section, 225 amp	2 Elec	44	.364	L.F.	171	21		192	220
0100	400 amp		36	.444		215	26		241	275
0150	600 amp		32	.500		297	29		326	370
0200	800 amp		26	.615		345	36		381	435
0250	1,000 amp		24	.667		430	39		469	535
0270	1,200 amp		23	.696		550	40.50		590.50	665
0300	1,350 amp		22	.727		283	42.50		325.50	375
0310	1,600 amp		18	.889		300	51.50		351.50	410
0320	2,000 amp		16	1		395	58		453	520
0330	2,500 amp		14	1.143		450	66.50		516.50	595
0340	3,000 amp		12	1.333		545	77.50		622.50	715
0350	Feeder, 600 amp		34	.471		110	27.50		137.50	162
0400	800 amp		28	.571		430	33.50		463.50	525
0450	1,000 amp		26	.615		520	36		556	625
0455	1,200 amp		25	.640		650	37.50		687.50	770
0500	1,350 amp		24	.667		230	39		269	310
0550	1,600 amp		20	.800		900	46.50		946.50	1,050
0600	2,000 amp		18	.889		1,100	51.50		1,151.50	1,275
0620	2,500 amp		14	1.143		1,350	66.50		1,416.50	1,575
0630	3,000 amp		12	1.333		1,575	77.50		1,652.50	1,850
0640	4,000 amp		10	1.600		665	93		758	870
0650	Elbow, 225 amp		4.40	3.636	Ea.	845	212		1,057	1,250
0700	400 amp		3.80	4.211		855	245		1,100	1,300
0750	600 amp		3.40	4.706		860	274		1,134	1,350
0800	800 amp		3	5.333		890	310		1,200	1,450
0850	1,000 amp		2.80	5.714		1,150	335		1,485	1,750
0870	1,200 amp		2.70	5.926		1,425	345		1,770	2,100
0900	1,350 amp		2.60	6.154		1,175	360		1,535	1,800
0950	1,600 amp		2.40	6.667		1,400	390		1,790	2,100
1000	2,000 amp		2	8		1,525	465		1,990	2,375
1020	2,500 amp		1.80	8.889		1,800	515		2,315	2,750
1030	3,000 amp		1.60	10		2,075	580		2,655	3,175
1040	4,000 amp		1.40	11.429		3,350	665		4,015	4,700
1100	Cable tap box end, 225 amp		3.60	4.444		1,650	259		1,909	2,200
1150	400 amp		3.20	5		1,700	291		1,991	2,300
1200	600 amp		2.60	6.154		1,500	360		1,860	2,175
1250	800 amp		2.20	7.273		1,775	425		2,200	2,575
1300	1,000 amp		2	8		3,400	465		3,865	4,450
1320	1,200 amp		2	8		1,775	465		2,240	2,650

R262513-10

For customer support on your Electrical Costs with RSMeans data, call 800.448.8182.

239

26 25 13.10 Aluminum Bus Duct		Crew	Daily Output	Labor-Hours	Unit	Material	2018 Bare Costs Labor	Equipment	Total	Total Incl O&P
1350	1,350 amp	2 Elec	1.60	10	Ea.	1,425	580		2,005	2,425
1400	1,600 amp		1.40	11.429		1,600	665		2,265	2,750
1450	2,000 amp		1.20	13.333		1,825	775		2,600	3,150
1460	2,500 amp		1	16		2,225	930		3,155	3,825
1470	3,000 amp		.80	20		2,450	1,175		3,625	4,450
1480	4,000 amp		.60	26.667		3,250	1,550		4,800	5,900
1500	Switchboard stub, 225 amp		5.80	2.759		1,625	161		1,786	2,025
1550	400 amp		5.40	2.963		1,650	172		1,822	2,050
1600	600 amp		4.60	3.478		1,675	202		1,877	2,150
1650	800 amp		4	4		1,700	233		1,933	2,225
1700	1,000 amp		3.20	5		1,725	291		2,016	2,300
1720	1,200 amp		3.10	5.161		1,750	300		2,050	2,375
1750	1,350 amp		3	5.333		1,600	310		1,910	2,250
1800	1,600 amp		2.60	6.154		1,800	360		2,160	2,500
1850	2,000 amp		2.40	6.667		1,800	390		2,190	2,550
1860	2,500 amp		2.20	7.273		1,975	425		2,400	2,800
1870	3,000 amp		2	8		2,050	465		2,515	2,950
1880	4,000 amp		1.80	8.889		2,225	515		2,740	3,225
1890	Tee fittings, 225 amp		3.20	5		875	291		1,166	1,400
1900	400 amp		2.80	5.714		875	335		1,210	1,450
1950	600 amp		2.60	6.154		875	360		1,235	1,500
2000	800 amp		2.40	6.667		930	390		1,320	1,600
2050	1,000 amp		2.20	7.273		985	425		1,410	1,700
2070	1,200 amp		2.10	7.619		1,150	445		1,595	1,950
2100	1,350 amp		2	8		1,600	465		2,065	2,450
2150	1,600 amp		1.60	10		1,925	580		2,505	2,975
2200	2,000 amp		1.20	13.333		2,125	775		2,900	3,475
2220	2,500 amp		1	16		2,525	930		3,455	4,200
2230	3,000 amp		.80	20		2,900	1,175		4,075	4,925
2240	4,000 amp		.60	26.667		4,800	1,550		6,350	7,625
2300	Wall flange, 600 amp		20	.800		263	46.50		309.50	360
2310	800 amp		16	1		263	58		321	375
2320	1,000 amp		13	1.231		263	71.50		334.50	395
2325	1,200 amp		12	1.333		263	77.50		340.50	405
2330	1,350 amp		10.80	1.481		263	86		349	420
2340	1,600 amp		9	1.778		263	103		366	445
2350	2,000 amp		8	2		263	116		379	465
2360	2,500 amp		6.60	2.424		263	141		404	500
2370	3,000 amp		5.40	2.963		263	172		435	545
2380	4,000 amp		4	4		380	233		613	770
2390	5,000 amp		3	5.333		380	310		690	885
2400	Vapor barrier		8	2		480	116		596	705
2420	Roof flange kit		4	4		760	233		993	1,175
2600	Expansion fitting, 225 amp		10	1.600		1,350	93		1,443	1,650
2610	400 amp		8	2		1,375	116		1,491	1,675
2620	600 amp		6	2.667		1,375	155		1,530	1,725
2630	800 amp		4.60	3.478		1,550	202		1,752	2,000
2640	1,000 amp		4	4		1,750	233		1,983	2,275
2650	1,350 amp		3.60	4.444		2,025	259		2,284	2,600
2660	1,600 amp		3.20	5		2,425	291		2,716	3,100
2670	2,000 amp		2.80	5.714		2,700	335		3,035	3,475
2680	2,500 amp		2.40	6.667		3,275	390		3,665	4,175
2690	3,000 amp		2	8		3,775	465		4,240	4,850

26 25 13.10 Aluminum Bus Duct		Crew	Daily Output	Labor-Hours	Unit	Material	2018 Bare Costs Labor	Equipment	Total	Total Incl O&P
2700	4,000 amp	2 Elec	1.60	10	Ea.	4,950	580		5,530	6,325
2800	Reducer nonfused, 400 amp		8	2		855	116		971	1,125
2810	600 amp		6	2.667		855	155		1,010	1,175
2820	800 amp		4.60	3.478		1,025	202		1,227	1,425
2830	1,000 amp		4	4		1,200	233		1,433	1,675
2840	1,350 amp		3.60	4.444		1,575	259		1,834	2,100
2850	1,600 amp		3.20	5		2,125	291		2,416	2,750
2860	2,000 amp		2.80	5.714		2,450	335		2,785	3,175
2870	2,500 amp		2.40	6.667		3,075	390		3,465	3,950
2880	3,000 amp		2	8		3,550	465		4,015	4,600
2890	4,000 amp		1.60	10		4,750	580		5,330	6,100
2950	Reducer fuse included, 225 amp		4.40	3.636		2,500	212		2,712	3,075
2960	400 amp		4.20	3.810		2,550	222		2,772	3,125
2970	600 amp		3.60	4.444		3,000	259		3,259	3,700
2980	800 amp		3.20	5		4,800	291		5,091	5,700
2990	1,000 amp		3	5.333		5,475	310		5,785	6,500
3000	1,200 amp		2.80	5.714		5,475	335		5,810	6,525
3010	1,600 amp		2.20	7.273		12,500	425		12,925	14,300
3020	2,000 amp		1.80	8.889		13,900	515		14,415	16,100
3100	Reducer circuit breaker, 225 amp		4.40	3.636		2,475	212		2,687	3,050
3110	400 amp		4.20	3.810		3,000	222		3,222	3,625
3120	600 amp		3.60	4.444		4,275	259		4,534	5,075
3130	800 amp		3.20	5		5,000	291		5,291	5,925
3140	1,000 amp		3	5.333		5,675	310		5,985	6,725
3150	1,200 amp		2.80	5.714		6,825	335		7,160	8,025
3160	1,600 amp		2.20	7.273		10,100	425		10,525	11,700
3170	2,000 amp		1.80	8.889		11,100	515		11,615	13,000
3250	Reducer circuit breaker, 75,000 AIC, 225 amp		4.40	3.636		3,875	212		4,087	4,575
3260	400 amp		4.20	3.810		3,875	222		4,097	4,575
3270	600 amp		3.60	4.444		5,175	259		5,434	6,050
3280	800 amp		3.20	5		5,675	291		5,966	6,675
3290	1,000 amp		3	5.333		9,125	310		9,435	10,500
3300	1,200 amp		2.80	5.714		9,125	335		9,460	10,500
3310	1,600 amp		2.20	7.273		10,100	425		10,525	11,700
3320	2,000 amp		1.80	8.889		11,100	515		11,615	13,000
3400	Reducer circuit breaker CLF 225 amp		4.40	3.636		3,975	212		4,187	4,700
3410	400 amp		4.20	3.810		4,725	222		4,947	5,525
3420	600 amp		3.60	4.444		7,075	259		7,334	8,150
3430	800 amp		3.20	5		7,375	291		7,666	8,525
3440	1,000 amp		3	5.333		7,675	310		7,985	8,925
3450	1,200 amp		2.80	5.714		10,000	335		10,335	11,500
3460	1,600 amp		2.20	7.273		10,100	425		10,525	11,700
3470	2,000 amp		1.80	8.889		11,100	515		11,615	13,000
3550	Ground bus added to bus duct, 225 amp		320	.050	L.F.	32.50	2.91		35.41	40.50
3560	400 amp		320	.050		32.50	2.91		35.41	40.50
3570	600 amp		280	.057		32.50	3.33		35.83	41
3580	800 amp		240	.067		32.50	3.88		36.38	42
3590	1,000 amp		200	.080		32.50	4.66		37.16	43
3600	1,350 amp		180	.089		32.50	5.15		37.65	44
3610	1,600 amp		160	.100		32.50	5.80		38.30	44.50
3620	2,000 amp		160	.100		32.50	5.80		38.30	44.50
3630	2,500 amp		140	.114		32.50	6.65		39.15	46
3640	3,000 amp		120	.133		32.50	7.75		40.25	47.50

26 25 13 – Low-Voltage Busways

26 25 13.10 Aluminum Bus Duct	Crew	Daily Output	Labor-Hours	Unit	Material	2018 Bare Costs Labor	2018 Bare Costs Equipment	Total	Total Incl O&P
3650	4,000 amp	2 Elec	100	.160	L.F.	32.50	9.30		41.80	50
3810	High short circuit, 400 amp		36	.444		205	26		231	265
3820	600 amp		32	.500		355	29		384	435
3830	800 amp		26	.615		415	36		451	510
3840	1,000 amp		24	.667		520	39		559	630
3850	1,350 amp		22	.727		184	42.50		226.50	266
3860	1,600 amp		18	.889		212	51.50		263.50	310
3870	2,000 amp		16	1		248	58		306	360
3880	2,500 amp		14	1.143		415	66.50		481.50	555
3890	3,000 amp		12	1.333		470	77.50		547.50	630
3920	Cross, 225 amp		5.60	2.857	Ea.	1,300	166		1,466	1,700
3930	400 amp		4.60	3.478		1,300	202		1,502	1,750
3940	600 amp		4	4		1,300	233		1,533	1,800
3950	800 amp		3.40	4.706		1,375	274		1,649	1,925
3960	1,000 amp		3	5.333		1,450	310		1,760	2,075
3970	1,350 amp		2.80	5.714		2,350	335		2,685	3,075
3980	1,600 amp		2.20	7.273		2,775	425		3,200	3,675
3990	2,000 amp		1.80	8.889		3,050	515		3,565	4,125
4000	2,500 amp		1.60	10		3,625	580		4,205	4,850
4010	3,000 amp		1.20	13.333		4,150	775		4,925	5,725
4020	4,000 amp		1	16		6,350	930		7,280	8,400
4040	Cable tap box center, 225 amp		3.60	4.444		3,025	259		3,284	3,700
4050	400 amp		3.20	5		3,475	291		3,766	4,250
4060	600 amp		2.60	6.154		4,625	360		4,985	5,625
4070	800 amp		2.20	7.273		5,125	425		5,550	6,250
4080	1,000 amp		2	8		6,500	465		6,965	7,850
4090	1,350 amp		1.60	10		1,425	580		2,005	2,425
4100	1,600 amp		1.40	11.429		1,600	665		2,265	2,750
4110	2,000 amp		1.20	13.333		1,825	775		2,600	3,150
4120	2,500 amp		1	16		2,225	930		3,155	3,825
4130	3,000 amp		.80	20		2,450	1,175		3,625	4,450
4140	4,000 amp		.60	26.667		3,250	1,550		4,800	5,900
4500	Weatherproof 3 pole 4 wire, feeder, 600 amp		30	.533	L.F.	132	31		163	192
4520	800 amp		24	.667		155	39		194	228
4540	1,000 amp		22	.727		177	42.50		219.50	258
4550	1,200 amp		21	.762		244	44.50		288.50	335
4560	1,350 amp		20	.800		276	46.50		322.50	375
4580	1,600 amp		17	.941		320	55		375	430
4600	2,000 amp		16	1		375	58		433	500
4620	2,500 amp		12	1.333		485	77.50		562.50	650
4640	3,000 amp		10	1.600		550	93		643	745
4660	4,000 amp		8	2		795	116		911	1,050
5000	Indoor 3 pole, 3 wire, feeder, 600 amp		40	.400		101	23.50		124.50	146
5010	800 amp		32	.500		355	29		384	435
5020	1,000 amp		30	.533		395	31		426	480
5025	1,200 amp		29	.552		535	32		567	635
5030	1,350 amp		28	.571		175	33.50		208.50	242
5040	1,600 amp		24	.667		755	39		794	890
5050	2,000 amp		20	.800		905	46.50		951.50	1,075
5060	2,500 amp		16	1		1,100	58		1,158	1,275
5070	3,000 amp		14	1.143		1,250	66.50		1,316.50	1,475
5080	4,000 amp		12	1.333		480	77.50		557.50	640
5200	Plug-in type, 225 amp		50	.320		138	18.60		156.60	180

26 25 13.10 Aluminum Bus Duct		Crew	Daily Output	Labor-Hours	Unit	Material	2018 Bare Costs Labor	Equipment	Total	Total Incl O&P
5210	400 amp	2 Elec	42	.381	L.F.	170	22		192	220
5220	600 amp		36	.444		211	26		237	271
5230	800 amp		30	.533		296	31		327	370
5240	1,000 amp		28	.571		335	33.50		368.50	420
5245	1,200 amp		27	.593		233	34.50		267.50	310
5250	1,350 amp		26	.615		184	36		220	256
5260	1,600 amp		20	.800		212	46.50		258.50	305
5270	2,000 amp		18	.889		248	51.50		299.50	350
5280	2,500 amp		16	1		340	58		398	460
5290	3,000 amp		14	1.143		405	66.50		471.50	545
5300	4,000 amp		12	1.333		490	77.50		567.50	650
5330	High short circuit, 400 amp		42	.381		204	22		226	257
5340	600 amp		36	.444		253	26		279	315
5350	800 amp		30	.533		355	31		386	435
5360	1,000 amp		28	.571		400	33.50		433.50	490
5370	1,350 amp		26	.615		184	36		220	256
5380	1,600 amp		20	.800		212	46.50		258.50	305
5390	2,000 amp		18	.889		248	51.50		299.50	350
5400	2,500 amp		16	1		390	58		448	515
5410	3,000 amp		14	1.143		405	66.50		471.50	545
5440	Elbow, 225 amp		5	3.200	Ea.	640	186		826	985
5450	400 amp		4.40	3.636		640	212		852	1,025
5460	600 amp		4	4		640	233		873	1,050
5470	800 amp		3.40	4.706		680	274		954	1,150
5480	1,000 amp		3.20	5		700	291		991	1,200
5485	1,200 amp		3.10	5.161		745	300		1,045	1,275
5490	1,350 amp		3	5.333		710	310		1,020	1,250
5500	1,600 amp		2.80	5.714		1,100	335		1,435	1,700
5510	2,000 amp		2.40	6.667		1,200	390		1,590	1,900
5520	2,500 amp		2	8		1,475	465		1,940	2,325
5530	3,000 amp		1.80	8.889		1,750	515		2,265	2,700
5540	4,000 amp		1.60	10		2,325	580		2,905	3,450
5560	Tee fittings, 225 amp		3.60	4.444		780	259		1,039	1,250
5570	400 amp		3.20	5		780	291		1,071	1,300
5580	600 amp		3	5.333		780	310		1,090	1,325
5590	800 amp		2.80	5.714		835	335		1,170	1,425
5600	1,000 amp		2.60	6.154		865	360		1,225	1,475
5605	1,200 amp		2.50	6.400		920	370		1,290	1,575
5610	1,350 amp		2.40	6.667		1,275	390		1,665	1,975
5620	1,600 amp		1.80	8.889		1,500	515		2,015	2,425
5630	2,000 amp		1.40	11.429		1,675	665		2,340	2,850
5640	2,500 amp		1.20	13.333		2,100	775		2,875	3,450
5650	3,000 amp		1	16		2,450	930		3,380	4,100
5660	4,000 amp		.70	22.857		3,600	1,325		4,925	5,950
5680	Cross, 225 amp		6.40	2.500		1,275	146		1,421	1,625
5690	400 amp		5.40	2.963		1,275	172		1,447	1,650
5700	600 amp		4.60	3.478		1,275	202		1,477	1,700
5710	800 amp		4	4		1,350	233		1,583	1,825
5720	1,000 amp		3.60	4.444		1,375	259		1,634	1,900
5730	1,350 amp		3.20	5		2,050	291		2,341	2,675
5740	1,600 amp		2.60	6.154		2,400	360		2,760	3,150
5750	2,000 amp		2.20	7.273		2,625	425		3,050	3,500
5760	2,500 amp		1.80	8.889		3,150	515		3,665	4,250

For customer support on your Electrical Costs with RSMeans data, call 800.448.8182.

243

26 25 Low-Voltage Enclosed Bus Assemblies

26 25 13 – Low-Voltage Busways

26 25 13.10 Aluminum Bus Duct	Crew	Daily Output	Labor-Hours	Unit	Material	2018 Bare Costs Labor	Equipment	Total	Total Incl O&P
5770	3,000 amp	2 Elec	1.40	11.429	Ea.	3,775	665		4,440	5,150
5780	4,000 amp		1.20	13.333		5,250	775		6,025	6,925
5800	Expansion fitting, 225 amp		11.60	1.379		1,000	80.50		1,080.50	1,225
5810	400 amp		9.20	1.739		1,000	101		1,101	1,250
5820	600 amp		7	2.286		1,000	133		1,133	1,300
5830	800 amp		5.20	3.077		1,175	179		1,354	1,575
5840	1,000 amp		4.60	3.478		1,300	202		1,502	1,725
5850	1,350 amp		4.20	3.810		1,450	222		1,672	1,900
5860	1,600 amp		3.60	4.444		1,800	259		2,059	2,350
5870	2,000 amp		3.20	5		2,125	291		2,416	2,750
5880	2,500 amp		2.80	5.714		2,400	335		2,735	3,125
5890	3,000 amp		2.40	6.667		2,800	390		3,190	3,650
5900	4,000 amp		1.80	8.889		3,825	515		4,340	5,000
5940	Reducer, nonfused, 400 amp		9.20	1.739		735	101		836	960
5950	600 amp		7	2.286		735	133		868	1,000
5960	800 amp		5.20	3.077		770	179		949	1,125
5970	1,000 amp		4.60	3.478		950	202		1,152	1,350
5980	1,350 amp		4.20	3.810		1,400	222		1,622	1,875
5990	1,600 amp		3.60	4.444		1,600	259		1,859	2,125
6000	2,000 amp		3.20	5		1,875	291		2,166	2,475
6010	2,500 amp		2.80	5.714		2,400	335		2,735	3,125
6020	3,000 amp		2.20	7.273		2,800	425		3,225	3,700
6030	4,000 amp		1.80	8.889		3,825	515		4,340	5,000
6050	Reducer, fuse included, 225 amp		5	3.200		1,925	186		2,111	2,400
6060	400 amp		4.80	3.333		2,550	194		2,744	3,100
6070	600 amp		4.20	3.810		3,275	222		3,497	3,925
6080	800 amp		3.60	4.444		5,075	259		5,334	5,950
6090	1,000 amp		3.40	4.706		5,525	274		5,799	6,475
6100	1,350 amp		3.20	5		10,100	291		10,391	11,500
6110	1,600 amp		2.60	6.154		12,000	360		12,360	13,600
6120	2,000 amp		2	8		13,900	465		14,365	16,000
6160	Reducer, circuit breaker, 225 amp		5	3.200		2,375	186		2,561	2,875
6170	400 amp		4.80	3.333		2,900	194		3,094	3,475
6180	600 amp		4.20	3.810		4,150	222		4,372	4,900
6190	800 amp		3.60	4.444		4,850	259		5,109	5,725
6200	1,000 amp		3.40	4.706		5,525	274		5,799	6,475
6210	1,350 amp		3.20	5		6,675	291		6,966	7,775
6220	1,600 amp		2.60	6.154		9,950	360		10,310	11,400
6230	2,000 amp		2	8		10,900	465		11,365	12,700
6270	Cable tap box center, 225 amp		4.20	3.810		1,000	222		1,222	1,425
6280	400 amp		3.60	4.444		1,000	259		1,259	1,475
6290	600 amp		3	5.333		1,000	310		1,310	1,575
6300	800 amp		2.60	6.154		1,100	360		1,460	1,725
6310	1,000 amp		2.40	6.667		1,150	390		1,540	1,850
6320	1,350 amp		1.80	8.889		1,225	515		1,740	2,125
6330	1,600 amp		1.60	10		1,375	580		1,955	2,400
6340	2,000 amp		1.40	11.429		1,575	665		2,240	2,750
6350	2,500 amp		1.20	13.333		1,975	775		2,750	3,325
6360	3,000 amp		1	16		2,225	930		3,155	3,850
6370	4,000 amp		.70	22.857		2,650	1,325		3,975	4,900
6390	Cable tap box end, 225 amp		4.20	3.810		610	222		832	1,000
6400	400 amp		3.60	4.444		610	259		869	1,050
6410	600 amp		3	5.333		610	310		920	1,150

26 25 13.10 Aluminum Bus Duct

		Crew	Daily Output	Labor-Hours	Unit	Material	2018 Bare Costs Labor	Equipment	Total	Total Incl O&P
6420	800 amp	2 Elec	2.60	6.154	Ea.	670	360		1,030	1,275
6430	1,000 amp		2.40	6.667		725	390		1,115	1,375
6435	1,200 amp		2.10	7.619		785	445		1,230	1,525
6440	1,350 amp		1.80	8.889		765	515		1,280	1,625
6450	1,600 amp		1.60	10		870	580		1,450	1,825
6460	2,000 amp		1.40	11.429		980	665		1,645	2,075
6470	2,500 amp		1.20	13.333		1,125	775		1,900	2,400
6480	3,000 amp		1	16		1,350	930		2,280	2,875
6490	4,000 amp		.70	22.857		1,625	1,325		2,950	3,775
7000	Weatherproof 3 pole 3 wire, feeder, 600 amp		34	.471	L.F.	131	27.50		158.50	185
7020	800 amp		28	.571		144	33.50		177.50	208
7040	1,000 amp		26	.615		155	36		191	224
7050	1,200 amp		25	.640		181	37.50		218.50	255
7060	1,350 amp		24	.667		210	39		249	289
7080	1,600 amp		20	.800		243	46.50		289.50	335
7100	2,000 amp		18	.889		287	51.50		338.50	395
7120	2,500 amp		14	1.143		395	66.50		461.50	535
7140	3,000 amp		12	1.333		475	77.50		552.50	640
7160	4,000 amp		10	1.600		575	93		668	770

26 25 13.20 Bus Duct

		Crew	Daily Output	Labor-Hours	Unit	Material	2018 Bare Costs Labor	Equipment	Total	Total Incl O&P
0010	**BUS DUCT** 100 amp and less, aluminum or copper, plug-in									
0080	Bus duct, 3 pole 3 wire, 100 amp	1 Elec	42	.190	L.F.	70.50	11.10		81.60	94
0110	Elbow		4	2	Ea.	131	116		247	320
0120	Tee		2	4		186	233		419	555
0130	Wall flange		8	1		24	58		82	113
0140	Ground kit		16	.500		61	29		90	111
0180	3 pole 4 wire, 100 amp		40	.200	L.F.	73.50	11.65		85.15	98.50
0200	Cable tap box		3.10	2.581	Ea.	223	150		373	470
0300	End closure		16	.500		29.50	29		58.50	76
0400	Elbow		4	2		186	116		302	380
0500	Tee		2	4		271	233		504	650
0600	Hangers		10	.800		17.85	46.50		64.35	89
0700	Circuit breakers, 15 to 50 amp, 1 pole		8	1		575	58		633	720
0800	15 to 60 amp, 2 pole		6.70	1.194		320	69.50		389.50	455
0900	3 pole		5.30	1.509		390	88		478	560
1000	60 to 100 amp, 1 pole		6.70	1.194		595	69.50		664.50	760
1100	70 to 100 amp, 2 pole		5.30	1.509		1,250	88		1,338	1,500
1200	3 pole		4.50	1.778		1,525	103		1,628	1,825
1220	Switch, nonfused, 3 pole, 4 wire		8	1		192	58		250	299
1240	Fused, 3 fuses, 4 wire, 30 amp		8	1		360	58		418	480
1260	60 amp		5.30	1.509		425	88		513	595
1280	100 amp		4.50	1.778		580	103		683	795
1300	Plug, fusible, 3 pole 250 volt, 30 amp		5.30	1.509		515	88		603	695
1310	60 amp		5.30	1.509		595	88		683	785
1320	100 amp		4.50	1.778		800	103		903	1,025
1330	3 pole 480 volt, 30 amp		5.30	1.509		530	88		618	715
1340	60 amp		5.30	1.509		570	88		658	760
1350	100 amp		4.50	1.778		825	103		928	1,075
1360	Circuit breaker, 3 pole 250 volt, 60 amp		5.30	1.509		940	88		1,028	1,150
1370	3 pole 480 volt, 100 amp		4.50	1.778		940	103		1,043	1,175
2000	Bus duct, 2 wire, 250 volt, 30 amp		60	.133	L.F.	7.25	7.75		15	19.60
2100	60 amp		50	.160		7.25	9.30		16.55	22

For customer support on your Electrical Costs with RSMeans data, call 800.448.8182.

245

26 25 13.20 Bus Duct		Crew	Daily Output	Labor-Hours	Unit	Material	2018 Bare Costs Labor	Equipment	Total	Total Incl O&P
2200	300 volt, 30 amp	1 Elec	60	.133	L.F.	7.25	7.75		15	19.60
2300	60 amp		50	.160		7.25	9.30		16.55	22
2400	3 wire, 250 volt, 30 amp		60	.133		9.80	7.75		17.55	22.50
2500	60 amp		50	.160		9.60	9.30		18.90	24.50
2600	480/277 volt, 30 amp		60	.133		9.80	7.75		17.55	22.50
2700	60 amp		50	.160		9.80	9.30		19.10	24.50
2750	End feed, 300 volt 2 wire max. 30 amp		6	1.333	Ea.	73	77.50		150.50	197
2800	60 amp		5.50	1.455		73	84.50		157.50	208
2850	30 amp miniature		6	1.333		73	77.50		150.50	197
2900	3 wire, 30 amp		6	1.333		92	77.50		169.50	217
2950	60 amp		5.50	1.455		92	84.50		176.50	228
3000	30 amp miniature		6	1.333		92	77.50		169.50	217
3050	Center feed, 300 volt 2 wire, 30 amp		6	1.333		101	77.50		178.50	227
3100	60 amp		5.50	1.455		101	84.50		185.50	238
3150	3 wire, 30 amp		6	1.333		115	77.50		192.50	243
3200	60 amp		5.50	1.455		115	84.50		199.50	254
3220	Elbow, 30 amp		6	1.333		39.50	77.50		117	160
3240	60 amp		5.50	1.455		39.50	84.50		124	171
3260	End cap		40	.200		8.90	11.65		20.55	27
3280	Strength beam, 10'		15	.533		23.50	31		54.50	72.50
3300	Hanger		24	.333		5.05	19.40		24.45	34.50
3320	Tap box, nonfusible		6.30	1.270		75.50	74		149.50	193
3340	Fusible switch 30 amp, 1 fuse		6	1.333		460	77.50		537.50	620
3360	2 fuse		6	1.333		485	77.50		562.50	645
3380	3 fuse		6	1.333		520	77.50		597.50	685
3400	Circuit breaker handle on cover, 1 pole		6	1.333		62.50	77.50		140	185
3420	2 pole		6	1.333		525	77.50		602.50	695
3440	3 pole		6	1.333		645	77.50		722.50	825
3460	Circuit breaker external operhandle, 1 pole		6	1.333		73.50	77.50		151	197
3480	2 pole		6	1.333		680	77.50		757.50	865
3500	3 pole		6	1.333		810	77.50		887.50	1,000
3520	Terminal plug only		16	.500		98.50	29		127.50	153
3540	Terminal with receptacle		16	.500		108	29		137	163
3560	Fixture plug		16	.500		73.50	29		102.50	124
4000	Copper bus duct, lighting, 2 wire 300 volt, 20 amp		70	.114	L.F.	6.90	6.65		13.55	17.55
4020	35 amp		60	.133		6.90	7.75		14.65	19.20
4040	50 amp		55	.145		6.90	8.45		15.35	20.50
4060	60 amp		50	.160		6.90	9.30		16.20	21.50
4080	3 wire 300 volt, 20 amp		70	.114		6.90	6.65		13.55	17.55
4100	35 amp		60	.133		6.75	7.75		14.50	19.05
4120	50 amp		55	.145		6.90	8.45		15.35	20.50
4140	60 amp		50	.160		6.90	9.30		16.20	21.50
4160	Feeder in box, end, 1 circuit		6	1.333	Ea.	94	77.50		171.50	220
4180	2 circuit		5.50	1.455		98	84.50		182.50	235
4200	Center, 1 circuit		6	1.333		130	77.50		207.50	259
4220	2 circuit		5.50	1.455		135	84.50		219.50	275
4240	End cap		40	.200		15.85	11.65		27.50	35
4260	Hanger, surface mount		24	.333		9.50	19.40		28.90	39.50
4280	Coupling		40	.200		12.05	11.65		23.70	30.50

246

For customer support on your Electrical Costs with RSMeans data, call 800.448.8182.

26 25 13.30 Copper Bus Duct	Crew	Daily Output	Labor-Hours	Unit	Material	2018 Bare Costs Labor	Equipment	Total	Total Incl O&P
0010 **COPPER BUS DUCT**									
0100 Weatherproof 3 pole 4 wire, feeder duct, 600 amp	2 Elec	24	.667	L.F.	315	39		354	405
0110 800 amp		18	.889		430	51.50		481.50	550
0120 1,000 amp		17	.941		480	55		535	605
0125 1,200 amp		16.50	.970		560	56.50		616.50	700
0130 1,350 amp		16	1		585	58		643	730
0140 1,600 amp		12	1.333		630	77.50		707.50	805
0150 2,000 amp		10	1.600		700	93		793	910
0160 2,500 amp		7	2.286		760	133		893	1,050
0170 3,000 amp		5	3.200		1,050	186		1,236	1,425
0180 4,000 amp		3.60	4.444		1,650	259		1,909	2,200
0200 Indoor 3 pole 4 wire, plug-in, bus duct high short circuit, 400 amp		32	.500		450	29		479	540
0210 600 amp		26	.615		510	36		546	615
0220 800 amp		20	.800		740	46.50		786.50	885
0230 1,000 amp		18	.889		870	51.50		921.50	1,025
0240 1,350 amp		16	1		450	58		508	580
0250 1,600 amp		12	1.333		510	77.50		587.50	675
0260 2,000 amp		10	1.600		645	93		738	845
0270 2,500 amp		8	2		795	116		911	1,050
0280 3,000 amp		6	2.667		905	155		1,060	1,225
0310 Cross, 225 amp		3	5.333	Ea.	2,975	310		3,285	3,750
0320 400 amp		2.80	5.714		2,975	335		3,310	3,775
0330 600 amp		2.60	6.154		2,975	360		3,335	3,800
0340 800 amp		2.20	7.273		3,225	425		3,650	4,175
0350 1,000 amp		2	8		3,625	465		4,090	4,675
0360 1,350 amp		1.80	8.889		3,875	515		4,390	5,050
0370 1,600 amp		1.70	9.412		4,325	550		4,875	5,575
0380 2,000 amp		1.60	10		7,100	580		7,680	8,675
0390 2,500 amp		1.40	11.429		8,575	665		9,240	10,400
0400 3,000 amp		1.20	13.333		9,375	775		10,150	11,500
0410 4,000 amp		1	16		12,200	930		13,130	14,800
0430 Expansion fitting, 225 amp		5.40	2.963		1,675	172		1,847	2,100
0440 400 amp		4.60	3.478		1,900	202		2,102	2,400
0450 600 amp		4	4		2,350	233		2,583	2,950
0460 800 amp		3.40	4.706		2,775	274		3,049	3,450
0470 1,000 amp		3	5.333		3,175	310		3,485	3,975
0480 1,350 amp		2.80	5.714		3,525	335		3,860	4,400
0490 1,600 amp		2.60	6.154		4,925	360		5,285	5,950
0500 2,000 amp		2.20	7.273		5,675	425		6,100	6,875
0510 2,500 amp		1.80	8.889		6,900	515		7,415	8,375
0520 3,000 amp		1.60	10		7,725	580		8,305	9,375
0530 4,000 amp		1.20	13.333		9,975	775		10,750	12,200
0550 Reducer nonfused, 225 amp		5.40	2.963		1,800	172		1,972	2,225
0560 400 amp		4.60	3.478		1,800	202		2,002	2,275
0570 600 amp		4	4		1,800	233		2,033	2,325
0580 800 amp		3.40	4.706		2,175	274		2,449	2,800
0590 1,000 amp		3	5.333		2,600	310		2,910	3,350
0600 1,350 amp		2.80	5.714		3,750	335		4,085	4,625
0610 1,600 amp		2.60	6.154		4,300	360		4,660	5,275
0620 2,000 amp		2.20	7.273		5,200	425		5,625	6,325
0630 2,500 amp		1.80	8.889		6,550	515		7,065	7,975
0640 3,000 amp		1.60	10		7,375	580		7,955	8,975

For customer support on your Electrical Costs with RSMeans data, call 800.448.8182.

247

26 25 13.30 Copper Bus Duct		Crew	Daily Output	Labor-Hours	Unit	Material	2018 Bare Costs Labor	Equipment	Total	Total Incl O&P
0650	4,000 amp	2 Elec	1.20	13.333	Ea.	9,600	775		10,375	11,800
0670	Reducer fuse included, 225 amp		4.40	3.636		3,825	212		4,037	4,550
0680	400 amp		4.20	3.810		4,800	222		5,022	5,600
0690	600 amp		3.60	4.444		5,900	259		6,159	6,875
0700	800 amp		3.20	5		8,350	291		8,641	9,600
0710	1,000 amp		3	5.333		10,500	310		10,810	12,000
0720	1,350 amp		2.80	5.714		16,300	335		16,635	18,400
0730	1,600 amp		2.20	7.273		22,000	425		22,425	24,800
0740	2,000 amp		1.80	8.889		24,800	515		25,315	28,100
0790	Reducer, circuit breaker, 225 amp		4.40	3.636		4,850	212		5,062	5,650
0800	400 amp		4.20	3.810		5,650	222		5,872	6,550
0810	600 amp		3.60	4.444		8,025	259		8,284	9,225
0820	800 amp		3.20	5		9,425	291		9,716	10,800
0830	1,000 amp		3	5.333		10,700	310		11,010	12,300
0840	1,350 amp		2.80	5.714		12,200	335		12,535	13,900
0850	1,600 amp		2.20	7.273		18,000	425		18,425	20,400
0860	2,000 amp		1.80	8.889		19,800	515		20,315	22,600
0910	Cable tap box, center, 225 amp		3.20	5		2,025	291		2,316	2,650
0920	400 amp		2.60	6.154		2,025	360		2,385	2,750
0930	600 amp		2.20	7.273		2,025	425		2,450	2,850
0940	800 amp		2	8		2,250	465		2,715	3,175
0950	1,000 amp		1.60	10		2,425	580		3,005	3,550
0960	1,350 amp		1.40	11.429		2,600	665		3,265	3,850
0970	1,600 amp		1.20	13.333		2,900	775		3,675	4,325
0980	2,000 amp		1	16		3,500	930		4,430	5,225
1040	2,500 amp		.80	20		4,125	1,175		5,300	6,300
1060	3,000 amp		.60	26.667		4,525	1,550		6,075	7,300
1080	4,000 amp		.40	40	↓	5,700	2,325		8,025	9,750
1800	Weatherproof 3 pole 3 wire, feeder duct, 600 amp		28	.571	L.F.	264	33.50		297.50	340
1820	800 amp		22	.727		320	42.50		362.50	420
1840	1,000 amp		20	.800		360	46.50		406.50	465
1850	1,200 amp		19	.842		370	49		419	485
1860	1,350 amp		18	.889		505	51.50		556.50	635
1880	1,600 amp		14	1.143		575	66.50		641.50	730
1900	2,000 amp		12	1.333		735	77.50		812.50	925
1920	2,500 amp		8	2		915	116		1,031	1,175
1940	3,000 amp		6	2.667		1,050	155		1,205	1,375
1960	4,000 amp		4	4		1,400	233		1,633	1,875
2000	Indoor 3 pole 3 wire, feeder duct, 600 amp		32	.500		220	29		249	286
2010	800 amp		26	.615		580	36		616	695
2020	1,000 amp		24	.667		299	39		338	390
2025	1,200 amp		22	.727		395	42.50		437.50	500
2030	1,350 amp		20	.800		420	46.50		466.50	530
2040	1,600 amp		16	1		1,075	58		1,133	1,250
2050	2,000 amp		14	1.143		1,375	66.50		1,441.50	1,600
2060	2,500 amp		10	1.600		765	93		858	980
2070	3,000 amp		8	2		880	116		996	1,150
2080	4,000 amp		6	2.667		1,150	155		1,305	1,500
2090	5,000 amp		5	3.200		1,425	186		1,611	1,825
2200	Indoor 3 pole 3 wire, bus duct plug-in, 225 amp		46	.348		171	20		191	219
2210	400 amp		36	.444		259	26		285	325
2220	600 amp		30	.533		330	31		361	410
2230	800 amp		24	.667		485	39		524	595

26 25 13.30 Copper Bus Duct		Crew	Daily Output	Labor-Hours	Unit	Material	2018 Bare Costs Labor	Equipment	Total	Total Incl O&P
2240	1,000 amp	2 Elec	20	.800	L.F.	525	46.50		571.50	645
2250	1,350 amp		18	.889		790	51.50		841.50	950
2260	1,600 amp		14	1.143		845	66.50		911.50	1,025
2270	2,000 amp		12	1.333		645	77.50		722.50	820
2280	2,500 amp		10	1.600		795	93		888	1,000
2290	3,000 amp		8	2		905	116		1,021	1,175
2330	High short circuit, 400 amp		36	.444		310	26		336	380
2340	600 amp		30	.533		400	31		431	480
2350	800 amp		24	.667		580	39		619	700
2360	1,000 amp		20	.800		630	46.50		676.50	760
2370	1,350 amp		18	.889		950	51.50		1,001.50	1,125
2380	1,600 amp		14	1.143		1,025	66.50		1,091.50	1,225
2390	2,000 amp		12	1.333		645	77.50		722.50	820
2400	2,500 amp		10	1.600		795	93		888	1,000
2410	3,000 amp		8	2		905	116		1,021	1,175
2440	Elbows, 225 amp		4.60	3.478	Ea.	1,425	202		1,627	1,850
2450	400 amp		4.20	3.810		1,425	222		1,647	1,875
2460	600 amp		3.60	4.444		1,425	259		1,684	1,925
2470	800 amp		3.20	5		1,525	291		1,816	2,100
2480	1,000 amp		3	5.333		1,600	310		1,910	2,225
2485	1,200 amp		2.90	5.517		1,525	320		1,845	2,150
2490	1,350 amp		2.80	5.714		1,825	335		2,160	2,500
2500	1,600 amp		2.60	6.154		1,950	360		2,310	2,675
2510	2,000 amp		2	8		2,100	465		2,565	3,025
2520	2,500 amp		1.80	8.889		3,200	515		3,715	4,300
2530	3,000 amp		1.60	10		3,500	580		4,080	4,725
2540	4,000 amp		1.40	11.429		4,475	665		5,140	5,925
2560	Tee fittings, 225 amp		2.80	5.714		1,475	335		1,810	2,125
2570	400 amp		2.40	6.667		1,475	390		1,865	2,200
2580	600 amp		2	8		1,475	465		1,940	2,325
2590	800 amp		1.80	8.889		1,625	515		2,140	2,550
2600	1,000 amp		1.60	10		1,800	580		2,380	2,875
2605	1,200 amp		1.50	10.667		1,925	620		2,545	3,050
2610	1,350 amp		1.40	11.429		2,025	665		2,690	3,225
2620	1,600 amp		1.20	13.333		2,350	775		3,125	3,725
2630	2,000 amp		1	16		3,900	930		4,830	5,675
2640	2,500 amp		.70	22.857		4,575	1,325		5,900	7,025
2650	3,000 amp		.60	26.667		5,025	1,550		6,575	7,850
2660	4,000 amp		.50	32		6,500	1,850		8,350	9,925
2680	Cross, 225 amp		3.60	4.444		2,275	259		2,534	2,875
2690	400 amp		3.20	5		2,275	291		2,566	2,925
2700	600 amp		3	5.333		2,275	310		2,585	2,975
2710	800 amp		2.60	6.154		2,700	360		3,060	3,500
2720	1,000 amp		2.40	6.667		2,825	390		3,215	3,700
2730	1,350 amp		2.20	7.273		3,225	425		3,650	4,175
2740	1,600 amp		2	8		3,475	465		3,940	4,525
2750	2,000 amp		1.80	8.889		5,500	515		6,015	6,825
2760	2,500 amp		1.60	10		6,400	580		6,980	7,900
2770	3,000 amp		1.40	11.429		7,000	665		7,665	8,700
2780	4,000 amp		1	16		8,950	930		9,880	11,300
2800	Expansion fitting, 225 amp		6.40	2.500		1,775	146		1,921	2,175
2810	400 amp		5.40	2.963		4,825	172		4,997	5,550
2820	600 amp		4.60	3.478		1,775	202		1,977	2,250

For customer support on your Electrical Costs with RSMeans data, call 800.448.8182.

249

26 25 13.30 Copper Bus Duct		Crew	Daily Output	Labor-Hours	Unit	Material	2018 Bare Costs Labor	Equipment	Total	Total Incl O&P
2830	800 amp	2 Elec	4	4	Ea.	2,125	233		2,358	2,675
2840	1,000 amp		3.60	4.444		2,325	259		2,584	2,925
2850	1,350 amp		3.20	5		2,750	291		3,041	3,450
2860	1,600 amp		3	5.333		3,025	310		3,335	3,800
2870	2,000 amp		2.60	6.154		3,725	360		4,085	4,625
2880	2,500 amp		2.20	7.273		5,200	425		5,625	6,350
2890	3,000 amp		1.80	8.889		5,825	515		6,340	7,175
2900	4,000 amp		1.40	11.429		7,425	665		8,090	9,175
2920	Reducer nonfused, 225 amp		6.40	2.500		1,475	146		1,621	1,850
2930	400 amp		5.40	2.963		1,475	172		1,647	1,875
2940	600 amp		4.60	3.478		1,475	202		1,677	1,925
2950	800 amp		4	4		1,725	233		1,958	2,225
2960	1,000 amp		3.60	4.444		1,925	259		2,184	2,500
2970	1,350 amp		3.20	5		2,425	291		2,716	3,100
2980	1,600 amp		3	5.333		2,750	310		3,060	3,500
2990	2,000 amp		2.60	6.154		3,300	360		3,660	4,150
3000	2,500 amp		2.20	7.273		4,825	425		5,250	5,925
3010	3,000 amp		1.80	8.889		5,400	515		5,915	6,725
3020	4,000 amp		1.40	11.429		7,025	665		7,690	8,725
3040	Reducer fuse included, 225 amp		5	3.200		3,475	186		3,661	4,100
3050	400 amp		4.80	3.333		4,625	194		4,819	5,400
3060	600 amp		4.20	3.810		5,525	222		5,747	6,400
3070	800 amp		3.60	4.444		7,875	259		8,134	9,025
3080	1,000 amp		3.40	4.706		9,325	274		9,599	10,600
3090	1,350 amp		3.20	5		9,025	291		9,316	10,400
3100	1,600 amp		2.60	6.154		21,000	360		21,360	23,600
3110	2,000 amp		2	8		23,800	465		24,265	26,900
3160	Reducer circuit breaker, 225 amp		5	3.200		4,450	186		4,636	5,175
3170	400 amp		4.80	3.333		5,450	194		5,644	6,300
3180	600 amp		4.20	3.810		7,825	222		8,047	8,925
3190	800 amp		3.60	4.444		9,150	259		9,409	10,500
3200	1,000 amp		3.40	4.706		10,400	274		10,674	11,800
3210	1,350 amp		3.20	5		11,900	291		12,191	13,500
3220	1,600 amp		2.60	6.154		17,800	360		18,160	20,100
3230	2,000 amp		2	8		19,400	465		19,865	22,100
3280	3 pole, 3 wire, cable tap box center, 225 amp		3.60	4.444		2,300	259		2,559	2,925
3290	400 amp		3	5.333		2,300	310		2,610	3,025
3300	600 amp		2.60	6.154		2,300	360		2,660	3,075
3310	800 amp		2.40	6.667		2,575	390		2,965	3,425
3320	1,000 amp		1.80	8.889		2,800	515		3,315	3,850
3330	1,350 amp		1.60	10		3,425	580		4,005	4,650
3340	1,600 amp		1.40	11.429		3,825	665		4,490	5,200
3350	2,000 amp		1.20	13.333		4,125	775		4,900	5,700
3360	2,500 amp		1	16		4,925	930		5,855	6,825
3370	3,000 amp		.70	22.857		5,425	1,325		6,750	7,975
3380	4,000 amp		.50	32		6,875	1,850		8,725	10,400
3400	Cable tap box end, 225 amp		3.60	4.444		1,250	259		1,509	1,750
3410	400 amp		3	5.333		1,250	310		1,560	1,850
3420	600 amp		2.60	6.154		1,400	360		1,760	2,075
3430	800 amp		2.40	6.667		1,400	390		1,790	2,125
3440	1,000 amp		1.80	8.889		1,550	515		2,065	2,475
3445	1,200 amp		1.70	9.412		1,550	550		2,100	2,525
3450	1,350 amp		1.60	10		1,825	580		2,405	2,900

26 25 13.30 Copper Bus Duct		Crew	Daily Output	Labor-Hours	Unit	Material	2018 Bare Costs Labor	Equipment	Total	Total Incl O&P
3460	1,600 amp	2 Elec	1.40	11.429	Ea.	2,100	665		2,765	3,325
3470	2,000 amp		1.20	13.333		2,150	775		2,925	3,500
3480	2,500 amp		1	16		2,550	930		3,480	4,225
3490	3,000 amp		.70	22.857		2,800	1,325		4,125	5,075
3500	4,000 amp		.50	32		3,550	1,850		5,400	6,675
4600	Plug-in, fusible switch w/3 fuses, 3 pole, 250 volt, 30 amp	1 Elec	4	2		435	116		551	655
4610	60 amp		3.60	2.222		570	129		699	820
4620	100 amp		2.70	2.963		815	172		987	1,150
4630	200 amp	2 Elec	3.20	5		1,375	291		1,666	1,925
4640	400 amp		1.40	11.429		3,575	665		4,240	4,925
4650	600 amp		.90	17.778		4,925	1,025		5,950	6,975
4700	4 pole, 120/208 volt, 30 amp	1 Elec	3.90	2.051		595	119		714	835
4710	60 amp		3.50	2.286		640	133		773	905
4720	100 amp		2.60	3.077		895	179		1,074	1,250
4730	200 amp	2 Elec	3	5.333		1,500	310		1,810	2,125
4740	400 amp		1.30	12.308		3,525	715		4,240	4,950
4750	600 amp		.80	20		4,950	1,175		6,125	7,175
4800	3 pole, 480 volt, 30 amp	1 Elec	4	2		445	116		561	665
4810	60 amp		3.60	2.222		470	129		599	715
4820	100 amp		2.70	2.963		795	172		967	1,125
4830	200 amp	2 Elec	3.20	5		1,375	291		1,666	1,925
4840	400 amp		1.40	11.429		3,200	665		3,865	4,500
4850	600 amp		.90	17.778		4,525	1,025		5,550	6,525
4860	800 amp		.66	24.242		16,100	1,400		17,500	19,800
4870	1,000 amp		.60	26.667		17,100	1,550		18,650	21,100
4880	1,200 amp		.50	32		18,000	1,850		19,850	22,600
4890	1,600 amp		.44	36.364		19,800	2,125		21,925	24,900
4900	4 pole, 277/480 volt, 30 amp	1 Elec	3.90	2.051		645	119		764	890
4910	60 amp		3.50	2.286		690	133		823	955
4920	100 amp		2.60	3.077		1,000	179		1,179	1,375
4930	200 amp	2 Elec	3	5.333		2,000	310		2,310	2,675
4940	400 amp		1.30	12.308		3,775	715		4,490	5,225
4950	600 amp		.80	20		5,150	1,175		6,325	7,425
5050	800 amp		.60	26.667		17,000	1,550		18,550	21,000
5060	1,000 amp		.56	28.571		19,500	1,675		21,175	24,000
5070	1,200 amp		.48	33.333		19,700	1,950		21,650	24,600
5080	1,600 amp		.42	38.095		21,700	2,225		23,925	27,200
5150	Fusible with starter, 3 pole 250 volt, 30 amp	1 Elec	3.50	2.286		3,100	133		3,233	3,625
5160	60 amp		3.20	2.500		3,300	146		3,446	3,850
5170	100 amp		2.50	3.200		3,725	186		3,911	4,375
5180	200 amp	2 Elec	2.80	5.714		6,150	335		6,485	7,250
5200	3 pole 480 volt, 30 amp	1 Elec	3.50	2.286		3,100	133		3,233	3,625
5210	60 amp		3.20	2.500		3,300	146		3,446	3,850
5220	100 amp		2.50	3.200		3,725	186		3,911	4,375
5230	200 amp	2 Elec	2.80	5.714		6,150	335		6,485	7,250
5300	Fusible with contactor, 3 pole 250 volt, 30 amp	1 Elec	3.50	2.286		3,025	133		3,158	3,525
5310	60 amp		3.20	2.500		3,850	146		3,996	4,475
5320	100 amp		2.50	3.200		5,400	186		5,586	6,225
5330	200 amp	2 Elec	2.80	5.714		6,175	335		6,510	7,300
5400	3 pole 480 volt, 30 amp	1 Elec	3.50	2.286		3,250	133		3,383	3,800
5410	60 amp		3.20	2.500		4,600	146		4,746	5,275
5420	100 amp		2.50	3.200		6,325	186		6,511	7,225
5430	200 amp	2 Elec	2.80	5.714		6,475	335		6,810	7,625

For customer support on your Electrical Costs with RSMeans data, call 800.448.8182.

251

26 25 13 – Low-Voltage Busways

26 25 13.30 Copper Bus Duct		Crew	Daily Output	Labor-Hours	Unit	Material	2018 Bare Costs Labor	Equipment	Total	Total Incl O&P
5450	Fusible with capacitor, 3 pole 250 volt, 30 amp	1 Elec	3	2.667	Ea.	7,875	155		8,030	8,875
5460	60 amp		2	4		9,175	233		9,408	10,500
5500	3 pole 480 volt, 30 amp		3	2.667		6,600	155		6,755	7,475
5510	60 amp		2	4		8,250	233		8,483	9,425
5600	Circuit breaker, 3 pole, 250 volt, 60 amp		4.50	1.778		610	103		713	830
5610	100 amp		3.20	2.500		750	146		896	1,050
5650	4 pole, 120/208 volt, 60 amp		4.40	1.818		690	106		796	920
5660	100 amp		3.10	2.581		820	150		970	1,125
5700	3 pole, 4 wire 277/480 volt, 60 amp		4.30	1.860		930	108		1,038	1,175
5710	100 amp		3	2.667		1,025	155		1,180	1,375
5720	225 amp	2 Elec	3.20	5		2,300	291		2,591	2,950
5730	400 amp		1.20	13.333		4,775	775		5,550	6,400
5740	600 amp		.96	16.667		6,400	970		7,370	8,500
5750	700 amp		.60	26.667		8,125	1,550		9,675	11,300
5760	800 amp		.60	26.667		8,125	1,550		9,675	11,300
5770	900 amp		.54	29.630		10,700	1,725		12,425	14,400
5780	1,000 amp		.54	29.630		10,700	1,725		12,425	14,400
5790	1,200 amp		.42	38.095		12,900	2,225		15,125	17,400
5810	Circuit breaker w/HIC fuses, 3 pole 480 volt, 60 amp	1 Elec	4.40	1.818		1,175	106		1,281	1,450
5820	100 amp	"	3.10	2.581		1,275	150		1,425	1,650
5830	225 amp	2 Elec	3.40	4.706		4,100	274		4,374	4,900
5840	400 amp		1.40	11.429		6,500	665		7,165	8,150
5850	600 amp		1	16		6,600	930		7,530	8,650
5860	700 amp		.64	25		8,750	1,450		10,200	11,800
5870	800 amp		.64	25		8,750	1,450		10,200	11,800
5880	900 amp		.56	28.571		18,900	1,675		20,575	23,300
5890	1,000 amp		.56	28.571		18,900	1,675		20,575	23,300
5950	3 pole 4 wire 277/480 volt, 60 amp	1 Elec	4.30	1.860		1,175	108		1,283	1,450
5960	100 amp	"	3	2.667		1,275	155		1,430	1,650
5970	225 amp	2 Elec	3	5.333		4,100	310		4,410	4,975
5980	400 amp		1.10	14.545		6,500	845		7,345	8,425
5990	600 amp		.94	17.021		6,600	990		7,590	8,725
6000	700 amp		.58	27.586		8,750	1,600		10,350	12,000
6010	800 amp		.58	27.586		8,750	1,600		10,350	12,000
6020	900 amp		.52	30.769		18,900	1,800		20,700	23,500
6030	1,000 amp		.52	30.769		18,900	1,800		20,700	23,500
6040	1,200 amp		.40	40		18,900	2,325		21,225	24,300
6100	Circuit breaker with starter, 3 pole 250 volt, 60 amp	1 Elec	3.20	2.500		1,875	146		2,021	2,275
6110	100 amp	"	2.50	3.200		2,600	186		2,786	3,125
6120	225 amp	2 Elec	3	5.333		3,325	310		3,635	4,150
6130	3 pole 480 volt, 60 amp	1 Elec	3.20	2.500		1,875	146		2,021	2,275
6140	100 amp	"	2.50	3.200		2,600	186		2,786	3,125
6150	225 amp	2 Elec	3	5.333		3,175	310		3,485	3,975
6200	Circuit breaker with contactor, 3 pole 250 volt, 60 amp	1 Elec	3.20	2.500		1,775	146		1,921	2,175
6210	100 amp	"	2.50	3.200		2,400	186		2,586	2,925
6220	225 amp	2 Elec	3	5.333		3,100	310		3,410	3,875
6250	3 pole 480 volt, 60 amp	1 Elec	3.20	2.500		1,775	146		1,921	2,175
6260	100 amp	"	2.50	3.200		2,400	186		2,586	2,925
6270	225 amp	2 Elec	3	5.333		2,950	310		3,260	3,725
6300	Circuit breaker with capacitor, 3 pole 250 volt, 60 amp	1 Elec	2	4		9,575	233		9,808	10,900
6310	3 pole 480 volt, 60 amp		2	4		9,600	233		9,833	11,000
6400	Add control transformer with pilot light to above starter		16	.500		530	29		559	630
6410	Switch, fusible, mechanically held contactor optional		16	.500		1,400	29		1,429	1,600

252

26 25 13 – Low-Voltage Busways

26 25 13.30 Copper Bus Duct		Crew	Daily Output	Labor-Hours	Unit	Material	2018 Bare Costs Labor	Equipment	Total	Total Incl O&P
6430	Circuit breaker, mechanically held contactor optional	1 Elec	16	.500	Ea.	1,400	29		1,429	1,600
6450	Ground neutralizer, 3 pole	↓	16	.500	↓	64.50	29		93.50	115

26 25 13.40 Copper Bus Duct		Crew	Daily Output	Labor-Hours	Unit	Material	2018 Bare Costs Labor	Equipment	Total	Total Incl O&P
0010	**COPPER BUS DUCT** 10' long									
0050	Indoor 3 pole 4 wire, plug-in, straight section, 225 amp	2 Elec	40	.400	L.F.	222	23.50		245.50	279
1000	400 amp		32	.500		375	29		404	455
1500	600 amp		26	.615		425	36		461	520
2400	800 amp		20	.800		620	46.50		666.50	750
2450	1,000 amp		18	.889		300	51.50		351.50	410
2470	1,200 amp		17	.941		395	55		450	510
2500	1,350 amp		16	1		415	58		473	540
2510	1,600 amp		12	1.333		470	77.50		547.50	630
2520	2,000 amp		10	1.600		595	93		688	795
2530	2,500 amp		8	2		730	116		846	980
2540	3,000 amp		6	2.667		840	155		995	1,150
2550	Feeder, 600 amp		28	.571		204	33.50		237.50	274
2600	800 amp		22	.727		750	42.50		792.50	890
2700	1,000 amp		20	.800		885	46.50		931.50	1,050
2750	1,200 amp		19	.842		1,050	49		1,099	1,250
2800	1,350 amp		18	.889		385	51.50		436.50	505
2900	1,600 amp		14	1.143		1,425	66.50		1,491.50	1,675
3000	2,000 amp		12	1.333		1,775	77.50		1,852.50	2,075
3010	2,500 amp		8	2		705	116		821	950
3020	3,000 amp		6	2.667		810	155		965	1,125
3030	4,000 amp		4	4		1,075	233		1,308	1,525
3040	5,000 amp		2	8	↓	1,300	465		1,765	2,150
3100	Elbows, 225 amp		4	4	Ea.	1,575	233		1,808	2,075
3200	400 amp		3.60	4.444		1,575	259		1,834	2,100
3300	600 amp		3.20	5		1,575	291		1,866	2,150
3400	800 amp		2.80	5.714		1,700	335		2,035	2,375
3500	1,000 amp		2.60	6.154		1,900	360		2,260	2,625
3550	1,200 amp		2.50	6.400		1,850	370		2,220	2,600
3600	1,350 amp		2.40	6.667		1,875	390		2,265	2,650
3700	1,600 amp		2.20	7.273		2,050	425		2,475	2,875
3800	2,000 amp		1.80	8.889		2,525	515		3,040	3,550
3810	2,500 amp		1.60	10		4,000	580		4,580	5,275
3820	3,000 amp		1.40	11.429		4,375	665		5,040	5,800
3830	4,000 amp		1.20	13.333		5,650	775		6,425	7,375
3840	5,000 amp		1	16		9,150	930		10,080	11,500
4000	End box, 225 amp		34	.471		170	27.50		197.50	228
4100	400 amp		32	.500		188	29		217	251
4200	600 amp		28	.571		188	33.50		221.50	257
4300	800 amp		26	.615		188	36		224	261
4400	1,000 amp		24	.667		179	39		218	255
4410	1,200 amp		23	.696		188	40.50		228.50	268
4500	1,350 amp		22	.727		179	42.50		221.50	260
4600	1,600 amp		20	.800		179	46.50		225.50	266
4700	2,000 amp		18	.889		219	51.50		270.50	320
4710	2,500 amp		16	1		219	58		277	330
4720	3,000 amp		14	1.143		208	66.50		274.50	330
4730	4,000 amp		12	1.333		251	77.50		328.50	390
4740	5,000 amp	↓	10	1.600		251	93		344	415

26 25 13.40 Copper Bus Duct		Crew	Daily Output	Labor-Hours	Unit	Material	2018 Bare Costs Labor	Equipment	Total	Total Incl O&P
4800	Cable tap box end, 225 amp	2 Elec	3.20	5	Ea.	1,175	291		1,466	1,725
5000	400 amp		2.60	6.154		1,200	360		1,560	1,850
5100	600 amp		2.20	7.273		1,625	425		2,050	2,400
5200	800 amp		2	8		1,700	465		2,165	2,575
5300	1,000 amp		1.60	10		1,725	580		2,305	2,775
5350	1,200 amp		1.50	10.667		2,000	620		2,620	3,125
5400	1,350 amp		1.40	11.429		2,250	665		2,915	3,475
5500	1,600 amp		1.20	13.333		2,500	775		3,275	3,900
5600	2,000 amp		1	16		2,800	930		3,730	4,475
5610	2,500 amp		.80	20		3,100	1,175		4,275	5,150
5620	3,000 amp		.60	26.667		3,625	1,550		5,175	6,300
5630	4,000 amp		.40	40		4,175	2,325		6,500	8,075
5640	5,000 amp		.20	80		4,925	4,650		9,575	12,400
5700	Switchboard stub, 225 amp		5.40	2.963		1,250	172		1,422	1,650
5800	400 amp		4.60	3.478		1,325	202		1,527	1,750
5900	600 amp		4	4		1,375	233		1,608	1,875
6000	800 amp		3.20	5		1,675	291		1,966	2,250
6100	1,000 amp		3	5.333		1,925	310		2,235	2,600
6150	1,200 amp		2.80	5.714		2,275	335		2,610	3,000
6200	1,350 amp		2.60	6.154		2,400	360		2,760	3,150
6300	1,600 amp		2.40	6.667		2,700	390		3,090	3,550
6400	2,000 amp		2	8		3,275	465		3,740	4,300
6410	2,500 amp		1.80	8.889		3,975	515		4,490	5,150
6420	3,000 amp		1.60	10		4,425	580		5,005	5,725
6430	4,000 amp		1.40	11.429		5,750	665		6,415	7,325
6440	5,000 amp		1.20	13.333		7,050	775		7,825	8,900
6490	Tee fittings, 225 amp		2.40	6.667		2,175	390		2,565	2,975
6500	400 amp		2	8		2,175	465		2,640	3,100
6600	600 amp		1.80	8.889		2,175	515		2,690	3,175
6700	800 amp		1.60	10		2,250	580		2,830	3,375
6750	1,000 amp		1.40	11.429		2,575	665		3,240	3,825
6770	1,200 amp		1.30	12.308		2,875	715		3,590	4,250
6800	1,350 amp		1.20	13.333		3,025	775		3,800	4,500
7000	1,600 amp		1	16		3,450	930		4,380	5,200
7100	2,000 amp		.80	20		4,100	1,175		5,275	6,250
7110	2,500 amp		.60	26.667		5,025	1,550		6,575	7,850
7120	3,000 amp		.50	32		5,550	1,850		7,400	8,875
7130	4,000 amp		.40	40		7,175	2,325		9,500	11,400
7140	5,000 amp		.20	80		8,500	4,650		13,150	16,300
7200	Plug-in fusible switches w/3 fuses, 600 volt, 3 pole, 30 amp	1 Elec	4	2		860	116		976	1,125
7300	60 amp		3.60	2.222		970	129		1,099	1,275
7400	100 amp		2.70	2.963		1,325	172		1,497	1,700
7500	200 amp	2 Elec	3.20	5		2,325	291		2,616	3,000
7600	400 amp		1.40	11.429		6,500	665		7,165	8,150
7700	600 amp		.90	17.778		7,850	1,025		8,875	10,200
7800	800 amp		.66	24.242		12,600	1,400		14,000	15,900
7900	1,200 amp		.50	32		23,600	1,850		25,450	28,700
7910	1,600 amp		.44	36.364		22,400	2,125		24,525	27,800
8000	Plug-in circuit breakers, molded case, 15 to 50 amp	1 Elec	4.40	1.818		815	106		921	1,050
8100	70 to 100 amp	"	3.10	2.581		905	150		1,055	1,225
8200	150 to 225 amp	2 Elec	3.40	4.706		2,450	274		2,724	3,100
8300	250 to 400 amp		1.40	11.429		4,300	665		4,965	5,750
8400	500 to 600 amp		1	16		5,825	930		6,755	7,800

For customer support on your Electrical Costs with RSMeans data, call 800.448.8182.

26 25 13.40 Copper Bus Duct		Crew	Daily Output	Labor-Hours	Unit	Material	2018 Bare Costs Labor	Equipment	Total	Total Incl O&P
8500	700 to 800 amp	2 Elec	.64	25	Ea.	7,175	1,450		8,625	10,100
8600	900 to 1,000 amp		.56	28.571		10,300	1,675		11,975	13,800
8700	1,200 amp		.44	36.364		12,400	2,125		14,525	16,800
8720	1,400 amp		.40	40		17,300	2,325		19,625	22,600
8730	1,600 amp		.40	40		19,000	2,325		21,325	24,400
8750	Circuit breakers, with current limiting fuse, 15 to 50 amp	1 Elec	4.40	1.818		1,625	106		1,731	1,950
8760	70 to 100 amp	"	3.10	2.581		1,925	150		2,075	2,350
8770	150 to 225 amp	2 Elec	3.40	4.706		4,150	274		4,424	4,975
8780	250 to 400 amp		1.40	11.429		6,425	665		7,090	8,075
8790	500 to 600 amp		1	16		7,400	930		8,330	9,550
8800	700 to 800 amp		.64	25		12,200	1,450		13,650	15,600
8810	900 to 1,000 amp		.56	28.571		13,900	1,675		15,575	17,800
8850	Combination starter FVNR, fusible switch, NEMA size 0, 30 amp	1 Elec	2	4		2,200	233		2,433	2,775
8860	NEMA size 1, 60 amp		1.80	4.444		2,350	259		2,609	2,950
8870	NEMA size 2, 100 amp		1.30	6.154		2,950	360		3,310	3,750
8880	NEMA size 3, 200 amp	2 Elec	2	8		4,675	465		5,140	5,850
8900	Circuit breaker, NEMA size 0, 30 amp	1 Elec	2	4		2,275	233		2,508	2,850
8910	NEMA size 1, 60 amp		1.80	4.444		2,350	259		2,609	2,975
8920	NEMA size 2, 100 amp		1.30	6.154		3,375	360		3,735	4,250
8930	NEMA size 3, 200 amp	2 Elec	2	8		4,300	465		4,765	5,425
8950	Combination contactor, fusible switch, NEMA size 0, 30 amp	1 Elec	2	4		1,275	233		1,508	1,775
8960	NEMA size 1, 60 amp		1.80	4.444		1,325	259		1,584	1,825
8970	NEMA size 2, 100 amp		1.30	6.154		1,975	360		2,335	2,700
8980	NEMA size 3, 200 amp	2 Elec	2	8		2,275	465		2,740	3,200
9000	Circuit breaker, NEMA size 0, 30 amp	1 Elec	2	4		1,475	233		1,708	1,975
9010	NEMA size 1, 60 amp		1.80	4.444		1,525	259		1,784	2,050
9020	NEMA size 2, 100 amp		1.30	6.154		2,325	360		2,685	3,100
9030	NEMA size 3, 200 amp	2 Elec	2	8		2,825	465		3,290	3,825
9050	Control transformer for above, NEMA size 0, 30 amp	1 Elec	8	1		250	58		308	360
9060	NEMA size 1, 60 amp		8	1		250	58		308	360
9070	NEMA size 2, 100 amp		7	1.143		350	66.50		416.50	485
9080	NEMA size 3, 200 amp	2 Elec	14	1.143		485	66.50		551.50	630
9100	Comb. fusible switch & lighting control, electrically held, 30 amp	1 Elec	2	4		1,025	233		1,258	1,475
9110	60 amp		1.80	4.444		1,475	259		1,734	2,000
9120	100 amp		1.30	6.154		1,900	360		2,260	2,600
9130	200 amp	2 Elec	2	8		4,650	465		5,115	5,825
9150	Mechanically held, 30 amp	1 Elec	2	4		1,275	233		1,508	1,750
9160	60 amp		1.80	4.444		1,900	259		2,159	2,475
9170	100 amp		1.30	6.154		2,450	360		2,810	3,225
9180	200 amp	2 Elec	2	8		5,000	465		5,465	6,200
9200	Ground bus added to bus duct, 225 amp		320	.050	L.F.	44.50	2.91		47.41	53
9210	400 amp		240	.067		44.50	3.88		48.38	54.50
9220	600 amp		240	.067		44.50	3.88		48.38	54.50
9230	800 amp		160	.100		51.50	5.80		57.30	65.50
9240	1,000 amp		160	.100		59	5.80		64.80	73.50
9250	1,350 amp		140	.114		84	6.65		90.65	102
9260	1,600 amp		120	.133		91	7.75		98.75	112
9270	2,000 amp		110	.145		119	8.45		127.45	144
9280	2,500 amp		100	.160		147	9.30		156.30	176
9290	3,000 amp		90	.178		166	10.35		176.35	198
9300	4,000 amp		80	.200		219	11.65		230.65	258
9310	5,000 amp		70	.229		266	13.30		279.30	310
9320	High short circuit bracing, add					18.55			18.55	20.50

26 25 13.60 Copper or Aluminum Bus Duct Fittings	Crew	Daily Output	Labor-Hours	Unit	Material	2018 Bare Costs Labor	Equipment	Total	Total Incl O&P
0010 **COPPER OR ALUMINUM BUS DUCT FITTINGS**									
0100 Flange, wall, with vapor barrier, 225 amp	2 Elec	6.20	2.581	Ea.	805	150		955	1,125
0110 400 amp		6	2.667		795	155		950	1,100
0120 600 amp		5.80	2.759		805	161		966	1,125
0130 800 amp		5.40	2.963		805	172		977	1,150
0140 1,000 amp		5	3.200		805	186		991	1,175
0145 1,200 amp		4.80	3.333		805	194		999	1,175
0150 1,350 amp		4.60	3.478		805	202		1,007	1,200
0160 1,600 amp		4.20	3.810		805	222		1,027	1,225
0170 2,000 amp		4	4		805	233		1,038	1,250
0180 2,500 amp		3.60	4.444		805	259		1,064	1,275
0190 3,000 amp		3.20	5		805	291		1,096	1,325
0200 4,000 amp		2.60	6.154		805	360		1,165	1,425
0300 Roof, 225 amp		6.20	2.581		930	150		1,080	1,250
0310 400 amp		6	2.667		930	155		1,085	1,250
0320 600 amp		5.80	2.759		930	161		1,091	1,275
0330 800 amp		5.40	2.963		930	172		1,102	1,275
0340 1,000 amp		5	3.200		930	186		1,116	1,300
0345 1,200 amp		4.80	3.333		930	194		1,124	1,325
0350 1,350 amp		4.60	3.478		930	202		1,132	1,325
0360 1,600 amp		4.20	3.810		930	222		1,152	1,350
0370 2,000 amp		4	4		930	233		1,163	1,375
0380 2,500 amp		3.60	4.444		930	259		1,189	1,400
0390 3,000 amp		3.20	5		930	291		1,221	1,450
0400 4,000 amp		2.60	6.154		930	360		1,290	1,550
0420 Support, floor mounted, 225 amp		20	.800		169	46.50		215.50	256
0430 400 amp		20	.800		169	46.50		215.50	256
0440 600 amp		18	.889		169	51.50		220.50	264
0450 800 amp		16	1		169	58		227	273
0460 1,000 amp		13	1.231		169	71.50		240.50	293
0465 1,200 amp		11.80	1.356		169	79		248	305
0470 1,350 amp		10.60	1.509		169	88		257	315
0480 1,600 amp		9.20	1.739		169	101		270	335
0490 2,000 amp		8	2		169	116		285	360
0500 2,500 amp		6.40	2.500		169	146		315	405
0510 3,000 amp		5.40	2.963		169	172		341	445
0520 4,000 amp		4	4		169	233		402	535
0540 Weather stop, 225 amp		12	1.333		520	77.50		597.50	690
0550 400 amp		10	1.600		520	93		613	715
0560 600 amp		9	1.778		520	103		623	730
0570 800 amp		8	2		520	116		636	750
0580 1,000 amp		6.40	2.500		520	146		666	795
0585 1,200 amp		5.90	2.712		520	158		678	810
0590 1,350 amp		5.40	2.963		520	172		692	835
0600 1,600 amp		4.60	3.478		520	202		722	880
0610 2,000 amp		4	4		520	233		753	925
0620 2,500 amp		3.20	5		520	291		811	1,000
0630 3,000 amp		2.60	6.154		520	360		880	1,100
0640 4,000 amp		2	8		520	465		985	1,275
0660 End closure, 225 amp		34	.471		194	27.50		221.50	254
0670 400 amp		32	.500		194	29		223	257
0680 600 amp		28	.571		194	33.50		227.50	263

26 25 13.60 Copper or Aluminum Bus Duct Fittings		Crew	Daily Output	Labor-Hours	Unit	Material	2018 Bare Costs Labor	Equipment	Total	Total Incl O&P
0690	800 amp	2 Elec	26	.615	Ea.	172	36		208	243
0700	1,000 amp		24	.667		194	39		233	271
0705	1,200 amp		23	.696		168	40.50		208.50	246
0710	1,350 amp		22	.727		168	42.50		210.50	249
0720	1,600 amp		20	.800		194	46.50		240.50	283
0730	2,000 amp		18	.889		269	51.50		320.50	375
0740	2,500 amp		16	1		249	58		307	360
0750	3,000 amp		14	1.143		249	66.50		315.50	375
0760	4,000 amp		12	1.333		249	77.50		326.50	390
0780	Switchboard stub, 3 pole 3 wire, 225 amp		6	2.667		1,000	155		1,155	1,325
0790	400 amp		5.20	3.077		1,000	179		1,179	1,375
0800	600 amp		4.60	3.478		1,000	202		1,202	1,400
0810	800 amp		3.60	4.444		1,225	259		1,484	1,725
0820	1,000 amp		3.40	4.706		1,425	274		1,699	1,975
0825	1,200 amp		3.20	5		1,425	291		1,716	2,000
0830	1,350 amp		3	5.333		1,725	310		2,035	2,350
0840	1,600 amp		2.80	5.714		2,025	335		2,360	2,725
0850	2,000 amp		2.40	6.667		2,350	390		2,740	3,175
0860	2,500 amp		2	8		2,850	465		3,315	3,850
0870	3,000 amp		1.80	8.889		3,275	515		3,790	4,375
0880	4,000 amp		1.60	10		4,225	580		4,805	5,525
0890	5,000 amp		1.40	11.429		5,200	665		5,865	6,725
0900	3 pole 4 wire, 225 amp		5.40	2.963		1,250	172		1,422	1,625
0910	400 amp		4.60	3.478		1,250	202		1,452	1,675
0920	600 amp		4	4		1,250	233		1,483	1,725
0930	800 amp		3.20	5		1,525	291		1,816	2,100
0940	1,000 amp		3	5.333		1,775	310		2,085	2,425
0950	1,350 amp		2.60	6.154		2,275	360		2,635	3,025
0960	1,600 amp		2.40	6.667		2,575	390		2,965	3,400
0970	2,000 amp		2	8		3,100	465		3,565	4,125
0980	2,500 amp		1.80	8.889		3,775	515		4,290	4,950
0990	3,000 amp		1.60	10		4,425	580		5,005	5,750
1000	4,000 amp		1.40	11.429		5,750	665		6,415	7,325
1050	Service head, weatherproof, 3 pole 3 wire, 225 amp		3	5.333		1,725	310		2,035	2,350
1060	400 amp		2.80	5.714		1,725	335		2,060	2,375
1070	600 amp		2.60	6.154		1,725	360		2,085	2,400
1080	800 amp		2.40	6.667		1,925	390		2,315	2,700
1090	1,000 amp		2	8		2,075	465		2,540	3,000
1100	1,350 amp		1.80	8.889		2,700	515		3,215	3,750
1110	1,600 amp		1.60	10		3,025	580		3,605	4,200
1120	2,000 amp		1.40	11.429		3,675	665		4,340	5,050
1130	2,500 amp		1.20	13.333		4,400	775		5,175	5,975
1140	3,000 amp		.90	17.778		5,125	1,025		6,150	7,175
1150	4,000 amp		.70	22.857		6,500	1,325		7,825	9,150
1200	3 pole 4 wire, 225 amp		2.60	6.154		1,925	360		2,285	2,625
1210	400 amp		2.40	6.667		1,925	390		2,315	2,675
1220	600 amp		2.20	7.273		1,925	425		2,350	2,725
1230	800 amp		2	8		2,225	465		2,690	3,150
1240	1,000 amp		1.70	9.412		2,575	550		3,125	3,650
1250	1,350 amp		1.50	10.667		3,050	620		3,670	4,275
1260	1,600 amp		1.40	11.429		3,325	665		3,990	4,650
1270	2,000 amp		1.20	13.333		4,450	775		5,225	6,050
1280	2,500 amp		1	16		5,475	930		6,405	7,425

For customer support on your Electrical Costs with RSMeans data, call 800.448.8182.

257

26 25 13.60 Copper or Aluminum Bus Duct Fittings		Crew	Daily Output	Labor-Hours	Unit	Material	2018 Bare Costs Labor	Equipment	Total	Total Incl O&P
1290	3,000 amp	2 Elec	.80	20	Ea.	6,400	1,175		7,575	8,800
1300	4,000 amp		.60	26.667		8,325	1,550		9,875	11,500
1350	Flanged end, 3 pole 3 wire, 225 amp		6	2.667		935	155		1,090	1,250
1360	400 amp		5.20	3.077		935	179		1,114	1,300
1370	600 amp		4.60	3.478		935	202		1,137	1,325
1380	800 amp		3.60	4.444		1,025	259		1,284	1,525
1390	1,000 amp		3.40	4.706		1,175	274		1,449	1,675
1395	1,200 amp		3.20	5		1,300	291		1,591	1,850
1400	1,350 amp		3	5.333		1,400	310		1,710	2,000
1410	1,600 amp		2.80	5.714		1,575	335		1,910	2,250
1420	2,000 amp		2.40	6.667		1,850	390		2,240	2,625
1430	2,500 amp		2	8		2,175	465		2,640	3,100
1440	3,000 amp		1.80	8.889		2,475	515		2,990	3,500
1450	4,000 amp		1.60	10		3,075	580		3,655	4,275
1500	3 pole 4 wire, 225 amp		5.40	2.963		1,075	172		1,247	1,425
1510	400 amp		4.60	3.478		1,075	202		1,277	1,475
1520	600 amp		4	4		1,075	233		1,308	1,525
1530	800 amp		3.20	5		1,275	291		1,566	1,825
1540	1,000 amp		3	5.333		1,400	310		1,710	2,025
1545	1,200 amp		2.80	5.714		1,600	335		1,935	2,250
1550	1,350 amp		2.60	6.154		1,725	360		2,085	2,425
1560	1,600 amp		2.40	6.667		1,975	390		2,365	2,750
1570	2,000 amp		2	8		2,325	465		2,790	3,250
1580	2,500 amp		1.80	8.889		2,775	515		3,290	3,825
1590	3,000 amp		1.60	10		3,150	580		3,730	4,350
1600	4,000 amp		1.40	11.429		4,100	665		4,765	5,500
1650	Hanger, standard, 225 amp		64	.250		26	14.55		40.55	50.50
1660	400 amp		48	.333		26	19.40		45.40	57.50
1670	600 amp		40	.400		26	23.50		49.50	63.50
1680	800 amp		32	.500		26	29		55	72
1690	1,000 amp		24	.667		26	39		65	86.50
1695	1,200 amp		22	.727		26	42.50		68.50	92
1700	1,350 amp		20	.800		26	46.50		72.50	98
1710	1,600 amp		20	.800		26	46.50		72.50	98
1720	2,000 amp		18	.889		26	51.50		77.50	106
1730	2,500 amp		16	1		26	58		84	116
1740	3,000 amp		16	1		26	58		84	116
1750	4,000 amp		16	1		26	58		84	116
1800	Spring type, 225 amp		16	1		88	58		146	184
1810	400 amp		14	1.143		88	66.50		154.50	196
1820	600 amp		14	1.143		88	66.50		154.50	196
1830	800 amp		14	1.143		88	66.50		154.50	196
1840	1,000 amp		14	1.143		88	66.50		154.50	196
1845	1,200 amp		14	1.143		88	66.50		154.50	196
1850	1,350 amp		14	1.143		88	66.50		154.50	196
1860	1,600 amp		12	1.333		88	77.50		165.50	213
1870	2,000 amp		12	1.333		88	77.50		165.50	213
1880	2,500 amp		12	1.333		88	77.50		165.50	213
1890	3,000 amp		10	1.600		88	93		181	236
1900	4,000 amp		10	1.600		88	93		181	236

26 25 13.70 Feedrail	Crew	Daily Output	Labor-Hours	Unit	Material	2018 Bare Costs Labor	Equipment	Total	Total Incl O&P
0010 **FEEDRAIL**, 12' mounting									
0050 Trolley busway, 3 pole									
0100 300 volt 60 amp, plain, 10' lengths	1 Elec	50	.160	L.F.	21	9.30		30.30	37
0300 Door track		50	.160		34	9.30		43.30	51.50
0500 Curved track	↓	30	.267	↓	14.95	15.50		30.45	39.50
0700 Coupling				Ea.	11.90			11.90	13.10
0900 Center feed	1 Elec	5.30	1.509		47	88		135	183
1100 End feed		5.30	1.509		48.50	88		136.50	184
1300 Hanger set		24	.333	↓	2.80	19.40		22.20	32
3000 600 volt 100 amp, plain, 10' lengths		35	.229	L.F.	60	13.30		73.30	86
3300 Door track	↓	35	.229	"	89	13.30		102.30	117
3700 Coupling				Ea.	52			52	57
4000 End cap	1 Elec	40	.200		63.50	11.65		75.15	87.50
4200 End feed		4	2		233	116		349	430
4500 Trolley, 600 volt, 20 amp		5.30	1.509		247	88		335	405
4700 30 amp		5.30	1.509		247	88		335	405
4900 Duplex, 40 amp		4	2		880	116		996	1,150
5000 60 amp		4	2		880	116		996	1,150
5300 Fusible, 20 amp		4	2		495	116		611	720
5500 30 amp		4	2		495	116		611	720
5900 300 volt, 20 amp		5.30	1.509		232	88		320	385
6000 30 amp		5.30	1.509		282	88		370	440
6300 Fusible, 20 amp		4.70	1.702		345	99		444	530
6500 30 amp		4.70	1.702	↓	480	99		579	680
7300 Busway, 250 volt 50 amp, 2 wire	↓	70	.114	L.F.	20.50	6.65		27.15	32.50
7330 Coupling				Ea.	42.50			42.50	47
7340 Center feed	1 Elec	6	1.333		495	77.50		572.50	660
7350 End feed		6	1.333		109	77.50		186.50	236
7360 End cap		40	.200		29	11.65		40.65	49.50
7370 Hanger set		24	.333	↓	2.80	19.40		22.20	32
7400 125/250 volt 50 amp, 3 wire		60	.133	L.F.	20.50	7.75		28.25	34
7430 Coupling		6	1.333	Ea.	61.50	77.50		139	184
7440 Center feed		6	1.333		495	77.50		572.50	660
7450 End feed		6	1.333		109	77.50		186.50	236
7460 End cap		40	.200		29	11.65		40.65	49.50
7470 Hanger set		24	.333		3.75	19.40		23.15	33
7480 Trolley, 250 volt, 2 pole, 20 amp		6	1.333		37	77.50		114.50	157
7490 30 amp		6	1.333		37	77.50		114.50	157
7500 125/250 volt, 3 pole, 20 amp		6	1.333		37	77.50		114.50	157
7510 30 amp	↓	6	1.333		37	77.50		114.50	157
8000 Cleaning tools, 300 volt, dust remover					107			107	118
8100 Bus bar cleaner					202			202	222
8300 600 volt, dust remover, 60 amp					297			297	325
8400 100 amp					650			650	715
8600 Bus bar cleaner, 60 amp					675			675	745
8700 100 amp				↓	800			800	880

For customer support on your Electrical Costs with RSMeans data, call 800.448.8182.

259

26 27 13.10 Meter Centers and Sockets	Crew	Daily Output	Labor-Hours	Unit	Material	2018 Bare Costs Labor	Equipment	Total	Total Incl O&P
0010 **METER CENTERS AND SOCKETS**									
0100 Sockets, single position, 4 terminal, 100 amp	1 Elec	3.20	2.500	Ea.	45	146		191	268
0200 150 amp		2.30	3.478		62	202		264	375
0300 200 amp		1.90	4.211		96	245		341	470
0400 Transformer rated, 20 amp		3.20	2.500		162	146		308	395
0500 Double position, 4 terminal, 100 amp		2.80	2.857		212	166		378	485
0600 150 amp		2.10	3.810		249	222		471	605
0700 200 amp		1.70	4.706		515	274		789	975
0800 Trans-socket, 13 terminal, 3 CT mounts, 400 amp		1	8		1,250	465		1,715	2,075
0900 800 amp	2 Elec	1.20	13.333		1,400	775		2,175	2,700
1100 Meter centers and sockets, three phase, single pos, 7 terminal, 100 amp	1 Elec	2.80	2.857		105	166		271	365
1200 200 amp		2.10	3.810		214	222		436	565
1400 400 amp		1.70	4.706		810	274		1,084	1,300
2000 Meter center, main fusible switch, 1P 3W 120/240 V									
2030 400 amp	2 Elec	1.60	10	Ea.	660	580		1,240	1,600
2040 600 amp		1.10	14.545		1,050	845		1,895	2,425
2050 800 amp		.90	17.778		1,225	1,025		2,250	2,900
2060 Rainproof 1P 3W 120/240 V, 400 A		1.60	10		1,675	580		2,255	2,700
2070 600 amp		1.10	14.545		2,900	845		3,745	4,475
2080 800 amp		.90	17.778		4,550	1,025		5,575	6,550
2100 3P 4W 120/208 V, 400 amp		1.60	10		675	580		1,255	1,625
2110 600 amp		1.10	14.545		1,150	845		1,995	2,525
2120 800 amp		.90	17.778		1,575	1,025		2,600	3,275
2130 Rainproof 3P 4W 120/208 V, 400 amp		1.60	10		1,900	580		2,480	2,975
2140 600 amp		1.10	14.545		3,575	845		4,420	5,200
2150 800 amp		.90	17.778		6,625	1,025		7,650	8,825
2170 Main circuit breaker, 1P 3W 120/240 V									
2180 400 amp	2 Elec	1.60	10	Ea.	1,400	580		1,980	2,425
2190 600 amp		1.10	14.545		1,725	845		2,570	3,175
2200 800 amp		.90	17.778		2,325	1,025		3,350	4,125
2210 1,000 amp		.80	20		3,075	1,175		4,250	5,150
2220 1,200 amp		.76	21.053		3,175	1,225		4,400	5,325
2230 1,600 amp		.68	23.529		18,700	1,375		20,075	22,600
2240 Rainproof 1P 3W 120/240 V, 400 amp		1.60	10		2,975	580		3,555	4,150
2250 600 amp		1.10	14.545		4,025	845		4,870	5,700
2260 800 amp		.90	17.778		4,675	1,025		5,700	6,700
2270 1,000 amp		.80	20		6,450	1,175		7,625	8,850
2280 1,200 amp		.76	21.053		8,700	1,225		9,925	11,400
2300 3P 4W 120/208 V, 400 amp		1.60	10		3,300	580		3,880	4,500
2310 600 amp		1.10	14.545		4,775	845		5,620	6,525
2320 800 amp		.90	17.778		5,675	1,025		6,700	7,800
2330 1,000 amp		.80	20		7,475	1,175		8,650	9,975
2340 1,200 amp		.76	21.053		9,550	1,225		10,775	12,300
2350 1,600 amp		.68	23.529		19,500	1,375		20,875	23,500
2360 Rainproof 3P 4W 120/208 V, 400 amp		1.60	10		3,400	580		3,980	4,625
2370 600 amp		1.10	14.545		4,775	845		5,620	6,525
2380 800 amp		.90	17.778		5,675	1,025		6,700	7,800
2390 1,000 amp		.76	21.053		7,475	1,225		8,700	10,100
2400 1,200 amp		.68	23.529		9,550	1,375		10,925	12,600
2420 Main lugs terminal box, 1P 3W 120/240 V									
2430 800 amp	2 Elec	.94	17.021	Ea.	450	990		1,440	1,975
2440 1,200 amp		.72	22.222		530	1,300		1,830	2,500
2450 Rainproof 1P 3W 120/240 V, 225 amp		2.40	6.667		380	390		770	1,000

For customer support on your Electrical Costs with RSMeans data, call 800.448.8182.

26 27 13.10 Meter Centers and Sockets		Crew	Daily Output	Labor-Hours	Unit	Material	2018 Bare Costs Labor	2018 Bare Costs Equipment	Total	Total Incl O&P
2460	800 amp	2 Elec	.94	17.021	Ea.	560	990		1,550	2,100
2470	1,200 amp		.72	22.222		1,125	1,300		2,425	3,150
2500	3P 4W 120/208 V, 800 amp		.94	17.021		620	990		1,610	2,150
2510	1,200 amp		.72	22.222		1,250	1,300		2,550	3,300
2520	Rainproof 3P 4W 120/208 V, 225 amp		2.40	6.667		380	390		770	1,000
2530	800 amp		.94	17.021		620	990		1,610	2,150
2540	1,200 amp		.72	22.222		1,250	1,300		2,550	3,300
2590	Basic meter device									
2600	1P 3W 120/240 V 4 jaw 125A sockets, 3 meter	2 Elec	1	16	Ea.	360	930		1,290	1,800
2610	4 meter		.90	17.778		450	1,025		1,475	2,050
2620	5 meter		.80	20		515	1,175		1,690	2,325
2630	6 meter		.60	26.667		560	1,550		2,110	2,950
2640	7 meter		.56	28.571		1,400	1,675		3,075	4,025
2650	8 meter		.52	30.769		1,525	1,800		3,325	4,375
2660	10 meter		.48	33.333		1,925	1,950		3,875	5,025
2680	Rainproof 1P 3W 120/240 V 4 jaw 125A sockets									
2690	3 meter	2 Elec	1	16	Ea.	640	930		1,570	2,100
2700	4 meter		.90	17.778		770	1,025		1,795	2,400
2710	6 meter		.60	26.667		1,100	1,550		2,650	3,550
2720	7 meter		.56	28.571		1,400	1,675		3,075	4,025
2730	8 meter		.52	30.769		1,525	1,800		3,325	4,375
2750	1P 3W 120/240 V 4 jaw sockets									
2760	with 125A circuit breaker, 3 meter	2 Elec	1	16	Ea.	1,200	930		2,130	2,725
2770	4 meter		.90	17.778		1,525	1,025		2,550	3,225
2780	5 meter		.80	20		1,900	1,175		3,075	3,825
2790	6 meter		.60	26.667		2,225	1,550		3,775	4,775
2800	7 meter		.56	28.571		2,725	1,675		4,400	5,450
2810	8 meter		.52	30.769		3,025	1,800		4,825	6,000
2820	10 meter		.48	33.333		3,800	1,950		5,750	7,075
2830	Rainproof 1P 3W 120/240 V 4 jaw sockets									
2840	with 125A circuit breaker, 3 meter	2 Elec	1	16	Ea.	1,200	930		2,130	2,725
2850	4 meter		.90	17.778		1,525	1,025		2,550	3,225
2870	6 meter		.60	26.667		2,225	1,550		3,775	4,775
2880	7 meter		.56	28.571		2,725	1,675		4,400	5,450
2890	8 meter		.52	30.769		3,025	1,800		4,825	6,000
2920	1P 3W on 3P 4W 120/208 V system 5 jaw									
2930	125A sockets, 3 meter	2 Elec	1	16	Ea.	640	930		1,570	2,100
2940	4 meter		.90	17.778		770	1,025		1,795	2,400
2950	5 meter		.80	20		960	1,175		2,135	2,800
2960	6 meter		.60	26.667		1,100	1,550		2,650	3,550
2970	7 meter		.56	28.571		1,400	1,675		3,075	4,025
2980	8 meter		.52	30.769		1,525	1,800		3,325	4,375
2990	10 meter		.48	33.333		1,925	1,950		3,875	5,025
3000	Rainproof 1P 3W on 3P 4W 120/208 V system									
3020	5 jaw 125A sockets, 3 meter	2 Elec	1	16	Ea.	640	930		1,570	2,100
3030	4 meter		.90	17.778		770	1,025		1,795	2,400
3050	6 meter		.60	26.667		1,100	1,550		2,650	3,550
3060	7 meter		.56	28.571		1,400	1,675		3,075	4,025
3070	8 meter		.52	30.769		1,525	1,800		3,325	4,375
3090	1P 3W on 3P 4W 120/208 V system 5 jaw sockets									
3100	With 125A circuit breaker, 3 meter	2 Elec	1	16	Ea.	1,200	930		2,130	2,725
3110	4 meter		.90	17.778		1,525	1,025		2,550	3,225
3120	5 meter		.80	20		1,900	1,175		3,075	3,825

For customer support on your Electrical Costs with RSMeans data, call 800.448.8182.

261

26 27 13.10 Meter Centers and Sockets		Crew	Daily Output	Labor-Hours	Unit	Material	2018 Bare Costs Labor	Equipment	Total	Total Incl O&P
3130	6 meter	2 Elec	.60	26.667	Ea.	2,225	1,550		3,775	4,775
3140	7 meter		.56	28.571		2,725	1,675		4,400	5,450
3150	8 meter		.52	30.769		3,025	1,800		4,825	6,000
3160	10 meter	↓	.48	33.333	↓	3,800	1,950		5,750	7,075
3170	Rainproof 1P 3W on 3P 4W 120/208 V system									
3180	5 jaw sockets w/125A circuit breaker, 3 meter	2 Elec	1	16	Ea.	1,200	930		2,130	2,725
3190	4 meter		.90	17.778		1,525	1,025		2,550	3,225
3210	6 meter		.60	26.667		2,225	1,550		3,775	4,775
3220	7 meter		.56	28.571		2,725	1,675		4,400	5,450
3230	8 meter	↓	.52	30.769	↓	3,025	1,800		4,825	6,000
3250	1P 3W 120/240 V 4 jaw sockets									
3260	with 200A circuit breaker, 3 meter	2 Elec	1	16	Ea.	1,800	930		2,730	3,375
3270	4 meter		.90	17.778		2,425	1,025		3,450	4,225
3290	6 meter		.60	26.667		3,600	1,550		5,150	6,275
3300	7 meter		.56	28.571		4,225	1,675		5,900	7,125
3310	8 meter		.56	28.571		4,875	1,675		6,550	7,825
3330	Rainproof 1P 3W 120/240 V 4 jaw sockets									
3350	with 200A circuit breaker, 3 meter	2 Elec	1	16	Ea.	1,800	930		2,730	3,375
3360	4 meter		.90	17.778		2,425	1,025		3,450	4,225
3380	6 meter		.60	26.667		3,600	1,550		5,150	6,275
3390	7 meter		.56	28.571		4,225	1,675		5,900	7,125
3400	8 meter	↓	.52	30.769	↓	4,875	1,800		6,675	8,025
3420	1P 3W on 3P 4W 120/208 V 5 jaw sockets									
3430	with 200A circuit breaker, 3 meter	2 Elec	1	16	Ea.	1,800	930		2,730	3,375
3440	4 meter		.90	17.778		2,425	1,025		3,450	4,225
3460	6 meter		.60	26.667		3,600	1,550		5,150	6,275
3470	7 meter		.56	28.571		4,225	1,675		5,900	7,125
3480	8 meter	↓	.52	30.769	↓	4,875	1,800		6,675	8,025
3500	Rainproof 1P 3W on 3P 4W 120/208 V 5 jaw socket									
3510	with 200A circuit breaker, 3 meter	2 Elec	1	16	Ea.	1,800	930		2,730	3,375
3520	4 meter		.90	17.778		2,425	1,025		3,450	4,225
3540	6 meter		.60	26.667		3,600	1,550		5,150	6,275
3550	7 meter		.56	28.571		4,225	1,675		5,900	7,125
3560	8 meter	↓	.52	30.769		4,875	1,800		6,675	8,025
3600	Automatic circuit closing, add						69.50		69.50	76.50
3610	Manual circuit closing, add				↓		79.50		79.50	87.50
3650	Branch meter device									
3660	3P 4W 208/120 or 240/120 V 7 jaw sockets									
3670	with 200A circuit breaker, 2 meter	2 Elec	.90	17.778	Ea.	3,000	1,025		4,025	4,850
3680	3 meter		.80	20		4,500	1,175		5,675	6,700
3690	4 meter		.70	22.857		6,025	1,325		7,350	8,625
3700	Main circuit breaker 42,000 rms, 400 amp		1.60	10		1,750	580		2,330	2,800
3710	600 amp		1.10	14.545		2,600	845		3,445	4,125
3720	800 amp		.90	17.778		3,300	1,025		4,325	5,175
3730	Rainproof main circ. breaker 42,000 rms, 400 amp		1.60	10		2,625	580		3,205	3,750
3740	600 amp		1.10	14.545		4,250	845		5,095	5,925
3750	800 amp		.90	17.778		5,675	1,025		6,700	7,800
3760	Main circuit breaker 65,000 rms, 400 amp		1.60	10		3,575	580		4,155	4,800
3770	600 amp		1.10	14.545		5,025	845		5,870	6,800
3780	800 amp		.90	17.778		5,675	1,025		6,700	7,800
3790	1,000 amp		.80	20		7,475	1,175		8,650	9,975
3800	1,200 amp		.76	21.053		9,550	1,225		10,775	12,300
3810	1,600 amp	↓	.68	23.529		19,500	1,375		20,875	23,500

For customer support on your Electrical Costs with RSMeans data, call 800.448.8182.

26 27 13 – Electricity Metering

26 27 13.10 Meter Centers and Sockets

		Crew	Daily Output	Labor-Hours	Unit	Material	2018 Bare Costs Labor	2018 Bare Costs Equipment	Total	Total Incl O&P
3820	Rainproof main circ. breaker 65,000 rms, 400 amp	2 Elec	1.60	10	Ea.	3,575	580		4,155	4,800
3830	600 amp		1.10	14.545		5,025	845		5,870	6,800
3840	800 amp		.90	17.778		5,675	1,025		6,700	7,800
3850	1,000 amp		.80	20		7,475	1,175		8,650	9,975
3860	1,200 amp		.76	21.053		9,550	1,225		10,775	12,300
3880	Main circuit breaker 100,000 rms, 400 amp		1.60	10		3,575	580		4,155	4,800
3890	600 amp		1.10	14.545		5,025	845		5,870	6,800
3900	800 amp		.90	17.778		5,900	1,025		6,925	8,025
3910	Rainproof main circ. breaker 100,000 rms, 400 amp		1.60	10		3,575	580		4,155	4,800
3920	600 amp		1.10	14.545		5,025	845		5,870	6,800
3930	800 amp		.90	17.778		5,900	1,025		6,925	8,025
3940	Main lugs terminal box, 800 amp		.94	17.021		620	990		1,610	2,150
3950	1,600 amp		.72	22.222		2,475	1,300		3,775	4,650
3960	Rainproof, 800 amp		.94	17.021		620	990		1,610	2,150
3970	1,600 amp		.72	22.222		2,475	1,300		3,775	4,650

26 27 16 – Electrical Cabinets and Enclosures

26 27 16.10 Cabinets

			Crew	Daily Output	Labor-Hours	Unit	Material	2018 Bare Costs Labor	2018 Bare Costs Equipment	Total	Total Incl O&P
0010	**CABINETS**	R260533-70									
7000	Cabinets, current transformer										
7050	Single door, 24" H x 24" W x 10" D	R262716-40	1 Elec	1.60	5	Ea.	165	291		456	615
7100	30" H x 24" W x 10" D			1.30	6.154		179	360		539	730
7150	36" H x 24" W x 10" D			1.10	7.273		380	425		805	1,050
7200	30" H x 30" W x 10" D			1	8		246	465		711	965
7250	36" H x 30" W x 10" D			.90	8.889		310	515		825	1,125
7300	36" H x 36" W x 10" D			.80	10		315	580		895	1,225
7500	Double door, 48" H x 36" W x 10" D			.60	13.333		575	775		1,350	1,775
7550	24" H x 24" W x 12" D			1	8		179	465		644	890
8000	NEMA 12, double door, floor mounted										
8020	54" H x 42" W x 8" D		2 Elec	6	2.667	Ea.	885	155		1,040	1,200
8040	60" H x 48" W x 8" D			5.40	2.963		1,175	172		1,347	1,550
8060	60" H x 48" W x 10" D			5.40	2.963		1,225	172		1,397	1,600
8080	60" H x 60" W x 10" D			5	3.200		1,350	186		1,536	1,775
8100	72" H x 60" W x 10" D			4	4		1,600	233		1,833	2,100
8120	72" H x 72" W x 10" D			3.40	4.706		1,775	274		2,049	2,350
8140	60" H x 48" W x 12" D			3.40	4.706		1,225	274		1,499	1,750
8160	60" H x 60" W x 12" D			3.20	5		1,400	291		1,691	1,975
8180	72" H x 60" W x 12" D			3	5.333		1,600	310		1,910	2,225
8200	72" H x 72" W x 12" D			3	5.333		1,775	310		2,085	2,425
8220	60" H x 48" W x 16" D			3.20	5		1,300	291		1,591	1,850
8240	72" H x 72" W x 16" D			2.60	6.154		1,875	360		2,235	2,575
8260	60" H x 48" W x 20" D			3	5.333		1,425	310		1,735	2,025
8280	72" H x 72" W x 20" D			2.20	7.273		2,000	425		2,425	2,825
8300	60" H x 48" W x 24" D			2.60	6.154		1,500	360		1,860	2,175
8320	72" H x 72" W x 24" D			2	8		2,125	465		2,590	3,050
8340	Pushbutton enclosure, oiltight										
8360	3-1/2" H x 3-1/4" W x 2-3/4" D, for 1 P.B.		1 Elec	12	.667	Ea.	39	39		78	101
8380	5-3/4" H x 3-1/4" W x 2-3/4" D, for 2 P.B.			11	.727		44	42.50		86.50	112
8400	8" H x 3-1/4" W x 2-3/4" D, for 3 P.B.			10.50	.762		49.50	44.50		94	121
8420	10-1/4" H x 3-1/4" W x 2-3/4" D, for 4 P.B.			10.50	.762		55	44.50		99.50	127
8460	12-1/2" H x 3-1/4" W x 3" D, for 5 P.B.			9	.889		92	51.50		143.50	179
8480	9-1/2" H x 6-1/4" W x 3" D, for 6 P.B.			8.50	.941		71.50	55		126.50	161
8500	9-1/2" H x 8-1/2" W x 3" D, for 9 P.B.			8	1		79	58		137	174

For customer support on your Electrical Costs with RSMeans data, call 800.448.8182.

263

26 27 16.10 Cabinets	Crew	Daily Output	Labor-Hours	Unit	Material	2018 Bare Costs Labor	Equipment	Total	Total Incl O&P
8510	11-3/4" H x 8-1/2" W x 3" D, for 12 P.B.	1 Elec	7	1.143	Ea.	86.50	66.50		153	195
8520	11-3/4" H x 10-3/4" W x 3" D, for 16 P.B.		6.50	1.231		94.50	71.50		166	211
8540	14" H x 10-3/4" W x 3" D, for 20 P.B.		5	1.600		153	93		246	310
8560	14" H x 13" W x 3" D, for 25 P.B.		4.50	1.778		167	103		270	340
8580	Sloping front pushbutton enclosures									
8600	3-1/2" H x 7-3/4" W x 4-7/8" D, for 3 P.B.	1 Elec	10	.800	Ea.	66	46.50		112.50	143
8620	7-1/4" H x 8-1/2" W x 6-3/4" D, for 6 P.B.		8	1		97	58		155	194
8640	9-1/2" H x 8-1/2" W x 7-7/8" D, for 9 P.B.		7	1.143		115	66.50		181.50	227
8660	11-1/4" H x 8-1/2" W x 9" D, for 12 P.B.		5	1.600		130	93		223	282
8680	11-3/4" H x 10" W x 9" D, for 16 P.B.		5	1.600		143	93		236	297
8700	11-3/4" H x 13" W x 9" D, for 20 P.B.		5	1.600		166	93		259	320
8720	14" H x 13" W x 10-1/8" D, for 25 P.B.		4.50	1.778		187	103		290	360
8740	Pedestals, not including P.B. enclosure or base									
8760	Straight column 4" x 4"	1 Elec	4.50	1.778	Ea.	175	103		278	350
8780	6" x 6"		4	2		273	116		389	475
8800	Angled column 4" x 4"		4.50	1.778		202	103		305	375
8820	6" x 6"		4	2		305	116		421	510
8840	Pedestal, base 18" x 18"		10	.800		104	46.50		150.50	184
8860	24" x 24"		9	.889		238	51.50		289.50	340
8900	Electronic rack enclosures									
8920	72" H x 19" W x 24" D	1 Elec	1.50	5.333	Ea.	2,000	310		2,310	2,650
8940	72" H x 23" W x 24" D		1.50	5.333		2,250	310		2,560	2,950
8960	72" H x 19" W x 30" D		1.30	6.154		2,375	360		2,735	3,150
8980	72" H x 19" W x 36" D		1.20	6.667		2,725	390		3,115	3,575
9000	72" H x 23" W x 36" D		1.20	6.667		3,025	390		3,415	3,900
9020	NEMA 12 & 4 enclosure panels									
9040	12" x 24"	1 Elec	20	.400	Ea.	41	23.50		64.50	80.50
9060	16" x 12"		20	.400		28.50	23.50		52	66.50
9080	20" x 16"		20	.400		42	23.50		65.50	81.50
9100	20" x 20"		19	.421		51	24.50		75.50	92.50
9120	24" x 20"		18	.444		65.50	26		91.50	111
9140	24" x 24"		17	.471		75.50	27.50		103	124
9160	30" x 20"		16	.500		105	29		134	159
9180	30" x 24"		16	.500		93	29		122	146
9200	36" x 24"		15	.533		111	31		142	170
9220	36" x 30"		15	.533		146	31		177	207
9240	42" x 24"		15	.533		127	31		158	187
9260	42" x 30"		14	.571		165	33.50		198.50	231
9280	42" x 36"		14	.571		198	33.50		231.50	268
9300	48" x 24"		14	.571		89	33.50		122.50	148
9320	48" x 30"		14	.571		115	33.50		148.50	177
9340	48" x 36"		13	.615		221	36		257	297
9360	60" x 36"		12	.667		272	39		311	355
9400	Wiring trough steel JIC, clamp cover									
9490	4" x 4", 12" long	1 Elec	12	.667	Ea.	61.50	39		100.50	126
9510	24" long		10	.800		77.50	46.50		124	155
9530	36" long		8	1		97	58		155	194
9540	48" long		7	1.143		112	66.50		178.50	224
9550	60" long		6	1.333		132	77.50		209.50	261
9560	6" x 6", 12" long		11	.727		81.50	42.50		124	153
9580	24" long		9	.889		108	51.50		159.50	197
9600	36" long		7	1.143		132	66.50		198.50	245
9610	48" long		6	1.333		161	77.50		238.50	293

For customer support on your Electrical Costs with RSMeans data, call 800.448.8182.

26 27 16.10 Cabinets		Crew	Daily Output	Labor-Hours	Unit	Material	2018 Bare Costs Labor	2018 Bare Costs Equipment	Total	Total Incl O&P
9620	60" long	1 Elec	5	1.600	Ea.	189	93		282	345
26 27 16.20 Cabinets and Enclosures										
0010	**CABINETS AND ENCLOSURES** Nonmetallic									
0080	Enclosures fiberglass NEMA 4X									
0100	Wall mount, quick release latch door, 20" H x 16" W x 6" D	1 Elec	4.80	1.667	Ea.	655	97		752	865
0110	20" H x 20" W x 6" D		4.50	1.778		780	103		883	1,000
0120	24" H x 20" W x 6" D		4.20	1.905		830	111		941	1,075
0130	20" H x 16" W x 8" D		4.50	1.778		745	103		848	975
0140	20" H x 20" W x 8" D		4.20	1.905		825	111		936	1,075
0150	24" H x 24" W x 8" D		3.80	2.105		935	123		1,058	1,200
0160	30" H x 24" W x 8" D		3.20	2.500		1,025	146		1,171	1,350
0170	36" H x 30" W x 8" D		3	2.667		1,425	155		1,580	1,800
0180	20" H x 16" W x 10" D		3.50	2.286		860	133		993	1,150
0190	20" H x 20" W x 10" D		3.20	2.500		920	146		1,066	1,225
0200	24" H x 20" W x 10" D		3	2.667		1,000	155		1,155	1,325
0210	30" H x 24" W x 10" D		2.80	2.857		1,125	166		1,291	1,475
0220	20" H x 16" W x 12" D		3	2.667		930	155		1,085	1,250
0230	20" H x 20" W x 12" D		2.80	2.857		970	166		1,136	1,325
0240	24" H x 24" W x 12" D		2.60	3.077		1,075	179		1,254	1,475
0250	30" H x 24" W x 12" D		2.40	3.333		1,200	194		1,394	1,625
0260	36" H x 30" W x 12" D		2.20	3.636		1,650	212		1,862	2,150
0270	36" H x 36" W x 12" D		2.10	3.810		1,825	222		2,047	2,325
0280	48" H x 36" W x 12" D		2	4		2,150	233		2,383	2,700
0290	60" H x 36" W x 12" D		1.80	4.444		1,475	259		1,734	2,000
0300	30" H x 24" W x 16" D		1.40	5.714		1,400	335		1,735	2,025
0310	48" H x 36" W x 16" D		1.20	6.667		2,325	390		2,715	3,125
0320	60" H x 36" W x 16" D		1	8		2,625	465		3,090	3,575
0480	Freestanding, one door, 72" H x 25" W x 25" D		.80	10		2,775	580		3,355	3,925
0490	Two doors with two panels, 72" H x 49" W x 24" D		.50	16		6,950	930		7,880	9,025
0500	Floor stand kits, for NEMA 4 & 12, 20" W or more, 6" H x 8" D		24	.333		106	19.40		125.40	146
0510	6" H x 10" D		24	.333		116	19.40		135.40	157
0520	6" H x 12" D		24	.333		130	19.40		149.40	172
0530	6" H x 18" D		24	.333		151	19.40		170.40	195
0540	12" H x 8" D		22	.364		134	21		155	180
0550	12" H x 10" D		22	.364		240	21		261	296
0560	12" H x 12" D		22	.364		256	21		277	315
0570	12" H x 16" D		22	.364		177	21		198	227
0580	12" H x 18" D		22	.364		181	21		202	231
0590	12" H x 20" D		22	.364		193	21		214	244
0600	18" H x 8" D		20	.400		160	23.50		183.50	211
0610	18" H x 10" D		20	.400		172	23.50		195.50	224
0620	18" H x 12" D		20	.400		184	23.50		207.50	238
0630	18" H x 16" D		20	.400		204	23.50		227.50	260
0640	24" H x 8" D		16	.500		198	29		227	262
0650	24" H x 10" D		16	.500		205	29		234	269
0660	24" H x 12" D		16	.500		221	29		250	287
0670	24" H x 16" D		16	.500		231	29		260	298
0680	Small, screw cover, 5-1/2" H x 4" W x 4-15/16" D		12	.667		101	39		140	170
0690	7-1/2" H x 4" W x 4-15/16" D		12	.667		66.50	39		105.50	132
0700	7-1/2" H x 6" W x 5-3/16" D		10	.800		124	46.50		170.50	207
0710	9-1/2" H x 6" W x 5-11/16" D		10	.800		129	46.50		175.50	211
0720	11-1/2" H x 8" W x 6-11/16" D		8	1		194	58		252	300

26 27 16.20 Cabinets and Enclosures	Crew	Daily Output	Labor-Hours	Unit	Material	2018 Bare Costs Labor	Equipment	Total	Total Incl O&P
0730	13-1/2" H x 10" W x 7-3/16" D	1 Elec	7	1.143	Ea.	227	66.50		293.50	350
0740	15-1/2" H x 12" W x 8-3/16" D		6	1.333		291	77.50		368.50	435
0750	17-1/2" H x 14" W x 8-11/16" D		5	1.600		345	93		438	520
0760	Screw cover with window, 6" H x 4" W x 5" D		12	.667		173	39		212	248
0770	8" H x 4" W x 5" D		11	.727		178	42.50		220.50	260
0780	8" H x 6" W x 5" D		11	.727		249	42.50		291.50	340
0790	10" H x 6" W x 6" D		10	.800		263	46.50		309.50	360
0800	12" H x 8" W x 7" D		8	1		330	58		388	450
0810	14" H x 10" W x 7" D		7	1.143		400	66.50		466.50	540
0820	16" H x 12" W x 8" D		6	1.333		475	77.50		552.50	640
0830	18" H x 14" W x 9" D		5	1.600		570	93		663	770
0840	Quick-release latch cover, 5-1/2" H x 4" W x 5" D		12	.667		132	39		171	203
0850	7-1/2" H x 4" W x 5" D		12	.667		134	39		173	206
0860	7-1/2" H x 6" W x 5-1/4" D		10	.800		158	46.50		204.50	244
0870	9-1/2" H x 6" W x 5-3/4" D		10	.800		108	46.50		154.50	189
0880	11-1/2" H x 8" W x 6-3/4" D		8	1		154	58		212	257
0890	13-1/2" H x 10" W x 7-1/4" D		7	1.143		193	66.50		259.50	310
0900	15-1/2" H x 12" W x 8-1/4" D		6	1.333		385	77.50		462.50	540
0910	17-1/2" H x 14" W x 8-3/4" D		5	1.600		460	93		553	650
0920	Pushbutton, 1 hole 5-1/2" H x 4" W x 4-15/16" D		12	.667		440	39		479	545
0930	2 hole 7-1/2" H x 4" W x 4-15/16" D		11	.727		615	42.50		657.50	740
0940	4 hole 7-1/2" H x 6" W x 5-3/16" D		10.50	.762		425	44.50		469.50	535
0950	6 hole 9-1/2" H x 6" W x 5-11/16" D		9	.889		515	51.50		566.50	650
0960	8 hole 11-1/2" H x 8" W x 6-11/16" D		8.50	.941		660	55		715	805
0970	12 hole 13-1/2" H x 10" W x 7-3/16" D		8	1		865	58		923	1,025
0980	20 hole 15-1/2" H x 12" W x 8-3/16" D		5	1.600		830	93		923	1,050
0990	30 hole 17-1/2" H x 14" W x 8-11/16" D	↓	4.50	1.778	↓	985	103		1,088	1,225
1450	Enclosures polyester NEMA 4X									
1460	Small, screw cover,									
1500	3-15/16" H x 3-15/16" W x 3-1/16" D	1 Elec	12	.667	Ea.	64	39		103	129
1510	5-3/16" H x 3-5/16" W x 3-1/16" D		12	.667		63	39		102	128
1520	5-7/8" H x 3-7/8" W x 4-3/16" D		12	.667		66.50	39		105.50	131
1530	5-7/8" H x 5-7/8" W x 4-3/16" D		12	.667		74.50	39		113.50	140
1540	7-5/8" H x 3-5/16" W x 3-1/16" D		12	.667		69	39		108	134
1550	10-3/16" H x 3-5/16" W x 3-1/16" D		10	.800		81.50	46.50		128	160
1560	Clear cover, 3-15/16" H x 3-15/16" W x 2-7/8" D		12	.667		72.50	39		111.50	138
1570	5-3/16" H x 3-5/16" W x 2-7/8" D		12	.667		77	39		116	143
1580	5-7/8" H x 3-7/8" W x 4" D		12	.667		95.50	39		134.50	163
1590	5-7/8" H x 5-7/8" W x 4" D		12	.667		118	39		157	188
1600	7-5/8" H x 3-5/16" W x 2-7/8" D		12	.667		86.50	39		125.50	153
1610	10-3/16" H x 3-5/16" W x 2-7/8" D		10	.800		109	46.50		155.50	189
1620	Pushbutton, 1 hole, 5-5/16" H x 3-5/16" W x 3-1/16" D		12	.667		56	39		95	120
1630	2 hole, 7-5/8" H x 3-5/16" W x 3-1/8" D		11	.727		63	42.50		105.50	133
1640	3 hole, 10-3/16" H x 3-5/16" W x 3-1/16" D		10.50	.762		74.50	44.50		119	149
8000	Wireway fiberglass, straight sect. screwcover, 12" L, 4" W x 4" D		40	.200		196	11.65		207.65	232
8010	6" W x 6" D		30	.267		266	15.50		281.50	315
8020	24" L, 4" W x 4" D		20	.400		253	23.50		276.50	315
8030	6" W x 6" D		15	.533		390	31		421	475
8040	36" L, 4" W x 4" D		13.30	.602		305	35		340	390
8050	6" W x 6" D		10	.800		490	46.50		536.50	610
8060	48" L, 4" W x 4" D		10	.800		360	46.50		406.50	465
8070	6" W x 6" D		7.50	1.067		585	62		647	735
8080	60" L, 4" W x 4" D	↓	8	1	↓	415	58		473	540

26 27 16 – Electrical Cabinets and Enclosures

26 27 16.20 Cabinets and Enclosures		Crew	Daily Output	Labor-Hours	Unit	Material	2018 Bare Costs Labor	2018 Bare Costs Equipment	Total	Total Incl O&P
8090	6" W x 6" D	1 Elec	6	1.333	Ea.	685	77.50		762.50	870
8100	Elbow, 90°, 4" W x 4" D		20	.400		198	23.50		221.50	253
8110	6" W x 6" D		18	.444		400	26		426	480
8120	Elbow, 45°, 4" W x 4" D		20	.400		200	23.50		223.50	255
8130	6" W x 6" D		18	.444		390	26		416	470
8140	Tee, 4" W x 4" D		16	.500		251	29		280	320
8150	6" W x 6" D		14	.571		455	33.50		488.50	555
8160	Cross, 4" W x 4" D		14	.571		233	33.50		266.50	305
8170	6" W x 6" D		12	.667		690	39		729	815
8180	Cut-off fitting, w/flange & adhesive, 4" W x 4" D		18	.444		152	26		178	207
8190	6" W x 6" D		16	.500		310	29		339	385
8200	Flexible ftng., hvy. neoprene coated nylon, 4" W x 4" D		20	.400		281	23.50		304.50	345
8210	6" W x 6" D		18	.444		420	26		446	505
8220	Closure plate, fiberglass, 4" W x 4" D		20	.400		54	23.50		77.50	94.50
8230	6" W x 6" D		18	.444		61.50	26		87.50	107
8240	Box connector, stainless steel type 304, 4" W x 4" D		20	.400		95.50	23.50		119	140
8250	6" W x 6" D		18	.444		106	26		132	156
8260	Hanger, 4" W x 4" D		100	.080		23.50	4.66		28.16	32.50
8270	6" W x 6" D		80	.100		28.50	5.80		34.30	40
8280	Straight tube section fiberglass, 4" W x 4" D, 12" long		40	.200		172	11.65		183.65	206
8290	24" long		20	.400		194	23.50		217.50	248
8300	36" long		13.30	.602		164	35		199	234
8310	48" long		10	.800		270	46.50		316.50	365
8320	60" long		8	1		310	58		368	425
8330	120" long		4	2		485	116		601	705

26 27 19 – Multi-Outlet Assemblies

26 27 19.10 Wiring Duct

		Crew	Daily Output	Labor-Hours	Unit	Material	2018 Bare Costs Labor	2018 Bare Costs Equipment	Total	Total Incl O&P
0010	**WIRING DUCT** Plastic									
1250	PVC, snap-in slots, adhesive backed									
1270	1-1/2" W x 2" H	2 Elec	120	.133	L.F.	4.86	7.75		12.61	16.95
1280	1-1/2" W x 3" H		120	.133		5.85	7.75		13.60	18
1290	1-1/2" W x 4" H		120	.133		6.90	7.75		14.65	19.20
1300	2" W x 1" H		120	.133		4.36	7.75		12.11	16.40
1310	2" W x 1-1/2" H		120	.133		4.73	7.75		12.48	16.80
1320	2" W x 2" H		120	.133		4.75	7.75		12.50	16.85
1340	2" W x 3" H		120	.133		6.20	7.75		13.95	18.40
1350	2" W x 4" H		120	.133		7.50	7.75		15.25	19.85
1360	2-1/2" W x 3" H		120	.133		6.65	7.75		14.40	18.95
1370	3" W x 1" H		110	.145		4.95	8.45		13.40	18.10
1390	3" W x 2" H		110	.145		5.90	8.45		14.35	19.15
1400	3" W x 3" H		110	.145		7.15	8.45		15.60	20.50
1410	3" W x 4" H		110	.145		8.95	8.45		17.40	22.50
1420	3" W x 5" H		110	.145		11.75	8.45		20.20	25.50
1430	4" W x 1-1/2" H		100	.160		5.90	9.30		15.20	20.50
1440	4" W x 2" H		100	.160		6.90	9.30		16.20	21.50
1450	4" W x 3" H		100	.160		7.80	9.30		17.10	22.50
1460	4" W x 4" H		100	.160		9.45	9.30		18.75	24.50
1470	4" W x 5" H		100	.160		13.30	9.30		22.60	28.50
1550	Cover, 1-1/2" W		200	.080		.99	4.66		5.65	8.05
1560	2" W		200	.080		1.07	4.66		5.73	8.15
1570	2-1/2" W		200	.080		1.53	4.66		6.19	8.65
1580	3" W		200	.080		1.62	4.66		6.28	8.75

26 27 19 – Multi-Outlet Assemblies

26 27 19.10 Wiring Duct	Crew	Daily Output	Labor-Hours	Unit	Material	2018 Bare Costs Labor	Equipment	Total	Total Incl O&P
1590	4" W	2 Elec	200	.080	L.F.	1.96	4.66		6.62	9.10

26 27 23 – Indoor Service Poles

26 27 23.40 Surface Raceway

26 27 23.40 Surface Raceway		Crew	Daily Output	Labor-Hours	Unit	Material	Labor	Equipment	Total	Total Incl O&P
0010	**SURFACE RACEWAY**									
0090	Metal, straight section									
0100	No. 500	1 Elec	100	.080	L.F.	1.07	4.66		5.73	8.15
0110	No. 700		100	.080		1.19	4.66		5.85	8.25
0200	No. 1000		90	.089		1.45	5.15		6.60	9.35
0400	No. 1500, small pancake		90	.089		2.20	5.15		7.35	10.15
0600	No. 2000, base & cover, blank		90	.089		2.24	5.15		7.39	10.20
0610	Receptacle, 6" OC		40	.200		20.50	11.65		32.15	40
0620	12" OC		44	.182		13.35	10.60		23.95	30.50
0630	18" OC		46	.174		11.05	10.10		21.15	27.50
0650	30" OC		50	.160		5.15	9.30		14.45	19.55
0660	60" OC		50	.160		4.75	9.30		14.05	19.15
0670	No. 2400, base & cover, blank		80	.100		1.95	5.80		7.75	10.85
0680	Receptacle, 6" OC		42	.190		40.50	11.10		51.60	61.50
0690	12" OC		53	.151		28.50	8.80		37.30	44
0700	18" OC		55	.145		20.50	8.45		28.95	35
0710	24" OC		57	.140		9.85	8.15		18	23
0720	30" OC		59	.136		6.75	7.90		14.65	19.25
0730	60" OC		61	.131		5.45	7.65		13.10	17.35
0800	No. 3000, base & cover, blank		75	.107		4.28	6.20		10.48	14
0810	Receptacle, 6" OC		45	.178		37.50	10.35		47.85	57
0820	12" OC		62	.129		21	7.50		28.50	35
0830	18" OC		64	.125		17.15	7.30		24.45	30
0840	24" OC		66	.121		12.85	7.05		19.90	24.50
0850	30" OC		68	.118		11.55	6.85		18.40	23
0860	60" OC		70	.114		8.45	6.65		15.10	19.25
1000	No. 4000, base & cover, blank		65	.123		6.95	7.15		14.10	18.35
1010	Receptacle, 6" OC		41	.195		50.50	11.35		61.85	72.50
1020	12" OC		52	.154		31.50	8.95		40.45	48
1030	18" OC		54	.148		25.50	8.60		34.10	41
1040	24" OC		56	.143		21	8.30		29.30	35.50
1050	30" OC		58	.138		19	8.05		27.05	33
1060	60" OC		60	.133		14.60	7.75		22.35	27.50
1200	No. 6000, base & cover, blank		50	.160		11.75	9.30		21.05	27
1210	Receptacle, 6" OC		30	.267		61.50	15.50		77	91
1220	12" OC		37	.216		41	12.60		53.60	64
1230	18" OC		39	.205		34.50	11.95		46.45	56
1240	24" OC		41	.195		28.50	11.35		39.85	48
1250	30" OC		43	.186		27	10.85		37.85	45.50
1260	60" OC		45	.178		21	10.35		31.35	38.50
2400	Fittings, elbows, No. 500		40	.200	Ea.	1.94	11.65		13.59	19.55
2800	Elbow cover, No. 2000		40	.200		3.64	11.65		15.29	21.50
2880	Tee, No. 500		42	.190		3.74	11.10		14.84	20.50
2900	No. 2000		27	.296		12.10	17.25		29.35	39.50
3000	Switch box, No. 500		16	.500		12.50	29		41.50	57.50
3400	Telephone outlet, No. 1500		16	.500		14.50	29		43.50	59.50
3600	Junction box, No. 1500		16	.500		10.30	29		39.30	55
3800	Plugmold wired sections, No. 2000									
4000	1 circuit, 6 outlets, 3' long	1 Elec	8	1	Ea.	38	58		96	129

268

For customer support on your Electrical Costs with RSMeans data, call 800.448.8182.

26 27 23.40 Surface Raceway

		Crew	Daily Output	Labor-Hours	Unit	Material	2018 Bare Costs Labor	2018 Bare Costs Equipment	Total	Total Incl O&P
4100	2 circuits, 8 outlets, 6' long	1 Elec	5.30	1.509	Ea.	62.50	88		150.50	200
4110	Tele-power pole, alum, w/2 recept, 10'		4	2		215	116		331	410
4120	12'		3.85	2.078		256	121		377	460
4130	15'		3.70	2.162		330	126		456	550
4140	Steel, w/2 recept, 10'		4	2		137	116		253	325
4150	One phone fitting, 10'		4	2		147	116		263	335
4160	Alum, 4 outlets, 10'		3.70	2.162		280	126		406	500
4300	Overhead distribution systems, 125 volt									
4800	No. 2000, entrance end fitting	1 Elec	20	.400	Ea.	5.60	23.50		29.10	41
5000	Blank end fitting		40	.200		2.25	11.65		13.90	19.90
5200	Supporting clip		40	.200		1.33	11.65		12.98	18.85
5800	No. 3000, entrance end fitting		20	.400		9.20	23.50		32.70	45
6000	Blank end fitting		40	.200		2.62	11.65		14.27	20.50
6020	Internal elbow		20	.400		12.10	23.50		35.60	48.50
6030	External elbow		20	.400		17.10	23.50		40.60	54
6040	Device bracket		53	.151		4.10	8.80		12.90	17.65
6400	Hanger clamp		32	.250		6.25	14.55		20.80	29
7000	No. 4000 base		90	.089	L.F.	4.40	5.15		9.55	12.60
7200	Divider		100	.080	"	.85	4.66		5.51	7.90
7400	Entrance end fitting		16	.500	Ea.	24.50	29		53.50	70.50
7600	Blank end fitting		40	.200		6.20	11.65		17.85	24
7610	Recpt. & tele. cover		53	.151		10.55	8.80		19.35	25
7620	External elbow		16	.500		34.50	29		63.50	81.50
7630	Coupling		53	.151		5.25	8.80		14.05	18.95
7640	Divider clip & coupling		80	.100		1.04	5.80		6.84	9.85
7650	Panel connector		16	.500		22	29		51	67.50
7800	Take off connector		16	.500		72.50	29		101.50	124
8000	No. 6000, take off connector		16	.500		86.50	29		115.50	139
8100	Take off fitting		16	.500		64	29		93	114
8200	Hanger clamp		32	.250		15.55	14.55		30.10	39
8230	Coupling					8.50			8.50	9.35
8240	One gang device plate	1 Elec	53	.151		8.30	8.80		17.10	22.50
8250	Two gang device plate		40	.200		10.15	11.65		21.80	28.50
8260	Blank end fitting		40	.200		8.80	11.65		20.45	27
8270	Combination elbow		14	.571		36	33.50		69.50	89
8300	Panel connector		16	.500		15.75	29		44.75	61
8500	Chan-L-Wire system installed in 1-5/8" x 1-5/8" strut. Strut									
8600	not incl., 30 amp, 4 wire, 3 phase	1 Elec	200	.040	L.F.	5	2.33		7.33	9
8700	Junction box		8	1	Ea.	33.50	58		91.50	124
8800	Insulating end cap		40	.200		8.90	11.65		20.55	27
8900	Strut splice plate		40	.200		11.80	11.65		23.45	30.50
9000	Tap		40	.200		23.50	11.65		35.15	43.50
9100	Fixture hanger		60	.133		10.15	7.75		17.90	23
9200	Pulling tool					90			90	99
9300	Non-metallic, straight section									
9310	7/16" x 7/8", base & cover, blank	1 Elec	160	.050	L.F.	1.72	2.91		4.63	6.25
9320	Base & cover w/adhesive		160	.050		1.45	2.91		4.36	5.95
9340	7/16" x 1-5/16", base & cover, blank		145	.055		1.88	3.21		5.09	6.85
9350	Base & cover w/adhesive		145	.055		2.21	3.21		5.42	7.25
9370	11/16" x 2-1/4", base & cover, blank		130	.062		2.66	3.58		6.24	8.30
9380	Base & cover w/adhesive		130	.062		3.07	3.58		6.65	8.75
9385	1-11/16" x 5-1/4", two compartment base & cover w/screws		80	.100		9.90	5.80		15.70	19.60
9400	Fittings, elbows, 7/16" x 7/8"		50	.160	Ea.	1.92	9.30		11.22	16

26 27 Low-Voltage Distribution Equipment

26 27 23 – Indoor Service Poles

26 27 23.40 Surface Raceway		Crew	Daily Output	Labor-Hours	Unit	Material	2018 Bare Costs Labor	Equipment	Total	Total Incl O&P
9410	7/16" x 1-5/16"	1 Elec	45	.178	Ea.	1.99	10.35		12.34	17.65
9420	11/16" x 2-1/4"		40	.200		2.15	11.65		13.80	19.75
9425	1-11/16" x 5-1/4"		28	.286		11.85	16.65		28.50	38
9430	Tees, 7/16" x 7/8"		35	.229		2.48	13.30		15.78	22.50
9440	7/16" x 1-5/16"		32	.250		2.54	14.55		17.09	25
9450	11/16" x 2-1/4"		30	.267		2.65	15.50		18.15	26
9455	1-11/16" x 5-1/4"		24	.333		18.85	19.40		38.25	49.50
9460	Cover clip, 7/16" x 7/8"		80	.100		.50	5.80		6.30	9.25
9470	7/16" x 1-5/16"		72	.111		.45	6.45		6.90	10.15
9480	11/16" x 2-1/4"		64	.125		.78	7.30		8.08	11.75
9484	1-11/16" x 5-1/4"		42	.190		2.77	11.10		13.87	19.60
9486	Wire clip, 1-11/16" x 5-1/4"		68	.118		.47	6.85		7.32	10.75
9490	Blank end, 7/16" x 7/8"		50	.160		.72	9.30		10.02	14.70
9500	7/16" x 1-5/16"		45	.178		.80	10.35		11.15	16.35
9510	11/16" x 2-1/4"		40	.200		1.20	11.65		12.85	18.70
9515	1-11/16" x 5-1/4"		38	.211		5.55	12.25		17.80	24.50
9520	Round fixture box, 5.5" diam. x 1"		25	.320		11.30	18.60		29.90	40.50
9530	Device box, 1 gang		30	.267		5.10	15.50		20.60	28.50
9540	2 gang		25	.320		7.65	18.60		26.25	36.50

26 27 26 – Wiring Devices

26 27 26.10 Low Voltage Switching

		Crew	Daily Output	Labor-Hours	Unit	Material	2018 Bare Costs Labor	Equipment	Total	Total Incl O&P
0010	**LOW VOLTAGE SWITCHING**									
3600	Relays, 120 V or 277 V standard	1 Elec	12	.667	Ea.	41	39		80	103
3800	Flush switch, standard		40	.200		11.50	11.65		23.15	30
4000	Interchangeable		40	.200		15.05	11.65		26.70	34
4100	Surface switch, standard		40	.200		8.20	11.65		19.85	26.50
4200	Transformer 115 V to 25 V		12	.667		128	39		167	199
4400	Master control, 12 circuit, manual		4	2		128	116		244	315
4500	25 circuit, motorized		4	2		141	116		257	330
4600	Rectifier, silicon		12	.667		46	39		85	109
4800	Switchplates, 1 gang, 1, 2 or 3 switch, plastic		80	.100		5	5.80		10.80	14.20
5000	Stainless steel		80	.100		11.35	5.80		17.15	21
5400	2 gang, 3 switch, stainless steel		53	.151		23	8.80		31.80	38
5500	4 switch, plastic		53	.151		10.45	8.80		19.25	24.50
5600	2 gang, 4 switch, stainless steel		53	.151		22	8.80		30.80	37
5700	6 switch, stainless steel		53	.151		43.50	8.80		52.30	61
5800	3 gang, 9 switch, stainless steel		32	.250		65.50	14.55		80.05	94
5900	Receptacle, triple, 1 return, 1 feed		26	.308		43	17.90		60.90	74.50
6000	2 feed		20	.400		43	23.50		66.50	82.50
6100	Relay gang boxes, flush or surface, 6 gang		5.30	1.509		90.50	88		178.50	231
6200	12 gang		4.70	1.702		102	99		201	260
6400	18 gang		4	2		113	116		229	298
6500	Frame, to hold up to 6 relays		12	.667		84	39		123	151
7200	Control wire, 2 conductor		6.30	1.270	C.L.F.	28	74		102	141
7400	3 conductor		5	1.600		39.50	93		132.50	183
7600	19 conductor		2.50	3.200		320	186		506	635
7800	26 conductor		2	4		435	233		668	830
8000	Weatherproof, 3 conductor		5	1.600		83	93		176	231

26 27 26.20 Wiring Devices Elements

		Crew	Daily Output	Labor-Hours	Unit	Material	2018 Bare Costs Labor	Equipment	Total	Total Incl O&P
0010	**WIRING DEVICES ELEMENTS** R262726-90									
0200	Toggle switch, quiet type, single pole, 15 amp	1 Elec	40	.200	Ea.	.49	11.65		12.14	17.95
0500	20 amp		27	.296		2.91	17.25		20.16	29

For customer support on your Electrical Costs with RSMeans data, call 800.448.8182.

26 27 26.20 Wiring Devices Elements		Crew	Daily Output	Labor-Hours	Unit	Material	2018 Bare Costs Labor	Equipment	Total	Total Incl O&P
0510	30 amp	1 Elec	23	.348	Ea.	24.50	20		44.50	57.50
0530	Lock handle, 20 amp		27	.296		27	17.25		44.25	55.50
0540	Security key, 20 amp		26	.308		87.50	17.90		105.40	123
0550	Rocker, 15 amp		40	.200		2.31	11.65		13.96	19.95
0560	20 amp		27	.296		6.45	17.25		23.70	33
0600	3 way, 15 amp		23	.348		1.52	20		21.52	32
0800	20 amp		18	.444		3.62	26		29.62	42.50
0810	30 amp		9	.889		24	51.50		75.50	104
0830	Lock handle, 20 amp		18	.444		29	26		55	70
0840	Security key, 20 amp		17	.471		114	27.50		141.50	166
0850	Rocker, 15 amp		23	.348		5.60	20		25.60	36.50
0860	20 amp		18	.444		8.55	26		34.55	48
0900	4 way, 15 amp		15	.533		10.70	31		41.70	58.50
1000	20 amp		11	.727		42.50	42.50		85	111
1020	Lock handle, 20 amp		11	.727		37.50	42.50		80	105
1030	Rocker, 15 amp		15	.533		8.75	31		39.75	56
1040	20 amp		11	.727		17.50	42.50		60	83
1100	Toggle switch, quiet type, double pole, 15 amp		15	.533		12.30	31		43.30	60
1200	20 amp		11	.727		17.10	42.50		59.60	82.50
1210	30 amp		9	.889		29.50	51.50		81	110
1230	Lock handle, 20 amp		11	.727		21	42.50		63.50	86.50
1250	Security key, 20 amp		10	.800		87	46.50		133.50	165
1420	Toggle switch quiet type, 1 pole, 2 throw center off, 15 amp		23	.348		40.50	20		60.50	75
1440	20 amp		18	.444		58	26		84	102
1460	2 pole, 2 throw center off, lock handle, 20 amp		11	.727		65.50	42.50		108	136
1480	1 pole, momentary contact, 15 amp		23	.348		17.65	20		37.65	50
1500	20 amp		18	.444		27	26		53	68
1520	Momentary contact, lock handle, 20 amp		18	.444		31	26		57	72.50
1650	Dimmer switch, 120 volt, incandescent, 600 watt, 1 pole	G	16	.500		22.50	29		51.50	68.50
1700	600 watt, 3 way	G	12	.667		10.30	39		49.30	69.50
1750	1,000 watt, 1 pole	G	16	.500		40	29		69	88
1800	1,000 watt, 3 way	G	12	.667		73	39		112	139
2000	1,500 watt, 1 pole	G	11	.727		84	42.50		126.50	156
2100	2,000 watt, 1 pole	G	8	1		129	58		187	229
2110	Fluorescent, 600 watt	G	15	.533		109	31		140	167
2120	1,000 watt	G	15	.533		133	31		164	193
2130	1,500 watt	G	10	.800		246	46.50		292.50	340
2160	Explosion proof, toggle switch, wall, single pole 20 amp		5.30	1.509		218	88		306	370
2180	Receptacle, single outlet, 20 amp		5.30	1.509		365	88		453	530
2190	30 amp		4	2		840	116		956	1,100
2290	60 amp		2.50	3.200		775	186		961	1,125
2360	Plug, 20 amp		16	.500		200	29		229	264
2370	30 amp		12	.667		355	39		394	455
2380	60 amp		8	1		505	58		563	645
2410	Furnace, thermal cutoff switch with plate		26	.308		15.90	17.90		33.80	44.50
2460	Receptacle, duplex, 120 volt, grounded, 15 amp		40	.200		1.30	11.65		12.95	18.85
2470	20 amp		27	.296		9.05	17.25		26.30	36
2480	Ground fault interrupting, 15 amp		27	.296		13.75	17.25		31	41
2482	20 amp		27	.296		40	17.25		57.25	70
2486	Clock receptacle, 15 amp		40	.200		27.50	11.65		39.15	47.50
2490	Dryer, 30 amp		15	.533		4.44	31		35.44	51.50
2500	Range, 50 amp		11	.727		10.60	42.50		53.10	75
2530	Surge suppressor receptacle, duplex, 20 amp		27	.296		38	17.25		55.25	68

For customer support on your Electrical Costs with RSMeans data, call 800.448.8182.

271

26 27 26 – Wiring Devices

26 27 26.20 Wiring Devices Elements	Crew	Daily Output	Labor-Hours	Unit	Material	2018 Bare Costs Labor	Equipment	Total	Total Incl O&P
2532	Quad, 20 amp	1 Elec	20	.400	Ea.	86.50	23.50		110	130
2540	Isolated ground receptacle, duplex, 20 amp		27	.296		23.50	17.25		40.75	52
2542	Quad, 20 amp		20	.400		30.50	23.50		54	68.50
2550	Simplex, 20 amp		27	.296		26.50	17.25		43.75	55
2560	Simplex, 30 amp		15	.533		31	31		62	80.50
2570	Cable reel w/receptacle 50' w/3#12, 120 V, 20 A	2 Elec	2.67	5.999		1,050	350		1,400	1,675
2600	Wall plates, stainless steel, 1 gang	1 Elec	80	.100		2.56	5.80		8.36	11.50
2800	2 gang		53	.151		4.33	8.80		13.13	17.90
3000	3 gang		32	.250		8.45	14.55		23	31.50
3100	4 gang		27	.296		11	17.25		28.25	38
3110	Brown plastic, 1 gang		80	.100		.38	5.80		6.18	9.10
3120	2 gang		53	.151		.75	8.80		9.55	14
3130	3 gang		32	.250		1.11	14.55		15.66	23
3140	4 gang		27	.296		2.99	17.25		20.24	29.50
3150	Brushed brass, 1 gang		80	.100		4.26	5.80		10.06	13.40
3160	Anodized aluminum, 1 gang		80	.100		3.01	5.80		8.81	12
3170	Switch cover, weatherproof, 1 gang		60	.133		5.25	7.75		13	17.40
3180	Vandal proof lock, 1 gang		60	.133		14.20	7.75		21.95	27
3200	Lampholder, keyless		26	.308		16.85	17.90		34.75	45.50
3400	Pullchain with receptacle		22	.364		20.50	21		41.50	54
3500	Pilot light, neon with jewel		27	.296		9.10	17.25		26.35	36
3600	Receptacle, 20 amp, 250 volt, NEMA 6		27	.296		21.50	17.25		38.75	49.50
3620	277 volt NEMA 7		27	.296		20.50	17.25		37.75	49
3640	125/250 volt NEMA 10		27	.296		20	17.25		37.25	48
3680	125/250 volt NEMA 14		25	.320		25.50	18.60		44.10	56
3700	3 pole, 250 volt NEMA 15		25	.320		28	18.60		46.60	59
3720	120/208 volt NEMA 18		25	.320		27.50	18.60		46.10	58
3740	30 amp, 125 volt NEMA 5		15	.533		19.70	31		50.70	68
3760	250 volt NEMA 6		15	.533		23	31		54	72
3780	277 volt NEMA 7		15	.533		27	31		58	76
3820	125/250 volt NEMA 14		14	.571		57	33.50		90.50	113
3840	3 pole, 250 volt NEMA 15		14	.571		51.50	33.50		85	106
3880	50 amp, 125 volt NEMA 5		11	.727		27.50	42.50		70	93.50
3900	250 volt NEMA 6		11	.727		27	42.50		69.50	93.50
3920	277 volt NEMA 7		11	.727		31	42.50		73.50	98
3960	125/250 volt NEMA 14		10	.800		73.50	46.50		120	150
3980	3 pole, 250 volt NEMA 15		10	.800		73.50	46.50		120	151
4020	60 amp, 125/250 volt, NEMA 14		8	1		79.50	58		137.50	175
4040	3 pole, 250 volt NEMA 15		8	1		84	58		142	180
4060	120/208 volt NEMA 18		8	1		97.50	58		155.50	194
4100	Receptacle locking, 20 amp, 125 volt NEMA L5		27	.296		19.55	17.25		36.80	47.50
4120	250 volt NEMA L6		27	.296		19.95	17.25		37.20	48
4140	277 volt NEMA L7		27	.296		19.70	17.25		36.95	47.50
4150	3 pole, 250 volt, NEMA L11		27	.296		24	17.25		41.25	52
4160	20 amp, 480 volt NEMA L8		27	.296		21.50	17.25		38.75	49.50
4180	600 volt NEMA L9		27	.296		31.50	17.25		48.75	60.50
4200	125/250 volt NEMA L10		27	.296		26.50	17.25		43.75	55
4230	125/250 volt NEMA L14		25	.320		27	18.60		45.60	57.50
4280	250 volt NEMA L15		25	.320		29	18.60		47.60	60
4300	480 volt NEMA L16		25	.320		23.50	18.60		42.10	53.50
4320	3 phase, 120/208 volt NEMA L18		25	.320		31.50	18.60		50.10	62.50
4340	277/480 volt NEMA L19		25	.320		30.50	18.60		49.10	61.50
4360	347/600 volt NEMA L20		25	.320		30.50	18.60		49.10	61.50

26 27 26.20 Wiring Devices Elements	Crew	Daily Output	Labor-Hours	Unit	Material	2018 Bare Costs Labor	Equipment	Total	Total Incl O&P
4380	120/208 volt NEMA L21	1 Elec	23	.348	Ea.	33	20		53	67
4400	277/480 volt NEMA L22		23	.348		35	20		55	69
4420	347/600 volt NEMA L23		23	.348		40.50	20		60.50	75
4440	30 amp, 125 volt NEMA L5		15	.533		27.50	31		58.50	76.50
4460	250 volt NEMA L6		15	.533		30	31		61	79.50
4480	277 volt NEMA L7		15	.533		30	31		61	79.50
4500	480 volt NEMA L8		15	.533		31	31		62	80.50
4520	600 volt NEMA L9		15	.533		35	31		66	84.50
4540	125/250 volt NEMA L10		15	.533		34.50	31		65.50	84.50
4560	3 phase, 250 volt NEMA L11		15	.533		27	31		58	76.50
4620	125/250 volt NEMA L14		14	.571		43	33.50		76.50	97
4640	250 volt NEMA L15		14	.571		43	33.50		76.50	97
4660	480 volt NEMA L16		14	.571		44	33.50		77.50	98
4680	600 volt NEMA L17		14	.571		43.50	33.50		77	97.50
4700	120/208 volt NEMA L18		14	.571		44	33.50		77.50	98
4720	277/480 volt NEMA L19		14	.571		45.50	33.50		79	99.50
4740	347/600 volt NEMA L20		14	.571		50.50	33.50		84	105
4760	120/208 volt NEMA L21		13	.615		45	36		81	103
4780	277/480 volt NEMA L22		13	.615		47.50	36		83.50	106
4800	347/600 volt NEMA L23		13	.615		55	36		91	114
4840	Receptacle, corrosion resistant, 15 or 20 amp, 125 volt NEMA L5		27	.296		26.50	17.25		43.75	55
4860	250 volt NEMA L6		27	.296		20	17.25		37.25	48
4870	Receptacle box assembly, cast aluminum, 60A, 4P5W, 3P, 120/208V, NEMA PR4		22	.364		590	21		611	675
4875	100A, 4P5W, 3P, 347/600V, NEMA PR4		22	.364		795	21		816	905
4880	200A, 4P5W, 3P, 347/600V, NEMA PR4		22	.364		1,700	21		1,721	1,900
4900	Receptacle, cover plate, phenolic plastic, NEMA 5 & 6		80	.100		.64	5.80		6.44	9.40
4910	NEMA 7-23		80	.100		.75	5.80		6.55	9.55
4920	Stainless steel, NEMA 5 & 6		80	.100		2.71	5.80		8.51	11.70
4930	NEMA 7-23		80	.100		3.64	5.80		9.44	12.70
4940	Brushed brass NEMA 5 & 6		80	.100		5.40	5.80		11.20	14.65
4950	NEMA 7-23		80	.100		5.85	5.80		11.65	15.15
4960	Anodized aluminum, NEMA 5 & 6		80	.100		3.51	5.80		9.31	12.55
4970	NEMA 7-23		80	.100		5.80	5.80		11.60	15.10
4980	Weatherproof NEMA 7-23		60	.133		36.50	7.75		44.25	51.50
5000	Duplex receptacle, combo 15A/125V, 3 wire w/2-5V 0.7A, port USB, AL		27	.296		15.55	17.25		32.80	43
5002	15A/125V, 3 wire w/2-5V 0.7A, port USB, BK		27	.296		18	17.25		35.25	46
5004	15A/125V, 3 wire w/2-5V 0.7A, port USB, LA		27	.296		18.25	17.25		35.50	46
5006	15A/125V, 3 wire w/2-5V 0.7A, port USB, IV		27	.296		18.50	17.25		35.75	46.50
5008	15A/125V, 3 wire w/2-5V 0.7A, port USB, WH		27	.296		18.85	17.25		36.10	47
5010	Duplex receptacle, combo 15A/125V, 3 wire w/2-5V 2.1A, port USB, AL		27	.296		41.50	17.25		58.75	72
5012	15A/125V, 3 wire w/2-5V 2.1A, port USB, BK		27	.296		45	17.25		62.25	75.50
5014	15A/125V, 3 wire w/2-5V 2.1A, port USB, LA		27	.296		33	17.25		50.25	62
5016	15A/125V, 3 wire w/2-5V 2.1A, port USB, IV		27	.296		32.50	17.25		49.75	62
5018	15A/125V, 3 wire w/2-5V 2.1A, port USB, WH		27	.296		30.50	17.25		47.75	59.50
5020	Duplex receptacle, combo 20A/125V, 3 wire w/2-5V 2.1A, port USB, AL		27	.296		37	17.25		54.25	66.50
5022	20A/125V, 3 wire w/2-5V 2.1A, port USB, BK		27	.296		49.50	17.25		66.75	80.50
5024	20A/125V, 3 wire w/2-5V 2.1A, port USB, LA		27	.296		34.50	17.25		51.75	64
5026	20A/125V, 3 wire w/2-5V 2.1A, port USB, IV		27	.296		34.50	17.25		51.75	63.50
5028	20A/125V, 3 wire w/2-5V 2.1A, port USB, WH		27	.296		35	17.25		52.25	64.50
5100	Plug, 20 amp, 250 volt NEMA 6		30	.267		18	15.50		33.50	43
5110	277 volt NEMA 7		30	.267		19.20	15.50		34.70	44
5120	3 pole, 120/250 volt NEMA 10		26	.308		21.50	17.90		39.40	50.50
5130	125/250 volt NEMA 14		26	.308		43.50	17.90		61.40	74.50

For customer support on your Electrical Costs with RSMeans data, call 800.448.8182.

273

26 27 26.20 Wiring Devices Elements	Crew	Daily Output	Labor-Hours	Unit	Material	2018 Bare Costs Labor	Equipment	Total	Total Incl O&P
5140	250 volt NEMA 15	1 Elec	26	.308	Ea.	45	17.90		62.90	76.50
5150	120/208 volt NEMA 8		26	.308		47.50	17.90		65.40	79.50
5160	30 amp, 125 volt NEMA 5		13	.615		50.50	36		86.50	109
5170	250 volt NEMA 6		13	.615		56.50	36		92.50	116
5180	277 volt NEMA 7		13	.615		44.50	36		80.50	102
5190	125/250 volt NEMA 14		13	.615		48	36		84	107
5200	3 pole, 250 volt NEMA 15		12	.667		51.50	39		90.50	115
5210	50 amp, 125 volt NEMA 5		9	.889		61.50	51.50		113	146
5220	250 volt NEMA 6		9	.889		64.50	51.50		116	149
5230	277 volt NEMA 7		9	.889		69	51.50		120.50	154
5240	125/250 volt NEMA 14		9	.889		61	51.50		112.50	145
5250	3 pole, 250 volt NEMA 15		8	1		61.50	58		119.50	155
5260	60 amp, 125/250 volt NEMA 14		7	1.143		68.50	66.50		135	175
5270	3 pole, 250 volt NEMA 15		7	1.143		73.50	66.50		140	180
5280	120/208 volt NEMA 18		7	1.143		82.50	66.50		149	191
5300	Plug angle, 20 amp, 250 volt NEMA 6		30	.267		24	15.50		39.50	49.50
5310	30 amp, 125 volt NEMA 5		13	.615		47.50	36		83.50	106
5320	250 volt NEMA 6		13	.615		51	36		87	110
5330	277 volt NEMA 7		13	.615		60	36		96	120
5340	125/250 volt NEMA 14		13	.615		53	36		89	112
5350	3 pole, 250 volt NEMA 15		12	.667		58	39		97	122
5360	50 amp, 125 volt NEMA 5		9	.889		57.50	51.50		109	141
5370	250 volt NEMA 6		9	.889		52.50	51.50		104	136
5380	277 volt NEMA 7		9	.889		67.50	51.50		119	152
5390	125/250 volt NEMA 14		9	.889		66.50	51.50		118	151
5400	3 pole, 250 volt NEMA 15		8	1		70	58		128	164
5410	60 amp, 125/250 volt NEMA 14		7	1.143		77.50	66.50		144	185
5420	3 pole, 250 volt NEMA 15		7	1.143		79	66.50		145.50	187
5430	120/208 volt NEMA 18		7	1.143		82.50	66.50		149	190
5500	Plug, locking, 20 amp, 125 volt NEMA L5		30	.267		15.50	15.50		31	40
5510	250 volt NEMA L6		30	.267		15.55	15.50		31.05	40
5520	277 volt NEMA L7		30	.267		15.20	15.50		30.70	40
5530	480 volt NEMA L8		30	.267		16.40	15.50		31.90	41
5540	600 volt NEMA L9		30	.267		18.35	15.50		33.85	43
5550	3 pole, 125/250 volt NEMA L10		26	.308		22.50	17.90		40.40	51.50
5560	250 volt NEMA L11		26	.308		22.50	17.90		40.40	51.50
5570	480 volt NEMA L12		26	.308		26	17.90		43.90	55.50
5580	125/250 volt NEMA L14		26	.308		23.50	17.90		41.40	52.50
5590	250 volt NEMA L15		26	.308		24.50	17.90		42.40	54
5600	480 volt NEMA L16		26	.308		26	17.90		43.90	56
5610	4 pole, 120/208 volt NEMA L18		24	.333		28	19.40		47.40	60
5620	277/480 volt NEMA L19		24	.333		29	19.40		48.40	61
5630	347/600 volt NEMA L20		24	.333		28	19.40		47.40	59.50
5640	120/208 volt NEMA L21		24	.333		30.50	19.40		49.90	62.50
5650	277/480 volt NEMA L22		24	.333		31.50	19.40		50.90	63.50
5660	347/600 volt NEMA L23		24	.333		35	19.40		54.40	67.50
5670	30 amp, 125 volt NEMA L5		13	.615		23.50	36		59.50	79.50
5680	250 volt NEMA L6		13	.615		24	36		60	80
5690	277 volt NEMA L7		13	.615		24	36		60	80
5700	480 volt NEMA L8		13	.615		25	36		61	81
5710	600 volt NEMA L9		13	.615		27	36		63	83
5720	3 pole, 125/250 volt NEMA L10		11	.727		24.50	42.50		67	90
5730	250 volt NEMA L11		11	.727		24.50	42.50		67	90

For customer support on your Electrical Costs with RSMeans data, call 800.448.8182.

26 27 26.20 Wiring Devices Elements		Crew	Daily Output	Labor-Hours	Unit	Material	2018 Bare Costs Labor	Equipment	Total	Total Incl O&P
5760	125/250 volt NEMA L14	1 Elec	11	.727	Ea.	31.50	42.50		74	98.50
5770	250 volt NEMA L15		11	.727		32	42.50		74.50	98.50
5780	480 volt NEMA L16		11	.727		32.50	42.50		75	99.50
5790	600 volt NEMA L17		11	.727		33.50	42.50		76	100
5800	4 pole, 120/208 volt NEMA L18		10	.800		35.50	46.50		82	109
5810	120/208 volt NEMA L19		10	.800		35.50	46.50		82	109
5820	347/600 volt NEMA L20		10	.800		37.50	46.50		84	111
5830	120/208 volt NEMA L21		10	.800		35	46.50		81.50	108
5840	277/480 volt NEMA L22		10	.800		36.50	46.50		83	110
5850	347/600 volt NEMA L23		10	.800		39	46.50		85.50	112
6000	Connector, 20 amp, 250 volt NEMA 6		30	.267		28	15.50		43.50	53.50
6010	277 volt NEMA 7		30	.267		31.50	15.50		47	57.50
6020	3 pole, 120/250 volt NEMA 10		26	.308		35.50	17.90		53.40	66
6030	125/250 volt NEMA 14		26	.308		35.50	17.90		53.40	66
6040	250 volt NEMA 15		26	.308		35.50	17.90		53.40	66.50
6050	120/208 volt NEMA 18		26	.308		39	17.90		56.90	69.50
6060	30 amp, 125 volt NEMA 5		13	.615		61.50	36		97.50	121
6070	250 volt NEMA 6		13	.615		61.50	36		97.50	121
6080	277 volt NEMA 7		13	.615		61.50	36		97.50	121
6110	50 amp, 125 volt NEMA 5		9	.889		84.50	51.50		136	171
6120	250 volt NEMA 6		9	.889		84.50	51.50		136	171
6130	277 volt NEMA 7		9	.889		84.50	51.50		136	171
6200	Connector, locking, 20 amp, 125 volt NEMA L5		30	.267		24	15.50		39.50	49.50
6210	250 volt NEMA L6		30	.267		23.50	15.50		39	49
6220	277 volt NEMA L7		30	.267		23.50	15.50		39	49
6230	480 volt NEMA L8		30	.267		26.50	15.50		42	52
6240	600 volt NEMA L9		30	.267		30	15.50		45.50	56
6250	3 pole, 125/250 volt NEMA L10		26	.308		34	17.90		51.90	64.50
6260	250 volt NEMA L11		26	.308		34	17.90		51.90	64.50
6280	125/250 volt NEMA L14		26	.308		32	17.90		49.90	62.50
6290	250 volt NEMA L15		26	.308		32	17.90		49.90	62.50
6300	480 volt NEMA L16		26	.308		35	17.90		52.90	65.50
6310	4 pole, 120/208 volt NEMA L18		24	.333		39	19.40		58.40	72
6320	277/480 volt NEMA L19		24	.333		44.50	19.40		63.90	78
6330	347/600 volt NEMA L20		24	.333		44	19.40		63.40	77
6340	120/208 volt NEMA L21		24	.333		48.50	19.40		67.90	82
6350	277/480 volt NEMA L22		24	.333		55	19.40		74.40	89.50
6360	347/600 volt NEMA L23		24	.333		60.50	19.40		79.90	95.50
6370	30 amp, 125 volt NEMA L5		13	.615		46.50	36		82.50	105
6380	250 volt NEMA L6		13	.615		47.50	36		83.50	106
6390	277 volt NEMA L7		13	.615		51.50	36		87.50	110
6400	480 volt NEMA L8		13	.615		49.50	36		85.50	108
6410	600 volt NEMA L9		13	.615		55.50	36		91.50	115
6420	3 pole, 125/250 volt NEMA L10		11	.727		66	42.50		108.50	136
6430	250 volt NEMA L11		11	.727		66	42.50		108.50	136
6460	125/250 volt NEMA L14		11	.727		64.50	42.50		107	135
6470	250 volt NEMA L15		11	.727		65	42.50		107.50	135
6480	480 volt NEMA L16		11	.727		69	42.50		111.50	140
6490	600 volt NEMA L17		11	.727		70	42.50		112.50	141
6500	4 pole, 120/208 volt NEMA L18		10	.800		77.50	46.50		124	155
6510	120/208 volt NEMA L19		10	.800		77.50	46.50		124	155
6520	347/600 volt NEMA L20		10	.800		77.50	46.50		124	155
6530	120/208 volt NEMA L21		10	.800		64	46.50		110.50	140

For customer support on your Electrical Costs with RSMeans data, call 800.448.8182.

275

26 27 26 – Wiring Devices

26 27 26.20 Wiring Devices Elements	Crew	Daily Output	Labor-Hours	Unit	Material	2018 Bare Costs Labor	Equipment	Total	Total Incl O&P
6540	277/480 volt NEMA L22	1 Elec	10	.800	Ea.	68	46.50		114.50	145
6550	347/600 volt NEMA L23		10	.800		68.50	46.50		115	145
7000	Receptacle computer, 250 volt, 15 amp, 3 pole 4 wire		8	1		82.50	58		140.50	178
7010	20 amp, 2 pole 3 wire		8	1		96.50	58		154.50	193
7020	30 amp, 2 pole 3 wire		6.50	1.231		154	71.50		225.50	276
7030	30 amp, 3 pole 4 wire		6.50	1.231		166	71.50		237.50	289
7040	60 amp, 3 pole 4 wire		4.50	1.778		270	103		373	450
7050	100 amp, 3 pole 4 wire		3	2.667		340	155		495	605
7100	Connector computer, 250 volt, 15 amp, 3 pole 4 wire		27	.296		130	17.25		147.25	169
7110	20 amp, 2 pole 3 wire		27	.296		133	17.25		150.25	172
7120	30 amp, 2 pole 3 wire		15	.533		197	31		228	264
7130	30 amp, 3 pole 4 wire		15	.533		205	31		236	272
7140	60 amp, 3 pole 4 wire		8	1		340	58		398	460
7150	100 amp, 3 pole 4 wire		4	2		475	116		591	700
7200	Plug, computer, 250 volt, 15 amp, 3 pole 4 wire		27	.296		116	17.25		133.25	154
7210	20 amp, 2 pole 3 wire		27	.296		119	17.25		136.25	157
7220	30 amp, 2 pole 3 wire		15	.533		190	31		221	256
7230	30 amp, 3 pole 4 wire		15	.533		184	31		215	249
7240	60 amp, 3 pole 4 wire		8	1		276	58		334	390
7250	100 amp, 3 pole 4 wire		4	2		345	116		461	555
7300	Connector adapter to flexible conduit, 1/2"		60	.133		3.10	7.75		10.85	15
7310	3/4"		50	.160		4.45	9.30		13.75	18.80
7320	1-1/4"		30	.267		12.35	15.50		27.85	36.50
7330	1-1/2"		23	.348		17.50	20		37.50	50
8100	Pin/sleeve, 20A, 480V, DSN1, male inlet, 3 pole, 3 phase, 4 wire		16	.500		43.50	29		72.50	91.50
8110	20A, 125V, DSN20, male inlet, 3 pole, 3 phase, 3 wire		16	.500		35	29		64	82
8120	60A, 480V, DS60, male inlet, 3 pole, 3 phase, 3 wire		16	.500		289	29		318	365
8130	Pin/sleeve, 20A, 480V, DSN10, female RCPT, 3P, 3 phase, 4 wire		16	.500		78.50	29		107.50	130
8140	30A, 480V, DS30, female RCPT, 3P, 3 phase, 4 wire		16	.500		310	29		339	385
8500	Wiring device terminal strip, 2 pole, 12-2AWG, 300VAC, CSA, 600VAC, 10A		36	.222		3.16	12.95		16.11	23
8505	4 pole, 12-2AWG, 300VAC, CSA, 600VAC, 10A		36	.222		4.46	12.95		17.41	24.50
8510	6 pole, 12-2AWG, 300VAC, CSA, 600VAC, 10A		36	.222		5.05	12.95		18	25
8515	8 pole, 12-2AWG, 300VAC, CSA, 600VAC, 10A		30	.267		5.65	15.50		21.15	29.50
8520	10 pole, 12-2AWG, 300VAC, CSA, 600VAC, 10A		30	.267		6.25	15.50		21.75	30
8525	12 pole, 12-2AWG, 300VAC, CSA, 600VAC, 10A		30	.267		6.90	15.50		22.40	30.50
8530	Wiring device terminal strip jumper, 10A		36	.222		.73	12.95		13.68	20
8535	Wiring device terminal strip, 2 pole, 12-22AWG, 300VAC, CSA, 600VAC, 20A		34	.235		3.42	13.70		17.12	24.50
8540	4 pole, 12-22AWG, 300VAC, CSA, 600VAC, 20A		34	.235		4.70	13.70		18.40	25.50
8545	6 pole, 12-22AWG, 300VAC, CSA, 600VAC, 20A		34	.235		5.30	13.70		19	26.50
8550	8 pole, 12-22AWG, 300VAC, CSA, 600VAC, 20A		30	.267		6	15.50		21.50	29.50
8555	10 pole, 12-22AWG, 300VAC, CSA, 600VAC, 20A		30	.267		6.60	15.50		22.10	30.50
8560	12 pole, 12-22AWG, 300VAC, CSA, 600VAC, 20A		30	.267		7.25	15.50		22.75	31
8565	Wiring device terminal strip jumper, 20A		34	.235		.89	13.70		14.59	21.50
8570	Wiring device terminal shorting type no cover, 4 pole, 6-10AWG, 600VAC, 50A		30	.267		16.55	15.50		32.05	41.50
8575	6 pole, 6-10AWG, 600VAC, 50A		30	.267		21.50	15.50		37	46.50
8580	8 pole, 6-10AWG, 600VAC, 50A		30	.267		24.50	15.50		40	50
8585	12 pole, 6-10AWG, 600VAC, 50A		30	.267		27	15.50		42.50	53
8590	Wiring device terminal 2 shorting pin 1/2 cover, 2 pole, 8-4AWG, 600V, 45A		30	.267		17.15	15.50		32.65	42
8595	4 pole, 8-4AWG, 600V, 45A		30	.267		21.50	15.50		37	46.50
8600	6 pole, 8-4AWG, 600V, 45A		30	.267		26.50	15.50		42	52.50
8605	8 pole, 8-4AWG, 600V, 45A		30	.267		38.50	15.50		54	65.50
8610	12 pole, 8-4AWG, 600V, 45A		30	.267		46	15.50		61.50	73.50
8615	Wiring device terminal 2 shorting pin W/cover, 4 pole, 8-4AWG, 600V, 45A		30	.267		29	15.50		44.50	55

26 27 Low-Voltage Distribution Equipment

26 27 26 – Wiring Devices

26 27 26.20 Wiring Devices Elements

		Crew	Daily Output	Labor-Hours	Unit	Material	2018 Bare Costs Labor	Equipment	Total	Total Incl O&P
8620	6 pole, 8-4AWG, 600V, 45A	1 Elec	30	.267	Ea.	34.50	15.50		50	61
8625	8 pole, 8-4AWG, 600V, 45A		30	.267		46.50	15.50		62	74.50
8630	12 pole, 8-4AWG, 600V, 45A		30	.267		55.50	15.50		71	84

26 27 33 – Power Distribution Units

26 27 33.10 Power Distribution Unit Cabinet

		Crew	Daily Output	Labor-Hours	Unit	Material	2018 Bare Costs Labor	Equipment	Total	Total Incl O&P
0010	**POWER DISTRIBUTION UNIT CABINET**									
0100	Power distribution unit, single cabinet, 50 kVA output, 480/208 input	1 Elec	2	4	Ea.	9,000	233		9,233	10,300
0110	75 kVA output, 480/208 input		2	4		9,775	233		10,008	11,200
0120	100 kVA output, 480/208 input		2	4		10,700	233		10,933	12,100
0130	125 kVA output, 480/208 input		1.50	5.333		11,700	310		12,010	13,400
0140	150 kVA output, 480/208 input		1.50	5.333		14,200	310		14,510	16,100
0150	200 kVA output, 480/208 input		1	8		15,700	465		16,165	17,900
0160	225 kVA output, 480/208 input		1	8		16,600	465		17,065	18,900

26 27 33.20 Power Distribution Unit

		Crew	Daily Output	Labor-Hours	Unit	Material	2018 Bare Costs Labor	Equipment	Total	Total Incl O&P
0010	**POWER DISTRIBUTION UNIT**									
0050	3PH, 60 kVA, PDU									
0100	208V-208V/120V	2 Elec	2	8	Ea.	13,900	465		14,365	15,900
0110	480V-208V/120V		2	8		13,900	465		14,365	15,900
0120	600V-208V/120V		2	8		13,900	465		14,365	15,900
0125	3PH, 80 kVA, PDU									
0130	208V-208V/120V	2 Elec	2	8	Ea.	27,500	465		27,965	31,000
0140	480V-208V/120V		2	8		27,500	465		27,965	31,000
0150	600V-208V/120V		2	8		27,500	465		27,965	31,000
0200	3PH, 15 kVA, 480-208/120V, 208-208/120V		3	5.333		11,700	310		12,010	13,400
0210	30 kVA, 400-208/120V, 208-208/120V		3	5.333		18,300	310		18,610	20,700
0250	50 kVA, 480-208/120V, 208-208/120V		2.50	6.400		7,825	370		8,195	9,175
0260	75 kVA, 480-208/120V, 208-208/120V		2.50	6.400		8,500	370		8,870	9,900
0270	100 kVA, 480-208/120V, 208-208/120V		2	8		9,275	465		9,740	10,900
0280	125 kVA, 480-208/120V, 208-208/120V		2	8		10,200	465		10,665	11,900
0290	150 kVA, 480-208/120V, 208-208/120V		2	8		12,400	465		12,865	14,300
0300	200 kVA, 480-208/120V, 208-208/120V		1.50	10.667		13,600	620		14,220	15,900
0310	225 kVA, 480-208/120V, 208-208/120V		1.50	10.667		14,400	620		15,020	16,700

26 27 73 – Door Chimes

26 27 73.10 Doorbell System

		Crew	Daily Output	Labor-Hours	Unit	Material	2018 Bare Costs Labor	Equipment	Total	Total Incl O&P
0010	**DOORBELL SYSTEM**, incl. transformer, button & signal									
0100	6" bell	1 Elec	4	2	Ea.	139	116		255	325
0200	Buzzer		4	2		109	116		225	294
1000	Door chimes, 2 notes		16	.500		30	29		59	76.50
1020	with ambient light		12	.667		120	39		159	190
1100	Tube type, 3 tube system		12	.667		226	39		265	305
1180	4 tube system		10	.800		480	46.50		526.50	600
1900	For transformer & button, add		5	1.600		14.20	93		107.20	155
3000	For push button only		24	.333		.94	19.40		20.34	30
3200	Bell transformer		16	.500		19.35	29		48.35	65

For customer support on your Electrical Costs with RSMeans data, call 800.448.8182.

277

26 28 13.10 Fuse Elements	Crew	Daily Output	Labor-Hours	Unit	Material	2018 Bare Costs Labor	Equipment	Total	Total Incl O&P
0010 **FUSE ELEMENTS**									
0020 Cartridge, nonrenewable									
0050 250 volt, 30 amp	1 Elec	50	.160	Ea.	2.35	9.30		11.65	16.50
0100 60 amp		50	.160		3.97	9.30		13.27	18.25
0150 100 amp		40	.200		17.05	11.65		28.70	36
0200 200 amp		36	.222		38	12.95		50.95	61
0250 400 amp		30	.267		75	15.50		90.50	106
0300 600 amp		24	.333		125	19.40		144.40	167
0400 600 volt, 30 amp		40	.200		10.50	11.65		22.15	29
0450 60 amp		40	.200		16.15	11.65		27.80	35
0500 100 amp		36	.222		33.50	12.95		46.45	56.50
0550 200 amp		30	.267		69	15.50		84.50	99
0600 400 amp		24	.333		142	19.40		161.40	185
0650 600 amp		20	.400		174	23.50		197.50	227
0800 Dual element, time delay, 250 volt, 30 amp		50	.160		9.95	9.30		19.25	25
0840 50 amp		50	.160		11.40	9.30		20.70	26.50
0850 60 amp		50	.160		11.60	9.30		20.90	26.50
0900 100 amp		40	.200		28	11.65		39.65	48.50
0950 200 amp		36	.222		61.50	12.95		74.45	87
1000 400 amp		30	.267		110	15.50		125.50	144
1050 600 amp		24	.333		181	19.40		200.40	228
1300 600 volt, 15 to 30 amp		40	.200		21.50	11.65		33.15	41
1350 35 to 60 amp		40	.200		31.50	11.65		43.15	52
1400 70 to 100 amp		36	.222		66.50	12.95		79.45	92.50
1450 110 to 200 amp		30	.267		129	15.50		144.50	165
1500 225 to 400 amp		24	.333		284	19.40		303.40	340
1550 600 amp		20	.400		355	23.50		378.50	430
1800 Class RK1, high capacity, 250 volt, 30 amp		50	.160		7.70	9.30		17	22.50
1850 60 amp		50	.160		14	9.30		23.30	29.50
1900 100 amp		40	.200		23.50	11.65		35.15	43.50
1950 200 amp		36	.222		75	12.95		87.95	102
2000 400 amp		30	.267		147	15.50		162.50	184
2050 600 amp		24	.333		194	19.40		213.40	242
2200 600 volt, 30 amp		40	.200		12.05	11.65		23.70	30.50
2250 60 amp		40	.200		20	11.65		31.65	39.50
2300 100 amp		36	.222		82	12.95		94.95	109
2350 200 amp		30	.267		100	15.50		115.50	133
2400 400 amp		24	.333		213	19.40		232.40	263
2450 600 amp		20	.400		375	23.50		398.50	450
2700 Class J, current limiting, 250 or 600 volt, 30 amp		40	.200		21	11.65		32.65	40.50
2750 60 amp		40	.200		33.50	11.65		45.15	54
2800 100 amp		36	.222		52	12.95		64.95	77
2850 200 amp		30	.267		97.50	15.50		113	130
2900 400 amp		24	.333		224	19.40		243.40	276
2950 600 amp		20	.400		293	23.50		316.50	360
3100 Class L, current limiting, 250 or 600 volt, 601 to 1,200 amp		16	.500		600	29		629	705
3150 1,500 to 1,600 amp		13	.615		695	36		731	820
3200 1,800 to 2,000 amp		10	.800		1,150	46.50		1,196.50	1,350
3250 2,500 amp		10	.800		1,300	46.50		1,346.50	1,500
3300 3,000 amp		8	1		1,600	58		1,658	1,825
3350 3,500 to 4,000 amp		8	1		2,075	58		2,133	2,350
3400 4,500 to 5,000 amp		6.70	1.194		1,900	69.50		1,969.50	2,200
3450 6,000 amp		5.70	1.404		4,750	81.50		4,831.50	5,350

For customer support on your Electrical Costs with RSMeans data, call 800.448.8182.

26 28 Low-Voltage Circuit Protective Devices

26 28 13 – Fuses

26 28 13.10 Fuse Elements

		Crew	Daily Output	Labor-Hours	Unit	Material	2018 Bare Costs Labor	Equipment	Total	Total Incl O&P
3600	Plug, 120 volt, 1 to 10 amp	1 Elec	50	.160	Ea.	3.05	9.30		12.35	17.25
3650	15 to 30 amp		50	.160		3.90	9.30		13.20	18.20
3700	Dual element 0.3 to 14 amp		50	.160		5.55	9.30		14.85	20
3750	15 to 30 amp		50	.160		5.75	9.30		15.05	20.50
3800	Fustat, 120 volt, 15 to 30 amp		50	.160		5.95	9.30		15.25	20.50
3850	0.3 to 14 amp		50	.160		5.95	9.30		15.25	20.50
3900	Adapters 0.3 to 10 amp		50	.160		5.45	9.30		14.75	19.90
3950	15 to 30 amp		50	.160		7.80	9.30		17.10	22.50
4000	F-frame current limiting fuse, 14 to 2 AWG, 3 ampere, 3P, aluminum terminal		40	.200		1,825	11.65		1,836.65	2,025
4010	7 ampere, 3P, aluminum terminal		40	.200		1,650	11.65		1,661.65	1,850
4020	15 AMP, 3P, aluminum terminal		40	.200		1,700	11.65		1,711.65	1,875
4030	30 AMP, 3P, aluminum terminal		40	.200		1,725	11.65		1,736.65	1,925
4040	50 AMP, 3P, aluminum terminal		40	.200		1,775	11.65		1,786.65	1,975
4050	F-frame current limiting fuse, 1 to 4/0 AWG, 100 AMP, 3P, aluminum terminal		40	.200		1,825	11.65		1,836.65	2,050
4060	150 AMP, 3P, aluminum terminal		40	.200		2,100	11.65		2,111.65	2,350

26 28 16 – Enclosed Switches and Circuit Breakers

26 28 16.10 Circuit Breakers

		Crew	Daily Output	Labor-Hours	Unit	Material	2018 Bare Costs Labor	Equipment	Total	Total Incl O&P
0010	**CIRCUIT BREAKERS** (in enclosure)									
0100	Enclosed (NEMA 1), 600 volt, 3 pole, 30 amp	1 Elec	3.20	2.500	Ea.	520	146		666	790
0200	60 amp		2.80	2.857		635	166		801	950
0400	100 amp		2.30	3.478		730	202		932	1,100
0500	200 amp		1.50	5.333		1,525	310		1,835	2,150
0600	225 amp		1.50	5.333		1,675	310		1,985	2,325
0700	400 amp	2 Elec	1.60	10		2,875	580		3,455	4,050
0800	600 amp		1.20	13.333		4,175	775		4,950	5,725
1000	800 amp		.94	17.021		5,425	990		6,415	7,450
1200	1,000 amp		.84	19.048		6,850	1,100		7,950	9,200
1220	1,200 amp		.80	20		8,775	1,175		9,950	11,400
1240	1,600 amp		.72	22.222		16,300	1,300		17,600	19,800
1260	2,000 amp		.64	25		17,600	1,450		19,050	21,600
1400	1,200 amp with ground fault		.80	20		14,100	1,175		15,275	17,300
1600	1,600 amp with ground fault		.72	22.222		18,400	1,300		19,700	22,200
1800	2,000 amp with ground fault		.64	25		19,800	1,450		21,250	24,000
2000	Disconnect, 240 volt 3 pole, 5 HP motor	1 Elec	3.20	2.500		445	146		591	710
2020	10 HP motor		3.20	2.500		445	146		591	710
2040	15 HP motor		2.80	2.857		445	166		611	740
2060	20 HP motor		2.30	3.478		540	202		742	900
2080	25 HP motor		2.30	3.478		540	202		742	900
2100	30 HP motor		2.30	3.478		540	202		742	900
2120	40 HP motor		2	4		925	233		1,158	1,375
2140	50 HP motor		1.50	5.333		925	310		1,235	1,500
2160	60 HP motor		1.50	5.333		2,125	310		2,435	2,825
2180	75 HP motor	2 Elec	2	8		2,125	465		2,590	3,050
2200	100 HP motor		1.60	10		2,125	580		2,705	3,225
2220	125 HP motor		1.60	10		2,125	580		2,705	3,225
2240	150 HP motor		1.20	13.333		4,175	775		4,950	5,725
2260	200 HP motor		1.20	13.333		5,425	775		6,200	7,125
2300	Enclosed (NEMA 7), explosion proof, 600 volt 3 pole, 50 amp	1 Elec	2.30	3.478		1,875	202		2,077	2,375
2350	100 amp		1.50	5.333		1,950	310		2,260	2,625
2400	150 amp		1	8		4,700	465		5,165	5,875
2450	250 amp	2 Elec	1.60	10		5,900	580		6,480	7,350
2500	400 amp	"	1.20	13.333		6,525	775		7,300	8,325

For customer support on your Electrical Costs with RSMeans data, call 800.448.8182.

279

26 28 16.13 Circuit Breakers

		Crew	Daily Output	Labor-Hours	Unit	Material	2018 Bare Costs Labor	Equipment	Total	Total Incl O&P
0010	**CIRCUIT BREAKERS**									
0100	Circuit breaker current limiter, 225 to 400, ampere, 1 pole	1 Elec	48	.167	Ea.	770	9.70		779.70	865
0200	Circuit breaker current limiter, 200 to 500, ampere, 1 pole, thermal magnet		46	.174		760	10.10		770.10	850
0300	300 to 500, ampere, 1 pole, thermal magnet		46	.174		1,200	10.10		1,210.10	1,325
0400	Circuit breaker current limiter, 600 to 1000, ampere, 1 pole, tri pac	▼	46	.174	▼	1,650	10.10		1,660.10	1,825

26 28 16.20 Safety Switches

		Crew	Daily Output	Labor-Hours	Unit	Material	2018 Bare Costs Labor	Equipment	Total	Total Incl O&P
0010	**SAFETY SWITCHES** R262816-80									
0100	General duty 240 volt, 3 pole NEMA 1, fusible, 30 amp	1 Elec	3.20	2.500	Ea.	55.50	146		201.50	279
0200	60 amp		2.30	3.478		94	202		296	410
0300	100 amp		1.90	4.211		159	245		404	540
0400	200 amp	▼	1.30	6.154		345	360		705	915
0500	400 amp	2 Elec	1.80	8.889		905	515		1,420	1,775
0600	600 amp	"	1.20	13.333		1,500	775		2,275	2,800
0610	Nonfusible, 30 amp	1 Elec	3.20	2.500		44.50	146		190.50	267
0650	60 amp		2.30	3.478		60.50	202		262.50	370
0700	100 amp		1.90	4.211		137	245		382	515
0750	200 amp	▼	1.30	6.154		249	360		609	810
0800	400 amp	2 Elec	1.80	8.889		580	515		1,095	1,425
0850	600 amp	"	1.20	13.333	▼	1,100	775		1,875	2,350
1100	Heavy duty, 600 volt, 3 pole NEMA 1 nonfused									
1110	30 amp	1 Elec	3.20	2.500	Ea.	81.50	146		227.50	310
1500	60 amp		2.30	3.478		138	202		340	455
1700	100 amp		1.90	4.211		214	245		459	600
1900	200 amp	▼	1.30	6.154		330	360		690	895
2100	400 amp	2 Elec	1.80	8.889		725	515		1,240	1,575
2300	600 amp		1.20	13.333		1,325	775		2,100	2,600
2500	800 amp		.94	17.021		2,250	990		3,240	3,950
2700	1,200 amp	▼	.80	20	▼	3,500	1,175		4,675	5,600
2900	Heavy duty, 240 volt, 3 pole NEMA 1 fusible									
2910	30 amp	1 Elec	3.20	2.500	Ea.	87.50	146		233.50	315
3000	60 amp		2.30	3.478		145	202		347	465
3300	100 amp		1.90	4.211		228	245		473	615
3500	200 amp		1.30	6.154		390	360		750	960
3700	400 amp	2 Elec	1.80	8.889		990	515		1,505	1,875
3900	600 amp		1.20	13.333		1,750	775		2,525	3,075
4100	800 amp		.94	17.021		3,950	990		4,940	5,825
4300	1,200 amp	▼	.80	20		5,400	1,175		6,575	7,700
4340	2 pole fusible, 30 amp	1 Elec	3.50	2.286		65	133		198	271
4350	600 volt, 3 pole, fusible, 30 amp		3.20	2.500		145	146		291	380
4380	60 amp		2.30	3.478		176	202		378	500
4400	100 amp		1.90	4.211		315	245		560	710
4420	200 amp	▼	1.30	6.154		450	360		810	1,025
4440	400 amp	2 Elec	1.80	8.889		1,250	515		1,765	2,150
4450	600 amp		1.20	13.333		1,975	775		2,750	3,325
4460	800 amp		.94	17.021		3,950	990		4,940	5,825
4480	1,200 amp	▼	.80	20	▼	5,400	1,175		6,575	7,700
4500	240 volt 3 pole NEMA 3R (no hubs), fusible									
4510	30 amp	1 Elec	3.10	2.581	Ea.	153	150		303	395
4700	60 amp		2.20	3.636		256	212		468	595
4900	100 amp		1.80	4.444		345	259		604	760
5100	200 amp	▼	1.20	6.667		470	390		860	1,100
5300	400 amp	2 Elec	1.60	10	▼	1,100	580		1,680	2,075

26 28 16.20 Safety Switches		Crew	Daily Output	Labor-Hours	Unit	Material	2018 Bare Costs Labor	Equipment	Total	Total Incl O&P
5500	600 amp	2 Elec	1	16	Ea.	2,200	930		3,130	3,825
5510	Heavy duty, 600 volt, 3 pole 3 ph. NEMA 3R fusible, 30 amp	1 Elec	3.10	2.581		245	150		395	495
5520	60 amp		2.20	3.636		310	212		522	655
5530	100 amp		1.80	4.444		435	259		694	865
5540	200 amp		1.20	6.667		630	390		1,020	1,275
5550	400 amp	2 Elec	1.60	10		1,525	580		2,105	2,550
5700	600 volt, 3 pole NEMA 3R nonfused									
5710	30 amp	1 Elec	3.10	2.581	Ea.	132	150		282	370
5900	60 amp		2.20	3.636		236	212		448	575
6100	100 amp		1.80	4.444		330	259		589	745
6300	200 amp		1.20	6.667		420	390		810	1,050
6500	400 amp	2 Elec	1.60	10		970	580		1,550	1,950
6700	600 amp	"	1	16		2,050	930		2,980	3,650
6900	600 volt, 6 pole NEMA 3R nonfused, 30 amp	1 Elec	2.70	2.963		830	172		1,002	1,175
7100	60 amp		2	4		1,175	233		1,408	1,625
7300	100 amp		1.50	5.333		1,475	310		1,785	2,100
7500	200 amp		1.20	6.667		3,300	390		3,690	4,200
7600	600 volt, 3 pole NEMA 7 explosion proof nonfused									
7610	30 amp	1 Elec	2.20	3.636	Ea.	1,450	212		1,662	1,900
7620	60 amp		1.80	4.444		1,725	259		1,984	2,275
7630	100 amp		1.20	6.667		2,100	390		2,490	2,875
7640	200 amp		.80	10		4,350	580		4,930	5,675
7710	600 volt 6 pole, NEMA 3R fusible, 30 amp		2.70	2.963		1,175	172		1,347	1,525
7900	60 amp		2	4		1,325	233		1,558	1,800
8100	100 amp		1.50	5.333		1,600	310		1,910	2,225
8110	240 volt 3 pole, NEMA 12 fusible, 30 amp		3.10	2.581		198	150		348	445
8120	60 amp		2.20	3.636		320	212		532	665
8130	100 amp		1.80	4.444		450	259		709	875
8140	200 amp		1.20	6.667		620	390		1,010	1,250
8150	400 amp	2 Elec	1.60	10		1,475	580		2,055	2,500
8160	600 amp	"	1	16		2,375	930		3,305	4,000
8180	600 volt 3 pole, NEMA 12 fusible, 30 amp	1 Elec	3.10	2.581		279	150		429	530
8190	60 amp		2.20	3.636		335	212		547	685
8200	100 amp		1.80	4.444		470	259		729	900
8210	200 amp		1.20	6.667		675	390		1,065	1,325
8220	400 amp	2 Elec	1.60	10		1,675	580		2,255	2,725
8230	600 amp	"	1	16		2,850	930		3,780	4,550
8240	600 volt 3 pole, NEMA 12 nonfused, 30 amp	1 Elec	3.10	2.581		176	150		326	420
8250	60 amp		2.20	3.636		246	212		458	585
8260	100 amp		1.80	4.444		375	259		634	800
8270	200 amp		1.20	6.667		435	390		825	1,050
8280	400 amp	2 Elec	1.60	10		1,225	580		1,805	2,225
8290	600 amp	"	1	16		2,250	930		3,180	3,875
8310	600 volt, 3 pole NEMA 4 fusible, 30 amp	1 Elec	3	2.667		660	155		815	955
8320	60 amp		2.20	3.636		775	212		987	1,175
8330	100 amp		1.80	4.444		1,550	259		1,809	2,100
8340	200 amp		1.20	6.667		2,025	390		2,415	2,825
8350	400 amp	2 Elec	1.60	10		4,075	580		4,655	5,375
8360	600 volt 3 pole NEMA 4 nonfused, 30 amp	1 Elec	3	2.667		560	155		715	850
8370	60 amp		2.20	3.636		665	212		877	1,050
8380	100 amp		1.80	4.444		1,400	259		1,659	1,900
8390	200 amp		1.20	6.667		1,825	390		2,215	2,575
8400	400 amp	2 Elec	1.60	10		3,550	580		4,130	4,775

For customer support on your Electrical Costs with RSMeans data, call 800.448.8182.

281

26 28 Low-Voltage Circuit Protective Devices

26 28 16 - Enclosed Switches and Circuit Breakers

26 28 16.20 Safety Switches		Crew	Daily Output	Labor-Hours	Unit	Material	2018 Bare Costs Labor	Equipment	Total	Total Incl O&P
8490	Motor starters, manual, single phase, NEMA 1	1 Elec	6.40	1.250	Ea.	84.50	73		157.50	202
8500	NEMA 4		4	2		239	116		355	435
8700	NEMA 7		4	2		300	116		416	505
8900	NEMA 1 with pilot		6.40	1.250		112	73		185	232
8920	3 pole, NEMA 1, 230/460 volt, 5 HP, size 0		3.50	2.286		207	133		340	425
8940	10 HP, size 1		2	4		245	233		478	620
9010	Disc. switch, 600 volt 3 pole fusible, 30 amp, to 10 HP motor		3.20	2.500		355	146		501	610
9050	60 amp, to 30 HP motor		2.30	3.478		815	202		1,017	1,200
9070	100 amp, to 60 HP motor		1.90	4.211		815	245		1,060	1,250
9100	200 amp, to 125 HP motor	↓	1.30	6.154		1,225	360		1,585	1,875
9110	400 amp, to 200 HP motor	2 Elec	1.80	8.889	↓	3,075	515		3,590	4,175

26 28 16.40 Time Switches

		Crew	Daily Output	Labor-Hours	Unit	Material	Labor	Equipment	Total	Total Incl O&P
0010	**TIME SWITCHES**									
0100	Single pole, single throw, 24 hour dial	1 Elec	4	2	Ea.	125	116		241	310
0200	24 hour dial with reserve power		3.60	2.222		565	129		694	815
0300	Astronomic dial		3.60	2.222		240	129		369	455
0400	Astronomic dial with reserve power		3.30	2.424		665	141		806	940
0500	7 day calendar dial		3.30	2.424		140	141		281	365
0600	7 day calendar dial with reserve power		3.20	2.500		253	146		399	495
0700	Photo cell 2,000 watt		8	1		27	58		85	117
1080	Load management device, 4 loads		2	4		1,150	233		1,383	1,600
1100	8 loads	↓	1	8	↓	2,325	465		2,790	3,250

26 28 16.50 Meter Socket Entry Hub

		Crew	Daily Output	Labor-Hours	Unit	Material	Labor	Equipment	Total	Total Incl O&P
0010	**METER SOCKET ENTRY HUB**									
0100	Meter socket entry hub closing cap, 3/4 to 4 inch	1 Elec	40	.200	Ea.	1.94	11.65		13.59	19.55
0110	Meter socket entry hub, 3/4 conduit		40	.200		15.40	11.65		27.05	34.50
0120	1 inch conduit		40	.200		15.35	11.65		27	34.50
0130	1.25 inch conduit		40	.200		15.60	11.65		27.25	34.50
0140	1.50 inch conduit		40	.200		15.65	11.65		27.30	34.50
0150	2 inch conduit		40	.200		19.10	11.65		30.75	38.50
0160	2.50 inch conduit		40	.200		50	11.65		61.65	73
0170	3 inch conduit	↓	40	.200	↓	90	11.65		101.65	116

26 29 Low-Voltage Controllers

26 29 13 - Enclosed Controllers

26 29 13.10 Contactors, AC

		Crew	Daily Output	Labor-Hours	Unit	Material	Labor	Equipment	Total	Total Incl O&P
0010	**CONTACTORS, AC** Enclosed (NEMA 1)									
0050	Lighting, 600 volt 3 pole, electrically held									
0100	20 amp	1 Elec	4	2	Ea.	380	116		496	590
0200	30 amp		3.60	2.222		250	129		379	470
0300	60 amp		3	2.667		535	155		690	820
0400	100 amp		2.50	3.200		835	186		1,021	1,200
0500	200 amp	↓	1.40	5.714		2,100	335		2,435	2,800
0600	300 amp	2 Elec	1.60	10		4,375	580		4,955	5,675
0800	600 volt 3 pole, mechanically held, 30 amp	1 Elec	3.60	2.222		605	129		734	860
0900	60 amp		3	2.667		1,225	155		1,380	1,575
1000	75 amp		2.80	2.857		1,450	166		1,616	1,850
1100	100 amp		2.50	3.200		1,550	186		1,736	2,000
1200	150 amp		2	4		3,475	233		3,708	4,175
1300	200 amp	↓	1.40	5.714	↓	5,325	335		5,660	6,375

26 29 Low-Voltage Controllers

26 29 13 – Enclosed Controllers

26 29 13.10 Contactors, AC

		Crew	Daily Output	Labor-Hours	Unit	Material	2018 Bare Costs Labor	Equipment	Total	Total Incl O&P
1500	Magnetic with auxiliary contact, size 00, 9 amp	1 Elec	4	2	Ea.	181	116		297	375
1600	Size 0, 18 amp		4	2		216	116		332	410
1700	Size 1, 27 amp		3.60	2.222		245	129		374	465
1800	Size 2, 45 amp		3	2.667		455	155		610	730
1900	Size 3, 90 amp		2.50	3.200		735	186		921	1,100
2000	Size 4, 135 amp		2.30	3.478		1,675	202		1,877	2,150
2100	Size 5, 270 amp	2 Elec	1.80	8.889		3,525	515		4,040	4,675
2200	Size 6, 540 amp		1.20	13.333		10,300	775		11,075	12,600
2300	Size 7, 810 amp		1	16		13,900	930		14,830	16,700
2310	Size 8, 1,215 amp		.80	20		21,600	1,175		22,775	25,600
2500	Magnetic, 240 volt, 1-2 pole, 0.75 HP motor	1 Elec	4	2		146	116		262	335
2520	2 HP motor		3.60	2.222		198	129		327	410
2540	5 HP motor		2.50	3.200		395	186		581	715
2560	10 HP motor		1.40	5.714		655	335		990	1,225
2600	240 volt or less, 3 pole, 0 .75 HP motor		4	2		146	116		262	335
2620	5 HP motor		3.60	2.222		181	129		310	390
2640	10 HP motor		3.60	2.222		210	129		339	425
2660	15 HP motor		2.50	3.200		420	186		606	740
2700	25 HP motor		2.50	3.200		420	186		606	740
2720	30 HP motor	2 Elec	2.80	5.714		700	335		1,035	1,275
2740	40 HP motor		2.80	5.714		700	335		1,035	1,275
2760	50 HP motor		1.60	10		700	580		1,280	1,650
2800	75 HP motor		1.60	10		1,650	580		2,230	2,675
2820	100 HP motor		1	16		1,650	930		2,580	3,200
2860	150 HP motor		1	16		3,500	930		4,430	5,250
2880	200 HP motor		1	16		3,500	930		4,430	5,250
3000	600 volt, 3 pole, 5 HP motor	1 Elec	4	2		181	116		297	375
3020	10 HP motor		3.60	2.222		210	129		339	425
3040	25 HP motor		3	2.667		420	155		575	690
3100	50 HP motor		2.50	3.200		700	186		886	1,050
3160	100 HP motor	2 Elec	2.80	5.714		1,650	335		1,985	2,300
3220	200 HP motor	"	1.60	10		3,500	580		4,080	4,725

26 29 13.20 Control Stations

		Crew	Daily Output	Labor-Hours	Unit	Material	2018 Bare Costs Labor	Equipment	Total	Total Incl O&P
0010	**CONTROL STATIONS**									
0050	NEMA 1, heavy duty, stop/start	1 Elec	8	1	Ea.	146	58		204	248
0100	Stop/start, pilot light		6.20	1.290		199	75		274	330
0200	Hand/off/automatic		6.20	1.290		108	75		183	231
0400	Stop/start/reverse		5.30	1.509		197	88		285	350
0500	NEMA 7, heavy duty, stop/start		6	1.333		530	77.50		607.50	700
0600	Stop/start, pilot light		4	2		645	116		761	885
0700	NEMA 7 or 9, 1 element		6	1.333		430	77.50		507.50	590
0800	2 element		6	1.333		555	77.50		632.50	725
0900	3 element		4	2		1,100	116		1,216	1,400
0910	Selector switch, 2 position		6	1.333		430	77.50		507.50	590
0920	3 position		4	2		430	116		546	650
0930	Oiltight, 1 element		8	1		110	58		168	208
0940	2 element		6.20	1.290		153	75		228	280
0950	3 element		5.30	1.509		143	88		231	288
0960	Selector switch, 2 position		6.20	1.290		111	75		186	234
0970	3 position		5.30	1.509		116	88		204	259

For customer support on your Electrical Costs with RSMeans data, call 800.448.8182.

283

26 29 13 – Enclosed Controllers

26 29 13.30 Control Switches		Crew	Daily Output	Labor-Hours	Unit	Material	2018 Bare Costs Labor	2018 Bare Costs Equipment	Total	Total Incl O&P
0010	**CONTROL SWITCHES** Field installed									
6000	Push button 600 V 10A, momentary contact									
6150	Standard operator with colored button	1 Elec	34	.235	Ea.	18.55	13.70		32.25	41
6160	With single block 1NO 1NC		18	.444		39	26		65	81.50
6170	With double block 2NO 2NC		15	.533		59.50	31		90.50	112
6180	Std operator w/mushroom button 1-9/16" diam.		34	.235		38	13.70		51.70	62.50
6190	Std operator w/mushroom button 2-1/4" diam.									
6200	With single block 1NO 1NC	1 Elec	18	.444	Ea.	55	26		81	99
6210	With double block 2NO 2NC		15	.533		73.50	31		104.50	128
6500	Maintained contact, selector operator		34	.235		55	13.70		68.70	81
6510	With single block 1NO 1NC		18	.444		73.50	26		99.50	120
6520	With double block 2NO 2NC		15	.533		92.50	31		123.50	149
6560	Spring-return selector operator		34	.235		55	13.70		68.70	81
6570	With single block 1NO 1NC		18	.444		73.50	26		99.50	120
6580	With double block 2NO 2NC		15	.533		92.50	31		123.50	149
6620	Transformer operator w/illuminated									
6630	button 6 V #12 lamp	1 Elec	32	.250	Ea.	92.50	14.55		107.05	124
6640	With single block 1NO 1NC w/guard		16	.500		112	29		141	167
6650	With double block 2NO 2NC w/guard		13	.615		130	36		166	197
6690	Combination operator		34	.235		55	13.70		68.70	81
6700	With single block 1NO 1NC		18	.444		73.50	26		99.50	120
6710	With double block 2NO 2NC		15	.533		92.50	31		123.50	149
9000	Indicating light unit, full voltage									
9010	110-125 V front mount	1 Elec	32	.250	Ea.	73.50	14.55		88.05	103
9020	130 V resistor type		32	.250		58	14.55		72.55	86
9030	6 V transformer type		32	.250		67.50	14.55		82.05	96

26 29 13.40 Relays

		Crew	Daily Output	Labor-Hours	Unit	Material	Labor	Equipment	Total	Total Incl O&P
0010	**RELAYS** Enclosed (NEMA 1)									
0050	600 volt AC, 1 pole, 12 amp	1 Elec	5.30	1.509	Ea.	86	88		174	226
0100	2 pole, 12 amp		5	1.600		86	93		179	234
0200	4 pole, 10 amp		4.50	1.778		115	103		218	281
0500	250 volt DC, 1 pole, 15 amp		5.30	1.509		113	88		201	255
0600	2 pole, 10 amp		5	1.600		108	93		201	258
0700	4 pole, 4 amp		4.50	1.778		142	103		245	310

26 29 23 – Variable-Frequency Motor Controllers

26 29 23.10 Variable Frequency Drives/Adj. Frequency Drives

			Crew	Daily Output	Labor-Hours	Unit	Material	Labor	Equipment	Total	Total Incl O&P
0010	**VARIABLE FREQUENCY DRIVES/ADJ. FREQUENCY DRIVES**										
0100	Enclosed (NEMA 1), 460 volt, for 3 HP motor size	G	1 Elec	.80	10	Ea.	1,950	580		2,530	3,025
0110	5 HP motor size	G		.80	10		2,225	580		2,805	3,325
0120	7.5 HP motor size	G		.67	11.940		2,700	695		3,395	4,025
0130	10 HP motor size	G		.67	11.940		2,975	695		3,670	4,325
0140	15 HP motor size	G	2 Elec	.89	17.978		3,825	1,050		4,875	5,775
0150	20 HP motor size	G		.89	17.978		4,525	1,050		5,575	6,550
0160	25 HP motor size	G		.67	23.881		5,550	1,400		6,950	8,200
0170	30 HP motor size	G		.67	23.881		6,950	1,400		8,350	9,725
0180	40 HP motor size	G		.67	23.881		8,025	1,400		9,425	10,900
0190	50 HP motor size	G		.53	30.189		10,300	1,750		12,050	13,900
0200	60 HP motor size	G	R-3	.56	35.714		12,400	2,075	231	14,706	17,100
0210	75 HP motor size	G		.56	35.714		14,400	2,075	231	16,706	19,200
0220	100 HP motor size	G		.50	40		18,500	2,325	259	21,084	24,200
0230	125 HP motor size	G		.50	40		20,100	2,325	259	22,684	25,900

26 29 Low-Voltage Controllers

26 29 23 – Variable-Frequency Motor Controllers

26 29 23.10 Variable Frequency Drives/Adj. Frequency Drives			Crew	Daily Output	Labor-Hours	Unit	Material	2018 Bare Costs Labor	Equipment	Total	Total Incl O&P
0240	150 HP motor size	G	R-3	.50	40	Ea.	24,300	2,325	259	26,884	30,500
0250	200 HP motor size	G	↓	.42	47.619		27,700	2,750	310	30,760	35,000
1100	Custom-engineered, 460 volt, for 3 HP motor size	G	1 Elec	.56	14.286		3,050	830		3,880	4,600
1110	5 HP motor size	G		.56	14.286		3,050	830		3,880	4,600
1120	7.5 HP motor size	G		.47	17.021		3,200	990		4,190	4,975
1130	10 HP motor size	G	↓	.47	17.021		3,350	990		4,340	5,150
1140	15 HP motor size	G	2 Elec	.62	25.806		4,175	1,500		5,675	6,850
1150	20 HP motor size	G		.62	25.806		4,700	1,500		6,200	7,425
1160	25 HP motor size	G		.47	34.043		5,500	1,975		7,475	9,000
1170	30 HP motor size	G		.47	34.043		6,875	1,975		8,850	10,500
1180	40 HP motor size	G		.47	34.043		8,125	1,975		10,100	11,900
1190	50 HP motor size	G	↓	.37	43.243		9,325	2,525		11,850	14,100
1200	60 HP motor size	G	R-3	.39	51.282		14,100	2,975	330	17,405	20,300
1210	75 HP motor size	G		.39	51.282		15,000	2,975	330	18,305	21,300
1220	100 HP motor size	G		.35	57.143		16,400	3,325	370	20,095	23,400
1230	125 HP motor size	G		.35	57.143		17,500	3,325	370	21,195	24,700
1240	150 HP motor size	G		.35	57.143		20,100	3,325	370	23,795	27,500
1250	200 HP motor size	G	↓	.29	68.966	↓	26,200	4,000	445	30,645	35,300
2000	For complex & special design systems to meet specific										
2010	requirements, obtain quote from vendor.										

26 31 Photovoltaic Collectors

26 31 13 – Photovoltaics

26 31 13.50 Solar Energy - Photovoltaics			Crew	Daily Output	Labor-Hours	Unit	Material	2018 Bare Costs Labor	Equipment	Total	Total Incl O&P
0010	**SOLAR ENERGY - PHOTOVOLTAICS**										
0220	Alt. energy source, photovoltaic module, 6 watt, 15 V	G	1 Elec	8	1	Ea.	50	58		108	143
0230	10 watt, 16.3 V	G		8	1		117	58		175	215
0240	20 watt, 14.5 V	G		8	1		180	58		238	285
0250	36 watt, 17 V	G		8	1		170	58		228	275
0260	55 watt, 17 V	G		8	1		238	58		296	350
0270	75 watt, 17 V	G		8	1		395	58		453	520
0280	130 watt, 33 V	G		8	1		620	58		678	765
0290	140 watt, 33 V	G		8	1		475	58		533	610
0300	150 watt, 33 V	G		8	1		400	58		458	520
0310	DC to AC inverter for, 12 V, 2,000 watt	G		4	2		1,725	116		1,841	2,075
0320	12 V, 2,500 watt	G		4	2		1,075	116		1,191	1,375
0330	24 V, 2,500 watt	G		4	2		1,500	116		1,616	1,825
0340	12 V, 3,000 watt	G		3	2.667		1,500	155		1,655	1,875
0350	24 V, 3,000 watt	G		3	2.667		2,375	155		2,530	2,850
0360	24 V, 4,000 watt	G		2	4		3,025	233		3,258	3,675
0370	48 V, 4,000 watt	G		2	4		2,525	233		2,758	3,125
0380	48 V, 5,500 watt	G		2	4		2,975	233		3,208	3,625
0390	PV components, combiner box, 10 lug, NEMA 3R enclosure	G		4	2		270	116		386	470
0400	Fuse, 15 A for combiner box	G		40	.200		21.50	11.65		33.15	41
0410	Battery charger controller w/temperature sensor	G		4	2		495	116		611	715
0420	Digital readout panel, displays hours, volts, amps, etc.	G		4	2		218	116		334	415
0430	Deep cycle solar battery, 6 V, 180 Ah (C/20)	G		8	1		305	58		363	420
0440	Battery interconn, 15" AWG #2/0, sealed w/copper ring lugs	G		16	.500		18.10	29		47.10	63.50
0442	Battery interconn, 24" AWG #2/0, sealed w/copper ring lugs	G		16	.500		23.50	29		52.50	69.50
0444	Battery interconn, 60" AWG #2/0, sealed w/copper ring lugs	G		16	.500		59.50	29		88.50	109
0446	Batt temp computer probe, RJ11 jack, 15' cord	G		16	.500		22	29		51	68

For customer support on your Electrical Costs with RSMeans data, call 800.448.8182.

285

26 31 13 – Photovoltaics

26 31 13.50 Solar Energy - Photovoltaics		Crew	Daily Output	Labor-Hours	Unit	Material	2018 Bare Costs Labor	Equipment	Total	Total Incl O&P
0450	System disconnect, DC 175 amp circuit breaker	G 1 Elec	8	1	Ea.	226	58		284	335
0460	Conduit box for inverter	G	8	1		62.50	58		120.50	156
0470	Low voltage disconnect	G	8	1		57	58		115	150
0480	Vented battery enclosure, wood	G 1 Carp	2	4		281	203		484	620
0490	PV rack system, roof, non-penetrating ballast, 1 panel	G R-1A	30.50	.525		965	27.50		992.50	1,100
0500	Penetrating surface mount, on steel framing, 1 panel		5.13	3.122		56	164		220	305
0510	On wood framing, 1 panel		11	1.455		54.50	76		130.50	174
0520	With standoff, 1 panel		11	1.455		64	76		140	185
0530	Ground, ballast, fixed, 3 panel		20.50	.780		1,025	41		1,066	1,175
0540	4 panel		20.50	.780		1,400	41		1,441	1,575
0550	5 panel		20.50	.780		1,850	41		1,891	2,075
0560	6 panel		20.50	.780		2,125	41		2,166	2,375
0570	Adjustable, 3 panel		20.50	.780		1,075	41		1,116	1,250
0580	4 panel		20.50	.780		1,475	41		1,516	1,675
0590	5 panel		20.50	.780		1,975	41		2,016	2,200
0600	6 panel		20.50	.780		2,250	41		2,291	2,525
0605	Top of Pole, see 26 56 13.10 3800+, 5400+ & 33 71 16.33 7400+ for poles									
0610	Passive tracking, 1 panel	R-1A	20.50	.780	Ea.	580	41		621	695
0620	2 panel		20.50	.780		1,175	41		1,216	1,350
0630	3 panel		20.50	.780		1,700	41		1,741	1,925
0640	4 panel		20.50	.780		1,850	41		1,891	2,075
0650	6 panel		20.50	.780		2,100	41		2,141	2,375
0660	8 panel		20.50	.780		3,125	41		3,166	3,475
1020	Photovoltaic module, Thin film	1 Elec	8	1		180	58		238	285
1040	Photovoltaic module, Cadium telluride		8	1		238	58		296	350
1060	Photovoltaic module, Polycrystalline		8	1		180	58		238	285
1080	Photovoltaic module, Monocrystalline		8	1		620	58		678	765

26 32 Packaged Generator Assemblies

26 32 13 – Engine Generators

26 32 13.13 Diesel-Engine-Driven Generator Sets

		Crew	Daily Output	Labor-Hours	Unit	Material	2018 Bare Costs Labor	Equipment	Total	Total Incl O&P
0010	**DIESEL-ENGINE-DRIVEN GENERATOR SETS**									
2000	Diesel engine, including battery, charger,									
2010	muffler, & day tank, 30 kW	R-3	.55	36.364	Ea.	11,000	2,100	236	13,336	15,500
2100	50 kW		.42	47.619		20,000	2,750	310	23,060	26,500
2110	60 kW		.39	51.282		20,800	2,975	330	24,105	27,700
2200	75 kW		.35	57.143		23,100	3,325	370	26,795	30,800
2300	100 kW		.31	64.516		27,900	3,750	420	32,070	36,800
2400	125 kW		.29	68.966		29,300	4,000	445	33,745	38,800
2500	150 kW		.26	76.923		37,600	4,450	500	42,550	48,600
2501	Generator set, dsl eng in alum encl, incl btry, chgr, muf & day tank,150 kW		.26	76.923		41,400	4,450	500	46,350	52,500
2600	175 kW		.25	80		41,900	4,650	520	47,070	53,500
2700	200 kW		.24	83.333		45,200	4,825	540	50,565	57,500
2800	250 kW		.23	86.957		48,400	5,050	565	54,015	61,500
2850	275 kW		.22	90.909		49,600	5,275	590	55,465	63,000
2900	300 kW		.22	90.909		51,500	5,275	590	57,365	65,000
3000	350 kW		.20	100		59,000	5,800	650	65,450	74,000
3100	400 kW		.19	105		72,000	6,100	680	78,780	89,000
3200	500 kW		.18	111		91,500	6,450	720	98,670	111,000
3220	600 kW		.17	118		118,500	6,825	760	126,085	141,500
3230	650 kW	R-13	.38	111		146,000	6,225	620	152,845	170,500

26 32 Packaged Generator Assemblies

26 32 13 – Engine Generators

26 32 13.13 Diesel-Engine-Driven Generator Sets		Crew	Daily Output	Labor-Hours	Unit	Material	Labor	Equipment	Total	Total Incl O&P
3240	750 kW	R-13	.38	111	Ea.	147,000	6,225	620	153,845	172,000
3250	800 kW		.36	117		154,500	6,575	655	161,730	180,500
3260	900 kW		.31	135		180,000	7,625	760	188,385	210,000
3270	1,000 kW		.27	156		186,000	8,775	870	195,645	218,500

26 32 13.16 Gas-Engine-Driven Generator Sets

		Crew	Daily Output	Labor-Hours	Unit	Material	Labor	Equipment	Total	Total Incl O&P
0010	**GAS-ENGINE-DRIVEN GENERATOR SETS**									
0020	Gas or gasoline operated, includes battery,									
0050	charger, & muffler									
0200	7.5 kW	R-3	.83	24.096	Ea.	8,550	1,400	156	10,106	11,700
0300	11.5 kW		.71	28.169		12,100	1,625	183	13,908	16,000
0400	20 kW		.63	31.746		14,300	1,850	206	16,356	18,700
0460	25 kW, steel enclosure		.63	31.746		21,000	1,850	206	23,056	26,100
0470	27 kW, aluminum enclosure		.63	31.746		26,200	1,850	206	28,256	31,900
0480	30 kW, steel enclosure		.63	31.746		23,700	1,850	206	25,756	29,100
0485	120/208 V, 30kW		.63	31.746		23,700	1,850	206	25,756	29,100
0500	35 kW		.55	36.364		17,000	2,100	236	19,336	22,100
0510	45 kW, steel enclosure		.50	40		34,000	2,325	259	36,584	41,300
0515	48 kW, aluminum enclosure		.50	40		40,400	2,325	259	42,984	48,200
0520	60 kW		.50	40		22,400	2,325	259	24,984	28,500
0550	70 kW, aluminum enclosure	R-13	.46	91.304		51,500	5,150	510	57,160	65,500
0600	80 kW		.40	105		27,900	5,925	590	34,415	40,200
0700	100 kW		.33	127		30,600	7,175	715	38,490	45,100
0800	125 kW		.28	150		62,500	8,450	840	71,790	82,500
0850	130 kW, aluminum enclosure		.28	150		73,000	8,450	840	82,290	93,500
0875	150 kW, aluminum enclosure		.25	168		85,500	9,475	940	95,915	109,000
0900	185 kW		.25	168		83,000	9,475	940	93,415	106,000

26 33 Battery Equipment

26 33 19 – Battery Units

26 33 19.10 Battery Units

		Crew	Daily Output	Labor-Hours	Unit	Material	Labor	Equipment	Total	Total Incl O&P
0010	**BATTERY UNITS**									
0500	Salt water battery, 2.1 kWA, 24V, 750W, 30A, wired in crate, IP22 rated	1 Elec	16.25	.492	Ea.	995	28.50		1,023.50	1,150
0510	2.2 kWh, 48V, 800W, 17A, wired in crate, IP22 rated		16.25	.492		1,200	28.50		1,228.50	1,375
0520	25.9 kWh, 48V, 11700W, 240A, wired in crate, IP2X rated		16.25	.492		15,000	28.50		15,028.50	16,500
0700	Nickel iron nife battery, deep cycle, 100Ah, 12V, 10 cells		14.25	.561		1,100	32.50		1,132.50	1,250
0710	24V, 10 cells		14.25	.561		2,200	32.50		2,232.50	2,450
0720	48V, 10 cells		14.25	.561		4,375	32.50		4,407.50	4,875
0730	Nickel iron nife battery, deep cycle, 300Ah, 12V, 10 cells		12.25	.653		3,175	38		3,213	3,550
0740	24V, 10 cells		12.25	.653		6,350	38		6,388	7,025
0750	48V, 10 cells		12.25	.653		12,700	38		12,738	14,100

26 33 43 – Battery Chargers

26 33 43.55 Electric Vehicle Charging

			Crew	Daily Output	Labor-Hours	Unit	Material	Labor	Equipment	Total	Total Incl O&P
0010	**ELECTRIC VEHICLE CHARGING**										
0020	Level 2, wall mounted										
2200	Heavy duty	G	R-1A	15.36	1.042	Ea.	1,575	54.50		1,629.50	1,825
2210	with RFID	G		12.29	1.302		3,175	68		3,243	3,600
2300	Free standing, single connector	G		10.24	1.563		1,800	82		1,882	2,125
2310	with RFID	G		8.78	1.822		3,625	95.50		3,720.50	4,125
2320	Double connector	G		7.68	2.083		3,350	109		3,459	3,850
2330	with RFID	G		6.83	2.343		6,725	123		6,848	7,575

For customer support on your Electrical Costs with RSMeans data, call 800.448.8182.

287

26 33 53 – Static Uninterruptible Power Supply

26 33 53.10 Uninterruptible Power Supply/Conditioner Trans.	Crew	Daily Output	Labor-Hours	Unit	Material	2018 Bare Costs Labor	Equipment	Total	Total Incl O&P
0010 **UNINTERRUPTIBLE POWER SUPPLY/CONDITIONER TRANSFORMERS**									
0100 Volt. regulating, isolating transf., w/invert. & 10 min. battery pack									
0110 Single-phase, 120 V, 0.35 kVA R263353-80	1 Elec	2.29	3.493	Ea.	1,175	203		1,378	1,600
0120 0.5 kVA		2	4		1,225	233		1,458	1,700
0130 For additional 55 min. battery, add to 0.35 kVA		2.29	3.493		690	203		893	1,075
0140 Add to 0.5 kVA		1.14	7.018		725	410		1,135	1,400
0150 Single-phase, 120 V, 0.75 kVA		.80	10		1,550	580		2,130	2,600
0160 1.0 kVA		.80	10		2,225	580		2,805	3,325
0170 1.5 kVA	2 Elec	1.14	14.035		3,825	815		4,640	5,425
0180 2 kVA	"	.89	17.978		4,150	1,050		5,200	6,150
0190 3 kVA	R-3	.63	31.746		5,025	1,850	206	7,081	8,525
0200 5 kVA		.42	47.619		7,300	2,750	310	10,360	12,500
0210 7.5 kVA		.33	60.606		9,325	3,525	395	13,245	15,900
0220 10 kVA		.28	71.429		12,100	4,150	465	16,715	20,000
0230 15 kVA		.22	90.909		15,300	5,275	590	21,165	25,400
0240 3 phase, 120/208 V input 120/208 V output, 20 kVA, incl 17 min. battery		.21	95.238		28,300	5,525	615	34,440	40,100
0242 30 kVA, incl 11 min. battery		.20	100		31,600	5,800	650	38,050	44,200
0250 40 kVA, incl 15 min. battery		.20	100		42,200	5,800	650	48,650	56,000
0260 480 V input 277/480 V output, 60 kVA, incl 6 min. battery		.19	105		47,500	6,100	680	54,280	62,000
0262 80 kVA, incl 4 min. battery		.19	105		56,500	6,100	680	63,280	72,000
0400 For additional 34 min./15 min. battery, add to 40 kVA		.71	28.011		14,700	1,625	182	16,507	18,800
0600 For complex & special design systems to meet specific									
0610 requirements, obtain quote from vendor									

26 35 Power Filters and Conditioners

26 35 13 – Capacitors

26 35 13.10 Capacitors Indoor

	Crew	Daily Output	Labor-Hours	Unit	Material	2018 Bare Costs Labor	Equipment	Total	Total Incl O&P
0010 **CAPACITORS INDOOR**									
0020 240 volts, single & 3 phase, 0.5 kVAR	1 Elec	2.70	2.963	Ea.	550	172		722	865
0100 1.0 kVAR		2.70	2.963		665	172		837	995
0150 2.5 kVAR		2	4		750	233		983	1,175
0200 5.0 kVAR		1.80	4.444		790	259		1,049	1,250
0250 7.5 kVAR		1.60	5		1,350	291		1,641	1,925
0300 10 kVAR		1.50	5.333		1,575	310		1,885	2,200
0350 15 kVAR		1.30	6.154		1,975	360		2,335	2,700
0400 20 kVAR		1.10	7.273		2,375	425		2,800	3,250
0450 25 kVAR		1	8		2,825	465		3,290	3,800
1000 480 volts, single & 3 phase, 1 kVAR		2.70	2.963		505	172		677	815
1050 2 kVAR		2.70	2.963		580	172		752	895
1100 5 kVAR		2	4		730	233		963	1,150
1150 7.5 kVAR		2	4		785	233		1,018	1,225
1200 10 kVAR		2	4		1,050	233		1,283	1,500
1250 15 kVAR		2	4		1,250	233		1,483	1,725
1300 20 kVAR		1.60	5		1,400	291		1,691	1,950
1350 30 kVAR		1.50	5.333		1,600	310		1,910	2,250
1400 40 kVAR		1.20	6.667		2,025	390		2,415	2,800
1450 50 kVAR		1.10	7.273		2,400	425		2,825	3,275
2000 600 volts, single & 3 phase, 1 kVAR		2.70	2.963		520	172		692	830
2050 2 kVAR		2.70	2.963		595	172		767	915
2100 5 kVAR		2	4		730	233		963	1,150

26 35 Power Filters and Conditioners

26 35 13 – Capacitors

26 35 13.10 Capacitors Indoor

		Crew	Daily Output	Labor-Hours	Unit	Material	2018 Bare Costs Labor	Equipment	Total	Total Incl O&P
2150	7.5 kVAR	1 Elec	2	4	Ea.	785	233		1,018	1,225
2200	10 kVAR		2	4		1,100	233		1,333	1,575
2250	15 kVAR		1.60	5		1,350	291		1,641	1,900
2300	20 kVAR		1.60	5		1,425	291		1,716	1,975
2350	25 kVAR		1.50	5.333		1,550	310		1,860	2,175
2400	35 kVAR		1.40	5.714		1,900	335		2,235	2,600
2450	50 kVAR		1.30	6.154		2,400	360		2,760	3,150

26 35 26 – Harmonic Filters

26 35 26.10 Computer Isolation Transformer

		Crew	Daily Output	Labor-Hours	Unit	Material	Labor	Equipment	Total	Total Incl O&P
0010	**COMPUTER ISOLATION TRANSFORMER**									
0100	Computer grade									
0110	Single-phase, 120/240 V, 0.5 kVA	1 Elec	4	2	Ea.	435	116		551	655
0120	1.0 kVA		2.67	2.996		620	174		794	940
0130	2.5 kVA		2	4		940	233		1,173	1,375
0140	5 kVA		1.14	7.018		1,075	410		1,485	1,775

26 35 26.20 Computer Regulator Transformer

		Crew	Daily Output	Labor-Hours	Unit	Material	Labor	Equipment	Total	Total Incl O&P
0010	**COMPUTER REGULATOR TRANSFORMER**									
0100	Ferro-resonant, constant voltage, variable transformer									
0110	Single-phase, 240 V, 0.5 kVA	1 Elec	2.67	2.996	Ea.	460	174		634	770
0120	1.0 kVA		2	4		635	233		868	1,050
0130	2.0 kVA		1	8		1,075	465		1,540	1,900
0210	Plug-in unit 120 V, 0.14 kVA		8	1		267	58		325	380
0220	0.25 kVA		8	1		310	58		368	430
0230	0.5 kVA		8	1		460	58		518	595
0240	1.0 kVA		5.33	1.501		635	87.50		722.50	830
0250	2.0 kVA		4	2		1,075	116		1,191	1,375

26 35 26.30 Power Conditioner Transformer

		Crew	Daily Output	Labor-Hours	Unit	Material	Labor	Equipment	Total	Total Incl O&P
0010	**POWER CONDITIONER TRANSFORMER**									
0100	Electronic solid state, buck-boost, transformer, w/tap switch									
0110	Single-phase, 115 V, 3.0 kVA, + or - 3% accuracy	2 Elec	1.60	10	Ea.	2,500	580		3,080	3,625
0120	208, 220, 230, or 240 V, 5.0 kVA, + or - 1.5% accuracy	3 Elec	1.60	15		3,225	875		4,100	4,850
0130	5.0 kVA, + or - 6% accuracy	2 Elec	1.14	14.035		2,900	815		3,715	4,400
0140	7.5 kVA, + or - 1.5% accuracy	3 Elec	1.50	16		4,100	930		5,030	5,900
0150	7.5 kVA, + or - 6% accuracy		1.60	15		3,425	875		4,300	5,050
0160	10.0 kVA, + or - 1.5% accuracy		1.33	18.045		5,450	1,050		6,500	7,575
0170	10.0 kVA, + or - 6% accuracy		1.41	17.021		4,650	990		5,640	6,600

26 35 26.40 Transient Voltage Suppressor Transformer

		Crew	Daily Output	Labor-Hours	Unit	Material	Labor	Equipment	Total	Total Incl O&P
0010	**TRANSIENT VOLTAGE SUPPRESSOR TRANSFORMER**									
0110	Single-phase, 120 V, 1.8 kVA	1 Elec	4	2	Ea.	1,575	116		1,691	1,900
0120	3.6 kVA		4	2		2,350	116		2,466	2,750
0130	7.2 kVA		3.20	2.500		2,775	146		2,921	3,275
0150	240 V, 3.6 kVA		4	2		3,475	116		3,591	3,975
0160	7.2 kVA		4	2		4,025	116		4,141	4,600
0170	14.4 kVA		3.20	2.500		4,575	146		4,721	5,275
0210	Plug-in unit, 120 V, 1.8 kVA		8	1		960	58		1,018	1,125

26 35 53 – Voltage Regulators

26 35 53.10 Automatic Voltage Regulators

		Crew	Daily Output	Labor-Hours	Unit	Material	Labor	Equipment	Total	Total Incl O&P
0010	**AUTOMATIC VOLTAGE REGULATORS**									
0100	Computer grade, solid state, variable transf. volt. regulator									
0110	Single-phase, 120 V, 8.6 kVA	2 Elec	1.33	12.030	Ea.	5,775	700		6,475	7,400
0120	17.3 kVA		1.14	14.035		6,825	815		7,640	8,725

For customer support on your Electrical Costs with RSMeans data, call 800.448.8182.

289

26 35 53 – Voltage Regulators

26 35 53.10 Automatic Voltage Regulators	Crew	Daily Output	Labor-Hours	Unit	Material	2018 Bare Costs Labor	Equipment	Total	Total Incl O&P
0130	208/240 V, 7.5/8.6 kVA	2 Elec	1.33	12.030	Ea.	5,775	700		6,475	7,400
0140	13.5/15.6 kVA		1.33	12.030		6,825	700		7,525	8,550
0150	27.0/31.2 kVA		1.14	14.035		8,625	815		9,440	10,700
0210	Two-phase, single control, 208/240 V, 15.0/17.3 kVA		1.14	14.035		6,825	815		7,640	8,725
0220	Individual phase control, 15.0/17.3 kVA		1.14	14.035		6,825	815		7,640	8,725
0230	30.0/34.6 kVA	3 Elec	1.33	18.045		8,625	1,050		9,675	11,100
0310	Three-phase single control, 208/240 V, 26/30 kVA	2 Elec	1	16		6,825	930		7,755	8,900
0320	380/480 V, 24/30 kVA	"	1	16		6,825	930		7,755	8,900
0330	43/54 kVA	3 Elec	1.33	18.045		12,400	1,050		13,450	15,200
0340	Individual phase control, 208 V, 26 kVA	"	1.33	18.045		6,825	1,050		7,875	9,075
0350	52 kVA	R-3	.91	21.978		8,625	1,275	142	10,042	11,500
0360	340/480 V, 24/30 kVA	2 Elec	1	16		6,825	930		7,755	8,900
0370	43/54 kVA	"	1	16		8,625	930		9,555	10,900
0380	48/60 kVA	3 Elec	1.33	18.045		12,500	1,050		13,550	15,300
0390	86/108 kVA	R-3	.91	21.978		13,900	1,275	142	15,317	17,400
0500	Standard grade, solid state, variable transformer volt. regulator									
0510	Single-phase, 115 V, 2.3 kVA	1 Elec	2.29	3.493	Ea.	2,125	203		2,328	2,625
0520	4.2 kVA		2	4		3,400	233		3,633	4,100
0530	6.6 kVA		1.14	7.018		4,175	410		4,585	5,175
0540	13.0 kVA		1.14	7.018		7,200	410		7,610	8,500
0550	16.6 kVA	2 Elec	1.23	13.008		8,500	755		9,255	10,500
0610	230 V, 8.3 kVA		1.33	12.030		7,200	700		7,900	8,950
0620	21.4 kVA		1.23	13.008		8,500	755		9,255	10,500
0630	29.9 kVA		1.23	13.008		8,500	755		9,255	10,500
0710	460 V, 9.2 kVA		1.33	12.030		7,200	700		7,900	8,950
0720	20.7 kVA		1.23	13.008		8,500	755		9,255	10,500
0810	Three-phase, 230 V, 13.1 kVA	3 Elec	1.41	17.021		7,200	990		8,190	9,375
0820	19.1 kVA		1.41	17.021		8,500	990		9,490	10,800
0830	25.1 kVA		1.23	19.512		8,500	1,125		9,625	11,100
0840	57.8 kVA	R-3	.95	21.053		15,400	1,225	136	16,761	19,000
0850	74.9 kVA	"	.91	21.978		15,400	1,275	142	16,817	19,100
0910	460 V, 14.3 kVA	3 Elec	1.41	17.021		7,200	990		8,190	9,375
0920	19.1 kVA		1.41	17.021		8,500	990		9,490	10,800
0930	27.9 kVA		1.23	19.512		8,500	1,125		9,625	11,100
0940	59.8 kVA	R-3	1	20		15,400	1,150	130	16,680	18,900
0950	79.7 kVA		.95	21.053		17,300	1,225	136	18,661	21,000
0960	118 kVA		.95	21.053		18,200	1,225	136	19,561	22,000
1000	Laboratory grade, precision, electronic voltage regulator									
1110	Single-phase, 115 V, 0.5 kVA	1 Elec	2.29	3.493	Ea.	1,425	203		1,628	1,850
1120	1.0 kVA		2	4		1,525	233		1,758	2,025
1130	3.0 kVA		.80	10		2,100	580		2,680	3,200
1140	6.0 kVA	2 Elec	1.46	10.959		3,900	640		4,540	5,250
1150	10.0 kVA	3 Elec	1	24		5,075	1,400		6,475	7,675
1160	15.0 kVA	"	1.50	16		5,800	930		6,730	7,775
1210	230 V, 3.0 kVA	1 Elec	.80	10		2,400	580		2,980	3,500
1220	6.0 kVA	2 Elec	1.46	10.959		4,000	640		4,640	5,350
1230	10.0 kVA	3 Elec	1.71	14.035		5,300	815		6,115	7,050
1240	15.0 kVA	"	1.60	15		6,000	875		6,875	7,900

26 35 53.30 Transient Suppressor/Voltage Regulator

		Crew	Daily Output	Labor-Hours	Unit	Material	Labor	Equipment	Total	Total Incl O&P
0010	**TRANSIENT SUPPRESSOR/VOLTAGE REGULATOR** (without isolation)									
0110	Single-phase, 115 V, 1.0 kVA	1 Elec	2.67	2.996	Ea.	1,150	174		1,324	1,500
0120	2.0 kVA		2.29	3.493		1,550	203		1,753	2,000

26 35 Power Filters and Conditioners

26 35 53 – Voltage Regulators

26 35 53.30 Transient Suppressor/Voltage Regulator	Crew	Daily Output	Labor-Hours	Unit	Material	2018 Bare Costs Labor	Equipment	Total	Total Incl O&P
0130	4.0 kVA	1 Elec	2.13	3.756	Ea.	1,925	219		2,144	2,425
0140	220 V, 1.0 kVA		2.67	2.996		1,150	174		1,324	1,500
0150	2.0 kVA		2.29	3.493		1,550	203		1,753	2,000
0160	4.0 kVA		2.13	3.756		2,025	219		2,244	2,550
0210	Plug-in unit, 120 V, 1.0 kVA		8	1		1,075	58		1,133	1,275
0220	2.0 kVA		8	1		1,525	58		1,583	1,750

26 36 Transfer Switches

26 36 13 – Manual Transfer Switches

26 36 13.10 Non-Automatic Transfer Switches

		Crew	Daily Output	Labor-Hours	Unit	Material	Labor	Equipment	Total	Total Incl O&P
0010	**NON-AUTOMATIC TRANSFER SWITCHES** enclosed									
0100	Manual operated, 480 volt 3 pole, 30 amp	1 Elec	2.30	3.478	Ea.	1,275	202		1,477	1,700
0150	60 amp		1.90	4.211		2,100	245		2,345	2,675
0200	100 amp		1.30	6.154		3,475	360		3,835	4,350
0250	200 amp	2 Elec	2	8		4,600	465		5,065	5,750
0300	400 amp		1.60	10		8,450	580		9,030	10,200
0350	600 amp		1	16		5,325	930		6,255	7,250
1000	250 volt 3 pole, 30 amp	1 Elec	2.30	3.478		1,225	202		1,427	1,650
1100	60 amp		1.90	4.211		1,675	245		1,920	2,225
1150	100 amp		1.30	6.154		2,850	360		3,210	3,650
1200	200 amp	2 Elec	2	8		4,300	465		4,765	5,425
1300	600 amp	"	1	16		9,825	930		10,755	12,200
1500	Electrically operated, 480 volt 3 pole, 60 amp	1 Elec	1.90	4.211		1,650	245		1,895	2,200
1600	100 amp	"	1.30	6.154		1,650	360		2,010	2,350
1650	200 amp	2 Elec	2	8		2,725	465		3,190	3,700
1700	400 amp		1.60	10		3,825	580		4,405	5,075
1750	600 amp		1	16		5,500	930		6,430	7,450
2000	250 volt 3 pole, 30 amp	1 Elec	2.30	3.478		1,800	202		2,002	2,300
2050	60 amp	"	1.90	4.211		2,175	245		2,420	2,775
2150	200 amp	2 Elec	2	8		3,575	465		4,040	4,625
2200	400 amp		1.60	10		5,000	580		5,580	6,375
2250	600 amp		1	16		7,200	930		8,130	9,325
2500	NEMA 3R, 480 volt 3 pole, 60 amp	1 Elec	1.80	4.444		2,550	259		2,809	3,200
2550	100 amp	"	1.20	6.667		4,150	390		4,540	5,125
2600	200 amp	2 Elec	1.80	8.889		5,050	515		5,565	6,350
2650	400 amp	"	1.40	11.429		4,175	665		4,840	5,600
2800	NEMA 3R, 250 volt 3 pole solid state, 100 amp	1 Elec	1.20	6.667		3,750	390		4,140	4,675
2850	150 amp	2 Elec	1.80	8.889		4,975	515		5,490	6,250
2900	250 volt 2 pole solid state, 100 amp	1 Elec	1.30	6.154		3,625	360		3,985	4,525
2950	150 amp	2 Elec	2	8		4,825	465		5,290	6,025

26 36 23 – Automatic Transfer Switches

26 36 23.10 Automatic Transfer Switch Devices

		Crew	Daily Output	Labor-Hours	Unit	Material	Labor	Equipment	Total	Total Incl O&P
0010	**AUTOMATIC TRANSFER SWITCH DEVICES** R263623-60									
0015	Switches, enclosed 120/240 volt, 2 pole, 30 amp	1 Elec	2.40	3.333	Ea.	1,650	194		1,844	2,125
0020	70 amp		2	4		1,650	233		1,883	2,175
0030	100 amp		1.35	5.926		1,650	345		1,995	2,350
0040	225 amp	2 Elec	2.10	7.619		2,450	445		2,895	3,375
0050	400 amp		1.70	9.412		3,725	550		4,275	4,925
0060	600 amp		1.06	15.094		8,200	880		9,080	10,400
0070	800 amp		.84	19.048		9,650	1,100		10,750	12,300

For customer support on your Electrical Costs with RSMeans data, call 800.448.8182.

291

26 36 Transfer Switches

26 36 23 – Automatic Transfer Switches

26 36 23.10 Automatic Transfer Switch Devices	Crew	Daily Output	Labor-Hours	Unit	Material	2018 Bare Costs Labor	Equipment	Total	Total Incl O&P
0100	Switches, enclosed 480 volt, 3 pole, 30 amp	1 Elec	2.30	3.478	Ea.	2,950	202		3,152	3,550
0200	60 amp		1.90	4.211		2,950	245		3,195	3,625
0300	100 amp		1.30	6.154		2,950	360		3,310	3,775
0400	150 amp	2 Elec	2.40	6.667		3,650	390		4,040	4,575
0500	225 amp		2	8		4,650	465		5,115	5,825
0600	260 amp		2	8		5,375	465		5,840	6,625
0700	400 amp		1.60	10		7,025	580		7,605	8,600
0800	600 amp		1	16		9,725	930		10,655	12,100
0900	800 amp		.80	20		11,400	1,175		12,575	14,400
1000	1,000 amp		.76	21.053		17,000	1,225		18,225	20,500
1100	1,200 amp		.70	22.857		23,300	1,325		24,625	27,600
1200	1,600 amp		.60	26.667		26,500	1,550		28,050	31,400
1300	2,000 amp		.50	32		29,600	1,850		31,450	35,300
1600	Accessories, time delay on engine starting					259			259	285
1700	Adjustable time delay on retransfer					259			259	285
1800	Shunt trips for customer connections					465			465	510
1900	Maintenance select switch					105			105	116
2000	Auxiliary contact when normal fails					122			122	134
2100	Pilot light-emergency					105			105	116
2200	Pilot light-normal					105			105	116
2300	Auxiliary contact-closed on normal					122			122	134
2400	Auxiliary contact-closed on emergency					122			122	134
2500	Emergency source sensing, frequency relay					535			535	590

26 41 Facility Lightning Protection

26 41 13 – Lightning Protection for Structures

26 41 13.13 Lightning Protection for Buildings

		Crew	Daily Output	Labor-Hours	Unit	Material	2018 Bare Costs Labor	Equipment	Total	Total Incl O&P
0010	**LIGHTNING PROTECTION FOR BUILDINGS**									
0200	Air terminals & base, copper									
0400	3/8" diameter x 10" (to 75' high)	1 Elec	8	1	Ea.	22.50	58		80.50	112
0500	1/2" diameter x 12" (over 75' high)		8	1		26	58		84	116
0520	1/2" diameter x 24"		7.30	1.096		35.50	64		99.50	135
0540	1/2" diameter x 60"		6.70	1.194		66.50	69.50		136	177
1000	Aluminum, 1/2" diameter x 12" (to 75' high)		8	1		17.85	58		75.85	107
1020	1/2" diameter x 24"		7.30	1.096		19.50	64		83.50	117
1040	1/2" diameter x 60"		6.70	1.194		26.50	69.50		96	134
1100	5/8" diameter x 12" (over 75' high)		8	1		19.10	58		77.10	108
2000	Cable, copper, 220 lb. per thousand ft. (to 75' high)		320	.025	L.F.	3.03	1.46		4.49	5.50
2100	375 lb. per thousand ft. (over 75' high)		230	.035		5.50	2.02		7.52	9.10
2500	Aluminum, 101 lb. per thousand ft. (to 75' high)		280	.029		.95	1.66		2.61	3.54
2600	199 lb. per thousand ft. (over 75' high)		240	.033		1.40	1.94		3.34	4.44
3000	Arrester, 175 volt AC to ground		8	1	Ea.	149	58		207	251
3100	650 volt AC to ground		6.70	1.194	"	149	69.50		218.50	268
4000	Air terminals, copper									
4010	3/8" x 10"	1 Elec	50	.160	Ea.	7.75	9.30		17.05	22.50
4020	1/2" x 12"		38	.211		11.30	12.25		23.55	31
4030	5/8" x 12"		33	.242		16	14.10		30.10	38.50
4040	1/2" x 24"		30	.267		21	15.50		36.50	46
4050	1/2" x 60"		19	.421		52	24.50		76.50	93.50
4060	Air terminals, aluminum									
4070	3/8" x 10"	1 Elec	50	.160	Ea.	2.79	9.30		12.09	16.95

26 41 13.13 Lightning Protection for Buildings		Crew	Daily Output	Labor-Hours	Unit	Material	2018 Bare Costs Labor	2018 Bare Costs Equipment	Total	Total Incl O&P
4080	1/2" x 12"	1 Elec	38	.211	Ea.	3.04	12.25		15.29	21.50
4090	5/8" x 12"		33	.242		4.32	14.10		18.42	26
4100	1/2" x 24"		30	.267		4.72	15.50		20.22	28
4110	1/2" x 60"		19	.421		11.90	24.50		36.40	49.50
4200	Air terminal bases, copper									
4210	Adhesive bases, 1/2"	1 Elec	15	.533	Ea.	16.45	31		47.45	64.50
4215	Adhesive base, lightning air terminal					16.45			16.45	18.05
4220	Bolted bases	1 Elec	9	.889		14.65	51.50		66.15	93.50
4230	Hinged bases		7	1.143		27.50	66.50		94	130
4240	Side mounted bases		9	.889		17.05	51.50		68.55	96.50
4250	Offset point support		9	.889		27	51.50		78.50	108
4260	Concealed base assembly		5	1.600		82.50	93		175.50	230
4270	Tee connector base		13	.615		46.50	36		82.50	105
4280	Tripod base, 36" for 60" air terminal		7	1.143		39	66.50		105.50	142
4290	Intermediate base, 1/2"		15	.533		28.50	31		59.50	78
4300	Air terminal bases, aluminum									
4310	Adhesive bases, 1/2"	1 Elec	15	.533	Ea.	8.85	31		39.85	56.50
4320	Bolted bases		9	.889		14.80	51.50		66.30	94
4330	Hinged bases		7	1.143		13.80	66.50		80.30	115
4340	Side mounted bases		9	.889		11.75	51.50		63.25	90.50
4350	Offset point support		9	.889		12.65	51.50		64.15	91.50
4360	Concealed base assembly		5	1.600		41.50	93		134.50	185
4370	Tee connector base		13	.615		14.20	36		50.20	69
4380	Tripod base, 36" for 60" air terminal		7	1.143		29	66.50		95.50	132
4390	Intermediate base, 1/2"		15	.533		13.40	31		44.40	61.50
4400	Connector cable, copper									
4410	Through wall connector	1 Elec	7	1.143	Ea.	81	66.50		147.50	189
4420	Through roof connector		5	1.600		97.50	93		190.50	246
4430	Beam connector		7	1.143		26	66.50		92.50	128
4440	Double bolt connector		38	.211		11.50	12.25		23.75	31
4450	Bar, copper		40	.200		12.80	11.65		24.45	31.50
4500	Connector cable, aluminum									
4510	Through wall connector	1 Elec	7	1.143	Ea.	32.50	66.50		99	136
4520	Through roof connector		5	1.600		36	93		129	179
4530	Beam connector		7	1.143		18	66.50		84.50	119
4540	Double bolt connector		38	.211		5.95	12.25		18.20	25
4600	Bonding plates, copper									
4610	I-beam, 8" square	1 Elec	7	1.143	Ea.	22.50	66.50		89	125
4620	Purlin, 8" square		7	1.143		26.50	66.50		93	129
4630	Large heavy duty, 16" square		5	1.600		31.50	93		124.50	174
4640	Aluminum, I-beam, 8" square		7	1.143		16.65	66.50		83.15	118
4650	Purlin, 8" square		7	1.143		16.50	66.50		83	118
4660	Large heavy duty, 16" square		5	1.600		17.95	93		110.95	159
4700	Cable support, copper, loop fastener		38	.211		1.10	12.25		13.35	19.50
4710	Adhesive cable holder		38	.211		1.43	12.25		13.68	19.85
4720	Aluminum, loop fastener		38	.211		.70	12.25		12.95	19.05
4730	Adhesive cable holder		38	.211		.77	12.25		13.02	19.15
4740	Bonding strap, copper, 3/4" x 9-1/2"		15	.533		11.30	31		42.30	59
4750	Aluminum, 3/4" x 9-1/2"		15	.533		4.83	31		35.83	52
4760	Swivel adapter, copper, 1/2"		73	.110		17	6.40		23.40	28.50
4770	Aluminum, 1/2"		73	.110		9	6.40		15.40	19.45

26 51 Interior Lighting

26 51 13 – Interior Lighting Fixtures, Lamps, and Ballasts

26 51 13.10 Fixture Hangers		Crew	Daily Output	Labor-Hours	Unit	Material	2018 Bare Costs Labor	Equipment	Total	Total Incl O&P
0010	**FIXTURE HANGERS**									
0220	Box hub cover	1 Elec	32	.250	Ea.	4.19	14.55		18.74	26.50
0240	Canopy		12	.667		9.95	39		48.95	69
0260	Connecting block		40	.200		2.78	11.65		14.43	20.50
0280	Cushion hanger		16	.500		23.50	29		52.50	69.50
0300	Box hanger, with mounting strap		8	1		10	58		68	98
0320	Connecting block		40	.200		2.34	11.65		13.99	19.95
0340	Flexible, 1/2" diameter, 4" long		12	.667		16.40	39		55.40	76
0360	6" long		12	.667		17.80	39		56.80	77.50
0380	8" long		12	.667		19.70	39		58.70	79.50
0400	10" long		12	.667		21	39		60	81
0420	12" long		12	.667		22	39		61	82.50
0440	15" long		12	.667		23.50	39		62.50	83.50
0460	18" long		12	.667		27.50	39		66.50	88
0480	3/4" diameter, 4" long		10	.800		21	46.50		67.50	92.50
0500	6" long		10	.800		23.50	46.50		70	95
0520	8" long		10	.800		23.50	46.50		70	95
0540	10" long		10	.800		26	46.50		72.50	98
0560	12" long		10	.800		27.50	46.50		74	100
0580	15" long		10	.800		31	46.50		77.50	104
0600	18" long		10	.800		35	46.50		81.50	108

26 51 13.40 Interior HID Fixtures		Crew	Daily Output	Labor-Hours	Unit	Material	2018 Bare Costs Labor	Equipment	Total	Total Incl O&P
0010	**INTERIOR HID FIXTURES** Incl. lamps and mounting hardware									
0700	High pressure sodium, recessed, round, 70 watt	1 Elec	3.50	2.286	Ea.	520	133		653	770
0720	100 watt		3.50	2.286		595	133		728	855
0740	150 watt		3.20	2.500		670	146		816	955
0760	Square, 70 watt		3.60	2.222		565	129		694	815
0780	100 watt		3.60	2.222		600	129		729	855
0820	250 watt		3	2.667		820	155		975	1,125
0840	1,000 watt	2 Elec	4.80	3.333		1,575	194		1,769	2,025
0860	Surface, round, 70 watt	1 Elec	3	2.667		800	155		955	1,100
0880	100 watt		3	2.667		820	155		975	1,125
0900	150 watt		2.70	2.963		860	172		1,032	1,200
0920	Square, 70 watt		3	2.667		760	155		915	1,075
0940	100 watt		3	2.667		800	155		955	1,100
0980	250 watt		2.50	3.200		665	186		851	1,025
1040	Pendent, round, 70 watt		3	2.667		830	155		985	1,150
1060	100 watt		3	2.667		800	155		955	1,100
1080	150 watt		2.70	2.963		815	172		987	1,150
1100	Square, 70 watt		3	2.667		865	155		1,020	1,175
1120	100 watt		3	2.667		875	155		1,030	1,200
1140	150 watt		2.70	2.963		900	172		1,072	1,250
1160	250 watt		2.50	3.200		1,225	186		1,411	1,625
1180	400 watt		2.40	3.333		1,275	194		1,469	1,700
1220	Wall, round, 70 watt		3	2.667		725	155		880	1,025
1240	100 watt		3	2.667		740	155		895	1,050
1260	150 watt		2.70	2.963		765	172		937	1,100
1300	Square, 70 watt		3	2.667		775	155		930	1,075
1320	100 watt		3	2.667		815	155		970	1,125
1340	150 watt		2.70	2.963		840	172		1,012	1,175
1360	250 watt		2.50	3.200		865	186		1,051	1,225
1380	400 watt	2 Elec	4.80	3.333		1,075	194		1,269	1,475

26 51 Interior Lighting

26 51 13 – Interior Lighting Fixtures, Lamps, and Ballasts

26 51 13.40 Interior HID Fixtures		Crew	Daily Output	Labor-Hours	Unit	Material	2018 Bare Costs Labor	Equipment	Total	Total Incl O&P
1400	1,000 watt	2 Elec	3.60	4.444	Ea.	1,525	259		1,784	2,050
1500	Metal halide, recessed, round, 175 watt	1 Elec	3.40	2.353		435	137		572	680
1520	250 watt	"	3.20	2.500		545	146		691	820
1540	400 watt	2 Elec	5.80	2.759		770	161		931	1,100
1580	Square, 175 watt	1 Elec	3.40	2.353		440	137		577	690
1640	Surface, round, 175 watt		2.90	2.759		605	161		766	905
1660	250 watt	↓	2.70	2.963		970	172		1,142	1,325
1680	400 watt	2 Elec	4.80	3.333		1,075	194		1,269	1,475
1720	Square, 175 watt	1 Elec	2.90	2.759		660	161		821	965
1800	Pendent, round, 175 watt		2.90	2.759		800	161		961	1,125
1820	250 watt	↓	2.70	2.963		1,050	172		1,222	1,400
1840	400 watt	2 Elec	4.80	3.333		1,150	194		1,344	1,575
1880	Square, 175 watt	1 Elec	2.90	2.759		475	161		636	760
1900	250 watt	"	2.70	2.963		685	172		857	1,025
1920	400 watt	2 Elec	4.80	3.333		1,100	194		1,294	1,525
1980	Wall, round, 175 watt	1 Elec	2.90	2.759		910	161		1,071	1,250
2000	250 watt	"	2.70	2.963		910	172		1,082	1,250
2020	400 watt	2 Elec	4.80	3.333		970	194		1,164	1,375
2060	Square, 175 watt	1 Elec	2.90	2.759		650	161		811	955
2080	250 watt	"	2.70	2.963		690	172		862	1,025
2100	400 watt	2 Elec	4.80	3.333		965	194		1,159	1,375
2800	High pressure sodium, vaporproof, recessed, 70 watt	1 Elec	3.50	2.286		790	133		923	1,075
2820	100 watt		3.50	2.286		810	133		943	1,100
2840	150 watt		3.20	2.500		830	146		976	1,125
2900	Surface, 70 watt		3	2.667		890	155		1,045	1,200
2920	100 watt		3	2.667		935	155		1,090	1,250
2940	150 watt		2.70	2.963		970	172		1,142	1,325
3000	Pendent, 70 watt		3	2.667		880	155		1,035	1,200
3020	100 watt		3	2.667		910	155		1,065	1,225
3040	150 watt		2.70	2.963		965	172		1,137	1,300
3100	Wall, 70 watt		3	2.667		950	155		1,105	1,275
3120	100 watt		3	2.667		995	155		1,150	1,325
3140	150 watt		2.70	2.963		1,025	172		1,197	1,375
3200	Metal halide, vaporproof, recessed, 175 watt		3.40	2.353		650	137		787	915
3220	250 watt	↓	3.20	2.500		745	146		891	1,050
3240	400 watt	2 Elec	5.80	2.759		935	161		1,096	1,275
3260	1,000 watt	"	4.80	3.333		1,700	194		1,894	2,175
3280	Surface, 175 watt	1 Elec	2.90	2.759		1,025	161		1,186	1,375
3300	250 watt	"	2.70	2.963		1,075	172		1,247	1,425
3320	400 watt	2 Elec	4.80	3.333		1,325	194		1,519	1,750
3340	1,000 watt	"	3.60	4.444		1,925	259		2,184	2,500
3360	Pendent, 175 watt	1 Elec	2.90	2.759		1,000	161		1,161	1,350
3380	250 watt	"	2.70	2.963		1,100	172		1,272	1,450
3400	400 watt	2 Elec	4.80	3.333		1,300	194		1,494	1,750
3420	1,000 watt	"	3.60	4.444		2,100	259		2,359	2,700
3440	Wall, 175 watt	1 Elec	2.90	2.759		1,050	161		1,211	1,400
3460	250 watt	"	2.70	2.963		1,150	172		1,322	1,525
3480	400 watt	2 Elec	4.80	3.333		1,375	194		1,569	1,800
3500	1,000 watt	"	3.60	4.444	↓	2,200	259		2,459	2,775

For customer support on your Electrical Costs with RSMeans data, call 800.448.8182.

295

26 51 13.50 Interior Lighting Fixtures		Crew	Daily Output	Labor-Hours	Unit	Material	2018 Bare Costs Labor	Equipment	Total	Total Incl O&P
0010	**INTERIOR LIGHTING FIXTURES** Including lamps, mounting R265113-40									
0030	hardware and connections									
0100	Fluorescent, C.W. lamps, troffer, recess mounted in grid, RS									
0130	Grid ceiling mount									
0200	Acrylic lens, 1' W x 4' L, two 40 watt	1 Elec	5.70	1.404	Ea.	51.50	81.50		133	179
0210	1' W x 4' L, three 40 watt		5.40	1.481		58.50	86		144.50	194
0300	2' W x 2' L, two U40 watt		5.70	1.404		55.50	81.50		137	183
0400	2' W x 4' L, two 40 watt		5.30	1.509		54	88		142	191
0500	2' W x 4' L, three 40 watt		5	1.600		59.50	93		152.50	205
0600	2' W x 4' L, four 40 watt		4.70	1.702		62	99		161	217
0700	4' W x 4' L, four 40 watt	2 Elec	6.40	2.500		276	146		422	525
0800	4' W x 4' L, six 40 watt		6.20	2.581		297	150		447	550
0900	4' W x 4' L, eight 40 watt		5.80	2.759		340	161		501	615
0910	Acrylic lens, 1' W x 4' L, two 32 watt T8 [G]	1 Elec	5.70	1.404		61.50	81.50		143	190
0930	2' W x 2' L, two U32 watt T8 [G]		5.70	1.404		96	81.50		177.50	227
0940	2' W x 4' L, two 32 watt T8 [G]		5.30	1.509		79	88		167	218
0950	2' W x 4' L, three 32 watt T8 [G]		5	1.600		70	93		163	216
0960	2' W x 4' L, four 32 watt T8 [G]		4.70	1.702		74.50	99		173.50	230
1000	Surface mounted, RS									
1030	Acrylic lens with hinged & latched door frame									
1100	1' W x 4' L, two 40 watt	1 Elec	7	1.143	Ea.	65.50	66.50		132	172
1110	1' W x 4' L, three 40 watt		6.70	1.194		68	69.50		137.50	179
1200	2' W x 2' L, two U40 watt		7	1.143		70.50	66.50		137	177
1300	2' W x 4' L, two 40 watt		6.20	1.290		80	75		155	200
1400	2' W x 4' L, three 40 watt		5.70	1.404		83	81.50		164.50	213
1500	2' W x 4' L, four 40 watt		5.30	1.509		85	88		173	225
1501	2' W x 4' L, six 40 watt T8		5.20	1.538		85	89.50		174.50	228
1600	4' W x 4' L, four 40 watt	2 Elec	7.20	2.222		420	129		549	655
1700	4' W x 4' L, six 40 watt		6.60	2.424		455	141		596	710
1800	4' W x 4' L, eight 40 watt		6.20	2.581		475	150		625	745
1900	2' W x 8' L, four 40 watt		6.40	2.500		167	146		313	400
2000	2' W x 8' L, eight 40 watt		6.20	2.581		179	150		329	420
2010	Acrylic wrap around lens									
2020	6" W x 4' L, one 40 watt	1 Elec	8	1	Ea.	68	58		126	162
2030	6" W x 8' L, two 40 watt	2 Elec	8	2		75	116		191	257
2040	11" W x 4' L, two 40 watt	1 Elec	7	1.143		45.50	66.50		112	150
2050	11" W x 8' L, four 40 watt	2 Elec	6.60	2.424		75.50	141		216.50	294
2060	16" W x 4' L, four 40 watt	1 Elec	5.30	1.509		75.50	88		163.50	214
2070	16" W x 8' L, eight 40 watt	2 Elec	6.40	2.500		164	146		310	400
2080	2' W x 2' L, two U40 watt	1 Elec	7	1.143		92	66.50		158.50	201
2100	Strip fixture									
2130	Surface mounted									
2200	4' long, one 40 watt, RS	1 Elec	8.50	.941	Ea.	31	55		86	116
2300	4' long, two 40 watt, RS		8	1		43	58		101	135
2310	4' long, two 32 watt T8, RS [G]		8	1		65.50	58		123.50	159
2400	4' long, one 40 watt, SL		8	1		53.50	58		111.50	146
2500	4' long, two 40 watt, SL		7	1.143		72.50	66.50		139	179
2580	8' long, one 60 watt T8, SL [G]	2 Elec	13.40	1.194		112	69.50		181.50	227
2590	8' long, two 60 watt T8, SL [G]		12.40	1.290		92.50	75		167.50	214
2600	8' long, one 75 watt, SL		13.40	1.194		55	69.50		124.50	165
2700	8' long, two 75 watt, SL		12.40	1.290		66.50	75		141.50	185
2800	4' long, two 60 watt, HO	1 Elec	6.70	1.194		107	69.50		176.50	222

26 51 13.50 Interior Lighting Fixtures		Crew	Daily Output	Labor-Hours	Unit	Material	2018 Bare Costs Labor	Equipment	Total	Total Incl O&P
2810	4' long, two 54 watt, T5HO **G**	1 Elec	6.70	1.194	Ea.	174	69.50		243.50	295
2900	8' long, two 110 watt, HO	2 Elec	10.60	1.509		108	88		196	250
2910	4' long, two 115 watt, VHO	1 Elec	6.50	1.231		142	71.50		213.50	263
2920	8' long, two 215 watt, VHO	2 Elec	10.40	1.538		153	89.50		242.50	305
2950	High bay pendent mounted, 16" W x 4' L, four 54 watt, T5HO **G**		8.90	1.798		244	105		349	425
2952	2' W x 4' L, six 54 watt, T5HO **G**		8.50	1.882		310	110		420	505
2954	2' W x 4' L, six 32 watt, T8 **G**		8.50	1.882		179	110		289	360
3000	Strip, pendent mounted, industrial, white porcelain enamel									
3100	4' long, two 40 watt, RS	1 Elec	5.70	1.404	Ea.	53	81.50		134.50	180
3110	4' long, two 32 watt T8, RS **G**		5.70	1.404		72	81.50		153.50	201
3200	4' long, two 60 watt, HO		5	1.600		83	93		176	231
3290	8' long, two 60 watt T8, SL **G**	2 Elec	8.80	1.818		118	106		224	288
3300	8' long, two 75 watt, SL		8.80	1.818		98.50	106		204.50	266
3400	8' long, two 110 watt, HO		8	2		126	116		242	315
3410	Acrylic finish, 4' long, two 40 watt, RS	1 Elec	5.70	1.404		86.50	81.50		168	217
3420	4' long, two 60 watt, HO		5	1.600		161	93		254	315
3430	4' long, two 115 watt, VHO		4.80	1.667		208	97		305	375
3440	8' long, two 75 watt, SL	2 Elec	8.80	1.818		170	106		276	345
3450	8' long, two 110 watt, HO		8	2		193	116		309	385
3460	8' long, two 215 watt, VHO		7.60	2.105		273	123		396	485
3470	Troffer, air handling, 2' W x 4' L with four 32 watt T8 **G**	1 Elec	4	2		116	116		232	300
3480	2' W x 2' L with two U32 watt T8 **G**		5.50	1.455		111	84.50		195.50	250
3490	Air connector insulated, 5" diameter		20	.400		68	23.50		91.50	110
3500	6" diameter		20	.400		69	23.50		92.50	111
3502	Troffer, direct/indirect, 2' W x 4' L with two 32 W T8 **G**		5.30	1.509		282	88		370	440
3510	Troffer parabolic lay-in, 1' W x 4' L with one 32 W T8 **G**		5.70	1.404		117	81.50		198.50	251
3520	1' W x 4' L with two 32 W T8 **G**		5.30	1.509		140	88		228	285
3525	2' W x 2' L with two U32 W T8 **G**		5.70	1.404		118	81.50		199.50	252
3530	2' W x 4' L with three 32 W T8 **G**		5	1.600		131	93		224	283
3531	Intr fxtr, fluor, troffer prismatic lay-in, 2' W x 4'l w/three 32 W T8		5	1.600		152	93		245	305
3535	Downlight, recess mounted **G**		8	1		148	58		206	249
3540	Wall wash reflector, recess mounted **G**		8	1		105	58		163	203
3550	Direct/indirect, 4' long, stl., pendent mtd. **G**		5	1.600		163	93		256	320
3560	4' long, alum., pendent mtd. **G**		5	1.600		335	93		428	510
3565	Prefabricated cove, 4' long, stl. continuous row **G**		5	1.600		204	93		297	365
3570	4' long, alum. continuous row **G**		5	1.600		350	93		443	525
3580	Wet location, recess mounted, 2' W x 4' L with two 32 watt T8 **G**		5.30	1.509		243	88		331	400
3590	Pendent mounted, 2' W x 4' L with two 32 watt T8 **G**		5.70	1.404		375	81.50		456.50	535
4000	Induction lamp, integral ballast, ceiling mounted									
4110	High bay, aluminum reflector, 160 watt	1 Elec	3.20	2.500	Ea.	860	146		1,006	1,175
4120	320 watt	"	3	2.667		1,550	155		1,705	1,925
4130	480 watt	2 Elec	5.80	2.759		2,350	161		2,511	2,825
4150	Low bay, aluminum reflector, 250 watt	1 Elec	3.20	2.500		920	146		1,066	1,225
4170	Garage, aluminum reflector, 80 watt		3.60	2.222		830	129		959	1,100
4180	Vandalproof, aluminum reflector, 100 watt		3.20	2.500		660	146		806	950
4220	Metal halide, integral ballast, ceiling, recess mounted									
4230	prismatic glass lens, floating door									
4240	2' W x 2' L, 250 watt	1 Elec	3.20	2.500	Ea.	255	146		401	500
4250	2' W x 2' L, 400 watt	2 Elec	5.80	2.759		365	161		526	645
4260	Surface mounted, 2' W x 2' L, 250 watt	1 Elec	2.70	2.963		345	172		517	640
4270	400 watt	2 Elec	4.80	3.333		405	194		599	740
4280	High bay, aluminum reflector,									
4290	Single unit, 400 watt	2 Elec	4.60	3.478	Ea.	415	202		617	760

26 51 13.50 Interior Lighting Fixtures		Crew	Daily Output	Labor-Hours	Unit	Material	2018 Bare Costs Labor	Equipment	Total	Total Incl O&P
4300	Single unit, 1,000 watt	2 Elec	4	4	Ea.	595	233		828	1,000
4310	Twin unit, 400 watt		3.20	5		825	291		1,116	1,350
4320	Low bay, aluminum reflector, 250W DX lamp	1 Elec	3.20	2.500		365	146		511	620
4330	400 watt lamp	2 Elec	5	3.200		530	186		716	860
4340	High pressure sodium integral ballast ceiling, recess mounted									
4350	prismatic glass lens, floating door									
4360	2' W x 2' L, 150 watt lamp	1 Elec	3.20	2.500	Ea.	390	146		536	645
4370	2' W x 2' L, 400 watt lamp	2 Elec	5.80	2.759		465	161		626	750
4380	Surface mounted, 2' W x 2' L, 150 watt lamp	1 Elec	2.70	2.963		475	172		647	780
4390	400 watt lamp	2 Elec	4.80	3.333		530	194		724	875
4400	High bay, aluminum reflector,									
4410	Single unit, 400 watt lamp	2 Elec	4.60	3.478	Ea.	380	202		582	725
4430	Single unit, 1,000 watt lamp	"	4	4		550	233		783	955
4440	Low bay, aluminum reflector, 150 watt lamp	1 Elec	3.20	2.500		330	146		476	580
4445	High bay H.I.D. quartz restrike	"	16	.500		164	29		193	224
4450	Incandescent, high hat can, round alzak reflector, prewired									
4470	100 watt	1 Elec	8	1	Ea.	64.50	58		122.50	158
4480	150 watt		8	1		105	58		163	202
4500	300 watt		6.70	1.194		242	69.50		311.50	370
4520	Round with reflector and baffles, 150 watt		8	1		51.50	58		109.50	144
4540	Round with concentric louver, 150 watt PAR		8	1		79	58		137	174
4600	Square glass lens with metal trim, prewired									
4630	100 watt	1 Elec	6.70	1.194	Ea.	55.50	69.50		125	165
4680	150 watt		6.70	1.194		98	69.50		167.50	212
4700	200 watt		6.70	1.194		98	69.50		167.50	212
4800	300 watt		5.70	1.404		146	81.50		227.50	282
4810	500 watt		5	1.600		286	93		379	455
4900	Ceiling/wall, surface mounted, metal cylinder, 75 watt		10	.800		50.50	46.50		97	126
4920	150 watt		10	.800		81	46.50		127.50	159
4930	300 watt		8	1		168	58		226	272
5000	500 watt		6.70	1.194		365	69.50		434.50	505
5010	Square, 100 watt		8	1		113	58		171	211
5020	150 watt		8	1		124	58		182	223
5030	300 watt		7	1.143		340	66.50		406.50	470
5040	500 watt		6	1.333		345	77.50		422.50	495
5200	Ceiling, surface mounted, opal glass drum									
5300	8", one 60 watt lamp	1 Elec	10	.800	Ea.	65.50	46.50		112	142
5400	10", two 60 watt lamps		8	1		72.50	58		130.50	167
5500	12", four 60 watt lamps		6.70	1.194		103	69.50		172.50	217
5510	Pendent, round, 100 watt		8	1		113	58		171	211
5520	150 watt		8	1		123	58		181	222
5530	300 watt		6.70	1.194		171	69.50		240.50	292
5540	500 watt		5.50	1.455		325	84.50		409.50	480
5550	Square, 100 watt		6.70	1.194		152	69.50		221.50	271
5560	150 watt		6.70	1.194		158	69.50		227.50	278
5570	300 watt		5.70	1.404		231	81.50		312.50	375
5580	500 watt		5	1.600		315	93		408	485
5600	Wall, round, 100 watt		8	1		65.50	58		123.50	159
5620	300 watt		8	1		143	58		201	244
5630	500 watt		6.70	1.194		380	69.50		449.50	520
5640	Square, 100 watt		8	1		104	58		162	201
5650	150 watt		8	1		106	58		164	203
5660	300 watt		7	1.143		166	66.50		232.50	283

For customer support on your Electrical Costs with RSMeans data, call 800.448.8182.

26 51 13.50 Interior Lighting Fixtures		Crew	Daily Output	Labor-Hours	Unit	Material	2018 Bare Costs Labor	Equipment	Total	Total Incl O&P
5670	500 watt	1 Elec	6	1.333	Ea.	283	77.50		360.50	425
6010	Vapor tight, incandescent, ceiling mounted, 200 watt		6.20	1.290		78.50	75		153.50	198
6020	Recessed, 200 watt		6.70	1.194		124	69.50		193.50	240
6030	Pendent, 200 watt		6.70	1.194		78.50	69.50		148	190
6040	Wall, 200 watt		8	1		80	58		138	175
6100	Fluorescent, surface mounted, 2 lamps, 4' L, RS, 40 watt		3.20	2.500		116	146		262	345
6110	Industrial, 2 lamps, 4' L in tandem, 430 MA		2.20	3.636		213	212		425	550
6130	2 lamps, 4' L, 800 MA		1.90	4.211		182	245		427	565
6160	Pendent, indust, 2 lamps, 4' L in tandem, 430 MA		1.90	4.211		248	245		493	640
6170	2 lamps, 4' L, 430 MA		2.30	3.478		166	202		368	490
6180	2 lamps, 4' L, 800 MA		1.70	4.706		209	274		483	640
6850	Vandalproof, surface mounted, fluorescent, two 32 watt T8 [G]		3.20	2.500		252	146		398	495
6860	Incandescent, one 150 watt		8	1		90	58		148	186
6900	Mirror light, fluorescent, RS, acrylic enclosure, two 40 watt		8	1		116	58		174	215
6910	One 40 watt		8	1		96	58		154	193
6920	One 20 watt		12	.667		82.50	39		121.50	149
7000	Low bay, aluminum reflector, 70 watt, high pressure sodium		4	2		275	116		391	480
7010	250 watt		3.20	2.500		365	146		511	625
7020	400 watt	2 Elec	5	3.200		385	186		571	700
7500	Ballast replacement, by weight of ballast, to 15' high									
7520	Indoor fluorescent, less than 2 lb.	1 Elec	10	.800	Ea.	23.50	46.50		70	95.50
7540	Two 40W, watt reducer, 2 to 5 lb.		9.40	.851		71	49.50		120.50	153
7560	Two F96 slimline, over 5 lb.		8	1		110	58		168	208
7580	Vaportite ballast, less than 2 lb.		9.40	.851		23.50	49.50		73	100
7600	2 lb. to 5 lb.		8.90	.899		71	52.50		123.50	157
7620	Over 5 lb.		7.60	1.053		110	61.50		171.50	213
7630	Electronic ballast for two tubes		8	1		43.50	58		101.50	135
7640	Dimmable ballast one-lamp [G]		8	1		108	58		166	206
7650	Dimmable ballast two-lamp [G]		7.60	1.053		108	61.50		169.50	210
7660	Dimmable ballast three-lamp		6.90	1.159		130	67.50		197.50	244
7690	Emergency ballast (factory installed in fixture)					179			179	197
7990	Decorator									
8000	Pendent RLM in colors, shallow dome, 12" diam., 100 watt	1 Elec	8	1	Ea.	80.50	58		138.50	176
8010	Regular dome, 12" diam., 100 watt		8	1		83	58		141	179
8020	16" diam., 200 watt		7	1.143		84.50	66.50		151	193
8030	18" diam., 300 watt		6	1.333		93	77.50		170.50	218
8100	Picture framing light		16	.500		96	29		125	150
8150	Miniature low voltage, recessed, pinhole		8	1		129	58		187	229
8160	Star		8	1		134	58		192	234
8170	Adjustable cone		8	1		166	58		224	270
8180	Eyeball		8	1		107	58		165	204
8190	Cone		8	1		130	58		188	230
8200	Coilex baffle		8	1		118	58		176	217
8210	Surface mounted, adjustable cylinder		8	1		125	58		183	224
8250	Chandeliers, incandescent									
8260	24" diam. x 42" high, 6 light candle	1 Elec	6	1.333	Ea.	420	77.50		497.50	580
8270	24" diam. x 42" high, 6 light candle w/glass shade		6	1.333		430	77.50		507.50	590
8280	17" diam. x 12" high, 8 light w/glass panels		8	1		257	58		315	370
8300	27" diam. x 29" high, 10 light bohemian lead crystal		4	2		560	116		676	790
8310	21" diam. x 9" high, 6 light sculptured ice crystal		8	1		400	58		458	525
8500	Accent lights, on floor or edge, 0.5 W low volt incandescent									
8520	incl. transformer & fastenings, based on 100' lengths									
8550	Lights in clear tubing, 12" OC	1 Elec	230	.035	L.F.	8.35	2.02		10.37	12.25

26 51 13 – Interior Lighting Fixtures, Lamps, and Ballasts

26 51 13.50 Interior Lighting Fixtures

		Crew	Daily Output	Labor-Hours	Unit	Material	2018 Bare Costs Labor	Equipment	Total	Total Incl O&P
8560	6" OC	1 Elec	160	.050	L.F.	10.90	2.91		13.81	16.35
8570	4" OC		130	.062		16.70	3.58		20.28	23.50
8580	3" OC		125	.064		18.50	3.72		22.22	26
8590	2" OC		100	.080		27	4.66		31.66	36.50
8600	Carpet, lights both sides 6" OC, in alum. extrusion		270	.030		25.50	1.72		27.22	30.50
8610	In bronze extrusion		270	.030		28.50	1.72		30.22	34
8620	Carpet-bare floor, lights 18" OC, in alum. extrusion		270	.030		20	1.72		21.72	24.50
8630	In bronze extrusion		270	.030		23.50	1.72		25.22	28.50
8640	Carpet edge-wall, lights 6" OC in alum. extrusion		270	.030		25.50	1.72		27.22	30.50
8650	In bronze extrusion		270	.030		28.50	1.72		30.22	34
8660	Bare floor, lights 18" OC, in alum. extrusion		300	.027		20	1.55		21.55	24.50
8670	In bronze extrusion		300	.027		23.50	1.55		25.05	28.50
8680	Bare floor conduit, alum. extrusion		300	.027		6.65	1.55		8.20	9.65
8690	In bronze extrusion		300	.027	↓	13.35	1.55		14.90	16.95
8700	Step edge to 36", lights 6" OC, in alum. extrusion		100	.080	Ea.	67	4.66		71.66	81
8710	In bronze extrusion		100	.080		70	4.66		74.66	84
8720	Step edge to 54", lights 6" OC, in alum. extrusion		100	.080		101	4.66		105.66	118
8730	In bronze extrusion		100	.080		106	4.66		110.66	124
8740	Step edge to 72", lights 6" OC, in alum. extrusion		100	.080		134	4.66		138.66	155
8750	In bronze extrusion		100	.080		146	4.66		150.66	168
8760	Connector, male		32	.250		2.48	14.55		17.03	24.50
8770	Female with pigtail		32	.250		5.20	14.55		19.75	28
8780	Clamps		400	.020		.49	1.16		1.65	2.28
8790	Transformers, 50 watt		8	1		89.50	58		147.50	186
8800	250 watt		4	2		285	116		401	490
8810	1,000 watt	↓	2.70	2.963	↓	325	172		497	620

26 51 13.55 Interior LED Fixtures

			Crew	Daily Output	Labor-Hours	Unit	Material	2018 Bare Costs Labor	Equipment	Total	Total Incl O&P
0010	**INTERIOR LED FIXTURES** Incl. lamps and mounting hardware										
0100	Downlight, recess mounted, 7.5" diameter, 25 watt	G	1 Elec	8	1	Ea.	335	58		393	450
0120	10" diameter, 36 watt	G		8	1		360	58		418	480
0160	cylinder, 10 watts	G		8	1		103	58		161	200
0180	20 watts	G		8	1		129	58		187	229
0900	Interior LED fixts, troffer, recess mounted, 2' x 2', 3,500K			8.50	.941		126	55		181	221
0910	2' x 2', 4,000K			8.50	.941		123	55		178	218
1000	Troffer, recess mounted, 2' x 4', 3,200 lumens	G		5.30	1.509		138	88		226	282
1010	4,800 lumens	G		5	1.600		179	93		272	335
1020	6,400 lumens	G		4.70	1.702		198	99		297	365
1100	Troffer retrofit lamp, 38 watt	G		21	.381		69.50	22		91.50	110
1110	60 watt	G		20	.400		141	23.50		164.50	191
1120	100 watt	G		18	.444		206	26		232	265
1200	Troffer, volumetric recess mounted, 2' x 2'	G		5.70	1.404		305	81.50		386.50	455
2000	Strip, surface mounted, one light bar 4' long, 3,500 K	G		8.50	.941		260	55		315	370
2010	5,000 K	G		8	1		260	58		318	375
2020	Two light bar 4' long, 5,000 K	G		7	1.143		410	66.50		476.50	550
3000	Linear, suspended mounted, one light bar 4' long, 37 watt	G	↓	6.70	1.194		153	69.50		222.50	272
3010	One light bar 8' long, 74 watt	G	2 Elec	12.20	1.311		283	76.50		359.50	425
3020	Two light bar 4' long, 74 watt	G	1 Elec	5.70	1.404		305	81.50		386.50	455
3030	Two light bar 8' long, 148 watt	G	2 Elec	8.80	1.818		350	106		456	545
4000	High bay, surface mounted, round, 150 watts	G		5.41	2.959		425	172		597	725
4010	2 bars, 164 watts	G		5.41	2.959		385	172		557	680
4020	3 bars, 246 watts	G		5.01	3.197		515	186		701	845
4030	4 bars, 328 watts	G	↓	4.60	3.478		715	202		917	1,100

26 51 13 – Interior Lighting Fixtures, Lamps, and Ballasts

26 51 13.55 Interior LED Fixtures

		Crew	Daily Output	Labor-Hours	Unit	Material	2018 Bare Costs Labor	Equipment	Total	Total Incl O&P
4040	5 bars, 410 watts [G]	3 Elec	4.20	5.716	Ea.	810	335		1,145	1,400
4050	6 bars, 492 watts [G]		3.80	6.324		910	370		1,280	1,550
4060	7 bars, 574 watts [G]		3.39	7.075		990	410		1,400	1,725
4070	8 bars, 656 watts [G]		2.99	8.029		1,050	465		1,515	1,850
5000	Track, lighthead, 6 watt [G]	1 Elec	32	.250		54.50	14.55		69.05	82
5010	9 watt [G]	"	32	.250		61.50	14.55		76.05	89.50
6000	Garage, surface mounted, 103 watts [G]	2 Elec	6.50	2.462		970	143		1,113	1,300
6100	pendent mounted, 80 watts [G]		6.50	2.462		660	143		803	940
6200	95 watts [G]		6.50	2.462		780	143		923	1,075
6300	125 watts [G]		6.50	2.462		820	143		963	1,125
7000	Downlight, recess mtd., low profile, 4" diam., 9W, 3,000K	1 Elec	8.50	.941		24.50	55		79.50	109
7010	4,000K		8.50	.941		23.50	55		78.50	108
7020	5,000K		8.50	.941		24	55		79	109

26 51 13.70 Residential Fixtures

		Crew	Daily Output	Labor-Hours	Unit	Material	2018 Bare Costs Labor	Equipment	Total	Total Incl O&P
0010	**RESIDENTIAL FIXTURES**									
0400	Fluorescent, interior, surface, circline, 32 watt & 40 watt	1 Elec	20	.400	Ea.	155	23.50		178.50	206
0500	2' x 2', two U-tube 32 watt T8		8	1		129	58		187	229
0700	Shallow under cabinet, two 20 watt		16	.500		71.50	29		100.50	122
0900	Wall mounted, 4' L, two 32 watt T8, with baffle		10	.800		133	46.50		179.50	216
2000	Incandescent, exterior lantern, wall mounted, 60 watt		16	.500		60.50	29		89.50	110
2100	Post light, 150 W, with 7' post		4	2		270	116		386	470
2500	Lamp holder, weatherproof with 150 W PAR		16	.500		33.50	29		62.50	80
2550	With reflector and guard		12	.667		59.50	39		98.50	124
2600	Interior pendent, globe with shade, 150 W		20	.400		201	23.50		224.50	256

26 51 13.90 Ballast, Replacement HID

		Crew	Daily Output	Labor-Hours	Unit	Material	2018 Bare Costs Labor	Equipment	Total	Total Incl O&P
0010	**BALLAST, REPLACEMENT HID**									
7510	Multi-tap 120/208/240/277 V									
7550	High pressure sodium, 70 watt	1 Elec	10	.800	Ea.	116	46.50		162.50	197
7560	100 watt		9.40	.851		82.50	49.50		132	165
7570	150 watt		9	.889		130	51.50		181.50	221
7580	250 watt		8.50	.941		193	55		248	295
7590	400 watt		7	1.143		171	66.50		237.50	288
7600	1,000 watt		6	1.333		183	77.50		260.50	320
7610	Metal halide, 175 watt		8	1		60.50	58		118.50	154
7620	250 watt		8	1		79	58		137	174
7630	400 watt		7	1.143		98.50	66.50		165	208
7640	1,000 watt		6	1.333		161	77.50		238.50	293
7650	1,500 watt		5	1.600		194	93		287	350

26 51 19 – LED Interior Lighting

26 51 19.10 LED Interior Lighting

		Crew	Daily Output	Labor-Hours	Unit	Material	2018 Bare Costs Labor	Equipment	Total	Total Incl O&P
0010	**LED INTERIOR LIGHTING**									
7000	Interior LED fixts, tape rope kit, 120V, 3' strip, 4W, w/5' cord, 3,000K	1 Elec	24	.333	Ea.	8.45	19.40		27.85	38.50
7010	5,000K		24	.333		12.05	19.40		31.45	42.50
7020	Interior LED fixts, tape rope kit, 120V, 6' strip, 8W, w/5' cord, 3,000K		23	.348		15.40	20		35.40	47.50
7030	5,000K		23	.348		15.40	20		35.40	47.50
7040	Interior LED fixts, tape rope kit, 120V, 13' strip, 15W, w/5' cord, 3,000K		22	.364		28.50	21		49.50	62.50
7050	5,000K		22	.364		28.50	21		49.50	62.50
7060	Interior LED fixts, tape rope kit, 120V, 20' strip, 22W, w/5' cord, 3,000K		21	.381		41	22		63	78.50
7070	5,000K		21	.381		41	22		63	78.50
7080	Interior LED fixts, tape rope kit, 120V, 33' strip, 37W, w/5' cord, 3,000K		20	.400		60.50	23.50		84	102
7090	5,000K		20	.400		60.50	23.50		84	102
7100	Interior LED fixts, tape rope kit, 120V, 50' strip, 55W, w/5' cord, 3,000K		19	.421		92	24.50		116.50	138

26 51 Interior Lighting

26 51 19 – LED Interior Lighting

26 51 19.10 LED Interior Lighting

	26 51 19.10 LED Interior Lighting	Crew	Daily Output	Labor-Hours	Unit	Material	2018 Bare Costs Labor	Equipment	Total	Total Incl O&P
7110	5,000K	1 Elec	19	.421	Ea.	131	24.50		155.50	182
7120	Interior LED tape light, starstrand, dimmable, 12V, super star, 2", 3,000K		25	.320		11.30	18.60		29.90	40.50
7130	Ultra star, 2", 3,000K		25	.320		14.10	18.60		32.70	43.50
7140	Rrainbow star, 2", 3,000K		25	.320		11.30	18.60		29.90	40.50
7150	Interior LED tape light, starstrand, dimmable, 24V, elite star, 4", 35K		25	.320		9.85	18.60		28.45	39
7160	12", 35K		25	.320		22.50	18.60		41.10	53
7170	60", 35K		23	.348		98.50	20		118.50	140
7180	120", 35K		21	.381		139	22		161	186
7190	240", 35K		18	.444		273	26		299	340
7200	Interior LED tape lighting channel , 36", AL flex, starstrand, elite star		48	.167		21	9.70		30.70	37.50

26 52 Safety Lighting

26 52 13 – Emergency and Exit Lighting

26 52 13.10 Emergency Lighting and Battery Units

	26 52 13.10 Emergency Lighting and Battery Units	Crew	Daily Output	Labor-Hours	Unit	Material	2018 Bare Costs Labor	Equipment	Total	Total Incl O&P
0010	**EMERGENCY LIGHTING AND BATTERY UNITS**									
0300	Emergency light units, battery operated									
0350	Twin sealed beam light, 25 W, 6 V each									
0500	Lead battery operated	1 Elec	4	2	Ea.	144	116		260	330
0700	Nickel cadmium battery operated		4	2		345	116		461	550
0780	Additional remote mount, sealed beam, 25 W 6 V		26.70	.300		29.50	17.45		46.95	58.50
0790	Twin sealed beam light, 25 W 6 V each		26.70	.300		62.50	17.45		79.95	94.50
0900	Self-contained fluorescent lamp pack		10	.800		191	46.50		237.50	280

26 52 13.16 Exit Signs

	26 52 13.16 Exit Signs		Crew	Daily Output	Labor-Hours	Unit	Material	2018 Bare Costs Labor	Equipment	Total	Total Incl O&P
0010	**EXIT SIGNS**										
0080	Exit light ceiling or wall mount, incandescent, single face		1 Elec	8	1	Ea.	70	58		128	164
0100	Double face			6.70	1.194		50	69.50		119.50	159
0120	Explosion proof			3.80	2.105		550	123		673	790
0150	Fluorescent, single face			8	1		81	58		139	176
0160	Double face			6.70	1.194		75	69.50		144.50	187
0200	LED standard, single face	G		8	1		47.50	58		105.50	139
0220	Double face	G		6.70	1.194		51	69.50		120.50	160
0230	LED vandal-resistant, single face	G		7.27	1.100		212	64		276	330
0240	LED w/battery unit, single face	G		4.40	1.818		184	106		290	360
0260	Double face	G		4	2		188	116		304	380
0262	LED w/battery unit, vandal-resistant, single face	G		4.40	1.818		245	106		351	430
0270	Combination emergency light units and exit sign			4	2		179	116		295	370
0290	LED retrofit kits	G		60	.133		50	7.75		57.75	66.50
1500	Exit sign, 12 V, 1 face, remote end mounted (Type 602A1)		R-19	18	1.111		69.50	65		134.50	174
1780	With emergency battery, explosion proof		"	7.70	2.597		4,375	151		4,526	5,050

26 54 Classified Location Lighting

26 54 13 – Incandescent Classified Location Lighting

26 54 13.20 Explosion Proof

		Crew	Daily Output	Labor-Hours	Unit	Material	2018 Bare Costs Labor	Equipment	Total	Total Incl O&P
0010	**EXPLOSION PROOF**, incl. lamps, mounting hardware and connections									
6310	Metal halide with ballast, ceiling, surface mounted, 175 watt	1 Elec	2.90	2.759	Ea.	1,300	161		1,461	1,675
6320	250 watt	"	2.70	2.963		1,575	172		1,747	1,975
6330	400 watt	2 Elec	4.80	3.333		1,675	194		1,869	2,150
6340	Ceiling, pendent mounted, 175 watt	1 Elec	2.60	3.077		1,250	179		1,429	1,650
6350	250 watt	"	2.40	3.333		1,500	194		1,694	1,950
6360	400 watt	2 Elec	4.20	3.810		1,625	222		1,847	2,100
6370	Wall, surface mounted, 175 watt	1 Elec	2.90	2.759		1,400	161		1,561	1,800
6380	250 watt	"	2.70	2.963		1,650	172		1,822	2,075
6390	400 watt	2 Elec	4.80	3.333		1,775	194		1,969	2,250
6400	High pressure sodium, ceiling surface mounted, 70 watt	1 Elec	3	2.667		2,200	155		2,355	2,650
6410	100 watt		3	2.667		2,275	155		2,430	2,725
6420	150 watt		2.70	2.963		2,350	172		2,522	2,825
6430	Pendent mounted, 70 watt		2.70	2.963		2,100	172		2,272	2,550
6440	100 watt		2.70	2.963		2,175	172		2,347	2,650
6450	150 watt		2.40	3.333		1,875	194		2,069	2,350
6460	Wall mounted, 70 watt		3	2.667		2,350	155		2,505	2,800
6470	100 watt		3	2.667		2,425	155		2,580	2,900
6480	150 watt		2.70	2.963		2,500	172		2,672	3,000
6510	Incandescent, ceiling mounted, 200 watt		4	2		1,550	116		1,666	1,875
6520	Pendent mounted, 200 watt		3.50	2.286		1,325	133		1,458	1,650
6530	Wall mounted, 200 watt		4	2		1,525	116		1,641	1,850
6600	Fluorescent, RS, 4' long, ceiling mounted, two 40 watt		2.70	2.963		4,075	172		4,247	4,725
6610	Three 40 watt		2.20	3.636		5,900	212		6,112	6,800
6620	Four 40 watt		1.90	4.211		7,575	245		7,820	8,700
6630	Pendent mounted, two 40 watt		2.30	3.478		4,725	202		4,927	5,500
6640	Three 40 watt		1.90	4.211		6,725	245		6,970	7,775
6650	Four 40 watt		1.70	4.706		8,850	274		9,124	10,200

26 55 Special Purpose Lighting

26 55 33 – Hazard Warning Lighting

26 55 33.10 Warning Beacons

		Crew	Daily Output	Labor-Hours	Unit	Material	2018 Bare Costs Labor	Equipment	Total	Total Incl O&P
0010	**WARNING BEACONS**									
0015	Surface mount with colored or clear lens									
0100	Rotating beacon									
0110	120V, 40 watt halogen	1 Elec	3.50	2.286	Ea.	106	133		239	315
0120	24V, 20 watt halogen	"	3.50	2.286	"	228	133		361	450
0200	Steady beacon									
0210	120V, 40 watt halogen	1 Elec	3.50	2.286	Ea.	108	133		241	320
0220	24V, 20 watt		3.50	2.286		112	133		245	320
0230	12V DC, incandescent		3.50	2.286		103	133		236	315
0300	Flashing beacon									
0310	120V, 40 watt halogen	1 Elec	3.50	2.286	Ea.	106	133		239	315
0320	24V, 20 watt halogen		3.50	2.286		106	133		239	315
0410	12V DC with two 6V lantern batteries		7	1.143		107	66.50		173.50	218

26 55 59 – Display Lighting

26 55 59.10 Track Lighting

		Crew	Daily Output	Labor-Hours	Unit	Material	2018 Bare Costs Labor	Equipment	Total	Total Incl O&P
0010	**TRACK LIGHTING**									
0080	Track, 1 circuit, 4' section	1 Elec	6.70	1.194	Ea.	42.50	69.50		112	151
0100	8' section	2 Elec	10.60	1.509		70.50	88		158.50	209

For customer support on your Electrical Costs with RSMeans data, call 800.448.8182.

303

26 55 59 – Display Lighting

26 55 59.10 Track Lighting	Crew	Daily Output	Labor-Hours	Unit	Material	2018 Bare Costs Labor	Equipment	Total	Total Incl O&P
0200	12' section	2 Elec	8.80	1.818	Ea.	105	106		211	273
0300	3 circuits, 4' section	1 Elec	6.70	1.194		94	69.50		163.50	208
0400	8' section	2 Elec	10.60	1.509		133	88		221	277
0500	12' section	"	8.80	1.818		159	106		265	335
1000	Feed kit, surface mounting	1 Elec	16	.500		14.80	29		43.80	60
1100	End cover		24	.333		5.55	19.40		24.95	35
1200	Feed kit, stem mounting, 1 circuit		16	.500		51.50	29		80.50	100
1300	3 circuit		16	.500		51.50	29		80.50	100
2000	Electrical joiner, for continuous runs, 1 circuit		32	.250		33.50	14.55		48.05	58.50
2100	3 circuit		32	.250		61	14.55		75.55	89
2200	Fixtures, spotlight, 75 W PAR halogen		16	.500		51	29		80	99.50
2210	50 W MR16 halogen		16	.500		171	29		200	232
3000	Wall washer, 250 W tungsten halogen		16	.500		132	29		161	189
3100	Low voltage, 25/50 W, 1 circuit		16	.500		133	29		162	190
3120	3 circuit		16	.500		198	29		227	262

26 55 61 – Theatrical Lighting

26 55 61.10 Lights

		Crew	Daily Output	Labor-Hours	Unit	Material	2018 Bare Costs Labor	Equipment	Total	Total Incl O&P
0010	**LIGHTS**									
2000	Lights, border, quartz, reflector, vented,									
2100	colored or white	1 Elec	20	.400	L.F.	182	23.50		205.50	235
2500	Spotlight, follow spot, with transformer, 2,100 watt	"	4	2	Ea.	3,450	116		3,566	3,950
2600	For no transformer, deduct					960			960	1,050
3000	Stationary spot, fresnel quartz, 6" lens	1 Elec	4	2		196	116		312	390
3100	8" lens		4	2		244	116		360	440
3500	Ellipsoidal quartz, 1,000 watt, 6" lens		4	2		360	116		476	575
3600	12" lens		4	2		655	116		771	895
4000	Strobe light, 1 to 15 flashes per second, quartz		3	2.667		845	155		1,000	1,150
4500	Color wheel, portable, five hole, motorized		4	2		217	116		333	415

26 55 63 – Detention Lighting

26 55 63.10 Detention Lighting Fixtures

		Crew	Daily Output	Labor-Hours	Unit	Material	2018 Bare Costs Labor	Equipment	Total	Total Incl O&P
0010	**DETENTION LIGHTING FIXTURES**									
1000	Surface mounted, cold rolled steel, 14 ga., 1' x 4'	1 Elec	5.40	1.481	Ea.	750	86		836	955
1010	2' x 2'		5.40	1.481		805	86		891	1,025
1020	2' x 4'		5	1.600		950	93		1,043	1,200
1100	12 ga., 1' x 4'		5.40	1.481		985	86		1,071	1,200
1110	2' x 2'		5.40	1.481		805	86		891	1,025
1120	2' x 4'		5	1.600		1,050	93		1,143	1,300
3000	Fluorescent vandal resistant light fixture, high abuse troffer		5.40	1.481		220	86		306	370
3010	50" L x 8" W x 4" H		5.40	1.481		203	86		289	350
3020	32W		5.40	1.481		278	86		364	435
3030	T8/32W		5.40	1.481		218	86		304	370
3040	No lamp		5.60	1.429		335	83		418	490
3050	Lamp included		5.40	1.481		385	86		471	555
3060	16 ga. cold rolled steel		5.40	1.481		242	86		328	395
3070	16 ga. cold rolled steel, ceiling/wall mount		5.40	1.481		284	86		370	440
3080	w/lamp, white		5.40	1.481		216	86		302	365
3090	wo/lamp, bronze		5.60	1.429		221	83		304	365
3100	w/lamp 6" L x 4-1/2" W x 8-1/2" H		5.40	1.481		56	86		142	191
3110	w/lamp 5-7/8" W x 4-1/2" D x 8-1/2" H clear		5.40	1.481		51.50	86		137.50	186
3120	w/lamp white		5.40	1.481		375	86		461	540
3130	Tall wallpack, w/lamp, bronze		5.40	1.481		111	86		197	251
3140	Tall wallpack, w/lamp, white		5.40	1.481		115	86		201	256

26 56 13.10 Lighting Poles	Crew	Daily Output	Labor-Hours	Unit	Material	2018 Bare Costs Labor	Equipment	Total	Total Incl O&P
0010 **LIGHTING POLES**									
0100 Exterior, light poles, concrete, 30' above 5' below, 13.5" Base, 5.5" Tip	2 Elec	4.60	3.478	Ea.	1,550	202		1,752	2,000
0110 39' above 6' below, 15.5" Base, 5.25" Tip		4.60	3.478		1,825	202		2,027	2,300
0120 43' above 7' below, 17.25" Base, 6.5" Tip		4.60	3.478		1,850	202		2,052	2,325
0130 43' above 7' below, 19.5" Base, 8.25" Tip	↓	4.60	3.478	↓	1,875	202		2,077	2,375
2800 Light poles, anchor base									
2820 not including concrete bases									
2840 Aluminum pole, 8' high	1 Elec	4	2	Ea.	785	116		901	1,025
2850 10' high		4	2		825	116		941	1,075
2860 12' high		3.80	2.105		860	123		983	1,125
2870 14' high		3.40	2.353		895	137		1,032	1,200
2880 16' high		3	2.667		980	155		1,135	1,300
3000 20' high	R-3	2.90	6.897		1,050	400	44.50	1,494.50	1,800
3200 30' high		2.60	7.692		1,975	445	50	2,470	2,900
3400 35' high		2.30	8.696		2,150	505	56.50	2,711.50	3,175
3600 40' high	↓	2	10		2,450	580	65	3,095	3,650
3800 Bracket arms, 1 arm	1 Elec	8	1		134	58		192	235
4000 2 arms		8	1		270	58		328	385
4200 3 arms		5.30	1.509		405	88		493	575
4400 4 arms		4.80	1.667		540	97		637	740
4500 Steel pole, galvanized, 8' high		3.80	2.105		680	123		803	930
4510 10' high		3.70	2.162		710	126		836	970
4520 12' high		3.40	2.353		765	137		902	1,050
4530 14' high		3.10	2.581		815	150		965	1,125
4540 16' high		2.90	2.759		865	161		1,026	1,200
4550 18' high	↓	2.70	2.963		910	172		1,082	1,250
4600 20' high	R-3	2.60	7.692		1,050	445	50	1,545	1,900
4800 30' high		2.30	8.696		1,450	505	56.50	2,011.50	2,425
5000 35' high		2.20	9.091		1,600	525	59	2,184	2,600
5200 40' high	↓	1.70	11.765		1,975	680	76	2,731	3,275
5400 Bracket arms, 1 arm	1 Elec	8	1		201	58		259	310
5600 2 arms		8	1		310	58		368	425
5800 3 arms		5.30	1.509		225	88		313	380
6000 4 arms	↓	5.30	1.509		320	88		408	480
6100 Fiberglass pole, 1 or 2 fixtures, 20' high	R-3	4	5		795	290	32.50	1,117.50	1,350
6200 30' high		3.60	5.556		985	320	36	1,341	1,600
6300 35' high		3.20	6.250		1,550	360	40.50	1,950.50	2,300
6400 40' high		2.80	7.143		1,800	415	46.50	2,261.50	2,650
6420 Wood pole, 4-1/2" x 5-1/8", 8' high	1 Elec	6	1.333		375	77.50		452.50	530
6430 10' high		6	1.333		430	77.50		507.50	585
6440 12' high		5.70	1.404		540	81.50		621.50	715
6450 15' high		5	1.600		630	93		723	835
6460 20' high	↓	4	2		765	116		881	1,025
6461 Light poles, anchor base, w/o conc base, pwdr ct stl, 16' H	2 Elec	3.10	5.161		865	300		1,165	1,400
6462 20' high	R-3	2.90	6.897		1,050	400	44.50	1,494.50	1,825
6463 30' high		2.30	8.696		1,450	505	56.50	2,011.50	2,425
6464 35' high		2.40	8.333		1,600	485	54	2,139	2,525
6465 25' high	↓	2.70	7.407	↓	1,050	430	48	1,528	1,875
7300 Transformer bases, not including concrete bases									
7320 Maximum pole size, steel, 40' high	1 Elec	2	4	Ea.	1,600	233		1,833	2,100
7340 Cast aluminum, 30' high		3	2.667		840	155		995	1,150
7350 40' high	↓	2.50	3.200		1,275	186		1,461	1,675
8000 Line cover protective devices									

26 56 13 – Lighting Poles and Standards

26 56 13.10 Lighting Poles		Crew	Daily Output	Labor-Hours	Unit	Material	2018 Bare Costs Labor	Equipment	Total	Total Incl O&P
8010	line cover									
8015	Refer to 26 01 02.20 for labor adjustment factors as they apply									
8100	MVLC 1" W x 1-1/2" H, 14/5'	1 Elec	60	.133	Ea.	4.56	7.75		12.31	16.60
8110	8'		58	.138		5.10	8.05		13.15	17.60
8120	MVLC 1" W x 1-1/2" H, 18/5'		60	.133		5.80	7.75		13.55	17.95
8130	8'		58	.138		6.45	8.05		14.50	19.05
8140	MVLC 1" W x 1-1/2" H, 38/5'		60	.133		10.40	7.75		18.15	23
8150	8'	↓	58	.138	↓	12.50	8.05		20.55	26

26 56 19 – LED Exterior Lighting

26 56 19.55 Roadway LED Luminaire

			Crew	Daily Output	Labor-Hours	Unit	Material	2018 Bare Costs Labor	Equipment	Total	Total Incl O&P
0010	**ROADWAY LED LUMINAIRE**										
0100	LED fixture, 72 LEDs, 120 V AC or 12 V DC, equal to 60 watt	G	1 Elec	2.70	2.963	Ea.	585	172		757	905
0110	108 LEDs, 120 V AC or 12 V DC, equal to 90 watt	G		2.70	2.963		690	172		862	1,025
0120	144 LEDs, 120 V AC or 12 V DC, equal to 120 watt	G	↓	2.70	2.963		845	172		1,017	1,200
0130	252 LEDs, 120 V AC or 12 V DC, equal to 210 watt	G	2 Elec	4.40	3.636		1,175	212		1,387	1,625
0140	Replaces high pressure sodium fixture, 75 watt	G	1 Elec	2.70	2.963		395	172		567	695
0150	125 watt	G		2.70	2.963		450	172		622	755
0160	150 watt	G		2.70	2.963		530	172		702	840
0170	175 watt	G		2.70	2.963		780	172		952	1,125
0180	200 watt	G	↓	2.70	2.963		745	172		917	1,075
0190	250 watt	G	2 Elec	4.40	3.636		850	212		1,062	1,250
0200	320 watt	G		4.40	3.636		930	212		1,142	1,350
2000	Exterior fixtures, LED roadway, type 2, 50W, neutral slipfitter bronze			5.20	3.077		545	179		724	870
2010	3, 50W, neutral slipfitter bronze			5.20	3.077		545	179		724	870
2020	4, 50W, neutral slipfitter bronze			5.20	3.077		545	179		724	870
2030	Exterior fixtures, LED roadway, type 2, 78W, neutral slipfitter bronze			5.20	3.077		545	179		724	870
2040	3, 78W, neutral slipfitter bronze			5.20	3.077		550	179		729	875
2050	4, 78W, neutral slipfitter bronze			5.20	3.077		545	179		724	870
2060	Exterior fixtures, LED roadway, type 2, 105W, neutral slipfitter bronze			5.20	3.077		645	179		824	980
2070	3, 105W, neutral slipfitter bronze			5.20	3.077		645	179		824	980
2080	4, 105W, neutral slipfitter bronze			5.20	3.077		645	179		824	980
2090	Exterior fixtures, LED roadway, type 2, 125W, neutral slipfitter bronze		↓	5.20	3.077	↓	645	179		824	980

26 56 19.60 Parking LED Lighting

			Crew	Daily Output	Labor-Hours	Unit	Material	2018 Bare Costs Labor	Equipment	Total	Total Incl O&P
0010	**PARKING LED LIGHTING**										
0100	Round pole mounting, 88 lamp watts	G	1 Elec	2	4	Ea.	1,150	233		1,383	1,600

26 56 21 – HID Exterior Lighting

26 56 21.20 Roadway Luminaire

		Crew	Daily Output	Labor-Hours	Unit	Material	2018 Bare Costs Labor	Equipment	Total	Total Incl O&P
0010	**ROADWAY LUMINAIRE**									
2650	Roadway area luminaire, low pressure sodium, 135 watt	1 Elec	2	4	Ea.	840	233		1,073	1,275
2700	180 watt	"	2	4		900	233		1,133	1,350
2750	Metal halide, 400 watt	2 Elec	4.40	3.636		650	212		862	1,025
2760	1,000 watt		4	4		730	233		963	1,150
2780	High pressure sodium, 400 watt		4.40	3.636		775	212		987	1,175
2790	1,000 watt	↓	4	4	↓	880	233		1,113	1,325

26 56 23 – Area Lighting

26 56 23.10 Exterior Fixtures

		Crew	Daily Output	Labor-Hours	Unit	Material	2018 Bare Costs Labor	Equipment	Total	Total Incl O&P
0010	**EXTERIOR FIXTURES** With lamps									
0200	Wall mounted, incandescent, 100 watt	1 Elec	8	1	Ea.	40.50	58		98.50	132
0400	Quartz, 500 watt		5.30	1.509		61.50	88		149.50	199
0420	1,500 watt		4.20	1.905		101	111		212	277
1100	Wall pack, low pressure sodium, 35 watt	↓	4	2	↓	191	116		307	385

For customer support on your Electrical Costs with RSMeans data, call 800.448.8182.

26 56 23 – Area Lighting

26 56 23.10 Exterior Fixtures		Crew	Daily Output	Labor-Hours	Unit	Material	2018 Bare Costs Labor	Equipment	Total	Total Incl O&P
1150	55 watt	1 Elec	4	2	Ea.	227	116		343	425
1160	High pressure sodium, 70 watt		4	2		187	116		303	380
1170	150 watt		4	2		205	116		321	400
1175	High pressure sodium, 250 watt		4	2		205	116		321	400
1180	Metal halide, 175 watt		4	2		210	116		326	405
1190	250 watt		4	2		236	116		352	435
1195	400 watt		4	2		415	116		531	630
1250	Induction lamp, 40 watt		4	2		470	116		586	690
1260	80 watt		4	2		560	116		676	790
1278	LED, poly lens, 26 watt		4	2		315	116		431	525
1280	110 watt		4	2		775	116		891	1,025
1500	LED, glass lens, 13 watt		4	2		315	116		431	525

26 56 23.55 Exterior LED Fixtures

			Crew	Daily Output	Labor-Hours	Unit	Material	Labor	Equipment	Total	Total Incl O&P
0010	**EXTERIOR LED FIXTURES**										
0100	Wall mounted, indoor/outdoor, 12 watt	G	1 Elec	10	.800	Ea.	216	46.50		262.50	310
0110	32 watt	G		10	.800		465	46.50		511.50	580
0120	66 watt	G		10	.800		470	46.50		516.50	585
0200	outdoor, 110 watt	G		10	.800		1,200	46.50		1,246.50	1,400
0210	220 watt	G		10	.800		1,750	46.50		1,796.50	2,000
0300	modular, type IV, 120 V, 50 lamp watts	G		9	.889		1,150	51.50		1,201.50	1,350
0310	101 lamp watts	G		9	.889		1,300	51.50		1,351.50	1,500
0320	126 lamp watts	G		9	.889		1,625	51.50		1,676.50	1,850
0330	202 lamp watts	G		9	.889		1,850	51.50		1,901.50	2,100
0340	240 V, 50 lamp watts	G		8	1		1,200	58		1,258	1,400
0350	101 lamp watts	G		8	1		1,350	58		1,408	1,550
0360	126 lamp watts	G		8	1		1,450	58		1,508	1,675
0370	202 lamp watts	G		8	1		1,900	58		1,958	2,150
0400	wall pack, glass, 13 lamp watts	G		4	2		445	116		561	665
0410	poly w/photocell, 26 lamp watts	G		4	2		340	116		456	550
0420	50 lamp watts	G		4	2		775	116		891	1,025
0430	replacement, 40 watts	G		4	2		390	116		506	605
0440	60 watts	G		4	2		400	116		516	615

26 56 26 – Landscape Lighting

26 56 26.20 Landscape Fixtures

		Crew	Daily Output	Labor-Hours	Unit	Material	Labor	Equipment	Total	Total Incl O&P
0010	**LANDSCAPE FIXTURES**									
7380	Landscape recessed uplight, incl. housing, ballast, transformer									
7390	& reflector, not incl. conduit, wire, trench									
7420	Incandescent, 250 watt	1 Elec	5	1.600	Ea.	655	93		748	860
7440	Quartz, 250 watt		5	1.600		620	93		713	820
7460	500 watt		4	2		635	116		751	875

26 56 26.50 Landscape LED Fixtures

		Crew	Daily Output	Labor-Hours	Unit	Material	Labor	Equipment	Total	Total Incl O&P
0010	**LANDSCAPE LED FIXTURES**									
0100	12 volt alum bullet hooded-BLK	1 Elec	5	1.600	Ea.	96	93		189	245
0200	12 volt alum bullet hooded-BRZ		5	1.600		96	93		189	245
0300	12 volt alum bullet hooded-GRN		5	1.600		96	93		189	245
1000	12 volt alum large bullet hooded-BLK		5	1.600		71	93		164	217
1100	12 volt alum large bullet hooded-BRZ		5	1.600		71	93		164	217
1200	12 volt alum large bullet hooded-GRN		5	1.600		71	93		164	217
2000	12 volt large bullet landscape light fixture		5	1.600		71	93		164	217
2100	12 volt alum light large bullet		5	1.600		71	93		164	217
2200	12 volt alum bullet light		5	1.600		71	93		164	217

26 56 33.10 Walkway Luminaire

26 56 33.10 Walkway Luminaire	Crew	Daily Output	Labor-Hours	Unit	Material	2018 Bare Costs Labor	Equipment	Total	Total Incl O&P
0010 **WALKWAY LUMINAIRE**									
6500 Bollard light, lamp & ballast, 42" high with polycarbonate lens									
6800 Metal halide, 175 watt	1 Elec	3	2.667	Ea.	915	155		1,070	1,225
6900 High pressure sodium, 70 watt		3	2.667		935	155		1,090	1,250
7000 100 watt		3	2.667		935	155		1,090	1,250
7100 150 watt		3	2.667		915	155		1,070	1,225
7200 Incandescent, 150 watt		3	2.667		670	155		825	965
7810 Walkway luminaire, square 16", metal halide 250 watt		2.70	2.963		715	172		887	1,050
7820 High pressure sodium, 70 watt		3	2.667		820	155		975	1,125
7830 100 watt		3	2.667		835	155		990	1,150
7840 150 watt		3	2.667		835	155		990	1,150
7850 200 watt		3	2.667		840	155		995	1,150
7910 Round 19", metal halide, 250 watt		2.70	2.963		1,050	172		1,222	1,400
7920 High pressure sodium, 70 watt		3	2.667		1,150	155		1,305	1,500
7930 100 watt		3	2.667		1,150	155		1,305	1,500
7940 150 watt		3	2.667		1,150	155		1,305	1,500
7950 250 watt		2.70	2.963		1,200	172		1,372	1,575
8000 Sphere 14" opal, incandescent, 200 watt		4	2		330	116		446	540
8020 Sphere 18" opal, incandescent, 300 watt		3.50	2.286		400	133		533	640
8040 Sphere 16" clear, high pressure sodium, 70 watt		3	2.667		695	155		850	995
8050 100 watt		3	2.667		740	155		895	1,050
8100 Cube 16" opal, incandescent, 300 watt		3.50	2.286		440	133		573	685
8120 High pressure sodium, 70 watt		3	2.667		645	155		800	940
8130 100 watt		3	2.667		660	155		815	955
8230 Lantern, high pressure sodium, 70 watt		3	2.667		575	155		730	865
8240 100 watt		3	2.667		620	155		775	910
8250 150 watt		3	2.667		580	155		735	870
8260 250 watt		2.70	2.963		810	172		982	1,150
8270 Incandescent, 300 watt		3.50	2.286		430	133		563	670
8330 Reflector 22" w/globe, high pressure sodium, 70 watt		3	2.667		555	155		710	840
8340 100 watt		3	2.667		560	155		715	845
8350 150 watt		3	2.667		570	155		725	855
8360 250 watt		2.70	2.963		725	172		897	1,050

26 56 33.55 Walkway LED Luminaire

26 56 33.55 Walkway LED Luminaire		Crew	Daily Output	Labor-Hours	Unit	Material	2018 Bare Costs Labor	Equipment	Total	Total Incl O&P
0010 **WALKWAY LED LUMINAIRE**										
0100 Pole mounted, 86 watts, 4,350 lumens	G	1 Elec	3	2.667	Ea.	1,325	155		1,480	1,700
0110 4,630 lumens	G		3	2.667		2,250	155		2,405	2,700
0120 80 watts, 4,000 lumens	G		3	2.667		2,025	155		2,180	2,450
0600 LED bollard pole, 6" diam. x 36" H, cast alum., surf. mtd., 8W, 6,000K, 120V			8.50	.941		380	55		435	500
0610 6" diam. x 42" H, cast alum., surf. mtd., 12W, 5,100K,120V			8.50	.941		505	55		560	635
0620 6" diam. x 42" H, cast alum., surf. mtd., 18W, 5,100K,120V			8.50	.941		515	55		570	645
0630 6" diam. x 42" H, cast alum., surf. mtd., 24W, 5,100K,120V			8.50	.941		500	55		555	630
0640 9.5" square x 42" H, cast resin, surf. mtd., A19, 9W, 120V			8.50	.941		625	55		680	770
0650 9.75" diam. x 42" H, cast resin, side mount, dome, 9W, 120V			8.50	.941		685	55		740	830
0660 9.75" diam. x 43" H, cast resin, burial base, dome, 9W, 120V			8.50	.941		685	55		740	830
0670 8" diam. x 36" H, concrete security base, w/rebar, 8W, HID			8	1		1,025	58		1,083	1,200
0680 10" diam. x 49" H, concrete base, w/Sch 40 Steel pipe, 8W			6.75	1.185		1,150	69		1,219	1,375
0690 12" diam. x 48" H, concrete base, w/Sch 40 Steel pipe, 8W			6.75	1.185		1,325	69		1,394	1,550
0700 8" diam. x 36" H, concrete, cast aluminum dome, 90 degree, 4.1W			7	1.143		1,025	66.50		1,091.50	1,225
0710 8" diam. x 36" H, concrete, cast aluminum dome, 180 degree, 4.1W			7	1.143		1,250	66.50		1,316.50	1,475
0720 8" diam. x 48" H, concrete, cast aluminum dome, 360 degree, 4.1W			7	1.143		1,550	66.50		1,616.50	1,800

26 56 36.20 Floodlights

		Crew	Daily Output	Labor-Hours	Unit	Material	2018 Bare Costs Labor	Equipment	Total	Total Incl O&P
0010	**FLOODLIGHTS** with ballast and lamp,									
1290	floor mtd, mount with swivel bracket									
1300	Induction lamp, 40 watt	1 Elec	3	2.667	Ea.	550	155		705	835
1310	80 watt		3	2.667		765	155		920	1,075
1320	150 watt	↓	3	2.667	↓	1,300	155		1,455	1,675
1400	Pole mounted, pole not included									
1950	Metal halide, 175 watt	1 Elec	2.70	2.963	Ea.	210	172		382	490
2000	400 watt	2 Elec	4.40	3.636		224	212		436	560
2200	1,000 watt		4	4		790	233		1,023	1,225
2210	1,500 watt	↓	3.70	4.324		420	252		672	835
2250	Low pressure sodium, 55 watt	1 Elec	2.70	2.963		425	172		597	725
2270	90 watt		2	4		575	233		808	985
2290	180 watt		2	4		635	233		868	1,050
2340	High pressure sodium, 70 watt		2.70	2.963		250	172		422	535
2360	100 watt		2.70	2.963		256	172		428	540
2380	150 watt	↓	2.70	2.963		310	172		482	600
2400	400 watt	2 Elec	4.40	3.636		315	212		527	665
2600	1,000 watt	"	4	4		580	233		813	985
9005	Solar powered floodlight, w/motion det, incl. batt pack for cloudy days G	1 Elec	8	1		60	58		118	153
9020	Replacement battery pack G	"	8	1	↓	19.90	58		77.90	109

26 56 36.55 LED Floodlights

		Crew	Daily Output	Labor-Hours	Unit	Material	2018 Bare Costs Labor	Equipment	Total	Total Incl O&P
0010	**LED FLOODLIGHTS** with ballast and lamp,									
0020	Pole mounted, pole not included									
0100	11 watt G	1 Elec	4	2	Ea.	435	116		551	655
0110	46 watt G		4	2		1,450	116		1,566	1,775
0120	90 watt G		4	2		2,000	116		2,116	2,375
0130	288 watt G	↓	4	2	↓	1,800	116		1,916	2,150

26 61 Lighting Systems and Accessories

26 61 13 – Lighting Accessories

26 61 13.30 Fixture Whips

		Crew	Daily Output	Labor-Hours	Unit	Material	2018 Bare Costs Labor	Equipment	Total	Total Incl O&P
0010	**FIXTURE WHIPS**									
0080	3/8" Greenfield, 2 connectors, 6' long									
0100	TFFN wire, three #18	1 Elec	32	.250	Ea.	6.20	14.55		20.75	29
0150	Four #18		28	.286		6.55	16.65		23.20	32.50
0200	Three #16		32	.250		6.60	14.55		21.15	29.50
0250	Four #16		28	.286		7.15	16.65		23.80	33
0300	THHN wire, three #14		32	.250		5.80	14.55		20.35	28.50
0350	Four #14		28	.286		8.25	16.65		24.90	34
0360	Three #12	↓	32	.250		8.50	14.55		23.05	31.50

26 61 23 – Lamps Applications

26 61 23.10 Lamps

		Crew	Daily Output	Labor-Hours	Unit	Material	2018 Bare Costs Labor	Equipment	Total	Total Incl O&P
0010	**LAMPS**									
0080	Fluorescent, rapid start, cool white, 2' long, 20 watt	1 Elec	1	8	C	305	465		770	1,025
0100	4' long, 40 watt		.90	8.889		256	515		771	1,050
0120	3' long, 30 watt		.90	8.889		365	515		880	1,175
0125	3' long, 25 watt energy saver R265723-10 G		.90	8.889		1,050	515		1,565	1,950
0150	U-40 watt		.80	10		1,175	580		1,755	2,175
0155	U-34 watt energy saver G		.80	10		12.05	580		592.05	885
0170	4' long, 34 watt energy saver G	↓	.90	8.889	↓	820	515		1,335	1,675

26 61 23.10 Lamps		Crew	Daily Output	Labor-Hours	Unit	Material	2018 Bare Costs Labor	Equipment	Total	Total Incl O&P
0176	2' long, T8, 17 watt energy saver Ⓖ	1 Elec	1	8	C	335	465		800	1,075
0178	3' long, T8, 25 watt energy saver Ⓖ		.90	8.889		365	515		880	1,175
0180	4' long, T8, 32 watt energy saver Ⓖ		.90	8.889		223	515		738	1,025
0200	Slimline, 4' long, 40 watt		.90	8.889		840	515		1,355	1,700
0210	4' long, 30 watt energy saver Ⓖ		.90	8.889		840	515		1,355	1,700
0300	8' long, 75 watt		.80	10		1,050	580		1,630	2,025
0350	8' long, 60 watt energy saver Ⓖ		.80	10		410	580		990	1,325
0400	High output, 4' long, 60 watt		.90	8.889		545	515		1,060	1,375
0410	8' long, 95 watt energy saver Ⓖ		.80	10		505	580		1,085	1,425
0500	8' long, 110 watt		.80	10		505	580		1,085	1,425
0512	2' long, T5, 14 watt energy saver Ⓖ		1	8		159	465		624	870
0514	3' long, T5, 21 watt energy saver Ⓖ		.90	8.889		180	515		695	975
0516	4' long, T5, 28 watt energy saver Ⓖ		.90	8.889		180	515		695	975
0517	4' long, T5, 54 watt energy saver Ⓖ		.90	8.889		555	515		1,070	1,375
0520	Very high output, 4' long, 110 watt		.90	8.889		1,225	515		1,740	2,125
0525	8' long, 195 watt energy saver Ⓖ		.70	11.429		1,150	665		1,815	2,275
0550	8' long, 215 watt		.70	11.429		1,100	665		1,765	2,200
0554	Full spectrum, 4' long, 60 watt ♿		.90	8.889		665	515		1,180	1,500
0556	6' long, 85 watt		.90	8.889		580	515		1,095	1,400
0558	8' long, 110 watt		.80	10		1,825	580		2,405	2,875
0560	Twin tube compact lamp Ⓖ		.90	8.889		450	515		965	1,275
0570	Double twin tube compact lamp Ⓖ		.80	10		985	580		1,565	1,950
0600	Mercury vapor, mogul base, deluxe white, 100 watt		.30	26.667		5,325	1,550		6,875	8,175
0650	175 watt		.30	26.667		3,250	1,550		4,800	5,900
0700	250 watt		.30	26.667		5,200	1,550		6,750	8,025
0800	400 watt		.30	26.667		5,200	1,550		6,750	8,050
0900	1,000 watt		.20	40		13,800	2,325		16,125	18,700
1000	Metal halide, mogul base, 175 watt		.30	26.667		1,075	1,550		2,625	3,525
1100	250 watt		.30	26.667		1,725	1,550		3,275	4,225
1200	400 watt		.30	26.667		1,900	1,550		3,450	4,400
1300	1,000 watt		.20	40		3,350	2,325		5,675	7,150
1320	1,000 watt, 125,000 initial lumens		.20	40		11,400	2,325		13,725	16,000
1330	1,500 watt		.20	40		3,700	2,325		6,025	7,550
1350	High pressure sodium, 70 watt		.30	26.667		1,650	1,550		3,200	4,125
1360	100 watt		.30	26.667		1,575	1,550		3,125	4,075
1370	150 watt		.30	26.667		1,675	1,550		3,225	4,175
1380	250 watt		.30	26.667		2,225	1,550		3,775	4,775
1400	400 watt		.30	26.667		1,675	1,550		3,225	4,150
1450	1,000 watt		.20	40		4,450	2,325		6,775	8,350
1500	Low pressure sodium, 35 watt		.30	26.667		16,500	1,550		18,050	20,400
1550	55 watt		.30	26.667		18,800	1,550		20,350	22,900
1600	90 watt		.30	26.667		6,500	1,550		8,050	9,475
1650	135 watt		.20	40		27,300	2,325		29,625	33,500
1700	180 watt		.20	40		30,900	2,325		33,225	37,500
1750	Quartz line, clear, 500 watt		1.10	7.273		580	425		1,005	1,275
1760	1,500 watt		.20	40		1,675	2,325		4,000	5,300
1762	Spot, MR 16, 50 watt		1.30	6.154		1,075	360		1,435	1,700
1770	Tungsten halogen, T4, 400 watt		1.10	7.273		3,975	425		4,400	5,000
1775	T3, 1,200 watt		.30	26.667		4,900	1,550		6,450	7,725
1778	PAR 30, 50 watt		1.30	6.154		1,125	360		1,485	1,775
1780	PAR 38, 90 watt		1.30	6.154		9,650	360		10,010	11,100
1800	Incandescent, interior, A21, 100 watt		1.60	5		2,800	291		3,091	3,500
1900	A21, 150 watt		1.60	5		16,600	291		16,891	18,700

26 61 23 – Lamps Applications

26 61 23.10 Lamps		Crew	Daily Output	Labor-Hours	Unit	Material	2018 Bare Costs Labor	Equipment	Total	Total Incl O&P
2000	A23, 200 watt	1 Elec	1.60	5	C	335	291		626	805
2200	PS 35, 300 watt		1.60	5		1,000	291		1,291	1,525
2210	PS 35, 500 watt		1.60	5		1,425	291		1,716	1,975
2230	PS 52, 1,000 watt		1.30	6.154		2,550	360		2,910	3,325
2240	PS 52, 1,500 watt		1.30	6.154		6,825	360		7,185	8,050
2300	R30, 75 watt		1.30	6.154		615	360		975	1,225
2400	R40, 100 watt		1.30	6.154		620	360		980	1,225
2500	Exterior, PAR 38, 75 watt		1.30	6.154		1,900	360		2,260	2,600
2600	PAR 38, 150 watt		1.30	6.154		2,200	360		2,560	2,925
2700	PAR 46, 200 watt		1.10	7.273		3,875	425		4,300	4,900
2800	PAR 56, 300 watt		1.10	7.273		2,650	425		3,075	3,550
3000	Guards, fluorescent lamp, 4' long		1	8		1,325	465		1,790	2,150
3200	8' long		.90	8.889	Ea.	13.20	515		528.20	790

26 61 23.55 LED Lamps

26 61 23.55 LED Lamps		Crew	Daily Output	Labor-Hours	Unit	Material	2018 Bare Costs Labor	Equipment	Total	Total Incl O&P
0010	**LED LAMPS**									
0100	LED lamp, interior, shape A60, equal to 60 W [G]	1 Elec	160	.050	Ea.	19	2.91		21.91	25.50
0110	7 W LED decorative c, ca, f, g shape		160	.050		34.50	2.91		37.41	42
0120	12V mini lamp LED		160	.050		5	2.91		7.91	9.85
0200	Globe frosted A60, equal to 60 W [G]		160	.050		11.40	2.91		14.31	16.90
0205	LED lamp, interior, globe		160	.050		16.15	2.91		19.06	22
0210	2.2 W LED LMP		160	.050		66	2.91		68.91	77.50
0220	2.2 W LED replacement decorative lamp		160	.050		8.70	2.91		11.61	13.90
0230	3.5 W LED replacement decorative lamp		160	.050		10.20	2.91		13.11	15.55
0240	4.5 W 120V LED replacement decorative lamp		160	.050		12.40	2.91		15.31	18
0250	4.5 W 120V LED, 2700k replacement decorative lamp		160	.050		10.10	2.91		13.01	15.45
0260	4.9 W 120V LED, 2700k replacement decorative lamp candelabra base		160	.050		10.05	2.91		12.96	15.40
0270	4.9 W 120V LED, 3000k replacement decorative lamp candelabra base		160	.050		9.40	2.91		12.31	14.70
0280	5 W LED PAR 20 parabolic reflector lamp FL		140	.057		60	3.33		63.33	71
0300	Globe earth, equal to 100 W [G]		140	.057		27.50	3.33		30.83	35.50
0305	7 W LED reflector lamp 3000k DIM		140	.057		58	3.33		61.33	69
0310	10 W omni LED warm white light bulb E26 medium base 120 volt card		140	.057		16.75	3.33		20.08	23.50
0315	7 W LED reflector lamp WFL 2700k DIM		140	.057		58	3.33		61.33	69
0320	9 W omni LED warm white light bulb E26 medium base 120 volt card		140	.057		9.05	3.33		12.38	14.90
0500	8 W LED, A19 lamp, equal to 40 W		140	.057		5.90	3.33		9.23	11.45
0505	9 W LED, A19 lamp, 5000K dimmable, equal to 60 W		140	.057		7.20	3.33		10.53	12.85
0510	10.5 W LED, A19 lamp, dimmable, equal to 60 W		140	.057		6.35	3.33		9.68	11.90
0515	10 W LED, A19 lamp, frosted, dimmable, equal to 60 W		140	.057		10.85	3.33		14.18	16.85
0520	10 W LED, A19 lamp, omni-directional, dimmable, equal to 60 W		140	.057		8.40	3.33		11.73	14.20
0525	12 W LED, A19 lamp, dimmable, equal to 75 W		140	.057		10.80	3.33		14.13	16.80
1100	MR16, 3 W, replacement of halogen lamp 25 W [G]		130	.062		19	3.58		22.58	26.50
1200	6 W replacement of halogen lamp 45 W [G]		130	.062		20	3.58		23.58	27.50
2100	10 W, PAR20, equal to 60 W [G]		130	.062		26.50	3.58		30.08	35
2200	15 W, PAR30, equal to 100 W [G]		130	.062		47.50	3.58		51.08	57.50
2210	LED lamp 50 W w/6ft pri		130	.062		610	3.58		613.58	675
2220	11 Watt reflector dimmable warm white LED light bulb with medium base		130	.062		41.50	3.58		45.08	51
2221	12 W A-Line LED lamp DIM		130	.062		78	3.58		81.58	91.50
2225	13 Watt reflector LED warm white e26 with medium base 120 volt box		130	.062		39	3.58		42.58	48.50
2226	13 W br30 LED lamp		130	.062		68	3.58		71.58	80
2227	15 W 120V br30 inc LED lamp, 2700k		130	.062		63.50	3.58		67.08	75
2228	15 W 120V br30 inc LED lamp, 4000k		130	.062		60	3.58		63.58	71.50
2230	3 Watt dimmable warm white decorative LED lamp with medium base		160	.050		12.60	2.91		15.51	18.20
2240	.43 W night light LED daylight bulb E12 candelabra base 120 volt 2 pack		160	.050		4.72	2.91		7.63	9.55

For customer support on your Electrical Costs with RSMeans data, call 800.448.8182.

311

26 61 23.55 LED Lamps		Crew	Daily Output	Labor-Hours	Unit	Material	2018 Bare Costs Labor	Equipment	Total	Total Incl O&P
2250	15 W omni-directional LED warm white e26 medium base 120 volt box	1 Elec	160	.050	Ea.	40	2.91		42.91	48.50
2251	11 W omni-directional LED warm white e26 medium base 120 volt box		160	.050		28	2.91		30.91	35
2252	10 W omni-directional LED warm white e26 medium base 120 volt box		160	.050		28	2.91		30.91	35
2253	7 W omni A19 LED warm white e26 medium base 120 volt box		160	.050		20.50	2.91		23.41	27
2255	8 PAR 20 parabolic reflector 2700 LED lamp		160	.050		18.80	2.91		21.71	25
2256	8 PAR 20 parabolic reflector 3000 LED lamp		160	.050		18.95	2.91		21.86	25.50
2260	12 W PAR 38 120V LED 15 degree directional lamp		160	.050		203	2.91		205.91	227
2270	12 W PAR 38 120V LED 25 degree directional lamp		160	.050		31.50	2.91		34.41	39
2280	12 W PAR 38 120V LED 40 degree directional lamp		160	.050		203	2.91		205.91	227
2285	16 W PAR 38 120V 2700k LED parabolic reflector lamp		160	.050		23	2.91		25.91	30
2290	16 W PAR 38 120V 3000k LED parabolic reflector lamp		160	.050		23	2.91		25.91	30
3000	15 W PAR 30 LED daylight E26 medium base 120V box		160	.050		60	2.91		62.91	70.50
3100	17 W LED 3000k PAR 38 100 W replacement		160	.050		47	2.91		49.91	56
3110	17 W LED PAR 38 100 W replacement		160	.050		35	2.91		37.91	43
3120	17 W LED PAR 38 100 W replacement parabolic reflector		160	.050		35	2.91		37.91	43
3130	24 W LED T8 PW straight fluorescent lamp		8.89	.900		289	52.50		341.50	400
3135	Linear fluorescent LED lamp 120V		8.89	.900		111	52.50		163.50	201
3200	30 W LED 2700K recessed 8055E PAR 38 high power		160	.050		105	2.91		107.91	120
3210	30 W LED 4200K 120 degree 8055E PAR 38 high power		160	.050		105	2.91		107.91	120
3220	30 W LED 5700K 8055E PAR 38 high power		160	.050		105	2.91		107.91	120
3230	50 W LED 2700K 8045M PAR 38 277V high power		160	.050		340	2.91		342.91	380
3240	50 W LED 4200K 8045M PAR 38 277V high power retro fit		160	.050		129	2.91		131.91	145
3250	50 W LED 5700K 8045M PAR 38 277V high power retro fit		160	.050		340	2.91		342.91	380
8000	10 W LED PAR 30/fl 10 pk		160	.050		96	2.91		98.91	110
8010	Gen 3 PAR 30 15 W short neck power LED 120 VAC E26 80 +cri 300k dimm		160	.050		12.95	2.91		15.86	18.60
8020	12 PAR 30 2700K parabolic reflector LED lamp		160	.050		43.50	2.91		46.41	52.50
8030	12 PAR 30 3000K parabolic reflector LED lamp		160	.050		28	2.91		30.91	35.50
8040	3500K LED advantage T8 9 W 800LM 2ft linear 2 BD frosted		69	.116		9	6.75		15.75	20
8050	3500K LED litespan T8 9 W 900LM 2ft linear frosted		69	.116		9.90	6.75		16.65	21
8060	4000K LED advantage T8 9 W 800LM 2ft linear 2BD frosted		69	.116		9	6.75		15.75	20
8070	4000K LED litespan T8 9 W 900LM 2ft linear frosted		69	.116		9.90	6.75		16.65	21
8080	5000K LED advantage T8 9 W 800LM 2ft linear 2BD frosted		69	.116		9	6.75		15.75	20
8090	5000K LED litespan T8 9 W 900LM 2ft linear frosted		69	.116		9.90	6.75		16.65	21
8100	3500K LED advantage T8 18 W 1600LM 4ft linear 2 BD frosted		69	.116		9.90	6.75		16.65	21
8105	18 W LED 4ft T8 4000K frost 1600L linear lamp		69	.116		18.50	6.75		25.25	30.50
8108	18 W LED 48 inch T8 4100K 1890LM linear lamp		69	.116		57	6.75		63.75	73
8110	3500K LED advantage T8 18 W 1800LM 4ft linear 2 BD frosted		69	.116		21	6.75		27.75	33
8120	4000K LED advantage T8 18 W 1600LM 4ft linear 2 BD frosted		69	.116		16.30	6.75		23.05	28
8130	4000K LED litespan T8 18 W 1800LM 4ft linear frosted		69	.116		21	6.75		27.75	33
8140	5000K LED advantage T8 18 W 1600LM 4ft linear 2BD frosted		69	.116		9.90	6.75		16.65	21
8150	5000K LED litespan T8 18 W 1800LM 4ft linear 2BD frosted		69	.116		21	6.75		27.75	33
8200	16.5 T8 3000 IF-6U U-shape fluorescent LED lamp		65	.123		21.50	7.15		28.65	34
8210	16.5 T8 3500 IF-6U U-shape fluorescent LED lamp		65	.123		21.50	7.15		28.65	34.50
8220	16.5 T8 4000 IF-6U U-shape fluorescent LED lamp		65	.123		21	7.15		28.15	33.50
8230	16.5 T8 5000 IF-6U U-shape fluorescent LED lamp		65	.123		21	7.15		28.15	34
8240	18 W 6 inch T8 4100K U-shape frosted fluorescent LED lamp		60	.133		11.70	7.75		19.45	24.50
8250	18 W 6 inch T8 5000K U-shape frosted fluorescent LED lamp		60	.133		11.70	7.75		19.45	24.50
8260	Circular 12 W linear fluorescent LED 2700K MOD lamp		62.40	.128		300	7.45		307.45	340
8270	Circular 12 W linear fluorescent LED 3000K MOD lamp		62.40	.128		300	7.45		307.45	340
8280	Circular 12 W linear fluorescent LED 3500K MOD lamp		62.40	.128		273	7.45		280.45	310
8290	Circular 18 W linear fluorescent LED 3000K MOD lamp		62.40	.128		375	7.45		382.45	420
8300	Circular 18 W linear fluorescent LED 3500K MOD lamp		62.40	.128		410	7.45		417.45	465
8310	Circular 18 W linear fluorescent LED 5000K MOD lamp		62.40	.128		375	7.45		382.45	420

26 61 Lighting Systems and Accessories

26 61 23 – Lamps Applications

		Crew	Daily Output	Labor-Hours	Unit	Material	2018 Bare Costs Labor	2018 Bare Costs Equipment	Total	Total Incl O&P
26 61 23.55 LED Lamps										
8320	Circular 18 W linear fluorescent LED 4100K MOD lamp	1 Elec	62.40	.128	Ea.	395	7.45		402.45	445

26 71 Electrical Machines

26 71 13 – Motors Applications

26 71 13.10 Handling

		Crew	Daily Output	Labor-Hours	Unit	Material	2018 Bare Costs Labor	2018 Bare Costs Equipment	Total	Total Incl O&P
0010	**HANDLING** Add to normal labor cost for restricted areas									
5000	Motors									
5100	1/2 HP, 23 pounds	1 Elec	4	2	Ea.		116		116	174
5110	3/4 HP, 28 pounds		4	2			116		116	174
5120	1 HP, 33 pounds		4	2			116		116	174
5130	1-1/2 HP, 44 pounds		3.20	2.500			146		146	218
5140	2 HP, 56 pounds		3	2.667			155		155	232
5150	3 HP, 71 pounds		2.30	3.478			202		202	305
5160	5 HP, 82 pounds		1.90	4.211			245		245	365
5170	7-1/2 HP, 124 pounds		1.50	5.333			310		310	465
5180	10 HP, 144 pounds		1.20	6.667			390		390	580
5190	15 HP, 185 pounds		1	8			465		465	695
5200	20 HP, 214 pounds	2 Elec	1.50	10.667			620		620	930
5210	25 HP, 266 pounds		1.40	11.429			665		665	995
5220	30 HP, 310 pounds		1.20	13.333			775		775	1,150
5230	40 HP, 400 pounds		1	16			930		930	1,400
5240	50 HP, 450 pounds		.90	17.778			1,025		1,025	1,550
5250	75 HP, 680 pounds		.80	20			1,175		1,175	1,750
5260	100 HP, 870 pounds	3 Elec	1	24			1,400		1,400	2,100
5270	125 HP, 940 pounds		.80	30			1,750		1,750	2,600
5280	150 HP, 1,200 pounds		.70	34.286			2,000		2,000	2,975
5290	175 HP, 1,300 pounds		.60	40			2,325		2,325	3,475
5300	200 HP, 1,400 pounds		.50	48			2,800		2,800	4,175

26 71 13.20 Motors

		Crew	Daily Output	Labor-Hours	Unit	Material	2018 Bare Costs Labor	2018 Bare Costs Equipment	Total	Total Incl O&P
0010	**MOTORS** 230/460 V, 60 HZ R263413-30									
0050	Dripproof, premium efficiency, 1.15 service factor									
0060	1,800 RPM, 1/4 HP	1 Elec	5.33	1.501	Ea.	261	87.50		348.50	420
0070	1/3 HP		5.33	1.501		241	87.50		328.50	395
0080	1/2 HP		5.33	1.501		201	87.50		288.50	350
0090	3/4 HP		5.33	1.501		281	87.50		368.50	440
0100	1 HP		4.50	1.778		300	103		403	485
0150	2 HP		4.50	1.778		360	103		463	550
0200	3 HP		4.50	1.778		720	103		823	950
0250	5 HP		4.50	1.778		640	103		743	860
0300	7.5 HP		4.20	1.905		970	111		1,081	1,250
0350	10 HP		4	2		1,200	116		1,316	1,475
0400	15 HP		3.20	2.500		1,675	146		1,821	2,050
0450	20 HP	2 Elec	5.20	3.077		2,025	179		2,204	2,500
0500	25 HP		5	3.200		2,650	186		2,836	3,175
0550	30 HP		4.80	3.333		2,875	194		3,069	3,475
0600	40 HP		4	4		3,750	233		3,983	4,475
0650	50 HP		3.20	5		3,825	291		4,116	4,650
0700	60 HP		2.80	5.714		4,250	335		4,585	5,175
0750	75 HP		2.40	6.667		4,975	390		5,365	6,025
0800	100 HP	3 Elec	2.70	8.889		6,375	515		6,890	7,800
0850	125 HP		2.10	11.429		6,675	665		7,340	8,325

For customer support on your Electrical Costs with RSMeans data, call 800.448.8182.

313

26 71 Electrical Machines

26 71 13 – Motors Applications

26 71 13.20 Motors

		Crew	Daily Output	Labor-Hours	Unit	Material	2018 Bare Costs Labor	2018 Bare Costs Equipment	Total	Total Incl O&P
0900	150 HP	3 Elec	1.80	13.333	Ea.	10,100	775		10,875	12,300
0950	200 HP	↓	1.50	16		12,300	930		13,230	15,000
1000	1,200 RPM, 1 HP	1 Elec	4.50	1.778		475	103		578	675
1050	2 HP		4.50	1.778		525	103		628	730
1100	3 HP		4.50	1.778		720	103		823	945
1150	5 HP		4.50	1.778		945	103		1,048	1,175
1200	3,600 RPM, 2 HP		4.50	1.778		600	103		703	815
1250	3 HP		4.50	1.778		620	103		723	840
1300	5 HP	↓	4.50	1.778	↓	540	103		643	750
1350	Totally enclosed, premium efficiency 1.15 service factor									
1360	1,800 RPM, 1/4 HP	1 Elec	5.33	1.501	Ea.	300	87.50		387.50	460
1370	1/3 HP		5.33	1.501		310	87.50		397.50	470
1380	1/2 HP		5.33	1.501		360	87.50		447.50	525
1390	3/4 HP		5.33	1.501		400	87.50		487.50	570
1400	1 HP		4.50	1.778		580	103		683	795
1450	2 HP		4.50	1.778		730	103		833	960
1500	3 HP		4.50	1.778		660	103		763	885
1550	5 HP		4.50	1.778		770	103		873	1,000
1600	7.5 HP		4.20	1.905		1,000	111		1,111	1,275
1650	10 HP		4	2		1,425	116		1,541	1,725
1700	15 HP		3.20	2.500		1,700	146		1,846	2,100
1750	20 HP	2 Elec	5.20	3.077		2,300	179		2,479	2,800
1800	25 HP		5	3.200		3,300	186		3,486	3,900
1850	30 HP		4.80	3.333		3,250	194		3,444	3,875
1900	40 HP		4	4		3,975	233		4,208	4,725
1950	50 HP		3.20	5		4,300	291		4,591	5,150
2000	60 HP		2.80	5.714		5,825	335		6,160	6,900
2050	75 HP	↓	2.40	6.667		7,550	390		7,940	8,900
2100	100 HP	3 Elec	2.70	8.889		8,925	515		9,440	10,600
2150	125 HP		2.10	11.429		11,600	665		12,265	13,800
2200	150 HP		1.80	13.333		13,600	775		14,375	16,200
2250	200 HP	↓	1.50	16		17,200	930		18,130	20,300
2300	1,200 RPM, 1 HP	1 Elec	4.50	1.778		480	103		583	685
2350	2 HP		4.50	1.778		545	103		648	755
2400	3 HP		4.50	1.778		790	103		893	1,025
2450	5 HP		4.50	1.778		940	103		1,043	1,175
2500	3,600 RPM, 2 HP		4.50	1.778		450	103		553	650
2550	3 HP		4.50	1.778		600	103		703	815
2600	5 HP	↓	4.50	1.778	↓	700	103		803	925

26 71 13.40 Motors Explosion Proof

		Crew	Daily Output	Labor-Hours	Unit	Material	2018 Bare Costs Labor	2018 Bare Costs Equipment	Total	Total Incl O&P
0010	**MOTORS EXPLOSION PROOF**, 208-230/460 V, 60 HZ									
0020	1,800 RPM, 1/4 HP	1 Elec	5	1.600	Ea.	565	93		658	765
0030	1/3 HP		5	1.600		272	93		365	440
0040	1/2 HP		5	1.600		440	93		533	625
0050	3/4 HP		5.33	1.501		350	87.50		437.50	515
0060	1 HP		4.20	1.905		555	111		666	775
0070	2 HP		4.20	1.905		645	111		756	875
0080	3 HP		4.20	1.905		725	111		836	960
0090	5 HP		4.20	1.905		890	111		1,001	1,150
0100	7.5 HP		4	2		1,100	116		1,216	1,375
0110	10 HP		3.70	2.162		1,300	126		1,426	1,625
0120	15 HP	↓	3.20	2.500		1,700	146		1,846	2,100

26 71 13.40 Motors Explosion Proof	Crew	Daily Output	Labor-Hours	Unit	Material	2018 Bare Costs Labor	Equipment	Total	Total Incl O&P
0130	20 HP	2 Elec	5	3.200	Ea.	2,125	186		2,311	2,600
0140	25 HP		4.80	3.333		2,550	194		2,744	3,100
0150	30 HP		4.40	3.636		2,950	212		3,162	3,575
0160	40 HP		3.80	4.211		3,775	245		4,020	4,525
0170	50 HP		3.20	5		4,125	291		4,416	4,950
0180	60 HP		2.60	6.154		6,000	360		6,360	7,125
0190	75 HP		2	8		7,400	465		7,865	8,825
0200	100 HP	3 Elec	2.50	9.600		9,425	560		9,985	11,200
0210	125 HP		2.20	10.909		12,300	635		12,935	14,500
0220	150 HP		1.60	15		14,500	875		15,375	17,200
0230	200 HP		1.20	20		17,800	1,175		18,975	21,400
1000	1,200 RPM, 1 HP	1 Elec	4.20	1.905		470	111		581	680
1010	2 HP		4.20	1.905		540	111		651	760
1020	3 HP		4.20	1.905		735	111		846	975
1030	5 HP		4.20	1.905		1,000	111		1,111	1,275
2000	3,600 RPM, 1/4 HP		5	1.600		405	93		498	585
2010	1/3 HP		5	1.600		425	93		518	605
2020	1/2 HP		5	1.600		320	93		413	490
2030	3/4 HP		5	1.600		430	93		523	615
2040	1 HP		4.20	1.905		610	111		721	835
2050	2 HP		4.20	1.905		670	111		781	900
2060	3 HP		4.20	1.905		730	111		841	970
2070	5 HP		4.20	1.905		975	111		1,086	1,250
2080	7.5 HP		4	2		1,100	116		1,216	1,375
2090	10 HP		3.70	2.162		1,275	126		1,401	1,600
2100	15 HP		3.20	2.500		1,725	146		1,871	2,125
2110	20 HP	2 Elec	5	3.200		2,150	186		2,336	2,625
2120	25 HP		4.80	3.333		2,650	194		2,844	3,225
2130	30 HP		4.40	3.636		3,100	212		3,312	3,725
2140	40 HP		3.80	4.211		3,825	245		4,070	4,575
2150	50 HP		3.20	5		4,125	291		4,416	4,950
2160	60 HP		2.60	6.154		6,050	360		6,410	7,175
2170	75 HP		2	8		7,400	465		7,865	8,825
2180	100 HP	3 Elec	2.50	9.600		9,425	560		9,985	11,200
2190	125 HP		2.20	10.909		12,700	635		13,335	15,000
2200	150 HP		1.60	15		14,500	875		15,375	17,200
2210	200 HP		1.20	20		18,300	1,175		19,475	21,900
2220	250 HP		1.20	20		18,300	1,175		19,475	21,900

For customer support on your Electrical Costs with RSMeans data, call 800.448.8182.

315

Division Notes

	CREW	DAILY OUTPUT	LABOR-HOURS	UNIT	BARE COSTS				TOTAL INCL O&P
					MAT.	LABOR	EQUIP.	TOTAL	

Estimating Tips

27 20 00 Data Communications
27 30 00 Voice Communications
27 40 00 Audio-Video Communications

- When estimating material costs for special systems, it is always prudent to obtain manufacturers' quotations for equipment prices and special installation requirements that may affect the total cost.

- For cost modifications for elevated tray installation, add the percentages to labor according to the height of the installation and only to the quantities exceeding the different height levels, not to the total tray quantities. Refer to 26 01 02.20 for labor adjustment factors.

- Do not overlook the costs for equipment used in the installation. If scissor lifts and boom lifts are available in the field, contractors may use them in lieu of the proposed ladders and rolling staging.

Reference Numbers

Reference numbers are shown at the beginning of some major classifications. These numbers refer to related items in the Reference Section. The reference information may be an estimating procedure, an alternate pricing method, or technical information.

Note: Not all subdivisions listed here necessarily appear. ■

27 01 Operation and Maintenance of Communications Systems

27 01 30 – Operation and Maintenance of Voice Communications

27 01 30.51 Operation and Maintenance of Voice Equipment	Crew	Daily Output	Labor-Hours	Unit	Material	2018 Bare Costs Labor	Equipment	Total	Total Incl O&P
0010	**OPERATION AND MAINTENANCE OF VOICE EQUIPMENT**								
3400	Remove and replace (reinstall), speaker	1 Elec	6	1.333	Ea.		77.50		77.50	116

27 05 Common Work Results for Communications

27 05 05 – Selective Demolition for Communications

27 05 05.20 Electrical Demolition, Communications

		Crew	Daily Output	Labor-Hours	Unit	Material	2018 Bare Costs Labor	Equipment	Total	Total Incl O&P
0010	**ELECTRICAL DEMOLITION, COMMUNICATIONS** R024119-10									
0100	Fiber optics									
0120	Cable R260105-30	1 Elec	2400	.003	L.F.		.19		.19	.29
0160	Multi-channel rack enclosure		6	1.333	Ea.		77.50		77.50	116
0180	Patch panel		18	.444	"		26		26	38.50
0200	Communication cables & fittings									
0220	Voice/data outlet	1 Elec	140	.057	Ea.		3.33		3.33	4.97
0240	Telephone cable		2800	.003	L.F.		.17		.17	.25
0260	Phone jack		135	.059	Ea.		3.45		3.45	5.15
0280	Data jack		135	.059	"		3.45		3.45	5.15
0300	High performance cable, 2 pair		3000	.003	L.F.		.16		.16	.23
0320	4 pair		2100	.004			.22		.22	.33
0340	25 pair		900	.009			.52		.52	.77
0400	Terminal cabinet		5	1.600	Ea.		93		93	139
1000	Nurse call system									
1020	Station	1 Elec	24	.333	Ea.		19.40		19.40	29
1040	Standard call button		24	.333			19.40		19.40	29
1060	Corridor dome light or zone indicator		24	.333			19.40		19.40	29
1080	Master control station	2 Elec	2	8			465		465	695

27 05 05.30 Electrical Demolition, Sound and Video

		Crew	Daily Output	Labor-Hours	Unit	Material	2018 Bare Costs Labor	Equipment	Total	Total Incl O&P
0010	**ELECTRICAL DEMOLITION, SOUND & VIDEO** R024119-10									
0100	Cables									
0120	TV antenna lead-in cable R260105-30	1 Elec	2100	.004	L.F.		.22		.22	.33
0140	Sound cable		2400	.003			.19		.19	.29
0160	Microphone cable		2400	.003			.19		.19	.29
0180	Coaxial cable		2400	.003			.19		.19	.29
0200	Doorbell system, not including wires, cables, and conduit		16	.500	Ea.		29		29	43.50
0220	Door chime or devices		36	.222	"		12.95		12.95	19.35
0300	Public address system, not including wires, cables, and conduit									
0320	Conventional office	1 Elec	16	.500	Speaker		29		29	43.50
0340	Conventional industrial	"	8	1	"		58		58	87
0352	PA cabinet/panel	2 Elec	5	3.200	Ea.		186		186	278
0400	Sound system, not including wires, cables, and conduit									
0410	Components	1 Elec	24	.333	Ea.		19.40		19.40	29
0412	Speaker		24	.333			19.40		19.40	29
0416	Volume control		24	.333			19.40		19.40	29
0420	Intercom, master station		6	1.333			77.50		77.50	116
0440	Remote station		24	.333			19.40		19.40	29
0460	Intercom outlets		24	.333			19.40		19.40	29
0480	Handset		12	.667			39		39	58
0500	Emergency call system, not including wires, cables, and conduit									
0520	Annunciator	1 Elec	4	2	Ea.		116		116	174
0540	Devices	"	16	.500			29		29	43.50
0600	Master door, buzzer type unit	2 Elec	1.60	10			580		580	870
0800	TV system, not including wires, cables, and conduit									

318

27 05 Common Work Results for Communications

27 05 05 – Selective Demolition for Communications

27 05 05.30 Electrical Demolition, Sound and Video		Crew	Daily Output	Labor-Hours	Unit	Material	2018 Bare Costs Labor	Equipment	Total	Total Incl O&P
0820	Master TV antenna system, per outlet	1 Elec	39	.205	Outlet		11.95		11.95	17.85
0840	School application, per outlet		16	.500	"		29		29	43.50
0860	Amplifier		12	.667	Ea.		39		39	58
0880	Antenna	↓	6	1.333	"		77.50		77.50	116
0900	One camera & one monitor	2 Elec	7.80	2.051	Total		119		119	178
0920	One camera	1 Elec	8	1	Ea.		58		58	87

27 05 29 – Hangers and Supports for Communications Systems

27 05 29.10 Cable Support

		Crew	Daily Output	Labor-Hours	Unit	Material	Labor	Equipment	Total	Total Incl O&P
0010	**CABLE SUPPORT**									
0110	J-hook, single tier, single sided, 1" diam.	1 Elec	68.57	.117	Ea.	1.89	6.80		8.69	12.25
0120	1-1/2" diam.		67.80	.118		2.42	6.85		9.27	12.90
0130	2" diam.		67.30	.119		2.92	6.90		9.82	13.55
0140	4" diam.		66.80	.120		5.40	6.95		12.35	16.35
0330	Double tier, single sided, 2" diam.		67	.119		11.40	6.95		18.35	23
0340	4" diam.		66.20	.121		14.95	7.05		22	27
0430	Double sided, 2" diam.		65.80	.122		17.50	7.10		24.60	30
0440	4" diam.		65.40	.122		23	7.10		30.10	35.50
0530	Triple tier, single sided, 2" diam.		66.40	.120		14.75	7		21.75	26.50
0540	4" diam.		65.85	.121		21.50	7.05		28.55	34
0630	Double sided, 2" diam.		65.10	.123		20.50	7.15		27.65	33.50
0640	4" diam.	↓	64.70	.124	↓	27	7.20		34.20	41

27 11 Communications Equipment Room Fittings

27 11 16 – Communications Cabinets, Racks, Frames and Enclosures

27 11 16.10 Public Phone

		Crew	Daily Output	Labor-Hours	Unit	Material	Labor	Equipment	Total	Total Incl O&P
0010	**PUBLIC PHONE**									
7600	Telephone with wood backboard									
7620	Single door, 12" H x 12" W x 4" D	1 Elec	5.30	1.509	Ea.	97	88		185	237
7650	18" H x 12" W x 4" D		4.70	1.702		111	99		210	270
7700	24" H x 12" W x 4" D		4.20	1.905		139	111		250	320
7720	18" H x 18" W x 4" D		4.20	1.905		117	111		228	294
7750	24" H x 18" W x 4" D		4	2		191	116		307	385
7780	36" H x 36" W x 4" D		3.60	2.222		216	129		345	430
7800	24" H x 24" W x 6" D		3.60	2.222		248	129		377	465
7820	30" H x 24" W x 6" D		3.20	2.500		276	146		422	525
7850	30" H x 30" W x 6" D		2.70	2.963		340	172		512	630
7880	36" H x 30" W x 6" D		2.50	3.200		370	186		556	690
7900	48" H x 36" W x 6" D		2.20	3.636		615	212		827	995
7920	Double door, 48" H x 36" W x 6" D	↓	2	4	↓	865	233		1,098	1,300

27 11 16.20 Rack Mount Cabinet

		Crew	Daily Output	Labor-Hours	Unit	Material	Labor	Equipment	Total	Total Incl O&P
0010	**RACK MOUNT CABINET**									
0100	80" H x 24" W x 40" D									
0110	No sides	1 Elec	9.60	.833	Ea.	1,200	48.50		1,248.50	1,400
0120	One side		8	1		1,275	58		1,333	1,475
0130	Two sides	↓	7.86	1.018	↓	1,350	59		1,409	1,575
0200	80" H x 24" W x 42" D									
0210	No sides	1 Elec	9.60	.833	Ea.	1,225	48.50		1,273.50	1,425
0220	One side		8	1		1,275	58		1,333	1,475
0230	Two sides	↓	7.86	1.018		1,300	59		1,359	1,525
0300	80" H x 24" W x 48" D									

For customer support on your Electrical Costs with RSMeans data, call 800.448.8182.

319

27 11 16.20 Rack Mount Cabinet		Crew	Daily Output	Labor-Hours	Unit	Material	2018 Bare Costs Labor	Equipment	Total	Total Incl O&P
0310	No sides	1 Elec	9.60	.833	Ea.	1,375	48.50		1,423.50	1,600
0320	One side		8	1		1,375	58		1,433	1,600
0330	Two sides	↓	7.86	1.018	↓	1,300	59		1,359	1,525
0400	80" H x 30" W x 40" D									
0410	No sides	1 Elec	9.60	.833	Ea.	1,400	48.50		1,448.50	1,600
0420	One side		8	1		1,375	58		1,433	1,600
0430	Two sides	↓	7.86	1.018	↓	1,375	59		1,434	1,625
0500	80" H x 30" W x 42" D									
0510	No sides	1 Elec	9.60	.833	Ea.	1,375	48.50		1,423.50	1,600
0520	One side		8	1		1,375	58		1,433	1,600
0530	Two sides	↓	7.86	1.018	↓	1,275	59		1,334	1,500
0600	80" H x 30" W x 48" D									
0610	No sides	1 Elec	9.60	.833	Ea.	1,375	48.50		1,423.50	1,600
0620	One side		8	1		1,375	58		1,433	1,600
0630	Two sides	↓	7.86	1.018	↓	1,375	59		1,434	1,625
0700	80" H x 32" W x 32" D									
0710	No sides	1 Elec	9.60	.833	Ea.	1,375	48.50		1,423.50	1,600
0720	One side		8	1		1,375	58		1,433	1,600
0730	Two sides	↓	7.86	1.018	↓	1,375	59		1,434	1,625
0800	80" H x 32" W x 40" D									
0810	No sides	1 Elec	9.60	.833	Ea.	1,375	48.50		1,423.50	1,600
0820	One side		8	1		1,375	58		1,433	1,600
0830	Two sides	↓	7.86	1.018	↓	1,375	59		1,434	1,600
0900	80" H x 32" W x 42" D									
0910	No sides	1 Elec	9.60	.833	Ea.	1,375	48.50		1,423.50	1,600
0920	One side		8	1		1,375	58		1,433	1,600
0930	Two sides	↓	7.86	1.018	↓	1,375	59		1,434	1,600
1000	80" H x 32" W x 48" D									
1010	No sides	1 Elec	9.60	.833	Ea.	1,375	48.50		1,423.50	1,600
1020	One side		8	1		1,375	58		1,433	1,600
1030	Two sides	↓	7.86	1.018	↓	1,350	59		1,409	1,600
1200	Freestanding cabinet with front and rear doors									
1300	84" H x 19" W x 24" D, gray									
1310	No sides	1 Elec	9.60	.833	Ea.	1,875	48.50		1,923.50	2,125
1320	One side		9.60	.833		2,225	48.50		2,273.50	2,500
1330	Two sides	↓	9.60	.833	↓	2,225	48.50		2,273.50	2,500
1400	84" H x 19" W x 24" D, white									
1410	No sides	1 Elec	9.60	.833	Ea.	1,875	48.50		1,923.50	2,125
1420	One side		9.60	.833		2,225	48.50		2,273.50	2,500
1430	Two sides	↓	9.60	.833	↓	2,225	48.50		2,273.50	2,500
1500	84" H x 19" W x 24" D, black									
1510	No sides	1 Elec	9.60	.833	Ea.	1,775	48.50		1,823.50	2,025
1520	One side		9.60	.833		2,100	48.50		2,148.50	2,400
1530	Two sides		9.60	.833		2,100	48.50		2,148.50	2,400
1600	84" H x 28" W x 34" D front glass, rear louvered door	↓	9.60	.833	↓	1,275	48.50		1,323.50	1,475

27 11 19 – Communications Termination Blocks and Patch Panels

27 11 19.10 Termination Blocks and Patch Panels

0010	TERMINATION BLOCKS AND PATCH PANELS									
2960	Patch panel, RJ-45/110 type, 24 ports	2 Elec	6	2.667	Ea.	199	155		354	450
3000	48 ports	3 Elec	6	4		279	233		512	655
3040	96 ports	"	4	6		505	350		855	1,075
3100	Punch down termination per port	1 Elec	107	.075	↓		4.35		4.35	6.50

27 13 23.13 Communications Optical Fiber		Crew	Daily Output	Labor-Hours	Unit	Material	2018 Bare Costs Labor	Equipment	Total	Total Incl O&P
0010	**COMMUNICATIONS OPTICAL FIBER**									
0040	Specialized tools & techniques cause installation costs to vary.									
0070	Fiber optic, cable, bulk simplex, single mode R271323-40	1 Elec	8	1	C.L.F.	22.50	58		80.50	112
0080	Multi mode		8	1		29.50	58		87.50	120
0090	4 strand, single mode		7.34	1.090		38	63.50		101.50	137
0095	Multi mode		7.34	1.090		50.50	63.50		114	151
0100	12 strand, single mode		6.67	1.199		72.50	70		142.50	184
0105	Multi mode		6.67	1.199		96.50	70		166.50	210
0150	Jumper				Ea.	33			33	36.50
0200	Pigtail					29			29	32
0300	Connector	1 Elec	24	.333		26	19.40		45.40	57.50
0350	Finger splice		32	.250		38	14.55		52.55	64
0400	Transceiver (low cost bi-directional)		8	1		430	58		488	560
0450	Rack housing, 4 rack spaces, 12 panels (144 fibers)		2	4		560	233		793	965
0500	Patch panel, 12 ports		6	1.333		300	77.50		377.50	445
1000	Cable, 62.5 microns, direct burial, 4 fiber	R-15	1200	.040	L.F.	.91	2.29	.24	3.44	4.68
1020	Indoor, 2 fiber	R-19	1000	.020		.42	1.17		1.59	2.20
1040	Outdoor, aerial/duct	"	1670	.012		.65	.70		1.35	1.76
1060	50 microns, direct burial, 8 fiber	R-22	4000	.009		1.27	.50		1.77	2.14
1080	12 fiber		4000	.009		2.08	.50		2.58	3.03
1100	Indoor, 12 fiber		759	.049		1.98	2.62		4.60	6.10
1120	Connectors, 62.5 micron cable, transmission	R-19	40	.500	Ea.	14.35	29		43.35	59.50
1140	Cable splice		40	.500		17.50	29		46.50	63
1160	125 micron cable, transmission		16	1.250		15.10	73		88.10	126
1180	Receiver, 1.2 mile range		20	1		240	58.50		298.50	350
1200	1.9 mile range		20	1		226	58.50		284.50	335
1220	6.2 mile range		5	4		282	233		515	660
1240	Transmitter, 1.2 mile range		20	1		264	58.50		322.50	375
1260	1.9 mile range		20	1		299	58.50		357.50	415
1280	6.2 mile range		5	4		380	233		613	770
1300	Modem, 1.2 mile range		5	4		173	233		406	540
1320	6.2 mile range		5	4		315	233		548	700
1340	1.9 mile range, 12 channel		5	4		2,000	233		2,233	2,550
1360	Repeater, 1.2 mile range		10	2		370	117		487	585
1380	1.9 mile range		10	2		475	117		592	700
1400	6.2 mile range		5	4		920	233		1,153	1,350
1420	1.2 mile range, digital		5	4		440	233		673	830
1440	Transceiver, 1.9 mile range		5	4		375	233		608	765
1460	1.2 mile range, digital		5	4		390	233		623	780
1480	Cable enclosure, interior NEMA 13		7	2.857		144	167		311	405
1500	Splice w/enclosure encapsulant		16	1.250		212	73		285	340
1510	1/4", black, non-metallic, flexible, tube, liquid tight		8	2.500	C.L.F.	149	146		295	380
1520	3/8"		8	2.500		101	146		247	330
1530	1/2"		7.34	2.725		110	159		269	360
1540	3/4"		7.34	2.725		198	159		357	455
1550	1"		7.34	2.725		278	159		437	540
1560	1-1/4"		6.67	2.999		515	175		690	825
1570	1-1/2"		6.67	2.999		585	175		760	900
1580	2"		6.35	3.150		880	184		1,064	1,250
2040	Fiber optic cable, 48 strand, single mode, steel armor, material		1.67	11.976	M.L.F.	5,050	700		5,750	6,600

For customer support on your Electrical Costs with RSMeans data, call 800.448.8182.

321

27 15 01 – Communications Horizontal Cabling Applications

27 15 01.19 Fire Alarm Communications Conductors & Cables	Crew	Daily Output	Labor-Hours	Unit	Material	2018 Bare Costs Labor	Equipment	Total	Total Incl O&P
0010 **FIRE ALARM COMMUNICATIONS CONDUCTORS AND CABLES**									
1500 Fire alarm FEP teflon 150 V to 200°C									
1550 #22, 1 pair	1 Elec	10	.800	C.L.F.	98	46.50		144.50	178
1600 2 pair		8	1		96	58		154	193
1650 4 pair		7	1.143		350	66.50		416.50	485
1700 6 pair		6	1.333		425	77.50		502.50	585
1750 8 pair		5.50	1.455		715	84.50		799.50	910
1800 10 pair		5	1.600		660	93		753	865
1850 #18, 1 pair	2 Elec	16	1		67.50	58		125.50	162
1900 2 pair		13	1.231		174	71.50		245.50	298
1950 4 pair		9.60	1.667		206	97		303	370
2000 6 pair		8	2		340	116		456	550
2050 8 pair		7	2.286		325	133		458	555
2100 10 pair		6	2.667		380	155		535	645

27 15 01.23 Audio-Video Communications Horizontal Cabling

	Crew	Daily Output	Labor-Hours	Unit	Material	Labor	Equipment	Total	Total Incl O&P
0010 **AUDIO-VIDEO COMMUNICATIONS HORIZONTAL CABLING**									
3000 S-video cable, MD4, 1M Pro interconnects, 3'	1 Elec	38	.211	Ea.	33	12.25		45.25	55
3010 6'		38	.211		37	12.25		49.25	59
3020 9'		38	.211		46	12.25		58.25	69
3030 16'		36	.222		49.50	12.95		62.45	74
3040 22'		36	.222		58	12.95		70.95	83.50
3300 HDMI cable, coaxial, HDMI/HDMI, Non plenum, 3', 24 kt terminations		40	.200		14.35	11.65		26	33
3310 6', 24 kt terminations		40	.200		16.95	11.65		28.60	36
3320 HDMI cable, coaxial, HDMI/HDMI metal, Non plenum, 3', 24 kt terminations		40	.200		16.40	11.65		28.05	35.50
3330 6', 24 kt terminations		40	.200		19.05	11.65		30.70	38.50
3340 15', 24 kt terminations		38	.211		26.50	12.25		38.75	47.50
3350 25', 24 kt terminations		38	.211		34.50	12.25		46.75	56.50

27 15 10 – Special Communications Cabling

27 15 10.23 Sound and Video Cables and Fittings

	Crew	Daily Output	Labor-Hours	Unit	Material	Labor	Equipment	Total	Total Incl O&P
0010 **SOUND AND VIDEO CABLES & FITTINGS**									
0900 TV antenna lead-in, 300 ohm, #20-2 conductor	1 Elec	7	1.143	C.L.F.	22.50	66.50		89	124
0950 Coaxial, feeder outlet		7	1.143		20.50	66.50		87	122
1000 Coaxial, main riser		6	1.333		13.75	77.50		91.25	131
1100 Sound, shielded with drain, #22-2 conductor		8	1		8.25	58		66.25	96
1150 #22-3 conductor		7.50	1.067		12.60	62		74.60	107
1200 #22-4 conductor		6.50	1.231		20.50	71.50		92	130
1250 Nonshielded, #22-2 conductor		10	.800		10.40	46.50		56.90	81
1300 #22-3 conductor		9	.889		20.50	51.50		72	101
1350 #22-4 conductor		8	1		23	58		81	113
1400 Microphone cable		8	1		30.50	58		88.50	121

27 15 13 – Communications Copper Horizontal Cabling

27 15 13.13 Communication Cables and Fittings

	Crew	Daily Output	Labor-Hours	Unit	Material	Labor	Equipment	Total	Total Incl O&P
0010 **COMMUNICATION CABLES AND FITTINGS**									
2200 Telephone twisted, PVC insulation, #22-2 conductor	1 Elec	10	.800	C.L.F.	10.40	46.50		56.90	81
2250 #22-3 conductor		9	.889		12.85	51.50		64.35	91.50
2300 #22-4 conductor		8	1		14.85	58		72.85	103
2350 #18-2 conductor		9	.889		20.50	51.50		72	100
2370 Telephone jack, eight pins		32	.250	Ea.	3.50	14.55		18.05	26
5000 High performance unshielded twisted pair (UTP)									
5100 Cable, category 3, #24, 2 pair solid, PVC jacket R271513-75	1 Elec	10	.800	C.L.F.	9.75	46.50		56.25	80
5200 4 pair solid, PVC jacket		7	1.143		13.75	66.50		80.25	115

27 15 13.13 Communication Cables and Fittings		Crew	Daily Output	Labor-Hours	Unit	Material	2018 Bare Costs Labor	Equipment	Total	Total Incl O&P
5300	25 pair solid, PVC jacket	1 Elec	3	2.667	C.L.F.	84.50	155		239.50	325
5400	2 pair solid, plenum		10	.800		11.15	46.50		57.65	82
5500	4 pair solid, plenum		7	1.143		13.05	66.50		79.55	114
5600	25 pair solid, plenum		3	2.667		114	155		269	360
5700	4 pair stranded, PVC jacket		7	1.143		12.05	66.50		78.55	113
7000	Category 5, #24, 4 pair solid, PVC jacket		7	1.143		10.20	66.50		76.70	111
7100	4 pair solid, plenum		7	1.143		14.65	66.50		81.15	116
7200	4 pair stranded, PVC jacket		7	1.143		21	66.50		87.50	123
7210	Category 5e, #24, 4 pair solid, PVC jacket		7	1.143		15.20	66.50		81.70	116
7212	4 pair solid, plenum		7	1.143		21.50	66.50		88	123
7214	4 pair stranded, PVC jacket		7	1.143		23.50	66.50		90	125
7240	Category 6, #24, 4 pair solid, PVC jacket		7	1.143		18.35	66.50		84.85	120
7242	4 pair solid, plenum		7	1.143		13.75	66.50		80.25	115
7244	4 pair stranded, PVC jacket		7	1.143		14.05	66.50		80.55	115
7300	Connector, RJ45, category 5		80	.100	Ea.	.75	5.80		6.55	9.55
7302	Shielded RJ45, category 5		72	.111		1.75	6.45		8.20	11.60
7310	Jack, UTP RJ45, category 3		72	.111		.45	6.45		6.90	10.15
7312	Category 5		65	.123		5.50	7.15		12.65	16.75
7314	Category 5e		65	.123		4.98	7.15		12.13	16.20
7316	Category 6		65	.123		3.23	7.15		10.38	14.25
7322	Jack, shielded RJ45, category 5		60	.133		6.45	7.75		14.20	18.70
7324	Category 5e		60	.133		6.45	7.75		14.20	18.70
7326	Category 6		60	.133		6.95	7.75		14.70	19.25
7400	Voice/data expansion module, category 5e		8	1		47	58		105	139
7401	Modular jack, cat 6 keystone, RJ45, office white		16.50	.485		6.50	28		34.50	49
7402	RJ45, green		16.50	.485		6	28		34	48.50
7404	RJ45, grey		16.50	.485		7.05	28		35.05	50
7408	RJ45, orange		16.50	.485		6.40	28		34.40	49
7420	5M, RJ45, 568B, cord set		16.50	.485		32.50	28		60.50	77.50
7422	3M, RJ45, 568B, cord set		16.50	.485		28	28		56	73
7424	1M, RJ45, 568B, cord set		16.50	.485		24.50	28		52.50	69
7450	M12 TO RJ45, 2M ultra-loch D-mode, cord set		15.50	.516		33.50	30		63.50	82
8000	Multipair unshielded non-plenum cable, 150 V PVC jacket									
8002	#22, 2 pair	1 Elec	8.40	.952	C.L.F.	15.75	55.50		71.25	100
8003	3 pair		8	1		24	58		82	114
8004	4 pair		7.30	1.096		50	64		114	151
8006	6 pair		6.20	1.290		35.50	75		110.50	151
8008	8 pair		5.70	1.404		56	81.50		137.50	184
8010	10 pair		5.30	1.509		63	88		151	201
8012	12 pair		4.20	1.905		160	111		271	340
8015	15 pair		3.80	2.105		209	123		332	415
8020	20 pair	2 Elec	6.60	2.424		190	141		331	420
8025	25 pair		6	2.667		221	155		376	475
8030	30 pair		5.60	2.857		365	166		531	655
8040	40 pair		5.20	3.077		430	179		609	740
8050	50 pair		4.90	3.265		600	190		790	945
8100	Multipair unshielded non-plenum cable, 300 V PVC jacket									
8102	#20, 2 pair	1 Elec	7.30	1.096	C.L.F.	22	64		86	120
8103	3 pair		6.70	1.194		29.50	69.50		99	137
8104	4 pair		6	1.333		33.50	77.50		111	153
8106	6 pair		5.30	1.509		49	88		137	185
8108	8 pair		4.40	1.818		80.50	106		186.50	247
8110	10 pair	2 Elec	7.30	2.192		89	128		217	289

For customer support on your Electrical Costs with RSMeans data, call 800.448.8182.

323

27 15 Communications Horizontal Cabling

27 15 13 – Communications Copper Horizontal Cabling

27 15 13.13 Communication Cables and Fittings		Crew	Daily Output	Labor-Hours	Unit	Material	2018 Bare Costs Labor	Equipment	Total	Total Incl O&P
8112	12 pair	2 Elec	6.20	2.581	C.L.F.	108	150		258	345
8115	15 pair	↓	5.90	2.712		136	158		294	385
8201	#18, 1 pair	1 Elec	8	1		71.50	58		129.50	166
8202	2 pair		6.50	1.231		78.50	71.50		150	194
8203	3 pair		5.60	1.429		114	83		197	249
8204	4 pair		5	1.600		149	93		242	305
8206	6 pair		4.40	1.818		250	106		356	435
8208	8 pair	↓	4	2		245	116		361	445
8215	15 pair	2 Elec	5	3.200	↓	645	186		831	990
8300	Multipair shielded non-plenum cable, 300 V PVC jacket									
8303	#22, 3 pair	1 Elec	6.70	1.194	C.L.F.	57.50	69.50		127	168
8306	6 pair		5.70	1.404		106	81.50		187.50	238
8309	9 pair	↓	5	1.600		141	93		234	294
8312	12 pair	2 Elec	8	2		405	116		521	620
8315	15 pair		7.30	2.192		495	128		623	735
8317	17 pair		6.80	2.353		600	137		737	865
8319	19 pair		6.40	2.500		600	146		746	880
8327	27 pair	↓	5.90	2.712		1,025	158		1,183	1,350
8402	#20, 2 pair	1 Elec	6.70	1.194		74	69.50		143.50	186
8403	3 pair		6.20	1.290		80.50	75		155.50	201
8406	6 pair		4.40	1.818		202	106		308	380
8409	9 pair	↓	3.60	2.222		278	129		407	500
8412	12 pair	2 Elec	5.90	2.712		510	158		668	795
8415	15 pair	"	5.70	2.807		470	163		633	760
8502	#18, 2 pair	1 Elec	6.20	1.290		110	75		185	233
8503	3 pair		5.30	1.509		172	88		260	320
8504	4 pair		4.70	1.702		200	99		299	370
8506	6 pair		4	2		320	116		436	525
8509	9 pair	↓	3.40	2.353		490	137		627	740
8515	15 pair	2 Elec	4.80	3.333	↓	815	194		1,009	1,200

27 15 33 – Communications Coaxial Horizontal Cabling

27 15 33.10 Coaxial Cable and Fittings

		Crew	Daily Output	Labor-Hours	Unit	Material	2018 Bare Costs Labor	Equipment	Total	Total Incl O&P
0010	**COAXIAL CABLE & FITTINGS**									
3500	Coaxial connectors, 50 ohm impedance quick disconnect									
3540	BNC plug, for RG A/U #58 cable	1 Elec	42	.190	Ea.	3.13	11.10		14.23	20
3550	RG A/U #59 cable		42	.190		3.10	11.10		14.20	19.95
3560	RG A/U #62 cable		42	.190		3.13	11.10		14.23	20
3600	BNC jack, for RG A/U #58 cable		42	.190		3.25	11.10		14.35	20
3610	RG A/U #59 cable		42	.190		2.62	11.10		13.72	19.45
3620	RG A/U #62 cable		42	.190		2.62	11.10		13.72	19.45
3660	BNC panel jack, for RG A/U #58 cable		40	.200		8	11.65		19.65	26
3670	RG A/U #59 cable		40	.200		6.60	11.65		18.25	24.50
3680	RG A/U #62 cable		40	.200		6.60	11.65		18.25	24.50
3720	BNC bulkhead jack, for RG A/U #58 cable		40	.200		5.95	11.65		17.60	24
3730	RG A/U #59 cable		40	.200		5.95	11.65		17.60	24
3740	RG A/U #62 cable		40	.200	↓	5.90	11.65		17.55	24
3850	Coaxial cable, RG A/U #58, 50 ohm		8	1	C.L.F.	49.50	58		107.50	142
3860	RG A/U #59, 75 ohm		8	1		38.50	58		96.50	130
3870	RG A/U #62, 93 ohm		8	1		46.50	58		104.50	138
3875	RG 6/U, 75 ohm		8	1		31	58		89	121
3950	Fire rated, RG A/U #58, 50 ohm		8	1		94.50	58		152.50	191
3960	RG A/U #59, 75 ohm		8	1		125	58		183	224

324

27 15 Communications Horizontal Cabling

27 15 33 – Communications Coaxial Horizontal Cabling

27 15 33.10 Coaxial Cable and Fittings	Crew	Daily Output	Labor-Hours	Unit	Material	2018 Bare Costs Labor	Equipment	Total	Total Incl O&P
3970 RG A/U #62, 93 ohm	1 Elec	8	1	C.L.F.	110	58		168	208

27 15 43 – Communications Faceplates and Connectors

27 15 43.13 Communication Outlets

		Crew	Daily Output	Labor-Hours	Unit	Material	2018 Bare Costs Labor	Equipment	Total	Total Incl O&P
0010	**COMMUNICATION OUTLETS**									
0100	Voice/data devices not included									
0120	Voice/data outlets, single opening	1 Elec	48	.167	Ea.	6.55	9.70		16.25	21.50
0140	Two jack openings		48	.167		2.26	9.70		11.96	17
0160	One jack & one 3/4" round opening		48	.167		7.45	9.70		17.15	22.50
0180	One jack & one twinaxial opening		48	.167		7.45	9.70		17.15	22.50
0200	One jack & one connector cabling opening		48	.167		7.45	9.70		17.15	22.50
0220	Two 3/8" coaxial openings		48	.167		7.45	9.70		17.15	22.50
0300	Data outlets, single opening		48	.167		7.45	9.70		17.15	22.50
0320	One 25-pin subminiature opening		48	.167		7.45	9.70		17.15	22.50
1000	Voice/data wall plate, plastic, 1 gang, 1-port		72	.111		2.01	6.45		8.46	11.85
1020	2-port		72	.111		2.97	6.45		9.42	12.90
1040	3-port		72	.111		2.08	6.45		8.53	11.95
1060	4-port		72	.111		2.48	6.45		8.93	12.40
1080	6-port		72	.111		2.08	6.45		8.53	11.95
1100	2 gang, 6-port		48	.167		6.95	9.70		16.65	22
1120	Voice/data wall plate, stainless steel, 1 gang, 1-port		72	.111		6.95	6.45		13.40	17.30
1140	2-port		72	.111		6.95	6.45		13.40	17.30
1160	3-port		72	.111		7.05	6.45		13.50	17.40
1180	4-port		72	.111		7	6.45		13.45	17.35
1200	2 gang, 6-port		48	.167		11.55	9.70		21.25	27.50

27 21 Data Communications Network Equipment

27 21 23 – Data Communications Switches and Hubs

27 21 23.10 Switching and Routing Equipment

		Crew	Daily Output	Labor-Hours	Unit	Material	2018 Bare Costs Labor	Equipment	Total	Total Incl O&P
0010	**SWITCHING AND ROUTING EQUIPMENT**									
1100	Network hub, dual speed, 24 ports, includes cabinet	3 Elec	.66	36.364	Ea.	860	2,125		2,985	4,125
1300	Network switch, 50/60 HZ, 8 port, multi-platform, analog KVM	1 Elec	6.85	1.168		1,000	68		1,068	1,200
1310	16 port, multi-platform, analog KVM		6	1.333		1,225	77.50		1,302.50	1,475
1320	Network switch, 0x2x16,CAT5, analog KVM		6	1.333		1,025	77.50		1,102.50	1,250
1330	2x1x16, digital KVM, w/VM		16	.500		3,225	29		3,254	3,600
1340	2x1x32, digital KVM, w/VM		16	.500		3,850	29		3,879	4,275
1350	8x1x32, digital KVM, w/VM		16	.500		4,300	29		4,329	4,800
1500	KVM,1 RMU 16 port, no keyboard		16	.500		1,250	29		1,279	1,425
1510	KVM,1 RMU, 17" LCD, 16 port, US keyboard		15.15	.528		2,400	30.50		2,430.50	2,700
1520	KVM,1 RMU, 17" LCD, 16 port, UK keyboard		15.15	.528		2,400	30.50		2,430.50	2,700
2000	10/100/1000 Mbps, 24 ports		14	.571		585	33.50		618.50	695
2040	10/100/1000 Mbps, 48 ports		12	.667		1,225	39		1,264	1,400
2050	10/100/Mbps, 5 port, industrial ethernet type		32	.250		273	14.55		287.55	320
2060	10/100/Mbps, 6 port, industrial ethernet type		32	.250		1,100	14.55		1,114.55	1,225
2070	10/100/1000 Mbps, 10 port, X307-3		28	.286		2,600	16.65		2,616.65	2,875
2080	10/100/1000 Mbps, 10 port, X308-2		28	.286		2,400	16.65		2,416.65	2,675
3000	10/100/1000 Mbps, 20 port, front ports		20	.400		3,775	23.50		3,798.50	4,175
3010	10/100/1000 Mbps, 20 port, rear ports		20	.400		3,775	23.50		3,798.50	4,175
3040	KVM, 10/100/1000/10000 Mbps, 28 port, rear ports		16	.500		7,800	29		7,829	8,650
3050	KVM, 10/100/1000/10000 Mbps, 28 port, front ports		16	.500		9,175	29		9,204	10,100
3070	KVM, 10/100/1000/10000 Mbps, 52 port, front ports	2 Elec	12	1.333		9,625	77.50		9,702.50	10,700

27 21 Data Communications Network Equipment

27 21 23 – Data Communications Switches and Hubs

27 21 23.10 Switching and Routing Equipment	Crew	Daily Output	Labor-Hours	Unit	Material	2018 Bare Costs Labor	Equipment	Total	Total Incl O&P
3080 KVM, 10/100/1000/10000 Mbps, 52 port, rear ports	2 Elec	12	1.333	Ea.	9,625	77.50		9,702.50	10,700

27 32 Voice Communications Terminal Equipment

27 32 26 – Ring-Down Emergency Telephones

27 32 26.10 Emergency Phone Stations

		Crew	Daily Output	Labor-Hours	Unit	Material	2018 Bare Costs Labor	Equipment	Total	Total Incl O&P
0010	**EMERGENCY PHONE STATIONS**									
0100	Hands free emergency speaker phone w/LED strobe	1 Elec	7.22	1.108	Ea.	850	64.50		914.50	1,025
0110	Wall/pole mounted emergency phone station w/LED strobe	"	4.60	1.739		1,525	101		1,626	1,825
2000	Call station, 30", pole mounted emergency phone, 120V/ VoIP, blue beacon	2 Elec	5	3.200		1,100	186		1,286	1,475
2010	Call station, 36"		5	3.200		1,575	186		1,761	2,000
2020	Call station, 5'8"		4.80	3.333		2,600	194		2,794	3,150
2030	Call station, 9'		4.80	3.333		2,700	194		2,894	3,275

27 32 36 – TTY Equipment

27 32 36.10 TTY Telephone Equipment

		Crew	Daily Output	Labor-Hours	Unit	Material	2018 Bare Costs Labor	Equipment	Total	Total Incl O&P
0010	**TTY TELEPHONE EQUIPMENT**									
1620	Telephone, TTY, compact, pocket type				Ea.	279			279	305
1630	Advanced, desk type	2 Elec	20	.800		500	46.50		546.50	620
1640	Full-featured public, wall type	"	4	4		700	233		933	1,125

27 41 Audio-Video Systems

27 41 33 – Master Antenna Television Systems

27 41 33.10 TV Systems

		Crew	Daily Output	Labor-Hours	Unit	Material	2018 Bare Costs Labor	Equipment	Total	Total Incl O&P
0010	**TV SYSTEMS,** not including rough-in wires, cables & conduits									
0100	Master TV antenna system									
0200	VHF reception & distribution, 12 outlets	1 Elec	6	1.333	Outlet	114	77.50		191.50	242
0400	30 outlets		10	.800		240	46.50		286.50	335
0600	100 outlets		13	.615		288	36		324	370
0800	VHF & UHF reception & distribution, 12 outlets		6	1.333		249	77.50		326.50	390
1000	30 outlets		10	.800		165	46.50		211.50	251
1200	100 outlets		13	.615		168	36		204	239
1400	School and deluxe systems, 12 outlets		2.40	3.333		330	194		524	650
1600	30 outlets		4	2		288	116		404	490
1800	80 outlets		5.30	1.509		276	88		364	435
1900	Amplifier		4	2	Ea.	825	116		941	1,075
5000	Antenna, small		6	1.333		55	77.50		132.50	177
5100	Large		4	2		231	116		347	430
5110	Rotor unit		8	1		63	58		121	157
5120	Single booster		8	1		31.50	58		89.50	122
5130	Antenna pole, 10'		3.20	2.500		20	146		166	240
6100	Satellite TV system	2 Elec	1	16		2,150	930		3,080	3,775
6110	Dish, mesh, 10' diam.	"	2.40	6.667		1,675	390		2,065	2,425
6111	Two way RF/IF tapeoff	1 Elec	36	.222		3.59	12.95		16.54	23.50
6112	Two way RF/IF splitter		24	.333		13.70	19.40		33.10	44
6113	Line amplifier		24	.333		13.10	19.40		32.50	43.50
6114	Line splitters		36	.222		4.95	12.95		17.90	25
6115	Line multi switches		8	1		655	58		713	805
6120	Motor unit		2.40	3.333		259	194		453	575
7000	Home theater, widescreen, 42", high definition, TV		10.24	.781		400	45.50		445.50	510
7050	Flat wall mount bracket		10.24	.781		76	45.50		121.50	152

For customer support on your Electrical Costs with RSMeans data, call 800.448.8182.

27 41 Audio-Video Systems

27 41 33 – Master Antenna Television Systems

27 41 33.10 TV Systems	Crew	Daily Output	Labor-Hours	Unit	Material	2018 Bare Costs Labor	Equipment	Total	Total Incl O&P
7100	7 channel home theater receiver	1 Elec	10.24	.781	Ea.	385	45.50		430.50	495
7200	Home theater speakers		10.24	.781	Set	216	45.50		261.50	305
7300	Home theater programmable remote		10.24	.781	Ea.	249	45.50		294.50	340
8000	Main video splitter		4	2		3,225	116		3,341	3,725
8010	Video distribution units		4	2		140	116		256	330

27 42 Electronic Digital Systems

27 42 13 – Point of Sale Systems

27 42 13.10 Bar Code Scanner

		Crew	Daily Output	Labor-Hours	Unit	Material	Labor	Equipment	Total	Total Incl O&P
0010	**BAR CODE SCANNER**									
0100	CR3600, w/B4 battery, handle	1 Elec	6	1.333	Outlet	905	77.50		982.50	1,100
0110	Palm		12	.667		875	39		914	1,025
0120	Batch, B4 batt, handle, chgstn, 3' USB		9	.889		875	51.50		926.50	1,050
0130	Palm, chgstn, 3' USB		12	.667		835	39		874	980
0140	CR2500, w/H2 handle, no cable		10	.800		615	46.50		661.50	745
0150	Core unit, palm, no cable		16	.500		535	29		564	635

27 51 Distributed Audio-Video Communications Systems

27 51 16 – Public Address Systems

27 51 16.10 Public Address System

		Crew	Daily Output	Labor-Hours	Unit	Material	Labor	Equipment	Total	Total Incl O&P
0010	**PUBLIC ADDRESS SYSTEM**									
0100	Conventional, office	1 Elec	5.33	1.501	Speaker	159	87.50		246.50	305
0200	Industrial	"	2.70	2.963	"	305	172		477	600

27 51 19 – Sound Masking Systems

27 51 19.10 Sound System

		Crew	Daily Output	Labor-Hours	Unit	Material	Labor	Equipment	Total	Total Incl O&P
0010	**SOUND SYSTEM**, not including rough-in wires, cables & conduits									
0100	Components, projector outlet	1 Elec	8	1	Ea.	46	58		104	138
0200	Microphone		4	2		95.50	116		211.50	279
0400	Speakers, ceiling or wall		8	1		139	58		197	240
0600	Trumpets		4	2		259	116		375	460
0800	Privacy switch		8	1		103	58		161	201
1000	Monitor panel		4	2		460	116		576	680
1200	Antenna, AM/FM		4	2		130	116		246	315
1400	Volume control		8	1		52	58		110	144
1600	Amplifier, 250 W		1	8		1,200	465		1,665	2,025
1800	Cabinets		1	8		1,000	465		1,465	1,800
2000	Intercom, 30 station capacity, master station	2 Elec	2	8		2,300	465		2,765	3,225
2020	10 station capacity	"	4	4		1,250	233		1,483	1,725
2200	Remote station	1 Elec	8	1		194	58		252	300
2400	Intercom outlets		8	1		114	58		172	212
2600	Handset		4	2		375	116		491	590
2800	Emergency call system, 12 zones, annunciator		1.30	6.154		1,125	360		1,485	1,775
3000	Bell		5.30	1.509		117	88		205	260
3200	Light or relay		8	1		58.50	58		116.50	152
3400	Transformer		4	2		257	116		373	455
3600	House telephone, talking station		1.60	5		555	291		846	1,050
3800	Press to talk, release to listen		5.30	1.509		129	88		217	273
4000	System-on button					77			77	85
4200	Door release	1 Elec	4	2		138	116		254	325

For customer support on your Electrical Costs with RSMeans data, call 800.448.8182.

327

27 51 Distributed Audio-Video Communications Systems

27 51 19 – Sound Masking Systems

27 51 19.10 Sound System	Crew	Daily Output	Labor-Hours	Unit	Material	2018 Bare Costs Labor	Equipment	Total	Total Incl O&P
4400	Combination speaker and microphone	1 Elec	8	1	Ea.	235	58		293	345
4600	Termination box		3.20	2.500		74	146		220	299
4800	Amplifier or power supply		5.30	1.509		850	88		938	1,075
5000	Vestibule door unit		16	.500	Name	156	29		185	215
5200	Strip cabinet		27	.296	Ea.	294	17.25		311.25	350
5400	Directory		16	.500		139	29		168	196
6000	Master door, button buzzer type, 100 unit	2 Elec	.54	29.630		1,400	1,725		3,125	4,125
6020	200 unit		.30	53.333		2,650	3,100		5,750	7,575
6040	300 unit		.20	80		4,050	4,650		8,700	11,400
6060	Transformer	1 Elec	8	1		36	58		94	127
6080	Door opener		5.30	1.509		51.50	88		139.50	188
6100	Buzzer with door release and plate		4	2		51.50	116		167.50	231
6200	Intercom type, 100 unit	2 Elec	.54	29.630		1,750	1,725		3,475	4,500
6220	200 unit		.30	53.333		3,425	3,100		6,525	8,425
6240	300 unit		.20	80		5,175	4,650		9,825	12,600
6260	Amplifier	1 Elec	2	4		257	233		490	635
6280	Speaker with door release	"	4	2		77	116		193	259

27 52 Healthcare Communications and Monitoring Systems

27 52 23 – Nurse Call/Code Blue Systems

27 52 23.10 Nurse Call Systems

		Crew	Daily Output	Labor-Hours	Unit	Material	2018 Bare Costs Labor	Equipment	Total	Total Incl O&P
0010	**NURSE CALL SYSTEMS**									
0100	Single bedside call station	1 Elec	8	1	Ea.	163	58		221	266
0200	Ceiling speaker station		8	1		83	58		141	178
0400	Emergency call station		8	1		89	58		147	185
0600	Pillow speaker		8	1		218	58		276	325
0800	Double bedside call station		4	2		176	116		292	370
1000	Duty station		4	2		131	116		247	320
1200	Standard call button		8	1		107	58		165	205
1400	Lights, corridor, dome or zone indicator		8	1		60.50	58		118.50	154
1600	Master control station for 20 stations	2 Elec	.65	24.615	Total	3,525	1,425		4,950	6,025

27 53 Distributed Systems

27 53 13 – Clock Systems

27 53 13.50 Clock Equipments

		Crew	Daily Output	Labor-Hours	Unit	Material	2018 Bare Costs Labor	Equipment	Total	Total Incl O&P
0010	**CLOCK EQUIPMENTS**, not including wires & conduits									
0100	Time system components, master controller	1 Elec	.33	24.242	Ea.	1,500	1,400		2,900	3,750
0200	Program bell		8	1		111	58		169	209
0400	Combination clock & speaker		3.20	2.500		198	146		344	435
0600	Frequency generator		2	4		2,425	233		2,658	3,025
0800	Job time automatic stamp recorder		4	2		565	116		681	795
1600	Master time clock system, clocks & bells, 20 room	4 Elec	.20	160		5,750	9,300		15,050	20,200
1800	50 room	"	.08	400		12,200	23,300		35,500	48,200
1900	Time clock	1 Elec	3.20	2.500		390	146		536	650
2000	100 cards in & out, 1 color					9			9	9.90
2200	2 colors					9			9	9.90
2800	Metal rack for 25 cards	1 Elec	7	1.143		46.50	66.50		113	151
4000	Wireless time systems component, master controller	"	2	4		640	233		873	1,050
4010	For transceiver and antenna, see Section 28 39 10.10									

27 53 13.50 Clock Equipments		Crew	Daily Output	Labor-Hours	Unit	Material	2018 Bare Costs			Total Incl O&P
							Labor	Equipment	Total	
4100	Wireless analog clock w/battery operated	1 Elec	8	1	Ea.	111	58		169	209
4200	Wireless digital clock w/battery operated	"	8	1	"	425	58		483	555

For customer support on your Electrical Costs with RSMeans data, call 800.448.8182.

329

Division Notes

	CREW	DAILY OUTPUT	LABOR-HOURS	UNIT	BARE COSTS				TOTAL INCL O&P
					MAT.	LABOR	EQUIP.	TOTAL	

Estimating Tips

- When estimating material costs for electronic safety and security systems, it is always prudent to obtain manufacturers' quotations for equipment prices and special installation requirements that may affect the total cost.

- Fire alarm systems consist of control panels, annunciator panels, batteries with rack, charger, and fire alarm actuating and indicating devices. Some fire alarm systems include speakers, telephone lines, door closer controls, and other components. Be careful not to overlook the costs related to installation for these items. Also be aware of costs for integrated automation instrumentation and terminal devices, control equipment, control wiring, and programming. Insurance underwriters may have specific requirements for the type of materials to be installed or design requirements based on the hazard to be protected. Local jurisdictions may have requirements not covered by code. It is advisable to be aware of any special conditions.

- Security equipment includes items such as CCTV, access control, and other detection and identification systems to perform alert and alarm functions. Be sure to consider the costs related to installation for this security equipment, such as for integrated automation instrumentation and terminal devices, control equipment, control wiring, and programming.

Reference Numbers

Reference numbers are shown at the beginning of some major classifications. These numbers refer to related items in the Reference Section. The reference information may be an estimating procedure, an alternate pricing method, or technical information.

Note: Not all subdivisions listed here necessarily appear. ∎

28 01 Operation and Maint. of Electronic Safety and Security

28 01 80 – Operation and Maintenance of Fire Detection and Alarm

28 01 80.51 Maintenance & Admin. of Fire Detection & Alarm	Crew	Daily Output	Labor-Hours	Unit	Material	2018 Bare Costs Labor	Equipment	Total	Total Incl O&P
0010 **MAINTENANCE AND ADMINISTRATION OF FIRE DETECTION AND ALARM**									
3300 Remove and replace (reinstall), fire alarm device	1 Elec	5.33	1.501	Ea.		87.50		87.50	131

28 05 Common Work Results for Electronic Safety and Security

28 05 05 – Selective Demolition for Electronic Safety and Security

28 05 05.10 Safety and Security Demolition

	Crew	Daily Output	Labor-Hours	Unit	Material	Labor	Equipment	Total	Total Incl O&P
0010 **SAFETY AND SECURITY DEMOLITION**									
1050 Finger print/card reader	1 Elec	8	1	Ea.		58		58	87
1051 Electrical dml, CA 3000 card reader sys w/25 users		.31	25.600			1,500		1,500	2,225
1060 Video camera		8	1			58		58	87
1070 Motion detector, multi-channel		7	1.143			66.50		66.50	99.50
1090 Infrared detector		7	1.143			66.50		66.50	99.50
1130 Video monitor, 19"		8	1			58		58	87
1210 Fire alarm horn and strobe light		16	.500			29		29	43.50
1215 Remote fire alarm indicator light		23	.348			20		20	30.50
1220 Flame detector		23	.348			20		20	30.50
1230 Duct detector		10	.800			46.50		46.50	69.50
1240 Smoke detector		23	.348			20		20	30.50
1241 Fire alarm duct smoke remote test station		21.60	.370			21.50		21.50	32
1250 Fire alarm pull station		23	.348			20		20	30.50
1260 Fire alarm control panel, 4 to 8 zone	2 Elec	4	4			233		233	350
1270 12 to 16 zone	"	2.67	5.993			350		350	520
1280 Fire alarm annunciation panel, 4 to 8 zone	1 Elec	6	1.333			77.50		77.50	116
1290 12 to 16 zone	2 Elec	6	2.667			155		155	232
1300 Electrical dml, tamper switch	1 Elec	38	.211			12.25		12.25	18.30
1310 Electrical dml, flow switch	"	38	.211			12.25		12.25	18.30

28 05 19 – Storage Appliances for Electronic Safety and Security

28 05 19.11 Digital Video Recorder (DVR)

	Crew	Daily Output	Labor-Hours	Unit	Material	Labor	Equipment	Total	Total Incl O&P
0010 **DIGITAL VIDEO RECORDERS**									
0100 Pentaplex hybrid, internet protocol, and hard drive									
0200 4 channel	1 Elec	1.33	6.015	Ea.	1,350	350		1,700	2,000
0300 8 channel		1	8		3,050	465		3,515	4,050
0400 16 channel		1	8		3,375	465		3,840	4,400

28 15 Access Control Hardware Devices

28 15 11 – Integrated Credential Readers and Field Entry Management

28 15 11.11 Standard Card Readers

	Crew	Daily Output	Labor-Hours	Unit	Material	Labor	Equipment	Total	Total Incl O&P
0010 **STANDARD CARD READERS**									
0015 Card key access									
0020 Computerized system, processor, proximity reader and cards									
0030 Does not include door hardware, lockset or wiring									
0040 Card key system for 1 door				Ea.	1,425			1,425	1,550
0060 Card key system for 2 doors					2,025			2,025	2,225
0080 Card key system for 4 doors					2,500			2,500	2,750
0100 Processor for card key access system					940			940	1,025
0160 Magnetic lock for electric access, 600 pound holding force					190			190	209
0170 Magnetic lock for electric access, 1200 pound holding force					190			190	209
0200 Proximity card reader					158			158	173
0210 Entrance card systems									

For customer support on your Electrical Costs with RSMeans data, call 800.448.8182.

28 15 Access Control Hardware Devices

28 15 11 – Integrated Credential Readers and Field Entry Management

28 15 11.11 Standard Card Readers

		Crew	Daily Output	Labor-Hours	Unit	Material	2018 Bare Costs Labor	2018 Bare Costs Equipment	Total	Total Incl O&P
0300	Entrance card, barium ferrite				Ea.	4.80			4.80	5.30
0320	Credential					7.85			7.85	8.60
0340	Proximity					3.43			3.43	3.77
0360	Weigand					7.55			7.55	8.30
0700	Entrance card reader, barium ferrite	R-19	4	5		335	292		627	805
0720	Credential		4	5		199	292		491	655
0740	Proximity		4	5		297	292		589	760
0760	Weigand		4	5		470	292		762	955
0850	Scanner, eye retina		4	5		2,500	292		2,792	3,175
0900	Gate opener, cantilever	R-18	3	8.667		2,300	445		2,745	3,175
0910	Light duty		3	8.667		1,650	445		2,095	2,475
0912	Custom		3	8.667		2,075	445		2,520	2,925
0950	Switch, tamper	R-19	6	3.333		34.50	194		228.50	330
1100	Accessories, electric door strike/bolt		5	4		75	233		308	435
1120	Electromagnetic lock		5	4		138	233		371	500
1140	Keypad for card reader		4	5		118	292		410	565

28 15 11.13 Keypads

		Crew	Daily Output	Labor-Hours	Unit	Material	2018 Bare Costs Labor	2018 Bare Costs Equipment	Total	Total Incl O&P
0010	**KEYPADS**									
0340	Digital keypad, int./ext., basic, excl. striker/power/wiring	1 Elec	3	2.667	Ea.	104	155		259	345
0350	Lockset, mechanical push-button type, complete, incl. hardware	1 Carp	4	2	"	405	101		506	600

28 15 11.15 Biometric Identity Devices

		Crew	Daily Output	Labor-Hours	Unit	Material	2018 Bare Costs Labor	2018 Bare Costs Equipment	Total	Total Incl O&P
0010	**BIOMETRIC IDENTITY DEVICES**									
0200	Fingerprint scanner unit, excl. striker/power supply	1 Elec	4	2	Ea.	1,575	116		1,691	1,900
0210	Fingerprint scanner unit, for computer keyboard access		8	1		1,200	58		1,258	1,400
0220	Hand geometry scanner, mem of 512 users, excl. striker/power		3	2.667		2,300	155		2,455	2,750
0230	Memory upgrade for, adds 9,700 user profiles		8	1		330	58		388	450
0240	Adds 32,500 user profiles		8	1		665	58		723	815
0250	Prison type, memory of 256 users, excl. striker/power		3	2.667		2,700	155		2,855	3,200
0260	Memory upgrade for, adds 3,300 user profiles		8	1		315	58		373	435
0270	Adds 9,700 user profiles		8	1		500	58		558	635
0280	Adds 27,900 user profiles		8	1		680	58		738	830
0290	All weather, mem of 512 users, excl. striker/power		3	2.667		4,075	155		4,230	4,700
0300	Facial & fingerprint scanner, combination unit, excl. striker/power		3	2.667		895	155		1,050	1,225
0310	Access for, for initial setup, excl. striker/power		3	2.667		1,200	155		1,355	1,550

28 15 11.19 Security Access Control Accessories

		Crew	Daily Output	Labor-Hours	Unit	Material	2018 Bare Costs Labor	2018 Bare Costs Equipment	Total	Total Incl O&P
0010	**SECURITY ACCESS CONTROL ACCESSORIES**									
0360	Scanner/reader access, power supply/transf, 110V to 12/24V	1 Elec	4	2	Ea.	215	116		331	410
0370	Elec/mag strikers, 12/24V		4	2		99	116		215	283
0380	1 hour battery backup power supply		4	2		46.50	116		162.50	225
0390	Deadbolt, digital, batt-operated, indoor/outdoor, complete, incl. hardware	1 Carp	4	2		360	101		461	550

For customer support on your Electrical Costs with RSMeans data, call 800.448.8182.

333

28 18 Security Access Detection Equipment

28 18 11 – Security Access Metal Detectors

28 18 11.13 Security Access Metal Detectors

		Crew	Daily Output	Labor-Hours	Unit	Material	2018 Bare Costs Labor	Equipment	Total	Total Incl O&P
0010	**SECURITY ACCESS METAL DETECTORS**				Ea.	129			129	138
0240	Metal detector, hand-held, wand type, unit only									
0250	Metal detector, walk through portal type, single zone	1 Elec	2	4		3,175	233		3,408	3,825
0260	Multi-zone	"	2	4		4,625	233		4,858	5,425

28 18 13 – Security Access X-Ray Equipment

28 18 13.16 Security Access X-Ray Equipment

		Crew	Daily Output	Labor-Hours	Unit	Material	2018 Bare Costs Labor	Equipment	Total	Total Incl O&P
0010	**SECURITY ACCESS X-RAY EQUIPMENT**				Ea.	3,475	116		3,591	4,000
0290	X-ray machine, desk top, for mail/small packages/letters	1 Elec	4	2						
0300	Conveyor type, incl. monitor		2	4		16,000	233		16,233	18,000
0310	Includes additional features		2	4		28,000	233		28,233	31,200
0320	X-ray machine, large unit, for airports, incl. monitor	2 Elec	1	16		40,000	930		40,930	45,400
0330	Full console	"	.50	32		68,500	1,850		70,350	78,500

28 18 15 – Security Access Explosive Detection Equipment

28 18 15.23 Security Access Explosive Detection Equipment

		Crew	Daily Output	Labor-Hours	Unit	Material	2018 Bare Costs Labor	Equipment	Total	Total Incl O&P
0010	**SECURITY ACCESS EXPLOSIVE DETECTION EQUIPMENT**									
0270	Explosives detector, walk through portal type	1 Elec	2	4	Ea.	44,200	233		44,433	49,000
0280	Hand-held, battery operated				"				25,500	28,100

28 23 Video Management System

28 23 13 – Video Management System Interfaces

28 23 13.10 Closed Circuit Television System

		Crew	Daily Output	Labor-Hours	Unit	Material	2018 Bare Costs Labor	Equipment	Total	Total Incl O&P
0010	**CLOSED CIRCUIT TELEVISION SYSTEM**									
2000	Surveillance, one station (camera & monitor)	2 Elec	2.60	6.154	Total	850	360		1,210	1,475
2200	For additional camera stations, add	1 Elec	2.70	2.963	Ea.	370	172		542	665
2400	Industrial quality, one station (camera & monitor)	2 Elec	2.60	6.154	Total	2,000	360		2,360	2,725
2600	For additional camera stations, add	1 Elec	2.70	2.963	Ea.	800	172		972	1,150
2610	For low light, add		2.70	2.963		600	172		772	920
2620	For very low light, add		2.70	2.963		3,900	172		4,072	4,550
2800	For weatherproof camera station, add		1.30	6.154		640	360		1,000	1,250
3000	For pan and tilt, add		1.30	6.154		1,875	360		2,235	2,600
3200	For zoom lens - remote control, add		2	4		2,200	233		2,433	2,775
3400	Extended zoom lens		2	4		6,000	233		6,233	6,950
3410	For automatic iris for low light, add		2	4		1,200	233		1,433	1,675
3600	Educational TV studio, basic 3 camera system, black & white,									
3800	electrical & electronic equip. only	4 Elec	.80	40	Total	10,000	2,325		12,325	14,500
4000	Full console		.28	114		29,700	6,650		36,350	42,700
4100	As above, but color system		.28	114		61,500	6,650		68,150	77,500
4120	Full console		.12	267		228,500	15,500		244,000	274,000
4200	For film chain, black & white, add	1 Elec	1	8	Ea.	15,400	465		15,865	17,600
4250	Color, add		.25	32		10,500	1,850		12,350	14,400
4400	For video recorders, add		1	8		3,000	465		3,465	4,000
4600	Premium	4 Elec	.40	80		19,400	4,650		24,050	28,300

28 23 23 – Video Surveillance Systems Infrastructure

28 23 23.50 Video Surveillance Equipments

		Crew	Daily Output	Labor-Hours	Unit	Material	2018 Bare Costs Labor	Equipment	Total	Total Incl O&P
0010	**VIDEO SURVEILLANCE EQUIPMENTS**									
0200	Video cameras, wireless, hidden in exit signs, clocks, etc., incl. receiver	1 Elec	3	2.667	Ea.	187	155		342	435
0210	Accessories for video recorder, single camera		3	2.667		183	155		338	435
0220	For multiple cameras		3	2.667		1,750	155		1,905	2,150
0230	Video cameras, wireless, for under vehicle searching, complete		2	4		16,000	233		16,233	18,000

28 23 Video Management System

28 23 23 – Video Surveillance Systems Infrastructure

28 23 23.50 Video Surveillance Equipments	Crew	Daily Output	Labor-Hours	Unit	Material	2018 Bare Costs Labor	Equipment	Total	Total Incl O&P
0234 Master monitor station, 3 doors x 5 color monitor with tilt feature	1 Elec	2	4	Ea.	800	233		1,033	1,225
0400 Internet protocol network camera, day/night, color & power supply		2.60	3.077		1,300	179		1,479	1,700
0500 Monitor, color flat screen, liquid crystal display (LCD), 15"		2.70	2.963		510	172		682	820
0520 17"		2.70	2.963		510	172		682	820
0540 19"		2.70	2.963		590	172		762	910

28 31 Intrusion Detection

28 31 16 – Intrusion Detection Systems Infrastructure

28 31 16.50 Intrusion Detection

		Crew	Daily Output	Labor-Hours	Unit	Material	2018 Bare Costs Labor	Equipment	Total	Total Incl O&P
0010	**INTRUSION DETECTION**, not including wires & conduits									
0100	Burglar alarm, battery operated, mechanical trigger	1 Elec	4	2	Ea.	280	116		396	485
0200	Electrical trigger		4	2		335	116		451	545
0400	For outside key control, add		8	1		87	58		145	183
0600	For remote signaling circuitry, add		8	1		138	58		196	239
0800	Card reader, flush type, standard		2.70	2.963		810	172		982	1,150
1000	Multi-code		2.70	2.963		1,200	172		1,372	1,575
1010	Card reader, proximity type		2.70	2.963		297	172		469	585
1200	Door switches, hinge switch		5.30	1.509		62.50	88		150.50	200
1400	Magnetic switch		5.30	1.509		100	88		188	241
1600	Exit control locks, horn alarm		4	2		219	116		335	415
1800	Flashing light alarm		4	2		247	116		363	445
2000	Indicating panels, 1 channel		2.70	2.963		271	172		443	555
2200	10 channel	2 Elec	3.20	5		1,075	291		1,366	1,600
2400	20 channel		2	8		2,500	465		2,965	3,425
2600	40 channel		1.14	14.035		4,500	815		5,315	6,175
2800	Ultrasonic motion detector, 12 V	1 Elec	2.30	3.478		194	202		396	520
3000	Infrared photoelectric detector		4	2		146	116		262	335
3200	Passive infrared detector		4	2		236	116		352	435
3400	Glass break alarm switch		8	1		99.50	58		157.50	196
3420	Switchmats, 30" x 5'		5.30	1.509		105	88		193	247
3440	30" x 25'		4	2		205	116		321	400
3460	Police connect panel		4	2		272	116		388	475
3480	Telephone dialer		5.30	1.509		390	88		478	555
3500	Alarm bell		4	2		105	116		221	290
3520	Siren		4	2		147	116		263	335
3540	Microwave detector, 10' to 200'		2	4		585	233		818	995
3560	10' to 350'		2	4		1,775	233		2,008	2,300

28 33 Security Monitoring and Control

28 33 11 – Electronic Structural Monitoring Systems

28 33 11.10 Load Cell Units

		Crew	Daily Output	Labor-Hours	Unit	Material	2018 Bare Costs Labor	Equipment	Total	Total Incl O&P
0010	**LOAD CELL UNITS**									
0100	Load cell, stainless steel, 0.5/1 ton rated, mounting unit	2 Elec	1.33	12.030	Ea.	675	700		1,375	1,800
0110	2/3 to 5/5 ton rated, mounting unit		1.20	13.333		705	775		1,480	1,925
0120	0.5/1/2/3.5/5 ton, self aligning bottom base		1.66	9.639		233	560		793	1,100
0130	Combo mounting unit, 10/25 ton rated		.86	18.605		1,850	1,075		2,925	3,650
0140	40/60 ton rated		.66	24.242		3,150	1,400		4,550	5,550
0150	60kg/3m, rated		1.84	8.696		1,575	505		2,080	2,475
0160	130kg/3m, rated		1.55	10.323		1,500	600		2,100	2,550

For customer support on your Electrical Costs with RSMeans data, call 800.448.8182.

335

28 33 Security Monitoring and Control

28 33 11 – Electronic Structural Monitoring Systems

28 33 11.10 Load Cell Units	Crew	Daily Output	Labor-Hours	Unit	Material	2018 Bare Costs Labor	Equipment	Total	Total Incl O&P
0170	280 kg, rated	2 Elec	1.22	13.115	Ea.	1,500	765		2,265	2,800

28 46 Fire Detection and Alarm

28 46 11 – Fire Sensors and Detectors

28 46 11.21 Carbon-Monoxide Detection Sensors

		Crew	Daily Output	Labor-Hours	Unit	Material	Labor	Equipment	Total	Total Incl O&P
0010	**CARBON-MONOXIDE DETECTION SENSORS**									
8400	Smoke and carbon monoxide alarm battery operated photoelectric low profile	1 Elec	24	.333	Ea.	54.50	19.40		73.90	88.50
8410	low profile photoelectric battery powered		24	.333		28.50	19.40		47.90	60.50
8420	photoelectric low profile sealed lithium		24	.333		52	19.40		71.40	86
8430	Photoelectric low profile sealed lithium smoke and CO with voice combo		24	.333		38	19.40		57.40	71
8500	Carbon monoxide sensor, wall mount 1Mod 1 relay output smoke & heat		24	.333		490	19.40		509.40	570
8510	1Mod, 1relay no display, smoke & heat		24	.333		480	19.40		499.40	555
8530	1Mod 1 relay no HW logo smoke & heat		24	.333		515	19.40		534.40	600
8700	Carbon monoxide detector, battery operated, wall mounted		16	.500		52	29		81	101
8710	Hardwired, wall and ceiling mounted		8	1		99.50	58		157.50	196
8720	Duct mounted		8	1		345	58		403	460
8730	Continuous air monitoring system, PC based, incl. remote sensors, min.	2 Elec	.25	64		31,200	3,725		34,925	39,900
8740	Maximum	"	.10	160		49,900	9,300		59,200	69,000

28 46 11.27 Other Sensors

		Crew	Daily Output	Labor-Hours	Unit	Material	Labor	Equipment	Total	Total Incl O&P
0010	**OTHER SENSORS**									
5200	Smoke detector, ceiling type	1 Elec	6.20	1.290	Ea.	120	75		195	244
5240	Smoke detector addressable type		6	1.333		224	77.50		301.50	360
5400	Duct type		3.20	2.500		330	146		476	585
5420	Duct addressable type		3.20	2.500		515	146		661	785
8300	Smoke alarm with integrated strobe light 120 V 16DB 60 fpm flash rate		16	.500		103	29		132	157
8310	Photoelectric smoke detector with strobe 120 V 90 DB ceiling mount		12	.667		158	39		197	232
8320	120 V, 90 DB wall mount		12	.667		166	39		205	240
8330	120 V, 9 VDC, 90 DB ceiling mount		12	.667		194	39		233	271
8340	120 V, 9 VDC backup 90 DB wall mount		12	.667		195	39		234	273
8440	Fire alarm beam detector, motorised reflective, infrared optical beam	2 Elec	1.60	10		800	580		1,380	1,750
8450	Fire alarm beam detector, motorised reflective, IR/UV optical beam	"	1.60	10		1,025	580		1,605	2,000

28 46 11.50 Fire and Heat Detectors

		Crew	Daily Output	Labor-Hours	Unit	Material	Labor	Equipment	Total	Total Incl O&P
0010	**FIRE & HEAT DETECTORS**									
5000	Detector, rate of rise	1 Elec	8	1	Ea.	51.50	58		109.50	144
5010	Heat addressable type		7.25	1.103		259	64		323	380
5100	Fixed temp fire alarm		7	1.143		51.50	66.50		118	156
5200	125 V SRF MT-135F		7.25	1.103		28	64		92	127
5300	MT-195F		7.25	1.103		31.50	64		95.50	131
5400	104/85 DB, indoor/outdoor, ceiling/wall, red		10.25	.780		63.50	45.50		109	138
5410	Ceiling/wall, white		10.25	.780		62	45.50		107.50	137
5420	102/98 DB, lumin 110cd, indoor/wall, red		10.25	.780		113	45.50		158.50	193
5430	Lumin15/75cd, indoor/wall, red		10.25	.780		110	45.50		155.50	189
5440	HI/LO DB, 24 V, fire marking, red		10.25	.780		60.50	45.50		106	135

28 46 21 – Fire Alarm

28 46 21.50 Alarm Panels and Devices

		Crew	Daily Output	Labor-Hours	Unit	Material	Labor	Equipment	Total	Total Incl O&P
0010	**ALARM PANELS AND DEVICES**, not including wires & conduits									
2200	Intercom remote station	1 Elec	8	1	Ea.	71	58		129	165
2400	Intercom outlet		8	1		53	58		111	146
2600	Sound system, intercom handset		8	1		475	58		533	610
3590	Signal device, beacon		8	1		199	58		257	305

28 46 Fire Detection and Alarm

28 46 21 – Fire Alarm

28 46 21.50 Alarm Panels and Devices

		Crew	Daily Output	Labor-Hours	Unit	Material	2018 Bare Costs Labor	2018 Bare Costs Equipment	Total	Total Incl O&P
3594	2 zone	1 Elec	1.35	5.926	Ea.	310	345		655	855
3600	4 zone	2 Elec	2	8		380	465		845	1,100
3610	3 zone		2	8		610	465		1,075	1,375
3800	8 zone		1	16		960	930		1,890	2,450
3810	5 zone		1.50	10.667		770	620		1,390	1,775
3900	10 zone		1.25	12.800		945	745		1,690	2,175
4000	12 zone		.67	23.988		2,100	1,400		3,500	4,375
4020	Alarm device, tamper, flow	1 Elec	8	1		240	58		298	350
4025	Fire alarm, loop expander card		16	.500		740	29		769	855
4050	Actuating device		8	1		340	58		398	460
4160	Alarm control panel, addressable w/o voice, up to 200 points	2 Elec	1.14	13.998		4,525	815		5,340	6,200
4170	addressable w/voice, up to 400 points	"	.73	22.008		9,525	1,275		10,800	12,400
4175	Addressable interface device	1 Elec	7.25	1.103		136	64		200	246
4200	Battery and rack		4	2		415	116		531	630
4400	Automatic charger		8	1		590	58		648	735
4600	Signal bell		8	1		124	58		182	224
4610	Fire alarm signal bell 10" red 20-24 V P		8	1		136	58		194	236
4800	Trouble buzzer or manual station		8	1		84	58		142	180
5425	Duct smoke and heat detector 2 wire		8	1		107	58		165	204
5430	Fire alarm duct detector controller		3	2.667		235	155		390	490
5435	Fire alarm duct detector sensor kit		8	1		75	58		133	170
5440	Remote test station for smoke detector duct type		5.30	1.509		65	88		153	203
5460	Remote fire alarm indicator light		5.30	1.509		18.55	88		106.55	152
5600	Strobe and horn		5.30	1.509		154	88		242	300
5610	Strobe and horn (ADA type)		5.30	1.509		154	88		242	300
5620	Visual alarm (ADA type)		6.70	1.194		104	69.50		173.50	218
5800	Fire alarm horn		6.70	1.194		61.50	69.50		131	172
6000	Door holder, electro-magnetic		4	2		104	116		220	288
6200	Combination holder and closer		3.20	2.500		124	146		270	355
6600	Drill switch		8	1		375	58		433	495
6800	Master box		2.70	2.963		6,475	172		6,647	7,375
7000	Break glass station		8	1		57.50	58		115.50	151
7010	Break glass station, addressable		7.25	1.103		149	64		213	260
7800	Remote annunciator, 8 zone lamp		1.80	4.444		213	259		472	620
8000	12 zone lamp	2 Elec	2.60	6.154		415	360		775	995
8200	16 zone lamp	"	2.20	7.273		375	425		800	1,050

28 47 Mass Notification

28 47 12 – Notification Systems

28 47 12.10 Mass Notification System

		Crew	Daily Output	Labor-Hours	Unit	Material	2018 Bare Costs Labor	2018 Bare Costs Equipment	Total	Total Incl O&P
0010	**MASS NOTIFICATION SYSTEM**									
0100	Wireless command center, 10,000 devices	2 Elec	1.33	12.030	Ea.	2,275	700		2,975	3,550
0200	Option, email notification					1,775			1,775	1,950
0210	Remote device supervision & monitor					2,525			2,525	2,800
0300	Antenna VHF or UHF, for medium range	1 Elec	4	2		87	116		203	270
0310	For high-power transmitter		2	4		1,100	233		1,333	1,550
0400	Transmitter, 25 watt		4	2		1,725	116		1,841	2,075
0410	40 watt		2.66	3.008		2,000	175		2,175	2,450
0420	100 watt		1.33	6.015		6,475	350		6,825	7,650
0500	Wireless receiver/control module for speaker		8	1		279	58		337	390
0600	Desktop paging controller, stand alone		4	2		835	116		951	1,100

For customer support on your Electrical Costs with RSMeans data, call 800.448.8182.

337

28 47 Mass Notification

28 47 12 – Notification Systems

28 47 12.10 Mass Notification System	Crew	Daily Output	Labor-Hours	Unit	Material	2018 Bare Costs Labor	Equipment	Total	Total Incl O&P
0700 Auxiliary level inputs	1 Elec	8.36	.957	Ea.	92	55.50		147.50	185

28 52 Detention Security Systems

28 52 11 – Detention Monitoring and Control Systems

28 52 11.10 Detention Control Systems

	Crew	Daily Output	Labor-Hours	Unit	Material	2018 Bare Costs Labor	Equipment	Total	Total Incl O&P
0010 **DETENTION CONTROL SYSTEMS**									
1000 Desk top control systems for 10 doors, 10 intercoms, and 10 lights	2 Elec	.30	53.333	Ea.	61,500	3,100		64,600	72,500
1020 Push button control panel systems for 10 doors, 10 intercoms, and 10 lights	"	.32	50	"	55,500	2,900		58,400	65,500

For customer support on your Electrical Costs with RSMeans data, call 800.448.8182.

Estimating Tips
31 05 00 Common Work Results for Earthwork

- Estimating the actual cost of performing earthwork requires careful consideration of the variables involved. This includes items such as type of soil, whether water will be encountered, dewatering, whether banks need bracing, disposal of excavated earth, and length of haul to fill or spoil sites, etc. If the project has large quantities of cut or fill, consider raising or lowering the site to reduce costs, while paying close attention to the effect on site drainage and utilities.

- If the project has large quantities of fill, creating a borrow pit on the site can significantly lower the costs.

- It is very important to consider what time of year the project is scheduled for completion. Bad weather can create large cost overruns from dewatering, site repair, and lost productivity from cold weather.

Reference Numbers

Reference numbers are shown at the beginning of some major classifications. These numbers refer to related items in the Reference Section. The reference information may be an estimating procedure, an alternate pricing method, or technical information.

Note: Not all subdivisions listed here necessarily appear. ■

31 23 16.13 Excavating, Trench

		Crew	Daily Output	Labor-Hours	Unit	Material	2018 Bare Costs Labor	2018 Bare Costs Equipment	Total	Total Incl O&P
0010	**EXCAVATING, TRENCH**									
0011	Or continuous footing									
0020	Common earth with no sheeting or dewatering included									
0050	1' to 4' deep, 3/8 C.Y. excavator	B-11C	150	.107	B.C.Y.		4.99	2.08	7.07	9.85
0060	1/2 C.Y. excavator	B-11M	200	.080			3.74	1.92	5.66	7.75
0062	3/4 C.Y. excavator	B-12F	270	.059			2.84	2.48	5.32	7.05
1400	By hand with pick and shovel 2' to 6' deep, light soil	1 Clab	8	1			40		40	60.50
1500	Heavy soil	"	4	2	↓		79.50		79.50	121
3000	Backfill trench, F.E. loader, wheel mtd., 1 C.Y. bucket									
3020	Minimal haul	B-10R	400	.030	L.C.Y.		1.47	.73	2.20	3.03
5020	Loam & sandy clay with no sheeting or dewatering included									
5050	1' to 4' deep, 3/8 C.Y. tractor loader/backhoe	B-11C	162	.099	B.C.Y.		4.62	1.93	6.55	9.10
5060	1/2 C.Y. excavator	B-11M	216	.074			3.47	1.78	5.25	7.20
5070	3/4 C.Y. excavator	B-12F	292	.055	↓		2.63	2.29	4.92	6.50
6020	Sand & gravel with no sheeting or dewatering included									
6050	1' to 4' deep, 3/8 C.Y. excavator	B-11C	165	.097	B.C.Y.		4.54	1.89	6.43	8.95
6060	1/2 C.Y. excavator	B-11M	220	.073			3.40	1.75	5.15	7.10
6070	3/4 C.Y. excavator	B-12F	297	.054	↓		2.58	2.25	4.83	6.40
7020	Dense hard clay with no sheeting or dewatering included									
7050	1' to 4' deep, 3/8 C.Y. excavator	B-11C	132	.121	B.C.Y.		5.65	2.37	8.02	11.20
7060	1/2 C.Y. excavator	B-11M	176	.091			4.25	2.19	6.44	8.85
7070	3/4 C.Y. excavator	B-12F	238	.067	↓		3.23	2.81	6.04	8

31 23 16.14 Excavating, Utility Trench

		Crew	Daily Output	Labor-Hours	Unit	Material	2018 Bare Costs Labor	2018 Bare Costs Equipment	Total	Total Incl O&P
0010	**EXCAVATING, UTILITY TRENCH**									
0011	Common earth									
0050	Trenching with chain trencher, 12 HP, operator walking									
0100	4" wide trench, 12" deep	B-53	800	.010	L.F.		.51	.08	.59	.86
0150	18" deep		750	.011			.55	.08	.63	.92
0200	24" deep		700	.011			.59	.09	.68	.98
0300	6" wide trench, 12" deep		650	.012			.63	.10	.73	1.05
0350	18" deep		600	.013			.68	.10	.78	1.14
0400	24" deep		550	.015			.75	.11	.86	1.25
0450	36" deep		450	.018			.91	.14	1.05	1.53
0600	8" wide trench, 12" deep		475	.017			.86	.13	.99	1.44
0650	18" deep		400	.020			1.03	.15	1.18	1.72
0700	24" deep		350	.023			1.17	.18	1.35	1.96
0750	36" deep	↓	300	.027	↓		1.37	.21	1.58	2.29
1000	Backfill by hand including compaction, add									
1050	4" wide trench, 12" deep	A-1G	800	.010	L.F.		.40	.07	.47	.68
1100	18" deep		530	.015			.60	.10	.70	1.03
1150	24" deep		400	.020			.80	.13	.93	1.35
1300	6" wide trench, 12" deep		540	.015			.59	.10	.69	1.01
1350	18" deep		405	.020			.79	.13	.92	1.34
1400	24" deep		270	.030			1.18	.19	1.37	2.01
1450	36" deep		180	.044			1.77	.29	2.06	3.02
1600	8" wide trench, 12" deep		400	.020			.80	.13	.93	1.35
1650	18" deep		265	.030			1.20	.20	1.40	2.05
1700	24" deep		200	.040			1.59	.26	1.85	2.72
1750	36" deep	↓	135	.059	↓		2.36	.39	2.75	4.02
2000	Chain trencher, 40 HP operator riding									
2050	6" wide trench and backfill, 12" deep	B-54	1200	.007	L.F.		.34	.28	.62	.83
2100	18" deep	↓	1000	.008	↓		.41	.34	.75	.99

31 23 Excavation and Fill

31 23 16 – Excavation

31 23 16.14 Excavating, Utility Trench

		Crew	Daily Output	Labor-Hours	Unit	Material	2018 Bare Costs Labor	Equipment	Total	Total Incl O&P
2150	24" deep	B-54	975	.008	L.F.		.42	.35	.77	1.02
2200	36" deep		900	.009			.46	.37	.83	1.10
2250	48" deep		750	.011			.55	.45	1	1.32
2300	60" deep		650	.012			.63	.52	1.15	1.52
2400	8" wide trench and backfill, 12" deep		1000	.008			.41	.34	.75	.99
2450	18" deep		950	.008			.43	.35	.78	1.04
2500	24" deep		900	.009			.46	.37	.83	1.10
2550	36" deep		800	.010			.51	.42	.93	1.23
2600	48" deep		650	.012			.63	.52	1.15	1.52
2700	12" wide trench and backfill, 12" deep		975	.008			.42	.35	.77	1.02
2750	18" deep		860	.009			.48	.39	.87	1.15
2800	24" deep		800	.010			.51	.42	.93	1.23
2850	36" deep		725	.011			.57	.46	1.03	1.36
3000	16" wide trench and backfill, 12" deep		835	.010			.49	.40	.89	1.18
3050	18" deep		750	.011			.55	.45	1	1.32
3100	24" deep	↓	700	.011	↓		.59	.48	1.07	1.41
3200	Compaction with vibratory plate, add								35%	35%
5100	Hand excavate and trim for pipe bells after trench excavation									
5200	8" pipe	1 Clab	155	.052	L.F.		2.06		2.06	3.13
5300	18" pipe	"	130	.062	"		2.45		2.45	3.74

31 23 16.15 Excavating, Utility Trench, Plow

		Crew	Daily Output	Labor-Hours	Unit	Material	2018 Bare Costs Labor	Equipment	Total	Total Incl O&P
0010	**EXCAVATING, UTILITY TRENCH, PLOW**									
0100	Single cable, plowed into fine material	B-54C	3800	.004	L.F.		.20	.29	.49	.62
0200	Two cable		3200	.005			.23	.34	.57	.73
0300	Single cable, plowed into coarse material	↓	2000	.008	↓		.37	.55	.92	1.17

31 23 16.16 Structural Excavation for Minor Structures

		Crew	Daily Output	Labor-Hours	Unit	Material	2018 Bare Costs Labor	Equipment	Total	Total Incl O&P
0010	**STRUCTURAL EXCAVATION FOR MINOR STRUCTURES**									
0015	Hand, pits to 6' deep, sandy soil	1 Clab	8	1	B.C.Y.		40		40	60.50
0100	Heavy soil or clay		4	2			79.50		79.50	121
0300	Pits 6' to 12' deep, sandy soil		5	1.600			64		64	97
0500	Heavy soil or clay		3	2.667			106		106	162
0700	Pits 12' to 18' deep, sandy soil		4	2			79.50		79.50	121
0900	Heavy soil or clay	↓	2	4			159		159	243
6030	Common earth, hydraulic backhoe, 1/2 C.Y. bucket	B-12E	55	.291			13.95	7.90	21.85	29.50
6035	3/4 C.Y. bucket	B-12F	90	.178			8.55	7.45	16	21
6040	1 C.Y. bucket	B-12A	108	.148			7.10	6.85	13.95	18.30
6050	1-1/2 C.Y. bucket	B-12B	144	.111			5.35	6.20	11.55	14.90
6060	2 C.Y. bucket	B-12C	200	.080			3.84	5.25	9.09	11.60
6070	Sand and gravel, 3/4 C.Y. bucket	B-12F	100	.160			7.70	6.70	14.40	18.95
6080	1 C.Y. bucket	B-12A	120	.133			6.40	6.20	12.60	16.50
6090	1-1/2 C.Y. bucket	B-12B	160	.100			4.80	5.60	10.40	13.40
6100	2 C.Y. bucket	B-12C	220	.073			3.49	4.78	8.27	10.55
6110	Clay, till, or blasted rock, 3/4 C.Y. bucket	B-12F	80	.200			9.60	8.35	17.95	24
6120	1 C.Y. bucket	B-12A	95	.168			8.10	7.80	15.90	21
6130	1-1/2 C.Y. bucket	B-12B	130	.123			5.90	6.85	12.75	16.50
6140	2 C.Y. bucket	B-12C	175	.091			4.39	6	10.39	13.25
6230	Sandy clay & loam, hydraulic backhoe, 1/2 C.Y. bucket	B-12E	60	.267			12.80	7.25	20.05	27.50
6235	3/4 C.Y. bucket	B-12F	98	.163			7.85	6.85	14.70	19.35
6240	1 C.Y. bucket	B-12A	116	.138			6.60	6.40	13	17.05
6250	1-1/2 C.Y. bucket	B-12B	156	.103	↓		4.92	5.75	10.67	13.75

For customer support on your Electrical Costs with RSMeans data, call 800.448.8182.

341

31 23 19 — Dewatering

31 23 19.20 Dewatering Systems

		Crew	Daily Output	Labor-Hours	Unit	Material	2018 Bare Costs Labor	Equipment	Total	Total Incl O&P
0010	**DEWATERING SYSTEMS**									
0020	Excavate drainage trench, 2' wide, 2' deep	B-11C	90	.178	C.Y.		8.30	3.47	11.77	16.40
0100	2' wide, 3' deep, with backhoe loader	"	135	.119			5.55	2.31	7.86	10.95
0200	Excavate sump pits by hand, light soil	1 Clab	7.10	1.127			45		45	68.50
0300	Heavy soil	"	3.50	2.286	↓		91		91	139
0500	Pumping 8 hrs., attended 2 hrs./day, incl. 20 L.F.									
0550	of suction hose & 100 L.F. discharge hose									
0600	2" diaphragm pump used for 8 hrs.	B-10H	4	3	Day		147	19.20	166.20	244
0650	4" diaphragm pump used for 8 hrs.	B-10I	4	3			147	30.50	177.50	257
0800	8 hrs. attended, 2" diaphragm pump	B-10H	1	12			590	77	667	975
0900	3" centrifugal pump	B-10J	1	12			590	85	675	985
1000	4" diaphragm pump	B-10I	1	12			590	123	713	1,025
1100	6" centrifugal pump	B-10K	1	12	↓		590	320	910	1,250

31 23 23 — Fill

31 23 23.13 Backfill

		Crew	Daily Output	Labor-Hours	Unit	Material	2018 Bare Costs Labor	Equipment	Total	Total Incl O&P
0010	**BACKFILL**									
0015	By hand, no compaction, light soil	1 Clab	14	.571	L.C.Y.		23		23	34.50
0100	Heavy soil		11	.727	"		29		29	44
0300	Compaction in 6" layers, hand tamp, add to above	↓	20.60	.388	E.C.Y.		15.50		15.50	23.50
0400	Roller compaction operator walking, add	B-10A	100	.120			5.90	1.80	7.70	10.90
0500	Air tamp, add	B-9D	190	.211			8.45	1.42	9.87	14.45
0600	Vibrating plate, add	A-1D	60	.133			5.30	.53	5.83	8.70
0800	Compaction in 12" layers, hand tamp, add to above	1 Clab	34	.235			9.40		9.40	14.30
0900	Roller compaction operator walking, add	B-10A	150	.080			3.93	1.20	5.13	7.25
1000	Air tamp, add	B-9	285	.140			5.65	.82	6.47	9.50
1100	Vibrating plate, add	A-1E	90	.089	↓		3.54	.45	3.99	5.90
1300	Dozer backfilling, bulk, up to 300' haul, no compaction	B-10B	1200	.010	L.C.Y.		.49	1.06	1.55	1.91
1400	Air tamped, add	B-11B	80	.200	E.C.Y.		9.10	3.56	12.66	17.70
1600	Compacting backfill, 6" to 12" lifts, vibrating roller	B-10C	800	.015			.74	2.12	2.86	3.44
1700	Sheepsfoot roller	B-10D	750	.016	↓		.79	2.27	3.06	3.68
1900	Dozer backfilling, trench, up to 300' haul, no compaction	B-10B	900	.013	L.C.Y.		.65	1.41	2.06	2.55
2000	Air tamped, add	B-11B	80	.200	E.C.Y.		9.10	3.56	12.66	17.70
2200	Compacting backfill, 6" to 12" lifts, vibrating roller	B-10C	700	.017			.84	2.42	3.26	3.93
2300	Sheepsfoot roller	B-10D	650	.018			.91	2.61	3.52	4.24

31 23 23.16 Fill By Borrow and Utility Bedding

		Crew	Daily Output	Labor-Hours	Unit	Material	2018 Bare Costs Labor	Equipment	Total	Total Incl O&P
0010	**FILL BY BORROW AND UTILITY BEDDING**									
0049	Utility bedding, for pipe & conduit, not incl. compaction									
0050	Crushed or screened bank run gravel	B-6	150	.160	L.C.Y.	21	7	2.08	30.08	36
0100	Crushed stone 3/4" to 1/2"		150	.160		28	7	2.08	37.08	43.50
0200	Sand, dead or bank	↓	150	.160	↓	18.15	7	2.08	27.23	33
0500	Compacting bedding in trench	A-1D	90	.089	E.C.Y.		3.54	.35	3.89	5.80
0610	See Section 31 23 23.20 for hauling mileage add.									

31 23 23.20 Hauling

		Crew	Daily Output	Labor-Hours	Unit	Material	2018 Bare Costs Labor	Equipment	Total	Total Incl O&P
0010	**HAULING**									
0011	Excavated or borrow, loose cubic yards									
0012	no loading equipment, including hauling, waiting, loading/dumping									
0013	time per cycle (wait, load, travel, unload or dump & return)									
0014	8 C.Y. truck, 15 MPH avg., cycle 0.5 miles, 10 min. wait/ld./uld.	B-34A	320	.025	L.C.Y.		1.15	1.06	2.21	2.89
0016	cycle 1 mile		272	.029			1.35	1.24	2.59	3.41
0018	cycle 2 miles		208	.038			1.77	1.63	3.40	4.46
0020	cycle 4 miles	↓	144	.056	↓		2.56	2.35	4.91	6.45

31 23 23.20 Hauling

		Crew	Daily Output	Labor-Hours	Unit	Material	2018 Bare Costs Labor	2018 Bare Costs Equipment	Total	Total Incl O&P
0022	cycle 6 miles	B-34A	112	.071	L.C.Y.		3.29	3.02	6.31	8.30
0024	cycle 8 miles		88	.091			4.18	3.85	8.03	10.55
0026	20 MPH avg., cycle 0.5 mile		336	.024			1.10	1.01	2.11	2.76
0028	cycle 1 mile		296	.027			1.24	1.14	2.38	3.13
0030	cycle 2 miles		240	.033			1.53	1.41	2.94	3.86
0032	cycle 4 miles		176	.045			2.09	1.92	4.01	5.25
0034	cycle 6 miles		136	.059			2.71	2.49	5.20	6.80
0036	cycle 8 miles		112	.071			3.29	3.02	6.31	8.30
0044	25 MPH avg., cycle 4 miles		192	.042			1.92	1.76	3.68	4.83
0046	cycle 6 miles		160	.050			2.30	2.12	4.42	5.80
0048	cycle 8 miles		128	.063			2.88	2.65	5.53	7.25
0050	30 MPH avg., cycle 4 miles		216	.037			1.70	1.57	3.27	4.29
0052	cycle 6 miles		176	.045			2.09	1.92	4.01	5.25
0054	cycle 8 miles		144	.056			2.56	2.35	4.91	6.45
0114	15 MPH avg., cycle 0.5 mile, 15 min. wait/ld./uld.		224	.036			1.64	1.51	3.15	4.13
0116	cycle 1 mile		200	.040			1.84	1.69	3.53	4.63
0118	cycle 2 miles		168	.048			2.19	2.02	4.21	5.50
0120	cycle 4 miles		120	.067			3.07	2.82	5.89	7.70
0122	cycle 6 miles		96	.083			3.83	3.53	7.36	9.65
0124	cycle 8 miles		80	.100			4.60	4.23	8.83	11.60
0126	20 MPH avg., cycle 0.5 mile		232	.034			1.59	1.46	3.05	4
0128	cycle 1 mile		208	.038			1.77	1.63	3.40	4.46
0130	cycle 2 miles		184	.043			2	1.84	3.84	5.05
0132	cycle 4 miles		144	.056			2.56	2.35	4.91	6.45
0134	cycle 6 miles		112	.071			3.29	3.02	6.31	8.30
0136	cycle 8 miles		96	.083			3.83	3.53	7.36	9.65
0144	25 MPH avg., cycle 4 miles		152	.053			2.42	2.23	4.65	6.10
0146	cycle 6 miles		128	.063			2.88	2.65	5.53	7.25
0148	cycle 8 miles		112	.071			3.29	3.02	6.31	8.30
0150	30 MPH avg., cycle 4 miles		168	.048			2.19	2.02	4.21	5.50
0152	cycle 6 miles		144	.056			2.56	2.35	4.91	6.45
0154	cycle 8 miles		120	.067			3.07	2.82	5.89	7.70
0214	15 MPH avg., cycle 0.5 mile, 20 min. wait/ld./uld.		176	.045			2.09	1.92	4.01	5.25
0216	cycle 1 mile		160	.050			2.30	2.12	4.42	5.80
0218	cycle 2 miles		136	.059			2.71	2.49	5.20	6.80
0220	cycle 4 miles		104	.077			3.54	3.26	6.80	8.95
0222	cycle 6 miles		88	.091			4.18	3.85	8.03	10.55
0224	cycle 8 miles		72	.111			5.10	4.70	9.80	12.85
0226	20 MPH avg., cycle 0.5 mile		176	.045			2.09	1.92	4.01	5.25
0228	cycle 1 mile		168	.048			2.19	2.02	4.21	5.50
0230	cycle 2 miles		144	.056			2.56	2.35	4.91	6.45
0232	cycle 4 miles		120	.067			3.07	2.82	5.89	7.70
0234	cycle 6 miles		96	.083			3.83	3.53	7.36	9.65
0236	cycle 8 miles		88	.091			4.18	3.85	8.03	10.55
0244	25 MPH avg., cycle 4 miles		128	.063			2.88	2.65	5.53	7.25
0246	cycle 6 miles		112	.071			3.29	3.02	6.31	8.30
0248	cycle 8 miles		96	.083			3.83	3.53	7.36	9.65
0250	30 MPH avg., cycle 4 miles		136	.059			2.71	2.49	5.20	6.80
0252	cycle 6 miles		120	.067			3.07	2.82	5.89	7.70
0254	cycle 8 miles		104	.077			3.54	3.26	6.80	8.95
0314	15 MPH avg., cycle 0.5 mile, 25 min. wait/ld./uld.		144	.056			2.56	2.35	4.91	6.45
0316	cycle 1 mile		128	.063			2.88	2.65	5.53	7.25
0318	cycle 2 miles		112	.071			3.29	3.02	6.31	8.30

31 23 23.20 Hauling		Crew	Daily Output	Labor-Hours	Unit	Material	2018 Bare Costs Labor	Equipment	Total	Total Incl O&P
0320	cycle 4 miles	B-34A	96	.083	L.C.Y.		3.83	3.53	7.36	9.65
0322	cycle 6 miles		80	.100			4.60	4.23	8.83	11.60
0324	cycle 8 miles		64	.125			5.75	5.30	11.05	14.45
0326	20 MPH avg., cycle 0.5 mile		144	.056			2.56	2.35	4.91	6.45
0328	cycle 1 mile		136	.059			2.71	2.49	5.20	6.80
0330	cycle 2 miles		120	.067			3.07	2.82	5.89	7.70
0332	cycle 4 miles		104	.077			3.54	3.26	6.80	8.95
0334	cycle 6 miles		88	.091			4.18	3.85	8.03	10.55
0336	cycle 8 miles		80	.100			4.60	4.23	8.83	11.60
0344	25 MPH avg., cycle 4 miles		112	.071			3.29	3.02	6.31	8.30
0346	cycle 6 miles		96	.083			3.83	3.53	7.36	9.65
0348	cycle 8 miles		88	.091			4.18	3.85	8.03	10.55
0350	30 MPH avg., cycle 4 miles		112	.071			3.29	3.02	6.31	8.30
0352	cycle 6 miles		104	.077			3.54	3.26	6.80	8.95
0354	cycle 8 miles		96	.083			3.83	3.53	7.36	9.65
0414	15 MPH avg., cycle 0.5 mile, 30 min. wait/ld./uld.		120	.067			3.07	2.82	5.89	7.70
0416	cycle 1 mile		112	.071			3.29	3.02	6.31	8.30
0418	cycle 2 miles		96	.083			3.83	3.53	7.36	9.65
0420	cycle 4 miles		80	.100			4.60	4.23	8.83	11.60
0422	cycle 6 miles		72	.111			5.10	4.70	9.80	12.85
0424	cycle 8 miles		64	.125			5.75	5.30	11.05	14.45
0426	20 MPH avg., cycle 0.5 mile		120	.067			3.07	2.82	5.89	7.70
0428	cycle 1 mile		112	.071			3.29	3.02	6.31	8.30
0430	cycle 2 miles		104	.077			3.54	3.26	6.80	8.95
0432	cycle 4 miles		88	.091			4.18	3.85	8.03	10.55
0434	cycle 6 miles		80	.100			4.60	4.23	8.83	11.60
0436	cycle 8 miles		72	.111			5.10	4.70	9.80	12.85
0444	25 MPH avg., cycle 4 miles		96	.083			3.83	3.53	7.36	9.65
0446	cycle 6 miles		88	.091			4.18	3.85	8.03	10.55
0448	cycle 8 miles		80	.100			4.60	4.23	8.83	11.60
0450	30 MPH avg., cycle 4 miles		96	.083			3.83	3.53	7.36	9.65
0452	cycle 6 miles		88	.091			4.18	3.85	8.03	10.55
0454	cycle 8 miles		80	.100			4.60	4.23	8.83	11.60
0514	15 MPH avg., cycle 0.5 mile, 35 min. wait/ld./uld.		104	.077			3.54	3.26	6.80	8.95
0516	cycle 1 mile		96	.083			3.83	3.53	7.36	9.65
0518	cycle 2 miles		88	.091			4.18	3.85	8.03	10.55
0520	cycle 4 miles		72	.111			5.10	4.70	9.80	12.85
0522	cycle 6 miles		64	.125			5.75	5.30	11.05	14.45
0524	cycle 8 miles		56	.143			6.55	6.05	12.60	16.55
0526	20 MPH avg., cycle 0.5 mile		104	.077			3.54	3.26	6.80	8.95
0528	cycle 1 mile		96	.083			3.83	3.53	7.36	9.65
0530	cycle 2 miles		96	.083			3.83	3.53	7.36	9.65
0532	cycle 4 miles		80	.100			4.60	4.23	8.83	11.60
0534	cycle 6 miles		72	.111			5.10	4.70	9.80	12.85
0536	cycle 8 miles		64	.125			5.75	5.30	11.05	14.45
0544	25 MPH avg., cycle 4 miles		88	.091			4.18	3.85	8.03	10.55
0546	cycle 6 miles		80	.100			4.60	4.23	8.83	11.60
0548	cycle 8 miles		72	.111			5.10	4.70	9.80	12.85
0550	30 MPH avg., cycle 4 miles		88	.091			4.18	3.85	8.03	10.55
0552	cycle 6 miles		80	.100			4.60	4.23	8.83	11.60
0554	cycle 8 miles		72	.111			5.10	4.70	9.80	12.85
1014	12 C.Y. truck, cycle 0.5 mile, 15 MPH avg., 15 min. wait/ld./uld.	B-34B	336	.024			1.10	1.62	2.72	3.43
1016	cycle 1 mile		300	.027			1.23	1.81	3.04	3.84

31 23 23.20 Hauling		Crew	Daily Output	Labor-Hours	Unit	Material	2018 Bare Costs Labor	Equipment	Total	Total Incl O&P
1018	cycle 2 miles	B-34B	252	.032	L.C.Y.		1.46	2.15	3.61	4.57
1020	cycle 4 miles		180	.044			2.04	3.02	5.06	6.40
1022	cycle 6 miles		144	.056			2.56	3.77	6.33	8
1024	cycle 8 miles		120	.067			3.07	4.52	7.59	9.60
1025	cycle 10 miles		96	.083			3.83	5.65	9.48	11.95
1026	20 MPH avg., cycle 0.5 mile		348	.023			1.06	1.56	2.62	3.31
1028	cycle 1 mile		312	.026			1.18	1.74	2.92	3.69
1030	cycle 2 miles		276	.029			1.33	1.97	3.30	4.17
1032	cycle 4 miles		216	.037			1.70	2.51	4.21	5.35
1034	cycle 6 miles		168	.048			2.19	3.23	5.42	6.85
1036	cycle 8 miles		144	.056			2.56	3.77	6.33	8
1038	cycle 10 miles		120	.067			3.07	4.52	7.59	9.60
1040	25 MPH avg., cycle 4 miles		228	.035			1.61	2.38	3.99	5.05
1042	cycle 6 miles		192	.042			1.92	2.83	4.75	6
1044	cycle 8 miles		168	.048			2.19	3.23	5.42	6.85
1046	cycle 10 miles		144	.056			2.56	3.77	6.33	8
1050	30 MPH avg., cycle 4 miles		252	.032			1.46	2.15	3.61	4.57
1052	cycle 6 miles		216	.037			1.70	2.51	4.21	5.35
1054	cycle 8 miles		180	.044			2.04	3.02	5.06	6.40
1056	cycle 10 miles		156	.051			2.36	3.48	5.84	7.40
1060	35 MPH avg., cycle 4 miles		264	.030			1.39	2.06	3.45	4.36
1062	cycle 6 miles		228	.035			1.61	2.38	3.99	5.05
1064	cycle 8 miles		204	.039			1.80	2.66	4.46	5.65
1066	cycle 10 miles		180	.044			2.04	3.02	5.06	6.40
1068	cycle 20 miles		120	.067			3.07	4.52	7.59	9.60
1069	cycle 30 miles		84	.095			4.38	6.45	10.83	13.70
1070	cycle 40 miles		72	.111			5.10	7.55	12.65	16
1072	40 MPH avg., cycle 6 miles		240	.033			1.53	2.26	3.79	4.80
1074	cycle 8 miles		216	.037			1.70	2.51	4.21	5.35
1076	cycle 10 miles		192	.042			1.92	2.83	4.75	6
1078	cycle 20 miles		120	.067			3.07	4.52	7.59	9.60
1080	cycle 30 miles		96	.083			3.83	5.65	9.48	11.95
1082	cycle 40 miles		72	.111			5.10	7.55	12.65	16
1084	cycle 50 miles		60	.133			6.15	9.05	15.20	19.20
1094	45 MPH avg., cycle 8 miles		216	.037			1.70	2.51	4.21	5.35
1096	cycle 10 miles		204	.039			1.80	2.66	4.46	5.65
1098	cycle 20 miles		132	.061			2.79	4.11	6.90	8.70
1100	cycle 30 miles		108	.074			3.41	5.05	8.46	10.70
1102	cycle 40 miles		84	.095			4.38	6.45	10.83	13.70
1104	cycle 50 miles		72	.111			5.10	7.55	12.65	16
1106	50 MPH avg., cycle 10 miles		216	.037			1.70	2.51	4.21	5.35
1108	cycle 20 miles		144	.056			2.56	3.77	6.33	8
1110	cycle 30 miles		108	.074			3.41	5.05	8.46	10.70
1112	cycle 40 miles		84	.095			4.38	6.45	10.83	13.70
1114	cycle 50 miles		72	.111			5.10	7.55	12.65	16
1214	15 MPH avg., cycle 0.5 mile, 20 min. wait/ld./uld.		264	.030			1.39	2.06	3.45	4.36
1216	cycle 1 mile		240	.033			1.53	2.26	3.79	4.80
1218	cycle 2 miles		204	.039			1.80	2.66	4.46	5.65
1220	cycle 4 miles		156	.051			2.36	3.48	5.84	7.40
1222	cycle 6 miles		132	.061			2.79	4.11	6.90	8.70
1224	cycle 8 miles		108	.074			3.41	5.05	8.46	10.70
1225	cycle 10 miles		96	.083			3.83	5.65	9.48	11.95
1226	20 MPH avg., cycle 0.5 mile		264	.030			1.39	2.06	3.45	4.36

31 23 23 – Fill

31 23 23.20 Hauling		Crew	Daily Output	Labor-Hours	Unit	Material	2018 Bare Costs Labor	Equipment	Total	Total Incl O&P
1228	cycle 1 mile	B-34B	252	.032	L.C.Y.		1.46	2.15	3.61	4.57
1230	cycle 2 miles		216	.037			1.70	2.51	4.21	5.35
1232	cycle 4 miles		180	.044			2.04	3.02	5.06	6.40
1234	cycle 6 miles		144	.056			2.56	3.77	6.33	8
1236	cycle 8 miles		132	.061			2.79	4.11	6.90	8.70
1238	cycle 10 miles		108	.074			3.41	5.05	8.46	10.70
1240	25 MPH avg., cycle 4 miles		192	.042			1.92	2.83	4.75	6
1242	cycle 6 miles		168	.048			2.19	3.23	5.42	6.85
1244	cycle 8 miles		144	.056			2.56	3.77	6.33	8
1246	cycle 10 miles		132	.061			2.79	4.11	6.90	8.70
1250	30 MPH avg., cycle 4 miles		204	.039			1.80	2.66	4.46	5.65
1252	cycle 6 miles		180	.044			2.04	3.02	5.06	6.40
1254	cycle 8 miles		156	.051			2.36	3.48	5.84	7.40
1256	cycle 10 miles		144	.056			2.56	3.77	6.33	8
1260	35 MPH avg., cycle 4 miles		216	.037			1.70	2.51	4.21	5.35
1262	cycle 6 miles		192	.042			1.92	2.83	4.75	6
1264	cycle 8 miles		168	.048			2.19	3.23	5.42	6.85
1266	cycle 10 miles		156	.051			2.36	3.48	5.84	7.40
1268	cycle 20 miles		108	.074			3.41	5.05	8.46	10.70
1269	cycle 30 miles		72	.111			5.10	7.55	12.65	16
1270	cycle 40 miles		60	.133			6.15	9.05	15.20	19.20
1272	40 MPH avg., cycle 6 miles		192	.042			1.92	2.83	4.75	6
1274	cycle 8 miles		180	.044			2.04	3.02	5.06	6.40
1276	cycle 10 miles		156	.051			2.36	3.48	5.84	7.40
1278	cycle 20 miles		108	.074			3.41	5.05	8.46	10.70
1280	cycle 30 miles		84	.095			4.38	6.45	10.83	13.70
1282	cycle 40 miles		72	.111			5.10	7.55	12.65	16
1284	cycle 50 miles		60	.133			6.15	9.05	15.20	19.20
1294	45 MPH avg., cycle 8 miles		180	.044			2.04	3.02	5.06	6.40
1296	cycle 10 miles		168	.048			2.19	3.23	5.42	6.85
1298	cycle 20 miles		120	.067			3.07	4.52	7.59	9.60
1300	cycle 30 miles		96	.083			3.83	5.65	9.48	11.95
1302	cycle 40 miles		72	.111			5.10	7.55	12.65	16
1304	cycle 50 miles		60	.133			6.15	9.05	15.20	19.20
1306	50 MPH avg., cycle 10 miles		180	.044			2.04	3.02	5.06	6.40
1308	cycle 20 miles		132	.061			2.79	4.11	6.90	8.70
1310	cycle 30 miles		96	.083			3.83	5.65	9.48	11.95
1312	cycle 40 miles		84	.095			4.38	6.45	10.83	13.70
1314	cycle 50 miles		72	.111			5.10	7.55	12.65	16
1414	15 MPH avg., cycle 0.5 mile, 25 min. wait/ld./uld.		204	.039			1.80	2.66	4.46	5.65
1416	cycle 1 mile		192	.042			1.92	2.83	4.75	6
1418	cycle 2 miles		168	.048			2.19	3.23	5.42	6.85
1420	cycle 4 miles		132	.061			2.79	4.11	6.90	8.70
1422	cycle 6 miles		120	.067			3.07	4.52	7.59	9.60
1424	cycle 8 miles		96	.083			3.83	5.65	9.48	11.95
1425	cycle 10 miles		84	.095			4.38	6.45	10.83	13.70
1426	20 MPH avg., cycle 0.5 mile		216	.037			1.70	2.51	4.21	5.35
1428	cycle 1 mile		204	.039			1.80	2.66	4.46	5.65
1430	cycle 2 miles		180	.044			2.04	3.02	5.06	6.40
1432	cycle 4 miles		156	.051			2.36	3.48	5.84	7.40
1434	cycle 6 miles		132	.061			2.79	4.11	6.90	8.70
1436	cycle 8 miles		120	.067			3.07	4.52	7.59	9.60
1438	cycle 10 miles		96	.083			3.83	5.65	9.48	11.95

31 23 23.20 Hauling		Crew	Daily Output	Labor-Hours	Unit	Material	2018 Bare Costs		Total	Total Incl O&P
							Labor	Equipment		
1440	25 MPH avg., cycle 4 miles	B-34B	168	.048	L.C.Y.		2.19	3.23	5.42	6.85
1442	cycle 6 miles		144	.056			2.56	3.77	6.33	8
1444	cycle 8 miles		132	.061			2.79	4.11	6.90	8.70
1446	cycle 10 miles		108	.074			3.41	5.05	8.46	10.70
1450	30 MPH avg., cycle 4 miles		168	.048			2.19	3.23	5.42	6.85
1452	cycle 6 miles		156	.051			2.36	3.48	5.84	7.40
1454	cycle 8 miles		132	.061			2.79	4.11	6.90	8.70
1456	cycle 10 miles		120	.067			3.07	4.52	7.59	9.60
1460	35 MPH avg., cycle 4 miles		180	.044			2.04	3.02	5.06	6.40
1462	cycle 6 miles		156	.051			2.36	3.48	5.84	7.40
1464	cycle 8 miles		144	.056			2.56	3.77	6.33	8
1466	cycle 10 miles		132	.061			2.79	4.11	6.90	8.70
1468	cycle 20 miles		96	.083			3.83	5.65	9.48	11.95
1469	cycle 30 miles		72	.111			5.10	7.55	12.65	16
1470	cycle 40 miles		60	.133			6.15	9.05	15.20	19.20
1472	40 MPH avg., cycle 6 miles		168	.048			2.19	3.23	5.42	6.85
1474	cycle 8 miles		156	.051			2.36	3.48	5.84	7.40
1476	cycle 10 miles		144	.056			2.56	3.77	6.33	8
1478	cycle 20 miles		96	.083			3.83	5.65	9.48	11.95
1480	cycle 30 miles		84	.095			4.38	6.45	10.83	13.70
1482	cycle 40 miles		60	.133			6.15	9.05	15.20	19.20
1484	cycle 50 miles		60	.133			6.15	9.05	15.20	19.20
1494	45 MPH avg., cycle 8 miles		156	.051			2.36	3.48	5.84	7.40
1496	cycle 10 miles		144	.056			2.56	3.77	6.33	8
1498	cycle 20 miles		108	.074			3.41	5.05	8.46	10.70
1500	cycle 30 miles		84	.095			4.38	6.45	10.83	13.70
1502	cycle 40 miles		72	.111			5.10	7.55	12.65	16
1504	cycle 50 miles		60	.133			6.15	9.05	15.20	19.20
1506	50 MPH avg., cycle 10 miles		156	.051			2.36	3.48	5.84	7.40
1508	cycle 20 miles		120	.067			3.07	4.52	7.59	9.60
1510	cycle 30 miles		96	.083			3.83	5.65	9.48	11.95
1512	cycle 40 miles		72	.111			5.10	7.55	12.65	16
1514	cycle 50 miles		60	.133			6.15	9.05	15.20	19.20
1614	15 MPH avg., cycle 0.5 mile, 30 min. wait/ld./uld.		180	.044			2.04	3.02	5.06	6.40
1616	cycle 1 mile		168	.048			2.19	3.23	5.42	6.85
1618	cycle 2 miles		144	.056			2.56	3.77	6.33	8
1620	cycle 4 miles		120	.067			3.07	4.52	7.59	9.60
1622	cycle 6 miles		108	.074			3.41	5.05	8.46	10.70
1624	cycle 8 miles		84	.095			4.38	6.45	10.83	13.70
1625	cycle 10 miles		84	.095			4.38	6.45	10.83	13.70
1626	20 MPH avg., cycle 0.5 mile		180	.044			2.04	3.02	5.06	6.40
1628	cycle 1 mile		168	.048			2.19	3.23	5.42	6.85
1630	cycle 2 miles		156	.051			2.36	3.48	5.84	7.40
1632	cycle 4 miles		132	.061			2.79	4.11	6.90	8.70
1634	cycle 6 miles		120	.067			3.07	4.52	7.59	9.60
1636	cycle 8 miles		108	.074			3.41	5.05	8.46	10.70
1638	cycle 10 miles		96	.083			3.83	5.65	9.48	11.95
1640	25 MPH avg., cycle 4 miles		144	.056			2.56	3.77	6.33	8
1642	cycle 6 miles		132	.061			2.79	4.11	6.90	8.70
1644	cycle 8 miles		108	.074			3.41	5.05	8.46	10.70
1646	cycle 10 miles		108	.074			3.41	5.05	8.46	10.70
1650	30 MPH avg., cycle 4 miles		144	.056			2.56	3.77	6.33	8
1652	cycle 6 miles		132	.061			2.79	4.11	6.90	8.70

31 23 23.20 Hauling		Crew	Daily Output	Labor-Hours	Unit	Material	2018 Bare Costs Labor	2018 Bare Costs Equipment	Total	Total Incl O&P
1654	cycle 8 miles	B-34B	120	.067	L.C.Y.		3.07	4.52	7.59	9.60
1656	cycle 10 miles		108	.074			3.41	5.05	8.46	10.70
1660	35 MPH avg., cycle 4 miles		156	.051			2.36	3.48	5.84	7.40
1662	cycle 6 miles		144	.056			2.56	3.77	6.33	8
1664	cycle 8 miles		132	.061			2.79	4.11	6.90	8.70
1666	cycle 10 miles		120	.067			3.07	4.52	7.59	9.60
1668	cycle 20 miles		84	.095			4.38	6.45	10.83	13.70
1669	cycle 30 miles		72	.111			5.10	7.55	12.65	16
1670	cycle 40 miles		60	.133			6.15	9.05	15.20	19.20
1672	40 MPH avg., cycle 6 miles		144	.056			2.56	3.77	6.33	8
1674	cycle 8 miles		132	.061			2.79	4.11	6.90	8.70
1676	cycle 10 miles		120	.067			3.07	4.52	7.59	9.60
1678	cycle 20 miles		96	.083			3.83	5.65	9.48	11.95
1680	cycle 30 miles		72	.111			5.10	7.55	12.65	16
1682	cycle 40 miles		60	.133			6.15	9.05	15.20	19.20
1684	cycle 50 miles		48	.167			7.65	11.30	18.95	24
1694	45 MPH avg., cycle 8 miles		144	.056			2.56	3.77	6.33	8
1696	cycle 10 miles		132	.061			2.79	4.11	6.90	8.70
1698	cycle 20 miles		96	.083			3.83	5.65	9.48	11.95
1700	cycle 30 miles		84	.095			4.38	6.45	10.83	13.70
1702	cycle 40 miles		60	.133			6.15	9.05	15.20	19.20
1704	cycle 50 miles		60	.133			6.15	9.05	15.20	19.20
1706	50 MPH avg., cycle 10 miles		132	.061			2.79	4.11	6.90	8.70
1708	cycle 20 miles		108	.074			3.41	5.05	8.46	10.70
1710	cycle 30 miles		84	.095			4.38	6.45	10.83	13.70
1712	cycle 40 miles		72	.111			5.10	7.55	12.65	16
1714	cycle 50 miles		60	.133			6.15	9.05	15.20	19.20
2000	Hauling, 8 C.Y. truck, small project cost per hour	B-34A	8	1	Hr.		46	42.50	88.50	116
2100	12 C.Y. truck	B-34B	8	1			46	68	114	144
2150	16.5 C.Y. truck	B-34C	8	1			46	76	122	154
2175	18 C.Y. 8 wheel truck	B-34I	8	1			46	88.50	134.50	167
2200	20 C.Y. truck	B-34D	8	1			46	78	124	156
2300	Grading at dump, or embankment if required, by dozer	B-10B	1000	.012	L.C.Y.		.59	1.27	1.86	2.29
2310	Spotter at fill or cut, if required	1 Clab	8	1	Hr.		40		40	60.50
9014	18 C.Y. truck, 8 wheels,15 min. wait/ld./uld.,15 MPH, cycle 0.5 mi.	B-34I	504	.016	L.C.Y.		.73	1.40	2.13	2.64
9016	cycle 1 mile		450	.018			.82	1.57	2.39	2.96
9018	cycle 2 miles		378	.021			.97	1.87	2.84	3.52
9020	cycle 4 miles		270	.030			1.36	2.61	3.97	4.93
9022	cycle 6 miles		216	.037			1.70	3.27	4.97	6.15
9024	cycle 8 miles		180	.044			2.04	3.92	5.96	7.40
9025	cycle 10 miles		144	.056			2.56	4.90	7.46	9.25
9026	20 MPH avg., cycle 0.5 mile		522	.015			.71	1.35	2.06	2.55
9028	cycle 1 mile		468	.017			.79	1.51	2.30	2.84
9030	cycle 2 miles		414	.019			.89	1.71	2.60	3.22
9032	cycle 4 miles		324	.025			1.14	2.18	3.32	4.11
9034	cycle 6 miles		252	.032			1.46	2.80	4.26	5.30
9036	cycle 8 miles		216	.037			1.70	3.27	4.97	6.15
9038	cycle 10 miles		180	.044			2.04	3.92	5.96	7.40
9040	25 MPH avg., cycle 4 miles		342	.023			1.08	2.06	3.14	3.89
9042	cycle 6 miles		288	.028			1.28	2.45	3.73	4.63
9044	cycle 8 miles		252	.032			1.46	2.80	4.26	5.30
9046	cycle 10 miles		216	.037			1.70	3.27	4.97	6.15
9050	30 MPH avg., cycle 4 miles		378	.021			.97	1.87	2.84	3.52

31 23 23.20 Hauling		Crew	Daily Output	Labor-Hours	Unit	Material	2018 Bare Costs Labor	Equipment	Total	Total Incl O&P
9052	cycle 6 miles	B-34I	324	.025	L.C.Y.		1.14	2.18	3.32	4.11
9054	cycle 8 miles		270	.030			1.36	2.61	3.97	4.93
9056	cycle 10 miles		234	.034			1.57	3.02	4.59	5.70
9060	35 MPH avg., cycle 4 miles		396	.020			.93	1.78	2.71	3.36
9062	cycle 6 miles		342	.023			1.08	2.06	3.14	3.89
9064	cycle 8 miles		288	.028			1.28	2.45	3.73	4.63
9066	cycle 10 miles		270	.030			1.36	2.61	3.97	4.93
9068	cycle 20 miles		162	.049			2.27	4.36	6.63	8.20
9070	cycle 30 miles		126	.063			2.92	5.60	8.52	10.55
9072	cycle 40 miles		90	.089			4.09	7.85	11.94	14.80
9074	40 MPH avg., cycle 6 miles		360	.022			1.02	1.96	2.98	3.70
9076	cycle 8 miles		324	.025			1.14	2.18	3.32	4.11
9078	cycle 10 miles		288	.028			1.28	2.45	3.73	4.63
9080	cycle 20 miles		180	.044			2.04	3.92	5.96	7.40
9082	cycle 30 miles		144	.056			2.56	4.90	7.46	9.25
9084	cycle 40 miles		108	.074			3.41	6.55	9.96	12.35
9086	cycle 50 miles		90	.089			4.09	7.85	11.94	14.80
9094	45 MPH avg., cycle 8 miles		324	.025			1.14	2.18	3.32	4.11
9096	cycle 10 miles		306	.026			1.20	2.31	3.51	4.35
9098	cycle 20 miles		198	.040			1.86	3.57	5.43	6.70
9100	cycle 30 miles		144	.056			2.56	4.90	7.46	9.25
9102	cycle 40 miles		126	.063			2.92	5.60	8.52	10.55
9104	cycle 50 miles		108	.074			3.41	6.55	9.96	12.35
9106	50 MPH avg., cycle 10 miles		324	.025			1.14	2.18	3.32	4.11
9108	cycle 20 miles		216	.037			1.70	3.27	4.97	6.15
9110	cycle 30 miles		162	.049			2.27	4.36	6.63	8.20
9112	cycle 40 miles		126	.063			2.92	5.60	8.52	10.55
9114	cycle 50 miles		108	.074			3.41	6.55	9.96	12.35
9214	20 min. wait/ld./uld.,15 MPH, cycle 0.5 mi.		396	.020			.93	1.78	2.71	3.36
9216	cycle 1 mile		360	.022			1.02	1.96	2.98	3.70
9218	cycle 2 miles		306	.026			1.20	2.31	3.51	4.35
9220	cycle 4 miles		234	.034			1.57	3.02	4.59	5.70
9222	cycle 6 miles		198	.040			1.86	3.57	5.43	6.70
9224	cycle 8 miles		162	.049			2.27	4.36	6.63	8.20
9225	cycle 10 miles		144	.056			2.56	4.90	7.46	9.25
9226	20 MPH avg., cycle 0.5 mile		396	.020			.93	1.78	2.71	3.36
9228	cycle 1 mile		378	.021			.97	1.87	2.84	3.52
9230	cycle 2 miles		324	.025			1.14	2.18	3.32	4.11
9232	cycle 4 miles		270	.030			1.36	2.61	3.97	4.93
9234	cycle 6 miles		216	.037			1.70	3.27	4.97	6.15
9236	cycle 8 miles		198	.040			1.86	3.57	5.43	6.70
9238	cycle 10 miles		162	.049			2.27	4.36	6.63	8.20
9240	25 MPH avg., cycle 4 miles		288	.028			1.28	2.45	3.73	4.63
9242	cycle 6 miles		252	.032			1.46	2.80	4.26	5.30
9244	cycle 8 miles		216	.037			1.70	3.27	4.97	6.15
9246	cycle 10 miles		198	.040			1.86	3.57	5.43	6.70
9250	30 MPH avg., cycle 4 miles		306	.026			1.20	2.31	3.51	4.35
9252	cycle 6 miles		270	.030			1.36	2.61	3.97	4.93
9254	cycle 8 miles		234	.034			1.57	3.02	4.59	5.70
9256	cycle 10 miles		216	.037			1.70	3.27	4.97	6.15
9260	35 MPH avg., cycle 4 miles		324	.025			1.14	2.18	3.32	4.11
9262	cycle 6 miles		288	.028			1.28	2.45	3.73	4.63
9264	cycle 8 miles		252	.032			1.46	2.80	4.26	5.30

For customer support on your Electrical Costs with RSMeans data, call 800.448.8182.

349

31 23 23.20 Hauling		Crew	Daily Output	Labor-Hours	Unit	Material	2018 Bare Costs Labor	2018 Bare Costs Equipment	Total	Total Incl O&P
9266	cycle 10 miles	B-34I	234	.034	L.C.Y.		1.57	3.02	4.59	5.70
9268	cycle 20 miles		162	.049			2.27	4.36	6.63	8.20
9270	cycle 30 miles		108	.074			3.41	6.55	9.96	12.35
9272	cycle 40 miles		90	.089			4.09	7.85	11.94	14.80
9274	40 MPH avg., cycle 6 miles		288	.028			1.28	2.45	3.73	4.63
9276	cycle 8 miles		270	.030			1.36	2.61	3.97	4.93
9278	cycle 10 miles		234	.034			1.57	3.02	4.59	5.70
9280	cycle 20 miles		162	.049			2.27	4.36	6.63	8.20
9282	cycle 30 miles		126	.063			2.92	5.60	8.52	10.55
9284	cycle 40 miles		108	.074			3.41	6.55	9.96	12.35
9286	cycle 50 miles		90	.089			4.09	7.85	11.94	14.80
9294	45 MPH avg., cycle 8 miles		270	.030			1.36	2.61	3.97	4.93
9296	cycle 10 miles		252	.032			1.46	2.80	4.26	5.30
9298	cycle 20 miles		180	.044			2.04	3.92	5.96	7.40
9300	cycle 30 miles		144	.056			2.56	4.90	7.46	9.25
9302	cycle 40 miles		108	.074			3.41	6.55	9.96	12.35
9304	cycle 50 miles		90	.089			4.09	7.85	11.94	14.80
9306	50 MPH avg., cycle 10 miles		270	.030			1.36	2.61	3.97	4.93
9308	cycle 20 miles		198	.040			1.86	3.57	5.43	6.70
9310	cycle 30 miles		144	.056			2.56	4.90	7.46	9.25
9312	cycle 40 miles		126	.063			2.92	5.60	8.52	10.55
9314	cycle 50 miles		108	.074			3.41	6.55	9.96	12.35
9414	25 min. wait/ld./uld.,15 MPH, cycle 0.5 mi.		306	.026			1.20	2.31	3.51	4.35
9416	cycle 1 mile		288	.028			1.28	2.45	3.73	4.63
9418	cycle 2 miles		252	.032			1.46	2.80	4.26	5.30
9420	cycle 4 miles		198	.040			1.86	3.57	5.43	6.70
9422	cycle 6 miles		180	.044			2.04	3.92	5.96	7.40
9424	cycle 8 miles		144	.056			2.56	4.90	7.46	9.25
9425	cycle 10 miles		126	.063			2.92	5.60	8.52	10.55
9426	20 MPH avg., cycle 0.5 mile		324	.025			1.14	2.18	3.32	4.11
9428	cycle 1 mile		306	.026			1.20	2.31	3.51	4.35
9430	cycle 2 miles		270	.030			1.36	2.61	3.97	4.93
9432	cycle 4 miles		234	.034			1.57	3.02	4.59	5.70
9434	cycle 6 miles		198	.040			1.86	3.57	5.43	6.70
9436	cycle 8 miles		180	.044			2.04	3.92	5.96	7.40
9438	cycle 10 miles		144	.056			2.56	4.90	7.46	9.25
9440	25 MPH avg., cycle 4 miles		252	.032			1.46	2.80	4.26	5.30
9442	cycle 6 miles		216	.037			1.70	3.27	4.97	6.15
9444	cycle 8 miles		198	.040			1.86	3.57	5.43	6.70
9446	cycle 10 miles		180	.044			2.04	3.92	5.96	7.40
9450	30 MPH avg., cycle 4 miles		252	.032			1.46	2.80	4.26	5.30
9452	cycle 6 miles		234	.034			1.57	3.02	4.59	5.70
9454	cycle 8 miles		198	.040			1.86	3.57	5.43	6.70
9456	cycle 10 miles		180	.044			2.04	3.92	5.96	7.40
9460	35 MPH avg., cycle 4 miles		270	.030			1.36	2.61	3.97	4.93
9462	cycle 6 miles		234	.034			1.57	3.02	4.59	5.70
9464	cycle 8 miles		216	.037			1.70	3.27	4.97	6.15
9466	cycle 10 miles		198	.040			1.86	3.57	5.43	6.70
9468	cycle 20 miles		144	.056			2.56	4.90	7.46	9.25
9470	cycle 30 miles		108	.074			3.41	6.55	9.96	12.35
9472	cycle 40 miles		90	.089			4.09	7.85	11.94	14.80
9474	40 MPH avg., cycle 6 miles		252	.032			1.46	2.80	4.26	5.30
9476	cycle 8 miles		234	.034			1.57	3.02	4.59	5.70

31 23 23.20 Hauling		Crew	Daily Output	Labor-Hours	Unit	Material	2018 Bare Costs Labor	2018 Bare Costs Equipment	Total	Total Incl O&P
9478	cycle 10 miles	B-34I	216	.037	L.C.Y.		1.70	3.27	4.97	6.15
9480	cycle 20 miles		144	.056			2.56	4.90	7.46	9.25
9482	cycle 30 miles		126	.063			2.92	5.60	8.52	10.55
9484	cycle 40 miles		90	.089			4.09	7.85	11.94	14.80
9486	cycle 50 miles		90	.089			4.09	7.85	11.94	14.80
9494	45 MPH avg., cycle 8 miles		234	.034			1.57	3.02	4.59	5.70
9496	cycle 10 miles		216	.037			1.70	3.27	4.97	6.15
9498	cycle 20 miles		162	.049			2.27	4.36	6.63	8.20
9500	cycle 30 miles		126	.063			2.92	5.60	8.52	10.55
9502	cycle 40 miles		108	.074			3.41	6.55	9.96	12.35
9504	cycle 50 miles		90	.089			4.09	7.85	11.94	14.80
9506	50 MPH avg., cycle 10 miles		234	.034			1.57	3.02	4.59	5.70
9508	cycle 20 miles		180	.044			2.04	3.92	5.96	7.40
9510	cycle 30 miles		144	.056			2.56	4.90	7.46	9.25
9512	cycle 40 miles		108	.074			3.41	6.55	9.96	12.35
9514	cycle 50 miles		90	.089			4.09	7.85	11.94	14.80
9614	30 min. wait/ld./uld.,15 MPH, cycle 0.5 mi.		270	.030			1.36	2.61	3.97	4.93
9616	cycle 1 mile		252	.032			1.46	2.80	4.26	5.30
9618	cycle 2 miles		216	.037			1.70	3.27	4.97	6.15
9620	cycle 4 miles		180	.044			2.04	3.92	5.96	7.40
9622	cycle 6 miles		162	.049			2.27	4.36	6.63	8.20
9624	cycle 8 miles		126	.063			2.92	5.60	8.52	10.55
9625	cycle 10 miles		126	.063			2.92	5.60	8.52	10.55
9626	20 MPH avg., cycle 0.5 mile		270	.030			1.36	2.61	3.97	4.93
9628	cycle 1 mile		252	.032			1.46	2.80	4.26	5.30
9630	cycle 2 miles		234	.034			1.57	3.02	4.59	5.70
9632	cycle 4 miles		198	.040			1.86	3.57	5.43	6.70
9634	cycle 6 miles		180	.044			2.04	3.92	5.96	7.40
9636	cycle 8 miles		162	.049			2.27	4.36	6.63	8.20
9638	cycle 10 miles		144	.056			2.56	4.90	7.46	9.25
9640	25 MPH avg., cycle 4 miles		216	.037			1.70	3.27	4.97	6.15
9642	cycle 6 miles		198	.040			1.86	3.57	5.43	6.70
9644	cycle 8 miles		180	.044			2.04	3.92	5.96	7.40
9646	cycle 10 miles		162	.049			2.27	4.36	6.63	8.20
9650	30 MPH avg., cycle 4 miles		216	.037			1.70	3.27	4.97	6.15
9652	cycle 6 miles		198	.040			1.86	3.57	5.43	6.70
9654	cycle 8 miles		180	.044			2.04	3.92	5.96	7.40
9656	cycle 10 miles		162	.049			2.27	4.36	6.63	8.20
9660	35 MPH avg., cycle 4 miles		234	.034			1.57	3.02	4.59	5.70
9662	cycle 6 miles		216	.037			1.70	3.27	4.97	6.15
9664	cycle 8 miles		198	.040			1.86	3.57	5.43	6.70
9666	cycle 10 miles		180	.044			2.04	3.92	5.96	7.40
9668	cycle 20 miles		126	.063			2.92	5.60	8.52	10.55
9670	cycle 30 miles		108	.074			3.41	6.55	9.96	12.35
9672	cycle 40 miles		90	.089			4.09	7.85	11.94	14.80
9674	40 MPH avg., cycle 6 miles		216	.037			1.70	3.27	4.97	6.15
9676	cycle 8 miles		198	.040			1.86	3.57	5.43	6.70
9678	cycle 10 miles		180	.044			2.04	3.92	5.96	7.40
9680	cycle 20 miles		144	.056			2.56	4.90	7.46	9.25
9682	cycle 30 miles		108	.074			3.41	6.55	9.96	12.35
9684	cycle 40 miles		90	.089			4.09	7.85	11.94	14.80
9686	cycle 50 miles		72	.111			5.10	9.80	14.90	18.50
9694	45 MPH avg., cycle 8 miles		216	.037			1.70	3.27	4.97	6.15

31 23 23.20 Hauling		Crew	Daily Output	Labor-Hours	Unit	Material	2018 Bare Costs Labor	2018 Bare Costs Equipment	Total	Total Incl O&P
9696	cycle 10 miles	B-34I	198	.040	L.C.Y.		1.86	3.57	5.43	6.70
9698	cycle 20 miles		144	.056			2.56	4.90	7.46	9.25
9700	cycle 30 miles		126	.063			2.92	5.60	8.52	10.55
9702	cycle 40 miles		108	.074			3.41	6.55	9.96	12.35
9704	cycle 50 miles		90	.089			4.09	7.85	11.94	14.80
9706	50 MPH avg., cycle 10 miles		198	.040			1.86	3.57	5.43	6.70
9708	cycle 20 miles		162	.049			2.27	4.36	6.63	8.20
9710	cycle 30 miles		126	.063			2.92	5.60	8.52	10.55
9712	cycle 40 miles		108	.074			3.41	6.55	9.96	12.35
9714	cycle 50 miles		90	.089			4.09	7.85	11.94	14.80
31 23 23.23 Compaction										
0010	**COMPACTION**									
7000	Walk behind, vibrating plate 18" wide, 6" lifts, 2 passes	A-1D	200	.040	E.C.Y.		1.59	.16	1.75	2.60
7020	3 passes		185	.043			1.72	.17	1.89	2.81
7040	4 passes		140	.057			2.28	.23	2.51	3.72

Estimating Tips

32 01 00 Operations and Maintenance of Exterior Improvements

- Recycling of asphalt pavement is becoming very popular and is an alternative to removal and replacement. It can be a good value engineering proposal if removed pavement can be recycled, either at the project site or at another site that is reasonably close to the project site. Sections on repair of flexible and rigid pavement are included.

32 10 00 Bases, Ballasts, and Paving

- When estimating paving, keep in mind the project schedule. Also note that prices for asphalt and concrete are generally higher in the cold seasons. Lines for pavement markings, including tactile warning systems and fence lines, are included.

32 90 00 Planting

- The timing of planting and guarantee specifications often dictate the costs for establishing tree and shrub growth and a stand of grass or ground cover. Establish the work performance schedule to coincide with the local planting season. Maintenance and growth guarantees can add from 20%–100% to the total landscaping cost and can be contractually cumbersome. The cost to replace trees and shrubs can be as high as 5% of the total cost, depending on the planting zone, soil conditions, and time of year.

Reference Numbers

Reference numbers are shown at the beginning of some major classifications. These numbers refer to related items in the Reference Section. The reference information may be an estimating procedure, an alternate pricing method, or technical information.

Note: Not all subdivisions listed here necessarily appear. ■

Did you know?

RSMeans data is available through our online application with 24/7 access:

- Search for unit prices by keyword
- Leverage the most up-to-date data
- Build and export estimates

Try it free for 30 days!
www.rsmeans.com/2018freetrial

32 31 Fences and Gates

32 31 13 – Chain Link Fences and Gates

32 31 13.30 Fence, Chain Link, Gates and Posts	Crew	Daily Output	Labor-Hours	Unit	Material	2018 Bare Costs Labor	Equipment	Total	Total Incl O&P
0010 **FENCE, CHAIN LINK, GATES & POSTS**									
0011 (1/3 post length in ground)									
0013 For Concrete, See Section 03 31									
6710 Corner post, galv. steel, 4" OD, set in conc., 4'	B-80	65	.492	Ea.	61	22	9.40	92.40	110
6715 6'	↓	63	.508		91.50	22.50	9.70	123.70	145
6720 7'		61	.525		108	23.50	10.05	141.55	165
6725 8'	↓	65	.492		122	22	9.40	153.40	177
7900 Auger fence post hole, 3' deep, medium soil, by hand	1 Clab	30	.267			10.65		10.65	16.20
7925 By machine	B-80	175	.183			8.10	3.50	11.60	16.15
7950 Rock, with jackhammer	B-9	32	1.250			50.50	7.30	57.80	84.50
7975 With rock drill	B-47C	65	.246	↓		11.20	23	34.20	42.50

Estimating Tips

33 10 00 Water Utilities
33 30 00 Sanitary Sewerage Utilities
33 40 00 Storm Drainage Utilities

- Never assume that the water, sewer, and drainage lines will go in at the early stages of the project. Consider the site access needs before dividing the site in half with open trenches, loose pipe, and machinery obstructions. Always inspect the site to establish that the site drawings are complete. Check off all existing utilities on your drawings as you locate them. Be especially careful with underground utilities because appurtenances are sometimes buried during regrading or repaving operations. If you find any discrepancies, mark up the site plan for further research. Differing site conditions can be very costly if discovered later in the project.

- See also Section 33 01 00 for restoration of pipe where removal/replacement may be undesirable. Use of new types of piping materials can reduce the overall project cost. Owners/design engineers should consider the installing contractor as a valuable source of current information on utility products and local conditions that could lead to significant cost savings.

Reference Numbers

Reference numbers are shown at the beginning of some major classifications. These numbers refer to related items in the Reference Section. The reference information may be an estimating procedure, an alternate pricing method, or technical information.

Note: Not all subdivisions listed here necessarily appear. ∎

Did you know?

RSMeans data is available through our online application with 24/7 access:

- Search for unit prices by keyword
- Leverage the most up-to-date data
- Build and export estimates

Try it free for 30 days!
www.rsmeans.com/2018freetrial

33 05 07 – Trenchless Installation of Utility Piping

33 05 07.13 Utility Directional Drilling

33 05 07.13 Utility Directional Drilling	Crew	Daily Output	Labor-Hours	Unit	Material	2018 Bare Costs Labor	2018 Bare Costs Equipment	Total	Total Incl O&P
0010 **UTILITY DIRECTIONAL DRILLING**									
0011 Excluding access and splice pits (if required) and conduit (required)									
0012 Drilled hole diameters shown should be 50% larger than conduit									
0013 Assume access to H20 & removal of spoil/drilling mud as non-hazard material									
0014 Actual production rates adj. to account for risk factors for each soil type									
0100 Sand, silt, clay, common earth									
0110 Mobilization or demobilization	B-82A	1	32	Ea.		1,450	2,250	3,700	4,675
0120 6" diameter		480	.067	L.F.		3.04	4.67	7.71	9.75
0130 12" diameter		270	.119			5.40	8.30	13.70	17.35
0140 18" diameter		180	.178			8.10	12.45	20.55	26
0150 24" diameter		135	.237			10.80	16.60	27.40	34.50
0200 Hard clay, cobble, random boulders									
0210 Mobilization or demobilization	B-82B	1	32	Ea.		1,450	2,500	3,950	4,950
0220 6" diameter		400	.080	L.F.		3.65	6.30	9.95	12.40
0230 12" diameter		220	.145			6.65	11.40	18.05	22.50
0240 18" diameter		145	.221			10.05	17.30	27.35	34.50
0250 24" diameter		105	.305			13.90	24	37.90	47.50
0260 30" diameter		80	.400			18.25	31.50	49.75	62
0270 36" diameter		60	.533			24.50	42	66.50	83
0300 Hard rock (solid bed, 24,000+ psi)									
0310 Mobilization or demobilization	B-82C	1	32	Ea.		1,450	2,800	4,250	5,275
0320 6" diameter		75	.427	L.F.		19.45	37.50	56.95	70.50
0330 12" diameter		40	.800			36.50	70	106.50	132
0340 18" diameter		25	1.280			58.50	112	170.50	212
0350 24" diameter		20	1.600			73	140	213	264
0360 30" diameter		15	2.133			97	187	284	355
0370 36" diameter		12	2.667			122	234	356	440

33 05 07.23 Utility Boring and Jacking

33 05 07.23 Utility Boring and Jacking	Crew	Daily Output	Labor-Hours	Unit	Material	Labor	Equipment	Total	Total Incl O&P
0010 **UTILITY BORING AND JACKING**									
0011 Casing only, 100' minimum,									
0020 not incl. jacking pits or dewatering									
0100 Roadwork, 1/2" thick wall, 24" diameter casing	B-42	20	3.200	L.F.	117	144	57	318	415
0200 36" diameter		16	4		215	180	71	466	590
0300 48" diameter		15	4.267		297	192	76	565	705
0500 Railroad work, 24" diameter		15	4.267		117	192	76	385	505
0600 36" diameter		14	4.571		215	206	81.50	502.50	640
0700 48" diameter		12	5.333		297	240	95	632	800
0900 For ledge, add								20%	20%
1000 Small diameter boring, 3", sandy soil	B-82	900	.018		21	.81	.09	21.90	24.50
1040 Rocky soil	"	500	.032		21	1.46	.16	22.62	25.50
1100 Prepare jacking pits, incl. mobilization & demobilization, minimum				Ea.				3,225	3,700
1101 Maximum				"				22,000	25,500

33 05 07.36 Microtunneling

33 05 07.36 Microtunneling	Crew	Daily Output	Labor-Hours	Unit	Material	Labor	Equipment	Total	Total Incl O&P
0010 **MICROTUNNELING**									
0011 Not including excavation, backfill, shoring,									
0020 or dewatering, average 50'/day, slurry method									
0100 24" to 48" outside diameter, minimum				L.F.				965	965
0110 Adverse conditions, add				"				500	500
1000 Rent microtunneling machine, average monthly lease				Month				97,500	107,000
1010 Operating technician				Day				630	705
1100 Mobilization and demobilization, minimum				Job				41,200	45,900
1110 Maximum				"				445,500	490,500

33 05 Common Work Results for Utilities

33 05 61 – Concrete Manholes

33 05 61.10 Storm Drainage Manholes, Frames and Covers

	33 05 61.10 Storm Drainage Manholes, Frames and Covers	Crew	Daily Output	Labor-Hours	Unit	Material	2018 Bare Costs Labor	Equipment	Total	Total Incl O&P
0010	**STORM DRAINAGE MANHOLES, FRAMES & COVERS**									
0020	Excludes footing, excavation, backfill (See line items for frame & cover)									
0050	Brick, 4' inside diameter, 4' deep	D-1	1	16	Ea.	590	715		1,305	1,750
0100	6' deep		.70	22.857		835	1,025		1,860	2,500
0150	8' deep		.50	32	▼	1,075	1,425		2,500	3,375
0200	For depths over 8', add		4	4	V.L.F.	88	179		267	370
0400	Concrete blocks (radial), 4' ID, 4' deep		1.50	10.667	Ea.	410	480		890	1,175
0500	6' deep		1	16		540	715		1,255	1,700
0600	8' deep		.70	22.857	▼	675	1,025		1,700	2,325
0700	For depths over 8', add	▼	5.50	2.909	V.L.F.	69	130		199	276
0800	Concrete, cast in place, 4' x 4', 8" thick, 4' deep	C-14H	2	24	Ea.	540	1,175	12.70	1,727.70	2,400
0900	6' deep		1.50	32		775	1,575	16.95	2,366.95	3,275
1000	8' deep		1	48	▼	1,125	2,375	25.50	3,525.50	4,850
1100	For depths over 8', add	▼	8	6	V.L.F.	123	296	3.18	422.18	590
1110	Precast, 4' ID, 4' deep	B-22	4.10	7.317	Ea.	845	345	47.50	1,237.50	1,500
1120	6' deep		3	10		1,050	470	65	1,585	1,925
1130	8' deep		2	15		1,200	705	97	2,002	2,500
1140	For depths over 8', add	▼	16	1.875	V.L.F.	128	88	12.15	228.15	288
1150	5' ID, 4' deep	B-6	3	8	Ea.	1,750	350	104	2,204	2,575
1160	6' deep		2	12		2,100	525	156	2,781	3,275
1170	8' deep		1.50	16	▼	2,550	700	208	3,458	4,075
1180	For depths over 8', add		12	2	V.L.F.	283	87.50	26	396.50	470
1190	6' ID, 4' deep		2	12	Ea.	2,500	525	156	3,181	3,725
1200	6' deep		1.50	16		2,950	700	208	3,858	4,525
1210	8' deep		1	24		3,625	1,050	310	4,985	5,950
1220	For depths over 8', add	▼	8	3	V.L.F.	385	131	39	555	660
1250	Slab tops, precast, 8" thick									
1300	4' diameter manhole	B-6	8	3	Ea.	275	131	39	445	540
1400	5' diameter manhole		7.50	3.200		450	140	41.50	631.50	755
1500	6' diameter manhole	▼	7	3.429		700	150	44.50	894.50	1,050
3800	Steps, heavyweight cast iron, 7" x 9"	1 Bric	40	.200		18.05	10.05		28.10	35.50
3900	8" x 9"		40	.200		21.50	10.05		31.55	39.50
3928	12" x 10-1/2"		40	.200		27	10.05		37.05	45
4000	Standard sizes, galvanized steel		40	.200		25	10.05		35.05	43
4100	Aluminum		40	.200		29	10.05		39.05	47.50
4150	Polyethylene	▼	40	.200	▼	29	10.05		39.05	47

33 05 63 – Concrete Vaults and Chambers

33 05 63.13 Precast Concrete Utility Structures

		Crew	Daily Output	Labor-Hours	Unit	Material	2018 Bare Costs Labor	Equipment	Total	Total Incl O&P
0010	**PRECAST CONCRETE UTILITY STRUCTURES**, 6" thick									
0050	5' x 10' x 6' high, ID	B-13	2	28	Ea.	1,800	1,225	295	3,320	4,150
0100	6' x 10' x 6' high, ID		2	28		1,875	1,225	295	3,395	4,250
0150	5' x 12' x 6' high, ID		2	28		1,975	1,225	295	3,495	4,350
0200	6' x 12' x 6' high, ID		1.80	31.111		2,225	1,350	330	3,905	4,875
0250	6' x 13' x 6' high, ID		1.50	37.333		2,925	1,625	395	4,945	6,100
0300	8' x 14' x 7' high, ID	▼	1	56	▼	3,150	2,450	590	6,190	7,850

33 05 97 – Identification and Signage for Utilities

33 05 97.05 Utility Connection

		Crew	Daily Output	Labor-Hours	Unit	Material	2018 Bare Costs Labor	Equipment	Total	Total Incl O&P
0010	**UTILITY CONNECTION**									
0020	Water, sanitary, stormwater, gas, single connection	B-14	1	48	Ea.	2,925	2,025	310	5,260	6,650
0030	Telecommunication	"	3	16	"	385	675	104	1,164	1,550

For customer support on your Electrical Costs with RSMeans data, call 800.448.8182.

357

33 05 Common Work Results for Utilities

33 05 97 – Identification and Signage for Utilities

33 05 97.10 Utility Accessories	Crew	Daily Output	Labor-Hours	Unit	Material	2018 Bare Costs Labor	Equipment	Total	Total Incl O&P
0010 **UTILITY ACCESSORIES**									
0400 Underground tape, detectable, reinforced, alum. foil core, 2"	1 Clab	150	.053	C.L.F.	9	2.13		11.13	13.15
0500 6"	"	140	.057	"	36	2.28		38.28	43

33 42 Stormwater Conveyance

33 42 33 – Stormwater Curbside Drains and Inlets

33 42 33.13 Catch Basins

	Crew	Daily Output	Labor-Hours	Unit	Material	2018 Bare Costs Labor	Equipment	Total	Total Incl O&P
0010 **CATCH BASINS**									
0011 Not including footing & excavation									
1600 Frames & grates, C.I., 24" square, 500 lb.	B-6	7.80	3.077	Ea.	355	134	40	529	640
1700 26" D shape, 600 lb.		7	3.429		620	150	44.50	814.50	955
1800 Light traffic, 18" diameter, 100 lb.		10	2.400		203	105	31.50	339.50	420
1900 24" diameter, 300 lb.		8.70	2.759		197	120	36	353	440
2000 36" diameter, 900 lb.		5.80	4.138		640	181	54	875	1,050
2100 Heavy traffic, 24" diameter, 400 lb.		7.80	3.077		263	134	40	437	540
2200 36" diameter, 1,150 lb.		3	8		875	350	104	1,329	1,600
2300 Mass. State standard, 26" diameter, 475 lb.		7	3.429		645	150	44.50	839.50	985
2400 30" diameter, 620 lb.		7	3.429		340	150	44.50	534.50	650
2500 Watertight, 24" diameter, 350 lb.		7.80	3.077		365	134	40	539	650
2600 26" diameter, 500 lb.		7	3.429		420	150	44.50	614.50	740
2700 32" diameter, 575 lb.	↓	6	4	↓	920	175	52	1,147	1,325
2800 3 piece cover & frame, 10" deep,									
2900 1,200 lb., for heavy equipment	B-6	3	8	Ea.	1,050	350	104	1,504	1,800
3000 Raised for paving 1-1/4" to 2" high									
3100 4 piece expansion ring									
3200 20" to 26" diameter	1 Clab	3	2.667	Ea.	178	106		284	355
3300 30" to 36" diameter	"	3	2.667	"	246	106		352	435
3320 Frames and covers, existing, raised for paving, 2", including									
3340 row of brick, concrete collar, up to 12" wide frame	B-6	18	1.333	Ea.	51.50	58	17.35	126.85	164
3360 20" to 26" wide frame		11	2.182		76.50	95.50	28.50	200.50	261
3380 30" to 36" wide frame	↓	9	2.667		95	116	34.50	245.50	320
3400 Inverts, single channel brick	D-1	3	5.333		104	239		343	480
3500 Concrete		5	3.200		122	143		265	355
3600 Triple channel, brick		2	8		166	360		526	735
3700 Concrete	↓	3	5.333	↓	157	239		396	535

33 71 Electrical Utility Transmission and Distribution

33 71 13 – Electrical Utility Towers

33 71 13.23 Steel Electrical Utility Towers

	Crew	Daily Output	Labor-Hours	Unit	Material	2018 Bare Costs Labor	Equipment	Total	Total Incl O&P
0010 **STEEL ELECTRICAL UTILITY TOWERS**									
0100 Excavation and backfill, earth	R-5	135.38	.650	C.Y.		33	8.95	41.95	60
0105 Rock		21.46	4.101	"		208	56.50	264.50	375
0200 Steel footings (grillage) in earth		3.91	22.506	Ton	2,200	1,150	310	3,660	4,500
0205 In rock	↓	3.20	27.500	"	2,200	1,400	380	3,980	4,950
0290 See also Section 33 05 23.19									
0300 Rock anchors	R-5	5.87	14.991	Ea.	575	760	207	1,542	2,000
0400 Concrete foundations	"	12.85	6.848	C.Y.	153	350	94.50	597.50	795
0490 See also Section 03 30 53.40									
0500 Towers: material handling and spotting	R-7	22.56	2.128	Ton		88	7.35	95.35	143

33 71 13.23 Steel Electrical Utility Towers

		Crew	Daily Output	Labor-Hours	Unit	Material	2018 Bare Costs Labor	2018 Bare Costs Equipment	Total	Total Incl O&P
0540	Steel tower erection	R-5	7.65	11.503	Ton	2,150	585	159	2,894	3,425
0550	Lace and box		7.10	12.394		2,150	630	171	2,951	3,525
0560	Painting total structure		1.47	59.864	Ea.	490	3,050	825	4,365	6,050
0570	Disposal of surplus material	R-7	20.87	2.300	Mile		95	7.95	102.95	154
0600	Special towers: material handling and spotting	"	12.31	3.899	Ton		161	13.50	174.50	262
0640	Special steel structure erection	R-6	6.52	13.497		2,700	685	500	3,885	4,525
0650	Special steel lace and box	"	6.29	13.990		2,700	710	520	3,930	4,600
0670	Disposal of surplus material	R-7	7.87	6.099	Mile		252	21	273	410

33 71 13.80 Transmission Line Right of Way

		Crew	Daily Output	Labor-Hours	Unit	Material	2018 Bare Costs Labor	2018 Bare Costs Equipment	Total	Total Incl O&P
0010	**TRANSMISSION LINE RIGHT OF WAY**									
0100	Clearing right of way	B-87	6.67	5.997	Acre		305	455	760	960
0200	Restoration & seeding	B-10D	4	3	"	440	147	425	1,012	1,175

33 71 16 – Electrical Utility Poles

33 71 16.23 Steel Electrical Utility Poles

		Crew	Daily Output	Labor-Hours	Unit	Material	2018 Bare Costs Labor	2018 Bare Costs Equipment	Total	Total Incl O&P
0010	**STEEL ELECTRICAL UTILITY POLES**									
0100	Steel, square, tapered shaft, 20'	R-15A	6	8	Ea.	410	410	47	867	1,125
0120	25'		5.35	8.972		440	460	53	953	1,225
0140	30'		4.80	10		495	510	59	1,064	1,375
0160	35'		4.35	11.034		530	565	65	1,160	1,500
0180	40'		4	12		735	610	70.50	1,415.50	1,800
0300	Galvanized, round, tapered shaft, 40'		4	12		2,150	610	70.50	2,830.50	3,350
0320	45'		3.70	12.973		2,425	660	76.50	3,161.50	3,750
0340	50'		3.45	13.913		2,600	710	82	3,392	4,050
0360	55'		3	16		2,850	815	94	3,759	4,450
0380	60'		2.65	18.113		3,075	925	107	4,107	4,900
0400	65'	R-13	2.20	19.091		3,925	1,075	107	5,107	6,050
0420	70'		1.90	22.105		4,300	1,250	124	5,674	6,750
0440	75'		1.70	24.706		5,650	1,400	138	7,188	8,450
0460	80'		1.50	28		5,100	1,575	157	6,832	8,175
0480	85'		1.40	30		5,675	1,700	168	7,543	8,925
0500	90'		1.25	33.600		5,725	1,900	188	7,813	9,325
0520	95'		1.15	36.522		6,350	2,050	205	8,605	10,300
0540	100'		1	42		7,050	2,375	235	9,660	11,600
0700	12 sided, tapered shaft, 40'	R-15A	4	12		1,250	610	70.50	1,930.50	2,375
0720	45'		3.70	12.973		1,375	660	76.50	2,111.50	2,600
0740	50'		3.45	13.913		1,600	710	82	2,392	2,925
0760	55'		3	16		1,825	815	94	2,734	3,325
0780	60'		2.65	18.113		1,900	925	107	2,932	3,625
0800	65'	R-13	2.20	19.091		2,225	1,075	107	3,407	4,150
0820	70'		1.90	22.105		2,400	1,250	124	3,774	4,650
0840	75'		1.70	24.706		2,625	1,400	138	4,163	5,100
0860	80'		1.50	28		4,500	1,575	157	6,232	7,500
0880	85'		1.40	30		7,200	1,700	168	9,068	10,600
0900	90'		1.25	33.600		9,425	1,900	188	11,513	13,400
0920	95'		1.15	36.522		11,800	2,050	205	14,055	16,300
0940	100'		1	42		14,400	2,375	235	17,010	19,700
0960	105'		.90	46.667		15,900	2,625	261	18,786	21,600
0980	Ladder clips					253			253	279
1000	Galvanized steel, round, tapered, w/one 6' arm, 20'	R-15A	9.20	5.217		1,050	266	30.50	1,346.50	1,575
1020	25'		7.65	6.275		1,125	320	37	1,482	1,775
1040	30'		6.55	7.328		1,250	375	43	1,668	1,975
1060	35'		5.75	8.348		1,350	425	49	1,824	2,175

For customer support on your Electrical Costs with RSMeans data, call 800.448.8182.

359

33 71 16 – Electrical Utility Poles

33 71 16.23 Steel Electrical Utility Poles	Crew	Daily Output	Labor-Hours	Unit	Material	2018 Bare Costs Labor	2018 Bare Costs Equipment	Total	Total Incl O&P
1080	40'	R-15A	5.15	9.320	Ea.	1,425	475	55	1,955	2,350
1200	Two 6' arms, 20'		6.55	7.328		1,350	375	43	1,768	2,075
1220	25'		5.75	8.348		1,375	425	49	1,849	2,200
1240	30'		5.15	9.320		1,425	475	55	1,955	2,350
1260	35'		4.60	10.435		1,625	530	61.50	2,216.50	2,650
1280	40'		4.20	11.429		1,875	585	67.50	2,527.50	3,000
1400	One 12' truss arm, 25'		7.65	6.275		1,300	320	37	1,657	1,950
1420	30'		6.55	7.328		1,350	375	43	1,768	2,075
1440	35'		5.75	8.348		1,450	425	49	1,924	2,300
1460	40'		5.15	9.320		1,625	475	55	2,155	2,575
1480	45'		4.60	10.435		1,875	530	61.50	2,466.50	2,950
1600	Two 12' truss arms, 25'		5.75	8.348		1,525	425	49	1,999	2,400
1620	30'		5.15	9.320		1,575	475	55	2,105	2,525
1640	35'		4.60	10.435		1,700	530	61.50	2,291.50	2,750
1660	40'		4.20	11.429		1,875	585	67.50	2,527.50	3,025
1680	45'		3.85	12.468		2,125	635	73.50	2,833.50	3,350
3400	Galvanized steel, tapered, 10'		15.50	3.097		425	158	18.25	601.25	725
3420	12'		11.50	4.174		500	213	24.50	737.50	895
3440	14'		9.20	5.217		370	266	30.50	666.50	840
3460	16'		8.45	5.680		385	290	33.50	708.50	895
3480	18'		7.65	6.275		395	320	37	752	955
3500	20'		7.10	6.761		380	345	40	765	985
5200	Galvanized steel, bracket type, 2", 6' long		18.50	2.595		161	132	15.30	308.30	395
5220	8' long		15.50	3.097		200	158	18.25	376.25	480
5240	10' long		12.50	3.840		335	196	22.50	553.50	690
6000	Digging holes in earth, average	R-5	25.14	3.500			178	48	226	320
6010	In rock, average	"	4.51	19.512			990	269	1,259	1,800
6020	Formed plate pole structure									
6030	Material handling and spotting	R-7	2.40	20	Ea.		825	69	894	1,350
6040	Erect steel plate pole	R-5	1.95	45.128		10,400	2,300	620	13,320	15,500
6050	Guys, anchors and hardware for pole, in earth		7.04	12.500		630	635	172	1,437	1,850
6060	In rock		17.96	4.900		750	249	67.50	1,066.50	1,275
6070	Foundations for line poles									
6080	Excavation, in earth	R-5	135.38	.650	C.Y.		33	8.95	41.95	60
6090	In rock		20	4.400			224	60.50	284.50	400
6110	Concrete foundations		11	8		153	405	110	668	905

33 71 16.33 Wood Electrical Utility Poles

		Crew	Daily Output	Labor-Hours	Unit	Material	2018 Bare Costs Labor	2018 Bare Costs Equipment	Total	Total Incl O&P
0010	**WOOD ELECTRICAL UTILITY POLES**									
0011	Excludes excavation, backfill and cast-in-place concrete									
1020	12" Ponderosa Pine Poles treated 0.40 ACQ, 16'	R-3	3.20	6.250	Ea.	875	360	40.50	1,275.50	1,550
5000	Wood, class 3 yellow pine, penta-treated, 25'	R-15A	8.60	5.581		278	285	33	596	770
5020	30'		7.70	6.234		320	320	36.50	676.50	870
5040	35'		5.80	8.276		400	420	48.50	868.50	1,125
5060	40'		5.30	9.057		475	460	53.50	988.50	1,275
5080	45'		4.70	10.213		515	520	60	1,095	1,425
5100	50'		4.20	11.429		635	585	67.50	1,287.50	1,650
5120	55'		3.80	12.632		745	645	74.50	1,464.50	1,875
5140	60'		3.50	13.714		995	700	81	1,776	2,250
5160	65'		3.20	15		1,300	765	88.50	2,153.50	2,675
5180	70'		3	16		1,825	815	94	2,734	3,325
5200	75'		2.80	17.143		2,350	875	101	3,326	4,000
6000	Wood, class 1 type C, CCA/ACA-treated, 25'		8.60	5.581		278	285	33	596	770

360

33 71 Electrical Utility Transmission and Distribution

33 71 16 – Electrical Utility Poles

33 71 16.33 Wood Electrical Utility Poles

		Crew	Daily Output	Labor-Hours	Unit	Material	2018 Bare Costs Labor	Equipment	Total	Total Incl O&P
6020	30'	R-15A	7.70	6.234	Ea.	340	320	36.50	696.50	895
6040	35'		5.80	8.276		475	420	48.50	943.50	1,225
6060	40'		5.30	9.057		555	460	53.50	1,068.50	1,375
6080	45'		4.70	10.213		675	520	60	1,255	1,600
6100	50'		4.20	11.429		740	585	67.50	1,392.50	1,775
6120	55'		3.80	12.632		835	645	74.50	1,554.50	1,975
6200	Electric & tel sitework, 20' high, treated wd., see Section 26 56 13.10	R-3	3.10	6.452		189	375	42	606	815
6400	25' high R337119-30		2.90	6.897		238	400	44.50	682.50	910
6600	30' high		2.60	7.692		340	445	50	835	1,100
6800	35' high		2.40	8.333		480	485	54	1,019	1,325
7000	40' high		2.30	8.696		640	505	56.50	1,201.50	1,525
7200	45' high		1.70	11.765		870	680	76	1,626	2,075
7400	Cross arms with hardware & insulators									
7600	4' long	1 Elec	2.50	3.200	Ea.	149	186		335	440
7800	5' long		2.40	3.333		165	194		359	470
8000	6' long		2.20	3.636		171	212		383	505
9000	Disposal of pole & hardware surplus material	R-7	20.87	2.300	Mile		95	7.95	102.95	154
9100	Disposal of crossarms & hardware surplus material	"	40	1.200	"		49.50	4.15	53.65	80.50

33 71 19 – Electrical Underground Ducts and Manholes

33 71 19.15 Underground Ducts and Manholes

		Crew	Daily Output	Labor-Hours	Unit	Material	2018 Bare Costs Labor	Equipment	Total	Total Incl O&P
0010	**UNDERGROUND DUCTS AND MANHOLES**									
0011	Not incl. excavation, backfill and concrete, in slab or duct bank									
1000	Direct burial R337119-30									
1010	PVC, schedule 40, w/coupling, 1/2" diameter	1 Elec	340	.024	L.F.	.30	1.37		1.67	2.38
1020	3/4" diameter		290	.028		.40	1.61		2.01	2.84
1030	1" diameter		260	.031		.66	1.79		2.45	3.41
1040	1-1/2" diameter		210	.038		.97	2.22		3.19	4.38
1050	2" diameter		180	.044		1.26	2.59		3.85	5.25
1060	3" diameter	2 Elec	240	.067		2.26	3.88		6.14	8.30
1070	4" diameter		160	.100		3.09	5.80		8.89	12.10
1080	5" diameter		120	.133		4.61	7.75		12.36	16.65
1090	6" diameter		90	.178		6.05	10.35		16.40	22
1110	Elbows, 1/2" diameter	1 Elec	48	.167	Ea.	.62	9.70		10.32	15.20
1120	3/4" diameter		38	.211		.78	12.25		13.03	19.15
1130	1" diameter		32	.250		1.30	14.55		15.85	23.50
1140	1-1/2" diameter		21	.381		2.22	22		24.22	35.50
1150	2" diameter		16	.500		2.91	29		31.91	46.50
1160	3" diameter		12	.667		10.50	39		49.50	69.50
1170	4" diameter		9	.889		17.05	51.50		68.55	96.50
1180	5" diameter		8	1		26	58		84	116
1190	6" diameter		5	1.600		36	93		129	179
1210	Adapters, 1/2" diameter		52	.154		.21	8.95		9.16	13.65
1220	3/4" diameter		43	.186		.34	10.85		11.19	16.55
1230	1" diameter		39	.205		.48	11.95		12.43	18.40
1240	1-1/2" diameter		35	.229		.79	13.30		14.09	21
1250	2" diameter		26	.308		1.11	17.90		19.01	28
1260	3" diameter		20	.400		2.63	23.50		26.13	38
1270	4" diameter		14	.571		4.40	33.50		37.90	54.50
1280	5" diameter		12	.667		8.30	39		47.30	67
1290	6" diameter		9	.889		19	51.50		70.50	98.50
1340	Bell end & cap, 1-1/2" diameter		35	.229		4.50	13.30		17.80	25
1350	Bell end & plug, 2" diameter		26	.308		6.10	17.90		24	34

33 71 19 – Electrical Underground Ducts and Manholes

33 71 19.15 Underground Ducts and Manholes	Crew	Daily Output	Labor-Hours	Unit	Material	2018 Bare Costs Labor	Equipment	Total	Total Incl O&P
1360	3" diameter	1 Elec	20	.400	Ea.	7.85	23.50		31.35	43.50
1370	4" diameter		14	.571		9.40	33.50		42.90	60
1380	5" diameter		12	.667		14.45	39		53.45	74
1390	6" diameter		9	.889		16.70	51.50		68.20	96
1450	Base spacer, 2" diameter		56	.143		1.21	8.30		9.51	13.80
1460	3" diameter		46	.174		1.57	10.10		11.67	16.90
1470	4" diameter		41	.195		1.77	11.35		13.12	18.95
1480	5" diameter		37	.216		1.98	12.60		14.58	21
1490	6" diameter		34	.235		2.35	13.70		16.05	23
1550	Intermediate spacer, 2" diameter		60	.133		1.19	7.75		8.94	12.90
1560	3" diameter		46	.174		1.72	10.10		11.82	17.05
1570	4" diameter		41	.195		1.54	11.35		12.89	18.70
1580	5" diameter		37	.216		1.82	12.60		14.42	21
1590	6" diameter		34	.235	↓	2.45	13.70		16.15	23
4010	PVC, schedule 80, w/coupling, 1/2" diameter		215	.037	L.F.	.88	2.17		3.05	4.21
4020	3/4" diameter		180	.044		1.20	2.59		3.79	5.20
4030	1" diameter		145	.055		1.64	3.21		4.85	6.60
4040	1-1/2" diameter		120	.067		2.58	3.88		6.46	8.65
4050	2" diameter		100	.080		3.56	4.66		8.22	10.85
4060	3" diameter	2 Elec	130	.123		7.40	7.15		14.55	18.80
4070	4" diameter		90	.178		10.45	10.35		20.80	27
4080	5" diameter		70	.229		17.10	13.30		30.40	38.50
4090	6" diameter		50	.320	↓	21	18.60		39.60	51.50
4110	Elbows, 1/2" diameter	1 Elec	29	.276	Ea.	2.36	16.05		18.41	26.50
4120	3/4" diameter		23	.348		5.40	20		25.40	36.50
4130	1" diameter		20	.400		6.55	23.50		30.05	42
4140	1-1/2" diameter		16	.500		10.85	29		39.85	55.50
4150	2" diameter		12	.667		14.15	39		53.15	73.50
4160	3" diameter		9	.889		36	51.50		87.50	118
4170	4" diameter		7	1.143		62.50	66.50		129	169
4180	5" diameter		6	1.333		170	77.50		247.50	305
4190	6" diameter		4	2		187	116		303	380
4210	Adapter, 1/2" diameter		39	.205		.21	11.95		12.16	18.10
4220	3/4" diameter		33	.242		.34	14.10		14.44	21.50
4230	1" diameter		29	.276		.48	16.05		16.53	24.50
4240	1-1/2" diameter		26	.308		.79	17.90		18.69	28
4250	2" diameter		23	.348		1.11	20		21.11	31.50
4260	3" diameter		18	.444		2.63	26		28.63	41.50
4270	4" diameter		13	.615		4.40	36		40.40	58.50
4280	5" diameter		11	.727		8.30	42.50		50.80	72.50
4290	6" diameter		8	1		19	58		77	108
4310	Bell end & cap, 1-1/2" diameter		26	.308		4.50	17.90		22.40	32
4320	Bell end & plug, 2" diameter		23	.348		6.10	20		26.10	37.50
4330	3" diameter		18	.444		7.85	26		33.85	47
4340	4" diameter		13	.615		9.40	36		45.40	64
4350	5" diameter		11	.727		14.45	42.50		56.95	79.50
4360	6" diameter		8	1		16.70	58		74.70	105
4370	Base spacer, 2" diameter		42	.190		1.21	11.10		12.31	17.90
4380	3" diameter		33	.242		1.57	14.10		15.67	22.50
4390	4" diameter		29	.276		1.77	16.05		17.82	26
4400	5" diameter		26	.308		1.98	17.90		19.88	29
4410	6" diameter		25	.320		2.35	18.60		20.95	30.50
4420	Intermediate spacer, 2" diameter	↓	45	.178	↓	1.19	10.35		11.54	16.75

33 71 19.15 Underground Ducts and Manholes

		Crew	Daily Output	Labor-Hours	Unit	Material	2018 Bare Costs Labor	Equipment	Total	Total Incl O&P
4430	3" diameter	1 Elec	34	.235	Ea.	1.72	13.70		15.42	22.50
4440	4" diameter		31	.258		1.54	15		16.54	24
4450	5" diameter		28	.286		1.82	16.65		18.47	27
4460	6" diameter		25	.320		2.45	18.60		21.05	30.50

33 71 19.17 Electric and Telephone Underground

		Crew	Daily Output	Labor-Hours	Unit	Material	2018 Bare Costs Labor	Equipment	Total	Total Incl O&P
0010	**ELECTRIC AND TELEPHONE UNDERGROUND**									
0011	Not including excavation									
0200	backfill and cast in place concrete									
0250	For bedding, see Section 31 23 23.16 R337119-30									
0400	Hand holes, precast concrete, with concrete cover									
0600	2' x 2' x 3' deep	R-3	2.40	8.333	Ea.	430	485	54	969	1,250
0800	3' x 3' x 3' deep		1.90	10.526		560	610	68	1,238	1,600
1000	4' x 4' x 4' deep		1.40	14.286		1,425	830	92.50	2,347.50	2,900
1200	Manholes, precast with iron racks & pulling irons, C.I. frame									
1400	and cover, 4' x 6' x 7' deep	B-13	2	28	Ea.	2,925	1,225	295	4,445	5,375
1600	6' x 8' x 7' deep		1.90	29.474		3,275	1,300	310	4,885	5,900
1800	6' x 10' x 7' deep		1.80	31.111		3,675	1,350	330	5,355	6,475
4200	Underground duct, banks ready for concrete fill, min. of 7.5"									
4400	between conduits, center to center									
4580	PVC, type EB, 1 @ 2" diameter	2 Elec	480	.033	L.F.	1.09	1.94		3.03	4.10
4600	2 @ 2" diameter		240	.067		2.18	3.88		6.06	8.20
4800	4 @ 2" diameter		120	.133		4.36	7.75		12.11	16.40
4900	1 @ 3" diameter		400	.040		1.82	2.33		4.15	5.50
5000	2 @ 3" diameter		200	.080		3.63	4.66		8.29	10.95
5200	4 @ 3" diameter		100	.160		7.25	9.30		16.55	22
5300	1 @ 4" diameter		320	.050		1.81	2.91		4.72	6.35
5400	2 @ 4" diameter		160	.100		3.61	5.80		9.41	12.70
5600	4 @ 4" diameter		80	.200		7.25	11.65		18.90	25.50
5800	6 @ 4" diameter		54	.296		10.85	17.25		28.10	38
5810	1 @ 5" diameter		260	.062		2.06	3.58		5.64	7.60
5820	2 @ 5" diameter		130	.123		4.12	7.15		11.27	15.25
5840	4 @ 5" diameter		70	.229		8.25	13.30		21.55	29
5860	6 @ 5" diameter		50	.320		12.35	18.60		30.95	41.50
5870	1 @ 6" diameter		200	.080		3.46	4.66		8.12	10.75
5880	2 @ 6" diameter		100	.160		6.90	9.30		16.20	21.50
5900	4 @ 6" diameter		50	.320		13.80	18.60		32.40	43
5920	6 @ 6" diameter		30	.533		20.50	31		51.50	69.50
6200	Rigid galvanized steel, 2 @ 2" diameter		180	.089		17.50	5.15		22.65	27
6400	4 @ 2" diameter		90	.178		35	10.35		45.35	54
6800	2 @ 3" diameter		100	.160		25	9.30		34.30	41.50
7000	4 @ 3" diameter		50	.320		50	18.60		68.60	83
7200	2 @ 4" diameter		70	.229		37	13.30		50.30	61
7400	4 @ 4" diameter		34	.471		74.50	27.50		102	123
7600	6 @ 4" diameter		22	.727		112	42.50		154.50	187
7620	2 @ 5" diameter		60	.267		70.50	15.50		86	101
7640	4 @ 5" diameter		30	.533		141	31		172	202
7660	6 @ 5" diameter		18	.889		211	51.50		262.50	310
7680	2 @ 6" diameter		40	.400		98	23.50		121.50	143
7700	4 @ 6" diameter		20	.800		196	46.50		242.50	286
7720	6 @ 6" diameter		14	1.143		294	66.50		360.50	425
7800	For cast-in-place concrete, add									
7810	Under 1 C.Y.	C-6	16	3	C.Y.	193	124	3.21	320.21	405

33 71 19 – Electrical Underground Ducts and Manholes

33 71 19.17 Electric and Telephone Underground

		Crew	Daily Output	Labor-Hours	Unit	Material	2018 Bare Costs Labor	Equipment	Total	Total Incl O&P
7820	1 C.Y. to 5 C.Y.	C-6	19.20	2.500	C.Y.	175	104	2.68	281.68	355
7830	Over 5 C.Y.	↓	24	2	↓	145	83	2.14	230.14	287
7850	For reinforcing rods, add									
7860	#4 to #7	2 Rodm	1.10	14.545	Ton	960	795		1,755	2,275
7870	#8 to #14	"	1.50	10.667	"	960	585		1,545	1,950
8000	Fittings, PVC type EB, elbow, 2" diameter	1 Elec	16	.500	Ea.	16.35	29		45.35	61.50
8200	3" diameter		14	.571		19.20	33.50		52.70	70.50
8400	4" diameter		12	.667		38.50	39		77.50	100
8420	5" diameter		10	.800		126	46.50		172.50	209
8440	6" diameter	↓	9	.889		164	51.50		215.50	259
8500	Coupling, 2" diameter					.82			.82	.90
8600	3" diameter					3.78			3.78	4.16
8700	4" diameter					4.24			4.24	4.66
8720	5" diameter					8.20			8.20	9
8740	6" diameter					21			21	23.50
8800	Adapter, 2" diameter	1 Elec	26	.308		.98	17.90		18.88	28
9000	3" diameter		20	.400		2.96	23.50		26.46	38.50
9200	4" diameter		16	.500		4.09	29		33.09	48
9220	5" diameter		13	.615		9.60	36		45.60	64
9240	6" diameter		10	.800		12.10	46.50		58.60	83
9400	End bell, 2" diameter		16	.500		2.13	29		31.13	46
9600	3" diameter		14	.571		3.67	33.50		37.17	53.50
9800	4" diameter		12	.667		4.81	39		43.81	63.50
9810	5" diameter		10	.800		8.35	46.50		54.85	78.50
9820	6" diameter		8	1		9.95	58		67.95	98
9830	5° angle coupling, 2" diameter		26	.308		7.65	17.90		25.55	35.50
9840	3" diameter		20	.400		29	23.50		52.50	66.50
9850	4" diameter		16	.500		18.05	29		47.05	63.50
9860	5" diameter		13	.615		15.70	36		51.70	71
9870	6" diameter		10	.800		25.50	46.50		72	97.50
9880	Expansion joint, 2" diameter		16	.500		34	29		63	81
9890	3" diameter		18	.444		64.50	26		90.50	110
9900	4" diameter		12	.667		98.50	39		137.50	167
9910	5" diameter		10	.800		179	46.50		225.50	267
9920	6" diameter	↓	8	1		162	58		220	265
9930	Heat bender, 2" diameter					400			400	440
9940	6" diameter					1,400			1,400	1,550
9950	Cement, quart				↓	14.70			14.70	16.15
9960	Nylon polyethylene pull rope, 1/4"	2 Elec	2000	.008	L.F.	.16	.47		.63	.88

33 71 23 – Insulators and Fittings

33 71 23.16 Post Insulators

		Crew	Daily Output	Labor-Hours	Unit	Material	2018 Bare Costs Labor	Equipment	Total	Total Incl O&P
0010	**POST INSULATORS**									
7400	Insulators, pedestal type	R-11	112	.500	Ea.		27.50	5.90	33.40	48
7490	See also line 33 71 39.13 1000									

33 71 26 – Transmission and Distribution Equipment

33 71 26.13 Capacitor Banks

		Crew	Daily Output	Labor-Hours	Unit	Material	2018 Bare Costs Labor	Equipment	Total	Total Incl O&P
0010	**CAPACITOR BANKS**									
1300	Station capacitors									
1350	Synchronous, 13 to 26 kV	R-11	3.11	18.006	MVAR	7,375	995	213	8,583	9,850
1360	46 kV		3.33	16.817		9,425	930	199	10,554	12,000
1370	69 kV		3.81	14.698		9,275	815	174	10,264	11,600
1380	161 kV	↓	6.51	8.602	↓	8,650	475	102	9,227	10,400

33 71 26 – Transmission and Distribution Equipment

33 71 26.13 Capacitor Banks

		Crew	Daily Output	Labor-Hours	Unit	Material	2018 Bare Costs Labor	Equipment	Total	Total Incl O&P
1390	500 kV	R-11	10.37	5.400	MVAR	7,525	299	64	7,888	8,800
1450	Static, 13 to 26 kV		3.11	18.006		6,250	995	213	7,458	8,600
1460	46 kV		3.01	18.605		7,900	1,025	220	9,145	10,500
1470	69 kV		3.81	14.698		7,675	815	174	8,664	9,875
1480	161 kV		6.51	8.602		7,125	475	102	7,702	8,675
1490	500 kV		10.37	5.400		6,500	299	64	6,863	7,675
1600	Voltage regulators, 13 to 26 kV		.75	74.667	Ea.	280,000	4,125	885	285,010	315,000

33 71 26.23 Current Transformers

		Crew	Daily Output	Labor-Hours	Unit	Material	2018 Bare Costs Labor	Equipment	Total	Total Incl O&P
0010	**CURRENT TRANSFORMERS**									
4050	Current transformers, 13 to 26 kV	R-11	14	4	Ea.	3,450	221	47.50	3,718.50	4,175
4060	46 kV		9.33	6.002		10,000	330	71	10,401	11,600
4070	69 kV		7	8		10,400	445	94.50	10,939.50	12,300
4080	161 kV		1.87	29.947		33,800	1,650	355	35,805	40,100

33 71 26.26 Potential Transformers

		Crew	Daily Output	Labor-Hours	Unit	Material	2018 Bare Costs Labor	Equipment	Total	Total Incl O&P
0010	**POTENTIAL TRANSFORMERS**									
4100	Potential transformers, 13 to 26 kV	R-11	11.20	5	Ea.	4,925	277	59	5,261	5,875
4110	46 kV		8	7		10,100	385	82.50	10,567.50	11,800
4120	69 kV		6.22	9.003		10,700	500	106	11,306	12,700
4130	161 kV		2.24	25		23,200	1,375	296	24,871	27,900
4140	500 kV		1.40	40		69,000	2,225	475	71,700	80,000

33 71 39 – High-Voltage Wiring

33 71 39.13 Overhead High-Voltage Wiring

		Crew	Daily Output	Labor-Hours	Unit	Material	2018 Bare Costs Labor	Equipment	Total	Total Incl O&P
0010	**OVERHEAD HIGH-VOLTAGE WIRING** R337116-60									
0100	Conductors, primary circuits									
0110	Material handling and spotting	R-5	9.78	8.998	W.Mile		455	124	579	825
0120	For river crossing, add		11	8			405	110	515	735
0150	Conductors, per wire, 210 to 636 kcmil		1.96	44.898		12,400	2,275	620	15,295	17,800
0160	795 to 954 kcmil		1.87	47.059		24,800	2,400	650	27,850	31,600
0170	1,000 to 1,600 kcmil		1.47	59.864		42,500	3,050	825	46,375	52,500
0180	Over 1,600 kcmil		1.35	65.185		59,500	3,325	900	63,725	71,500
0200	For river crossing, add, 210 to 636 kcmil		1.24	70.968			3,600	980	4,580	6,525
0220	795 to 954 kcmil		1.09	80.734			4,100	1,125	5,225	7,425
0230	1,000 to 1,600 kcmil		.97	90.722			4,600	1,250	5,850	8,325
0240	Over 1,600 kcmil		.87	101			5,150	1,400	6,550	9,275
0300	Joints and dead ends	R-8	6	8	Ea.	1,650	410	47	2,107	2,500
0400	Sagging	R-5	7.33	12.001	W.Mile		610	165	775	1,100
0500	Clipping, per structure, 69 kV	R-10	9.60	5	Ea.		274	62.50	336.50	480
0510	161 kV		5.33	9.006			495	113	608	865
0520	345 to 500 kV		2.53	18.972			1,050	237	1,287	1,800
0600	Make and install jumpers, per structure, 69 kV	R-8	3.20	15		450	770	88	1,308	1,775
0620	161 kV		1.20	40		895	2,050	234	3,179	4,350
0640	345 to 500 kV		.32	150		1,500	7,725	880	10,105	14,200
0700	Spacers	R-10	68.57	.700		89.50	38.50	8.75	136.75	166
0720	For river crossings, add	"	60	.800			44	10	54	77
0800	Installing pulling line (500 kV only)	R-9	1.45	44.138	W.Mile	840	2,125	194	3,159	4,350
0810	Disposal of surplus material, high voltage conductors	R-7	6.96	6.897	Mile		285	24	309	460
0820	With trailer mounted reel stands	"	13.71	3.501	"		145	12.10	157.10	234
0900	Insulators and hardware, primary circuits									
0920	Material handling and spotting, 69 kV	R-7	480	.100	Ea.		4.13	.35	4.48	6.75
0930	161 kV		685.71	.070			2.89	.24	3.13	4.70
0950	345 to 500 kV		960	.050			2.06	.17	2.23	3.35
1000	Disk insulators, 69 kV	R-5	880	.100		91.50	5.10	1.38	97.98	109

For customer support on your Electrical Costs with RSMeans data, call 800.448.8182.

365

33 71 39.13 Overhead High-Voltage Wiring		Crew	Daily Output	Labor-Hours	Unit	Material	2018 Bare Costs Labor	2018 Bare Costs Equipment	Total	Total Incl O&P
1020	161 kV	R-5	977.78	.090	Ea.	104	4.57	1.24	109.81	123
1040	345 to 500 kV	↓	1100	.080	↓	104	4.07	1.10	109.17	122
1060	See Section 33 71 23.16 for pin or pedestal insulator									
1100	Install disk insulator at river crossing, add									
1110	69 kV	R-5	586.67	.150	Ea.		7.60	2.07	9.67	13.75
1120	161 kV		880	.100			5.10	1.38	6.48	9.15
1140	345 to 500 kV	↓	880	.100	↓		5.10	1.38	6.48	9.15
1150	Disposal of surplus material, high voltage insulators	R-7	41.74	1.150	Mile		47.50	3.98	51.48	77
1300	Overhead ground wire installation									
1320	Material handling and spotting	R-7	5.65	8.496	W.Mile		350	29.50	379.50	570
1340	Overhead ground wire	R-5	1.76	50		3,300	2,550	690	6,540	8,200
1350	At river crossing, add		1.17	75.214	↓		3,825	1,025	4,850	6,925
1360	Disposal of surplus material, grounding wire	↓	41.74	2.108	Mile		107	29	136	194
1400	Installing conductors, underbuilt circuits									
1420	Material handling and spotting	R-7	5.65	8.496	W.Mile		350	29.50	379.50	570
1440	Conductors, per wire, 210 to 636 kcmil	R-5	1.96	44.898		12,400	2,275	620	15,295	17,800
1450	795 to 954 kcmil		1.87	47.059		24,800	2,400	650	27,850	31,600
1460	1,000 to 1,600 kcmil		1.47	59.864		42,500	3,050	825	46,375	52,500
1470	Over 1,600 kcmil		1.35	65.185	↓	59,500	3,325	900	63,725	71,500
1500	Joints and dead ends	R-8	6	8	Ea.	1,650	410	47	2,107	2,500
1550	Sagging	R-5	8.80	10	W.Mile		510	138	648	915
1600	Clipping, per structure, 69 kV	R-10	9.60	5	Ea.		274	62.50	336.50	480
1620	161 kV		5.33	9.006			495	113	608	865
1640	345 to 500 kV	↓	2.53	18.972			1,050	237	1,287	1,800
1700	Making and installing jumpers, per structure, 69 kV	R-8	5.87	8.177		450	420	48	918	1,175
1720	161 kV		.96	50		895	2,575	293	3,763	5,175
1740	345 to 500 kV	↓	.32	150		1,500	7,725	880	10,105	14,200
1800	Spacers	R-10	96	.500	↓	89.50	27.50	6.25	123.25	146
1810	Disposal of surplus material, conductors & hardware	R-7	6.96	6.897	Mile		285	24	309	460
2000	Insulators and hardware for underbuilt circuits									
2100	Material handling and spotting	R-7	1200	.040	Ea.		1.65	.14	1.79	2.68
2150	Disk insulators, 69 kV	R-8	600	.080		91.50	4.12	.47	96.09	107
2160	161 kV		686	.070		104	3.60	.41	108.01	121
2170	345 to 500 kV	↓	800	.060	↓	104	3.09	.35	107.44	120
2180	Disposal of surplus material, insulators & hardware	R-7	41.74	1.150	Mile		47.50	3.98	51.48	77
2300	Sectionalizing switches, 69 kV	R-5	1.26	69.841	Ea.	23,500	3,550	960	28,010	32,300
2310	161 kV		.80	110		26,600	5,600	1,525	33,725	39,400
2500	Protective devices	↓	5.50	16	↓	7,700	815	220	8,735	9,925
2600	Clearance poles, 8 poles per mile									
2650	In earth, 69 kV	R-5	1.16	75.862	Mile	6,325	3,850	1,050	11,225	13,900
2660	161 kV	"	.64	138		10,400	7,000	1,900	19,300	24,000
2670	345 to 500 kV	R-6	.48	183		12,500	9,325	6,825	28,650	35,300
2800	In rock, 69 kV	R-5	.69	128		6,325	6,475	1,750	14,550	18,700
2820	161 kV	"	.35	251		10,400	12,800	3,475	26,675	34,500
2840	345 to 500 kV	R-6	.24	367	↓	12,500	18,600	13,600	44,700	57,000

33 72 Utility Substations

33 72 26 – Substation Bus Assemblies

33 72 26.13 Aluminum Substation Bus Assemblies	Crew	Daily Output	Labor-Hours	Unit	Material	2018 Bare Costs Labor	Equipment	Total	Total Incl O&P
0010 **ALUMINUM SUBSTATION BUS ASSEMBLIES**									
7300 Bus	R-11	590	.095	Lb.	3.30	5.25	1.12	9.67	12.75

33 72 33 – Control House Equipment

33 72 33.43 Substation Backup Batteries

	Crew	Daily Output	Labor-Hours	Unit	Material	Labor	Equipment	Total	Total Incl O&P
0010 **SUBSTATION BACKUP BATTERIES**									
9120 Battery chargers	R-11	11.20	5	Ea.	4,200	277	59	4,536	5,100
9200 Control batteries	"	14	4	K.A.H.	535	221	47.50	803.50	965

33 73 Utility Transformers

33 73 23 – Dry-Type Utility Transformers

33 73 23.20 Primary Transformers

	Crew	Daily Output	Labor-Hours	Unit	Material	Labor	Equipment	Total	Total Incl O&P
0010 **PRIMARY TRANSFORMERS**									
1000 Main conversion equipment									
1050 Power transformers, 13 to 26 kV	R-11	1.72	32.558	MVA	26,400	1,800	385	28,585	32,200
1060 46 kV		3.50	16		24,800	885	189	25,874	28,800
1070 69 kV		3.11	18.006		21,100	995	213	22,308	25,000
1080 110 kV		3.29	17.021		20,100	940	201	21,241	23,700
1090 161 kV		4.31	12.993		18,700	720	154	19,574	21,700
1100 500 kV		7	8		18,600	445	94.50	19,139.50	21,200
1200 Grounding transformers		3.11	18.006	Ea.	117,000	995	213	118,208	130,000

33 75 High-Voltage Switchgear and Protection Devices

33 75 13 – Air High-Voltage Circuit Breaker

33 75 13.13 High-Voltage Circuit Breaker, Air

	Crew	Daily Output	Labor-Hours	Unit	Material	Labor	Equipment	Total	Total Incl O&P
0010 **HIGH-VOLTAGE CIRCUIT BREAKER, AIR**									
2100 Air circuit breakers, 13 to 26 kV	R-11	.56	100	Ea.	65,500	5,525	1,175	72,200	81,500
2110 161 kV	"	.24	233	"	281,500	12,900	2,750	297,150	332,000

33 75 16 – Oil High-Voltage Circuit Breaker

33 75 16.13 Oil Circuit Breaker

	Crew	Daily Output	Labor-Hours	Unit	Material	Labor	Equipment	Total	Total Incl O&P
0010 **OIL CIRCUIT BREAKER**									
2000 Power circuit breakers									
2050 Oil circuit breakers, 13 to 26 kV	R-11	1.12	50	Ea.	63,000	2,775	590	66,365	74,500
2060 46 kV		.75	74.667		92,000	4,125	885	97,010	108,000
2070 69 kV		.45	124		213,000	6,900	1,475	221,375	246,500
2080 161 kV		.16	350		324,500	19,400	4,125	348,025	390,500
2090 500 kV		.06	933		1,213,500	51,500	11,000	1,276,000	1,424,500

33 75 19 – Gas High-Voltage Circuit Breaker

33 75 19.13 Gas Circuit Breaker

	Crew	Daily Output	Labor-Hours	Unit	Material	Labor	Equipment	Total	Total Incl O&P
0010 **GAS CIRCUIT BREAKER**									
2000 Power circuit breakers									
2150 Gas circuit breakers, 13 to 26 kV	R-11	.56	100	Ea.	246,000	5,525	1,175	252,700	280,000
2160 161 kV		.08	700		323,000	38,700	8,275	369,975	422,000
2170 500 kV		.04	1400		1,136,000	77,500	16,500	1,230,000	1,383,500

For customer support on your Electrical Costs with RSMeans data, call 800.448.8182.

367

33 75 23 – Vacuum High-Voltage Circuit Breaker

33 75 23.13 Vacuum Circuit Breaker	Crew	Daily Output	Labor-Hours	Unit	Material	2018 Bare Costs Labor	Equipment	Total	Total Incl O&P
0010 **VACUUM CIRCUIT BREAKER**									
2000 Power circuit breakers									
2200 Vacuum circuit breakers, 13 to 26 kV	R-11	.56	100	Ea.	56,000	5,525	1,175	62,700	71,000

33 75 36 – High-Voltage Utility Fuses

33 75 36.13 Fuses

	Crew	Daily Output	Labor-Hours	Unit	Material	Labor	Equipment	Total	Total Incl O&P
0010 **FUSES**									
8000 Protective equipment									
8250 Fuses, 13 to 26 kV	R-11	18.67	2.999	Ea.	2,300	166	35.50	2,501.50	2,825
8260 46 kV		11.20	5		2,625	277	59	2,961	3,350
8270 69 kV		8	7		2,800	385	82.50	3,267.50	3,750
8280 161 kV		4.67	11.991		3,450	665	142	4,257	4,950

33 75 39 – High-Voltage Surge Arresters

33 75 39.13 Surge Arresters

	Crew	Daily Output	Labor-Hours	Unit	Material	Labor	Equipment	Total	Total Incl O&P
0010 **SURGE ARRESTERS**									
8000 Protective equipment									
8050 Lightning arresters, 13 to 26 kV	R-11	18.67	2.999	Ea.	1,725	166	35.50	1,926.50	2,200
8060 46 kV		14	4		4,600	221	47.50	4,868.50	5,425
8070 69 kV		11.20	5		5,900	277	59	6,236	6,975
8080 161 kV		5.60	10		8,200	555	118	8,873	9,975
8090 500 kV		1.40	40		27,900	2,225	475	30,600	34,500

33 75 43 – Shunt Reactors

33 75 43.13 Reactors

	Crew	Daily Output	Labor-Hours	Unit	Material	Labor	Equipment	Total	Total Incl O&P
0010 **REACTORS**									
8000 Protective equipment									
8150 Reactors and resistors, 13 to 26 kV	R-11	28	2	Ea.	4,100	111	23.50	4,234.50	4,700
8160 46 kV		4.31	12.993		12,300	720	154	13,174	14,700
8170 69 kV		2.80	20		19,700	1,100	236	21,036	23,500
8180 161 kV		2.24	25		22,100	1,375	296	23,771	26,800
8190 500 kV		.08	700		87,000	38,700	8,275	133,975	162,500

33 75 53 – High-Voltage Switches

33 75 53.13 Switches

	Crew	Daily Output	Labor-Hours	Unit	Material	Labor	Equipment	Total	Total Incl O&P
0010 **SWITCHES**									
3000 Disconnecting switches									
3050 Gang operated switches									
3060 Manual operation, 13 to 26 kV	R-11	1.65	33.939	Ea.	15,600	1,875	400	17,875	20,400
3070 46 kV		1.12	50		26,200	2,775	590	29,565	33,700
3080 69 kV		.80	70		28,700	3,875	825	33,400	38,300
3090 161 kV		.56	100		34,400	5,525	1,175	41,100	47,500
3100 500 kV		.14	400		92,500	22,100	4,725	119,325	140,500
3110 Motor operation, 161 kV		.51	110		52,000	6,075	1,300	59,375	67,500
3120 500 kV		.28	200		141,000	11,100	2,375	154,475	174,000
3250 Circuit switches, 161 kV		.41	137		101,500	7,550	1,625	110,675	125,000
3300 Single pole switches									
3350 Disconnecting switches, 13 to 26 kV	R-11	28	2	Ea.	13,900	111	23.50	14,034.50	15,500
3360 46 kV		8	7		24,600	385	82.50	25,067.50	27,800
3370 69 kV		5.60	10		26,600	555	118	27,273	30,200
3380 161 kV		2.80	20		99,000	1,100	236	100,336	111,000
3390 500 kV		.22	255		282,000	14,100	3,000	299,100	335,000
3450 Grounding switches, 46 kV		5.60	10		36,900	555	118	37,573	41,600
3460 69 kV		3.73	15.013		37,700	830	177	38,707	42,900

33 75 High-Voltage Switchgear and Protection Devices

33 75 53 – High-Voltage Switches

33 75 53.13 Switches

		Crew	Daily Output	Labor-Hours	Unit	Material	2018 Bare Costs Labor	Equipment	Total	Total Incl O&P
3470	161 kV	R-11	2.24	25	Ea.	40,200	1,375	296	41,871	46,600
3480	500 kV	↓	.62	90.323	↓	50,500	5,000	1,075	56,575	64,000

33 78 Substation Converter Stations

33 78 33 – Converter Stations

33 78 33.46 Substation Converter Stations

		Crew	Daily Output	Labor-Hours	Unit	Material	2018 Bare Costs Labor	Equipment	Total	Total Incl O&P
0010	**SUBSTATION CONVERTER STATIONS**									
9000	Station service equipment									
9100	Conversion equipment									
9110	Station service transformers	R-11	5.60	10	Ea.	101,500	555	118	102,173	112,500

33 79 Site Grounding

33 79 83 – Site Grounding Conductors

33 79 83.13 Grounding Wire, Bar, and Rod

		Crew	Daily Output	Labor-Hours	Unit	Material	2018 Bare Costs Labor	Equipment	Total	Total Incl O&P
0010	**GROUNDING WIRE, BAR, AND ROD**									
7500	Grounding systems	R-11	280	.200	Lb.	11.55	11.05	2.36	24.96	32

33 81 Communications Structures

33 81 13 – Communications Transmission Towers

33 81 13.10 Radio Towers

		Crew	Daily Output	Labor-Hours	Unit	Material	2018 Bare Costs Labor	Equipment	Total	Total Incl O&P
0010	**RADIO TOWERS**									
0020	Guyed, 50' H, 40 lb. section, 70 MPH basic wind speed	2 Sswk	1	16	Ea.	2,775	875		3,650	4,475
0100	Wind load 90 MPH basic wind speed	"	1	16		3,675	875		4,550	5,450
0300	190' high, 40 lb. section, wind load 70 MPH basic wind speed	K-2	.33	72.727		9,525	3,775	720	14,020	17,300
0400	200' high, 70 lb. section, wind load 90 MPH basic wind speed		.33	72.727		15,800	3,775	720	20,295	24,100
0600	300' high, 70 lb. section, wind load 70 MPH basic wind speed		.20	120		25,300	6,225	1,200	32,725	39,100
0700	270' high, 90 lb. section, wind load 90 MPH basic wind speed		.20	120		27,400	6,225	1,200	34,825	41,400
0800	400' high, 100 lb. section, wind load 70 MPH basic wind speed		.14	171		38,200	8,900	1,700	48,800	58,000
0900	Self-supporting, 60' high, wind load 70 MPH basic wind speed		.80	30		4,500	1,550	298	6,348	7,775
0910	60' high, wind load 90 MPH basic wind speed		.45	53.333		5,250	2,775	530	8,555	10,800
1000	120' high, wind load 70 MPH basic wind speed		.40	60		9,925	3,125	595	13,645	16,500
1200	190' high, wind load 90 MPH basic wind speed	↓	.20	120		28,300	6,225	1,200	35,725	42,400
2000	For states west of Rocky Mountains, add for shipping				↓	10%				

For customer support on your Electrical Costs with RSMeans data, call 800.448.8182.

369

Division Notes

	CREW	DAILY OUTPUT	LABOR-HOURS	UNIT	BARE COSTS				TOTAL INCL O&P
					MAT.	LABOR	EQUIP.	TOTAL	

Estimating Tips
34 11 00 Rail Tracks
This subdivision includes items that may involve either repair of existing or construction of new railroad tracks. Additional preparation work, such as the roadbed earthwork, would be found in Division 31. Additional new construction siding and turnouts are found in Subdivision 34 72. Maintenance of railroads is found under 34 01 23 Operation and Maintenance of Railways.

34 40 00 Traffic Signals
This subdivision includes traffic signal systems. Other traffic control devices such as traffic signs are found in Subdivision 10 14 53 Traffic Signage.

34 70 00 Vehicle Barriers
This subdivision includes security vehicle barriers, guide and guard rails, crash barriers, and delineators. The actual maintenance and construction of concrete and asphalt pavement are found in Division 32.

Reference Numbers
Reference numbers are shown at the beginning of some major classifications. These numbers refer to related items in the Reference Section. The reference information may be an estimating procedure, an alternate pricing method, or technical information.

Note: Not all subdivisions listed here necessarily appear. ∎

34 43 13 – Airfield Signals and Lighting

34 43 13.16 Airfield Runway and Taxiway Inset Lighting	Crew	Daily Output	Labor-Hours	Unit	Material	2018 Bare Costs Labor	Equipment	Total	Total Incl O&P
0010	**AIRFIELD RUNWAY AND TAXIWAY INSET LIGHTING**									
0100	Runway centerline, bidir., semi-flush, 200 W, w/shallow insert base	R-22	12.40	3.006	Ea.	1,975	160		2,135	2,425
0110	for mounting in base housing		18.60	2.004		1,025	107		1,132	1,300
0120	Flush, 200 W, w/shallow insert base		12.40	3.006		1,950	160		2,110	2,400
0130	for mounting in base housing		18.64	2		1,025	107		1,132	1,275
0140	45 W, for mounting in base housing		18.64	2		1,025	107		1,132	1,275
0150	Touchdown zone light, unidirectional, 200 W, w/shallow insert base		12.40	3.006		1,800	160		1,960	2,225
0160	115 W		12.40	3.006		1,725	160		1,885	2,150
0170	Bidirectional, 62 W		12.40	3.006		1,825	160		1,985	2,250
0180	Unidirectional, 200 W, for mounting in base housing		18.60	2.004		805	107		912	1,050
0190	115 W		18.60	2.004		865	107		972	1,100
0200	Bidirectional, 62 W		18.60	2.004		965	107		1,072	1,200
0210	Runway edge & threshold light, bidir., 200 W, for base housing		9.36	3.983		1,075	212		1,287	1,500
0220	300 W		9.36	3.983		1,500	212		1,712	1,975
0230	499 W		7.44	5.011		2,525	267		2,792	3,175
0240	Threshold & approach light, unidir., 200 W, for base housing		9.36	3.983		625	212		837	1,000
0250	499 W		7.44	5.011		760	267		1,027	1,250
0260	Runway edge, bidirectional, 2-115 W, for base housing		12.40	3.006		1,650	160		1,810	2,075
0270	2-185 W		12.40	3.006		1,825	160		1,985	2,275
0280	Runway threshold & end, bidir., 2-115 W, for base housing		12.40	3.006		1,825	160		1,985	2,250
0370	45 W, flush, for mounting in base housing		18.64	2		890	107		997	1,150
0380	115 W		18.64	2		915	107		1,022	1,150
0400	Transformer, 45 W		37.20	1.002		196	53.50		249.50	295
0410	65 W		37.20	1.002		201	53.50		254.50	300
0420	100 W		37.20	1.002		248	53.50		301.50	355
0430	200 W, 6.6/6.6 A		37.20	1.002		244	53.50		297.50	350
0440	20/6.6 A		24.80	1.503		261	80		341	405
0450	500 W		24.80	1.503		340	80		420	495
0460	Base housing, 24" deep, 12" diameter		18.64	2		550	107		657	765
0470	15" diameter		18.64	2		1,200	107		1,307	1,475
0480	Connection kit for 1 conductor cable		24.80	1.503		39.50	80		119.50	163
0500	Elevated runway marker, high intensity quartz, edge, 115 W		14.88	2.505		440	133		573	685
0510	175 W		14.88	2.505		455	133		588	700
0520	Threshold, 115 W		14.88	2.505		495	133		628	745
0530	Edge, 200 W, w/base and transformer		11.85	3.146		1,225	168		1,393	1,600
0540	Elevated runway light, quartz, runway edge, 45 W		14.88	2.505		259	133		392	485
0550	Threshold, 115 W		14.88	2.505		430	133		563	670
0560	Taxiway, 30 W		14.88	2.505		237	133		370	460
0570	Fixture, 200 W, w/base and transformer		11.85	3.146		1,225	168		1,393	1,600
0580	Elevated threshold marker, quartz, 115 W		14.88	2.505		540	133		673	795
0590	175 W		14.88	2.505		540	133		673	795

34 43 23 – Weather Observation Equipment

34 43 23.16 Airfield Wind Cones	Crew	Daily Output	Labor-Hours	Unit	Material	2018 Bare Costs Labor	Equipment	Total	Total Incl O&P
0010	**AIRFIELD WIND CONES**									
1200	Wind cone, 12' lighted assembly, rigid, w/obstruction light	R-21	1.36	24.118	Ea.	10,600	1,400	49	12,049	13,800
1210	Without obstruction light		1.52	21.579		8,850	1,250	44	10,144	11,600
1220	Unlighted assembly, w/obstruction light		1.68	19.524		9,650	1,125	40	10,815	12,300
1230	Without obstruction light		1.84	17.826		8,850	1,050	36.50	9,936.50	11,300
1240	Wind cone slip fitter, 2-1/2" pipe		21.84	1.502		117	87.50	3.06	207.56	262
1250	Wind cone sock, 12' x 3', cotton		6.56	5		640	291	10.20	941.20	1,150
1260	Nylon		6.56	5		650	291	10.20	951.20	1,150

Estimating Tips

- When estimating costs for the installation of electrical power generation equipment, factors to review include access to the job site, access and setting up at the installation site, required connections, uncrating pads, anchors, leveling, final assembly of the components, and temporary protection from physical damage, such as environmental exposure.

- Be aware of the costs of equipment supports, concrete pads, and vibration isolators. Cross-reference them against other trades' specifications. Also, review site and structural drawings for items that must be included in the estimates.

- It is important to include items that are not documented in the plans and specifications but must be priced. These items include, but are not limited to, testing, dust protection, roof penetration, core drilling concrete floors and walls, patching, cleanup, and final adjustments. Add a contingency or allowance for utility company fees for power hookups, if needed.

- The project size and scope of electrical power generation equipment will have a significant impact on cost. The intent of RSMeans cost data is to provide a benchmark cost so that owners, engineers, and electrical contractors will have a comfortable number with which to start a project. Additionally, there are many websites available to use for research and to obtain a vendor's quote to finalize costs.

Reference Numbers

Reference numbers are shown at the beginning of some major classifications. These numbers refer to related items in the Reference Section. The reference information may be an estimating procedure, an alternate pricing method, or technical information.

Note: Not all subdivisions listed here necessarily appear. ■

Did you know?

RSMeans data is available through our online application with 24/7 access:

- Search for unit prices by keyword
- Leverage the most up-to-date data
- Build and export estimates

Try it free for 30 days!
www.rsmeans.com/2018freetrial

48 13 Hydroelectric Plant Electrical Power Generation Equipment

48 13 13 – Hydroelectric Power Plant Water Turbines

48 13 13.10 Water Turbines and Components	Crew	Daily Output	Labor-Hours	Unit	Material	2018 Bare Costs Labor	Equipment	Total	Total Incl O&P
0010 **WATER TURBINES & COMPONENTS**									
0500 Water turbine, 100-500 watt, 3P, 48VDC, 54"W x 24"H, 260LBS	2 Elec	2	8	Ea.	9,875	465		10,340	11,600
0600 Water turbine, 1500-2000 watt, 3P, 130-200VAC to DC, 95LBS	"	3	5.333	"	3,900	310		4,210	4,775

48 15 Wind Energy Electrical Power Generation Equipment

48 15 13 – Wind Turbines

48 15 13.50 Wind Turbines and Components

		Crew	Daily Output	Labor-Hours	Unit	Material	2018 Bare Costs Labor	Equipment	Total	Total Incl O&P
0010	**WIND TURBINES & COMPONENTS**									
0500	Complete system, grid connected									
0501	Wind turbine marine rated, 160 W, 12 V, DC	2 Elec	2.20	7.273	Ea.	1,150	425		1,575	1,875
0510	160 W, 24 V, DC		2.20	7.273		1,150	425		1,575	1,875
0520	160 W, 48 V, DC		2.20	7.273		1,150	425		1,575	1,875
1000	20 kW, 31' diam., incl. labor & material G				System				49,900	49,900
1001	Wind turbine, 300 W, 12 V, DC	2 Elec	2.20	7.273	Ea.	855	425		1,280	1,575
1010	Enhanced G				System				92,000	92,000
1011	300 W, 24 V, DC	2 Elec	2.20	7.273	Ea.	855	425		1,280	1,575
1020	300 W, 48 V, DC		2.20	7.273		855	425		1,280	1,575
1030	Wind turbine, 400 W, 12 V, DC		2.20	7.273		855	425		1,280	1,575
1040	400 W, 24 V, DC		2.20	7.273		855	425		1,280	1,575
1050	400 W, 48 V, DC		2.20	7.273		855	425		1,280	1,575
1500	10 kW, 23' diam., incl. labor & material G				System				74,000	74,000
2000	2.4 kW, 12' diam., incl. labor & material G				"				18,000	18,000
2900	Component system									
3200	1,000 W, 9' diam. G	1 Elec	2.05	3.902	Ea.	1,500	227		1,727	2,000
3300	Wind turbine/charge control, 2,900-7,900 W, 5.5" H x 11' W, 180 lb.	2 Elec	2	8		15,900	465		16,365	18,200
3320	Wind turbine/charge control, 4,000-10,800 W, 5.5" H x 13' W, 240 lb.	"	2	8		26,400	465		26,865	29,700
3400	Mounting hardware									
3401	Wind turbine tower, 2 stage steel, 8' H x 6' W	2 Elec	2	8	Ea.	2,000	465		2,465	2,900
3500	30' guyed tower kit G	2 Clab	5.12	3.125		395	125		520	620
3505	3' galvanized helical earth screw G	1 Clab	8	1		48	40		88	113
3510	Attic mount kit G	1 Rofc	2.56	3.125		202	137		339	455
3520	Roof mount kit G	1 Clab	3.41	2.346		292	93.50		385.50	460
4000	Anemometer, standard AC or 9V, 2-digit LCD, wood case, +/-2% of Input	2 Elec	2	8		160	465		625	870
8900	Equipment									
9100	DC to AC inverter for, 48 V, 4,000 W G	1 Elec	2	4	Ea.	2,275	233		2,508	2,850

48 18 Fuel Cell Electrical Power Generation Equipment

48 18 13 – Electrical Power Generation Fuel Cells

48 18 13.10 Fuel Cells

					Unit	Material	Labor	Equipment	Total	Total Incl O&P
0010	**FUEL CELLS**									
2001	Complete system, natural gas, installed, 200 kW G				System				1,150,000	1,150,000
2002	400 kW G								2,168,000	2,168,000
2003	600 kW G								3,060,000	3,060,000
2004	800 kW G								4,000,000	4,000,000
2005	1,000 kW G								5,005,000	5,005,000
2010	Comm type, for battery charging, uses hydrogen, 100 W, 12 V G				Ea.	2,350			2,350	2,575
2020	500 W, 12 V G					4,575			4,575	5,025
2030	1,000 W, 48 V G					7,250			7,250	7,975
2040	Small, for demonstration or battery charging units, 12 V G					159			159	175

48 18 Fuel Cell Electrical Power Generation Equipment

48 18 13 – Electrical Power Generation Fuel Cells

48 18 13.10 Fuel Cells		Crew	Daily Output	Labor-Hours	Unit	Material	2018 Bare Costs Labor	Equipment	Total	Total Incl O&P
2050	Spare anode set for unit	G				Ea.	39			39	43
2060	3 W, uses hydrogen gas	G					299			299	330
2070	6 W, uses hydrogen gas	G					520			520	570
2080	10 W, uses hydrogen gas	G					1,325			1,325	1,450

For customer support on your Electrical Costs with RSMeans data, call 800.448.8182.

375

Division Notes

		CREW	DAILY OUTPUT	LABOR-HOURS	UNIT	BARE COSTS				TOTAL INCL O&P
						MAT.	LABOR	EQUIP.	TOTAL	

Assemblies Section

Table of Contents

RSMeans data: Assemblies— How They Work

Assemblies estimating provides a fast and reasonably accurate way to develop construction costs. An assembly is the grouping of individual work items—with appropriate quantities—to provide a cost for a major construction component in a convenient unit of measure.

An assemblies estimate is often used during early stages of design development to compare the cost impact of various design alternatives on total building cost.

Assemblies estimates are also used as an efficient tool to verify construction estimates.

Assemblies estimates do not require a completed design or detailed drawings. Instead, they are based on the general size of the structure and other known parameters of the project. The degree of accuracy of an assemblies estimate is generally within +/- 15%.

Most assemblies consist of three major elements: a graphic, the system components, and the cost data itself. The **Graphic** is a visual representation showing the typical appearance of the assembly

① Unique 12-character Identifier

Our assemblies are identified by a **unique 12-character identifier**. The assemblies are numbered using UNIFORMAT II, ASTM Standard E1557. The first 5 characters represent this system to Level 3. The last 7 characters represent further breakdown by in order to arrange items in understandable groups of similar tasks. Line numbers are consistent across all our publications, so a line number in any assemblies data set will always refer to the same item.

② Reference Box

Information is available in the Reference Section to assist the estimator with estimating procedures, alternate pricing methods, and additional technical information.

The **Reference Box** indicates the exact location of this information in the Reference Section. The "R" stands for "reference," and the remaining characters are the line numbers.

③ Narrative Descriptions

Our assemblies descriptions appear in two formats: narrative and table. **Narrative descriptions** are shown in a hierarchical structure to make them readable. In order to read a complete description, read up through the indents to the top of the section. Include everything that is above and to the left that is not contradicted by information below.

D20 Plumbing

D2020 Domestic Water Distribution

Installation includes piping and fittings within 10' of heater. Electric water heaters do not require venting.

1 Kilowatt hour will raise:

Gallons of Water	Degrees F	Gallons of Water	Degrees F
4.1	100°	6.8	60°
4.5	90°	8.2	50°
5.1	80°	10.0	40°
5.9	70°		

System Components	QUANTITY	UNIT	COST EACH MAT.	COST EACH INST.	COST EACH TOTAL
SYSTEM D2020 210 1780					
ELECTRIC WATER HEATER, RESIDENTIAL, 100°F RISE					
10 GALLON TANK, 7 GPH					
Water heater, residential electric, glass lined tank, 10 gal.	1.000	Ea.	470	325	795
Copper tubing, type L, solder joint, hanger 10' OC 1/2" diam.	30.000	L.F.	110.40	277.50	387.90
Wrought copper 90° elbow for solder joints 1/2" diam.	4.000	Ea.	4.96	150	154.96
Wrought copper Tee for solder joints, 1/2" diam.	2.000	Ea.	4.22	115	119.22
Union, wrought copper, 1/2" diam.	2.000	Ea.	42	79	121
Valve, gate, bronze, 125 lb, NRS, soldered 1/2" diam.	2.000	Ea.	123	62	185
Relief valve, bronze, press & temp, self-close, 3/4" IPS	1.000	Ea.	230	26.50	256.50
Wrought copper adapter, CTS to MPT 3/4" IPS	1.000	Ea.	5.95	44	49.95
Copper tubing, type L, solder joints, 3/4" diam.	1.000	L.F.	4.94	9.85	14.79
Wrought copper 90° elbow for solder joints 3/4" diam.	1.000	Ea.	2.77	39.50	42.27
TOTAL			998.24	1,128.35	2,126.59

D2020 210	Electric Water Heaters - Residential Systems		MAT.	INST.	TOTAL
1760	Electric water heater, residential, 100°F rise				
1780	10 gallon tank, 7 GPH		1,000	1,125	2,125
1820	20 gallon tank, 7 GPH	② R224000 -20	1,250	1,200	2,450
1860	③ 30 gallon tank, 7 GPH		1,900	1,275	3,175
1900	40 gallon tank, 8 GPH		2,150	1,400	3,550
1940	52 gallon tank, 10 GPH		2,300	1,400	3,700
1980	66 gallon tank, 13 GPH		3,600	1,600	5,200
2020	80 gallon tank, 16 GPH		3,875	1,675	5,550
2060	120 gallon tank, 23 GPH		5,475	1,950	7,425

For supplemental customizable square foot estimating forms, visit: **www.RSMeans.com/2018books**

in question. It is frequently accompanied by additional explanatory technical information describing the class of items. The **System Components** is a listing of the individual tasks that make up the assembly, including the quantity and unit of measure for each item, along with the cost of material and installation. The **Assemblies**

Data below lists prices for other similar systems with dimensional and/or size variations.

All of our assemblies costs represent the cost for the installing contractor. An allowance for profit has been added to all material, labor, and equipment rental costs. A markup for labor burdens, including workers' compensation, fixed

overhead, and business overhead, is included with installation costs.

The information in RSMeans cost data represents a "national average" cost. This data should be modified to the project location using the **City Cost Indexes** or **Location Factors** tables found in the Reference Section.

D20 Plumbing

D2020 Domestic Water Distribution

Installation includes piping and fittings within 10' of heater. Electric water heaters do not require venting.

1 Kilowatt hour will raise:			
Gallons of Water	Degrees F	Gallons of Water	Degrees F
4.1	100°	6.8	60°
4.5	90°	8.2	50°
5.1	80°	10.0	40°
5.9	70°		

4 Unit of Measure

All RSMeans data: Assemblies include a typical **Unit of Measure** used for estimating that item. For instance, for continuous footings or foundation walls the unit is linear feet (L.F.). For spread footings the unit is each (Ea.). The estimator needs to take special care that the unit in the data matches the unit in the takeoff. Abbreviations and unit conversions can be found in the Reference Section.

5 System Components

System components are listed separately to detail what is included in the development of the total system price.

System Components 5

	QUANTITY	UNIT	COST EACH		
			MAT.	INST.	TOTAL
SYSTEM D2020 210 1780					
ELECTRIC WATER HEATER, RESIDENTIAL, 100°F RISE					
10 GALLON TANK, 7 GPH					
Water heater, residential electric, glass lined tank, 10 gal.	1.000	Ea.	470	325	795
Copper tubing, type L, solder joint, hanger 10' OC 1/2" diam.	30.000	L.F.	110.40	277.50	387.90
Wrought copper 90° elbow for solder joints 1/2" diam.	4.000	Ea.	4.96	150	154.96
Wrought copper Tee for solder joints, 1/2" diam.	2.000	Ea.	4.22	115	119.22
Union, wrought copper, 1/2" diam.	2.000	Ea.	42	79	121
Valve, gate, bronze, 125 lb, NRS, soldered 1/2" diam.	2.000	Ea.	123	62	185
Relief valve, bronze, press & temp, self-close, 3/4" IPS	1.000	Ea.	230	26.50	256.50
Wrought copper adapter, CTS to MPT 3/4" IPS	1.000	Ea.	5.95	44	49.95
Copper tubing, type L, solder joints, 3/4" diam.	1.000	L.F.	4.94	9.85	14.79
Wrought copper 90° elbow for solder joints 3/4" diam.	1.000	Ea.	2.77	39.50	42.27
TOTAL			998.24	1,128.35	2,126.59

D2020 210	Electric Water Heaters - Residential Systems		COST EACH		
			MAT.	INST.	TOTAL
1760	Electric water heater, residential, 100°F rise				
1780	10 gallon tank, 7 GPH		1,000	1,125	2,125
1820	20 gallon tank, 7 GPH		1,250	1,200	2,450
1860	30 gallon tank, 7 GPH	R224000 -20	1,900	1,275	3,175
1900	40 gallon tank, 8 GPH		2,150	1,400	3,550
1940	52 gallon tank, 10 GPH		2,300	1,400	3,700
1980	66 gallon tank, 13 GPH		3,600	1,600	5,200
2020	80 gallon tank, 16 GPH		3,875	1,675	5,550
2060	120 gallon tank, 23 GPH		5,475	1,950	7,425

RSMeans data: Assemblies—How They Work (Continued)

Sample Estimate

This sample demonstrates the elements of an estimate, including a tally of the RSMeans data lines. Published assemblies costs include all markups for labor burden and profit for the installing contractor. This estimate adds a summary of the markups applied by a general contractor on the installing contractor's work. These figures represent the total cost to the owner. The location factor with RSMeans data is applied at the bottom of the estimate to adjust the cost of the work to a specific location.

Project Name:	Interior Fit-out, ABC Office			
Location:	**Anywhere, USA**	**Date: 1/1/2018**		**STD**
Assembly Number	**Description**	**Qty.**	**Unit**	**Subtotal**
❶ C1010 124 1200	Wood partition, 2 x 4 @ 16" OC w/5/8" FR gypsum board	560.000	S.F.	$2,856.00
C1020 114 1800	Metal door & frame, flush hollow core, 3'-0" x 7'-0"	2.000	Ea.	$2,510.00
C3010 230 0080	Painting, brushwork, primer & 2 coats	1,120.000	S.F.	$1,422.40
C3020 410 0140	Carpet, tufted, nylon, roll goods, 12' wide, 26 oz	240.000	S.F.	$842.40
C3030 210 6000	Acoustic ceilings, 24" x 48" tile, tee grid suspension	200.000	S.F.	$1,260.00
D5020 125 0560	Receptacles incl plate, box, conduit, wire, 20 A duplex	8.000	Ea.	$2,320.00
D5020 125 0720	Light switch incl plate, box, conduit, wire, 20 A single pole	2.000	Ea.	$566.00
D5020 210 0560	Fluorescent fixtures, recess mounted, 20 per 1000 SF	200.000	S.F.	$2,226.00
	Assembly Subtotal			**$14,002.80**
	Sales Tax @ ❷	5 %		$ 350.07
	General Requirements @ ❸	7 %		$ 980.20
	Subtotal A			**$15,333.07**
	GC Overhead @ ❹	5 %		$ 766.65
	Subtotal B			**$16,099.72**
	GC Profit @ ❺	5 %		$ 804.99
	Subtotal C			**$16,904.71**
	Adjusted by Location Factor ❻	115.2		$ 19,474.22
	Architects Fee @ ❼	8 %		$ 1,557.94
	Contingency @ ❽	15 %		$ 2,921.13
	Project Total Cost			**$ 23,953.29**

This estimate is based on an interactive spreadsheet. You are free to download it and adjust it to your methodology.
A copy of this spreadsheet is available at **www.RSMeans.com/2018books.**

1 Work Performed

The body of the estimate shows the RSMeans data selected, including line numbers, a brief description of each item, its takeoff quantity and unit, and the total installed cost, including the installing contractor's overhead and profit.

2 Sales Tax

If the work is subject to state or local sales taxes, the amount must be added to the estimate. In a conceptual estimate it can be assumed that one half of the total represents material costs. Therefore, apply the sales tax rate to 50% of the assembly subtotal.

3 General Requirements

This item covers project-wide needs provided by the general contractor. These items vary by project but may include temporary facilities and utilities, security, testing, project cleanup, etc. In assemblies estimates a percentage is used—typically between 5% and 15% of project cost.

4 General Contractor Overhead

This entry represents the general contractor's markup on all work to cover project administration costs.

5 General Contractor Profit

This entry represents the GC's profit on all work performed. The value included here can vary widely by project and is influenced by the GC's perception of the project's financial risk and market conditions.

6 Location Factor

RSMeans published data are based on national average costs. If necessary, adjust the total cost of the project using a location factor from the "Location Factor" table or the "City Cost Indexes" table found in the Reference Section. Use location factors if the work is general, covering the work of multiple trades. If the work is by a single trade (e.g., masonry) use the more specific data found in the City Cost Indexes.

To adjust costs by location factors, multiply the base cost by the factor and divide by 100.

7 Architect's Fee

If appropriate, add the design cost to the project estimate. These fees vary based on project complexity and size. Typical design and engineering fees can be found in the Reference Section.

8 Contingency

A factor for contingency may be added to any estimate to represent the cost of unknowns that may occur between the time that the estimate is performed and the time the project is constructed. The amount of the allowance will depend on the stage of design at which the estimate is done, and the contractor's assessment of the risk involved.

D2020 Domestic Water Distribution

Installation includes piping and fittings within 10' of heater. Electric water heaters do not require venting.

1 Kilowatt hour will raise:			
Gallons of Water	Degrees F	Gallons of Water	Degrees F
4.1	100°	6.8	60°
4.5	90°	8.2	50°
5.1	80°	10.0	40°
5.9	70°		

System Components	QUANTITY	UNIT	COST EACH		
			MAT.	INST.	TOTAL
SYSTEM D2020 210 1780					
ELECTRIC WATER HEATER, RESIDENTIAL, 100°F RISE					
10 GALLON TANK, 7 GPH					
Water heater, residential electric, glass lined tank, 10 gal.	1.000	Ea.	470	325	795
Copper tubing, type L, solder joint, hanger 10' OC 1/2" diam.	30.000	L.F.	110.40	277.50	387.90
Wrought copper 90° elbow for solder joints 1/2" diam.	4.000	Ea.	4.96	150	154.96
Wrought copper Tee for solder joints, 1/2" diam.	2.000	Ea.	4.22	115	119.22
Union, wrought copper, 1/2" diam.	2.000	Ea.	42	79	121
Valve, gate, bronze, 125 lb, NRS, soldered 1/2" diam.	2.000	Ea.	123	62	185
Relief valve, bronze, press & temp, self-close, 3/4" IPS	1.000	Ea.	230	26.50	256.50
Wrought copper adapter, CTS to MPT 3/4" IPS	1.000	Ea.	5.95	44	49.95
Copper tubing, type L, solder joints, 3/4" diam.	1.000	L.F.	4.94	9.85	14.79
Wrought copper 90° elbow for solder joints 3/4" diam.	1.000	Ea.	2.77	39.50	42.27
TOTAL			998.24	1,128.35	2,126.59

D2020 210	Electric Water Heaters - Residential Systems	COST EACH		
		MAT.	INST.	TOTAL
1760	Electric water heater, residential, 100°F rise			
1780	10 gallon tank, 7 GPH	1,000	1,125	2,125
1820	20 gallon tank, 7 GPH	1,250	1,200	2,450
1860	30 gallon tank, 7 GPH	1,900	1,275	3,175
1900	40 gallon tank, 8 GPH	2,150	1,400	3,550
1940	52 gallon tank, 10 GPH	2,300	1,400	3,700
1980	66 gallon tank, 13 GPH	3,600	1,600	5,200
2020	80 gallon tank, 16 GPH	3,875	1,675	5,550
2060	120 gallon tank, 23 GPH	5,475	1,950	7,425

(Note: R224000-20 appears near the 1820/1860 rows)

D2020   Domestic Water Distribution

Systems below include piping and fittings within 10' of heater. Electric water heaters do not require venting.

System Components	QUANTITY	UNIT	COST EACH		
			MAT.	INST.	TOTAL
SYSTEM D2020 240 1820					
ELECTRIC WATER HEATER, COMMERCIAL, 100°F RISE					
50 GALLON TANK, 9 KW, 37 GPH					
Water heater, commercial, electric, 50 Gal, 9 KW, 37 GPH	1.000	Ea.	8,000	415	8,415
Copper tubing, type L, solder joint, hanger 10' OC, 3/4" diam	34.000	L.F.	167.96	334.90	502.86
Wrought copper 90° elbow for solder joints 3/4" diam	5.000	Ea.	13.85	197.50	211.35
Wrought copper Tee for solder joints, 3/4" diam	2.000	Ea.	14	125	139
Wrought copper union for soldered joints, 3/4" diam.	2.000	Ea.	53	83	136
Valve, gate, bronze, 125 lb, NRS, soldered 3/4" diam	2.000	Ea.	140	75	215
Relief valve, bronze, press & temp, self-close, 3/4" IPS	1.000	Ea.	230	26.50	256.50
Wrought copper adapter, copper tubing to male, 3/4" IPS	1.000	Ea.	5.95	44	49.95
TOTAL			8,624.76	1,300.90	9,925.66

D2020 240	Electric Water Heaters - Commercial Systems		COST EACH		
			MAT.	INST.	TOTAL
1800	Electric water heater, commercial, 100°F rise				
1820	50 gallon tank, 9 KW 37 GPH		8,625	1,300	9,925
1860	80 gal, 12 KW 49 GPH		11,000	1,600	12,600
1900	36 KW 147 GPH	R224000 -10	14,900	1,750	16,650
1940	120 gal, 36 KW 147 GPH		16,100	1,875	17,975
1980	150 gal, 120 KW 490 GPH	R224000 -20	50,000	2,000	52,000
2020	200 gal, 120 KW 490 GPH		53,000	2,075	55,075
2060	250 gal, 150 KW 615 GPH		58,500	2,425	60,925
2100	300 gal, 180 KW 738 GPH		63,500	2,525	66,025
2140	350 gal, 30 KW 123 GPH		47,200	2,750	49,950
2180	180 KW 738 GPH		65,500	2,750	68,250
2220	500 gal, 30 KW 123 GPH		61,500	3,200	64,700
2260	240 KW 984 GPH		99,500	3,200	102,700
2300	700 gal, 30 KW 123 GPH		74,500	3,675	78,175
2340	300 KW 1230 GPH		110,500	3,675	114,175
2380	1000 gal, 60 KW 245 GPH		90,000	5,125	95,125
2420	480 KW 1970 GPH		146,500	5,125	151,625
2460	1500 gal, 60 KW 245 GPH		131,500	6,325	137,825
2500	480 KW 1970 GPH		181,000	6,325	187,325

For customer support on your Electrical Costs with RSMeans data, call 800.448.8182.

385

D3020 Heat Generating Systems

Boiler

Baseboard Radiation

**Small Electric Boiler
System Considerations:**
1. Terminal units are fin tube baseboard radiation rated at 720 BTU/hr with 200° water temperature or 820 BTU/hr steam.
2. Primary use being for residential or smaller supplementary areas, the floor levels are based on 7-1/2' ceiling heights.
3. All distribution piping is copper for boilers through 205 MBH. All piping for larger systems is steel pipe.

System Components	QUANTITY	UNIT	COST EACH		
			MAT.	INST.	TOTAL
SYSTEM D3020 102 1120					
SMALL HEATING SYSTEM, HYDRONIC, ELECTRIC BOILER					
1,480 S.F., 61 MBH, STEAM, 1 FLOOR					
Boiler, electric steam, std cntrls, trim, ftngs and valves, 18 KW, 61.4 MBH	1.000	Ea.	5,225	1,897.50	7,122.50
Copper tubing type L, solder joint, hanger 10'OC, 1-1/4" diam	160.000	L.F.	2,016	2,064	4,080
Radiation, 3/4" copper tube w/alum fin baseboard pkg 7" high	60.000	L.F.	519	1,410	1,929
Rough in baseboard panel or fin tube with valves & traps	10.000	Set	2,670	7,150	9,820
Pipe covering, calcium silicate w/cover, 1" wall 1-1/4" diam	160.000	L.F.	707.20	1,224	1,931.20
Low water cut-off, quick hookup, in gage glass tappings	1.000	Ea.	310	47.50	357.50
TOTAL			11,447.20	13,793	25,240.20
COST PER S.F.			7.73	9.32	17.05

D3020 102	Small Heating Systems, Hydronic, Electric Boilers	COST PER S.F.		
		MAT.	INST.	TOTAL
1100	Small heating systems, hydronic, electric boilers			
1120	Steam, 1 floor, 1480 S.F., 61 M.B.H.	7.70	9.32	17.02
1160	3,000 S.F., 123 M.B.H.	5.95	8.20	14.15
1200	5,000 S.F., 205 M.B.H.	4.85	7.55	12.40
1240	2 floors, 12,400 S.F., 512 M.B.H.	3.94	7.50	11.44
1280	3 floors, 24,800 S.F., 1023 M.B.H.	4.57	7.40	11.97
1320	34,750 S.F., 1,433 M.B.H.	4.19	7.20	11.39
1360	Hot water, 1 floor, 1,000 S.F., 41 M.B.H.	13.15	5.15	18.30
1400	2,500 S.F., 103 M.B.H.	8.90	9.30	18.20
1440	2 floors, 4,850 S.F., 205 M.B.H.	8.20	11.20	19.40
1480	3 floors, 9,700 S.F., 410 M.B.H.	8.80	11.60	20.40

D3020 Heat Generating Systems

Boiler

Unit Heater

Large Electric Boiler System Considerations:

1. Terminal units are all unit heaters of the same size. Quantities are varied to accommodate total requirements.
2. All air is circulated through the heaters a minimum of three times per hour.
3. As the capacities are adequate for commercial use, floor levels are based on 10' ceiling heights.
4. All distribution piping is black steel pipe.

System Components	QUANTITY	UNIT	COST EACH		
			MAT.	INST.	TOTAL
SYSTEM D3020 104 1240					
LARGE HEATING SYSTEM, HYDRONIC, ELECTRIC BOILER					
9,280 S.F., 135 KW, 461 MBH, 1 FLOOR					
Boiler, electric hot water, std ctrls, trim, ftngs, valves, 135 KW, 461 MBH	1.000	Ea.	12,450	4,125	16,575
Expansion tank, painted steel, 60 Gal capacity ASME	1.000	Ea.	4,875	228	5,103
Circulating pump, CI, close cpld, 50 GPM, 2 HP, 2" pipe conn	1.000	Ea.	3,275	450	3,725
Unit heater, 1 speed propeller, horizontal, 200° EWT, 72.7 MBH	7.000	Ea.	6,475	1,736	8,211
Unit heater piping hookup with controls	7.000	Set	5,810	11,375	17,185
Pipe, steel, black, schedule 40, welded, 2-1/2" diam	380.000	L.F.	6,308	11,358.20	17,666.20
Pipe covering, calcium silicate w/cover, 1" wall, 2-1/2" diam	380.000	L.F.	2,090	2,983	5,073
TOTAL			41,283	32,255.20	73,538.20
COST PER S.F.			4.45	3.48	7.93

D3020 104	Large Heating Systems, Hydronic, Electric Boilers	COST PER S.F.		
		MAT.	INST.	TOTAL
1230	Large heating systems, hydronic, electric boilers			
1240	9,280 S.F., 135 K.W., 461 M.B.H., 1 floor	4.45	3.47	7.92
1280	14,900 S.F., 240 K.W., 820 M.B.H., 2 floors	5.85	5.65	11.50
1320	18,600 S.F., 296 K.W., 1,010 M.B.H., 3 floors	5.80	6.10	11.90
1360	26,100 S.F., 420 K.W., 1,432 M.B.H., 4 floors	5.70	6	11.70
1400	39,100 S.F., 666 K.W., 2,273 M.B.H., 4 floors	4.87	5	9.87
1440	57,700 S.F., 900 K.W., 3,071 M.B.H., 5 floors	4.63	4.96	9.59
1480	111,700 S.F., 1,800 K.W., 6,148 M.B.H., 6 floors	4.10	4.26	8.36
1520	149,000 S.F., 2,400 K.W., 8,191 M.B.H., 8 floors	4.07	4.25	8.32
1560	223,300 S.F., 3,600 K.W., 12,283 M.B.H., 14 floors	4.34	4.83	9.17

D4090 Other Fire Protection Systems

General: Automatic fire protection (suppression) systems other than water sprinklers may be desired for special environments, high risk areas, isolated locations or unusual hazards. Some typical applications would include:

Paint dip tanks
Securities vaults
Electronic data processing
Tape and data storage
Transformer rooms
Spray booths
Petroleum storage
High rack storage

Piping and wiring costs are dependent on the individual application and must be added to the component costs shown below.

All areas are assumed to be open.

D4090 910	Fire Suppression Unit Components	COST EACH		
		MAT.	INST.	TOTAL
0020	Detectors with brackets			
0040	Fixed temperature heat detector	56.50	99.50	156
0060	Rate of temperature rise detector	56.50	87	143.50
0080	Ion detector (smoke) detector	132	112	244
0200	Extinguisher agent			
0240	200 lb FM200, container	7,275	335	7,610
0280	75 lb carbon dioxide cylinder	1,450	225	1,675
0320	Dispersion nozzle			
0340	FM200 1-1/2" dispersion nozzle	200	53.50	253.50
0380	Carbon dioxide 3" x 5" dispersion nozzle	158	41.50	199.50
0420	Control station			
0440	Single zone control station with batteries	1,025	695	1,720
0470	Multizone (4) control station with batteries	3,050	1,400	4,450
0490				
0500	Electric mechanical release	1,025	360	1,385
0520				
0550	Manual pull station	87	125	212
0570				
0640	Battery standby power 10" x 10" x 17"	455	174	629
0700				
0740	Bell signalling device	137	87	224

D4090 920	FM200 Systems	COST PER C.F.		
		MAT.	INST.	TOTAL
0820	Average FM200 system, minimum			1.94
0840	Maximum			3.85

D5010 Electrical Service/Distribution

System Components			COST PER L.F.		
	QUANTITY	UNIT	MAT.	INST.	TOTAL
SYSTEM D5010 110 0200					
HIGH VOLTAGE CABLE, NEUTRAL AND CONDUIT INCLUDED, COPPER #2, 5 kV					
Shielded cable, no splice/termn, copper, XLP shielding, 5 kV, #2	.030	C.L.F.	8.64	10.50	19.14
Wire, 600 volt, type THW, copper, stranded, #8	.010	C.L.F.	.38	.87	1.25
Rigid galv steel conduit to 15' H, 2" diam, w/term, ftng & support	1.000	L.F.	10.95	15.45	26.40
TOTAL			19.97	26.82	46.79

D5010 110	High Voltage Shielded Conductors	COST PER L.F.		
		MAT.	INST.	TOTAL
0200	High voltage cable, neutral & conduit included, copper #2, 5 kV	19.95	27	46.95
0240	Copper #1, 5 kV	24	31.50	55.50
0280	15 kV	25	31.50	56.50
0320	Copper 1/0, 5 kV	27	40	67
0360	15 kV	29.50	40	69.50
0400	25 kV	34.50	40.50	75
0440	35 kV	36	41.50	77.50
0480	Copper 2/0, 5 kV	29.50	41	70.50
0520	15 kV	32	41	73
0560	25 kV	36.50	41.50	78
0600	35 kV	43.50	46	89.50
0640	Copper 4/0, 5 kV	39	46	85
0680	15 kV	41.50	46	87.50
0720	25 kV	46.50	47	93.50
0760	35 kV	50	48	98
0800	Copper 250 kcmil, 5 kV	42.50	47	89.50
0840	15 kV	43.50	47	90.50
0880	25 kV	52.50	48	100.50
0920	35 kV	57.50	52.50	110
0960	Copper 350 kcmil, 5 kV	52	52.50	104.50
1000	15 kV	52.50	52.50	105
1040	25 kV	61.50	54	115.50
1080	35 kV	65	55.50	120.50
1120	Copper 500 kcmil, 5 kV	81.50	66	147.50
1160	15 kV	80	66	146
1200	25 kV	89	67.50	156.50
1240	35 kV	93.50	69.50	163

For customer support on your Electrical Costs with RSMeans data, call 800.448.8182.

389

D5010 Electrical Service/Distribution

System Components	QUANTITY	UNIT	COST EACH		
			MAT.	INST.	TOTAL
SYSTEM D5010 120 0220					
SERVICE INSTALLATION, INCLUDES BREAKERS, METERING, 20' CONDUIT & WIRE					
3 PHASE, 4 WIRE, 60 A					
Wire, 600 volt, copper type XHHW, stranded #6	.600	C.L.F.	33.90	64.20	98.10
Rigid galvanized steel conduit, 3/4" including fittings	20.000	L.F.	101	174	275
Service entrance cap, 3/4" diameter	1.000	Ea.	6.55	53.50	60.05
Conduit LB fitting with cover, 3/4" diameter	1.000	Ea.	9.15	53.50	62.65
Meter socket, three phase, 100 A	1.000	Ea.	116	249	365
Safety switches, heavy duty, 240 volt, 3 pole NEMA 1 fusible, 60 amp	1.000	Ea.	160	305	465
Grounding, wire ground bare armored, #8-1 conductor	.200	C.L.F.	15.30	70	85.30
Grounding, clamp, bronze, 3/4" diameter	1.000	Ea.	6.15	22	28.15
Grounding, rod, copper clad, 8' long, 3/4" diameter	1.000	Ea.	38.50	131	169.50
Wireway w/fittings, 2-1/2" x 2-1/2"	1.000	L.F.	13.45	15.45	28.90
TOTAL			500	1,137.65	1,637.65

D5010 120	Overhead Electric Service, 3 Phase - 4 Wire	COST EACH		
		MAT.	INST.	TOTAL
0200	Service installation, includes breakers, metering, 20' conduit & wire			
0220	3 phase, 4 wire, 120/208 volts, 60 A	500	1,150	1,650
0240	100 A	675	1,275	1,950
0245	100 A w/circuit breaker	1,475	1,600	3,075
0280	200 A	1,175	1,775	2,950
0285	200 A w/circuit breaker	2,850	2,225	5,075
0320	400 A	2,350	3,525	5,875
0325	400 A w/circuit breaker	5,525	4,400	9,925
0360	600 A	3,850	5,200	9,050
0365	600 A, w/switchboard	8,250	6,600	14,850
0400	800 A	6,100	6,250	12,350
0405	800 A, w/switchboard	10,500	7,825	18,325
0440	1000 A	7,425	7,700	15,125
0445	1000 A, w/switchboard	12,700	9,450	22,150
0480	1200 A	9,700	8,775	18,475
0485	1200 A, w/groundfault switchboard	32,500	10,900	43,400
0520	1600 A	12,400	11,500	23,900
0525	1600 A, w/groundfault switchboard	37,100	13,700	50,800

D5010 Electrical Service/Distribution

D5010 120	Overhead Electric Service, 3 Phase - 4 Wire	COST EACH		
		MAT.	INST.	TOTAL
0560	2000 A	16,000	13,900	29,900
0565	2000 A, w/groundfault switchboard	43,800	17,200	61,000
0610	1 phase, 3 wire, 120/240 volts, 100 A (no safety switch)	160	630	790
0615	100 A w/load center	330	1,200	1,530
0620	200 A	360	860	1,220
0625	200 A w/load center	660	1,825	2,485

For customer support on your Electrical Costs with RSMeans data, call 800.448.8182.

391

System Components	QUANTITY	UNIT	COST EACH		
			MAT.	INST.	TOTAL
SYSTEM D5010 130 1000					
2000 AMP UNDERGROUND ELECTRIC SERVICE WITH SAFETY SWITCH					
INCLUDING EXCAVATION, BACKFILL AND COMPACTION					
Excavate Trench	44.440	B.C.Y.		312.86	312.86
4 inch conduit bank	108.000	L.F.	1,290.60	2,808	4,098.60
4 inch fitting	8.000	Ea.	336	464	800
4 inch bells	4.000	Ea.	21.20	232	253.20
Concrete material	16.580	C.Y.	2,205.14		2,205.14
Concrete placement	16.580	C.Y.		422.29	422.29
Backfill trench	33.710	L.C.Y.		102.14	102.14
Compact fill material in trench	25.930	E.C.Y.		134.31	134.31
Dispose of excess fill material on-site	24.070	L.C.Y.		291.24	291.24
500 kcmil power cable	18.000	C.L.F.	13,680	7,830	21,510
Wire, 600 volt, type THW, copper, stranded, 1/0	6.000	C.L.F.	1,428	1,266	2,694
Saw cutting, concrete walls, plain, per inch of depth	64.000	L.F.	3.20	485.12	488.32
Meter centers and sockets, single pos, 4 terminal, 400 amp	5.000	Ea.	4,450	2,050	6,500
Safety switch, 400 amp	5.000	Ea.	5,500	3,875	9,375
Ground rod clamp	6.000	Ea.	36.90	132	168.90
Wireway	1.000	L.F.	13.45	15.45	28.90
600 volt stranded copper wire, 1/0	1.200	C.L.F.	285.60	253.20	538.80
Flexible metallic conduit	120.000	L.F.	76.80	522	598.80
Ground rod	6.000	Ea.	324	1,044	1,368
TOTAL			29,650.89	22,239.61	51,890.50

D5010 130	Underground Electric Service	COST EACH		
		MAT.	INST.	TOTAL
0950	Underground electric service including excavation, backfill and compaction			
1000	3 phase, 4 wire, 277/480 volts, 2000 A	29,700	22,200	51,900
1050	2000 A w/groundfault switchboard	56,500	24,700	81,200
1100	1600 A	23,300	17,900	41,200
1150	1600 A w/groundfault switchboard	48,000	20,200	68,200
1200	1200 A	18,100	14,300	32,400
1250	1200 A w/groundfault switchboard	40,900	16,400	57,300
1400	800 A	12,800	11,800	24,600

D50 Electrical

D5010 Electrical Service/Distribution

D5010 130	Underground Electric Service	COST EACH		
		MAT.	INST.	TOTAL
1450	800 A w/switchboard	17,200	13,400	30,600
1500	600 A	8,350	10,800	19,150
1550	600 A w/switchboard	12,700	12,200	24,900
1600	1 phase, 3 wire, 120/240 volts, 200 A	2,950	3,675	6,625
1650	200 A w/load center	3,250	4,625	7,875
1700	100 A	2,350	2,975	5,325
1750	100 A w/load center	2,650	3,950	6,600

For customer support on your Electrical Costs with RSMeans data, call 800.448.8182.

393

D5010 Electrical Service/Distribution

System Components	QUANTITY	UNIT	COST PER L.F.		
			MAT.	INST.	TOTAL
SYSTEM D5010 230 0200					
FEEDERS, INCLUDING STEEL CONDUIT & WIRE, 60 A					
Rigid galvanized steel conduit, 3/4", including fittings	1.000	L.F.	5.05	8.70	13.75
Wire 600 volt, type XHHW copper stranded #6	.030	C.L.F.	1.70	3.21	4.91
Wire 600 volt, type XHHW copper stranded #8	.010	C.L.F.	.34	.87	1.21
TOTAL			7.09	12.78	19.87

D5010 230	Feeder Installation	COST PER L.F.		
		MAT.	INST.	TOTAL
0200	Feeder installation 600 V, including RGS conduit and XHHW wire, 60 A	7.10	12.80	19.90
0240	100 A	8.80	16.65	25.45
0280	200 A	22.50	25	47.50
0320	400 A	45	50	95
0360	600 A	72.50	82	154.50
0400	800 A	94	97	191
0440	1000 A	117	128	245
0480	1200 A	145	164	309
0520	1600 A	188	194	382
0560	2000 A	235	256	491
1200	Branch installation 600 V, including EMT conduit and THW wire, 15 A	1.25	6.25	7.50
1240	20 A	1.45	6.65	8.10
1280	30 A	1.79	6.85	8.64
1320	50 A	3.53	9.55	13.08
1360	65 A	3.85	10.15	14
1400	85 A	5.95	11.75	17.70
1440	100 A	7.10	12	19.10
1480	130 A	8.05	14.05	22.10
1520	150 A	11.85	16.10	27.95
1560	200 A	12.75	18.35	31.10

D5010 Electrical Service/Distribution

System Components	QUANTITY	UNIT	COST EACH		
			MAT.	INST.	TOTAL
SYSTEM D5010 240 0240					
SWITCHGEAR INSTALLATION, INCL SWBD, PANELS & CIRC BREAKERS, 600 A					
Switchboards, 120/208 V, 600 amp	1.000	Ea.	4,400	1,400	5,800
Aluminum bus bars, 120/208 V, 600 amp	1.000	Ea.	1,675	1,400	3,075
Feeder section, circuit breakers, KA frame, 70 to 225 amp	1.000	Ea.	1,700	218	1,918
Feeder section, circuit breakers, LA frame, 125 to 400 amp	1.000	Ea.	3,875	305	4,180
TOTAL			11,650	3,323	14,973

D5010 240	Switchgear	COST PER EACH		
		MAT.	INST.	TOTAL
0190	Switchgear installation, including switchboard, panels, & circuit breaker			
0200	120/208 V, 3 phase, 400 A	8,950	2,975	11,925
0240	600 A	11,700	3,325	15,025
0280	800 A	15,000	3,800	18,800
0300	1000 A	18,600	4,250	22,850
0320	1200 A	19,400	4,775	24,175
0360	1600 A	31,700	5,500	37,200
0400	2000 A	38,200	6,025	44,225
0500	277/480 V, 3 phase, 400 A	12,200	5,525	17,725
0520	600 A	16,700	6,350	23,050
0540	800 A	19,200	6,950	26,150
0560	1000 A	24,500	7,750	32,250
0580	1200 A	26,300	8,625	34,925
0600	1600 A	39,000	9,500	48,500
0620	2000 A	49,200	10,500	59,700

For customer support on your Electrical Costs with RSMeans data, call 800.448.8182.

395

D5010 Electrical Service/Distribution

System Components	QUANTITY	UNIT	COST EACH		
			MAT.	INST.	TOTAL
SYSTEM D5010 250 1020					
PANELBOARD INSTALLATION, INCLUDES PANELBOARD, CONDUCTOR, & CONDUIT					
Conduit, galvanized steel, 1-1/4"dia.	37.000	L.F.	192.40	429.20	621.60
Panelboards, NQOD, 4 wire, 120/208 volts, 100 amp main, 24 circuits	1.000	Ea.	2,100	1,475	3,575
Wire, 600 volt, type THW, copper, stranded, #3	1.480	C.L.F.	170.20	205.72	375.92
TOTAL			2,462.60	2,109.92	4,572.52

D5010 250	Panelboard	COST EACH		
		MAT.	INST.	TOTAL
0900	Panelboards, NQOD, 4 wire, 120/208 volts w/conductor & conduit			
1000	100 A, 0 stories, 0' horizontal	2,100	1,475	3,575
1020	1 stories, 25' horizontal	2,475	2,100	4,575
1040	5 stories, 50' horizontal	3,175	3,375	6,550
1060	10 stories, 75' horizontal	4,000	4,825	8,825
1080	225A, 0 stories, 0' horizontal	3,800	1,925	5,725
2000	1 stories, 25' horizontal	4,850	3,075	7,925
2020	5 stories, 50' horizontal	6,925	5,325	12,250
2040	10 stories, 75' horizontal	9,325	7,950	17,275
2060	400A, 0 stories, 0' horizontal	5,600	2,900	8,500
2080	1 stories, 25' horizontal	7,050	4,750	11,800
3000	5 stories, 50' horizontal	9,875	8,425	18,300
3020	10 stories, 75' horizontal	16,500	16,500	33,000
3040	600 A, 0 stories, 0' horizontal	8,250	3,475	11,725
3060	1 stories, 25' horizontal	10,600	6,525	17,125
3080	5 stories, 50' horizontal	15,200	12,500	27,700
4000	10 stories, 75' horizontal	26,900	24,600	51,500
4010	Panelboards, NEHB, 4 wire, 277/480 volts w/conductor, conduit, & safety switch			
4020	100 A, 0 stories, 0' horizontal, includes safety switch	3,150	2,025	5,175
4040	1 stories, 25' horizontal	3,525	2,650	6,175
4060	5 stories, 50' horizontal	4,225	3,900	8,125
4080	10 stories, 75' horizontal	5,050	5,350	10,400
5000	225 A, 0 stories, 0' horizontal	4,700	2,450	7,150
5020	1 stories, 25' horizontal	5,750	3,600	9,350
5040	5 stories, 50' horizontal	7,825	5,850	13,675
5060	10 stories, 75' horizontal	10,200	8,475	18,675

D5010 Electrical Service/Distribution

D5010 250	Panelboard	COST EACH		
		MAT.	INST.	TOTAL
5080	400 A, 0 stories, 0' horizontal	7,425	3,800	11,225
6000	1 stories, 25' horizontal	8,875	5,650	14,525
6020	5 stories, 50' horizontal	11,700	9,325	21,025
6040	10 stories, 75' horizontal	18,300	17,400	35,700
6060	600 A, 0 stories, 0' horizontal	10,600	4,825	15,425
6080	1 stories, 25' horizontal	12,900	7,875	20,775
7000	5 stories, 50' horizontal	17,500	13,900	31,400
7020	10 stories, 75' horizontal	29,300	26,000	55,300

For customer support on your Electrical Costs with RSMeans data, call 800.448.8182.

397

D5020 Lighting and Branch Wiring

Duplex Receptacle

System Components	QUANTITY	UNIT	COST PER S.F.		
			MAT.	INST.	TOTAL
SYSTEM D5020 110 0200					
RECEPTACLES INCL. PLATE, BOX, CONDUIT, WIRE & TRANS. WHEN REQUIRED					
2.5 PER 1000 S.F., .3 WATTS PER S.F.					
Steel intermediate conduit, (IMC) 1/2″ diam	167.000	L.F.	.31	1.16	1.47
Wire 600V type THWN-THHN, copper solid #12	3.340	C.L.F.	.04	.21	.25
Wiring device, receptacle, duplex, 120V grounded, 15 amp	2.500	Ea.		.04	.04
Wall plate, 1 gang, brown plastic	2.500	Ea.		.02	.02
Steel outlet box 4″ square	2.500	Ea.	.01	.09	.10
Steel outlet box 4″ plaster rings	2.500	Ea.		.03	.03
TOTAL			.36	1.55	1.91

D5020 110	Receptacle (by Wattage)	COST PER S.F.		
		MAT.	INST.	TOTAL
0190	Receptacles include plate, box, conduit, wire & transformer when required			
0200	2.5 per 1000 S.F., .3 watts per S.F.	.36	1.55	1.91
0240	With transformer	.45	1.63	2.08
0280	4 per 1000 S.F., .5 watts per S.F.	.42	1.80	2.22
0320	With transformer	.57	1.92	2.49
0360	5 per 1000 S.F., .6 watts per S.F.	.14	.71	.85
0400	With transformer	.69	2.29	2.98
0440	8 per 1000 S.F., .9 watts per S.F.	.52	2.38	2.90
0480	With transformer	.82	2.63	3.45
0520	10 per 1000 S.F., 1.2 watts per S.F.	.56	2.57	3.13
0560	With transformer	.93	2.88	3.81
0600	16.5 per 1000 S.F., 2.0 watts per S.F.	.67	3.21	3.88
0640	With transformer	1.29	3.72	5.01
0680	20 per 1000 S.F., 2.4 watts per S.F.	.70	3.50	4.20
0720	With transformer	1.45	4.12	5.57

D5020 Lighting and Branch Wiring

Underfloor Receptacle System

Description: Table D5020 115 includes installed costs of raceways and copper wire from panel to and including receptacle.

National Electrical Code prohibits use of undercarpet system in residential, school or hospital buildings. Can only be used with carpet squares.

Low density = (1) Outlet per 259 S.F. of floor area.

High density = (1) Outlet per 127 S.F. of floor area.

System Components	QUANTITY	UNIT	COST PER S.F.		
			MAT.	INST.	TOTAL
SYSTEM D5020 115 0200					
RECEPTACLE SYSTEMS, UNDERFLOOR DUCT, 5' ON CENTER, LOW DENSITY					
Underfloor duct 3-1/8" x 7/8" w/insert 24" on center	.190	L.F.	4.18	1.89	6.07
Underfloor duct junction box, single duct, 3-1/8"	.003	Ea.	1.49	.52	2.01
Underfloor junction box carpet pan	.003	Ea.	1.25	.03	1.28
Underfloor duct outlet, high tension receptacle	.004	Ea.	1.25	.03	1.28
Wire 600V type THWN-THHN copper solid #12	.010	C.L.F.	.48	.35	.83
Vertical elbow for underfloor duct, 3-1/8", included			.11	.64	.75
Underfloor duct conduit adapter, 2" x 1-1/4", included					
TOTAL			7.51	3.43	10.94

D5020 115	Receptacles, Floor	COST PER S.F.		
		MAT.	INST.	TOTAL
0200	Receptacle systems, underfloor duct, 5' on center, low density	7.50	3.43	10.93
0240	High density	8.10	4.41	12.51
0280	7' on center, low density	5.95	2.95	8.90
0320	High density	6.50	3.93	10.43
0400	Poke thru fittings, low density	1.32	1.67	2.99
0440	High density	2.62	3.32	5.94
0520	Telepoles, using Romex, low density	1.27	1.03	2.30
0560	High density	2.53	2.06	4.59
0600	Using EMT, low density	1.34	1.37	2.71
0640	High density	2.70	2.72	5.42
0720	Conduit system with floor boxes, low density	1.18	1.17	2.35
0760	High density	2.37	2.34	4.71
0840	Undercarpet power system, 3 conductor with 5 conductor feeder, low density	1.53	.41	1.94
0880	High density	3.03	.85	3.88

D5020 Lighting and Branch Wiring

Duplex Receptacle

Wall Switch

System Components			COST PER S.F.		
	QUANTITY	UNIT	MAT.	INST.	TOTAL
SYSTEM D5020 120 0520					
RECEPTACLES AND WALL SWITCHES					
4 RECEPTACLES PER 400 S.F.					
Steel intermediate conduit, (IMC), 1/2" diam	.220	L.F.	.41	1.53	1.94
Wire, 600 volt, type THWN-THHN, copper, solid #12	.005	C.L.F.	.05	.32	.37
Steel outlet box 4" square	.010	Ea.	.06	.35	.41
Steel outlet box, 4" square, plaster rings	.010	Ea.	.02	.11	.13
Receptacle, duplex, 120 volt grounded, 15 amp	.010	Ea.	.01	.17	.18
Wall plate, 1 gang, brown plastic	.010	Ea.		.09	.09
TOTAL			.55	2.57	3.12

D5020 120	Receptacles & Wall Switches	COST PER S.F.		
		MAT.	INST.	TOTAL
0520	Receptacles and wall switches, 400 S.F., 4 receptacles	.55	2.57	3.12
0560	6 receptacles	.63	3	3.63
0600	8 receptacles	.69	3.48	4.17
0640	1 switch	.12	.58	.70
0680	600 S.F., 6 receptacles	.55	2.57	3.12
0720	8 receptacles	.61	2.91	3.52
0760	10 receptacles	.66	3.28	3.94
0800	2 switches	.16	.77	.93
0840	1000 S.F., 10 receptacles	.54	2.56	3.10
0880	12 receptacles	.56	2.69	3.25
0920	14 receptacles	.60	2.91	3.51
0960	2 switches	.08	.41	.49
1000	1600 S.F., 12 receptacles	.48	2.28	2.76
1040	14 receptacles	.52	2.49	3.01
1080	16 receptacles	.54	2.56	3.10
1120	4 switches	.09	.48	.57
1160	2000 S.F., 14 receptacles	.50	2.34	2.84
1200	16 receptacles	.50	2.35	2.85
1240	18 receptacles	.52	2.49	3.01
1280	4 switches	.08	.41	.49
1320	3000 S.F., 12 receptacles	.40	1.78	2.18
1360	18 receptacles	.48	2.20	2.68
1400	24 receptacles	.50	2.35	2.85
1440	6 switches	.08	.41	.49
1480	3600 S.F., 20 receptacles	.46	2.13	2.59
1520	24 receptacles	.47	2.21	2.68
1560	28 receptacles	.49	2.28	2.77
1600	8 switches	.09	.48	.57
1640	4000 S.F., 16 receptacles	.40	1.78	2.18
1680	24 receptacles	.48	2.20	2.68
1720	30 receptacles	.48	2.28	2.76
1760	8 switches	.08	.41	.49
1800	5000 S.F., 20 receptacles	.40	1.78	2.18
1840	26 receptacles	.48	2.13	2.61
1880	30 receptacles	.48	2.20	2.68
1920	10 switches	.08	.41	.49

D5020   Lighting and Branch Wiring

Duplex Receptacle   **Wall Switch**

System Components	QUANTITY	UNIT	COST PER EACH		
			MAT.	INST.	TOTAL
SYSTEM D5020 125 0520					
RECEPTACLES AND WALL SWITCHES, RECEPTICLE DUPLEX 120 V GROUNDED, 15 A					
Electric metallic tubing conduit, (EMT), 3/4" diam	22.000	L.F.	27.28	117.70	144.98
Wire, 600 volt, type THWN-THHN, copper, solid #12	.630	C.L.F.	6.65	40.01	46.66
Steel outlet box 4" square	1.000	Ea.	5.80	35	40.80
Steel outlet box, 4" square, plaster rings	1.000	Ea.	1.69	10.90	12.59
Receptacle, duplex, 120 volt grounded, 15 amp	1.000	Ea.	1.43	17.40	18.83
Wall plate, 1 gang, brown plastic	1.000	Ea.	.42	8.70	9.12
TOTAL			43.27	229.71	272.98

D5020 125	Receptacles & Switches by Each	COST PER EACH		
		MAT.	INST.	TOTAL
0460	Receptacles & Switches, with box, plate, 3/4" EMT conduit & wire			
0520	Receptacle duplex 120 V grounded, 15 A	43.50	230	273.50
0560	20 A	52	238	290
0600	Receptacle duplex ground fault interrupting, 15 A	57	238	295
0640	20 A	86	238	324
0680	Toggle switch single, 15 A	42.50	230	272.50
0720	20 A	45	238	283
0760	3 way switch, 15 A	43.50	243	286.50
0800	20 A	46	251	297
0840	4 way switch, 15 A	53.50	259	312.50
0880	20 A	89	276	365

For customer support on your Electrical Costs with RSMeans data, call 800.448.8182.

401

D5020 Lighting and Branch Wiring

Description: Table D5020 130 includes the cost for switch, plate, box, conduit in slab or EMT exposed and copper wire. Add 20% for exposed conduit.

No power required for switches.

Federal energy guidelines recommend the maximum lighting area controlled per switch shall not exceed 1000 S.F. and that areas over 500 S.F. shall be so controlled that total illumination can be reduced by at least 50%.

System Components			COST PER S.F.		
	QUANTITY	UNIT	MAT.	INST.	TOTAL
SYSTEM D5020 130 0360					
WALL SWITCHES, 5.0 PER 1000 S.F.					
Steel, intermediate conduit (IMC), 1/2" diameter	88.000	L.F.	.16	.61	.77
Wire, 600V type THWN-THHN, copper solid #12	1.710	C.L.F.	.02	.11	.13
Toggle switch, single pole, 15 amp	5.000	Ea.		.09	.09
Wall plate, 1 gang, brown plastic	5.000	Ea.		.04	.04
Steel outlet box 4" plaster rings	5.000	Ea.	.03	.18	.21
Plaster rings	5.000	Ea.	.01	.05	.06
TOTAL			.22	1.08	1.30

D5020 130	Wall Switch by Sq. Ft.	COST PER S.F.		
		MAT.	INST.	TOTAL
0200	Wall switches, 1.0 per 1000 S.F.	.05	.26	.31
0240	1.2 per 1000 S.F.	.05	.27	.32
0280	2.0 per 1000 S.F.	.08	.40	.48
0320	2.5 per 1000 S.F.	.09	.50	.59
0360	5.0 per 1000 S.F.	.22	1.08	1.30
0400	10.0 per 1000 S.F.	.46	2.19	2.65

D5020 Lighting and Branch Wiring

System D5020 135 includes all wiring and connections.

System Components	QUANTITY	UNIT	COST PER S.F.		
			MAT.	INST.	TOTAL
SYSTEM D5020 135 0200					
MISCELLANEOUS POWER, TO .5 WATTS					
Steel intermediate conduit, (IMC) 1/2" diam	15.000	L.F.	.03	.10	.13
Wire 600V type THWN-THHN, copper solid #12	.325	C.L.F.		.02	.02
TOTAL			.03	.12	.15

D5020 135	Miscellaneous Power	COST PER S.F.		
		MAT.	INST.	TOTAL
0200	Miscellaneous power, to .5 watts	.03	.12	.15
0240	.8 watts	.04	.18	.22
0280	1 watt	.06	.24	.30
0320	1.2 watts	.07	.28	.35
0360	1.5 watts	.08	.33	.41
0400	1.8 watts	.10	.38	.48
0440	2 watts	.11	.45	.56
0480	2.5 watts	.13	.55	.68
0520	3 watts	.17	.65	.82

D5020 Lighting and Branch Wiring

System D5020 140 includes all wiring and connections for central air conditioning units.

System Components	QUANTITY	UNIT	COST PER S.F.		
			MAT.	INST.	TOTAL
SYSTEM D5020 140 0200					
CENTRAL AIR CONDITIONING POWER, 1 WATT					
Steel intermediate conduit, 1/2" diam.	.030	L.F.	.06	.21	.27
Wire 600V type THWN-THHN, copper solid #12	.001	C.L.F.	.01	.06	.07
TOTAL			.07	.27	.34

D5020 140	Central A. C. Power (by Wattage)	COST PER S.F.		
		MAT.	INST.	TOTAL
0200	Central air conditioning power, 1 watt	.07	.27	.34
0220	2 watts	.08	.31	.39
0240	3 watts	.15	.47	.62
0280	4 watts	.16	.47	.63
0320	6 watts	.24	.63	.87
0360	8 watts	.34	.67	1.01
0400	10 watts	.48	.79	1.27

D50 Electrical

D5020 Lighting and Branch Wiring

| Power System | Switch | Motor Starter | Switch | Motor Connection / Motor |

System D5020 145 installed cost of motor wiring using 50' of rigid conduit and copper wire. **Cost and setting of motor not included.**

System Components			COST EACH		
	QUANTITY	UNIT	MAT.	INST.	TOTAL
SYSTEM D5020 145 0200					
MOTOR INST., SINGLE PHASE, 115V, TO AND INCLUDING 1/3 HP MOTOR SIZE					
Wire 600V type THWN-THHN, copper solid #12	1.250	C.L.F.	13.19	79.38	92.57
Steel intermediate conduit, (IMC) 1/2" diam.	50.000	L.F.	92.50	347.50	440
Magnetic FVNR, 115V, 1/3 HP, size 00 starter	1.000	Ea.	197	174	371
Safety switch, fused, heavy duty, 240V 2P 30 amp	1.000	Ea.	71.50	199	270.50
Safety switch, non fused, heavy duty, 600V, 3 phase, 30 A	1.000	Ea.	89.50	218	307.50
Flexible metallic conduit, Greenfield 1/2" diam.	1.500	L.F.	.69	5.22	5.91
Connectors for flexible metallic conduit Greenfield 1/2" diam.	1.000	Ea.	1.63	8.70	10.33
Coupling for Greenfield to conduit 1/2" diam. flexible metallic conduit	1.000	Ea.	.87	13.90	14.77
Fuse cartridge nonrenewable, 250V 30 amp	1.000	Ea.	2.59	13.90	16.49
TOTAL			469.47	1,059.60	1,529.07

D5020 145	Motor Installation	COST EACH		
		MAT.	INST.	TOTAL
0200	Motor installation, single phase, 115V, 1/3 HP motor size	470	1,050	1,520
0240	1 HP motor size	490	1,050	1,540
0280	2 HP motor size	540	1,125	1,665
0320	3 HP motor size	620	1,150	1,770
0360	230V, 1 HP motor size	470	1,075	1,545
0400	2 HP motor size	510	1,075	1,585
0440	3 HP motor size	570	1,150	1,720
0520	Three phase, 200V, 1-1/2 HP motor size	545	1,175	1,720
0560	3 HP motor size	655	1,275	1,930
0600	5 HP motor size	660	1,425	2,085
0640	7-1/2 HP motor size	685	1,450	2,135
0680	10 HP motor size	1,050	1,825	2,875
0720	15 HP motor size	1,475	2,050	3,525
0760	20 HP motor size	1,775	2,350	4,125
0800	25 HP motor size	1,850	2,350	4,200
0840	30 HP motor size	2,975	2,750	5,725
0880	40 HP motor size	3,625	3,275	6,900
0920	50 HP motor size	6,600	3,825	10,425
0960	60 HP motor size	6,850	4,025	10,875
1000	75 HP motor size	8,350	4,625	12,975
1040	100 HP motor size	24,700	5,425	30,125
1080	125 HP motor size	25,400	5,975	31,375
1120	150 HP motor size	28,500	7,025	35,525
1160	200 HP motor size	29,300	8,250	37,550
1240	230V, 1-1/2 HP motor size	525	1,175	1,700
1280	3 HP motor size	635	1,275	1,910
1320	5 HP motor size	635	1,425	2,060
1360	7-1/2 HP motor size	635	1,425	2,060
1400	10 HP motor size	990	1,725	2,715
1440	15 HP motor size	1,125	1,900	3,025
1480	20 HP motor size	1,675	2,275	3,950
1520	25 HP motor size	1,775	2,350	4,125

D5020 Lighting and Branch Wiring

D5020 145	Motor Installation	COST EACH		
		MAT.	INST.	TOTAL
1560	30 HP motor size	1,825	2,350	4,175
1600	40 HP motor size	3,475	3,175	6,650
1640	50 HP motor size	3,700	3,375	7,075
1680	60 HP motor size	6,600	3,825	10,425
1720	75 HP motor size	7,775	4,350	12,125
1760	100 HP motor size	8,450	4,850	13,300
1800	125 HP motor size	25,600	5,600	31,200
1840	150 HP motor size	26,400	6,350	32,750
1880	200 HP motor size	27,000	7,075	34,075
1960	460V, 2 HP motor size	650	1,175	1,825
2000	5 HP motor size	760	1,275	2,035
2040	10 HP motor size	730	1,425	2,155
2080	15 HP motor size	995	1,625	2,620
2120	20 HP motor size	1,050	1,750	2,800
2160	25 HP motor size	1,125	1,825	2,950
2200	30 HP motor size	1,475	1,975	3,450
2240	40 HP motor size	1,825	2,125	3,950
2280	50 HP motor size	2,000	2,350	4,350
2320	60 HP motor size	3,150	2,725	5,875
2360	75 HP motor size	3,650	3,025	6,675
2400	100 HP motor size	3,975	3,375	7,350
2440	125 HP motor size	6,900	3,850	10,750
2480	150 HP motor size	8,575	4,300	12,875
2520	200 HP motor size	9,300	4,850	14,150
2600	575V, 2 HP motor size	650	1,175	1,825
2640	5 HP motor size	760	1,275	2,035
2680	10 HP motor size	730	1,425	2,155
2720	20 HP motor size	995	1,625	2,620
2760	25 HP motor size	1,050	1,750	2,800
2800	30 HP motor size	1,475	1,975	3,450
2840	50 HP motor size	1,575	2,050	3,625
2880	60 HP motor size	3,100	2,725	5,825
2920	75 HP motor size	3,150	2,725	5,875
2960	100 HP motor size	3,650	3,025	6,675
3000	125 HP motor size	6,650	3,775	10,425
3040	150 HP motor size	6,900	3,850	10,750
3080	200 HP motor size	8,625	4,375	13,000

D5020 Lighting and Branch Wiring

System Components	QUANTITY	UNIT	COST PER L.F.		
			MAT.	INST.	TOTAL
SYSTEM D5020 155 0200					
MOTOR FEEDER SYSTEMS, SINGLE PHASE, UP TO 115V, 1HP OR 230V, 2HP					
Steel intermediate conduit, (IMC) 1/2" diam	1.000	L.F.	1.85	6.95	8.80
Wire 600V type THWN-THHN, copper solid #12	.020	C.L.F.	.21	1.27	1.48
TOTAL			2.06	8.22	10.28

D5020 155	Motor Feeder	COST PER L.F.		
		MAT.	INST.	TOTAL
0200	Motor feeder systems, single phase, feed up to 115V 1HP or 230V 2 HP	2.06	8.20	10.26
0240	115V 2HP, 230V 3HP	2.16	8.35	10.51
0280	115V 3HP	2.47	8.70	11.17
0360	Three phase, feed to 200V 3HP, 230V 5HP, 460V 10HP, 575V 10HP	2.17	8.85	11.02
0440	200V 5HP, 230V 7.5HP, 460V 15HP, 575V 20HP	2.32	9.05	11.37
0520	200V 10HP, 230V 10HP, 460V 30HP, 575V 30HP	2.78	9.55	12.33
0600	200V 15HP, 230V 15HP, 460V 40HP, 575V 50HP	3.84	10.95	14.79
0680	200V 20HP, 230V 25HP, 460V 50HP, 575V 60HP	5.10	13.90	19
0760	200V 25HP, 230V 30HP, 460V 60HP, 575V 75HP	5.95	14.10	20.05
0840	200V 30HP	6.70	14.60	21.30
0920	230V 40HP, 460V 75HP, 575V 100HP	8.90	15.90	24.80
1000	200V 40HP	10.40	17.05	27.45
1080	230V 50HP, 460V 100HP, 575V 125HP	12.55	18.80	31.35
1160	200V 50HP, 230V 60HP, 460V 125HP, 575V 150HP	14.55	19.95	34.50
1240	200V 60HP, 460V 150HP	18.30	23.50	41.80
1320	230V 75HP, 575V 200HP	21	24.50	45.50
1400	200V 75HP	23.50	25	48.50
1480	230V 100HP, 460V 200HP	29.50	29	58.50
1560	200V 100HP	41	36	77
1640	230V 125HP	41	36	77
1720	200V 125HP, 230V 150HP	46.50	45	91.50
1800	200V 150HP	57.50	49	106.50
1880	200V 200HP	74	52	126
1960	230V 200HP	68.50	54	122.50

For customer support on your Electrical Costs with RSMeans data, call 800.448.8182.

407

D5020 Lighting and Branch Wiring

Starters are full voltage, type NEMA 1 for general purpose indoor application with motor overload protection and include mounting and wire connections.

System Components	QUANTITY	UNIT	COST EACH		
			MAT.	INST.	TOTAL
SYSTEM D5020 160 0200					
MAGNETIC STARTER, SIZE 00 TO 1/3 HP, 1 PHASE 115V OR 1 HP 230V					
Magnetic starter, size 00, to 1/3 HP, 1 phase, 115V or 1 HP 230V	1.000	Ea.	197	174	371
TOTAL			197	174	371

D5020 160	Magnetic Starter	COST EACH		
		MAT.	INST.	TOTAL
0200	Magnetic starter, size 00, to 1/3 HP, 1 phase, 115V or 1 HP 230V	197	174	371
0280	Size 00, to 1-1/2 HP, 3 phase, 200-230V or 2 HP 460-575V	216	199	415
0360	Size 0, to 1 HP, 1 phase, 115V or 2 HP 230V	219	174	393
0440	Size 0, to 3 HP, 3 phase, 200-230V or 5 HP 460-575V	325	305	630
0520	Size 1, to 2 HP, 1 phase, 115V or 3 HP 230V	252	232	484
0600	Size 1, to 7-1/2 HP, 3 phase, 200-230V or 10 HP 460-575V	294	435	729
0680	Size 2, to 10 HP, 3 phase, 200V, 15 HP-230V or 25 HP 460-575V	550	635	1,185
0760	Size 3, to 25 HP, 3 phase, 200V, 30 HP-230V or 50 HP 460-575V	900	775	1,675
0840	Size 4, to 40 HP, 3 phase, 200V, 50 HP-230V or 100 HP 460-575V	2,000	1,150	3,150
0920	Size 5, to 75 HP, 3 phase, 200V, 100 HP-230V or 200 HP 460-575V	4,675	1,550	6,225
1000	Size 6, to 150 HP, 3 phase, 200V, 200 HP-230V or 400 HP 460-575V	20,200	1,750	21,950

D5020 Lighting and Branch Wiring

Safety switches are type NEMA 1 for general purpose indoor application, and include time delay fuses, insulation and wire terminations.

System Components	QUANTITY	UNIT	COST EACH		
			MAT.	INST.	TOTAL
SYSTEM D5020 165 0200					
SAFETY SWITCH, 30A FUSED, 1 PHASE, 115V OR 230V					
Safety switch fused, hvy duty, 240V 2p 30 amp	1.000	Ea.	71.50	199	270.50
Fuse, dual element time delay 250V, 30 amp	2.000	Ea.	21.90	27.80	49.70
TOTAL			93.40	226.80	320.20

D5020 165	Safety Switches	COST EACH		
		MAT.	INST.	TOTAL
0200	Safety switch, 30 A fused, 1 phase, 2 HP 115 V or 3 HP, 230 V	93.50	227	320.50
0280	3 phase, 5 HP, 200 V or 7 1/2 HP, 230 V	129	260	389
0360	15 HP, 460 V or 20 HP, 575 V	231	270	501
0440	60 A fused, 3 phase, 15 HP 200 V or 15 HP 230 V	198	345	543
0520	30 HP 460 V or 40 HP 575 V	298	355	653
0600	100 A fused, 3 phase, 20 HP 200 V or 25 HP 230 V	345	415	760
0680	50 HP 460 V or 60 HP 575 V	565	425	990
0760	200 A fused, 3 phase, 50 HP 200 V or 60 HP 230 V	630	595	1,225
0840	125 HP 460 V or 150 HP 575 V	920	605	1,525
0920	400 A fused, 3 phase, 100 HP 200 V or 125 HP 230 V	1,475	845	2,320
1000	250 HP 460 V or 350 HP 575 V	2,300	860	3,160
1020	600 A fused, 3 phase, 150 HP 200 V or 200 HP 230 V	2,525	1,225	3,750
1040	400 HP 460 V	3,350	1,250	4,600

D5020 Lighting and Branch Wiring

Straight Connector

Angle Connector

Flexible Conduit

Table below includes costs for the flexible conduit. Not included are wire terminations and testing motor for correct rotation.

System Components	QUANTITY	UNIT	COST EACH		
			MAT.	INST.	TOTAL
SYSTEM D5020 170 0200 **MOTOR CONNECTIONS, SINGLE PHASE, 115V/230V UP TO 1 HP** Motor connection, flexible conduit & fittings, 1 HP motor 115V	1.000	Ea.	5.35	87	92.35
TOTAL			5.35	87	92.35

D5020 170	Motor Connections	COST EACH		
		MAT.	INST.	TOTAL
0200	Motor connections, single phase, 115/230V, up to 1 HP	5.35	87	92.35
0240	Up to 3 HP	10.25	107	117.25
0280	Three phase, 200/230/460/575V, up to 3 HP	6.50	103	109.50
0320	Up to 5 HP	5.50	127	132.50
0360	Up to 7-1/2 HP	9.45	151	160.45
0400	Up to 10 HP	17.50	166	183.50
0440	Up to 15 HP	17.50	211	228.50
0480	Up to 25 HP	26.50	258	284.50
0520	Up to 50 HP	53.50	315	368.50
0560	Up to 100 HP	113	465	578

D5020 Lighting and Branch Wiring

Manual Starter

Magnetic Starter

Induction Motor

For 230/460 Volt A.C., 3 phase, 60 cycle ball bearing squirrel cage induction motors, NEMA Class B standard line. Installation included.

No conduit, wire, or terminations included.

System Components		QUANTITY	UNIT	COST EACH		
				MAT.	INST.	TOTAL
SYSTEM D5020 175 0220						
MOTOR, DRIPPROOF CLASS B INSULATION, 1.15 SERVICE FACTOR, WITH STARTER						
1 H.P., 1200 RPM WITH MANUAL STARTER						
Motor, dripproof, class B insul, 1.15 serv fact, 1,200 RPM, 1 HP		1.000	Ea.	520	155	675
Motor starter, manual, 3 phase, 1 HP motor		1.000	Ea.	228	199	427
	TOTAL			748	354	1,102

D5020 175	Motor & Starter	COST EACH		
		MAT.	INST.	TOTAL
0190	Motor, dripproof, premium efficient, 1.15 service factor			
0200	1 HP, 1200 RPM, motor only	520	155	675
0220	With manual starter	750	355	1,105
0240	With magnetic starter	735	355	1,090
0260	1800 RPM, motor only	330	155	485
0280	With manual starter	560	355	915
0300	With magnetic starter	545	355	900
0320	2 HP, 1200 RPM, motor only	575	155	730
0340	With manual starter	805	355	1,160
0360	With magnetic starter	900	460	1,360
0380	1800 RPM, motor only	395	155	550
0400	With manual starter	625	355	980
0420	With magnetic starter	720	460	1,180
0440	3600 RPM, motor only	660	155	815
0460	With manual starter	890	355	1,245
0480	With magnetic starter	985	460	1,445
0500	3 HP, 1200 RPM, motor only	790	155	945
0520	With manual starter	1,025	355	1,380
0540	With magnetic starter	1,125	460	1,585
0560	1800 RPM, motor only	795	155	950
0580	With manual starter	1,025	355	1,380
0600	With magnetic starter	1,125	460	1,585
0620	3600 RPM, motor only	685	155	840
0640	With manual starter	915	355	1,270
0660	With magnetic starter	1,000	460	1,460
0680	5 HP, 1200 RPM, motor only	1,025	155	1,180
0700	With manual starter	1,300	505	1,805
0720	With magnetic starter	1,325	590	1,915
0740	1800 RPM, motor only	705	155	860
0760	With manual starter	975	505	1,480
0780	With magnetic starter	1,000	590	1,590
0800	3600 RPM, motor only	595	155	750

For customer support on your Electrical Costs with RSMeans data, call 800.448.8182.

411

D5020 Lighting and Branch Wiring

D5020 175	Motor & Starter	COST EACH		
		MAT.	INST.	TOTAL
0820	With manual starter	865	505	1,370
0840	With magnetic starter	890	590	1,480
0860	7.5 HP, 1800 RPM, motor only	1,075	166	1,241
0880	With manual starter	1,350	515	1,865
0900	With magnetic starter	1,625	800	2,425
0920	10 HP, 1800 RPM, motor only	1,300	174	1,474
0940	With manual starter	1,575	525	2,100
0960	With magnetic starter	1,850	810	2,660
0980	15 HP, 1800 RPM, motor only	1,825	218	2,043
1000	With magnetic starter	2,375	855	3,230
1040	20 HP, 1800 RPM, motor only	2,225	268	2,493
1060	With magnetic starter	3,125	1,050	4,175
1100	25 HP, 1800 RPM, motor only	2,900	278	3,178
1120	With magnetic starter	3,800	1,050	4,850
1160	30 HP, 1800 RPM, motor only	3,175	290	3,465
1180	With magnetic starter	4,075	1,075	5,150
1220	40 HP, 1800 RPM, motor only	4,125	350	4,475
1240	With magnetic starter	6,125	1,500	7,625
1280	50 HP, 1800 RPM, motor only	4,225	435	4,660
1300	With magnetic starter	6,225	1,575	7,800
1340	60 HP, 1800 RPM, motor only	4,675	495	5,170
1360	With magnetic starter	9,350	2,050	11,400
1400	75 HP, 1800 RPM, motor only	5,450	580	6,030
1420	With magnetic starter	10,100	2,125	12,225
1460	100 HP, 1800 RPM, motor only	7,025	775	7,800
1480	With magnetic starter	11,700	2,325	14,025
1520	125 HP, 1800 RPM, motor only	7,325	995	8,320
1540	With magnetic starter	27,500	2,750	30,250
1580	150 HP, 1800 RPM, motor only	11,100	1,150	12,250
1600	With magnetic starter	31,300	2,900	34,200
1640	200 HP, 1800 RPM, motor only	13,600	1,400	15,000
1660	With magnetic starter	33,800	3,150	36,950
1680	Totally encl, premium efficient, 1.0 ser. fac., 1HP, 1200 RPM, motor only	530	155	685
1700	With manual starter	760	355	1,115
1720	With magnetic starter	745	355	1,100
1740	1800 RPM, motor only	640	155	795
1760	With manual starter	870	355	1,225
1780	With magnetic starter	855	355	1,210
1800	2 HP, 1200 RPM, motor only	600	155	755
1820	With manual starter	830	355	1,185
1840	With magnetic starter	925	460	1,385
1860	1800 RPM, motor only	805	155	960
1880	With manual starter	1,025	355	1,380
1900	With magnetic starter	1,125	460	1,585
1920	3600 RPM, motor only	495	155	650
1940	With manual starter	725	355	1,080
1960	With magnetic starter	820	460	1,280
1980	3 HP, 1200 RPM, motor only	870	155	1,025
2000	With manual starter	1,100	355	1,455
2020	With magnetic starter	1,200	460	1,660
2040	1800 RPM, motor only	730	155	885
2060	With manual starter	960	355	1,315
2080	With magnetic starter	1,050	460	1,510
2100	3600 RPM, motor only	660	155	815
2120	With manual starter	890	355	1,245
2140	With magnetic starter	985	460	1,445
2160	5 HP, 1200 RPM, motor only	1,025	155	1,180
2180	With manual starter	1,300	505	1,805

D5020 Lighting and Branch Wiring

D5020 175	Motor & Starter	MAT.	INST.	TOTAL
2200	With magnetic starter	1,325	590	1,915
2220	1800 RPM, motor only	850	155	1,005
2240	With manual starter	1,125	505	1,630
2260	With magnetic starter	1,150	590	1,740
2280	3600 RPM, motor only	770	155	925
2300	With manual starter	1,050	505	1,555
2320	With magnetic starter	1,075	590	1,665
2340	7.5 HP, 1800 RPM, motor only	1,100	166	1,266
2360	With manual starter	1,375	515	1,890
2380	With magnetic starter	1,650	800	2,450
2400	10 HP, 1800 RPM, motor only	1,550	174	1,724
2420	With manual starter	1,825	525	2,350
2440	With magnetic starter	2,100	810	2,910
2460	15 HP, 1800 RPM, motor only	1,875	218	2,093
2480	With magnetic starter	2,425	855	3,280
2500	20 HP, 1800 RPM, motor only	2,525	268	2,793
2520	With magnetic starter	3,425	1,050	4,475
2540	25 HP, 1800 RPM, motor only	3,625	278	3,903
2560	With magnetic starter	4,525	1,050	5,575
2580	30 HP, 1800 RPM, motor only	3,575	290	3,865
2600	With magnetic starter	4,475	1,075	5,550
2620	40 HP, 1800 RPM, motor only	4,375	350	4,725
2640	With magnetic starter	6,375	1,500	7,875
2660	50 HP, 1800 RPM, motor only	4,725	435	5,160
2680	With magnetic starter	6,725	1,575	8,300
2700	60 HP, 1800 RPM, motor only	6,400	495	6,895
2720	With magnetic starter	11,100	2,050	13,150
2740	75 HP, 1800 RPM, motor only	8,325	580	8,905
2760	With magnetic starter	13,000	2,125	15,125
2780	100 HP, 1800 RPM, motor only	9,825	775	10,600
2800	With magnetic starter	14,500	2,325	16,825
2820	125 HP, 1800 RPM, motor only	12,800	995	13,795
2840	With magnetic starter	33,000	2,750	35,750
2860	150 HP, 1800 RPM, motor only	15,000	1,150	16,150
2880	With magnetic starter	35,200	2,900	38,100
2900	200 HP, 1800 RPM, motor only	18,900	1,400	20,300
2920	With magnetic starter	39,100	3,150	42,250

General: The cost of the lighting portion of the electrical costs is dependent upon:
1. The footcandle requirement of the proposed building.
2. The type of fixtures required.
3. The ceiling heights of the building.
4. Reflectance value of ceilings, walls and floors.
5. Fixture efficiencies and spacing vs. mounting height ratios.

Footcandle Requirements: See Table D5020-204 for Footcandle and Watts per S.F. determination.

Table D5020-201   IESNA* Recommended Illumination Levels in Footcandles

Commercial Buildings			Industrial Buildings		
Type	Description	Footcandles	Type	Description	Footcandles
Bank	Lobby	50	Assembly Areas	Rough bench & machine work	50
	Customer Areas	70		Medium bench & machine work	100
	Teller Stations	150		Fine bench & machine work	500
	Accounting Areas	150	Inspection Areas	Ordinary	50
Offices	Routine Work	100		Difficult	100
	Accounting	150		Highly Difficult	200
	Drafting	200	Material Handling	Loading	20
	Corridors, Halls, Washrooms	30		Stock Picking	30
Schools	Reading or Writing	70		Packing, Wrapping	50
	Drafting, Labs, Shops	100	Stairways	Service Areas	20
	Libraries	70	Washrooms	Service Areas	20
	Auditoriums, Assembly	15	Storage Areas	Inactive	5
	Auditoriums, Exhibition	30		Active, Rough, Bulky	10
Stores	Circulation Areas	30		Active, Medium	20
	Stock Rooms	30		Active, Fine	50
	Merchandise Areas, Service	100	Garages	Active Traffic Areas	20
	Self-Service Areas	200		Service & Repair	100

*IESNA - Illuminating Engineering Society of North America

Table D5020-202 General Lighting Loads by Occupancies

Type of Occupancy	Unit Load per S.F. (Watts)
Armories and Auditoriums	1
Banks	5
Barber Shops and Beauty Parlors	3
Churches	1
Clubs	2
Court Rooms	2
*Dwelling Units	3
Garages — Commercial (storage)	½
Hospitals	2
*Hotels and Motels, including apartment houses without provisions for cooking by tenants	2
Industrial Commercial (Loft) Buildings	2
Lodge Rooms	1½
Office Buildings	5
Restaurants	2
Schools	3
Stores	3
Warehouses (storage)	¼
*In any of the above occupancies except one-family dwellings and individual dwelling units of multi-family dwellings:	
Assembly Halls and Auditoriums	1
Halls, Corridors, Closets	½
Storage Spaces	¼

Table D5020-203 Lighting Limit (Connected Load) for Listed Occupancies: New Building Proposed Energy Conservation Guideline

Type of Use	Maximum Watts per S.F.
Interior	
Category A: Classrooms, office areas, automotive mechanical areas, museums, conference rooms, drafting rooms, clerical areas, laboratories, merchandising areas, kitchens, examining rooms, book stacks, athletic facilities.	3.00
Category B: Auditoriums, waiting areas, spectator areas, restrooms, dining areas, transportation terminals, working corridors in prisons and hospitals, book storage areas, active inventory storage, hospital bedrooms, hotel and motel bedrooms, enclosed shopping mall concourse areas, stairways.	1.00
Category C: Corridors, lobbies, elevators, inactive storage areas.	0.50
Category D: Indoor parking.	0.25
Exterior	
Category E: Building perimeter: wall-wash, facade, canopy.	5.00 (per linear foot)
Category F: Outdoor parking.	0.10

For customer support on your Electrical Costs with RSMeans data, call 800.448.8182.

415

Table D5020-204 Procedure for Calculating Footcandles and Watts Per Square Foot

1. Initial footcandles = No. of fixtures × lamps per fixture × lumens per lamp × coefficient of utilization ÷ square feet
2. Maintained footcandles = initial footcandles × maintenance factor
3. Watts per square foot = No. of fixtures × lamps × (lamp watts + ballast watts) ÷ square feet

Example: To find footcandles and watts per S.F. for an office 20′ x 20′ with 11 fluorescent fixtures each having 4–40 watt C.W. lamps.

Based on good reflectance and clean conditions:

Lumens per lamp = 40 watt cool white at 3150 lumens per lamp

Coefficient of utilization = .42 (varies from .62 for light colored areas to .27 for dark)

Maintenance factor = .75 (varies from .80 for clean areas with good maintenance to .50 for poor)

Ballast loss = 8 watts per lamp (Varies with manufacturer. See manufacturers' catalog.)

1. Initial footcandles:

$$\frac{11 \times 4 \times 3150 \times .42}{400} = \frac{58,212}{400} = 145 \text{ footcandles}$$

2. Maintained footcandles:

$$145 \times .75 = 109 \text{ footcandles}$$

3. Watts per S.F.

$$\frac{11 \times 4 (40 + 8)}{400} = \frac{2,112}{400} = 5.3 \text{ watts per S.F.}$$

Table D5020-205 Approximate Watts Per Square Foot for Popular Fixture Types

Due to the many variables involved, use for preliminary estimating only:
  a. Fluorescent – industrial System D5020 208
  b. Fluorescent – lens unit System D5020 208 Fixture types B & C
  c. Fluorescent – louvered unit
  d. Incandescent – open reflector System D5020 214, Type D
  e. Incandescent – lens unit System D5020 214, Type A
  f. Incandescent – down light System D5020 214, Type B

D5020 Lighting and Branch Wiring

A. Strip Fixture

B. Surface Mounted

C. Recessed

D. Pendent Mounted

Design Assumptions:

1. A 100 footcandle average maintained level of illumination.
2. Ceiling heights range from 9' to 11'.
3. Average reflectance values are assumed for ceilings, walls and floors.
4. Cool white (CW) fluorescent lamps with 3150 lumens for 40 watt lamps and 6300 lumens for 8' slimline lamps.
5. Four 40 watt lamps per 4' fixture and two 8' lamps per 8' fixture.
6. Average fixture efficiency values and spacing to mounting height ratios.
7. Installation labor is average U.S. rate as of January 1.

System Components	QUANTITY	UNIT	COST PER S.F.		
			MAT.	INST.	TOTAL
SYSTEM D5020 208 0520					
FLUORESCENT FIXTURES MOUNTED 9'-11" ABOVE FLOOR, 100 FC					
TYPE A, 8 FIXTURES PER 400 S.F.					
Conduit, steel intermediate, 1/2" diam.	.185	L.F.	.34	1.29	1.63
Wire, 600V, type THWN-THHN, copper, solid, #12	.004	C.L.F.	.04	.23	.27
Fluorescent strip fixture 8' long, surface mounted, two 75W SL	.020	Ea.	1.46	2.24	3.70
Steel outlet box 4" concrete	.020	Ea.	.19	.70	.89
Steel outlet box plate with stud, 4" concrete	.020	Ea.	.20	.17	.37
Fixture hangers, flexible, 1/2" diameter, 4" long	.040	Ea.	.72	2.32	3.04
Fixture whip, THHN wire, three #12, 3/8" Greenfield	.040	Ea.	.37	.88	1.25
TOTAL			**3.32**	**7.83**	**11.15**

D5020 208	Fluorescent Fixtures (by Type)	COST PER S.F.		
		MAT.	INST.	TOTAL
0520	Fluorescent fixtures, type A, 8 fixtures per 400 S.F.	3.32	7.85	11.17
0560	11 fixtures per 600 S.F.	3.03	7.15	10.18
0600	17 fixtures per 1000 S.F.	2.82	6.65	9.47
0640	23 fixtures per 1600 S.F.	2.37	5.60	7.97
0680	28 fixtures per 2000 S.F.	2.34	5.50	7.84
0720	41 fixtures per 3000 S.F.	2.30	5.40	7.70
0800	53 fixtures per 4000 S.F.	2.21	5.20	7.41
0840	64 fixtures per 5000 S.F.	2.17	5.10	7.27
0880	Type B, 11 fixtures per 400 S.F.	5.15	11.15	16.30
0920	15 fixtures per 600 S.F.	4.62	10.05	14.67
0960	24 fixtures per 1000 S.F.	4.45	9.70	14.15
1000	35 fixtures per 1600 S.F.	4.06	8.85	12.91
1040	42 fixtures per 2000 S.F.	3.88	8.45	12.33
1080	61 fixtures per 3000 S.F.	3.72	8.15	11.87
1160	80 fixtures per 4000 S.F.	2.70	5.15	7.85
1200	98 fixtures per 5000 S.F.	3.65	7.90	11.55
1240	Type C, 11 fixtures per 400 S.F.	4.44	11.60	16.04
1280	14 fixtures per 600 S.F.	3.71	9.75	13.46
1320	23 fixtures per 1000 S.F.	3.68	9.65	13.33
1360	34 fixtures per 1600 S.F.	3.39	8.90	12.29
1400	43 fixtures per 2000 S.F.	3.49	9.10	12.59
1440	63 fixtures per 3000 S.F.	3.36	8.80	12.16
1520	81 fixtures per 4000 S.F.	3.22	8.45	11.67
1560	101 fixtures per 5000 S.F.	3.19	8.40	11.59
1600	Type D, 8 fixtures per 400 S.F.	4.02	8.75	12.77
1640	12 fixtures per 600 S.F.	4.02	8.75	12.77
1680	19 fixtures per 1000 S.F.	3.84	8.30	12.14
1720	27 fixtures per 1600 S.F.	3.42	7.45	10.87
1760	34 fixtures per 2000 S.F.	3.42	7.45	10.87
1800	48 fixtures per 3000 S.F.	3.23	7	10.23
1880	64 fixtures per 4000 S.F.	3.23	7	10.23
1920	79 fixtures per 5000 S.F.	3.23	7	10.23

D5020 Lighting and Branch Wiring

Type C. Recessed, mounted on grid ceiling suspension system, 2' x 4', four 40 watt lamps, acrylic prismatic diffusers.

5.3 watts per S.F. for 100 footcandles.

3 watts per S.F. for 57 footcandles.

System Components	QUANTITY	UNIT	COST PER S.F.		
			MAT.	INST.	TOTAL
SYSTEM D5020 210 0200					
FLUORESCENT FIXTURES RECESS MOUNTED IN CEILING					
1 WATT PER S.F., 20 FC, 5 FIXTURES PER 1000 S.F.					
Steel intermediate conduit, (IMC) 1/2" diam.	.128	L.F.	.24	.89	1.13
Wire, 600 volt, type THW, copper, solid, #12	.003	C.L.F.	.03	.19	.22
Fluorescent fixture, recessed, 2'x 4', four 40W, w/lens, for grid ceiling	.005	Ea.	.34	.74	1.08
Steel outlet box 4" square	.005	Ea.	.05	.18	.23
Fixture whip, Greenfield w/#12 THHN wire	.005	Ea.	.05	.04	.09
TOTAL			.71	2.04	2.75

D5020 210	Fluorescent Fixtures (by Wattage)	COST PER S.F.		
		MAT.	INST.	TOTAL
0190	Fluorescent fixtures recess mounted in ceiling			
0195	T12, standard 40 watt lamps			
0200	1 watt per S.F., 20 FC, 5 fixtures @40 watts per 1000 S.F.	.71	2.04	2.75
0240	2 watt per S.F., 40 FC, 10 fixtures @40 watt per 1000 S.F.	1.41	4.02	5.43
0280	3 watt per S.F., 60 FC, 15 fixtures @40 watt per 1000 S.F	2.12	6.05	8.17
0320	4 watt per S.F., 80 FC, 20 fixtures @40 watt per 1000 S.F.	2.82	8.05	10.87
0400	5 watt per S.F., 100 FC, 25 fixtures @40 watt per 1000 S.F.	3.52	10.10	13.62
0450	T8, energy saver 32 watt lamps			
0500	0.8 watt per S.F., 20 FC, 5 fixtures @32 watt per 1000 S.F.	.78	2.04	2.82
0520	1.6 watt per S.F., 40 FC, 10 fixtures @32 watt per 1000 S.F.	1.54	4.02	5.56
0540	2.4 watt per S.F., 60 FC, 15 fixtures @ 32 watt per 1000 S.F	2.31	6.05	8.36
0560	3.2 watt per S.F., 80 FC, 20 fixtures @32 watt per 1000 S.F.	3.08	8.05	11.13
0580	4 watt per S.F., 100 FC, 25 fixtures @32 watt per 1000 S.F.	3.85	10.10	13.95

D5020 Lighting and Branch Wiring

Type A

Type B

Type C

Type D

Type A. Recessed wide distribution reflector with flat glass lens 150 W.

Maximum spacing = 1.2 x mounting height.

13 watts per S.F. for 100 footcandles.

Type B. Recessed reflector down light with baffles 150 W.

Maximum spacing = 0.8 x mounting height.

18 watts per S.F. for 100 footcandles.

Type C. Recessed PAR–38 flood lamp with concentric louver 150 W.

Maximum spacing = 0.5 x mounting height.

19 watts per S.F. for 100 footcandles.

Type D. Recessed R–40 flood lamp with reflector skirt.

Maximum spacing = 0.7 x mounting height.

15 watts per S.F. for 100 footcandles.

System Components	QUANTITY	UNIT	COST PER S.F.		
			MAT.	INST.	TOTAL
SYSTEM D5020 214 0400					
INCANDESCENT FIXTURE RECESS MOUNTED, 100 FC					
TYPE A, 34 FIXTURES PER 400 S.F.					
Steel intermediate conduit, (IMC) 1/2" diam	1.060	L.F.	1.96	7.37	9.33
Wire, 600V, type THWN-THHN, copper, solid, #12	.033	C.L.F.	.35	2.10	2.45
Steel outlet box 4" square	.085	Ea.	.82	2.98	3.80
Fixture whip, Greenfield w/#12 THHN wire	.085	Ea.	.84	.74	1.58
Incandescent fixture, recessed, w/lens, prewired, square trim, 200W	.085	Ea.	9.18	8.84	18.02
TOTAL			13.15	22.03	35.18

D5020 214	Incandescent Fixture (by Type)	COST PER S.F.		
		MAT.	INST.	TOTAL
0380	Incandescent fixture recess mounted, 100 FC			
0400	Type A, 34 fixtures per 400 S.F.	13.15	22	35.15
0440	49 fixtures per 600 S.F.	12.80	21.50	34.30
0480	63 fixtures per 800 S.F.	12.45	21.50	33.95
0520	90 fixtures per 1200 S.F.	12	21	33
0560	116 fixtures per 1600 S.F.	11.75	21	32.75
0600	143 fixtures per 2000 S.F.	11.65	20.50	32.15
0640	Type B, 47 fixtures per 400 S.F.	12	28	40
0680	66 fixtures per 600 S.F.	11.50	27	38.50
0720	88 fixtures per 800 S.F.	11.50	27	38.50
0760	127 fixtures per 1200 S.F.	11.25	27	38.25
0800	160 fixtures per 1600 S.F.	11.05	26.50	37.55
0840	206 fixtures per 2000 S.F.	11	26.50	37.50
0880	Type C, 51 fixtures per 400 S.F.	16.65	29	45.65
0920	74 fixtures per 600 S.F.	16.15	28.50	44.65
0960	97 fixtures per 800 S.F.	15.95	28.50	44.45
1000	142 fixtures per 1200 S.F.	15.65	28	43.65
1040	186 fixtures per 1600 S.F.	15.45	28	43.45
1080	230 fixtures per 2000 S.F.	15.40	28	43.40
1120	Type D, 39 fixtures per 400 S.F.	15.65	23	38.65
1160	57 fixtures per 600 S.F.	15.25	22.50	37.75
1200	75 fixtures per 800 S.F.	15.15	22.50	37.65
1240	109 fixtures per 1200 S.F.	14.80	22.50	37.30
1280	143 fixtures per 1600 S.F.	14.50	22	36.50
1320	176 fixtures per 2000 S.F.	14.40	22	36.40

D5020 Lighting and Branch Wiring

Type A. Recessed, wide distribution reflector with flat glass lens.

150 watt inside frost—2500 lumens per lamp.

PS–25 extended service lamp.

Maximum spacing = 1.2 x mounting height.

13 watts per S.F. for 100 footcandles.

System Components	QUANTITY	UNIT	COST PER S.F.		
			MAT.	INST.	TOTAL
SYSTEM D5020 216 0200					
INCANDESCENT FIXTURE RECESS MOUNTED, TYPE A					
1 WATT PER S.F., 8 FC, 6 FIXT PER 1000 S.F.					
Steel intermediate conduit, (IMC) 1/2" diam	.091	L.F.	.17	.63	.80
Wire, 600V, type THWN-THHN, copper, solid, #12	.002	C.L.F.	.02	.13	.15
Incandescent fixture, recessed, w/lens, prewired, square trim, 200W	.006	Ea.	.65	.62	1.27
Steel outlet box 4" square	.006	Ea.	.06	.21	.27
Fixture whip, Greenfield w/#12 THHN wire	.006	Ea.	.06	.05	.11
TOTAL			.96	1.64	2.60

D5020 216	Incandescent Fixture (by Wattage)	COST PER S.F.		
		MAT.	INST.	TOTAL
0190	Incandescent fixture recess mounted, type A			
0200	1 watt per S.F., 8 FC, 6 fixtures per 1000 S.F.	.96	1.64	2.60
0240	2 watt per S.F., 16 FC, 12 fixtures per 1000 S.F.	1.92	3.28	5.20
0280	3 watt per S.F., 24 FC, 18 fixtures, per 1000 S.F.	2.85	4.88	7.73
0320	4 watt per S.F., 32 FC, 24 fixtures per 1000 S.F.	3.80	6.50	10.30
0400	5 watt per S.F., 40 FC, 30 fixtures per 1000 S.F.	4.76	8.15	12.91

D5020 Lighting and Branch Wiring

High Bay Fluorescent Fixtures
Four T5 HO/54 watt lamps

System Components	QUANTITY	UNIT	COST PER S.F.		
			MAT.	INST.	TOTAL
SYSTEM D5020 218 0200					
FLUORESCENT HIGH BAY FIXTURE, 4 LAMP, 8'-10' ABOVE WORK PLANE					
0.5 WATT/S.F., 29 FC, 2 FIXTURES/1000 S.F.					
Steel intermediate conduit, (IMC) 1/2" diam	.100	L.F.	.19	.70	.89
Wire, 600V, type THWN-THHN, copper, solid, #10	.002	C.L.F.	.03	.14	.17
Steel outlet box 4" concrete	.002	Ea.	.02	.07	.09
Steel outlet box plate with stud, 4" concrete	.002	Ea.	.02	.02	.04
Flourescent , hi bay, 4 lamp, T-5 fixture	.002	Ea.	.54	.31	.85
TOTAL			.80	1.24	2.04

D5020 218	Fluorescent Fixture, High Bay, 8'-10' (by Wattage)	COST PER S.F.		
		MAT.	INST.	TOTAL
0190	Fluorescent high bay-4 lamp fixture, 8'-10' above work plane			
0200	.5 watt/SF, 29 FC, 2 fixtures per 1000 S.F.	.80	1.24	2.04
0400	1 watt/SF, 59 FC, 4 fixtures per 1000 S.F.	1.55	2.36	3.91
0600	1.5 watt/SF, 103 FC, 7 fixtures per 1000 S.F.	2.61	3.70	6.31
0800	2 watt/SF, 133 FC, 9 fixtures per 1000 S.F.	3.39	4.88	8.27
1000	2.5 watt/SF, 162 FC, 11 fixtures per 1000 S.F.	4.17	6.05	10.22

For customer support on your Electrical Costs with RSMeans data, call 800.448.8182.

421

D5020 Lighting and Branch Wiring

HIGH BAY FIXTURES

B. Metal halide 400 watt

C. High pressure sodium 400 watt

E. Metal halide 1000 watt

F. High pressure sodium 1000 watt

G. Metal halide 1000 watt
125,000 lumen lamp

System Components	QUANTITY	UNIT	COST PER S.F.		
			MAT.	INST.	TOTAL
SYSTEM D5020 220 0880					
HIGH INTENSITY DISCHARGE FIXTURE, 8'-10' ABOVE WORK PLANE, 100 FC					
TYPE B, 8 FIXTURES PER 900 S.F.					
Steel intermediate conduit, (IMC) 1/2" diam	.460	L.F.	.85	3.20	4.05
Wire, 600V, type THWN-THHN, copper, solid, #10	.009	C.L.F.	.14	.63	.77
Steel outlet box 4" concrete	.009	Ea.	.09	.32	.41
Steel outlet box plate with stud 4" concrete	.009	Ea.	.09	.08	.17
Metal halide, hi bay, aluminum reflector, 400 W lamp	.009	Ea.	4.10	2.75	6.85
TOTAL			5.27	6.98	12.25

D5020 220	H.I.D. Fixture, High Bay, 8'-10' (by Type)	COST PER S.F.		
		MAT.	INST.	TOTAL
0500	High intensity discharge fixture, 8'-10' above work plane, 100 FC			
0880	Type B, 8 fixtures per 900 S.F.	5.25	7	12.25
0920	15 fixtures per 1800 S.F.	4.81	6.70	11.51
0960	24 fixtures per 3000 S.F.	4.81	6.70	11.51
1000	31 fixtures per 4000 S.F.	4.81	6.70	11.51
1040	38 fixtures per 5000 S.F.	4.81	6.70	11.51
1080	60 fixtures per 8000 S.F.	4.81	6.70	11.51
1120	72 fixtures per 10000 S.F.	4.30	6.20	10.50
1160	115 fixtures per 16000 S.F.	4.30	6.20	10.50
1200	230 fixtures per 32000 S.F.	4.30	6.20	10.50
1240	Type C, 4 fixtures per 900 S.F.	2.49	4.24	6.73
1280	8 fixtures per 1800 S.F.	2.49	4.27	6.76
1320	13 fixtures per 3000 S.F.	2.49	4.24	6.73
1360	17 fixtures per 4000 S.F.	2.49	4.24	6.73
1400	21 fixtures per 5000 S.F.	2.41	3.90	6.31
1440	33 fixtures per 8000 S.F.	2.40	3.83	6.23
1480	40 fixtures per 10000 S.F.	2.34	3.62	5.96
1520	63 fixtures per 16000 S.F.	2.34	3.62	5.96
1560	126 fixtures per 32000 S.F.	2.34	3.62	5.96

D5020 Lighting and Branch Wiring

HIGH BAY FIXTURES

B. Metal halide 400 watt

C. High pressure sodium 400 watt

E. Metal halide 1000 watt

F. High pressure sodium 1000 watt

G. Metal halide 1000 watt
125,000 lumen lamp

System Components	QUANTITY	UNIT	COST PER S.F.		
			MAT.	INST.	TOTAL
SYSTEM D5020 222 0240					
HIGH INTENSITY DISCHARGE FIXTURE, 8'-10' ABOVE WORK PLANE					
1 WATT/S.F., TYPE B, 29 FC, 2 FIXTURES/1000 S.F.					
Steel intermediate conduit, (IMC) 1/2" diam	.100	L.F.	.19	.70	.89
Wire, 600V, type THWN-THHN, copper, solid, #10	.002	C.L.F.	.03	.14	.17
Steel outlet box 4" concrete	.002	Ea.	.02	.07	.09
Steel outlet box plate with stud, 4" concrete	.002	Ea.	.02	.02	.04
Metal halide, hi bay, aluminum reflector, 400 W lamp	.002	Ea.	.91	.61	1.52
TOTAL			1.17	1.54	2.71

D5020 222	H.I.D. Fixture, High Bay, 8'-10' (by Wattage)	COST PER S.F.		
		MAT.	INST.	TOTAL
0190	High intensity discharge fixture, 8'-10' above work plane			
0240	1 watt/S.F., type B, 29 FC, 2 fixtures/1000 S.F.	1.17	1.54	2.71
0280	Type C, 54 FC, 2 fixtures/1000 S.F.	1.01	1.17	2.18
0400	2 watt/S.F., type B, 59 FC, 4 fixtures/1000 S.F.	2.30	2.96	5.26
0440	Type C, 108 FC, 4 fixtures/1000 S.F.	1.98	2.26	4.24
0560	3 watt/S.F., type B, 103 FC, 7 fixtures/1000 S.F.	3.92	4.75	8.67
0600	Type C, 189 FC, 6 fixtures/1000 S.F.	3.20	2.94	6.14
0720	4 watt/S.F., type B, 133 FC, 9 fixtures/1000 S.F.	5.10	6.25	11.35
0760	Type C, 243 FC, 9 fixtures/1000 S.F.	4.45	5.05	9.50
0880	5 watt/S.F., type B, 162 FC, 11 fixtures/1000 S.F.	6.25	7.70	13.95
0920	Type C, 297 FC, 11 fixtures/1000 S.F.	5.45	6.25	11.70

D5020 Lighting and Branch Wiring

HIGH BAY FIXTURES
B. Metal halide 400 watt
C. High pressure sodium 400 watt
E. Metal halide 1000 watt
F. High pressure sodium 1000 watt
G. Metal halide 1000 watt
   125,000 lumen lamp

System Components	QUANTITY	UNIT	COST PER S.F.		
			MAT.	INST.	TOTAL
SYSTEM D5020 224 1240					
HIGH INTENSITY DISCHARGE FIXTURE, 16′ ABOVE WORK PLANE, 100 FC					
TYPE C, 5 FIXTURES PER 900 S.F.					
Steel intermediate conduit, (IMC) 1/2″ diam	.260	L.F.	.48	1.81	2.29
Wire, 600V, type THWN-THHN, copper, solid, #10	.007	C.L.F.	.11	.49	.60
Steel outlet box 4″ concrete	.006	Ea.	.06	.21	.27
Steel outlet box plate with stud, 4″ concrete	.006	Ea.	.06	.05	.11
High pressure sodium, hi bay, aluminum reflector, 400 W lamp	.006	Ea.	2.52	1.83	4.35
TOTAL			3.23	4.39	7.62

D5020 224	H.I.D. Fixture, High Bay, 16′ (by Type)	COST PER S.F.		
		MAT.	INST.	TOTAL
0510	High intensity discharge fixture, 16′ above work plane, 100 FC			
1240	Type C, 5 fixtures per 900 S.F.	3.23	4.39	7.62
1280	9 fixtures per 1800 S.F.	2.94	4.60	7.54
1320	15 fixtures per 3000 S.F.	2.94	4.60	7.54
1360	18 fixtures per 4000 S.F.	2.81	4.12	6.93
1400	22 fixtures per 5000 S.F.	2.81	4.12	6.93
1440	36 fixtures per 8000 S.F.	2.81	4.12	6.93
1480	42 fixtures per 10,000 S.F.	2.50	4.24	6.74
1520	65 fixtures per 16,000 S.F.	2.50	4.24	6.74
1600	Type G, 4 fixtures per 900 S.F.	4.09	6.90	10.99
1640	6 fixtures per 1800 S.F.	3.39	6.45	9.84
1720	9 fixtures per 4000 S.F.	2.77	6.30	9.07
1760	11 fixtures per 5000 S.F.	2.77	6.30	9.07
1840	21 fixtures per 10,000 S.F.	2.61	5.65	8.26
1880	33 fixtures per 16,000 S.F.	2.61	5.65	8.26

D5020 Lighting and Branch Wiring

HIGH BAY FIXTURES

B. Metal halide 400 watt

C. High pressure sodium 400 watt

E. Metal halide 1000 watt

F. High pressure sodium 1000 watt

G. Metal halide 1000 watt
125,000 lumen lamp

System Components	QUANTITY	UNIT	COST PER S.F.		
			MAT.	INST.	TOTAL
SYSTEM D5020 226 0240					
HIGH INTENSITY DISCHARGE FIXTURE, 16' ABOVE WORK PLANE					
1 WATT/S.F., TYPE E, 42 FC, 1 FIXTURE/1000 S.F.					
Steel intermediate conduit, (IMC) 1/2" diam	.160	L.F.	.30	1.11	1.41
Wire, 600V, type THWN-THHN, copper, solid, #10	.003	C.L.F.	.05	.21	.26
Steel outlet box 4" concrete	.001	Ea.	.01	.04	.05
Steel outlet box plate with stud, 4" concrete	.001	Ea.	.01	.01	.02
Metal halide, hi bay, aluminum reflector, 1000 W lamp	.001	Ea.	.66	.35	1.01
TOTAL			1.03	1.72	2.75

D5020 226	H.I.D. Fixture, High Bay, 16' (by Wattage)	COST PER S.F.		
		MAT.	INST.	TOTAL
0190	High intensity discharge fixture, 16' above work plane			
0240	1 watt/S.F., type E, 42 FC, 1 fixture/1000 S.F.	1.03	1.72	2.75
0280	Type G, 52 FC, 1 fixture/1000 S.F.	1.03	1.72	2.75
0320	Type C, 54 FC, 2 fixture/1000 S.F.	1.20	1.92	3.12
0440	2 watt/S.F., type E, 84 FC, 2 fixture/1000 S.F.	2.05	3.50	5.55
0480	Type G, 105 FC, 2 fixture/1000 S.F.	2.05	3.50	5.55
0520	Type C, 108 FC, 4 fixture/1000 S.F.	2.39	3.84	6.23
0640	3 watt/S.F., type E, 126 FC, 3 fixture/1000 S.F.	3.08	5.25	8.33
0680	Type G, 157 FC, 3 fixture/1000 S.F.	3.08	5.25	8.33
0720	Type C, 162 FC, 6 fixture/1000 S.F.	3.59	5.75	9.34
0840	4 watt/S.F., type E, 168 FC, 4 fixture/1000 S.F.	4.10	7	11.10
0880	Type G, 210 FC, 4 fixture/1000 S.F.	4.10	7	11.10
0920	Type C, 243 FC, 9 fixture/1000 S.F.	5.25	8.05	13.30
1040	5 watt/S.F., type E, 210 FC, 5 fixture/1000 S.F.	5.15	8.70	13.85
1080	Type G, 262 FC, 5 fixture/1000 S.F.	5.15	8.70	13.85
1120	Type C, 297 FC, 11 fixture/1000 S.F.	6.45	9.95	16.40

For customer support on your Electrical Costs with RSMeans data, call 800.448.8182.

425

D5020 Lighting and Branch Wiring

HIGH BAY FIXTURES

B. Metal halide 400 watt

C. High pressure sodium 400 watt

E. Metal halide 1000 watt

F. High pressure sodium 1000 watt

G. Metal halide 1000 watt
125,000 lumen lamp

System Components	QUANTITY	UNIT	COST PER S.F.		
			MAT.	INST.	TOTAL
SYSTEM D5020 228 1240					
HIGH INTENSITY DISCHARGE FIXTURE, 20' ABOVE WORK PLANE, 100 FC					
TYPE C, 6 FIXTURES PER 900 S.F.					
Steel intermediate conduit, (IMC) 1/2" diam	.350	L.F.	.65	2.43	3.08
Wire, 600V, type THWN-THHN, copper, solid, #10.	.011	C.L.F.	.17	.76	.93
Steel outlet box 4" concrete	.007	Ea.	.07	.25	.32
Steel outlet box plate with stud, 4" concrete	.007	Ea.	.07	.06	.13
High pressure sodium, hi bay, aluminum reflector, 400 W lamp	.007	Ea.	2.94	2.14	5.08
TOTAL			3.90	5.64	9.54

D5020 228	H.I.D. Fixture, High Bay, 20' (by Type)	COST PER S.F.		
		MAT.	INST.	TOTAL
0510	High intensity discharge fixture 20' above work plane, 100 FC			
1240	Type C, 6 fixtures per 900 S.F.	3.90	5.65	9.55
1280	10 fixtures per 1800 S.F.	3.46	5.30	8.76
1320	16 fixtures per 3000 S.F.	3.06	5.10	8.16
1360	20 fixtures per 4000 S.F.	3.06	5.10	8.16
1400	24 fixtures per 5000 S.F.	3.06	5.10	8.16
1440	38 fixtures per 8000 S.F.	3	4.87	7.87
1520	68 fixtures per 16000 S.F.	2.78	5.35	8.13
1560	132 fixtures per 32000 S.F.	2.78	5.35	8.13
1600	Type G, 4 fixtures per 900 S.F.	4.04	6.80	10.84
1640	6 fixtures per 1800 S.F.	3.38	6.45	9.83
1680	7 fixtures per 3000 S.F.	3.33	6.25	9.58
1720	10 fixtures per 4000 S.F.	3.32	6.20	9.52
1760	11 fixtures per 5000 S.F.	2.85	6.65	9.50
1800	18 fixtures per 8000 S.F.	2.85	6.65	9.50
1840	22 fixtures per 10000 S.F.	2.85	6.65	9.50
1880	34 fixtures per 16000 S.F.	2.80	6.40	9.20
1920	66 fixtures per 32000 S.F.	2.65	5.85	8.50

D5020 Lighting and Branch Wiring

HIGH BAY FIXTURES

B. Metal halide 400 watt

C. High pressure sodium 400 watt

E. Metal halide 1000 watt

F. High pressure sodium 1000 watt

G. Metal halide 1000 watt
   125,000 lumen lamp

System Components	QUANTITY	UNIT	COST PER S.F.		
			MAT.	INST.	TOTAL
SYSTEM D5020 230 0240					
HIGH INTENSITY DISCHARGE FIXTURE, 20' ABOVE WORK PLANE					
1 WATT/S.F., TYPE E, 40 FC, 1 FIXTURE 1000 S.F.					
Steel intermediate conduit, (IMC) 1/2" diam	.160	L.F.	.30	1.11	1.41
Wire, 600V, type THWN-THHN, copper, solid, #10	.005	C.L.F.	.08	.35	.43
Steel outlet box 4" concrete	.001	Ea.	.01	.04	.05
Steel outlet box plate with stud, 4" concrete	.001	Ea.	.01	.01	.02
Metal halide, hi bay, aluminum reflector, 1000 W lamp	.001	Ea.	.66	.35	1.01
TOTAL			1.06	1.86	2.92

D5020 230	H.I.D. Fixture, High Bay, 20' (by Wattage)	COST PER S.F.		
		MAT.	INST.	TOTAL
0190	High intensity discharge fixture, 20' above work plane			
0240	1 watt/S.F., type E, 40 FC, 1 fixture/1000 S.F.	1.06	1.86	2.92
0280	Type G, 50 FC, 1 fixture/1000 S.F.	1.06	1.86	2.92
0320	Type C, 52 FC, 2 fixtures/1000 S.F.	1.25	2.15	3.40
0440	2 watt/S.F., type E, 81 FC, 2 fixtures/1000 S.F.	2.10	3.71	5.81
0480	Type G, 101 FC, 2 fixtures/1000 S.F.	2.10	3.71	5.81
0520	Type C, 104 FC, 4 fixtures/1000 S.F.	2.48	4.22	6.70
0640	3 watt/S.F., type E, 121 FC, 3 fixtures/1000 S.F.	3.16	5.55	8.71
0680	Type G, 151 FC, 3 fixtures/1000 S.F.	3.16	5.55	8.71
0720	Type C, 155 FC, 6 fixtures/1000 S.F.	3.74	6.35	10.09
0840	4 watt/S.F., type E, 161 FC, 4 fixtures/1000 S.F.	4.19	7.40	11.59
0880	Type G, 202 FC, 4 fixtures/1000 S.F.	4.19	7.40	11.59
0920	Type C, 233 FC, 9 fixtures/1000 S.F.	5.40	8.80	14.20
1040	5 watt/S.F., type E, 202 FC, 5 fixtures/1000 S.F.	5.25	9.25	14.50
1080	Type G, 252 FC, 5 fixtures/1000 S.F.	5.25	9.25	14.50
1120	Type C, 285 FC, 11 fixtures/1000 S.F.	6.65	10.95	17.60

D5020 Lighting and Branch Wiring

HIGH BAY FIXTURES

B. Metal halide 400 watt

C. High pressure sodium 400 watt

E. Metal halide 1000 watt

F. High pressure sodium 1000 watt

G. Metal halide 1000 watt
   125,000 lumen lamp

System Components	QUANTITY	UNIT	COST PER S.F.		
			MAT.	INST.	TOTAL
SYSTEM D5020 232 1240					
HIGH INTENSITY DISCHARGE FIXTURE, 30' ABOVE WORK PLANE, 100 FC					
TYPE F, 4 FIXTURES PER 900 S.F.					
Steel intermediate conduit, (IMC) 1/2" diam	.580	L.F.	1.07	4.03	5.10
Wire, 600V, type THWN-THHN, copper, solid, #10	.018	C.L.F.	.28	1.25	1.53
Steel outlet box 4" concrete	.004	Ea.	.04	.14	.18
Steel outlet box plate with stud, 4" concrete	.004	Ea.	.04	.03	.07
High pressure sodium, hi bay, aluminum, 1000 W lamp	.004	Ea.	2.42	1.40	3.82
TOTAL			3.85	6.85	10.70

D5020 232	H.I.D. Fixture, High Bay, 30' (by Type)	COST PER S.F.		
		MAT.	INST.	TOTAL
0510	High intensity discharge fixture, 30' above work plane, 100 FC			
1240	Type F, 4 fixtures per 900 S.F.	3.85	6.85	10.70
1280	6 fixtures per 1800 S.F.	3.25	6.55	9.80
1320	8 fixtures per 3000 S.F.	3.25	6.55	9.80
1360	9 fixtures per 4000 S.F.	2.64	6.20	8.84
1400	10 fixtures per 5000 S.F.	2.64	6.20	8.84
1440	17 fixtures per 8000 S.F.	2.64	6.20	8.84
1480	18 fixtures per 10,000 S.F.	2.51	5.70	8.21
1520	27 fixtures per 16,000 S.F.	2.51	5.70	8.21
1560	52 fixtures per 32000 S.F.	2.50	5.65	8.15
1600	Type G, 4 fixtures per 900 S.F.	4.17	7.25	11.42
1640	6 fixtures per 1800 S.F.	3.43	6.60	10.03
1680	9 fixtures per 3000 S.F.	3.37	6.40	9.77
1720	11 fixtures per 4000 S.F.	3.32	6.20	9.52
1760	13 fixtures per 5000 S.F.	3.32	6.20	9.52
1800	21 fixtures per 8000 S.F.	3.32	6.20	9.52
1840	23 fixtures per 10,000 S.F.	2.81	6.55	9.36
1880	36 fixtures per 16,000 S.F.	2.81	6.55	9.36
1920	70 fixtures per 32,000 S.F.	2.81	6.55	9.36

D5020 Lighting and Branch Wiring

HIGH BAY FIXTURES

B. Metal halide 400 watt

C. High pressure sodium 400 watt

E. Metal halide 1000 watt

F. High pressure sodium 1000 watt

G. Metal halide 1000 watt
  125,000 lumen lamp

System Components	QUANTITY	UNIT	COST PER S.F.		
			MAT.	INST.	TOTAL
SYSTEM D5020 234 0240					
HIGH INTENSITY DISCHARGE FIXTURE, 30′ ABOVE WORK PLANE					
1 WATT/S.F., TYPE E, 37 FC, 1 FIXTURE/1000 S.F.					
Steel intermediate conduit, (IMC) 1/2″ diam	.196	L.F.	.36	1.36	1.72
Wire, 600V type THWN-THHN, copper, solid, #10	.006	C.L.F.	.09	.42	.51
Steel outlet box 4″ concrete	.001	Ea.	.01	.04	.05
Steel outlet box plate with stud, 4″ concrete	.001	Ea.	.01	.01	.02
Metal halide, hi bay, aluminum reflector, 1000 W lamp	.001	Ea.	.66	.35	1.01
TOTAL			1.13	2.18	3.31

D5020 234	H.I.D. Fixture, High Bay, 30′ (by Wattage)	COST PER S.F.		
		MAT.	INST.	TOTAL
0190	High intensity discharge fixture, 30′ above work plane			
0240	1 watt/S.F., type E, 37 FC, 1 fixture/1000 S.F.	1.13	2.18	3.31
0280	Type G, 45 FC, 1 fixture/1000 S.F.	1.13	2.18	3.31
0320	Type F, 50 FC, 1 fixture/1000 S.F.	.96	1.71	2.67
0440	2 watt/S.F., type E, 74 FC, 2 fixtures/1000 S.F.	2.27	4.34	6.61
0480	Type G, 92 FC, 2 fixtures/1000 S.F.	2.27	4.34	6.61
0520	Type F, 100 FC, 2 fixtures/1000 S.F.	1.94	3.48	5.42
0640	3 watt/S.F., type E, 110 FC, 3 fixtures/1000 S.F.	3.42	6.60	10.02
0680	Type G, 138 FC, 3 fixtures/1000 S.F.	3.42	6.60	10.02
0720	Type F, 150 FC, 3 fixtures/1000 S.F.	2.90	5.20	8.10
0840	4 watt/S.F., type E, 148 FC, 4 fixtures/1000 S.F.	4.54	8.75	13.29
0880	Type G, 185 FC, 4 fixtures/1000 S.F.	4.54	8.75	13.29
0920	Type F, 200 FC, 4 fixtures/1000 S.F.	3.88	6.95	10.83
1040	5 watt/S.F., type E, 185 FC, 5 fixtures/1000 S.F.	5.70	10.95	16.65
1080	Type G, 230 FC, 5 fixtures/1000 S.F.	5.70	10.95	16.65
1120	Type F, 250 FC, 5 fixtures/1000 S.F.	4.85	8.65	13.50

D5020 Lighting and Branch Wiring

LOW BAY FIXTURES

J. Metal halide 250 watt

K. High pressure sodium 150 watt

System Components	QUANTITY	UNIT	COST PER S.F.		
			MAT.	INST.	TOTAL
SYSTEM D5020 236 0920					
HIGH INTENSITY DISCHARGE FIXTURE, 8'-10' ABOVE WORK PLANE, 50 FC					
TYPE J, 13 FIXTURES PER 1800 S.F.					
Steel intermediate conduit, (IMC) 1/2" diam	.550	L.F.	1.02	3.82	4.84
Wire, 600V, type THWN-THHN, copper, solid, #10	.012	C.L.F.	.19	.83	1.02
Steel outlet box 4" concrete	.007	Ea.	.07	.25	.32
Steel outlet box plate with stud, 4" concrete	.007	Ea.	.07	.06	.13
Metal halide, lo bay, aluminum reflector, 250 W DX lamp	.007	Ea.	2.80	1.53	4.33
TOTAL			4.15	6.49	10.64

D5020 236	H.I.D. Fixture, Low Bay, 8'-10' (by Type)	COST PER S.F.		
		MAT.	INST.	TOTAL
0510	High intensity discharge fixture, 8'-10' above work plane, 50 FC			
0880	Type J, 7 fixtures per 900 S.F.	4.51	6.55	11.06
0920	13 fixtures per 1800 S.F.	4.15	6.50	10.65
0960	21 fixtures per 3000 S.F.	4.15	6.50	10.65
1000	28 fixtures per 4000 S.F.	4.15	6.50	10.65
1040	35 fixtures per 5000 S.F.	4.15	6.50	10.65
1120	62 fixtures per 10,000 S.F.	3.73	6.20	9.93
1160	99 fixtures per 16,000 S.F.	3.73	6.20	9.93
1200	199 fixtures per 32,000 S.F.	3.73	6.20	9.93
1240	Type K, 9 fixtures per 900 S.F.	4.58	5.60	10.18
1280	16 fixtures per 1800 S.F.	4.22	5.40	9.62
1320	26 fixtures per 3000 S.F.	4.20	5.35	9.55
1360	31 fixtures per 4000 S.F.	3.85	5.20	9.05
1400	39 fixtures per 5000 S.F.	3.85	5.20	9.05
1440	62 fixtures per 8000 S.F.	3.85	5.20	9.05
1480	78 fixtures per 10,000 S.F.	3.85	5.20	9.05
1520	124 fixtures per 16,000 S.F.	3.82	5.10	8.92
1560	248 fixtures per 32,000 S.F.	3.65	5.65	9.30

D5020 Lighting and Branch Wiring

LOW BAY FIXTURES

J. Metal halide 250 watt

K. High pressure sodium 150 watt

System Components	QUANTITY	UNIT	COST PER S.F.		
			MAT.	INST.	TOTAL
SYSTEM D5020 238 0240					
HIGH INTENSITY DISCHARGE FIXTURE, 8'-10' ABOVE WORK PLANE					
1 WATT/S.F., TYPE J, 30 FC, 4 FIXTURES/1000 S.F.					
Steel intermediate conduit, (IMC) 1/2" diam	.280	L.F.	.52	1.95	2.47
Wire, 600V, type THWN-THHN, copper, solid, #10	.008	C.L.F.	.13	.56	.69
Steel outlet box 4" concrete	.004	Ea.	.04	.14	.18
Steel outlet box plate with stud, 4" concrete	.004	Ea.	.04	.03	.07
Metal halide, lo bay, aluminum reflector, 250 W DX lamp	.004	Ea.	1.60	.87	2.47
TOTAL			2.33	3.55	5.88

D5020 238	H.I.D. Fixture, Low Bay, 8'-10' (by Wattage)	COST PER S.F.		
		MAT.	INST.	TOTAL
0190	High intensity discharge fixture, 8'-10' above work plane			
0240	1 watt/S.F., type J, 30 FC, 4 fixtures/1000 S.F.	2.33	3.55	5.88
0280	Type K, 29 FC, 5 fixtures/1000 S.F.	2.38	3.19	5.57
0400	2 watt/S.F., type J, 52 FC, 7 fixtures/1000 S.F.	4.14	6.50	10.64
0440	Type K, 63 FC, 11 fixtures/1000 S.F.	5.15	6.70	11.85
0560	3 watt/S.F., type J, 81 FC, 11 fixtures/1000 S.F.	6.40	9.85	16.25
0600	Type K, 92 FC, 16 fixtures/1000 S.F.	7.55	9.90	17.45
0720	4 watt/S.F., type J, 103 FC, 14 fixtures/1000 S.F.	8.25	12.95	21.20
0760	Type K, 127 FC, 22 fixtures/1000 S.F.	10.30	13.40	23.70
0880	5 watt/S.F., type J, 133 FC, 18 fixtures/1000 S.F.	10.55	16.30	26.85
0920	Type K, 155 FC, 27 fixtures/1000 S.F.	12.70	16.60	29.30

D5020 Lighting and Branch Wiring

LOW BAY FIXTURES

J. Metal halide 250 watt

K. High pressure sodium 150 watt

System Components	QUANTITY	UNIT	COST PER S.F.		
			MAT.	INST.	TOTAL
SYSTEM D5020 240 0880					
HIGH INTENSITY DISCHARGE FIXTURE, 16' ABOVE WORK PLANE, 50 FC					
TYPE J, 9 FIXTURES PER 900 S.F.					
Steel intermediate conduit, (IMC) 1/2" diam	.630	L.F.	1.17	4.38	5.55
Wire, 600V type, THWN-THHN, copper, solid, #10	.012	C.L.F.	.19	.83	1.02
Steel outlet box 4" concrete	.010	Ea.	.10	.35	.45
Steel outlet box plate with stud, 4" concrete	.010	Ea.	.10	.09	.19
Metal halide, lo bay, aluminum reflector, 250 W DX lamp	.010	Ea.	4	2.18	6.18
TOTAL			5.56	7.83	13.39

D5020 240	H.I.D. Fixture, Low Bay, 16' (by Type)	COST PER S.F.		
		MAT.	INST.	TOTAL
0510	High intensity discharge fixture, 16' above work plane, 50 FC			
0880	Type J, 9 fixtures per 900 S.F.	5.55	7.85	13.40
0920	14 fixtures per 1800 S.F.	4.73	7.35	12.08
0960	24 fixtures per 3000 S.F.	4.76	7.50	12.26
1000	32 fixtures per 4000 S.F.	4.76	7.50	12.26
1040	35 fixtures per 5000 S.F.	4.36	7.35	11.71
1080	56 fixtures per 8000 S.F.	4.36	7.35	11.71
1120	70 fixtures per 10,000 S.F.	4.36	7.35	11.71
1160	111 fixtures per 16,000 S.F.	4.36	7.35	11.71
1200	222 fixtures per 32,000 S.F.	4.36	7.35	11.71
1240	Type K, 11 fixtures per 900 S.F.	5.65	7.30	12.95
1280	20 fixtures per 1800 S.F.	5.20	6.80	12
1320	29 fixtures per 3000 S.F.	4.85	6.65	11.50
1360	39 fixtures per 4000 S.F.	4.83	6.60	11.43
1400	44 fixtures per 5000 S.F.	4.48	6.45	10.93
1440	62 fixtures per 8000 S.F.	4.48	6.45	10.93
1480	87 fixtures per 10,000 S.F.	4.48	6.45	10.93
1520	138 fixtures per 16,000 S.F.	4.48	6.45	10.93

D5020 Lighting and Branch Wiring

LOW BAY FIXTURES

J. Metal halide 250 watt

K. High pressure sodium 150 watt

System Components	QUANTITY	UNIT	COST PER S.F.		
			MAT.	INST.	TOTAL
SYSTEM D5020 242 0240					
HIGH INTENSITY DISCHARGE FIXTURE, 16' ABOVE WORK PLANE					
1 WATT/S.F., TYPE J, 28 FC, 4 FIXTURES/1000 S.F.					
Steel intermediate conduit, (IMC) 1/2" diam	.328	L.F.	.61	2.28	2.89
Wire, 600V, type THWN-THHN, copper, solid, #10	.010	C.L.F.	.16	.70	.86
Steel outlet box 4" concrete	.004	Ea.	.04	.14	.18
Steel outlet box plate with stud, 4" concrete	.004	Ea.	.04	.03	.07
Metal halide, lo bay, aluminum reflector, 250 W DX lamp	.004	Ea.	1.60	.87	2.47
TOTAL			2.45	4.02	6.47

D5020 242	H.I.D. Fixture, Low Bay, 16' (by Wattage)	COST PER S.F.		
		MAT.	INST.	TOTAL
0190	High intensity discharge fixture, mounted 16' above work plane			
0240	1 watt/S.F., type J, 28 FC, 4 fixt./1000 S.F.	2.45	4.02	6.47
0280	Type K, 27 FC, 5 fixt./1000 S.F.	2.72	4.50	7.22
0400	2 watt/S.F., type J, 48 FC, 7 fixt/1000 S.F.	4.46	7.80	12.26
0440	Type K, 58 FC, 11 fixt/1000 S.F.	5.80	9.20	15
0560	3 watt/S.F., type J, 75 FC, 11 fixt/1000 S.F.	6.90	11.80	18.70
0600	Type K, 85 FC, 16 fixt/1000 S.F.	8.50	13.70	22.20
0720	4 watt/S.F., type J, 95 FC, 14 fixt/1000 S.F.	8.95	15.55	24.50
0760	Type K, 117 FC, 22 fixt/1000 S.F.	11.60	18.40	30
0880	5 watt/S.F., type J, 122 FC, 18 fixt/1000 S.F.	11.35	19.60	30.95
0920	Type K, 143 FC, 27 fixt/1000 S.F.	14.30	23	37.30

For customer support on your Electrical Costs with RSMeans data, call 800.448.8182.

433

D5020 Lighting and Branch Wiring

Daylight Dimming System

System Components	QUANTITY	UNIT	COST PER S.F.		
			MAT.	INST.	TOTAL
SYSTEM D5020 290 0800					
DAYLIGHT DIMMING CONTROL SYSTEM					
5 FIXTURES PER 1000 S.F.					
Tray cable, type TC, copper #16-4 conductor	.300	C.L.F.	.01	.03	.04
Wire, 600 volt, type THWN-THHN, copper, solid, #12	.320	C.L.F.		.02	.02
Conduit (EMT) , to 10' H, incl 2 termn,2 elb&11 bm clp per 100', 3/4"	10.000	L.F.	.01	.05	.06
Cabinet, hinged, steel, NEMA 1, 12"W x 12"H x 4"D	.200	Ea.	.01	.02	.03
Lighting control module	.200	Ea.	.07	.07	.14
Dimmable ballast three-lamp	5.000	Ea.	.72	.51	1.23
Daylight level sensor, wall mounted, on/off or dimming	.200	Ea.	.04	.02	.06
Automatic wall switches	.200	Ea.	.02	.01	.03
Remote power pack	.100	Ea.		.01	.01
TOTAL			.88	.74	1.62

D5020 290	Daylight Dimming System	COST PER S.F.		
		MAT.	INST.	TOTAL
0500	Daylight Dimming System (no fixtures or fixture power)			
0800	5 fixtures per 1000 S.F.	.88	.74	1.62
1000	10 fixtures per 1000 S.F.	1.75	1.38	3.13
2000	12 fixtures per 1000 S.F.	2.11	1.65	3.76
3000	15 fixtures per 1000 S.F.	2.62	2.05	4.67
4000	20 fixtures per 1000 S.F.	3.49	2.69	6.18
5000	25 fixtures per 1000 S.F.	4.36	3.36	7.72
6000	50 fixtures per 1000 S.F.	8.70	6.65	15.35

D5020 Lighting and Branch Wiring

Lighting AC Control System

System Components	QUANTITY	UNIT	COST PER S.F.		
			MAT.	INST.	TOTAL
SYSTEM D5020 295 0800					
LIGHTING ON/OFF CONTROL SYSTEM					
5 FIXTURES PER 1000 S.F.					
Tray cable, type TC, copper #16-4 conductor	.600	C.L.F.	.03	.06	.09
Wire, 600 volt, type THWN-THHN, copper, solid, #12	.320	C.L.F.		.02	.02
Conduit (EMT) , to 10' H, incl 2 termn,2 elb&11 bm clp per 100', 3/4"	10.000	L.F.	.01	.05	.06
Cabinet, hinged, steel, NEMA 1, 12"W x 12"H x 4"D	.400	Ea.	.01	.05	.06
Relays, 120 V or 277 V standard	.800	Ea.	.04	.05	.09
24 hour dial with reserve power	.400	Ea.	.25	.08	.33
Lighting control module	.200	Ea.	.07	.07	.14
Occupancy sensors, passive infrared ceiling mounted	.400	Ea.	.03	.04	.07
Automatic wall switches	.400	Ea.	.03	.01	.04
Remote power pack	.200	Ea.	.01	.01	.02
TOTAL			.48	.44	.92

D5020 295	Lighting On/Off Control System	COST PER S.F.		
		MAT.	INST.	TOTAL
0500	Includes occupancy and time switching (no fixtures or fixture power)			
0800	5 fixtures per 1000 SF	.48	.44	.92
1000	10 fixtures per 1000 S.F.	.55	.51	1.06
2000	12 fixtures per 1000 S.F.	.67	.60	1.27
3000	15 fixtures per 1000 S.F.	.81	.73	1.54
4000	20 fixtures per 1000 S.F.	1.10	.92	2.02
5000	25 fixtures per 1000 S.F.	1.35	1.15	2.50
6000	50 fixtures per 1000 S.F.	2.71	2.24	4.95

For customer support on your Electrical Costs with RSMeans data, call 800.448.8182.

435

D5030 Communications and Security

Description: System below includes telephone fitting installed. Does not include cable.

When poke thru fittings and telepoles are used for power, they can also be used for telephones at a negligible additional cost.

System Components	QUANTITY	UNIT	COST PER S.F.		
			MAT.	INST.	TOTAL
SYSTEM D5030 310 0200					
TELEPHONE SYSTEMS, UNDERFLOOR DUCT, 5' ON CENTER, LOW DENSITY					
Underfloor duct 7-1/4" w/insert 2' O.C. 1-3/8" x 7-1/4" super duct	.190	L.F.	7.32	2.64	9.96
Vertical elbow for underfloor superduct, 7-1/4", included					
Underfloor duct conduit adapter, 2" x 1-1/4", included					
Underfloor duct junction box, single duct, 7-1/4" x 3 1/8"	.003	Ea.	1.85	.52	2.37
Underfloor junction box carpet pan	.003	Ea.	1.25	.03	1.28
Underfloor duct outlet, low tension	.004	Ea.	.46	.35	.81
TOTAL			10.88	3.54	14.42

D5030 310	Telephone Systems	COST PER S.F.		
		MAT.	INST.	TOTAL
0200	Telephone systems, underfloor duct, 5' on center, low density	10.90	3.54	14.44
0240	5' on center, high density	11.35	3.89	15.24
0280	7' on center, low density	8.70	2.94	11.64
0320	7' on center, high density	9.15	3.29	12.44
0400	Poke thru fittings, low density	1.24	1.25	2.49
0440	High density	2.48	2.50	4.98
0520	Telepoles, low density	1.24	.75	1.99
0560	High density	2.48	1.50	3.98
0640	Conduit system with floor boxes, low density	1.49	1.23	2.72
0680	High density	2.96	2.44	5.40
1020	Telephone wiring for offices & laboratories, 8 jacks/M.S.F.	.37	1.92	2.29

D5030 810	Security & Detection Systems	COST EACH		
		MAT.	INST.	TOTAL
0200	Security system, head end equipment	29,800	4,650	34,450
0240	Security system, door/window contact biased, box, conduit & cable	181	525	706
0280	Security system, door/window contact balanced, box, conduit & cable	123	525	648
0440	Security system, proximity card reader, box, conduit & cable	445	695	1,140
1600	Card control entrance system, to 6 zone, including hardware for 100 doors	44,100	32,300	76,400

D5030 Communications and Security

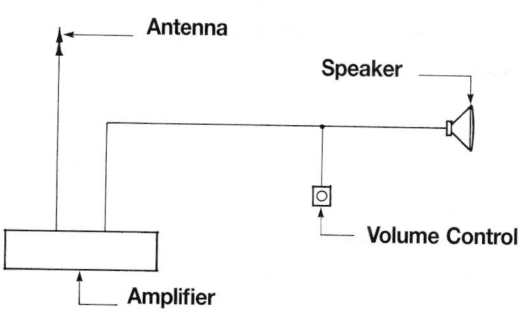

Antenna

Speaker

Volume Control

Amplifier

Sound System Includes AM–FM antenna, outlets, rigid conduit, and copper wire.
Fire Detection System Includes pull stations, signals, smoke and heat detectors, rigid conduit, and copper wire.
Intercom System Includes master and remote stations, rigid conduit, and copper wire.
Master Clock System Includes clocks, bells, rigid conduit, and copper wire.
Master TV Antenna Includes antenna, VHF–UHF reception and distribution, rigid conduit, and copper wire.

System Components	QUANTITY	UNIT	COST EACH		
			MAT.	INST.	TOTAL
SYSTEM D5030 910 0220					
SOUND SYSTEM, INCLUDES OUTLETS, BOXES, CONDUIT & WIRE					
Steel intermediate conduit, (IMC) 1/2" diam	1200.000	L.F.	2,220	8,340	10,560
Wire sound shielded w/drain, #22-2 conductor	15.500	C.L.F.	141.05	1,348.50	1,489.55
Sound system speakers ceiling or wall	12.000	Ea.	1,836	1,044	2,880
Sound system volume control	12.000	Ea.	684	1,044	1,728
Sound system amplifier, 250 W	1.000	Ea.	1,325	695	2,020
Sound system antenna, AM FM	1.000	Ea.	143	174	317
Sound system monitor panel	1.000	Ea.	505	174	679
Sound system cabinet	1.000	Ea.	1,100	695	1,795
Steel outlet box 4" square	12.000	Ea.	69.60	420	489.60
Steel outlet box 4" plaster rings	12.000	Ea.	20.28	130.80	151.08
TOTAL			8,043.93	14,065.30	22,109.23

D5030 910	Communication & Alarm Systems	COST EACH		
		MAT.	INST.	TOTAL
0200	Communication & alarm systems, includes outlets, boxes, conduit & wire			
0210	Sound system, 6 outlets	5,650	8,800	14,450
0220	12 outlets	8,050	14,100	22,150
0240	30 outlets	14,100	26,600	40,700
0280	100 outlets	42,100	89,000	131,100
0320	Fire detection systems, non-addressable, 12 detectors	3,350	7,300	10,650
0360	25 detectors	5,900	12,300	18,200
0400	50 detectors	11,600	24,400	36,000
0440	100 detectors	21,700	44,400	66,100
0450	Addressable type, 12 detectors	4,900	7,350	12,250
0452	25 detectors	8,950	12,400	21,350
0454	50 detectors	17,200	24,400	41,600
0456	100 detectors	33,500	44,900	78,400
0458	Fire alarm control panel, 8 zone, excluding wire and conduit	1,050	1,400	2,450
0459	12 zone	2,300	2,075	4,375
0460	Fire alarm command center, addressable without voice, excl. wire & conduit	4,975	1,225	6,200
0462	Addressable with voice	10,500	1,925	12,425
0480	Intercom systems, 6 stations	4,000	6,025	10,025
0520	12 stations	7,775	12,000	19,775
0560	25 stations	13,200	23,000	36,200
0600	50 stations	25,600	43,300	68,900
0640	100 stations	50,500	84,500	135,000
0680	Master clock systems, 6 rooms	4,650	9,800	14,450
0720	12 rooms	7,450	16,700	24,150
0760	20 rooms	10,500	23,700	34,200
0800	30 rooms	17,200	43,600	60,800

D50 Electrical

D5030 Communications and Security

D5030 910	Communication & Alarm Systems	COST EACH		
		MAT.	INST.	TOTAL
0840	50 rooms	28,000	73,500	101,500
0880	100 rooms	54,500	146,000	200,500
0920	Master TV antenna systems, 6 outlets	2,250	6,200	8,450
0960	12 outlets	4,200	11,600	15,800
1000	30 outlets	14,300	26,800	41,100
1040	100 outlets	52,500	87,500	140,000

D5030 920	Data Communication	COST PER M.S.F.		
		MAT.	INST.	TOTAL
0100	Data communication system, incl data/voice outlets, boxes, conduit & cable			
0102	Data and voice system, 2 data/voice outlets per 1000 S.F.	135	515	650
0104	4 data/voice outlets per 1000 S.F.	255	1,025	1,280
0106	6 data/voice outlets per 1000 S.F.	360	1,500	1,860
0110	8 data/voice outlets per 1000 S.F.	465	1,975	2,440

D5090 Other Electrical Systems

Muffler · Sleeve · Transfer Switch · Engine · Charger · Battery

Description: System below tabulates the installed cost for generators by kW. Included in costs are battery, charger, muffler, and transfer switch.

No conduit, wire, or terminations included.

System Components	QUANTITY	UNIT	COST PER kW		
			MAT.	INST.	TOTAL
SYSTEM D5090 210 0200 GENERATOR SET, INCL. BATTERY, CHARGER, MUFFLER & TRANSFER SWITCH GAS/GASOLINE OPER., 3 PHASE, 4 WIRE, 277/480V, 7.5 kW					
Generator set, gas or gasoline operated, 3 ph 4 W, 277/480 V, 7.5 kW	.133	Ea.	1,256.67	302.93	1,559.60
TOTAL			1,256.67	302.93	1,559.60

D5090 210	Generators (by kW)	COST PER kW		
		MAT.	INST.	TOTAL
0190	Generator sets, include battery, charger, muffler & transfer switch			
0200	Gas/gasoline operated, 3 phase, 4 wire, 277/480 volt, 7.5 kW	1,250	305	1,555
0240	11.5 kW	1,150	231	1,381
0280	20 kW	785	149	934
0320	35 kW	535	97.50	632.50
0360	80 kW	385	119	504
0400	100 kW	335	115	450
0440	125 kW	550	109	659
0480	185 kW	490	82.50	572.50
0560	Diesel engine with fuel tank, 30 kW	405	114	519
0600	50 kW	440	89.50	529.50
0640	75 kW	340	71.50	411.50
0680	100 kW	305	60.50	365.50
0720	125 kW	258	52	310
0760	150 kW	276	48	324
0800	175 kW	263	43	306
0840	200 kW	249	39	288
0880	250 kW	214	32.50	246.50
0920	300 kW	188	28.50	216.50
0960	350 kW	184	27	211
1000	400 kW	198	25	223
1040	500 kW	201	21	222
1200	750 kW	216	13.35	229.35
1400	1000 kW	205	14.05	219.05

For customer support on your Electrical Costs with RSMeans data, call 800.448.8182.

439

D50 Electrical

D5090 Other Electrical Systems

PV power system, stand alone, AC and DC loads

System Components	QUANTITY	UNIT	COST EACH		
			MAT.	INST.	TOTAL
SYSTEM D5090 420 0100					
PHOTOVOLTAIC POWER SYSTEM, STAND ALONE					
Alternative energy sources, photovoltaic module, 75 watt, 17 V	12.000	Ea.	5,220	1,044	6,264
PV rack system, roof, penetrating surface mount, on wood framing, 1 panel	12.000	Ea.	720	1,368	2,088
DC to AC inverter for, 24 V, 2,500 watt	1.000	Ea.	1,650	174	1,824
PV components, combiner box, 10 lug, NEMA 3R enclosure	1.000	Ea.	297	174	471
Alternative energy sources, PV components, fuse, 15 A for combiner box	10.000	Ea.	235	174	409
Battery charger controller w/temperature sensor	1.000	Ea.	540	174	714
Digital readout panel, displays hours, volts, amps, etc.	1.000	Ea.	240	174	414
Deep cycle solar battery, 6 V, 180 Ah (C/20)	4.000	Ea.	1,340	348	1,688
Battery interconnection, 15" AWG #2/0, sealed w/copper ring lugs	3.000	Ea.	59.70	130.50	190.20
Battery interconn, 24" AWG #2/0, sealed w/copper ring lugs	2.000	Ea.	52	87	139
Battery interconn, 60" AWG #2/0, sealed w/copper ring lugs	2.000	Ea.	130	87	217
Batt temp computer probe, RJ11 jack, 15' cord	1.000	Ea.	24.50	43.50	68
System disconnect, DC 175 amp circuit breaker	1.000	Ea.	248	87	335
Conduit box for inverter	1.000	Ea.	69	87	156
Low voltage disconnect	1.000	Ea.	63	87	150
Vented battery enclosure, wood	1.000	Ea.	310	310	620
Grounding, rod, copper clad, 8' long, 5/8" diameter	1.000	Ea.	23.50	127	150.50
Grounding, clamp, bronze, 5/8" dia	1.000	Ea.	5.30	22	27.30
Bare copper wire, stranded, #8	1.000	C.L.F.	47.50	63.50	111
Wire, 600 volt, type THW, copper, stranded, #12	3.600	C.L.F.	52.20	228.60	280.80
Wire, 600 volt, type THW, copper, stranded, #10	1.050	C.L.F.	24.15	72.98	97.13
Cond to 10' H,incl 2 termn,2 elb&11 bm CLP per 100',galv stl,1/2" dia	120.000	L.F.	304.80	930	1,234.80
Conduit,to 10' H,incl 2 termn,2 elb&11 bm clp per 100',(EMT), 1" dia	30.000	L.F.	60.90	181.50	242.40
Lightning surge suppressor	1.000	Ea.	84	22	106
General duty 240 volt, 2 pole, nonfusible, NEMA 3R, 60 amp	1.000	Ea.	137	305	442
Load centers, 3 wire, 120/240V, 100 amp main lugs, indoor, 8 circuits	2.000	Ea.	188	990	1,178
Circuit breaker, Plug-in, 120/240 volt, to 60 amp, 1 pole	5.000	Ea.	33.75	290	323.75
Fuses, dual element, time delay, 250 volt, 50 amp	2.000	Ea.	25.10	27.80	52.90
TOTAL			12,184.40	7,809.38	19,993.78

D5090 Other Electrical Systems

D5090 420	Photovoltaic Power System, Stand Alone	COST EACH		
		MAT.	INST.	TOTAL
0050	24 V capacity, 900 W (~102 SF)			
0100	Roof mounted, on wood framing	12,200	7,800	20,000
0150	with standoff	12,300	7,800	20,100
0200	Ground placement, ballast, fixed	17,100	8,325	25,425
0250	Adjustable	17,400	8,325	25,725
0300	Top of pole, passive tracking	17,100	8,325	25,425
0350	1.8 kW (~204 SF)			
0400	Roof mounted, on wood framing	19,700	12,400	32,100
0450	with standoff	20,000	12,400	32,400
0500	Ground placement, ballast, fixed	28,600	11,700	40,300
0550	Adjustable	29,200	11,700	40,900
0600	Top of pole, passive tracking	29,600	11,700	41,300
0650	3.2 kW (~408 SF)			
0700	Roof mounted, on wood framing	35,800	21,300	57,100
0750	with standoff	36,300	21,300	57,600
0800	Ground placement, ballast, fixed	52,500	18,200	70,700
0850	Adjustable	54,000	18,200	72,200
0900	Top of pole, passive tracking	54,500	18,000	72,500

For customer support on your Electrical Costs with RSMeans data, call 800.448.8182.

441

D5090 Other Electrical Systems

Photovoltaic Modules →

Photovoltaic power system, grid connected 10 kW

Utility Connection →

Combiner Box

Isolation Transformer

AC Disconnect

Inverter

DC Disconnect

System Components	QUANTITY	UNIT	COST EACH		
			MAT.	INST.	TOTAL
SYSTEM D5090 430 0100					
PHOTOVOLTAIC POWER SYSTEM, GRID CONNECTED 10 kW					
Photovoltaic module, 150 watt, 33 V	60.000	Ea.	26,100	5,220	31,320
PV rack system, roof, non-penetrating ballast, 1 panel	60.000	Ea.	63,000	2,460	65,460
DC to AC inverter for, 24 V, 4,000 watt	3.000	Ea.	9,975	1,050	11,025
Combiner box 10 lug, NEMA 3R	1.000	Ea.	297	174	471
15 amp fuses	10.000	Ea.	235	174	409
Safety switch, 60 amp	1.000	Ea.	281	315	596
Safety switch, 100 amp	1.000	Ea.	480	385	865
Utility connection, 3 pole breaker	1.000	Ea.	253	112	365
Fuse, 60 A	3.000	Ea.	38.40	41.70	80.10
Fuse, 100 A	3.000	Ea.	93	52.20	145.20
10 kVA isolation tranformer	1.000	Ea.	1,950	870	2,820
EMT Conduit w/fittings & support	1.000	L.F.	365.40	1,089	1,454.40
RGS Conduit w/fittings & support	267.000	L.F.	1,989.15	2,856.90	4,846.05
Enclosure 24" x 24" x 10", NEMA 4	1.000	Ea.	5,400	1,400	6,800
Wire, 600 volt, type THWN-THHN, copper, stranded, #6	8.000	C.L.F.	424	856	1,280
Wire, 600 volt, copper type XLPE-USE(RHW), stranded, #12	18.000	C.L.F.	360	1,143	1,503
Grounding, bare copper wire stranded, 4/0	2.000	C.L.F.	970	488	1,458
Grounding, exothermic weld, 4/0 wire to building steel	4.000	Ea.	42.20	398	440.20
Grounding, brazed connections, #6 wire	2.000	Ea.	35.60	116	151.60
Insulated ground wire, copper, #6	.400	C.L.F.	21.20	42.80	64
TOTAL			112,309.95	19,243.60	131,553.55

D5090 430	Photovoltaic Power System, Grid Connected	COST EACH		
		MAT.	INST.	TOTAL
0050	10 kW			
0100	Roof mounted, non-penetrating ballast	112,500	19,200	131,700
0200	Penetrating surface mount, on steel framing	53,000	31,400	84,400
0300	on wood framing	53,000	23,600	76,600
0400	with standoff	53,500	23,700	77,200
0500	Ground placement, ballast, fixed	73,500	19,100	92,600
0600	adjustable	75,000	19,100	94,100
0700	Top of pole, passive tracking	75,000	19,500	94,500
1050	20 kW			
1100	Roof mounted, non-penetrating ballast	217,000	32,200	249,200
1200	Penetrating surface mount, on steel framing	98,500	56,500	155,000
1300	on wood framing	98,500	41,000	139,500
1400	with standoff	99,500	41,000	140,500
1500	Ground placement, ballast, fixed	139,000	31,000	170,000
1600	adjustable	142,000	31,000	173,000
1700	Top of pole, passive tracking	142,000	31,900	173,900

D5090 Other Electrical Systems

System Components	QUANTITY	UNIT	COST EACH		
			MAT.	INST.	TOTAL
SYSTEM D5090 480 1100					
ELECTRICAL, SINGLE PHASE, 1 METER					
#18 twisted shielded pair in 1/2" EMT conduit	.050	C.L.F.	7.85	11.60	19.45
Wire, 600 volt, type THW, copper, solid, #12	.150	C.L.F.	1.59	9.53	11.12
Conduit (EMT) , to 10' H, incl 2 termn,2 elb&11 bm clp per 100', 3/4"	5.000	L.F.	6.20	26.75	32.95
Outlet boxes, pressed steel, handy box	1.000	Ea.	2.19	26	28.19
Outlet boxes, pressed steel, handy box, covers, device	1.000	Ea.	1.10	10.90	12
Wiring devices, receptacle, duplex, 120 volt, ground, 20 amp	1.000	Ea.	10	26	36
Single phase, 277 volt, 200 amp	1.000	Ea.	430	79.50	509.50
Data recorder, 8 meters	1.000	Ea.	1,575	63.50	1,638.50
Software package, per meter, premium	1.000	Ea.	735		735
TOTAL			2,768.93	253.78	3,022.71

D5090 480	Energy Monitoring Systems	COST EACH		
		MAT.	INST.	TOTAL
1000	Electrical			
1100	Single phase, 1 meter	2,775	254	3,029
1110	4 meters	6,925	1,925	8,850
1120	8 meters	13,900	3,950	17,850
1200	Three phase, 1 meter	3,250	320	3,570
1210	5 meters	10,800	3,000	13,800
1220	10 meters	22,200	6,150	28,350
1230	25 meters	52,500	12,200	64,700
2000	Mechanical			
2100	BTU, 1 meter	3,550	925	4,475
2110	w/1 duct sensor	3,750	1,150	4,900
2120	& 1 space sensor	4,250	1,300	5,550
2130	& 5 space sensors	6,275	1,925	8,200
2140	& 10 space sensors	11,200	2,700	13,900
2200	BTU, 3 meters	7,525	2,650	10,175
2210	w/3 duct sensors	8,125	3,300	11,425
2220	& 3 space sensors	9,625	3,775	13,400
2230	& 15 space sensors	15,700	5,625	21,325
2240	& 30 space sensors	25,600	7,975	33,575
9000	Front end display	525	136	661
9100	Computer workstation	1,225	1,800	3,025

D5090 Other Electrical Systems

Low Density and Medium Density Baseboard Radiation
The costs shown in Table below are based on the following system considerations:

1. The heat loss per square foot is based on approximately 34 BTU/hr. per S.F. of floor or 10 watts per S.F. of floor.
2. Baseboard radiation is based on the low watt density type rated 187 watts per L.F. and the medium density type rated 250 watts per L.F.
3. Thermostat is not included.
4. Wiring costs include branch circuit wiring.

System Components	QUANTITY	UNIT	COST PER S.F.		
			MAT.	INST.	TOTAL
SYSTEM D5090 510 1000					
ELECTRIC BASEBOARD RADIATION, LOW DENSITY, 900 S.F., 31 MBH, 9 kW					
Electric baseboard radiator, 5' long 935 watt	.011	Ea.	.56	1.34	1.90
Steel intermediate conduit, (IMC) 1/2" diam	.170	L.F.	.31	1.18	1.49
Wire 600 volt, type THW, copper, solid, #12	.005	C.L.F.	.05	.32	.37
TOTAL			.92	2.84	3.76

D5090 510	Electric Baseboard Radiation (Low Density)	COST PER S.F.		
		MAT.	INST.	TOTAL
1000	Electric baseboard radiation, low density, 900 S.F., 31 MBH, 9 kW	.92	2.84	3.76
1200	1500 S.F., 51 MBH, 15 kW	.91	2.77	3.68
1400	2100 S.F., 72 MBH, 21 kW	.82	2.48	3.30
1600	3000 S.F., 102 MBH, 30 kW	.74	2.26	3
2000	Medium density, 900 S.F., 31 MBH, 9 kW	.82	2.60	3.42
2200	1500 S.F., 51 MBH, 15 kW	.81	2.53	3.34
2400	2100 S.F., 72 MBH, 21 kW	.77	2.36	3.13
2600	3000 S.F., 102 MBH, 30 kW	.69	2.14	2.83

D5090 Other Electrical Systems

Commercial Duty Baseboard Radiation
The costs shown in Table below are based on the following system considerations:

1. The heat loss per square foot is based on approximately 41 BTU/hr. per S.F. of floor or 12 watts per S.F.
2. The baseboard radiation is of the commercial duty type rated 250 watts per L.F. served by 277 volt, single phase power.
3. Thermostat is not included.
4. Wiring costs include branch circuit wiring.

System Components	QUANTITY	UNIT	COST PER S.F.		
			MAT.	INST.	TOTAL
SYSTEM D5090 520 1000					
ELECTRIC BASEBOARD RADIATION, MEDIUM DENSITY, 1230 S.F., 51 MBH, 15 kW					
Electric baseboard radiator, 5' long	.013	Ea.	.66	1.59	2.25
Steel intermediate conduit, (IMC) 1/2" diam	.154	L.F.	.28	1.07	1.35
Wire 600 volt, type THW, copper, solid, #12	.004	C.L.F.	.04	.25	.29
TOTAL			.98	2.91	3.89

D5090 520	Electric Baseboard Radiation (Medium Density)	COST PER S.F.		
		MAT.	INST.	TOTAL
1000	Electric baseboard radiation, medium density, 1230 SF, 51 MBH, 15 kW	.98	2.91	3.89
1200	2500 S.F. floor area, 106 MBH, 31 kW	.92	2.72	3.64
1400	3700 S.F. floor area, 157 MBH, 46 kW	.90	2.64	3.54
1600	4800 S.F. floor area, 201 MBH, 59 kW	.83	2.44	3.27
1800	11,300 S.F. floor area, 464 MBH, 136 kW	.78	2.25	3.03
2000	30,000 S.F. floor area, 1229 MBH, 360 kW	.77	2.19	2.96

For customer support on your Electrical Costs with RSMeans data, call 800.448.8182.

445

G10 Site Preparation

G1030 Site Earthwork

Trenching Systems are shown on a cost per linear foot basis. The systems include: excavation; backfill and removal of spoil; and compaction for various depths and trench bottom widths. The backfill has been reduced to accommodate a pipe of suitable diameter and bedding.

The slope for trench sides varies from none to 1:1.

The Expanded System Listing shows Trenching Systems that range from 2' to 12' in width. Depths range from 2' to 25'.

System Components	QUANTITY	UNIT	COST PER L.F.		
			EQUIP.	LABOR	TOTAL
SYSTEM G1030 805 1310					
TRENCHING, COMMON EARTH, NO SLOPE, 2' WIDE, 2' DP, 3/8 C.Y. BUCKET					
Excavation, trench, hyd. backhoe, track mtd., 3/8 C.Y. bucket	.148	B.C.Y.	.34	1.12	1.46
Backfill and load spoil, from stockpile	.153	L.C.Y.	.12	.34	.46
Compaction by vibrating plate, 6" lifts, 4 passes	.118	E.C.Y.	.03	.41	.44
Remove excess spoil, 8 C.Y. dump truck, 2 mile roundtrip	.040	L.C.Y.	.12	.18	.30
TOTAL			.61	2.05	2.66

G1030 805	Trenching Common Earth	COST PER L.F.		
		EQUIP.	LABOR	TOTAL
1310	Trenching, common earth, no slope, 2' wide, 2' deep, 3/8 C.Y. bucket	.61	2.05	2.66
1320	3' deep, 3/8 C.Y. bucket	.88	3.09	3.97
1330	4' deep, 3/8 C.Y. bucket	1.15	4.10	5.25
1340	6' deep, 3/8 C.Y. bucket	1.59	5.35	6.94
1350	8' deep, 1/2 C.Y. bucket	2.12	7.05	9.17
1360	10' deep, 1 C.Y. bucket	3.45	8.45	11.90
1400	4' wide, 2' deep, 3/8 C.Y. bucket	1.39	4.07	5.46
1410	3' deep, 3/8 C.Y. bucket	1.91	6.15	8.06
1420	4' deep, 1/2 C.Y. bucket	2.38	6.85	9.23
1430	6' deep, 1/2 C.Y. bucket	3.92	11	14.92
1440	8' deep, 1/2 C.Y. bucket	6.50	14.25	20.75
1450	10' deep, 1 C.Y. bucket	7.95	17.70	25.65
1460	12' deep, 1 C.Y. bucket	10.25	22.50	32.75
1470	15' deep, 1-1/2 C.Y. bucket	8.60	20	28.60
1480	18' deep, 2-1/2 C.Y. bucket	12.20	28	40.20
1520	6' wide, 6' deep, 5/8 C.Y. bucket w/trench box	8.35	16.55	24.90
1530	8' deep, 3/4 C.Y. bucket	11.05	21.50	32.55
1540	10' deep, 1 C.Y. bucket	10.75	22.50	33.25
1550	12' deep, 1-1/2 C.Y. bucket	11.30	24	35.30
1560	16' deep, 2-1/2 C.Y. bucket	15.55	30	45.55
1570	20' deep, 3-1/2 C.Y. bucket	20.50	36	56.50
1580	24' deep, 3-1/2 C.Y. bucket	24.50	43	67.50
1640	8' wide, 12' deep, 1-1/2 C.Y. bucket w/trench box	15.65	30.50	46.15
1650	15' deep, 1-1/2 C.Y. bucket	20.50	40	60.50
1660	18' deep, 2-1/2 C.Y. bucket	22.50	40	62.50
1680	24' deep, 3-1/2 C.Y. bucket	33	55.50	88.50
1730	10' wide, 20' deep, 3-1/2 C.Y. bucket w/trench box	26.50	53	79.50
1740	24' deep, 3-1/2 C.Y. bucket	39.50	63.50	103
1780	12' wide, 20' deep, 3-1/2 C.Y. bucket w/trench box	41.50	67.50	109
1790	25' deep, bucket	51.50	85.50	137
1800	1/2 to 1 slope, 2' wide, 2' deep, 3/8 C.Y. bucket	.88	3.09	3.97
1810	3' deep, 3/8 C.Y. bucket	1.47	5.40	6.87

448

For customer support on your Electrical Costs with RSMeans data, call 800.448.8182.

G1030 Site Earthwork

G1030 805	Trenching Common Earth	COST PER L.F.		
		EQUIP.	LABOR	TOTAL
1820	4' deep, 3/8 C.Y. bucket	2.21	8.25	10.46
1840	6' deep, 3/8 C.Y. bucket	3.87	13.35	17.22
1860	8' deep, 1/2 C.Y. bucket	6.15	21.50	27.65
1880	10' deep, 1 C.Y. bucket	11.90	30	41.90
2300	4' wide, 2' deep, 3/8 C.Y. bucket	1.64	5.10	6.74
2310	3' deep, 3/8 C.Y. bucket	2.51	8.45	10.96
2320	4' deep, 1/2 C.Y. bucket	3.38	10.40	13.78
2340	6' deep, 1/2 C.Y. bucket	6.65	19.55	26.20
2360	8' deep, 1/2 C.Y. bucket	12.70	29	41.70
2380	10' deep, 1 C.Y. bucket	17.70	40.50	58.20
2400	12' deep, 1 C.Y. bucket	23	55	78
2430	15' deep, 1-1/2 C.Y. bucket	24.50	59	83.50
2460	18' deep, 2-1/2 C.Y. bucket	43	92	135
2840	6' wide, 6' deep, 5/8 C.Y. bucket w/trench box	12.35	24.50	36.85
2860	8' deep, 3/4 C.Y. bucket	18	36.50	54.50
2880	10' deep, 1 C.Y. bucket	17.05	37.50	54.55
2900	12' deep, 1-1/2 C.Y. bucket	21.50	49	70.50
2940	16' deep, 2-1/2 C.Y. bucket	35.50	71.50	107
2980	20' deep, 3-1/2 C.Y. bucket	51.50	96.50	148
3020	24' deep, 3-1/2 C.Y. bucket	72.50	132	204.50
3100	8' wide, 12' deep, 1-1/2 C.Y. bucket w/trench box	26.50	55.50	82
3120	15' deep, 1-1/2 C.Y. bucket	38.50	80.50	119
3140	18' deep, 2-1/2 C.Y. bucket	49	95.50	144.50
3180	24' deep, 3-1/2 C.Y. bucket	81	144	225
3270	10' wide, 20' deep, 3-1/2 C.Y. bucket w/trench box	52.50	111	163.50
3280	24' deep, 3-1/2 C.Y. bucket	89.50	158	247.50
3370	12' wide, 20' deep, 3-1/2 C.Y. bucket w/trench box	75	129	204
3380	25' deep, 3-1/2 C.Y. bucket	104	184	288
3500	1 to 1 slope, 2' wide, 2' deep, 3/8 C.Y. bucket	1.15	4.11	5.26
3520	3' deep, 3/8 C.Y. bucket	3.19	9.40	12.59
3540	4' deep, 3/8 C.Y. bucket	3.27	12.35	15.62
3560	6' deep, 1/2 C.Y. bucket	3.86	13.35	17.21
3580	8' deep, 1/2 C.Y. bucket	7.65	26.50	34.15
3600	10' deep, 1 C.Y. bucket	20.50	51	71.50
3800	4' wide, 2' deep, 3/8 C.Y. bucket	1.91	6.15	8.06
3820	3' deep, 3/8 C.Y. bucket	3.11	10.80	13.91
3840	4' deep, 1/2 C.Y. bucket	4.39	14	18.39
3860	6' deep, 1/2 C.Y. bucket	9.40	28	37.40
3880	8' deep, 1/2 C.Y. bucket	18.95	43.50	62.45
3900	10' deep, 1 C.Y. bucket	27.50	63	90.50
3920	12' deep, 1 C.Y. bucket	40.50	91.50	132
3940	15' deep, 1-1/2 C.Y. bucket	40.50	97.50	138
3960	18' deep, 2-1/2 C.Y. bucket	59	126	185
4030	6' wide, 6' deep, 5/8 C.Y. bucket w/trench box	16.30	33	49.30
4040	8' deep, 3/4 C.Y. bucket	23.50	45.50	69
4050	10' deep, 1 C.Y. bucket	24.50	55	79.50
4060	12' deep, 1-1/2 C.Y. bucket	33	74.50	107.50
4070	16' deep, 2-1/2 C.Y. bucket	56	113	169
4080	20' deep, 3-1/2 C.Y. bucket	83	157	240
4090	24' deep, 3-1/2 C.Y. bucket	121	221	342
4500	8' wide, 12' deep, 1-1/2 C.Y. bucket w/trench box	37.50	81	118.50
4550	15' deep, 1-1/2 C.Y. bucket	57	121	178
4600	18' deep, 2-1/2 C.Y. bucket	74.50	148	222.50
4650	24' deep, 3-1/2 C.Y. bucket	129	233	362
4800	10' wide, 20' deep, 3-1/2 C.Y. bucket w/trench box	78	169	247
4850	24' deep, 3-1/2 C.Y. bucket	138	246	384
4950	12' wide, 20' deep, 3-1/2 C.Y. bucket w/trench box	108	190	298
4980	25' deep, 3-1/2 C.Y. bucket	156	279	435

G1030 Site Earthwork

Trenching Systems are shown on a cost per linear foot basis. The systems include: excavation; backfill and removal of spoil; and compaction for various depths and trench bottom widths. The backfill has been reduced to accommodate a pipe of suitable diameter and bedding.

The slope for trench sides varies from none to 1:1.

The Expanded System Listing shows Trenching Systems that range from 2' to 12' in width. Depths range from 2' to 25'.

System Components			COST PER L.F.		
	QUANTITY	UNIT	EQUIP.	LABOR	TOTAL
SYSTEM G1030 806 1310					
TRENCHING, LOAM & SANDY CLAY, NO SLOPE, 2' WIDE, 2' DP, 3/8 C.Y. BUCKET					
Excavation, trench, hyd. backhoe, track mtd., 3/8 C.Y. bucket	.148	B.C.Y.	.31	1.04	1.35
Backfill and load spoil, from stockpile	.165	L.C.Y.	.13	.37	.50
Compaction by vibrating plate 18" wide, 6" lifts, 4 passes	.118	E.C.Y.	.03	.41	.44
Remove excess spoil, 8 C.Y. dump truck, 2 mile roundtrip	.042	L.C.Y.	.13	.19	.32
TOTAL			.60	2.01	2.61

G1030 806	Trenching Loam & Sandy Clay	COST PER L.F.		
		EQUIP.	LABOR	TOTAL
1310	Trenching, loam & sandy clay, no slope, 2' wide, 2' deep, 3/8 C.Y. bucket	.60	2.01	2.61
1320	3' deep, 3/8 C.Y. bucket	.95	3.30	4.25
1330	4' deep, 3/8 C.Y. bucket	1.13	4.01	5.15
1340	6' deep, 3/8 C.Y. bucket	1.71	4.79	6.50
1350	8' deep, 1/2 C.Y. bucket	2.27	6.35	8.60
1360	10' deep, 1 C.Y. bucket	2.49	6.75	9.25
1400	4' wide, 2' deep, 3/8 C.Y. bucket	1.38	4	5.40
1410	3' deep, 3/8 C.Y. bucket	1.89	6.05	7.95
1420	4' deep, 1/2 C.Y. bucket	2.35	6.75	9.10
1430	6' deep, 1/2 C.Y. bucket	4.18	9.95	14.15
1440	8' deep, 1/2 C.Y. bucket	6.35	14.10	20.50
1450	10' deep, 1 C.Y. bucket	6.05	14.40	20.50
1460	12' deep, 1 C.Y. bucket	7.60	17.95	25.50
1470	15' deep, 1-1/2 C.Y. bucket	9	21	30
1480	18' deep, 2-1/2 C.Y. bucket	10.65	23	33.50
1520	6' wide, 6' deep, 5/8 C.Y. bucket w/trench box	8.05	16.25	24.50
1530	8' deep, 3/4 C.Y. bucket	10.65	21.50	32
1540	10' deep, 1 C.Y. bucket	9.90	21.50	31.50
1550	12' deep, 1-1/2 C.Y. bucket	11.10	24	35
1560	16' deep, 2-1/2 C.Y. bucket	15.25	30.50	46
1570	20' deep, 3-1/2 C.Y. bucket	18.85	36	55
1580	24' deep, 3-1/2 C.Y. bucket	23.50	43.50	67
1640	8' wide, 12' deep, 1-1/4 C.Y. bucket w/trench box	15.50	30.50	46
1650	15' deep, 1-1/2 C.Y. bucket	19	39	58
1660	18' deep, 2-1/2 C.Y. bucket	23	43.50	66.50
1680	24' deep, 3-1/2 C.Y. bucket	32	56.50	88.50
1730	10' wide, 20' deep, 3-1/2 C.Y. bucket w/trench box	32	57	89
1740	24' deep, 3-1/2 C.Y. bucket	40	70	110
1780	12' wide, 20' deep, 3-1/2 C.Y. bucket w/trench box	39	67.50	107
1790	25' deep, 3-1/2 C.Y. bucket	50.50	87	138
1800	1/2:1 slope, 2' wide, 2' deep, 3/8 C.Y. bucket	.87	3.01	3.88
1810	3' deep, 3/8 C.Y. bucket	1.44	5.30	6.75
1820	4' deep, 3/8 C.Y. bucket	2.16	8.05	10.20
1840	6' deep, 3/8 C.Y. bucket	4.16	12	16.15

G1030 Site Earthwork

G1030 806	Trenching Loam & Sandy Clay	COST PER L.F.		
		EQUIP.	LABOR	TOTAL
1860	8' deep, 1/2 C.Y. bucket	6.65	19.15	26
1880	10' deep, 1 C.Y. bucket	8.50	24	32.50
2300	4' wide, 2' deep, 3/8 C.Y. bucket	1.63	5	6.65
2310	3' deep, 3/8 C.Y. bucket	2.47	8.30	10.75
2320	4' deep, 1/2 C.Y. bucket	3.34	10.25	13.60
2340	6' deep, 1/2 C.Y. bucket	7.10	17.65	25
2360	8' deep, 1/2 C.Y. bucket	12.35	28.50	41
2380	10' deep, 1 C.Y. bucket	13.35	33	46.50
2400	12' deep, 1 C.Y. bucket	22.50	54.50	77
2430	15' deep, 1-1/2 C.Y. bucket	25.50	61	86.50
2460	18' deep, 2-1/2 C.Y. bucket	42	93	135
2840	6' wide, 6' deep, 5/8 C.Y. bucket w/trench box	11.80	24.50	36.50
2860	8' deep, 3/4 C.Y. bucket	17.30	36	53.50
2880	10' deep, 1 C.Y. bucket	17.80	40.50	58.50
2900	12' deep, 1-1/2 C.Y. bucket	22	49.50	71.50
2940	16' deep, 2-1/2 C.Y. bucket	35	72.50	108
2980	20' deep, 3-1/2 C.Y. bucket	50	97.50	148
3020	24' deep, 3-1/2 C.Y. bucket	70.50	134	205
3100	8' wide, 12' deep, 1-1/2 C.Y. bucket w/trench box	26	56	82
3120	15' deep, 1-1/2 C.Y. bucket	35.50	78.50	114
3140	18' deep, 2-1/2 C.Y. bucket	48	97	145
3180	24' deep, 3-1/2 C.Y. bucket	79	147	226
3270	10' wide, 20' deep, 3-1/2 C.Y. bucket w/trench box	63	119	182
3280	24' deep, 3-1/2 C.Y. bucket	87	160	247
3320	12' wide, 20' deep, 3-1/2 C.Y. bucket w/trench box	70	129	199
3380	25' deep, 3-1/2 C.Y. bucket w/trench box	95.50	173	269
3500	1:1 slope, 2' wide, 2' deep, 3/8 C.Y. bucket	1.13	4.01	5.15
3520	3' deep, 3/8 C.Y. bucket	2.03	7.55	9.60
3540	4' deep, 3/8 C.Y. bucket	3.18	12.05	15.25
3560	6' deep, 1/2 C.Y. bucket	4.16	12	16.15
3580	8' deep, 1/2 C.Y. bucket	10.95	32	43
3600	10' deep, 1 C.Y. bucket	14.55	41	55.50
3800	4' wide, 2' deep, 3/8 C.Y. bucket	1.89	6.05	7.95
3820	3' deep, 1/2 C.Y. bucket	3.06	10.55	13.60
3840	4' deep, 1/2 C.Y. bucket	4.33	13.75	18.10
3860	6' deep, 1/2 C.Y. bucket	10	25.50	35.50
3880	8' deep, 1/2 C.Y. bucket	18.40	43	61.50
3900	10' deep, 1 C.Y. bucket	20.50	51.50	72
3920	12' deep, 1 C.Y. bucket	30.	73	103
3940	15' deep, 1-1/2 C.Y. bucket	42	101	143
3960	18' deep, 2-1/2 C.Y. bucket	58	128	186
4030	6' wide, 6' deep, 5/8 C.Y. bucket w/trench box	15.55	33	48.50
4040	8' deep, 3/4 C.Y. bucket	24	50.50	74.50
4050	10' deep, 1 C.Y. bucket	25.50	59.50	85
4060	12' deep, 1-1/2 C.Y. bucket	32.50	74.50	107
4070	16' deep, 2-1/2 C.Y. bucket	54.50	114	169
4080	20' deep, 3-1/2 C.Y. bucket	81	160	241
4090	24' deep, 3-1/2 C.Y. bucket	117	224	340
4500	8' wide, 12' deep, 1-1/4 C.Y. bucket w/trench box	37.	81	118
4550	15' deep, 1-1/2 C.Y. bucket	52.50	118	171
4600	18' deep, 2-1/2 C.Y. bucket	73	150	223
4650	24' deep, 3-1/2 C.Y. bucket	126	237	365
4800	10' wide, 20' deep, 3-1/2 C.Y. bucket w/trench box	94	181	275
4850	24' deep, 3-1/2 C.Y. bucket	134	250	385
4950	12' wide, 20' deep, 3-1/2 C.Y. bucket w/trench box	101	191	292
4980	25' deep, 3-1/2 C.Y. bucket	152	282	435

For customer support on your Electrical Costs with RSMeans data, call 800.448.8182.

451

Trenching Systems are shown on a cost per linear foot basis. The systems include: excavation; backfill and removal of spoil; and compaction for various depths and trench bottom widths. The backfill has been reduced to accommodate a pipe of suitable diameter and bedding.

The slope for trench sides varies from none to 1:1.

The Expanded System Listing shows Trenching Systems that range from 2' to 12' in width. Depths range from 2' to 25'.

System Components				COST PER L.F.		
		QUANTITY	UNIT	EQUIP.	LABOR	TOTAL
SYSTEM G1030 807 1310						
TRENCHING, SAND & GRAVEL, NO SLOPE, 2' WIDE, 2' DEEP, 3/8 C.Y. BUCKET						
Excavation, trench, hyd. backhoe, track mtd., 3/8 C.Y. bucket		.148	B.C.Y.	.31	1.01	1.32
Backfill and load spoil, from stockpile		.140	L.C.Y.	.11	.31	.42
Compaction by vibrating plate 18" wide, 6" lifts, 4 passes		.118	E.C.Y.	.03	.41	.44
Remove excess spoil, 8 C.Y. dump truck, 2 mile roundtrip		.035	L.C.Y.	.11	.16	.27
	TOTAL			.56	1.89	2.45

G1030 807	Trenching Sand & Gravel	COST PER L.F.		
		EQUIP.	LABOR	TOTAL
1310	Trenching, sand & gravel, no slope, 2' wide, 2' deep, 3/8 C.Y. bucket	.56	1.89	2.45
1320	3' deep, 3/8 C.Y. bucket	.90	3.17	4.07
1330	4' deep, 3/8 C.Y. bucket	1.05	3.81	4.86
1340	6' deep, 3/8 C.Y. bucket	1.63	4.57	6.20
1350	8' deep, 1/2 C.Y. bucket	2.16	6.05	8.20
1360	10' deep, 1 C.Y. bucket	2.33	6.40	8.75
1400	4' wide, 2' deep, 3/8 C.Y. bucket	1.26	3.75	5
1410	3' deep, 3/8 C.Y. bucket	1.73	5.65	7.40
1420	4' deep, 1/2 C.Y. bucket	2.16	6.40	8.55
1430	6' deep, 1/2 C.Y. bucket	3.96	9.50	13.45
1440	8' deep, 1/2 C.Y. bucket	6	13.35	19.35
1450	10' deep, 1 C.Y. bucket	5.70	13.60	19.30
1460	12' deep, 1 C.Y. bucket	7.15	16.95	24
1470	15' deep, 1-1/2 C.Y. bucket	8.45	19.65	28
1480	18' deep, 2-1/2 C.Y. bucket	10.05	21.50	31.50
1520	6' wide, 6' deep, 5/8 C.Y. bucket w/trench box	7.65	15.35	23
1530	8' deep, 3/4 C.Y. bucket	10.10	20	30
1540	10' deep, 1 C.Y. bucket	9.35	20.50	30
1550	12' deep, 1-1/2 C.Y. bucket	10.45	22.50	33
1560	16' deep, 2 C.Y. bucket	14.50	28.50	43
1570	20' deep, 3-1/2 C.Y. bucket	17.85	33.50	51.50
1580	24' deep, 3-1/2 C.Y. bucket	22.50	41	63.50
1640	8' wide, 12' deep, 1-1/2 C.Y. bucket w/trench box	14.55	29	43.50
1650	15' deep, 1-1/2 C.Y. bucket	17.75	36.50	54.50
1660	18' deep, 2-1/2 C.Y. bucket	21.50	41	62.50
1680	24' deep, 3-1/2 C.Y. bucket	30	53	83
1730	10' wide, 20' deep, 3-1/2 C.Y. bucket w/trench box	30	53	83
1740	24' deep, 3-1/2 C.Y. bucket	38	65.50	104
1780	12' wide, 20' deep, 3-1/2 C.Y. bucket w/trench box	36.50	63	99.50
1790	25' deep, 3-1/2 C.Y. bucket	47.50	81.50	129
1800	1/2:1 slope, 2' wide, 2' deep, 3/8 C.Y. bucket	.80	2.86	3.66
1810	3' deep, 3/8 C.Y. bucket	1.35	5	6.35
1820	4' deep, 3/8 C.Y. bucket	2.01	7.65	9.65
1840	6' deep, 3/8 C.Y. bucket	3.96	11.45	15.40

G1030 Site Earthwork

G1030 807	Trenching Sand & Gravel	COST PER L.F.		
		EQUIP.	LABOR	TOTAL
1860	8' deep, 1/2 C.Y. bucket	6.30	18.30	24.50
1880	10' deep, 1 C.Y. bucket	8	22.50	30.50
2300	4' wide, 2' deep, 3/8 C.Y. bucket	1.49	4.72	6.20
2310	3' deep, 3/8 C.Y. bucket	2.28	7.85	10.15
2320	4' deep, 1/2 C.Y. bucket	3.09	9.70	12.80
2340	6' deep, 1/2 C.Y. bucket	6.75	16.90	23.50
2360	8' deep, 1/2 C.Y. bucket	11.75	27	39
2380	10' deep, 1 C.Y. bucket	12.65	31	43.50
2400	12' deep, 1 C.Y. bucket	21	52	73
2430	15' deep, 1-1/2 C.Y. bucket	24	57.50	81.50
2460	18' deep, 2-1/2 C.Y. bucket	40	87.50	128
2840	6' wide, 6' deep, 5/8 C.Y. bucket w/trench box	11.25	23	34.50
2860	8' deep, 3/4 C.Y. bucket	16.50	34	50.50
2880	10' deep, 1 C.Y. bucket	16.85	38	55
2900	12' deep, 1-1/2 C.Y. bucket	20.50	46.50	67
2940	16' deep, 2 C.Y. bucket	33.50	68.50	102
2980	20' deep, 3-1/2 C.Y. bucket	47	91.50	139
3020	24' deep, 3-1/2 C.Y. bucket	67	126	193
3100	8' wide, 12' deep, 1-1/4 C.Y. bucket w/trench box	25	53	78
3120	15' deep, 1-1/2 C.Y. bucket	33.50	74	108
3140	18' deep, 2-1/2 C.Y. bucket	45.50	91	137
3180	24' deep, 3-1/2 C.Y. bucket	74.50	138	213
3270	10' wide, 20' deep, 3-1/2 C.Y. bucket w/trench box	59.50	111	171
3280	24' deep, 3-1/2 C.Y. bucket	82.50	150	233
3370	12' wide, 20' deep, 3-1/2 C.Y. bucket w/trench box	66.50	124	191
3380	25' deep, 3-1/2 C.Y. bucket	95.50	173	269
3500	1:1 slope, 2' wide, 2' deep, 3/8 C.Y. bucket	1.73	4.79	6.50
3520	3' deep, 3/8 C.Y. bucket	1.90	7.15	9.05
3540	4' deep, 3/8 C.Y. bucket	2.98	11.50	14.50
3560	6' deep, 3/8 C.Y. bucket	3.97	11.45	15.40
3580	8' deep, 1/2 C.Y. bucket	10.45	30.50	41
3600	10' deep, 1 C.Y. bucket	13.65	39	52.50
3800	4' wide, 2' deep, 3/8 C.Y. bucket	1.73	5.65	7.40
3820	3' deep, 3/8 C.Y. bucket	2.83	10	12.85
3840	4' deep, 1/2 C.Y. bucket	4.02	13.05	17.05
3860	6' deep, 1/2 C.Y. bucket	9.55	24.50	34
3880	8' deep, 1/2 C.Y. bucket	17.50	40.50	58
3900	10' deep, 1 C.Y. bucket	19.55	49	68.50
3920	12' deep, 1 C.Y. bucket	28	69.50	97.50
3940	15' deep, 1-1/2 C.Y. bucket	39.50	95.50	135
3960	18' deep, 2-1/2 C.Y. bucket	55	120	175
4030	6' wide, 6' deep, 5/8 C.Y. bucket w/trench box	14.80	31	46
4040	8' deep, 3/4 C.Y. bucket	23	48	71
4050	10' deep, 1 C.Y. bucket	24.50	56	80.50
4060	12' deep, 1-1/2 C.Y. bucket	30.50	70.50	101
4070	16' deep, 2 C.Y. bucket	52.50	108	161
4080	20' deep, 3-1/2 C.Y. bucket	76.50	150	227
4090	24' deep, 3-1/2 C.Y. bucket	111	210	320
4500	8' wide, 12' deep, 1-1/2 C.Y. bucket w/trench box	35	76.50	112
4550	15' deep, 1-1/2 C.Y. bucket	49.50	111	161
4600	18' deep, 2-1/2 C.Y. bucket	69.50	141	211
4650	24' deep, 3-1/2 C.Y. bucket	119	222	340
4800	10' wide, 20' deep, 3-1/2 C.Y. bucket w/trench box	89	169	258
4850	24' deep, 3-1/2 C.Y. bucket	127	235	360
4950	12' wide, 20' deep, 3-1/2 C.Y. bucket w/trench box	95	179	274
4980	25' deep, 3-1/2 C.Y. bucket	144	265	410

For customer support on your Electrical Costs with RSMeans data, call 800.448.8182.

453

G4010 Electrical Distribution

UNDERGROUND ELECTRICAL CONDUIT

System Components	QUANTITY	UNIT	COST EACH		
			MAT.	INST.	TOTAL
SYSTEM G4010 312 1000					
2000 AMP UNDERGROUND SERVICE INCLUDING COMMON EARTH EXCAVATION,					
CONCRETE, BACKFILL AND COMPACTION					
Excavate Trench	44.440	B.C.Y.		312.86	312.86
4 inch conduit bank	108.000	L.F.	1,290.60	2,808	4,098.60
4 inch fitting	8.000	Ea.	336	464	800
4 inch bells	4.000	Ea.	21.20	232	253.20
Concrete material	16.580	C.Y.	2,205.14		2,205.14
Concrete placement	16.580	C.Y.		422.29	422.29
Backfill trench	33.710	L.C.Y.		102.14	102.14
Compact fill material in trench	25.930	E.C.Y.		134.31	134.31
Dispose of excess fill material on-site	24.070	L.C.Y.		291.24	291.24
500 kcmil power cable	18.000	C.L.F.	13,680	7,830	21,510
Wire, 600 volt stranded copper wire, type THW, 1/0	6.000	C.L.F.	1,428	1,266	2,694
TOTAL			18,960.94	13,862.84	32,823.78

G4010 312	Underground Power Feed	COST EACH		
		MAT.	INST.	TOTAL
0900	Underground electrical			
0950	Including common earth excavation, concrete, backfill and compaction			
1000	2000 Amp service, 100' length, 4' depth	19,000	13,900	32,900
1100	1600 Amp service	14,700	10,700	25,400
1200	1200 Amp service	11,700	8,850	20,550
1300	1000 Amp service	9,775	8,125	17,900
1400	800 Amp service	8,425	7,600	16,025
1500	600 Amp service	6,150	7,050	13,200

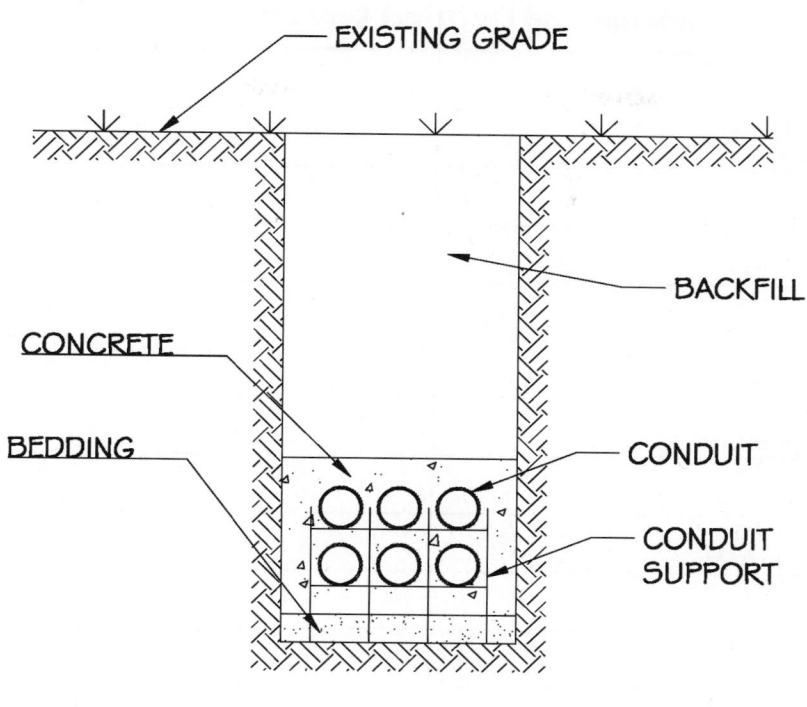

System Components	QUANTITY	UNIT	COST L.F.		
			MAT.	INST.	TOTAL
SYSTEM G4010 320 1012					
UNDERGROUND ELECTRICAL CONDUIT, 2″ DIAMETER, INCLUDING EXCAVATION,					
CONCRETE, BEDDING, BACKFILL AND COMPACTION					
Excavate Trench	.150	B.C.Y.		1.06	1.06
Base spacer	.200	Ea.	.27	2.49	2.76
Conduit	1.000	L.F.	1.39	3.87	5.26
Concrete material	.050	C.Y.	6.65		6.65
Concrete placement	.050	C.Y.		1.27	1.27
Backfill trench	.120	L.C.Y.		.37	.37
Compact fill material in trench	.100	E.C.Y.		.51	.51
Dispose of excess fill material on-site	.120	L.C.Y.		1.45	1.45
Marking tape	.010	C.L.F.	.40	.03	.43
TOTAL			8.71	11.05	19.76

G4010 320	Underground Electrical Conduit	COST L.F.		
		MAT.	INST.	TOTAL
0950	Underground electrical conduit, including common earth excavation,			
0952	Concrete, bedding, backfill and compaction			
1012	2″ dia. Schedule 40 PVC, 2′ deep	8.70	11.05	19.75
1014	4′ deep	8.70	15.70	24.40
1016	6′ deep	8.70	20	28.70
1022	2 @ 2″ dia. Schedule 40 PVC, 2′ deep	10.35	17.40	27.75
1024	4′ deep	10.35	22	32.35
1026	6′ deep	10.35	26.50	36.85
1032	3 @ 2″ dia. Schedule 40 PVC, 2′ deep	12	24	36
1034	4′ deep	12	28.50	40.50
1036	6′ deep	12	33	45
1042	4 @ 2″ dia. Schedule 40 PVC, 2′ deep	16.30	30	46.30
1044	4′ deep	16.30	34	50.30
1046	6′ deep	16.30	39	55.30

For customer support on your Electrical Costs with RSMeans data, call 800.448.8182.

455

G4010 Electrical Distribution

G4010 320	Underground Electrical Conduit	COST L.F.		
		MAT.	INST.	TOTAL
1062	6 @ 2" dia. Schedule 40 PVC, 2' deep	19.65	42.50	62.15
1064	4' deep	19.65	47	66.65
1066	6' deep	19.65	51.50	71.15
1082	8 @ 2" dia. Schedule 40 PVC, 2' deep	23	55	78
1084	4' deep	23	59.50	82.50
1086	6' deep	23	64	87
1092	9 @ 2" dia. Schedule 40 PVC, 2' deep	24.50	61	85.50
1094	4' deep	27.50	65.50	93
1096	6' deep	27.50	70	97.50
1212	3" dia. Schedule 40 PVC, 2' deep	11.20	13.55	24.75
1214	4' deep	11.20	18.30	29.50
1216	6' deep	11.20	23	34.20
1222	2 @ 3" dia. Schedule 40 PVC, 2' deep	12.70	22	34.70
1224	4' deep	12.70	27	39.70
1226	6' deep	12.70	31.50	44.20
1232	3 @ 3" dia. Schedule 40 PVC, 2' deep	15.55	31	46.55
1234	4' deep	15.55	36	51.55
1236	6' deep	15.55	40	55.55
1242	4 @ 3" dia. Schedule 40 PVC, 2' deep	22.50	40	62.50
1244	4' deep	22.50	44.50	67
1246	6' deep	22.50	49	71.50
1262	6 @ 3" dia. Schedule 40 PVC, 2' deep	27	57.50	84.50
1264	4' deep	27	62	89
1266	6' deep	27	66.50	93.50
1282	8 @ 3" dia. Schedule 40 PVC, 2' deep	32.50	75.50	108
1284	4' deep	32.50	79.50	112
1286	6' deep	32.50	85	117.50
1292	9 @ 3" dia. Schedule 40 PVC, 2' deep	38	85	123
1294	4' deep	39.50	88.50	128
1296	6' deep	39.50	92.50	132
1312	4" dia. Schedule 40 PVC, 2' deep	12.15	16.65	28.80
1314	4' deep	12.15	21.50	33.65
1316	6' deep	12.15	26	38.15
1322	2 @ 4" dia. Schedule 40 PVC, 2' deep	15.95	29	44.95
1324	4' deep	15.95	33	48.95
1326	6' deep	15.95	38	53.95
1332	3 @ 4" dia. Schedule 40 PVC, 2' deep	18.40	40.50	58.90
1334	4' deep	18.40	45	63.40
1336	6' deep	18.40	50	68.40
1342	4 @ 4" dia. Schedule 40 PVC, 2' deep	27.50	52.50	80
1344	4' deep	27.50	57.50	85
1346	6' deep	27.50	62.50	90
1362	6 @ 4" dia. Schedule 40 PVC, 2' deep	33.50	76.50	110
1364	4' deep	33.50	81.50	115
1366	6' deep	33.50	86.50	120
1382	8 @ 4" dia. Schedule 40 PVC, 2' deep	42.50	102	144.50
1384	4' deep	42.50	107	149.50
1386	6' deep	42.50	113	155.50
1392	9 @ 4" dia. Schedule 40 PVC, 2' deep	47.50	114	161.50
1394	4' deep	47.50	118	165.50
1396	6' deep	47.50	122	169.50
1412	5" dia. Schedule 40 PVC, 2' deep	13.85	19.90	33.75
1414	4' deep	13.85	24.50	38.35
1416	6' deep	13.85	29	42.85
1422	2 @ 5" dia. Schedule 40 PVC, 2' deep	19.35	35.50	54.85
1424	4' deep	19.35	39.50	58.85
1426	6' deep	19.35	44.50	63.85
1432	3 @ 5" dia. Schedule 40 PVC, 2' deep	25	51	76

G4010 Electrical Distribution

G4010 320	Underground Electrical Conduit	COST L.F.		
		MAT.	INST.	TOTAL
1434	4' deep	25	56	81
1436	6' deep	25	60.50	85.50
1442	4 @ 5" dia. Schedule 40 PVC, 2' deep	34	65.50	99.50
1444	4' deep	34	70.50	104.50
1446	6' deep	34	74.50	108.50
1462	6 @ 5" dia. Schedule 40 PVC, 2' deep	44	96	140
1464	4' deep	44	101	145
1466	6' deep	44	106	150
1482	8 @ 5" dia. Schedule 40 PVC, 2' deep	57.50	128	185.50
1484	4' deep	57.50	135	192.50
1486	6' deep	57.50	140	197.50
1492	9 @ 5" dia. Schedule 40 PVC, 2' deep	64	144	208
1494	4' deep	64	147	211
1496	6' deep	64	152	216
1512	6" dia. Schedule 40 PVC, 2' deep	16.90	24.50	41.40
1514	4' deep	16.90	28.50	45.40
1516	6' deep	16.90	33.50	50.40
1522	2 @ 6" dia. Schedule 40 PVC, 2' deep	22.50	43.50	66
1524	4' deep	22.50	48	70.50
1526	6' deep	22.50	52.50	75
1532	3 @ 6" dia. Schedule 40 PVC, 2' deep	31	64	95
1534	4' deep	31	69.50	100.50
1536	6' deep	31	75	106
1542	4 @ 6" dia. Schedule 40 PVC, 2' deep	41	82	123
1544	4' deep	41	87	128
1546	6' deep	41	91	132
1562	6 @ 6" dia. Schedule 40 PVC, 2' deep	57	122	179
1564	4' deep	57	127	184
1566	6' deep	57	133	190
1582	8 @ 6" dia. Schedule 40 PVC, 2' deep	75	162	237
1584	4' deep	75	169	244
1586	6' deep	75	176	251
1592	9 @ 6" dia. Schedule 40 PVC, 2' deep	83.50	182	265.50
1594	4' deep	83.50	186	269.50
1596	6' deep	83.50	191	274.50

G4020 Site Lighting

The Site Lighting System includes the complete unit from foundation to electrical fixtures. Each system includes: excavation; concrete base; backfill by hand; compaction with a plate compacter; pole of specified material; all fixtures; and lamps.

The Expanded System Listing shows Site Lighting Systems that use one of two types of lamps: high pressure sodium; and metal halide. Systems are listed for 400-watt and 1000-watt lamps. Pole height varies from 20′ to 40′. There are four types of poles possibly listed: aluminum, fiberglass, steel and wood.

System Components	QUANTITY	UNIT	COST EACH		
			MAT.	INST.	TOTAL
SYSTEM G4020 110 5820					
SITE LIGHTING, METAL HALIDE, 400 WATT, ALUMINUM POLE, 20′ HIGH					
Excavating, by hand, pits to 6′ deep, heavy soil	2.368	C.Y.		286.53	286.53
Concrete in place incl. forms and reinf. stl. spread footings under 1 CY	.465	C.Y.	94.86	136.25	231.11
Backfill	1.903	C.Y.		83.73	83.73
Compact, vibrating plate	1.903	C.Y.		11.23	11.23
Aluminum light pole, 20′, no concrete base	1.000	Ea.	1,150	649	1,799
Road area luminaire wire	.300	C.L.F.	9.30	26.10	35.40
Roadway area luminaire, metal halide 400 W	1.000	Ea.	715	315	1,030
TOTAL			1,969.16	1,507.84	3,477

G4020 110	Site Lighting	COST EACH		
		MAT.	INST.	TOTAL
2320	Site lighting, high pressure sodium, 400 watt, aluminum pole, 20′ high	2,100	1,500	3,600
2340	30′ high	3,150	1,850	5,000
2360	40′ high	3,700	2,375	6,075
2520	Fiberglass pole, 20′ high	1,825	1,325	3,150
2540	30′ high	2,050	1,650	3,700
2560	40′ high	2,975	2,100	5,075
2720	Steel pole, 20′ high	2,125	1,575	3,700
2740	30′ high	2,575	1,950	4,525
2760	40′ high	3,175	2,525	5,700
2920	Wood pole, 20′ high	1,150	1,475	2,625
2940	30′ high	1,350	1,850	3,200
2960	40′ high	1,725	2,275	4,000
3120	1000 watt, aluminum pole, 20′ high	2,225	1,550	3,775
3140	30′ high	3,275	1,875	5,150
3160	40′ high	3,825	2,425	6,250
3320	Fiberglass pole, 20′ high	1,950	1,350	3,300
3340	30′ high	2,175	1,675	3,850
3360	40′ high	3,100	2,150	5,250
3420	Steel pole, 20′ high	2,250	1,600	3,850
3440	30′ high	2,700	1,975	4,675

G4020 Site Lighting

G4020 110	Site Lighting	COST EACH		
		MAT.	INST.	TOTAL
3460	40' high	3,300	2,575	5,875
3520	Wood pole, 20' high	1,275	1,500	2,775
3540	30' high	1,475	1,875	3,350
3560	40' high	1,825	2,300	4,125
5820	Metal halide, 400 watt, aluminum pole, 20' high	1,975	1,500	3,475
5840	30' high	3,025	1,850	4,875
5860	40' high	3,575	2,375	5,950
6120	Fiberglass pole, 20' high	1,700	1,325	3,025
6140	30' high	1,925	1,650	3,575
6160	40' high	2,850	2,100	4,950
6320	Steel pole, 20' high	2,000	1,575	3,575
6340	30' high	2,450	1,950	4,400
6360	40' high	3,050	2,525	5,575
7620	1000 watt, aluminum pole, 20' high	2,050	1,550	3,600
7640	30' high	3,100	1,875	4,975
7660	40' high	3,675	2,425	6,100
7900	Fiberglass pole, 20' high	1,775	1,350	3,125
7920	30' high	2,000	1,675	3,675
7940	40' high	2,950	2,150	5,100
8120	Steel pole, 20' high	2,075	1,600	3,675
8140	30' high	2,525	1,975	4,500
8160	40' high	3,150	2,575	5,725

G4020 Site Lighting

Table G4020 210 Procedure for Calculating Floodlights Required for Various Footcandles
Poles should not be spaced more than 4 times the fixture mounting height for good light distribution.

Estimating Chart
Select Lamp type.

Determine total square feet.

Chart will show quantity of fixtures to provide 1 footcandle initial, at intersection of lines. Multiply fixture quantity by desired footcandle level.

Chart based on use of wide beam luminaires in an area whose dimensions are large compared to mounting height and is approximate only.

To maintain 1 footcandle over a large area use these watts per square foot:

Incandescent	0.15
Metal Halide	0.032
Mercury Vapor	0.05
High Pressure Sodium	0.024

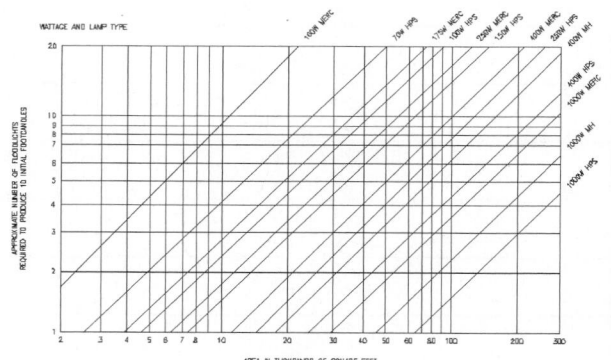

System Components				COST EACH		
		QUANTITY	UNIT	MAT.	INST.	TOTAL
SYSTEM G4020 210 0200						
LIGHT POLES, ALUMINUM, 20' HIGH, 1 ARM BRACKET						
Aluminum light pole, 20', no concrete base		1.000	Ea.	1,150	649	1,799
Bracket arm for aluminum light pole		1.000	Ea.	148	87	235
Excavation by hand, pits to 6' deep, heavy soil or clay		2.368	C.Y.		286.53	286.53
Footing, concrete incl forms, reinforcing, spread, under 1 C.Y.		.465	C.Y.	94.86	136.25	231.11
Backfill by hand		1.903	C.Y.		83.73	83.73
Compaction vibrating plate		1.903	C.Y.		11.23	11.23
	TOTAL			1,392.86	1,253.74	2,646.60

G4020 210	Light Pole (Installed)	COST EACH		
		MAT.	INST.	TOTAL
0200	Light pole, aluminum, 20' high, 1 arm bracket	1,400	1,250	2,650
0240	2 arm brackets	1,550	1,250	2,800
0280	3 arm brackets	1,700	1,300	3,000
0320	4 arm brackets	1,850	1,300	3,150
0360	30' high, 1 arm bracket	2,450	1,575	4,025
0400	2 arm brackets	2,600	1,575	4,175
0440	3 arm brackets	2,750	1,625	4,375
0480	4 arm brackets	2,900	1,625	4,525
0680	40' high, 1 arm bracket	3,000	2,100	5,100
0720	2 arm brackets	3,150	2,100	5,250
0760	3 arm brackets	3,275	2,150	5,425
0800	4 arm brackets	3,425	2,175	5,600
0840	Steel, 20' high, 1 arm bracket	1,500	1,325	2,825
0880	2 arm brackets	1,600	1,325	2,925
0920	3 arm brackets	1,525	1,375	2,900
0960	4 arm brackets	1,625	1,375	3,000
1000	30' high, 1 arm bracket	1,950	1,700	3,650
1040	2 arm brackets	2,050	1,700	3,750
1080	3 arm brackets	1,975	1,725	3,700
1120	4 arm brackets	2,075	1,725	3,800
1320	40' high, 1 arm bracket	2,550	2,275	4,825
1360	2 arm brackets	2,650	2,275	4,925
1400	3 arm brackets	2,575	2,300	4,875
1440	4 arm brackets	2,675	2,300	4,975

Reference Section

All the reference information is in one section, making it easy to find what you need to know and easy to use the data set on a daily basis. This section is visually identified by a vertical black bar on the page edges.

In this Reference Section, we've included Equipment Rental Costs, a listing of rental and operating costs; Crew Listings, a full listing of all crews, equipment, and their costs; Historical Cost Indexes for cost comparisons over time; City Cost Indexes and Location Factors for adjusting costs to the region you are in; Reference Tables, where you will find explanations, estimating information and procedures, and technical data; and an explanation of all the Abbreviations in the data set.

Table of Contents

461

Estimating Tips

- This section contains the average costs to rent and operate hundreds of pieces of construction equipment. This is useful information when estimating the time and material requirements of any particular operation in order to establish a unit or total cost. Bare equipment costs shown on a unit cost line include not only rental, but also operating costs for equipment under normal use.

Rental Costs

- Equipment rental rates are obtained from the following industry sources throughout North America: contractors, suppliers, dealers, manufacturers, and distributors.
- Rental rates vary throughout the country, with larger cities generally having lower rates. Lease plans for new equipment are available for periods in excess of six months, with a percentage of payments applying toward purchase.
- Monthly rental rates vary from 2% to 5% of the purchase price of the equipment depending on the anticipated life of the equipment and its wearing parts.
- Weekly rental rates are about 1/3 of the monthly rates, and daily rental rates are about 1/3 of the weekly rate.
- Rental rates can also be treated as reimbursement costs for contractor-owned equipment. Owned equipment costs include depreciation, loan payments, interest, taxes, insurance, storage, and major repairs.

Operating Costs

- The operating costs include parts and labor for routine servicing, such as the repair and replacement of pumps, filters, and worn lines. Normal operating expendables, such as fuel, lubricants, tires, and electricity (where applicable), are also included.
- Extraordinary operating expendables with highly variable wear patterns, such as diamond bits and blades, are excluded. These costs can be found as material costs in the Unit Price section.
- The hourly operating costs listed do not include the operator's wages.

Equipment Cost/Day

- Any power equipment required by a crew is shown in the Crew Listings with a daily cost.
- This daily cost of equipment needed by a crew includes both the rental cost and the operating cost and is based on dividing the weekly rental rate by 5 (number of working days in the week), and then adding the hourly operating cost times 8 (the number of hours in a day). This "Equipment Cost/Day" is shown in the far right column of the Equipment Rental section.
- If equipment is needed for only one or two days, it is best to develop your own cost by including components for daily rent and hourly operating costs. This is important when the listed Crew for a task does not contain the equipment needed, such as a crane for lifting mechanical heating/cooling equipment up onto a roof.
- If the quantity of work is less than the crew's Daily Output shown for a Unit Price line item that includes a bare unit equipment cost, the recommendation is to estimate one day's rental cost and operating cost for equipment shown in the Crew Listing for that line item.

Mobilization, Demobilization Costs

- The cost to move construction equipment from an equipment yard or rental company to the job site and back again is not included in equipment rental costs listed in the Reference Section, nor in the bare equipment cost of any unit price line item, nor in any equipment costs shown in the Crew Listings.
- Mobilization (to the site) and demobilization (from the site) costs can be found in the Unit Price section.
- If a piece of equipment is already at the job site, it is not appropriate to utilize mobilization, demobilization costs again in an estimate. ■

Did you know?

RSMeans data is available through our online application with 24/7 access:

- Search for unit prices by keyword
- Leverage the most up-to-date data
- Build and export estimates

Try it free for 30 days!
www.rsmeans.com/2018freetrial

---

01 54 33 | Equipment Rental

		UNIT	HOURLY OPER. COST	RENT PER DAY	RENT PER WEEK	RENT PER MONTH	EQUIPMENT COST/DAY	
0010	CONCRETE EQUIPMENT RENTAL without operators	R015433 -10		.85	24.50	73	219	21.40	10
0200	Bucket, concrete lightweight, 1/2 C.Y.		Ea.						
0300	1 C.Y.			.95	28.50	85	255	24.60	
0400	1-1/2 C.Y.			1.20	38.50	115	345	32.60	
0500	2 C.Y.			1.30	46.50	140	420	38.40	
0580	8 C.Y.			6.35	263	790	2,375	208.80	
0600	Cart, concrete, self-propelled, operator walking, 10 C.F.			2.85	58.50	175	525	57.80	
0700	Operator riding, 18 C.F.			4.80	98.50	295	885	97.40	
0800	Conveyer for concrete, portable, gas, 16" wide, 26' long			10.60	130	390	1,175	162.80	
0900	46' long			11.00	155	465	1,400	181	
1000	56' long			11.15	163	490	1,475	187.20	
1100	Core drill, electric, 2-1/2 H.P., 1" to 8" bit diameter			1.56	58.50	175	525	47.50	
1150	11 H.P., 8" to 18" cores			5.40	115	345	1,025	112.20	
1200	Finisher, concrete floor, gas, riding trowel, 96" wide			9.65	148	445	1,325	166.20	
1300	Gas, walk-behind, 3 blade, 36" trowel			2.15	23	69	207	31	
1400	4 blade, 48" trowel			3.05	28	84	252	41.20	
1500	Float, hand-operated (Bull float), 48" wide			.08	13.65	41	123	8.85	
1570	Curb builder, 14 H.P., gas, single screw			13.90	275	825	2,475	276.20	
1590	Double screw			14.80	330	995	2,975	317.40	
1600	Floor grinder, concrete and terrazzo, electric, 22" path			2.95	183	550	1,650	133.60	
1700	Edger, concrete, electric, 7" path			1.12	56.50	170	510	42.95	
1750	Vacuum pick-up system for floor grinders, wet/dry			1.62	90	270	810	66.95	
1800	Mixer, powered, mortar and concrete, gas, 6 C.F., 18 H.P.			7.40	123	370	1,100	133.20	
1900	10 C.F., 25 H.P.			9.00	148	445	1,325	161	
2000	16 C.F.			9.35	172	515	1,550	177.80	
2100	Concrete, stationary, tilt drum, 2 C.Y.			7.40	242	725	2,175	204.20	
2120	Pump, concrete, truck mounted, 4" line, 80' boom			29.20	1,075	3,240	9,725	881.60	
2140	5" line, 110' boom			36.60	1,375	4,115	12,300	1,116	
2160	Mud jack, 50 C.F. per hr.			6.45	128	385	1,150	128.60	
2180	225 C.F. per hr.			8.80	147	440	1,325	158.40	
2190	Shotcrete pump rig, 12 C.Y./hr.			13.95	225	675	2,025	246.60	
2200	35 C.Y./hr.			15.65	242	725	2,175	270.20	
2600	Saw, concrete, manual, gas, 18 H.P.			5.40	46.50	140	420	71.20	
2650	Self-propelled, gas, 30 H.P.			7.55	70	210	630	102.40	
2675	V-groove crack chaser, manual, gas, 6 H.P.			1.80	18.35	55	165	25.40	
2700	Vibrators, concrete, electric, 60 cycle, 2 H.P.			.47	9	27	81	9.15	
2800	3 H.P.			.60	11.65	35	105	11.80	
2900	Gas engine, 5 H.P.			1.50	16.35	49	147	21.80	
3000	8 H.P.			2.00	16	48	144	25.60	
3050	Vibrating screed, gas engine, 8 H.P.			2.82	88	264	790	75.35	
3120	Concrete transit mixer, 6 x 4, 250 H.P., 8 C.Y., rear discharge			48.75	600	1,805	5,425	751	
3200	Front discharge			56.60	735	2,205	6,625	893.80	
3300	6 x 6, 285 H.P., 12 C.Y., rear discharge			55.65	700	2,095	6,275	864.20	
3400	Front discharge			57.95	735	2,210	6,625	905.60	
0010	EARTHWORK EQUIPMENT RENTAL without operators	R015433 -10		2.60	26.50	80	240	36.80	20
0040	Aggregate spreader, push type, 8' to 12' wide		Ea.						
0045	Tailgate type, 8' wide			2.55	33.50	100	300	40.40	
0055	Earth auger, truck mounted, for fence & sign posts, utility poles			12.60	455	1,365	4,100	373.80	
0060	For borings and monitoring wells			42.60	695	2,090	6,275	758.80	
0070	Portable, trailer mounted			2.30	33	99	297	38.20	
0075	Truck mounted, for caissons, water wells			85.75	2,925	8,790	26,400	2,444	
0080	Horizontal boring machine, 12" to 36" diameter, 45 H.P.			22.70	197	590	1,775	299.60	
0090	12" to 48" diameter, 65 H.P.			30.60	340	1,020	3,050	448.80	
0095	Auger, for fence posts, gas engine, hand held			.45	6.35	19	57	7.40	
0100	Excavator, diesel hydraulic, crawler mounted, 1/2 C.Y. cap.			21.30	440	1,325	3,975	435.40	
0120	5/8 C.Y. capacity			28.60	580	1,740	5,225	576.80	
0140	3/4 C.Y. capacity			31.95	690	2,070	6,200	669.60	
0150	1 C.Y. capacity			39.90	705	2,115	6,350	742.20	

01 54 33 | Equipment Rental

		UNIT	HOURLY OPER. COST	RENT PER DAY	RENT PER WEEK	RENT PER MONTH	EQUIPMENT COST/DAY
0200	1-1/2 C.Y. capacity	Ea.	47.45	855	2,570	7,700	893.60
0300	2 C.Y. capacity		54.15	1,025	3,095	9,275	1,052
0320	2-1/2 C.Y. capacity		78.90	1,300	3,900	11,700	1,411
0325	3-1/2 C.Y. capacity		113.65	2,150	6,430	19,300	2,195
0330	4-1/2 C.Y. capacity		143.70	2,650	7,925	23,800	2,735
0335	6 C.Y. capacity		181.40	3,400	10,180	30,500	3,487
0340	7 C.Y. capacity		175.20	3,125	9,365	28,100	3,275
0342	Excavator attachments, bucket thumbs		3.35	252	755	2,275	177.80
0345	Grapples		3.05	220	660	1,975	156.40
0346	Hydraulic hammer for boom mounting, 4000 ft lb.		13.20	365	1,095	3,275	324.60
0347	5000 ft lb.		15.50	450	1,350	4,050	394
0348	8000 ft lb.		22.85	655	1,970	5,900	576.80
0349	12,000 ft lb.		24.90	785	2,350	7,050	669.20
0350	Gradall type, truck mounted, 3 ton @ 15' radius, 5/8 C.Y.		44.95	850	2,550	7,650	869.60
0370	1 C.Y. capacity		60.20	1,275	3,805	11,400	1,243
0400	Backhoe-loader, 40 to 45 H.P., 5/8 C.Y. capacity		12.25	248	745	2,225	247
0450	45 H.P. to 60 H.P., 3/4 C.Y. capacity		17.80	283	850	2,550	312.40
0460	80 H.P., 1-1/4 C.Y. capacity		19.50	380	1,145	3,425	385
0470	112 H.P., 1-1/2 C.Y. capacity		31.90	605	1,820	5,450	619.20
0482	Backhoe-loader attachment, compactor, 20,000 lb.		6.25	148	445	1,325	139
0485	Hydraulic hammer, 750 ft lb.		3.55	102	305	915	89.40
0486	Hydraulic hammer, 1200 ft lb.		6.55	208	625	1,875	177.40
0500	Brush chipper, gas engine, 6" cutter head, 35 H.P.		9.15	110	330	990	139.20
0550	Diesel engine, 12" cutter head, 130 H.P.		24.00	335	1,005	3,025	393
0600	15" cutter head, 165 H.P.		26.05	400	1,200	3,600	448.40
0750	Bucket, clamshell, general purpose, 3/8 C.Y.		1.40	40	120	360	35.20
0800	1/2 C.Y.		1.50	48.50	145	435	41
0850	3/4 C.Y.		1.65	56.50	170	510	47.20
0900	1 C.Y.		1.70	61.50	185	555	50.60
0950	1-1/2 C.Y.		2.75	85	255	765	73
1000	2 C.Y.		2.90	93.50	280	840	79.20
1010	Bucket, dragline, medium duty, 1/2 C.Y.		.80	24.50	73	219	21
1020	3/4 C.Y.		.80	25.50	77	231	21.80
1030	1 C.Y.		.85	27	81	243	23
1040	1-1/2 C.Y.		1.30	41.50	125	375	35.40
1050	2 C.Y.		1.35	45	135	405	37.80
1070	3 C.Y.		2.10	65	195	585	55.80
1200	Compactor, manually guided 2-drum vibratory smooth roller, 7.5 H.P.		7.20	203	610	1,825	179.60
1250	Rammer/tamper, gas, 8"		2.25	46.50	140	420	46
1260	15"		2.50	53.50	160	480	52
1300	Vibratory plate, gas, 18" plate, 3000 lb. blow		2.15	24.50	73	219	31.80
1350	21" plate, 5000 lb. blow		2.60	33	99	297	40.60
1370	Curb builder/extruder, 14 H.P., gas, single screw		13.90	275	825	2,475	276.20
1390	Double screw		14.80	330	995	2,975	317.40
1500	Disc harrow attachment, for tractor		.47	78.50	235	705	50.75
1810	Feller buncher, shearing & accumulating trees, 100 H.P.		40.55	810	2,430	7,300	810.40
1860	Grader, self-propelled, 25,000 lb.		32.75	745	2,235	6,700	709
1910	30,000 lb.		32.35	640	1,925	5,775	643.80
1920	40,000 lb.		54.75	1,250	3,765	11,300	1,191
1930	55,000 lb.		67.80	1,650	4,975	14,900	1,537
1950	Hammer, pavement breaker, self-propelled, diesel, 1000 to 1250 lb.		28.40	460	1,380	4,150	503.20
2000	1300 to 1500 lb.		42.60	920	2,760	8,275	892.80
2050	Pile driving hammer, steam or air, 4150 ft lb. @ 225 bpm		11.80	565	1,695	5,075	433.40
2100	8750 ft lb. @ 145 bpm		14.30	805	2,410	7,225	596.40
2150	15,000 ft lb. @ 60 bpm		14.65	845	2,530	7,600	623.20
2200	24,450 ft lb. @ 111 bpm		15.65	935	2,800	8,400	685.20
2250	Leads, 60' high for pile driving hammers up to 20,000 ft lb.		3.60	84.50	253	760	79.40
2300	90' high for hammers over 20,000 ft lb.		5.35	148	444	1,325	131.60

01 54 33 | Equipment Rental

		UNIT	HOURLY OPER. COST	RENT PER DAY	RENT PER WEEK	RENT PER MONTH	EQUIPMENT COST/DAY	
2350	Diesel type hammer, 22,400 ft lb.	Ea.	18.25	470	1,405	4,225	427	20
2400	41,300 ft lb.		26.50	600	1,800	5,400	572	
2450	141,000 ft lb.		41.80	945	2,840	8,525	902.40	
2500	Vib. elec. hammer/extractor, 200 kW diesel generator, 34 H.P.		40.35	695	2,080	6,250	738.80	
2550	80 H.P.		70.10	1,000	3,000	9,000	1,161	
2600	150 H.P.		129.40	1,925	5,780	17,300	2,191	
2800	Log chipper, up to 22" diameter, 600 H.P.		46.10	660	1,980	5,950	764.80	
2850	Logger, for skidding & stacking logs, 150 H.P.		44.05	830	2,485	7,450	849.40	
2860	Mulcher, diesel powered, trailer mounted		17.10	220	660	1,975	268.80	
2900	Rake, spring tooth, with tractor		14.54	360	1,075	3,225	331.30	
3000	Roller, vibratory, tandem, smooth drum, 20 H.P.		7.65	150	450	1,350	151.20	
3050	35 H.P.		10.05	252	755	2,275	231.40	
3100	Towed type vibratory compactor, smooth drum, 50 H.P.		25.25	370	1,105	3,325	423	
3150	Sheepsfoot, 50 H.P.		25.35	370	1,115	3,350	425.80	
3170	Landfill compactor, 220 H.P.		71.35	1,575	4,710	14,100	1,513	
3200	Pneumatic tire roller, 80 H.P.		12.90	390	1,175	3,525	338.20	
3250	120 H.P.		19.45	645	1,930	5,800	541.60	
3300	Sheepsfoot vibratory roller, 240 H.P.		62.05	1,375	4,100	12,300	1,316	
3320	340 H.P.		83.10	2,025	6,075	18,200	1,880	
3350	Smooth drum vibratory roller, 75 H.P.		23.30	650	1,950	5,850	576.40	
3400	125 H.P.		27.60	730	2,195	6,575	659.80	
3410	Rotary mower, brush, 60", with tractor		19.00	340	1,025	3,075	357	
3420	Rototiller, walk-behind, gas, 5 H.P.		2.21	76.50	229	685	63.50	
3422	8 H.P.		2.80	87	261	785	74.60	
3440	Scrapers, towed type, 7 C.Y. capacity		6.30	122	365	1,100	123.40	
3450	10 C.Y. capacity		7.10	165	495	1,475	155.80	
3500	15 C.Y. capacity		7.60	192	575	1,725	175.80	
3525	Self-propelled, single engine, 14 C.Y. capacity		131.25	2,400	7,200	21,600	2,490	
3550	Dual engine, 21 C.Y. capacity		137.25	2,275	6,850	20,600	2,468	
3600	31 C.Y. capacity		184.55	3,500	10,500	31,500	3,576	
3640	44 C.Y. capacity		225.40	4,425	13,270	39,800	4,457	
3650	Elevating type, single engine, 11 C.Y. capacity		61.05	1,100	3,335	10,000	1,155	
3700	22 C.Y. capacity		114.65	2,275	6,835	20,500	2,284	
3710	Screening plant, 110 H.P. w/5' x 10' screen		20.55	385	1,160	3,475	396.40	
3720	5' x 16' screen		25.96	495	1,480	4,450	503.70	
3850	Shovel, crawler-mounted, front-loading, 7 C.Y. capacity		204.00	3,750	11,240	33,700	3,880	
3855	12 C.Y. capacity		332.00	5,200	15,567	46,700	5,769	
3860	Shovel/backhoe bucket, 1/2 C.Y.		2.70	71.50	215	645	64.60	
3870	3/4 C.Y.		2.75	78.50	235	705	69	
3880	1 C.Y.		2.85	88.50	265	795	75.80	
3890	1-1/2 C.Y.		3.00	102	305	915	85	
3910	3 C.Y.		3.35	137	410	1,225	108.80	
3950	Stump chipper, 18" deep, 30 H.P.		6.93	214	643	1,925	184.05	
4110	Dozer, crawler, torque converter, diesel 80 H.P.		25.05	440	1,320	3,950	464.40	
4150	105 H.P.		34.25	555	1,670	5,000	608	
4200	140 H.P.		42.20	845	2,540	7,625	845.60	
4260	200 H.P.		61.65	1,300	3,900	11,700	1,273	
4310	300 H.P.		82.45	1,950	5,845	17,500	1,829	
4360	410 H.P.		109.35	2,350	7,080	21,200	2,291	
4370	500 H.P.		137.25	2,875	8,600	25,800	2,818	
4380	700 H.P.		229.75	5,275	15,805	47,400	4,999	
4400	Loader, crawler, torque conv., diesel, 1-1/2 C.Y., 80 H.P.		29.60	570	1,715	5,150	579.80	
4450	1-1/2 to 1-3/4 C.Y., 95 H.P.		30.55	670	2,005	6,025	645.40	
4510	1-3/4 to 2-1/4 C.Y., 130 H.P.		47.50	1,000	2,995	8,975	979	
4530	2-1/2 to 3-1/4 C.Y., 190 H.P.		57.55	1,250	3,720	11,200	1,204	
4560	3-1/2 to 5 C.Y., 275 H.P.		71.55	1,475	4,435	13,300	1,459	
4610	Front end loader, 4WD, articulated frame, diesel, 1 to 1-1/4 C.Y., 70 H.P.		16.10	270	810	2,425	290.80	
4620	1-1/2 to 1-3/4 C.Y., 95 H.P.		19.35	315	940	2,825	342.80	

01 54 33 | Equipment Rental

		UNIT	HOURLY OPER. COST	RENT PER DAY	RENT PER WEEK	RENT PER MONTH	EQUIPMENT COST/DAY
20	4650	1-3/4 to 2 C.Y., 130 H.P.	Ea.	20.50	380	1,140	3,425	392
	4710	2-1/2 to 3-1/2 C.Y., 145 H.P.		29.15	485	1,450	4,350	523.20
	4730	3 to 4-1/2 C.Y., 185 H.P.		32.20	530	1,585	4,750	574.60
	4760	5-1/4 to 5-3/4 C.Y., 270 H.P.		53.15	930	2,785	8,350	982.20
	4810	7 to 9 C.Y., 475 H.P.		90.60	1,750	5,275	15,800	1,780
	4870	9 to 11 C.Y., 620 H.P.		131.70	2,625	7,880	23,600	2,630
	4880	Skid-steer loader, wheeled, 10 C.F., 30 H.P. gas		9.50	163	490	1,475	174
	4890	1 C.Y., 78 H.P., diesel		18.35	365	1,090	3,275	364.80
	4892	Skid-steer attachment, auger		.82	136	408	1,225	88.15
	4893	Backhoe		.74	123	369	1,100	79.70
	4894	Broom		.71	118	355	1,075	76.70
	4895	Forks		.16	27	81	243	17.50
	4896	Grapple		.69	115	346	1,050	74.70
	4897	Concrete hammer		1.05	175	526	1,575	113.60
	4898	Tree spade		.60	100	300	900	64.80
	4899	Trencher		.69	115	344	1,025	74.30
	4900	Trencher, chain, boom type, gas, operator walking, 12 H.P.		4.10	48.50	145	435	61.80
	4910	Operator riding, 40 H.P.		16.30	345	1,030	3,100	336.40
	5000	Wheel type, diesel, 4' deep, 12" wide		69.35	900	2,700	8,100	1,095
	5100	6' deep, 20" wide		87.50	2,100	6,285	18,900	1,957
	5150	Chain type, diesel, 5' deep, 8" wide		16.30	345	1,030	3,100	336.40
	5200	Diesel, 8' deep, 16" wide		95.25	2,150	6,430	19,300	2,048
	5202	Rock trencher, wheel type, 6" wide x 18" deep		43.65	925	2,775	8,325	904.20
	5206	Chain type, 18" wide x 7' deep		106.10	3,050	9,160	27,500	2,681
	5210	Tree spade, self-propelled		14.24	390	1,167	3,500	347.30
	5250	Truck, dump, 2-axle, 12 ton, 8 C.Y. payload, 220 H.P.		23.95	245	735	2,200	338.60
	5300	Three axle dump, 16 ton, 12 C.Y. payload, 400 H.P.		41.60	350	1,050	3,150	542.80
	5310	Four axle dump, 25 ton, 18 C.Y. payload, 450 H.P.		50.00	510	1,530	4,600	706
	5350	Dump trailer only, rear dump, 16-1/2 C.Y.		5.75	145	435	1,300	133
	5400	20 C.Y.		6.20	163	490	1,475	147.60
	5450	Flatbed, single axle, 1-1/2 ton rating		18.30	70	210	630	188.40
	5500	3 ton rating		22.25	100	300	900	238
	5550	Off highway rear dump, 25 ton capacity		61.80	1,425	4,245	12,700	1,343
	5600	35 ton capacity		66.10	1,525	4,595	13,800	1,448
	5610	50 ton capacity		81.20	1,675	5,060	15,200	1,662
	5620	65 ton capacity		84.95	1,950	5,820	17,500	1,844
	5630	100 ton capacity		119.60	2,850	8,560	25,700	2,669
	6000	Vibratory plow, 25 H.P., walking		6.80	61.50	185	555	91.40
40	0010	**GENERAL EQUIPMENT RENTAL** without operators						
	0020	Aerial lift, scissor type, to 20' high, 1200 lb. capacity, electric	R015433 -10	Ea.	3.40	51.50	155	465	58.20
	0030	To 30' high, 1200 lb. capacity		3.85	68.50	205	615	71.80
	0040	Over 30' high, 1500 lb. capacity		5.15	122	365	1,100	114.20
	0070	Articulating boom, to 45' high, 500 lb. capacity, diesel	R015433 -15		9.70	273	820	2,450	241.60
	0075	To 60' high, 500 lb. capacity		13.70	470	1,410	4,225	391.60
	0080	To 80' high, 500 lb. capacity		16.10	560	1,685	5,050	465.80
	0085	To 125' high, 500 lb. capacity		18.40	780	2,335	7,000	614.20
	0100	Telescoping boom to 40' high, 500 lb. capacity, diesel		11.60	315	950	2,850	282.80
	0105	To 45' high, 500 lb. capacity		12.40	320	958	2,875	290.80
	0110	To 60' high, 500 lb. capacity		16.20	540	1,625	4,875	454.60
	0115	To 80' high, 500 lb. capacity		21.70	625	1,875	5,625	548.60
	0120	To 100' high, 500 lb. capacity		28.80	805	2,420	7,250	714.40
	0125	To 120' high, 500 lb. capacity		29.25	845	2,530	7,600	740
	0195	Air compressor, portable, 6.5 CFM, electric		.91	13	39	117	15.10
	0196	Gasoline		.66	19.65	59	177	17.10
	0200	Towed type, gas engine, 60 CFM		9.30	51.50	155	465	105.40
	0300	160 CFM		10.65	53.50	160	480	117.20
	0400	Diesel engine, rotary screw, 250 CFM		12.05	118	355	1,075	167.40
	0500	365 CFM		15.80	142	425	1,275	211.40

For customer support on your Electrical Costs with RSMeans data, call 800.448.8182.

467

01 54 33 | Equipment Rental

	UNIT	HOURLY OPER. COST	RENT PER DAY	RENT PER WEEK	RENT PER MONTH	EQUIPMENT COST/DAY	
0550	450 CFM	Ea.	19.70	177	530	1,600	263.60	40
0600	600 CFM		33.90	243	730	2,200	417.20	
0700	750 CFM		34.10	252	755	2,275	423.80	
0930	Air tools, breaker, pavement, 60 lb.		.55	10.35	31	93	10.60	
0940	80 lb.		.55	10.65	32	96	10.80	
0950	Drills, hand (jackhammer), 65 lb.		.65	17.65	53	159	15.80	
0960	Track or wagon, swing boom, 4" drifter		54.80	925	2,775	8,325	993.40	
0970	5" drifter		63.45	1,100	3,325	9,975	1,173	
0975	Track mounted quarry drill, 6" diameter drill		104.15	1,650	4,945	14,800	1,822	
0980	Dust control per drill		1.04	24.50	74	222	23.10	
0990	Hammer, chipping, 12 lb.		.60	27	81	243	21	
1000	Hose, air with couplings, 50' long, 3/4" diameter		.07	11.35	34	102	7.35	
1100	1" diameter		.08	13	39	117	8.45	
1200	1-1/2" diameter		.21	35	105	315	22.70	
1300	2" diameter		.24	40	120	360	25.90	
1400	2-1/2" diameter		.35	58.50	175	525	37.80	
1410	3" diameter		.40	66.50	200	600	43.20	
1450	Drill, steel, 7/8" x 2'		.08	13.65	41	123	8.85	
1460	7/8" x 6'		.11	17.65	53	159	11.50	
1520	Moil points		.03	4.67	14	42	3.05	
1525	Pneumatic nailer w/accessories		.48	32	96	288	23.05	
1530	Sheeting driver for 60 lb. breaker		.04	7.35	22	66	4.70	
1540	For 90 lb. breaker		.15	9.65	29	87	7	
1550	Spade, 25 lb.		.50	7.35	22	66	8.40	
1560	Tamper, single, 35 lb.		.59	39.50	118	355	28.30	
1570	Triple, 140 lb.		.89	59	177	530	42.50	
1580	Wrenches, impact, air powered, up to 3/4" bolt		.45	13	39	117	11.40	
1590	Up to 1-1/4" bolt		.55	23.50	71	213	18.60	
1600	Barricades, barrels, reflectorized, 1 to 99 barrels		.03	5.35	16	48	3.45	
1610	100 to 200 barrels		.02	4.13	12.40	37	2.65	
1620	Barrels with flashers, 1 to 99 barrels		.04	6	18	54	3.90	
1630	100 to 200 barrels		.03	4.80	14.40	43	3.10	
1640	Barrels with steady burn type C lights		.05	8	24	72	5.20	
1650	Illuminated board, trailer mounted, with generator		3.30	133	400	1,200	106.40	
1670	Portable barricade, stock, with flashers, 1 to 6 units		.04	6	18	54	3.90	
1680	25 to 50 units		.03	5.60	16.80	50.50	3.60	
1685	Butt fusion machine, wheeled, 1.5 H.P. electric, 2" - 8" diameter pipe		2.63	167	500	1,500	121.05	
1690	Tracked, 20 H.P. diesel, 4" - 12" diameter pipe		11.21	560	1,680	5,050	425.70	
1695	83 H.P. diesel, 8" - 24" diameter pipe		49.46	2,525	7,560	22,700	1,908	
1700	Carts, brick, gas engine, 1000 lb. capacity		2.95	61.50	185	555	60.60	
1800	1500 lb., 7-1/2' lift		3.00	65	195	585	63	
1822	Dehumidifier, medium, 6 lb./hr., 150 CFM		1.16	72.50	218	655	52.90	
1824	Large, 18 lb./hr., 600 CFM		2.20	138	413	1,250	100.20	
1830	Distributor, asphalt, trailer mounted, 2000 gal., 38 H.P. diesel		10.75	350	1,050	3,150	296	
1840	3000 gal., 38 H.P. diesel		12.35	380	1,140	3,425	326.80	
1850	Drill, rotary hammer, electric		1.12	27	81	243	25.15	
1860	Carbide bit, 1-1/2" diameter, add to electric rotary hammer		.03	5	15	45	3.25	
1865	Rotary, crawler, 250 H.P.		136.15	2,225	6,690	20,100	2,427	
1870	Emulsion sprayer, 65 gal., 5 H.P. gas engine		2.77	103	309	925	83.95	
1880	200 gal., 5 H.P. engine		7.30	172	515	1,550	161.40	
1900	Floor auto-scrubbing machine, walk-behind, 28" path		5.41	350	1,055	3,175	254.30	
1930	Floodlight, mercury vapor, or quartz, on tripod, 1000 watt		.46	22	66	198	16.90	
1940	2000 watt		.63	27.50	82	246	21.45	
1950	Floodlights, trailer mounted with generator, 1 - 300 watt light		3.60	76.50	230	690	74.80	
1960	2 - 1000 watt lights		4.50	102	305	915	97	
2000	4 - 300 watt lights		4.25	96.50	290	870	92	
2005	Foam spray rig, incl. box trailer, compressor, generator, proportioner		23.78	515	1,545	4,625	499.25	
2015	Forklift, pneumatic tire, rough terr, straight mast, 5000 lb, 12' lift, gas		19.00	212	635	1,900	279	

01 54 33 | Equipment Rental

		UNIT	HOURLY OPER. COST	RENT PER DAY	RENT PER WEEK	RENT PER MONTH	EQUIPMENT COST/DAY	
2025	8000 lb., 12' lift	Ea.	22.75	283	850	2,550	352	40
2030	5000 lb., 12' lift, diesel		15.70	237	710	2,125	267.60	
2035	8000 lb., 12' lift, diesel		16.75	268	805	2,425	295	
2045	All terrain, telescoping boom, diesel, 5000 lb., 10' reach, 19' lift		17.25	325	980	2,950	334	
2055	6600 lb., 29' reach, 42' lift		21.10	380	1,140	3,425	396.80	
2065	10,000 lb., 31' reach, 45' lift		23.65	490	1,475	4,425	484.20	
2070	Cushion tire, smooth floor, gas, 5000 lb. capacity		8.25	76.50	230	690	112	
2075	8000 lb. capacity		11.40	95	285	855	148.20	
2085	Diesel, 5000 lb. capacity		7.75	83.50	250	750	112	
2090	12,000 lb. capacity		12.05	130	390	1,175	174.40	
2095	20,000 lb. capacity		17.00	165	495	1,475	235	
2100	Generator, electric, gas engine, 1.5 kW to 3 kW		2.70	11.35	34	102	28.40	
2200	5 kW		3.35	14.35	43	129	35.40	
2300	10 kW		6.25	35	105	315	71	
2400	25 kW		7.60	86.50	260	780	112.80	
2500	Diesel engine, 20 kW		9.20	76.50	230	690	119.60	
2600	50 kW		15.85	100	300	900	186.80	
2700	100 kW		28.20	137	410	1,225	307.60	
2800	250 kW		56.05	260	780	2,350	604.40	
2850	Hammer, hydraulic, for mounting on boom, to 500 ft lb.		2.85	86.50	260	780	74.80	
2860	1000 ft lb.		4.70	133	400	1,200	117.60	
2900	Heaters, space, oil or electric, 50 MBH		1.47	8	24	72	16.55	
3000	100 MBH		2.73	11.35	34	102	28.65	
3100	300 MBH		7.84	40	120	360	86.70	
3150	500 MBH		12.75	45	135	405	129	
3200	Hose, water, suction with coupling, 20' long, 2" diameter		.02	3	9	27	1.95	
3210	3" diameter		.03	4.33	13	39	2.85	
3220	4" diameter		.03	5	15	45	3.25	
3230	6" diameter		.11	17.65	53	159	11.50	
3240	8" diameter		.28	46.50	140	420	30.25	
3250	Discharge hose with coupling, 50' long, 2" diameter		.01	1.33	4	12	.90	
3260	3" diameter		.01	2.33	7	21	1.50	
3270	4" diameter		.02	3.67	11	33	2.35	
3280	6" diameter		.06	9.35	28	84	6.10	
3290	8" diameter		.24	40	120	360	25.90	
3295	Insulation blower		.83	6	18	54	10.25	
3300	Ladders, extension type, 16' to 36' long		.18	30	90	270	19.45	
3400	40' to 60' long		.67	112	335	1,000	72.35	
3405	Lance for cutting concrete		2.23	58.50	176	530	53.05	
3407	Lawn mower, rotary, 22", 5 H.P.		1.15	25	75	225	24.20	
3408	48" self-propelled		2.86	90	270	810	76.90	
3410	Level, electronic, automatic, with tripod and leveling rod		1.05	70	210	630	50.40	
3430	Laser type, for pipe and sewer line and grade		2.13	142	425	1,275	102.05	
3440	Rotating beam for interior control		.90	60	180	540	43.20	
3460	Builder's optical transit, with tripod and rod		.10	16.35	49	147	10.60	
3500	Light towers, towable, with diesel generator, 2000 watt		4.25	96.50	290	870	92	
3600	4000 watt		4.50	102	305	915	97	
3700	Mixer, powered, plaster and mortar, 6 C.F., 7 H.P.		2.05	20.50	62	186	28.80	
3800	10 C.F., 9 H.P.		2.20	33.50	100	300	37.60	
3850	Nailer, pneumatic		.48	32	96	288	23.05	
3900	Paint sprayers complete, 8 CFM		.94	62.50	188	565	45.10	
4000	17 CFM		1.69	112	337	1,000	80.90	
4020	Pavers, bituminous, rubber tires, 8' wide, 50 H.P., diesel		31.55	550	1,645	4,925	581.40	
4030	10' wide, 150 H.P.		96.50	1,875	5,655	17,000	1,903	
4050	Crawler, 8' wide, 100 H.P., diesel		87.60	2,025	6,105	18,300	1,922	
4060	10' wide, 150 H.P.		104.50	2,350	7,015	21,000	2,239	
4070	Concrete paver, 12' to 24' wide, 250 H.P.		87.45	1,625	4,875	14,600	1,675	
4080	Placer-spreader-trimmer, 24' wide, 300 H.P.		117.20	2,375	7,115	21,300	2,361	

For customer support on your Electrical Costs with RSMeans data, call 800.448.8182.

469

01 54 | Construction Aids

01 54 33 | Equipment Rental

		UNIT	HOURLY OPER. COST	RENT PER DAY	RENT PER WEEK	RENT PER MONTH	EQUIPMENT COST/DAY
40	4100	Pump, centrifugal gas pump, 1-1/2" diameter, 65 GPM	Ea.	3.90	53.50	160	480	63.20
	4200	2" diameter, 130 GPM		5.00	63.50	190	570	78
	4300	3" diameter, 250 GPM		5.15	63.50	190	570	79.20
	4400	6" diameter, 1500 GPM		22.30	197	590	1,775	296.40
	4500	Submersible electric pump, 1-1/4" diameter, 55 GPM		.41	17.65	53	159	13.90
	4600	1-1/2" diameter, 83 GPM		.45	20.50	61	183	15.80
	4700	2" diameter, 120 GPM		1.65	25.50	76	228	28.40
	4800	3" diameter, 300 GPM		2.94	45	135	405	50.50
	4900	4" diameter, 560 GPM		14.70	167	500	1,500	217.60
	5000	6" diameter, 1590 GPM		21.94	218	655	1,975	306.50
	5100	Diaphragm pump, gas, single, 1-1/2" diameter		1.12	54.50	164	490	41.75
	5200	2" diameter		4.00	68.50	205	615	73
	5300	3" diameter		4.05	68.50	205	615	73.40
	5400	Double, 4" diameter		5.85	113	340	1,025	114.80
	5450	Pressure washer 5 GPM, 3000 psi		3.95	53.50	160	480	63.60
	5460	7 GPM, 3000 psi		4.90	63.50	190	570	77.20
	5500	Trash pump, self-priming, gas, 2" diameter		3.80	23.50	70	210	44.40
	5600	Diesel, 4" diameter		6.95	95	285	855	112.60
	5650	Diesel, 6" diameter		16.90	167	500	1,500	235.20
	5655	Grout pump		19.50	275	825	2,475	321
	5700	Salamanders, L.P. gas fired, 100,000 BTU		2.93	14	42	126	31.85
	5705	50,000 BTU		1.67	11.35	34	102	20.15
	5720	Sandblaster, portable, open top, 3 C.F. capacity		.60	27	81	243	21
	5730	6 C.F. capacity		1.00	40	120	360	32
	5740	Accessories for above		.14	22.50	68	204	14.70
	5750	Sander, floor		.77	17.65	53	159	16.75
	5760	Edger		.52	15	45	135	13.15
	5800	Saw, chain, gas engine, 18" long		1.80	22.50	67	201	27.80
	5900	Hydraulic powered, 36" long		.80	66.50	200	600	46.40
	5950	60" long		.80	68.50	205	615	47.40
	6000	Masonry, table mounted, 14" diameter, 5 H.P.		1.32	56.50	170	510	44.55
	6050	Portable cut-off, 8 H.P.		1.85	33.50	100	300	34.80
	6100	Circular, hand held, electric, 7-1/4" diameter		.23	5	15	45	4.85
	6200	12" diameter		.23	8	24	72	6.65
	6250	Wall saw, w/hydraulic power, 10 H.P.		3.30	33.50	100	300	46.40
	6275	Shot blaster, walk-behind, 20" wide		4.85	293	880	2,650	214.80
	6280	Sidewalk broom, walk-behind		2.39	85	255	765	70.10
	6300	Steam cleaner, 100 gallons per hour		3.35	80	240	720	74.80
	6310	200 gallons per hour		4.40	96.50	290	870	93.20
	6340	Tar kettle/pot, 400 gallons		15.15	76.50	230	690	167.20
	6350	Torch, cutting, acetylene-oxygen, 150' hose, excludes gases		.45	15	45	135	12.60
	6360	Hourly operating cost includes tips and gas		21.00				168
	6410	Toilet, portable chemical		.13	22	66	198	14.25
	6420	Recycle flush type		.16	27	81	243	17.50
	6430	Toilet, fresh water flush, garden hose,		.19	32.50	97	291	20.90
	6440	Hoisted, non-flush, for high rise		.16	26.50	79	237	17.10
	6465	Tractor, farm with attachment		17.80	340	1,025	3,075	347.40
	6480	Trailers, platform, flush deck, 2 axle, 3 ton capacity		1.60	21	63	189	25.40
	6500	25 ton capacity		6.25	138	415	1,250	133
	6600	40 ton capacity		8.00	193	580	1,750	180
	6700	3 axle, 50 ton capacity		8.65	215	645	1,925	198.20
	6800	75 ton capacity		10.90	285	855	2,575	258.20
	6810	Trailer mounted cable reel for high voltage line work		5.79	276	827	2,475	211.70
	6820	Trailer mounted cable tensioning rig		11.48	545	1,640	4,925	419.85
	6830	Cable pulling rig		72.98	3,075	9,210	27,600	2,426
	6850	Portable cable/wire puller, 8000 lb. max pulling capacity		3.72	167	502	1,500	130.15
	6900	Water tank trailer, engine driven discharge, 5000 gallons		7.20	150	450	1,350	147.60
	6925	10,000 gallons		9.70	207	620	1,850	201.60

01 54 33 | Equipment Rental

		Description	UNIT	HOURLY OPER. COST	RENT PER DAY	RENT PER WEEK	RENT PER MONTH	EQUIPMENT COST/DAY
40	6950	Water truck, off highway, 6000 gallons	Ea.	70.16	805	2,420	7,250	1,045
	7010	Tram car for high voltage line work, powered, 2 conductor		6.85	150	449	1,350	144.60
	7020	Transit (builder's level) with tripod		.10	16.35	49	147	10.60
	7030	Trench box, 3000 lb., 6' x 8'		.56	93.50	280	840	60.50
	7040	7200 lb., 6' x 20'		.75	125	375	1,125	81
	7050	8000 lb., 8' x 16'		1.08	180	540	1,625	116.65
	7060	9500 lb., 8' x 20'		1.21	201	603	1,800	130.30
	7065	11,000 lb., 8' x 24'		1.27	211	633	1,900	136.75
	7070	12,000 lb., 10' x 20'		1.50	251	752	2,250	162.40
	7100	Truck, pickup, 3/4 ton, 2 wheel drive		9.90	60	180	540	115.20
	7200	4 wheel drive		10.20	75	225	675	126.60
	7250	Crew carrier, 9 passenger		14.00	90	270	810	166
	7290	Flat bed truck, 20,000 lb. GVW		14.90	130	390	1,175	197.20
	7300	Tractor, 4 x 2, 220 H.P.		21.00	203	610	1,825	290
	7410	330 H.P.		30.80	280	840	2,525	414.40
	7500	6 x 4, 380 H.P.		35.15	325	975	2,925	476.20
	7600	450 H.P.		43.30	395	1,185	3,550	583.40
	7610	Tractor, with A frame, boom and winch, 225 H.P.		24.10	282	845	2,525	361.80
	7620	Vacuum truck, hazardous material, 2500 gallons		12.85	305	910	2,725	284.80
	7625	5000 gallons		13.11	425	1,270	3,800	358.90
	7650	Vacuum, HEPA, 16 gallon, wet/dry		.90	18	54	162	18
	7655	55 gallon, wet/dry		.81	27	81	243	22.70
	7660	Water tank, portable		.74	123	370	1,100	79.90
	7690	Sewer/catch basin vacuum, 14 C.Y., 1500 gallons		17.59	635	1,910	5,725	522.70
	7700	Welder, electric, 200 amp		3.99	16.35	49	147	41.70
	7800	300 amp		5.90	20	60	180	59.20
	7900	Gas engine, 200 amp		9.10	24.50	74	222	87.60
	8000	300 amp		10.35	26	78	234	98.40
	8100	Wheelbarrow, any size		.06	10.65	32	96	6.90
	8200	Wrecking ball, 4000 lb.		2.45	71.50	215	645	62.60
50	0010	**HIGHWAY EQUIPMENT RENTAL** without operators						
	0050	Asphalt batch plant, portable drum mixer, 100 ton/hr.	Ea.	85.49	1,500	4,505	13,500	1,585
	0060	200 ton/hr.		97.81	1,600	4,800	14,400	1,742
	0070	300 ton/hr.		116.21	1,875	5,625	16,900	2,055
	0100	Backhoe attachment, long stick, up to 185 H.P., 10.5' long		.37	24.50	73	219	17.55
	0140	Up to 250 H.P., 12' long		.41	27	81	243	19.50
	0180	Over 250 H.P., 15' long		.56	37	111	335	26.70
	0200	Special dipper arm, up to 100 H.P., 32' long		1.14	75.50	227	680	54.50
	0240	Over 100 H.P., 33' long		1.42	94.50	284	850	68.15
	0280	Catch basin/sewer cleaning truck, 3 ton, 9 C.Y., 1000 gal.		35.10	405	1,210	3,625	522.80
	0300	Concrete batch plant, portable, electric, 200 C.Y./hr.		24.34	545	1,630	4,900	520.70
	0520	Grader/dozer attachment, ripper/scarifier, rear mounted, up to 135 H.P.		3.15	61.50	185	555	62.20
	0540	Up to 180 H.P.		4.10	91.50	275	825	87.80
	0580	Up to 250 H.P.		5.70	145	435	1,300	132.60
	0700	Pvmt. removal bucket, for hyd. excavator, up to 90 H.P.		2.10	56.50	170	510	50.80
	0740	Up to 200 H.P.		2.25	71.50	215	645	61
	0780	Over 200 H.P.		2.45	88.50	265	795	72.60
	0900	Aggregate spreader, self-propelled, 187 H.P.		50.00	730	2,185	6,550	837
	1000	Chemical spreader, 3 C.Y.		3.15	45	135	405	52.20
	1900	Hammermill, traveling, 250 H.P.		68.23	2,200	6,620	19,900	1,870
	2000	Horizontal borer, 3" diameter, 13 H.P. gas driven		5.50	56.50	170	510	78
	2150	Horizontal directional drill, 20,000 lb. thrust, 78 H.P. diesel		27.50	680	2,045	6,125	629
	2160	30,000 lb. thrust, 115 H.P.		33.65	1,050	3,135	9,400	896.20
	2170	50,000 lb. thrust, 170 H.P.		48.35	1,325	4,005	12,000	1,188
	2190	Mud trailer for HDD, 1500 gallons, 175 H.P., gas		24.10	158	475	1,425	287.80
	2200	Hydromulcher, diesel, 3000 gallon, for truck mounting		16.35	253	760	2,275	282.80
	2300	Gas, 600 gallon		7.40	103	310	930	121.20
	2400	Joint & crack cleaner, walk behind, 25 H.P.		3.10	51.50	155	465	55.80

R015433 -10

For customer support on your Electrical Costs with RSMeans data, call 800.448.8182.

471

01 54 33 | Equipment Rental

		UNIT	HOURLY OPER. COST	RENT PER DAY	RENT PER WEEK	RENT PER MONTH	EQUIPMENT COST/DAY	
2500	Filler, trailer mounted, 400 gallons, 20 H.P.	Ea.	8.40	218	655	1,975	198.20	50
3000	Paint striper, self-propelled, 40 gallon, 22 H.P.		6.75	162	485	1,450	151	
3100	120 gallon, 120 H.P.		18.90	405	1,220	3,650	395.20	
3200	Post drivers, 6" I-Beam frame, for truck mounting		12.45	390	1,175	3,525	334.60	
3400	Road sweeper, self-propelled, 8' wide, 90 H.P.		35.95	670	2,005	6,025	688.60	
3450	Road sweeper, vacuum assisted, 4 C.Y., 220 gallons		56.05	655	1,960	5,875	840.40	
4000	Road mixer, self-propelled, 130 H.P.		45.95	800	2,405	7,225	848.60	
4100	310 H.P.		75.55	2,150	6,425	19,300	1,889	
4220	Cold mix paver, incl. pug mill and bitumen tank, 165 H.P.		94.60	2,300	6,915	20,700	2,140	
4240	Pavement brush, towed		3.40	96.50	290	870	85.20	
4250	Paver, asphalt, wheel or crawler, 130 H.P., diesel		94.25	2,275	6,845	20,500	2,123	
4300	Paver, road widener, gas, 1' to 6', 67 H.P.		46.65	940	2,825	8,475	938.20	
4400	Diesel, 2' to 14', 88 H.P.		56.75	1,125	3,355	10,100	1,125	
4600	Slipform pavers, curb and gutter, 2 track, 75 H.P.		56.30	1,200	3,615	10,800	1,173	
4700	4 track, 165 H.P.		36.95	825	2,470	7,400	789.60	
4800	Median barrier, 215 H.P.		57.45	1,275	3,805	11,400	1,221	
4901	Trailer, low bed, 75 ton capacity		11.05	268	805	2,425	249.40	
5000	Road planer, walk behind, 10" cutting width, 10 H.P.		2.50	33.50	100	300	40	
5100	Self-propelled, 12" cutting width, 64 H.P.		8.00	115	345	1,025	133	
5120	Traffic line remover, metal ball blaster, truck mounted, 115 H.P.		46.70	800	2,395	7,175	852.60	
5140	Grinder, truck mounted, 115 H.P.		51.05	850	2,555	7,675	919.40	
5160	Walk-behind, 11 H.P.		3.55	55	165	495	61.40	
5200	Pavement profiler, 4' to 6' wide, 450 H.P.		218.90	3,450	10,350	31,100	3,821	
5300	8' to 10' wide, 750 H.P.		336.50	4,550	13,635	40,900	5,419	
5400	Roadway plate, steel, 1" x 8' x 20'		.09	14.35	43	129	9.30	
5600	Stabilizer, self-propelled, 150 H.P.		41.50	680	2,045	6,125	741	
5700	310 H.P.		77.65	1,825	5,485	16,500	1,718	
5800	Striper, truck mounted, 120 gallon paint, 460 H.P.		48.50	505	1,510	4,525	690	
5900	Thermal paint heating kettle, 115 gallons		7.73	26.50	80	240	77.85	
6000	Tar kettle, 330 gallon, trailer mounted		11.65	60	180	540	129.20	
7000	Tunnel locomotive, diesel, 8 to 12 ton		29.90	600	1,800	5,400	599.20	
7005	Electric, 10 ton		28.40	685	2,060	6,175	639.20	
7010	Muck cars, 1/2 C.Y. capacity		2.25	26	78	234	33.60	
7020	1 C.Y. capacity		2.45	33.50	100	300	39.60	
7030	2 C.Y. capacity		2.60	38.50	115	345	43.80	
7040	Side dump, 2 C.Y. capacity		2.80	46.50	140	420	50.40	
7050	3 C.Y. capacity		3.80	51.50	155	465	61.40	
7060	5 C.Y. capacity		5.50	66.50	200	600	84	
7100	Ventilating blower for tunnel, 7-1/2 H.P.		2.16	51.50	155	465	48.30	
7110	10 H.P.		2.39	53.50	160	480	51.10	
7120	20 H.P.		3.56	69.50	208	625	70.10	
7140	40 H.P.		6.02	80	240	720	96.15	
7160	60 H.P.		8.75	98.50	295	885	129	
7175	75 H.P.		10.31	153	460	1,375	174.50	
7180	200 H.P.		20.73	305	920	2,750	349.85	
7800	Windrow loader, elevating		54.10	1,350	4,045	12,100	1,242	
0010	LIFTING AND HOISTING EQUIPMENT RENTAL without operators (R015433-10)	Ea.						60
0150	Crane, flatbed mounted, 3 ton capacity		14.10	205	615	1,850	235.80	
0200	Crane, climbing, 106' jib, 6000 lb. capacity, 410 fpm (R312316-45)		41.13	1,750	5,260	15,800	1,381	
0300	101' jib, 10,250 lb. capacity, 270 fpm		48.18	2,225	6,670	20,000	1,719	
0500	Tower, static, 130' high, 106' jib, 6200 lb. capacity at 400 fpm		45.23	2,025	6,080	18,200	1,578	
0520	Mini crawler spider crane, up to 24" wide, 1990 lb. lifting capacity		12.54	755	2,265	6,800	553.30	
0525	Up to 30" wide, 6450 lb. lifting capacity		14.57	840	2,520	7,550	620.55	
0530	Up to 52" wide, 6680 lb. lifting capacity		23.17	1,325	3,960	11,900	977.35	
0535	Up to 55" wide, 8920 lb. lifting capacity		25.87	1,500	4,500	13,500	1,107	
0540	Up to 66" wide, 13,350 lb. lifting capacity		35.03	2,050	6,120	18,400	1,504	
0600	Crawler mounted, lattice boom, 1/2 C.Y., 15 tons at 12' radius		37.10	885	2,660	7,975	828.80	
0700	3/4 C.Y., 20 tons at 12' radius		49.46	1,100	3,320	9,950	1,060	

01 54 33 | Equipment Rental

		UNIT	HOURLY OPER. COST	RENT PER DAY	RENT PER WEEK	RENT PER MONTH	EQUIPMENT COST/DAY
0800	1 C.Y., 25 tons at 12' radius	Ea.	65.95	1,375	4,100	12,300	1,348
0900	1-1/2 C.Y., 40 tons at 12' radius		66.45	1,400	4,190	12,600	1,370
1000	2 C.Y., 50 tons at 12' radius		88.90	2,050	6,145	18,400	1,940
1100	3 C.Y., 75 tons at 12' radius		76.00	1,825	5,500	16,500	1,708
1200	100 ton capacity, 60' boom		86.05	1,975	5,920	17,800	1,872
1300	165 ton capacity, 60' boom		105.00	2,325	6,970	20,900	2,234
1400	200 ton capacity, 70' boom		140.85	3,125	9,385	28,200	3,004
1500	350 ton capacity, 80' boom		182.50	4,125	12,375	37,100	3,935
1600	Truck mounted, lattice boom, 6 x 4, 20 tons at 10' radius		37.84	1,300	3,900	11,700	1,083
1700	25 tons at 10' radius		40.92	1,425	4,240	12,700	1,175
1800	8 x 4, 30 tons at 10' radius		44.29	1,500	4,520	13,600	1,258
1900	40 tons at 12' radius		47.09	1,575	4,720	14,200	1,321
2000	60 tons at 15' radius		52.56	1,675	5,000	15,000	1,420
2050	82 tons at 15' radius		58.46	1,775	5,340	16,000	1,536
2100	90 tons at 15' radius		65.33	1,950	5,820	17,500	1,687
2200	115 tons at 15' radius		73.60	2,175	6,500	19,500	1,889
2300	150 tons at 18' radius		80.95	2,275	6,845	20,500	2,017
2350	165 tons at 18' radius		85.95	2,425	7,260	21,800	2,140
2400	Truck mounted, hydraulic, 12 ton capacity		31.00	415	1,240	3,725	496
2500	25 ton capacity		37.45	485	1,455	4,375	590.60
2550	33 ton capacity		50.70	890	2,675	8,025	940.60
2560	40 ton capacity		51.00	905	2,710	8,125	950
2600	55 ton capacity		56.60	900	2,705	8,125	993.80
2700	80 ton capacity		77.85	1,500	4,475	13,400	1,518
2720	100 ton capacity		76.80	1,525	4,595	13,800	1,533
2740	120 ton capacity		101.90	1,825	5,460	16,400	1,907
2760	150 ton capacity		107.85	1,975	5,960	17,900	2,055
2800	Self-propelled, 4 x 4, with telescoping boom, 5 ton		15.00	232	695	2,075	259
2900	12-1/2 ton capacity		20.45	335	1,000	3,000	363.60
3000	15 ton capacity		33.95	520	1,560	4,675	583.60
3050	20 ton capacity		25.40	615	1,840	5,525	571.20
3100	25 ton capacity		37.30	615	1,850	5,550	668.40
3150	40 ton capacity		44.10	635	1,910	5,725	734.80
3200	Derricks, guy, 20 ton capacity, 60' boom, 75' mast		23.07	430	1,288	3,875	442.15
3300	100' boom, 115' mast		36.55	735	2,210	6,625	734.40
3400	Stiffleg, 20 ton capacity, 70' boom, 37' mast		25.74	555	1,670	5,000	539.90
3500	100' boom, 47' mast		39.84	895	2,680	8,050	854.70
3550	Helicopter, small, lift to 1250 lb. maximum, w/pilot		97.10	3,475	10,400	31,200	2,857
3600	Hoists, chain type, overhead, manual, 3/4 ton		.15	.33	1	3	1.40
3900	10 ton		.80	6	18	54	10
4000	Hoist and tower, 5000 lb. cap., portable electric, 40' high		5.19	247	742	2,225	189.90
4100	For each added 10' section, add		.12	19.35	58	174	12.55
4200	Hoist and single tubular tower, 5000 lb. electric, 100' high		7.03	345	1,036	3,100	263.45
4300	For each added 6'-6" section, add		.20	33.50	101	305	21.80
4400	Hoist and double tubular tower, 5000 lb., 100' high		7.56	380	1,141	3,425	288.70
4500	For each added 6'-6" section, add		.22	37	111	335	23.95
4550	Hoist and tower, mast type, 6000 lb., 100' high		8.14	395	1,183	3,550	301.70
4570	For each added 10' section, add		.14	22.50	68	204	14.70
4600	Hoist and tower, personnel, electric, 2000 lb., 100' @ 125 fpm		17.23	1,050	3,150	9,450	767.85
4700	3000 lb., 100' @ 200 fpm		19.70	1,200	3,570	10,700	871.60
4800	3000 lb., 150' @ 300 fpm		21.85	1,325	4,000	12,000	974.80
4900	4000 lb., 100' @ 300 fpm		22.62	1,350	4,080	12,200	996.95
5000	6000 lb., 100' @ 275 fpm		24.32	1,425	4,270	12,800	1,049
5100	For added heights up to 500', add	L.F.	.01	1.67	5	15	1.10
5200	Jacks, hydraulic, 20 ton	Ea.	.05	2	6	18	1.60
5500	100 ton		.40	12	36	108	10.40
6100	Jacks, hydraulic, climbing w/50' jackrods, control console, 30 ton cap.		2.13	142	426	1,275	102.25
6150	For each added 10' jackrod section, add		.05	3.33	10	30	2.40

For customer support on your Electrical Costs with RSMeans data, call 800.448.8182.

473

01 54 33 | Equipment Rental

		UNIT	HOURLY OPER. COST	RENT PER DAY	RENT PER WEEK	RENT PER MONTH	EQUIPMENT COST/DAY	
60								**60**
6300	50 ton capacity	Ea.	3.43	228	685	2,050	164.45	
6350	For each added 10' jackrod section, add		.06	4	12	36	2.90	
6500	125 ton capacity		8.95	595	1,790	5,375	429.60	
6550	For each added 10' jackrod section, add		.61	40.50	121	365	29.10	
6600	Cable jack, 10 ton capacity with 200' cable		1.79	119	357	1,075	85.70	
6650	For each added 50' of cable, add		.22	14.35	43	129	10.35	
70								**70**
0010	**WELLPOINT EQUIPMENT RENTAL** without operators R015433 -10							
0020	Based on 2 months rental	Ea.	15.83	350	1,057	3,175	338.05	
0100	Combination jetting & wellpoint pump, 60 H.P. diesel	"	34.42	300	903	2,700	455.95	
0200	High pressure gas jet pump, 200 H.P., 300 psi	L.F.	.01	.57	1.71	5.15	.40	
0300	Discharge pipe, 8" diameter		.01	.84	2.53	7.60	.60	
0350	12" diameter		.01	.52	1.56	4.68	.40	
0400	Header pipe, flows up to 150 GPM, 4" diameter		.01	.61	1.83	5.50	.45	
0500	400 GPM, 6" diameter		.01	.84	2.53	7.60	.60	
0600	800 GPM, 8" diameter		.01	.89	2.66	8	.60	
0700	1500 GPM, 10" diameter		.03	1.68	5.03	15.10	1.25	
0800	2500 GPM, 12" diameter		.03	2.15	6.44	19.30	1.55	
0900	4500 GPM, 16" diameter		.03	2.22	6.67	20	1.55	
0950	For quick coupling aluminum and plastic pipe, add	Ea.	.07	4.44	13.31	40	3.20	
1100	Wellpoint, 25' long, with fittings & riser pipe, 1-1/2" or 2" diameter		7.07	203	609	1,825	178.35	
1200	Wellpoint pump, diesel powered, 4" suction, 20 H.P.		9.51	252	756	2,275	227.30	
1300	6" suction, 30 H.P.		12.87	345	1,036	3,100	310.15	
1400	8" suction, 40 H.P.		19.01	405	1,211	3,625	394.30	
1500	10" suction, 75 H.P.		27.56	645	1,930	5,800	606.50	
1600	12" suction, 100 H.P.		39.50	710	2,130	6,400	742	
1700	12" suction, 175 H.P.							
80								**80**
0010	**MARINE EQUIPMENT RENTAL** without operators R015433 -10	Ea.	18.05	1,150	3,455	10,400	835.40	
0200	Barge, 400 ton, 30' wide x 90' long		21.95	1,425	4,240	12,700	1,024	
0240	800 ton, 45' wide x 90' long		28.80	228	685	2,050	367.40	
2000	Tugboat, diesel, 100 H.P.		54.40	410	1,225	3,675	680.20	
2040	250 H.P.		122.45	1,225	3,685	11,100	1,717	
2080	380 H.P.		12.50	63.50	190	570	138	
3000	Small work boat, gas, 16-foot, 50 H.P.		74.90	1,300	3,930	11,800	1,385	
4000	Large, diesel, 48-foot, 200 H.P.							

Crew No.	Bare Costs		Incl. Subs O&P		Cost Per Labor-Hour	
Crew A-1	Hr.	Daily	Hr.	Daily	Bare Costs	Incl. O&P
1 Building Laborer	$39.85	$318.80	$60.70	$485.60	$39.85	$60.70
1 Concrete Saw, Gas Manual		71.20		78.32	8.90	9.79
8 L.H., Daily Totals		$390.00		$563.92	$48.75	$70.49
Crew A-1A	Hr.	Daily	Hr.	Daily	Bare Costs	Incl. O&P
1 Skilled Worker	$52.35	$418.80	$80.15	$641.20	$52.35	$80.15
1 Shot Blaster, 20"		214.80		236.28	26.85	29.54
8 L.H., Daily Totals		$633.60		$877.48	$79.20	$109.69
Crew A-1B	Hr.	Daily	Hr.	Daily	Bare Costs	Incl. O&P
1 Building Laborer	$39.85	$318.80	$60.70	$485.60	$39.85	$60.70
1 Concrete Saw		102.40		112.64	12.80	14.08
8 L.H., Daily Totals		$421.20		$598.24	$52.65	$74.78
Crew A-1C	Hr.	Daily	Hr.	Daily	Bare Costs	Incl. O&P
1 Building Laborer	$39.85	$318.80	$60.70	$485.60	$39.85	$60.70
1 Chain Saw, Gas, 18"		27.80		30.58	3.48	3.82
8 L.H., Daily Totals		$346.60		$516.18	$43.33	$64.52
Crew A-1D	Hr.	Daily	Hr.	Daily	Bare Costs	Incl. O&P
1 Building Laborer	$39.85	$318.80	$60.70	$485.60	$39.85	$60.70
1 Vibrating Plate, Gas, 18"		31.80		34.98	3.98	4.37
8 L.H., Daily Totals		$350.60		$520.58	$43.83	$65.07
Crew A-1E	Hr.	Daily	Hr.	Daily	Bare Costs	Incl. O&P
1 Building Laborer	$39.85	$318.80	$60.70	$485.60	$39.85	$60.70
1 Vibrating Plate, Gas, 21"		40.60		44.66	5.08	5.58
8 L.H., Daily Totals		$359.40		$530.26	$44.92	$66.28
Crew A-1F	Hr.	Daily	Hr.	Daily	Bare Costs	Incl. O&P
1 Building Laborer	$39.85	$318.80	$60.70	$485.60	$39.85	$60.70
1 Rammer/Tamper, Gas, 8"		46.00		50.60	5.75	6.33
8 L.H., Daily Totals		$364.80		$536.20	$45.60	$67.03
Crew A-1G	Hr.	Daily	Hr.	Daily	Bare Costs	Incl. O&P
1 Building Laborer	$39.85	$318.80	$60.70	$485.60	$39.85	$60.70
1 Rammer/Tamper, Gas, 15"		52.00		57.20	6.50	7.15
8 L.H., Daily Totals		$370.80		$542.80	$46.35	$67.85
Crew A-1H	Hr.	Daily	Hr.	Daily	Bare Costs	Incl. O&P
1 Building Laborer	$39.85	$318.80	$60.70	$485.60	$39.85	$60.70
1 Exterior Steam Cleaner		74.80		82.28	9.35	10.29
8 L.H., Daily Totals		$393.60		$567.88	$49.20	$70.98
Crew A-1J	Hr.	Daily	Hr.	Daily	Bare Costs	Incl. O&P
1 Building Laborer	$39.85	$318.80	$60.70	$485.60	$39.85	$60.70
1 Cultivator, Walk-Behind, 5 H.P.		63.50		69.85	7.94	8.73
8 L.H., Daily Totals		$382.30		$555.45	$47.79	$69.43
Crew A-1K	Hr.	Daily	Hr.	Daily	Bare Costs	Incl. O&P
1 Building Laborer	$39.85	$318.80	$60.70	$485.60	$39.85	$60.70
1 Cultivator, Walk-Behind, 8 H.P.		74.60		82.06	9.32	10.26
8 L.H., Daily Totals		$393.40		$567.66	$49.17	$70.96
Crew A-1M	Hr.	Daily	Hr.	Daily	Bare Costs	Incl. O&P
1 Building Laborer	$39.85	$318.80	$60.70	$485.60	$39.85	$60.70
1 Snow Blower, Walk-Behind		70.10		77.11	8.76	9.64
8 L.H., Daily Totals		$388.90		$562.71	$48.61	$70.34

Crew No.	Bare Costs		Incl. Subs O&P		Cost Per Labor-Hour	
Crew A-2	Hr.	Daily	Hr.	Daily	Bare Costs	Incl. O&P
2 Laborers	$39.85	$637.60	$60.70	$971.20	$41.40	$62.80
1 Truck Driver (light)	44.50	356.00	67.00	536.00		
1 Flatbed Truck, Gas, 1.5 Ton		188.40		207.24	7.85	8.63
24 L.H., Daily Totals		$1182.00		$1714.44	$49.25	$71.44
Crew A-2A	Hr.	Daily	Hr.	Daily	Bare Costs	Incl. O&P
2 Laborers	$39.85	$637.60	$60.70	$971.20	$41.40	$62.80
1 Truck Driver (light)	44.50	356.00	67.00	536.00		
1 Flatbed Truck, Gas, 1.5 Ton		188.40		207.24		
1 Concrete Saw		102.40		112.64	12.12	13.33
24 L.H., Daily Totals		$1284.40		$1827.08	$53.52	$76.13
Crew A-2B	Hr.	Daily	Hr.	Daily	Bare Costs	Incl. O&P
1 Truck Driver (light)	$44.50	$356.00	$67.00	$536.00	$44.50	$67.00
1 Flatbed Truck, Gas, 1.5 Ton		188.40		207.24	23.55	25.91
8 L.H., Daily Totals		$544.40		$743.24	$68.05	$92.91
Crew A-3A	Hr.	Daily	Hr.	Daily	Bare Costs	Incl. O&P
1 Equip. Oper. (light)	$51.30	$410.40	$77.35	$618.80	$51.30	$77.35
1 Pickup Truck, 4x4, 3/4 Ton		126.60		139.26	15.82	17.41
8 L.H., Daily Totals		$537.00		$758.06	$67.13	$94.76
Crew A-3B	Hr.	Daily	Hr.	Daily	Bare Costs	Incl. O&P
1 Equip. Oper. (medium)	$53.75	$430.00	$81.05	$648.40	$49.88	$75.17
1 Truck Driver (heavy)	46.00	368.00	69.30	554.40		
1 Dump Truck, 12 C.Y., 400 H.P.		542.80		597.08		
1 F.E. Loader, W.M., 2.5 C.Y.		523.20		575.52	66.63	73.29
16 L.H., Daily Totals		$1864.00		$2375.40	$116.50	$148.46
Crew A-3C	Hr.	Daily	Hr.	Daily	Bare Costs	Incl. O&P
1 Equip. Oper. (light)	$51.30	$410.40	$77.35	$618.80	$51.30	$77.35
1 Loader, Skid Steer, 78 H.P.		364.80		401.28	45.60	50.16
8 L.H., Daily Totals		$775.20		$1020.08	$96.90	$127.51
Crew A-3D	Hr.	Daily	Hr.	Daily	Bare Costs	Incl. O&P
1 Truck Driver (light)	$44.50	$356.00	$67.00	$536.00	$44.50	$67.00
1 Pickup Truck, 4x4, 3/4 Ton		126.60		139.26		
1 Flatbed Trailer, 25 Ton		133.00		146.30	32.45	35.70
8 L.H., Daily Totals		$615.60		$821.56	$76.95	$102.69
Crew A-3E	Hr.	Daily	Hr.	Daily	Bare Costs	Incl. O&P
1 Equip. Oper. (crane)	$56.10	$448.80	$84.60	$676.80	$51.05	$76.95
1 Truck Driver (heavy)	46.00	368.00	69.30	554.40		
1 Pickup Truck, 4x4, 3/4 Ton		126.60		139.26	7.91	8.70
16 L.H., Daily Totals		$943.40		$1370.46	$58.96	$85.65
Crew A-3F	Hr.	Daily	Hr.	Daily	Bare Costs	Incl. O&P
1 Equip. Oper. (crane)	$56.10	$448.80	$84.60	$676.80	$51.05	$76.95
1 Truck Driver (heavy)	46.00	368.00	69.30	554.40		
1 Pickup Truck, 4x4, 3/4 Ton		126.60		139.26		
1 Truck Tractor, 6x4, 380 H.P.		476.20		523.82		
1 Lowbed Trailer, 75 Ton		249.40		274.34	53.26	58.59
16 L.H., Daily Totals		$1669.00		$2168.62	$104.31	$135.54

Crew No.	Bare Costs		Incl. Subs O&P		Cost Per Labor-Hour	

Crew A-3G	Hr.	Daily	Hr.	Daily	Bare Costs	Incl. O&P
1 Equip. Oper. (crane)	$56.10	$448.80	$84.60	$676.80	$51.05	$76.95
1 Truck Driver (heavy)	46.00	368.00	69.30	554.40		
1 Pickup Truck, 4x4, 3/4 Ton		126.60		139.26		
1 Truck Tractor, 6x4, 450 H.P.		583.40		641.74		
1 Lowbed Trailer, 75 Ton		249.40		274.34	59.96	65.96
16 L.H., Daily Totals		$1776.20		$2286.54	$111.01	$142.91

Crew A-3H	Hr.	Daily	Hr.	Daily	Bare Costs	Incl. O&P
1 Equip. Oper. (crane)	$56.10	$448.80	$84.60	$676.80	$56.10	$84.60
1 Hyd. Crane, 12 Ton (Daily)		628.60		691.46	78.58	86.43
8 L.H., Daily Totals		$1077.40		$1368.26	$134.68	$171.03

Crew A-3I	Hr.	Daily	Hr.	Daily	Bare Costs	Incl. O&P
1 Equip. Oper. (crane)	$56.10	$448.80	$84.60	$676.80	$56.10	$84.60
1 Hyd. Crane, 25 Ton (Daily)		759.40		835.34	94.92	104.42
8 L.H., Daily Totals		$1208.20		$1512.14	$151.03	$189.02

Crew A-3J	Hr.	Daily	Hr.	Daily	Bare Costs	Incl. O&P
1 Equip. Oper. (crane)	$56.10	$448.80	$84.60	$676.80	$56.10	$84.60
1 Hyd. Crane, 40 Ton (Daily)		1313.00		1444.30	164.13	180.54
8 L.H., Daily Totals		$1761.80		$2121.10	$220.22	$265.14

Crew A-3K	Hr.	Daily	Hr.	Daily	Bare Costs	Incl. O&P
1 Equip. Oper. (crane)	$56.10	$448.80	$84.60	$676.80	$52.35	$78.95
1 Equip. Oper. (oiler)	48.60	388.80	73.30	586.40		
1 Hyd. Crane, 55 Ton (Daily)		1353.00		1488.30		
1 P/U, 3/4 Ton (Daily)		139.20		153.12	93.26	102.59
16 L.H., Daily Totals		$2329.80		$2904.62	$145.61	$181.54

Crew A-3L	Hr.	Daily	Hr.	Daily	Bare Costs	Incl. O&P
1 Equip. Oper. (crane)	$56.10	$448.80	$84.60	$676.80	$52.35	$78.95
1 Equip. Oper. (oiler)	48.60	388.80	73.30	586.40		
1 Hyd. Crane, 80 Ton (Daily)		2113.00		2324.30		
1 P/U Truck, 3/4 Ton (Daily)		139.20		153.12	140.76	154.84
16 L.H., Daily Totals		$3089.80		$3740.62	$193.11	$233.79

Crew A-3M	Hr.	Daily	Hr.	Daily	Bare Costs	Incl. O&P
1 Equip. Oper. (crane)	$56.10	$448.80	$84.60	$676.80	$52.35	$78.95
1 Equip. Oper. (oiler)	48.60	388.80	73.30	586.40		
1 Hyd. Crane, 100 Ton (Daily)		2144.00		2358.40		
1 P/U Truck, 3/4 Ton (Daily)		139.20		153.12	142.70	156.97
16 L.H., Daily Totals		$3120.80		$3774.72	$195.05	$235.92

Crew A-3N	Hr.	Daily	Hr.	Daily	Bare Costs	Incl. O&P
1 Equip. Oper. (crane)	$56.10	$448.80	$84.60	$676.80	$56.10	$84.60
1 Tower Crane (monthly)		1180.00		1298.00	147.50	162.25
8 L.H., Daily Totals		$1628.80		$1974.80	$203.60	$246.85

Crew A-3P	Hr.	Daily	Hr.	Daily	Bare Costs	Incl. O&P
1 Equip. Oper. (light)	$51.30	$410.40	$77.35	$618.80	$51.30	$77.35
1 A.T. Forklift, 31' reach, 45' lift		484.20		532.62	60.52	66.58
8 L.H., Daily Totals		$894.60		$1151.42	$111.83	$143.93

Crew A-3Q	Hr.	Daily	Hr.	Daily	Bare Costs	Incl. O&P
1 Equip. Oper. (light)	$51.30	$410.40	$77.35	$618.80	$51.30	$77.35
1 Pickup Truck, 4x4, 3/4 Ton		126.60		139.26		
1 Flatbed Trailer, 3 Ton		25.40		27.94	19.00	20.90
8 L.H., Daily Totals		$562.40		$786.00	$70.30	$98.25

Crew A-3R	Hr.	Daily	Hr.	Daily	Bare Costs	Incl. O&P
1 Equip. Oper. (light)	$51.30	$410.40	$77.35	$618.80	$51.30	$77.35
1 Forklift, Smooth Floor, 8,000 Lb.		148.20		163.02	18.52	20.38
8 L.H., Daily Totals		$558.60		$781.82	$69.83	$97.73

Crew A-4	Hr.	Daily	Hr.	Daily	Bare Costs	Incl. O&P
2 Carpenters	$50.70	$811.20	$77.20	$1235.20	$47.98	$72.82
1 Painter, Ordinary	42.55	340.40	64.05	512.40		
24 L.H., Daily Totals		$1151.60		$1747.60	$47.98	$72.82

Crew A-5	Hr.	Daily	Hr.	Daily	Bare Costs	Incl. O&P
2 Laborers	$39.85	$637.60	$60.70	$971.20	$40.37	$61.40
.25 Truck Driver (light)	44.50	89.00	67.00	134.00		
.25 Flatbed Truck, Gas, 1.5 Ton		47.10		51.81	2.62	2.88
18 L.H., Daily Totals		$773.70		$1157.01	$42.98	$64.28

Crew A-6	Hr.	Daily	Hr.	Daily	Bare Costs	Incl. O&P
1 Instrument Man	$52.35	$418.80	$80.15	$641.20	$50.42	$76.80
1 Rodman/Chainman	48.50	388.00	73.45	587.60		
1 Level, Electronic		50.40		55.44	3.15	3.46
16 L.H., Daily Totals		$857.20		$1284.24	$53.58	$80.27

Crew A-7	Hr.	Daily	Hr.	Daily	Bare Costs	Incl. O&P
1 Chief of Party	$63.65	$509.20	$96.45	$771.60	$54.83	$83.35
1 Instrument Man	52.35	418.80	80.15	641.20		
1 Rodman/Chainman	48.50	388.00	73.45	587.60		
1 Level, Electronic		50.40		55.44	2.10	2.31
24 L.H., Daily Totals		$1366.40		$2055.84	$56.93	$85.66

Crew A-8	Hr.	Daily	Hr.	Daily	Bare Costs	Incl. O&P
1 Chief of Party	$63.65	$509.20	$96.45	$771.60	$53.25	$80.88
1 Instrument Man	52.35	418.80	80.15	641.20		
2 Rodmen/Chainmen	48.50	776.00	73.45	1175.20		
1 Level, Electronic		50.40		55.44	1.58	1.73
32 L.H., Daily Totals		$1754.40		$2643.44	$54.83	$82.61

Crew A-9	Hr.	Daily	Hr.	Daily	Bare Costs	Incl. O&P
1 Asbestos Foreman	$56.80	$454.40	$88.15	$705.20	$56.36	$87.49
7 Asbestos Workers	56.30	3152.80	87.40	4894.40		
64 L.H., Daily Totals		$3607.20		$5599.60	$56.36	$87.49

Crew A-10A	Hr.	Daily	Hr.	Daily	Bare Costs	Incl. O&P
1 Asbestos Foreman	$56.80	$454.40	$88.15	$705.20	$56.47	$87.65
2 Asbestos Workers	56.30	900.80	87.40	1398.40		
24 L.H., Daily Totals		$1355.20		$2103.60	$56.47	$87.65

Crew A-10B	Hr.	Daily	Hr.	Daily	Bare Costs	Incl. O&P
1 Asbestos Foreman	$56.80	$454.40	$88.15	$705.20	$56.42	$87.59
3 Asbestos Workers	56.30	1351.20	87.40	2097.60		
32 L.H., Daily Totals		$1805.60		$2802.80	$56.42	$87.59

Crew A-10C	Hr.	Daily	Hr.	Daily	Bare Costs	Incl. O&P
3 Asbestos Workers	$56.30	$1351.20	$87.40	$2097.60	$56.30	$87.40
1 Flatbed Truck, Gas, 1.5 Ton		188.40		207.24	7.85	8.63
24 L.H., Daily Totals		$1539.60		$2304.84	$64.15	$96.03

Crew A-10D

Crew No.	Bare Costs Hr.	Daily	Incl. Subs O&P Hr.	Daily	Bare Costs	Incl. O&P
2 Asbestos Workers	$56.30	$900.80	$87.40	$1398.40	$54.33	$83.17
1 Equip. Oper. (crane)	56.10	448.80	84.60	676.80		
1 Equip. Oper. (oiler)	48.60	388.80	73.30	586.40		
1 Hydraulic Crane, 33 Ton		940.60		1034.66	29.39	32.33
32 L.H., Daily Totals		$2679.00		$3696.26	$83.72	$115.51

Crew A-11

Crew No.	Bare Costs Hr.	Daily	Incl. Subs O&P Hr.	Daily	Bare Costs	Incl. O&P
1 Asbestos Foreman	$56.80	$454.40	$88.15	$705.20	$56.36	$87.49
7 Asbestos Workers	56.30	3152.80	87.40	4894.40		
2 Chip. Hammers, 12 Lb., Elec.		42.00		46.20	.66	.72
64 L.H., Daily Totals		$3649.20		$5645.80	$57.02	$88.22

Crew A-12

Crew No.	Bare Costs Hr.	Daily	Incl. Subs O&P Hr.	Daily	Bare Costs	Incl. O&P
1 Asbestos Foreman	$56.80	$454.40	$88.15	$705.20	$56.36	$87.49
7 Asbestos Workers	56.30	3152.80	87.40	4894.40		
1 Trk-Mtd Vac, 14 CY, 1500 Gal.		522.70		574.97		
1 Flatbed Truck, 20,000 GVW		197.20		216.92	11.25	12.37
64 L.H., Daily Totals		$4327.10		$6391.49	$67.61	$99.87

Crew A-13

Crew No.	Bare Costs Hr.	Daily	Incl. Subs O&P Hr.	Daily	Bare Costs	Incl. O&P
1 Equip. Oper. (light)	$51.30	$410.40	$77.35	$618.80	$51.30	$77.35
1 Trk-Mtd Vac, 14 CY, 1500 Gal.		522.70		574.97		
1 Flatbed Truck, 20,000 GVW		197.20		216.92	89.99	98.99
8 L.H., Daily Totals		$1130.30		$1410.69	$141.29	$176.34

Crew B-1

Crew No.	Bare Costs Hr.	Daily	Incl. Subs O&P Hr.	Daily	Bare Costs	Incl. O&P
1 Labor Foreman (outside)	$41.85	$334.80	$63.75	$510.00	$40.52	$61.72
2 Laborers	39.85	637.60	60.70	971.20		
24 L.H., Daily Totals		$972.40		$1481.20	$40.52	$61.72

Crew B-1A

Crew No.	Bare Costs Hr.	Daily	Incl. Subs O&P Hr.	Daily	Bare Costs	Incl. O&P
1 Labor Foreman (outside)	$41.85	$334.80	$63.75	$510.00	$40.52	$61.72
2 Laborers	39.85	637.60	60.70	971.20		
2 Cutting Torches		25.20		27.72		
2 Sets of Gases		336.00		369.60	15.05	16.56
24 L.H., Daily Totals		$1333.60		$1878.52	$55.57	$78.27

Crew B-1B

Crew No.	Bare Costs Hr.	Daily	Incl. Subs O&P Hr.	Daily	Bare Costs	Incl. O&P
1 Labor Foreman (outside)	$41.85	$334.80	$63.75	$510.00	$44.41	$67.44
2 Laborers	39.85	637.60	60.70	971.20		
1 Equip. Oper. (crane)	56.10	448.80	84.60	676.80		
2 Cutting Torches		25.20		27.72		
2 Sets of Gases		336.00		369.60		
1 Hyd. Crane, 12 Ton		496.00		545.60	26.79	29.47
32 L.H., Daily Totals		$2278.40		$3100.92	$71.20	$96.90

Crew B-1C

Crew No.	Bare Costs Hr.	Daily	Incl. Subs O&P Hr.	Daily	Bare Costs	Incl. O&P
1 Labor Foreman (outside)	$41.85	$334.80	$63.75	$510.00	$40.52	$61.72
2 Laborers	39.85	637.60	60.70	971.20		
1 Telescoping Boom Lift, to 60'		454.60		500.06	18.94	20.84
24 L.H., Daily Totals		$1427.00		$1981.26	$59.46	$82.55

Crew B-1D

Crew No.	Bare Costs Hr.	Daily	Incl. Subs O&P Hr.	Daily	Bare Costs	Incl. O&P
2 Laborers	$39.85	$637.60	$60.70	$971.20	$39.85	$60.70
1 Small Work Boat, Gas, 50 H.P.		138.00		151.80		
1 Pressure Washer, 7 GPM		77.20		84.92	13.45	14.80
16 L.H., Daily Totals		$852.80		$1207.92	$53.30	$75.50

Crew B-1E

Crew No.	Bare Costs Hr.	Daily	Incl. Subs O&P Hr.	Daily	Bare Costs	Incl. O&P
1 Labor Foreman (outside)	$41.85	$334.80	$63.75	$510.00	$40.35	$61.46
3 Laborers	39.85	956.40	60.70	1456.80		
1 Work Boat, Diesel, 200 H.P.		1385.00		1523.50		
2 Pressure Washers, 7 GPM		154.40		169.84	48.11	52.92
32 L.H., Daily Totals		$2830.60		$3660.14	$88.46	$114.38

Crew B-1F

Crew No.	Bare Costs Hr.	Daily	Incl. Subs O&P Hr.	Daily	Bare Costs	Incl. O&P
2 Skilled Workers	$52.35	$837.60	$80.15	$1282.40	$48.18	$73.67
1 Laborer	39.85	318.80	60.70	485.60		
1 Small Work Boat, Gas, 50 H.P.		138.00		151.80		
1 Pressure Washer, 7 GPM		77.20		84.92	8.97	9.86
24 L.H., Daily Totals		$1371.60		$2004.72	$57.15	$83.53

Crew B-1G

Crew No.	Bare Costs Hr.	Daily	Incl. Subs O&P Hr.	Daily	Bare Costs	Incl. O&P
2 Laborers	$39.85	$637.60	$60.70	$971.20	$39.85	$60.70
1 Small Work Boat, Gas, 50 H.P.		138.00		151.80	8.63	9.49
16 L.H., Daily Totals		$775.60		$1123.00	$48.48	$70.19

Crew B-1H

Crew No.	Bare Costs Hr.	Daily	Incl. Subs O&P Hr.	Daily	Bare Costs	Incl. O&P
2 Skilled Workers	$52.35	$837.60	$80.15	$1282.40	$48.18	$73.67
1 Laborer	39.85	318.80	60.70	485.60		
1 Small Work Boat, Gas, 50 H.P.		138.00		151.80	5.75	6.33
24 L.H., Daily Totals		$1294.40		$1919.80	$53.93	$79.99

Crew B-1J

Crew No.	Bare Costs Hr.	Daily	Incl. Subs O&P Hr.	Daily	Bare Costs	Incl. O&P
1 Labor Foreman (inside)	$40.35	$322.80	$61.45	$491.60	$40.10	$61.08
1 Laborer	39.85	318.80	60.70	485.60		
16 L.H., Daily Totals		$641.60		$977.20	$40.10	$61.08

Crew B-1K

Crew No.	Bare Costs Hr.	Daily	Incl. Subs O&P Hr.	Daily	Bare Costs	Incl. O&P
1 Carpenter Foreman (inside)	$51.20	$409.60	$78.00	$624.00	$50.95	$77.60
1 Carpenter	50.70	405.60	77.20	617.60		
16 L.H., Daily Totals		$815.20		$1241.60	$50.95	$77.60

Crew B-2

Crew No.	Bare Costs Hr.	Daily	Incl. Subs O&P Hr.	Daily	Bare Costs	Incl. O&P
1 Labor Foreman (outside)	$41.85	$334.80	$63.75	$510.00	$40.25	$61.31
4 Laborers	39.85	1275.20	60.70	1942.40		
40 L.H., Daily Totals		$1610.00		$2452.40	$40.25	$61.31

Crew B-2A

Crew No.	Bare Costs Hr.	Daily	Incl. Subs O&P Hr.	Daily	Bare Costs	Incl. O&P
1 Labor Foreman (outside)	$41.85	$334.80	$63.75	$510.00	$40.52	$61.72
2 Laborers	39.85	637.60	60.70	971.20		
1 Telescoping Boom Lift, to 60'		454.60		500.06	18.94	20.84
24 L.H., Daily Totals		$1427.00		$1981.26	$59.46	$82.55

Crew B-3

Crew No.	Bare Costs Hr.	Daily	Incl. Subs O&P Hr.	Daily	Bare Costs	Incl. O&P
1 Labor Foreman (outside)	$41.85	$334.80	$63.75	$510.00	$44.55	$67.47
2 Laborers	39.85	637.60	60.70	971.20		
1 Equip. Oper. (medium)	53.75	430.00	81.05	648.40		
2 Truck Drivers (heavy)	46.00	736.00	69.30	1108.80		
1 Crawler Loader, 3 C.Y.		1204.00		1324.40		
2 Dump Trucks, 12 C.Y., 400 H.P.		1085.60		1194.16	47.70	52.47
48 L.H., Daily Totals		$4428.00		$5756.96	$92.25	$119.94

Crew B-3A

Crew No.	Bare Costs Hr.	Daily	Incl. Subs O&P Hr.	Daily	Bare Costs	Incl. O&P
4 Laborers	$39.85	$1275.20	$60.70	$1942.40	$42.63	$64.77
1 Equip. Oper. (medium)	53.75	430.00	81.05	648.40		
1 Hyd. Excavator, 1.5 C.Y.		893.60		982.96	22.34	24.57
40 L.H., Daily Totals		$2598.80		$3573.76	$64.97	$89.34

For customer support on your Electrical Costs with RSMeans data, call 800.448.8182.

477

| Crew No. | | Bare Costs | | Incl. Subs O&P | | Cost Per Labor-Hour | |

Crew B-3B

Crew B-3B	Hr.	Daily	Hr.	Daily	Bare Costs	Incl. O&P
2 Laborers	$39.85	$637.60	$60.70	$971.20	$44.86	$67.94
1 Equip. Oper. (medium)	53.75	430.00	81.05	648.40		
1 Truck Driver (heavy)	46.00	368.00	69.30	554.40		
1 Backhoe Loader, 80 H.P.		385.00		423.50		
1 Dump Truck, 12 C.Y., 400 H.P.		542.80		597.08	28.99	31.89
32 L.H., Daily Totals		$2363.40		$3194.58	$73.86	$99.83

Crew B-3C

Crew B-3C	Hr.	Daily	Hr.	Daily	Bare Costs	Incl. O&P
3 Laborers	$39.85	$956.40	$60.70	$1456.80	$43.33	$65.79
1 Equip. Oper. (medium)	53.75	430.00	81.05	648.40		
1 Crawler Loader, 4 C.Y.		1459.00		1604.90	45.59	50.15
32 L.H., Daily Totals		$2845.40		$3710.10	$88.92	$115.94

Crew B-4

Crew B-4	Hr.	Daily	Hr.	Daily	Bare Costs	Incl. O&P
1 Labor Foreman (outside)	$41.85	$334.80	$63.75	$510.00	$41.21	$62.64
4 Laborers	39.85	1275.20	60.70	1942.40		
1 Truck Driver (heavy)	46.00	368.00	69.30	554.40		
1 Truck Tractor, 220 H.P.		290.00		319.00		
1 Flatbed Trailer, 40 Ton		180.00		198.00	9.79	10.77
48 L.H., Daily Totals		$2448.00		$3523.80	$51.00	$73.41

Crew B-5

Crew B-5	Hr.	Daily	Hr.	Daily	Bare Costs	Incl. O&P
1 Labor Foreman (outside)	$41.85	$334.80	$63.75	$510.00	$44.11	$66.95
4 Laborers	39.85	1275.20	60.70	1942.40		
2 Equip. Oper. (medium)	53.75	860.00	81.05	1296.80		
1 Air Compressor, 250 cfm		167.40		184.14		
2 Breakers, Pavement, 60 lb.		21.20		23.32		
2 -50' Air Hoses, 1.5"		45.40		49.94		
1 Crawler Loader, 3 C.Y.		1204.00		1324.40	25.68	28.25
56 L.H., Daily Totals		$3908.00		$5331.00	$69.79	$95.20

Crew B-5A

Crew B-5A	Hr.	Daily	Hr.	Daily	Bare Costs	Incl. O&P
1 Labor Foreman (outside)	$41.85	$334.80	$63.75	$510.00	$44.31	$67.17
6 Laborers	39.85	1912.80	60.70	2913.60		
2 Equip. Oper. (medium)	53.75	860.00	81.05	1296.80		
1 Equip. Oper. (light)	51.30	410.40	77.35	618.80		
2 Truck Drivers (heavy)	46.00	736.00	69.30	1108.80		
1 Air Compressor, 365 cfm		211.40		232.54		
2 Breakers, Pavement, 60 lb.		21.20		23.32		
8 -50' Air Hoses, 1"		67.60		74.36		
2 Dump Trucks, 8 C.Y., 220 H.P.		677.20		744.92	10.18	11.20
96 L.H., Daily Totals		$5231.40		$7523.14	$54.49	$78.37

Crew B-5B

Crew B-5B	Hr.	Daily	Hr.	Daily	Bare Costs	Incl. O&P
1 Powderman	$52.35	$418.80	$80.15	$641.20	$49.64	$75.03
2 Equip. Oper. (medium)	53.75	860.00	81.05	1296.80		
3 Truck Drivers (heavy)	46.00	1104.00	69.30	1663.20		
1 F.E. Loader, W.M., 2.5 C.Y.		523.20		575.52		
3 Dump Trucks, 12 C.Y., 400 H.P.		1628.40		1791.24		
1 Air Compressor, 365 cfm		211.40		232.54	49.23	54.15
48 L.H., Daily Totals		$4745.80		$6200.50	$98.87	$129.18

Crew B-5C

Crew B-5C	Hr.	Daily	Hr.	Daily	Bare Costs	Incl. O&P
3 Laborers	$39.85	$956.40	$60.70	$1456.80	$46.25	$69.96
1 Equip. Oper. (medium)	53.75	430.00	81.05	648.40		
2 Truck Drivers (heavy)	46.00	736.00	69.30	1108.80		
1 Equip. Oper. (crane)	56.10	448.80	84.60	676.80		
1 Equip. Oper. (oiler)	48.60	388.80	73.30	586.40		
2 Dump Trucks, 12 C.Y., 400 H.P.		1085.60		1194.16		
1 Crawler Loader, 4 C.Y.		1459.00		1604.90		
1 S.P. Crane, 4x4, 25 Ton		668.40		735.24	50.20	55.22
64 L.H., Daily Totals		$6173.00		$8011.50	$96.45	$125.18

Crew B-5D

Crew B-5D	Hr.	Daily	Hr.	Daily	Bare Costs	Incl. O&P
1 Labor Foreman (outside)	$41.85	$334.80	$63.75	$510.00	$44.34	$67.24
4 Laborers	39.85	1275.20	60.70	1942.40		
2 Equip. Oper. (medium)	53.75	860.00	81.05	1296.80		
1 Truck Driver (heavy)	46.00	368.00	69.30	554.40		
1 Air Compressor, 250 cfm		167.40		184.14		
2 Breakers, Pavement, 60 lb.		21.20		23.32		
2 -50' Air Hoses, 1.5"		45.40		49.94		
1 Crawler Loader, 3 C.Y.		1204.00		1324.40		
1 Dump Truck, 12 C.Y., 400 H.P.		542.80		597.08	30.95	34.05
64 L.H., Daily Totals		$4818.80		$6482.48	$75.29	$101.29

Crew B-6

Crew B-6	Hr.	Daily	Hr.	Daily	Bare Costs	Incl. O&P
2 Laborers	$39.85	$637.60	$60.70	$971.20	$43.67	$66.25
1 Equip. Oper. (light)	51.30	410.40	77.35	618.80		
1 Backhoe Loader, 48 H.P.		312.40		343.64	13.02	14.32
24 L.H., Daily Totals		$1360.40		$1933.64	$56.68	$80.57

Crew B-6A

Crew B-6A	Hr.	Daily	Hr.	Daily	Bare Costs	Incl. O&P
.5 Labor Foreman (outside)	$41.85	$167.40	$63.75	$255.00	$45.81	$69.45
1 Laborer	39.85	318.80	60.70	485.60		
1 Equip. Oper. (medium)	53.75	430.00	81.05	648.40		
1 Vacuum Truck, 5000 Gal.		358.90		394.79	17.95	19.74
20 L.H., Daily Totals		$1275.10		$1783.79	$63.76	$89.19

Crew B-6B

Crew B-6B	Hr.	Daily	Hr.	Daily	Bare Costs	Incl. O&P
2 Labor Foremen (outside)	$41.85	$669.60	$63.75	$1020.00	$40.52	$61.72
4 Laborers	39.85	1275.20	60.70	1942.40		
1 S.P. Crane, 4x4, 5 Ton		259.00		284.90		
1 Flatbed Truck, Gas, 1.5 Ton		188.40		207.24		
1 Butt Fusion Mach., 4"-12" diam.		425.70		468.27	18.19	20.01
48 L.H., Daily Totals		$2817.90		$3922.81	$58.71	$81.73

Crew B-6C

Crew B-6C	Hr.	Daily	Hr.	Daily	Bare Costs	Incl. O&P
2 Labor Foremen (outside)	$41.85	$669.60	$63.75	$1020.00	$40.52	$61.72
4 Laborers	39.85	1275.20	60.70	1942.40		
1 S.P. Crane, 4x4, 12 Ton		363.60		399.96		
1 Flatbed Truck, Gas, 3 Ton		238.00		261.80		
1 Butt Fusion Mach., 8"-24" diam.		1908.00		2098.80	52.28	57.51
48 L.H., Daily Totals		$4454.40		$5722.96	$92.80	$119.23

Crew B-7

Crew B-7	Hr.	Daily	Hr.	Daily	Bare Costs	Incl. O&P
1 Labor Foreman (outside)	$41.85	$334.80	$63.75	$510.00	$42.50	$64.60
4 Laborers	39.85	1275.20	60.70	1942.40		
1 Equip. Oper. (medium)	53.75	430.00	81.05	648.40		
1 Brush Chipper, 12", 130 H.P.		393.00		432.30		
1 Crawler Loader, 3 C.Y.		1204.00		1324.40		
2 Chain Saws, Gas, 36" Long		92.80		102.08	35.20	38.72
48 L.H., Daily Totals		$3729.80		$4959.58	$77.70	$103.32

Crew No.	Bare Costs		Incl. Subs O&P		Cost Per Labor-Hour	

Crew B-7A	Hr.	Daily	Hr.	Daily	Bare Costs	Incl. O&P
2 Laborers	$39.85	$637.60	$60.70	$971.20	$43.67	$66.25
1 Equip. Oper. (light)	51.30	410.40	77.35	618.80		
1 Rake w/Tractor		331.30		364.43		
2 Chain Saws, Gas, 18"		55.60		61.16	16.12	17.73
24 L.H., Daily Totals		$1434.90		$2015.59	$59.79	$83.98

Crew B-7B	Hr.	Daily	Hr.	Daily	Bare Costs	Incl. O&P
1 Labor Foreman (outside)	$41.85	$334.80	$63.75	$510.00	$43.00	$65.27
4 Laborers	39.85	1275.20	60.70	1942.40		
1 Equip. Oper. (medium)	53.75	430.00	81.05	648.40		
1 Truck Driver (heavy)	46.00	368.00	69.30	554.40		
1 Brush Chipper, 12", 130 H.P.		393.00		432.30		
1 Crawler Loader, 3 C.Y.		1204.00		1324.40		
2 Chain Saws, Gas, 36" Long		92.80		102.08		
1 Dump Truck, 8 C.Y., 220 H.P.		338.60		372.46	36.22	39.84
56 L.H., Daily Totals		$4436.40		$5886.44	$79.22	$105.11

Crew B-7C	Hr.	Daily	Hr.	Daily	Bare Costs	Incl. O&P
1 Labor Foreman (outside)	$41.85	$334.80	$63.75	$510.00	$43.00	$65.27
4 Laborers	39.85	1275.20	60.70	1942.40		
1 Equip. Oper. (medium)	53.75	430.00	81.05	648.40		
1 Truck Driver (heavy)	46.00	368.00	69.30	554.40		
1 Brush Chipper, 12", 130 H.P.		393.00		432.30		
1 Crawler Loader, 3 C.Y.		1204.00		1324.40		
2 Chain Saws, Gas, 36" Long		92.80		102.08		
1 Dump Truck, 12 C.Y., 400 H.P.		542.80		597.08	39.87	43.85
56 L.H., Daily Totals		$4640.60		$6111.06	$82.87	$109.13

Crew B-8	Hr.	Daily	Hr.	Daily	Bare Costs	Incl. O&P
1 Labor Foreman (outside)	$41.85	$334.80	$63.75	$510.00	$46.21	$69.89
2 Laborers	39.85	637.60	60.70	971.20		
2 Equip. Oper. (medium)	53.75	860.00	81.05	1296.80		
1 Equip. Oper. (oiler)	48.60	388.80	73.30	586.40		
2 Truck Drivers (heavy)	46.00	736.00	69.30	1108.80		
1 Hyd. Crane, 25 Ton		590.60		649.66		
1 Crawler Loader, 3 C.Y.		1204.00		1324.40		
2 Dump Trucks, 12 C.Y., 400 H.P.		1085.60		1194.16	45.00	49.50
64 L.H., Daily Totals		$5837.40		$7641.42	$91.21	$119.40

Crew B-9	Hr.	Daily	Hr.	Daily	Bare Costs	Incl. O&P
1 Labor Foreman (outside)	$41.85	$334.80	$63.75	$510.00	$40.25	$61.31
4 Laborers	39.85	1275.20	60.70	1942.40		
1 Air Compressor, 250 cfm		167.40		184.14		
2 Breakers, Pavement, 60 lb.		21.20		23.32		
2 -50' Air Hoses, 1.5"		45.40		49.94	5.85	6.43
40 L.H., Daily Totals		$1844.00		$2709.80	$46.10	$67.75

Crew B-9A	Hr.	Daily	Hr.	Daily	Bare Costs	Incl. O&P
2 Laborers	$39.85	$637.60	$60.70	$971.20	$41.90	$63.57
1 Truck Driver (heavy)	46.00	368.00	69.30	554.40		
1 Water Tank Trailer, 5000 Gal.		147.60		162.36		
1 Truck Tractor, 220 H.P.		290.00		319.00		
2 -50' Discharge Hoses, 3"		3.00		3.30	18.36	20.19
24 L.H., Daily Totals		$1446.20		$2010.26	$60.26	$83.76

Crew B-9B	Hr.	Daily	Hr.	Daily	Bare Costs	Incl. O&P
2 Laborers	$39.85	$637.60	$60.70	$971.20	$41.90	$63.57
1 Truck Driver (heavy)	46.00	368.00	69.30	554.40		
2 -50' Discharge Hoses, 3"		3.00		3.30		
1 Water Tank Trailer, 5000 Gal.		147.60		162.36		
1 Truck Tractor, 220 H.P.		290.00		319.00		
1 Pressure Washer		63.60		69.96	21.01	23.11
24 L.H., Daily Totals		$1509.80		$2080.22	$62.91	$86.68

Crew B-9D	Hr.	Daily	Hr.	Daily	Bare Costs	Incl. O&P
1 Labor Foreman (outside)	$41.85	$334.80	$63.75	$510.00	$40.25	$61.31
4 Common Laborers	39.85	1275.20	60.70	1942.40		
1 Air Compressor, 250 cfm		167.40		184.14		
2 -50' Air Hoses, 1.5"		45.40		49.94		
2 Air Powered Tampers		56.60		62.26	6.74	7.41
40 L.H., Daily Totals		$1879.40		$2748.74	$46.98	$68.72

Crew B-10	Hr.	Daily	Hr.	Daily	Bare Costs	Incl. O&P
1 Equip. Oper. (medium)	$53.75	$430.00	$81.05	$648.40	$49.12	$74.27
.5 Laborer	39.85	159.40	60.70	242.80		
12 L.H., Daily Totals		$589.40		$891.20	$49.12	$74.27

Crew B-10A	Hr.	Daily	Hr.	Daily	Bare Costs	Incl. O&P
1 Equip. Oper. (medium)	$53.75	$430.00	$81.05	$648.40	$49.12	$74.27
.5 Laborer	39.85	159.40	60.70	242.80		
1 Roller, 2-Drum, W.B., 7.5 H.P.		179.60		197.56	14.97	16.46
12 L.H., Daily Totals		$769.00		$1088.76	$64.08	$90.73

Crew B-10B	Hr.	Daily	Hr.	Daily	Bare Costs	Incl. O&P
1 Equip. Oper. (medium)	$53.75	$430.00	$81.05	$648.40	$49.12	$74.27
.5 Laborer	39.85	159.40	60.70	242.80		
1 Dozer, 200 H.P.		1273.00		1400.30	106.08	116.69
12 L.H., Daily Totals		$1862.40		$2291.50	$155.20	$190.96

Crew B-10C	Hr.	Daily	Hr.	Daily	Bare Costs	Incl. O&P
1 Equip. Oper. (medium)	$53.75	$430.00	$81.05	$648.40	$49.12	$74.27
.5 Laborer	39.85	159.40	60.70	242.80		
1 Dozer, 200 H.P.		1273.00		1400.30		
1 Vibratory Roller, Towed, 23 Ton		423.00		465.30	141.33	155.47
12 L.H., Daily Totals		$2285.40		$2756.80	$190.45	$229.73

Crew B-10D	Hr.	Daily	Hr.	Daily	Bare Costs	Incl. O&P
1 Equip. Oper. (medium)	$53.75	$430.00	$81.05	$648.40	$49.12	$74.27
.5 Laborer	39.85	159.40	60.70	242.80		
1 Dozer, 200 H.P.		1273.00		1400.30		
1 Sheepsft. Roller, Towed		425.80		468.38	141.57	155.72
12 L.H., Daily Totals		$2288.20		$2759.88	$190.68	$229.99

Crew B-10E	Hr.	Daily	Hr.	Daily	Bare Costs	Incl. O&P
1 Equip. Oper. (medium)	$53.75	$430.00	$81.05	$648.40	$49.12	$74.27
.5 Laborer	39.85	159.40	60.70	242.80		
1 Tandem Roller, 5 Ton		151.20		166.32	12.60	13.86
12 L.H., Daily Totals		$740.60		$1057.52	$61.72	$88.13

Crew B-10F	Hr.	Daily	Hr.	Daily	Bare Costs	Incl. O&P
1 Equip. Oper. (medium)	$53.75	$430.00	$81.05	$648.40	$49.12	$74.27
.5 Laborer	39.85	159.40	60.70	242.80		
1 Tandem Roller, 10 Ton		231.40		254.54	19.28	21.21
12 L.H., Daily Totals		$820.80		$1145.74	$68.40	$95.48

479

For customer support on your Electrical Costs with RSMeans data, call 800.448.8182.

Crew No.	Bare Costs		Incl. Subs O&P		Cost Per Labor-Hour	

Crew B-10G	Hr.	Daily	Hr.	Daily	Bare Costs	Incl. O&P
1 Equip. Oper. (medium)	$53.75	$430.00	$81.05	$648.40	$49.12	$74.27
.5 Laborer	39.85	159.40	60.70	242.80		
1 Sheepsfoot Roller, 240 H.P.		1316.00		1447.60	109.67	120.63
12 L.H., Daily Totals		$1905.40		$2338.80	$158.78	$194.90

Crew B-10H	Hr.	Daily	Hr.	Daily	Bare Costs	Incl. O&P
1 Equip. Oper. (medium)	$53.75	$430.00	$81.05	$648.40	$49.12	$74.27
.5 Laborer	39.85	159.40	60.70	242.80		
1 Diaphragm Water Pump, 2"		73.00		80.30		
1 -20' Suction Hose, 2"		1.95		2.15		
2 -50' Discharge Hoses, 2"		1.80		1.98	6.40	7.04
12 L.H., Daily Totals		$666.15		$975.63	$55.51	$81.30

Crew B-10I	Hr.	Daily	Hr.	Daily	Bare Costs	Incl. O&P
1 Equip. Oper. (medium)	$53.75	$430.00	$81.05	$648.40	$49.12	$74.27
.5 Laborer	39.85	159.40	60.70	242.80		
1 Diaphragm Water Pump, 4"		114.80		126.28		
1 -20' Suction Hose, 4"		3.25		3.58		
2 -50' Discharge Hoses, 4"		4.70		5.17	10.23	11.25
12 L.H., Daily Totals		$712.15		$1026.22	$59.35	$85.52

Crew B-10J	Hr.	Daily	Hr.	Daily	Bare Costs	Incl. O&P
1 Equip. Oper. (medium)	$53.75	$430.00	$81.05	$648.40	$49.12	$74.27
.5 Laborer	39.85	159.40	60.70	242.80		
1 Centrifugal Water Pump, 3"		79.20		87.12		
1 -20' Suction Hose, 3"		2.85		3.13		
2 -50' Discharge Hoses, 3"		3.00		3.30	7.09	7.80
12 L.H., Daily Totals		$674.45		$984.76	$56.20	$82.06

Crew B-10K	Hr.	Daily	Hr.	Daily	Bare Costs	Incl. O&P
1 Equip. Oper. (medium)	$53.75	$430.00	$81.05	$648.40	$49.12	$74.27
.5 Laborer	39.85	159.40	60.70	242.80		
1 Centr. Water Pump, 6"		296.40		326.04		
1 -20' Suction Hose, 6"		11.50		12.65		
2 -50' Discharge Hoses, 6"		12.20		13.42	26.68	29.34
12 L.H., Daily Totals		$909.50		$1243.31	$75.79	$103.61

Crew B-10L	Hr.	Daily	Hr.	Daily	Bare Costs	Incl. O&P
1 Equip. Oper. (medium)	$53.75	$430.00	$81.05	$648.40	$49.12	$74.27
.5 Laborer	39.85	159.40	60.70	242.80		
1 Dozer, 80 H.P.		464.40		510.84	38.70	42.57
12 L.H., Daily Totals		$1053.80		$1402.04	$87.82	$116.84

Crew B-10M	Hr.	Daily	Hr.	Daily	Bare Costs	Incl. O&P
1 Equip. Oper. (medium)	$53.75	$430.00	$81.05	$648.40	$49.12	$74.27
.5 Laborer	39.85	159.40	60.70	242.80		
1 Dozer, 300 H.P.		1829.00		2011.90	152.42	167.66
12 L.H., Daily Totals		$2418.40		$2903.10	$201.53	$241.93

Crew B-10N	Hr.	Daily	Hr.	Daily	Bare Costs	Incl. O&P
1 Equip. Oper. (medium)	$53.75	$430.00	$81.05	$648.40	$49.12	$74.27
.5 Laborer	39.85	159.40	60.70	242.80		
1 F.E. Loader, T.M., 1.5 C.Y.		579.80		637.78	48.32	53.15
12 L.H., Daily Totals		$1169.20		$1528.98	$97.43	$127.42

Crew B-10O	Hr.	Daily	Hr.	Daily	Bare Costs	Incl. O&P
1 Equip. Oper. (medium)	$53.75	$430.00	$81.05	$648.40	$49.12	$74.27
.5 Laborer	39.85	159.40	60.70	242.80		
1 F.E. Loader, T.M., 2.25 C.Y.		979.00		1076.90	81.58	89.74
12 L.H., Daily Totals		$1568.40		$1968.10	$130.70	$164.01

Crew B-10P	Hr.	Daily	Hr.	Daily	Bare Costs	Incl. O&P
1 Equip. Oper. (medium)	$53.75	$430.00	$81.05	$648.40	$49.12	$74.27
.5 Laborer	39.85	159.40	60.70	242.80		
1 Crawler Loader, 3 C.Y.		1204.00		1324.40	100.33	110.37
12 L.H., Daily Totals		$1793.40		$2215.60	$149.45	$184.63

Crew B-10Q	Hr.	Daily	Hr.	Daily	Bare Costs	Incl. O&P
1 Equip. Oper. (medium)	$53.75	$430.00	$81.05	$648.40	$49.12	$74.27
.5 Laborer	39.85	159.40	60.70	242.80		
1 Crawler Loader, 4 C.Y.		1459.00		1604.90	121.58	133.74
12 L.H., Daily Totals		$2048.40		$2496.10	$170.70	$208.01

Crew B-10R	Hr.	Daily	Hr.	Daily	Bare Costs	Incl. O&P
1 Equip. Oper. (medium)	$53.75	$430.00	$81.05	$648.40	$49.12	$74.27
.5 Laborer	39.85	159.40	60.70	242.80		
1 F.E. Loader, W.M., 1 C.Y.		290.80		319.88	24.23	26.66
12 L.H., Daily Totals		$880.20		$1211.08	$73.35	$100.92

Crew B-10S	Hr.	Daily	Hr.	Daily	Bare Costs	Incl. O&P
1 Equip. Oper. (medium)	$53.75	$430.00	$81.05	$648.40	$49.12	$74.27
.5 Laborer	39.85	159.40	60.70	242.80		
1 F.E. Loader, W.M., 1.5 C.Y.		342.80		377.08	28.57	31.42
12 L.H., Daily Totals		$932.20		$1268.28	$77.68	$105.69

Crew B-10T	Hr.	Daily	Hr.	Daily	Bare Costs	Incl. O&P
1 Equip. Oper. (medium)	$53.75	$430.00	$81.05	$648.40	$49.12	$74.27
.5 Laborer	39.85	159.40	60.70	242.80		
1 F.E. Loader, W.M., 2.5 C.Y.		523.20		575.52	43.60	47.96
12 L.H., Daily Totals		$1112.60		$1466.72	$92.72	$122.23

Crew B-10U	Hr.	Daily	Hr.	Daily	Bare Costs	Incl. O&P
1 Equip. Oper. (medium)	$53.75	$430.00	$81.05	$648.40	$49.12	$74.27
.5 Laborer	39.85	159.40	60.70	242.80		
1 F.E. Loader, W.M., 5.5 C.Y.		982.20		1080.42	81.85	90.03
12 L.H., Daily Totals		$1571.60		$1971.62	$130.97	$164.30

Crew B-10V	Hr.	Daily	Hr.	Daily	Bare Costs	Incl. O&P
1 Equip. Oper. (medium)	$53.75	$430.00	$81.05	$648.40	$49.12	$74.27
.5 Laborer	39.85	159.40	60.70	242.80		
1 Dozer, 700 H.P.		4999.00		5498.90	416.58	458.24
12 L.H., Daily Totals		$5588.40		$6390.10	$465.70	$532.51

Crew B-10W	Hr.	Daily	Hr.	Daily	Bare Costs	Incl. O&P
1 Equip. Oper. (medium)	$53.75	$430.00	$81.05	$648.40	$49.12	$74.27
.5 Laborer	39.85	159.40	60.70	242.80		
1 Dozer, 105 H.P.		608.00		668.80	50.67	55.73
12 L.H., Daily Totals		$1197.40		$1560.00	$99.78	$130.00

Crew B-10X	Hr.	Daily	Hr.	Daily	Bare Costs	Incl. O&P
1 Equip. Oper. (medium)	$53.75	$430.00	$81.05	$648.40	$49.12	$74.27
.5 Laborer	39.85	159.40	60.70	242.80		
1 Dozer, 410 H.P.		2291.00		2520.10	190.92	210.01
12 L.H., Daily Totals		$2880.40		$3411.30	$240.03	$284.27

Crew B-10Y	Hr.	Daily	Hr.	Daily	Bare Costs	Incl. O&P
1 Equip. Oper. (medium)	$53.75	$430.00	$81.05	$648.40	$49.12	$74.27
.5 Laborer	39.85	159.40	60.70	242.80		
1 Vibr. Roller, Towed, 12 Ton		576.40		634.04	48.03	52.84
12 L.H., Daily Totals		$1165.80		$1525.24	$97.15	$127.10

Crews - Standard

Crew B-11A

Crew B-11A	Hr.	Daily	Hr.	Daily	Bare Costs	Incl. O&P
1 Equipment Oper. (med.)	$53.75	$430.00	$81.05	$648.40	$46.80	$70.88
1 Laborer	39.85	318.80	60.70	485.60		
1 Dozer, 200 H.P.		1273.00		1400.30	79.56	87.52
16 L.H., Daily Totals		$2021.80		$2534.30	$126.36	$158.39

Crew B-11B

Crew B-11B	Hr.	Daily	Hr.	Daily	Bare Costs	Incl. O&P
1 Equipment Oper. (light)	$51.30	$410.40	$77.35	$618.80	$45.58	$69.03
1 Laborer	39.85	318.80	60.70	485.60		
1 Air Powered Tamper		28.30		31.13		
1 Air Compressor, 365 cfm		211.40		232.54		
2 -50' Air Hoses, 1.5"		45.40		49.94	17.82	19.60
16 L.H., Daily Totals		$1014.30		$1418.01	$63.39	$88.63

Crew B-11C

Crew B-11C	Hr.	Daily	Hr.	Daily	Bare Costs	Incl. O&P
1 Equipment Oper. (med.)	$53.75	$430.00	$81.05	$648.40	$46.80	$70.88
1 Laborer	39.85	318.80	60.70	485.60		
1 Backhoe Loader, 48 H.P.		312.40		343.64	19.52	21.48
16 L.H., Daily Totals		$1061.20		$1477.64	$66.33	$92.35

Crew B-11J

Crew B-11J	Hr.	Daily	Hr.	Daily	Bare Costs	Incl. O&P
1 Equipment Oper. (med.)	$53.75	$430.00	$81.05	$648.40	$46.80	$70.88
1 Laborer	39.85	318.80	60.70	485.60		
1 Grader, 30,000 Lbs.		643.80		708.18		
1 Ripper, Beam & 1 Shank		87.80		96.58	45.73	50.30
16 L.H., Daily Totals		$1480.40		$1938.76	$92.53	$121.17

Crew B-11K

Crew B-11K	Hr.	Daily	Hr.	Daily	Bare Costs	Incl. O&P
1 Equipment Oper. (med.)	$53.75	$430.00	$81.05	$648.40	$46.80	$70.88
1 Laborer	39.85	318.80	60.70	485.60		
1 Trencher, Chain Type, 8' D		2048.00		2252.80	128.00	140.80
16 L.H., Daily Totals		$2796.80		$3386.80	$174.80	$211.68

Crew B-11L

Crew B-11L	Hr.	Daily	Hr.	Daily	Bare Costs	Incl. O&P
1 Equipment Oper. (med.)	$53.75	$430.00	$81.05	$648.40	$46.80	$70.88
1 Laborer	39.85	318.80	60.70	485.60		
1 Grader, 30,000 Lbs.		643.80		708.18	40.24	44.26
16 L.H., Daily Totals		$1392.60		$1842.18	$87.04	$115.14

Crew B-11M

Crew B-11M	Hr.	Daily	Hr.	Daily	Bare Costs	Incl. O&P
1 Equipment Oper. (med.)	$53.75	$430.00	$81.05	$648.40	$46.80	$70.88
1 Laborer	39.85	318.80	60.70	485.60		
1 Backhoe Loader, 80 H.P.		385.00		423.50	24.06	26.47
16 L.H., Daily Totals		$1133.80		$1557.50	$70.86	$97.34

Crew B-11N

Crew B-11N	Hr.	Daily	Hr.	Daily	Bare Costs	Incl. O&P
1 Labor Foreman (outside)	$41.85	$334.80	$63.75	$510.00	$47.26	$71.29
2 Equipment Operators (med.)	53.75	860.00	81.05	1296.80		
6 Truck Drivers (heavy)	46.00	2208.00	69.30	3326.40		
1 F.E. Loader, W.M., 5.5 C.Y.		982.20		1080.42		
1 Dozer, 410 H.P.		2291.00		2520.10		
6 Dump Trucks, Off Hwy., 50 Ton		9972.00		10969.20	183.96	202.36
72 L.H., Daily Totals		$16648.00		$19702.92	$231.22	$273.65

Crew B-11Q

Crew B-11Q	Hr.	Daily	Hr.	Daily	Bare Costs	Incl. O&P
1 Equipment Operator (med.)	$53.75	$430.00	$81.05	$648.40	$49.12	$74.27
.5 Laborer	39.85	159.40	60.70	242.80		
1 Dozer, 140 H.P.		845.60		930.16	70.47	77.51
12 L.H., Daily Totals		$1435.00		$1821.36	$119.58	$151.78

Crew B-11R

Crew B-11R	Hr.	Daily	Hr.	Daily	Bare Costs	Incl. O&P
1 Equipment Operator (med.)	$53.75	$430.00	$81.05	$648.40	$49.12	$74.27
.5 Laborer	39.85	159.40	60.70	242.80		
1 Dozer, 200 H.P.		1273.00		1400.30	106.08	116.69
12 L.H., Daily Totals		$1862.40		$2291.50	$155.20	$190.96

Crew B-11S

Crew B-11S	Hr.	Daily	Hr.	Daily	Bare Costs	Incl. O&P
1 Equipment Operator (med.)	$53.75	$430.00	$81.05	$648.40	$49.12	$74.27
.5 Laborer	39.85	159.40	60.70	242.80		
1 Dozer, 300 H.P.		1829.00		2011.90		
1 Ripper, Beam & 1 Shank		87.80		96.58	159.73	175.71
12 L.H., Daily Totals		$2506.20		$2999.68	$208.85	$249.97

Crew B-11T

Crew B-11T	Hr.	Daily	Hr.	Daily	Bare Costs	Incl. O&P
1 Equipment Operator (med.)	$53.75	$430.00	$81.05	$648.40	$49.12	$74.27
.5 Laborer	39.85	159.40	60.70	242.80		
1 Dozer, 410 H.P.		2291.00		2520.10		
1 Ripper, Beam & 2 Shanks		132.60		145.86	201.97	222.16
12 L.H., Daily Totals		$3013.00		$3557.16	$251.08	$296.43

Crew B-11U

Crew B-11U	Hr.	Daily	Hr.	Daily	Bare Costs	Incl. O&P
1 Equipment Operator (med.)	$53.75	$430.00	$81.05	$648.40	$49.12	$74.27
.5 Laborer	39.85	159.40	60.70	242.80		
1 Dozer, 520 H.P.		2818.00		3099.80	234.83	258.32
12 L.H., Daily Totals		$3407.40		$3991.00	$283.95	$332.58

Crew B-11V

Crew B-11V	Hr.	Daily	Hr.	Daily	Bare Costs	Incl. O&P
3 Laborers	$39.85	$956.40	$60.70	$1456.80	$39.85	$60.70
1 Roller, 2-Drum, W.B., 7.5 H.P.		179.60		197.56	7.48	8.23
24 L.H., Daily Totals		$1136.00		$1654.36	$47.33	$68.93

Crew B-11W

Crew B-11W	Hr.	Daily	Hr.	Daily	Bare Costs	Incl. O&P
1 Equipment Operator (med.)	$53.75	$430.00	$81.05	$648.40	$46.13	$69.56
1 Common Laborer	39.85	318.80	60.70	485.60		
10 Truck Drivers (heavy)	46.00	3680.00	69.30	5544.00		
1 Dozer, 200 H.P.		1273.00		1400.30		
1 Vibratory Roller, Towed, 23 Ton		423.00		465.30		
10 Dump Trucks, 8 C.Y., 220 H.P.		3386.00		3724.60	52.94	58.23
96 L.H., Daily Totals		$9510.80		$12268.20	$99.07	$127.79

Crew B-11Y

Crew B-11Y	Hr.	Daily	Hr.	Daily	Bare Costs	Incl. O&P
1 Labor Foreman (outside)	$41.85	$334.80	$63.75	$510.00	$44.71	$67.82
5 Common Laborers	39.85	1594.00	60.70	2428.00		
3 Equipment Operators (med.)	53.75	1290.00	81.05	1945.20		
1 Dozer, 80 H.P.		464.40		510.84		
2 Rollers, 2-Drums, W.B., 7.5 H.P.		359.20		395.12		
4 Vibrating Plates, Gas, 21"		162.40		178.64	13.69	15.06
72 L.H., Daily Totals		$4204.80		$5967.80	$58.40	$82.89

Crew B-12A

Crew B-12A	Hr.	Daily	Hr.	Daily	Bare Costs	Incl. O&P
1 Equip. Oper. (crane)	$56.10	$448.80	$84.60	$676.80	$47.98	$72.65
1 Laborer	39.85	318.80	60.70	485.60		
1 Hyd. Excavator, 1 C.Y.		742.20		816.42	46.39	51.03
16 L.H., Daily Totals		$1509.80		$1978.82	$94.36	$123.68

Crew B-12B

Crew B-12B	Hr.	Daily	Hr.	Daily	Bare Costs	Incl. O&P
1 Equip. Oper. (crane)	$56.10	$448.80	$84.60	$676.80	$47.98	$72.65
1 Laborer	39.85	318.80	60.70	485.60		
1 Hyd. Excavator, 1.5 C.Y.		893.60		982.96	55.85	61.44
16 L.H., Daily Totals		$1661.20		$2145.36	$103.83	$134.09

For customer support on your Electrical Costs with RSMeans data, call 800.448.8182.

481

Crew No.	Bare Costs Hr.	Bare Costs Daily	Incl. Subs O&P Hr.	Incl. Subs O&P Daily	Cost Per Labor-Hour Bare Costs	Cost Per Labor-Hour Incl. O&P
Crew B-12C					Bare Costs	Incl. O&P
1 Equip. Oper. (crane)	$56.10	$448.80	$84.60	$676.80	$47.98	$72.65
1 Laborer	39.85	318.80	60.70	485.60		
1 Hyd. Excavator, 2 C.Y.		1052.00		1157.20	65.75	72.33
16 L.H., Daily Totals		$1819.60		$2319.60	$113.72	$144.97
Crew B-12D	Hr.	Daily	Hr.	Daily	Bare Costs	Incl. O&P
1 Equip. Oper. (crane)	$56.10	$448.80	$84.60	$676.80	$47.98	$72.65
1 Laborer	39.85	318.80	60.70	485.60		
1 Hyd. Excavator, 3.5 C.Y.		2195.00		2414.50	137.19	150.91
16 L.H., Daily Totals		$2962.60		$3576.90	$185.16	$223.56
Crew B-12E	Hr.	Daily	Hr.	Daily	Bare Costs	Incl. O&P
1 Equip. Oper. (crane)	$56.10	$448.80	$84.60	$676.80	$47.98	$72.65
1 Laborer	39.85	318.80	60.70	485.60		
1 Hyd. Excavator, .5 C.Y.		435.40		478.94	27.21	29.93
16 L.H., Daily Totals		$1203.00		$1641.34	$75.19	$102.58
Crew B-12F	Hr.	Daily	Hr.	Daily	Bare Costs	Incl. O&P
1 Equip. Oper. (crane)	$56.10	$448.80	$84.60	$676.80	$47.98	$72.65
1 Laborer	39.85	318.80	60.70	485.60		
1 Hyd. Excavator, .75 C.Y.		669.60		736.56	41.85	46.03
16 L.H., Daily Totals		$1437.20		$1898.96	$89.83	$118.69
Crew B-12G	Hr.	Daily	Hr.	Daily	Bare Costs	Incl. O&P
1 Equip. Oper. (crane)	$56.10	$448.80	$84.60	$676.80	$47.98	$72.65
1 Laborer	39.85	318.80	60.70	485.60		
1 Crawler Crane, 15 Ton		828.80		911.68		
1 Clamshell Bucket, .5 C.Y.		41.00		45.10	54.36	59.80
16 L.H., Daily Totals		$1637.40		$2119.18	$102.34	$132.45
Crew B-12H	Hr.	Daily	Hr.	Daily	Bare Costs	Incl. O&P
1 Equip. Oper. (crane)	$56.10	$448.80	$84.60	$676.80	$47.98	$72.65
1 Laborer	39.85	318.80	60.70	485.60		
1 Crawler Crane, 25 Ton		1348.00		1482.80		
1 Clamshell Bucket, 1 C.Y.		50.60		55.66	87.41	96.15
16 L.H., Daily Totals		$2166.20		$2700.86	$135.39	$168.80
Crew B-12I	Hr.	Daily	Hr.	Daily	Bare Costs	Incl. O&P
1 Equip. Oper. (crane)	$56.10	$448.80	$84.60	$676.80	$47.98	$72.65
1 Laborer	39.85	318.80	60.70	485.60		
1 Crawler Crane, 20 Ton		1060.00		1166.00		
1 Dragline Bucket, .75 C.Y.		21.80		23.98	67.61	74.37
16 L.H., Daily Totals		$1849.40		$2352.38	$115.59	$147.02
Crew B-12J	Hr.	Daily	Hr.	Daily	Bare Costs	Incl. O&P
1 Equip. Oper. (crane)	$56.10	$448.80	$84.60	$676.80	$47.98	$72.65
1 Laborer	39.85	318.80	60.70	485.60		
1 Gradall, 5/8 C.Y.		869.60		956.56	54.35	59.78
16 L.H., Daily Totals		$1637.20		$2118.96	$102.33	$132.44
Crew B-12K	Hr.	Daily	Hr.	Daily	Bare Costs	Incl. O&P
1 Equip. Oper. (crane)	$56.10	$448.80	$84.60	$676.80	$47.98	$72.65
1 Laborer	39.85	318.80	60.70	485.60		
1 Gradall, 3 Ton, 1 C.Y.		1243.00		1367.30	77.69	85.46
16 L.H., Daily Totals		$2010.60		$2529.70	$125.66	$158.11

Crew No.	Bare Costs Hr.	Bare Costs Daily	Incl. Subs O&P Hr.	Incl. Subs O&P Daily	Cost Per Labor-Hour Bare Costs	Cost Per Labor-Hour Incl. O&P
Crew B-12L	Hr.	Daily	Hr.	Daily	Bare Costs	Incl. O&P
1 Equip. Oper. (crane)	$56.10	$448.80	$84.60	$676.80	$47.98	$72.65
1 Laborer	39.85	318.80	60.70	485.60		
1 Crawler Crane, 15 Ton		828.80		911.68		
1 F.E. Attachment, .5 C.Y.		64.60		71.06	55.84	61.42
16 L.H., Daily Totals		$1661.00		$2145.14	$103.81	$134.07
Crew B-12M	Hr.	Daily	Hr.	Daily	Bare Costs	Incl. O&P
1 Equip. Oper. (crane)	$56.10	$448.80	$84.60	$676.80	$47.98	$72.65
1 Laborer	39.85	318.80	60.70	485.60		
1 Crawler Crane, 20 Ton		1060.00		1166.00		
1 F.E. Attachment, .75 C.Y.		69.00		75.90	70.56	77.62
16 L.H., Daily Totals		$1896.60		$2404.30	$118.54	$150.27
Crew B-12N	Hr.	Daily	Hr.	Daily	Bare Costs	Incl. O&P
1 Equip. Oper. (crane)	$56.10	$448.80	$84.60	$676.80	$47.98	$72.65
1 Laborer	39.85	318.80	60.70	485.60		
1 Crawler Crane, 25 Ton		1348.00		1482.80		
1 F.E. Attachment, 1 C.Y.		75.80		83.38	88.99	97.89
16 L.H., Daily Totals		$2191.40		$2728.58	$136.96	$170.54
Crew B-12O	Hr.	Daily	Hr.	Daily	Bare Costs	Incl. O&P
1 Equip. Oper. (crane)	$56.10	$448.80	$84.60	$676.80	$47.98	$72.65
1 Laborer	39.85	318.80	60.70	485.60		
1 Crawler Crane, 40 Ton		1370.00		1507.00		
1 F.E. Attachment, 1.5 C.Y.		85.00		93.50	90.94	100.03
16 L.H., Daily Totals		$2222.60		$2762.90	$138.91	$172.68
Crew B-12P	Hr.	Daily	Hr.	Daily	Bare Costs	Incl. O&P
1 Equip. Oper. (crane)	$56.10	$448.80	$84.60	$676.80	$47.98	$72.65
1 Laborer	39.85	318.80	60.70	485.60		
1 Crawler Crane, 40 Ton		1370.00		1507.00		
1 Dragline Bucket, 1.5 C.Y.		35.40		38.94	87.84	96.62
16 L.H., Daily Totals		$2173.00		$2708.34	$135.81	$169.27
Crew B-12Q	Hr.	Daily	Hr.	Daily	Bare Costs	Incl. O&P
1 Equip. Oper. (crane)	$56.10	$448.80	$84.60	$676.80	$47.98	$72.65
1 Laborer	39.85	318.80	60.70	485.60		
1 Hyd. Excavator, 5/8 C.Y.		576.80		634.48	36.05	39.66
16 L.H., Daily Totals		$1344.40		$1796.88	$84.03	$112.31
Crew B-12S	Hr.	Daily	Hr.	Daily	Bare Costs	Incl. O&P
1 Equip. Oper. (crane)	$56.10	$448.80	$84.60	$676.80	$47.98	$72.65
1 Laborer	39.85	318.80	60.70	485.60		
1 Hyd. Excavator, 2.5 C.Y.		1411.00		1552.10	88.19	97.01
16 L.H., Daily Totals		$2178.60		$2714.50	$136.16	$169.66
Crew B-12T	Hr.	Daily	Hr.	Daily	Bare Costs	Incl. O&P
1 Equip. Oper. (crane)	$56.10	$448.80	$84.60	$676.80	$47.98	$72.65
1 Laborer	39.85	318.80	60.70	485.60		
1 Crawler Crane, 75 Ton		1708.00		1878.80		
1 F.E. Attachment, 3 C.Y.		108.80		119.68	113.55	124.91
16 L.H., Daily Totals		$2584.40		$3160.88	$161.53	$197.56
Crew B-12V	Hr.	Daily	Hr.	Daily	Bare Costs	Incl. O&P
1 Equip. Oper. (crane)	$56.10	$448.80	$84.60	$676.80	$47.98	$72.65
1 Laborer	39.85	318.80	60.70	485.60		
1 Crawler Crane, 75 Ton		1708.00		1878.80		
1 Dragline Bucket, 3 C.Y.		55.80		61.38	110.24	121.26
16 L.H., Daily Totals		$2531.40		$3102.58	$158.21	$193.91

Crew No.	Bare Costs Hr.	Daily	Incl. Subs O&P Hr.	Daily	Cost Per Labor-Hour Bare Costs	Incl. O&P
Crew B-12Y	Hr.	Daily	Hr.	Daily	Bare Costs	Incl. O&P
1 Equip. Oper. (crane)	$56.10	$448.80	$84.60	$676.80	$45.27	$68.67
2 Laborers	39.85	637.60	60.70	971.20		
1 Hyd. Excavator, 3.5 C.Y.		2195.00		2414.50	91.46	100.60
24 L.H., Daily Totals		$3281.40		$4062.50	$136.72	$169.27
Crew B-12Z	Hr.	Daily	Hr.	Daily	Bare Costs	Incl. O&P
1 Equip. Oper. (crane)	$56.10	$448.80	$84.60	$676.80	$45.27	$68.67
2 Laborers	39.85	637.60	60.70	971.20		
1 Hyd. Excavator, 2.5 C.Y.		1411.00		1552.10	58.79	64.67
24 L.H., Daily Totals		$2497.40		$3200.10	$104.06	$133.34
Crew B-13	Hr.	Daily	Hr.	Daily	Bare Costs	Incl. O&P
1 Labor Foreman (outside)	$41.85	$334.80	$63.75	$510.00	$43.71	$66.35
4 Laborers	39.85	1275.20	60.70	1942.40		
1 Equip. Oper. (crane)	56.10	448.80	84.60	676.80		
1 Equip. Oper. (oiler)	48.60	388.80	73.30	586.40		
1 Hyd. Crane, 25 Ton		590.60		649.66	10.55	11.60
56 L.H., Daily Totals		$3038.20		$4365.26	$54.25	$77.95
Crew B-13A	Hr.	Daily	Hr.	Daily	Bare Costs	Incl. O&P
1 Labor Foreman (outside)	$41.85	$334.80	$63.75	$510.00	$45.86	$69.41
2 Laborers	39.85	637.60	60.70	971.20		
2 Equipment Operators (med.)	53.75	860.00	81.05	1296.80		
2 Truck Drivers (heavy)	46.00	736.00	69.30	1108.80		
1 Crawler Crane, 75 Ton		1708.00		1878.80		
1 Crawler Loader, 4 C.Y.		1459.00		1604.90		
2 Dump Trucks, 8 C.Y., 220 H.P.		677.20		744.92	68.65	75.51
56 L.H., Daily Totals		$6412.60		$8115.42	$114.51	$144.92
Crew B-13B	Hr.	Daily	Hr.	Daily	Bare Costs	Incl. O&P
1 Labor Foreman (outside)	$41.85	$334.80	$63.75	$510.00	$43.71	$66.35
4 Laborers	39.85	1275.20	60.70	1942.40		
1 Equip. Oper. (crane)	56.10	448.80	84.60	676.80		
1 Equip. Oper. (oiler)	48.60	388.80	73.30	586.40		
1 Hyd. Crane, 55 Ton		993.80		1093.18	17.75	19.52
56 L.H., Daily Totals		$3441.40		$4808.78	$61.45	$85.87
Crew B-13C	Hr.	Daily	Hr.	Daily	Bare Costs	Incl. O&P
1 Labor Foreman (outside)	$41.85	$334.80	$63.75	$510.00	$43.71	$66.35
4 Laborers	39.85	1275.20	60.70	1942.40		
1 Equip. Oper. (crane)	56.10	448.80	84.60	676.80		
1 Equip. Oper. (oiler)	48.60	388.80	73.30	586.40		
1 Crawler Crane, 100 Ton		1872.00		2059.20	33.43	36.77
56 L.H., Daily Totals		$4319.60		$5774.80	$77.14	$103.12
Crew B-13D	Hr.	Daily	Hr.	Daily	Bare Costs	Incl. O&P
1 Laborer	$39.85	$318.80	$60.70	$485.60	$47.98	$72.65
1 Equip. Oper. (crane)	56.10	448.80	84.60	676.80		
1 Hyd. Excavator, 1 C.Y.		742.20		816.42		
1 Trench Box		81.00		89.10	51.45	56.59
16 L.H., Daily Totals		$1590.80		$2067.92	$99.42	$129.25
Crew B-13E	Hr.	Daily	Hr.	Daily	Bare Costs	Incl. O&P
1 Laborer	$39.85	$318.80	$60.70	$485.60	$47.98	$72.65
1 Equip. Oper. (crane)	56.10	448.80	84.60	676.80		
1 Hyd. Excavator, 1.5 C.Y.		893.60		982.96		
1 Trench Box		81.00		89.10	60.91	67.00
16 L.H., Daily Totals		$1742.20		$2234.46	$108.89	$139.65

Crew No.	Bare Costs Hr.	Daily	Incl. Subs O&P Hr.	Daily	Cost Per Labor-Hour Bare Costs	Incl. O&P
Crew B-13F	Hr.	Daily	Hr.	Daily	Bare Costs	Incl. O&P
1 Laborer	$39.85	$318.80	$60.70	$485.60	$47.98	$72.65
1 Equip. Oper. (crane)	56.10	448.80	84.60	676.80		
1 Hyd. Excavator, 3.5 C.Y.		2195.00		2414.50		
1 Trench Box		81.00		89.10	142.25	156.47
16 L.H., Daily Totals		$3043.60		$3666.00	$190.22	$229.13
Crew B-13G	Hr.	Daily	Hr.	Daily	Bare Costs	Incl. O&P
1 Laborer	$39.85	$318.80	$60.70	$485.60	$47.98	$72.65
1 Equip. Oper. (crane)	56.10	448.80	84.60	676.80		
1 Hyd. Excavator, .75 C.Y.		669.60		736.56		
1 Trench Box		81.00		89.10	46.91	51.60
16 L.H., Daily Totals		$1518.20		$1988.06	$94.89	$124.25
Crew B-13H	Hr.	Daily	Hr.	Daily	Bare Costs	Incl. O&P
1 Laborer	$39.85	$318.80	$60.70	$485.60	$47.98	$72.65
1 Equip. Oper. (crane)	56.10	448.80	84.60	676.80		
1 Gradall, 5/8 C.Y.		869.60		956.56		
1 Trench Box		81.00		89.10	59.41	65.35
16 L.H., Daily Totals		$1718.20		$2208.06	$107.39	$138.00
Crew B-13I	Hr.	Daily	Hr.	Daily	Bare Costs	Incl. O&P
1 Laborer	$39.85	$318.80	$60.70	$485.60	$47.98	$72.65
1 Equip. Oper. (crane)	56.10	448.80	84.60	676.80		
1 Gradall, 3 Ton, 1 C.Y.		1243.00		1367.30		
1 Trench Box		81.00		89.10	82.75	91.03
16 L.H., Daily Totals		$2091.60		$2618.80	$130.72	$163.68
Crew B-13J	Hr.	Daily	Hr.	Daily	Bare Costs	Incl. O&P
1 Laborer	$39.85	$318.80	$60.70	$485.60	$47.98	$72.65
1 Equip. Oper. (crane)	56.10	448.80	84.60	676.80		
1 Hyd. Excavator, 2.5 C.Y.		1411.00		1552.10		
1 Trench Box		81.00		89.10	93.25	102.58
16 L.H., Daily Totals		$2259.60		$2803.60	$141.22	$175.22
Crew B-13K	Hr.	Daily	Hr.	Daily	Bare Costs	Incl. O&P
2 Equip. Opers. (crane)	$56.10	$897.60	$84.60	$1353.60	$56.10	$84.60
1 Hyd. Excavator, .75 C.Y.		669.60		736.56		
1 Hyd. Hammer, 4000 ft-lb		324.60		357.06		
1 Hyd. Excavator, .75 C.Y.		669.60		736.56	103.99	114.39
16 L.H., Daily Totals		$2561.40		$3183.78	$160.09	$198.99
Crew B-13L	Hr.	Daily	Hr.	Daily	Bare Costs	Incl. O&P
2 Equip. Opers. (crane)	$56.10	$897.60	$84.60	$1353.60	$56.10	$84.60
1 Hyd. Excavator, 1.5 C.Y.		893.60		982.96		
1 Hyd. Hammer, 5000 ft-lb		394.00		433.40		
1 Hyd. Excavator, .75 C.Y.		669.60		736.56	122.33	134.56
16 L.H., Daily Totals		$2854.80		$3506.52	$178.43	$219.16
Crew B-13M	Hr.	Daily	Hr.	Daily	Bare Costs	Incl. O&P
2 Equip. Opers. (crane)	$56.10	$897.60	$84.60	$1353.60	$56.10	$84.60
1 Hyd. Excavator, 2.5 C.Y.		1411.00		1552.10		
1 Hyd. Hammer, 8000 ft-lb		576.80		634.48		
1 Hyd. Excavator, 1.5 C.Y.		893.60		982.96	180.09	198.10
16 L.H., Daily Totals		$3779.00		$4523.14	$236.19	$282.70

Crews - Standard

Crew No.	Bare Costs Hr.	Daily	Incl. Subs O&P Hr.	Daily	Cost Per Labor-Hour Bare Costs	Incl. O&P
Crew B-13N						
2 Equip. Opers. (crane)	$56.10	$897.60	$84.60	$1353.60	$56.10	$84.60
1 Hyd. Excavator, 3.5 C.Y.		2195.00		2414.50		
1 Hyd. Hammer, 12,000 ft-lb		669.20		736.12		
1 Hyd. Excavator, 1.5 C.Y.		893.60		982.96	234.86	258.35
16 L.H., Daily Totals		$4655.40		$5487.18	$290.96	$342.95
Crew B-14						
1 Labor Foreman (outside)	$41.85	$334.80	$63.75	$510.00	$42.09	$63.98
4 Laborers	39.85	1275.20	60.70	1942.40		
1 Equip. Oper. (light)	51.30	410.40	77.35	618.80		
1 Backhoe Loader, 48 H.P.		312.40		343.64	6.51	7.16
48 L.H., Daily Totals		$2332.80		$3414.84	$48.60	$71.14
Crew B-14A						
1 Equip. Oper. (crane)	$56.10	$448.80	$84.60	$676.80	$50.68	$76.63
.5 Laborer	39.85	159.40	60.70	242.80		
1 Hyd. Excavator, 4.5 C.Y.		2735.00		3008.50	227.92	250.71
12 L.H., Daily Totals		$3343.20		$3928.10	$278.60	$327.34
Crew B-14B						
1 Equip. Oper. (crane)	$56.10	$448.80	$84.60	$676.80	$50.68	$76.63
.5 Laborer	39.85	159.40	60.70	242.80		
1 Hyd. Excavator, 6 C.Y.		3487.00		3835.70	290.58	319.64
12 L.H., Daily Totals		$4095.20		$4755.30	$341.27	$396.27
Crew B-14C						
1 Equip. Oper. (crane)	$56.10	$448.80	$84.60	$676.80	$50.68	$76.63
.5 Laborer	39.85	159.40	60.70	242.80		
1 Hyd. Excavator, 7 C.Y.		3275.00		3602.50	272.92	300.21
12 L.H., Daily Totals		$3883.20		$4522.10	$323.60	$376.84
Crew B-14F						
1 Equip. Oper. (crane)	$56.10	$448.80	$84.60	$676.80	$50.68	$76.63
.5 Laborer	39.85	159.40	60.70	242.80		
1 Hyd. Shovel, 7 C.Y.		3880.00		4268.00	323.33	355.67
12 L.H., Daily Totals		$4488.20		$5187.60	$374.02	$432.30
Crew B-14G						
1 Equip. Oper. (crane)	$56.10	$448.80	$84.60	$676.80	$50.68	$76.63
.5 Laborer	39.85	159.40	60.70	242.80		
1 Hyd. Shovel, 12 C.Y.		5769.00		6345.90	480.75	528.83
12 L.H., Daily Totals		$6377.20		$7265.50	$531.43	$605.46
Crew B-14J						
1 Equip. Oper. (medium)	$53.75	$430.00	$81.05	$648.40	$49.12	$74.27
.5 Laborer	39.85	159.40	60.70	242.80		
1 F.E. Loader, 8 C.Y.		1780.00		1958.00	148.33	163.17
12 L.H., Daily Totals		$2369.40		$2849.20	$197.45	$237.43
Crew B-14K						
1 Equip. Oper. (medium)	$53.75	$430.00	$81.05	$648.40	$49.12	$74.27
.5 Laborer	39.85	159.40	60.70	242.80		
1 F.E. Loader, 10 C.Y.		2630.00		2893.00	219.17	241.08
12 L.H., Daily Totals		$3219.40		$3784.20	$268.28	$315.35
Crew B-15						
1 Equipment Oper. (med.)	$53.75	$430.00	$81.05	$648.40	$47.34	$71.43
.5 Laborer	39.85	159.40	60.70	242.80		
2 Truck Drivers (heavy)	46.00	736.00	69.30	1108.80		
2 Dump Trucks, 12 C.Y., 400 H.P.		1085.60		1194.16		
1 Dozer, 200 H.P.		1273.00		1400.30	84.24	92.66
28 L.H., Daily Totals		$3684.00		$4594.46	$131.57	$164.09
Crew B-16						
1 Labor Foreman (outside)	$41.85	$334.80	$63.75	$510.00	$41.89	$63.61
2 Laborers	39.85	637.60	60.70	971.20		
1 Truck Driver (heavy)	46.00	368.00	69.30	554.40		
1 Dump Truck, 12 C.Y., 400 H.P.		542.80		597.08	16.96	18.66
32 L.H., Daily Totals		$1883.20		$2632.68	$58.85	$82.27
Crew B-17						
2 Laborers	$39.85	$637.60	$60.70	$971.20	$44.25	$67.01
1 Equip. Oper. (light)	51.30	410.40	77.35	618.80		
1 Truck Driver (heavy)	46.00	368.00	69.30	554.40		
1 Backhoe Loader, 48 H.P.		312.40		343.64		
1 Dump Truck, 8 C.Y., 220 H.P.		338.60		372.46	20.34	22.38
32 L.H., Daily Totals		$2067.00		$2860.50	$64.59	$89.39
Crew B-17A						
2 Labor Foremen (outside)	$41.85	$669.60	$63.75	$1020.00	$42.95	$65.50
6 Laborers	39.85	1912.80	60.70	2913.60		
1 Skilled Worker Foreman (out)	54.35	434.80	83.20	665.60		
1 Skilled Worker	52.35	418.80	80.15	641.20		
80 L.H., Daily Totals		$3436.00		$5240.40	$42.95	$65.50
Crew B-17B						
2 Laborers	$39.85	$637.60	$60.70	$971.20	$44.25	$67.01
1 Equip. Oper. (light)	51.30	410.40	77.35	618.80		
1 Truck Driver (heavy)	46.00	368.00	69.30	554.40		
1 Backhoe Loader, 48 H.P.		312.40		343.64		
1 Dump Truck, 12 C.Y., 400 H.P.		542.80		597.08	26.73	29.40
32 L.H., Daily Totals		$2271.20		$3085.12	$70.97	$96.41
Crew B-18						
1 Labor Foreman (outside)	$41.85	$334.80	$63.75	$510.00	$40.52	$61.72
2 Laborers	39.85	637.60	60.70	971.20		
1 Vibrating Plate, Gas, 21"		40.60		44.66	1.69	1.86
24 L.H., Daily Totals		$1013.00		$1525.86	$42.21	$63.58
Crew B-19						
1 Pile Driver Foreman (outside)	$53.30	$426.40	$84.05	$672.40	$52.41	$81.27
4 Pile Drivers	51.30	1641.60	80.90	2588.80		
2 Equip. Oper. (crane)	56.10	897.60	84.60	1353.60		
1 Equip. Oper. (oiler)	48.60	388.80	73.30	586.40		
1 Crawler Crane, 40 Ton		1370.00		1507.00		
1 Lead, 90' High		131.60		144.76		
1 Hammer, Diesel, 22k ft-lb		427.00		469.70	30.13	33.15
64 L.H., Daily Totals		$5283.00		$7322.66	$82.55	$114.42

Crew No.	Bare Costs		Incl. Subs O&P		Cost Per Labor-Hour	
Crew B-19A	Hr.	Daily	Hr.	Daily	Bare Costs	Incl. O&P
1 Pile Driver Foreman (outside)	$53.30	$426.40	$84.05	$672.40	$52.41	$81.27
4 Pile Drivers	51.30	1641.60	80.90	2588.80		
2 Equip. Oper. (crane)	56.10	897.60	84.60	1353.60		
1 Equip. Oper. (oiler)	48.60	388.80	73.30	586.40		
1 Crawler Crane, 75 Ton		1708.00		1878.80		
1 Lead, 90' High		131.60		144.76		
1 Hammer, Diesel, 41k ft-lb		572.00		629.20	37.68	41.45
64 L.H., Daily Totals		$5766.00		$7853.96	$90.09	$122.72

Crew No.	Bare Costs		Incl. Subs O&P		Cost Per Labor-Hour	
Crew B-19B	Hr.	Daily	Hr.	Daily	Bare Costs	Incl. O&P
1 Pile Driver Foreman (outside)	$53.30	$426.40	$84.05	$672.40	$52.41	$81.27
4 Pile Drivers	51.30	1641.60	80.90	2588.80		
2 Equip. Oper. (crane)	56.10	897.60	84.60	1353.60		
1 Equip. Oper. (oiler)	48.60	388.80	73.30	586.40		
1 Crawler Crane, 40 Ton		1370.00		1507.00		
1 Lead, 90' High		131.60		144.76		
1 Hammer, Diesel, 22k ft-lb		427.00		469.70		
1 Barge, 400 Ton		835.40		918.94	43.19	47.51
64 L.H., Daily Totals		$6118.40		$8241.60	$95.60	$128.78

Crew No.	Bare Costs		Incl. Subs O&P		Cost Per Labor-Hour	
Crew B-19C	Hr.	Daily	Hr.	Daily	Bare Costs	Incl. O&P
1 Pile Driver Foreman (outside)	$53.30	$426.40	$84.05	$672.40	$52.41	$81.27
4 Pile Drivers	51.30	1641.60	80.90	2588.80		
2 Equip. Oper. (crane)	56.10	897.60	84.60	1353.60		
1 Equip. Oper. (oiler)	48.60	388.80	73.30	586.40		
1 Crawler Crane, 75 Ton		1708.00		1878.80		
1 Lead, 90' High		131.60		144.76		
1 Hammer, Diesel, 41k ft-lb		572.00		629.20		
1 Barge, 400 Ton		835.40		918.94	50.73	55.81
64 L.H., Daily Totals		$6601.40		$8772.90	$103.15	$137.08

Crew No.	Bare Costs		Incl. Subs O&P		Cost Per Labor-Hour	
Crew B-20	Hr.	Daily	Hr.	Daily	Bare Costs	Incl. O&P
1 Labor Foreman (outside)	$41.85	$334.80	$63.75	$510.00	$44.68	$68.20
1 Skilled Worker	52.35	418.80	80.15	641.20		
1 Laborer	39.85	318.80	60.70	485.60		
24 L.H., Daily Totals		$1072.40		$1636.80	$44.68	$68.20

Crew No.	Bare Costs		Incl. Subs O&P		Cost Per Labor-Hour	
Crew B-20A	Hr.	Daily	Hr.	Daily	Bare Costs	Incl. O&P
1 Labor Foreman (outside)	$41.85	$334.80	$63.75	$510.00	$48.39	$73.22
1 Laborer	39.85	318.80	60.70	485.60		
1 Plumber	62.15	497.20	93.60	748.80		
1 Plumber Apprentice	49.70	397.60	74.85	598.80		
32 L.H., Daily Totals		$1548.40		$2343.20	$48.39	$73.22

Crew No.	Bare Costs		Incl. Subs O&P		Cost Per Labor-Hour	
Crew B-21	Hr.	Daily	Hr.	Daily	Bare Costs	Incl. O&P
1 Labor Foreman (outside)	$41.85	$334.80	$63.75	$510.00	$46.31	$70.54
1 Skilled Worker	52.35	418.80	80.15	641.20		
1 Laborer	39.85	318.80	60.70	485.60		
.5 Equip. Oper. (crane)	56.10	224.40	84.60	338.40		
.5 S.P. Crane, 4x4, 5 Ton		129.50		142.45	4.63	5.09
28 L.H., Daily Totals		$1426.30		$2117.65	$50.94	$75.63

Crew No.	Bare Costs		Incl. Subs O&P		Cost Per Labor-Hour	
Crew B-21A	Hr.	Daily	Hr.	Daily	Bare Costs	Incl. O&P
1 Labor Foreman (outside)	$41.85	$334.80	$63.75	$510.00	$49.93	$75.50
1 Laborer	39.85	318.80	60.70	485.60		
1 Plumber	62.15	497.20	93.60	748.80		
1 Plumber Apprentice	49.70	397.60	74.85	598.80		
1 Equip. Oper. (crane)	56.10	448.80	84.60	676.80		
1 S.P. Crane, 4x4, 12 Ton		363.60		399.96	9.09	10.00
40 L.H., Daily Totals		$2360.80		$3419.96	$59.02	$85.50

Crew No.	Bare Costs		Incl. Subs O&P		Cost Per Labor-Hour	
Crew B-21B	Hr.	Daily	Hr.	Daily	Bare Costs	Incl. O&P
1 Labor Foreman (outside)	$41.85	$334.80	$63.75	$510.00	$43.50	$66.09
3 Laborers	39.85	956.40	60.70	1456.80		
1 Equip. Oper. (crane)	56.10	448.80	84.60	676.80		
1 Hyd. Crane, 12 Ton		496.00		545.60	12.40	13.64
40 L.H., Daily Totals		$2236.00		$3189.20	$55.90	$79.73

Crew No.	Bare Costs		Incl. Subs O&P		Cost Per Labor-Hour	
Crew B-21C	Hr.	Daily	Hr.	Daily	Bare Costs	Incl. O&P
1 Labor Foreman (outside)	$41.85	$334.80	$63.75	$510.00	$43.71	$66.35
4 Laborers	39.85	1275.20	60.70	1942.40		
1 Equip. Oper. (crane)	56.10	448.80	84.60	676.80		
1 Equip. Oper. (oiler)	48.60	388.80	73.30	586.40		
2 Cutting Torches		25.20		27.72		
2 Sets of Gases		336.00		369.60		
1 Lattice Boom Crane, 90 Ton		1687.00		1855.70	36.58	40.23
56 L.H., Daily Totals		$4495.80		$5968.62	$80.28	$106.58

Crew No.	Bare Costs		Incl. Subs O&P		Cost Per Labor-Hour	
Crew B-22	Hr.	Daily	Hr.	Daily	Bare Costs	Incl. O&P
1 Labor Foreman (outside)	$41.85	$334.80	$63.75	$510.00	$46.97	$71.48
1 Skilled Worker	52.35	418.80	80.15	641.20		
1 Laborer	39.85	318.80	60.70	485.60		
.75 Equip. Oper. (crane)	56.10	336.60	84.60	507.60		
.75 S.P. Crane, 4x4, 5 Ton		194.25		213.68	6.47	7.12
30 L.H., Daily Totals		$1603.25		$2358.07	$53.44	$78.60

Crew No.	Bare Costs		Incl. Subs O&P		Cost Per Labor-Hour	
Crew B-22A	Hr.	Daily	Hr.	Daily	Bare Costs	Incl. O&P
1 Labor Foreman (outside)	$41.85	$334.80	$63.75	$510.00	$46.00	$69.98
1 Skilled Worker	52.35	418.80	80.15	641.20		
2 Laborers	39.85	637.60	60.70	971.20		
1 Equipment Operator, Crane	56.10	448.80	84.60	676.80		
1 S.P. Crane, 4x4, 5 Ton		259.00		284.90		
1 Butt Fusion Mach., 4"-12" diam.		425.70		468.27	17.12	18.83
40 L.H., Daily Totals		$2524.70		$3552.37	$63.12	$88.81

Crew No.	Bare Costs		Incl. Subs O&P		Cost Per Labor-Hour	
Crew B-22B	Hr.	Daily	Hr.	Daily	Bare Costs	Incl. O&P
1 Labor Foreman (outside)	$41.85	$334.80	$63.75	$510.00	$46.00	$69.98
1 Skilled Worker	52.35	418.80	80.15	641.20		
2 Laborers	39.85	637.60	60.70	971.20		
1 Equip. Oper. (crane)	56.10	448.80	84.60	676.80		
1 S.P. Crane, 4x4, 5 Ton		259.00		284.90		
1 Butt Fusion Mach., 8"-24" diam.		1908.00		2098.80	54.17	59.59
40 L.H., Daily Totals		$4007.00		$5182.90	$100.18	$129.57

Crew No.	Bare Costs		Incl. Subs O&P		Cost Per Labor-Hour	
Crew B-22C	Hr.	Daily	Hr.	Daily	Bare Costs	Incl. O&P
1 Skilled Worker	$52.35	$418.80	$80.15	$641.20	$46.10	$70.42
1 Laborer	39.85	318.80	60.70	485.60		
1 Butt Fusion Mach., 2"-8" diam.		121.05		133.16	7.57	8.32
16 L.H., Daily Totals		$858.65		$1259.95	$53.67	$78.75

Crew No.	Bare Costs		Incl. Subs O&P		Cost Per Labor-Hour	
Crew B-23	Hr.	Daily	Hr.	Daily	Bare Costs	Incl. O&P
1 Labor Foreman (outside)	$41.85	$334.80	$63.75	$510.00	$40.25	$61.31
4 Laborers	39.85	1275.20	60.70	1942.40		
1 Drill Rig, Truck-Mounted		2444.00		2688.40		
1 Flatbed Truck, Gas, 3 Ton		238.00		261.80	67.05	73.75
40 L.H., Daily Totals		$4292.00		$5402.60	$107.30	$135.07

For customer support on your Electrical Costs with RSMeans data, call 800.448.8182.

485

Crew No.	Bare Costs		Incl. Subs O&P		Cost Per Labor-Hour	
					Bare Costs	Incl. O&P
Crew B-23A	Hr.	Daily	Hr.	Daily		
1 Labor Foreman (outside)	$41.85	$334.80	$63.75	$510.00	$45.15	$68.50
1 Laborer	39.85	318.80	60.70	485.60		
1 Equip. Oper. (medium)	53.75	430.00	81.05	648.40		
1 Drill Rig, Truck-Mounted		2444.00		2688.40		
1 Pickup Truck, 3/4 Ton		115.20		126.72	106.63	117.30
24 L.H., Daily Totals		$3642.80		$4459.12	$151.78	$185.80

Crew No.	Bare Costs		Incl. Subs O&P		Cost Per Labor-Hour	
					Bare Costs	Incl. O&P
Crew B-23B	Hr.	Daily	Hr.	Daily		
1 Labor Foreman (outside)	$41.85	$334.80	$63.75	$510.00	$45.15	$68.50
1 Laborer	39.85	318.80	60.70	485.60		
1 Equip. Oper. (medium)	53.75	430.00	81.05	648.40		
1 Drill Rig, Truck-Mounted		2444.00		2688.40		
1 Pickup Truck, 3/4 Ton		115.20		126.72		
1 Centr. Water Pump, 6"		296.40		326.04	118.98	130.88
24 L.H., Daily Totals		$3939.20		$4785.16	$164.13	$199.38

Crew B-24	Hr.	Daily	Hr.	Daily	Bare Costs	Incl. O&P
1 Cement Finisher	$47.55	$380.40	$70.45	$563.60	$46.03	$69.45
1 Laborer	39.85	318.80	60.70	485.60		
1 Carpenter	50.70	405.60	77.20	617.60		
24 L.H., Daily Totals		$1104.80		$1666.80	$46.03	$69.45

Crew B-25	Hr.	Daily	Hr.	Daily	Bare Costs	Incl. O&P
1 Labor Foreman (outside)	$41.85	$334.80	$63.75	$510.00	$43.82	$66.53
7 Laborers	39.85	2231.60	60.70	3399.20		
3 Equip. Oper. (medium)	53.75	1290.00	81.05	1945.20		
1 Asphalt Paver, 130 H.P.		2123.00		2335.30		
1 Tandem Roller, 10 Ton		231.40		254.54		
1 Roller, Pneum. Whl., 12 Ton		338.20		372.02	30.60	33.66
88 L.H., Daily Totals		$6549.00		$8816.26	$74.42	$100.18

Crew B-25B	Hr.	Daily	Hr.	Daily	Bare Costs	Incl. O&P
1 Labor Foreman (outside)	$41.85	$334.80	$63.75	$510.00	$44.65	$67.74
7 Laborers	39.85	2231.60	60.70	3399.20		
4 Equip. Oper. (medium)	53.75	1720.00	81.05	2593.60		
1 Asphalt Paver, 130 H.P.		2123.00		2335.30		
2 Tandem Rollers, 10 Ton		462.80		509.08		
1 Roller, Pneum. Whl., 12 Ton		338.20		372.02	30.46	33.50
96 L.H., Daily Totals		$7210.40		$9719.20	$75.11	$101.24

Crew B-25C	Hr.	Daily	Hr.	Daily	Bare Costs	Incl. O&P
1 Labor Foreman (outside)	$41.85	$334.80	$63.75	$510.00	$44.82	$67.99
3 Laborers	39.85	956.40	60.70	1456.80		
2 Equip. Oper. (medium)	53.75	860.00	81.05	1296.80		
1 Asphalt Paver, 130 H.P.		2123.00		2335.30		
1 Tandem Roller, 10 Ton		231.40		254.54	49.05	53.95
48 L.H., Daily Totals		$4505.60		$5853.44	$93.87	$121.95

Crew B-25D	Hr.	Daily	Hr.	Daily	Bare Costs	Incl. O&P
1 Labor Foreman (outside)	$41.85	$334.80	$63.75	$510.00	$45.02	$68.28
3 Laborers	39.85	956.40	60.70	1456.80		
2.125 Equip. Oper. (medium)	53.75	913.75	81.05	1377.85		
.125 Truck Driver (heavy)	46.00	46.00	69.30	69.30		
.125 Truck Tractor, 6x4, 380 H.P.		59.52		65.48		
.125 Dist. Tanker, 3000 Gallon		40.85		44.94		
1 Asphalt Paver, 130 H.P.		2123.00		2335.30		
1 Tandem Roller, 10 Ton		231.40		254.54	49.10	54.01
50 L.H., Daily Totals		$4705.73		$6114.20	$94.11	$122.28

Crew B-25E	Hr.	Daily	Hr.	Daily	Bare Costs	Incl. O&P
1 Labor Foreman (outside)	$41.85	$334.80	$63.75	$510.00	$45.21	$68.54
3 Laborers	39.85	956.40	60.70	1456.80		
2.250 Equip. Oper. (medium)	53.75	967.50	81.05	1458.90		
.25 Truck Driver (heavy)	46.00	92.00	69.30	138.60		
.25 Truck Tractor, 6x4, 380 H.P.		119.05		130.96		
.25 Dist. Tanker, 3000 Gallon		81.70		89.87		
1 Asphalt Paver, 130 H.P.		2123.00		2335.30		
1 Tandem Roller, 10 Ton		231.40		254.54	49.14	54.05
52 L.H., Daily Totals		$4905.85		$6374.97	$94.34	$122.60

Crew B-26	Hr.	Daily	Hr.	Daily	Bare Costs	Incl. O&P
1 Labor Foreman (outside)	$41.85	$334.80	$63.75	$510.00	$44.60	$67.63
6 Laborers	39.85	1912.80	60.70	2913.60		
2 Equip. Oper. (medium)	53.75	860.00	81.05	1296.80		
1 Rodman (reinf.)	54.65	437.20	83.45	667.60		
1 Cement Finisher	47.55	380.40	70.45	563.60		
1 Grader, 30,000 Lbs.		643.80		708.18		
1 Paving Mach. & Equip.		2361.00		2597.10	34.15	37.56
88 L.H., Daily Totals		$6930.00		$9256.88	$78.75	$105.19

Crew B-26A	Hr.	Daily	Hr.	Daily	Bare Costs	Incl. O&P
1 Labor Foreman (outside)	$41.85	$334.80	$63.75	$510.00	$44.60	$67.63
6 Laborers	39.85	1912.80	60.70	2913.60		
2 Equip. Oper. (medium)	53.75	860.00	81.05	1296.80		
1 Rodman (reinf.)	54.65	437.20	83.45	667.60		
1 Cement Finisher	47.55	380.40	70.45	563.60		
1 Grader, 30,000 Lbs.		643.80		708.18		
1 Paving Mach. & Equip.		2361.00		2597.10		
1 Concrete Saw		102.40		112.64	35.31	38.84
88 L.H., Daily Totals		$7032.40		$9369.52	$79.91	$106.47

Crew B-26B	Hr.	Daily	Hr.	Daily	Bare Costs	Incl. O&P
1 Labor Foreman (outside)	$41.85	$334.80	$63.75	$510.00	$45.37	$68.75
6 Laborers	39.85	1912.80	60.70	2913.60		
3 Equip. Oper. (medium)	53.75	1290.00	81.05	1945.20		
1 Rodman (reinf.)	54.65	437.20	83.45	667.60		
1 Cement Finisher	47.55	380.40	70.45	563.60		
1 Grader, 30,000 Lbs.		643.80		708.18		
1 Paving Mach. & Equip.		2361.00		2597.10		
1 Concrete Pump, 110' Boom		1116.00		1227.60	42.92	47.22
96 L.H., Daily Totals		$8476.00		$11132.88	$88.29	$115.97

Crew B-26C	Hr.	Daily	Hr.	Daily	Bare Costs	Incl. O&P
1 Labor Foreman (outside)	$41.85	$334.80	$63.75	$510.00	$43.69	$66.29
6 Laborers	39.85	1912.80	60.70	2913.60		
1 Equip. Oper. (medium)	53.75	430.00	81.05	648.40		
1 Rodman (reinf.)	54.65	437.20	83.45	667.60		
1 Cement Finisher	47.55	380.40	70.45	563.60		
1 Paving Mach. & Equip.		2361.00		2597.10		
1 Concrete Saw		102.40		112.64	30.79	33.87
80 L.H., Daily Totals		$5958.60		$8012.94	$74.48	$100.16

Crew B-27	Hr.	Daily	Hr.	Daily	Bare Costs	Incl. O&P
1 Labor Foreman (outside)	$41.85	$334.80	$63.75	$510.00	$40.35	$61.46
3 Laborers	39.85	956.40	60.70	1456.80		
1 Berm Machine		317.40		349.14	9.92	10.91
32 L.H., Daily Totals		$1608.60		$2315.94	$50.27	$72.37

Crew No.	Bare Costs		Incl. Subs O&P		Cost Per Labor-Hour	
Crew B-28	Hr.	Daily	Hr.	Daily	Bare Costs	Incl. O&P
2 Carpenters	$50.70	$811.20	$77.20	$1235.20	$47.08	$71.70
1 Laborer	39.85	318.80	60.70	485.60		
24 L.H., Daily Totals		$1130.00		$1720.80	$47.08	$71.70
Crew B-29	Hr.	Daily	Hr.	Daily	Bare Costs	Incl. O&P
1 Labor Foreman (outside)	$41.85	$334.80	$63.75	$510.00	$43.71	$66.35
4 Laborers	39.85	1275.20	60.70	1942.40		
1 Equip. Oper. (crane)	56.10	448.80	84.60	676.80		
1 Equip. Oper. (oiler)	48.60	388.80	73.30	586.40		
1 Gradall, 5/8 C.Y.		869.60		956.56	15.53	17.08
56 L.H., Daily Totals		$3317.20		$4672.16	$59.24	$83.43
Crew B-30	Hr.	Daily	Hr.	Daily	Bare Costs	Incl. O&P
1 Equip. Oper. (medium)	$53.75	$430.00	$81.05	$648.40	$48.58	$73.22
2 Truck Drivers (heavy)	46.00	736.00	69.30	1108.80		
1 Hyd. Excavator, 1.5 C.Y.		893.60		982.96		
2 Dump Trucks, 12 C.Y., 400 H.P.		1085.60		1194.16	82.47	90.71
24 L.H., Daily Totals		$3145.20		$3934.32	$131.05	$163.93
Crew B-31	Hr.	Daily	Hr.	Daily	Bare Costs	Incl. O&P
1 Labor Foreman (outside)	$41.85	$334.80	$63.75	$510.00	$42.42	$64.61
3 Laborers	39.85	956.40	60.70	1456.80		
1 Carpenter	50.70	405.60	77.20	617.60		
1 Air Compressor, 250 cfm		167.40		184.14		
1 Sheeting Driver		7.00		7.70		
2 -50' Air Hoses, 1.5"		45.40		49.94	5.50	6.04
40 L.H., Daily Totals		$1916.60		$2826.18	$47.91	$70.65
Crew B-32	Hr.	Daily	Hr.	Daily	Bare Costs	Incl. O&P
1 Laborer	$39.85	$318.80	$60.70	$485.60	$50.27	$75.96
3 Equip. Oper. (medium)	53.75	1290.00	81.05	1945.20		
1 Grader, 30,000 Lbs.		643.80		708.18		
1 Tandem Roller, 10 Ton		231.40		254.54		
1 Dozer, 200 H.P.		1273.00		1400.30	67.13	73.84
32 L.H., Daily Totals		$3757.00		$4793.82	$117.41	$149.81
Crew B-32A	Hr.	Daily	Hr.	Daily	Bare Costs	Incl. O&P
1 Laborer	$39.85	$318.80	$60.70	$485.60	$49.12	$74.27
2 Equip. Oper. (medium)	53.75	860.00	81.05	1296.80		
1 Grader, 30,000 Lbs.		643.80		708.18		
1 Roller, Vibratory, 25 Ton		659.80		725.78	54.32	59.75
24 L.H., Daily Totals		$2482.40		$3216.36	$103.43	$134.01
Crew B-32B	Hr.	Daily	Hr.	Daily	Bare Costs	Incl. O&P
1 Laborer	$39.85	$318.80	$60.70	$485.60	$49.12	$74.27
2 Equip. Oper. (medium)	53.75	860.00	81.05	1296.80		
1 Dozer, 200 H.P.		1273.00		1400.30		
1 Roller, Vibratory, 25 Ton		659.80		725.78	80.53	88.59
24 L.H., Daily Totals		$3111.60		$3908.48	$129.65	$162.85
Crew B-32C	Hr.	Daily	Hr.	Daily	Bare Costs	Incl. O&P
1 Labor Foreman (outside)	$41.85	$334.80	$63.75	$510.00	$47.13	$71.38
2 Laborers	39.85	637.60	60.70	971.20		
3 Equip. Oper. (medium)	53.75	1290.00	81.05	1945.20		
1 Grader, 30,000 Lbs.		643.80		708.18		
1 Tandem Roller, 10 Ton		231.40		254.54		
1 Dozer, 200 H.P.		1273.00		1400.30	44.75	49.23
48 L.H., Daily Totals		$4410.60		$5789.42	$91.89	$120.61

Crew No.	Bare Costs		Incl. Subs O&P		Cost Per Labor-Hour	
Crew B-33A	Hr.	Daily	Hr.	Daily	Bare Costs	Incl. O&P
1 Equip. Oper. (medium)	$53.75	$430.00	$81.05	$648.40	$49.78	$75.24
.5 Laborer	39.85	159.40	60.70	242.80		
.25 Equip. Oper. (medium)	53.75	107.50	81.05	162.10		
1 Scraper, Towed, 7 C.Y.		123.40		135.74		
1.25 Dozers, 300 H.P.		2286.25		2514.88	172.12	189.33
14 L.H., Daily Totals		$3106.55		$3703.92	$221.90	$264.57
Crew B-33B	Hr.	Daily	Hr.	Daily	Bare Costs	Incl. O&P
1 Equip. Oper. (medium)	$53.75	$430.00	$81.05	$648.40	$49.78	$75.24
.5 Laborer	39.85	159.40	60.70	242.80		
.25 Equip. Oper. (medium)	53.75	107.50	81.05	162.10		
1 Scraper, Towed, 10 C.Y.		155.80		171.38		
1.25 Dozers, 300 H.P.		2286.25		2514.88	174.43	191.88
14 L.H., Daily Totals		$3138.95		$3739.55	$224.21	$267.11
Crew B-33C	Hr.	Daily	Hr.	Daily	Bare Costs	Incl. O&P
1 Equip. Oper. (medium)	$53.75	$430.00	$81.05	$648.40	$49.78	$75.24
.5 Laborer	39.85	159.40	60.70	242.80		
.25 Equip. Oper. (medium)	53.75	107.50	81.05	162.10		
1 Scraper, Towed, 15 C.Y.		175.80		193.38		
1.25 Dozers, 300 H.P.		2286.25		2514.88	175.86	193.45
14 L.H., Daily Totals		$3158.95		$3761.55	$225.64	$268.68
Crew B-33D	Hr.	Daily	Hr.	Daily	Bare Costs	Incl. O&P
1 Equip. Oper. (medium)	$53.75	$430.00	$81.05	$648.40	$49.78	$75.24
.5 Laborer	39.85	159.40	60.70	242.80		
.25 Equip. Oper. (medium)	53.75	107.50	81.05	162.10		
1 S.P. Scraper, 14 C.Y.		2490.00		2739.00		
.25 Dozer, 300 H.P.		457.25		502.98	210.52	231.57
14 L.H., Daily Totals		$3644.15		$4295.27	$260.30	$306.81
Crew B-33E	Hr.	Daily	Hr.	Daily	Bare Costs	Incl. O&P
1 Equip. Oper. (medium)	$53.75	$430.00	$81.05	$648.40	$49.78	$75.24
.5 Laborer	39.85	159.40	60.70	242.80		
.25 Equip. Oper. (medium)	53.75	107.50	81.05	162.10		
1 S.P. Scraper, 21 C.Y.		2468.00		2714.80		
.25 Dozer, 300 H.P.		457.25		502.98	208.95	229.84
14 L.H., Daily Totals		$3622.15		$4271.07	$258.73	$305.08
Crew B-33F	Hr.	Daily	Hr.	Daily	Bare Costs	Incl. O&P
1 Equip. Oper. (medium)	$53.75	$430.00	$81.05	$648.40	$49.78	$75.24
.5 Laborer	39.85	159.40	60.70	242.80		
.25 Equip. Oper. (medium)	53.75	107.50	81.05	162.10		
1 Elev. Scraper, 11 C.Y.		1155.00		1270.50		
.25 Dozer, 300 H.P.		457.25		502.98	115.16	126.68
14 L.H., Daily Totals		$2309.15		$2826.78	$164.94	$201.91
Crew B-33G	Hr.	Daily	Hr.	Daily	Bare Costs	Incl. O&P
1 Equip. Oper. (medium)	$53.75	$430.00	$81.05	$648.40	$49.78	$75.24
.5 Laborer	39.85	159.40	60.70	242.80		
.25 Equip. Oper. (medium)	53.75	107.50	81.05	162.10		
1 Elev. Scraper, 22 C.Y.		2284.00		2512.40		
.25 Dozer, 300 H.P.		457.25		502.98	195.80	215.38
14 L.H., Daily Totals		$3438.15		$4068.68	$245.58	$290.62

Crew No.	Bare Costs		Incl. Subs O&P		Cost Per Labor-Hour	
Crew B-33H	Hr.	Daily	Hr.	Daily	Bare Costs	Incl. O&P
.5 Laborer	$39.85	$159.40	$60.70	$242.80	$49.78	$75.24
1 Equipment Operator (med.)	53.75	430.00	81.05	648.40		
.25 Equipment Operator (med.)	53.75	107.50	81.05	162.10		
1 S.P. Scraper, 44 C.Y.		4457.00		4902.70		
.25 Dozer, 410 H.P.		572.75		630.02	359.27	395.19
14 L.H., Daily Totals		$5726.65		$6586.02	$409.05	$470.43
Crew B-33J	Hr.	Daily	Hr.	Daily	Bare Costs	Incl. O&P
1 Equipment Operator (med.)	$53.75	$430.00	$81.05	$648.40	$53.75	$81.05
1 S.P. Scraper, 14 C.Y.		2490.00		2739.00	311.25	342.38
8 L.H., Daily Totals		$2920.00		$3387.40	$365.00	$423.43
Crew B-33K	Hr.	Daily	Hr.	Daily	Bare Costs	Incl. O&P
1 Equipment Operator (med.)	$53.75	$430.00	$81.05	$648.40	$49.78	$75.24
.25 Equipment Operator (med.)	53.75	107.50	81.05	162.10		
.5 Laborer	39.85	159.40	60.70	242.80		
1 S.P. Scraper, 31 C.Y.		3576.00		3933.60		
.25 Dozer, 410 H.P.		572.75		630.02	296.34	325.97
14 L.H., Daily Totals		$4845.65		$5616.93	$346.12	$401.21
Crew B-34A	Hr.	Daily	Hr.	Daily	Bare Costs	Incl. O&P
1 Truck Driver (heavy)	$46.00	$368.00	$69.30	$554.40	$46.00	$69.30
1 Dump Truck, 8 C.Y., 220 H.P.		338.60		372.46	42.33	46.56
8 L.H., Daily Totals		$706.60		$926.86	$88.33	$115.86
Crew B-34B	Hr.	Daily	Hr.	Daily	Bare Costs	Incl. O&P
1 Truck Driver (heavy)	$46.00	$368.00	$69.30	$554.40	$46.00	$69.30
1 Dump Truck, 12 C.Y., 400 H.P.		542.80		597.08	67.85	74.64
8 L.H., Daily Totals		$910.80		$1151.48	$113.85	$143.94
Crew B-34C	Hr.	Daily	Hr.	Daily	Bare Costs	Incl. O&P
1 Truck Driver (heavy)	$46.00	$368.00	$69.30	$554.40	$46.00	$69.30
1 Truck Tractor, 6x4, 380 H.P.		476.20		523.82		
1 Dump Trailer, 16.5 C.Y.		133.00		146.30	76.15	83.77
8 L.H., Daily Totals		$977.20		$1224.52	$122.15	$153.07
Crew B-34D	Hr.	Daily	Hr.	Daily	Bare Costs	Incl. O&P
1 Truck Driver (heavy)	$46.00	$368.00	$69.30	$554.40	$46.00	$69.30
1 Truck Tractor, 6x4, 380 H.P.		476.20		523.82		
1 Dump Trailer, 20 C.Y.		147.60		162.36	77.97	85.77
8 L.H., Daily Totals		$991.80		$1240.58	$123.97	$155.07
Crew B-34E	Hr.	Daily	Hr.	Daily	Bare Costs	Incl. O&P
1 Truck Driver (heavy)	$46.00	$368.00	$69.30	$554.40	$46.00	$69.30
1 Dump Truck, Off Hwy., 25 Ton		1343.00		1477.30	167.88	184.66
8 L.H., Daily Totals		$1711.00		$2031.70	$213.88	$253.96
Crew B-34F	Hr.	Daily	Hr.	Daily	Bare Costs	Incl. O&P
1 Truck Driver (heavy)	$46.00	$368.00	$69.30	$554.40	$46.00	$69.30
1 Dump Truck, Off Hwy., 35 Ton		1448.00		1592.80	181.00	199.10
8 L.H., Daily Totals		$1816.00		$2147.20	$227.00	$268.40
Crew B-34G	Hr.	Daily	Hr.	Daily	Bare Costs	Incl. O&P
1 Truck Driver (heavy)	$46.00	$368.00	$69.30	$554.40	$46.00	$69.30
1 Dump Truck, Off Hwy., 50 Ton		1662.00		1828.20	207.75	228.53
8 L.H., Daily Totals		$2030.00		$2382.60	$253.75	$297.82

Crew No.	Bare Costs		Incl. Subs O&P		Cost Per Labor-Hour	
Crew B-34H	Hr.	Daily	Hr.	Daily	Bare Costs	Incl. O&P
1 Truck Driver (heavy)	$46.00	$368.00	$69.30	$554.40	$46.00	$69.30
1 Dump Truck, Off Hwy., 65 Ton		1844.00		2028.40	230.50	253.55
8 L.H., Daily Totals		$2212.00		$2582.80	$276.50	$322.85
Crew B-34I	Hr.	Daily	Hr.	Daily	Bare Costs	Incl. O&P
1 Truck Driver (heavy)	$46.00	$368.00	$69.30	$554.40	$46.00	$69.30
1 Dump Truck, 18 C.Y., 450 H.P.		706.00		776.60	88.25	97.08
8 L.H., Daily Totals		$1074.00		$1331.00	$134.25	$166.38
Crew B-34J	Hr.	Daily	Hr.	Daily	Bare Costs	Incl. O&P
1 Truck Driver (heavy)	$46.00	$368.00	$69.30	$554.40	$46.00	$69.30
1 Dump Truck, Off Hwy., 100 Ton		2669.00		2935.90	333.63	366.99
8 L.H., Daily Totals		$3037.00		$3490.30	$379.63	$436.29
Crew B-34K	Hr.	Daily	Hr.	Daily	Bare Costs	Incl. O&P
1 Truck Driver (heavy)	$46.00	$368.00	$69.30	$554.40	$46.00	$69.30
1 Truck Tractor, 6x4, 450 H.P.		583.40		641.74		
1 Lowbed Trailer, 75 Ton		249.40		274.34	104.10	114.51
8 L.H., Daily Totals		$1200.80		$1470.48	$150.10	$183.81
Crew B-34L	Hr.	Daily	Hr.	Daily	Bare Costs	Incl. O&P
1 Equip. Oper. (light)	$51.30	$410.40	$77.35	$618.80	$51.30	$77.35
1 Flatbed Truck, Gas, 1.5 Ton		188.40		207.24	23.55	25.91
8 L.H., Daily Totals		$598.80		$826.04	$74.85	$103.26
Crew B-34M	Hr.	Daily	Hr.	Daily	Bare Costs	Incl. O&P
1 Equip. Oper. (light)	$51.30	$410.40	$77.35	$618.80	$51.30	$77.35
1 Flatbed Truck, Gas, 3 Ton		238.00		261.80	29.75	32.73
8 L.H., Daily Totals		$648.40		$880.60	$81.05	$110.08
Crew B-34N	Hr.	Daily	Hr.	Daily	Bare Costs	Incl. O&P
1 Truck Driver (heavy)	$46.00	$368.00	$69.30	$554.40	$49.88	$75.17
1 Equip. Oper. (medium)	53.75	430.00	81.05	648.40		
1 Truck Tractor, 6x4, 380 H.P.		476.20		523.82		
1 Flatbed Trailer, 40 Ton		180.00		198.00	41.01	45.11
16 L.H., Daily Totals		$1454.20		$1924.62	$90.89	$120.29
Crew B-34P	Hr.	Daily	Hr.	Daily	Bare Costs	Incl. O&P
1 Pipe Fitter	$63.00	$504.00	$94.90	$759.20	$53.75	$80.98
1 Truck Driver (light)	44.50	356.00	67.00	536.00		
1 Equip. Oper. (medium)	53.75	430.00	81.05	648.40		
1 Flatbed Truck, Gas, 3 Ton		238.00		261.80		
1 Backhoe Loader, 48 H.P.		312.40		343.64	22.93	25.23
24 L.H., Daily Totals		$1840.40		$2549.04	$76.68	$106.21
Crew B-34Q	Hr.	Daily	Hr.	Daily	Bare Costs	Incl. O&P
1 Pipe Fitter	$63.00	$504.00	$94.90	$759.20	$54.53	$82.17
1 Truck Driver (light)	44.50	356.00	67.00	536.00		
1 Equip. Oper. (crane)	56.10	448.80	84.60	676.80		
1 Flatbed Trailer, 25 Ton		133.00		146.30		
1 Dump Truck, 8 C.Y., 220 H.P.		338.60		372.46		
1 Hyd. Crane, 25 Ton		590.60		649.66	44.26	48.68
24 L.H., Daily Totals		$2371.00		$3140.42	$98.79	$130.85

Crew No.	Bare Costs		Incl. Subs O&P		Cost Per Labor-Hour	
Crew B-34R	Hr.	Daily	Hr.	Daily	Bare Costs	Incl. O&P
1 Pipe Fitter	$63.00	$504.00	$94.90	$759.20	$54.53	$82.17
1 Truck Driver (light)	44.50	356.00	67.00	536.00		
1 Equip. Oper. (crane)	56.10	448.80	84.60	676.80		
1 Flatbed Trailer, 25 Ton		133.00		146.30		
1 Dump Truck, 8 C.Y., 220 H.P.		338.60		372.46		
1 Hyd. Crane, 25 Ton		590.60		649.66		
1 Hyd. Excavator, 1 C.Y.		742.20		816.42	75.18	82.70
24 L.H., Daily Totals		$3113.20		$3956.84	$129.72	$164.87
Crew B-34S	Hr.	Daily	Hr.	Daily	Bare Costs	Incl. O&P
2 Pipe Fitters	$63.00	$1008.00	$94.90	$1518.40	$57.02	$85.92
1 Truck Driver (heavy)	46.00	368.00	69.30	554.40		
1 Equip. Oper. (crane)	56.10	448.80	84.60	676.80		
1 Flatbed Trailer, 40 Ton		180.00		198.00		
1 Truck Tractor, 6x4, 380 H.P.		476.20		523.82		
1 Hyd. Crane, 80 Ton		1518.00		1669.80		
1 Hyd. Excavator, 2 C.Y.		1052.00		1157.20	100.82	110.90
32 L.H., Daily Totals		$5051.00		$6298.42	$157.84	$196.83
Crew B-34T	Hr.	Daily	Hr.	Daily	Bare Costs	Incl. O&P
2 Pipe Fitters	$63.00	$1008.00	$94.90	$1518.40	$57.02	$85.92
1 Truck Driver (heavy)	46.00	368.00	69.30	554.40		
1 Equip. Oper. (crane)	56.10	448.80	84.60	676.80		
1 Flatbed Trailer, 40 Ton		180.00		198.00		
1 Truck Tractor, 6x4, 380 H.P.		476.20		523.82		
1 Hyd. Crane, 80 Ton		1518.00		1669.80	67.94	74.74
32 L.H., Daily Totals		$3999.00		$5141.22	$124.97	$160.66
Crew B-34U	Hr.	Daily	Hr.	Daily	Bare Costs	Incl. O&P
1 Truck Driver (heavy)	$46.00	$368.00	$69.30	$554.40	$48.65	$73.33
1 Equip. Oper. (light)	51.30	410.40	77.35	618.80		
1 Truck Tractor, 220 H.P.		290.00		319.00		
1 Flatbed Trailer, 25 Ton		133.00		146.30	26.44	29.08
16 L.H., Daily Totals		$1201.40		$1638.50	$75.09	$102.41
Crew B-34V	Hr.	Daily	Hr.	Daily	Bare Costs	Incl. O&P
1 Truck Driver (heavy)	$46.00	$368.00	$69.30	$554.40	$51.13	$77.08
1 Equip. Oper. (crane)	56.10	448.80	84.60	676.80		
1 Equip. Oper. (light)	51.30	410.40	77.35	618.80		
1 Truck Tractor, 6x4, 450 H.P.		583.40		641.74		
1 Equipment Trailer, 50 Ton		198.20		218.02		
1 Pickup Truck, 4x4, 3/4 Ton		126.60		139.26	37.84	41.63
24 L.H., Daily Totals		$2135.40		$2849.02	$88.97	$118.71
Crew B-34W	Hr.	Daily	Hr.	Daily	Bare Costs	Incl. O&P
5 Truck Drivers (heavy)	$46.00	$1840.00	$69.30	$2772.00	$48.71	$73.48
2 Equip. Opers. (crane)	56.10	897.60	84.60	1353.60		
1 Equip. Oper. (mechanic)	56.30	450.40	84.90	679.20		
1 Laborer	39.85	318.80	60.70	485.60		
4 Truck Tractors, 6x4, 380 H.P.		1904.80		2095.28		
2 Equipment Trailers, 50 Ton		396.40		436.04		
2 Flatbed Trailers, 40 Ton		360.00		396.00		
1 Pickup Truck, 4x4, 3/4 Ton		126.60		139.26		
1 S.P. Crane, 4x4, 20 Ton		571.20		628.32	46.65	51.32
72 L.H., Daily Totals		$6865.80		$8985.30	$95.36	$124.80

Crew No.	Bare Costs		Incl. Subs O&P		Cost Per Labor-Hour	
Crew B-35	Hr.	Daily	Hr.	Daily	Bare Costs	Incl. O&P
1 Labor Foreman (outside)	$41.85	$334.80	$63.75	$510.00	$50.15	$76.02
1 Skilled Worker	52.35	418.80	80.15	641.20		
1 Welder (plumber)	62.15	497.20	93.60	748.80		
1 Laborer	39.85	318.80	60.70	485.60		
1 Equip. Oper. (crane)	56.10	448.80	84.60	676.80		
1 Equip. Oper. (oiler)	48.60	388.80	73.30	586.40		
1 Welder, Electric, 300 amp		59.20		65.12		
1 Hyd. Excavator, .75 C.Y.		669.60		736.56	15.18	16.70
48 L.H., Daily Totals		$3136.00		$4450.48	$65.33	$92.72
Crew B-35A	Hr.	Daily	Hr.	Daily	Bare Costs	Incl. O&P
1 Labor Foreman (outside)	$41.85	$334.80	$63.75	$510.00	$48.68	$73.83
2 Laborers	39.85	637.60	60.70	971.20		
1 Skilled Worker	52.35	418.80	80.15	641.20		
1 Welder (plumber)	62.15	497.20	93.60	748.80		
1 Equip. Oper. (crane)	56.10	448.80	84.60	676.80		
1 Equip. Oper. (oiler)	48.60	388.80	73.30	586.40		
1 Welder, Gas Engine, 300 amp		98.40		108.24		
1 Crawler Crane, 75 Ton		1708.00		1878.80	32.26	35.48
56 L.H., Daily Totals		$4532.40		$6121.44	$80.94	$109.31
Crew B-36	Hr.	Daily	Hr.	Daily	Bare Costs	Incl. O&P
1 Labor Foreman (outside)	$41.85	$334.80	$63.75	$510.00	$45.81	$69.45
2 Laborers	39.85	637.60	60.70	971.20		
2 Equip. Oper. (medium)	53.75	860.00	81.05	1296.80		
1 Dozer, 200 H.P.		1273.00		1400.30		
1 Aggregate Spreader		36.80		40.48		
1 Tandem Roller, 10 Ton		231.40		254.54	38.53	42.38
40 L.H., Daily Totals		$3373.60		$4473.32	$84.34	$111.83
Crew B-36A	Hr.	Daily	Hr.	Daily	Bare Costs	Incl. O&P
1 Labor Foreman (outside)	$41.85	$334.80	$63.75	$510.00	$48.08	$72.76
2 Laborers	39.85	637.60	60.70	971.20		
4 Equip. Oper. (medium)	53.75	1720.00	81.05	2593.60		
1 Dozer, 200 H.P.		1273.00		1400.30		
1 Aggregate Spreader		36.80		40.48		
1 Tandem Roller, 10 Ton		231.40		254.54		
1 Roller, Pneum. Whl., 12 Ton		338.20		372.02	33.56	36.92
56 L.H., Daily Totals		$4571.80		$6142.14	$81.64	$109.68
Crew B-36B	Hr.	Daily	Hr.	Daily	Bare Costs	Incl. O&P
1 Labor Foreman (outside)	$41.85	$334.80	$63.75	$510.00	$47.82	$72.33
2 Laborers	39.85	637.60	60.70	971.20		
4 Equip. Oper. (medium)	53.75	1720.00	81.05	2593.60		
1 Truck Driver (heavy)	46.00	368.00	69.30	554.40		
1 Grader, 30,000 Lbs.		643.80		708.18		
1 F.E. Loader, Crl, 1.5 C.Y.		645.40		709.94		
1 Dozer, 300 H.P.		1829.00		2011.90		
1 Roller, Vibratory, 25 Ton		659.80		725.78		
1 Truck Tractor, 6x4, 450 H.P.		583.40		641.74		
1 Water Tank Trailer, 5000 Gal.		147.60		162.36	70.45	77.50
64 L.H., Daily Totals		$7569.40		$9589.10	$118.27	$149.83

Crews - Standard

Crew No.	Bare Costs		Incl. Subs O&P		Cost Per Labor-Hour	

Crew B-36C

	Hr.	Daily	Hr.	Daily	Bare Costs	Incl. O&P
1 Labor Foreman (outside)	$41.85	$334.80	$63.75	$510.00	$49.82	$75.24
3 Equip. Oper. (medium)	53.75	1290.00	81.05	1945.20		
1 Truck Driver (heavy)	46.00	368.00	69.30	554.40		
1 Grader, 30,000 Lbs.		643.80		708.18		
1 Dozer, 300 H.P.		1829.00		2011.90		
1 Roller, Vibratory, 25 Ton		659.80		725.78		
1 Truck Tractor, 6x4, 450 H.P.		583.40		641.74		
1 Water Tank Trailer, 5000 Gal.		147.60		162.36	96.59	106.25
40 L.H., Daily Totals		$5856.40		$7259.56	$146.41	$181.49

Crew B-36D

	Hr.	Daily	Hr.	Daily	Bare Costs	Incl. O&P
1 Labor Foreman (outside)	$41.85	$334.80	$63.75	$510.00	$50.77	$76.72
3 Equip. Oper. (medium)	53.75	1290.00	81.05	1945.20		
1 Grader, 30,000 Lbs.		643.80		708.18		
1 Dozer, 300 H.P.		1829.00		2011.90		
1 Roller, Vibratory, 25 Ton		659.80		725.78	97.89	107.68
32 L.H., Daily Totals		$4757.40		$5901.06	$148.67	$184.41

Crew B-37

	Hr.	Daily	Hr.	Daily	Bare Costs	Incl. O&P
1 Labor Foreman (outside)	$41.85	$334.80	$63.75	$510.00	$42.09	$63.98
4 Laborers	39.85	1275.20	60.70	1942.40		
1 Equip. Oper. (light)	51.30	410.40	77.35	618.80		
1 Tandem Roller, 5 Ton		151.20		166.32	3.15	3.46
48 L.H., Daily Totals		$2171.60		$3237.52	$45.24	$67.45

Crew B-37A

	Hr.	Daily	Hr.	Daily	Bare Costs	Incl. O&P
2 Laborers	$39.85	$637.60	$60.70	$971.20	$41.40	$62.80
1 Truck Driver (light)	44.50	356.00	67.00	536.00		
1 Flatbed Truck, Gas, 1.5 Ton		188.40		207.24		
1 Tar Kettle, T.M.		129.20		142.12	13.23	14.56
24 L.H., Daily Totals		$1311.20		$1856.56	$54.63	$77.36

Crew B-37B

	Hr.	Daily	Hr.	Daily	Bare Costs	Incl. O&P
3 Laborers	$39.85	$956.40	$60.70	$1456.80	$41.01	$62.27
1 Truck Driver (light)	44.50	356.00	67.00	536.00		
1 Flatbed Truck, Gas, 1.5 Ton		188.40		207.24		
1 Tar Kettle, T.M.		129.20		142.12	9.93	10.92
32 L.H., Daily Totals		$1630.00		$2342.16	$50.94	$73.19

Crew B-37C

	Hr.	Daily	Hr.	Daily	Bare Costs	Incl. O&P
2 Laborers	$39.85	$637.60	$60.70	$971.20	$42.17	$63.85
2 Truck Drivers (light)	44.50	712.00	67.00	1072.00		
2 Flatbed Trucks, Gas, 1.5 Ton		376.80		414.48		
1 Tar Kettle, T.M.		129.20		142.12	15.81	17.39
32 L.H., Daily Totals		$1855.60		$2599.80	$57.99	$81.24

Crew B-37D

	Hr.	Daily	Hr.	Daily	Bare Costs	Incl. O&P
1 Laborer	$39.85	$318.80	$60.70	$485.60	$42.17	$63.85
1 Truck Driver (light)	44.50	356.00	67.00	536.00		
1 Pickup Truck, 3/4 Ton		115.20		126.72	7.20	7.92
16 L.H., Daily Totals		$790.00		$1148.32	$49.38	$71.77

Crew B-37E

	Hr.	Daily	Hr.	Daily	Bare Costs	Incl. O&P
3 Laborers	$39.85	$956.40	$60.70	$1456.80	$44.80	$67.79
1 Equip. Oper. (light)	51.30	410.40	77.35	618.80		
1 Equip. Oper. (medium)	53.75	430.00	81.05	648.40		
2 Truck Drivers (light)	44.50	712.00	67.00	1072.00		
4 Barrels w/Flasher		15.60		17.16		
1 Concrete Saw		102.40		112.64		
1 Rotary Hammer Drill		25.15		27.66		
1 Hammer Drill Bit		3.25		3.58		
1 Loader, Skid Steer, 30 H.P.		174.00		191.40		
1 Conc. Hammer Attach.		113.60		124.96		
1 Vibrating Plate, Gas, 18"		31.80		34.98		
2 Flatbed Trucks, Gas, 1.5 Ton		376.80		414.48	15.05	16.55
56 L.H., Daily Totals		$3351.40		$4722.86	$59.85	$84.34

Crew B-37F

	Hr.	Daily	Hr.	Daily	Bare Costs	Incl. O&P
3 Laborers	$39.85	$956.40	$60.70	$1456.80	$41.01	$62.27
1 Truck Driver (light)	44.50	356.00	67.00	536.00		
4 Barrels w/Flasher		15.60		17.16		
1 Concrete Mixer, 10 C.F.		161.00		177.10		
1 Air Compressor, 60 cfm		105.40		115.94		
1 -50' Air Hose, 3/4"		7.35		8.09		
1 Spade (Chipper)		8.40		9.24		
1 Flatbed Truck, Gas, 1.5 Ton		188.40		207.24	15.19	16.71
32 L.H., Daily Totals		$1798.55		$2527.57	$56.20	$78.99

Crew B-37G

	Hr.	Daily	Hr.	Daily	Bare Costs	Incl. O&P
1 Labor Foreman (outside)	$41.85	$334.80	$63.75	$510.00	$42.09	$63.98
4 Laborers	39.85	1275.20	60.70	1942.40		
1 Equip. Oper. (light)	51.30	410.40	77.35	618.80		
1 Berm Machine		317.40		349.14		
1 Tandem Roller, 5 Ton		151.20		166.32	9.76	10.74
48 L.H., Daily Totals		$2489.00		$3586.66	$51.85	$74.72

Crew B-37H

	Hr.	Daily	Hr.	Daily	Bare Costs	Incl. O&P
1 Labor Foreman (outside)	$41.85	$334.80	$63.75	$510.00	$42.09	$63.98
4 Laborers	39.85	1275.20	60.70	1942.40		
1 Equip. Oper. (light)	51.30	410.40	77.35	618.80		
1 Tandem Roller, 5 Ton		151.20		166.32		
1 Flatbed Truck, Gas, 1.5 Ton		188.40		207.24		
1 Tar Kettle, T.M.		129.20		142.12	9.77	10.74
48 L.H., Daily Totals		$2489.20		$3586.88	$51.86	$74.73

Crew B-37I

	Hr.	Daily	Hr.	Daily	Bare Costs	Incl. O&P
3 Laborers	$39.85	$956.40	$60.70	$1456.80	$44.80	$67.79
1 Equip. Oper. (light)	51.30	410.40	77.35	618.80		
1 Equip. Oper. (medium)	53.75	430.00	81.05	648.40		
2 Truck Drivers (light)	44.50	712.00	67.00	1072.00		
4 Barrels w/Flasher		15.60		17.16		
1 Concrete Saw		102.40		112.64		
1 Rotary Hammer Drill		25.15		27.66		
1 Hammer Drill Bit		3.25		3.58		
1 Air Compressor, 60 cfm		105.40		115.94		
1 -50' Air Hose, 3/4"		7.35		8.09		
1 Spade (Chipper)		8.40		9.24		
1 Loader, Skid Steer, 30 H.P.		174.00		191.40		
1 Conc. Hammer Attach.		113.60		124.96		
1 Concrete Mixer, 10 C.F.		161.00		177.10		
1 Vibrating Plate, Gas, 18"		31.80		34.98		
2 Flatbed Trucks, Gas, 1.5 Ton		376.80		414.48	20.08	22.09
56 L.H., Daily Totals		$3633.55		$5033.23	$64.88	$89.88

Crews - Standard

Crew No.	Bare Costs Hr.	Bare Costs Daily	Incl. Subs O&P Hr.	Incl. Subs O&P Daily	Cost Per Labor-Hour Bare Costs	Cost Per Labor-Hour Incl. O&P
Crew B-37J					Bare Costs	Incl. O&P
1 Labor Foreman (outside)	$41.85	$334.80	$63.75	$510.00	$42.09	$63.98
4 Laborers	39.85	1275.20	60.70	1942.40		
1 Equip. Oper. (light)	51.30	410.40	77.35	618.80		
1 Air Compressor, 60 cfm		105.40		115.94		
1 -50' Air Hose, 3/4"		7.35		8.09		
2 Concrete Mixers, 10 C.F.		322.00		354.20		
2 Flatbed Trucks, Gas, 1.5 Ton		376.80		414.48		
1 Shot Blaster, 20"		214.80		236.28	21.38	23.52
48 L.H., Daily Totals		$3046.75		$4200.19	$63.47	$87.50
Crew B-37K	Hr.	Daily	Hr.	Daily	Bare Costs	Incl. O&P
1 Labor Foreman (outside)	$41.85	$334.80	$63.75	$510.00	$42.09	$63.98
4 Laborers	39.85	1275.20	60.70	1942.40		
1 Equip. Oper. (light)	51.30	410.40	77.35	618.80		
1 Air Compressor, 60 cfm		105.40		115.94		
1 -50' Air Hose, 3/4"		7.35		8.09		
2 Flatbed Trucks, Gas, 1.5 Ton		376.80		414.48		
1 Shot Blaster, 20"		214.80		236.28	14.67	16.14
48 L.H., Daily Totals		$2724.75		$3845.99	$56.77	$80.12
Crew B-38	Hr.	Daily	Hr.	Daily	Bare Costs	Incl. O&P
1 Labor Foreman (outside)	$41.85	$334.80	$63.75	$510.00	$45.32	$68.71
2 Laborers	39.85	637.60	60.70	971.20		
1 Equip. Oper. (light)	51.30	410.40	77.35	618.80		
1 Equip. Oper. (medium)	53.75	430.00	81.05	648.40		
1 Backhoe Loader, 48 H.P.		312.40		343.64		
1 Hyd. Hammer (1200 lb.)		177.40		195.14		
1 F.E. Loader, W.M., 4 C.Y.		574.60		632.06		
1 Pvmt. Rem. Bucket		61.00		67.10	28.14	30.95
40 L.H., Daily Totals		$2938.20		$3986.34	$73.45	$99.66
Crew B-39	Hr.	Daily	Hr.	Daily	Bare Costs	Incl. O&P
1 Labor Foreman (outside)	$41.85	$334.80	$63.75	$510.00	$42.09	$63.98
4 Laborers	39.85	1275.20	60.70	1942.40		
1 Equip. Oper. (light)	51.30	410.40	77.35	618.80		
1 Air Compressor, 250 cfm		167.40		184.14		
2 Breakers, Pavement, 60 lb.		21.20		23.32		
2 -50' Air Hoses, 1.5"		45.40		49.94	4.88	5.36
48 L.H., Daily Totals		$2254.40		$3328.60	$46.97	$69.35
Crew B-40	Hr.	Daily	Hr.	Daily	Bare Costs	Incl. O&P
1 Pile Driver Foreman (outside)	$53.30	$426.40	$84.05	$672.40	$52.41	$81.27
4 Pile Drivers	51.30	1641.60	80.90	2588.80		
2 Equip. Oper. (crane)	56.10	897.60	84.60	1353.60		
1 Equip. Oper. (oiler)	48.60	388.80	73.30	586.40		
1 Crawler Crane, 40 Ton		1370.00		1507.00		
1 Vibratory Hammer & Gen.		2191.00		2410.10	55.64	61.20
64 L.H., Daily Totals		$6915.40		$9118.30	$108.05	$142.47
Crew B-40B	Hr.	Daily	Hr.	Daily	Bare Costs	Incl. O&P
1 Labor Foreman (outside)	$41.85	$334.80	$63.75	$510.00	$44.35	$67.29
3 Laborers	39.85	956.40	60.70	1456.80		
1 Equip. Oper. (crane)	56.10	448.80	84.60	676.80		
1 Equip. Oper. (oiler)	48.60	388.80	73.30	586.40		
1 Lattice Boom Crane, 40 Ton		1321.00		1453.10	27.52	30.27
48 L.H., Daily Totals		$3449.80		$4683.10	$71.87	$97.56

Crew No.	Bare Costs Hr.	Bare Costs Daily	Incl. Subs O&P Hr.	Incl. Subs O&P Daily	Cost Per Labor-Hour Bare Costs	Cost Per Labor-Hour Incl. O&P
Crew B-41	Hr.	Daily	Hr.	Daily	Bare Costs	Incl. O&P
1 Labor Foreman (outside)	$41.85	$334.80	$63.75	$510.00	$41.35	$62.91
4 Laborers	39.85	1275.20	60.70	1942.40		
.25 Equip. Oper. (crane)	56.10	112.20	84.60	169.20		
.25 Equip. Oper. (oiler)	48.60	97.20	73.30	146.60		
.25 Crawler Crane, 40 Ton		342.50		376.75	7.78	8.56
44 L.H., Daily Totals		$2161.90		$3144.95	$49.13	$71.48
Crew B-42	Hr.	Daily	Hr.	Daily	Bare Costs	Incl. O&P
1 Labor Foreman (outside)	$41.85	$334.80	$63.75	$510.00	$45.08	$69.22
4 Laborers	39.85	1275.20	60.70	1942.40		
1 Equip. Oper. (crane)	56.10	448.80	84.60	676.80		
1 Equip. Oper. (oiler)	48.60	388.80	73.30	586.40		
1 Welder	54.65	437.20	89.35	714.80		
1 Hyd. Crane, 25 Ton		590.60		649.66		
1 Welder, Gas Engine, 300 amp		98.40		108.24		
1 Horz. Boring Csg. Mch.		448.80		493.68	17.78	19.56
64 L.H., Daily Totals		$4022.60		$5681.98	$62.85	$88.78
Crew B-43	Hr.	Daily	Hr.	Daily	Bare Costs	Incl. O&P
1 Labor Foreman (outside)	$41.85	$334.80	$63.75	$510.00	$44.35	$67.29
3 Laborers	39.85	956.40	60.70	1456.80		
1 Equip. Oper. (crane)	56.10	448.80	84.60	676.80		
1 Equip. Oper. (oiler)	48.60	388.80	73.30	586.40		
1 Drill Rig, Truck-Mounted		2444.00		2688.40	50.92	56.01
48 L.H., Daily Totals		$4572.80		$5918.40	$95.27	$123.30
Crew B-44	Hr.	Daily	Hr.	Daily	Bare Costs	Incl. O&P
1 Pile Driver Foreman (outside)	$53.30	$426.40	$84.05	$672.40	$51.32	$79.69
4 Pile Drivers	51.30	1641.60	80.90	2588.80		
2 Equip. Oper. (crane)	56.10	897.60	84.60	1353.60		
1 Laborer	39.85	318.80	60.70	485.60		
1 Crawler Crane, 40 Ton		1370.00		1507.00		
1 Lead, 60' High		79.40		87.34		
1 Hammer, Diesel, 15K ft.-lbs.		623.20		685.52	32.38	35.62
64 L.H., Daily Totals		$5357.00		$7380.26	$83.70	$115.32
Crew B-45	Hr.	Daily	Hr.	Daily	Bare Costs	Incl. O&P
1 Equip. Oper. (medium)	$53.75	$430.00	$81.05	$648.40	$49.88	$75.17
1 Truck Driver (heavy)	46.00	368.00	69.30	554.40		
1 Dist. Tanker, 3000 Gallon		326.80		359.48		
1 Truck Tractor, 6x4, 380 H.P.		476.20		523.82	50.19	55.21
16 L.H., Daily Totals		$1601.00		$2086.10	$100.06	$130.38
Crew B-46	Hr.	Daily	Hr.	Daily	Bare Costs	Incl. O&P
1 Pile Driver Foreman (outside)	$53.30	$426.40	$84.05	$672.40	$45.91	$71.33
2 Pile Drivers	51.30	820.80	80.90	1294.40		
3 Laborers	39.85	956.40	60.70	1456.80		
1 Chain Saw, Gas, 36" Long		46.40		51.04	.97	1.06
48 L.H., Daily Totals		$2250.00		$3474.64	$46.88	$72.39
Crew B-47	Hr.	Daily	Hr.	Daily	Bare Costs	Incl. O&P
1 Blast Foreman (outside)	$41.85	$334.80	$63.75	$510.00	$44.33	$67.27
1 Driller	39.85	318.80	60.70	485.60		
1 Equip. Oper. (light)	51.30	410.40	77.35	618.80		
1 Air Track Drill, 4"		993.40		1092.74		
1 Air Compressor, 600 cfm		417.20		458.92		
2 -50' Air Hoses, 3"		86.40		95.04	62.38	68.61
24 L.H., Daily Totals		$2561.00		$3261.10	$106.71	$135.88

For customer support on your Electrical Costs with RSMeans data, call 800.448.8182.

491

Crew No.	Bare Costs Hr.	Daily	Incl. Subs O&P Hr.	Daily	Cost Per Labor-Hour Bare Costs	Incl. O&P
Crew B-47A						
1 Drilling Foreman (outside)	$41.85	$334.80	$63.75	$510.00	$48.85	$73.88
1 Equip. Oper. (heavy)	56.10	448.80	84.60	676.80		
1 Equip. Oper. (oiler)	48.60	388.80	73.30	586.40		
1 Air Track Drill, 5"		1173.00		1290.30	48.88	53.76
24 L.H., Daily Totals		$2345.40		$3063.50	$97.72	$127.65
Crew B-47C						
1 Laborer	$39.85	$318.80	$60.70	$485.60	$45.58	$69.03
1 Equip. Oper. (light)	51.30	410.40	77.35	618.80		
1 Air Compressor, 750 cfm		423.80		466.18		
2 -50' Air Hoses, 3"		86.40		95.04		
1 Air Track Drill, 4"		993.40		1092.74	93.97	103.37
16 L.H., Daily Totals		$2232.80		$2758.36	$139.55	$172.40
Crew B-47E						
1 Labor Foreman (outside)	$41.85	$334.80	$63.75	$510.00	$40.35	$61.46
3 Laborers	39.85	956.40	60.70	1456.80		
1 Flatbed Truck, Gas, 3 Ton		238.00		261.80	7.44	8.18
32 L.H., Daily Totals		$1529.20		$2228.60	$47.79	$69.64
Crew B-47G						
1 Labor Foreman (outside)	$41.85	$334.80	$63.75	$510.00	$43.21	$65.63
2 Laborers	39.85	637.60	60.70	971.20		
1 Equip. Oper. (light)	51.30	410.40	77.35	618.80		
1 Air Track Drill, 4"		993.40		1092.74		
1 Air Compressor, 600 cfm		417.20		458.92		
2 -50' Air Hoses, 3"		86.40		95.04		
1 Gunite Pump Rig		321.00		353.10	56.81	62.49
32 L.H., Daily Totals		$3200.80		$4099.80	$100.03	$128.12
Crew B-47H						
1 Skilled Worker Foreman (out)	$54.35	$434.80	$83.20	$665.60	$52.85	$80.91
3 Skilled Workers	52.35	1256.40	80.15	1923.60		
1 Flatbed Truck, Gas, 3 Ton		238.00		261.80	7.44	8.18
32 L.H., Daily Totals		$1929.20		$2851.00	$60.29	$89.09
Crew B-48						
1 Labor Foreman (outside)	$41.85	$334.80	$63.75	$510.00	$45.34	$68.73
3 Laborers	39.85	956.40	60.70	1456.80		
1 Equip. Oper. (crane)	56.10	448.80	84.60	676.80		
1 Equip. Oper. (oiler)	48.60	388.80	73.30	586.40		
1 Equip. Oper. (light)	51.30	410.40	77.35	618.80		
1 Centr. Water Pump, 6"		296.40		326.04		
1 -20' Suction Hose, 6"		11.50		12.65		
1 -50' Discharge Hose, 6"		6.10		6.71		
1 Drill Rig, Truck-Mounted		2444.00		2688.40	49.25	54.17
56 L.H., Daily Totals		$5297.20		$6882.60	$94.59	$122.90
Crew B-49						
1 Labor Foreman (outside)	$41.85	$334.80	$63.75	$510.00	$47.70	$72.80
3 Laborers	39.85	956.40	60.70	1456.80		
2 Equip. Oper. (crane)	56.10	897.60	84.60	1353.60		
2 Equip. Oper. (oilers)	48.60	777.60	73.30	1172.80		
1 Equip. Oper. (light)	51.30	410.40	77.35	618.80		
2 Pile Drivers	51.30	820.80	80.90	1294.40		
1 Hyd. Crane, 25 Ton		590.60		649.66		
1 Centr. Water Pump, 6"		296.40		326.04		
1 -20' Suction Hose, 6"		11.50		12.65		
1 -50' Discharge Hose, 6"		6.10		6.71		
1 Drill Rig, Truck-Mounted		2444.00		2688.40	38.05	41.86
88 L.H., Daily Totals		$7546.20		$10089.86	$85.75	$114.66

Crew No.	Bare Costs Hr.	Daily	Incl. Subs O&P Hr.	Daily	Cost Per Labor-Hour Bare Costs	Incl. O&P
Crew B-50						
2 Pile Driver Foremen (outside)	$53.30	$852.80	$84.05	$1344.80	$49.63	$77.01
6 Pile Drivers	51.30	2462.40	80.90	3883.20		
2 Equip. Oper. (crane)	56.10	897.60	84.60	1353.60		
1 Equip. Oper. (oiler)	48.60	388.80	73.30	586.40		
3 Laborers	39.85	956.40	60.70	1456.80		
1 Crawler Crane, 40 Ton		1370.00		1507.00		
1 Lead, 60' High		79.40		87.34		
1 Hammer, Diesel, 15K ft.-lbs.		623.20		685.52		
1 Air Compressor, 600 cfm		417.20		458.92		
2 -50' Air Hoses, 3"		86.40		95.04		
1 Chain Saw, Gas, 36" Long		46.40		51.04	23.42	25.76
112 L.H., Daily Totals		$8180.60		$11509.66	$73.04	$102.76
Crew B-51						
1 Labor Foreman (outside)	$41.85	$334.80	$63.75	$510.00	$40.96	$62.26
4 Laborers	39.85	1275.20	60.70	1942.40		
1 Truck Driver (light)	44.50	356.00	67.00	536.00		
1 Flatbed Truck, Gas, 1.5 Ton		188.40		207.24	3.92	4.32
48 L.H., Daily Totals		$2154.40		$3195.64	$44.88	$66.58
Crew B-52						
1 Carpenter Foreman (outside)	$52.70	$421.60	$80.25	$642.00	$46.39	$70.32
1 Carpenter	50.70	405.60	77.20	617.60		
3 Laborers	39.85	956.40	60.70	1456.80		
1 Cement Finisher	47.55	380.40	70.45	563.60		
.5 Rodman (reinf.)	54.65	218.60	83.45	333.80		
.5 Equip. Oper. (medium)	53.75	215.00	81.05	324.20		
.5 Crawler Loader, 3 C.Y.		602.00		662.20	10.75	11.82
56 L.H., Daily Totals		$3199.60		$4600.20	$57.14	$82.15
Crew B-53						
1 Equip. Oper. (light)	$51.30	$410.40	$77.35	$618.80	$51.30	$77.35
1 Trencher, Chain, 12 H.P.		61.80		67.98	7.72	8.50
8 L.H., Daily Totals		$472.20		$686.78	$59.02	$85.85
Crew B-54						
1 Equip. Oper. (light)	$51.30	$410.40	$77.35	$618.80	$51.30	$77.35
1 Trencher, Chain, 40 H.P.		336.40		370.04	42.05	46.26
8 L.H., Daily Totals		$746.80		$988.84	$93.35	$123.61
Crew B-54A						
.17 Labor Foreman (outside)	$41.85	$56.92	$63.75	$86.70	$52.02	$78.54
1 Equipment Operator (med.)	53.75	430.00	81.05	648.40		
1 Wheel Trencher, 67 H.P.		1095.00		1204.50	116.99	128.69
9.36 L.H., Daily Totals		$1581.92		$1939.60	$169.01	$207.22
Crew B-54B						
.25 Labor Foreman (outside)	$41.85	$83.70	$63.75	$127.50	$51.37	$77.59
1 Equipment Operator (med.)	53.75	430.00	81.05	648.40		
1 Wheel Trencher, 150 H.P.		1957.00		2152.70	195.70	215.27
10 L.H., Daily Totals		$2470.70		$2928.60	$247.07	$292.86
Crew B-54C						
1 Laborer	$39.85	$318.80	$60.70	$485.60	$46.80	$70.88
1 Equipment Operator (med.)	53.75	430.00	81.05	648.40		
1 Wheel Trencher, 67 H.P.		1095.00		1204.50	68.44	75.28
16 L.H., Daily Totals		$1843.80		$2338.50	$115.24	$146.16

Crew B-54D

Crew No.	Bare Costs Hr.	Daily	Incl. Subs O&P Hr.	Daily	Cost Per Labor-Hour Bare Costs	Incl. O&P
1 Laborer	$39.85	$318.80	$60.70	$485.60	$46.80	$70.88
1 Equipment Operator (med.)	53.75	430.00	81.05	648.40		
1 Rock Trencher, 6" Width		904.20		994.62	56.51	62.16
16 L.H., Daily Totals		$1653.00		$2128.62	$103.31	$133.04

Crew B-54E

Crew No.	Bare Costs Hr.	Daily	Incl. Subs O&P Hr.	Daily	Cost Per Labor-Hour Bare Costs	Incl. O&P
1 Laborer	$39.85	$318.80	$60.70	$485.60	$46.80	$70.88
1 Equipment Operator (med.)	53.75	430.00	81.05	648.40		
1 Rock Trencher, 18" Width		2681.00		2949.10	167.56	184.32
16 L.H., Daily Totals		$3429.80		$4083.10	$214.36	$255.19

Crew B-55

Crew No.	Bare Costs Hr.	Daily	Incl. Subs O&P Hr.	Daily	Cost Per Labor-Hour Bare Costs	Incl. O&P
2 Laborers	$39.85	$637.60	$60.70	$971.20	$41.40	$62.80
1 Truck Driver (light)	44.50	356.00	67.00	536.00		
1 Truck-Mounted Earth Auger		758.80		834.68		
1 Flatbed Truck, Gas, 3 Ton		238.00		261.80	41.53	45.69
24 L.H., Daily Totals		$1990.40		$2603.68	$82.93	$108.49

Crew B-56

Crew No.	Bare Costs Hr.	Daily	Incl. Subs O&P Hr.	Daily	Cost Per Labor-Hour Bare Costs	Incl. O&P
1 Laborer	$39.85	$318.80	$60.70	$485.60	$45.58	$69.03
1 Equip. Oper. (light)	51.30	410.40	77.35	618.80		
1 Air Track Drill, 4"		993.40		1092.74		
1 Air Compressor, 600 cfm		417.20		458.92		
1 -50' Air Hose, 3"		43.20		47.52	90.86	99.95
16 L.H., Daily Totals		$2183.00		$2703.58	$136.44	$168.97

Crew B-57

Crew No.	Bare Costs Hr.	Daily	Incl. Subs O&P Hr.	Daily	Cost Per Labor-Hour Bare Costs	Incl. O&P
1 Labor Foreman (outside)	$41.85	$334.80	$63.75	$510.00	$46.26	$70.07
2 Laborers	39.85	637.60	60.70	971.20		
1 Equip. Oper. (crane)	56.10	448.80	84.60	676.80		
1 Equip. Oper. (light)	51.30	410.40	77.35	618.80		
1 Equip. Oper. (oiler)	48.60	388.80	73.30	586.40		
1 Crawler Crane, 25 Ton		1348.00		1482.80		
1 Clamshell Bucket, 1 C.Y.		50.60		55.66		
1 Centr. Water Pump, 6"		296.40		326.04		
1 -20' Suction Hose, 6"		11.50		12.65		
20 -50' Discharge Hoses, 6"		122.00		134.20	38.09	41.90
48 L.H., Daily Totals		$4048.90		$5374.55	$84.35	$111.97

Crew B-58

Crew No.	Bare Costs Hr.	Daily	Incl. Subs O&P Hr.	Daily	Cost Per Labor-Hour Bare Costs	Incl. O&P
2 Laborers	$39.85	$637.60	$60.70	$971.20	$43.67	$66.25
1 Equip. Oper. (light)	51.30	410.40	77.35	618.80		
1 Backhoe Loader, 48 H.P.		312.40		343.64		
1 Small Helicopter, w/ Pilot		2857.00		3142.70	132.06	145.26
24 L.H., Daily Totals		$4217.40		$5076.34	$175.72	$211.51

Crew B-59

Crew No.	Bare Costs Hr.	Daily	Incl. Subs O&P Hr.	Daily	Cost Per Labor-Hour Bare Costs	Incl. O&P
1 Truck Driver (heavy)	$46.00	$368.00	$69.30	$554.40	$46.00	$69.30
1 Truck Tractor, 220 H.P.		290.00		319.00		
1 Water Tank Trailer, 5000 Gal.		147.60		162.36	54.70	60.17
8 L.H., Daily Totals		$805.60		$1035.76	$100.70	$129.47

Crew B-59A

Crew No.	Bare Costs Hr.	Daily	Incl. Subs O&P Hr.	Daily	Cost Per Labor-Hour Bare Costs	Incl. O&P
2 Laborers	$39.85	$637.60	$60.70	$971.20	$41.90	$63.57
1 Truck Driver (heavy)	46.00	368.00	69.30	554.40		
1 Water Tank Trailer, 5000 Gal.		147.60		162.36		
1 Truck Tractor, 220 H.P.		290.00		319.00	18.23	20.06
24 L.H., Daily Totals		$1443.20		$2006.96	$60.13	$83.62

Crew B-60

Crew No.	Bare Costs Hr.	Daily	Incl. Subs O&P Hr.	Daily	Cost Per Labor-Hour Bare Costs	Incl. O&P
1 Labor Foreman (outside)	$41.85	$334.80	$63.75	$510.00	$46.98	$71.11
2 Laborers	39.85	637.60	60.70	971.20		
1 Equip. Oper. (crane)	56.10	448.80	84.60	676.80		
2 Equip. Oper. (light)	51.30	820.80	77.35	1237.60		
1 Equip. Oper. (oiler)	48.60	388.80	73.30	586.40		
1 Crawler Crane, 40 Ton		1370.00		1507.00		
1 Lead, 60' High		79.40		87.34		
1 Hammer, Diesel, 15K ft.-lbs.		623.20		685.52		
1 Backhoe Loader, 48 H.P.		312.40		343.64	42.59	46.85
56 L.H., Daily Totals		$5015.80		$6605.50	$89.57	$117.96

Crew B-61

Crew No.	Bare Costs Hr.	Daily	Incl. Subs O&P Hr.	Daily	Cost Per Labor-Hour Bare Costs	Incl. O&P
1 Labor Foreman (outside)	$41.85	$334.80	$63.75	$510.00	$42.54	$64.64
3 Laborers	39.85	956.40	60.70	1456.80		
1 Equip. Oper. (light)	51.30	410.40	77.35	618.80		
1 Cement Mixer, 2 C.Y.		204.20		224.62		
1 Air Compressor, 160 cfm		117.20		128.92	8.04	8.84
40 L.H., Daily Totals		$2023.00		$2939.14	$50.58	$73.48

Crew B-62

Crew No.	Bare Costs Hr.	Daily	Incl. Subs O&P Hr.	Daily	Cost Per Labor-Hour Bare Costs	Incl. O&P
2 Laborers	$39.85	$637.60	$60.70	$971.20	$43.67	$66.25
1 Equip. Oper. (light)	51.30	410.40	77.35	618.80		
1 Loader, Skid Steer, 30 H.P.		174.00		191.40	7.25	7.97
24 L.H., Daily Totals		$1222.00		$1781.40	$50.92	$74.22

Crew B-62A

Crew No.	Bare Costs Hr.	Daily	Incl. Subs O&P Hr.	Daily	Cost Per Labor-Hour Bare Costs	Incl. O&P
2 Laborers	$39.85	$637.60	$60.70	$971.20	$43.67	$66.25
1 Equip. Oper. (light)	51.30	410.40	77.35	618.80		
1 Loader, Skid Steer, 30 H.P.		174.00		191.40		
1 Trencher Attachment		74.30		81.73	10.35	11.38
24 L.H., Daily Totals		$1296.30		$1863.13	$54.01	$77.63

Crew B-63

Crew No.	Bare Costs Hr.	Daily	Incl. Subs O&P Hr.	Daily	Cost Per Labor-Hour Bare Costs	Incl. O&P
4 Laborers	$39.85	$1275.20	$60.70	$1942.40	$42.14	$64.03
1 Equip. Oper. (light)	51.30	410.40	77.35	618.80		
1 Loader, Skid Steer, 30 H.P.		174.00		191.40	4.35	4.79
40 L.H., Daily Totals		$1859.60		$2752.60	$46.49	$68.81

Crew B-63B

Crew No.	Bare Costs Hr.	Daily	Incl. Subs O&P Hr.	Daily	Cost Per Labor-Hour Bare Costs	Incl. O&P
1 Labor Foreman (inside)	$40.35	$322.80	$61.45	$491.60	$42.84	$65.05
2 Laborers	39.85	637.60	60.70	971.20		
1 Equip. Oper. (light)	51.30	410.40	77.35	618.80		
1 Loader, Skid Steer, 78 H.P.		364.80		401.28	11.40	12.54
32 L.H., Daily Totals		$1735.60		$2482.88	$54.24	$77.59

Crew B-64

Crew No.	Bare Costs Hr.	Daily	Incl. Subs O&P Hr.	Daily	Cost Per Labor-Hour Bare Costs	Incl. O&P
1 Laborer	$39.85	$318.80	$60.70	$485.60	$42.17	$63.85
1 Truck Driver (light)	44.50	356.00	67.00	536.00		
1 Power Mulcher (small)		139.20		153.12		
1 Flatbed Truck, Gas, 1.5 Ton		188.40		207.24	20.48	22.52
16 L.H., Daily Totals		$1002.40		$1381.96	$62.65	$86.37

Crew B-65

Crew No.	Bare Costs Hr.	Daily	Incl. Subs O&P Hr.	Daily	Cost Per Labor-Hour Bare Costs	Incl. O&P
1 Laborer	$39.85	$318.80	$60.70	$485.60	$42.17	$63.85
1 Truck Driver (light)	44.50	356.00	67.00	536.00		
1 Power Mulcher (Large)		268.80		295.68		
1 Flatbed Truck, Gas, 1.5 Ton		188.40		207.24	28.57	31.43
16 L.H., Daily Totals		$1132.00		$1524.52	$70.75	$95.28

For customer support on your Electrical Costs with RSMeans data, call 800.448.8182.

493

Crew No.	Bare Costs Hr.	Bare Costs Daily	Incl. Subs O&P Hr.	Incl. Subs O&P Daily	Cost Per Labor-Hour Bare Costs	Cost Per Labor-Hour Incl. O&P
Crew B-66	Hr.	Daily	Hr.	Daily	Bare Costs	Incl. O&P
1 Equip. Oper. (light)	$51.30	$410.40	$77.35	$618.80	$51.30	$77.35
1 Loader-Backhoe, 40 H.P.		247.00		271.70	30.88	33.96
8 L.H., Daily Totals		$657.40		$890.50	$82.17	$111.31
Crew B-67	Hr.	Daily	Hr.	Daily	Bare Costs	Incl. O&P
1 Millwright	$52.75	$422.00	$76.95	$615.60	$52.02	$77.15
1 Equip. Oper. (light)	51.30	410.40	77.35	618.80		
1 R.T. Forklift, 5,000 Lb., diesel		267.60		294.36	16.73	18.40
16 L.H., Daily Totals		$1100.00		$1528.76	$68.75	$95.55
Crew B-67B	Hr.	Daily	Hr.	Daily	Bare Costs	Incl. O&P
1 Millwright Foreman (inside)	$53.25	$426.00	$77.70	$621.60	$53.00	$77.33
1 Millwright	52.75	422.00	76.95	615.60		
16 L.H., Daily Totals		$848.00		$1237.20	$53.00	$77.33
Crew B-68	Hr.	Daily	Hr.	Daily	Bare Costs	Incl. O&P
2 Millwrights	$52.75	$844.00	$76.95	$1231.20	$52.27	$77.08
1 Equip. Oper. (light)	51.30	410.40	77.35	618.80		
1 R.T. Forklift, 5,000 Lb., diesel		267.60		294.36	11.15	12.27
24 L.H., Daily Totals		$1522.00		$2144.36	$63.42	$89.35
Crew B-68A	Hr.	Daily	Hr.	Daily	Bare Costs	Incl. O&P
1 Millwright Foreman (inside)	$53.25	$426.00	$77.70	$621.60	$52.92	$77.20
2 Millwrights	52.75	844.00	76.95	1231.20		
1 Forklift, Smooth Floor, 8,000 Lb.		148.20		163.02	6.17	6.79
24 L.H., Daily Totals		$1418.20		$2015.82	$59.09	$83.99
Crew B-68B	Hr.	Daily	Hr.	Daily	Bare Costs	Incl. O&P
1 Millwright Foreman (inside)	$53.25	$426.00	$77.70	$621.60	$57.06	$84.69
2 Millwrights	52.75	844.00	76.95	1231.20		
2 Electricians	58.20	931.20	87.00	1392.00		
2 Plumbers	62.15	994.40	93.60	1497.60		
1 R.T. Forklift, 5,000 Lb., gas		279.00		306.90	4.98	5.48
56 L.H., Daily Totals		$3474.60		$5049.30	$62.05	$90.17
Crew B-68C	Hr.	Daily	Hr.	Daily	Bare Costs	Incl. O&P
1 Millwright Foreman (inside)	$53.25	$426.00	$77.70	$621.60	$56.59	$83.81
1 Millwright	52.75	422.00	76.95	615.60		
1 Electrician	58.20	465.60	87.00	696.00		
1 Plumber	62.15	497.20	93.60	748.80		
1 R.T. Forklift, 5,000 Lb., gas		279.00		306.90	8.72	9.59
32 L.H., Daily Totals		$2089.80		$2988.90	$65.31	$93.40
Crew B-68D	Hr.	Daily	Hr.	Daily	Bare Costs	Incl. O&P
1 Labor Foreman (inside)	$40.35	$322.80	$61.45	$491.60	$43.83	$66.50
1 Laborer	39.85	318.80	60.70	485.60		
1 Equip. Oper. (light)	51.30	410.40	77.35	618.80		
1 R.T. Forklift, 5,000 Lb., gas		279.00		306.90	11.63	12.79
24 L.H., Daily Totals		$1331.00		$1902.90	$55.46	$79.29
Crew B-68E	Hr.	Daily	Hr.	Daily	Bare Costs	Incl. O&P
1 Struc. Steel Foreman (inside)	$55.15	$441.20	$90.15	$721.20	$54.75	$89.51
3 Struc. Steel Workers	54.65	1311.60	89.35	2144.40		
1 Welder	54.65	437.20	89.35	714.80		
1 Forklift, Smooth Floor, 8,000 Lb.		148.20		163.02	3.71	4.08
40 L.H., Daily Totals		$2338.20		$3743.42	$58.45	$93.59

Crew No.	Bare Costs Hr.	Bare Costs Daily	Incl. Subs O&P Hr.	Incl. Subs O&P Daily	Cost Per Labor-Hour Bare Costs	Cost Per Labor-Hour Incl. O&P
Crew B-68F	Hr.	Daily	Hr.	Daily	Bare Costs	Incl. O&P
1 Skilled Worker Foreman (out)	$54.35	$434.80	$83.20	$665.60	$53.02	$81.17
2 Skilled Workers	52.35	837.60	80.15	1282.40		
1 R.T. Forklift, 5,000 Lb., gas		279.00		306.90	11.63	12.79
24 L.H., Daily Totals		$1551.40		$2254.90	$64.64	$93.95
Crew B-68G	Hr.	Daily	Hr.	Daily	Bare Costs	Incl. O&P
2 Structural Steel Workers	$54.65	$874.40	$89.35	$1429.60	$54.65	$89.35
1 R.T. Forklift, 5,000 Lb., gas		279.00		306.90	17.44	19.18
16 L.H., Daily Totals		$1153.40		$1736.50	$72.09	$108.53
Crew B-69	Hr.	Daily	Hr.	Daily	Bare Costs	Incl. O&P
1 Labor Foreman (outside)	$41.85	$334.80	$63.75	$510.00	$44.35	$67.29
3 Laborers	39.85	956.40	60.70	1456.80		
1 Equip. Oper. (crane)	56.10	448.80	84.60	676.80		
1 Equip. Oper. (oiler)	48.60	388.80	73.30	586.40		
1 Hyd. Crane, 80 Ton		1518.00		1669.80	31.63	34.79
48 L.H., Daily Totals		$3646.80		$4899.80	$75.97	$102.08
Crew B-69A	Hr.	Daily	Hr.	Daily	Bare Costs	Incl. O&P
1 Labor Foreman (outside)	$41.85	$334.80	$63.75	$510.00	$43.78	$66.22
3 Laborers	39.85	956.40	60.70	1456.80		
1 Equip. Oper. (medium)	53.75	430.00	81.05	648.40		
1 Concrete Finisher	47.55	380.40	70.45	563.60		
1 Curb/Gutter Paver, 2-Track		1173.00		1290.30	24.44	26.88
48 L.H., Daily Totals		$3274.60		$4469.10	$68.22	$93.11
Crew B-69B	Hr.	Daily	Hr.	Daily	Bare Costs	Incl. O&P
1 Labor Foreman (outside)	$41.85	$334.80	$63.75	$510.00	$43.78	$66.22
3 Laborers	39.85	956.40	60.70	1456.80		
1 Equip. Oper. (medium)	53.75	430.00	81.05	648.40		
1 Cement Finisher	47.55	380.40	70.45	563.60		
1 Curb/Gutter Paver, 4-Track		789.60		868.56	16.45	18.09
48 L.H., Daily Totals		$2891.20		$4047.36	$60.23	$84.32
Crew B-70	Hr.	Daily	Hr.	Daily	Bare Costs	Incl. O&P
1 Labor Foreman (outside)	$41.85	$334.80	$63.75	$510.00	$46.09	$69.86
3 Laborers	39.85	956.40	60.70	1456.80		
3 Equip. Oper. (medium)	53.75	1290.00	81.05	1945.20		
1 Grader, 30,000 Lbs.		643.80		708.18		
1 Ripper, Beam & 1 Shank		87.80		96.58		
1 Road Sweeper, S.P., 8' wide		688.60		757.46		
1 F.E. Loader, W.M., 1.5 C.Y.		342.80		377.08	31.48	34.63
56 L.H., Daily Totals		$4344.20		$5851.30	$77.58	$104.49
Crew B-70A	Hr.	Daily	Hr.	Daily	Bare Costs	Incl. O&P
1 Laborer	$39.85	$318.80	$60.70	$485.60	$50.97	$76.98
4 Equip. Oper. (medium)	53.75	1720.00	81.05	2593.60		
1 Grader, 40,000 Lbs.		1191.00		1310.10		
1 F.E. Loader, W.M., 2.5 C.Y.		523.20		575.52		
1 Dozer, 80 H.P.		464.40		510.84		
1 Roller, Pneum. Whl., 12 Ton		338.20		372.02	62.92	69.21
40 L.H., Daily Totals		$4555.60		$5847.68	$113.89	$146.19
Crew B-71	Hr.	Daily	Hr.	Daily	Bare Costs	Incl. O&P
1 Labor Foreman (outside)	$41.85	$334.80	$63.75	$510.00	$46.09	$69.86
3 Laborers	39.85	956.40	60.70	1456.80		
3 Equip. Oper. (medium)	53.75	1290.00	81.05	1945.20		
1 Pvmt. Profiler, 750 H.P.		5419.00		5960.90		
1 Road Sweeper, S.P., 8' wide		688.60		757.46		
1 F.E. Loader, W.M., 1.5 C.Y.		342.80		377.08	115.19	126.70
56 L.H., Daily Totals		$9031.60		$11007.44	$161.28	$196.56

Crew No.	Bare Costs		Incl. Subs O&P		Cost Per Labor-Hour	

Crew B-72

	Hr.	Daily	Hr.	Daily	Bare Costs	Incl. O&P
1 Labor Foreman (outside)	$41.85	$334.80	$63.75	$510.00	$47.05	$71.26
3 Laborers	39.85	956.40	60.70	1456.80		
4 Equip. Oper. (medium)	53.75	1720.00	81.05	2593.60		
1 Pvmt. Profiler, 750 H.P.		5419.00		5960.90		
1 Hammermill, 250 H.P.		1870.00		2057.00		
1 Windrow Loader		1242.00		1366.20		
1 Mix Paver, 165 H.P.		2140.00		2354.00		
1 Roller, Pneum. Whl., 12 Ton		338.20		372.02	172.02	189.22
64 L.H., Daily Totals		$14020.40		$16670.52	$219.07	$260.48

Crew B-73

	Hr.	Daily	Hr.	Daily	Bare Costs	Incl. O&P
1 Labor Foreman (outside)	$41.85	$334.80	$63.75	$510.00	$48.79	$73.80
2 Laborers	39.85	637.60	60.70	971.20		
5 Equip. Oper. (medium)	53.75	2150.00	81.05	3242.00		
1 Road Mixer, 310 H.P.		1889.00		2077.90		
1 Tandem Roller, 10 Ton		231.40		254.54		
1 Hammermill, 250 H.P.		1870.00		2057.00		
1 Grader, 30,000 Lbs.		643.80		708.18		
.5 F.E. Loader, W.M., 1.5 C.Y.		171.40		188.54		
.5 Truck Tractor, 220 H.P.		145.00		159.50		
.5 Water Tank Trailer, 5000 Gal.		73.80		81.18	78.51	86.36
64 L.H., Daily Totals		$8146.80		$10250.04	$127.29	$160.16

Crew B-74

	Hr.	Daily	Hr.	Daily	Bare Costs	Incl. O&P
1 Labor Foreman (outside)	$41.85	$334.80	$63.75	$510.00	$48.59	$73.41
1 Laborer	39.85	318.80	60.70	485.60		
4 Equip. Oper. (medium)	53.75	1720.00	81.05	2593.60		
2 Truck Drivers (heavy)	46.00	736.00	69.30	1108.80		
1 Grader, 30,000 Lbs.		643.80		708.18		
1 Ripper, Beam & 1 Shank		87.80		96.58		
2 Stabilizers, 310 H.P.		3436.00		3779.60		
1 Flatbed Truck, Gas, 3 Ton		238.00		261.80		
1 Chem. Spreader, Towed		52.20		57.42		
1 Roller, Vibratory, 25 Ton		659.80		725.78		
1 Water Tank Trailer, 5000 Gal.		147.60		162.36		
1 Truck Tractor, 220 H.P.		290.00		319.00	86.80	95.48
64 L.H., Daily Totals		$8664.80		$10808.72	$135.39	$168.89

Crew B-75

	Hr.	Daily	Hr.	Daily	Bare Costs	Incl. O&P
1 Labor Foreman (outside)	$41.85	$334.80	$63.75	$510.00	$48.96	$73.99
1 Laborer	39.85	318.80	60.70	485.60		
4 Equip. Oper. (medium)	53.75	1720.00	81.05	2593.60		
1 Truck Driver (heavy)	46.00	368.00	69.30	554.40		
1 Grader, 30,000 Lbs.		643.80		708.18		
1 Ripper, Beam & 1 Shank		87.80		96.58		
2 Stabilizers, 310 H.P.		3436.00		3779.60		
1 Dist. Tanker, 3000 Gallon		326.80		359.48		
1 Truck Tractor, 6x4, 380 H.P.		476.20		523.82		
1 Roller, Vibratory, 25 Ton		659.80		725.78	100.54	110.60
56 L.H., Daily Totals		$8372.00		$10337.04	$149.50	$184.59

Crew B-76

	Hr.	Daily	Hr.	Daily	Bare Costs	Incl. O&P
1 Dock Builder Foreman (outside)	$53.30	$426.40	$84.05	$672.40	$52.29	$81.23
5 Dock Builders	51.30	2052.00	80.90	3236.00		
2 Equip. Oper. (crane)	56.10	897.60	84.60	1353.60		
1 Equip. Oper. (oiler)	48.60	388.80	73.30	586.40		
1 Crawler Crane, 50 Ton		1940.00		2134.00		
1 Barge, 400 Ton		835.40		918.94		
1 Hammer, Diesel, 15K ft.-lbs.		623.20		685.52		
1 Lead, 60' High		79.40		87.34		
1 Air Compressor, 600 cfm		417.20		458.92		
2 -50' Air Hoses, 3"		86.40		95.04	55.30	60.83
72 L.H., Daily Totals		$7746.40		$10228.16	$107.59	$142.06

Crew B-76A

	Hr.	Daily	Hr.	Daily	Bare Costs	Incl. O&P
1 Labor Foreman (outside)	$41.85	$334.80	$63.75	$510.00	$43.23	$65.64
5 Laborers	39.85	1594.00	60.70	2428.00		
1 Equip. Oper. (crane)	56.10	448.80	84.60	676.80		
1 Equip. Oper. (oiler)	48.60	388.80	73.30	586.40		
1 Crawler Crane, 50 Ton		1940.00		2134.00		
1 Barge, 400 Ton		835.40		918.94	43.37	47.70
64 L.H., Daily Totals		$5541.80		$7254.14	$86.59	$113.35

Crew B-77

	Hr.	Daily	Hr.	Daily	Bare Costs	Incl. O&P
1 Labor Foreman (outside)	$41.85	$334.80	$63.75	$510.00	$41.18	$62.57
3 Laborers	39.85	956.40	60.70	1456.80		
1 Truck Driver (light)	44.50	356.00	67.00	536.00		
1 Crack Cleaner, 25 H.P.		55.80		61.38		
1 Crack Filler, Trailer Mtd.		198.20		218.02		
1 Flatbed Truck, Gas, 3 Ton		238.00		261.80	12.30	13.53
40 L.H., Daily Totals		$2139.20		$3044.00	$53.48	$76.10

Crew B-78

	Hr.	Daily	Hr.	Daily	Bare Costs	Incl. O&P
1 Labor Foreman (outside)	$41.85	$334.80	$63.75	$510.00	$40.96	$62.26
4 Laborers	39.85	1275.20	60.70	1942.40		
1 Truck Driver (light)	44.50	356.00	67.00	536.00		
1 Paint Striper, S.P., 40 Gallon		151.00		166.10		
1 Flatbed Truck, Gas, 3 Ton		238.00		261.80		
1 Pickup Truck, 3/4 Ton		115.20		126.72	10.50	11.55
48 L.H., Daily Totals		$2470.20		$3543.02	$51.46	$73.81

Crew B-78A

	Hr.	Daily	Hr.	Daily	Bare Costs	Incl. O&P
1 Equip. Oper. (light)	$51.30	$410.40	$77.35	$618.80	$51.30	$77.35
1 Line Rem. (Metal Balls) 115 H.P.		852.60		937.86	106.58	117.23
8 L.H., Daily Totals		$1263.00		$1556.66	$157.88	$194.58

Crew B-78B

	Hr.	Daily	Hr.	Daily	Bare Costs	Incl. O&P
2 Laborers	$39.85	$637.60	$60.70	$971.20	$41.12	$62.55
.25 Equip. Oper. (light)	51.30	102.60	77.35	154.70		
1 Pickup Truck, 3/4 Ton		115.20		126.72		
1 Line Rem.,11 H.P.,Walk Behind		61.40		67.54		
.25 Road Sweeper, S.P., 8' wide		172.15		189.37	19.38	21.31
18 L.H., Daily Totals		$1088.95		$1509.53	$60.50	$83.86

Crew B-78C

	Hr.	Daily	Hr.	Daily	Bare Costs	Incl. O&P
1 Labor Foreman (outside)	$41.85	$334.80	$63.75	$510.00	$40.96	$62.26
4 Laborers	39.85	1275.20	60.70	1942.40		
1 Truck Driver (light)	44.50	356.00	67.00	536.00		
1 Paint Striper, T.M., 120 Gal.		690.00		759.00		
1 Flatbed Truck, Gas, 3 Ton		238.00		261.80		
1 Pickup Truck, 3/4 Ton		115.20		126.72	21.73	23.91
48 L.H., Daily Totals		$3009.20		$4135.92	$62.69	$86.17

For customer support on your Electrical Costs with RSMeans data, call 800.448.8182.

495

Crew No.	Bare Costs			Incl. Subs O&P	Cost Per Labor-Hour	

Crew B-78D	Hr.	Daily	Hr.	Daily	Bare Costs	Incl. O&P
2 Labor Foremen (outside)	$41.85	$669.60	$63.75	$1020.00	$40.72	$61.94
7 Laborers	39.85	2231.60	60.70	3399.20		
1 Truck Driver (light)	44.50	356.00	67.00	536.00		
1 Paint Striper, T.M., 120 Gal.		690.00		759.00		
1 Flatbed Truck, Gas, 3 Ton		238.00		261.80		
3 Pickup Trucks, 3/4 Ton		345.60		380.16		
1 Air Compressor, 60 cfm		105.40		115.94		
1 -50' Air Hose, 3/4"		7.35		8.09		
1 Breaker, Pavement, 60 lb.		10.60		11.66	17.46	19.21
80 L.H., Daily Totals		$4654.15		$6491.85	$58.18	$81.15

Crew B-78E	Hr.	Daily	Hr.	Daily	Bare Costs	Incl. O&P
2 Labor Foremen (outside)	$41.85	$669.60	$63.75	$1020.00	$40.57	$61.73
9 Laborers	39.85	2869.20	60.70	4370.40		
1 Truck Driver (light)	44.50	356.00	67.00	536.00		
1 Paint Striper, T.M., 120 Gal.		690.00		759.00		
1 Flatbed Truck, Gas, 3 Ton		238.00		261.80		
4 Pickup Trucks, 3/4 Ton		460.80		506.88		
2 Air Compressors, 60 cfm		210.80		231.88		
2 -50' Air Hoses, 3/4"		14.70		16.17		
2 Breakers, Pavement, 60 lb.		21.20		23.32	17.04	18.74
96 L.H., Daily Totals		$5530.30		$7725.45	$57.61	$80.47

Crew B-78F	Hr.	Daily	Hr.	Daily	Bare Costs	Incl. O&P
2 Labor Foremen (outside)	$41.85	$669.60	$63.75	$1020.00	$40.47	$61.59
11 Laborers	39.85	3506.80	60.70	5341.60		
1 Truck Driver (light)	44.50	356.00	67.00	536.00		
1 Paint Striper, T.M., 120 Gal.		690.00		759.00		
1 Flatbed Truck, Gas, 3 Ton		238.00		261.80		
7 Pickup Trucks, 3/4 Ton		806.40		887.04		
3 Air Compressors, 60 cfm		316.20		347.82		
3 -50' Air Hoses, 3/4"		22.05		24.25		
3 Breakers, Pavement, 60 lb.		31.80		34.98	18.79	20.67
112 L.H., Daily Totals		$6636.85		$9212.50	$59.26	$82.25

Crew B-79	Hr.	Daily	Hr.	Daily	Bare Costs	Incl. O&P
1 Labor Foreman (outside)	$41.85	$334.80	$63.75	$510.00	$41.18	$62.57
3 Laborers	39.85	956.40	60.70	1456.80		
1 Truck Driver (light)	44.50	356.00	67.00	536.00		
1 Paint Striper, T.M., 120 Gal.		690.00		759.00		
1 Heating Kettle, 115 Gallon		77.85		85.64		
1 Flatbed Truck, Gas, 3 Ton		238.00		261.80		
2 Pickup Trucks, 3/4 Ton		230.40		253.44	30.91	34.00
40 L.H., Daily Totals		$2883.45		$3862.68	$72.09	$96.57

Crew B-79A	Hr.	Daily	Hr.	Daily	Bare Costs	Incl. O&P
1.5 Equip. Oper. (light)	$51.30	$615.60	$77.35	$928.20	$51.30	$77.35
.5 Line Remov. (Grinder) 115 H.P.		459.70		505.67		
1 Line Rem. (Metal Balls) 115 H.P.		852.60		937.86	109.36	120.29
12 L.H., Daily Totals		$1927.90		$2371.73	$160.66	$197.64

Crew B-79B	Hr.	Daily	Hr.	Daily	Bare Costs	Incl. O&P
1 Laborer	$39.85	$318.80	$60.70	$485.60	$39.85	$60.70
1 Set of Gases		168.00		184.80	21.00	23.10
8 L.H., Daily Totals		$486.80		$670.40	$60.85	$83.80

Crew B-79C	Hr.	Daily	Hr.	Daily	Bare Costs	Incl. O&P
1 Labor Foreman (outside)	$41.85	$334.80	$63.75	$510.00	$40.80	$62.04
5 Laborers	39.85	1594.00	60.70	2428.00		
1 Truck Driver (light)	44.50	356.00	67.00	536.00		
1 Paint Striper, T.M., 120 Gal.		690.00		759.00		
1 Heating Kettle, 115 Gallon		77.85		85.64		
1 Flatbed Truck, Gas, 3 Ton		238.00		261.80		
3 Pickup Trucks, 3/4 Ton		345.60		380.16		
1 Air Compressor, 60 cfm		105.40		115.94		
1 -50' Air Hose, 3/4"		7.35		8.09		
1 Breaker, Pavement, 60 lb.		10.60		11.66	26.34	28.97
56 L.H., Daily Totals		$3759.60		$5096.28	$67.14	$91.00

Crew B-79D	Hr.	Daily	Hr.	Daily	Bare Costs	Incl. O&P
2 Labor Foremen (outside)	$41.85	$669.60	$63.75	$1020.00	$40.93	$62.25
5 Laborers	39.85	1594.00	60.70	2428.00		
1 Truck Driver (light)	44.50	356.00	67.00	536.00		
1 Paint Striper, T.M., 120 Gal.		690.00		759.00		
1 Heating Kettle, 115 Gallon		77.85		85.64		
1 Flatbed Truck, Gas, 3 Ton		238.00		261.80		
4 Pickup Trucks, 3/4 Ton		460.80		506.88		
1 Air Compressor, 60 cfm		105.40		115.94		
1 -50' Air Hose, 3/4"		7.35		8.09		
1 Breaker, Pavement, 60 lb.		10.60		11.66	24.84	27.33
64 L.H., Daily Totals		$4209.60		$5733.00	$65.78	$89.58

Crew B-79E	Hr.	Daily	Hr.	Daily	Bare Costs	Incl. O&P
2 Labor Foremen (outside)	$41.85	$669.60	$63.75	$1020.00	$40.72	$61.94
7 Laborers	39.85	2231.60	60.70	3399.20		
1 Truck Driver (light)	44.50	356.00	67.00	536.00		
1 Paint Striper, T.M., 120 Gal.		690.00		759.00		
1 Heating Kettle, 115 Gallon		77.85		85.64		
1 Flatbed Truck, Gas, 3 Ton		238.00		261.80		
5 Pickup Trucks, 3/4 Ton		576.00		633.60		
2 Air Compressors, 60 cfm		210.80		231.88		
2 -50' Air Hoses, 3/4"		14.70		16.17		
2 Breakers, Pavement, 60 lb.		21.20		23.32	22.86	25.14
80 L.H., Daily Totals		$5085.75		$6966.60	$63.57	$87.08

Crew B-80	Hr.	Daily	Hr.	Daily	Bare Costs	Incl. O&P
1 Labor Foreman (outside)	$41.85	$334.80	$63.75	$510.00	$44.38	$67.20
1 Laborer	39.85	318.80	60.70	485.60		
1 Truck Driver (light)	44.50	356.00	67.00	536.00		
1 Equip. Oper. (light)	51.30	410.40	77.35	618.80		
1 Flatbed Truck, Gas, 3 Ton		238.00		261.80		
1 Earth Auger, Truck-Mtd.		373.80		411.18	19.12	21.03
32 L.H., Daily Totals		$2031.80		$2823.38	$63.49	$88.23

Crew B-80A	Hr.	Daily	Hr.	Daily	Bare Costs	Incl. O&P
3 Laborers	$39.85	$956.40	$60.70	$1456.80	$39.85	$60.70
1 Flatbed Truck, Gas, 3 Ton		238.00		261.80	9.92	10.91
24 L.H., Daily Totals		$1194.40		$1718.60	$49.77	$71.61

Crew B-80B	Hr.	Daily	Hr.	Daily	Bare Costs	Incl. O&P
3 Laborers	$39.85	$956.40	$60.70	$1456.80	$42.71	$64.86
1 Equip. Oper. (light)	51.30	410.40	77.35	618.80		
1 Crane, Flatbed Mounted, 3 Ton		235.80		259.38	7.37	8.11
32 L.H., Daily Totals		$1602.60		$2334.98	$50.08	$72.97

Crew B-80C

Crew No.	Bare Costs Hr.	Bare Costs Daily	Incl. Subs O&P Hr.	Incl. Subs O&P Daily	Cost Per Labor-Hour Bare Costs	Cost Per Labor-Hour Incl. O&P
2 Laborers	$39.85	$637.60	$60.70	$971.20	$41.40	$62.80
1 Truck Driver (light)	44.50	356.00	67.00	536.00		
1 Flatbed Truck, Gas, 1.5 Ton		188.40		207.24		
1 Manual Fence Post Auger, Gas		7.40		8.14	8.16	8.97
24 L.H., Daily Totals		$1189.40		$1722.58	$49.56	$71.77

Crew B-81

Crew No.	Bare Costs Hr.	Bare Costs Daily	Incl. Subs O&P Hr.	Incl. Subs O&P Daily	Bare Costs	Incl. O&P
1 Laborer	$39.85	$318.80	$60.70	$485.60	$46.53	$70.35
1 Equip. Oper. (medium)	53.75	430.00	81.05	648.40		
1 Truck Driver (heavy)	46.00	368.00	69.30	554.40		
1 Hydromulcher, T.M., 3000 Gal.		282.80		311.08		
1 Truck Tractor, 220 H.P.		290.00		319.00	23.87	26.25
24 L.H., Daily Totals		$1689.60		$2318.48	$70.40	$96.60

Crew B-81A

Crew No.	Bare Costs Hr.	Bare Costs Daily	Incl. Subs O&P Hr.	Incl. Subs O&P Daily	Bare Costs	Incl. O&P
1 Laborer	$39.85	$318.80	$60.70	$485.60	$42.17	$63.85
1 Truck Driver (light)	44.50	356.00	67.00	536.00		
1 Hydromulcher, T.M., 600 Gal.		121.20		133.32		
1 Flatbed Truck, Gas, 3 Ton		238.00		261.80	22.45	24.70
16 L.H., Daily Totals		$1034.00		$1416.72	$64.63	$88.55

Crew B-82

Crew No.	Bare Costs Hr.	Bare Costs Daily	Incl. Subs O&P Hr.	Incl. Subs O&P Daily	Bare Costs	Incl. O&P
1 Laborer	$39.85	$318.80	$60.70	$485.60	$45.58	$69.03
1 Equip. Oper. (light)	51.30	410.40	77.35	618.80		
1 Horiz. Borer, 6 H.P.		78.00		85.80	4.88	5.36
16 L.H., Daily Totals		$807.20		$1190.20	$50.45	$74.39

Crew B-82A

Crew No.	Bare Costs Hr.	Bare Costs Daily	Incl. Subs O&P Hr.	Incl. Subs O&P Daily	Bare Costs	Incl. O&P
2 Laborers	$39.85	$637.60	$60.70	$971.20	$45.58	$69.03
2 Equip. Opers. (light)	51.30	820.80	77.35	1237.60		
2 Dump Trucks, 8 C.Y., 220 H.P.		677.20		744.92		
1 Flatbed Trailer, 25 Ton		133.00		146.30		
1 Horiz. Dir. Drill, 20k lb. Thrust		629.00		691.90		
1 Mud Trailer for HDD, 1500 Gal.		287.80		316.58		
1 Pickup Truck, 4x4, 3/4 Ton		126.60		139.26		
1 Flatbed Trailer, 3 Ton		25.40		27.94		
1 Loader, Skid Steer, 78 H.P.		364.80		401.28	70.12	77.13
32 L.H., Daily Totals		$3702.20		$4676.98	$115.69	$146.16

Crew B-82B

Crew No.	Bare Costs Hr.	Bare Costs Daily	Incl. Subs O&P Hr.	Incl. Subs O&P Daily	Bare Costs	Incl. O&P
2 Laborers	$39.85	$637.60	$60.70	$971.20	$45.58	$69.03
2 Equip. Opers. (light)	51.30	820.80	77.35	1237.60		
2 Dump Trucks, 8 C.Y., 220 H.P.		677.20		744.92		
1 Flatbed Trailer, 25 Ton		133.00		146.30		
1 Horiz. Dir. Drill, 30k lb. Thrust		896.20		985.82		
1 Mud Trailer for HDD, 1500 Gal.		287.80		316.58		
1 Pickup Truck, 4x4, 3/4 Ton		126.60		139.26		
1 Flatbed Trailer, 3 Ton		25.40		27.94		
1 Loader, Skid Steer, 78 H.P.		364.80		401.28	78.47	86.32
32 L.H., Daily Totals		$3969.40		$4970.90	$124.04	$155.34

Crew B-82C

Crew No.	Bare Costs Hr.	Bare Costs Daily	Incl. Subs O&P Hr.	Incl. Subs O&P Daily	Bare Costs	Incl. O&P
2 Laborers	$39.85	$637.60	$60.70	$971.20	$45.58	$69.03
2 Equip. Opers. (light)	51.30	820.80	77.35	1237.60		
2 Dump Trucks, 8 C.Y., 220 H.P.		677.20		744.92		
1 Flatbed Trailer, 25 Ton		133.00		146.30		
1 Horiz. Dir. Drill, 50k lb. Thrust		1188.00		1306.80		
1 Mud Trailer for HDD, 1500 Gal.		287.80		316.58		
1 Pickup Truck, 4x4, 3/4 Ton		126.60		139.26		
1 Flatbed Trailer, 3 Ton		25.40		27.94		
1 Loader, Skid Steer, 78 H.P.		364.80		401.28	87.59	96.35
32 L.H., Daily Totals		$4261.20		$5291.88	$133.16	$165.37

Crew B-82D

Crew No.	Bare Costs Hr.	Bare Costs Daily	Incl. Subs O&P Hr.	Incl. Subs O&P Daily	Bare Costs	Incl. O&P
1 Equip. Oper. (light)	$51.30	$410.40	$77.35	$618.80	$51.30	$77.35
1 Mud Trailer for HDD, 1500 Gal.		287.80		316.58	35.98	39.57
8 L.H., Daily Totals		$698.20		$935.38	$87.28	$116.92

Crew B-83

Crew No.	Bare Costs Hr.	Bare Costs Daily	Incl. Subs O&P Hr.	Incl. Subs O&P Daily	Bare Costs	Incl. O&P
1 Tugboat Captain	$53.75	$430.00	$81.05	$648.40	$46.80	$70.88
1 Tugboat Hand	39.85	318.80	60.70	485.60		
1 Tugboat, 250 H.P.		680.20		748.22	42.51	46.76
16 L.H., Daily Totals		$1429.00		$1882.22	$89.31	$117.64

Crew B-84

Crew No.	Bare Costs Hr.	Bare Costs Daily	Incl. Subs O&P Hr.	Incl. Subs O&P Daily	Bare Costs	Incl. O&P
1 Equip. Oper. (medium)	$53.75	$430.00	$81.05	$648.40	$53.75	$81.05
1 Rotary Mower/Tractor		357.00		392.70	44.63	49.09
8 L.H., Daily Totals		$787.00		$1041.10	$98.38	$130.14

Crew B-85

Crew No.	Bare Costs Hr.	Bare Costs Daily	Incl. Subs O&P Hr.	Incl. Subs O&P Daily	Bare Costs	Incl. O&P
3 Laborers	$39.85	$956.40	$60.70	$1456.80	$43.86	$66.49
1 Equip. Oper. (medium)	53.75	430.00	81.05	648.40		
1 Truck Driver (heavy)	46.00	368.00	69.30	554.40		
1 Telescoping Boom Lift, to 80'		548.60		603.46		
1 Brush Chipper, 12", 130 H.P.		393.00		432.30		
1 Pruning Saw, Rotary		6.65		7.32	23.71	26.08
40 L.H., Daily Totals		$2702.65		$3702.68	$67.57	$92.57

Crew B-86

Crew No.	Bare Costs Hr.	Bare Costs Daily	Incl. Subs O&P Hr.	Incl. Subs O&P Daily	Bare Costs	Incl. O&P
1 Equip. Oper. (medium)	$53.75	$430.00	$81.05	$648.40	$53.75	$81.05
1 Stump Chipper, S.P.		184.05		202.46	23.01	25.31
8 L.H., Daily Totals		$614.05		$850.86	$76.76	$106.36

Crew B-86A

Crew No.	Bare Costs Hr.	Bare Costs Daily	Incl. Subs O&P Hr.	Incl. Subs O&P Daily	Bare Costs	Incl. O&P
1 Equip. Oper. (medium)	$53.75	$430.00	$81.05	$648.40	$53.75	$81.05
1 Grader, 30,000 Lbs.		643.80		708.18	80.47	88.52
8 L.H., Daily Totals		$1073.80		$1356.58	$134.22	$169.57

Crew B-86B

Crew No.	Bare Costs Hr.	Bare Costs Daily	Incl. Subs O&P Hr.	Incl. Subs O&P Daily	Bare Costs	Incl. O&P
1 Equip. Oper. (medium)	$53.75	$430.00	$81.05	$648.40	$53.75	$81.05
1 Dozer, 200 H.P.		1273.00		1400.30	159.13	175.04
8 L.H., Daily Totals		$1703.00		$2048.70	$212.88	$256.09

Crew B-87

Crew No.	Bare Costs Hr.	Bare Costs Daily	Incl. Subs O&P Hr.	Incl. Subs O&P Daily	Bare Costs	Incl. O&P
1 Laborer	$39.85	$318.80	$60.70	$485.60	$50.97	$76.98
4 Equip. Oper. (medium)	53.75	1720.00	81.05	2593.60		
2 Feller Bunchers, 100 H.P.		1620.80		1782.88		
1 Log Chipper, 22" Tree		764.80		841.28		
1 Dozer, 105 H.P.		608.00		668.80		
1 Chain Saw, Gas, 36" Long		46.40		51.04	76.00	83.60
40 L.H., Daily Totals		$5078.80		$6423.20	$126.97	$160.58

Crew No.	Bare Costs		Incl. Subs O&P		Cost Per Labor-Hour	

Crew B-88

Crew B-88	Hr.	Daily	Hr.	Daily	Bare Costs	Incl. O&P
1 Laborer	$39.85	$318.80	$60.70	$485.60	$51.76	$78.14
6 Equip. Oper. (medium)	53.75	2580.00	81.05	3890.40		
2 Feller Bunchers, 100 H.P.		1620.80		1782.88		
1 Log Chipper, 22" Tree		764.80		841.28		
2 Log Skidders, 50 H.P.		1698.80		1868.68		
1 Dozer, 105 H.P.		608.00		668.80		
1 Chain Saw, Gas, 36" Long		46.40		51.04	84.62	93.08
56 L.H., Daily Totals		$7637.60		$9588.68	$136.39	$171.23

Crew B-89	Hr.	Daily	Hr.	Daily	Bare Costs	Incl. O&P
1 Equip. Oper. (light)	$51.30	$410.40	$77.35	$618.80	$47.90	$72.17
1 Truck Driver (light)	44.50	356.00	67.00	536.00		
1 Flatbed Truck, Gas, 3 Ton		238.00		261.80		
1 Concrete Saw		102.40		112.64		
1 Water Tank, 65 Gal.		79.90		87.89	26.27	28.90
16 L.H., Daily Totals		$1186.70		$1617.13	$74.17	$101.07

Crew B-89A	Hr.	Daily	Hr.	Daily	Bare Costs	Incl. O&P
1 Skilled Worker	$52.35	$418.80	$80.15	$641.20	$46.10	$70.42
1 Laborer	39.85	318.80	60.70	485.60		
1 Core Drill (Large)		112.20		123.42	7.01	7.71
16 L.H., Daily Totals		$849.80		$1250.22	$53.11	$78.14

Crew B-89B	Hr.	Daily	Hr.	Daily	Bare Costs	Incl. O&P
1 Equip. Oper. (light)	$51.30	$410.40	$77.35	$618.80	$47.90	$72.17
1 Truck Driver (light)	44.50	356.00	67.00	536.00		
1 Wall Saw, Hydraulic, 10 H.P.		46.40		51.04		
1 Generator, Diesel, 100 kW		307.60		338.36		
1 Water Tank, 65 Gal.		79.90		87.89		
1 Flatbed Truck, Gas, 3 Ton		238.00		261.80	41.99	46.19
16 L.H., Daily Totals		$1438.30		$1893.89	$89.89	$118.37

Crew B-90	Hr.	Daily	Hr.	Daily	Bare Costs	Incl. O&P
1 Labor Foreman (outside)	$41.85	$334.80	$63.75	$510.00	$44.50	$67.39
3 Laborers	39.85	956.40	60.70	1456.80		
2 Equip. Oper. (light)	51.30	820.80	77.35	1237.60		
2 Truck Drivers (heavy)	46.00	736.00	69.30	1108.80		
1 Road Mixer, 310 H.P.		1889.00		2077.90		
1 Dist. Truck, 2000 Gal.		296.00		325.60	34.14	37.55
64 L.H., Daily Totals		$5033.00		$6716.70	$78.64	$104.95

Crew B-90A	Hr.	Daily	Hr.	Daily	Bare Costs	Incl. O&P
1 Labor Foreman (outside)	$41.85	$334.80	$63.75	$510.00	$48.08	$72.76
2 Laborers	39.85	637.60	60.70	971.20		
4 Equip. Oper. (medium)	53.75	1720.00	81.05	2593.60		
2 Graders, 30,000 Lbs.		1287.60		1416.36		
1 Tandem Roller, 10 Ton		231.40		254.54		
1 Roller, Pneum. Whl., 12 Ton		338.20		372.02	33.16	36.48
56 L.H., Daily Totals		$4549.60		$6117.72	$81.24	$109.25

Crew B-90B	Hr.	Daily	Hr.	Daily	Bare Costs	Incl. O&P
1 Labor Foreman (outside)	$41.85	$334.80	$63.75	$510.00	$47.13	$71.38
2 Laborers	39.85	637.60	60.70	971.20		
3 Equip. Oper. (medium)	53.75	1290.00	81.05	1945.20		
1 Roller, Pneum. Whl., 12 Ton		338.20		372.02		
1 Road Mixer, 310 H.P.		1889.00		2077.90	46.40	51.04
48 L.H., Daily Totals		$4489.60		$5876.32	$93.53	$122.42

Crew B-90C	Hr.	Daily	Hr.	Daily	Bare Costs	Incl. O&P
1 Labor Foreman (outside)	$41.85	$334.80	$63.75	$510.00	$45.50	$68.87
4 Laborers	39.85	1275.20	60.70	1942.40		
3 Equip. Oper. (medium)	53.75	1290.00	81.05	1945.20		
3 Truck Drivers (heavy)	46.00	1104.00	69.30	1663.20		
3 Road Mixers, 310 H.P.		5667.00		6233.70	64.40	70.84
88 L.H., Daily Totals		$9671.00		$12294.50	$109.90	$139.71

Crew B-90D	Hr.	Daily	Hr.	Daily	Bare Costs	Incl. O&P
1 Labor Foreman (outside)	$41.85	$334.80	$63.75	$510.00	$44.63	$67.62
6 Laborers	39.85	1912.80	60.70	2913.60		
3 Equip. Oper. (medium)	53.75	1290.00	81.05	1945.20		
3 Truck Drivers (heavy)	46.00	1104.00	69.30	1663.20		
3 Road Mixers, 310 H.P.		5667.00		6233.70	54.49	59.94
104 L.H., Daily Totals		$10308.60		$13265.70	$99.12	$127.55

Crew B-90E	Hr.	Daily	Hr.	Daily	Bare Costs	Incl. O&P
1 Labor Foreman (outside)	$41.85	$334.80	$63.75	$510.00	$45.39	$68.78
4 Laborers	39.85	1275.20	60.70	1942.40		
3 Equip. Oper. (medium)	53.75	1290.00	81.05	1945.20		
1 Truck Driver (heavy)	46.00	368.00	69.30	554.40		
1 Road Mixer, 310 H.P.		1889.00		2077.90	26.24	28.86
72 L.H., Daily Totals		$5157.00		$7029.90	$71.63	$97.64

Crew B-91	Hr.	Daily	Hr.	Daily	Bare Costs	Incl. O&P
1 Labor Foreman (outside)	$41.85	$334.80	$63.75	$510.00	$47.82	$72.33
2 Laborers	39.85	637.60	60.70	971.20		
4 Equip. Oper. (medium)	53.75	1720.00	81.05	2593.60		
1 Truck Driver (heavy)	46.00	368.00	69.30	554.40		
1 Dist. Tanker, 3000 Gallon		326.80		359.48		
1 Truck Tractor, 6x4, 380 H.P.		476.20		523.82		
1 Aggreg. Spreader, S.P.		837.00		920.70		
1 Roller, Pneum. Whl., 12 Ton		338.20		372.02		
1 Tandem Roller, 10 Ton		231.40		254.54	34.52	37.98
64 L.H., Daily Totals		$5270.00		$7059.76	$82.34	$110.31

Crew B-91B	Hr.	Daily	Hr.	Daily	Bare Costs	Incl. O&P
1 Laborer	$39.85	$318.80	$60.70	$485.60	$46.80	$70.88
1 Equipment Oper. (med.)	53.75	430.00	81.05	648.40		
1 Road Sweeper, Vac. Assist.		840.40		924.44	52.52	57.78
16 L.H., Daily Totals		$1589.20		$2058.44	$99.33	$128.65

Crew B-91C	Hr.	Daily	Hr.	Daily	Bare Costs	Incl. O&P
1 Laborer	$39.85	$318.80	$60.70	$485.60	$42.17	$63.85
1 Truck Driver (light)	44.50	356.00	67.00	536.00		
1 Catch Basin Cleaning Truck		522.80		575.08	32.67	35.94
16 L.H., Daily Totals		$1197.60		$1596.68	$74.85	$99.79

Crew B-91D	Hr.	Daily	Hr.	Daily	Bare Costs	Incl. O&P
1 Labor Foreman (outside)	$41.85	$334.80	$63.75	$510.00	$46.30	$70.08
5 Laborers	39.85	1594.00	60.70	2428.00		
5 Equip. Oper. (medium)	53.75	2150.00	81.05	3242.00		
2 Truck Drivers (heavy)	46.00	736.00	69.30	1108.80		
1 Aggreg. Spreader, S.P.		837.00		920.70		
2 Truck Tractors, 6x4, 380 H.P.		952.40		1047.64		
2 Dist. Tankers, 3000 Gallon		653.60		718.96		
2 Pavement Brushes, Towed		170.40		187.44		
2 Rollers Pneum. Whl., 12 Ton		676.40		744.04	31.63	34.80
104 L.H., Daily Totals		$8104.60		$10907.58	$77.93	$104.88

Crew No.	Bare Costs		Incl. Subs O&P		Cost Per Labor-Hour	
Crew B-92	Hr.	Daily	Hr.	Daily	Bare Costs	Incl. O&P
1 Labor Foreman (outside)	$41.85	$334.80	$63.75	$510.00	$40.35	$61.46
3 Laborers	39.85	956.40	60.70	1456.80		
1 Crack Cleaner, 25 H.P.		55.80		61.38		
1 Air Compressor, 60 cfm		105.40		115.94		
1 Tar Kettle, T.M.		129.20		142.12		
1 Flatbed Truck, Gas, 3 Ton		238.00		261.80	16.51	18.16
32 L.H., Daily Totals		$1819.60		$2548.04	$56.86	$79.63
Crew B-93	Hr.	Daily	Hr.	Daily	Bare Costs	Incl. O&P
1 Equip. Oper. (medium)	$53.75	$430.00	$81.05	$648.40	$53.75	$81.05
1 Feller Buncher, 100 H.P.		810.40		891.44	101.30	111.43
8 L.H., Daily Totals		$1240.40		$1539.84	$155.05	$192.48
Crew B-94A	Hr.	Daily	Hr.	Daily	Bare Costs	Incl. O&P
1 Laborer	$39.85	$318.80	$60.70	$485.60	$39.85	$60.70
1 Diaphragm Water Pump, 2"		73.00		80.30		
1 -20' Suction Hose, 2"		1.95		2.15		
2 -50' Discharge Hoses, 2"		1.80		1.98	9.59	10.55
8 L.H., Daily Totals		$395.55		$570.02	$49.44	$71.25
Crew B-94B	Hr.	Daily	Hr.	Daily	Bare Costs	Incl. O&P
1 Laborer	$39.85	$318.80	$60.70	$485.60	$39.85	$60.70
1 Diaphragm Water Pump, 4"		114.80		126.28		
1 -20' Suction Hose, 4"		3.25		3.58		
2 -50' Discharge Hoses, 4"		4.70		5.17	15.34	16.88
8 L.H., Daily Totals		$441.55		$620.63	$55.19	$77.58
Crew B-94C	Hr.	Daily	Hr.	Daily	Bare Costs	Incl. O&P
1 Laborer	$39.85	$318.80	$60.70	$485.60	$39.85	$60.70
1 Centrifugal Water Pump, 3"		79.20		87.12		
1 -20' Suction Hose, 3"		2.85		3.13		
2 -50' Discharge Hoses, 3"		3.00		3.30	10.63	11.69
8 L.H., Daily Totals		$403.85		$579.15	$50.48	$72.39
Crew B-94D	Hr.	Daily	Hr.	Daily	Bare Costs	Incl. O&P
1 Laborer	$39.85	$318.80	$60.70	$485.60	$39.85	$60.70
1 Centr. Water Pump, 6"		296.40		326.04		
1 -20' Suction Hose, 6"		11.50		12.65		
2 -50' Discharge Hoses, 6"		12.20		13.42	40.01	44.01
8 L.H., Daily Totals		$638.90		$837.71	$79.86	$104.71
Crew C-1	Hr.	Daily	Hr.	Daily	Bare Costs	Incl. O&P
3 Carpenters	$50.70	$1216.80	$77.20	$1852.80	$47.99	$73.08
1 Laborer	39.85	318.80	60.70	485.60		
32 L.H., Daily Totals		$1535.60		$2338.40	$47.99	$73.08
Crew C-2	Hr.	Daily	Hr.	Daily	Bare Costs	Incl. O&P
1 Carpenter Foreman (outside)	$52.70	$421.60	$80.25	$642.00	$49.23	$74.96
4 Carpenters	50.70	1622.40	77.20	2470.40		
1 Laborer	39.85	318.80	60.70	485.60		
48 L.H., Daily Totals		$2362.80		$3598.00	$49.23	$74.96
Crew C-2A	Hr.	Daily	Hr.	Daily	Bare Costs	Incl. O&P
1 Carpenter Foreman (outside)	$52.70	$421.60	$80.25	$642.00	$48.70	$73.83
3 Carpenters	50.70	1216.80	77.20	1852.80		
1 Cement Finisher	47.55	380.40	70.45	563.60		
1 Laborer	39.85	318.80	60.70	485.60		
48 L.H., Daily Totals		$2337.60		$3544.00	$48.70	$73.83

Crew No.	Bare Costs		Incl. Subs O&P		Cost Per Labor-Hour	
Crew C-3	Hr.	Daily	Hr.	Daily	Bare Costs	Incl. O&P
1 Rodman Foreman (outside)	$56.65	$453.20	$86.50	$692.00	$50.78	$77.38
4 Rodmen (reinf.)	54.65	1748.80	83.45	2670.40		
1 Equip. Oper. (light)	51.30	410.40	77.35	618.80		
2 Laborers	39.85	637.60	60.70	971.20		
3 Stressing Equipment		31.20		34.32		
.5 Grouting Equipment		79.20		87.12	1.73	1.90
64 L.H., Daily Totals		$3360.40		$5073.84	$52.51	$79.28
Crew C-4	Hr.	Daily	Hr.	Daily	Bare Costs	Incl. O&P
1 Rodman Foreman (outside)	$56.65	$453.20	$86.50	$692.00	$55.15	$84.21
3 Rodmen (reinf.)	54.65	1311.60	83.45	2002.80		
3 Stressing Equipment		31.20		34.32	.97	1.07
32 L.H., Daily Totals		$1796.00		$2729.12	$56.13	$85.28
Crew C-4A	Hr.	Daily	Hr.	Daily	Bare Costs	Incl. O&P
2 Rodmen (reinf.)	$54.65	$874.40	$83.45	$1335.20	$54.65	$83.45
4 Stressing Equipment		41.60		45.76	2.60	2.86
16 L.H., Daily Totals		$916.00		$1380.96	$57.25	$86.31
Crew C-5	Hr.	Daily	Hr.	Daily	Bare Costs	Incl. O&P
1 Rodman Foreman (outside)	$56.65	$453.20	$86.50	$692.00	$54.28	$82.60
4 Rodmen (reinf.)	54.65	1748.80	83.45	2670.40		
1 Equip. Oper. (crane)	56.10	448.80	84.60	676.80		
1 Equip. Oper. (oiler)	48.60	388.80	73.30	586.40		
1 Hyd. Crane, 25 Ton		590.60		649.66	10.55	11.60
56 L.H., Daily Totals		$3630.20		$5275.26	$64.83	$94.20
Crew C-6	Hr.	Daily	Hr.	Daily	Bare Costs	Incl. O&P
1 Labor Foreman (outside)	$41.85	$334.80	$63.75	$510.00	$41.47	$62.83
4 Laborers	39.85	1275.20	60.70	1942.40		
1 Cement Finisher	47.55	380.40	70.45	563.60		
2 Gas Engine Vibrators		51.20		56.32	1.07	1.17
48 L.H., Daily Totals		$2041.60		$3072.32	$42.53	$64.01
Crew C-7	Hr.	Daily	Hr.	Daily	Bare Costs	Incl. O&P
1 Labor Foreman (outside)	$41.85	$334.80	$63.75	$510.00	$43.44	$65.78
5 Laborers	39.85	1594.00	60.70	2428.00		
1 Cement Finisher	47.55	380.40	70.45	563.60		
1 Equip. Oper. (medium)	53.75	430.00	81.05	648.40		
1 Equip. Oper. (oiler)	48.60	388.80	73.30	586.40		
2 Gas Engine Vibrators		51.20		56.32		
1 Concrete Bucket, 1 C.Y.		24.60		27.06		
1 Hyd. Crane, 55 Ton		993.80		1093.18	14.86	16.34
72 L.H., Daily Totals		$4197.60		$5912.96	$58.30	$82.12
Crew C-7A	Hr.	Daily	Hr.	Daily	Bare Costs	Incl. O&P
1 Labor Foreman (outside)	$41.85	$334.80	$63.75	$510.00	$41.64	$63.23
5 Laborers	39.85	1594.00	60.70	2428.00		
2 Truck Drivers (heavy)	46.00	736.00	69.30	1108.80		
2 Conc. Transit Mixers		1811.20		1992.32	28.30	31.13
64 L.H., Daily Totals		$4476.00		$6039.12	$69.94	$94.36
Crew C-7B	Hr.	Daily	Hr.	Daily	Bare Costs	Incl. O&P
1 Labor Foreman (outside)	$41.85	$334.80	$63.75	$510.00	$43.23	$65.64
5 Laborers	39.85	1594.00	60.70	2428.00		
1 Equipment Operator, Crane	56.10	448.80	84.60	676.80		
1 Equipment Oiler	48.60	388.80	73.30	586.40		
1 Conc. Bucket, 2 C.Y.		38.40		42.24		
1 Lattice Boom Crane, 165 Ton		2140.00		2354.00	34.04	37.44
64 L.H., Daily Totals		$4944.80		$6597.44	$77.26	$103.09

Crews - Standard

Crew C-7C

Crew C-7C	Bare Costs Hr.	Bare Costs Daily	Incl. Subs O&P Hr.	Incl. Subs O&P Daily	Cost Per LH Bare Costs	Cost Per LH Incl. O&P
1 Labor Foreman (outside)	$41.85	$334.80	$63.75	$510.00	$43.58	$66.17
5 Laborers	39.85	1594.00	60.70	2428.00		
2 Equipment Operators (med.)	53.75	860.00	81.05	1296.80		
2 F.E. Loaders, W.M., 4 C.Y.		1149.20		1264.12	17.96	19.75
64 L.H., Daily Totals		$3938.00		$5498.92	$61.53	$85.92

Crew C-7D

Crew C-7D	Bare Costs Hr.	Bare Costs Daily	Incl. Subs O&P Hr.	Incl. Subs O&P Daily	Cost Per LH Bare Costs	Cost Per LH Incl. O&P
1 Labor Foreman (outside)	$41.85	$334.80	$63.75	$510.00	$42.12	$64.04
5 Laborers	39.85	1594.00	60.70	2428.00		
1 Equip. Oper. (medium)	53.75	430.00	81.05	648.40		
1 Concrete Conveyer		187.20		205.92	3.34	3.68
56 L.H., Daily Totals		$2546.00		$3792.32	$45.46	$67.72

Crew C-8

Crew C-8	Bare Costs Hr.	Bare Costs Daily	Incl. Subs O&P Hr.	Incl. Subs O&P Daily	Cost Per LH Bare Costs	Cost Per LH Incl. O&P
1 Labor Foreman (outside)	$41.85	$334.80	$63.75	$510.00	$44.32	$66.83
3 Laborers	39.85	956.40	60.70	1456.80		
2 Cement Finishers	47.55	760.80	70.45	1127.20		
1 Equip. Oper. (medium)	53.75	430.00	81.05	648.40		
1 Concrete Pump (Small)		881.60		969.76	15.74	17.32
56 L.H., Daily Totals		$3363.60		$4712.16	$60.06	$84.15

Crew C-8A

Crew C-8A	Bare Costs Hr.	Bare Costs Daily	Incl. Subs O&P Hr.	Incl. Subs O&P Daily	Cost Per LH Bare Costs	Cost Per LH Incl. O&P
1 Labor Foreman (outside)	$41.85	$334.80	$63.75	$510.00	$42.75	$64.46
3 Laborers	39.85	956.40	60.70	1456.80		
2 Cement Finishers	47.55	760.80	70.45	1127.20		
48 L.H., Daily Totals		$2052.00		$3094.00	$42.75	$64.46

Crew C-8B

Crew C-8B	Bare Costs Hr.	Bare Costs Daily	Incl. Subs O&P Hr.	Incl. Subs O&P Daily	Cost Per LH Bare Costs	Cost Per LH Incl. O&P
1 Labor Foreman (outside)	$41.85	$334.80	$63.75	$510.00	$43.03	$65.38
3 Laborers	39.85	956.40	60.70	1456.80		
1 Equip. Oper. (medium)	53.75	430.00	81.05	648.40		
1 Vibrating Power Screed		75.35		82.89		
1 Roller, Vibratory, 25 Ton		659.80		725.78		
1 Dozer, 200 H.P.		1273.00		1400.30	50.20	55.22
40 L.H., Daily Totals		$3729.35		$4824.17	$93.23	$120.60

Crew C-8C

Crew C-8C	Bare Costs Hr.	Bare Costs Daily	Incl. Subs O&P Hr.	Incl. Subs O&P Daily	Cost Per LH Bare Costs	Cost Per LH Incl. O&P
1 Labor Foreman (outside)	$41.85	$334.80	$63.75	$510.00	$43.78	$66.22
3 Laborers	39.85	956.40	60.70	1456.80		
1 Cement Finisher	47.55	380.40	70.45	563.60		
1 Equip. Oper. (medium)	53.75	430.00	81.05	648.40		
1 Shotcrete Rig, 12 C.Y./hr		246.60		271.26		
1 Air Compressor, 160 cfm		117.20		128.92		
4 -50' Air Hoses, 1"		33.80		37.18		
4 -50' Air Hoses, 2"		103.60		113.96	10.44	11.49
48 L.H., Daily Totals		$2602.80		$3730.12	$54.23	$77.71

Crew C-8D

Crew C-8D	Bare Costs Hr.	Bare Costs Daily	Incl. Subs O&P Hr.	Incl. Subs O&P Daily	Cost Per LH Bare Costs	Cost Per LH Incl. O&P
1 Labor Foreman (outside)	$41.85	$334.80	$63.75	$510.00	$45.14	$68.06
1 Laborer	39.85	318.80	60.70	485.60		
1 Cement Finisher	47.55	380.40	70.45	563.60		
1 Equipment Oper. (light)	51.30	410.40	77.35	618.80		
1 Air Compressor, 250 cfm		167.40		184.14		
2 -50' Air Hoses, 1"		16.90		18.59	5.76	6.34
32 L.H., Daily Totals		$1628.70		$2380.73	$50.90	$74.40

Crew C-8E

Crew C-8E	Bare Costs Hr.	Bare Costs Daily	Incl. Subs O&P Hr.	Incl. Subs O&P Daily	Cost Per LH Bare Costs	Cost Per LH Incl. O&P
1 Labor Foreman (outside)	$41.85	$334.80	$63.75	$510.00	$43.38	$65.61
3 Laborers	39.85	956.40	60.70	1456.80		
1 Cement Finisher	47.55	380.40	70.45	563.60		
1 Equipment Oper. (light)	51.30	410.40	77.35	618.80		
1 Shotcrete Rig, 35 C.Y./hr		270.20		297.22		
1 Air Compressor, 250 cfm		167.40		184.14		
4 -50' Air Hoses, 1"		33.80		37.18		
4 -50' Air Hoses, 2"		103.60		113.96	11.98	13.18
48 L.H., Daily Totals		$2657.00		$3781.70	$55.35	$78.79

Crew C-10

Crew C-10	Bare Costs Hr.	Bare Costs Daily	Incl. Subs O&P Hr.	Incl. Subs O&P Daily	Cost Per LH Bare Costs	Cost Per LH Incl. O&P
1 Laborer	$39.85	$318.80	$60.70	$485.60	$44.98	$67.20
2 Cement Finishers	47.55	760.80	70.45	1127.20		
24 L.H., Daily Totals		$1079.60		$1612.80	$44.98	$67.20

Crew C-10B

Crew C-10B	Bare Costs Hr.	Bare Costs Daily	Incl. Subs O&P Hr.	Incl. Subs O&P Daily	Cost Per LH Bare Costs	Cost Per LH Incl. O&P
3 Laborers	$39.85	$956.40	$60.70	$1456.80	$42.93	$64.60
2 Cement Finishers	47.55	760.80	70.45	1127.20		
1 Concrete Mixer, 10 C.F.		161.00		177.10		
2 Trowels, 48" Walk-Behind		82.40		90.64	6.09	6.69
40 L.H., Daily Totals		$1960.60		$2851.74	$49.02	$71.29

Crew C-10C

Crew C-10C	Bare Costs Hr.	Bare Costs Daily	Incl. Subs O&P Hr.	Incl. Subs O&P Daily	Cost Per LH Bare Costs	Cost Per LH Incl. O&P
1 Laborer	$39.85	$318.80	$60.70	$485.60	$44.98	$67.20
2 Cement Finishers	47.55	760.80	70.45	1127.20		
1 Trowel, 48" Walk-Behind		41.20		45.32	1.72	1.89
24 L.H., Daily Totals		$1120.80		$1658.12	$46.70	$69.09

Crew C-10D

Crew C-10D	Bare Costs Hr.	Bare Costs Daily	Incl. Subs O&P Hr.	Incl. Subs O&P Daily	Cost Per LH Bare Costs	Cost Per LH Incl. O&P
1 Laborer	$39.85	$318.80	$60.70	$485.60	$44.98	$67.20
2 Cement Finishers	47.55	760.80	70.45	1127.20		
1 Vibrating Power Screed		75.35		82.89		
1 Trowel, 48" Walk-Behind		41.20		45.32	4.86	5.34
24 L.H., Daily Totals		$1196.15		$1741.01	$49.84	$72.54

Crew C-10E

Crew C-10E	Bare Costs Hr.	Bare Costs Daily	Incl. Subs O&P Hr.	Incl. Subs O&P Daily	Cost Per LH Bare Costs	Cost Per LH Incl. O&P
1 Laborer	$39.85	$318.80	$60.70	$485.60	$44.98	$67.20
2 Cement Finishers	47.55	760.80	70.45	1127.20		
1 Vibrating Power Screed		75.35		82.89		
1 Cement Trowel, 96" Ride-On		166.20		182.82	10.06	11.07
24 L.H., Daily Totals		$1321.15		$1878.51	$55.05	$78.27

Crew C-10F

Crew C-10F	Bare Costs Hr.	Bare Costs Daily	Incl. Subs O&P Hr.	Incl. Subs O&P Daily	Cost Per LH Bare Costs	Cost Per LH Incl. O&P
1 Laborer	$39.85	$318.80	$60.70	$485.60	$44.98	$67.20
2 Cement Finishers	47.55	760.80	70.45	1127.20		
1 Telescoping Boom Lift, to 60'		454.60		500.06	18.94	20.84
24 L.H., Daily Totals		$1534.20		$2112.86	$63.92	$88.04

Crew C-11

Crew C-11	Bare Costs Hr.	Bare Costs Daily	Incl. Subs O&P Hr.	Incl. Subs O&P Daily	Cost Per LH Bare Costs	Cost Per LH Incl. O&P
1 Struc. Steel Foreman (outside)	$56.65	$453.20	$92.60	$740.80	$54.36	$87.40
6 Struc. Steel Workers	54.65	2623.20	89.35	4288.80		
1 Equip. Oper. (crane)	56.10	448.80	84.60	676.80		
1 Equip. Oper. (oiler)	48.60	388.80	73.30	586.40		
1 Lattice Boom Crane, 150 Ton		2017.00		2218.70	28.01	30.82
72 L.H., Daily Totals		$5931.00		$8511.50	$82.38	$118.22

Crew C-12

Crew No.	Bare Costs Hr.	Daily	Incl. Subs O&P Hr.	Daily	Cost Per Labor-Hour Bare Costs	Incl. O&P
1 Carpenter Foreman (outside)	$52.70	$421.60	$80.25	$642.00	$50.13	$76.19
3 Carpenters	50.70	1216.80	77.20	1852.80		
1 Laborer	39.85	318.80	60.70	485.60		
1 Equip. Oper. (crane)	56.10	448.80	84.60	676.80		
1 Hyd. Crane, 12 Ton		496.00		545.60	10.33	11.37
48 L.H., Daily Totals		$2902.00		$4202.80	$60.46	$87.56

Crew C-13

Crew No.	Bare Costs Hr.	Daily	Incl. Subs O&P Hr.	Daily	Cost Per Labor-Hour Bare Costs	Incl. O&P
1 Struc. Steel Worker	$54.65	$437.20	$89.35	$714.80	$53.33	$85.30
1 Welder	54.65	437.20	89.35	714.80		
1 Carpenter	50.70	405.60	77.20	617.60		
1 Welder, Gas Engine, 300 amp		98.40		108.24	4.10	4.51
24 L.H., Daily Totals		$1378.40		$2155.44	$57.43	$89.81

Crew C-14

Crew No.	Bare Costs Hr.	Daily	Incl. Subs O&P Hr.	Daily	Cost Per Labor-Hour Bare Costs	Incl. O&P
1 Carpenter Foreman (outside)	$52.70	$421.60	$80.25	$642.00	$49.11	$74.54
5 Carpenters	50.70	2028.00	77.20	3088.00		
4 Laborers	39.85	1275.20	60.70	1942.40		
4 Rodmen (reinf.)	54.65	1748.80	83.45	2670.40		
2 Cement Finishers	47.55	760.80	70.45	1127.20		
1 Equip. Oper. (crane)	56.10	448.80	84.60	676.80		
1 Equip. Oper. (oiler)	48.60	388.80	73.30	586.40		
1 Hyd. Crane, 80 Ton		1518.00		1669.80	10.54	11.60
144 L.H., Daily Totals		$8590.00		$12403.00	$59.65	$86.13

Crew C-14A

Crew No.	Bare Costs Hr.	Daily	Incl. Subs O&P Hr.	Daily	Cost Per Labor-Hour Bare Costs	Incl. O&P
1 Carpenter Foreman (outside)	$52.70	$421.60	$80.25	$642.00	$50.54	$76.89
16 Carpenters	50.70	6489.60	77.20	9881.60		
4 Rodmen (reinf.)	54.65	1748.80	83.45	2670.40		
2 Laborers	39.85	637.60	60.70	971.20		
1 Cement Finisher	47.55	380.40	70.45	563.60		
1 Equip. Oper. (medium)	53.75	430.00	81.05	648.40		
1 Gas Engine Vibrator		25.60		28.16		
1 Concrete Pump (Small)		881.60		969.76	4.54	4.99
200 L.H., Daily Totals		$11015.20		$16375.12	$55.08	$81.88

Crew C-14B

Crew No.	Bare Costs Hr.	Daily	Incl. Subs O&P Hr.	Daily	Cost Per Labor-Hour Bare Costs	Incl. O&P
1 Carpenter Foreman (outside)	$52.70	$421.60	$80.25	$642.00	$50.42	$76.64
16 Carpenters	50.70	6489.60	77.20	9881.60		
4 Rodmen (reinf.)	54.65	1748.80	83.45	2670.40		
2 Laborers	39.85	637.60	60.70	971.20		
2 Cement Finishers	47.55	760.80	70.45	1127.20		
1 Equip. Oper. (medium)	53.75	430.00	81.05	648.40		
1 Gas Engine Vibrator		25.60		28.16		
1 Concrete Pump (Small)		881.60		969.76	4.36	4.80
208 L.H., Daily Totals		$11395.60		$16938.72	$54.79	$81.44

Crew C-14C

Crew No.	Bare Costs Hr.	Daily	Incl. Subs O&P Hr.	Daily	Cost Per Labor-Hour Bare Costs	Incl. O&P
1 Carpenter Foreman (outside)	$52.70	$421.60	$80.25	$642.00	$48.08	$73.11
6 Carpenters	50.70	2433.60	77.20	3705.60		
2 Rodmen (reinf.)	54.65	874.40	83.45	1335.20		
4 Laborers	39.85	1275.20	60.70	1942.40		
1 Cement Finisher	47.55	380.40	70.45	563.60		
1 Gas Engine Vibrator		25.60		28.16	.23	.25
112 L.H., Daily Totals		$5410.80		$8216.96	$48.31	$73.37

Crew C-14D

Crew No.	Bare Costs Hr.	Daily	Incl. Subs O&P Hr.	Daily	Cost Per Labor-Hour Bare Costs	Incl. O&P
1 Carpenter Foreman (outside)	$52.70	$421.60	$80.25	$642.00	$50.22	$76.39
18 Carpenters	50.70	7300.80	77.20	11116.80		
2 Rodmen (reinf.)	54.65	874.40	83.45	1335.20		
2 Laborers	39.85	637.60	60.70	971.20		
1 Cement Finisher	47.55	380.40	70.45	563.60		
1 Equip. Oper. (medium)	53.75	430.00	81.05	648.40		
1 Gas Engine Vibrator		25.60		28.16		
1 Concrete Pump (Small)		881.60		969.76	4.54	4.99
200 L.H., Daily Totals		$10952.00		$16275.12	$54.76	$81.38

Crew C-14E

Crew No.	Bare Costs Hr.	Daily	Incl. Subs O&P Hr.	Daily	Cost Per Labor-Hour Bare Costs	Incl. O&P
1 Carpenter Foreman (outside)	$52.70	$421.60	$80.25	$642.00	$49.07	$74.64
2 Carpenters	50.70	811.20	77.20	1235.20		
4 Rodmen (reinf.)	54.65	1748.80	83.45	2670.40		
3 Laborers	39.85	956.40	60.70	1456.80		
1 Cement Finisher	47.55	380.40	70.45	563.60		
1 Gas Engine Vibrator		25.60		28.16	.29	.32
88 L.H., Daily Totals		$4344.00		$6596.16	$49.36	$74.96

Crew C-14F

Crew No.	Bare Costs Hr.	Daily	Incl. Subs O&P Hr.	Daily	Cost Per Labor-Hour Bare Costs	Incl. O&P
1 Labor Foreman (outside)	$41.85	$334.80	$63.75	$510.00	$45.21	$67.54
2 Laborers	39.85	637.60	60.70	971.20		
6 Cement Finishers	47.55	2282.40	70.45	3381.60		
1 Gas Engine Vibrator		25.60		28.16	.36	.39
72 L.H., Daily Totals		$3280.40		$4890.96	$45.56	$67.93

Crew C-14G

Crew No.	Bare Costs Hr.	Daily	Incl. Subs O&P Hr.	Daily	Cost Per Labor-Hour Bare Costs	Incl. O&P
1 Labor Foreman (outside)	$41.85	$334.80	$63.75	$510.00	$44.54	$66.71
2 Laborers	39.85	637.60	60.70	971.20		
4 Cement Finishers	47.55	1521.60	70.45	2254.40		
1 Gas Engine Vibrator		25.60		28.16	.46	.50
56 L.H., Daily Totals		$2519.60		$3763.76	$44.99	$67.21

Crew C-14H

Crew No.	Bare Costs Hr.	Daily	Incl. Subs O&P Hr.	Daily	Cost Per Labor-Hour Bare Costs	Incl. O&P
1 Carpenter Foreman (outside)	$52.70	$421.60	$80.25	$642.00	$49.36	$74.88
2 Carpenters	50.70	811.20	77.20	1235.20		
1 Rodman (reinf.)	54.65	437.20	83.45	667.60		
1 Laborer	39.85	318.80	60.70	485.60		
1 Cement Finisher	47.55	380.40	70.45	563.60		
1 Gas Engine Vibrator		25.60		28.16	.53	.59
48 L.H., Daily Totals		$2394.80		$3622.16	$49.89	$75.46

Crew C-14L

Crew No.	Bare Costs Hr.	Daily	Incl. Subs O&P Hr.	Daily	Cost Per Labor-Hour Bare Costs	Incl. O&P
1 Carpenter Foreman (outside)	$52.70	$421.60	$80.25	$642.00	$46.99	$71.39
6 Carpenters	50.70	2433.60	77.20	3705.60		
4 Laborers	39.85	1275.20	60.70	1942.40		
1 Cement Finisher	47.55	380.40	70.45	563.60		
1 Gas Engine Vibrator		25.60		28.16	.27	.29
96 L.H., Daily Totals		$4536.40		$6881.76	$47.25	$71.69

Crew C-14M

Crew No.	Bare Costs Hr.	Daily	Incl. Subs O&P Hr.	Daily	Cost Per Labor-Hour Bare Costs	Incl. O&P
1 Carpenter Foreman (outside)	$52.70	$421.60	$80.25	$642.00	$48.72	$73.88
2 Carpenters	50.70	811.20	77.20	1235.20		
1 Rodman (reinf.)	54.65	437.20	83.45	667.60		
2 Laborers	39.85	637.60	60.70	971.20		
1 Cement Finisher	47.55	380.40	70.45	563.60		
1 Equip. Oper. (medium)	53.75	430.00	81.05	648.40		
1 Gas Engine Vibrator		25.60		28.16		
1 Concrete Pump (Small)		881.60		969.76	14.18	15.59
64 L.H., Daily Totals		$4025.20		$5725.92	$62.89	$89.47

For customer support on your Electrical Costs with RSMeans data, call 800.448.8182.

501

Crew No.	Bare Costs		Incl. Subs O&P		Cost Per Labor-Hour	
	Hr.	Daily	Hr.	Daily	Bare Costs	Incl. O&P
Crew C-15	Hr.	Daily	Hr.	Daily	Bare Costs	Incl. O&P
1 Carpenter Foreman (outside)	$52.70	$421.60	$80.25	$642.00	$47.04	$71.23
2 Carpenters	50.70	811.20	77.20	1235.20		
3 Laborers	39.85	956.40	60.70	1456.80		
2 Cement Finishers	47.55	760.80	70.45	1127.20		
1 Rodman (reinf.)	54.65	437.20	83.45	667.60		
72 L.H., Daily Totals		$3387.20		$5128.80	$47.04	$71.23
Crew C-16	Hr.	Daily	Hr.	Daily	Bare Costs	Incl. O&P
1 Labor Foreman (outside)	$41.85	$334.80	$63.75	$510.00	$44.32	$66.83
3 Laborers	39.85	956.40	60.70	1456.80		
2 Cement Finishers	47.55	760.80	70.45	1127.20		
1 Equip. Oper. (medium)	53.75	430.00	81.05	648.40		
1 Gunite Pump Rig		321.00		353.10		
2 -50' Air Hoses, 3/4"		14.70		16.17		
2 -50' Air Hoses, 2"		51.80		56.98	6.92	7.61
56 L.H., Daily Totals		$2869.50		$4168.65	$51.24	$74.44
Crew C-16A	Hr.	Daily	Hr.	Daily	Bare Costs	Incl. O&P
1 Laborer	$39.85	$318.80	$60.70	$485.60	$47.17	$70.66
2 Cement Finishers	47.55	760.80	70.45	1127.20		
1 Equip. Oper. (medium)	53.75	430.00	81.05	648.40		
1 Gunite Pump Rig		321.00		353.10		
2 -50' Air Hoses, 3/4"		14.70		16.17		
2 -50' Air Hoses, 2"		51.80		56.98		
1 Telescoping Boom Lift, to 60'		454.60		500.06	26.32	28.95
32 L.H., Daily Totals		$2351.70		$3187.51	$73.49	$99.61
Crew C-17	Hr.	Daily	Hr.	Daily	Bare Costs	Incl. O&P
2 Skilled Worker Foremen (out)	$54.35	$869.60	$83.20	$1331.20	$52.75	$80.76
8 Skilled Workers	52.35	3350.40	80.15	5129.60		
80 L.H., Daily Totals		$4220.00		$6460.80	$52.75	$80.76
Crew C-17A	Hr.	Daily	Hr.	Daily	Bare Costs	Incl. O&P
2 Skilled Worker Foremen (out)	$54.35	$869.60	$83.20	$1331.20	$52.79	$80.81
8 Skilled Workers	52.35	3350.40	80.15	5129.60		
.125 Equip. Oper. (crane)	56.10	56.10	84.60	84.60		
.125 Hyd. Crane, 80 Ton		189.75		208.72	2.34	2.58
81 L.H., Daily Totals		$4465.85		$6754.13	$55.13	$83.38
Crew C-17B	Hr.	Daily	Hr.	Daily	Bare Costs	Incl. O&P
2 Skilled Worker Foremen (out)	$54.35	$869.60	$83.20	$1331.20	$52.83	$80.85
8 Skilled Workers	52.35	3350.40	80.15	5129.60		
.25 Equip. Oper. (crane)	56.10	112.20	84.60	169.20		
.25 Hyd. Crane, 80 Ton		379.50		417.45		
.25 Trowel, 48" Walk-Behind		10.30		11.33	4.75	5.23
82 L.H., Daily Totals		$4722.00		$7058.78	$57.59	$86.08
Crew C-17C	Hr.	Daily	Hr.	Daily	Bare Costs	Incl. O&P
2 Skilled Worker Foremen (out)	$54.35	$869.60	$83.20	$1331.20	$52.87	$80.90
8 Skilled Workers	52.35	3350.40	80.15	5129.60		
.375 Equip. Oper. (crane)	56.10	168.30	84.60	253.80		
.375 Hyd. Crane, 80 Ton		569.25		626.17	6.86	7.54
83 L.H., Daily Totals		$4957.55		$7340.77	$59.73	$88.44
Crew C-17D	Hr.	Daily	Hr.	Daily	Bare Costs	Incl. O&P
2 Skilled Worker Foremen (out)	$54.35	$869.60	$83.20	$1331.20	$52.91	$80.94
8 Skilled Workers	52.35	3350.40	80.15	5129.60		
.5 Equip. Oper. (crane)	56.10	224.40	84.60	338.40		
.5 Hyd. Crane, 80 Ton		759.00		834.90	9.04	9.94
84 L.H., Daily Totals		$5203.40		$7634.10	$61.95	$90.88

Crew No.	Bare Costs		Incl. Subs O&P		Cost Per Labor-Hour	
Crew C-17E	Hr.	Daily	Hr.	Daily	Bare Costs	Incl. O&P
2 Skilled Worker Foremen (out)	$54.35	$869.60	$83.20	$1331.20	$52.75	$80.76
8 Skilled Workers	52.35	3350.40	80.15	5129.60		
1 Hyd. Jack with Rods		102.25		112.47	1.28	1.41
80 L.H., Daily Totals		$4322.25		$6573.27	$54.03	$82.17
Crew C-18	Hr.	Daily	Hr.	Daily	Bare Costs	Incl. O&P
.125 Labor Foreman (outside)	$41.85	$41.85	$63.75	$63.75	$40.07	$61.04
1 Laborer	39.85	318.80	60.70	485.60		
1 Concrete Cart, 10 C.F.		57.80		63.58	6.42	7.06
9 L.H., Daily Totals		$418.45		$612.93	$46.49	$68.10
Crew C-19	Hr.	Daily	Hr.	Daily	Bare Costs	Incl. O&P
.125 Labor Foreman (outside)	$41.85	$41.85	$63.75	$63.75	$40.07	$61.04
1 Laborer	39.85	318.80	60.70	485.60		
1 Concrete Cart, 18 C.F.		97.40		107.14	10.82	11.90
9 L.H., Daily Totals		$458.05		$656.49	$50.89	$72.94
Crew C-20	Hr.	Daily	Hr.	Daily	Bare Costs	Incl. O&P
1 Labor Foreman (outside)	$41.85	$334.80	$63.75	$510.00	$42.80	$64.84
5 Laborers	39.85	1594.00	60.70	2428.00		
1 Cement Finisher	47.55	380.40	70.45	563.60		
1 Equip. Oper. (medium)	53.75	430.00	81.05	648.40		
2 Gas Engine Vibrators		51.20		56.32		
1 Concrete Pump (Small)		881.60		969.76	14.57	16.03
64 L.H., Daily Totals		$3672.00		$5176.08	$57.38	$80.88
Crew C-21	Hr.	Daily	Hr.	Daily	Bare Costs	Incl. O&P
1 Labor Foreman (outside)	$41.85	$334.80	$63.75	$510.00	$42.80	$64.84
5 Laborers	39.85	1594.00	60.70	2428.00		
1 Cement Finisher	47.55	380.40	70.45	563.60		
1 Equip. Oper. (medium)	53.75	430.00	81.05	648.40		
2 Gas Engine Vibrators		51.20		56.32		
1 Concrete Conveyer		187.20		205.92	3.73	4.10
64 L.H., Daily Totals		$2977.60		$4412.24	$46.52	$68.94
Crew C-22	Hr.	Daily	Hr.	Daily	Bare Costs	Incl. O&P
1 Rodman Foreman (outside)	$56.65	$453.20	$86.50	$692.00	$54.92	$83.82
4 Rodmen (reinf.)	54.65	1748.80	83.45	2670.40		
.125 Equip. Oper. (crane)	56.10	56.10	84.60	84.60		
.125 Equip. Oper. (oiler)	48.60	48.60	73.30	73.30		
.125 Hyd. Crane, 25 Ton		73.83		81.21	1.76	1.93
42 L.H., Daily Totals		$2380.53		$3601.51	$56.68	$85.75
Crew C-23	Hr.	Daily	Hr.	Daily	Bare Costs	Incl. O&P
2 Skilled Worker Foremen (out)	$54.35	$869.60	$83.20	$1331.20	$52.75	$80.52
6 Skilled Workers	52.35	2512.80	80.15	3847.20		
1 Equip. Oper. (crane)	56.10	448.80	84.60	676.80		
1 Equip. Oper. (oiler)	48.60	388.80	73.30	586.40		
1 Lattice Boom Crane, 90 Ton		1687.00		1855.70	21.09	23.20
80 L.H., Daily Totals		$5907.00		$8297.30	$73.84	$103.72
Crew C-23A	Hr.	Daily	Hr.	Daily	Bare Costs	Incl. O&P
1 Labor Foreman (outside)	$41.85	$334.80	$63.75	$510.00	$45.25	$68.61
2 Laborers	39.85	637.60	60.70	971.20		
1 Equip. Oper. (crane)	56.10	448.80	84.60	676.80		
1 Equip. Oper. (oiler)	48.60	388.80	73.30	586.40		
1 Crawler Crane, 100 Ton		1872.00		2059.20		
3 Conc. Buckets, 8 C.Y.		626.40		689.04	62.46	68.71
40 L.H., Daily Totals		$4308.40		$5492.64	$107.71	$137.32

Crew C-24

	Bare Costs Hr.	Daily	Incl. Subs O&P Hr.	Daily	Cost Per LH Bare Costs	Incl. O&P
2 Skilled Worker Foremen (out)	$54.35	$869.60	$83.20	$1331.20	$52.75	$80.52
6 Skilled Workers	52.35	2512.80	80.15	3847.20		
1 Equip. Oper. (crane)	56.10	448.80	84.60	676.80		
1 Equip. Oper. (oiler)	48.60	388.80	73.30	586.40		
1 Lattice Boom Crane, 150 Ton		2017.00		2218.70	25.21	27.73
80 L.H., Daily Totals		$6237.00		$8660.30	$77.96	$108.25

Crew C-25

	Bare Costs Hr.	Daily	Incl. Subs O&P Hr.	Daily	Cost Per LH Bare Costs	Incl. O&P
2 Rodmen (reinf.)	$54.65	$874.40	$83.45	$1335.20	$43.75	$69.38
2 Rodmen Helpers	32.85	525.60	55.30	884.80		
32 L.H., Daily Totals		$1400.00		$2220.00	$43.75	$69.38

Crew C-27

	Bare Costs Hr.	Daily	Incl. Subs O&P Hr.	Daily	Cost Per LH Bare Costs	Incl. O&P
2 Cement Finishers	$47.55	$760.80	$70.45	$1127.20	$47.55	$70.45
1 Concrete Saw		102.40		112.64	6.40	7.04
16 L.H., Daily Totals		$863.20		$1239.84	$53.95	$77.49

Crew C-28

	Bare Costs Hr.	Daily	Incl. Subs O&P Hr.	Daily	Cost Per LH Bare Costs	Incl. O&P
1 Cement Finisher	$47.55	$380.40	$70.45	$563.60	$47.55	$70.45
1 Portable Air Compressor, Gas		17.10		18.81	2.14	2.35
8 L.H., Daily Totals		$397.50		$582.41	$49.69	$72.80

Crew C-29

	Bare Costs Hr.	Daily	Incl. Subs O&P Hr.	Daily	Cost Per LH Bare Costs	Incl. O&P
1 Laborer	$39.85	$318.80	$60.70	$485.60	$39.85	$60.70
1 Pressure Washer		63.60		69.96	7.95	8.74
8 L.H., Daily Totals		$382.40		$555.56	$47.80	$69.44

Crew C-30

	Bare Costs Hr.	Daily	Incl. Subs O&P Hr.	Daily	Cost Per LH Bare Costs	Incl. O&P
1 Laborer	$39.85	$318.80	$60.70	$485.60	$39.85	$60.70
1 Concrete Mixer, 10 C.F.		161.00		177.10	20.13	22.14
8 L.H., Daily Totals		$479.80		$662.70	$59.98	$82.84

Crew C-31

	Bare Costs Hr.	Daily	Incl. Subs O&P Hr.	Daily	Cost Per LH Bare Costs	Incl. O&P
1 Cement Finisher	$47.55	$380.40	$70.45	$563.60	$47.55	$70.45
1 Grout Pump		321.00		353.10	40.13	44.14
8 L.H., Daily Totals		$701.40		$916.70	$87.67	$114.59

Crew C-32

	Bare Costs Hr.	Daily	Incl. Subs O&P Hr.	Daily	Cost Per LH Bare Costs	Incl. O&P
1 Cement Finisher	$47.55	$380.40	$70.45	$563.60	$43.70	$65.58
1 Laborer	39.85	318.80	60.70	485.60		
1 Crack Chaser Saw, Gas, 6 H.P.		25.40		27.94		
1 Vacuum Pick-Up System		66.95		73.64	5.77	6.35
16 L.H., Daily Totals		$791.55		$1150.79	$49.47	$71.92

Crew D-1

	Bare Costs Hr.	Daily	Incl. Subs O&P Hr.	Daily	Cost Per LH Bare Costs	Incl. O&P
1 Bricklayer	$50.30	$402.40	$77.00	$616.00	$44.80	$68.58
1 Bricklayer Helper	39.30	314.40	60.15	481.20		
16 L.H., Daily Totals		$716.80		$1097.20	$44.80	$68.58

Crew D-2

	Bare Costs Hr.	Daily	Incl. Subs O&P Hr.	Daily	Cost Per LH Bare Costs	Incl. O&P
3 Bricklayers	$50.30	$1207.20	$77.00	$1848.00	$46.34	$70.89
2 Bricklayer Helpers	39.30	628.80	60.15	962.40		
.5 Carpenter	50.70	202.80	77.20	308.80		
44 L.H., Daily Totals		$2038.80		$3119.20	$46.34	$70.89

Crew D-3

	Bare Costs Hr.	Daily	Incl. Subs O&P Hr.	Daily	Cost Per LH Bare Costs	Incl. O&P
3 Bricklayers	$50.30	$1207.20	$77.00	$1848.00	$46.13	$70.59
2 Bricklayer Helpers	39.30	628.80	60.15	962.40		
.25 Carpenter	50.70	101.40	77.20	154.40		
42 L.H., Daily Totals		$1937.40		$2964.80	$46.13	$70.59

Crew D-4

	Bare Costs Hr.	Daily	Incl. Subs O&P Hr.	Daily	Cost Per LH Bare Costs	Incl. O&P
1 Bricklayer	$50.30	$402.40	$77.00	$616.00	$45.05	$68.66
2 Bricklayer Helpers	39.30	628.80	60.15	962.40		
1 Equip. Oper. (light)	51.30	410.40	77.35	618.80		
1 Grout Pump, 50 C.F./hr.		128.60		141.46	4.02	4.42
32 L.H., Daily Totals		$1570.20		$2338.66	$49.07	$73.08

Crew D-5

	Bare Costs Hr.	Daily	Incl. Subs O&P Hr.	Daily	Cost Per LH Bare Costs	Incl. O&P
1 Bricklayer	50.30	402.40	77.00	616.00	50.30	77.00
8 L.H., Daily Totals		$402.40		$616.00	$50.30	$77.00

Crew D-6

	Bare Costs Hr.	Daily	Incl. Subs O&P Hr.	Daily	Cost Per LH Bare Costs	Incl. O&P
3 Bricklayers	$50.30	$1207.20	$77.00	$1848.00	$45.04	$68.92
3 Bricklayer Helpers	39.30	943.20	60.15	1443.60		
.25 Carpenter	50.70	101.40	77.20	154.40		
50 L.H., Daily Totals		$2251.80		$3446.00	$45.04	$68.92

Crew D-7

	Bare Costs Hr.	Daily	Incl. Subs O&P Hr.	Daily	Cost Per LH Bare Costs	Incl. O&P
1 Tile Layer	$46.85	$374.80	$69.35	$554.80	$42.02	$62.20
1 Tile Layer Helper	37.20	297.60	55.05	440.40		
16 L.H., Daily Totals		$672.40		$995.20	$42.02	$62.20

Crew D-8

	Bare Costs Hr.	Daily	Incl. Subs O&P Hr.	Daily	Cost Per LH Bare Costs	Incl. O&P
3 Bricklayers	$50.30	$1207.20	$77.00	$1848.00	$45.90	$70.26
2 Bricklayer Helpers	39.30	628.80	60.15	962.40		
40 L.H., Daily Totals		$1836.00		$2810.40	$45.90	$70.26

Crew D-9

	Bare Costs Hr.	Daily	Incl. Subs O&P Hr.	Daily	Cost Per LH Bare Costs	Incl. O&P
3 Bricklayers	$50.30	$1207.20	$77.00	$1848.00	$44.80	$68.58
3 Bricklayer Helpers	39.30	943.20	60.15	1443.60		
48 L.H., Daily Totals		$2150.40		$3291.60	$44.80	$68.58

Crew D-10

	Bare Costs Hr.	Daily	Incl. Subs O&P Hr.	Daily	Cost Per LH Bare Costs	Incl. O&P
1 Bricklayer Foreman (outside)	$52.30	$418.40	$80.05	$640.40	$49.50	$75.45
1 Bricklayer	50.30	402.40	77.00	616.00		
1 Bricklayer Helper	39.30	314.40	60.15	481.20		
1 Equip. Oper. (crane)	56.10	448.80	84.60	676.80		
1 S.P. Crane, 4x4, 12 Ton		363.60		399.96	11.36	12.50
32 L.H., Daily Totals		$1947.60		$2814.36	$60.86	$87.95

Crew D-11

	Bare Costs Hr.	Daily	Incl. Subs O&P Hr.	Daily	Cost Per LH Bare Costs	Incl. O&P
1 Bricklayer Foreman (outside)	$52.30	$418.40	$80.05	$640.40	$47.30	$72.40
1 Bricklayer	50.30	402.40	77.00	616.00		
1 Bricklayer Helper	39.30	314.40	60.15	481.20		
24 L.H., Daily Totals		$1135.20		$1737.60	$47.30	$72.40

Crew D-12

	Bare Costs Hr.	Daily	Incl. Subs O&P Hr.	Daily	Cost Per LH Bare Costs	Incl. O&P
1 Bricklayer Foreman (outside)	$52.30	$418.40	$80.05	$640.40	$45.30	$69.34
1 Bricklayer	50.30	402.40	77.00	616.00		
2 Bricklayer Helpers	39.30	628.80	60.15	962.40		
32 L.H., Daily Totals		$1449.60		$2218.80	$45.30	$69.34

For customer support on your Electrical Costs with RSMeans data, call 800.448.8182.

503

Crew No.	Bare Costs		Incl. Subs O&P		Cost Per Labor-Hour	
	Hr.	Daily	Hr.	Daily	Bare Costs	Incl. O&P
Crew D-13	Hr.	Daily	Hr.	Daily	Bare Costs	Incl. O&P
1 Bricklayer Foreman (outside)	$52.30	$418.40	$80.05	$640.40	$48.00	$73.19
1 Bricklayer	50.30	402.40	77.00	616.00		
2 Bricklayer Helpers	39.30	628.80	60.15	962.40		
1 Carpenter	50.70	405.60	77.20	617.60		
1 Equip. Oper. (crane)	56.10	448.80	84.60	676.80		
1 S.P. Crane, 4x4, 12 Ton		363.60		399.96	7.58	8.33
48 L.H., Daily Totals		$2667.60		$3913.16	$55.58	$81.52

Crew D-14	Hr.	Daily	Hr.	Daily	Bare Costs	Incl. O&P
3 Bricklayers	$50.30	$1207.20	$77.00	$1848.00	$47.55	$72.79
1 Bricklayer Helper	39.30	314.40	60.15	481.20		
32 L.H., Daily Totals		$1521.60		$2329.20	$47.55	$72.79

Crew E-1	Hr.	Daily	Hr.	Daily	Bare Costs	Incl. O&P
1 Welder Foreman (outside)	$56.65	$453.20	$92.60	$740.80	$54.20	$86.43
1 Welder	54.65	437.20	89.35	714.80		
1 Equip. Oper. (light)	51.30	410.40	77.35	618.80		
1 Welder, Gas Engine, 300 amp		98.40		108.24	4.10	4.51
24 L.H., Daily Totals		$1399.20		$2182.64	$58.30	$90.94

Crew E-2	Hr.	Daily	Hr.	Daily	Bare Costs	Incl. O&P
1 Struc. Steel Foreman (outside)	$56.65	$453.20	$92.60	$740.80	$54.28	$86.84
4 Struc. Steel Workers	54.65	1748.80	89.35	2859.20		
1 Equip. Oper. (crane)	56.10	448.80	84.60	676.80		
1 Equip. Oper. (oiler)	48.60	388.80	73.30	586.40		
1 Lattice Boom Crane, 90 Ton		1687.00		1855.70	30.13	33.14
56 L.H., Daily Totals		$4726.60		$6718.90	$84.40	$119.98

Crew E-3	Hr.	Daily	Hr.	Daily	Bare Costs	Incl. O&P
1 Struc. Steel Foreman (outside)	$56.65	$453.20	$92.60	$740.80	$55.32	$90.43
1 Struc. Steel Worker	54.65	437.20	89.35	714.80		
1 Welder	54.65	437.20	89.35	714.80		
1 Welder, Gas Engine, 300 amp		98.40		108.24	4.10	4.51
24 L.H., Daily Totals		$1426.00		$2278.64	$59.42	$94.94

Crew E-3A	Hr.	Daily	Hr.	Daily	Bare Costs	Incl. O&P
1 Struc. Steel Foreman (outside)	$56.65	$453.20	$92.60	$740.80	$55.32	$90.43
1 Struc. Steel Worker	54.65	437.20	89.35	714.80		
1 Welder	54.65	437.20	89.35	714.80		
1 Welder, Gas Engine, 300 amp		98.40		108.24		
1 Telescoping Boom Lift, to 40'		282.80		311.08	15.88	17.47
24 L.H., Daily Totals		$1708.80		$2589.72	$71.20	$107.91

Crew E-4	Hr.	Daily	Hr.	Daily	Bare Costs	Incl. O&P
1 Struc. Steel Foreman (outside)	$56.65	$453.20	$92.60	$740.80	$55.15	$90.16
3 Struc. Steel Workers	54.65	1311.60	89.35	2144.40		
1 Welder, Gas Engine, 300 amp		98.40		108.24	3.08	3.38
32 L.H., Daily Totals		$1863.20		$2993.44	$58.23	$93.55

Crew E-5	Hr.	Daily	Hr.	Daily	Bare Costs	Incl. O&P
2 Struc. Steel Foremen (outside)	$56.65	$906.40	$92.60	$1481.60	$54.59	$87.92
5 Struc. Steel Workers	54.65	2186.00	89.35	3574.00		
1 Equip. Oper. (crane)	56.10	448.80	84.60	676.80		
1 Welder	54.65	437.20	89.35	714.80		
1 Equip. Oper. (oiler)	48.60	388.80	73.30	586.40		
1 Lattice Boom Crane, 90 Ton		1687.00		1855.70		
1 Welder, Gas Engine, 300 amp		98.40		108.24	22.32	24.55
80 L.H., Daily Totals		$6152.60		$8997.54	$76.91	$112.47

Crew E-6	Hr.	Daily	Hr.	Daily	Bare Costs	Incl. O&P
3 Struc. Steel Foremen (outside)	$56.65	$1359.60	$92.60	$2222.40	$54.53	$87.91
9 Struc. Steel Workers	54.65	3934.80	89.35	6433.20		
1 Equip. Oper. (crane)	56.10	448.80	84.60	676.80		
1 Welder	54.65	437.20	89.35	714.80		
1 Equip. Oper. (oiler)	48.60	388.80	73.30	586.40		
1 Equip. Oper. (light)	51.30	410.40	77.35	618.80		
1 Lattice Boom Crane, 90 Ton		1687.00		1855.70		
1 Welder, Gas Engine, 300 amp		98.40		108.24		
1 Air Compressor, 160 cfm		117.20		128.92		
2 Impact Wrenches		37.20		40.92	15.15	16.67
128 L.H., Daily Totals		$8919.40		$13386.18	$69.68	$104.58

Crew E-7	Hr.	Daily	Hr.	Daily	Bare Costs	Incl. O&P
1 Struc. Steel Foreman (outside)	$56.65	$453.20	$92.60	$740.80	$54.59	$87.92
4 Struc. Steel Workers	54.65	1748.80	89.35	2859.20		
1 Equip. Oper. (crane)	56.10	448.80	84.60	676.80		
1 Equip. Oper. (oiler)	48.60	388.80	73.30	586.40		
1 Welder Foreman (outside)	56.65	453.20	92.60	740.80		
2 Welders	54.65	874.40	89.35	1429.60		
1 Lattice Boom Crane, 90 Ton		1687.00		1855.70		
2 Welders, Gas Engine, 300 amp		196.80		216.48	23.55	25.90
80 L.H., Daily Totals		$6251.00		$9105.78	$78.14	$113.82

Crew E-8	Hr.	Daily	Hr.	Daily	Bare Costs	Incl. O&P
1 Struc. Steel Foreman (outside)	$56.65	$453.20	$92.60	$740.80	$54.35	$87.33
4 Struc. Steel Workers	54.65	1748.80	89.35	2859.20		
1 Welder Foreman (outside)	56.65	453.20	92.60	740.80		
4 Welders	54.65	1748.80	89.35	2859.20		
1 Equip. Oper. (crane)	56.10	448.80	84.60	676.80		
1 Equip. Oper. (oiler)	48.60	388.80	73.30	586.40		
1 Equip. Oper. (light)	51.30	410.40	77.35	618.80		
1 Lattice Boom Crane, 90 Ton		1687.00		1855.70		
4 Welders, Gas Engine, 300 amp		393.60		432.96	20.01	22.01
104 L.H., Daily Totals		$7732.60		$11370.66	$74.35	$109.33

Crew E-9	Hr.	Daily	Hr.	Daily	Bare Costs	Incl. O&P
2 Struc. Steel Foremen (outside)	$56.65	$906.40	$92.60	$1481.60	$54.53	$87.91
5 Struc. Steel Workers	54.65	2186.00	89.35	3574.00		
1 Welder Foreman (outside)	56.65	453.20	92.60	740.80		
5 Welders	54.65	2186.00	89.35	3574.00		
1 Equip. Oper. (crane)	56.10	448.80	84.60	676.80		
1 Equip. Oper. (oiler)	48.60	388.80	73.30	586.40		
1 Equip. Oper. (light)	51.30	410.40	77.35	618.80		
1 Lattice Boom Crane, 90 Ton		1687.00		1855.70		
5 Welders, Gas Engine, 300 amp		492.00		541.20	17.02	18.73
128 L.H., Daily Totals		$9158.60		$13649.30	$71.55	$106.64

Crew E-10	Hr.	Daily	Hr.	Daily	Bare Costs	Incl. O&P
1 Welder Foreman (outside)	$56.65	$453.20	$92.60	$740.80	$55.65	$90.97
1 Welder	54.65	437.20	89.35	714.80		
1 Welder, Gas Engine, 300 amp		98.40		108.24		
1 Flatbed Truck, Gas, 3 Ton		238.00		261.80	21.02	23.13
16 L.H., Daily Totals		$1226.80		$1825.64	$76.67	$114.10

Crew E-11	Hr.	Daily	Hr.	Daily	Bare Costs	Incl. O&P
2 Painters, Struc. Steel	$43.50	$696.00	$72.45	$1159.20	$44.54	$70.74
1 Building Laborer	39.85	318.80	60.70	485.60		
1 Equip. Oper. (light)	51.30	410.40	77.35	618.80		
1 Air Compressor, 250 cfm		167.40		184.14		
1 Sandblaster, Portable, 3 C.F.		21.00		23.10		
1 Set Sand Blasting Accessories		14.70		16.17	6.35	6.98
32 L.H., Daily Totals		$1628.30		$2487.01	$50.88	$77.72

Crew No.	Bare Costs		Incl. Subs O&P		Cost Per Labor-Hour	

Crew E-11A	Hr.	Daily	Hr.	Daily	Bare Costs	Incl. O&P
2 Painters, Struc. Steel	$43.50	$696.00	$72.45	$1159.20	$44.54	$70.74
1 Building Laborer	39.85	318.80	60.70	485.60		
1 Equip. Oper. (light)	51.30	410.40	77.35	618.80		
1 Air Compressor, 250 cfm		167.40		184.14		
1 Sandblaster, Portable, 3 C.F.		21.00		23.10		
1 Set Sand Blasting Accessories		14.70		16.17		
1 Telescoping Boom Lift, to 60'		454.60		500.06	20.55	22.61
32 L.H., Daily Totals		$2082.90		$2987.07	$65.09	$93.35

Crew E-11B	Hr.	Daily	Hr.	Daily	Bare Costs	Incl. O&P
2 Painters, Struc. Steel	$43.50	$696.00	$72.45	$1159.20	$42.28	$68.53
1 Building Laborer	39.85	318.80	60.70	485.60		
2 Paint Sprayer, 8 C.F.M.		90.20		99.22		
1 Telescoping Boom Lift, to 60'		454.60		500.06	22.70	24.97
24 L.H., Daily Totals		$1559.60		$2244.08	$64.98	$93.50

Crew E-12	Hr.	Daily	Hr.	Daily	Bare Costs	Incl. O&P
1 Welder Foreman (outside)	$56.65	$453.20	$92.60	$740.80	$53.98	$84.97
1 Equip. Oper. (light)	51.30	410.40	77.35	618.80		
1 Welder, Gas Engine, 300 amp		98.40		108.24	6.15	6.76
16 L.H., Daily Totals		$962.00		$1467.84	$60.13	$91.74

Crew E-13	Hr.	Daily	Hr.	Daily	Bare Costs	Incl. O&P
1 Welder Foreman (outside)	$56.65	$453.20	$92.60	$740.80	$54.87	$87.52
.5 Equip. Oper. (light)	51.30	205.20	77.35	309.40		
1 Welder, Gas Engine, 300 amp		98.40		108.24	8.20	9.02
12 L.H., Daily Totals		$756.80		$1158.44	$63.07	$96.54

Crew E-14	Hr.	Daily	Hr.	Daily	Bare Costs	Incl. O&P
1 Welder Foreman (outside)	$56.65	$453.20	$92.60	$740.80	$56.65	$92.60
1 Welder, Gas Engine, 300 amp		98.40		108.24	12.30	13.53
8 L.H., Daily Totals		$551.60		$849.04	$68.95	$106.13

Crew E-16	Hr.	Daily	Hr.	Daily	Bare Costs	Incl. O&P
1 Welder Foreman (outside)	$56.65	$453.20	$92.60	$740.80	$55.65	$90.97
1 Welder	54.65	437.20	89.35	714.80		
1 Welder, Gas Engine, 300 amp		98.40		108.24	6.15	6.76
16 L.H., Daily Totals		$988.80		$1563.84	$61.80	$97.74

Crew E-17	Hr.	Daily	Hr.	Daily	Bare Costs	Incl. O&P
1 Struc. Steel Foreman (outside)	$56.65	$453.20	$92.60	$740.80	$55.65	$90.97
1 Structural Steel Worker	54.65	437.20	89.35	714.80		
16 L.H., Daily Totals		$890.40		$1455.60	$55.65	$90.97

Crew E-18	Hr.	Daily	Hr.	Daily	Bare Costs	Incl. O&P
1 Struc. Steel Foreman (outside)	$56.65	$453.20	$92.60	$740.80	$54.87	$88.34
3 Structural Steel Workers	54.65	1311.60	89.35	2144.40		
1 Equipment Operator (med.)	53.75	430.00	81.05	648.40		
1 Lattice Boom Crane, 20 Ton		1083.00		1191.30	27.07	29.78
40 L.H., Daily Totals		$3277.80		$4724.90	$81.94	$118.12

Crew E-19	Hr.	Daily	Hr.	Daily	Bare Costs	Incl. O&P
1 Struc. Steel Foreman (outside)	$56.65	$453.20	$92.60	$740.80	$54.20	$86.43
1 Structural Steel Worker	54.65	437.20	89.35	714.80		
1 Equip. Oper. (light)	51.30	410.40	77.35	618.80		
1 Lattice Boom Crane, 20 Ton		1083.00		1191.30	45.13	49.64
24 L.H., Daily Totals		$2383.80		$3265.70	$99.33	$136.07

Crew E-20	Hr.	Daily	Hr.	Daily	Bare Costs	Incl. O&P
1 Struc. Steel Foreman (outside)	$56.65	$453.20	$92.60	$740.80	$54.33	$87.16
5 Structural Steel Workers	54.65	2186.00	89.35	3574.00		
1 Equip. Oper. (crane)	56.10	448.80	84.60	676.80		
1 Equip. Oper. (oiler)	48.60	388.80	73.30	586.40		
1 Lattice Boom Crane, 40 Ton		1321.00		1453.10	20.64	22.70
64 L.H., Daily Totals		$4797.80		$7031.10	$74.97	$109.86

Crew E-22	Hr.	Daily	Hr.	Daily	Bare Costs	Incl. O&P
1 Skilled Worker Foreman (out)	$54.35	$434.80	$83.20	$665.60	$53.02	$81.17
2 Skilled Workers	52.35	837.60	80.15	1282.40		
24 L.H., Daily Totals		$1272.40		$1948.00	$53.02	$81.17

Crew E-24	Hr.	Daily	Hr.	Daily	Bare Costs	Incl. O&P
3 Structural Steel Workers	$54.65	$1311.60	$89.35	$2144.40	$54.42	$87.28
1 Equipment Operator (med.)	53.75	430.00	81.05	648.40		
1 Hyd. Crane, 25 Ton		590.60		649.66	18.46	20.30
32 L.H., Daily Totals		$2332.20		$3442.46	$72.88	$107.58

Crew E-25	Hr.	Daily	Hr.	Daily	Bare Costs	Incl. O&P
1 Welder Foreman (outside)	$56.65	$453.20	$92.60	$740.80	$56.65	$92.60
1 Cutting Torch		12.60		13.86	1.58	1.73
8 L.H., Daily Totals		$465.80		$754.66	$58.23	$94.33

Crew E-26	Hr.	Daily	Hr.	Daily	Bare Costs	Incl. O&P
1 Struc. Steel Foreman (outside)	$56.65	$453.20	$92.60	$740.80	$56.01	$90.41
1 Struc. Steel Worker	54.65	437.20	89.35	714.80		
1 Welder	54.65	437.20	89.35	714.80		
.25 Electrician	58.20	116.40	87.00	174.00		
.25 Plumber	62.15	124.30	93.60	187.20		
1 Welder, Gas Engine, 300 amp		98.40		108.24	3.51	3.87
28 L.H., Daily Totals		$1666.70		$2639.84	$59.52	$94.28

Crew E-27	Hr.	Daily	Hr.	Daily	Bare Costs	Incl. O&P
1 Struc. Steel Foreman (outside)	$56.65	$453.20	$92.60	$740.80	$54.33	$87.16
5 Struc. Steel Workers	54.65	2186.00	89.35	3574.00		
1 Equip. Oper. (crane)	56.10	448.80	84.60	676.80		
1 Equip. Oper. (oiler)	48.60	388.80	73.30	586.40		
1 Hyd. Crane, 12 Ton		496.00		545.60		
1 Hyd. Crane, 80 Ton		1518.00		1669.80	31.47	34.62
64 L.H., Daily Totals		$5490.80		$7793.40	$85.79	$121.77

Crew F-3	Hr.	Daily	Hr.	Daily	Bare Costs	Incl. O&P
4 Carpenters	$50.70	$1622.40	$77.20	$2470.40	$51.78	$78.68
1 Equip. Oper. (crane)	56.10	448.80	84.60	676.80		
1 Hyd. Crane, 12 Ton		496.00		545.60	12.40	13.64
40 L.H., Daily Totals		$2567.20		$3692.80	$64.18	$92.32

Crew F-4	Hr.	Daily	Hr.	Daily	Bare Costs	Incl. O&P
4 Carpenters	$50.70	$1622.40	$77.20	$2470.40	$51.25	$77.78
1 Equip. Oper. (crane)	56.10	448.80	84.60	676.80		
1 Equip. Oper. (oiler)	48.60	388.80	73.30	586.40		
1 Hyd. Crane, 55 Ton		993.80		1093.18	20.70	22.77
48 L.H., Daily Totals		$3453.80		$4826.78	$71.95	$100.56

Crew F-5	Hr.	Daily	Hr.	Daily	Bare Costs	Incl. O&P
1 Carpenter Foreman (outside)	$52.70	$421.60	$80.25	$642.00	$51.20	$77.96
3 Carpenters	50.70	1216.80	77.20	1852.80		
32 L.H., Daily Totals		$1638.40		$2494.80	$51.20	$77.96

Crew No.	Bare Costs		Incl. Subs O&P		Cost Per Labor-Hour	
	Hr.	Daily	Hr.	Daily	Bare Costs	Incl. O&P
Crew F-6					$47.44	$72.08
2 Carpenters	$50.70	$811.20	$77.20	$1235.20		
2 Building Laborers	39.85	637.60	60.70	971.20		
1 Equip. Oper. (crane)	56.10	448.80	84.60	676.80		
1 Hyd. Crane, 12 Ton		496.00		545.60	12.40	13.64
40 L.H., Daily Totals		$2393.60		$3428.80	$59.84	$85.72
Crew F-7	Hr.	Daily	Hr.	Daily	Bare Costs	Incl. O&P
2 Carpenters	$50.70	$811.20	$77.20	$1235.20	$45.27	$68.95
2 Building Laborers	39.85	637.60	60.70	971.20		
32 L.H., Daily Totals		$1448.80		$2206.40	$45.27	$68.95
Crew G-1	Hr.	Daily	Hr.	Daily	Bare Costs	Incl. O&P
1 Roofer Foreman (outside)	$45.95	$367.60	$77.35	$618.80	$41.06	$69.11
4 Roofers Composition	43.95	1406.40	73.95	2366.40		
2 Roofer Helpers	32.85	525.60	55.30	884.80		
1 Application Equipment		181.00		199.10		
1 Tar Kettle/Pot		167.20		183.92		
1 Crew Truck		166.00		182.60	9.18	10.10
56 L.H., Daily Totals		$2813.80		$4435.62	$50.25	$79.21
Crew G-2	Hr.	Daily	Hr.	Daily	Bare Costs	Incl. O&P
1 Plasterer	$46.45	$371.60	$69.70	$557.60	$42.02	$63.35
1 Plasterer Helper	39.75	318.00	59.65	477.20		
1 Building Laborer	39.85	318.80	60.70	485.60		
1 Grout Pump, 50 C.F./hr.		128.60		141.46	5.36	5.89
24 L.H., Daily Totals		$1137.00		$1661.86	$47.38	$69.24
Crew G-2A	Hr.	Daily	Hr.	Daily	Bare Costs	Incl. O&P
1 Roofer Composition	$43.95	$351.60	$73.95	$591.60	$38.88	$63.32
1 Roofer Helper	32.85	262.80	55.30	442.40		
1 Building Laborer	39.85	318.80	60.70	485.60		
1 Foam Spray Rig, Trailer-Mtd.		499.25		549.17		
1 Pickup Truck, 3/4 Ton		115.20		126.72	25.60	28.16
24 L.H., Daily Totals		$1547.65		$2195.49	$64.49	$91.48
Crew G-3	Hr.	Daily	Hr.	Daily	Bare Costs	Incl. O&P
2 Sheet Metal Workers	$59.80	$956.80	$91.25	$1460.00	$49.83	$75.97
2 Building Laborers	39.85	637.60	60.70	971.20		
32 L.H., Daily Totals		$1594.40		$2431.20	$49.83	$75.97
Crew G-4	Hr.	Daily	Hr.	Daily	Bare Costs	Incl. O&P
1 Labor Foreman (outside)	$41.85	$334.80	$63.75	$510.00	$40.52	$61.72
2 Building Laborers	39.85	637.60	60.70	971.20		
1 Flatbed Truck, Gas, 1.5 Ton		188.40		207.24		
1 Air Compressor, 160 cfm		117.20		128.92	12.73	14.01
24 L.H., Daily Totals		$1278.00		$1817.36	$53.25	$75.72
Crew G-5	Hr.	Daily	Hr.	Daily	Bare Costs	Incl. O&P
1 Roofer Foreman (outside)	$45.95	$367.60	$77.35	$618.80	$39.91	$67.17
2 Roofers Composition	43.95	703.20	73.95	1183.20		
2 Roofer Helpers	32.85	525.60	55.30	884.80		
1 Application Equipment		181.00		199.10	4.53	4.98
40 L.H., Daily Totals		$1777.40		$2885.90	$44.44	$72.15
Crew G-6A	Hr.	Daily	Hr.	Daily	Bare Costs	Incl. O&P
2 Roofers Composition	$43.95	$703.20	$73.95	$1183.20	$43.95	$73.95
1 Small Compressor, Electric		15.10		16.61		
2 Pneumatic Nailers		46.10		50.71	3.83	4.21
16 L.H., Daily Totals		$764.40		$1250.52	$47.77	$78.16

Crew No.	Bare Costs		Incl. Subs O&P		Cost Per Labor-Hour	
Crew G-7	Hr.	Daily	Hr.	Daily	Bare Costs	Incl. O&P
1 Carpenter	$50.70	$405.60	$77.20	$617.60	$50.70	$77.20
1 Small Compressor, Electric		15.10		16.61		
1 Pneumatic Nailer		23.05		25.36	4.77	5.25
8 L.H., Daily Totals		$443.75		$659.57	$55.47	$82.45
Crew H-1	Hr.	Daily	Hr.	Daily	Bare Costs	Incl. O&P
2 Glaziers	$48.50	$776.00	$73.45	$1175.20	$51.58	$81.40
2 Struc. Steel Workers	54.65	874.40	89.35	1429.60		
32 L.H., Daily Totals		$1650.40		$2604.80	$51.58	$81.40
Crew H-2	Hr.	Daily	Hr.	Daily	Bare Costs	Incl. O&P
2 Glaziers	$48.50	$776.00	$73.45	$1175.20	$45.62	$69.20
1 Building Laborer	39.85	318.80	60.70	485.60		
24 L.H., Daily Totals		$1094.80		$1660.80	$45.62	$69.20
Crew H-3	Hr.	Daily	Hr.	Daily	Bare Costs	Incl. O&P
1 Glazier	$48.50	$388.00	$73.45	$587.60	$43.15	$65.90
1 Helper	37.80	302.40	58.35	466.80		
16 L.H., Daily Totals		$690.40		$1054.40	$43.15	$65.90
Crew H-4	Hr.	Daily	Hr.	Daily	Bare Costs	Incl. O&P
1 Carpenter	$50.70	$405.60	$77.20	$617.60	$47.04	$71.62
1 Carpenter Helper	37.80	302.40	58.35	466.80		
.5 Electrician	58.20	232.80	87.00	348.00		
20 L.H., Daily Totals		$940.80		$1432.40	$47.04	$71.62
Crew J-1	Hr.	Daily	Hr.	Daily	Bare Costs	Incl. O&P
3 Plasterers	$46.45	$1114.80	$69.70	$1672.80	$43.77	$65.68
2 Plasterer Helpers	39.75	636.00	59.65	954.40		
1 Mixing Machine, 6 C.F.		133.20		146.52	3.33	3.66
40 L.H., Daily Totals		$1884.00		$2773.72	$47.10	$69.34
Crew J-2	Hr.	Daily	Hr.	Daily	Bare Costs	Incl. O&P
3 Plasterers	$46.45	$1114.80	$69.70	$1672.80	$44.67	$66.84
2 Plasterer Helpers	39.75	636.00	59.65	954.40		
1 Lather	49.15	393.20	72.65	581.20		
1 Mixing Machine, 6 C.F.		133.20		146.52	2.77	3.05
48 L.H., Daily Totals		$2277.20		$3354.92	$47.44	$69.89
Crew J-3	Hr.	Daily	Hr.	Daily	Bare Costs	Incl. O&P
1 Terrazzo Worker	$47.00	$376.00	$69.55	$556.40	$43.05	$63.70
1 Terrazzo Helper	39.10	312.80	57.85	462.80		
1 Floor Grinder, 22" Path		133.60		146.96		
1 Terrazzo Mixer		177.80		195.58	19.46	21.41
16 L.H., Daily Totals		$1000.20		$1361.74	$62.51	$85.11
Crew J-4	Hr.	Daily	Hr.	Daily	Bare Costs	Incl. O&P
2 Cement Finishers	$47.55	$760.80	$70.45	$1127.20	$44.98	$67.20
1 Laborer	39.85	318.80	60.70	485.60		
1 Floor Grinder, 22" Path		133.60		146.96		
1 Floor Edger, 7" Path		42.95		47.24		
1 Vacuum Pick-Up System		66.95		73.64	10.15	11.16
24 L.H., Daily Totals		$1323.10		$1880.65	$55.13	$78.36

Crews - Standard

Crew No.	Bare Costs Hr.	Daily	Incl. Subs O&P Hr.	Daily	Cost Per Labor-Hour Bare Costs	Incl. O&P
Crew J-4A	Hr.	Daily	Hr.	Daily	Bare Costs	Incl. O&P
2 Cement Finishers	$47.55	$760.80	$70.45	$1127.20	$43.70	$65.58
2 Laborers	39.85	637.60	60.70	971.20		
1 Floor Grinder, 22" Path		133.60		146.96		
1 Floor Edger, 7" Path		42.95		47.24		
1 Vacuum Pick-Up System		66.95		73.64		
1 Floor Auto Scrubber		254.30		279.73	15.56	17.11
32 L.H., Daily Totals		$1896.20		$2645.98	$59.26	$82.69
Crew J-4B	Hr.	Daily	Hr.	Daily	Bare Costs	Incl. O&P
1 Laborer	$39.85	$318.80	$60.70	$485.60	$39.85	$60.70
1 Floor Auto Scrubber		254.30		279.73	31.79	34.97
8 L.H., Daily Totals		$573.10		$765.33	$71.64	$95.67
Crew J-6	Hr.	Daily	Hr.	Daily	Bare Costs	Incl. O&P
2 Painters	$42.55	$680.80	$64.05	$1024.80	$44.06	$66.54
1 Building Laborer	39.85	318.80	60.70	485.60		
1 Equip. Oper. (light)	51.30	410.40	77.35	618.80		
1 Air Compressor, 250 cfm		167.40		184.14		
1 Sandblaster, Portable, 3 C.F.		21.00		23.10		
1 Set Sand Blasting Accessories		14.70		16.17	6.35	6.98
32 L.H., Daily Totals		$1613.10		$2352.61	$50.41	$73.52
Crew J-7	Hr.	Daily	Hr.	Daily	Bare Costs	Incl. O&P
2 Painters	$42.55	$680.80	$64.05	$1024.80	$42.55	$64.05
1 Floor Belt Sander		16.75		18.43		
1 Floor Sanding Edger		13.15		14.47	1.87	2.06
16 L.H., Daily Totals		$710.70		$1057.69	$44.42	$66.11
Crew K-1	Hr.	Daily	Hr.	Daily	Bare Costs	Incl. O&P
1 Carpenter	$50.70	$405.60	$77.20	$617.60	$47.60	$72.10
1 Truck Driver (light)	44.50	356.00	67.00	536.00		
1 Flatbed Truck, Gas, 3 Ton		238.00		261.80	14.88	16.36
16 L.H., Daily Totals		$999.60		$1415.40	$62.48	$88.46
Crew K-2	Hr.	Daily	Hr.	Daily	Bare Costs	Incl. O&P
1 Struc. Steel Foreman (outside)	$56.65	$453.20	$92.60	$740.80	$51.93	$82.98
1 Struc. Steel Worker	54.65	437.20	89.35	714.80		
1 Truck Driver (light)	44.50	356.00	67.00	536.00		
1 Flatbed Truck, Gas, 3 Ton		238.00		261.80	9.92	10.91
24 L.H., Daily Totals		$1484.40		$2253.40	$61.85	$93.89
Crew L-1	Hr.	Daily	Hr.	Daily	Bare Costs	Incl. O&P
1 Electrician	$58.20	$465.60	$87.00	$696.00	$60.17	$90.30
1 Plumber	62.15	497.20	93.60	748.80		
16 L.H., Daily Totals		$962.80		$1444.80	$60.17	$90.30
Crew L-2	Hr.	Daily	Hr.	Daily	Bare Costs	Incl. O&P
1 Carpenter	$50.70	$405.60	$77.20	$617.60	$44.25	$67.78
1 Carpenter Helper	37.80	302.40	58.35	466.80		
16 L.H., Daily Totals		$708.00		$1084.40	$44.25	$67.78
Crew L-3	Hr.	Daily	Hr.	Daily	Bare Costs	Incl. O&P
1 Carpenter	$50.70	$405.60	$77.20	$617.60	$54.85	$83.16
.5 Electrician	58.20	232.80	87.00	348.00		
.5 Sheet Metal Worker	59.80	239.20	91.25	365.00		
16 L.H., Daily Totals		$877.60		$1330.60	$54.85	$83.16

Crew No.	Bare Costs Hr.	Daily	Incl. Subs O&P Hr.	Daily	Cost Per Labor-Hour Bare Costs	Incl. O&P
Crew L-3A	Hr.	Daily	Hr.	Daily	Bare Costs	Incl. O&P
1 Carpenter Foreman (outside)	$52.70	$421.60	$80.25	$642.00	$55.07	$83.92
.5 Sheet Metal Worker	59.80	239.20	91.25	365.00		
12 L.H., Daily Totals		$660.80		$1007.00	$55.07	$83.92
Crew L-4	Hr.	Daily	Hr.	Daily	Bare Costs	Incl. O&P
2 Skilled Workers	$52.35	$837.60	$80.15	$1282.40	$47.50	$72.88
1 Helper	37.80	302.40	58.35	466.80		
24 L.H., Daily Totals		$1140.00		$1749.20	$47.50	$72.88
Crew L-5	Hr.	Daily	Hr.	Daily	Bare Costs	Incl. O&P
1 Struc. Steel Foreman (outside)	$56.65	$453.20	$92.60	$740.80	$55.14	$89.14
5 Struc. Steel Workers	54.65	2186.00	89.35	3574.00		
1 Equip. Oper. (crane)	56.10	448.80	84.60	676.80		
1 Hyd. Crane, 25 Ton		590.60		649.66	10.55	11.60
56 L.H., Daily Totals		$3678.60		$5641.26	$65.69	$100.74
Crew L-5A	Hr.	Daily	Hr.	Daily	Bare Costs	Incl. O&P
1 Struc. Steel Foreman (outside)	$56.65	$453.20	$92.60	$740.80	$55.51	$88.97
2 Structural Steel Workers	54.65	874.40	89.35	1429.60		
1 Equip. Oper. (crane)	56.10	448.80	84.60	676.80		
1 S.P. Crane, 4x4, 25 Ton		668.40		735.24	20.89	22.98
32 L.H., Daily Totals		$2444.80		$3582.44	$76.40	$111.95
Crew L-5B	Hr.	Daily	Hr.	Daily	Bare Costs	Incl. O&P
1 Struc. Steel Foreman (outside)	$56.65	$453.20	$92.60	$740.80	$57.01	$88.11
2 Structural Steel Workers	54.65	874.40	89.35	1429.60		
2 Electricians	58.20	931.20	87.00	1392.00		
2 Steamfitters/Pipefitters	63.00	1008.00	94.90	1518.40		
1 Equip. Oper. (crane)	56.10	448.80	84.60	676.80		
1 Equip. Oper. (oiler)	48.60	388.80	73.30	586.40		
1 Hyd. Crane, 80 Ton		1518.00		1669.80	21.08	23.19
72 L.H., Daily Totals		$5622.40		$8013.80	$78.09	$111.30
Crew L-6	Hr.	Daily	Hr.	Daily	Bare Costs	Incl. O&P
1 Plumber	$62.15	$497.20	$93.60	$748.80	$60.83	$91.40
.5 Electrician	58.20	232.80	87.00	348.00		
12 L.H., Daily Totals		$730.00		$1096.80	$60.83	$91.40
Crew L-7	Hr.	Daily	Hr.	Daily	Bare Costs	Incl. O&P
2 Carpenters	$50.70	$811.20	$77.20	$1235.20	$48.67	$73.89
1 Building Laborer	39.85	318.80	60.70	485.60		
.5 Electrician	58.20	232.80	87.00	348.00		
28 L.H., Daily Totals		$1362.80		$2068.80	$48.67	$73.89
Crew L-8	Hr.	Daily	Hr.	Daily	Bare Costs	Incl. O&P
2 Carpenters	$50.70	$811.20	$77.20	$1235.20	$52.99	$80.48
.5 Plumber	62.15	248.60	93.60	374.40		
20 L.H., Daily Totals		$1059.80		$1609.60	$52.99	$80.48
Crew L-9	Hr.	Daily	Hr.	Daily	Bare Costs	Incl. O&P
1 Labor Foreman (inside)	$40.35	$322.80	$61.45	$491.60	$45.29	$70.16
2 Building Laborers	39.85	637.60	60.70	971.20		
1 Struc. Steel Worker	54.65	437.20	89.35	714.80		
.5 Electrician	58.20	232.80	87.00	348.00		
36 L.H., Daily Totals		$1630.40		$2525.60	$45.29	$70.16

For customer support on your Electrical Costs with RSMeans data, call 800.448.8182.

Crew No.	Bare Costs Hr.	Daily	Incl. Subs O&P Hr.	Daily	Cost Per Labor-Hour Bare Costs	Incl. O&P
Crew L-10	Hr.	Daily	Hr.	Daily	Bare Costs	Incl. O&P
1 Struc. Steel Foreman (outside)	$56.65	$453.20	$92.60	$740.80	$55.80	$88.85
1 Structural Steel Worker	54.65	437.20	89.35	714.80		
1 Equip. Oper. (crane)	56.10	448.80	84.60	676.80		
1 Hyd. Crane, 12 Ton		496.00		545.60	20.67	22.73
24 L.H., Daily Totals		$1835.20		$2678.00	$76.47	$111.58
Crew L-11	Hr.	Daily	Hr.	Daily	Bare Costs	Incl. O&P
2 Wreckers	$39.85	$637.60	$62.70	$1003.20	$46.77	$71.84
1 Equip. Oper. (crane)	56.10	448.80	84.60	676.80		
1 Equip. Oper. (light)	51.30	410.40	77.35	618.80		
1 Hyd. Excavator, 2.5 C.Y.		1411.00		1552.10		
1 Loader, Skid Steer, 78 H.P.		364.80		401.28	55.49	61.04
32 L.H., Daily Totals		$3272.60		$4252.18	$102.27	$132.88
Crew M-1	Hr.	Daily	Hr.	Daily	Bare Costs	Incl. O&P
3 Elevator Constructors	$82.20	$1972.80	$122.50	$2940.00	$78.09	$116.36
1 Elevator Apprentice	65.75	526.00	97.95	783.60		
5 Hand Tools		50.00		55.00	1.56	1.72
32 L.H., Daily Totals		$2548.80		$3778.60	$79.65	$118.08
Crew M-3	Hr.	Daily	Hr.	Daily	Bare Costs	Incl. O&P
1 Electrician Foreman (outside)	$60.20	$481.60	$90.00	$720.00	$61.51	$92.10
1 Common Laborer	39.85	318.80	60.70	485.60		
.25 Equipment Operator (med.)	53.75	107.50	81.05	162.10		
1 Elevator Constructor	82.20	657.60	122.50	980.00		
1 Elevator Apprentice	65.75	526.00	97.95	783.60		
.25 S.P. Crane, 4x4, 20 Ton		142.80		157.08	4.20	4.62
34 L.H., Daily Totals		$2234.30		$3288.38	$65.71	$96.72
Crew M-4	Hr.	Daily	Hr.	Daily	Bare Costs	Incl. O&P
1 Electrician Foreman (outside)	$60.20	$481.60	$90.00	$720.00	$60.93	$91.25
1 Common Laborer	39.85	318.80	60.70	485.60		
.25 Equipment Operator, Crane	56.10	112.20	84.60	169.20		
.25 Equip. Oper. (oiler)	48.60	97.20	73.30	146.60		
1 Elevator Constructor	82.20	657.60	122.50	980.00		
1 Elevator Apprentice	65.75	526.00	97.95	783.60		
.25 S.P. Crane, 4x4, 40 Ton		183.70		202.07	5.10	5.61
36 L.H., Daily Totals		$2377.10		$3487.07	$66.03	$96.86
Crew Q-1	Hr.	Daily	Hr.	Daily	Bare Costs	Incl. O&P
1 Plumber	$62.15	$497.20	$93.60	$748.80	$55.92	$84.22
1 Plumber Apprentice	49.70	397.60	74.85	598.80		
16 L.H., Daily Totals		$894.80		$1347.60	$55.92	$84.22
Crew Q-1A	Hr.	Daily	Hr.	Daily	Bare Costs	Incl. O&P
.25 Plumber Foreman (outside)	$64.15	$128.30	$96.60	$193.20	$62.55	$94.20
1 Plumber	62.15	497.20	93.60	748.80		
10 L.H., Daily Totals		$625.50		$942.00	$62.55	$94.20
Crew Q-1C	Hr.	Daily	Hr.	Daily	Bare Costs	Incl. O&P
1 Plumber	$62.15	$497.20	$93.60	$748.80	$55.20	$83.17
1 Plumber Apprentice	49.70	397.60	74.85	598.80		
1 Equip. Oper. (medium)	53.75	430.00	81.05	648.40		
1 Trencher, Chain Type, 8' D		2048.00		2252.80	85.33	93.87
24 L.H., Daily Totals		$3372.80		$4248.80	$140.53	$177.03
Crew Q-2	Hr.	Daily	Hr.	Daily	Bare Costs	Incl. O&P
2 Plumbers	$62.15	$994.40	$93.60	$1497.60	$58.00	$87.35
1 Plumber Apprentice	49.70	397.60	74.85	598.80		
24 L.H., Daily Totals		$1392.00		$2096.40	$58.00	$87.35

Crew No.	Bare Costs Hr.	Daily	Incl. Subs O&P Hr.	Daily	Cost Per Labor-Hour Bare Costs	Incl. O&P
Crew Q-3	Hr.	Daily	Hr.	Daily	Bare Costs	Incl. O&P
1 Plumber Foreman (inside)	$62.65	$501.20	$94.35	$754.80	$59.16	$89.10
2 Plumbers	62.15	994.40	93.60	1497.60		
1 Plumber Apprentice	49.70	397.60	74.85	598.80		
32 L.H., Daily Totals		$1893.20		$2851.20	$59.16	$89.10
Crew Q-4	Hr.	Daily	Hr.	Daily	Bare Costs	Incl. O&P
1 Plumber Foreman (inside)	$62.65	$501.20	$94.35	$754.80	$59.16	$89.10
1 Plumber	62.15	497.20	93.60	748.80		
1 Welder (plumber)	62.15	497.20	93.60	748.80		
1 Plumber Apprentice	49.70	397.60	74.85	598.80		
1 Welder, Electric, 300 amp		59.20		65.12	1.85	2.04
32 L.H., Daily Totals		$1952.40		$2916.32	$61.01	$91.14
Crew Q-5	Hr.	Daily	Hr.	Daily	Bare Costs	Incl. O&P
1 Steamfitter	$63.00	$504.00	$94.90	$759.20	$56.70	$85.40
1 Steamfitter Apprentice	50.40	403.20	75.90	607.20		
16 L.H., Daily Totals		$907.20		$1366.40	$56.70	$85.40
Crew Q-6	Hr.	Daily	Hr.	Daily	Bare Costs	Incl. O&P
2 Steamfitters	$63.00	$1008.00	$94.90	$1518.40	$58.80	$88.57
1 Steamfitter Apprentice	50.40	403.20	75.90	607.20		
24 L.H., Daily Totals		$1411.20		$2125.60	$58.80	$88.57
Crew Q-7	Hr.	Daily	Hr.	Daily	Bare Costs	Incl. O&P
1 Steamfitter Foreman (inside)	$63.50	$508.00	$95.65	$765.20	$59.98	$90.34
2 Steamfitters	63.00	1008.00	94.90	1518.40		
1 Steamfitter Apprentice	50.40	403.20	75.90	607.20		
32 L.H., Daily Totals		$1919.20		$2890.80	$59.98	$90.34
Crew Q-8	Hr.	Daily	Hr.	Daily	Bare Costs	Incl. O&P
1 Steamfitter Foreman (inside)	$63.50	$508.00	$95.65	$765.20	$59.98	$90.34
1 Steamfitter	63.00	504.00	94.90	759.20		
1 Welder (steamfitter)	63.00	504.00	94.90	759.20		
1 Steamfitter Apprentice	50.40	403.20	75.90	607.20		
1 Welder, Electric, 300 amp		59.20		65.12	1.85	2.04
32 L.H., Daily Totals		$1978.40		$2955.92	$61.83	$92.37
Crew Q-9	Hr.	Daily	Hr.	Daily	Bare Costs	Incl. O&P
1 Sheet Metal Worker	$59.80	$478.40	$91.25	$730.00	$53.83	$82.13
1 Sheet Metal Apprentice	47.85	382.80	73.00	584.00		
16 L.H., Daily Totals		$861.20		$1314.00	$53.83	$82.13
Crew Q-10	Hr.	Daily	Hr.	Daily	Bare Costs	Incl. O&P
2 Sheet Metal Workers	$59.80	$956.80	$91.25	$1460.00	$55.82	$85.17
1 Sheet Metal Apprentice	47.85	382.80	73.00	584.00		
24 L.H., Daily Totals		$1339.60		$2044.00	$55.82	$85.17
Crew Q-11	Hr.	Daily	Hr.	Daily	Bare Costs	Incl. O&P
1 Sheet Metal Foreman (inside)	$60.30	$482.40	$92.00	$736.00	$56.94	$86.88
2 Sheet Metal Workers	59.80	956.80	91.25	1460.00		
1 Sheet Metal Apprentice	47.85	382.80	73.00	584.00		
32 L.H., Daily Totals		$1822.00		$2780.00	$56.94	$86.88
Crew Q-12	Hr.	Daily	Hr.	Daily	Bare Costs	Incl. O&P
1 Sprinkler Installer	$59.90	$479.20	$90.45	$723.60	$53.90	$81.40
1 Sprinkler Apprentice	47.90	383.20	72.35	578.80		
16 L.H., Daily Totals		$862.40		$1302.40	$53.90	$81.40

For customer support on your Electrical Costs with RSMeans data, call 800.448.8182.

Crew No.	Bare Costs		Incl. Subs O&P		Cost Per Labor-Hour	

Crew Q-13

	Hr.	Daily	Hr.	Daily	Bare Costs	Incl. O&P
1 Sprinkler Foreman (inside)	$60.40	$483.20	$91.20	$729.60	$57.02	$86.11
2 Sprinkler Installers	59.90	958.40	90.45	1447.20		
1 Sprinkler Apprentice	47.90	383.20	72.35	578.80		
32 L.H., Daily Totals		$1824.80		$2755.60	$57.02	$86.11

Crew Q-14

	Hr.	Daily	Hr.	Daily	Bare Costs	Incl. O&P
1 Asbestos Worker	$56.30	$450.40	$87.40	$699.20	$50.67	$78.65
1 Asbestos Apprentice	45.05	360.40	69.90	559.20		
16 L.H., Daily Totals		$810.80		$1258.40	$50.67	$78.65

Crew Q-15

	Hr.	Daily	Hr.	Daily	Bare Costs	Incl. O&P
1 Plumber	$62.15	$497.20	$93.60	$748.80	$55.92	$84.22
1 Plumber Apprentice	49.70	397.60	74.85	598.80		
1 Welder, Electric, 300 amp		59.20		65.12	3.70	4.07
16 L.H., Daily Totals		$954.00		$1412.72	$59.63	$88.30

Crew Q-16

	Hr.	Daily	Hr.	Daily	Bare Costs	Incl. O&P
2 Plumbers	$62.15	$994.40	$93.60	$1497.60	$58.00	$87.35
1 Plumber Apprentice	49.70	397.60	74.85	598.80		
1 Welder, Electric, 300 amp		59.20		65.12	2.47	2.71
24 L.H., Daily Totals		$1451.20		$2161.52	$60.47	$90.06

Crew Q-17

	Hr.	Daily	Hr.	Daily	Bare Costs	Incl. O&P
1 Steamfitter	$63.00	$504.00	$94.90	$759.20	$56.70	$85.40
1 Steamfitter Apprentice	50.40	403.20	75.90	607.20		
1 Welder, Electric, 300 amp		59.20		65.12	3.70	4.07
16 L.H., Daily Totals		$966.40		$1431.52	$60.40	$89.47

Crew Q-17A

	Hr.	Daily	Hr.	Daily	Bare Costs	Incl. O&P
1 Steamfitter	$63.00	$504.00	$94.90	$759.20	$56.50	$85.13
1 Steamfitter Apprentice	50.40	403.20	75.90	607.20		
1 Equip. Oper. (crane)	56.10	448.80	84.60	676.80		
1 Hyd. Crane, 12 Ton		496.00		545.60		
1 Welder, Electric, 300 amp		59.20		65.12	23.13	25.45
24 L.H., Daily Totals		$1911.20		$2653.92	$79.63	$110.58

Crew Q-18

	Hr.	Daily	Hr.	Daily	Bare Costs	Incl. O&P
2 Steamfitters	$63.00	$1008.00	$94.90	$1518.40	$58.80	$88.57
1 Steamfitter Apprentice	50.40	403.20	75.90	607.20		
1 Welder, Electric, 300 amp		59.20		65.12	2.47	2.71
24 L.H., Daily Totals		$1470.40		$2190.72	$61.27	$91.28

Crew Q-19

	Hr.	Daily	Hr.	Daily	Bare Costs	Incl. O&P
1 Steamfitter	$63.00	$504.00	$94.90	$759.20	$57.20	$85.93
1 Steamfitter Apprentice	50.40	403.20	75.90	607.20		
1 Electrician	58.20	465.60	87.00	696.00		
24 L.H., Daily Totals		$1372.80		$2062.40	$57.20	$85.93

Crew Q-20

	Hr.	Daily	Hr.	Daily	Bare Costs	Incl. O&P
1 Sheet Metal Worker	$59.80	$478.40	$91.25	$730.00	$54.70	$83.10
1 Sheet Metal Apprentice	47.85	382.80	73.00	584.00		
.5 Electrician	58.20	232.80	87.00	348.00		
20 L.H., Daily Totals		$1094.00		$1662.00	$54.70	$83.10

Crew Q-21

	Hr.	Daily	Hr.	Daily	Bare Costs	Incl. O&P
2 Steamfitters	$63.00	$1008.00	$94.90	$1518.40	$58.65	$88.17
1 Steamfitter Apprentice	50.40	403.20	75.90	607.20		
1 Electrician	58.20	465.60	87.00	696.00		
32 L.H., Daily Totals		$1876.80		$2821.60	$58.65	$88.17

Crew Q-22

	Hr.	Daily	Hr.	Daily	Bare Costs	Incl. O&P
1 Plumber	$62.15	$497.20	$93.60	$748.80	$55.92	$84.22
1 Plumber Apprentice	49.70	397.60	74.85	598.80		
1 Hyd. Crane, 12 Ton		496.00		545.60	31.00	34.10
16 L.H., Daily Totals		$1390.80		$1893.20	$86.92	$118.33

Crew Q-22A

	Hr.	Daily	Hr.	Daily	Bare Costs	Incl. O&P
1 Plumber	$62.15	$497.20	$93.60	$748.80	$51.95	$78.44
1 Plumber Apprentice	49.70	397.60	74.85	598.80		
1 Laborer	39.85	318.80	60.70	485.60		
1 Equip. Oper. (crane)	56.10	448.80	84.60	676.80		
1 Hyd. Crane, 12 Ton		496.00		545.60	15.50	17.05
32 L.H., Daily Totals		$2158.40		$3055.60	$67.45	$95.49

Crew Q-23

	Hr.	Daily	Hr.	Daily	Bare Costs	Incl. O&P
1 Plumber Foreman (outside)	$64.15	$513.20	$96.60	$772.80	$60.02	$90.42
1 Plumber	62.15	497.20	93.60	748.80		
1 Equip. Oper. (medium)	53.75	430.00	81.05	648.40		
1 Lattice Boom Crane, 20 Ton		1083.00		1191.30	45.13	49.64
24 L.H., Daily Totals		$2523.40		$3361.30	$105.14	$140.05

Crew R-1

	Hr.	Daily	Hr.	Daily	Bare Costs	Incl. O&P
1 Electrician Foreman	$58.70	$469.60	$87.75	$702.00	$54.40	$81.33
3 Electricians	58.20	1396.80	87.00	2088.00		
2 Electrician Apprentices	46.55	744.80	69.60	1113.60		
48 L.H., Daily Totals		$2611.20		$3903.60	$54.40	$81.33

Crew R-1A

	Hr.	Daily	Hr.	Daily	Bare Costs	Incl. O&P
1 Electrician	$58.20	$465.60	$87.00	$696.00	$52.38	$78.30
1 Electrician Apprentice	46.55	372.40	69.60	556.80		
16 L.H., Daily Totals		$838.00		$1252.80	$52.38	$78.30

Crew R-1B

	Hr.	Daily	Hr.	Daily	Bare Costs	Incl. O&P
1 Electrician	$58.20	$465.60	$87.00	$696.00	$50.43	$75.40
2 Electrician Apprentices	46.55	744.80	69.60	1113.60		
24 L.H., Daily Totals		$1210.40		$1809.60	$50.43	$75.40

Crew R-1C

	Hr.	Daily	Hr.	Daily	Bare Costs	Incl. O&P
2 Electricians	$58.20	$931.20	$87.00	$1392.00	$52.38	$78.30
2 Electrician Apprentices	46.55	744.80	69.60	1113.60		
1 Portable Cable Puller, 8000 lb.		130.15		143.16	4.07	4.47
32 L.H., Daily Totals		$1806.15		$2648.76	$56.44	$82.77

Crew R-2

	Hr.	Daily	Hr.	Daily	Bare Costs	Incl. O&P
1 Electrician Foreman	$58.70	$469.60	$87.75	$702.00	$54.64	$81.79
3 Electricians	58.20	1396.80	87.00	2088.00		
2 Electrician Apprentices	46.55	744.80	69.60	1113.60		
1 Equip. Oper. (crane)	56.10	448.80	84.60	676.80		
1 S.P. Crane, 4x4, 5 Ton		259.00		284.90	4.63	5.09
56 L.H., Daily Totals		$3319.00		$4865.30	$59.27	$86.88

Crew R-3

	Hr.	Daily	Hr.	Daily	Bare Costs	Incl. O&P
1 Electrician Foreman	$58.70	$469.60	$87.75	$702.00	$57.98	$86.82
1 Electrician	58.20	465.60	87.00	696.00		
.5 Equip. Oper. (crane)	56.10	224.40	84.60	338.40		
.5 S.P. Crane, 4x4, 5 Ton		129.50		142.45	6.47	7.12
20 L.H., Daily Totals		$1289.10		$1878.85	$64.45	$93.94

For customer support on your Electrical Costs with RSMeans data, call 800.448.8182.

509

Crew No.	Bare Costs		Incl. Subs O&P		Cost Per Labor-Hour	
Crew R-4	Hr.	Daily	Hr.	Daily	Bare Costs	Incl. O&P
1 Struc. Steel Foreman (outside)	$56.65	$453.20	$92.60	$740.80	$55.76	$89.53
3 Struc. Steel Workers	54.65	1311.60	89.35	2144.40		
1 Electrician	58.20	465.60	87.00	696.00		
1 Welder, Gas Engine, 300 amp		98.40		108.24	2.46	2.71
40 L.H., Daily Totals		$2328.80		$3689.44	$58.22	$92.24
Crew R-5	Hr.	Daily	Hr.	Daily	Bare Costs	Incl. O&P
1 Electrician Foreman	$58.70	$469.60	$87.75	$702.00	$50.83	$76.65
4 Electrician Linemen	58.20	1862.40	87.00	2784.00		
2 Electrician Operators	58.20	931.20	87.00	1392.00		
4 Electrician Groundmen	37.80	1209.60	58.35	1867.20		
1 Crew Truck		166.00		182.60		
1 Flatbed Truck, 20,000 GVW		197.20		216.92		
1 Pickup Truck, 3/4 Ton		115.20		126.72		
.2 Hyd. Crane, 55 Ton		198.76		218.64		
.2 Hyd. Crane, 12 Ton		99.20		109.12		
.2 Earth Auger, Truck-Mtd.		74.76		82.24		
1 Tractor w/Winch		361.80		397.98	13.78	15.16
88 L.H., Daily Totals		$5685.72		$8079.41	$64.61	$91.81
Crew R-6	Hr.	Daily	Hr.	Daily	Bare Costs	Incl. O&P
1 Electrician Foreman	$58.70	$469.60	$87.75	$702.00	$50.83	$76.65
4 Electrician Linemen	58.20	1862.40	87.00	2784.00		
2 Electrician Operators	58.20	931.20	87.00	1392.00		
4 Electrician Groundmen	37.80	1209.60	58.35	1867.20		
1 Crew Truck		166.00		182.60		
1 Flatbed Truck, 20,000 GVW		197.20		216.92		
1 Pickup Truck, 3/4 Ton		115.20		126.72		
.2 Hyd. Crane, 55 Ton		198.76		218.64		
.2 Hyd. Crane, 12 Ton		99.20		109.12		
.2 Earth Auger, Truck-Mtd.		74.76		82.24		
1 Tractor w/Winch		361.80		397.98		
3 Cable Trailers		635.10		698.61		
.5 Tensioning Rig		209.93		230.92		
.5 Cable Pulling Rig		1213.00		1334.30	37.17	40.89
88 L.H., Daily Totals		$7743.74		$10343.24	$88.00	$117.54
Crew R-7	Hr.	Daily	Hr.	Daily	Bare Costs	Incl. O&P
1 Electrician Foreman	$58.70	$469.60	$87.75	$702.00	$41.28	$63.25
5 Electrician Groundmen	37.80	1512.00	58.35	2334.00		
1 Crew Truck		166.00		182.60	3.46	3.80
48 L.H., Daily Totals		$2147.60		$3218.60	$44.74	$67.05
Crew R-8	Hr.	Daily	Hr.	Daily	Bare Costs	Incl. O&P
1 Electrician Foreman	$58.70	$469.60	$87.75	$702.00	$51.48	$77.58
3 Electrician Linemen	58.20	1396.80	87.00	2088.00		
2 Electrician Groundmen	37.80	604.80	58.35	933.60		
1 Pickup Truck, 3/4 Ton		115.20		126.72		
1 Crew Truck		166.00		182.60	5.86	6.44
48 L.H., Daily Totals		$2752.40		$4032.92	$57.34	$84.02
Crew R-9	Hr.	Daily	Hr.	Daily	Bare Costs	Incl. O&P
1 Electrician Foreman	$58.70	$469.60	$87.75	$702.00	$48.06	$72.77
1 Electrician Lineman	58.20	465.60	87.00	696.00		
2 Electrician Operators	58.20	931.20	87.00	1392.00		
4 Electrician Groundmen	37.80	1209.60	58.35	1867.20		
1 Pickup Truck, 3/4 Ton		115.20		126.72		
1 Crew Truck		166.00		182.60	4.39	4.83
64 L.H., Daily Totals		$3357.20		$4966.52	$52.46	$77.60
Crew R-10	Hr.	Daily	Hr.	Daily	Bare Costs	Incl. O&P
1 Electrician Foreman	$58.70	$469.60	$87.75	$702.00	$54.88	$82.35
4 Electrician Linemen	58.20	1862.40	87.00	2784.00		
1 Electrician Groundman	37.80	302.40	58.35	466.80		
1 Crew Truck		166.00		182.60		
3 Tram Cars		433.80		477.18	12.50	13.75
48 L.H., Daily Totals		$3234.20		$4612.58	$67.38	$96.10
Crew R-11	Hr.	Daily	Hr.	Daily	Bare Costs	Incl. O&P
1 Electrician Foreman	$58.70	$469.60	$87.75	$702.00	$55.35	$83.01
4 Electricians	58.20	1862.40	87.00	2784.00		
1 Equip. Oper. (crane)	56.10	448.80	84.60	676.80		
1 Common Laborer	39.85	318.80	60.70	485.60		
1 Crew Truck		166.00		182.60		
1 Hyd. Crane, 12 Ton		496.00		545.60	11.82	13.00
56 L.H., Daily Totals		$3761.60		$5376.60	$67.17	$96.01
Crew R-12	Hr.	Daily	Hr.	Daily	Bare Costs	Incl. O&P
1 Carpenter Foreman (inside)	$51.20	$409.60	$78.00	$624.00	$47.44	$72.73
4 Carpenters	50.70	1622.40	77.20	2470.40		
4 Common Laborers	39.85	1275.20	60.70	1942.40		
1 Equip. Oper. (medium)	53.75	430.00	81.05	648.40		
1 Steel Worker	54.65	437.20	89.35	714.80		
1 Dozer, 200 H.P.		1273.00		1400.30		
1 Pickup Truck, 3/4 Ton		115.20		126.72	15.78	17.35
88 L.H., Daily Totals		$5562.60		$7927.02	$63.21	$90.08
Crew R-13	Hr.	Daily	Hr.	Daily	Bare Costs	Incl. O&P
1 Electrician Foreman	$58.70	$469.60	$87.75	$702.00	$56.37	$84.42
3 Electricians	58.20	1396.80	87.00	2088.00		
.25 Equip. Oper. (crane)	56.10	112.20	84.60	169.20		
1 Equipment Oiler	48.60	388.80	73.30	586.40		
.25 Hydraulic Crane, 33 Ton		235.15		258.67	5.60	6.16
42 L.H., Daily Totals		$2602.55		$3804.26	$61.97	$90.58
Crew R-15	Hr.	Daily	Hr.	Daily	Bare Costs	Incl. O&P
1 Electrician Foreman	$58.70	$469.60	$87.75	$702.00	$57.13	$85.52
4 Electricians	58.20	1862.40	87.00	2784.00		
1 Equipment Oper. (light)	51.30	410.40	77.35	618.80		
1 Telescoping Boom Lift, to 40'		282.80		311.08	5.89	6.48
48 L.H., Daily Totals		$3025.20		$4415.88	$63.02	$92.00
Crew R-15A	Hr.	Daily	Hr.	Daily	Bare Costs	Incl. O&P
1 Electrician Foreman	$58.70	$469.60	$87.75	$702.00	$51.02	$76.75
2 Electricians	58.20	931.20	87.00	1392.00		
2 Common Laborers	39.85	637.60	60.70	971.20		
1 Equip. Oper. (light)	51.30	410.40	77.35	618.80		
1 Telescoping Boom Lift, to 40'		282.80		311.08	5.89	6.48
48 L.H., Daily Totals		$2731.60		$3995.08	$56.91	$83.23
Crew R-18	Hr.	Daily	Hr.	Daily	Bare Costs	Incl. O&P
.25 Electrician Foreman	$58.70	$117.40	$87.75	$175.50	$51.07	$76.35
1 Electrician	58.20	465.60	87.00	696.00		
2 Electrician Apprentices	46.55	744.80	69.60	1113.60		
26 L.H., Daily Totals		$1327.80		$1985.10	$51.07	$76.35
Crew R-19	Hr.	Daily	Hr.	Daily	Bare Costs	Incl. O&P
.5 Electrician Foreman	$58.70	$234.80	$87.75	$351.00	$58.30	$87.15
2 Electricians	58.20	931.20	87.00	1392.00		
20 L.H., Daily Totals		$1166.00		$1743.00	$58.30	$87.15

For customer support on your Electrical Costs with RSMeans data, call 800.448.8182.

Crew No.	Bare Costs		Incl. Subs O & P		Cost Per Labor-Hour	
Crew R-21	**Hr.**	**Daily**	**Hr.**	**Daily**	**Bare Costs**	**Incl. O&P**
1 Electrician Foreman	$58.70	$469.60	$87.75	$702.00	$58.21	$87.04
3 Electricians	58.20	1396.80	87.00	2088.00		
.1 Equip. Oper. (medium)	53.75	43.00	81.05	64.84		
.1 S.P. Crane, 4x4, 25 Ton		66.84		73.52	2.04	2.24
32.8 L.H., Daily Totals		$1976.24		$2928.36	$60.25	$89.28
Crew R-22	**Hr.**	**Daily**	**Hr.**	**Daily**	**Bare Costs**	**Incl. O&P**
.66 Electrician Foreman	$58.70	$309.94	$87.75	$463.32	$53.27	$79.64
2 Electricians	58.20	931.20	87.00	1392.00		
2 Electrician Apprentices	46.55	744.80	69.60	1113.60		
37.28 L.H., Daily Totals		$1985.94		$2968.92	$53.27	$79.64
Crew R-30	**Hr.**	**Daily**	**Hr.**	**Daily**	**Bare Costs**	**Incl. O&P**
.25 Electrician Foreman (outside)	$60.20	$120.40	$90.00	$180.00	$47.06	$71.05
1 Electrician	58.20	465.60	87.00	696.00		
2 Laborers (Semi-Skilled)	39.85	637.60	60.70	971.20		
26 L.H., Daily Totals		$1223.60		$1847.20	$47.06	$71.05
Crew R-31	**Hr.**	**Daily**	**Hr.**	**Daily**	**Bare Costs**	**Incl. O&P**
1 Electrician	$58.20	$465.60	$87.00	$696.00	$58.20	$87.00
1 Core Drill, Electric, 2.5 H.P.		47.50		52.25	5.94	6.53
8 L.H., Daily Totals		$513.10		$748.25	$64.14	$93.53
Crew W-41E	**Hr.**	**Daily**	**Hr.**	**Daily**	**Bare Costs**	**Incl. O&P**
.5 Plumber Foreman (outside)	$64.15	$256.60	$96.60	$386.40	$53.63	$81.04
1 Plumber	62.15	497.20	93.60	748.80		
1 Laborer	39.85	318.80	60.70	485.60		
20 L.H., Daily Totals		$1072.60		$1620.80	$53.63	$81.04

The table below lists both the RSMeans® historical cost index based on Jan. 1, 1993 = 100 as well as the computed value of an index based on Jan. 1, 2018 costs. Since the Jan. 1, 2018 figure is estimated, space is left to write in the actual index figures as they become available through the quarterly RSMeans Construction Cost Indexes.

To compute the actual index based on Jan. 1, 2018 = 100, divide the historical cost index for a particular year by the actual Jan. 1, 2018 construction cost index. Space has been left to advance the index figures as the year progresses.

Year	Historical Cost Index Jan. 1, 1993 = 100 Est.	Historical Cost Index Jan. 1, 1993 = 100 Actual	Current Index Based on Jan. 1, 2018 = 100 Est.	Current Index Based on Jan. 1, 2018 = 100 Actual
Oct 2018*				
July 2018*				
April 2018*				
Jan 2018*	215.8		100.0	100.0
July 2017		213.6	99.0	
2016		207.3	96.1	
2015		206.2	95.6	
2014		204.9	94.9	
2013		201.2	93.2	
2012		194.6	90.2	
2011		191.2	88.6	
2010		183.5	85.0	
2009		180.1	83.5	
2008		180.4	83.6	
2007		169.4	78.5	
2006		162.0	75.1	
2005		151.6	70.3	
2004		143.7	66.6	

Year	Historical Cost Index Jan. 1, 1993 = 100 Actual	Current Index Based on Jan. 1, 2018 = 100 Est.	Current Index Based on Jan. 1, 2018 = 100 Actual
July 2003	132.0	61.2	
2002	128.7	59.6	
2001	125.1	58.0	
2000	120.9	56.0	
1999	117.6	54.5	
1998	115.1	53.3	
1997	112.8	52.3	
1996	110.2	51.1	
1995	107.6	49.9	
1994	104.4	48.4	
1993	101.7	47.1	
1992	99.4	46.1	
1991	96.8	44.9	
1990	94.3	43.7	
1989	92.1	42.7	
1988	89.9	41.6	
1987	87.7	40.6	
1986	84.2	39.0	

Year	Historical Cost Index Jan. 1, 1993 = 100 Actual	Current Index Based on Jan. 1, 2018 = 100 Est.	Current Index Based on Jan. 1, 2018 = 100 Actual
July 1985	82.6	38.3	
1984	82.0	38.0	
1983	80.2	37.1	
1982	76.1	35.3	
1981	70.0	32.4	
1980	62.9	29.1	
1979	57.8	26.8	
1978	53.5	24.8	
1977	49.5	22.9	
1976	46.9	21.7	
1975	44.8	20.8	
1974	41.4	19.2	
1973	37.7	17.5	
1972	34.8	16.1	
1971	32.1	14.9	
1970	28.7	13.3	
1969	26.9	12.5	
1968	24.9	11.5	

Adjustments to Costs

The "Historical Cost Index" can be used to convert national average building costs at a particular time to the approximate building costs for some other time.

Example:

Estimate and compare construction costs for different years in the same city.

To estimate the national average construction cost of a building in 1970, knowing that it cost $900,000 in 2018:

INDEX in 1970 = 28.7

INDEX in 2018 = 215.8

Note: The city cost indexes for Canada can be used to convert U.S. national averages to local costs in Canadian dollars.

Time Adjustment Using the Historical Cost Indexes:

$$\frac{\text{Index for Year A}}{\text{Index for Year B}} \times \text{Cost in Year B} = \text{Cost in Year A}$$

$$\frac{\text{INDEX 1970}}{\text{INDEX 2018}} \times \text{Cost 2018} = \text{Cost 1970}$$

$$\frac{28.7}{215.8} \times \$900,000 = .133 \times \$900,000 = \$119,694$$

The construction cost of the building in 1970 was $119,694.

Example:

To estimate and compare the cost of a building in Toronto, ON in 2018 with the known cost of $600,000 (US$) in New York, NY in 2018:

INDEX Toronto = 110.8

INDEX New York = 134.6

$$\frac{\text{INDEX Toronto}}{\text{INDEX New York}} \times \text{Cost New York} = \text{Cost Toronto}$$

$$\frac{110.8}{134.6} \times \$600,000 = .823 \times \$600,000 = \$493,908$$

The construction cost of the building in Toronto is $493,908 (CN$).

*Historical Cost Index updates and other resources are provided on the following website:
http://info.thegordiangroup.com/RSMeans.html

How to Use the City Cost Indexes

What you should know before you begin

RSMeans City Cost Indexes (CCI) are an extremely useful tool for when you want to compare costs from city to city and region to region.

This publication contains average construction cost indexes for 731 U.S. and Canadian cities covering over 930 three-digit zip code locations, as listed directly under each city.

Keep in mind that a City Cost Index number is a percentage ratio of a specific city's cost to the national average cost of the same item at a stated time period.

In other words, these index figures represent relative construction factors (or, if you prefer, multipliers) for material and installation costs, as well as the weighted average for Total In Place costs for each CSI MasterFormat division. Installation costs include both labor and equipment rental costs. When estimating equipment rental rates only for a specific location, use 01 54 33 EQUIPMENT RENTAL COSTS in the Reference Section.

The 30 City Average Index is the average of 30 major U.S. cities and serves as a national average.

Index figures for both material and installation are based on the 30 major city average of 100 and represent the cost relationship as of July 1, 2017. The index for each division is computed from representative material and labor quantities for that division. The weighted average for each city is a weighted total of the components listed above it. It does not include relative productivity between trades or cities.

As changes occur in local material prices, labor rates, and equipment rental rates (including fuel costs), the impact of these changes should be accurately measured by the change in the City Cost Index for each particular city (as compared to the 30 city average).

Therefore, if you know (or have estimated) building costs in one city today, you can easily convert those costs to expected building costs in another city.

In addition, by using the Historical Cost Index, you can easily convert national average building costs at a particular time to the approximate building costs for some other time. The City Cost Indexes can then be applied to calculate the costs for a particular city.

Quick calculations

Location Adjustment Using the City Cost Indexes:

$$\frac{\text{Index for City A}}{\text{Index for City B}} \times \text{Cost in City B} = \text{Cost in City A}$$

Time Adjustment for the National Average Using the Historical Cost Index:

$$\frac{\text{Index for Year A}}{\text{Index for Year B}} \times \text{Cost in Year B} = \text{Cost in Year A}$$

Adjustment from the National Average:

$$\frac{\text{Index for City A}}{100} \times \text{National Average Cost} = \text{Cost in City A}$$

Since each of the other RSMeans data sets contains many different items, any *one* item multiplied by the particular city index may give incorrect results. However, the larger the number of items compiled, the closer the results should be to actual costs for that particular city.

The City Cost Indexes for Canadian cities are calculated using Canadian material and equipment prices and labor rates in Canadian dollars. Therefore, indexes for Canadian cities can be used to convert U.S. national average prices to local costs in Canadian dollars.

How to use this section

1. Compare costs from city to city.

In using the RSMeans Indexes, remember that an index number is not a fixed number but a ratio: It's a percentage ratio of a building component's cost at any stated time to the national average cost of that same component at the same time period. Put in the form of an equation:

$$\frac{\text{Specific City Cost}}{\text{National Average Cost}} \times 100 = \text{City Index Number}$$

Therefore, when making cost comparisons between cities, do not subtract one city's index number from the index number of another city and read the result as a percentage difference. Instead, divide one city's index number by that of the other city. The resulting number may then be used as a multiplier to calculate cost differences from city to city.

The formula used to find cost differences between cities for the purpose of comparison is as follows:

$$\frac{\text{City A Index}}{\text{City B Index}} \times \text{City B Cost (Known)} = \text{City A Cost (Unknown)}$$

In addition, you can use RSMeans CCI to calculate and compare costs division by division between cities using the same basic formula. (Just be sure that you're comparing similar divisions.)

2. Compare a specific city's construction costs with the national average.

When you're studying construction location feasibility, it's advisable to compare a prospective project's cost index with an index of the national average cost.

For example, divide the weighted average index of construction costs of a specific city by that of the 30 City Average, which = 100.

$$\frac{\text{City Index}}{100} = \% \text{ of National Average}$$

As a result, you get a ratio that indicates the relative cost of construction in that city in comparison with the national average.

3. Convert U.S. national average to actual costs in Canadian City.

$$\frac{\text{Index for Canadian City}}{100} \times \text{National Average Cost} = \text{Cost in Canadian City in \$ CAN}$$

For customer support on your Electrical Costs with RSMeans data, call 800.448.8182.

513

4. Adjust construction cost data based on a national average.

When you use a source of construction cost data which is based on a national average (such as RSMeans cost data), it is necessary to adjust those costs to a specific location.

$$\frac{City\ Index}{100} \times \begin{array}{c} Cost\ Based\ on \\ National\ Average\ Costs \end{array} = \begin{array}{c} City\ Cost \\ (Unknown) \end{array}$$

5. When applying the City Cost Indexes to demolition projects, use the appropriate division installation index. For example, for removal of existing doors and windows, use the Division 8 (Openings) index.

What you might like to know about how we developed the Indexes

The information presented in the CCI is organized according to the Construction Specifications Institute (CSI) MasterFormat 2014 classification system.

To create a reliable index, RSMeans researched the building type most often constructed in the United States and Canada. Because it was concluded that no one type of building completely represented the building construction industry, nine different types of buildings were combined to create a composite model.

The exact material, labor, and equipment quantities are based on detailed analyses of these nine building types, and then each quantity is weighted in proportion to expected usage. These various material items, labor hours, and equipment rental rates are thus combined to form a composite building representing as closely as possible the actual usage of materials, labor, and equipment in the North American building construction industry.

The following structures were chosen to make up that composite model:

1. Factory, 1 story
2. Office, 2–4 stories
3. Store, Retail
4. Town Hall, 2–3 stories
5. High School, 2–3 stories
6. Hospital, 4–8 stories
7. Garage, Parking
8. Apartment, 1–3 stories
9. Hotel/Motel, 2–3 stories

For the purposes of ensuring the timeliness of the data, the components of the index for the composite model have been streamlined. They currently consist of:

- specific quantities of 66 commonly used construction materials;
- specific labor-hours for 21 building construction trades; and
- specific days of equipment rental for 6 types of construction equipment (normally used to install the 66 material items by the 21 trades.) Fuel costs and routine maintenance costs are included in the equipment cost.

Material and equipment price quotations are gathered quarterly from cities in the United States and Canada. These prices and the latest negotiated labor wage rates for 21 different building trades are used to compile the quarterly update of the City Cost Index.

The 30 major U.S. cities used to calculate the national average are:

Atlanta, GA	Memphis, TN
Baltimore, MD	Milwaukee, WI
Boston, MA	Minneapolis, MN
Buffalo, NY	Nashville, TN
Chicago, IL	New Orleans, LA
Cincinnati, OH	New York, NY
Cleveland, OH	Philadelphia, PA
Columbus, OH	Phoenix, AZ
Dallas, TX	Pittsburgh, PA
Denver, CO	St. Louis, MO
Detroit, MI	San Antonio, TX
Houston, TX	San Diego, CA
Indianapolis, IN	San Francisco, CA
Kansas City, MO	Seattle, WA
Los Angeles, CA	Washington, DC

What the CCI does not indicate

The weighted average for each city is a total of the divisional components weighted to reflect typical usage. It does not include the productivity variations between trades or cities.

In addition, the CCI does not take into consideration factors such as the following:

- managerial efficiency
- competitive conditions
- automation
- restrictive union practices
- unique local requirements
- regional variations due to specific building codes

City Cost Indexes

DIVISION		UNITED STATES 30 CITY AVERAGE			ALABAMA ANNISTON 362			BIRMINGHAM 350 - 352			BUTLER 369			DECATUR 356			DOTHAN 363		
		MAT.	INST.	TOTAL	MAT.	INST.	TOTAL	MAT.	INST.	TOTAL	MAT.	INST.	TOTAL	MAT.	INST.	TOTAL	MAT.	INST.	TOTAL
015433	CONTRACTOR EQUIPMENT		100.0	100.0		104.4	104.4		104.5	104.5		101.9	101.9		104.4	104.4		101.9	101.9
0241, 31 - 34	SITE & INFRASTRUCTURE, DEMOLITION	100.0	100.0	100.0	86.8	92.4	90.7	93.0	92.8	92.8	99.8	87.9	91.5	86.4	91.9	90.3	97.5	88.2	91.0
0310	Concrete Forming & Accessories	100.0	100.0	100.0	88.7	66.5	69.6	92.5	67.5	70.9	85.3	67.1	69.6	95.0	62.6	67.1	93.9	68.2	71.7
0320	Concrete Reinforcing	100.0	100.0	100.0	87.2	72.0	79.5	95.5	71.9	83.6	92.3	72.2	82.2	89.4	68.9	79.1	92.3	71.6	81.9
0330	Cast-in-Place Concrete	100.0	100.0	100.0	91.6	67.5	82.5	102.4	68.4	89.5	89.3	67.9	81.2	100.5	67.4	87.9	89.3	67.6	81.1
03	CONCRETE	100.0	100.0	100.0	92.3	69.6	81.9	92.6	70.3	82.4	93.2	70.0	82.6	92.4	67.2	80.9	89.3	67.6	81.1
04	MASONRY	100.0	100.0	100.0	92.9	70.6	79.1	90.4	68.0	76.5	97.6	70.6	80.9	88.3	70.1	77.0	99.0	60.0	74.8
05	METALS	100.0	100.0	100.0	92.1	95.2	93.1	92.2	95.4	93.2	91.1	96.3	92.7	94.2	94.2	94.2	91.2	95.7	92.6
06	WOOD, PLASTICS & COMPOSITES	100.0	100.0	100.0	88.9	67.1	76.8	92.2	67.1	78.2	83.7	67.1	74.5	102.1	61.5	79.5	95.3	68.6	80.5
07	THERMAL & MOISTURE PROTECTION	100.0	100.0	100.0	95.4	65.2	82.6	98.3	67.3	85.2	95.4	67.8	83.8	97.3	67.0	84.5	95.4	64.9	82.5
08	OPENINGS	100.0	100.0	100.0	95.8	68.8	89.6	101.0	68.8	93.6	95.8	68.8	89.6	107.7	64.8	97.9	95.8	69.6	89.8
0920	Plaster & Gypsum Board	100.0	100.0	100.0	88.0	66.7	73.7	87.9	66.7	73.7	85.0	66.7	72.7	93.9	60.9	71.8	95.2	68.3	77.1
0950, 0980	Ceilings & Acoustic Treatment	100.0	100.0	100.0	77.6	66.7	70.2	82.6	66.7	71.9	77.6	66.7	70.2	82.9	60.9	68.1	77.6	68.3	71.3
0960	Flooring	100.0	100.0	100.0	80.5	82.5	81.0	92.3	75.9	87.8	84.9	82.5	84.2	87.0	82.5	85.7	90.0	84.1	88.3
0970, 0990	Wall Finishes & Painting/Coating	100.0	100.0	100.0	86.2	60.0	70.9	87.8	60.0	71.6	86.2	48.9	64.5	83.5	62.6	71.3	86.2	80.5	82.9
09	FINISHES	100.0	100.0	100.0	79.2	69.2	73.7	87.3	68.2	76.9	81.9	68.0	74.4	83.4	66.0	74.0	84.6	72.5	78.1
COVERS	DIVS. 10 - 14, 25, 28, 41, 43, 44, 46	100.0	100.0	100.0	100.0	71.6	93.7	100.0	84.9	96.6	100.0	78.5	95.2	100.0	70.8	93.5	100.0	73.0	94.0
21, 22, 23	FIRE SUPPRESSION, PLUMBING & HVAC	100.0	100.0	100.0	100.9	51.3	79.7	100.0	63.1	84.2	98.2	66.9	84.8	100.0	65.9	85.5	98.2	65.3	84.1
26, 27, 3370	ELECTRICAL, COMMUNICATIONS & UTIL.	100.0	100.0	100.0	96.4	60.2	77.6	98.1	61.5	79.1	98.5	67.4	82.3	93.9	66.0	79.4	97.2	76.1	86.3
MF2016	WEIGHTED AVERAGE	100.0	100.0	100.0	94.5	68.5	83.1	95.9	71.4	85.2	94.8	72.8	85.2	96.0	71.3	85.2	94.9	72.9	85.3

DIVISION		ALABAMA EVERGREEN 364			GADSDEN 359			HUNTSVILLE 357 - 358			JASPER 355			MOBILE 365 - 366			MONTGOMERY 360 - 361		
		MAT.	INST.	TOTAL	MAT.	INST.	TOTAL	MAT.	INST.	TOTAL	MAT.	INST.	TOTAL	MAT.	INST.	TOTAL	MAT.	INST.	TOTAL
015433	CONTRACTOR EQUIPMENT		101.9	101.9		104.4	104.4		104.4	104.4		104.4	104.4		101.9	101.9		101.9	101.9
0241, 31 - 34	SITE & INFRASTRUCTURE, DEMOLITION	100.2	87.9	91.6	91.9	92.4	92.3	86.1	92.3	90.4	91.8	92.5	92.3	92.9	89.0	90.1	91.3	88.2	89.1
0310	Concrete Forming & Accessories	82.1	68.0	69.9	87.2	66.9	69.7	95.0	64.3	68.5	92.3	67.2	70.7	93.0	68.3	71.4	94.6	67.0	70.8
0320	Concrete Reinforcing	92.3	72.2	82.2	95.0	71.9	83.3	89.4	71.1	80.2	89.4	72.1	80.7	89.8	71.6	80.6	91.7	71.5	84.5
0330	Cast-in-Place Concrete	89.3	67.6	81.1	100.6	67.5	88.0	97.9	66.6	86.0	111.5	68.3	95.1	93.7	67.6	83.7	90.6	67.7	81.9
03	CONCRETE	93.5	70.3	82.9	96.8	69.7	84.4	91.2	68.1	80.6	100.5	70.2	86.6	89.1	70.2	80.5	87.4	69.8	79.4
04	MASONRY	97.6	70.1	80.6	86.8	66.4	74.2	89.6	65.0	74.4	84.5	71.7	76.6	95.8	60.0	73.6	93.1	60.5	72.9
05	METALS	91.2	96.2	92.7	92.2	95.5	93.2	94.2	94.9	94.4	92.1	95.5	93.2	93.1	95.5	93.8	92.2	95.0	93.1
06	WOOD, PLASTICS & COMPOSITES	80.3	68.6	73.8	92.9	67.1	78.5	102.1	64.3	81.1	99.2	67.1	81.4	93.9	68.6	79.9	92.3	67.1	78.3
07	THERMAL & MOISTURE PROTECTION	95.4	67.8	83.7	97.5	66.6	84.4	97.2	65.7	83.9	97.5	63.7	83.2	95.0	64.9	82.3	93.6	65.0	81.5
08	OPENINGS	95.8	69.6	89.8	104.2	68.8	96.0	107.4	67.0	98.2	104.1	68.8	96.0	98.7	69.6	92.0	97.5	68.8	90.9
0920	Plaster & Gypsum Board	84.4	68.3	73.6	86.7	66.7	73.3	93.9	63.8	73.7	90.7	66.7	74.6	92.1	68.3	76.1	88.8	66.7	74.0
0950, 0980	Ceilings & Acoustic Treatment	77.6	68.3	71.3	80.3	66.7	71.1	84.8	63.8	70.7	80.3	66.7	71.1	82.9	68.3	73.0	84.2	66.7	72.4
0960	Flooring	82.9	82.5	82.8	83.3	77.5	81.7	87.0	77.5	84.3	85.3	82.5	84.5	89.5	65.1	82.8	87.8	65.1	81.5
0970, 0990	Wall Finishes & Painting/Coating	86.2	48.9	64.4	83.5	60.0	69.8	83.5	62.7	71.4	83.5	60.0	69.8	89.4	48.9	65.8	86.5	60.0	71.1
09	FINISHES	81.3	68.8	74.5	81.3	68.1	74.1	83.8	66.4	74.3	82.3	69.5	75.3	84.7	65.2	74.1	85.7	65.7	74.8
COVERS	DIVS. 10 - 14, 25, 28, 41, 43, 44, 46	100.0	71.4	93.6	100.0	84.4	96.5	100.0	83.7	96.4	100.0	72.8	93.9	100.0	85.4	96.8	100.0	84.6	96.6
21, 22, 23	FIRE SUPPRESSION, PLUMBING & HVAC	98.2	66.7	84.7	102.1	62.9	85.3	100.0	62.2	83.8	102.1	63.0	85.4	99.8	66.7	85.6	99.9	65.5	85.2
26, 27, 3370	ELECTRICAL, COMMUNICATIONS & UTIL.	95.8	67.4	81.0	93.8	61.5	77.0	94.8	66.0	79.8	93.6	60.2	76.2	99.1	58.0	77.7	99.6	76.1	87.4
MF2016	WEIGHTED AVERAGE	94.5	72.6	84.9	96.1	71.0	85.1	96.0	70.7	84.9	96.5	71.2	85.5	95.4	70.2	84.4	94.8	72.3	85.0

DIVISION		ALABAMA PHENIX CITY 368			SELMA 367			TUSCALOOSA 354			ALASKA ANCHORAGE 995 - 996			FAIRBANKS 997			JUNEAU 998		
		MAT.	INST.	TOTAL	MAT.	INST.	TOTAL	MAT.	INST.	TOTAL	MAT.	INST.	TOTAL	MAT.	INST.	TOTAL	MAT.	INST.	TOTAL
015433	CONTRACTOR EQUIPMENT		101.9	101.9		101.9	101.9		104.4	104.4		115.3	115.3		115.3	115.3		115.3	115.3
0241, 31 - 34	SITE & INFRASTRUCTURE, DEMOLITION	103.9	88.2	92.9	97.4	88.2	90.9	86.6	92.5	90.7	127.0	130.4	129.4	126.5	130.4	129.2	147.3	130.4	135.5
0310	Concrete Forming & Accessories	88.7	67.2	70.1	86.4	67.1	69.8	94.9	67.2	71.0	125.4	119.5	120.3	130.1	118.0	119.7	131.0	119.5	121.1
0320	Concrete Reinforcing	92.3	65.9	78.9	92.3	72.1	82.1	89.4	71.9	80.6	158.3	118.8	138.4	152.5	118.8	135.5	137.9	118.8	128.3
0330	Cast-in-Place Concrete	89.3	67.8	81.2	89.3	67.8	81.1	102.0	67.8	89.0	108.1	119.1	112.3	117.0	119.4	117.9	127.8	119.1	124.5
03	CONCRETE	96.2	68.9	83.7	92.0	70.0	81.9	93.1	69.9	82.5	117.6	118.4	118.0	110.2	117.9	113.7	123.5	118.4	121.2
04	MASONRY	97.6	70.6	80.8	100.9	70.6	82.1	88.5	66.8	75.1	175.5	121.0	141.7	185.3	121.0	145.4	165.8	121.0	138.0
05	METALS	91.1	93.6	91.9	91.1	95.8	92.6	93.5	95.6	94.1	120.0	104.9	115.4	121.2	105.0	116.2	117.0	104.9	113.3
06	WOOD, PLASTICS & COMPOSITES	88.6	67.0	76.6	85.7	67.1	75.4	102.1	67.1	82.6	117.8	117.9	117.9	129.1	115.8	121.7	124.7	117.9	120.9
07	THERMAL & MOISTURE PROTECTION	95.8	67.8	84.0	95.3	67.8	83.7	97.3	66.8	84.4	169.2	117.9	147.5	180.8	116.9	153.8	183.4	117.9	155.7
08	OPENINGS	95.8	67.1	89.2	95.8	68.8	89.6	107.4	68.8	98.6	128.9	117.3	126.3	134.8	116.4	130.6	129.7	117.3	126.8
0920	Plaster & Gypsum Board	89.3	66.6	74.1	87.3	66.7	73.5	93.9	66.7	75.6	143.2	118.3	126.5	166.9	116.1	132.8	147.9	118.3	128.0
0950, 0980	Ceilings & Acoustic Treatment	77.6	66.6	70.2	77.6	66.7	70.2	84.8	66.7	72.6	131.3	118.3	122.5	130.9	116.1	120.9	143.6	118.3	125.9
0960	Flooring	86.8	82.5	85.6	85.4	82.5	84.6	87.0	77.5	84.3	119.0	125.1	120.7	117.5	125.1	119.6	125.7	125.1	125.5
0970, 0990	Wall Finishes & Painting/Coating	86.2	80.5	82.9	86.2	60.0	70.9	83.5	60.0	69.8	103.4	115.3	110.3	106.4	119.9	114.3	108.9	115.3	112.6
09	FINISHES	83.5	71.4	76.9	82.1	69.2	75.1	83.8	68.2	75.3	125.6	120.4	122.8	126.1	119.7	122.6	128.4	120.4	124.1
COVERS	DIVS. 10 - 14, 25, 28, 41, 43, 44, 46	100.0	84.6	96.6	100.0	72.1	93.8	100.0	84.6	96.6	100.0	111.1	102.5	100.0	110.9	102.4	100.0	111.1	102.5
21, 22, 23	FIRE SUPPRESSION, PLUMBING & HVAC	98.2	65.4	84.2	98.2	65.5	84.2	100.0	68.3	86.5	100.8	106.4	103.2	100.6	110.1	104.7	100.6	106.4	103.1
26, 27, 3370	ELECTRICAL, COMMUNICATIONS & UTIL.	97.9	67.7	82.2	97.0	76.1	86.1	94.3	61.5	77.3	114.4	111.5	112.9	118.2	111.5	114.7	103.1	111.5	107.5
MF2016	WEIGHTED AVERAGE	95.4	72.7	85.4	94.6	73.6	85.4	96.0	72.3	85.6	118.9	114.8	117.1	120.1	115.3	118.0	118.8	114.8	117.0

DIVISION		ALASKA — KETCHIKAN 999 MAT.	INST.	TOTAL	CHAMBERS 865 MAT.	INST.	TOTAL	FLAGSTAFF 860 MAT.	INST.	TOTAL	GLOBE 855 MAT.	INST.	TOTAL	KINGMAN 864 MAT.	INST.	TOTAL	MESA/TEMPE 852 MAT.	INST.	TOTAL
015433	CONTRACTOR EQUIPMENT		115.3	115.3		90.7	90.7		90.7	90.7		92.0	92.0		90.7	90.7		92.0	92.0
0241, 31 - 34	SITE & INFRASTRUCTURE, DEMOLITION	183.9	130.4	146.4	72.9	94.9	88.3	92.6	94.9	94.2	105.2	95.9	98.7	72.9	94.8	88.3	94.5	95.9	95.5
0310	Concrete Forming & Accessories	121.3	119.5	119.7	97.2	68.7	72.6	102.4	68.7	73.3	93.4	68.7	72.1	95.5	68.6	72.3	96.5	68.8	72.6
0320	Concrete Reinforcing	113.8	118.8	116.3	100.0	81.8	90.9	99.8	81.8	90.8	110.6	81.9	96.1	100.2	81.8	90.9	111.3	81.9	96.5
0330	Cast-in-Place Concrete	236.9	119.1	192.1	89.7	68.4	81.6	89.8	68.4	81.6	82.7	68.1	77.2	89.4	68.4	81.4	83.4	68.1	77.6
03	CONCRETE	187.2	118.4	155.7	92.0	70.8	82.3	111.6	70.8	93.0	103.7	70.8	88.6	91.7	70.8	82.1	94.6	70.8	83.8
04	MASONRY	192.8	121.0	148.2	95.7	59.3	73.1	95.8	58.7	72.8	103.6	59.2	76.0	95.7	59.3	73.1	103.7	59.2	76.1
05	METALS	121.3	104.9	116.3	101.6	73.5	92.9	102.1	73.5	93.3	98.7	74.4	91.3	102.2	73.4	93.4	99.0	74.5	91.5
06	WOOD, PLASTICS & COMPOSITES	119.9	117.9	118.8	100.1	70.5	83.6	105.8	70.5	86.2	97.0	70.7	82.4	95.3	70.5	81.5	99.7	68.5	86.5
07	THERMAL & MOISTURE PROTECTION	185.6	117.9	157.0	93.6	70.1	83.6	95.4	69.8	84.6	99.9	68.8	86.7	93.5	69.9	83.5	96.5	70.9	90.6
08	OPENINGS	130.4	117.3	127.4	105.9	69.9	97.3	106.0	73.0	98.4	96.4	68.8	90.1	106.1	70.8	98.0	96.5	70.9	90.6
0920	Plaster & Gypsum Board	155.6	118.3	130.6	88.6	69.9	76.0	91.8	69.9	77.1	89.0	69.9	76.2	81.2	69.9	73.6	91.3	69.9	76.9
0950, 0980	Ceilings & Acoustic Treatment	124.2	118.3	120.2	105.7	69.9	81.6	106.6	69.9	81.9	95.0	69.9	78.1	106.6	69.9	81.9	95.0	69.9	78.1
0960	Flooring	117.5	125.1	119.6	91.1	60.1	82.5	93.5	60.1	84.3	102.7	60.1	90.9	89.8	60.1	81.6	104.6	60.1	92.3
0970, 0990	Wall Finishes & Painting/Coating	106.4	115.3	111.6	88.8	64.0	74.3	88.8	64.0	74.3	94.1	64.0	76.5	88.8	64.0	74.3	94.1	64.0	76.5
09	FINISHES	127.1	120.4	123.5	90.7	66.4	77.5	93.9	66.4	78.9	97.7	66.6	80.8	89.5	66.4	76.9	97.5	66.6	80.7
COVERS	DIVS. 10 - 14, 25, 28, 41, 43, 44, 46	100.0	111.1	102.5	100.0	84.1	96.5	100.0	84.0	96.4	100.0	84.4	96.5	100.0	84.0	96.4	100.0	84.4	96.5
21, 22, 23	FIRE SUPPRESSION, PLUMBING & HVAC	98.7	106.4	102.0	98.0	77.2	89.1	100.2	78.2	90.8	97.0	77.3	88.6	98.0	78.0	89.4	100.1	77.4	90.4
26, 27, 3370	ELECTRICAL, COMMUNICATIONS & UTIL.	118.2	111.5	114.7	103.2	86.2	94.4	102.2	63.1	81.9	99.0	66.4	82.0	103.2	66.4	84.0	95.9	66.4	80.5
MF2016	WEIGHTED AVERAGE	130.8	114.8	123.8	97.9	75.0	87.9	101.7	72.1	88.8	99.2	72.4	87.4	97.9	72.5	86.8	98.2	72.5	87.0

ARIZONA / ARKANSAS

DIVISION		PHOENIX 850, 853 MAT.	INST.	TOTAL	PRESCOTT 863 MAT.	INST.	TOTAL	SHOW LOW 859 MAT.	INST.	TOTAL	TUCSON 856 - 857 MAT.	INST.	TOTAL	BATESVILLE 725 MAT.	INST.	TOTAL	CAMDEN 717 MAT.	INST.	TOTAL
015433	CONTRACTOR EQUIPMENT		93.0	93.0		90.7	90.7		92.0	92.0		92.0	92.0		92.2	92.2		92.2	92.2
0241, 31 - 34	SITE & INFRASTRUCTURE, DEMOLITION	95.1	94.5	94.7	79.8	94.8	90.3	107.5	95.9	99.4	90.1	95.9	94.2	73.6	90.1	85.2	82.5	90.1	87.8
0310	Concrete Forming & Accessories	101.4	72.9	76.9	98.7	68.7	72.8	100.5	68.8	73.1	96.9	68.8	72.7	83.8	57.4	61.1	81.0	57.4	60.7
0320	Concrete Reinforcing	109.2	79.3	94.2	99.8	81.8	90.8	111.3	81.9	96.5	92.4	81.8	87.1	90.9	65.5	78.1	94.9	65.2	79.9
0330	Cast-in-Place Concrete	83.0	70.7	78.3	89.7	68.3	81.6	82.8	68.1	77.2	85.9	68.1	79.1	68.9	75.8	71.5	80.4	75.8	78.7
03	CONCRETE	97.2	73.3	86.2	97.4	70.8	85.2	106.1	70.8	89.9	93.2	70.8	83.0	73.3	66.1	70.0	79.5	66.0	73.4
04	MASONRY	96.1	62.9	75.5	95.8	59.3	73.1	103.6	59.2	76.1	90.2	58.6	70.6	85.6	43.2	62.0	109.8	37.7	65.1
05	METALS	100.5	75.7	92.9	102.1	73.4	93.3	98.5	74.4	91.1	99.8	74.5	92.0	99.1	73.7	91.3	100.9	73.3	92.4
06	WOOD, PLASTICS & COMPOSITES	103.4	74.2	87.1	101.4	70.7	84.3	104.6	70.7	85.7	100.6	70.7	83.9	93.1	58.2	73.7	93.0	58.2	73.6
07	THERMAL & MOISTURE PROTECTION	100.9	70.4	88.0	94.1	70.2	84.0	100.1	68.6	86.8	100.8	68.3	87.1	103.1	59.0	84.4	96.4	57.5	79.9
08	OPENINGS	97.7	74.4	92.4	106.0	68.8	97.5	95.7	70.9	90.0	92.9	73.1	88.4	99.7	54.2	89.2	102.6	56.1	91.9
0920	Plaster & Gypsum Board	99.2	73.3	81.8	88.7	70.1	76.2	93.3	69.9	77.6	96.2	69.9	78.5	82.0	57.5	65.5	85.1	57.5	66.5
0950, 0980	Ceilings & Acoustic Treatment	105.6	73.3	83.8	104.9	70.1	81.4	95.0	69.9	78.1	95.8	69.9	78.4	86.7	57.5	67.0	82.1	57.5	65.5
0960	Flooring	106.7	60.1	93.8	92.0	60.1	83.2	106.6	60.1	93.7	95.7	60.1	85.8	84.5	54.2	76.1	92.0	36.3	76.6
0970, 0990	Wall Finishes & Painting/Coating	98.9	66.2	79.8	88.8	64.0	74.3	94.1	64.0	76.5	94.8	64.0	76.8	90.5	36.0	58.7	93.1	50.5	68.3
09	FINISHES	102.2	69.8	84.6	91.4	66.5	77.9	99.7	66.6	81.7	95.5	66.6	79.8	78.4	53.9	65.1	80.9	52.3	65.3
COVERS	DIVS. 10 - 14, 25, 28, 41, 43, 44, 46	100.0	86.5	97.0	100.0	84.1	96.5	100.0	84.4	96.5	100.0	84.4	96.5	100.0	67.2	92.7	100.0	67.2	92.7
21, 22, 23	FIRE SUPPRESSION, PLUMBING & HVAC	99.9	80.2	91.5	100.2	77.2	90.4	97.0	77.3	88.6	100.0	77.5	90.4	97.1	53.2	78.3	97.0	57.9	80.3
26, 27, 3370	ELECTRICAL, COMMUNICATIONS & UTIL.	101.4	63.1	81.5	101.9	66.4	83.4	96.2	66.4	80.7	98.1	61.0	78.8	95.5	62.6	78.4	94.4	59.0	76.0
MF2016	WEIGHTED AVERAGE	99.5	74.1	88.4	99.3	72.2	87.5	99.4	72.5	87.6	97.1	71.8	86.0	92.6	60.8	78.7	95.0	60.5	79.9

ARKANSAS

DIVISION		FAYETTEVILLE 727 MAT.	INST.	TOTAL	FORT SMITH 729 MAT.	INST.	TOTAL	HARRISON 726 MAT.	INST.	TOTAL	HOT SPRINGS 719 MAT.	INST.	TOTAL	JONESBORO 724 MAT.	INST.	TOTAL	LITTLE ROCK 720 - 722 MAT.	INST.	TOTAL
015433	CONTRACTOR EQUIPMENT		92.2	92.2		92.2	92.2		92.2	92.2		92.2	92.2		116.6	116.6		92.2	92.2
0241, 31 - 34	SITE & INFRASTRUCTURE, DEMOLITION	73.0	90.1	85.0	78.1	90.1	86.5	78.5	90.1	86.6	85.8	90.1	88.8	97.6	108.2	105.0	86.4	90.3	89.1
0310	Concrete Forming & Accessories	79.4	57.6	60.6	97.9	57.9	63.5	88.1	57.3	61.5	78.6	57.6	60.5	87.2	57.6	61.7	92.2	59.5	64.0
0320	Concrete Reinforcing	91.0	62.8	76.7	92.1	65.2	78.6	90.6	65.6	78.0	93.1	65.5	79.2	87.8	65.7	76.6	97.4	68.4	82.7
0330	Cast-in-Place Concrete	68.9	75.9	71.6	78.8	76.1	77.7	76.4	75.8	76.2	82.3	75.9	79.9	75.1	77.2	75.9	76.3	76.2	76.3
03	CONCRETE	73.0	65.7	69.7	80.1	66.3	73.8	79.7	66.0	73.4	82.8	66.2	75.2	77.3	67.7	72.9	81.0	67.6	74.9
04	MASONRY	84.0	42.7	58.4	90.9	63.4	73.8	93.0	44.3	62.8	82.3	42.9	57.9	85.2	40.8	57.6	86.7	67.2	74.6
05	METALS	99.1	72.9	91.0	101.4	75.2	93.3	100.2	73.6	92.1	100.9	73.8	92.6	95.6	89.1	93.6	101.6	76.9	94.0
06	WOOD, PLASTICS & COMPOSITES	89.7	58.2	72.2	109.4	58.2	80.9	98.9	58.2	76.3	90.4	58.2	72.5	97.3	58.6	75.7	93.9	59.7	74.9
07	THERMAL & MOISTURE PROTECTION	103.9	58.9	84.8	104.2	64.7	87.5	103.4	59.3	84.7	96.7	58.9	80.7	108.7	58.3	87.4	98.2	65.9	84.2
08	OPENINGS	99.7	56.1	89.7	101.7	56.6	91.4	100.5	53.1	89.6	102.6	56.0	91.9	105.4	57.2	94.3	95.9	58.2	87.2
0920	Plaster & Gypsum Board	81.4	57.5	65.3	88.3	57.5	67.6	87.3	57.5	67.3	83.8	57.5	66.1	96.1	57.5	70.1	95.4	59.0	71.0
0950, 0980	Ceilings & Acoustic Treatment	86.7	57.5	67.0	88.4	57.5	67.5	88.4	57.5	67.5	82.1	57.5	65.5	90.7	57.5	68.3	90.5	59.0	69.3
0960	Flooring	81.7	53.1	73.8	91.0	75.0	86.6	86.7	54.2	77.7	90.9	53.1	80.4	61.2	48.0	57.5	90.9	83.9	88.9
0970, 0990	Wall Finishes & Painting/Coating	90.5	48.2	65.8	90.5	50.8	67.3	90.5	31.1	55.9	93.1	50.5	68.3	80.1	44.1	59.1	88.8	50.5	66.5
09	FINISHES	77.5	54.9	65.2	81.8	60.1	70.0	80.5	53.3	65.7	80.7	55.3	66.9	76.2	53.7	64.0	87.8	62.8	74.2
COVERS	DIVS. 10 - 14, 25, 28, 41, 43, 44, 46	100.0	67.2	92.7	100.0	79.3	95.4	100.0	68.3	92.9	100.0	67.2	92.7	100.0	66.3	92.5	100.0	79.5	95.4
21, 22, 23	FIRE SUPPRESSION, PLUMBING & HVAC	97.2	62.3	82.3	100.3	49.2	78.4	97.1	49.4	76.7	97.0	53.5	78.4	100.6	53.2	80.4	99.9	53.6	80.1
26, 27, 3370	ELECTRICAL, COMMUNICATIONS & UTIL.	90.0	54.7	71.6	93.0	59.0	75.4	94.2	54.7	73.6	96.3	68.8	82.0	98.8	62.8	80.1	99.7	68.8	83.7
MF2016	WEIGHTED AVERAGE	91.5	61.6	78.5	95.0	63.1	81.0	93.9	58.9	78.6	94.3	62.0	80.1	94.5	63.7	81.0	95.3	66.6	82.7

ARKANSAS / CALIFORNIA

DIVISION		PINE BLUFF 716 MAT.	INST.	TOTAL	RUSSELLVILLE 728 MAT.	INST.	TOTAL	TEXARKANA 718 MAT.	INST.	TOTAL	WEST MEMPHIS 723 MAT.	INST.	TOTAL	ALHAMBRA 917-918 MAT.	INST.	TOTAL	ANAHEIM 928 MAT.	INST.	TOTAL
015433	CONTRACTOR EQUIPMENT		92.2	92.2		92.2	92.2		93.5	93.5		116.6	116.6		95.8	95.8		101.0	101.0
0241, 31 - 34	SITE & INFRASTRUCTURE, DEMOLITION	87.8	90.2	89.4	74.9	90.1	85.5	98.9	92.3	94.3	104.6	108.5	107.3	101.0	109.1	106.7	102.0	109.9	107.5
0310	Concrete Forming & Accessories	78.4	58.2	61.0	84.4	57.3	61.0	84.1	57.5	61.1	92.6	57.9	62.7	116.4	134.6	132.1	104.9	138.5	133.9
0320	Concrete Reinforcing	94.8	68.1	81.3	91.5	65.4	78.4	94.4	65.5	79.8	87.8	64.0	75.7	104.8	128.4	116.7	94.8	128.1	111.6
0330	Cast-in-Place Concrete	82.3	76.2	80.0	72.2	75.8	73.6	89.7	75.9	84.5	78.7	77.3	78.2	82.8	129.0	100.4	87.9	132.7	104.9
03	CONCRETE	83.5	67.0	75.9	76.1	65.9	71.5	81.9	66.1	74.7	83.8	67.5	76.4	94.6	130.3	110.9	95.6	133.4	112.9
04	MASONRY	116.6	63.4	83.6	90.2	41.8	60.2	96.5	42.2	62.8	73.9	40.6	53.3	112.0	141.0	130.0	78.0	136.0	114.0
05	METALS	101.7	76.4	93.9	99.1	73.4	91.2	93.8	73.6	87.6	94.6	88.7	92.8	86.5	109.6	93.6	104.9	111.5	106.9
06	WOOD, PLASTICS & COMPOSITES	89.9	58.2	72.3	94.4	58.2	74.2	97.4	58.2	75.6	102.9	58.6	78.2	99.6	130.3	116.7	99.7	135.3	119.5
07	THERMAL & MOISTURE PROTECTION	96.7	64.6	83.1	104.1	58.6	84.8	97.4	58.7	81.0	109.2	58.2	87.6	98.9	130.1	112.1	106.2	134.4	118.1
08	OPENINGS	103.8	57.4	93.2	99.7	55.3	89.5	108.9	55.3	96.6	102.8	56.0	92.1	91.2	131.2	100.4	107.1	134.0	113.3
0920	Plaster & Gypsum Board	83.5	57.5	66.0	82.0	57.5	65.5	86.5	57.5	67.0	98.4	57.5	70.9	96.2	131.4	119.9	105.7	136.3	126.2
0950, 0980	Ceilings & Acoustic Treatment	82.1	57.5	65.5	86.7	57.5	67.0	85.5	57.5	66.6	88.8	57.5	67.7	110.3	131.4	124.5	111.1	136.3	128.1
0960	Flooring	90.6	75.0	86.3	84.1	54.2	75.8	92.8	45.8	79.8	63.4	48.0	59.2	102.9	116.0	106.5	97.4	118.6	103.3
0970, 0990	Wall Finishes & Painting/Coating	93.1	50.8	68.4	90.5	33.2	57.1	93.1	41.7	63.1	80.1	50.5	62.9	108.0	120.9	115.5	89.3	118.5	106.4
09	FINISHES	80.6	60.1	69.5	78.5	53.6	65.0	82.9	53.1	66.7	77.5	54.6	65.1	102.3	129.3	117.0	97.4	132.8	116.7
COVERS	DIVS. 10 - 14, 25, 28, 41, 43, 44, 46	100.0	79.3	95.4	100.0	67.2	92.7	100.0	67.2	92.7	100.0	68.1	92.9	100.0	119.7	104.4	100.0	120.9	104.6
21, 22, 23	FIRE SUPPRESSION, PLUMBING & HVAC	100.2	53.5	80.2	97.2	52.9	78.2	100.2	56.7	81.6	97.5	65.4	83.8	96.9	125.5	109.1	100.0	133.2	114.2
26, 27, 3370	ELECTRICAL, COMMUNICATIONS & UTIL.	94.6	64.4	78.9	93.0	54.7	73.1	96.2	56.4	75.5	100.3	64.6	81.7	117.8	126.8	122.5	91.1	108.6	100.2
MF2016	WEIGHTED AVERAGE	96.9	65.0	82.9	92.7	59.4	78.1	95.7	60.7	80.3	94.2	66.6	82.1	98.0	125.9	110.2	99.1	125.9	110.8

CALIFORNIA

DIVISION		BAKERSFIELD 932-933 MAT.	INST.	TOTAL	BERKELEY 947 MAT.	INST.	TOTAL	EUREKA 955 MAT.	INST.	TOTAL	FRESNO 936-938 MAT.	INST.	TOTAL	INGLEWOOD 903-905 MAT.	INST.	TOTAL	LONG BEACH 906-908 MAT.	INST.	TOTAL
015433	CONTRACTOR EQUIPMENT		98.9	98.9		102.7	102.7		98.7	98.7		98.9	98.9		97.8	97.8		97.8	97.8
0241, 31 - 34	SITE & INFRASTRUCTURE, DEMOLITION	99.6	107.6	105.2	107.4	109.4	108.8	113.2	106.2	108.3	104.1	107.0	106.1	91.7	105.8	101.6	99.1	105.9	103.8
0310	Concrete Forming & Accessories	104.7	134.4	130.3	115.2	165.0	158.2	113.6	150.3	145.2	102.1	149.9	143.3	108.3	134.9	131.2	103.3	130.5	130.6
0320	Concrete Reinforcing	90.0	128.0	109.2	90.6	130.1	110.5	102.9	129.5	116.3	75.5	129.1	102.5	94.7	128.4	111.7	93.9	128.4	111.3
0330	Cast-in-Place Concrete	89.3	131.6	105.4	117.4	133.0	123.3	95.4	129.0	108.2	96.9	128.8	109.0	80.8	131.6	100.1	92.0	131.7	107.1
03	CONCRETE	88.1	131.1	107.8	102.9	145.7	122.5	105.9	137.6	120.4	92.1	137.3	112.8	88.9	131.4	108.3	98.3	131.5	113.4
04	MASONRY	92.0	135.4	118.9	118.6	155.8	141.7	103.0	150.3	132.4	96.7	145.1	126.7	74.2	141.1	115.7	83.2	141.1	119.1
05	METALS	102.6	110.5	105.1	106.6	114.8	109.1	104.7	112.8	107.2	103.0	111.9	105.8	96.3	111.4	100.9	96.2	111.5	100.9
06	WOOD, PLASTICS & COMPOSITES	96.1	130.7	115.4	110.2	170.1	143.5	113.7	153.2	135.7	103.1	153.2	130.9	100.0	130.7	117.1	94.2	130.7	114.5
07	THERMAL & MOISTURE PROTECTION	102.6	124.5	111.9	102.7	153.7	124.3	110.0	145.8	125.1	94.0	133.0	110.5	103.4	130.2	114.7	103.7	131.8	115.6
08	OPENINGS	95.1	130.0	103.1	96.2	158.5	110.5	106.3	140.2	114.1	96.7	142.4	107.2	91.0	131.4	100.3	91.0	131.4	100.3
0920	Plaster & Gypsum Board	95.3	131.4	119.5	107.3	171.5	150.4	110.7	154.5	140.1	92.8	154.5	134.2	103.7	131.4	122.3	100.1	131.4	121.1
0950, 0980	Ceilings & Acoustic Treatment	97.9	131.4	120.5	100.5	171.5	148.4	114.5	154.5	141.5	95.3	154.5	135.2	109.2	131.4	124.2	100.1	131.4	121.1
0960	Flooring	100.9	116.0	105.1	116.9	136.3	122.3	101.2	130.6	109.4	103.5	114.8	106.6	109.0	116.0	110.9	105.7	116.0	108.6
0970, 0990	Wall Finishes & Painting/Coating	91.8	108.5	101.5	108.5	165.7	141.9	90.7	127.2	112.0	103.6	120.9	113.7	107.3	120.9	115.3	107.3	120.9	115.3
09	FINISHES	94.9	128.3	113.1	106.8	161.6	136.6	102.3	146.0	126.0	94.3	142.7	120.6	106.1	129.5	118.8	105.1	129.5	118.4
COVERS	DIVS. 10 - 14, 25, 28, 41, 43, 44, 46	100.0	112.1	102.7	100.0	131.3	107.0	100.0	128.1	106.2	100.0	128.0	106.2	100.0	120.6	104.6	100.0	120.6	104.6
21, 22, 23	FIRE SUPPRESSION, PLUMBING & HVAC	100.0	123.0	109.9	97.1	161.4	124.6	96.9	122.8	108.0	100.2	128.4	112.3	96.6	122.9	107.8	96.6	125.6	109.0
26, 27, 3370	ELECTRICAL, COMMUNICATIONS & UTIL.	106.3	105.1	105.6	102.5	159.0	131.9	97.6	120.7	109.6	96.5	108.5	102.7	100.5	126.8	114.2	100.2	126.8	114.1
MF2016	WEIGHTED AVERAGE	98.3	121.3	108.3	102.5	148.9	122.8	102.4	130.0	114.4	98.2	128.2	111.3	95.5	125.5	108.6	97.1	126.1	109.8

CALIFORNIA

DIVISION		LOS ANGELES 900-902 MAT.	INST.	TOTAL	MARYSVILLE 959 MAT.	INST.	TOTAL	MODESTO 953 MAT.	INST.	TOTAL	MOJAVE 935 MAT.	INST.	TOTAL	OAKLAND 946 MAT.	INST.	TOTAL	OXNARD 930 MAT.	INST.	TOTAL
015433	CONTRACTOR EQUIPMENT		103.6	103.6		98.7	98.7		98.7	98.7		98.9	98.9		102.7	102.7		97.8	97.8
0241, 31 - 34	SITE & INFRASTRUCTURE, DEMOLITION	97.7	111.1	107.1	109.7	106.3	107.3	104.3	106.5	105.8	97.3	107.6	104.5	112.3	109.4	110.3	104.3	105.7	105.3
0310	Concrete Forming & Accessories	106.3	138.9	134.4	103.8	150.0	143.7	100.3	149.9	143.1	113.8	134.3	131.5	104.4	165.0	156.7	105.3	138.6	134.0
0320	Concrete Reinforcing	95.4	128.6	112.1	102.9	128.9	116.0	106.6	128.8	116.0	90.4	128.0	109.4	92.7	130.1	111.6	88.8	128.1	108.6
0330	Cast-in-Place Concrete	83.4	132.6	102.1	106.6	128.9	115.1	95.5	128.8	108.1	83.8	131.6	102.0	111.3	133.0	119.6	97.6	132.0	110.7
03	CONCRETE	95.8	133.7	113.1	106.7	137.4	120.7	97.9	137.3	115.9	84.1	131.1	105.6	102.9	145.7	122.5	92.2	133.2	110.9
04	MASONRY	89.0	142.4	122.1	104.0	140.4	126.5	101.3	140.4	125.5	94.8	135.4	120.0	126.1	155.8	144.5	97.8	134.6	120.6
05	METALS	102.8	113.6	106.2	104.2	112.1	106.6	101.0	111.9	104.3	100.6	110.4	103.6	101.9	114.8	105.9	98.3	111.2	102.3
06	WOOD, PLASTICS & COMPOSITES	103.9	135.7	121.6	101.2	153.2	130.1	96.9	153.2	128.2	102.2	130.7	118.1	98.8	170.1	138.5	97.4	135.4	118.5
07	THERMAL & MOISTURE PROTECTION	99.5	134.1	114.1	109.4	135.1	120.3	108.9	132.4	118.9	99.9	122.2	109.3	101.4	153.7	123.6	103.2	132.5	115.6
08	OPENINGS	100.8	134.9	108.6	105.5	143.1	114.1	104.2	143.6	113.3	91.4	130.0	100.3	96.2	158.5	110.5	94.1	134.0	100.3
0920	Plaster & Gypsum Board	99.3	136.3	124.1	103.3	154.5	137.7	105.7	154.5	138.5	102.6	131.4	121.9	101.9	171.5	148.7	96.6	136.3	123.2
0950, 0980	Ceilings & Acoustic Treatment	116.2	136.3	129.7	113.6	154.5	141.2	111.1	154.5	140.4	96.7	131.4	120.1	103.3	171.5	149.3	99.2	136.3	124.2
0960	Flooring	106.9	118.6	110.1	96.9	121.2	103.6	97.3	121.0	103.9	104.5	116.0	107.7	109.7	136.3	117.1	96.7	118.6	102.7
0970, 0990	Wall Finishes & Painting/Coating	103.9	120.9	113.8	90.7	134.7	116.4	90.7	131.9	114.7	88.6	107.0	99.3	108.5	165.7	141.9	88.6	115.4	104.2
09	FINISHES	107.1	133.1	121.3	99.3	145.2	124.3	98.8	144.9	123.9	94.6	128.0	112.8	104.8	161.6	135.7	92.1	132.5	114.1
COVERS	DIVS. 10 - 14, 25, 28, 41, 43, 44, 46	100.0	118.7	104.2	100.0	128.1	106.2	100.0	128.0	106.2	100.0	112.1	102.7	100.0	131.3	107.0	100.0	121.1	104.7
21, 22, 23	FIRE SUPPRESSION, PLUMBING & HVAC	100.0	130.1	112.9	96.9	125.4	109.1	100.0	124.1	110.3	96.9	122.4	107.8	100.2	161.4	126.4	100.0	133.2	114.2
26, 27, 3370	ELECTRICAL, COMMUNICATIONS & UTIL.	98.9	130.6	115.4	94.5	118.3	106.9	96.8	107.3	102.3	95.0	107.9	101.7	101.7	159.0	131.5	100.4	117.2	109.1
MF2016	WEIGHTED AVERAGE	99.9	129.3	112.8	101.6	128.8	113.5	100.4	126.9	112.0	95.2	121.4	106.7	102.5	148.9	122.8	97.6	126.5	110.2

For customer support on your Electrical Costs with RSMeans data, call 800.448.8182.

517

CALIFORNIA

DIVISION		PALM SPRINGS 922			PALO ALTO 943			PASADENA 910 - 912			REDDING 960			RICHMOND 948			RIVERSIDE 925		
		MAT.	INST.	TOTAL	MAT.	INST.	TOTAL	MAT.	INST.	TOTAL	MAT.	INST.	TOTAL	MAT.	INST.	TOTAL	MAT.	INST.	TOTAL
015433	CONTRACTOR EQUIPMENT		99.8	99.8		102.7	102.7		95.8	95.8		98.7	98.7		102.7	102.7		99.8	99.8
0241, 31 - 34	SITE & INFRASTRUCTURE, DEMOLITION	93.4	107.9	103.5	103.7	109.4	107.7	97.7	109.1	105.7	133.9	106.3	114.5	111.6	109.4	110.1	105.4	107.9	105.7
0310	Concrete Forming & Accessories	101.7	134.9	130.3	102.6	164.7	156.1	105.4	134.5	130.5	104.3	149.7	143.5	117.8	164.6	158.2	105.4	138.5	133.9
0320	Concrete Reinforcing	108.3	128.2	118.3	90.6	129.6	110.3	105.8	128.4	117.2	127.7	128.9	128.3	90.6	129.6	110.3	105.2	128.1	116.7
0330	Cast-in-Place Concrete	83.8	132.6	102.4	99.3	132.9	112.1	78.6	128.9	97.7	115.6	128.9	120.6	114.2	132.8	121.3	91.2	132.7	106.9
03	CONCRETE	90.1	131.7	109.1	92.6	145.5	116.8	90.4	130.2	108.6	117.1	137.2	126.3	104.8	145.4	123.4	96.0	133.4	113.1
04	MASONRY	76.0	138.5	114.8	100.7	148.3	130.2	97.5	141.0	124.5	132.5	140.4	137.4	118.4	148.3	136.9	77.0	135.7	113.4
05	METALS	105.5	111.5	107.3	99.4	114.0	103.9	86.6	109.5	93.6	106.4	112.1	108.2	99.5	113.8	103.9	104.9	111.5	106.9
06	WOOD, PLASTICS & COMPOSITES	94.7	130.6	114.7	96.4	169.9	137.3	86.5	130.3	110.9	111.3	152.7	134.3	113.4	170.1	145.0	99.7	135.3	119.5
07	THERMAL & MOISTURE PROTECTION	105.5	133.6	117.4	100.6	152.2	122.4	98.5	130.1	111.9	130.6	135.1	132.5	101.3	149.3	121.6	105.7	134.0	112.2
08	OPENINGS	102.8	131.4	109.3	96.2	157.4	110.3	91.2	131.2	100.4	119.0	142.9	124.5	96.2	157.5	110.3	105.2	134.0	112.2
0920	Plaster & Gypsum Board	100.9	131.4	121.4	100.3	171.3	148.0	90.3	131.4	117.9	104.9	154.1	137.9	108.5	171.5	150.8	104.9	136.3	126.0
0950, 0980	Ceilings & Acoustic Treatment	107.7	131.4	123.7	101.4	171.3	148.5	110.3	131.4	124.5	141.6	154.1	150.0	101.4	171.5	148.7	116.3	136.3	129.7
0960	Flooring	99.4	122.3	105.7	108.2	136.3	116.0	96.4	116.0	101.8	91.7	124.9	100.9	119.5	136.3	124.2	101.0	118.6	105.9
0970, 0990	Wall Finishes & Painting/Coating	87.8	118.5	105.7	108.5	165.7	141.9	108.0	120.9	115.5	100.4	134.7	120.4	108.5	165.7	141.9	87.8	118.5	105.7
09	FINISHES	96.0	130.3	114.7	103.2	161.4	134.9	99.3	129.3	115.6	104.6	145.6	126.9	108.3	161.6	137.3	99.1	132.8	117.4
COVERS	DIVS. 10 - 14, 25, 28, 41, 43, 44, 46	100.0	120.4	104.5	100.0	131.3	107.0	100.0	119.7	104.4	100.0	128.0	106.2	100.0	131.2	106.9	100.0	120.9	104.6
21, 22, 23	FIRE SUPPRESSION, PLUMBING & HVAC	96.9	125.3	109.1	97.1	155.6	122.1	96.9	122.8	108.0	100.5	125.4	111.1	97.1	154.1	121.5	100.0	133.2	114.2
26, 27, 3370	ELECTRICAL, COMMUNICATIONS & UTIL.	94.1	110.8	102.8	101.6	169.0	136.6	114.7	126.8	121.0	99.0	118.3	109.1	102.2	136.0	119.8	90.9	108.0	99.8
MF2016	WEIGHTED AVERAGE	97.2	123.9	108.9	98.5	148.0	120.2	96.0	125.3	108.8	109.1	128.8	117.7	101.7	143.0	119.8	99.0	125.6	110.7

CALIFORNIA

DIVISION		SACRAMENTO 942, 956 - 958			SALINAS 939			SAN BERNARDINO 923 - 924			SAN DIEGO 919 - 921			SAN FRANCISCO 940 - 941			SAN JOSE 951		
		MAT.	INST.	TOTAL	MAT.	INST.	TOTAL	MAT.	INST.	TOTAL	MAT.	INST.	TOTAL	MAT.	INST.	TOTAL	MAT.	INST.	TOTAL
015433	CONTRACTOR EQUIPMENT		101.4	101.4		98.9	98.9		99.8	99.8		100.8	100.8		109.7	109.7		100.5	100.5
0241, 31 - 34	SITE & INFRASTRUCTURE, DEMOLITION	93.6	115.5	108.9	118.0	107.1	110.4	79.5	107.9	99.4	106.2	106.5	106.4	113.9	113.0	113.3	135.4	101.4	111.6
0310	Concrete Forming & Accessories	102.5	152.8	145.9	108.4	153.1	147.0	109.2	138.4	134.4	105.3	126.0	123.1	108.4	165.8	157.9	106.2	164.9	156.8
0320	Concrete Reinforcing	86.2	129.3	107.9	89.2	129.4	109.5	105.2	128.3	116.8	102.4	128.3	115.4	105.6	130.3	118.1	94.2	129.7	112.1
0330	Cast-in-Place Concrete	94.2	129.9	107.7	96.4	129.3	108.9	63.0	132.6	89.5	90.0	119.2	101.1	124.3	133.3	127.7	110.9	132.2	119.0
03	CONCRETE	93.9	138.7	114.4	101.1	139.0	118.4	71.0	133.4	99.5	100.6	123.2	110.9	114.2	146.7	129.1	103.9	145.7	123.0
04	MASONRY	101.6	146.1	129.2	94.9	146.5	126.9	83.3	138.5	117.5	89.8	132.5	116.3	135.9	156.5	148.7	131.2	149.8	142.7
05	METALS	97.6	108.5	101.0	103.2	113.2	106.3	104.9	111.6	107.0	102.8	113.0	106.0	107.4	122.2	112.0	99.3	119.1	105.4
06	WOOD, PLASTICS & COMPOSITES	92.2	156.5	128.0	101.8	156.3	132.1	103.2	135.3	121.1	97.9	122.6	111.6	101.1	170.1	139.5	109.7	169.9	143.2
07	THERMAL & MOISTURE PROTECTION	110.6	139.1	122.7	100.5	143.3	118.6	104.5	135.1	117.5	103.6	118.2	109.8	106.5	156.0	127.4	105.4	151.6	125.0
08	OPENINGS	109.6	145.4	117.8	95.3	150.9	108.1	102.8	134.6	110.1	100.1	125.8	106.0	102.2	158.5	115.2	95.7	158.4	110.1
0920	Plaster & Gypsum Board	97.6	157.7	137.9	97.6	157.7	138.0	106.3	136.3	126.4	96.2	123.0	114.2	99.9	171.5	148.0	101.8	171.5	148.6
0950, 0980	Ceilings & Acoustic Treatment	101.4	157.7	139.3	96.7	157.7	137.8	111.1	136.3	128.1	120.8	123.0	122.3	110.6	171.5	151.7	111.2	171.5	151.9
0960	Flooring	108.4	125.7	113.2	98.7	137.9	109.6	102.9	118.6	107.3	106.2	118.6	109.6	110.2	137.9	117.9	90.7	136.3	103.4
0970, 0990	Wall Finishes & Painting/Coating	105.2	134.7	122.4	89.4	165.7	133.9	87.8	120.9	107.1	105.6	120.9	114.5	106.4	175.4	146.7	91.0	165.7	134.6
09	FINISHES	102.1	148.0	127.1	94.0	153.3	126.3	97.4	132.9	116.7	107.0	123.7	116.1	106.4	162.9	137.2	97.8	161.4	132.4
COVERS	DIVS. 10 - 14, 25, 28, 41, 43, 44, 46	100.0	128.9	106.4	100.0	128.4	106.3	100.0	118.1	104.0	100.0	114.5	103.2	100.0	131.4	107.0	100.0	130.8	106.8
21, 22, 23	FIRE SUPPRESSION, PLUMBING & HVAC	100.1	128.1	112.1	96.9	134.5	113.0	96.9	129.9	111.1	99.9	126.5	111.3	100.0	189.7	138.4	100.0	166.0	128.3
26, 27, 3370	ELECTRICAL, COMMUNICATIONS & UTIL.	97.4	122.0	110.2	96.0	128.7	113.0	94.1	111.6	103.2	101.3	103.1	102.2	101.8	174.7	139.7	99.7	170.0	136.3
MF2016	WEIGHTED AVERAGE	100.0	131.6	113.8	98.8	134.8	114.6	94.8	125.7	108.4	101.0	119.5	109.1	106.3	158.5	129.1	102.4	150.4	123.4

CALIFORNIA

DIVISION		SAN LUIS OBISPO 934			SAN MATEO 944			SAN RAFAEL 949			SANTA ANA 926 - 927			SANTA BARBARA 931			SANTA CRUZ 950		
		MAT.	INST.	TOTAL	MAT.	INST.	TOTAL	MAT.	INST.	TOTAL	MAT.	INST.	TOTAL	MAT.	INST.	TOTAL	MAT.	INST.	TOTAL
015433	CONTRACTOR EQUIPMENT		98.9	98.9		102.7	102.7		101.8	101.8		99.8	99.8		98.9	98.9		100.5	100.5
0241, 31 - 34	SITE & INFRASTRUCTURE, DEMOLITION	110.0	107.7	108.4	109.5	109.4	109.4	106.3	115.2	112.5	91.8	107.9	103.0	104.3	107.7	106.7	135.0	101.3	111.4
0310	Concrete Forming & Accessories	115.5	136.8	133.9	108.4	164.9	157.1	112.9	164.9	157.7	109.4	134.7	131.2	106.1	138.6	134.1	106.2	153.5	147.0
0320	Concrete Reinforcing	90.4	128.1	109.4	90.6	129.8	110.4	91.4	129.9	110.8	108.9	128.2	118.6	88.8	128.1	108.6	116.2	129.4	122.9
0330	Cast-in-Place Concrete	103.6	131.8	114.3	110.6	132.9	119.1	128.6	132.0	129.9	80.5	132.5	100.2	97.2	131.9	110.4	110.2	131.1	118.1
03	CONCRETE	99.5	132.3	114.5	101.6	145.6	121.7	122.8	145.1	133.0	87.7	131.6	107.8	92.0	133.1	110.8	106.4	140.1	121.8
04	MASONRY	96.5	135.6	120.7	118.1	151.6	138.8	96.9	152.5	131.4	73.1	138.8	113.8	95.1	135.6	120.2	134.9	146.7	142.2
05	METALS	101.3	111.0	104.3	99.3	114.3	103.9	100.6	110.8	103.7	105.0	111.4	106.9	98.9	111.2	102.7	106.5	117.8	110.0
06	WOOD, PLASTICS & COMPOSITES	104.3	133.2	120.4	103.6	170.1	140.6	99.4	169.9	138.6	105.1	130.6	119.3	97.4	135.4	118.5	109.7	156.4	135.7
07	THERMAL & MOISTURE PROTECTION	100.7	131.7	113.8	101.0	152.4	122.7	105.3	151.9	125.0	105.8	133.7	117.6	100.1	132.6	113.9	97.0	150.9	109.3
08	OPENINGS	93.4	131.4	102.1	96.2	157.5	110.3	107.1	158.0	118.7	102.1	131.4	108.8	95.1	134.4	104.1	97.0	150.9	109.3
0920	Plaster & Gypsum Board	103.5	134.0	124.0	105.2	171.5	149.7	106.9	171.5	150.3	108.0	131.4	123.7	96.6	136.3	123.2	110.4	157.7	142.1
0950, 0980	Ceilings & Acoustic Treatment	96.7	134.0	121.9	101.4	171.5	148.7	108.2	171.5	150.9	111.1	131.4	124.8	99.2	136.3	124.2	111.9	157.7	142.8
0960	Flooring	105.4	116.0	108.4	112.3	136.3	118.9	124.7	130.6	126.3	103.5	116.0	107.0	97.8	118.6	103.6	94.7	137.9	106.7
0970, 0990	Wall Finishes & Painting/Coating	88.6	110.9	101.6	108.5	165.7	141.9	104.6	164.8	139.7	87.8	118.5	105.7	88.6	115.4	104.2	91.2	165.7	134.7
09	FINISHES	96.0	129.9	114.4	105.5	161.6	136.0	108.9	160.3	136.9	98.9	129.2	115.4	92.6	132.3	114.2	100.4	153.4	129.2
COVERS	DIVS. 10 - 14, 25, 28, 41, 43, 44, 46	100.0	127.6	106.1	100.0	131.3	107.0	100.0	130.6	106.8	100.0	120.4	104.5	100.0	118.3	104.1	100.0	128.7	106.4
21, 22, 23	FIRE SUPPRESSION, PLUMBING & HVAC	96.9	127.2	109.9	97.1	157.3	122.9	97.1	185.0	134.7	96.9	107.6	101.5	93.9	133.2	114.2	100.0	134.9	115.0
26, 27, 3370	ELECTRICAL, COMMUNICATIONS & UTIL.	95.0	112.5	104.1	101.6	161.1	132.6	98.8	123.8	111.8	94.1	110.0	102.4	93.9	114.8	104.8	98.9	128.7	114.4
MF2016	WEIGHTED AVERAGE	98.1	124.4	109.6	100.8	147.7	121.3	103.8	148.4	123.3	96.8	122.9	108.3	96.9	126.3	109.8	104.3	135.1	117.8

CALIFORNIA / COLORADO

DIVISION		SANTA ROSA 954			STOCKTON 952			SUSANVILLE 961			VALLEJO 945			VAN NUYS 913 - 916			ALAMOSA 811		
		MAT.	INST.	TOTAL	MAT.	INST.	TOTAL	MAT.	INST.	TOTAL	MAT.	INST.	TOTAL	MAT.	INST.	TOTAL	MAT.	INST.	TOTAL
015433	CONTRACTOR EQUIPMENT		99.3	99.3		98.7	98.7		98.7	98.7		101.8	101.8		95.8	95.8		92.8	92.8
0241, 31 - 34	SITE & INFRASTRUCTURE, DEMOLITION	106.1	106.5	106.4	104.0	106.5	105.8	142.3	106.2	117.0	94.8	115.3	109.1	116.1	109.1	111.2	143.5	88.4	104.9
0310	Concrete Forming & Accessories	102.6	164.2	155.7	104.2	152.2	145.6	105.4	141.9	136.9	103.6	163.8	155.5	112.1	134.5	131.4	102.5	63.8	69.1
0320	Concrete Reinforcing	103.8	130.0	117.0	106.6	128.8	117.8	127.7	128.8	128.3	92.5	129.8	111.3	105.8	128.4	117.2	107.9	67.8	87.7
0330	Cast-in-Place Concrete	104.7	130.5	114.5	93.0	128.8	106.6	105.1	128.8	114.1	102.5	131.1	113.4	82.9	128.9	100.4	107.9	73.6	89.9
03	CONCRETE	106.6	144.6	124.0	96.9	138.3	115.9	119.5	133.7	126.0	98.7	144.2	119.5	104.7	130.2	116.4	110.6	68.6	91.4
04	MASONRY	101.8	153.3	133.7	101.2	140.4	125.5	131.1	140.4	136.8	73.7	151.4	121.9	112.0	141.0	130.0	130.4	60.9	87.3
05	METALS	105.3	115.2	108.3	101.2	111.9	104.5	105.2	111.9	107.3	100.5	110.1	103.5	85.7	109.5	93.0	103.7	78.2	95.9
06	WOOD, PLASTICS & COMPOSITES	96.1	169.6	137.0	102.3	156.3	132.3	113.0	142.3	129.3	89.3	169.9	134.1	94.6	130.3	114.5	97.8	64.1	79.0
07	THERMAL & MOISTURE PROTECTION	106.3	152.0	125.6	109.4	134.0	119.8	132.3	131.9	132.1	103.3	151.8	123.8	99.7	130.1	112.6	104.3	69.0	89.4
08	OPENINGS	103.7	158.3	116.3	104.2	143.2	113.2	119.9	137.1	123.9	108.9	156.3	119.8	91.0	131.2	100.3	96.4	67.0	89.6
0920	Plaster & Gypsum Board	102.7	171.5	148.9	105.7	157.7	140.6	105.7	143.4	131.0	101.6	171.5	148.6	94.3	131.4	119.2	78.0	63.0	67.9
0950, 0980	Ceilings & Acoustic Treatment	111.1	171.5	151.8	119.7	157.7	145.3	134.1	143.4	140.3	110.1	171.5	151.5	107.7	131.4	123.7	102.3	63.0	75.8
0960	Flooring	99.9	128.7	107.9	97.3	121.0	103.9	92.1	124.9	101.2	118.0	136.3	123.1	99.8	116.0	104.3	109.3	73.0	99.2
0970, 0990	Wall Finishes & Painting/Coating	87.8	155.6	127.4	90.7	128.3	112.6	100.4	134.7	120.4	105.6	164.8	140.1	108.0	120.9	115.5	103.1	61.9	79.1
09	FINISHES	98.0	158.5	130.9	100.5	146.3	125.4	104.4	139.5	123.5	105.4	161.0	135.6	101.7	129.2	116.7	99.6	64.8	80.7
COVERS	DIVS. 10 - 14, 25, 28, 41, 43, 44, 46	100.0	129.7	106.6	100.0	128.4	106.3	100.0	126.9	106.0	100.0	130.2	106.7	100.0	119.7	104.4	100.0	84.4	96.5
21, 22, 23	FIRE SUPPRESSION, PLUMBING & HVAC	96.9	182.2	133.4	100.0	124.1	110.3	97.4	125.4	109.4	100.2	143.2	118.5	96.9	122.8	108.0	96.9	65.5	83.5
26, 27, 3370	ELECTRICAL, COMMUNICATIONS & UTIL.	94.4	124.4	110.0	96.8	110.8	104.1	99.4	118.3	109.2	95.0	126.8	111.5	114.7	126.8	121.0	96.9	65.1	80.8
MF2016	WEIGHTED AVERAGE	101.1	147.3	121.3	100.5	127.7	112.4	108.8	127.0	116.8	99.5	139.6	117.0	99.1	125.3	110.6	103.3	69.0	88.3

COLORADO

DIVISION		BOULDER 803			COLORADO SPRINGS 808 - 809			DENVER 800 - 802			DURANGO 813			FORT COLLINS 805			FORT MORGAN 807		
		MAT.	INST.	TOTAL	MAT.	INST.	TOTAL	MAT.	INST.	TOTAL	MAT.	INST.	TOTAL	MAT.	INST.	TOTAL	MAT.	INST.	TOTAL
015433	CONTRACTOR EQUIPMENT		94.7	94.7		92.6	92.6		99.1	99.1		92.8	92.8		94.7	94.7		94.7	94.7
0241, 31 - 34	SITE & INFRASTRUCTURE, DEMOLITION	97.3	95.7	96.2	99.1	90.7	93.2	104.2	102.4	102.9	136.6	88.4	102.8	109.6	95.6	99.8	99.5	95.4	96.6
0310	Concrete Forming & Accessories	103.0	74.9	78.8	93.8	60.3	64.9	102.0	63.8	69.1	108.6	63.6	69.8	100.6	69.9	74.1	103.5	53.3	60.2
0320	Concrete Reinforcing	97.7	67.9	82.7	96.9	67.9	82.3	96.9	68.0	82.3	107.9	67.8	87.7	97.8	67.9	82.7	97.9	67.9	82.8
0330	Cast-in-Place Concrete	106.0	74.2	93.9	108.9	73.3	95.4	113.6	72.7	98.0	115.0	73.5	99.2	119.7	72.8	101.9	104.0	72.7	92.1
03	CONCRETE	103.2	73.8	89.7	106.8	66.9	88.6	109.4	68.3	90.6	112.8	68.4	92.5	114.2	71.0	94.5	101.6	63.5	84.2
04	MASONRY	97.3	64.7	77.1	98.9	62.7	76.4	100.1	65.1	78.4	117.1	60.9	82.2	115.2	62.2	82.3	112.5	62.2	81.3
05	METALS	98.1	77.1	91.7	101.3	77.4	93.9	103.8	78.4	96.0	103.7	78.0	95.8	99.4	77.1	92.5	97.8	76.9	91.4
06	WOOD, PLASTICS & COMPOSITES	105.7	77.7	90.1	96.1	59.1	75.5	103.0	63.6	81.1	106.7	64.1	83.0	103.3	72.2	86.0	105.7	49.9	74.6
07	THERMAL & MOISTURE PROTECTION	98.9	73.5	88.1	99.6	70.7	87.4	97.4	71.6	86.5	104.3	69.0	89.4	99.2	71.8	87.6	98.8	69.3	86.3
08	OPENINGS	98.3	74.5	92.8	102.6	64.2	93.8	105.3	66.7	96.4	103.4	67.0	95.0	98.2	71.5	92.1	98.2	59.1	89.2
0920	Plaster & Gypsum Board	116.2	77.5	90.2	100.0	58.2	71.9	111.9	63.0	79.1	90.8	63.0	72.2	110.6	71.8	84.6	116.2	48.9	71.0
0950, 0980	Ceilings & Acoustic Treatment	90.2	77.5	81.6	97.9	58.2	71.1	102.5	63.0	75.9	102.3	63.0	75.8	90.2	71.8	77.8	90.2	48.9	62.4
0960	Flooring	111.8	73.0	101.0	102.8	73.0	94.5	109.0	73.0	99.0	114.7	73.0	103.1	107.7	73.0	98.1	112.3	73.0	101.4
0970, 0990	Wall Finishes & Painting/Coating	101.5	61.9	78.4	101.2	61.9	78.3	108.2	61.9	81.2	103.1	61.9	79.1	101.5	61.9	78.4	101.5	61.9	78.4
09	FINISHES	102.8	73.4	86.8	99.4	61.8	79.0	104.8	64.6	82.9	102.1	64.8	81.8	101.4	69.5	84.0	102.9	56.3	77.6
COVERS	DIVS. 10 - 14, 25, 28, 41, 43, 44, 46	100.0	85.8	96.8	100.0	83.2	96.3	100.0	83.5	96.3	100.0	84.4	96.5	100.0	84.4	96.5	100.0	81.9	96.0
21, 22, 23	FIRE SUPPRESSION, PLUMBING & HVAC	96.9	73.3	86.8	100.1	80.6	91.8	99.9	75.5	89.5	96.9	57.3	80.0	100.0	72.0	88.0	96.9	74.8	87.4
26, 27, 3370	ELECTRICAL, COMMUNICATIONS & UTIL.	95.2	79.5	87.1	98.2	75.8	86.5	99.8	79.6	89.3	97.2	53.1	74.3	95.2	79.5	87.1	95.5	79.4	87.2
MF2016	WEIGHTED AVERAGE	98.8	75.9	88.8	101.0	73.2	88.8	102.8	74.6	90.5	103.7	65.5	87.0	102.1	74.3	89.9	99.4	71.2	87.0

COLORADO

DIVISION		GLENWOOD SPRINGS 816			GOLDEN 804			GRAND JUNCTION 815			GREELEY 806			MONTROSE 814			PUEBLO 810		
		MAT.	INST.	TOTAL	MAT.	INST.	TOTAL	MAT.	INST.	TOTAL	MAT.	INST.	TOTAL	MAT.	INST.	TOTAL	MAT.	INST.	TOTAL
015433	CONTRACTOR EQUIPMENT		96.0	96.0		94.7	94.7		96.0	96.0		94.7	94.7		94.4	94.4		92.8	92.8
0241, 31 - 34	SITE & INFRASTRUCTURE, DEMOLITION	153.2	96.4	113.4	110.6	95.7	100.2	135.9	96.6	108.4	95.9	95.6	95.7	146.3	92.4	108.5	127.3	88.4	100.0
0310	Concrete Forming & Accessories	99.5	53.2	59.6	96.1	59.6	64.6	107.5	74.3	78.9	98.5	73.9	77.3	99.0	53.5	59.8	104.7	60.3	66.4
0320	Concrete Reinforcing	106.7	67.8	87.1	97.9	67.9	82.8	107.0	67.9	87.3	97.7	67.9	82.7	106.5	67.8	87.0	103.1	67.9	85.3
0330	Cast-in-Place Concrete	99.9	72.4	89.5	104.1	74.1	92.7	110.7	73.9	96.7	100.0	72.8	89.7	100.0	73.0	89.7	99.2	73.6	89.5
03	CONCRETE	115.7	63.4	91.8	112.7	66.8	91.7	109.3	73.5	92.9	98.3	72.8	86.6	106.8	63.7	87.1	99.5	67.0	84.7
04	MASONRY	101.7	62.1	77.2	115.6	64.7	84.0	137.4	65.4	92.7	109.0	62.2	80.0	110.2	60.9	79.6	97.9	62.0	75.7
05	METALS	103.4	77.7	95.5	98.0	77.0	91.6	105.1	78.1	96.8	99.3	77.1	92.5	102.5	78.0	95.0	106.8	78.3	98.0
06	WOOD, PLASTICS & COMPOSITES	93.1	50.0	69.2	98.3	57.1	75.4	104.4	77.0	89.2	100.6	77.6	87.8	94.4	50.2	69.8	100.3	59.4	77.5
07	THERMAL & MOISTURE PROTECTION	104.2	67.9	88.8	99.9	71.2	87.8	103.3	72.2	90.1	98.5	72.4	87.4	104.4	67.5	88.8	102.8	69.1	88.5
08	OPENINGS	102.4	59.2	92.5	98.2	63.1	90.2	103.1	74.1	96.5	98.2	74.5	92.8	103.5	59.3	93.4	98.2	64.4	90.5
0920	Plaster & Gypsum Board	120.4	48.9	72.4	108.3	56.4	73.4	132.9	76.7	95.1	109.0	77.4	87.8	77.3	48.9	58.2	82.0	58.2	66.0
0950, 0980	Ceilings & Acoustic Treatment	101.5	48.9	66.0	90.2	56.4	67.4	101.5	76.7	84.7	90.2	77.4	81.6	102.3	48.9	66.3	110.0	58.2	75.1
0960	Flooring	108.4	73.0	98.6	105.2	73.0	96.3	114.0	73.0	102.6	106.5	73.0	97.2	112.0	73.0	101.1	110.6	73.0	100.1
0970, 0990	Wall Finishes & Painting/Coating	103.1	61.9	79.1	101.5	61.9	78.4	103.1	61.9	79.1	101.5	61.9	78.4	103.1	61.9	79.1	103.1	61.9	79.1
09	FINISHES	105.3	56.5	78.8	101.0	61.2	79.4	106.9	73.0	88.5	100.0	72.7	85.2	100.3	56.6	76.5	100.2	62.0	79.4
COVERS	DIVS. 10 - 14, 25, 28, 41, 43, 44, 46	100.0	82.2	96.0	100.0	83.5	96.3	100.0	86.0	96.9	100.0	85.0	96.7	100.0	82.5	96.1	100.0	83.9	96.4
21, 22, 23	FIRE SUPPRESSION, PLUMBING & HVAC	96.9	66.6	83.9	96.9	73.3	86.8	99.9	76.4	89.9	100.0	72.0	88.0	96.9	77.1	88.4	99.9	66.6	85.7
26, 27, 3370	ELECTRICAL, COMMUNICATIONS & UTIL.	94.7	53.1	73.1	95.5	79.5	87.2	96.9	53.1	74.1	95.2	79.5	87.1	96.9	53.1	74.1	97.7	66.2	81.3
MF2016	WEIGHTED AVERAGE	103.5	65.9	87.0	101.0	72.6	88.6	105.5	73.0	91.3	99.3	75.1	88.7	102.4	67.8	87.3	101.3	68.8	87.1

For customer support on your Electrical Costs with RSMeans data, call 800.448.8182.

519

COLORADO / CONNECTICUT

DIVISION		SALIDA 812 MAT.	INST.	TOTAL	BRIDGEPORT 066 MAT.	INST.	TOTAL	BRISTOL 060 MAT.	INST.	TOTAL	HARTFORD 061 MAT.	INST.	TOTAL	MERIDEN 064 MAT.	INST.	TOTAL	NEW BRITAIN 060 MAT.	INST.	TOTAL
015433	CONTRACTOR EQUIPMENT		94.4	94.4		98.8	98.8		98.8	98.8		98.8	98.8		99.3	99.3		98.8	98.8
0241, 31 - 34	SITE & INFRASTRUCTURE, DEMOLITION	136.3	92.4	105.5	105.9	103.2	104.0	105.0	103.2	103.7	100.3	103.2	102.3	101.9	103.9	103.6	102.6	118.6	116.4
0310	Concrete Forming & Accessories	107.4	53.3	60.8	102.2	118.7	116.5	102.2	118.6	116.3	102.1	118.6	116.3	101.9	118.5	116.3	102.6	118.6	116.4
0320	Concrete Reinforcing	106.2	67.8	86.8	117.4	142.1	129.8	117.4	142.1	129.8	112.6	142.1	127.5	117.4	142.1	129.8	117.4	142.1	129.8
0330	Cast-in-Place Concrete	114.5	72.9	98.7	98.7	129.3	110.3	92.5	129.2	106.4	96.9	129.2	109.2	89.0	129.2	104.3	94.0	129.2	107.4
03	CONCRETE	107.8	63.6	87.6	96.2	125.6	109.7	93.3	125.5	108.0	93.5	125.6	108.2	91.6	125.5	107.1	94.0	125.5	108.4
04	MASONRY	138.9	60.9	90.5	113.9	130.1	124.0	105.1	130.1	120.6	104.4	130.1	120.3	104.8	130.1	120.5	107.5	130.1	121.5
05	METALS	102.2	77.8	94.7	96.2	116.4	102.4	96.2	116.2	102.4	101.2	116.3	105.8	93.6	116.2	100.6	92.7	116.3	100.0
06	WOOD, PLASTICS & COMPOSITES	101.3	50.2	72.9	110.3	116.7	113.9	110.3	116.7	113.9	96.7	116.7	107.8	110.3	116.7	113.9	110.3	116.7	113.9
07	THERMAL & MOISTURE PROTECTION	103.2	67.5	88.1	100.1	124.8	110.6	100.3	121.4	109.2	105.4	121.4	112.1	100.3	121.3	109.2	100.3	121.6	109.3
08	OPENINGS	96.5	59.3	87.9	95.5	122.4	101.7	95.5	122.4	101.7	94.0	122.4	100.5	97.8	122.4	103.4	95.5	122.4	101.7
0920	Plaster & Gypsum Board	77.6	48.9	58.3	115.6	116.8	116.4	115.6	116.8	116.4	103.9	116.8	112.5	116.8	116.8	116.8	115.6	116.8	116.4
0950, 0980	Ceilings & Acoustic Treatment	102.3	48.9	66.3	95.3	116.8	109.8	95.3	116.8	109.8	93.2	116.8	109.1	98.5	116.8	110.8	95.3	116.8	109.8
0960	Flooring	117.6	73.0	105.2	87.3	131.6	99.6	87.3	124.4	97.6	90.2	131.6	101.7	87.3	124.4	97.6	87.3	124.4	97.6
0970, 0990	Wall Finishes & Painting/Coating	103.1	61.9	79.1	92.3	128.8	113.6	92.3	128.8	113.6	94.4	128.8	114.5	92.3	128.8	113.6	92.3	128.8	113.6
09	FINISHES	101.0	56.6	76.8	91.4	121.4	107.7	91.4	120.2	107.0	93.6	121.4	108.7	92.2	120.2	107.4	91.4	120.2	107.0
COVERS	DIVS. 10 - 14, 25, 28, 41, 43, 44, 46	100.0	82.6	96.1	100.0	113.2	102.9	100.0	113.2	102.9	100.0	113.2	102.9	100.0	113.2	102.9	100.0	113.2	102.9
21, 22, 23	FIRE SUPPRESSION, PLUMBING & HVAC	96.9	62.9	82.3	100.1	118.8	108.1	100.1	118.7	108.1	100.0	118.8	108.0	97.0	118.7	106.3	100.1	118.7	108.1
26, 27, 3370	ELECTRICAL, COMMUNICATIONS & UTIL.	97.1	60.8	78.2	99.0	108.3	103.8	99.0	106.3	102.8	98.5	110.5	104.8	98.9	105.7	102.5	99.0	106.3	102.8
MF2016	WEIGHTED AVERAGE	103.0	65.8	86.7	98.5	118.4	107.2	97.7	117.9	106.5	98.4	118.6	107.3	96.6	117.8	105.9	97.4	117.9	106.3

CONNECTICUT

DIVISION		NEW HAVEN 065 MAT.	INST.	TOTAL	NEW LONDON 063 MAT.	INST.	TOTAL	NORWALK 068 MAT.	INST.	TOTAL	STAMFORD 069 MAT.	INST.	TOTAL	WATERBURY 067 MAT.	INST.	TOTAL	WILLIMANTIC 062 MAT.	INST.	TOTAL
015433	CONTRACTOR EQUIPMENT		99.3	99.3		99.3	99.3		98.8	98.8		98.8	98.8		98.8	98.8		98.8	98.8
0241, 31 - 34	SITE & INFRASTRUCTURE, DEMOLITION	105.1	103.7	104.1	97.4	103.7	101.8	105.6	103.1	103.9	106.3	103.2	104.1	105.7	103.2	103.9	105.6	102.9	103.7
0310	Concrete Forming & Accessories	101.9	118.6	116.3	101.9	118.6	116.3	102.2	118.7	116.4	102.2	119.1	116.7	102.2	118.6	116.4	102.2	118.5	116.2
0320	Concrete Reinforcing	117.4	142.1	129.8	92.0	142.1	117.3	117.4	142.1	129.8	117.4	142.2	129.9	117.4	142.1	129.8	117.4	142.1	129.8
0330	Cast-in-Place Concrete	95.6	129.2	108.4	81.5	129.2	99.6	97.1	129.3	109.3	98.7	129.4	110.4	98.7	129.2	110.3	92.2	129.2	106.2
03	CONCRETE	106.6	125.5	115.3	82.3	125.5	102.1	95.4	125.6	109.2	96.2	125.8	109.8	96.2	125.6	109.6	93.1	125.5	107.9
04	MASONRY	105.4	130.1	120.7	103.7	130.1	120.1	104.8	130.1	120.5	105.6	130.1	120.8	105.6	130.1	120.6	96.0	116.2	102.2
05	METALS	92.9	116.3	100.1	92.7	116.3	99.9	96.2	116.4	102.4	96.2	116.8	102.5	96.2	116.3	102.4	96.0	116.2	102.2
06	WOOD, PLASTICS & COMPOSITES	110.3	116.7	113.9	110.3	116.7	113.9	110.3	116.7	113.9	110.3	116.7	113.9	110.3	116.7	113.9	110.3	116.7	113.9
07	THERMAL & MOISTURE PROTECTION	100.4	121.7	109.4	100.2	121.4	109.1	100.3	124.8	110.7	100.3	124.8	110.7	100.3	121.8	109.4	100.5	121.0	109.2
08	OPENINGS	95.5	122.4	101.7	97.9	122.4	103.5	95.5	122.4	101.7	95.5	122.4	101.7	95.5	122.4	101.7	97.9	122.4	103.5
0920	Plaster & Gypsum Board	115.6	116.8	116.4	115.6	116.8	116.4	115.6	116.8	116.4	115.6	116.8	116.4	115.6	116.8	116.4	115.6	116.8	116.4
0950, 0980	Ceilings & Acoustic Treatment	95.3	116.8	109.8	93.4	116.8	109.2	95.3	116.8	109.8	95.3	116.8	109.8	95.3	116.8	109.8	93.4	116.8	109.2
0960	Flooring	87.3	131.6	99.6	87.3	131.6	99.6	87.3	124.4	97.6	87.3	131.6	99.6	87.3	129.1	98.9	87.3	128.0	98.6
0970, 0990	Wall Finishes & Painting/Coating	92.3	128.8	113.6	92.3	128.8	113.6	92.3	128.8	113.6	92.3	128.8	113.6	92.3	128.8	113.6	92.3	128.8	113.6
09	FINISHES	91.4	121.4	107.7	90.6	121.4	107.3	91.4	120.2	107.1	91.5	121.4	107.7	91.3	121.0	107.4	91.1	120.8	107.3
COVERS	DIVS. 10 - 14, 25, 28, 41, 43, 44, 46	100.0	113.2	102.9	100.0	113.2	102.9	100.0	113.2	102.9	100.0	113.4	103.0	100.0	113.2	102.9	100.0	113.2	102.9
21, 22, 23	FIRE SUPPRESSION, PLUMBING & HVAC	100.1	118.8	108.1	97.0	118.7	106.3	100.1	118.8	108.1	100.1	118.8	108.1	100.1	118.6	108.0	100.1	118.6	108.0
26, 27, 3370	ELECTRICAL, COMMUNICATIONS & UTIL.	98.9	112.0	105.7	96.0	109.6	103.1	99.0	110.7	105.1	99.0	158.5	130.0	98.5	111.4	105.2	97.9	110.5	105.0
MF2016	WEIGHTED AVERAGE	98.9	118.9	107.6	94.7	118.5	105.1	98.0	118.6	107.0	98.1	125.5	110.1	98.1	118.7	107.1	97.9	118.5	106.9

D.C. / DELAWARE / FLORIDA

DIVISION		WASHINGTON 200 - 205 MAT.	INST.	TOTAL	DOVER 199 MAT.	INST.	TOTAL	NEWARK 197 MAT.	INST.	TOTAL	WILMINGTON 198 MAT.	INST.	TOTAL	DAYTONA BEACH 321 MAT.	INST.	TOTAL	FORT LAUDERDALE 333 MAT.	INST.	TOTAL
015433	CONTRACTOR EQUIPMENT		106.4	106.4		122.9	122.9		122.9	122.9		123.1	123.1		101.9	101.9		94.8	94.8
0241, 31 - 34	SITE & INFRASTRUCTURE, DEMOLITION	104.0	96.5	98.7	104.1	114.8	111.6	103.8	114.8	111.5	105.0	115.2	112.1	102.4	88.7	92.8	94.2	76.1	81.5
0310	Concrete Forming & Accessories	99.2	72.3	76.0	95.9	102.3	101.4	98.3	102.3	101.8	96.7	102.3	101.5	97.3	63.6	68.3	95.8	56.5	61.9
0320	Concrete Reinforcing	100.6	87.1	93.8	105.3	114.3	109.9	100.8	114.3	107.6	105.3	114.3	109.9	92.6	65.9	79.2	92.3	57.5	74.7
0330	Cast-in-Place Concrete	109.9	81.0	98.9	101.0	107.7	103.5	88.3	107.7	95.7	98.2	107.7	101.8	88.8	65.8	80.0	88.6	62.7	78.7
03	CONCRETE	107.2	79.0	94.3	96.2	107.4	101.3	91.2	107.4	98.6	95.0	107.4	100.7	86.4	66.5	77.3	87.4	60.7	75.2
04	MASONRY	101.8	82.1	89.6	101.0	100.5	100.7	103.1	100.5	101.5	101.3	100.5	100.8	89.7	59.6	71.0	94.7	63.5	75.4
05	METALS	101.9	96.6	100.3	102.0	124.3	108.9	103.5	124.3	109.9	102.0	124.3	108.9	96.0	90.9	94.5	94.8	87.8	92.7
06	WOOD, PLASTICS & COMPOSITES	99.7	70.2	83.3	91.3	100.3	96.3	96.8	100.3	98.8	87.3	100.3	94.5	100.5	62.9	79.6	90.3	56.4	71.4
07	THERMAL & MOISTURE PROTECTION	99.8	85.6	93.8	100.7	113.2	106.0	105.0	113.2	108.5	100.1	113.2	105.6	97.2	65.8	83.9	98.4	62.2	83.1
08	OPENINGS	99.8	74.6	94.0	90.6	110.5	95.2	91.5	110.5	95.9	88.7	110.5	93.7	95.3	62.4	87.7	97.6	56.3	87.0
0920	Plaster & Gypsum Board	106.8	69.3	81.6	101.9	100.3	100.8	98.9	100.3	99.8	101.3	100.3	100.6	83.2	62.4	69.2	87.2	55.7	65.9
0950, 0980	Ceilings & Acoustic Treatment	99.2	69.3	79.0	94.2	100.3	98.3	96.1	100.3	98.9	91.6	100.3	97.5	83.2	62.4	69.2	87.2	55.7	65.9
0960	Flooring	97.4	79.0	92.3	87.5	111.5	94.1	86.5	111.5	93.4	88.0	111.5	94.5	96.8	58.0	86.0	94.1	78.5	89.8
0970, 0990	Wall Finishes & Painting/Coating	103.1	74.9	86.7	88.0	110.6	101.2	86.7	110.6	100.6	83.9	110.6	99.5	94.7	62.4	75.8	87.8	63.1	73.3
09	FINISHES	96.3	72.8	83.5	92.2	104.0	98.6	88.9	104.0	97.1	91.7	104.0	98.4	88.2	61.9	73.9	88.7	60.9	73.6
COVERS	DIVS. 10 - 14, 25, 28, 41, 43, 44, 46	100.0	95.9	99.1	100.0	106.2	101.4	100.0	106.2	101.4	100.0	106.2	101.4	100.0	82.6	96.1	100.0	81.6	95.9
21, 22, 23	FIRE SUPPRESSION, PLUMBING & HVAC	100.0	89.9	95.7	100.0	122.0	109.4	100.2	122.0	109.5	100.0	122.0	109.4	99.9	76.4	89.8	99.9	58.9	82.4
26, 27, 3370	ELECTRICAL, COMMUNICATIONS & UTIL.	100.6	102.8	101.7	97.3	112.7	105.3	98.9	112.7	106.1	97.3	112.7	105.3	92.7	62.2	76.8	94.6	69.4	81.4
MF2016	WEIGHTED AVERAGE	101.1	87.6	95.2	98.0	112.4	104.3	97.9	112.4	104.3	97.6	112.5	104.1	94.9	71.0	84.4	95.1	66.0	82.5

City Cost Indexes

FLORIDA

DIVISION		FORT MYERS 339, 341			GAINESVILLE 326, 344			JACKSONVILLE 320, 322			LAKELAND 338			MELBOURNE 329			MIAMI 330 - 332, 340		
		MAT.	INST.	TOTAL	MAT.	INST.	TOTAL	MAT.	INST.	TOTAL	MAT.	INST.	TOTAL	MAT.	INST.	TOTAL	MAT.	INST.	TOTAL
015433	CONTRACTOR EQUIPMENT		101.9	101.9		101.9	101.9		101.9	101.9		101.9	101.9		101.9	101.9		94.8	94.8
0241, 31 - 34	SITE & INFRASTRUCTURE, DEMOLITION	105.9	88.2	93.5	110.0	88.5	95.0	102.4	88.5	92.7	107.8	88.7	94.4	109.1	88.9	94.9	95.5	76.1	81.9
0310	Concrete Forming & Accessories	91.7	56.5	61.3	92.5	56.6	61.5	97.1	56.6	62.1	88.0	62.5	66.0	93.7	63.8	67.9	100.0	56.2	62.3
0320	Concrete Reinforcing	93.3	56.0	74.5	98.2	58.4	78.1	92.6	58.2	75.3	95.6	76.5	86.0	93.6	65.9	79.7	99.1	56.9	77.8
0330	Cast-in-Place Concrete	92.4	62.6	81.1	101.9	64.0	87.5	89.6	63.9	79.9	94.5	65.3	83.4	107.0	65.9	91.4	85.6	62.5	76.8
03	CONCRETE	88.0	60.4	75.4	97.0	61.4	80.7	86.8	61.3	75.2	89.6	67.6	79.6	97.5	66.6	83.4	85.8	60.5	74.2
04	MASONRY	88.4	63.5	73.0	103.0	56.6	74.2	89.4	54.1	67.6	104.5	58.8	76.2	87.5	59.6	70.2	95.0	54.4	69.8
05	METALS	96.8	87.2	93.8	94.9	87.9	92.8	94.6	87.7	92.5	96.7	94.3	95.9	104.1	91.0	100.1	95.2	86.8	92.6
06	WOOD, PLASTICS & COMPOSITES	87.2	56.4	70.0	94.4	55.5	72.8	100.5	55.5	75.5	82.4	62.1	71.1	96.1	62.9	77.6	96.2	56.4	74.0
07	THERMAL & MOISTURE PROTECTION	98.2	62.5	83.1	97.6	60.9	82.1	97.5	60.2	81.7	98.1	62.6	83.0	97.7	64.6	83.7	98.2	59.7	81.9
08	OPENINGS	97.4	56.0	87.9	94.9	56.2	86.0	95.3	56.2	86.3	97.3	64.0	89.7	94.5	56.4	87.6	98.8	56.3	89.0
0920	Plaster & Gypsum Board	93.8	55.7	68.2	85.5	54.8	64.9	88.7	54.8	65.9	90.5	61.6	71.1	85.5	62.4	70.0	87.7	55.7	66.2
0950, 0980	Ceilings & Acoustic Treatment	82.7	55.7	64.5	77.6	54.8	62.2	83.2	54.8	64.0	82.7	61.6	68.5	82.3	62.4	68.9	86.8	55.7	65.8
0960	Flooring	91.1	76.7	87.1	94.2	58.0	84.2	96.8	58.0	86.0	88.9	58.0	80.3	94.4	58.0	84.3	93.0	58.0	83.3
0970, 0990	Wall Finishes & Painting/Coating	91.7	63.0	75.0	94.7	62.4	75.8	94.7	63.0	76.2	91.7	63.0	75.0	94.7	83.5	88.2	85.8	63.1	72.5
09	FINISHES	88.4	60.5	73.2	86.6	56.7	70.4	88.2	56.8	71.1	87.4	61.2	73.2	87.4	64.2	74.8	87.5	56.7	70.7
COVERS	DIVS. 10 - 14, 25, 28, 41, 43, 44, 46	100.0	78.4	95.2	100.0	80.8	95.7	100.0	79.0	95.3	100.0	80.5	95.7	100.0	82.6	96.1	100.0	81.6	95.9
21, 22, 23	FIRE SUPPRESSION, PLUMBING & HVAC	98.2	57.0	80.6	98.7	61.9	82.9	99.9	61.9	83.6	98.2	59.5	81.7	99.9	76.4	89.8	99.9	57.3	81.7
26, 27, 3370	ELECTRICAL, COMMUNICATIONS & UTIL.	96.4	60.7	79.0	92.9	59.0	75.3	92.4	64.1	77.7	94.8	63.4	78.5	93.7	61.0	76.7	98.2	73.3	85.2
MF2016	WEIGHTED AVERAGE	95.4	65.4	82.3	96.4	65.0	82.7	94.7	65.3	81.8	96.2	67.8	83.7	97.6	71.2	86.0	95.6	64.6	82.0

FLORIDA

DIVISION		ORLANDO 327 - 328, 347			PANAMA CITY 324			PENSACOLA 325			SARASOTA 342			ST. PETERSBURG 337			TALLAHASSEE 323		
		MAT.	INST.	TOTAL	MAT.	INST.	TOTAL	MAT.	INST.	TOTAL	MAT.	INST.	TOTAL	MAT.	INST.	TOTAL	MAT.	INST.	TOTAL
015433	CONTRACTOR EQUIPMENT		101.9	101.9		101.9	101.9		101.9	101.9		101.9	101.9		101.9	101.9		101.9	101.9
0241, 31 - 34	SITE & INFRASTRUCTURE, DEMOLITION	99.7	88.7	92.0	113.1	88.0	95.5	113.8	88.5	96.0	113.3	88.7	96.1	109.4	88.3	94.6	101.7	88.5	92.5
0310	Concrete Forming & Accessories	100.1	63.3	68.4	96.4	61.4	66.2	94.5	64.7	68.8	95.4	62.5	67.0	95.2	60.3	65.2	97.1	56.6	62.2
0320	Concrete Reinforcing	96.2	65.9	80.9	96.7	69.1	82.8	99.2	68.6	83.7	93.3	76.5	84.8	95.6	76.5	86.0	99.8	58.4	78.9
0330	Cast-in-Place Concrete	105.9	65.7	90.7	94.2	59.9	81.2	116.4	65.1	96.9	103.2	65.3	88.8	95.5	62.1	82.8	91.4	64.0	81.0
03	CONCRETE	94.1	66.3	81.4	95.1	64.0	80.9	104.9	67.2	87.7	95.2	67.6	82.6	91.1	65.5	79.4	88.3	61.4	76.0
04	MASONRY	95.3	59.6	73.2	94.1	58.6	72.1	114.3	58.7	79.8	92.0	58.8	71.4	144.1	53.2	87.7	91.7	56.6	69.9
05	METALS	95.2	90.6	93.8	95.8	92.1	94.6	96.8	91.4	95.1	98.7	94.3	97.4	97.6	93.2	96.3	92.3	87.9	91.0
06	WOOD, PLASTICS & COMPOSITES	89.8	62.9	74.9	99.3	65.4	80.4	97.3	65.4	79.5	98.1	62.1	78.1	91.7	62.1	75.2	96.0	55.5	73.5
07	THERMAL & MOISTURE PROTECTION	100.0	65.8	85.5	97.8	59.9	81.8	97.7	63.0	83.0	98.2	62.6	83.1	98.3	60.0	82.1	96.9	60.8	81.7
08	OPENINGS	98.1	62.0	89.8	93.3	64.0	86.6	93.3	64.0	86.6	99.5	64.0	91.3	96.1	63.6	88.6	100.1	56.2	90.0
0920	Plaster & Gypsum Board	94.7	62.4	73.0	87.8	64.9	72.4	95.5	64.9	75.0	94.7	61.6	72.5	96.1	61.6	72.9	93.3	54.8	67.4
0950, 0980	Ceilings & Acoustic Treatment	89.4	62.4	71.2	82.3	64.9	70.6	82.3	64.9	70.6	87.3	61.6	69.9	84.6	61.6	69.1	87.9	54.8	65.6
0960	Flooring	91.3	58.0	82.1	96.3	76.7	90.9	92.2	58.0	82.7	101.4	58.0	89.4	93.0	58.0	83.3	92.9	58.0	83.2
0970, 0990	Wall Finishes & Painting/Coating	93.4	63.0	75.7	94.7	62.4	75.8	94.7	62.4	75.8	96.3	63.0	76.9	91.7	63.6	75.3	91.9	63.0	75.1
09	FINISHES	90.4	61.9	74.9	88.8	64.3	75.5	88.4	63.1	74.6	94.3	61.2	76.3	89.9	59.9	73.6	89.8	56.8	71.8
COVERS	DIVS. 10 - 14, 25, 28, 41, 43, 44, 46	100.0	82.6	96.1	100.0	78.0	95.1	100.0	81.1	95.8	100.0	80.5	95.7	100.0	78.6	95.2	100.0	78.6	95.2
21, 22, 23	FIRE SUPPRESSION, PLUMBING & HVAC	100.0	57.8	81.9	99.9	51.9	79.4	99.9	63.0	84.1	99.9	59.5	82.6	99.9	56.6	81.4	99.9	67.8	86.2
26, 27, 3370	ELECTRICAL, COMMUNICATIONS & UTIL.	96.6	62.7	78.9	91.6	59.0	74.6	94.8	53.5	73.3	95.7	63.9	79.2	94.8	64.0	78.8	96.7	59.0	77.1
MF2016	WEIGHTED AVERAGE	96.8	67.0	83.8	96.2	65.1	82.5	98.8	67.1	84.9	98.1	67.9	84.9	99.0	65.9	84.5	95.6	66.2	82.7

FLORIDA / GEORGIA

DIVISION		TAMPA (FLORIDA) 335 - 336, 346			WEST PALM BEACH (FLORIDA) 334, 349			ALBANY (GEORGIA) 317, 398			ATHENS (GEORGIA) 306			ATLANTA (GEORGIA) 300 - 303, 399			AUGUSTA (GEORGIA) 308 - 309		
		MAT.	INST.	TOTAL	MAT.	INST.	TOTAL	MAT.	INST.	TOTAL	MAT.	INST.	TOTAL	MAT.	INST.	TOTAL	MAT.	INST.	TOTAL
015433	CONTRACTOR EQUIPMENT		101.9	101.9		94.8	94.8		95.7	95.7		94.5	94.5		96.4	96.4		94.5	94.5
0241, 31 - 34	SITE & INFRASTRUCTURE, DEMOLITION	109.9	88.7	95.1	91.1	76.1	80.6	104.3	79.0	86.6	102.6	95.1	97.3	99.4	95.3	96.5	95.6	95.4	95.5
0310	Concrete Forming & Accessories	98.3	62.4	67.4	99.4	56.3	62.2	94.0	68.9	72.4	91.4	45.1	51.5	96.0	71.7	75.1	92.5	72.9	75.6
0320	Concrete Reinforcing	92.3	76.5	84.3	94.9	56.9	75.7	88.1	71.9	79.9	102.0	65.6	83.6	101.2	72.0	86.5	102.4	68.1	85.1
0330	Cast-in-Place Concrete	93.4	65.3	82.7	84.2	62.5	76.0	88.3	69.0	81.0	109.3	69.9	94.3	112.7	71.4	97.0	103.3	70.2	90.7
03	CONCRETE	89.8	67.6	79.7	84.4	60.5	73.5	86.4	71.2	79.5	104.9	58.6	83.7	107.1	72.4	91.2	97.3	71.8	85.6
04	MASONRY	95.7	58.8	72.8	94.3	54.4	69.5	92.2	69.2	77.9	78.5	79.4	79.1	91.9	69.4	77.9	92.2	69.3	78.0
05	METALS	96.7	94.3	96.0	93.9	86.8	91.7	97.8	98.0	97.9	93.1	80.0	89.1	93.9	85.1	91.2	92.8	81.3	89.3
06	WOOD, PLASTICS & COMPOSITES	95.7	62.1	77.0	95.3	56.4	73.6	85.9	69.6	76.8	97.0	37.4	63.9	99.2	73.1	84.7	98.5	75.3	85.6
07	THERMAL & MOISTURE PROTECTION	98.6	62.6	83.3	98.1	61.3	82.5	97.3	68.9	85.3	97.9	70.2	86.2	99.4	72.4	88.0	97.7	71.7	86.7
08	OPENINGS	97.4	64.0	89.7	95.7	56.3	86.7	87.7	71.6	84.0	93.1	52.0	83.6	97.6	73.5	92.1	93.1	73.8	88.7
0920	Plaster & Gypsum Board	98.4	61.6	73.7	102.1	55.7	70.9	103.6	69.3	80.5	93.6	36.1	55.0	96.0	72.5	80.2	94.9	75.0	81.6
0950, 0980	Ceilings & Acoustic Treatment	87.2	61.6	69.9	82.7	55.7	64.5	83.8	69.3	74.0	98.8	36.1	56.6	91.6	72.5	78.7	99.8	75.0	83.1
0960	Flooring	94.1	58.0	84.1	95.9	58.0	85.4	96.9	67.4	88.7	95.8	86.6	93.3	98.8	67.4	90.1	96.0	67.4	88.1
0970, 0990	Wall Finishes & Painting/Coating	91.7	63.0	75.0	87.8	62.4	72.9	89.1	95.9	93.1	97.5	95.9	96.6	101.3	95.9	98.2	97.5	80.4	87.5
09	FINISHES	91.1	61.2	74.9	88.7	56.7	71.3	92.2	71.1	80.7	96.2	56.0	74.4	96.3	73.2	83.7	96.0	72.8	83.4
COVERS	DIVS. 10 - 14, 25, 28, 41, 43, 44, 46	100.0	80.5	95.7	100.0	81.6	95.9	100.0	84.9	96.6	100.0	81.2	95.8	100.0	86.0	96.9	100.0	85.3	96.7
21, 22, 23	FIRE SUPPRESSION, PLUMBING & HVAC	99.9	59.5	82.6	98.2	56.8	80.5	99.9	71.1	87.6	97.0	68.5	84.8	100.0	70.1	87.2	100.1	69.2	86.9
26, 27, 3370	ELECTRICAL, COMMUNICATIONS & UTIL.	94.5	62.2	77.7	95.6	69.4	81.9	95.0	63.6	78.7	99.1	67.5	82.7	98.5	71.4	84.4	99.7	70.5	84.5
MF2016	WEIGHTED AVERAGE	96.6	67.6	83.9	94.2	64.0	81.0	95.0	73.4	85.5	96.6	69.1	84.5	98.7	75.0	88.3	96.8	74.2	86.9

For customer support on your Electrical Costs with RSMeans data, call 800.448.8182.

521

GEORGIA

DIVISION		COLUMBUS 318-319 MAT.	INST.	TOTAL	DALTON 307 MAT.	INST.	TOTAL	GAINESVILLE 305 MAT.	INST.	TOTAL	MACON 310-312 MAT.	INST.	TOTAL	SAVANNAH 313-314 MAT.	INST.	TOTAL	STATESBORO 304 MAT.	INST.	TOTAL
015433	CONTRACTOR EQUIPMENT		95.7	95.7		110.4	110.4		94.5	94.5		105.7	105.7		96.6	96.6		97.6	97.6
0241, 31 - 34	SITE & INFRASTRUCTURE, DEMOLITION	104.2	79.1	86.6	102.0	100.6	101.0	102.5	95.0	97.2	106.0	94.5	97.9	104.0	80.6	87.6	103.2	81.3	87.8
0310	Concrete Forming & Accessories	93.9	69.0	72.4	84.4	66.7	69.1	94.9	42.5	49.7	93.6	69.1	72.5	96.1	71.8	75.2	79.2	53.1	56.7
0320	Concrete Reinforcing	87.9	71.9	79.9	101.4	62.7	81.9	101.8	65.5	83.4	89.1	72.0	80.4	94.7	68.2	81.3	101.0	36.7	68.5
0330	Cast-in-Place Concrete	88.0	69.1	80.8	106.2	68.5	91.9	114.9	69.7	97.7	86.8	70.4	80.6	91.7	68.9	83.1	109.1	64.8	93.7
03	CONCRETE	86.2	71.3	79.4	104.1	68.5	87.8	106.7	57.3	84.1	85.8	71.8	79.4	87.7	71.9	80.4	104.5	57.8	83.2
04	MASONRY	92.2	69.2	77.9	79.6	78.7	79.1	87.0	79.4	82.3	104.5	69.3	82.7	87.2	69.2	76.0	81.6	79.4	80.2
05	METALS	97.4	98.3	97.7	94.1	94.7	94.3	92.4	79.4	88.4	93.1	98.4	94.7	94.2	96.4	94.9	97.6	84.9	93.7
06	WOOD, PLASTICS & COMPOSITES	85.9	69.6	76.8	79.7	68.3	73.3	100.8	34.7	64.0	92.6	69.6	79.8	90.3	73.6	81.0	73.6	49.3	60.1
07	THERMAL & MOISTURE PROTECTION	97.2	70.5	85.9	100.0	72.4	88.3	97.9	69.9	86.0	95.7	72.7	85.9	95.3	70.1	84.6	98.6	69.5	86.3
08	OPENINGS	87.7	71.6	84.0	94.1	68.3	88.2	93.1	50.5	83.3	87.5	71.6	83.8	94.4	72.9	89.4	95.0	50.5	84.8
0920	Plaster & Gypsum Board	103.6	69.3	80.5	82.4	67.8	72.6	95.9	33.3	53.9	109.5	69.3	82.5	98.9	73.4	81.8	82.3	48.4	59.6
0950, 0980	Ceilings & Acoustic Treatment	83.8	69.3	74.0	113.1	67.8	82.6	98.8	33.3	54.7	79.0	69.3	72.4	90.3	73.4	78.9	109.4	48.4	68.3
0960	Flooring	96.9	67.4	88.7	96.6	86.6	93.8	97.4	86.6	94.4	76.4	67.4	73.9	92.5	67.4	85.6	113.6	86.6	106.1
0970, 0990	Wall Finishes & Painting/Coating	89.1	95.9	93.1	87.6	73.2	79.2	97.5	95.9	96.6	91.1	95.9	93.9	88.3	87.9	88.1	95.3	73.2	82.4
09	FINISHES	92.1	71.1	80.7	105.3	71.6	87.0	96.9	54.4	73.8	81.9	71.1	76.0	92.1	72.6	81.5	108.8	60.6	82.6
COVERS	DIVS. 10 - 14, 25, 28, 41, 43, 44, 46	100.0	85.0	96.7	100.0	28.8	84.2	100.0	36.8	85.9	100.0	85.0	96.7	100.0	85.0	96.7	100.0	42.9	87.3
21, 22, 23	FIRE SUPPRESSION, PLUMBING & HVAC	100.0	67.4	86.1	97.1	63.2	82.6	97.0	68.3	84.7	100.0	70.5	87.4	100.1	68.3	86.5	97.6	69.1	85.4
26, 27, 3370	ELECTRICAL, COMMUNICATIONS & UTIL.	95.2	71.5	82.9	107.6	65.7	85.8	99.1	71.4	84.7	93.9	66.2	79.5	98.7	72.4	85.0	99.3	55.7	76.6
MF2016	WEIGHTED AVERAGE	94.9	73.8	85.7	98.3	72.2	86.9	97.2	67.6	84.2	93.8	75.1	85.6	95.4	74.4	86.2	98.7	66.3	84.5

DIVISION		GEORGIA VALDOSTA 316 MAT.	INST.	TOTAL	WAYCROSS 315 MAT.	INST.	TOTAL	HAWAII HILO 967 MAT.	INST.	TOTAL	HONOLULU 968 MAT.	INST.	TOTAL	STATES & POSS., GUAM 969 MAT.	INST.	TOTAL	IDAHO BOISE 836 - 837 MAT.	INST.	TOTAL
015433	CONTRACTOR EQUIPMENT		95.7	95.7		95.7	95.7		99.3	99.3		99.3	99.3		165.6	165.6		97.4	97.4
0241, 31 - 34	SITE & INFRASTRUCTURE, DEMOLITION	113.8	79.0	89.4	110.6	80.3	89.4	156.3	106.9	121.7	164.8	106.9	124.2	201.3	102.9	132.3	88.9	94.6	94.5
0310	Concrete Forming & Accessories	84.4	42.8	48.5	86.4	65.9	68.7	107.6	122.5	120.4	118.4	122.5	121.9	110.0	53.9	61.7	98.7	81.3	83.7
0320	Concrete Reinforcing	90.0	61.4	75.5	90.0	61.6	75.7	133.0	113.9	123.4	154.7	113.9	134.1	243.0	28.2	134.7	101.6	77.6	89.9
0330	Cast-in-Place Concrete	86.5	68.9	79.8	97.7	68.7	86.7	194.4	122.0	166.9	156.3	122.0	143.3	168.5	100.5	142.7	90.5	84.8	88.3
03	CONCRETE	91.1	57.5	75.7	94.1	67.7	82.0	151.7	119.8	137.1	142.6	119.8	132.2	153.7	66.6	113.9	96.1	82.0	89.7
04	MASONRY	97.6	79.4	86.3	98.5	79.4	86.7	153.6	121.4	133.6	140.0	121.4	128.5	217.7	36.2	105.2	125.6	89.0	102.9
05	METALS	97.0	94.1	96.1	96.0	89.7	94.1	112.9	102.3	109.6	125.9	102.3	118.7	145.3	77.1	124.4	108.1	82.1	100.1
06	WOOD, PLASTICS & COMPOSITES	74.1	34.5	52.0	75.7	66.4	70.5	114.9	122.4	119.1	129.4	122.4	125.5	126.7	56.3	87.5	92.4	79.9	85.5
07	THERMAL & MOISTURE PROTECTION	97.5	68.6	85.3	97.2	71.1	86.2	130.8	119.0	125.8	148.7	119.0	136.1	153.9	60.9	114.5	94.3	86.3	90.9
08	OPENINGS	84.3	49.2	76.2	84.6	64.2	79.9	117.0	118.9	117.5	131.5	118.9	128.6	121.6	45.9	104.3	97.1	74.3	91.9
0920	Plaster & Gypsum Board	95.7	33.2	53.7	95.7	66.0	75.8	108.1	122.9	118.0	151.3	122.9	132.2	216.4	44.6	101.0	92.4	79.4	83.7
0950, 0980	Ceilings & Acoustic Treatment	81.3	33.2	48.8	79.4	66.0	70.3	131.5	122.9	125.7	140.9	122.9	128.8	246.8	44.6	110.5	102.4	79.4	86.9
0960	Flooring	90.9	88.2	90.2	92.0	86.6	90.5	107.3	136.7	115.5	125.7	136.7	128.8	126.6	41.8	103.1	93.8	81.4	90.4
0970, 0990	Wall Finishes & Painting/Coating	89.1	95.9	93.1	89.1	73.2	79.8	95.6	137.2	119.9	103.4	137.2	123.1	100.4	33.0	61.1	93.6	41.3	63.1
09	FINISHES	89.8	54.6	70.6	89.3	70.6	79.2	108.7	127.1	118.7	124.8	127.1	126.1	178.9	50.4	109.0	92.7	77.2	84.3
COVERS	DIVS. 10 - 14, 25, 28, 41, 43, 44, 46	100.0	80.6	95.7	100.0	50.7	89.0	100.0	112.1	102.7	100.0	112.1	102.7	100.0	68.5	93.0	100.0	90.4	97.9
21, 22, 23	FIRE SUPPRESSION, PLUMBING & HVAC	100.0	70.6	87.4	98.1	65.4	84.1	100.5	109.4	104.3	100.5	109.4	104.3	103.5	34.9	74.2	100.1	74.9	89.3
26, 27, 3370	ELECTRICAL, COMMUNICATIONS & UTIL.	93.5	54.6	73.2	97.3	55.7	75.6	105.3	121.7	113.9	106.9	121.7	114.6	151.8	37.9	92.6	97.9	67.8	82.2
MF2016	WEIGHTED AVERAGE	95.1	67.3	82.9	95.3	69.5	84.0	116.9	116.0	116.5	121.1	116.0	118.9	139.1	53.4	101.6	100.5	79.8	91.4

DIVISION		IDAHO COEUR D'ALENE 838 MAT.	INST.	TOTAL	IDAHO FALLS 834 MAT.	INST.	TOTAL	LEWISTON 835 MAT.	INST.	TOTAL	POCATELLO 832 MAT.	INST.	TOTAL	TWIN FALLS 833 MAT.	INST.	TOTAL	ILLINOIS BLOOMINGTON 617 MAT.	INST.	TOTAL
015433	CONTRACTOR EQUIPMENT		92.2	92.2		97.4	97.4		92.2	92.2		97.4	97.4		97.4	97.4		105.2	105.2
0241, 31 - 34	SITE & INFRASTRUCTURE, DEMOLITION	87.9	90.9	90.0	86.5	96.8	93.7	94.7	91.8	92.7	89.9	96.8	94.7	97.0	96.6	96.7	95.7	100.5	99.1
0310	Concrete Forming & Accessories	108.9	79.6	83.6	92.8	80.1	81.8	113.9	81.6	86.0	98.8	80.9	83.4	100.0	79.9	82.7	83.4	119.9	114.9
0320	Concrete Reinforcing	110.0	99.2	104.6	103.8	77.6	90.6	110.0	99.3	104.6	102.1	77.6	89.7	104.1	77.6	90.7	96.1	105.6	100.9
0330	Cast-in-Place Concrete	97.8	82.6	92.0	86.2	83.2	85.0	101.7	86.0	95.7	93.0	84.7	89.9	95.5	83.1	90.8	97.7	117.5	105.2
03	CONCRETE	102.6	84.2	94.2	88.3	80.9	84.9	106.0	86.2	97.0	95.3	81.8	89.1	102.9	80.8	92.9	93.2	117.2	104.2
04	MASONRY	127.3	85.4	101.3	120.6	86.3	99.4	127.7	87.7	102.9	123.1	89.0	102.0	125.9	86.4	101.4	115.2	122.8	119.9
05	METALS	101.6	89.5	97.9	116.4	81.9	105.8	101.0	89.8	97.6	116.5	81.8	105.8	116.5	81.7	105.8	99.9	125.4	107.7
06	WOOD, PLASTICS & COMPOSITES	97.5	78.4	86.8	86.7	79.9	82.9	102.8	79.4	89.8	92.4	79.9	85.5	93.5	79.9	85.9	86.7	118.5	104.4
07	THERMAL & MOISTURE PROTECTION	147.3	83.8	120.4	93.5	76.4	86.3	147.6	85.5	121.4	94.0	77.2	86.9	94.9	83.3	90.0	96.9	115.8	104.9
08	OPENINGS	114.3	76.9	105.7	100.1	74.3	94.1	108.0	80.0	101.8	97.8	69.2	91.2	100.9	64.9	92.6	94.5	124.8	101.5
0920	Plaster & Gypsum Board	166.3	77.9	107.0	77.5	79.4	78.8	168.0	79.1	108.2	79.4	79.4	79.4	81.0	79.4	79.9	92.8	119.1	110.4
0950, 0980	Ceilings & Acoustic Treatment	136.6	77.9	97.1	103.2	79.4	87.1	136.6	79.0	97.8	110.0	79.4	89.4	105.7	79.4	88.0	89.2	119.1	109.3
0960	Flooring	132.6	81.4	118.4	93.4	81.4	90.1	136.1	81.4	120.9	97.1	81.4	92.7	98.3	81.4	93.7	84.6	126.3	96.2
0970, 0990	Wall Finishes & Painting/Coating	112.8	75.9	91.3	93.6	41.3	63.1	112.8	75.9	91.3	93.5	41.3	63.0	93.6	41.3	63.1	85.6	140.9	117.8
09	FINISHES	155.8	79.2	114.1	90.3	76.5	82.8	157.2	80.4	115.4	93.4	77.2	84.6	93.9	76.5	84.5	86.9	123.7	106.9
COVERS	DIVS. 10 - 14, 25, 28, 41, 43, 44, 46	100.0	95.0	98.9	100.0	89.5	97.7	100.0	95.9	99.1	100.0	90.4	97.9	100.0	89.5	97.7	100.0	107.6	101.7
21, 22, 23	FIRE SUPPRESSION, PLUMBING & HVAC	99.6	83.2	92.6	100.9	72.3	88.7	100.9	88.1	95.4	99.9	74.8	89.2	99.9	70.1	87.2	96.9	107.5	101.4
26, 27, 3370	ELECTRICAL, COMMUNICATIONS & UTIL.	90.2	86.1	88.1	89.9	68.2	78.6	88.4	82.6	85.4	95.4	69.7	82.0	91.3	68.2	79.3	95.6	95.4	95.5
MF2016	WEIGHTED AVERAGE	108.1	84.7	97.8	100.0	78.5	90.6	108.2	86.3	98.6	101.5	79.5	91.9	102.8	77.8	91.8	96.8	113.0	103.9

For customer support on your Electrical Costs with RSMeans data, call 800.448.8182.

City Cost Indexes

ILLINOIS

DIVISION		CARBONDALE 629 MAT.	INST.	TOTAL	CENTRALIA 628 MAT.	INST.	TOTAL	CHAMPAIGN 618 - 619 MAT.	INST.	TOTAL	CHICAGO 606 - 608 MAT.	INST.	TOTAL	DECATUR 625 MAT.	INST.	TOTAL	EAST ST. LOUIS 620 - 622 MAT.	INST.	TOTAL
015433	CONTRACTOR EQUIPMENT		114.0	114.0		114.0	114.0		106.1	106.1		100.2	100.2		106.1	106.1		114.0	114.0
0241, 31 - 34	SITE & INFRASTRUCTURE, DEMOLITION	100.6	101.6	101.3	100.9	102.3	101.9	104.7	101.6	102.5	101.9	104.4	103.6	97.6	101.8	100.5	102.9	102.4	102.5
0310	Concrete Forming & Accessories	90.5	106.1	104.0	92.1	110.8	108.3	89.6	117.8	113.9	99.8	160.4	152.0	92.6	118.2	114.7	88.0	111.9	108.6
0320	Concrete Reinforcing	98.3	102.6	100.5	98.3	102.8	100.6	96.1	100.9	98.6	104.7	150.8	128.0	96.3	101.2	98.8	98.2	103.1	100.7
0330	Cast-in-Place Concrete	90.8	103.6	95.7	91.3	117.2	101.1	113.3	112.9	113.2	123.6	153.6	135.0	99.2	116.5	105.8	92.8	118.0	102.4
03	CONCRETE	83.1	106.3	93.7	83.5	113.2	97.1	105.5	113.8	109.3	109.9	155.7	130.8	93.5	115.2	103.4	84.6	114.0	98.0
04	MASONRY	78.1	109.9	97.8	78.1	112.4	99.4	139.8	121.5	128.5	102.6	164.9	141.2	73.8	122.8	104.2	78.3	117.2	102.4
05	METALS	97.8	130.5	107.9	97.9	132.9	108.8	99.9	121.7	106.6	97.9	146.9	113.0	101.6	121.2	107.6	99.0	133.3	109.5
06	WOOD, PLASTICS & COMPOSITES	90.9	102.7	97.5	93.3	108.3	101.6	93.2	117.1	106.5	104.3	160.0	135.3	91.2	117.1	105.6	88.4	108.9	99.8
07	THERMAL & MOISTURE PROTECTION	93.9	101.4	97.1	93.9	108.0	99.9	97.7	114.6	104.9	96.7	150.2	119.3	99.6	113.2	105.4	94.0	107.6	99.7
08	OPENINGS	89.4	115.7	95.4	89.4	118.8	96.1	95.1	119.4	100.7	101.9	170.3	117.6	100.5	119.5	104.9	89.4	119.1	96.2
0920	Plaster & Gypsum Board	97.3	102.9	101.1	98.3	108.6	105.2	94.7	117.6	110.1	103.3	161.7	142.5	99.7	117.6	111.7	96.3	109.2	105.0
0950, 0980	Ceilings & Acoustic Treatment	87.5	102.9	97.9	87.5	108.6	101.7	89.2	117.6	108.3	102.2	161.7	142.3	93.4	117.6	109.7	87.5	109.2	102.1
0960	Flooring	107.2	116.4	109.8	108.0	112.1	109.2	87.8	117.3	96.0	94.4	163.4	113.5	96.1	117.8	102.2	106.2	112.1	107.9
0970, 0990	Wall Finishes & Painting/Coating	94.5	104.8	100.5	94.5	115.2	106.6	85.6	117.6	104.4	101.4	162.5	137.0	86.5	111.6	101.1	94.5	115.2	106.6
09	FINISHES	91.6	106.6	99.7	92.0	111.1	102.4	88.8	118.3	104.8	99.3	162.1	133.4	91.4	118.1	105.9	91.2	112.4	102.7
COVERS	DIVS. 10 - 14, 25, 28, 41, 43, 44, 46	100.0	103.4	100.8	100.0	104.2	100.9	100.0	106.6	101.5	100.0	127.1	106.0	100.0	107.1	101.6	100.0	104.6	101.0
21, 22, 23	FIRE SUPPRESSION, PLUMBING & HVAC	96.9	105.8	100.7	96.9	94.1	95.7	96.9	105.1	100.4	100.0	138.4	116.4	100.1	98.2	99.3	100.0	97.7	99.0
26, 27, 3370	ELECTRICAL, COMMUNICATIONS & UTIL.	94.9	107.4	101.4	96.2	107.4	102.0	98.7	95.4	97.0	96.4	135.5	116.7	98.6	92.6	95.5	95.8	100.5	98.2
MF2016	WEIGHTED AVERAGE	93.2	108.8	100.0	93.4	108.8	100.1	100.3	110.6	104.8	100.8	145.8	120.5	97.3	109.0	102.4	94.4	109.4	100.9

ILLINOIS

DIVISION		EFFINGHAM 624 MAT.	INST.	TOTAL	GALESBURG 614 MAT.	INST.	TOTAL	JOLIET 604 MAT.	INST.	TOTAL	KANKAKEE 609 MAT.	INST.	TOTAL	LA SALLE 613 MAT.	INST.	TOTAL	NORTH SUBURBAN 600 - 603 MAT.	INST.	TOTAL
015433	CONTRACTOR EQUIPMENT		106.1	106.1		105.2	105.2		96.6	96.6		96.6	96.6		96.6	96.6		105.2	105.2
0241, 31 - 34	SITE & INFRASTRUCTURE, DEMOLITION	101.3	100.6	100.8	98.1	100.1	99.5	97.6	102.8	101.3	91.6	102.0	98.9	97.5	101.1	100.0	96.9	102.6	100.9
0310	Concrete Forming & Accessories	97.1	116.3	113.7	89.5	118.9	114.9	99.1	158.3	150.1	92.7	141.6	134.9	103.1	122.5	119.8	98.4	153.7	146.1
0320	Concrete Reinforcing	99.5	92.1	95.8	95.6	102.7	99.2	104.7	141.0	123.0	105.6	134.9	120.4	95.8	133.8	115.0	104.7	146.1	125.6
0330	Cast-in-Place Concrete	98.8	111.0	103.4	100.6	106.8	103.0	112.4	152.0	127.4	104.8	139.2	117.8	100.5	126.2	110.3	112.5	150.1	126.7
03	CONCRETE	94.3	110.6	101.6	96.0	112.6	103.6	102.8	152.2	125.4	96.7	139.1	116.1	96.8	126.5	110.4	102.8	150.4	124.6
04	MASONRY	82.0	115.2	102.6	115.4	120.6	118.6	101.8	163.4	140.0	98.1	147.8	128.9	115.4	145.7	134.2	98.6	159.1	136.2
05	METALS	98.7	114.1	103.4	99.9	122.9	107.0	95.9	138.4	108.9	95.9	133.9	107.5	100.0	143.1	113.2	97.0	140.1	110.2
06	WOOD, PLASTICS & COMPOSITES	93.5	117.1	106.6	93.0	118.6	107.3	102.9	157.3	133.2	96.1	139.8	120.4	107.7	119.5	114.3	101.3	152.6	129.8
07	THERMAL & MOISTURE PROTECTION	99.1	110.2	103.8	97.1	110.0	102.6	101.1	149.2	121.5	100.2	138.3	116.3	97.3	124.0	108.6	101.4	144.3	119.6
08	OPENINGS	95.3	116.0	100.0	94.5	117.3	99.7	99.5	165.7	114.7	92.7	153.2	106.6	94.5	133.5	103.5	99.6	163.8	114.3
0920	Plaster & Gypsum Board	99.6	117.6	111.7	94.7	119.3	111.2	96.1	159.0	138.3	94.1	141.0	125.6	101.7	120.1	114.1	99.1	154.1	136.0
0950, 0980	Ceilings & Acoustic Treatment	87.5	117.6	107.8	89.2	119.3	109.4	101.8	159.0	140.3	101.8	141.0	128.2	89.2	120.1	110.1	101.8	154.1	137.1
0960	Flooring	97.2	110.9	101.0	87.6	126.3	98.3	92.4	158.3	110.7	89.3	159.3	108.7	94.2	125.7	102.9	92.9	158.3	111.0
0970, 0990	Wall Finishes & Painting/Coating	86.5	109.3	99.8	85.6	99.9	93.9	93.5	170.4	138.4	93.5	140.9	121.2	85.6	139.1	116.8	95.5	159.5	132.9
09	FINISHES	90.6	115.7	104.3	88.2	119.2	105.1	94.7	160.6	130.6	93.2	145.2	121.4	91.1	123.5	108.7	95.4	156.0	128.3
COVERS	DIVS. 10 - 14, 25, 28, 41, 43, 44, 46	100.0	78.1	95.1	100.0	102.4	100.5	100.0	126.9	106.0	100.0	116.9	103.8	100.0	103.1	100.7	100.0	120.6	104.6
21, 22, 23	FIRE SUPPRESSION, PLUMBING & HVAC	97.0	104.2	100.1	96.9	105.2	100.5	100.0	135.4	115.2	96.9	129.7	111.0	96.9	123.8	108.4	99.9	133.2	114.2
26, 27, 3370	ELECTRICAL, COMMUNICATIONS & UTIL.	96.5	107.3	102.2	96.5	86.4	91.2	95.6	127.2	112.0	91.0	126.4	109.5	93.7	126.5	110.8	95.5	132.0	114.5
MF2016	WEIGHTED AVERAGE	95.8	108.7	101.5	97.4	108.9	102.5	98.8	142.0	117.7	95.6	133.5	112.2	97.6	126.5	110.3	98.9	140.6	117.1

ILLINOIS

DIVISION		PEORIA 615 - 616 MAT.	INST.	TOTAL	QUINCY 623 MAT.	INST.	TOTAL	ROCK ISLAND 612 MAT.	INST.	TOTAL	ROCKFORD 610 - 611 MAT.	INST.	TOTAL	SOUTH SUBURBAN 605 MAT.	INST.	TOTAL	SPRINGFIELD 626 - 627 MAT.	INST.	TOTAL
015433	CONTRACTOR EQUIPMENT		105.2	105.2		106.1	106.1		105.2	105.2		105.2	105.2		96.6	96.6		106.1	106.1
0241, 31 - 34	SITE & INFRASTRUCTURE, DEMOLITION	98.4	100.6	99.9	100.2	101.1	100.8	96.2	99.6	98.6	97.9	102.0	100.8	96.9	102.3	100.7	101.2	101.6	101.5
0310	Concrete Forming & Accessories	92.3	120.3	116.4	95.0	114.5	111.8	91.0	102.7	101.1	96.5	132.2	127.3	98.4	153.7	146.1	93.3	117.8	114.4
0320	Concrete Reinforcing	93.2	103.0	98.1	99.1	84.5	91.7	95.6	96.2	95.9	88.2	128.6	108.6	104.7	146.1	125.6	99.6	101.2	100.4
0330	Cast-in-Place Concrete	97.6	120.0	106.1	99.0	101.6	100.0	98.4	102.1	99.8	99.9	131.2	111.8	112.5	150.1	126.7	94.1	110.7	100.4
03	CONCRETE	93.3	117.8	104.5	93.9	105.5	99.2	94.0	102.5	97.8	94.0	131.7	111.2	102.8	150.4	124.6	91.7	113.1	101.4
04	MASONRY	114.5	123.6	120.2	105.4	111.0	108.9	115.2	101.9	106.9	89.1	139.1	120.1	98.6	159.1	136.2	84.6	121.6	107.5
05	METALS	102.6	124.2	109.3	98.7	112.4	102.9	99.9	118.7	105.7	102.6	140.4	114.2	97.0	140.1	110.2	99.1	121.3	105.9
06	WOOD, PLASTICS & COMPOSITES	100.5	118.6	110.6	91.1	117.1	105.5	94.6	102.2	98.8	100.5	129.6	116.7	101.3	152.6	129.8	92.2	117.1	106.0
07	THERMAL & MOISTURE PROTECTION	97.8	116.6	105.8	99.1	115.4	101.8	97.1	100.7	98.6	100.2	132.0	113.7	101.4	144.3	119.6	101.6	114.6	107.1
08	OPENINGS	100.6	124.1	106.0	96.0	113.6	100.0	94.5	106.7	97.3	100.6	140.6	109.7	99.6	163.8	114.3	98.8	118.8	103.4
0920	Plaster & Gypsum Board	98.5	119.3	112.4	98.3	117.6	111.3	94.7	102.4	99.9	98.5	130.5	120.0	99.1	154.1	136.0	100.8	117.6	112.1
0950, 0980	Ceilings & Acoustic Treatment	94.3	119.3	111.1	87.5	117.6	107.8	89.2	102.4	98.1	94.3	130.5	118.7	101.8	154.1	137.1	94.7	117.6	110.2
0960	Flooring	91.1	126.3	100.9	96.1	114.1	101.1	88.7	98.2	91.4	91.1	125.7	100.7	92.9	158.3	111.0	99.3	117.8	104.5
0970, 0990	Wall Finishes & Painting/Coating	85.6	140.9	117.8	86.5	115.2	103.2	85.6	99.9	93.9	85.6	149.2	122.7	95.5	159.5	132.9	84.7	115.2	102.5
09	FINISHES	90.8	123.9	108.8	90.1	115.7	104.0	88.4	101.4	95.5	90.8	132.6	113.6	95.4	156.0	128.3	96.9	118.1	108.4
COVERS	DIVS. 10 - 14, 25, 28, 41, 43, 44, 46	100.0	107.8	101.7	100.0	79.3	95.4	100.0	100.7	100.1	100.0	116.0	103.6	100.0	120.6	104.6	100.0	106.6	101.5
21, 22, 23	FIRE SUPPRESSION, PLUMBING & HVAC	100.0	102.7	101.1	97.0	101.5	98.9	96.9	100.1	98.3	100.1	118.6	108.0	99.9	133.2	114.2	100.0	103.0	101.3
26, 27, 3370	ELECTRICAL, COMMUNICATIONS & UTIL.	97.5	96.6	97.0	94.1	86.8	90.3	89.0	93.8	91.5	97.7	132.6	115.9	95.5	132.0	114.5	101.1	86.8	93.7
MF2016	WEIGHTED AVERAGE	99.3	112.2	104.9	96.7	103.9	99.8	96.4	101.9	98.8	98.2	128.3	111.4	98.9	140.6	117.1	97.8	108.8	102.6

For customer support on your Electrical Costs with RSMeans data, call 800.448.8182.

523

INDIANA

DIVISION		ANDERSON 460 MAT.	INST.	TOTAL	BLOOMINGTON 474 MAT.	INST.	TOTAL	COLUMBUS 472 MAT.	INST.	TOTAL	EVANSVILLE 476-477 MAT.	INST.	TOTAL	FORT WAYNE 467-468 MAT.	INST.	TOTAL	GARY 463-464 MAT.	INST.	TOTAL
015433	CONTRACTOR EQUIPMENT		96.7	96.7		83.4	83.4		83.4	83.4		113.9	113.9		96.7	96.7		96.7	96.7
0241, 31 - 34	SITE & INFRASTRUCTURE, DEMOLITION	97.9	93.4	94.7	87.8	92.1	90.8	84.4	91.9	89.7	93.4	121.8	113.3	98.9	93.2	94.9	98.5	97.1	97.5
0310	Concrete Forming & Accessories	94.6	81.2	83.0	99.0	81.6	84.0	93.2	79.1	81.0	92.6	82.3	83.7	92.7	74.8	77.3	94.7	114.8	112.1
0320	Concrete Reinforcing	90.1	85.7	87.8	87.1	85.3	86.2	87.5	85.4	86.4	95.4	78.6	86.9	90.1	77.1	83.5	90.1	113.1	101.7
0330	Cast-in-Place Concrete	100.6	78.7	92.3	99.1	77.8	91.0	98.7	75.7	89.9	94.7	87.6	92.0	106.9	75.0	94.8	105.1	116.3	109.4
03	CONCRETE	89.8	81.6	86.0	97.8	80.5	89.9	97.1	78.6	88.6	97.9	83.6	91.4	92.6	76.0	85.0	91.9	114.7	102.3
04	MASONRY	88.8	77.2	81.6	91.1	73.5	80.2	91.0	74.6	80.8	86.5	80.6	82.8	91.8	74.5	81.0	90.2	111.7	103.5
05	METALS	96.3	89.9	94.3	98.2	74.8	91.0	98.2	74.2	90.9	91.5	83.2	88.9	96.3	86.6	93.3	96.3	106.3	99.4
06	WOOD, PLASTICS & COMPOSITES	96.5	81.3	88.5	111.0	82.0	94.9	106.1	78.8	90.9	91.8	81.4	86.0	96.3	74.4	84.1	93.8	113.1	104.5
07	THERMAL & MOISTURE PROTECTION	110.1	77.9	96.5	96.6	78.9	89.1	96.0	78.5	88.6	100.9	84.8	94.1	109.8	74.0	94.7	108.5	108.2	108.4
08	OPENINGS	93.5	79.7	90.3	100.3	80.1	95.6	96.4	78.3	92.3	94.2	78.3	90.5	93.5	72.0	88.5	98.9	116.6	104.8
0920	Plaster & Gypsum Board	105.8	81.1	89.2	99.0	82.3	87.8	96.4	78.9	84.7	94.8	80.5	85.2	105.1	74.0	84.2	98.9	113.7	108.8
0950, 0980	Ceilings & Acoustic Treatment	89.7	81.1	83.9	79.8	82.3	81.5	79.8	78.9	79.2	83.6	80.5	81.5	89.7	74.0	79.1	89.7	113.7	105.9
0960	Flooring	94.0	79.3	89.9	98.5	83.7	94.4	93.3	83.7	90.6	93.0	75.5	88.1	94.0	73.2	88.2	94.0	116.8	100.3
0970, 0990	Wall Finishes & Painting/Coating	93.1	68.8	78.9	86.0	79.8	82.4	86.0	79.8	82.4	91.1	88.2	89.4	93.1	73.4	81.6	93.1	123.7	110.9
09	FINISHES	91.2	79.7	85.0	90.6	81.9	85.9	88.6	80.0	83.9	89.2	81.5	85.0	91.0	74.4	82.0	90.2	116.2	104.3
COVERS	DIVS. 10 - 14, 25, 28, 41, 43, 44, 46	100.0	91.0	98.0	100.0	87.5	97.2	100.0	87.2	97.1	100.0	95.1	98.9	100.0	90.5	97.9	100.0	107.2	101.6
21, 22, 23	FIRE SUPPRESSION, PLUMBING & HVAC	100.0	80.2	91.5	99.7	79.3	91.0	96.6	80.1	89.6	100.0	79.3	91.2	100.0	73.3	88.6	100.0	107.1	103.0
26, 27, 3370	ELECTRICAL, COMMUNICATIONS & UTIL.	87.9	87.8	87.9	99.9	88.1	93.8	99.2	87.7	93.2	96.0	86.0	90.8	88.6	76.5	82.3	99.4	114.1	107.0
MF2016	WEIGHTED AVERAGE	95.1	83.3	89.9	97.9	81.3	90.6	96.3	80.8	89.5	95.5	85.5	91.1	95.7	77.7	87.8	96.5	110.4	102.6

INDIANA

DIVISION		INDIANAPOLIS 461-462 MAT.	INST.	TOTAL	KOKOMO 469 MAT.	INST.	TOTAL	LAFAYETTE 479 MAT.	INST.	TOTAL	LAWRENCEBURG 470 MAT.	INST.	TOTAL	MUNCIE 473 MAT.	INST.	TOTAL	NEW ALBANY 471 MAT.	INST.	TOTAL
015433	CONTRACTOR EQUIPMENT		84.3	84.3		96.7	96.7		83.4	83.4		103.6	103.6		95.3	95.3		93.1	93.1
0241, 31 - 34	SITE & INFRASTRUCTURE, DEMOLITION	99.0	90.4	93.0	94.3	93.3	93.6	85.3	92.1	90.0	83.1	107.7	100.3	87.7	92.6	91.1	80.1	94.4	90.1
0310	Concrete Forming & Accessories	99.3	84.6	86.6	97.7	77.5	80.2	90.9	82.5	83.6	89.8	78.1	79.7	90.7	80.6	82.0	88.3	78.8	80.1
0320	Concrete Reinforcing	91.1	85.7	88.4	81.4	85.5	83.5	87.1	85.6	86.3	86.4	77.5	81.9	96.4	85.6	90.9	87.7	81.4	84.5
0330	Cast-in-Place Concrete	98.8	86.0	93.9	99.6	82.5	93.1	99.2	81.7	92.6	92.8	75.0	86.0	104.2	77.8	94.2	95.8	74.7	87.8
03	CONCRETE	95.7	84.7	90.7	86.9	81.2	84.3	97.3	82.3	90.4	90.5	77.5	84.6	96.2	81.1	89.3	95.8	78.1	87.7
04	MASONRY	90.5	79.2	83.5	88.5	75.2	80.3	96.5	77.1	84.5	75.7	73.3	74.2	92.9	77.3	83.2	82.3	69.5	74.4
05	METALS	97.1	75.3	90.4	92.7	89.6	91.8	96.6	75.1	90.0	93.3	85.3	90.8	100.0	89.9	96.9	95.2	81.8	91.1
06	WOOD, PLASTICS & COMPOSITES	98.8	85.1	91.2	99.5	76.2	86.5	103.3	83.0	92.0	90.1	77.9	83.3	105.1	80.8	91.6	92.0	80.1	85.4
07	THERMAL & MOISTURE PROTECTION	101.3	81.2	92.8	109.0	77.7	95.8	95.9	81.0	89.6	101.7	77.7	91.5	99.1	79.3	90.7	88.0	72.9	81.6
08	OPENINGS	99.7	81.8	95.6	89.0	76.9	86.2	94.9	80.6	91.6	96.1	74.7	91.2	93.7	79.4	90.4	93.7	78.8	90.3
0920	Plaster & Gypsum Board	94.7	85.0	88.2	110.4	75.8	87.2	94.0	83.3	86.8	73.6	77.9	76.5	94.8	81.1	85.6	92.5	80.0	84.1
0950, 0980	Ceilings & Acoustic Treatment	90.9	85.0	86.9	89.7	75.8	80.4	76.4	83.3	81.0	87.0	77.9	80.9	79.8	81.1	80.7	83.6	80.0	81.2
0960	Flooring	96.2	83.7	92.7	98.2	87.4	95.2	92.2	83.7	89.8	68.4	83.7	72.6	92.3	79.3	88.7	90.4	57.1	81.2
0970, 0990	Wall Finishes & Painting/Coating	98.6	79.8	87.6	93.1	71.0	80.2	86.0	85.1	85.4	86.8	74.5	79.6	86.0	68.8	76.0	91.1	68.2	77.8
09	FINISHES	94.6	84.1	88.9	93.0	78.4	85.0	87.3	83.1	85.0	79.2	79.0	79.1	87.8	79.1	83.1	88.3	73.7	80.4
COVERS	DIVS. 10 - 14, 25, 28, 41, 43, 44, 46	100.0	91.9	98.2	100.0	87.8	97.3	100.0	90.2	97.8	100.0	85.7	96.8	100.0	89.9	97.8	100.0	85.2	96.7
21, 22, 23	FIRE SUPPRESSION, PLUMBING & HVAC	99.9	80.8	91.7	96.9	80.2	89.8	96.6	79.4	89.3	97.6	75.9	88.3	91.8	78.2	84.7	94.5	77.1	85.5
26, 27, 3370	ELECTRICAL, COMMUNICATIONS & UTIL.	102.2	87.8	94.7	92.2	81.5	86.7	98.6	82.0	90.5	93.9	73.7	83.4	96.3	81.7	89.9	94.0	78.0	87.0
MF2016	WEIGHTED AVERAGE	98.3	83.2	91.7	93.5	81.7	88.3	96.0	81.6	89.7	92.8	79.6	87.0	96.3	81.7	89.9	94.0	78.0	87.0

INDIANA / IOWA

DIVISION		SOUTH BEND 465-466 MAT.	INST.	TOTAL	TERRE HAUTE 478 MAT.	INST.	TOTAL	WASHINGTON 475 MAT.	INST.	TOTAL	BURLINGTON 526 MAT.	INST.	TOTAL	CARROLL 514 MAT.	INST.	TOTAL	CEDAR RAPIDS 522-524 MAT.	INST.	TOTAL
015433	CONTRACTOR EQUIPMENT		108.4	108.4		113.9	113.9		113.9	113.9		101.5	101.5		101.5	101.5		98.2	98.2
0241, 31 - 34	SITE & INFRASTRUCTURE, DEMOLITION	97.4	93.9	94.9	95.2	122.2	114.1	95.0	122.4	114.2	99.5	96.6	97.5	88.3	97.4	94.7	98.7	87.9	89.4
0310	Concrete Forming & Accessories	94.5	79.6	81.6	93.4	80.8	82.6	94.3	82.5	84.1	93.0	82.5	83.9	81.3	80.6	80.7	92.4	82.7	87.5
0320	Concrete Reinforcing	89.4	86.2	87.8	95.4	85.7	90.5	88.2	85.1	86.7	91.7	86.1	88.9	92.4	82.9	87.6	92.4	82.7	87.5
0330	Cast-in-Place Concrete	99.1	81.5	92.4	91.7	82.4	88.2	99.8	87.8	95.3	107.6	57.4	88.5	107.6	58.9	89.1	107.9	85.3	99.3
03	CONCRETE	88.3	83.0	85.9	100.8	82.4	92.4	106.6	84.9	96.7	94.6	75.2	85.7	93.5	74.3	84.7	94.7	86.7	91.0
04	MASONRY	93.4	77.2	83.4	94.2	76.4	83.2	86.7	80.3	82.8	100.4	75.0	84.6	102.1	74.2	84.8	106.2	83.4	92.0
05	METALS	96.3	105.9	99.2	92.2	86.2	90.3	86.7	86.1	86.5	90.2	95.0	91.7	90.2	94.0	91.4	95.3	87.6	91.1
06	WOOD, PLASTICS & COMPOSITES	94.3	78.6	85.6	93.8	80.4	86.3	94.2	81.3	87.0	89.0	81.4	84.8	76.4	84.4	80.9	95.3	87.6	91.1
07	THERMAL & MOISTURE PROTECTION	101.2	82.0	93.1	101.0	83.0	93.4	100.9	84.5	93.9	104.2	78.5	93.3	104.5	73.4	91.4	105.2	83.5	96.0
08	OPENINGS	95.2	78.5	91.4	94.7	79.2	91.1	91.5	80.1	88.9	93.5	84.9	91.5	97.9	82.6	94.4	98.4	85.4	95.4
0920	Plaster & Gypsum Board	95.1	78.3	83.9	94.8	79.4	84.5	95.1	80.4	85.2	109.4	81.0	90.3	104.8	84.1	90.9	113.9	87.7	96.3
0950, 0980	Ceilings & Acoustic Treatment	91.6	78.3	82.7	83.6	79.4	80.8	79.4	80.4	80.1	98.0	81.0	86.5	98.0	84.1	88.7	107.9	89.1	102.7
0960	Flooring	91.6	91.1	91.5	93.0	80.0	89.4	93.9	81.1	90.3	94.1	73.0	88.3	88.2	84.1	87.1	96.4	74.0	83.3
0970, 0990	Wall Finishes & Painting/Coating	86.9	86.3	86.5	91.1	83.4	86.6	91.1	88.2	89.4	94.8	76.1	83.9	90.2	81.6	85.5	99.4	86.8	92.6
09	FINISHES	91.1	82.2	86.2	89.2	80.8	84.6	88.9	82.6	85.5	94.1	81.6	87.3	90.2	81.6	85.5	99.4	86.8	92.6
COVERS	DIVS. 10 - 14, 25, 28, 41, 43, 44, 46	100.0	92.1	98.2	100.0	92.4	98.3	100.0	95.1	98.9	100.0	91.7	98.1	100.0	64.9	92.2	100.0	93.7	98.6
21, 22, 23	FIRE SUPPRESSION, PLUMBING & HVAC	99.9	78.1	90.6	100.0	79.7	91.3	96.9	80.9	90.0	97.1	82.6	90.9	97.1	73.8	87.2	100.2	84.5	93.5
26, 27, 3370	ELECTRICAL, COMMUNICATIONS & UTIL.	100.1	88.6	94.1	94.3	88.1	91.1	94.8	86.0	90.2	100.8	76.3	88.0	101.4	81.2	90.9	98.4	79.6	88.6
MF2016	WEIGHTED AVERAGE	96.3	85.0	91.4	96.3	85.4	91.5	94.8	86.5	91.1	96.0	82.3	90.0	95.8	79.9	88.8	98.2	86.5	93.1

IOWA

DIVISION		COUNCIL BLUFFS 515			CRESTON 508			DAVENPORT 527 - 528			DECORAH 521			DES MOINES 500 - 503, 509			DUBUQUE 520		
		MAT.	INST.	TOTAL	MAT.	INST.	TOTAL	MAT.	INST.	TOTAL	MAT.	INST.	TOTAL	MAT.	INST.	TOTAL	MAT.	INST.	TOTAL
015433	CONTRACTOR EQUIPMENT		97.5	97.5		101.5	101.5		101.5	101.5		101.5	101.5		103.4	103.4		97.0	97.0
0241, 31 - 34	SITE & INFRASTRUCTURE, DEMOLITION	103.5	92.5	95.8	98.0	98.3	98.2	99.8	99.6	99.7	98.1	96.3	96.8	104.4	101.2	102.2	98.5	93.2	94.8
0310	Concrete Forming & Accessories	80.8	76.2	76.8	78.5	83.7	83.0	98.2	100.5	100.2	90.7	74.7	76.9	96.4	86.0	87.4	82.0	85.4	85.0
0320	Concrete Reinforcing	94.3	80.0	87.1	91.5	82.4	86.9	92.4	108.9	100.7	91.7	71.0	81.3	96.0	83.4	89.6	91.1	82.5	86.8
0330	Cast-in-Place Concrete	112.1	81.9	100.7	108.8	63.4	91.6	103.9	100.3	102.5	104.7	79.7	95.2	92.6	93.3	92.9	105.6	87.2	98.6
03	CONCRETE	96.9	79.8	89.1	93.9	77.2	86.3	92.7	102.2	97.1	92.6	76.7	85.3	87.5	89.0	88.2	91.6	86.2	89.2
04	MASONRY	107.2	77.5	88.8	101.9	72.4	83.6	102.8	96.1	98.7	122.3	69.7	89.7	89.0	86.0	87.1	107.1	71.8	85.2
05	METALS	97.7	91.5	95.8	94.5	93.5	94.2	92.7	106.3	96.9	90.3	88.6	89.8	100.5	100.7	100.6	91.2	94.0	92.1
06	WOOD, PLASTICS & COMPOSITES	75.3	75.6	75.5	72.1	84.4	78.9	95.3	99.9	97.9	86.1	74.3	79.5	91.4	84.4	87.5	76.9	86.2	82.0
07	THERMAL & MOISTURE PROTECTION	104.6	76.9	92.9	107.5	80.1	95.9	104.6	95.7	100.9	104.4	72.7	91.0	99.1	86.7	93.9	104.9	80.7	94.6
08	OPENINGS	97.4	78.7	93.1	105.0	84.4	100.3	98.4	100.5	98.9	96.5	77.4	92.1	96.9	87.5	94.7	97.5	86.6	95.0
0920	Plaster & Gypsum Board	104.8	75.3	85.0	102.5	84.1	90.2	113.9	100.0	104.6	108.1	73.7	85.0	94.5	84.1	87.5	104.8	86.2	92.3
0950, 0980	Ceilings & Acoustic Treatment	98.0	75.3	82.7	89.8	84.1	86.0	100.6	100.0	100.2	98.0	73.7	81.6	91.5	84.1	86.5	98.0	86.2	90.0
0960	Flooring	86.8	89.1	87.4	78.6	73.0	77.1	96.6	94.5	96.0	93.5	73.0	87.8	87.5	94.1	89.3	98.8	73.0	91.7
0970, 0990	Wall Finishes & Painting/Coating	90.9	74.0	81.0	84.3	77.7	80.4	94.8	94.6	94.7	94.8	85.0	89.1	85.8	85.0	85.3	95.6	82.0	87.7
09	FINISHES	90.8	78.2	84.0	83.4	81.5	82.4	95.9	98.8	97.5	93.7	75.6	83.9	89.5	87.0	88.1	94.7	82.9	88.3
COVERS	DIVS. 10 - 14, 25, 28, 41, 43, 44, 46	100.0	90.6	97.9	100.0	67.5	92.8	100.0	97.4	99.4	100.0	81.1	95.8	100.0	93.6	98.6	100.0	92.5	98.3
21, 22, 23	FIRE SUPPRESSION, PLUMBING & HVAC	100.2	74.8	89.3	96.9	81.7	90.4	100.2	97.8	99.2	97.1	73.1	86.9	99.9	83.9	93.0	100.2	78.7	91.0
26, 27, 3370	ELECTRICAL, COMMUNICATIONS & UTIL.	103.6	82.8	92.8	93.3	81.2	87.0	96.6	92.5	94.4	98.4	49.9	73.2	104.3	83.8	93.6	102.2	77.1	89.1
MF2016	WEIGHTED AVERAGE	99.0	81.1	91.2	96.2	82.1	90.0	97.3	98.6	97.9	96.8	74.0	86.8	97.2	88.6	93.5	97.3	82.8	91.0

IOWA

DIVISION		FORT DODGE 505			MASON CITY 504			OTTUMWA 525			SHENANDOAH 516			SIBLEY 512			SIOUX CITY 510 - 511		
		MAT.	INST.	TOTAL	MAT.	INST.	TOTAL	MAT.	INST.	TOTAL	MAT.	INST.	TOTAL	MAT.	INST.	TOTAL	MAT.	INST.	TOTAL
015433	CONTRACTOR EQUIPMENT		101.5	101.5		101.5	101.5		97.0	97.0		97.5	97.5		101.5	101.5		101.5	101.5
0241, 31 - 34	SITE & INFRASTRUCTURE, DEMOLITION	107.1	95.1	98.7	107.1	96.2	99.5	99.0	90.9	93.3	102.1	92.9	95.7	109.7	96.3	100.3	111.4	97.5	101.6
0310	Concrete Forming & Accessories	79.1	75.2	75.8	83.1	74.6	75.7	88.7	74.9	76.8	82.3	57.9	61.3	82.7	38.9	44.9	98.7	75.7	78.9
0320	Concrete Reinforcing	91.5	69.0	80.2	91.4	82.4	86.9	91.7	86.1	88.9	94.3	69.4	81.7	94.3	68.9	81.5	92.4	79.6	85.9
0330	Cast-in-Place Concrete	102.2	43.5	79.9	102.2	74.7	91.7	108.3	66.7	92.5	108.5	79.3	97.4	106.3	58.0	87.9	106.9	73.3	94.2
03	CONCRETE	89.5	64.2	77.9	89.8	77.0	83.9	94.2	75.0	85.4	94.4	68.8	82.7	93.4	52.7	74.8	94.0	76.5	86.0
04	MASONRY	100.8	51.7	70.3	114.2	68.4	85.8	103.6	55.1	73.5	106.8	73.2	86.0	126.0	52.9	80.1	100.0	69.6	81.1
05	METALS	94.6	87.6	92.4	94.6	93.4	94.2	90.2	94.8	91.6	96.7	88.1	94.1	90.4	87.1	89.4	92.7	92.5	92.6
06	WOOD, PLASTICS & COMPOSITES	72.5	84.4	79.1	76.3	74.3	75.2	83.5	81.2	82.2	76.9	52.8	63.5	77.6	35.7	54.3	95.3	75.7	84.4
07	THERMAL & MOISTURE PROTECTION	106.8	67.0	90.0	106.3	73.3	92.3	105.0	69.5	90.0	103.8	67.4	88.4	104.1	54.6	83.2	104.6	73.0	91.3
08	OPENINGS	99.0	70.3	92.4	91.4	80.5	88.9	97.9	81.7	94.2	89.1	59.0	82.2	94.6	43.3	82.8	98.4	77.1	93.5
0920	Plaster & Gypsum Board	102.5	84.1	90.2	102.5	73.7	83.2	105.7	81.0	89.1	104.8	51.9	69.2	104.8	34.1	57.3	113.9	75.2	87.9
0950, 0980	Ceilings & Acoustic Treatment	89.8	84.1	86.0	89.8	73.7	79.0	98.0	81.0	86.5	98.0	51.9	66.9	98.0	34.1	54.9	100.6	75.2	83.5
0960	Flooring	79.8	73.0	77.9	81.7	73.0	79.3	102.0	73.0	93.9	87.5	78.1	84.9	89.1	73.0	84.7	96.6	75.7	90.8
0970, 0990	Wall Finishes & Painting/Coating	84.3	85.0	84.7	84.3	85.0	84.7	95.6	85.0	89.4	90.9	63.7	75.0	94.8	59.8	74.4	94.8	69.5	80.0
09	FINISHES	85.3	77.1	80.8	85.9	75.3	80.2	95.9	76.0	85.1	90.9	60.9	74.6	93.4	45.6	67.4	97.4	74.9	85.2
COVERS	DIVS. 10 - 14, 25, 28, 41, 43, 44, 46	100.0	84.1	96.5	100.0	88.5	97.4	100.0	84.5	96.5	100.0	61.7	91.5	100.0	75.9	94.6	100.0	90.6	97.9
21, 22, 23	FIRE SUPPRESSION, PLUMBING & HVAC	96.9	70.6	85.7	96.9	79.5	89.5	97.1	72.7	86.7	97.1	87.1	92.8	97.1	66.3	84.0	100.2	79.5	91.4
26, 27, 3370	ELECTRICAL, COMMUNICATIONS & UTIL.	99.3	73.2	85.7	98.4	49.9	73.2	100.6	74.9	87.2	98.4	79.8	88.7	98.4	49.9	73.2	98.4	72.6	85.0
MF2016	WEIGHTED AVERAGE	95.9	73.0	85.9	95.8	76.0	87.1	96.7	76.2	87.7	96.4	76.5	87.7	97.1	61.0	81.3	97.9	79.2	89.7

DIVISION		IOWA SPENCER 513			WATERLOO 506 - 507			KANSAS BELLEVILLE 669			COLBY 677			DODGE CITY 678			EMPORIA 668		
		MAT.	INST.	TOTAL	MAT.	INST.	TOTAL	MAT.	INST.	TOTAL	MAT.	INST.	TOTAL	MAT.	INST.	TOTAL	MAT.	INST.	TOTAL
015433	CONTRACTOR EQUIPMENT		101.5	101.5		101.5	101.5		107.1	107.1		107.1	107.1		107.1	107.1		105.2	105.2
0241, 31 - 34	SITE & INFRASTRUCTURE, DEMOLITION	109.8	95.0	99.5	112.5	97.4	101.9	110.7	97.2	101.3	109.2	97.9	101.3	111.7	96.8	101.3	102.5	94.7	97.0
0310	Concrete Forming & Accessories	88.7	38.7	45.6	94.3	71.6	74.7	93.9	56.2	61.4	99.6	63.6	68.6	92.9	63.3	67.3	85.1	68.2	70.5
0320	Concrete Reinforcing	94.3	68.9	81.5	92.1	82.9	87.5	96.8	63.4	80.0	104.6	63.5	83.9	102.0	63.3	83.8	95.5	63.9	79.6
0330	Cast-in-Place Concrete	106.3	68.7	92.0	109.4	83.4	99.5	120.9	87.7	108.3	115.6	91.6	106.5	117.7	91.3	107.6	116.9	91.7	107.3
03	CONCRETE	93.8	56.3	76.7	95.3	78.7	87.7	111.5	70.0	92.5	108.9	74.7	93.3	110.3	74.4	93.9	104.1	76.9	91.6
04	MASONRY	126.0	52.0	80.1	101.5	77.7	86.7	92.0	55.2	69.2	102.2	61.5	77.0	112.5	59.9	79.9	97.9	68.9	79.9
05	METALS	90.4	87.0	89.3	97.0	94.7	96.3	98.7	86.4	95.0	99.1	87.3	95.5	100.6	86.0	96.1	98.5	87.9	95.2
06	WOOD, PLASTICS & COMPOSITES	83.4	35.7	56.9	89.3	67.4	77.1	98.2	53.6	73.4	107.0	60.1	80.9	99.0	60.1	77.4	89.6	65.9	76.4
07	THERMAL & MOISTURE PROTECTION	105.1	55.2	84.0	106.5	79.9	95.3	93.7	63.5	80.9	95.4	66.9	83.3	95.3	66.1	83.0	91.8	78.4	86.1
08	OPENINGS	105.9	43.3	91.5	91.7	76.7	88.3	98.2	55.9	88.5	103.4	63.3	94.2	103.4	59.5	93.3	96.1	65.2	89.0
0920	Plaster & Gypsum Board	105.7	34.1	57.6	111.0	66.7	81.2	95.4	52.4	66.5	100.6	59.1	72.7	95.0	59.1	70.9	92.7	65.1	74.2
0950, 0980	Ceilings & Acoustic Treatment	98.0	34.1	54.9	92.4	66.7	75.0	84.0	52.4	62.7	81.5	59.1	66.4	81.5	59.1	66.4	84.0	65.1	71.3
0960	Flooring	91.9	73.0	86.7	86.7	83.4	85.8	88.5	70.8	83.6	87.4	70.8	82.8	83.6	70.8	80.1	83.8	70.8	80.2
0970, 0990	Wall Finishes & Painting/Coating	94.8	59.8	74.4	84.3	85.0	84.7	84.5	57.6	68.8	91.7	57.6	71.8	91.7	57.6	71.8	84.5	57.6	68.8
09	FINISHES	94.3	44.4	67.2	89.4	74.3	81.2	86.6	57.8	71.0	86.3	63.3	73.8	84.5	63.3	72.9	83.8	66.7	74.5
COVERS	DIVS. 10 - 14, 25, 28, 41, 43, 44, 46	100.0	74.6	94.4	100.0	91.0	98.0	100.0	85.3	96.7	100.0	88.2	97.4	100.0	88.2	97.4	100.0	88.8	97.5
21, 22, 23	FIRE SUPPRESSION, PLUMBING & HVAC	97.1	70.7	85.9	100.0	83.1	92.8	96.9	71.7	86.1	96.9	72.5	86.5	100.0	72.5	88.2	96.9	75.9	87.9
26, 27, 3370	ELECTRICAL, COMMUNICATIONS & UTIL.	100.1	49.9	74.0	95.1	65.0	79.5	102.2	66.2	83.5	99.6	71.7	85.1	96.8	71.7	83.7	99.5	70.0	84.2
MF2016	WEIGHTED AVERAGE	98.7	62.1	82.7	97.2	80.2	89.7	99.1	69.9	86.3	99.8	73.5	88.3	100.9	72.9	88.7	97.4	75.8	87.9

KANSAS

DIVISION		FORT SCOTT 667 MAT.	INST.	TOTAL	HAYS 676 MAT.	INST.	TOTAL	HUTCHINSON 675 MAT.	INST.	TOTAL	INDEPENDENCE 673 MAT.	INST.	TOTAL	KANSAS CITY 660-662 MAT.	INST.	TOTAL	LIBERAL 679 MAT.	INST.	TOTAL
015433	CONTRACTOR EQUIPMENT		106.1	106.1		107.1	107.1		107.1	107.1		107.1	107.1		103.6	103.6		107.1	107.1
0241, 31-34	SITE & INFRASTRUCTURE, DEMOLITION	99.3	94.7	96.1	114.1	97.4	102.4	93.1	97.9	96.5	112.8	97.9	102.4	93.3	94.9	94.4	93.4	60.7	65.2
0310	Concrete Forming & Accessories	101.8	83.7	86.2	97.1	61.1	66.1	87.6	58.6	62.6	108.3	70.4	75.6	98.5	98.7	98.7	103.5	63.3	83.2
0320	Concrete Reinforcing	94.9	101.7	98.3	102.0	63.4	82.5	102.0	63.3	82.5	101.4	64.9	83.0	92.0	104.4	98.3	91.2	87.6	89.8
0330	Cast-in-Place Concrete	108.4	86.8	100.2	91.2	88.0	90.0	84.5	91.3	87.1	118.2	91.5	108.0	92.8	99.9	95.5	91.2	87.6	89.8
03	CONCRETE	99.3	88.6	94.4	100.4	72.3	87.6	83.7	72.3	78.5	111.4	77.9	96.1	91.4	100.6	95.6	102.3	71.9	88.4
04	MASONRY	99.3	55.9	72.4	111.6	55.3	76.6	102.1	59.9	75.9	99.3	64.6	77.8	99.8	98.2	98.8	110.3	53.6	75.1
05	METALS	98.4	100.3	99.0	98.7	87.0	95.1	98.5	86.0	94.7	98.4	86.5	94.8	106.0	106.3	106.1	99.0	85.8	94.9
06	WOOD, PLASTICS & COMPOSITES	108.2	90.6	98.4	103.9	60.1	79.5	94.1	53.8	71.7	117.3	69.0	90.4	104.1	98.6	101.0	99.6	60.1	77.6
07	THERMAL & MOISTURE PROTECTION	92.8	75.2	85.4	95.7	64.2	82.4	94.2	65.4	82.0	95.4	77.8	88.0	92.5	100.6	95.9	95.9	63.2	82.0
08	OPENINGS	96.1	87.5	94.1	103.3	59.5	93.3	103.3	56.0	92.4	101.1	64.0	92.6	97.4	98.0	97.6	103.4	59.5	93.3
0920	Plaster & Gypsum Board	97.7	90.4	92.8	97.9	59.1	71.8	94.0	52.6	66.2	107.5	68.2	81.1	91.4	98.6	96.2	95.6	59.1	71.1
0950, 0980	Ceilings & Acoustic Treatment	84.0	90.4	88.3	81.5	59.1	66.4	81.5	52.6	62.0	81.5	68.2	72.6	84.0	98.6	93.9	81.5	59.1	66.4
0960	Flooring	97.9	70.7	90.3	86.2	70.8	81.9	80.8	70.8	78.0	91.6	70.7	85.8	78.8	99.7	84.6	83.9	69.7	79.9
0970, 0990	Wall Finishes & Painting/Coating	86.1	81.0	83.1	91.7	57.6	71.8	91.7	57.6	71.8	91.7	57.6	71.8	91.5	102.4	97.8	91.7	57.6	71.8
09	FINISHES	89.0	82.2	85.3	86.0	61.7	72.8	82.0	59.6	69.8	88.6	69.1	78.0	84.1	99.3	92.4	85.3	61.4	72.3
COVERS	DIVS. 10-14, 25, 28, 41, 43, 44, 46	100.0	88.5	97.5	100.0	86.1	96.9	100.0	87.5	97.2	100.0	89.2	97.6	100.0	95.9	99.1	100.0	86.1	96.9
21, 22, 23	FIRE SUPPRESSION, PLUMBING & HVAC	96.9	69.0	85.0	96.9	69.3	85.1	96.9	70.4	85.6	96.9	72.9	86.6	99.9	98.8	99.4	96.9	67.9	84.5
26, 27, 3370	ELECTRICAL, COMMUNICATIONS & UTIL.	98.8	71.7	84.7	98.6	71.7	84.6	94.1	64.4	78.6	96.1	72.3	83.7	104.4	98.4	101.3	96.8	71.7	83.7
MF2016	WEIGHTED AVERAGE	97.3	79.1	89.3	99.1	71.3	86.9	95.0	70.5	84.3	99.6	75.6	89.1	98.4	99.3	98.8	99.0	70.6	86.6

KANSAS / KENTUCKY

DIVISION		SALINA 674 MAT.	INST.	TOTAL	TOPEKA 664-666 MAT.	INST.	TOTAL	WICHITA 670-672 MAT.	INST.	TOTAL	ASHLAND 411-412 MAT.	INST.	TOTAL	BOWLING GREEN 421-422 MAT.	INST.	TOTAL	CAMPTON 413-414 MAT.	INST.	TOTAL
015433	CONTRACTOR EQUIPMENT		107.1	107.1		105.2	105.2		107.1	107.1		100.3	100.3		93.1	93.1		99.5	99.5
0241, 31-34	SITE & INFRASTRUCTURE, DEMOLITION	102.0	97.4	98.8	96.9	94.5	95.2	98.0	98.6	98.4	115.9	82.6	92.6	80.2	94.2	90.0	89.3	95.5	93.6
0310	Concrete Forming & Accessories	89.4	56.3	60.9	97.2	69.0	72.8	95.7	56.1	61.6	85.8	95.9	94.5	84.7	79.2	79.9	87.7	83.2	83.8
0320	Concrete Reinforcing	101.4	63.4	82.2	91.6	90.6	91.1	99.5	101.3	100.4	89.2	94.7	92.0	86.5	79.1	82.8	87.3	95.2	91.3
0330	Cast-in-Place Concrete	102.1	88.0	96.8	97.5	89.9	94.6	95.5	76.3	88.2	87.2	97.1	91.0	86.5	72.5	81.2	96.5	70.6	86.7
03	CONCRETE	97.4	70.2	85.0	93.2	81.2	87.7	93.2	72.5	83.7	91.7	97.2	94.2	95.4	69.8	79.5	92.1	57.0	70.4
04	MASONRY	127.8	55.3	82.8	93.5	67.0	77.1	99.3	51.5	69.7	93.3	94.3	93.9	95.9	84.4	92.4	95.2	90.8	93.8
05	METALS	100.5	87.0	96.3	102.6	97.7	101.1	102.6	99.3	101.6	94.3	109.5	99.0	95.9	84.4	92.4	95.2	90.8	93.8
06	WOOD, PLASTICS & COMPOSITES	95.6	53.6	72.2	101.0	68.1	82.7	104.0	53.8	76.1	74.4	95.5	86.2	86.9	79.8	83.0	85.5	90.6	88.3
07	THERMAL & MOISTURE PROTECTION	94.8	63.5	81.6	96.5	77.7	88.6	94.1	60.6	79.9	92.0	91.7	91.9	88.0	79.2	84.2	101.0	70.9	88.3
08	OPENINGS	103.3	55.9	92.4	103.3	74.0	96.6	105.7	64.7	96.2	93.0	94.4	93.3	93.7	81.2	90.8	95.0	88.7	93.6
0920	Plaster & Gypsum Board	94.0	52.4	66.1	100.8	67.3	78.3	93.7	52.6	66.1	61.6	95.6	84.4	88.2	79.7	82.5	88.2	89.9	89.4
0950, 0980	Ceilings & Acoustic Treatment	81.5	52.4	61.9	88.7	67.3	74.3	85.3	52.6	63.3	78.9	95.6	90.2	83.6	79.7	81.0	83.6	89.9	87.9
0960	Flooring	82.2	70.8	79.0	89.6	70.8	84.4	92.6	70.8	86.6	73.6	85.9	77.0	88.3	64.4	81.7	90.4	65.7	83.5
0970, 0990	Wall Finishes & Painting/Coating	91.7	57.6	71.8	88.8	70.8	78.3	92.0	57.6	71.9	93.2	91.7	92.3	91.1	89.2	90.0	91.1	56.3	70.8
09	FINISHES	83.2	57.8	69.4	92.1	68.5	79.3	90.3	57.5	72.4	76.5	93.9	85.9	87.0	77.3	81.7	87.8	78.1	82.5
COVERS	DIVS. 10-14, 25, 28, 41, 43, 44, 46	100.0	86.1	96.9	100.0	83.3	96.3	100.0	85.4	96.8	100.0	91.7	98.1	100.0	87.4	97.2	100.0	49.5	88.8
21, 22, 23	FIRE SUPPRESSION, PLUMBING & HVAC	100.0	70.3	87.3	100.0	76.0	89.8	99.8	69.8	87.0	96.7	87.6	92.8	100.0	80.2	91.5	96.9	76.8	88.3
26, 27, 3370	ELECTRICAL, COMMUNICATIONS & UTIL.	96.5	74.1	84.8	103.2	73.7	87.8	100.2	74.1	86.6	92.3	93.2	92.8	94.8	78.9	86.5	92.4	93.1	92.8
MF2016	WEIGHTED AVERAGE	99.6	70.8	87.0	99.1	78.1	89.9	99.2	72.1	87.3	93.5	93.4	93.5	94.6	79.9	88.2	94.7	80.2	88.3

KENTUCKY

DIVISION		CORBIN 407-409 MAT.	INST.	TOTAL	COVINGTON 410 MAT.	INST.	TOTAL	ELIZABETHTOWN 427 MAT.	INST.	TOTAL	FRANKFORT 406 MAT.	INST.	TOTAL	HAZARD 417-418 MAT.	INST.	TOTAL	HENDERSON 424 MAT.	INST.	TOTAL
015433	CONTRACTOR EQUIPMENT		99.5	99.5		103.6	103.6		93.1	93.1		99.5	99.5		99.5	99.5		113.9	113.9
0241, 31-34	SITE & INFRASTRUCTURE, DEMOLITION	95.1	96.0	95.7	84.6	107.6	100.7	74.6	93.9	88.1	93.0	96.4	95.4	84.3	83.8	83.9	90.7	80.5	81.9
0310	Concrete Forming & Accessories	83.2	74.4	75.6	83.3	73.1	74.5	79.6	75.4	75.9	94.0	76.0	78.4	87.7	62.3	74.9	86.6	77.8	82.2
0320	Concrete Reinforcing	90.9	62.1	76.4	86.0	81.1	83.5	86.9	82.7	84.8	95.8	80.5	88.1	92.8	72.9	85.2	76.5	88.8	81.2
0330	Cast-in-Place Concrete	91.8	75.2	85.5	92.3	76.3	86.2	78.2	69.9	75.0	90.9	77.4	85.8	90.6	76.6	84.2	87.3	83.2	85.4
03	CONCRETE	87.1	73.1	80.7	92.1	76.5	84.9	82.3	75.2	79.1	89.3	77.7	84.0	90.6	76.6	84.2	87.3	83.2	85.4
04	MASONRY	90.0	62.6	73.0	106.9	75.1	87.1	79.2	64.4	70.0	85.5	73.3	77.9	90.8	59.2	71.2	99.0	82.1	88.5
05	METALS	93.9	78.4	89.1	93.2	89.9	92.2	95.1	85.1	92.0	96.6	86.2	93.4	95.2	78.8	90.2	86.4	84.6	85.9
06	WOOD, PLASTICS & COMPOSITES	72.1	75.0	73.7	83.7	70.2	76.1	82.4	77.5	79.7	89.6	75.0	81.5	82.7	90.6	87.1	89.5	79.4	83.9
07	THERMAL & MOISTURE PROTECTION	105.1	74.2	91.2	101.9	75.1	90.5	87.4	71.1	80.5	102.6	75.7	91.2	100.9	72.7	89.0	100.2	80.7	91.9
08	OPENINGS	89.4	62.5	83.2	97.0	74.1	91.7	93.7	75.6	89.6	95.3	77.2	91.2	95.4	80.7	92.0	91.9	79.8	89.1
0920	Plaster & Gypsum Board	93.1	73.9	80.2	70.8	70.0	70.2	87.5	77.3	80.6	92.5	73.9	80.0	87.5	89.9	89.1	91.5	78.4	82.7
0950, 0980	Ceilings & Acoustic Treatment	80.1	73.9	75.9	86.2	70.0	75.2	83.6	77.3	79.4	88.7	73.9	78.7	83.6	89.9	87.9	79.4	78.4	78.7
0960	Flooring	85.0	65.7	79.6	65.9	85.9	71.4	85.5	78.0	83.5	90.5	65.7	83.6	88.5	65.7	82.2	92.0	80.4	88.8
0970, 0990	Wall Finishes & Painting/Coating	90.9	64.8	75.7	86.8	74.4	79.5	91.1	70.1	78.9	93.2	71.2	80.4	91.1	56.3	70.8	91.1	90.7	90.9
09	FINISHES	83.2	71.9	77.0	78.1	75.0	76.4	85.7	75.0	79.9	89.9	73.6	81.0	87.0	78.6	82.4	87.1	81.8	84.2
COVERS	DIVS. 10-14, 25, 28, 41, 43, 44, 46	100.0	90.7	97.9	100.0	89.3	97.6	100.0	72.8	94.0	100.0	59.0	90.9	100.0	50.2	88.9	100.0	58.0	90.6
21, 22, 23	FIRE SUPPRESSION, PLUMBING & HVAC	97.0	76.4	88.2	97.7	79.0	89.7	97.0	77.1	88.5	100.0	80.2	91.5	96.9	77.6	88.7	97.0	79.0	89.3
26, 27, 3370	ELECTRICAL, COMMUNICATIONS & UTIL.	92.6	93.1	92.9	95.8	76.5	85.8	92.2	78.9	85.3	101.1	75.5	87.8	92.4	93.1	92.8	94.3	75.8	84.7
MF2016	WEIGHTED AVERAGE	92.7	77.7	86.2	94.7	80.5	88.5	91.5	77.2	85.3	96.0	78.5	88.3	94.1	78.7	87.4	92.4	83.0	88.3

City Cost Indexes

KENTUCKY

DIVISION		LEXINGTON 403 - 405			LOUISVILLE 400 - 402			OWENSBORO 423			PADUCAH 420			PIKEVILLE 415 - 416			SOMERSET 425 - 426		
		MAT.	INST.	TOTAL	MAT.	INST.	TOTAL	MAT.	INST.	TOTAL	MAT.	INST.	TOTAL	MAT.	INST.	TOTAL	MAT.	INST.	TOTAL
015433	CONTRACTOR EQUIPMENT		99.5	99.5		93.1	93.1		113.9	113.9		113.9	113.9		100.3	100.3		99.5	99.5
0241, 31 - 34	SITE & INFRASTRUCTURE, DEMOLITION	97.2	97.8	97.6	91.6	94.2	93.4	93.3	122.0	113.4	85.9	121.2	110.6	127.4	81.8	95.5	79.6	96.0	91.1
0310	Concrete Forming & Accessories	94.2	76.7	79.1	94.7	78.4	80.7	89.2	80.5	81.7	87.2	82.3	82.9	94.5	78.1	80.3	85.3	74.8	76.3
0320	Concrete Reinforcing	99.8	83.0	91.3	95.8	82.9	89.3	86.6	79.8	83.1	87.1	81.0	84.0	89.7	94.5	92.1	86.9	62.3	74.5
0330	Cast-in-Place Concrete	93.9	80.8	88.9	90.9	71.7	83.6	89.2	88.3	88.9	81.5	84.5	82.6	95.9	92.3	94.5	76.5	92.1	82.4
03	CONCRETE	89.9	79.6	85.2	89.3	77.3	83.8	99.1	83.4	91.9	91.7	83.1	87.8	105.0	87.4	97.0	77.5	79.1	78.2
04	MASONRY	88.2	73.3	79.0	86.5	69.4	75.9	91.3	82.0	85.5	94.1	78.8	84.6	90.8	86.8	88.3	85.7	65.4	73.1
05	METALS	96.4	86.9	93.5	97.5	85.5	93.8	87.9	86.5	87.5	84.9	86.7	85.4	94.2	108.0	98.5	95.1	79.0	90.2
06	WOOD, PLASTICS & COMPOSITES	86.0	75.0	79.8	89.7	79.8	84.2	87.5	79.4	83.0	85.3	82.2	83.6	83.3	77.0	79.8	83.2	75.0	78.6
07	THERMAL & MOISTURE PROTECTION	105.3	76.5	93.1	102.4	74.2	90.5	100.9	80.1	92.1	100.3	84.7	93.7	92.8	82.6	88.5	100.0	73.2	88.8
08	OPENINGS	89.6	77.1	86.8	85.9	76.9	83.9	91.9	80.2	89.2	91.2	82.6	89.2	93.6	76.6	89.7	94.4	69.3	88.7
0920	Plaster & Gypsum Board	102.4	73.9	83.3	92.8	79.7	84.0	90.2	78.4	82.3	89.2	81.3	83.9	64.5	76.6	72.6	87.5	73.9	78.4
0950, 0980	Ceilings & Acoustic Treatment	83.5	73.9	77.0	88.7	79.7	82.6	79.4	78.4	78.7	79.4	81.3	80.7	78.9	76.6	77.3	83.6	73.9	77.1
0960	Flooring	89.9	65.7	83.2	89.8	64.4	82.7	91.3	64.4	83.9	90.3	80.4	87.5	77.9	65.7	74.5	88.9	65.7	82.4
0970, 0990	Wall Finishes & Painting/Coating	90.9	80.9	85.0	92.5	70.1	79.5	91.1	90.7	90.9	91.1	76.5	82.6	93.2	69.9	79.6	91.1	70.1	78.9
09	FINISHES	86.7	74.6	80.1	89.6	74.9	81.6	87.3	78.2	82.3	86.5	81.3	83.7	79.0	74.6	76.6	86.4	72.5	78.8
COVERS	DIVS. 10 - 14, 25, 28, 41, 43, 44, 46	100.0	91.8	98.2	100.0	89.3	97.6	100.0	96.9	99.3	100.0	88.1	97.3	100.0	48.3	88.5	100.0	90.7	97.9
21, 22, 23	FIRE SUPPRESSION, PLUMBING & HVAC	100.1	79.0	91.1	100.0	78.7	90.9	100.0	80.3	91.6	97.0	79.8	90.7	96.7	82.8	90.8	97.0	75.8	88.0
26, 27, 3370	ELECTRICAL, COMMUNICATIONS & UTIL.	95.2	78.9	86.7	101.1	78.9	89.5	94.3	75.8	84.7	96.5	78.1	87.0	95.1	66.7	80.3	92.8	93.1	93.0
MF2016	WEIGHTED AVERAGE	94.8	80.3	88.5	95.1	79.1	88.1	94.7	84.3	90.2	92.6	84.5	89.1	95.9	81.4	89.6	91.9	79.2	86.4

LOUISIANA

DIVISION		ALEXANDRIA 713 - 714			BATON ROUGE 707 - 708			HAMMOND 704			LAFAYETTE 705			LAKE CHARLES 706			MONROE 712		
		MAT.	INST.	TOTAL	MAT.	INST.	TOTAL	MAT.	INST.	TOTAL	MAT.	INST.	TOTAL	MAT.	INST.	TOTAL	MAT.	INST.	TOTAL
015433	CONTRACTOR EQUIPMENT		93.5	93.5		91.4	91.4		91.9	91.9		91.9	91.9		91.4	91.4		93.5	93.5
0241, 31 - 34	SITE & INFRASTRUCTURE, DEMOLITION	105.3	92.4	96.3	102.3	91.1	94.5	98.9	90.4	93.0	99.9	92.7	94.9	100.7	91.8	94.5	105.3	92.3	96.2
0310	Concrete Forming & Accessories	80.2	62.9	65.3	95.5	74.7	77.5	77.9	57.3	60.1	95.0	70.9	74.2	95.8	71.0	74.4	79.8	62.2	64.7
0320	Concrete Reinforcing	96.3	65.1	80.5	91.7	55.6	73.5	93.0	53.7	73.2	94.3	54.0	74.0	94.3	54.3	74.1	95.1	65.0	80.0
0330	Cast-in-Place Concrete	93.8	68.2	84.0	93.5	77.2	87.3	90.1	65.7	80.8	89.7	69.5	82.0	94.4	69.5	84.9	93.8	67.4	83.7
03	CONCRETE	87.1	65.8	77.3	88.8	72.5	81.4	86.9	60.4	74.8	87.8	67.9	78.7	90.1	68.0	80.0	86.9	65.2	77.0
04	MASONRY	114.2	62.9	82.4	87.1	64.9	73.4	91.7	54.7	68.8	91.7	65.3	75.4	91.1	69.7	77.8	108.7	61.6	79.5
05	METALS	92.5	75.1	87.2	96.6	70.4	88.5	87.6	69.1	81.9	86.9	69.8	81.7	86.9	70.2	81.8	92.5	75.0	87.3
06	WOOD, PLASTICS & COMPOSITES	92.9	61.6	75.5	102.1	76.5	87.9	85.6	56.8	69.6	105.4	71.3	86.4	103.6	71.3	85.6	92.3	61.6	75.2
07	THERMAL & MOISTURE PROTECTION	97.9	67.4	85.0	95.3	69.9	84.5	97.1	63.7	83.0	97.6	69.5	85.7	97.3	70.5	86.0	97.9	66.6	84.6
08	OPENINGS	110.5	63.8	99.8	99.7	70.4	93.0	95.5	57.0	86.6	99.1	64.9	91.3	99.1	64.9	91.3	110.5	62.2	99.4
0920	Plaster & Gypsum Board	84.0	60.9	68.5	99.3	76.3	83.9	101.2	56.0	70.9	109.0	70.9	83.4	109.0	70.9	83.4	83.7	60.9	68.4
0950, 0980	Ceilings & Acoustic Treatment	83.0	60.9	68.1	94.8	76.3	82.3	101.6	56.0	70.9	99.9	70.9	80.3	100.9	70.9	80.7	83.0	60.9	68.1
0960	Flooring	90.9	64.9	83.7	86.9	60.4	79.6	88.1	58.4	79.8	96.8	64.3	87.8	96.8	67.4	88.6	90.5	58.2	81.5
0970, 0990	Wall Finishes & Painting/Coating	93.1	62.5	75.2	89.4	62.5	73.7	94.9	62.5	76.0	94.9	62.5	76.0	94.9	62.5	76.0	93.1	62.5	75.2
09	FINISHES	82.0	62.8	71.6	89.7	71.2	79.6	91.4	57.6	73.0	94.8	69.0	80.8	95.0	69.7	81.2	81.8	61.1	70.6
COVERS	DIVS. 10 - 14, 25, 28, 41, 43, 44, 46	100.0	80.6	95.7	100.0	87.2	97.2	100.0	80.4	95.6	100.0	86.8	97.1	100.0	86.9	97.1	100.0	80.1	95.6
21, 22, 23	FIRE SUPPRESSION, PLUMBING & HVAC	100.2	65.4	85.3	100.0	66.4	85.6	97.1	63.0	82.5	100.2	66.6	85.8	100.2	66.6	85.8	100.2	64.6	85.0
26, 27, 3370	ELECTRICAL, COMMUNICATIONS & UTIL.	93.2	64.9	78.5	97.9	59.8	78.1	93.7	70.5	81.7	94.8	67.4	80.5	94.4	69.6	81.5	94.8	60.5	77.0
MF2016	WEIGHTED AVERAGE	96.9	68.2	84.3	96.3	70.1	84.8	93.2	65.1	80.9	94.9	70.1	84.0	95.1	70.9	84.5	96.7	66.9	83.7

DIVISION		LOUISIANA NEW ORLEANS 700 - 701			SHREVEPORT 710 - 711			THIBODAUX 703			MAINE AUGUSTA 043			BANGOR 044			BATH 045		
		MAT.	INST.	TOTAL	MAT.	INST.	TOTAL	MAT.	INST.	TOTAL	MAT.	INST.	TOTAL	MAT.	INST.	TOTAL	MAT.	INST.	TOTAL
015433	CONTRACTOR EQUIPMENT		87.5	87.5		93.5	93.5		91.9	91.9		98.8	98.8		98.8	98.8		98.8	98.8
0241, 31 - 34	SITE & INFRASTRUCTURE, DEMOLITION	99.9	93.5	95.4	107.6	92.3	96.9	101.2	92.4	95.1	90.4	99.3	96.6	93.6	100.7	98.6	91.6	99.3	97.0
0310	Concrete Forming & Accessories	97.3	69.3	73.1	94.8	62.8	67.2	89.4	67.2	70.3	101.1	77.7	80.9	93.8	79.5	81.5	89.7	77.9	79.5
0320	Concrete Reinforcing	96.2	55.0	75.4	96.3	53.8	74.9	93.0	67.5	80.1	102.3	79.0	90.5	92.1	80.8	86.4	91.1	80.2	85.6
0330	Cast-in-Place Concrete	91.1	69.3	82.8	96.7	67.5	85.6	96.8	67.4	85.6	86.7	112.4	96.5	70.0	113.9	86.7	70.0	112.5	86.1
03	CONCRETE	93.1	66.7	81.0	90.5	63.6	78.2	91.7	67.6	80.7	88.1	90.5	89.2	81.7	92.1	86.5	81.7	90.8	85.9
04	MASONRY	96.9	60.6	74.4	98.1	58.9	73.8	114.7	53.9	77.0	102.7	66.9	80.5	118.3	95.5	104.1	125.5	86.9	101.5
05	METALS	96.8	62.7	86.3	96.6	70.8	88.7	87.6	68.9	81.9	101.1	89.5	97.6	91.0	92.7	91.5	89.4	91.2	89.9
06	WOOD, PLASTICS & COMPOSITES	103.7	71.4	85.7	99.1	62.4	78.6	92.8	68.7	79.4	98.6	76.5	86.3	97.9	76.5	86.0	92.5	76.5	83.6
07	THERMAL & MOISTURE PROTECTION	97.9	68.3	85.4	95.6	66.0	83.1	97.0	65.1	83.5	106.2	70.1	90.9	102.8	84.7	95.1	102.8	76.3	91.5
08	OPENINGS	103.1	65.3	94.5	105.4	59.0	94.8	100.0	59.4	90.7	100.6	80.3	96.0	98.0	80.3	94.0	98.0	80.3	93.9
0920	Plaster & Gypsum Board	98.6	70.9	80.0	93.5	61.7	72.2	102.5	68.2	79.5	108.2	75.4	86.2	112.7	75.4	87.7	108.5	75.4	86.3
0950, 0980	Ceilings & Acoustic Treatment	105.8	70.9	82.3	89.7	61.7	70.8	101.6	68.2	79.1	104.0	75.4	84.7	86.6	75.4	79.1	85.6	75.4	78.8
0960	Flooring	103.8	60.4	91.8	94.9	58.2	84.7	94.1	40.3	79.2	88.5	48.4	77.4	85.0	104.4	90.4	83.3	48.4	73.6
0970, 0990	Wall Finishes & Painting/Coating	99.0	62.5	77.7	85.6	62.5	72.1	96.1	62.5	76.5	90.5	93.0	91.9	90.7	93.0	92.0	90.7	93.0	92.0
09	FINISHES	101.3	67.1	82.7	88.7	61.7	74.0	93.7	61.6	76.3	93.1	72.7	82.0	87.4	84.8	86.0	86.1	72.7	78.8
COVERS	DIVS. 10 - 14, 25, 28, 41, 43, 44, 46	100.0	85.6	96.8	100.0	85.7	96.8	100.0	85.2	96.7	100.0	100.9	100.2	100.0	104.0	100.9	100.0	100.9	100.2
21, 22, 23	FIRE SUPPRESSION, PLUMBING & HVAC	100.1	64.0	84.7	99.9	64.8	84.9	97.1	64.5	83.2	100.0	71.9	88.0	100.2	73.0	88.5	97.1	72.0	86.4
26, 27, 3370	ELECTRICAL, COMMUNICATIONS & UTIL.	98.4	70.5	83.9	100.2	64.5	81.6	92.5	70.5	81.1	101.5	77.6	89.1	99.7	73.2	86.0	97.8	77.6	87.3
MF2016	WEIGHTED AVERAGE	98.7	68.4	85.5	97.9	66.7	84.2	95.6	67.4	83.3	98.4	79.8	90.3	95.8	85.0	91.1	94.8	82.2	89.3

For customer support on your Electrical Costs with RSMeans data, call 800.448.8182.

527

City Cost Indexes

MAINE

DIVISION		HOULTON 047 MAT.	INST.	TOTAL	KITTERY 039 MAT.	INST.	TOTAL	LEWISTON 042 MAT.	INST.	TOTAL	MACHIAS 046 MAT.	INST.	TOTAL	PORTLAND 040-041 MAT.	INST.	TOTAL	ROCKLAND 048 MAT.	INST.	TOTAL
015433	CONTRACTOR EQUIPMENT		98.8	98.8		98.8	98.8		98.8	98.8		98.8	98.8		98.8	98.8		98.8	98.8
0241, 31-34	SITE & INFRASTRUCTURE, DEMOLITION	93.4	99.3	97.5	82.9	99.3	94.4	91.3	100.7	97.9	92.8	99.3	97.3	91.2	100.7	97.8	89.3	99.3	96.3
0310	Concrete Forming & Accessories	97.6	77.7	80.4	87.8	78.2	79.6	99.4	79.6	82.3	94.7	77.7	80.0	101.1	79.6	82.5	95.7	77.7	80.2
0320	Concrete Reinforcing	92.1	78.7	85.3	85.3	80.3	82.8	113.1	80.8	96.8	92.1	78.7	85.3	102.3	80.8	91.4	92.1	78.7	85.3
0330	Cast-in-Place Concrete	70.0	112.4	86.1	73.3	112.6	88.2	71.4	113.9	87.6	70.0	112.4	86.1	86.7	113.9	97.0	71.5	112.4	87.0
03	CONCRETE	82.6	90.4	86.2	78.8	91.0	84.4	82.1	92.1	86.7	82.1	90.4	85.9	87.4	92.1	89.6	79.8	90.4	84.7
04	MASONRY	100.2	62.4	76.8	115.6	86.9	97.8	100.8	95.5	97.5	100.2	62.4	76.8	106.1	95.5	99.5	93.7	62.4	74.3
05	METALS	89.6	89.1	89.5	85.8	91.5	87.6	94.3	92.7	93.8	89.6	89.1	89.5	101.2	92.7	98.6	89.5	89.1	89.4
06	WOOD, PLASTICS & COMPOSITES	101.8	76.5	87.7	95.8	76.5	85.0	103.6	76.5	88.5	98.8	76.5	86.4	97.3	76.5	85.7	99.6	76.5	86.7
07	THERMAL & MOISTURE PROTECTION	102.9	69.9	88.9	102.4	75.4	91.0	102.6	84.7	95.0	102.8	69.1	88.6	106.5	84.7	97.2	102.5	69.1	88.4
08	OPENINGS	98.1	80.3	94.0	97.4	84.0	94.3	101.3	80.3	96.5	98.1	80.3	94.0	99.9	80.3	95.4	98.0	80.3	94.0
0920	Plaster & Gypsum Board	115.0	75.4	88.4	104.8	75.4	85.1	118.0	75.4	89.4	113.4	75.4	87.9	108.6	75.4	86.3	113.4	75.4	87.9
0950, 0980	Ceilings & Acoustic Treatment	85.6	75.4	78.8	99.5	75.4	83.3	95.1	75.4	81.8	85.6	75.4	78.8	102.2	75.4	84.1	85.6	75.4	78.8
0960	Flooring	86.2	45.2	74.9	85.4	50.4	75.7	87.8	104.4	92.4	85.4	45.2	74.2	89.2	104.4	93.4	85.8	45.2	74.5
0970, 0990	Wall Finishes & Painting/Coating	90.7	93.0	92.0	84.8	106.8	97.6	90.7	93.0	92.0	90.7	93.0	92.0	93.2	93.0	93.1	90.7	93.0	92.0
09	FINISHES	88.0	72.1	79.3	87.8	74.5	80.6	90.3	84.8	87.3	87.4	72.1	79.1	93.7	84.8	88.8	87.2	72.1	79.0
COVERS	DIVS. 10-14, 25, 28, 41, 43, 44, 46	100.0	100.9	100.2	100.0	100.9	100.2	100.0	104.1	100.9	100.0	100.9	100.2	100.0	104.1	100.9	100.0	100.9	100.2
21, 22, 23	FIRE SUPPRESSION, PLUMBING & HVAC	97.1	71.9	86.3	97.0	82.6	90.9	100.2	73.0	88.5	97.1	71.9	86.3	100.0	73.0	88.5	97.1	71.9	86.3
26, 27, 3370	ELECTRICAL, COMMUNICATIONS & UTIL.	101.7	77.6	89.2	96.0	77.6	86.4	101.7	77.6	89.2	101.7	77.6	89.2	103.8	77.6	90.2	101.6	77.6	89.1
MF2016	WEIGHTED AVERAGE	94.4	79.2	87.8	93.0	84.9	89.5	96.3	85.6	91.6	94.2	79.2	87.7	98.7	85.6	93.0	93.5	79.2	87.2

MAINE / MARYLAND

DIVISION		WATERVILLE 049 MAT.	INST.	TOTAL	ANNAPOLIS 214 MAT.	INST.	TOTAL	BALTIMORE 210-212 MAT.	INST.	TOTAL	COLLEGE PARK 207-208 MAT.	INST.	TOTAL	CUMBERLAND 215 MAT.	INST.	TOTAL	EASTON 216 MAT.	INST.	TOTAL
015433	CONTRACTOR EQUIPMENT		98.8	98.8		105.7	105.7		105.6	105.6		111.0	111.0		105.7	105.7		105.7	105.7
0241, 31-34	SITE & INFRASTRUCTURE, DEMOLITION	93.3	99.3	97.5	105.3	94.1	97.5	102.5	98.2	99.4	101.3	98.6	99.4	95.9	95.3	95.5	103.1	92.4	95.6
0310	Concrete Forming & Accessories	89.2	77.7	79.3	97.4	75.7	78.7	98.8	78.3	81.2	85.6	73.8	75.4	90.9	85.8	86.5	89.1	74.1	76.2
0320	Concrete Reinforcing	92.1	79.0	85.5	104.8	90.0	97.3	100.6	90.1	95.3	100.6	89.1	94.8	91.6	85.8	88.7	90.9	87.7	89.3
0330	Cast-in-Place Concrete	70.0	112.4	86.1	110.5	74.5	96.8	109.9	76.3	97.1	116.0	78.2	101.7	94.8	87.3	91.9	105.3	68.1	91.1
03	CONCRETE	83.1	90.5	86.5	98.2	79.3	89.5	107.2	80.7	95.1	106.9	79.7	94.4	89.6	87.6	88.7	97.6	75.8	87.6
04	MASONRY	110.8	66.9	83.6	99.5	70.6	81.6	101.8	70.6	82.5	115.6	76.5	91.4	101.0	87.9	92.9	115.2	61.1	81.6
05	METALS	89.6	89.5	89.6	102.1	105.0	103.0	101.7	99.9	101.1	88.5	107.3	94.3	98.1	105.0	100.2	98.4	101.0	99.2
06	WOOD, PLASTICS & COMPOSITES	92.0	76.5	83.4	94.2	77.4	84.8	101.2	80.7	89.8	83.4	71.7	76.9	90.2	84.9	87.2	88.3	80.9	84.1
07	THERMAL & MOISTURE PROTECTION	102.9	70.1	89.0	103.1	79.8	93.2	101.1	80.8	92.5	100.6	82.9	93.1	99.9	85.6	93.8	100.1	75.8	89.8
08	OPENINGS	98.1	80.3	94.0	97.3	83.0	94.0	99.8	84.8	96.4	95.2	79.6	91.6	100.1	87.8	97.3	98.3	84.3	95.1
0920	Plaster & Gypsum Board	108.5	75.4	86.3	100.1	77.0	84.6	104.3	80.1	88.1	98.7	71.0	80.1	105.7	84.7	91.6	105.7	80.6	88.9
0950, 0980	Ceilings & Acoustic Treatment	85.6	75.4	78.8	91.7	77.0	81.8	100.0	80.1	86.6	101.6	71.0	81.0	103.0	84.7	90.7	103.0	80.6	87.9
0960	Flooring	83.0	48.4	73.4	86.2	80.0	84.5	99.3	80.0	93.9	89.3	80.0	86.7	87.9	96.3	90.2	87.0	80.0	85.1
0970, 0990	Wall Finishes & Painting/Coating	90.7	93.0	92.0	87.3	76.5	81.0	102.9	76.5	87.5	104.1	74.3	86.7	97.3	88.5	92.2	97.3	74.3	83.9
09	FINISHES	86.2	72.7	78.8	86.1	76.2	80.7	99.8	78.2	88.1	92.8	74.1	82.6	94.1	88.0	90.8	94.3	75.7	84.2
COVERS	DIVS. 10-14, 25, 28, 41, 43, 44, 46	100.0	100.7	100.2	100.0	87.9	97.3	100.0	89.1	97.6	100.0	84.0	96.4	100.0	92.5	98.3	100.0	70.9	93.5
21, 22, 23	FIRE SUPPRESSION, PLUMBING & HVAC	97.1	71.9	86.3	100.1	82.1	92.4	100.1	82.2	92.3	97.0	86.1	92.3	96.8	74.9	87.5	96.8	75.8	87.8
26, 27, 3370	ELECTRICAL, COMMUNICATIONS & UTIL.	101.7	77.6	89.1	100.1	90.9	95.3	100.7	90.9	95.6	100.2	98.6	99.4	98.1	80.6	89.0	97.6	64.1	80.2
MF2016	WEIGHTED AVERAGE	94.7	79.8	88.2	98.9	84.2	92.4	101.4	84.7	94.1	97.9	86.8	93.0	96.9	86.3	92.3	98.6	76.6	88.9

MARYLAND / MASSACHUSETTS

DIVISION		ELKTON 219 MAT.	INST.	TOTAL	HAGERSTOWN 217 MAT.	INST.	TOTAL	SALISBURY 218 MAT.	INST.	TOTAL	SILVER SPRING 209 MAT.	INST.	TOTAL	WALDORF 206 MAT.	INST.	TOTAL	BOSTON 020-022, 024 MAT.	INST.	TOTAL
015433	CONTRACTOR EQUIPMENT		105.7	105.7		105.7	105.7		105.7	105.7		102.4	102.4		102.4	102.4		103.0	103.0
0241, 31-34	SITE & INFRASTRUCTURE, DEMOLITION	89.8	93.9	92.6	94.0	95.2	94.8	103.1	92.4	95.6	89.7	90.0	89.9	96.1	90.1	91.9	101.6	102.1	100.1
0310	Concrete Forming & Accessories	94.8	90.7	91.3	90.0	82.5	83.5	102.8	52.6	59.5	94.1	73.1	75.9	101.3	70.5	74.8	101.6	139.0	133.8
0320	Concrete Reinforcing	90.9	113.2	102.2	91.6	85.8	88.7	90.9	67.3	79.0	99.4	89.0	94.1	100.0	89.0	94.4	101.1	147.6	124.6
0330	Cast-in-Place Concrete	85.3	74.0	81.0	90.3	87.3	89.2	105.3	66.1	90.4	118.8	78.6	103.5	133.1	78.6	112.4	96.0	139.2	112.4
03	CONCRETE	82.3	89.9	85.8	86.0	86.0	86.0	98.4	61.8	81.7	104.8	79.3	93.1	115.7	78.1	98.5	100.9	139.9	118.7
04	MASONRY	99.6	69.9	81.2	106.6	87.9	95.0	114.9	57.5	79.3	115.1	76.9	91.4	98.5	76.9	85.1	112.0	148.2	134.5
05	METALS	98.4	113.6	103.1	98.3	104.9	100.3	98.4	93.0	96.8	92.9	103.1	96.1	92.9	103.1	96.1	101.1	133.3	111.0
06	WOOD, PLASTICS & COMPOSITES	95.3	98.1	96.9	89.5	80.4	84.4	105.3	53.8	76.6	89.9	70.9	79.3	96.8	67.5	80.5	101.9	139.5	122.8
07	THERMAL & MOISTURE PROTECTION	99.6	81.6	92.0	100.3	86.7	94.5	100.3	71.1	87.9	104.3	88.0	97.4	104.8	87.6	97.5	106.0	137.6	119.3
08	OPENINGS	98.3	100.6	98.9	98.3	82.5	94.7	98.6	63.7	90.6	87.4	79.1	85.5	88.1	77.3	85.6	98.7	145.2	109.3
0920	Plaster & Gypsum Board	109.3	98.4	102.0	105.7	80.1	88.5	116.2	52.8	73.6	104.3	71.0	81.9	107.3	67.5	80.6	108.8	140.4	130.0
0950, 0980	Ceilings & Acoustic Treatment	103.0	98.4	99.9	103.9	80.1	87.9	103.0	52.8	69.2	110.1	71.0	83.7	110.1	67.5	81.4	109.4	140.4	130.3
0960	Flooring	89.5	80.0	86.9	87.5	96.3	89.9	93.4	80.0	89.7	95.4	80.0	91.2	99.2	80.0	93.8	95.3	163.7	114.3
0970, 0990	Wall Finishes & Painting/Coating	97.3	74.3	83.9	97.3	74.3	83.9	97.3	74.3	83.9	111.7	74.3	89.9	111.7	74.3	89.9	97.1	158.1	132.7
09	FINISHES	94.7	88.1	91.1	94.0	83.8	88.5	97.7	58.8	76.5	93.8	73.6	82.8	95.6	71.6	82.5	102.7	146.3	126.4
COVERS	DIVS. 10-14, 25, 28, 41, 43, 44, 46	100.0	58.1	90.7	100.0	92.0	98.2	100.0	79.7	95.5	100.0	84.4	96.5	100.0	84.0	96.4	100.0	116.7	103.7
21, 22, 23	FIRE SUPPRESSION, PLUMBING & HVAC	96.8	80.5	89.8	99.9	86.5	94.2	96.8	73.7	86.9	97.0	85.9	92.3	97.0	85.8	92.2	100.1	126.1	111.2
26, 27, 3370	ELECTRICAL, COMMUNICATIONS & UTIL.	99.3	89.1	94.0	97.9	80.6	88.9	96.4	63.5	79.3	97.5	98.6	98.1	95.1	98.6	96.9	101.4	129.8	116.2
MF2016	WEIGHTED AVERAGE	95.8	87.4	92.1	97.2	87.7	93.0	99.0	69.8	86.2	97.2	85.8	92.2	97.9	85.2	92.3	101.2	133.1	115.2

City Cost Indexes

MASSACHUSETTS

DIVISION		BROCKTON 023 MAT.	INST.	TOTAL	BUZZARDS BAY 025 MAT.	INST.	TOTAL	FALL RIVER 027 MAT.	INST.	TOTAL	FITCHBURG 014 MAT.	INST.	TOTAL	FRAMINGHAM 017 MAT.	INST.	TOTAL	GREENFIELD 013 MAT.	INST.	TOTAL
015433	CONTRACTOR EQUIPMENT		101.4	101.4		101.4	101.4		102.3	102.3		98.8	98.8		100.7	100.7		98.8	98.8
0241, 31 - 34	SITE & INFRASTRUCTURE, DEMOLITION	94.0	103.8	100.9	84.2	103.8	97.9	93.0	103.9	100.6	85.3	103.7	98.2	82.0	103.6	97.1	88.9	98.8	95.4
0310	Concrete Forming & Accessories	98.6	125.5	121.8	96.3	124.9	120.9	98.6	125.1	121.4	92.9	125.1	120.7	100.3	125.5	122.0	91.3	109.4	106.9
0320	Concrete Reinforcing	102.0	147.3	124.9	81.8	124.4	103.3	102.0	124.5	113.4	84.6	147.0	116.1	84.6	147.1	116.1	88.3	120.1	104.3
0330	Cast-in-Place Concrete	89.4	138.6	108.1	74.3	138.4	98.6	86.5	138.9	106.4	78.7	138.5	101.4	78.7	138.7	101.5	80.9	122.3	96.6
03	CONCRETE	92.2	133.3	111.0	78.1	129.0	101.4	90.8	129.2	108.4	76.7	132.9	102.3	79.2	133.3	103.9	80.0	115.4	96.2
04	MASONRY	103.0	144.1	128.5	96.0	142.3	124.7	103.7	142.3	127.6	102.0	140.7	126.0	108.7	140.8	128.6	106.9	122.8	116.8
05	METALS	97.0	128.7	106.8	91.9	118.8	100.1	97.0	119.2	103.8	94.0	125.0	103.5	94.1	128.3	104.6	96.5	111.5	101.1
06	WOOD, PLASTICS & COMPOSITES	101.4	122.2	113.0	98.3	122.2	111.6	101.4	122.4	113.1	99.7	122.2	112.2	105.9	122.0	114.8	97.8	107.1	103.0
07	THERMAL & MOISTURE PROTECTION	100.1	132.9	114.0	99.0	129.5	111.9	100.0	127.8	111.8	100.8	126.7	111.7	100.9	131.5	113.8	100.8	110.2	104.8
08	OPENINGS	98.8	133.2	106.7	94.6	122.4	101.0	98.8	122.2	104.2	100.1	133.2	107.7	90.9	133.1	100.6	100.5	113.1	103.4
0920	Plaster & Gypsum Board	92.1	122.4	112.5	87.9	122.4	111.1	92.1	122.4	112.5	108.3	122.4	117.8	111.0	122.4	118.6	109.3	106.9	107.7
0950, 0980	Ceilings & Acoustic Treatment	108.7	122.4	117.9	93.3	122.4	112.9	108.7	122.4	117.9	93.1	122.4	112.9	93.1	122.4	112.9	101.6	106.9	105.2
0960	Flooring	90.8	163.7	111.0	88.7	163.7	109.5	89.9	163.7	110.4	86.6	163.7	108.0	88.3	163.7	109.2	85.8	140.4	101.0
0970, 0990	Wall Finishes & Painting/Coating	90.5	137.4	117.9	90.5	137.4	117.9	90.5	137.4	117.9	87.8	137.4	116.7	88.7	137.4	117.1	87.8	114.3	103.3
09	FINISHES	92.8	133.9	115.2	88.0	133.9	112.9	92.6	134.1	115.1	87.5	133.9	112.7	88.2	133.7	113.0	89.4	115.7	103.7
COVERS	DIVS. 10 - 14, 25, 28, 41, 43, 44, 46	100.0	115.1	103.4	100.0	115.1	103.4	100.0	115.7	103.5	100.0	109.7	102.2	100.0	114.7	103.3	100.0	105.3	101.2
21, 22, 23	FIRE SUPPRESSION, PLUMBING & HVAC	100.3	109.7	104.3	97.2	109.2	102.3	100.3	109.3	104.1	97.4	110.4	102.9	97.4	123.4	108.5	97.4	102.4	99.5
26, 27, 3370	ELECTRICAL, COMMUNICATIONS & UTIL.	100.1	103.2	101.7	97.5	103.2	100.4	100.0	103.2	101.6	101.4	102.5	101.9	98.1	129.3	114.3	101.4	98.2	99.7
MF2016	WEIGHTED AVERAGE	97.9	121.8	108.4	92.8	119.5	104.5	97.7	119.6	107.3	94.4	120.8	105.9	93.7	127.9	108.7	95.7	109.0	101.5

MASSACHUSETTS

DIVISION		HYANNIS 026 MAT.	INST.	TOTAL	LAWRENCE 019 MAT.	INST.	TOTAL	LOWELL 018 MAT.	INST.	TOTAL	NEW BEDFORD 027 MAT.	INST.	TOTAL	PITTSFIELD 012 MAT.	INST.	TOTAL	SPRINGFIELD 010 - 011 MAT.	INST.	TOTAL
015433	CONTRACTOR EQUIPMENT		101.4	101.4		101.4	101.4		98.8	98.8		102.3	102.3		98.8	98.8		98.8	98.8
0241, 31 - 34	SITE & INFRASTRUCTURE, DEMOLITION	90.3	103.8	99.8	94.1	103.8	100.9	93.0	103.8	100.5	91.4	103.9	100.2	94.0	102.2	99.7	93.5	102.5	99.8
0310	Concrete Forming & Accessories	90.9	124.9	120.2	101.5	125.8	122.4	98.0	125.7	121.9	98.6	125.2	121.5	98.0	107.8	106.4	98.2	109.4	107.9
0320	Concrete Reinforcing	81.8	124.4	103.3	104.7	139.0	122.0	105.6	138.7	122.3	102.0	124.5	113.4	87.7	120.8	104.4	105.6	120.1	112.9
0330	Cast-in-Place Concrete	81.6	138.4	103.2	91.0	138.8	109.2	82.8	138.7	104.0	76.2	138.9	100.0	90.3	119.5	101.4	86.1	122.3	99.9
03	CONCRETE	84.0	129.0	104.6	92.4	132.0	110.5	84.7	131.7	106.2	86.0	129.3	105.8	86.0	113.8	98.7	86.3	115.4	99.9
04	MASONRY	102.0	142.3	127.0	114.6	144.9	133.4	100.9	140.7	125.6	101.9	142.3	126.9	101.5	113.8	109.2	101.2	122.8	114.6
05	METALS	93.4	118.8	101.2	96.9	125.7	105.7	96.9	122.2	104.6	97.0	119.3	103.9	96.7	111.5	101.2	99.5	111.5	103.2
06	WOOD, PLASTICS & COMPOSITES	92.1	122.2	108.8	106.1	122.2	115.0	105.6	122.2	114.8	101.4	122.4	113.1	105.6	107.5	106.7	105.6	107.1	106.4
07	THERMAL & MOISTURE PROTECTION	99.5	129.5	112.2	101.5	133.1	114.8	101.2	131.9	114.2	99.9	127.8	111.7	101.3	106.9	103.7	101.2	110.2	105.0
08	OPENINGS	95.2	122.4	101.4	94.8	130.9	103.1	101.6	130.9	108.3	98.8	127.1	105.3	101.6	113.6	104.3	101.6	113.1	104.2
0920	Plaster & Gypsum Board	83.7	122.4	109.7	113.6	122.4	119.5	113.6	122.4	119.5	92.1	122.4	112.5	113.6	107.3	109.4	113.6	106.9	109.1
0950, 0980	Ceilings & Acoustic Treatment	101.9	122.4	115.7	103.5	122.4	116.3	103.5	122.4	116.3	108.7	122.4	117.9	103.5	107.3	106.1	103.5	106.9	105.8
0960	Flooring	86.0	163.7	107.6	88.8	163.7	109.6	88.8	163.7	109.6	89.9	163.7	110.4	89.2	132.9	101.4	88.2	140.4	102.7
0970, 0990	Wall Finishes & Painting/Coating	90.5	137.4	117.9	87.9	137.4	116.8	87.8	137.4	116.7	90.5	137.4	117.9	87.8	114.3	103.3	89.0	114.3	103.7
09	FINISHES	88.6	133.9	113.2	91.4	133.9	114.5	91.4	133.9	114.5	92.4	134.1	115.1	91.5	113.8	103.3	91.3	115.7	104.6
COVERS	DIVS. 10 - 14, 25, 28, 41, 43, 44, 46	100.0	115.1	103.4	100.0	115.2	103.4	100.0	115.2	103.4	100.0	115.7	103.5	100.0	103.7	100.8	100.0	105.3	101.2
21, 22, 23	FIRE SUPPRESSION, PLUMBING & HVAC	100.3	109.2	104.1	100.1	125.9	111.1	100.1	125.8	111.1	100.3	109.3	104.1	100.1	98.1	99.2	100.1	102.4	101.1
26, 27, 3370	ELECTRICAL, COMMUNICATIONS & UTIL.	97.9	103.2	100.6	100.5	129.8	115.7	100.9	126.0	114.0	100.9	103.2	102.1	100.9	98.2	99.5	101.0	102.4	101.1
MF2016	WEIGHTED AVERAGE	95.1	119.5	105.8	98.0	128.5	111.4	97.1	127.1	110.3	97.0	119.8	107.0	97.3	106.5	101.3	97.8	109.0	102.7

MASSACHUSETTS / MICHIGAN

DIVISION		WORCESTER 015 - 016 MAT.	INST.	TOTAL	ANN ARBOR 481 MAT.	INST.	TOTAL	BATTLE CREEK 490 MAT.	INST.	TOTAL	BAY CITY 487 MAT.	INST.	TOTAL	DEARBORN 481 MAT.	INST.	TOTAL	DETROIT 482 MAT.	INST.	TOTAL
015433	CONTRACTOR EQUIPMENT		98.8	98.8		112.1	112.1		98.9	98.9		112.1	112.1		112.1	112.1		96.3	96.3
0241, 31 - 34	SITE & INFRASTRUCTURE, DEMOLITION	93.4	103.7	100.6	82.2	96.7	92.3	92.1	85.8	87.7	73.6	96.3	89.5	81.9	96.8	92.3	97.3	100.7	99.7
0310	Concrete Forming & Accessories	98.6	125.1	121.5	96.4	106.3	105.0	95.2	80.9	82.9	96.5	82.6	84.5	96.3	107.0	105.5	99.3	107.0	105.9
0320	Concrete Reinforcing	105.6	147.0	126.5	98.4	104.5	101.5	88.6	80.9	84.7	98.4	103.7	101.1	98.4	104.6	101.5	99.3	106.5	102.9
0330	Cast-in-Place Concrete	85.6	138.5	105.7	88.4	101.5	93.3	85.3	95.8	89.3	84.6	87.7	85.7	86.4	102.3	92.5	103.2	101.9	102.7
03	CONCRETE	86.1	132.9	107.5	89.2	105.2	96.5	84.8	85.8	85.3	87.4	89.6	88.4	88.3	105.8	96.3	101.3	104.5	102.8
04	MASONRY	100.7	140.7	125.5	97.8	100.5	99.5	97.3	81.7	87.7	97.4	81.0	87.2	97.7	102.4	100.6	96.1	102.5	100.0
05	METALS	99.5	125.0	107.3	97.2	117.5	103.4	99.2	84.1	94.6	97.8	115.4	103.2	97.2	117.7	103.5	98.5	96.3	97.8
06	WOOD, PLASTICS & COMPOSITES	106.0	122.2	115.0	90.9	107.9	100.4	89.2	79.2	83.6	90.9	82.1	86.0	90.9	107.9	100.4	95.7	107.8	102.4
07	THERMAL & MOISTURE PROTECTION	101.2	126.7	112.0	106.1	102.7	104.6	97.5	81.9	90.9	103.3	84.5	95.4	104.3	106.2	105.1	103.5	106.7	104.9
08	OPENINGS	101.6	133.2	108.8	96.6	103.3	98.1	91.7	77.3	88.4	96.6	85.1	93.9	96.6	102.7	98.0	98.3	103.7	99.5
0920	Plaster & Gypsum Board	113.6	122.4	119.5	105.4	107.8	107.0	98.0	75.2	82.7	105.4	81.3	89.2	105.4	107.8	107.0	100.6	107.8	105.4
0950, 0980	Ceilings & Acoustic Treatment	103.5	122.4	116.3	88.9	107.8	101.6	82.4	75.2	77.5	89.8	81.3	84.0	88.9	107.8	101.6	96.9	107.8	104.3
0960	Flooring	88.8	163.7	109.6	90.1	113.4	96.6	88.3	74.5	84.4	90.1	81.3	87.7	89.6	107.4	94.6	97.6	107.4	100.3
0970, 0990	Wall Finishes & Painting/Coating	87.8	137.4	116.7	82.0	99.9	92.4	86.7	79.7	82.6	82.0	82.7	82.4	82.0	98.1	91.4	92.3	99.8	96.7
09	FINISHES	91.4	133.9	114.5	90.3	107.3	99.5	85.7	79.0	82.0	90.1	81.7	85.5	90.2	106.5	99.0	97.4	106.6	102.4
COVERS	DIVS. 10 - 14, 25, 28, 41, 43, 44, 46	100.0	109.7	102.2	100.0	97.1	99.4	100.0	97.6	99.5	100.0	91.2	98.0	100.0	97.6	99.5	100.0	102.2	100.5
21, 22, 23	FIRE SUPPRESSION, PLUMBING & HVAC	100.1	110.4	104.5	100.2	95.2	98.0	100.2	86.0	94.1	100.2	81.9	92.4	100.2	104.2	101.9	100.0	104.2	101.8
26, 27, 3370	ELECTRICAL, COMMUNICATIONS & UTIL.	101.0	102.5	101.7	96.3	100.9	98.7	93.5	82.1	87.6	95.5	87.1	91.1	96.3	101.5	99.0	99.0	101.4	100.3
MF2016	WEIGHTED AVERAGE	97.7	120.8	107.8	96.2	102.4	98.9	94.8	83.7	89.9	95.7	88.3	92.5	96.1	104.7	99.8	99.2	103.0	100.9

For customer support on your Electrical Costs with RSMeans data, call 800.448.8182.

529

MICHIGAN

DIVISION		FLINT 484 - 485			GAYLORD 497			GRAND RAPIDS 493, 495			IRON MOUNTAIN 498 - 499			JACKSON 492			KALAMAZOO 491		
		MAT.	INST.	TOTAL	MAT.	INST.	TOTAL	MAT.	INST.	TOTAL	MAT.	INST.	TOTAL	MAT.	INST.	TOTAL	MAT.	INST.	TOTAL
015433	CONTRACTOR EQUIPMENT		112.1	112.1		106.8	106.8		98.9	98.9		93.6	93.6		106.8	106.8		98.9	98.9
0241, 31 - 34	SITE & INFRASTRUCTURE, DEMOLITION	71.5	96.2	88.8	86.6	82.6	83.8	91.0	85.7	87.3	95.1	91.6	92.7	109.1	85.1	92.3	92.4	85.8	87.7
0310	Concrete Forming & Accessories	99.2	85.8	87.6	93.9	73.2	76.0	93.8	79.4	81.4	85.7	79.8	80.6	90.5	84.1	85.0	95.2	80.8	82.8
0320	Concrete Reinforcing	98.4	104.1	101.3	82.6	89.0	85.8	93.0	80.7	86.8	82.5	85.7	84.1	80.2	104.0	92.2	88.6	78.7	83.6
0330	Cast-in-Place Concrete	89.0	90.2	89.4	85.1	82.2	84.0	89.5	94.0	91.2	100.7	70.8	89.3	84.9	92.1	87.6	87.0	95.8	90.3
03	CONCRETE	89.7	91.9	90.7	81.7	81.0	81.4	88.6	84.5	86.7	90.1	78.4	84.8	77.0	91.6	83.7	87.9	85.4	86.7
04	MASONRY	97.9	89.6	92.7	108.0	75.7	88.0	90.2	77.2	82.2	93.2	80.8	85.5	87.5	87.5	87.5	95.8	81.7	87.1
05	METALS	97.2	118.1	103.0	100.6	113.0	104.4	96.3	83.6	92.4	100.0	91.8	97.5	100.8	114.1	104.9	99.2	83.2	94.3
06	WOOD, PLASTICS & COMPOSITES	94.3	84.5	88.9	82.6	71.9	76.6	93.2	77.6	84.5	78.2	79.8	79.1	81.4	82.1	81.8	89.2	79.2	83.6
07	THERMAL & MOISTURE PROTECTION	103.7	88.7	97.3	96.2	70.8	85.4	98.8	74.1	88.3	100.0	76.4	90.0	95.4	89.0	92.7	97.5	81.9	90.9
08	OPENINGS	96.6	86.3	94.2	92.1	80.0	89.3	101.3	77.0	95.8	99.0	71.4	92.7	91.2	88.5	90.6	91.7	77.2	88.4
0920	Plaster & Gypsum Board	107.4	83.8	91.5	97.8	70.1	79.2	101.4	73.6	82.7	52.7	79.8	70.9	96.2	80.6	85.7	98.0	75.2	82.7
0950, 0980	Ceilings & Acoustic Treatment	88.9	83.8	85.4	81.5	70.1	73.8	95.1	73.6	80.0	80.5	79.8	80.0	81.5	80.6	80.9	82.4	75.2	77.5
0960	Flooring	90.1	90.9	90.3	81.8	90.5	84.2	90.3	80.5	87.6	100.0	92.4	97.9	80.5	83.1	81.2	88.3	74.5	84.4
0970, 0990	Wall Finishes & Painting/Coating	82.0	81.9	81.9	83.1	80.9	81.9	87.9	80.4	83.5	100.4	67.9	81.5	83.1	98.1	91.9	86.7	79.7	82.6
09	FINISHES	89.8	85.7	87.5	85.2	76.5	80.5	91.3	79.1	84.7	85.6	80.9	83.1	86.1	84.6	85.3	85.7	79.0	82.0
COVERS	DIVS. 10 - 14, 25, 28, 41, 43, 44, 46	100.0	92.5	98.3	100.0	80.5	95.7	100.0	97.3	99.4	100.0	89.0	97.6	100.0	93.7	98.6	100.0	97.6	99.5
21, 22, 23	FIRE SUPPRESSION, PLUMBING & HVAC	100.2	86.8	94.5	97.3	75.3	87.9	100.0	82.9	92.7	97.2	82.1	90.7	97.3	86.9	92.8	100.2	81.5	92.2
26, 27, 3370	ELECTRICAL, COMMUNICATIONS & UTIL.	96.3	94.1	95.2	91.7	75.9	83.5	98.5	87.6	92.8	97.4	81.3	89.0	95.2	99.8	97.6	93.4	78.3	85.5
MF2016	WEIGHTED AVERAGE	95.9	92.3	94.4	94.1	80.7	88.2	96.4	82.9	90.5	96.0	82.4	90.0	93.4	91.8	92.7	95.1	82.1	89.4

MICHIGAN / MINNESOTA

DIVISION		LANSING 488 - 489			MUSKEGON 494			ROYAL OAK 480, 483			SAGINAW 486			TRAVERSE CITY 496			BEMIDJI 566		
		MAT.	INST.	TOTAL	MAT.	INST.	TOTAL	MAT.	INST.	TOTAL	MAT.	INST.	TOTAL	MAT.	INST.	TOTAL	MAT.	INST.	TOTAL
015433	CONTRACTOR EQUIPMENT		112.1	112.1		98.9	98.9		91.7	91.7		112.1	112.1		93.6	93.6		100.3	100.3
0241, 31 - 34	SITE & INFRASTRUCTURE, DEMOLITION	94.1	96.5	95.8	90.1	85.7	87.0	86.1	95.8	92.9	74.6	96.2	89.8	80.8	91.1	88.0	96.1	99.6	98.6
0310	Concrete Forming & Accessories	93.3	78.4	80.5	95.5	79.2	81.5	91.7	106.6	104.6	96.4	83.4	85.2	85.7	72.9	74.7	85.0	85.2	85.1
0320	Concrete Reinforcing	101.2	103.8	102.5	89.2	80.6	84.9	89.2	90.0	89.6	98.4	103.6	101.1	83.7	77.9	80.8	94.4	98.7	96.6
0330	Cast-in-Place Concrete	99.3	89.0	95.4	85.0	93.9	88.4	77.3	102.0	86.7	87.3	87.6	87.4	78.5	78.5	78.5	102.1	107.1	104.0
03	CONCRETE	92.1	88.2	90.3	83.2	84.3	83.7	76.7	101.5	88.0	88.7	89.9	89.3	73.9	76.6	75.1	89.9	96.4	92.9
04	MASONRY	96.9	88.1	91.5	94.4	77.2	83.8	91.5	100.8	97.3	99.3	81.0	87.9	91.4	76.9	82.4	100.1	100.9	100.6
05	METALS	96.3	115.5	102.2	96.9	83.4	92.8	100.4	91.8	97.8	97.2	115.3	102.8	99.9	88.6	96.4	93.5	116.4	100.6
06	WOOD, PLASTICS & COMPOSITES	89.9	74.6	81.4	86.0	77.6	81.3	86.7	107.9	98.5	87.3	83.5	85.2	78.2	72.6	75.1	66.8	80.7	74.5
07	THERMAL & MOISTURE PROTECTION	101.7	86.1	95.1	96.4	74.4	87.1	101.9	102.6	102.2	104.6	84.8	96.2	99.0	68.4	86.0	106.1	91.9	100.1
08	OPENINGS	100.4	80.6	95.9	91.0	77.6	87.9	96.4	103.2	98.0	94.6	85.9	92.6	99.0	66.1	91.5	98.8	101.8	99.5
0920	Plaster & Gypsum Board	99.3	73.6	82.0	77.1	73.6	74.7	103.1	107.8	106.2	105.4	82.6	90.1	52.7	72.5	66.0	108.9	80.5	89.9
0950, 0980	Ceilings & Acoustic Treatment	88.7	73.6	78.5	83.2	73.6	76.7	88.1	107.8	101.4	88.9	82.6	84.7	80.5	72.5	75.1	125.3	80.5	95.1
0960	Flooring	91.9	83.1	89.5	87.0	76.4	84.1	87.2	107.4	92.8	90.1	81.3	87.7	100.0	90.5	97.4	90.6	90.4	90.6
0970, 0990	Wall Finishes & Painting/Coating	89.6	80.4	84.2	84.9	79.1	81.5	83.6	98.1	92.0	82.0	82.7	82.4	100.4	41.2	65.9	90.9	108.5	101.1
09	FINISHES	89.4	78.3	83.4	82.0	77.5	79.6	89.2	106.2	98.4	90.0	82.5	85.9	84.6	72.7	78.1	97.8	88.1	92.5
COVERS	DIVS. 10 - 14, 25, 28, 41, 43, 44, 46	100.0	91.7	98.2	100.0	97.3	99.4	100.0	102.4	100.5	100.0	91.4	98.1	100.0	87.2	97.2	100.0	95.3	99.0
21, 22, 23	FIRE SUPPRESSION, PLUMBING & HVAC	100.0	87.1	94.5	99.9	82.9	92.7	97.3	99.9	98.4	100.2	81.4	92.2	97.2	79.0	89.4	97.3	82.8	91.1
26, 27, 3370	ELECTRICAL, COMMUNICATIONS & UTIL.	97.9	92.5	95.1	93.8	77.0	85.1	98.3	101.4	99.9	94.7	89.2	91.8	93.2	75.9	84.2	104.4	104.2	104.3
MF2016	WEIGHTED AVERAGE	97.0	90.1	94.0	93.6	81.2	88.2	94.3	100.5	97.0	95.6	88.7	92.6	92.9	78.4	86.6	97.0	96.0	96.6

MINNESOTA

DIVISION		BRAINERD 564			DETROIT LAKES 565			DULUTH 556 - 558			MANKATO 560			MINNEAPOLIS 553 - 555			ROCHESTER 559		
		MAT.	INST.	TOTAL	MAT.	INST.	TOTAL	MAT.	INST.	TOTAL	MAT.	INST.	TOTAL	MAT.	INST.	TOTAL	MAT.	INST.	TOTAL
015433	CONTRACTOR EQUIPMENT		103.0	103.0		100.3	100.3		103.4	103.4		103.0	103.0		107.9	107.9		103.4	103.4
0241, 31 - 34	SITE & INFRASTRUCTURE, DEMOLITION	98.6	104.6	102.8	94.3	99.8	98.2	99.8	102.9	102.0	95.2	104.3	101.6	97.9	108.3	105.2	98.8	102.6	101.4
0310	Concrete Forming & Accessories	85.9	85.7	85.7	82.1	84.9	84.5	99.5	96.7	97.1	93.6	95.3	95.1	98.1	114.7	112.4	98.1	98.9	98.8
0320	Concrete Reinforcing	93.2	98.6	95.9	94.4	98.5	96.5	95.2	99.0	97.2	93.0	106.0	99.6	94.9	106.5	100.8	92.9	106.2	99.6
0330	Cast-in-Place Concrete	111.0	111.6	111.2	99.2	110.0	103.3	102.2	101.6	102.0	102.3	99.2	101.1	101.5	115.9	106.9	99.7	99.9	99.8
03	CONCRETE	93.8	98.2	95.8	87.6	97.3	92.0	95.1	99.9	97.3	89.4	99.6	94.1	98.0	114.3	105.5	91.5	101.6	96.2
04	MASONRY	125.1	106.4	113.5	124.5	103.7	111.6	99.2	110.1	106.0	112.9	104.3	107.6	105.4	118.5	113.5	98.8	107.8	104.4
05	METALS	94.7	116.0	101.2	93.5	115.7	100.3	101.2	118.8	106.6	94.5	119.7	102.2	101.2	124.0	108.2	100.9	122.0	107.4
06	WOOD, PLASTICS & COMPOSITES	83.6	77.8	80.4	64.1	78.0	71.8	93.1	93.6	93.3	92.5	93.3	93.0	95.7	112.1	104.8	95.7	97.4	96.6
07	THERMAL & MOISTURE PROTECTION	104.4	101.8	103.3	105.9	100.6	103.7	105.1	104.0	104.6	104.8	92.8	99.7	103.4	116.7	109.0	110.0	93.5	103.0
08	OPENINGS	85.8	100.2	89.1	98.8	100.3	99.1	105.8	106.0	105.8	90.3	112.2	95.3	101.3	122.6	106.2	100.7	113.5	103.6
0920	Plaster & Gypsum Board	94.8	77.8	83.3	107.9	77.8	87.7	92.7	93.9	93.5	99.4	93.7	95.6	100.8	112.7	108.8	103.6	97.8	99.7
0950, 0980	Ceilings & Acoustic Treatment	57.8	77.8	71.3	125.3	77.8	93.3	94.3	93.9	94.0	57.8	93.7	82.0	99.6	112.7	108.4	96.3	97.8	97.3
0960	Flooring	89.5	90.4	89.7	89.4	90.4	89.7	92.7	122.4	100.9	91.4	86.5	90.0	100.8	116.9	105.3	93.6	86.5	91.6
0970, 0990	Wall Finishes & Painting/Coating	85.3	108.5	98.8	90.9	87.2	88.7	85.5	109.2	99.3	96.2	102.6	100.0	100.9	129.5	117.6	85.7	102.6	95.6
09	FINISHES	81.8	88.0	85.2	97.1	85.5	90.8	90.5	102.2	96.8	83.6	94.7	89.6	99.8	116.7	109.0	91.4	97.5	94.7
COVERS	DIVS. 10 - 14, 25, 28, 41, 43, 44, 46	100.0	96.9	99.3	100.0	96.7	99.3	100.0	97.9	99.5	100.0	97.1	99.4	100.0	105.6	101.2	100.0	98.1	99.6
21, 22, 23	FIRE SUPPRESSION, PLUMBING & HVAC	96.5	86.3	92.1	97.3	85.3	92.2	100.0	96.3	98.4	96.5	86.4	92.2	99.9	108.4	103.6	100.0	93.9	97.4
26, 27, 3370	ELECTRICAL, COMMUNICATIONS & UTIL.	102.1	99.5	100.8	104.2	70.3	86.6	103.1	99.5	101.2	108.5	92.0	99.9	101.2	104.4	102.9	101.6	92.0	96.6
MF2016	WEIGHTED AVERAGE	95.9	97.5	96.6	97.8	92.0	95.3	99.8	102.7	101.1	96.0	98.0	96.9	100.4	113.0	105.9	98.8	100.9	99.9

DIVISION		MINNESOTA																MISSISSIPPI		
		SAINT PAUL			ST. CLOUD			THIEF RIVER FALLS			WILLMAR			WINDOM			BILOXI		
		550 - 551			563			567			562			561			395		
		MAT.	INST.	TOTAL	MAT.	INST.	TOTAL	MAT.	INST.	TOTAL	MAT.	INST.	TOTAL	MAT.	INST.	TOTAL	MAT.	INST.	TOTAL
015433	CONTRACTOR EQUIPMENT		103.4	103.4		103.0	103.0		100.3	100.3		103.0	103.0		103.0	103.0		102.5	102.5
0241, 31 - 34	SITE & INFRASTRUCTURE, DEMOLITION	96.0	103.6	101.3	93.7	105.6	102.1	95.1	99.4	98.1	93.2	104.6	101.2	87.0	103.9	98.8	102.9	89.7	93.6
0310	Concrete Forming & Accessories	97.1	114.4	112.0	83.3	113.7	109.5	85.6	84.2	84.4	83.1	89.8	88.9	87.3	85.8	86.0	93.4	66.9	70.5
0320	Concrete Reinforcing	95.5	106.5	101.1	93.2	105.8	99.6	94.8	98.4	96.6	92.8	105.6	99.3	92.8	105.0	99.0	83.1	50.1	66.4
0330	Cast-in-Place Concrete	101.2	114.4	106.2	98.0	114.1	104.1	101.2	83.5	94.5	99.5	82.2	92.9	86.1	87.3	86.5	114.5	67.1	96.5
03	CONCRETE	94.6	113.7	103.3	85.5	113.1	98.1	88.7	87.8	88.3	85.5	91.2	88.1	76.7	91.1	83.3	98.1	65.7	83.3
04	MASONRY	97.7	118.6	110.7	109.0	108.7	108.8	100.1	98.4	99.0	113.0	102.9	106.7	124.2	85.2	100.0	97.5	61.5	75.2
05	METALS	101.2	123.5	108.1	95.3	120.3	103.0	93.7	114.9	100.2	94.4	119.2	102.1	94.4	117.9	101.6	89.4	84.8	88.0
06	WOOD, PLASTICS & COMPOSITES	97.1	111.8	105.3	81.2	111.8	98.2	67.7	80.7	75.0	80.9	83.7	82.5	84.9	83.7	84.2	102.6	67.6	83.1
07	THERMAL & MOISTURE PROTECTION	106.0	116.3	110.4	104.5	106.0	105.1	106.9	87.4	98.6	104.3	97.9	101.6	104.3	83.6	95.5	94.8	63.9	81.7
08	OPENINGS	100.3	121.5	105.2	90.7	121.5	97.8	98.8	101.8	99.5	87.8	106.0	91.9	91.3	106.0	94.7	100.8	57.8	90.9
0920	Plaster & Gypsum Board	102.0	112.7	109.1	94.8	112.7	106.8	108.6	80.5	89.7	94.8	83.8	87.4	94.8	83.8	87.4	97.0	67.2	77.0
0950, 0980	Ceilings & Acoustic Treatment	95.8	112.7	107.2	57.8	112.7	94.8	125.3	80.5	95.1	57.8	83.8	75.4	57.8	83.8	75.4	90.2	67.2	74.7
0960	Flooring	87.7	116.9	95.8	86.3	90.4	87.4	90.3	90.4	90.3	87.6	90.4	88.4	90.2	90.4	90.2	88.6	57.5	80.0
0970, 0990	Wall Finishes & Painting/Coating	88.1	123.2	108.6	96.2	123.2	112.0	90.9	87.2	88.7	90.9	87.2	88.7	90.9	102.6	97.7	81.9	49.0	62.7
09	FINISHES	91.5	115.8	104.7	81.2	111.2	97.5	97.6	85.7	91.1	81.2	89.1	85.5	81.5	88.4	85.2	85.7	63.6	73.7
COVERS	DIVS. 10 - 14, 25, 28, 41, 43, 44, 46	100.0	102.8	100.6	100.0	102.6	100.6	100.0	95.2	98.9	100.0	97.5	99.4	100.0	94.3	98.7	100.0	72.7	93.9
21, 22, 23	FIRE SUPPRESSION, PLUMBING & HVAC	99.9	111.3	104.8	99.6	108.1	103.2	97.3	82.4	91.0	96.5	101.9	98.8	96.5	82.3	90.4	99.9	57.6	81.8
26, 27, 3370	ELECTRICAL, COMMUNICATIONS & UTIL.	101.6	113.2	107.6	102.1	113.2	107.9	101.6	70.2	85.3	102.1	79.7	90.5	108.5	92.0	99.9	100.0	56.7	77.5
MF2016	WEIGHTED AVERAGE	98.9	114.1	105.6	95.2	111.3	102.3	96.6	89.2	93.4	94.3	97.4	95.6	94.6	92.3	93.6	96.7	65.5	83.1

DIVISION		MISSISSIPPI																		
		CLARKSDALE			COLUMBUS			GREENVILLE			GREENWOOD			JACKSON			LAUREL		
		386			397			387			389			390 - 392			394		
		MAT.	INST.	TOTAL	MAT.	INST.	TOTAL	MAT.	INST.	TOTAL	MAT.	INST.	TOTAL	MAT.	INST.	TOTAL	MAT.	INST.	TOTAL
015433	CONTRACTOR EQUIPMENT		102.5	102.5		102.5	102.5		102.5	102.5		102.5	102.5		102.5	102.5		102.5	102.5
0241, 31 - 34	SITE & INFRASTRUCTURE, DEMOLITION	97.3	88.3	91.0	101.7	89.2	92.9	102.8	89.5	93.5	100.1	88.1	91.7	98.0	89.5	92.1	106.6	88.4	93.9
0310	Concrete Forming & Accessories	84.9	45.7	51.1	82.3	47.6	52.4	81.7	64.5	66.9	94.0	45.5	52.2	90.7	65.6	69.1	82.4	61.8	64.6
0320	Concrete Reinforcing	100.7	64.8	82.6	89.0	35.9	62.2	101.2	51.5	76.1	100.7	44.0	72.1	92.2	51.1	71.5	89.6	32.9	61.0
0330	Cast-in-Place Concrete	102.2	50.8	82.7	116.8	56.1	93.7	105.3	57.2	87.0	109.9	50.4	87.3	99.8	65.7	86.8	114.2	52.1	90.6
03	CONCRETE	93.7	52.9	75.1	99.5	50.8	77.2	99.1	61.5	81.9	99.9	49.1	76.7	92.4	64.8	79.8	102.0	55.2	80.6
04	MASONRY	92.5	41.0	60.6	124.2	47.2	76.4	138.0	69.1	95.3	93.2	40.9	60.8	103.3	59.2	75.9	120.1	43.8	72.8
05	METALS	89.3	87.2	88.6	86.5	79.3	84.3	90.4	85.7	88.9	89.3	77.4	85.7	95.7	85.2	92.5	86.6	76.0	83.3
06	WOOD, PLASTICS & COMPOSITES	85.2	46.4	63.6	88.6	47.3	65.6	82.0	65.8	73.0	97.6	46.4	69.1	97.2	67.0	80.4	89.7	67.6	77.4
07	THERMAL & MOISTURE PROTECTION	95.8	49.5	76.2	94.8	54.5	77.7	96.3	63.8	82.5	96.2	51.8	77.4	93.1	62.4	80.1	94.9	55.5	78.2
08	OPENINGS	96.7	49.8	85.9	100.4	43.3	87.3	96.4	57.1	87.4	96.7	43.6	84.5	101.2	57.8	91.3	97.3	54.4	87.4
0920	Plaster & Gypsum Board	87.5	45.5	59.3	88.0	46.4	60.0	87.2	65.3	72.5	98.1	45.5	62.7	85.8	66.6	72.9	88.0	67.2	74.0
0950, 0980	Ceilings & Acoustic Treatment	84.7	45.5	58.2	84.9	46.4	58.9	88.5	65.3	72.9	98.1	45.5	62.7	90.8	66.6	74.5	84.9	67.2	72.9
0960	Flooring	95.7	47.6	82.3	82.9	53.4	74.7	94.0	47.6	81.1	101.4	47.6	86.5	85.9	57.5	78.1	81.6	47.6	72.1
0970, 0990	Wall Finishes & Painting/Coating	92.1	49.0	67.0	81.9	49.0	62.7	92.1	49.0	67.0	92.1	49.0	67.0	82.6	49.0	63.0	81.9	49.0	62.7
09	FINISHES	88.7	46.0	65.5	81.8	48.4	63.6	89.4	60.0	73.4	92.2	46.0	67.1	85.7	62.7	73.2	81.9	59.2	69.5
COVERS	DIVS. 10 - 14, 25, 28, 41, 43, 44, 46	100.0	49.8	88.8	100.0	50.8	89.1	100.0	71.8	93.7	100.0	49.8	88.8	100.0	71.9	93.7	100.0	35.4	85.6
21, 22, 23	FIRE SUPPRESSION, PLUMBING & HVAC	98.6	53.3	79.2	98.2	54.9	79.7	100.0	58.9	82.4	98.6	53.5	79.3	99.9	58.9	82.4	98.2	43.4	74.8
26, 27, 3370	ELECTRICAL, COMMUNICATIONS & UTIL.	95.6	43.9	68.7	97.5	46.6	71.0	95.6	56.7	75.4	95.6	41.2	67.3	101.5	56.7	78.2	99.0	56.7	77.0
MF2016	WEIGHTED AVERAGE	94.7	55.2	77.4	96.6	53.9	78.7	98.3	64.8	83.9	96.0	53.2	77.3	97.3	63.3	83.3	96.7	56.3	79.0

DIVISION		MISSISSIPPI									MISSOURI								
		MCCOMB			MERIDIAN			TUPELO			BOWLING GREEN			CAPE GIRARDEAU			CHILLICOTHE		
		396			393			388			633			637			646		
		MAT.	INST.	TOTAL	MAT.	INST.	TOTAL	MAT.	INST.	TOTAL	MAT.	INST.	TOTAL	MAT.	INST.	TOTAL	MAT.	INST.	TOTAL
015433	CONTRACTOR EQUIPMENT		102.5	102.5		102.5	102.5		102.5	102.5		110.5	110.5		110.5	110.5		104.8	104.8
0241, 31 - 34	SITE & INFRASTRUCTURE, DEMOLITION	94.8	88.2	90.2	98.8	89.7	92.4	94.8	88.2	90.2	89.1	94.9	93.1	90.7	94.8	93.6	105.8	92.9	96.8
0310	Concrete Forming & Accessories	82.3	47.0	51.9	79.6	65.3	67.3	82.2	47.4	52.2	93.6	91.5	91.8	86.4	78.3	79.4	82.9	96.1	94.2
0320	Concrete Reinforcing	90.1	34.5	62.1	89.0	51.1	69.9	98.5	44.1	71.0	106.8	96.7	101.7	108.1	78.4	93.1	98.0	101.3	99.7
0330	Cast-in-Place Concrete	101.3	50.3	81.9	108.4	66.7	92.6	102.2	71.0	90.3	91.4	98.5	94.1	90.4	88.1	89.6	94.8	89.0	92.6
03	CONCRETE	89.0	48.1	70.3	93.5	65.0	80.5	93.2	57.0	76.7	92.8	96.4	94.5	92.0	83.7	88.2	98.6	95.3	97.1
04	MASONRY	125.4	40.4	72.7	97.1	60.8	74.6	126.7	43.6	75.2	113.0	93.7	101.0	109.9	79.8	91.2	103.0	92.1	96.3
05	METALS	86.7	75.3	83.2	87.6	85.2	86.8	89.2	77.6	85.7	97.5	119.0	104.1	98.6	110.2	102.2	92.4	110.9	98.1
06	WOOD, PLASTICS & COMPOSITES	88.6	48.9	66.6	86.1	65.8	74.8	82.5	47.3	62.9	99.6	91.5	95.1	92.6	75.3	83.0	93.0	97.5	95.5
07	THERMAL & MOISTURE PROTECTION	94.3	51.9	76.3	94.5	63.0	81.1	95.8	53.1	77.7	100.1	99.3	99.8	99.6	85.4	93.6	92.5	94.5	93.3
08	OPENINGS	100.4	43.3	87.3	100.1	57.1	90.2	96.6	47.0	85.3	102.9	98.4	101.9	102.9	74.4	96.4	92.1	97.6	93.3
0920	Plaster & Gypsum Board	88.0	48.0	61.2	88.0	65.3	72.8	87.2	46.4	59.8	103.9	91.6	95.6	103.2	75.0	84.3	100.6	97.3	98.4
0950, 0980	Ceilings & Acoustic Treatment	84.9	48.0	60.1	86.8	65.3	72.3	84.7	46.4	58.8	91.4	91.6	91.5	91.4	75.0	80.3	91.1	97.3	95.3
0960	Flooring	82.9	47.6	73.1	81.5	57.5	74.9	94.2	47.6	81.3	95.6	98.5	96.4	92.2	87.2	90.8	94.7	104.7	97.5
0970, 0990	Wall Finishes & Painting/Coating	81.9	49.0	62.7	81.9	49.0	62.7	92.1	47.3	66.0	96.0	105.8	101.8	96.0	70.1	80.9	95.5	94.2	94.8
09	FINISHES	81.2	47.4	62.8	81.4	62.4	71.1	88.2	47.0	65.8	97.7	94.0	95.7	96.5	77.7	86.3	97.8	97.8	97.8
COVERS	DIVS. 10 - 14, 25, 28, 41, 43, 44, 46	100.0	52.8	89.5	100.0	72.3	93.8	100.0	50.8	89.1	100.0	82.7	96.2	100.0	83.4	96.3	100.0	83.4	96.3
21, 22, 23	FIRE SUPPRESSION, PLUMBING & HVAC	98.2	52.8	78.8	99.9	59.8	82.7	98.7	54.5	79.8	96.9	95.9	96.4	99.9	99.4	99.7	97.0	100.9	98.7
26, 27, 3370	ELECTRICAL, COMMUNICATIONS & UTIL.	96.1	46.8	70.5	99.0	58.7	78.1	95.3	46.7	70.0	101.8	79.6	90.3	101.7	101.0	101.3	95.9	77.9	86.5
MF2016	WEIGHTED AVERAGE	95.0	53.7	76.9	95.1	65.9	82.3	96.2	55.9	78.6	98.7	95.0	97.1	99.1	91.5	95.8	96.5	95.1	95.9

For customer support on your Electrical Costs with RSMeans data, call 800.448.8182.

531

MISSOURI

DIVISION		COLUMBIA 652			FLAT RIVER 636			HANNIBAL 634			HARRISONVILLE 647			JEFFERSON CITY 650 - 651			JOPLIN 648		
		MAT.	INST.	TOTAL	MAT.	INST.	TOTAL	MAT.	INST.	TOTAL	MAT.	INST.	TOTAL	MAT.	INST.	TOTAL	MAT.	INST.	TOTAL
015433	CONTRACTOR EQUIPMENT		114.0	114.0		110.5	110.5		110.5	110.5		104.8	104.8		114.0	114.0		108.3	108.3
0241, 31 - 34	SITE & INFRASTRUCTURE, DEMOLITION	95.1	99.0	97.8	91.8	94.7	93.9	87.0	94.9	92.5	97.1	94.2	95.1	94.7	99.0	97.7	106.3	97.8	100.4
0310	Concrete Forming & Accessories	83.2	81.6	81.8	99.9	86.0	87.9	91.9	83.2	84.4	80.4	102.0	99.0	94.4	81.6	83.3	93.4	74.5	77.1
0320	Concrete Reinforcing	87.7	96.2	92.0	108.1	103.5	105.8	106.2	96.7	101.4	97.6	106.8	102.3	94.7	96.2	95.5	101.2	86.9	94.0
0330	Cast-in-Place Concrete	83.5	86.5	84.6	94.4	94.3	94.3	86.5	98.4	91.0	97.1	103.1	99.4	88.4	86.5	87.7	102.7	78.1	93.3
03	CONCRETE	80.7	87.8	83.9	95.7	93.7	94.8	89.2	92.7	90.8	94.6	103.9	98.9	87.4	87.8	87.6	97.4	79.2	89.1
04	MASONRY	144.4	87.5	109.1	110.2	78.5	90.5	104.7	93.7	97.9	97.4	102.3	100.4	102.0	87.5	93.0	96.3	81.3	87.0
05	METALS	102.6	118.1	107.4	97.4	120.9	104.6	97.5	118.6	104.0	92.9	114.6	99.5	102.2	118.1	107.1	95.6	101.0	97.2
06	WOOD, PLASTICS & COMPOSITES	85.6	79.1	82.0	107.7	85.2	95.2	97.9	80.8	88.4	89.9	101.6	96.4	97.8	79.1	87.4	103.5	73.0	86.5
07	THERMAL & MOISTURE PROTECTION	93.6	87.7	91.1	100.3	93.1	97.3	90.6	96.2	98.4	91.7	104.2	97.0	100.8	87.7	95.2	91.6	83.1	88.0
08	OPENINGS	99.6	84.1	96.0	102.9	97.0	101.6	102.9	85.0	98.8	92.1	104.5	95.0	97.6	84.1	94.5	93.2	78.0	89.7
0920	Plaster & Gypsum Board	93.1	78.6	83.3	109.8	85.1	93.2	103.6	80.6	88.1	96.2	101.5	99.8	96.8	78.6	84.6	107.0	72.1	83.6
0950, 0980	Ceilings & Acoustic Treatment	93.0	78.6	83.3	91.4	85.1	87.2	91.4	80.6	84.1	91.1	101.5	98.1	92.5	78.6	83.1	91.9	72.1	78.6
0960	Flooring	88.5	75.5	84.9	98.9	87.2	95.6	94.9	100.4	96.5	90.3	106.5	94.8	93.9	75.5	88.8	119.6	75.5	107.4
0970, 0990	Wall Finishes & Painting/Coating	92.4	81.7	86.2	96.0	75.0	83.8	96.0	98.4	97.4	99.8	107.6	104.3	86.0	81.7	83.5	95.1	79.3	85.9
09	FINISHES	85.9	79.9	82.7	99.7	83.9	91.1	97.3	86.7	91.6	95.5	103.4	99.8	91.2	79.9	85.1	104.7	74.9	88.8
COVERS	DIVS. 10 - 14, 25, 28, 41, 43, 44, 46	100.0	96.6	99.3	100.0	93.5	98.5	100.0	81.6	95.9	100.0	85.8	96.8	100.0	96.6	99.3	100.0	82.2	96.0
21, 22, 23	FIRE SUPPRESSION, PLUMBING & HVAC	99.9	99.9	99.9	96.9	99.0	97.8	96.9	99.9	98.2	96.9	102.7	99.4	99.9	100.0	100.0	100.1	71.8	88.0
26, 27, 3370	ELECTRICAL, COMMUNICATIONS & UTIL.	95.4	83.8	89.4	106.3	101.0	103.6	100.5	79.6	89.6	102.1	101.5	101.8	100.6	83.8	91.8	93.9	69.7	81.3
MF2016	WEIGHTED AVERAGE	98.1	92.5	95.7	99.6	95.7	97.9	97.6	93.6	95.9	95.9	102.8	98.9	97.8	92.5	95.5	97.8	79.6	89.8

MISSOURI

DIVISION		KANSAS CITY 640 - 641			KIRKSVILLE 635			POPLAR BLUFF 639			ROLLA 654 - 655			SEDALIA 653			SIKESTON 638		
		MAT.	INST.	TOTAL	MAT.	INST.	TOTAL	MAT.	INST.	TOTAL	MAT.	INST.	TOTAL	MAT.	INST.	TOTAL	MAT.	INST.	TOTAL
015433	CONTRACTOR EQUIPMENT		105.7	105.7		100.2	100.2		102.6	102.6		114.0	114.0		103.6	103.6		102.6	102.6
0241, 31 - 34	SITE & INFRASTRUCTURE, DEMOLITION	98.5	98.4	98.4	90.9	89.6	90.0	78.0	93.5	88.8	94.0	99.4	97.8	93.7	94.1	94.0	81.4	94.1	90.3
0310	Concrete Forming & Accessories	93.3	102.0	100.8	84.6	80.8	81.3	84.8	79.9	80.6	90.2	96.5	95.7	88.0	80.7	81.7	85.9	76.7	78.0
0320	Concrete Reinforcing	96.1	96.9	96.5	107.0	85.9	96.3	110.2	78.3	94.1	88.1	96.4	92.3	86.9	106.2	96.6	109.5	78.3	93.8
0330	Cast-in-Place Concrete	100.6	104.0	101.9	94.3	86.4	91.3	72.6	86.4	77.8	85.4	98.0	90.2	89.2	85.3	87.7	77.5	86.5	80.9
03	CONCRETE	97.4	102.4	99.7	107.7	84.9	97.3	82.9	83.1	83.0	82.3	98.5	89.7	96.0	88.0	92.3	86.6	81.7	84.4
04	MASONRY	103.4	102.3	102.7	116.8	87.6	98.7	108.2	76.1	88.3	117.4	88.2	99.3	123.8	84.1	99.2	107.9	76.1	88.2
05	METALS	102.6	109.1	104.6	97.2	102.8	98.9	97.8	99.5	98.3	102.1	118.7	107.2	100.9	111.9	104.3	98.2	99.7	98.7
06	WOOD, PLASTICS & COMPOSITES	98.5	101.5	100.2	85.3	78.8	81.7	84.4	79.9	81.9	92.9	98.1	95.8	86.0	79.1	82.1	85.9	75.4	80.1
07	THERMAL & MOISTURE PROTECTION	91.8	104.7	97.3	106.8	89.2	99.3	104.9	83.4	95.8	93.8	95.8	94.7	99.7	90.6	95.9	105.1	84.5	96.4
08	OPENINGS	100.1	101.9	100.5	108.2	82.2	102.2	109.3	76.9	101.9	99.6	94.6	98.4	104.5	88.7	100.9	109.3	74.4	101.3
0920	Plaster & Gypsum Board	102.5	101.5	101.8	98.9	78.6	85.3	99.3	79.7	86.1	95.1	98.2	97.2	88.8	78.6	81.9	100.0	75.0	83.5
0950, 0980	Ceilings & Acoustic Treatment	95.1	101.5	99.4	89.7	78.6	82.2	91.4	79.7	83.5	93.0	98.2	96.5	93.0	78.6	83.3	91.4	75.0	80.3
0960	Flooring	96.5	106.5	99.3	74.1	75.5	74.5	87.2	87.2	87.2	92.1	75.5	87.5	71.8	75.5	72.8	87.7	87.2	87.6
0970, 0990	Wall Finishes & Painting/Coating	98.4	107.6	103.7	92.0	81.9	86.1	91.2	70.1	78.9	92.4	94.3	93.5	92.4	81.9	86.3	91.2	70.1	78.9
09	FINISHES	99.1	103.4	101.4	96.4	79.8	87.4	96.2	79.6	87.2	87.4	92.7	90.3	84.7	79.1	81.7	96.8	77.3	86.2
COVERS	DIVS. 10 - 14, 25, 28, 41, 43, 44, 46	100.0	100.3	100.1	100.0	81.4	95.9	100.0	91.9	98.2	100.0	100.3	100.1	100.0	91.4	98.1	100.0	91.6	98.1
21, 22, 23	FIRE SUPPRESSION, PLUMBING & HVAC	99.9	102.6	101.1	96.9	99.3	97.9	96.9	96.8	96.8	96.9	100.4	98.4	96.8	97.4	97.0	96.9	96.8	96.9
26, 27, 3370	ELECTRICAL, COMMUNICATIONS & UTIL.	103.8	101.5	102.6	100.7	77.9	88.8	100.9	100.9	100.9	94.0	83.8	88.7	95.0	101.5	98.4	100.0	100.9	100.5
MF2016	WEIGHTED AVERAGE	100.3	102.7	101.4	101.1	85.3	95.6	97.5	89.7	94.1	96.2	96.9	96.5	98.6	93.2	96.2	98.0	89.1	94.1

DIVISION		MISSOURI									MONTANA								
		SPRINGFIELD 656 - 658			ST. JOSEPH 644 - 645			ST. LOUIS 630 - 631			BILLINGS 590 - 591			BUTTE 597			GREAT FALLS 594		
		MAT.	INST.	TOTAL	MAT.	INST.	TOTAL	MAT.	INST.	TOTAL	MAT.	INST.	TOTAL	MAT.	INST.	TOTAL	MAT.	INST.	TOTAL
015433	CONTRACTOR EQUIPMENT		106.1	106.1		104.8	104.8		111.1	111.1		100.6	100.6		100.3	100.3		100.3	100.3
0241, 31 - 34	SITE & INFRASTRUCTURE, DEMOLITION	96.4	96.6	96.5	100.8	92.0	94.6	96.0	99.4	98.4	90.7	97.6	95.6	95.9	96.4	96.3	99.5	97.4	98.0
0310	Concrete Forming & Accessories	96.1	76.6	79.3	92.2	90.2	90.5	97.1	104.1	103.2	94.7	67.7	71.4	82.8	67.5	69.6	94.6	66.9	70.7
0320	Concrete Reinforcing	84.6	95.8	90.2	94.9	106.4	100.7	98.8	104.9	101.9	95.4	80.9	88.1	103.5	80.8	92.1	95.4	80.8	88.1
0330	Cast-in-Place Concrete	90.7	77.5	85.7	95.4	100.1	97.1	102.1	102.8	102.3	108.9	71.3	94.7	119.9	71.1	101.4	126.7	71.2	105.6
03	CONCRETE	92.2	81.5	87.3	93.6	97.4	95.4	98.6	105.0	101.5	92.4	72.2	83.1	95.9	72.0	85.0	100.6	71.8	87.4
04	MASONRY	94.1	81.7	86.4	99.3	91.8	94.7	94.1	111.1	104.7	127.7	75.4	95.3	123.6	75.4	93.7	127.9	75.4	95.3
05	METALS	107.5	104.0	106.4	98.9	113.4	103.3	103.4	121.5	109.0	105.2	88.1	99.9	99.4	87.9	95.9	102.7	88.0	98.2
06	WOOD, PLASTICS & COMPOSITES	93.4	75.3	83.4	103.5	89.1	95.5	99.9	102.6	101.4	86.5	64.5	74.3	74.6	64.5	69.0	87.6	63.6	74.3
07	THERMAL & MOISTURE PROTECTION	97.9	78.4	89.6	92.1	91.6	91.9	97.9	107.1	101.8	107.6	71.1	92.2	107.4	71.0	92.0	108.0	70.6	92.2
08	OPENINGS	107.0	86.8	102.4	97.3	97.6	97.4	101.5	107.4	102.8	96.8	66.4	89.8	95.1	66.4	88.5	98.0	65.9	90.6
0920	Plaster & Gypsum Board	96.1	74.7	81.7	108.8	88.6	95.3	108.2	103.0	104.7	113.2	63.9	80.1	112.3	63.9	79.8	121.9	62.9	82.3
0950, 0980	Ceilings & Acoustic Treatment	93.0	74.7	80.7	99.6	88.6	92.2	93.1	103.0	99.8	91.9	63.9	73.0	97.8	63.9	74.9	99.5	62.9	74.9
0960	Flooring	91.1	75.5	86.8	99.9	104.7	101.2	98.0	100.4	98.7	93.7	79.7	89.8	91.1	79.7	87.9	97.9	79.7	92.9
0970, 0990	Wall Finishes & Painting/Coating	86.9	103.9	96.8	95.5	107.6	102.5	97.0	108.5	103.7	95.2	70.8	81.0	94.1	70.8	80.5	94.1	70.8	80.5
09	FINISHES	89.6	79.6	84.1	101.5	94.5	97.7	100.1	103.7	102.1	91.6	69.6	79.7	91.6	69.6	79.7	95.6	69.1	81.2
COVERS	DIVS. 10 - 14, 25, 28, 41, 43, 44, 46	100.0	94.0	98.7	100.0	97.4	99.4	100.0	102.2	100.5	100.0	92.2	98.3	100.0	92.2	98.3	100.0	92.1	98.2
21, 22, 23	FIRE SUPPRESSION, PLUMBING & HVAC	100.0	72.3	88.1	100.1	89.0	95.4	99.9	105.8	102.4	100.1	74.4	89.1	100.2	70.8	87.6	100.2	70.8	87.6
26, 27, 3370	ELECTRICAL, COMMUNICATIONS & UTIL.	99.1	71.7	84.8	102.1	77.9	89.5	104.3	101.0	102.6	98.6	70.8	84.8	105.4	71.2	87.6	98.0	71.2	84.1
MF2016	WEIGHTED AVERAGE	99.6	81.7	91.7	98.9	92.8	96.2	100.5	106.2	103.0	100.0	76.4	89.6	99.8	75.3	89.1	101.3	75.3	89.9

MONTANA

DIVISION		HAVRE 595			HELENA 596			KALISPELL 599			MILES CITY 593			MISSOULA 598			WOLF POINT 592		
		MAT.	INST.	TOTAL	MAT.	INST.	TOTAL	MAT.	INST.	TOTAL	MAT.	INST.	TOTAL	MAT.	INST.	TOTAL	MAT.	INST.	TOTAL
015433	CONTRACTOR EQUIPMENT		100.3	100.3		100.3	100.3		100.3	100.3		100.3	100.3		100.3	100.3		100.3	100.3
0241, 31 - 34	SITE & INFRASTRUCTURE, DEMOLITION	102.8	96.5	98.4	90.0	96.4	94.5	86.6	96.3	93.4	92.5	96.4	95.2	79.6	96.2	91.2	108.8	96.5	100.2
0310	Concrete Forming & Accessories	76.4	67.4	68.6	97.5	67.4	71.6	85.7	67.5	70.0	93.2	67.5	71.0	85.7	67.4	69.9	86.4	66.8	69.5
0320	Concrete Reinforcing	104.3	80.8	92.5	110.2	79.6	94.8	106.2	85.6	95.8	104.0	80.8	92.3	105.2	85.6	95.3	105.4	80.1	92.7
0330	Cast-in-Place Concrete	129.1	71.1	107.1	93.7	71.1	85.1	104.1	71.1	91.6	114.0	71.1	97.7	88.4	71.1	81.8	127.7	71.1	106.2
03	CONCRETE	103.3	72.0	89.0	89.2	71.8	81.2	86.4	72.9	80.2	93.0	72.0	83.4	76.0	72.8	74.5	106.4	71.6	90.5
04	MASONRY	124.6	75.4	94.1	115.4	75.4	90.6	122.6	75.8	93.6	129.9	75.4	96.1	149.3	75.8	103.7	131.0	75.4	96.5
05	METALS	95.7	87.8	93.3	101.4	87.5	97.1	95.5	89.6	93.7	94.8	87.9	92.7	96.1	89.4	94.0	94.9	87.6	92.7
06	WOOD, PLASTICS & COMPOSITES	67.1	64.5	65.7	90.0	64.5	75.8	77.5	64.5	70.3	85.0	64.5	73.6	77.5	64.5	70.3	77.3	63.6	69.7
07	THERMAL & MOISTURE PROTECTION	107.8	66.2	90.2	102.4	71.0	89.1	107.0	73.2	92.7	107.3	66.3	90.0	106.6	70.6	91.3	108.3	66.2	90.5
08	OPENINGS	95.6	66.4	88.9	99.5	66.2	91.9	95.6	67.5	89.1	95.1	66.4	88.5	95.1	67.5	88.8	95.1	65.8	88.4
0920	Plaster & Gypsum Board	108.0	63.9	78.4	106.4	63.9	77.8	112.3	63.9	79.8	121.1	63.9	82.7	112.3	63.9	79.8	115.9	62.9	80.3
0950, 0980	Ceilings & Acoustic Treatment	97.8	63.9	74.9	96.8	63.9	74.6	97.8	63.9	74.9	96.1	63.9	74.4	97.8	63.9	74.9	96.1	62.9	73.7
0960	Flooring	88.6	79.7	86.1	98.9	79.7	93.6	92.8	79.7	89.2	97.7	79.7	92.8	92.8	79.7	89.2	94.3	79.7	90.2
0970, 0990	Wall Finishes & Painting/Coating	94.1	70.8	80.5	94.4	70.8	80.6	94.1	70.8	80.5	94.1	70.8	80.5	94.1	56.0	71.9	94.1	70.8	80.5
09	FINISHES	90.9	69.6	79.3	97.2	69.6	82.2	91.6	69.6	79.7	94.4	69.6	80.9	91.1	68.1	78.6	93.8	69.1	80.4
COVERS	DIVS. 10 - 14, 25, 28, 41, 43, 44, 46	100.0	92.2	98.3	100.0	92.2	98.3	100.0	92.2	98.3	100.0	92.2	98.3	100.0	92.2	98.3	100.0	92.1	98.3
21, 22, 23	FIRE SUPPRESSION, PLUMBING & HVAC	97.1	70.8	85.8	100.2	70.8	87.6	97.1	68.8	85.0	97.1	74.4	87.4	100.2	68.8	86.8	97.1	74.4	87.4
26, 27, 3370	ELECTRICAL, COMMUNICATIONS & UTIL.	98.0	71.2	84.1	104.6	71.2	87.2	102.2	68.7	84.8	98.0	76.6	86.8	103.2	67.7	84.8	98.0	76.6	86.8
MF2016	WEIGHTED AVERAGE	98.9	75.2	88.5	99.6	75.3	88.9	96.8	75.0	87.2	97.8	76.7	88.6	97.4	74.5	87.4	99.9	76.5	89.7

NEBRASKA

DIVISION		ALLIANCE 693			COLUMBUS 686			GRAND ISLAND 688			HASTINGS 689			LINCOLN 683 - 685			MCCOOK 690		
		MAT.	INST.	TOTAL	MAT.	INST.	TOTAL	MAT.	INST.	TOTAL	MAT.	INST.	TOTAL	MAT.	INST.	TOTAL	MAT.	INST.	TOTAL
015433	CONTRACTOR EQUIPMENT		98.2	98.2		105.2	105.2		105.2	105.2		105.2	105.2		105.2	105.2		105.2	105.2
0241, 31 - 34	SITE & INFRASTRUCTURE, DEMOLITION	99.4	101.2	100.6	100.8	95.2	96.9	105.6	95.2	98.3	104.4	95.2	97.9	91.1	95.2	94.0	102.7	95.2	97.4
0310	Concrete Forming & Accessories	86.3	55.8	60.0	95.2	74.8	77.6	94.8	71.2	74.4	98.0	74.3	77.6	93.7	76.4	78.8	91.6	55.9	60.9
0320	Concrete Reinforcing	111.2	88.0	99.5	97.2	86.9	92.0	96.6	76.3	86.4	96.6	73.4	84.9	95.9	76.5	86.2	103.4	69.7	86.4
0330	Cast-in-Place Concrete	107.8	77.3	96.2	110.5	80.9	99.2	117.0	80.3	103.0	117.0	62.2	96.2	91.8	80.4	87.5	116.6	58.6	94.6
03	CONCRETE	115.6	69.7	94.6	100.3	80.2	91.1	105.0	76.5	92.0	105.2	71.2	89.7	88.5	78.9	84.1	105.1	61.0	85.0
04	MASONRY	110.3	75.1	88.5	116.5	78.6	93.0	109.1	75.1	88.0	118.6	75.1	91.6	96.9	78.6	85.6	105.5	75.1	86.6
05	METALS	104.0	84.1	97.9	98.3	98.4	98.3	100.1	94.2	98.3	101.0	92.8	98.5	102.1	94.5	99.8	98.5	90.8	96.2
06	WOOD, PLASTICS & COMPOSITES	86.6	50.1	66.3	99.2	74.1	85.2	98.5	69.4	82.3	102.0	74.1	86.5	98.3	76.3	86.1	95.2	50.1	70.1
07	THERMAL & MOISTURE PROTECTION	102.5	65.9	87.0	101.7	80.5	92.7	101.8	79.1	92.2	101.9	78.7	92.1	98.2	80.9	90.9	97.5	76.1	88.4
08	OPENINGS	94.1	59.7	86.2	93.8	74.2	89.3	93.8	69.7	88.3	93.8	70.9	88.6	101.3	70.3	94.1	94.6	55.1	85.5
0920	Plaster & Gypsum Board	82.7	48.9	60.0	94.6	73.5	80.4	93.9	68.7	77.0	95.6	73.5	80.7	102.7	75.8	84.6	94.7	48.9	63.9
0950, 0980	Ceilings & Acoustic Treatment	93.2	48.9	63.3	85.9	73.5	77.5	85.9	68.7	74.3	85.9	73.5	77.5	95.1	75.8	82.1	89.2	48.9	62.0
0960	Flooring	89.7	83.4	88.0	82.4	89.7	84.5	82.2	83.4	82.5	83.5	83.4	83.5	93.7	89.7	92.6	88.5	83.4	87.1
0970, 0990	Wall Finishes & Painting/Coating	150.3	52.9	93.5	73.3	61.6	66.5	73.3	65.3	68.7	73.3	61.6	66.5	91.4	81.3	85.5	85.6	46.1	62.6
09	FINISHES	91.4	59.5	74.1	83.6	76.0	79.4	83.7	72.3	77.5	84.3	74.7	79.1	94.2	79.4	86.2	89.1	58.8	72.6
COVERS	DIVS. 10 - 14, 25, 28, 41, 43, 44, 46	100.0	60.9	91.3	100.0	84.1	96.5	100.0	89.0	97.5	100.0	84.1	96.5	100.0	89.7	97.7	100.0	60.3	91.2
21, 22, 23	FIRE SUPPRESSION, PLUMBING & HVAC	97.1	73.4	86.9	97.0	76.3	88.2	100.1	80.6	91.8	97.0	76.0	88.0	100.0	80.6	91.7	96.9	76.1	88.0
26, 27, 3370	ELECTRICAL, COMMUNICATIONS & UTIL.	93.5	65.6	79.0	93.7	80.2	86.7	92.4	65.7	78.5	91.8	81.4	86.4	101.4	65.7	82.8	95.5	65.6	80.0
MF2016	WEIGHTED AVERAGE	100.4	71.9	87.9	97.3	81.4	90.3	98.5	78.4	89.7	98.5	79.0	89.9	98.3	80.2	90.4	98.0	71.5	86.4

NEBRASKA / NEVADA

DIVISION		NORFOLK 687			NORTH PLATTE 691			OMAHA 680 - 681			VALENTINE 692			CARSON CITY 897			ELKO 898		
		MAT.	INST.	TOTAL	MAT.	INST.	TOTAL	MAT.	INST.	TOTAL	MAT.	INST.	TOTAL	MAT.	INST.	TOTAL	MAT.	INST.	TOTAL
015433	CONTRACTOR EQUIPMENT		94.4	94.4		105.2	105.2		94.4	94.4		97.9	97.9		97.4	97.4		97.4	97.4
0241, 31 - 34	SITE & INFRASTRUCTURE, DEMOLITION	83.6	94.3	91.1	103.9	95.2	97.8	90.9	94.3	93.3	87.3	100.3	96.4	89.2	97.8	95.2	71.8	96.1	88.8
0310	Concrete Forming & Accessories	81.9	75.5	76.4	94.1	70.3	73.6	93.3	76.1	78.5	82.7	55.3	59.1	105.0	79.7	83.2	110.2	77.4	82.0
0320	Concrete Reinforcing	97.3	64.6	80.8	102.8	73.3	87.9	95.9	76.5	86.1	103.4	64.2	83.6	99.0	113.3	106.2	106.2	91.7	98.9
0330	Cast-in-Place Concrete	111.1	61.5	92.2	116.6	63.3	96.3	92.1	78.2	86.8	102.8	56.3	85.1	102.5	82.7	94.9	99.2	74.1	89.7
03	CONCRETE	98.9	69.3	85.4	105.2	69.7	89.0	88.6	77.4	83.5	102.5	58.3	82.3	96.7	86.7	92.1	94.4	79.0	87.4
04	MASONRY	123.3	78.6	95.5	93.6	75.1	82.1	97.9	78.6	85.9	105.4	75.0	86.6	115.7	72.4	88.8	122.3	67.3	88.2
05	METALS	101.9	79.4	95.0	97.7	92.2	96.0	102.1	84.3	96.7	110.3	78.2	100.4	105.7	94.4	102.2	109.1	86.6	102.2
06	WOOD, PLASTICS & COMPOSITES	82.5	75.8	78.8	97.1	69.4	81.7	94.5	75.8	84.1	81.0	49.7	63.6	92.3	77.8	84.2	101.7	77.8	88.4
07	THERMAL & MOISTURE PROTECTION	101.6	79.2	92.1	97.4	78.2	89.3	98.1	80.3	90.6	98.3	75.0	88.4	105.7	80.4	95.0	102.3	71.8	89.4
08	OPENINGS	95.4	69.7	89.5	93.9	68.4	88.0	99.3	76.4	94.0	96.3	55.2	86.9	98.7	80.0	94.4	101.5	75.3	95.5
0920	Plaster & Gypsum Board	94.7	75.8	82.0	94.7	68.7	77.2	101.0	75.8	84.0	96.1	48.9	64.4	99.3	77.3	84.5	101.3	77.3	85.2
0950, 0980	Ceilings & Acoustic Treatment	99.8	75.8	83.6	89.2	68.7	75.4	97.2	75.8	82.8	105.6	48.9	67.4	101.4	77.3	85.1	100.1	77.3	84.7
0960	Flooring	102.0	89.7	98.6	89.6	83.4	87.9	94.3	89.7	93.0	114.6	83.4	105.9	96.0	71.0	89.0	101.0	71.0	92.6
0970, 0990	Wall Finishes & Painting/Coating	127.6	61.6	89.1	85.6	61.6	71.6	101.4	75.4	86.2	150.8	63.8	100.0	93.9	80.4	86.0	93.4	80.4	85.8
09	FINISHES	99.9	76.9	87.4	89.4	71.9	79.9	96.1	78.4	86.5	109.6	60.4	82.8	96.1	77.5	86.0	92.8	76.5	84.0
COVERS	DIVS. 10 - 14, 25, 28, 41, 43, 44, 46	100.0	86.1	96.9	100.0	62.4	91.6	100.0	88.7	97.5	100.0	58.3	90.7	100.0	95.2	98.9	100.0	90.5	97.9
21, 22, 23	FIRE SUPPRESSION, PLUMBING & HVAC	96.7	76.7	88.1	100.0	75.6	89.6	100.0	76.9	90.1	96.5	76.2	87.8	100.1	80.1	91.6	98.6	77.7	89.7
26, 27, 3370	ELECTRICAL, COMMUNICATIONS & UTIL.	92.7	83.9	88.1	93.8	65.6	79.2	102.7	83.9	92.9	91.2	66.1	78.2	103.2	93.2	98.0	100.4	93.2	96.7
MF2016	WEIGHTED AVERAGE	98.8	78.6	90.0	97.8	75.2	87.9	98.4	80.8	90.7	100.5	70.5	87.4	101.0	84.9	93.9	100.6	81.2	92.1

For customer support on your Electrical Costs with RSMeans data, call 800.448.8182.

533

City Cost Indexes

DIVISION		NEVADA ELY 893 MAT.	INST.	TOTAL	LAS VEGAS 889-891 MAT.	INST.	TOTAL	RENO 894-895 MAT.	INST.	TOTAL	NEW HAMPSHIRE CHARLESTON 036 MAT.	INST.	TOTAL	CLAREMONT 037 MAT.	INST.	TOTAL	CONCORD 032-033 MAT.	INST.	TOTAL
015433	CONTRACTOR EQUIPMENT		97.4	97.4		97.4	97.4		97.4	97.4		98.8	98.8		98.8	98.8		98.8	98.8
0241, 31 - 34	SITE & INFRASTRUCTURE, DEMOLITION	77.9	97.6	91.7	81.2	100.1	94.4	77.6	97.8	91.8	84.9	99.8	95.3	78.8	99.8	93.5	90.2	101.2	97.9
0310	Concrete Forming & Accessories	103.4	99.8	100.3	104.3	108.8	108.1	99.7	79.8	82.5	85.5	78.7	79.6	91.2	78.7	80.4	95.9	93.5	93.9
0320	Concrete Reinforcing	105.0	91.9	98.4	96.5	120.6	108.7	99.2	119.4	109.4	85.3	83.3	84.3	85.3	83.3	84.3	99.6	83.7	91.6
0330	Cast-in-Place Concrete	106.5	100.1	104.1	103.1	106.7	104.5	112.7	82.7	101.3	87.7	104.5	94.1	80.6	104.5	94.1	103.4	115.5	108.0
03	CONCRETE	102.6	98.3	100.6	97.9	109.5	103.2	102.5	87.9	95.8	87.3	88.7	88.0	80.2	88.7	84.1	94.6	99.3	96.7
04	MASONRY	127.9	78.1	97.0	114.8	99.2	105.1	121.5	72.4	91.0	95.6	77.4	84.3	95.9	77.4	84.4	104.0	100.1	101.6
05	METALS	109.1	89.7	103.1	117.6	101.6	112.7	110.8	96.4	106.4	93.0	89.3	91.9	93.0	89.3	91.9	99.6	90.6	96.9
06	WOOD, PLASTICS & COMPOSITES	92.6	101.0	97.3	90.7	106.9	99.7	87.0	77.8	81.9	94.2	83.2	88.1	100.0	83.2	90.7	94.9	92.7	93.7
07	THERMAL & MOISTURE PROTECTION	102.9	95.8	99.9	115.7	100.3	109.2	102.4	80.4	93.1	101.9	97.6	100.1	101.7	97.6	100.0	106.1	108.3	107.0
08	OPENINGS	101.4	88.2	98.4	100.4	111.9	103.0	99.2	81.4	95.1	97.3	82.9	94.0	98.4	82.9	94.8	99.2	88.1	94.0
0920	Plaster & Gypsum Board	96.7	101.1	99.7	91.9	107.2	102.1	87.0	77.3	80.5	104.8	82.4	89.7	106.1	82.4	89.7	107.3	92.1	97.1
0950, 0980	Ceilings & Acoustic Treatment	100.1	101.1	100.8	107.8	107.2	107.4	105.2	77.3	86.4	99.5	82.4	87.9	99.5	82.4	87.9	98.6	92.1	94.2
0960	Flooring	98.4	71.0	90.8	90.1	102.0	93.5	95.0	71.0	88.4	83.9	106.7	90.2	86.4	106.7	92.0	92.5	106.7	96.4
0970, 0990	Wall Finishes & Painting/Coating	93.4	117.1	107.2	95.9	114.9	107.0	93.4	80.4	85.8	84.8	94.1	90.2	84.8	94.1	90.2	88.8	94.1	91.9
09	FINISHES	92.0	96.9	94.7	90.5	108.3	100.1	90.5	77.5	83.4	86.6	85.4	85.9	87.1	85.4	86.2	90.6	95.7	93.3
COVERS	DIVS. 10 - 14, 25, 28, 41, 43, 44, 46	100.0	72.7	93.9	100.0	103.9	100.9	100.0	95.2	98.9	100.0	85.5	96.8	100.0	85.5	96.8	100.0	104.8	101.1
21, 22, 23	FIRE SUPPRESSION, PLUMBING & HVAC	98.6	99.5	99.0	100.3	100.7	100.5	100.2	80.1	91.6	97.0	60.1	81.2	97.0	60.1	81.2	100.2	85.3	93.7
26, 27, 3370	ELECTRICAL, COMMUNICATIONS & UTIL.	100.7	98.2	99.4	104.9	113.8	109.5	101.1	93.2	97.0	97.2	49.1	72.2	97.2	49.1	72.2	98.0	79.5	88.4
MF2016	WEIGHTED AVERAGE	101.9	94.1	98.5	103.1	105.2	104.0	101.8	85.3	94.6	94.3	76.3	86.4	93.5	76.3	86.0	97.9	92.4	95.5

DIVISION		NEW HAMPSHIRE KEENE 034 MAT.	INST.	TOTAL	LITTLETON 035 MAT.	INST.	TOTAL	MANCHESTER 031 MAT.	INST.	TOTAL	NASHUA 030 MAT.	INST.	TOTAL	PORTSMOUTH 038 MAT.	INST.	TOTAL	NEW JERSEY ATLANTIC CITY 082, 084 MAT.	INST.	TOTAL
015433	CONTRACTOR EQUIPMENT		98.8	98.8		98.8	98.8		98.8	98.8		98.8	98.8		98.8	98.8		96.3	96.3
0241, 31 - 34	SITE & INFRASTRUCTURE, DEMOLITION	91.9	100.1	97.6	79.0	100.1	93.7	92.1	101.2	98.5	93.4	101.0	98.7	86.9	101.6	97.2	89.7	104.6	100.2
0310	Concrete Forming & Accessories	89.8	80.3	81.6	102.0	80.3	83.3	95.9	93.7	94.2	97.8	93.6	94.2	86.9	93.5	92.6	110.1	143.9	139.3
0320	Concrete Reinforcing	85.3	83.4	84.3	86.0	83.4	84.7	99.6	83.7	91.6	106.5	83.7	95.0	85.3	83.7	84.5	75.8	142.1	109.2
0330	Cast-in-Place Concrete	88.1	106.5	95.1	79.0	106.6	89.5	103.4	115.6	108.0	83.4	115.4	95.5	79.1	116.0	93.1	77.9	142.7	102.5
03	CONCRETE	87.0	90.2	88.4	79.7	90.2	84.5	94.5	99.4	96.8	87.0	99.3	92.6	79.8	99.5	88.8	82.1	141.5	109.2
04	MASONRY	98.4	80.7	87.4	108.5	80.7	91.2	100.7	100.1	100.4	100.7	100.0	100.3	96.0	100.3	98.7	111.3	140.6	129.5
05	METALS	93.7	89.7	92.5	93.8	89.7	92.5	101.2	90.8	98.0	99.3	90.3	96.6	95.3	91.5	94.2	96.5	115.7	102.4
06	WOOD, PLASTICS & COMPOSITES	98.4	83.2	89.9	109.2	83.2	94.7	95.0	92.7	93.7	107.6	92.7	99.3	95.4	92.7	93.9	118.0	144.6	132.8
07	THERMAL & MOISTURE PROTECTION	102.4	99.1	101.0	101.8	99.1	100.7	106.3	108.3	107.2	102.8	108.3	105.1	102.4	108.3	104.9	101.0	137.4	119.6
08	OPENINGS	96.0	86.5	93.8	99.3	82.9	95.5	96.7	91.8	95.5	100.6	91.1	98.4	100.9	81.2	96.4	96.8	140.3	106.8
0920	Plaster & Gypsum Board	105.1	82.4	89.8	118.9	82.4	94.4	107.3	92.1	97.1	114.5	92.1	99.5	104.8	92.1	96.3	116.0	145.4	135.7
0950, 0980	Ceilings & Acoustic Treatment	99.5	82.4	87.9	99.5	82.4	87.9	99.6	92.1	94.5	111.5	92.1	98.4	100.4	92.1	94.8	91.4	145.4	127.8
0960	Flooring	86.0	106.7	91.8	96.7	105.6	99.2	89.8	106.7	94.5	89.7	106.7	94.4	84.1	106.7	90.4	92.5	168.1	113.4
0970, 0990	Wall Finishes & Painting/Coating	84.8	108.1	98.4	84.8	94.1	90.2	90.8	108.1	100.9	84.8	106.9	97.7	84.8	106.9	97.7	83.0	145.9	119.6
09	FINISHES	88.4	87.8	88.0	91.9	86.0	88.7	91.9	97.2	94.8	93.1	97.1	95.3	87.4	97.1	92.7	89.7	149.9	122.4
COVERS	DIVS. 10 - 14, 25, 28, 41, 43, 44, 46	100.0	93.2	98.5	100.0	98.5	99.7	100.0	104.8	101.1	100.0	104.8	101.1	100.0	104.8	101.1	100.0	119.2	104.3
21, 22, 23	FIRE SUPPRESSION, PLUMBING & HVAC	97.0	63.3	82.6	97.0	69.8	85.4	100.1	85.2	93.7	100.1	85.2	93.7	100.1	85.2	93.7	99.8	131.7	113.4
26, 27, 3370	ELECTRICAL, COMMUNICATIONS & UTIL.	97.2	58.6	77.2	98.2	51.9	74.1	98.1	79.5	88.4	99.4	79.4	89.0	97.7	79.5	88.2	93.2	144.6	119.9
MF2016	WEIGHTED AVERAGE	94.8	79.6	88.1	94.9	79.8	88.3	98.3	92.7	95.9	97.7	92.6	95.5	95.1	92.4	93.9	95.7	134.7	112.8

DIVISION		NEW JERSEY CAMDEN 081 MAT.	INST.	TOTAL	DOVER 078 MAT.	INST.	TOTAL	ELIZABETH 072 MAT.	INST.	TOTAL	HACKENSACK 076 MAT.	INST.	TOTAL	JERSEY CITY 073 MAT.	INST.	TOTAL	LONG BRANCH 077 MAT.	INST.	TOTAL
015433	CONTRACTOR EQUIPMENT		96.3	96.3		98.8	98.8		98.8	98.8		98.8	98.8		96.3	96.3		95.9	95.9
0241, 31 - 34	SITE & INFRASTRUCTURE, DEMOLITION	90.9	104.6	100.5	96.5	106.3	103.4	100.3	106.3	104.5	97.5	106.3	103.6	88.7	106.2	100.9	92.5	105.9	101.9
0310	Concrete Forming & Accessories	101.3	143.8	137.9	99.4	146.5	140.0	111.6	146.5	141.7	99.4	146.4	139.9	103.7	146.4	140.6	104.2	145.6	139.9
0320	Concrete Reinforcing	99.5	132.6	116.2	78.1	150.5	114.6	78.1	150.5	114.6	78.1	150.5	114.6	101.3	150.5	126.1	78.1	150.5	114.6
0330	Cast-in-Place Concrete	75.6	140.8	100.3	85.2	136.1	104.5	73.2	136.1	97.1	83.2	136.1	103.3	66.4	136.1	92.9	73.8	140.8	99.3
03	CONCRETE	82.6	139.1	108.4	84.2	142.1	110.7	80.8	142.1	110.8	82.7	142.1	109.9	78.6	141.9	105.7	82.3	143.1	110.1
04	MASONRY	101.4	140.6	125.7	93.5	143.0	124.2	111.1	143.0	130.9	97.6	143.0	125.8	89.3	143.0	122.6	103.2	141.1	126.7
05	METALS	102.3	112.3	105.3	94.3	123.7	103.4	95.9	123.7	104.4	94.4	123.7	103.4	100.2	120.6	106.4	94.4	120.4	102.4
06	WOOD, PLASTICS & COMPOSITES	106.7	144.6	127.8	102.4	146.2	126.7	117.3	146.2	133.3	102.4	146.2	126.7	103.2	146.2	127.1	104.2	146.1	127.5
07	THERMAL & MOISTURE PROTECTION	100.8	138.4	116.8	108.7	136.4	120.4	108.9	136.4	120.5	108.4	135.6	119.9	108.1	136.4	120.1	108.3	135.2	119.7
08	OPENINGS	99.1	138.1	108.0	104.2	144.2	113.4	102.6	144.2	112.1	102.0	144.2	111.7	100.7	144.2	110.7	96.6	144.2	107.6
0920	Plaster & Gypsum Board	111.5	145.4	134.3	107.2	147.0	133.9	114.4	147.0	136.3	107.2	147.0	133.9	110.2	147.0	134.9	109.1	147.0	134.6
0950, 0980	Ceilings & Acoustic Treatment	101.0	145.4	130.9	87.8	147.0	127.7	89.7	147.0	128.3	87.8	147.0	127.7	97.3	147.0	130.8	87.8	147.0	127.7
0960	Flooring	88.6	168.1	110.6	81.8	191.9	112.4	86.9	191.9	116.0	81.8	191.9	112.4	82.9	191.9	113.1	83.1	187.0	111.9
0970, 0990	Wall Finishes & Painting/Coating	83.0	145.9	119.6	84.0	145.9	120.1	84.0	145.9	120.1	84.0	145.9	120.1	84.1	145.9	120.1	84.1	145.9	120.1
09	FINISHES	89.7	149.9	122.4	85.4	155.4	123.5	88.8	155.4	125.0	85.3	155.4	123.4	87.8	155.4	124.6	86.3	154.2	123.2
COVERS	DIVS. 10 - 14, 25, 28, 41, 43, 44, 46	100.0	119.2	104.3	100.0	131.8	107.1	100.0	131.8	107.1	100.0	131.8	107.1	100.0	131.8	107.1	100.0	119.4	104.3
21, 22, 23	FIRE SUPPRESSION, PLUMBING & HVAC	100.0	131.6	113.5	99.9	137.5	116.0	100.1	137.5	116.1	99.9	137.5	116.0	100.1	137.5	116.1	99.9	136.5	115.6
26, 27, 3370	ELECTRICAL, COMMUNICATIONS & UTIL.	97.7	139.5	119.4	94.8	142.9	119.8	95.4	142.9	120.1	94.8	140.9	118.8	99.5	140.9	121.0	94.4	137.2	116.7
MF2016	WEIGHTED AVERAGE	97.0	133.3	112.9	95.7	138.2	114.3	96.8	138.2	114.9	95.5	137.9	114.0	95.9	137.6	114.1	95.0	136.2	113.1

City Cost Indexes

NEW JERSEY

DIVISION		NEW BRUNSWICK 088-089 MAT.	INST.	TOTAL	NEWARK 070-071 MAT.	INST.	TOTAL	PATERSON 074-075 MAT.	INST.	TOTAL	POINT PLEASANT 087 MAT.	INST.	TOTAL	SUMMIT 079 MAT.	INST.	TOTAL	TRENTON 085-086 MAT.	INST.	TOTAL
015433	CONTRACTOR EQUIPMENT		95.9	95.9		98.8	98.8		98.8	98.8		95.9	95.9		98.8	98.8		95.9	95.9
0241, 31 - 34	SITE & INFRASTRUCTURE, DEMOLITION	102.2	106.0	104.9	100.9	106.3	104.7	99.4	106.3	104.2	103.7	105.9	105.3	98.1	106.3	103.9	89.0	105.9	100.9
0310	Concrete Forming & Accessories	104.3	146.4	140.6	104.4	146.5	140.7	101.4	146.4	140.2	98.8	145.5	139.1	102.2	146.5	140.4	100.5	145.3	139.1
0320	Concrete Reinforcing	76.7	150.5	113.9	99.7	150.5	125.3	101.3	150.5	126.1	76.7	150.5	113.9	78.1	150.5	114.6	100.7	119.9	110.4
0330	Cast-in-Place Concrete	96.3	141.9	113.6	93.9	136.1	109.9	84.7	136.1	104.2	96.3	142.9	114.0	70.7	136.1	95.6	94.8	140.6	112.2
03	CONCRETE	97.5	143.8	118.7	91.6	142.1	114.7	87.1	142.1	112.2	97.2	143.8	118.5	78.2	142.1	107.4	91.9	137.6	112.8
04	MASONRY	109.7	143.0	130.4	96.6	143.0	125.4	94.0	143.0	124.4	98.0	141.1	124.7	96.6	143.0	125.4	101.9	141.1	126.2
05	METALS	96.6	120.5	103.9	102.0	123.7	108.7	95.5	123.7	104.2	96.6	120.3	103.9	94.3	123.7	103.4	102.0	109.1	104.2
06	WOOD, PLASTICS & COMPOSITES	111.5	146.1	130.8	100.0	146.2	125.7	105.0	146.2	127.5	104.1	146.1	127.5	106.1	146.2	128.4	93.5	146.1	122.8
07	THERMAL & MOISTURE PROTECTION	101.3	134.8	115.5	109.8	136.4	121.1	108.7	135.6	120.1	101.3	137.6	116.7	109.1	136.4	120.6	101.7	138.6	117.3
08	OPENINGS	92.0	144.2	104.0	100.6	144.2	110.6	107.3	144.2	115.8	93.8	144.2	105.3	108.3	144.2	116.5	98.5	136.7	107.3
0920	Plaster & Gypsum Board	113.3	147.0	136.0	102.5	147.0	132.4	110.2	147.0	134.9	108.4	147.0	134.3	109.1	147.0	134.6	103.4	147.0	132.7
0950, 0980	Ceilings & Acoustic Treatment	91.4	147.0	128.9	98.8	147.0	131.3	97.3	147.0	130.8	91.4	147.0	128.9	87.8	147.0	127.7	97.9	147.0	131.0
0960	Flooring	90.0	191.9	118.3	87.4	191.9	116.3	82.9	191.9	113.1	87.5	168.1	109.8	83.1	191.9	113.3	91.9	187.0	118.2
0970, 0990	Wall Finishes & Painting/Coating	83.0	145.9	120.6	85.4	145.9	120.6	84.0	145.9	120.1	83.0	145.9	119.6	84.0	145.9	120.1	86.6	145.9	121.2
09	FINISHES	89.6	155.3	125.4	90.7	155.4	125.9	88.0	155.4	124.6	88.2	150.9	122.3	86.4	155.4	123.9	92.4	154.2	126.0
COVERS	DIVS. 10 - 14, 25, 28, 41, 43, 44, 46	100.0	131.7	107.0	100.0	131.8	107.1	100.0	131.8	107.1	100.0	116.4	103.7	100.0	131.8	107.1	100.0	119.4	104.3
21, 22, 23	FIRE SUPPRESSION, PLUMBING & HVAC	99.8	137.5	115.9	100.0	137.5	116.1	100.1	137.5	116.1	99.8	136.5	115.5	99.9	137.5	116.0	100.2	136.2	115.6
26, 27, 3370	ELECTRICAL, COMMUNICATIONS & UTIL.	93.9	141.2	118.5	103.1	140.9	122.8	99.5	142.9	122.1	93.2	137.2	116.1	95.4	142.9	120.1	101.2	135.9	119.2
MF2016	WEIGHTED AVERAGE	97.4	137.8	115.1	99.1	137.9	116.1	97.4	138.2	115.3	96.8	135.8	113.9	95.8	138.2	114.3	98.5	133.9	114.0

NEW JERSEY / NEW MEXICO

DIVISION		VINELAND 080, 083 MAT.	INST.	TOTAL	ALBUQUERQUE 870-872 MAT.	INST.	TOTAL	CARRIZOZO 883 MAT.	INST.	TOTAL	CLOVIS 881 MAT.	INST.	TOTAL	FARMINGTON 874 MAT.	INST.	TOTAL	GALLUP 873 MAT.	INST.	TOTAL
015433	CONTRACTOR EQUIPMENT		96.3	96.3		111.0	111.0		111.0	111.0		111.0	111.0		111.0	111.0		111.0	111.0
0241, 31 - 34	SITE & INFRASTRUCTURE, DEMOLITION	93.9	104.7	101.4	92.3	102.4	99.4	112.0	102.4	105.3	98.8	102.4	101.3	98.8	102.4	101.3	108.5	102.4	104.2
0310	Concrete Forming & Accessories	96.0	144.0	137.4	99.0	64.1	68.9	97.1	64.1	68.6	97.1	64.0	68.5	99.1	64.1	68.9	99.1	64.1	68.9
0320	Concrete Reinforcing	75.8	135.4	105.9	97.0	71.0	83.9	109.7	71.0	90.2	110.9	71.0	90.8	106.1	71.0	88.4	101.5	71.0	86.1
0330	Cast-in-Place Concrete	84.0	140.9	105.6	93.6	70.1	84.7	94.7	70.1	85.3	94.6	70.0	85.3	94.5	70.1	85.2	88.9	70.1	81.8
03	CONCRETE	86.3	139.8	110.8	93.3	68.8	82.1	114.3	68.8	93.5	102.6	68.7	87.1	96.7	68.8	83.9	102.9	68.8	87.3
04	MASONRY	98.6	141.1	125.0	107.3	60.0	78.0	106.9	60.0	77.8	106.9	60.0	77.8	116.9	60.0	81.6	101.6	60.0	75.8
05	METALS	96.5	113.9	101.8	109.2	91.0	103.6	105.2	91.0	100.8	104.8	90.8	100.5	106.8	91.0	101.9	105.9	91.0	101.3
06	WOOD, PLASTICS & COMPOSITES	101.0	144.6	125.2	94.7	64.6	77.9	92.4	64.6	77.0	92.4	64.6	77.0	94.8	64.6	78.0	94.8	64.6	78.0
07	THERMAL & MOISTURE PROTECTION	100.8	137.6	116.4	96.9	70.8	85.8	99.5	70.8	87.3	98.2	70.8	86.6	97.1	70.8	86.0	98.2	70.8	86.6
08	OPENINGS	93.3	139.1	103.8	98.5	65.8	91.0	96.6	65.8	89.5	96.7	65.8	89.7	100.8	65.8	92.8	100.9	65.8	92.8
0920	Plaster & Gypsum Board	106.8	145.4	132.7	100.6	63.4	75.6	77.6	63.4	68.1	77.6	63.4	68.1	92.7	63.4	73.0	92.7	63.4	73.0
0950, 0980	Ceilings & Acoustic Treatment	91.4	145.4	127.8	99.3	63.4	75.1	102.3	63.4	76.1	102.3	63.4	76.1	97.1	63.4	74.4	97.1	63.4	74.4
0960	Flooring	86.6	168.1	109.2	87.7	69.5	82.7	98.5	69.5	90.4	98.5	69.5	90.4	89.3	69.5	83.8	89.3	69.5	83.8
0970, 0990	Wall Finishes & Painting/Coating	83.0	145.9	119.6	95.3	52.8	70.5	93.6	52.8	69.8	93.6	52.8	69.8	89.9	52.8	68.3	89.9	52.8	68.3
09	FINISHES	87.0	150.0	121.3	89.4	63.6	75.4	93.9	63.6	77.4	92.6	63.6	76.8	88.3	63.6	74.9	89.6	63.6	75.4
COVERS	DIVS. 10 - 14, 25, 28, 41, 43, 44, 46	100.0	119.4	104.3	100.0	84.0	96.5	100.0	84.0	96.5	100.0	84.0	96.5	100.0	84.0	96.5	100.0	84.0	96.5
21, 22, 23	FIRE SUPPRESSION, PLUMBING & HVAC	99.8	131.9	113.5	100.2	69.2	86.9	98.1	69.2	85.7	98.1	68.8	85.6	100.0	69.2	86.8	98.1	69.2	85.7
26, 27, 3370	ELECTRICAL, COMMUNICATIONS & UTIL.	93.2	144.6	119.9	88.1	88.5	88.3	91.7	88.5	90.0	89.4	88.5	88.9	86.2	88.5	87.4	85.6	88.5	87.1
MF2016	WEIGHTED AVERAGE	95.0	134.4	112.2	98.5	75.1	88.3	101.1	75.1	89.7	99.0	75.0	88.4	99.1	75.1	88.6	98.9	75.1	88.5

NEW MEXICO

DIVISION		LAS CRUCES 880 MAT.	INST.	TOTAL	LAS VEGAS 877 MAT.	INST.	TOTAL	ROSWELL 882 MAT.	INST.	TOTAL	SANTA FE 875 MAT.	INST.	TOTAL	SOCORRO 878 MAT.	INST.	TOTAL	TRUTH/CONSEQUENCES 879 MAT.	INST.	TOTAL
015433	CONTRACTOR EQUIPMENT		85.7	85.7		111.0	111.0		111.0	111.0		111.0	111.0		111.0	111.0		85.7	85.7
0241, 31 - 34	SITE & INFRASTRUCTURE, DEMOLITION	99.0	80.8	86.3	97.9	102.4	101.1	101.0	102.4	102.0	102.2	102.4	102.3	94.4	102.4	100.0	114.9	80.9	91.0
0310	Concrete Forming & Accessories	93.8	63.0	67.2	99.1	64.1	68.9	97.1	64.1	68.6	97.8	64.1	68.7	99.1	64.1	68.9	96.8	63.0	67.6
0320	Concrete Reinforcing	106.4	70.9	88.5	103.3	71.0	87.0	110.9	71.0	90.8	96.5	71.0	83.7	105.3	71.0	88.0	98.4	70.9	84.5
0330	Cast-in-Place Concrete	89.4	62.6	79.2	91.9	70.1	83.6	94.6	70.1	85.3	97.5	70.1	87.1	90.0	70.1	82.5	98.6	62.6	85.0
03	CONCRETE	81.3	65.3	74.0	94.3	68.8	82.6	103.4	68.8	87.6	93.5	68.8	82.2	93.1	68.8	82.0	87.1	65.3	77.1
04	MASONRY	102.2	59.6	75.8	101.9	60.0	75.9	118.3	60.0	82.1	94.9	60.0	73.2	101.8	60.0	75.9	99.0	59.6	74.6
05	METALS	103.6	83.0	97.2	105.6	91.0	101.1	106.1	91.0	101.4	102.7	91.0	99.1	105.9	91.0	101.3	105.5	83.0	98.6
06	WOOD, PLASTICS & COMPOSITES	81.4	63.5	71.4	94.8	64.6	78.0	92.4	64.6	77.0	93.9	64.6	77.6	94.8	64.6	78.0	85.9	63.5	73.5
07	THERMAL & MOISTURE PROTECTION	85.7	65.9	77.3	96.7	70.8	85.7	98.4	70.8	86.7	99.2	70.8	87.2	96.6	70.8	85.7	85.6	65.9	77.2
08	OPENINGS	92.3	65.2	86.1	97.3	65.8	90.1	96.6	65.8	89.5	100.0	65.8	92.2	97.1	65.8	89.9	90.7	65.2	84.9
0920	Plaster & Gypsum Board	75.9	63.4	67.5	92.7	63.4	73.0	77.6	63.4	68.1	107.8	63.4	78.0	92.7	63.4	73.0	94.1	63.4	73.5
0950, 0980	Ceilings & Acoustic Treatment	88.0	63.4	71.4	97.1	63.4	74.4	102.3	63.4	76.1	97.7	63.4	74.6	97.1	63.4	74.4	85.9	63.4	70.7
0960	Flooring	129.6	69.5	112.9	89.3	69.5	83.8	98.5	69.5	90.4	97.0	69.5	89.4	89.3	69.5	83.8	118.0	69.5	104.5
0970, 0990	Wall Finishes & Painting/Coating	82.8	52.8	65.3	89.9	52.8	68.3	93.6	52.8	69.8	95.9	52.8	70.8	89.9	52.8	68.3	82.3	52.8	65.1
09	FINISHES	102.6	62.7	80.9	88.2	63.6	74.8	92.7	63.6	76.9	96.6	63.6	78.6	88.1	63.6	74.8	100.0	62.8	79.8
COVERS	DIVS. 10 - 14, 25, 28, 41, 43, 44, 46	100.0	81.5	95.9	100.0	84.0	96.5	100.0	84.0	96.5	100.0	84.0	96.5	100.0	84.0	96.5	100.0	81.5	95.9
21, 22, 23	FIRE SUPPRESSION, PLUMBING & HVAC	100.4	68.9	86.9	98.1	69.2	85.7	99.9	69.2	86.8	100.1	69.2	86.9	98.1	69.2	85.7	98.1	68.9	85.6
26, 27, 3370	ELECTRICAL, COMMUNICATIONS & UTIL.	91.4	88.5	89.9	87.7	88.5	88.1	90.8	88.5	89.6	99.8	88.5	93.9	86.0	88.5	87.3	89.6	88.5	89.0
MF2016	WEIGHTED AVERAGE	96.4	71.7	85.6	97.2	75.1	87.5	100.3	75.1	89.3	99.1	75.1	88.6	96.8	75.1	87.3	96.6	71.8	85.7

For customer support on your Electrical Costs with RSMeans data, call 800.448.8182.

535

DIVISION		NEW MEXICO TUCUMCARI 884			NEW YORK ALBANY 120 - 122			NEW YORK BINGHAMTON 137 - 139			NEW YORK BRONX 104			NEW YORK BROOKLYN 112			NEW YORK BUFFALO 140 - 142		
		MAT.	INST.	TOTAL	MAT.	INST.	TOTAL	MAT.	INST.	TOTAL	MAT.	INST.	TOTAL	MAT.	INST.	TOTAL	MAT.	INST.	TOTAL
015433	CONTRACTOR EQUIPMENT		111.0	111.0		115.8	115.8		121.0	121.0		106.9	106.9		112.3	112.3		98.9	98.9
0241, 31 - 34	SITE & INFRASTRUCTURE, DEMOLITION	98.4	104.2	101.2	83.9	104.7	98.5	96.2	93.2	94.1	92.6	113.7	107.4	120.9	125.5	124.1	99.6	101.6	101.0
0310	Concrete Forming & Accessories	97.1	64.0	68.5	99.3	105.9	105.0	100.6	90.7	92.0	99.1	193.1	180.2	106.7	182.5	172.1	102.3	120.3	117.8
0320	Concrete Reinforcing	108.7	71.0	89.7	103.7	119.5	111.7	100.4	104.2	102.3	96.4	172.4	134.8	101.8	222.6	162.8	108.6	115.3	111.9
0330	Cast-in-Place Concrete	94.6	70.0	85.3	79.7	116.1	93.5	105.1	107.6	106.1	86.4	173.4	119.5	107.2	171.7	131.7	108.3	125.3	114.8
03	CONCRETE	101.9	68.7	86.7	84.1	112.9	97.3	93.4	101.3	97.0	89.4	181.7	131.6	104.6	184.0	140.9	105.1	120.6	112.2
04	MASONRY	118.3	60.0	82.1	90.5	116.0	106.3	106.8	104.1	105.1	90.1	182.8	147.6	117.2	182.8	157.9	113.5	123.6	119.7
05	METALS	104.8	90.8	100.5	101.2	128.8	109.7	94.2	134.4	106.6	90.8	170.3	115.2	102.1	167.5	122.2	99.4	108.2	102.1
06	WOOD, PLASTICS & COMPOSITES	92.4	64.6	77.0	94.4	102.3	98.8	105.6	86.6	95.0	98.0	197.8	153.6	107.2	183.4	149.6	100.9	119.7	111.4
07	THERMAL & MOISTURE PROTECTION	98.2	70.8	86.6	106.6	110.3	108.2	107.1	95.0	102.0	104.4	168.9	131.7	107.3	166.6	132.4	104.8	113.3	108.4
08	OPENINGS	96.5	65.8	89.5	95.4	104.5	97.5	91.9	92.8	92.1	92.8	200.7	117.6	89.4	190.6	112.7	99.8	114.0	103.1
0920	Plaster & Gypsum Board	77.6	63.4	68.1	102.5	102.3	102.3	107.2	85.9	92.9	96.7	200.3	166.3	103.4	185.8	158.7	105.4	120.1	115.3
0950, 0980	Ceilings & Acoustic Treatment	102.3	63.4	76.1	99.6	102.3	101.4	94.1	85.9	88.6	85.2	200.3	162.8	89.9	185.8	154.5	97.3	120.1	112.7
0960	Flooring	98.5	69.5	90.4	90.6	111.1	96.3	99.4	99.2	99.4	92.6	189.9	119.6	106.3	189.9	129.5	95.2	118.5	101.7
0970, 0990	Wall Finishes & Painting/Coating	93.6	52.8	69.8	93.8	102.3	98.8	87.4	99.6	94.5	103.8	159.9	136.5	114.2	159.9	140.8	97.8	115.8	108.3
09	FINISHES	92.5	63.6	76.8	90.3	105.9	98.8	91.4	92.2	91.8	92.6	190.5	145.8	104.3	181.9	146.5	100.1	120.1	111.0
COVERS	DIVS. 10 - 14, 25, 28, 41, 43, 44, 46	100.0	84.0	96.5	100.0	102.0	100.4	100.0	98.8	99.7	100.0	137.9	108.4	100.0	135.5	107.9	100.0	108.1	101.8
21, 22, 23	FIRE SUPPRESSION, PLUMBING & HVAC	98.1	68.8	85.6	100.2	109.0	103.9	100.6	99.6	100.2	100.2	170.4	130.2	99.7	170.2	129.9	100.0	103.4	101.5
26, 27, 3370	ELECTRICAL, COMMUNICATIONS & UTIL.	91.7	88.5	90.0	95.5	110.9	103.5	99.4	100.8	100.1	94.2	181.5	139.6	99.1	181.5	141.9	98.9	106.0	102.6
MF2016	WEIGHTED AVERAGE	99.5	75.0	88.8	95.8	111.2	102.5	97.2	101.7	99.2	94.7	173.4	129.2	101.6	172.5	132.7	101.2	111.6	105.8

DIVISION		NEW YORK ELMIRA 148 - 149			NEW YORK FAR ROCKAWAY 116			NEW YORK FLUSHING 113			NEW YORK GLENS FALLS 128			NEW YORK HICKSVILLE 115, 117, 118			NEW YORK JAMAICA 114		
		MAT.	INST.	TOTAL	MAT.	INST.	TOTAL	MAT.	INST.	TOTAL	MAT.	INST.	TOTAL	MAT.	INST.	TOTAL	MAT.	INST.	TOTAL
015433	CONTRACTOR EQUIPMENT		123.0	123.0		112.3	112.3		112.3	112.3		115.8	115.8		112.3	112.3		112.3	112.3
0241, 31 - 34	SITE & INFRASTRUCTURE, DEMOLITION	102.2	92.8	95.6	124.2	125.5	125.1	124.2	125.5	125.1	74.3	104.2	95.3	113.8	124.6	121.3	118.4	125.5	123.4
0310	Concrete Forming & Accessories	83.5	96.0	94.3	93.1	193.0	179.2	97.0	193.0	179.8	84.8	96.3	94.8	89.6	162.1	152.1	97.0	193.0	179.8
0320	Concrete Reinforcing	108.1	108.3	108.2	101.8	222.7	162.8	103.5	222.7	163.6	99.3	111.3	105.4	101.8	222.5	162.7	101.8	222.7	162.8
0330	Cast-in-Place Concrete	98.3	106.7	101.5	115.9	171.8	137.1	115.9	171.8	137.1	77.0	110.8	89.8	96.6	173.4	131.7	104.0	188.8	142.7
03	CONCRETE	89.5	104.3	96.2	110.7	188.8	146.4	111.2	188.8	146.7	78.3	105.3	90.6	96.6	173.4	131.7	104.0	188.8	142.7
04	MASONRY	105.1	105.1	105.1	122.0	182.8	159.7	115.9	182.8	157.4	96.6	108.1	103.7	111.7	178.0	152.8	119.9	182.8	158.9
05	METALS	95.2	137.2	108.1	102.1	167.5	122.2	102.1	167.5	122.2	94.6	125.1	104.0	103.6	164.4	122.3	102.1	167.5	122.2
06	WOOD, PLASTICS & COMPOSITES	84.7	94.3	90.0	91.1	197.5	150.3	95.6	197.5	152.3	87.2	92.9	90.4	87.8	159.5	127.7	95.6	197.5	152.3
07	THERMAL & MOISTURE PROTECTION	103.2	95.6	100.0	107.2	168.2	133.0	107.2	168.2	133.0	99.7	105.3	102.1	106.9	161.4	130.0	107.0	168.2	132.9
08	OPENINGS	98.3	97.3	98.0	88.1	198.4	113.5	88.1	198.4	113.5	90.4	98.6	92.3	88.5	177.4	108.9	88.1	198.4	113.5
0920	Plaster & Gypsum Board	101.8	94.1	96.6	92.5	200.3	164.9	94.8	200.3	165.6	95.0	92.6	93.4	92.1	161.2	138.5	94.8	200.3	165.6
0950, 0980	Ceilings & Acoustic Treatment	98.5	94.1	95.5	79.7	200.3	161.0	79.7	200.3	161.0	85.6	92.6	90.3	78.7	161.2	134.4	79.7	200.3	161.0
0960	Flooring	86.5	102.4	90.9	101.1	189.9	125.7	102.6	189.9	126.9	82.3	111.1	90.3	100.0	176.4	121.2	102.6	189.9	126.9
0970, 0990	Wall Finishes & Painting/Coating	94.8	91.5	92.9	114.2	159.9	140.8	114.2	159.9	140.8	89.8	97.3	94.2	114.2	159.9	140.8	114.2	159.9	140.8
09	FINISHES	92.0	96.5	94.5	99.6	190.2	148.9	100.4	190.2	149.3	82.4	98.2	91.0	98.2	164.1	134.1	100.0	190.2	149.1
COVERS	DIVS. 10 - 14, 25, 28, 41, 43, 44, 46	100.0	97.9	99.5	100.0	137.1	108.3	100.0	137.1	108.3	100.0	98.1	99.6	100.0	124.9	105.5	100.0	137.1	108.3
21, 22, 23	FIRE SUPPRESSION, PLUMBING & HVAC	97.1	96.5	96.9	96.6	170.2	128.1	96.6	170.2	128.1	97.1	104.4	100.2	99.7	159.5	125.3	97.6	181.5	141.2
26, 27, 3370	ELECTRICAL, COMMUNICATIONS & UTIL.	97.7	103.8	100.9	105.9	181.5	145.2	105.9	181.5	145.2	90.7	110.9	101.2	98.6	142.4	121.4	100.2	174.8	132.9
MF2016	WEIGHTED AVERAGE	96.5	103.0	99.3	102.0	174.8	133.9	101.9	174.8	133.8	91.4	106.4	98.0	99.6	158.9	125.6	100.2	174.8	132.9

DIVISION		NEW YORK JAMESTOWN 147			NEW YORK KINGSTON 124			NEW YORK LONG ISLAND CITY 111			NEW YORK MONTICELLO 127			NEW YORK MOUNT VERNON 105			NEW YORK NEW ROCHELLE 108		
		MAT.	INST.	TOTAL	MAT.	INST.	TOTAL	MAT.	INST.	TOTAL	MAT.	INST.	TOTAL	MAT.	INST.	TOTAL	MAT.	INST.	TOTAL
015433	CONTRACTOR EQUIPMENT		91.9	91.9		112.3	112.3		112.3	112.3		112.3	112.3		106.9	106.9		106.9	106.9
0241, 31 - 34	SITE & INFRASTRUCTURE, DEMOLITION	103.7	93.5	96.6	145.6	120.7	128.1	122.0	125.5	124.5	93.3	128.1	123.3	89.4	140.1	133.1	104.5	140.1	135.2
0310	Concrete Forming & Accessories	83.6	89.1	88.3	86.2	128.1	122.4	101.3	193.0	180.4	99.0	156.7	128.1	95.2	170.9	133.4	95.3	170.9	133.4
0320	Concrete Reinforcing	108.3	114.5	111.4	99.8	156.7	128.5	101.8	222.7	162.8	99.0	156.7	128.1	95.2	170.9	133.4	95.3	170.9	133.4
0330	Cast-in-Place Concrete	102.0	85.3	95.7	103.9	145.2	119.6	110.6	171.8	133.8	97.3	145.2	115.5	96.5	150.7	117.1	96.5	150.7	117.1
03	CONCRETE	92.4	92.3	92.4	98.3	138.6	116.7	107.0	188.8	144.4	93.6	138.6	114.2	95.9	154.7	132.4	95.9	154.7	132.4
04	MASONRY	114.6	101.4	106.4	111.3	150.7	135.7	114.0	182.8	156.7	104.2	150.7	133.0	90.8	162.4	112.8	90.8	162.4	112.8
05	METALS	92.7	102.6	95.7	102.9	134.2	112.5	102.1	167.5	122.2	102.9	134.2	112.5	105.4	134.9	121.8	105.4	134.9	121.8
06	WOOD, PLASTICS & COMPOSITES	83.3	86.1	84.8	88.4	120.0	107.1	101.4	197.5	154.9	95.2	122.0	110.1	88.6	134.9	114.3	105.4	146.6	122.9
07	THERMAL & MOISTURE PROTECTION	102.7	94.0	99.0	123.4	143.5	131.9	107.2	168.2	133.0	123.1	143.5	131.7	105.4	146.6	122.8	105.4	146.6	122.9
08	OPENINGS	98.1	94.8	97.3	92.6	140.8	103.7	88.1	198.4	113.4	88.3	140.8	100.3	92.8	165.9	109.6	92.9	165.9	109.6
0920	Plaster & Gypsum Board	88.9	85.6	86.7	97.2	122.8	114.4	99.4	200.3	167.2	97.9	122.8	114.6	91.7	135.6	121.2	104.9	135.6	125.5
0950, 0980	Ceilings & Acoustic Treatment	95.1	85.6	88.7	76.3	122.8	107.6	79.7	200.3	161.0	76.3	122.8	107.6	83.5	135.6	118.6	83.5	135.6	118.6
0960	Flooring	89.0	102.4	92.7	98.7	162.9	116.5	104.3	189.9	128.0	101.1	162.9	118.2	84.6	173.4	109.3	92.0	173.4	114.6
0970, 0990	Wall Finishes & Painting/Coating	96.4	95.1	95.6	118.7	129.0	124.7	114.2	159.9	140.8	114.2	159.9	140.8	102.2	159.9	135.9	102.2	159.9	135.9
09	FINISHES	90.6	91.4	91.0	96.0	134.0	116.7	101.3	190.2	149.7	96.4	134.0	116.9	89.6	147.5	121.1	93.5	147.5	122.8
COVERS	DIVS. 10 - 14, 25, 28, 41, 43, 44, 46	100.0	96.6	99.3	100.0	115.9	103.5	100.0	137.1	108.3	100.0	115.9	103.5	100.0	119.9	104.4	100.0	116.8	103.7
21, 22, 23	FIRE SUPPRESSION, PLUMBING & HVAC	97.0	89.0	93.6	97.1	125.7	109.3	99.7	170.2	129.9	97.1	121.8	107.6	97.2	140.6	115.8	97.2	140.6	115.8
26, 27, 3370	ELECTRICAL, COMMUNICATIONS & UTIL.	96.7	93.8	95.2	92.0	121.0	107.1	98.1	181.5	141.4	92.0	121.0	107.1	92.4	149.6	122.2	92.4	149.6	122.2
MF2016	WEIGHTED AVERAGE	96.6	93.9	95.4	100.0	131.6	113.8	101.2	174.8	133.5	98.5	130.8	112.7	95.1	145.6	117.2	95.5	145.6	117.4

City Cost Indexes

For customer support on your Electrical Costs with RSMeans data, call 800.448.8182.

537

NEW YORK

DIVISION		NEW YORK 100 - 102			NIAGARA FALLS 143			PLATTSBURGH 129			POUGHKEEPSIE 125 - 126			QUEENS 110			RIVERHEAD 119		
		MAT.	INST.	TOTAL	MAT.	INST.	TOTAL	MAT.	INST.	TOTAL	MAT.	INST.	TOTAL	MAT.	INST.	TOTAL	MAT.	INST.	TOTAL
015433	CONTRACTOR EQUIPMENT		104.9	104.9		91.9	91.9		.96.6	96.6		112.3	112.3		112.3	112.3		112.3	112.3
0241, 31 - 34	SITE & INFRASTRUCTURE, DEMOLITION	101.4	111.7	108.6	106.0	95.0	98.3	113.2	101.7	105.2	141.7	121.1	127.3	117.1	125.5	123.0	115.0	124.7	121.8
0310	Concrete Forming & Accessories	108.5	196.3	184.2	83.5	118.6	113.8	90.5	95.8	95.1	86.2	175.7	163.3	89.9	193.0	178.8	94.2	162.2	152.8
0320	Concrete Reinforcing	102.0	236.2	169.7	106.8	115.6	111.2	103.8	115.6	109.8	99.8	156.8	128.5	103.5	222.7	163.6	103.7	222.5	163.6
0330	Cast-in-Place Concrete	99.9	181.8	131.0	105.5	130.7	115.1	94.2	106.0	98.7	100.7	142.7	116.6	101.8	171.8	128.4	100.1	168.6	126.1
03	CONCRETE	103.3	195.8	145.6	94.6	121.5	106.9	90.9	102.5	96.2	95.8	159.3	124.8	99.8	188.8	140.5	97.2	173.6	132.1
04	MASONRY	103.5	189.8	157.0	121.8	130.5	127.2	90.9	101.5	97.5	103.8	145.5	129.7	108.1	182.8	154.4	117.5	178.4	155.3
05	METALS	102.0	169.5	122.8	95.3	104.3	98.0	98.8	101.6	99.7	102.9	135.3	112.9	102.1	167.5	122.2	104.1	164.5	122.7
06	WOOD, PLASTICS & COMPOSITES	102.4	197.9	155.5	83.2	113.4	100.0	93.6	92.4	93.0	88.4	188.0	143.9	87.9	197.5	148.9	92.7	159.5	129.9
07	THERMAL & MOISTURE PROTECTION	107.4	173.4	135.3	102.8	115.4	108.1	117.8	102.0	111.1	123.4	148.4	133.9	106.8	168.2	132.8	107.8	161.6	130.6
08	OPENINGS	95.8	201.1	119.9	98.1	110.0	100.9	98.1	99.5	98.5	92.7	177.3	112.1	88.1	198.4	113.4	88.5	177.4	108.9
0920	Plaster & Gypsum Board	102.2	200.3	168.1	88.9	113.7	105.6	114.8	91.7	99.3	97.2	190.6	159.9	92.1	200.3	164.8	93.3	161.2	138.9
0950, 0980	Ceilings & Acoustic Treatment	100.5	200.3	167.8	95.1	113.7	107.7	104.7	91.7	95.9	76.3	190.6	153.3	79.7	200.3	161.0	79.6	161.2	134.6
0960	Flooring	93.9	189.9	120.5	89.0	118.5	97.2	104.6	111.1	106.4	98.7	167.7	117.8	100.0	189.9	125.0	101.1	176.4	122.0
0970, 0990	Wall Finishes & Painting/Coating	100.5	164.8	138.0	96.4	116.5	108.1	114.4	99.5	105.7	118.7	129.0	124.7	114.2	159.9	140.8	114.2	159.9	140.8
09	FINISHES	97.5	192.8	149.3	90.7	118.6	105.9	94.7	98.2	96.6	95.8	172.5	137.5	98.6	190.2	148.4	98.8	163.9	134.2
COVERS	DIVS. 10 - 14, 25, 28, 41, 43, 44, 46	100.0	143.3	109.6	100.0	107.0	101.6	100.0	96.6	99.2	100.0	121.4	104.8	100.0	137.1	108.3	100.0	125.0	105.6
21, 22, 23	FIRE SUPPRESSION, PLUMBING & HVAC	100.1	174.1	131.8	97.0	107.7	101.6	97.1	104.5	100.3	97.1	125.3	109.1	99.7	170.2	129.9	99.9	159.6	125.5
26, 27, 3370	ELECTRICAL, COMMUNICATIONS & UTIL.	100.9	188.1	146.3	95.4	104.4	100.1	88.8	96.9	93.0	92.0	128.0	110.8	98.6	181.5	141.7	100.1	142.4	122.1
MF2016	WEIGHTED AVERAGE	100.6	178.2	134.6	97.6	111.9	103.9	96.6	100.0	98.5	99.2	142.4	118.1	99.6	174.8	132.6	100.4	159.0	126.1

NEW YORK

DIVISION		ROCHESTER 144 - 146			SCHENECTADY 123			STATEN ISLAND 103			SUFFERN 109			SYRACUSE 130 - 132			UTICA 133 - 135		
		MAT.	INST.	TOTAL	MAT.	INST.	TOTAL	MAT.	INST.	TOTAL	MAT.	INST.	TOTAL	MAT.	INST.	TOTAL	MAT.	INST.	TOTAL
015433	CONTRACTOR EQUIPMENT		119.8	119.8		115.8	115.8		106.9	106.9		106.9	106.9		115.8	115.8		115.8	115.8
0241, 31 - 34	SITE & INFRASTRUCTURE, DEMOLITION	94.1	108.8	104.4	84.6	104.7	98.7	101.5	113.7	110.1	94.5	109.2	104.8	94.8	103.5	100.9	73.2	103.1	94.1
0310	Concrete Forming & Accessories	99.8	101.5	101.3	101.1	105.9	105.3	88.9	182.9	169.9	97.8	147.6	140.7	99.7	93.4	94.2	100.8	90.5	91.9
0320	Concrete Reinforcing	110.7	108.4	109.5	98.3	119.5	109.0	96.4	222.7	160.1	95.3	156.9	126.4	101.4	104.2	102.8	101.4	103.5	102.5
0330	Cast-in-Place Concrete	97.7	107.7	101.5	92.9	116.1	101.7	96.5	173.5	125.8	93.6	146.5	113.7	97.6	107.7	101.4	89.3	106.3	95.8
03	CONCRETE	94.3	106.1	99.7	91.2	112.9	101.2	100.5	184.9	139.1	95.3	147.9	119.3	96.0	101.6	98.6	94.0	99.7	96.6
04	MASONRY	99.6	107.8	104.7	93.3	116.0	107.4	102.0	182.8	152.1	95.4	148.3	128.2	99.3	105.3	103.0	90.4	103.7	98.6
05	METALS	101.4	123.2	108.1	98.7	128.8	108.0	88.9	170.5	113.9	88.9	135.0	103.1	97.7	120.4	104.7	95.8	120.0	103.2
06	WOOD, PLASTICS & COMPOSITES	98.0	100.0	99.1	105.3	102.3	103.6	87.1	183.7	140.9	97.8	149.4	126.5	101.9	90.1	95.4	101.9	86.7	93.4
07	THERMAL & MOISTURE PROTECTION	102.5	103.6	102.9	101.2	110.3	105.1	104.8	167.3	131.3	105.3	145.0	122.1	101.7	98.8	100.5	90.1	98.7	93.7
08	OPENINGS	102.3	101.6	102.1	96.1	104.5	98.0	92.8	192.9	115.8	92.9	155.9	107.3	94.2	92.8	93.9	96.9	90.8	95.5
0920	Plaster & Gypsum Board	105.9	100.1	102.0	104.3	102.3	103.0	91.8	185.8	154.9	96.0	150.5	132.6	98.5	89.8	92.7	96.9	86.2	90.3
0950, 0980	Ceilings & Acoustic Treatment	97.0	100.1	99.1	92.4	102.3	99.1	85.2	185.8	153.0	83.5	150.5	128.7	94.1	89.8	91.2	94.1	86.2	88.8
0960	Flooring	90.1	111.0	95.9	89.5	111.1	95.5	88.1	189.9	116.3	88.1	181.4	114.0	88.9	97.6	91.3	86.9	97.7	89.9
0970, 0990	Wall Finishes & Painting/Coating	94.7	102.9	99.5	89.6	102.3	97.1	103.8	159.9	136.5	102.2	132.7	120.0	91.2	103.2	98.2	84.8	103.2	95.5
09	FINISHES	95.2	103.5	99.7	87.8	105.9	97.6	91.3	182.1	140.7	90.9	153.2	124.8	90.5	94.5	92.7	88.9	92.3	90.7
COVERS	DIVS. 10 - 14, 25, 28, 41, 43, 44, 46	100.0	102.0	100.5	100.0	102.0	100.4	100.0	136.3	108.1	100.0	119.5	104.3	100.0	99.0	99.8	100.0	93.5	98.6
21, 22, 23	FIRE SUPPRESSION, PLUMBING & HVAC	100.1	91.8	96.5	100.2	109.0	104.0	100.2	170.4	130.2	97.2	128.8	110.7	100.2	96.7	98.7	100.2	96.4	98.6
26, 27, 3370	ELECTRICAL, COMMUNICATIONS & UTIL.	102.7	94.9	98.6	94.7	110.9	103.1	94.2	181.5	139.6	99.0	121.0	110.5	99.4	106.7	103.2	97.5	106.7	102.3
MF2016	WEIGHTED AVERAGE	99.5	102.7	100.9	96.1	111.2	102.7	96.4	172.2	129.6	95.1	136.0	113.0	97.6	102.0	99.5	95.7	100.8	97.9

		NEW YORK									NORTH CAROLINA								
DIVISION		WATERTOWN 136			WHITE PLAINS 106			YONKERS 107			ASHEVILLE 287 - 288			CHARLOTTE 281 - 282			DURHAM 277		
		MAT.	INST.	TOTAL	MAT.	INST.	TOTAL	MAT.	INST.	TOTAL	MAT.	INST.	TOTAL	MAT.	INST.	TOTAL	MAT.	INST.	TOTAL
015433	CONTRACTOR EQUIPMENT		115.8	115.8		106.9	106.9		106.9	106.9		101.9	101.9		101.9	101.9		107.3	107.3
0241, 31 - 34	SITE & INFRASTRUCTURE, DEMOLITION	80.9	103.6	96.8	92.4	109.9	104.7	99.2	110.0	106.7	97.4	81.6	86.3	99.1	81.8	87.0	101.5	90.7	93.9
0310	Concrete Forming & Accessories	85.8	97.5	95.9	102.8	140.1	135.0	103.0	151.3	144.7	91.1	63.4	67.2	96.1	62.9	67.5	95.2	63.4	67.7
0320	Concrete Reinforcing	102.1	104.2	103.2	95.3	170.9	133.4	99.1	171.0	135.4	94.9	68.0	81.3	99.2	67.0	82.9	98.2	67.0	82.0
0330	Cast-in-Place Concrete	104.0	110.2	106.3	85.7	150.7	110.4	95.9	150.9	116.8	109.3	71.3	94.9	111.5	70.6	95.9	98.2	71.3	88.0
03	CONCRETE	106.4	104.4	105.5	89.1	149.4	116.7	98.1	154.6	124.0	99.5	68.6	85.4	99.2	68.0	84.9	94.2	68.4	82.4
04	MASONRY	91.6	109.3	102.6	94.8	154.7	131.9	98.6	154.7	133.4	90.1	63.7	73.7	93.9	62.3	74.3	86.6	63.7	72.4
05	METALS	95.8	120.4	103.4	90.5	162.5	112.6	98.6	163.0	118.4	97.8	91.7	95.9	98.6	91.4	96.4	113.7	91.3	106.8
06	WOOD, PLASTICS & COMPOSITES	84.4	94.2	89.9	103.1	134.9	120.8	103.0	149.4	128.8	90.7	61.8	74.7	91.2	61.8	74.8	94.9	61.8	76.5
07	THERMAL & MOISTURE PROTECTION	90.4	102.0	95.3	105.1	146.6	122.7	105.4	149.5	124.1	101.0	65.3	85.9	95.4	64.6	82.4	107.4	65.3	89.6
08	OPENINGS	96.9	97.3	97.0	92.9	165.9	109.6	96.5	173.7	114.2	93.7	62.9	86.6	98.6	62.7	90.3	100.8	62.7	92.0
0920	Plaster & Gypsum Board	89.7	94.0	92.6	98.9	135.6	123.6	102.3	150.5	134.7	106.6	60.6	75.7	101.6	60.6	74.1	98.7	60.6	73.1
0950, 0980	Ceilings & Acoustic Treatment	94.1	94.0	94.0	83.5	135.6	118.6	99.7	150.5	133.9	83.8	60.6	68.2	87.8	60.6	69.5	86.3	60.6	69.0
0960	Flooring	80.0	97.7	84.9	90.5	173.4	113.5	90.0	189.9	117.7	92.8	63.5	84.6	92.1	63.5	84.1	96.8	63.5	87.6
0970, 0990	Wall Finishes & Painting/Coating	84.8	97.2	92.0	102.2	159.9	135.9	102.2	159.9	135.9	104.5	59.7	78.4	97.0	59.7	75.3	98.8	59.7	76.0
09	FINISHES	86.2	97.0	92.1	91.7	147.5	122.0	95.7	159.3	110.3	90.1	62.7	75.2	90.0	62.4	75.0	89.7	62.7	75.0
COVERS	DIVS. 10 - 14, 25, 28, 41, 43, 44, 46	100.0	100.5	100.1	100.0	119.9	104.4	100.0	129.1	106.5	100.0	85.2	96.7	100.0	84.8	96.6	100.0	85.2	96.7
21, 22, 23	FIRE SUPPRESSION, PLUMBING & HVAC	100.2	91.4	96.5	100.4	140.6	117.6	100.4	140.8	117.7	100.5	62.6	84.3	99.9	63.1	84.1	100.5	62.6	84.3
26, 27, 3370	ELECTRICAL, COMMUNICATIONS & UTIL.	99.3	95.2	97.2	92.4	149.6	122.2	99.1	164.7	133.2	100.5	61.1	80.0	99.7	62.3	80.3	98.0	60.6	78.5
MF2016	WEIGHTED AVERAGE	97.3	100.7	98.8	94.6	145.6	116.9	98.9	150.8	121.6	97.6	68.3	84.8	98.1	68.3	85.0	100.2	68.9	86.5

NORTH CAROLINA

DIVISION		ELIZABETH CITY 279			FAYETTEVILLE 283			GASTONIA 280			GREENSBORO 270, 272 - 274			HICKORY 286			KINSTON 285		
		MAT.	INST.	TOTAL	MAT.	INST.	TOTAL	MAT.	INST.	TOTAL	MAT.	INST.	TOTAL	MAT.	INST.	TOTAL	MAT.	INST.	TOTAL
015433	CONTRACTOR EQUIPMENT		111.5	111.5		107.3	107.3		101.9	101.9		107.3	107.3		107.3	107.3		107.3	107.3
0241, 31 - 34	SITE & INFRASTRUCTURE, DEMOLITION	105.8	92.5	96.5	96.8	90.6	92.5	97.3	81.9	86.5	101.3	90.8	94.0	96.4	89.4	91.5	95.2	89.3	91.1
0310	Concrete Forming & Accessories	81.6	63.6	66.0	90.8	62.8	66.7	97.3	63.6	68.2	95.0	63.3	67.7	87.7	63.4	66.7	84.1	62.6	65.6
0320	Concrete Reinforcing	96.2	71.9	84.0	98.6	67.0	82.6	95.3	67.0	81.0	97.1	67.0	81.9	94.9	67.0	80.8	94.4	66.9	80.5
0330	Cast-in-Place Concrete	98.4	73.0	88.8	114.7	70.5	97.9	106.9	71.4	93.4	97.5	71.3	87.6	109.3	71.3	94.9	105.5	70.4	92.2
03	CONCRETE	94.4	70.0	83.2	101.4	67.9	86.1	98.0	68.6	84.5	93.6	68.4	82.1	99.3	68.4	85.2	96.1	67.8	83.2
04	MASONRY	98.8	62.3	76.2	93.2	62.3	74.1	94.5	63.7	75.4	82.5	63.7	70.9	78.9	63.7	69.5	85.8	62.3	71.2
05	METALS	100.0	94.0	98.2	117.8	91.3	109.7	98.3	91.6	96.3	106.2	91.3	101.7	97.8	91.3	95.8	96.6	91.1	94.9
06	WOOD, PLASTICS & COMPOSITES	80.1	62.5	70.3	90.0	61.8	74.3	98.5	61.8	78.1	94.6	61.8	76.4	85.9	61.8	72.5	83.0	61.8	71.2
07	THERMAL & MOISTURE PROTECTION	106.6	64.5	88.8	100.4	64.6	85.3	101.2	65.3	86.0	107.1	65.3	89.4	101.3	65.3	86.1	101.2	64.6	85.7
08	OPENINGS	97.6	64.1	89.9	93.8	62.7	86.7	97.5	62.7	89.5	100.8	62.7	92.0	93.8	62.7	86.6	93.9	62.7	86.7
0920	Plaster & Gypsum Board	92.3	60.6	71.0	110.8	60.6	77.1	112.7	60.6	77.7	100.2	60.6	73.6	106.6	60.6	75.7	106.4	60.6	75.7
0950, 0980	Ceilings & Acoustic Treatment	86.3	60.6	69.0	85.5	60.6	68.8	87.3	60.6	69.3	86.3	60.6	69.0	83.8	60.6	68.2	87.3	60.6	69.3
0960	Flooring	89.2	81.6	87.1	93.0	63.5	84.8	96.1	81.6	92.1	96.8	63.5	87.6	92.7	81.6	89.6	90.0	81.6	87.7
0970, 0990	Wall Finishes & Painting/Coating	98.8	59.7	76.0	104.5	59.7	78.4	104.5	59.7	78.4	98.8	59.7	76.0	104.5	59.7	78.4	104.5	59.7	78.4
09	FINISHES	86.9	66.5	75.8	91.1	62.4	75.5	92.6	66.4	78.3	90.0	62.4	75.1	90.3	66.4	77.3	89.9	66.0	76.9
COVERS	DIVS. 10 - 14, 25, 28, 41, 43, 44, 46	100.0	82.7	96.1	100.0	84.7	96.6	100.0	85.2	96.7	100.0	82.4	96.1	100.0	85.2	96.7	100.0	84.7	96.6
21, 22, 23	FIRE SUPPRESSION, PLUMBING & HVAC	97.3	60.5	81.6	100.2	61.9	83.9	100.5	61.5	83.8	100.4	62.6	84.2	97.4	61.5	82.0	97.4	60.3	81.5
26, 27, 3370	ELECTRICAL, COMMUNICATIONS & UTIL.	97.7	68.2	82.4	100.3	60.6	79.6	100.0	62.3	80.4	97.1	61.1	78.4	98.4	62.3	79.6	98.2	58.9	77.8
MF2016	WEIGHTED AVERAGE	97.2	70.4	85.5	101.3	68.4	86.9	98.4	68.7	85.4	98.6	68.9	85.6	96.1	69.3	84.4	95.8	68.2	83.7

NORTH CAROLINA / NORTH DAKOTA

DIVISION		MURPHY 289			RALEIGH 275 - 276			ROCKY MOUNT 278			WILMINGTON 284			WINSTON-SALEM 271			BISMARCK 585		
		MAT.	INST.	TOTAL	MAT.	INST.	TOTAL	MAT.	INST.	TOTAL	MAT.	INST.	TOTAL	MAT.	INST.	TOTAL	MAT.	INST.	TOTAL
015433	CONTRACTOR EQUIPMENT		101.9	101.9		107.3	107.3		107.3	107.3		101.9	101.9		107.3	107.3		100.3	100.3
0241, 31 - 34	SITE & INFRASTRUCTURE, DEMOLITION	98.7	80.1	85.6	101.2	90.7	93.8	103.9	90.7	94.7	98.5	81.5	86.6	101.7	90.8	94.1	102.1	98.6	99.7
0310	Concrete Forming & Accessories	97.9	62.7	67.5	95.5	62.8	67.3	87.5	62.9	66.3	92.5	62.8	66.9	96.8	63.4	68.0	102.8	74.9	78.7
0320	Concrete Reinforcing	94.4	66.3	80.2	99.2	67.0	82.9	96.2	67.0	81.5	95.6	67.0	81.2	97.1	67.0	81.9	93.8	94.8	94.3
0330	Cast-in-Place Concrete	113.1	70.5	96.9	101.1	70.5	89.5	96.3	70.6	86.5	108.9	70.5	94.3	99.9	71.3	89.1	102.9	85.8	96.4
03	CONCRETE	102.4	67.7	86.6	92.5	67.9	81.3	95.1	68.0	82.7	99.4	67.9	85.0	94.9	68.4	82.8	92.5	82.8	88.1
04	MASONRY	82.0	62.3	69.8	80.9	62.3	69.4	76.1	62.3	67.6	79.4	62.3	68.8	82.8	63.7	71.0	103.6	83.3	91.0
05	METALS	95.6	91.1	94.2	98.7	91.3	96.5	99.2	91.4	96.8	97.3	91.3	95.5	103.4	91.3	99.7	103.2	93.7	100.3
06	WOOD, PLASTICS & COMPOSITES	99.1	61.7	78.3	92.7	61.8	75.5	86.6	61.8	72.8	92.5	61.8	75.5	94.6	61.8	76.4	97.2	70.7	82.5
07	THERMAL & MOISTURE PROTECTION	101.2	64.6	85.7	101.8	64.6	86.1	107.1	64.6	89.1	101.0	64.6	85.6	107.1	65.3	89.4	105.0	84.7	96.4
08	OPENINGS	93.7	62.4	86.5	99.1	62.7	90.7	96.8	62.7	89.0	93.9	62.7	86.7	100.8	62.7	92.0	106.5	81.0	100.7
0920	Plaster & Gypsum Board	111.9	60.5	77.4	92.6	60.6	71.1	94.2	60.6	71.7	108.7	60.6	76.4	100.2	60.6	73.6	102.8	70.3	81.0
0950, 0980	Ceilings & Acoustic Treatment	83.8	60.5	68.1	83.3	60.6	68.0	84.6	60.6	68.4	85.5	60.6	68.8	86.3	60.6	69.0	110.2	70.3	83.3
0960	Flooring	96.4	81.6	92.3	90.4	63.5	82.9	92.6	81.6	89.6	93.5	63.5	85.2	96.8	63.5	87.6	86.0	52.6	76.7
0970, 0990	Wall Finishes & Painting/Coating	104.5	59.7	78.4	93.8	59.7	73.9	98.8	59.7	76.0	104.5	59.7	78.4	98.8	59.7	76.0	91.2	60.1	73.1
09	FINISHES	92.1	65.9	77.9	88.4	62.4	74.2	88.0	66.0	76.0	91.0	62.4	75.4	90.0	62.7	75.1	96.0	69.6	81.7
COVERS	DIVS. 10 - 14, 25, 28, 41, 43, 44, 46	100.0	84.7	96.6	100.0	84.7	96.6	100.0	84.7	96.6	100.0	84.7	96.6	100.0	85.2	96.7	100.0	90.9	98.0
21, 22, 23	FIRE SUPPRESSION, PLUMBING & HVAC	97.4	60.7	81.7	99.9	61.9	83.7	97.3	60.6	81.6	100.5	61.9	84.0	100.4	62.6	84.2	100.1	76.9	90.2
26, 27, 3370	ELECTRICAL, COMMUNICATIONS & UTIL.	101.4	58.9	79.3	100.5	59.5	79.1	99.5	60.6	79.3	100.8	58.9	79.0	97.1	58.9	77.3	99.5	76.7	87.6
MF2016	WEIGHTED AVERAGE	96.9	67.5	84.0	96.9	68.3	84.4	96.3	68.6	84.2	97.2	67.5	84.2	98.4	68.7	85.4	100.3	81.4	92.0

NORTH DAKOTA

DIVISION		DEVILS LAKE 583			DICKINSON 586			FARGO 580 - 581			GRAND FORKS 582			JAMESTOWN 584			MINOT 587		
		MAT.	INST.	TOTAL	MAT.	INST.	TOTAL	MAT.	INST.	TOTAL	MAT.	INST.	TOTAL	MAT.	INST.	TOTAL	MAT.	INST.	TOTAL
015433	CONTRACTOR EQUIPMENT		100.3	100.3		100.3	100.3		100.3	100.3		100.3	100.3		100.3	100.3		100.3	100.3
0241, 31 - 34	SITE & INFRASTRUCTURE, DEMOLITION	106.9	98.6	101.1	114.4	98.6	103.3	97.6	98.6	98.3	110.4	98.6	102.1	105.9	98.6	100.8	108.0	98.6	101.4
0310	Concrete Forming & Accessories	99.3	74.8	78.1	89.1	74.8	76.8	96.9	75.0	78.0	92.8	74.8	77.3	90.7	74.7	76.9	88.8	74.8	76.8
0320	Concrete Reinforcing	95.3	94.7	95.0	96.2	94.8	95.5	95.4	95.1	95.3	93.9	94.7	94.3	95.9	95.1	95.5	97.2	94.8	96.0
0330	Cast-in-Place Concrete	123.1	85.8	108.9	111.4	85.8	101.6	100.8	85.8	95.1	111.3	85.8	101.6	121.5	85.7	107.9	111.3	85.8	101.6
03	CONCRETE	101.6	82.7	93.0	100.3	82.8	92.3	93.4	82.9	88.6	97.9	82.8	91.0	100.2	82.8	92.2	96.6	82.8	90.3
04	MASONRY	120.5	82.6	97.0	122.1	83.3	98.0	107.1	89.0	95.9	114.1	82.6	94.6	133.8	89.0	106.0	113.2	83.3	94.7
05	METALS	98.9	93.5	97.3	98.8	93.6	97.2	101.2	94.2	99.1	98.9	93.5	97.2	98.9	93.9	97.4	99.2	93.7	97.5
06	WOOD, PLASTICS & COMPOSITES	92.5	70.7	80.4	80.7	70.7	75.1	92.5	70.7	80.4	84.7	70.7	76.9	82.4	70.7	75.9	80.4	70.7	75.0
07	THERMAL & MOISTURE PROTECTION	105.7	84.5	96.7	106.3	84.7	97.1	105.9	86.7	97.8	105.9	84.5	96.9	105.5	86.7	97.6	105.7	84.7	96.8
08	OPENINGS	100.2	81.0	95.8	100.1	81.0	95.7	101.4	81.0	96.7	98.7	81.0	94.7	100.1	81.0	95.7	98.9	81.0	94.8
0920	Plaster & Gypsum Board	121.8	70.3	87.2	112.6	70.3	84.2	101.0	70.3	80.4	113.9	70.3	84.6	113.6	70.3	84.5	112.6	70.3	84.2
0950, 0980	Ceilings & Acoustic Treatment	108.4	70.3	82.7	108.4	70.3	82.7	99.9	70.3	79.9	108.4	70.3	82.7	108.4	70.3	82.7	108.4	70.3	82.7
0960	Flooring	92.6	52.6	81.5	86.1	52.6	76.8	96.1	52.6	84.1	88.0	52.6	78.2	86.8	52.6	77.3	85.8	52.6	76.6
0970, 0990	Wall Finishes & Painting/Coating	89.6	58.9	71.7	89.6	58.9	71.7	90.8	70.4	78.9	89.6	69.0	77.6	89.6	58.9	71.7	89.6	60.1	72.4
09	FINISHES	96.2	69.4	81.6	93.8	69.4	80.5	96.9	70.6	82.6	94.0	70.5	81.2	93.2	69.4	80.2	93.0	69.5	80.2
COVERS	DIVS. 10 - 14, 25, 28, 41, 43, 44, 46	100.0	90.8	98.0	100.0	90.9	98.0	100.0	90.9	98.0	100.0	90.9	98.0	100.0	90.9	98.0	100.0	90.9	98.0
21, 22, 23	FIRE SUPPRESSION, PLUMBING & HVAC	97.2	81.5	90.5	97.2	75.2	87.8	100.1	77.0	90.2	100.3	74.4	89.2	97.2	72.8	86.8	100.3	75.0	89.5
26, 27, 3370	ELECTRICAL, COMMUNICATIONS & UTIL.	97.4	48.0	71.7	105.8	71.7	88.0	103.5	73.3	87.8	100.8	63.0	81.1	97.4	48.0	71.7	103.8	76.7	89.7
MF2016	WEIGHTED AVERAGE	100.1	78.2	90.5	100.8	80.3	91.8	100.1	81.8	92.1	100.1	79.0	90.8	100.2	77.1	90.1	100.0	81.0	91.7

City Cost Indexes

NORTH DAKOTA / OHIO

DIVISION		WILLISTON 588 MAT.	INST.	TOTAL	AKRON 442-443 MAT.	INST.	TOTAL	ATHENS 457 MAT.	INST.	TOTAL	CANTON 446-447 MAT.	INST.	TOTAL	CHILLICOTHE 456 MAT.	INST.	TOTAL	CINCINNATI 451-452 MAT.	INST.	TOTAL
015433	CONTRACTOR EQUIPMENT		100.3	100.3		91.6	91.6		87.4	87.4		91.6	91.6		98.5	98.5		96.0	96.0
0241, 31 - 34	SITE & INFRASTRUCTURE, DEMOLITION	108.4	96.3	99.9	97.4	99.2	98.7	115.8	88.9	97.0	97.6	99.0	98.6	101.6	99.8	100.4	96.9	99.1	98.4
0310	Concrete Forming & Accessories	94.5	74.5	77.3	100.5	86.0	88.0	92.6	87.6	88.3	100.5	76.9	80.2	95.0	80.2	82.2	98.6	82.5	84.7
0320	Concrete Reinforcing	98.1	94.7	96.4	103.4	90.2	96.8	89.1	83.5	86.3	103.4	78.6	90.9	86.0	78.3	82.1	91.2	79.4	85.2
0330	Cast-in-Place Concrete	111.3	85.7	101.6	98.9	90.3	95.6	109.4	86.2	100.6	99.8	88.5	95.5	99.3	83.4	93.3	94.8	76.6	87.9
03	CONCRETE	97.8	82.6	90.9	98.3	87.6	93.4	101.1	85.9	94.1	98.7	80.9	90.6	95.5	81.4	89.1	94.6	80.1	88.0
04	MASONRY	107.4	83.3	92.5	93.5	91.6	92.3	80.1	81.2	80.8	94.1	79.1	84.8	87.3	91.2	89.7	90.5	78.0	82.7
05	METALS	99.0	93.3	97.3	96.8	80.1	91.7	103.2	77.5	95.3	96.8	74.8	90.0	95.2	85.7	92.3	97.5	83.8	93.3
06	WOOD, PLASTICS & COMPOSITES	86.2	70.7	77.6	107.2	85.0	94.8	86.4	90.6	88.8	107.5	75.2	89.6	99.7	77.0	87.1	99.4	83.9	90.7
07	THERMAL & MOISTURE PROTECTION	105.9	84.7	96.9	103.3	92.7	98.8	98.5	87.5	93.9	104.4	87.6	97.3	100.4	85.8	94.3	98.6	81.5	91.4
08	OPENINGS	100.2	81.0	95.8	109.6	85.3	104.0	97.1	84.9	94.3	103.3	73.2	96.4	88.7	74.4	85.4	96.6	80.1	92.8
0920	Plaster & Gypsum Board	113.9	70.3	84.6	103.3	84.5	90.7	95.6	90.2	92.0	104.3	74.4	84.2	98.3	76.7	83.8	97.8	83.8	88.4
0950, 0980	Ceilings & Acoustic Treatment	108.4	70.3	82.7	93.4	84.5	87.4	108.9	90.2	96.3	93.4	74.4	80.6	102.5	76.7	85.1	96.2	83.8	87.8
0960	Flooring	88.9	52.6	78.8	95.1	86.2	92.6	120.1	78.2	108.5	95.2	73.1	89.1	98.1	78.2	92.6	99.4	80.8	94.3
0970, 0990	Wall Finishes & Painting/Coating	89.6	58.9	71.7	98.2	95.0	96.3	100.7	88.8	93.7	98.2	72.2	83.0	97.9	80.1	87.5	97.5	74.6	84.1
09	FINISHES	94.2	69.4	80.7	97.4	86.6	91.5	101.5	86.5	93.3	97.6	75.0	85.3	98.4	79.3	88.0	97.7	81.6	88.9
COVERS	DIVS. 10 - 14, 25, 28, 41, 43, 44, 46	100.0	90.9	98.0	100.0	93.6	98.6	100.0	88.0	97.3	100.0	91.6	98.1	100.0	87.9	97.3	100.0	88.9	97.5
21, 22, 23	FIRE SUPPRESSION, PLUMBING & HVAC	97.2	75.2	87.8	100.0	89.0	95.3	96.9	76.3	88.1	100.0	78.0	90.6	97.4	94.6	96.2	99.9	77.1	90.2
26, 27, 3370	ELECTRICAL, COMMUNICATIONS & UTIL.	101.1	72.6	86.3	98.3	85.0	91.4	100.8	90.6	95.5	97.6	85.6	91.4	100.1	76.5	87.8	99.1	74.8	86.5
MF2016	WEIGHTED AVERAGE	99.3	80.2	90.9	99.7	88.2	94.7	99.1	83.7	92.4	99.1	81.0	91.1	96.2	86.1	91.7	97.7	80.9	90.3

OHIO

DIVISION		CLEVELAND 441 MAT.	INST.	TOTAL	COLUMBUS 430-432 MAT.	INST.	TOTAL	DAYTON 453-454 MAT.	INST.	TOTAL	HAMILTON 450 MAT.	INST.	TOTAL	LIMA 458 MAT.	INST.	TOTAL	LORAIN 440 MAT.	INST.	TOTAL
015433	CONTRACTOR EQUIPMENT		92.7	92.7		91.4	91.4		91.5	91.5		98.5	98.5		90.8	90.8		91.6	91.6
0241, 31 - 34	SITE & INFRASTRUCTURE, DEMOLITION	96.0	97.5	97.1	101.7	91.9	94.8	97.1	99.2	98.5	97.0	99.5	98.8	109.1	89.0	95.0	96.8	99.5	98.7
0310	Concrete Forming & Accessories	100.1	90.7	92.0	97.7	77.4	80.1	96.8	73.2	76.5	96.8	71.9	75.3	92.6	79.2	81.1	100.6	77.7	80.8
0320	Concrete Reinforcing	103.9	92.0	97.9	96.2	80.3	88.2	91.2	81.5	86.3	91.2	79.4	85.2	89.1	81.6	85.3	103.4	90.5	96.9
0330	Cast-in-Place Concrete	95.8	99.3	97.1	96.9	82.2	91.3	85.1	78.8	82.7	91.2	77.2	85.9	100.5	88.0	95.8	94.1	93.7	94.0
03	CONCRETE	98.5	93.4	96.2	96.2	79.6	88.6	86.9	76.6	82.2	89.8	75.7	83.4	94.0	82.7	88.8	96.0	85.2	91.1
04	MASONRY	100.0	99.2	99.5	91.9	86.5	88.5	86.3	77.2	80.6	86.7	79.4	82.2	108.7	78.6	90.0	90.2	94.6	92.9
05	METALS	98.4	84.9	94.3	97.5	80.8	92.4	96.7	77.8	90.9	96.8	86.9	93.7	103.2	80.7	96.3	97.4	81.5	92.5
06	WOOD, PLASTICS & COMPOSITES	101.0	88.3	93.9	99.4	76.1	86.4	103.5	71.7	85.8	102.1	69.0	83.7	86.4	78.7	82.1	107.2	73.4	88.4
07	THERMAL & MOISTURE PROTECTION	100.2	98.4	99.5	100.4	85.8	94.2	104.8	79.1	93.9	100.5	80.5	92.1	98.0	85.0	92.5	104.3	94.2	100.0
08	OPENINGS	97.9	87.5	95.5	99.7	74.4	93.9	95.8	72.8	90.5	93.5	71.5	88.4	97.1	75.9	92.2	103.3	78.9	97.7
0920	Plaster & Gypsum Board	103.2	87.7	92.8	97.2	75.5	82.6	100.1	71.3	80.8	100.1	68.5	78.9	95.6	78.0	83.8	103.3	78.9	82.7
0950, 0980	Ceilings & Acoustic Treatment	89.0	87.7	88.1	90.9	75.5	80.5	104.3	71.3	82.0	103.3	68.5	79.8	108.0	78.0	87.7	93.4	72.6	79.4
0960	Flooring	96.5	92.6	95.4	92.8	78.2	88.8	101.8	71.9	93.5	99.2	80.8	94.1	119.0	77.9	107.6	95.2	90.8	94.0
0970, 0990	Wall Finishes & Painting/Coating	101.1	95.0	97.5	100.4	80.1	88.5	97.9	71.6	82.6	97.9	71.0	82.2	100.7	71.9	83.9	98.2	95.0	96.3
09	FINISHES	97.8	91.2	94.2	93.9	77.3	84.8	99.9	72.3	84.8	98.9	72.8	84.7	100.5	78.0	88.3	97.4	81.0	88.5
COVERS	DIVS. 10 - 14, 25, 28, 41, 43, 44, 46	100.0	98.0	99.6	100.0	89.0	97.6	100.0	87.2	97.1	100.0	87.4	97.2	100.0	87.1	97.1	100.0	94.5	98.8
21, 22, 23	FIRE SUPPRESSION, PLUMBING & HVAC	100.0	93.0	97.0	99.9	85.7	93.9	100.8	75.7	90.1	100.5	77.5	90.7	96.9	83.1	91.0	100.0	90.0	95.7
26, 27, 3370	ELECTRICAL, COMMUNICATIONS & UTIL.	98.0	95.7	96.8	99.0	84.7	91.6	97.8	76.4	86.6	98.0	76.7	86.9	101.1	76.4	88.3	97.7	80.8	89.0
MF2016	WEIGHTED AVERAGE	98.9	93.4	96.5	98.1	83.3	91.6	96.7	77.9	88.5	96.6	79.4	89.1	99.4	81.1	91.4	98.6	86.9	93.5

OHIO

DIVISION		MANSFIELD 448-449 MAT.	INST.	TOTAL	MARION 433 MAT.	INST.	TOTAL	SPRINGFIELD 455 MAT.	INST.	TOTAL	STEUBENVILLE 439 MAT.	INST.	TOTAL	TOLEDO 434-436 MAT.	INST.	TOTAL	YOUNGSTOWN 444-445 MAT.	INST.	TOTAL
015433	CONTRACTOR EQUIPMENT		91.6	91.6		91.2	91.2		91.5	91.5		95.1	95.1		95.1	95.1		91.6	91.6
0241, 31 - 34	SITE & INFRASTRUCTURE, DEMOLITION	93.4	95.0	94.5	96.3	95.0	95.4	97.4	99.1	98.6	141.4	103.5	114.8	99.9	95.5	96.8	97.3	99.0	98.5
0310	Concrete Forming & Accessories	89.8	76.3	78.2	95.6	77.0	79.5	96.8	73.3	76.5	96.8	82.3	84.3	99.2	88.1	89.6	100.5	79.3	82.2
0320	Concrete Reinforcing	94.1	79.3	86.7	88.6	80.5	84.5	91.2	81.5	86.3	86.2	96.8	91.5	96.2	84.5	90.3	103.4	86.4	94.8
0330	Cast-in-Place Concrete	91.6	83.0	88.3	84.0	81.1	82.9	87.4	78.7	84.1	91.2	88.6	90.2	92.0	91.7	91.9	97.9	88.6	94.4
03	CONCRETE	90.6	78.9	85.2	83.4	78.9	81.4	88.0	76.6	82.8	87.2	86.5	86.9	91.0	88.7	89.9	97.8	83.4	91.2
04	MASONRY	92.6	91.7	92.0	93.5	89.2	90.8	86.5	76.9	80.5	80.8	91.3	87.3	99.9	90.7	94.2	93.8	87.3	89.7
05	METALS	97.6	76.0	91.0	96.5	78.9	91.1	96.7	77.9	90.9	93.1	82.2	89.7	97.3	85.9	93.8	96.8	78.2	91.1
06	WOOD, PLASTICS & COMPOSITES	94.9	73.4	83.0	96.6	76.1	85.2	105.0	71.7	86.5	92.1	80.0	85.4	100.5	88.0	93.5	107.2	78.2	91.1
07	THERMAL & MOISTURE PROTECTION	102.5	90.4	97.4	96.9	89.1	93.6	104.6	78.9	93.8	108.7	89.1	100.4	97.0	91.9	94.9	104.5	85.8	96.6
08	OPENINGS	103.7	73.2	96.7	93.4	71.5	88.4	94.0	72.8	89.1	93.7	82.0	91.0	96.2	85.1	93.6	103.3	78.6	97.6
0920	Plaster & Gypsum Board	96.9	72.6	80.6	98.5	75.5	83.1	100.1	71.3	80.8	97.6	79.1	85.1	100.5	87.7	91.9	103.3	77.5	86.0
0950, 0980	Ceilings & Acoustic Treatment	94.3	72.6	79.7	97.1	75.5	82.5	104.3	71.3	82.0	94.2	79.1	84.0	97.1	87.7	90.8	93.4	77.5	82.7
0960	Flooring	89.8	92.2	90.4	92.2	92.2	92.2	101.8	71.9	93.5	119.4	92.8	112.0	92.5	88.8	91.5	95.2	83.4	91.9
0970, 0990	Wall Finishes & Painting/Coating	98.2	76.5	85.5	104.9	76.5	88.3	97.9	71.6	82.6	118.5	88.8	101.2	104.9	87.3	94.6	98.2	82.0	88.7
09	FINISHES	94.8	78.6	86.0	95.3	79.5	86.7	99.9	72.2	84.8	112.9	84.3	97.4	96.0	88.0	91.6	97.5	79.9	87.9
COVERS	DIVS. 10 - 14, 25, 28, 41, 43, 44, 46	100.0	91.7	98.2	100.0	86.1	96.9	100.0	87.1	97.1	100.0	89.0	97.5	100.0	94.8	98.8	100.0	91.8	98.2
21, 22, 23	FIRE SUPPRESSION, PLUMBING & HVAC	96.9	89.0	93.5	96.9	85.1	91.9	100.8	81.1	92.4	97.3	93.3	95.6	100.0	93.5	97.2	100.0	83.8	93.1
26, 27, 3370	ELECTRICAL, COMMUNICATIONS & UTIL.	95.6	70.6	82.6	93.5	70.6	81.6	97.8	84.7	91.0	88.3	107.1	98.1	99.1	109.2	104.4	97.7	77.1	87.0
MF2016	WEIGHTED AVERAGE	96.8	82.8	90.7	94.4	81.6	88.8	96.7	80.2	89.5	96.2	91.8	94.3	97.5	93.1	95.6	98.9	83.3	92.1

For customer support on your Electrical Costs with RSMeans data, call 800.448.8182.

539

City Cost Indexes

OHIO / OKLAHOMA

DIVISION		OHIO ZANESVILLE 437-438 MAT.	INST.	TOTAL	OKLAHOMA ARDMORE 734 MAT.	INST.	TOTAL	CLINTON 736 MAT.	INST.	TOTAL	DURANT 747 MAT.	INST.	TOTAL	ENID 737 MAT.	INST.	TOTAL	GUYMON 739 MAT.	INST.	TOTAL
015433	CONTRACTOR EQUIPMENT		91.2	91.2		82.8	82.8		81.8	81.8		81.8	81.8		81.8	81.8		81.8	81.8
0241, 31 - 34	SITE & INFRASTRUCTURE, DEMOLITION	99.4	95.0	96.3	101.6	97.4	98.6	103.1	95.8	98.0	95.5	93.4	94.0	104.7	95.8	98.5	107.6	95.5	99.1
0310	Concrete Forming & Accessories	92.7	76.9	79.1	91.6	57.8	62.4	90.2	57.8	62.2	84.2	57.4	61.1	93.7	57.9	62.8	96.9	57.6	63.0
0320	Concrete Reinforcing	88.1	93.0	90.6	81.1	66.5	73.7	81.6	66.5	74.0	92.1	65.6	78.7	81.0	66.5	73.7	81.6	62.7	72.0
0330	Cast-in-Place Concrete	88.5	80.9	85.6	103.5	73.9	92.2	100.0	73.9	90.1	92.0	73.8	85.1	100.0	74.0	90.1	100.0	73.6	90.0
03	CONCRETE	87.0	81.1	84.3	89.8	64.7	78.3	89.1	64.7	78.0	85.7	64.3	75.9	89.6	64.8	78.3	92.1	63.9	79.2
04	MASONRY	91.2	77.7	82.8	95.8	57.9	72.3	119.7	57.9	81.4	88.7	64.8	73.8	101.5	57.9	74.5	98.0	56.6	72.4
05	METALS	98.0	83.6	93.6	102.0	61.3	89.5	102.1	61.3	89.6	91.3	60.9	82.0	103.6	61.4	90.7	102.6	59.1	89.3
06	WOOD, PLASTICS & COMPOSITES	92.4	76.1	83.3	104.1	57.0	77.9	103.1	57.0	77.5	91.5	57.0	72.3	106.5	57.0	79.0	110.0	57.0	80.5
07	THERMAL & MOISTURE PROTECTION	97.0	82.8	91.0	105.3	65.2	88.3	105.4	65.2	88.4	99.3	66.2	85.2	105.5	65.2	88.5	105.9	64.8	88.5
08	OPENINGS	93.4	78.1	89.9	103.2	57.0	92.6	103.2	57.0	92.6	96.9	56.7	87.7	104.4	57.3	93.6	103.3	56.1	92.5
0920	Plaster & Gypsum Board	95.2	75.5	82.0	88.9	56.4	67.0	88.5	56.4	66.9	77.9	56.4	63.4	89.2	56.4	67.1	89.2	56.4	67.1
0950, 0980	Ceilings & Acoustic Treatment	97.1	75.5	82.5	86.0	56.4	66.0	86.0	56.4	66.0	82.3	56.4	64.8	86.0	56.4	66.0	86.0	56.4	66.0
0960	Flooring	90.3	78.2	87.0	88.5	60.6	80.7	87.4	60.6	79.9	90.8	54.6	80.8	89.1	74.8	85.1	90.6	60.6	82.3
0970, 0990	Wall Finishes & Painting/Coating	104.9	80.1	90.4	84.0	45.1	61.3	84.0	45.1	61.3	90.9	45.1	64.2	84.0	45.1	61.3	84.0	44.2	60.8
09	FINISHES	94.5	77.0	85.0	81.8	55.9	67.7	81.6	55.9	67.6	82.5	54.8	67.4	82.2	58.8	69.5	83.1	57.2	69.0
COVERS	DIVS. 10 - 14, 25, 28, 41, 43, 44, 46	100.0	82.1	96.0	100.0	79.1	95.3	100.0	79.1	95.3	100.0	79.0	95.3	100.0	79.1	95.3	100.0	79.0	95.3
21, 22, 23	FIRE SUPPRESSION, PLUMBING & HVAC	96.9	90.9	94.3	97.2	67.8	84.7	97.2	67.8	84.7	97.2	67.8	84.6	100.3	67.9	86.4	97.2	65.6	83.7
26, 27, 3370	ELECTRICAL, COMMUNICATIONS & UTIL.	93.6	74.0	83.4	93.9	71.2	82.1	94.8	71.2	82.5	96.2	62.1	78.4	94.8	71.2	82.5	96.4	57.5	76.2
MF2016	WEIGHTED AVERAGE	95.0	82.5	89.5	96.7	66.7	83.6	97.9	66.6	84.2	93.2	65.6	81.1	98.3	67.0	84.6	97.8	63.8	82.9

OKLAHOMA

DIVISION		LAWTON 735 MAT.	INST.	TOTAL	MCALESTER 745 MAT.	INST.	TOTAL	MIAMI 743 MAT.	INST.	TOTAL	MUSKOGEE 744 MAT.	INST.	TOTAL	OKLAHOMA CITY 730-731 MAT.	INST.	TOTAL	PONCA CITY 746 MAT.	INST.	TOTAL
015433	CONTRACTOR EQUIPMENT		82.8	82.8		81.8	81.8		93.5	93.5		93.5	93.5		83.1	83.1		81.8	81.8
0241, 31 - 34	SITE & INFRASTRUCTURE, DEMOLITION	100.6	97.4	98.3	88.7	95.5	93.5	90.0	92.8	91.9	90.3	92.7	92.0	99.0	97.9	98.2	96.0	95.8	95.9
0310	Concrete Forming & Accessories	96.9	57.9	63.3	82.3	43.2	48.6	95.1	57.6	62.7	99.5	57.4	63.2	94.3	57.9	62.9	90.6	57.7	62.3
0320	Concrete Reinforcing	81.2	66.5	73.8	91.8	65.2	78.4	90.3	66.5	78.3	91.2	64.9	77.9	89.7	66.5	78.0	91.2	66.5	78.7
0330	Cast-in-Place Concrete	96.7	74.0	88.1	80.7	73.5	78.0	84.6	74.9	80.9	85.6	74.8	81.5	96.7	74.0	88.1	94.5	73.9	86.7
03	CONCRETE	86.0	64.8	76.3	76.8	57.7	68.1	81.1	66.0	74.2	82.7	65.6	74.9	88.8	64.8	77.9	87.8	64.7	77.2
04	MASONRY	97.8	57.9	73.1	106.4	57.8	76.3	91.6	58.0	70.7	108.3	51.3	73.0	100.3	57.9	74.0	84.4	57.9	68.0
05	METALS	107.5	61.4	93.3	91.2	60.0	81.6	91.2	76.7	86.7	92.7	75.3	87.4	100.4	61.4	88.4	91.2	61.2	82.0
06	WOOD, PLASTICS & COMPOSITES	109.0	57.0	80.1	89.1	37.9	60.6	103.7	57.1	77.8	107.9	57.1	79.7	94.9	57.0	73.8	99.4	57.0	75.8
07	THERMAL & MOISTURE PROTECTION	105.3	65.2	88.3	98.9	62.1	83.3	99.3	65.2	84.8	99.6	62.9	84.1	96.2	65.5	83.2	99.4	65.5	85.1
08	OPENINGS	106.2	57.3	95.0	96.9	46.1	85.2	96.9	57.1	87.7	98.1	56.8	88.6	101.5	57.3	91.4	96.9	57.0	87.7
0920	Plaster & Gypsum Board	90.7	56.4	67.7	76.9	36.7	49.9	83.8	56.4	65.4	85.8	56.4	66.0	96.0	56.4	69.4	82.8	56.4	65.1
0950, 0980	Ceilings & Acoustic Treatment	92.8	56.4	68.2	82.3	36.7	51.6	82.3	56.4	64.8	90.9	56.4	67.6	92.6	56.4	68.2	82.3	56.4	64.8
0960	Flooring	90.9	74.8	86.5	89.7	60.6	81.6	97.0	54.6	85.2	99.4	35.1	81.6	88.2	74.8	84.5	93.9	60.6	84.7
0970, 0990	Wall Finishes & Painting/Coating	84.0	45.1	61.3	90.9	44.2	63.7	90.9	44.2	63.7	90.9	44.2	63.7	87.1	45.1	62.6	90.9	45.1	64.2
09	FINISHES	83.9	58.8	70.3	81.5	44.5	61.4	84.6	54.8	68.3	87.3	51.3	67.7	86.5	58.8	71.4	84.2	56.9	69.4
COVERS	DIVS. 10 - 14, 25, 28, 41, 43, 44, 46	100.0	79.1	95.3	100.0	77.0	94.9	100.0	79.5	95.4	100.0	79.5	95.4	100.0	79.1	95.3	100.0	79.1	95.3
21, 22, 23	FIRE SUPPRESSION, PLUMBING & HVAC	100.3	67.9	86.4	97.2	62.4	82.3	97.2	62.5	82.4	100.3	62.4	84.1	100.1	67.9	86.3	97.2	62.5	82.4
26, 27, 3370	ELECTRICAL, COMMUNICATIONS & UTIL.	96.4	69.4	82.4	94.7	64.5	79.0	96.0	64.6	79.7	94.3	70.9	82.1	101.7	71.2	85.8	94.3	67.6	80.4
MF2016	WEIGHTED AVERAGE	98.7	66.9	84.8	92.6	61.1	78.8	92.9	65.8	81.0	95.1	65.3	82.1	97.7	67.2	84.4	93.3	65.1	81.0

OKLAHOMA / OREGON

DIVISION		POTEAU 749 MAT.	INST.	TOTAL	SHAWNEE 748 MAT.	INST.	TOTAL	TULSA 740-741 MAT.	INST.	TOTAL	WOODWARD 738 MAT.	INST.	TOTAL	BEND 977 MAT.	INST.	TOTAL	EUGENE 974 MAT.	INST.	TOTAL
015433	CONTRACTOR EQUIPMENT		92.2	92.2		81.8	81.8		93.5	93.5		81.8	81.8		98.9	98.9		98.9	98.9
0241, 31 - 34	SITE & INFRASTRUCTURE, DEMOLITION	77.2	88.1	84.9	99.1	95.8	96.8	96.7	91.8	93.3	103.5	95.8	98.1	109.4	102.5	104.6	99.4	102.6	101.6
0310	Concrete Forming & Accessories	88.6	57.3	61.6	84.1	57.7	61.3	99.6	57.6	63.4	90.3	45.5	51.6	100.8	96.2	97.9	105.3	96.3	97.5
0320	Concrete Reinforcing	92.2	66.5	79.2	91.2	62.8	76.8	91.4	66.5	78.9	81.0	66.5	73.7	100.0	109.8	104.9	104.4	109.8	107.1
0330	Cast-in-Place Concrete	84.6	74.8	80.9	97.6	73.9	88.6	93.2	75.0	86.3	100.0	73.9	90.1	105.4	101.3	103.9	102.1	101.4	101.8
03	CONCRETE	83.2	65.8	75.3	89.4	64.0	77.8	87.9	66.1	77.9	89.4	59.2	75.6	103.8	99.9	102.0	95.6	99.9	97.6
04	MASONRY	91.9	58.0	70.9	107.8	57.9	76.8	92.5	58.0	71.1	91.0	57.9	70.5	104.0	100.8	102.0	101.0	100.8	100.9
05	METALS	91.2	76.4	86.7	91.1	59.9	81.5	95.8	76.8	89.9	102.2	61.2	89.6	96.3	94.5	95.7	97.0	94.6	96.3
06	WOOD, PLASTICS & COMPOSITES	96.1	57.1	74.4	91.3	57.0	72.2	107.1	57.1	79.3	103.3	40.5	68.3	101.8	95.1	98.1	97.8	95.1	96.3
07	THERMAL & MOISTURE PROTECTION	99.4	65.2	84.9	99.4	64.9	84.8	99.6	66.1	85.4	105.5	63.6	87.8	115.4	99.4	108.6	114.5	102.5	109.4
08	OPENINGS	96.9	57.1	87.7	96.9	56.1	87.5	99.8	57.1	90.0	103.2	47.9	90.5	101.2	98.9	100.6	101.4	98.9	100.9
0920	Plaster & Gypsum Board	80.9	56.4	64.4	77.9	56.4	63.4	85.8	56.4	66.0	88.5	39.4	55.5	115.2	94.8	101.5	113.2	94.8	100.8
0950, 0980	Ceilings & Acoustic Treatment	82.3	56.4	64.8	82.3	56.4	64.8	90.9	56.4	67.6	86.0	39.4	54.6	90.8	94.8	93.5	91.7	94.8	93.8
0960	Flooring	93.3	54.6	82.6	90.8	60.6	82.4	98.3	62.5	88.4	87.4	57.7	79.1	98.4	105.4	100.3	96.6	105.4	99.1
0970, 0990	Wall Finishes & Painting/Coating	90.9	45.1	64.2	90.9	44.2	63.7	90.9	44.2	63.7	84.0	45.1	61.3	90.9	72.8	80.3	90.9	72.8	80.3
09	FINISHES	82.3	54.9	67.4	82.7	55.8	68.1	87.1	56.3	70.4	81.6	45.6	62.0	96.1	95.1	95.6	94.5	95.1	94.8
COVERS	DIVS. 10 - 14, 25, 28, 41, 43, 44, 46	100.0	79.3	95.4	100.0	79.1	95.3	100.0	79.5	95.4	100.0	77.3	94.9	100.0	99.7	99.9	100.0	99.7	99.9
21, 22, 23	FIRE SUPPRESSION, PLUMBING & HVAC	97.2	62.4	82.4	97.2	67.8	84.6	100.3	64.3	84.9	97.2	67.8	84.7	97.0	102.2	99.2	100.1	102.2	101.0
26, 27, 3370	ELECTRICAL, COMMUNICATIONS & UTIL.	94.4	64.6	78.9	96.2	71.2	83.2	96.2	64.6	79.8	96.3	71.2	83.2	101.0	94.5	97.6	99.6	94.5	97.0
MF2016	WEIGHTED AVERAGE	92.5	65.4	80.6	94.7	66.3	82.3	96.0	66.4	83.0	96.7	63.8	82.3	100.0	98.7	99.4	99.1	98.8	99.0

OREGON

DIVISION		KLAMATH FALLS 976 MAT.	INST.	TOTAL	MEDFORD 975 MAT.	INST.	TOTAL	PENDLETON 978 MAT.	INST.	TOTAL	PORTLAND 970-972 MAT.	INST.	TOTAL	SALEM 973 MAT.	INST.	TOTAL	VALE 979 MAT.	INST.	TOTAL
015433	CONTRACTOR EQUIPMENT		98.9	98.9		98.9	98.9		96.4	96.4		98.9	98.9		98.9	98.9		96.4	96.4
0241, 31 - 34	SITE & INFRASTRUCTURE, DEMOLITION	113.4	102.5	105.8	107.1	102.5	103.9	106.8	95.5	98.9	102.1	102.6	102.4	93.9	102.5	99.9	94.0	95.4	95.0
0310	Concrete Forming & Accessories	101.7	95.9	96.7	100.8	95.9	96.5	102.2	96.2	97.1	106.4	96.3	97.7	105.9	96.2	97.5	108.7	95.1	97.0
0320	Concrete Reinforcing	100.0	109.7	104.9	101.7	109.7	105.8	99.2	109.8	104.5	105.2	109.8	107.5	113.1	109.8	111.4	96.7	109.6	103.2
0330	Cast-in-Place Concrete	105.4	101.2	103.8	105.4	101.2	103.8	106.2	100.2	103.9	104.9	101.4	103.6	96.4	101.3	98.3	83.6	102.2	90.6
03	CONCRETE	106.4	99.7	103.4	101.4	99.7	100.6	88.6	99.5	93.6	97.1	99.9	98.4	93.3	99.9	96.3	74.1	99.6	85.8
04	MASONRY	118.1	100.8	107.4	98.0	100.8	99.7	108.5	100.9	103.8	102.7	100.8	101.5	107.2	100.8	103.2	106.6	100.9	103.1
05	METALS	96.3	94.3	95.7	96.6	94.2	95.8	103.0	95.0	100.5	98.1	94.6	97.0	103.7	94.5	100.9	102.9	93.9	100.1
06	WOOD, PLASTICS & COMPOSITES	92.9	95.1	94.1	91.8	95.1	93.6	94.6	95.2	94.9	98.7	95.1	96.7	92.5	95.1	93.9	102.9	95.2	98.6
07	THERMAL & MOISTURE PROTECTION	115.6	95.2	107.0	115.2	95.2	106.8	108.0	95.8	102.9	114.4	99.4	108.1	110.5	99.4	105.8	107.4	93.4	101.5
08	OPENINGS	101.2	98.9	100.7	104.3	98.9	103.0	97.7	99.0	98.0	99.1	98.9	99.1	102.7	98.9	101.8	97.7	89.1	95.7
0920	Plaster & Gypsum Board	110.0	94.8	99.8	109.4	94.8	99.6	96.7	94.8	95.4	113.1	94.8	100.8	110.9	94.8	100.1	102.9	94.8	97.5
0950, 0980	Ceilings & Acoustic Treatment	98.4	94.8	96.0	105.1	94.8	98.2	65.0	94.8	85.1	93.6	94.8	94.4	99.6	94.8	96.4	65.0	94.8	85.1
0960	Flooring	95.2	105.4	98.1	94.6	105.4	97.6	66.1	105.4	77.0	94.5	105.4	97.5	98.0	105.4	100.0	68.2	105.4	78.5
0970, 0990	Wall Finishes & Painting/Coating	90.9	68.2	77.7	90.9	68.2	77.7	83.5	75.0	78.6	90.7	75.0	81.6	90.6	75.0	81.5	83.5	72.8	77.2
09	FINISHES	96.4	94.6	95.4	96.7	94.6	95.5	68.1	95.4	83.0	94.2	95.3	94.8	96.1	95.3	95.6	68.7	95.1	83.1
COVERS	DIVS. 10 - 14, 25, 28, 41, 43, 44, 46	100.0	99.6	99.9	100.0	99.6	99.9	100.0	99.5	99.9	100.0	99.7	99.9	100.0	99.7	99.9	100.0	99.5	99.9
21, 22, 23	FIRE SUPPRESSION, PLUMBING & HVAC	97.0	106.1	100.9	100.1	102.2	101.0	98.9	108.7	103.1	100.1	106.2	102.7	100.0	102.2	101.0	98.9	73.5	88.1
26, 27, 3370	ELECTRICAL, COMMUNICATIONS & UTIL.	99.7	77.5	88.2	103.1	77.5	89.8	92.2	96.3	94.3	99.8	103.5	101.7	107.2	94.5	100.6	92.2	67.7	79.4
MF2016	WEIGHTED AVERAGE	101.0	96.9	99.2	100.6	96.1	98.6	96.0	99.7	97.6	99.4	100.8	100.0	100.9	98.7	100.0	93.8	87.7	91.1

PENNSYLVANIA

DIVISION		ALLENTOWN 181 MAT.	INST.	TOTAL	ALTOONA 166 MAT.	INST.	TOTAL	BEDFORD 155 MAT.	INST.	TOTAL	BRADFORD 167 MAT.	INST.	TOTAL	BUTLER 160 MAT.	INST.	TOTAL	CHAMBERSBURG 172 MAT.	INST.	TOTAL
015433	CONTRACTOR EQUIPMENT		115.8	115.8		115.8	115.8		114.1	114.1		115.8	115.8		115.8	115.8		115.0	115.0
0241, 31 - 34	SITE & INFRASTRUCTURE, DEMOLITION	93.5	102.2	99.6	96.3	102.1	100.3	105.9	99.9	101.7	92.5	101.0	98.4	87.6	102.8	98.3	91.3	99.4	97.0
0310	Concrete Forming & Accessories	99.1	109.4	108.0	84.2	83.2	83.3	82.6	79.9	80.3	86.3	101.0	99.0	85.6	88.9	88.4	86.9	80.0	81.0
0320	Concrete Reinforcing	101.4	117.0	109.3	98.3	104.2	101.2	96.4	102.0	99.2	100.4	104.0	102.2	99.0	120.4	109.8	99.0	111.9	105.5
0330	Cast-in-Place Concrete	88.5	105.2	94.8	98.6	91.3	95.8	111.9	86.3	102.2	94.1	95.3	94.6	87.1	99.6	91.8	91.3	97.7	93.7
03	CONCRETE	90.8	110.1	99.6	86.8	91.2	88.8	103.3	87.6	96.1	92.7	100.6	96.3	78.9	99.5	88.3	94.5	93.3	93.9
04	MASONRY	94.3	97.1	96.0	97.8	87.0	91.1	105.7	83.7	92.0	95.2	82.1	87.1	100.0	96.7	97.9	100.0	82.3	89.0
05	METALS	98.0	123.2	105.7	92.0	115.6	99.3	97.2	114.0	102.3	95.8	115.4	101.8	91.7	123.4	101.5	95.5	117.8	102.4
06	WOOD, PLASTICS & COMPOSITES	101.3	111.6	107.0	80.5	81.3	80.9	84.0	79.1	81.3	86.9	106.1	97.6	81.9	85.6	84.0	88.2	79.2	83.2
07	THERMAL & MOISTURE PROTECTION	101.7	112.1	106.1	100.7	92.8	97.4	99.6	89.0	95.1	101.5	91.5	97.3	100.2	96.3	98.5	95.7	86.1	91.6
08	OPENINGS	94.2	109.9	97.8	87.9	85.6	87.4	96.3	84.5	93.6	94.1	99.4	95.3	87.9	96.1	89.8	91.4	84.7	89.9
0920	Plaster & Gypsum Board	96.8	111.8	106.9	88.5	80.7	83.2	100.5	78.5	85.8	89.6	106.2	100.7	88.5	85.2	86.3	104.3	78.5	87.0
0950, 0980	Ceilings & Acoustic Treatment	86.4	111.8	103.5	89.0	80.7	83.4	98.8	78.5	85.2	89.1	106.2	100.6	89.9	85.2	86.7	88.5	78.5	81.8
0960	Flooring	88.9	106.5	92.1	82.8	104.6	88.9	94.7	82.4	91.3	83.2	104.6	89.1	83.7	111.3	91.4	87.3	82.4	85.9
0970, 0990	Wall Finishes & Painting/Coating	91.2	106.5	100.1	86.8	111.0	100.9	96.7	111.0	105.1	91.2	105.3	99.4	86.8	111.0	100.9	88.6	105.3	98.3
09	FINISHES	88.8	108.0	99.2	86.5	89.2	88.0	99.4	83.3	90.6	86.6	102.9	95.5	86.5	94.4	90.8	86.5	82.7	84.5
COVERS	DIVS. 10 - 14, 25, 28, 41, 43, 44, 46	100.0	103.2	100.7	100.0	97.9	99.5	100.0	96.4	99.2	100.0	100.1	100.0	100.0	100.0	100.0	100.0	95.6	99.0
21, 22, 23	FIRE SUPPRESSION, PLUMBING & HVAC	100.2	112.0	105.3	99.7	85.6	93.7	96.9	82.3	90.6	97.2	94.4	96.0	96.7	97.1	96.8	97.2	88.1	93.3
26, 27, 3370	ELECTRICAL, COMMUNICATIONS & UTIL.	98.7	98.0	98.3	89.5	110.6	100.5	95.6	110.6	103.4	92.8	110.6	102.1	90.1	109.1	100.0	89.2	86.3	87.7
MF2016	WEIGHTED AVERAGE	96.5	107.7	101.4	93.2	95.1	94.0	98.7	92.2	95.8	94.8	100.2	97.2	91.4	101.5	95.8	94.4	90.8	92.8

PENNSYLVANIA

DIVISION		DOYLESTOWN 189 MAT.	INST.	TOTAL	DUBOIS 158 MAT.	INST.	TOTAL	ERIE 164 - 165 MAT.	INST.	TOTAL	GREENSBURG 156 MAT.	INST.	TOTAL	HARRISBURG 170 - 171 MAT.	INST.	TOTAL	HAZLETON 182 MAT.	INST.	TOTAL
015433	CONTRACTOR EQUIPMENT		93.5	93.5		114.1	114.1		115.8	115.8		114.1	114.1		115.0	115.0		115.8	115.8
0241, 31 - 34	SITE & INFRASTRUCTURE, DEMOLITION	107.5	89.2	94.7	110.8	100.0	103.2	93.2	102.4	99.6	101.8	101.9	101.9	91.4	100.3	97.6	86.9	101.5	97.1
0310	Concrete Forming & Accessories	83.3	126.4	120.5	82.1	80.7	80.9	98.4	87.0	88.5	88.8	92.8	92.2	98.6	87.5	89.0	80.9	86.7	85.9
0320	Concrete Reinforcing	98.1	124.0	111.2	95.7	104.1	99.9	100.4	105.2	102.8	95.7	120.1	108.0	108.4	114.5	111.5	98.4	113.0	105.8
0330	Cast-in-Place Concrete	83.6	133.8	102.6	107.9	94.7	102.9	96.9	93.3	95.5	103.8	99.2	102.0	89.8	100.9	94.0	83.7	97.0	89.5
03	CONCRETE	86.5	127.8	105.4	105.1	91.1	98.7	85.8	93.8	89.4	98.2	101.0	99.5	89.9	98.3	93.7	83.7	97.0	89.8
04	MASONRY	97.7	126.7	115.7	106.2	84.1	92.5	86.9	90.6	89.2	116.1	92.9	101.7	94.2	87.4	90.0	107.4	91.5	97.6
05	METALS	95.5	114.2	101.3	97.2	114.2	102.4	92.2	116.2	99.6	97.1	121.7	104.6	102.0	121.2	107.9	97.7	119.9	104.5
06	WOOD, PLASTICS & COMPOSITES	82.6	126.6	107.1	83.1	79.1	80.9	97.4	84.5	90.2	90.1	91.3	90.7	99.8	86.4	92.3	81.3	83.8	82.7
07	THERMAL & MOISTURE PROTECTION	99.2	127.8	111.3	99.9	91.2	96.2	101.2	92.1	97.4	99.5	95.8	97.9	98.6	103.9	100.8	101.0	100.8	100.9
08	OPENINGS	96.5	129.8	104.2	96.3	84.5	93.6	88.1	89.0	88.3	96.3	99.2	97.0	97.8	89.3	95.8	94.7	93.1	94.4
0920	Plaster & Gypsum Board	87.5	127.3	114.2	99.5	78.5	85.4	96.8	83.9	88.2	101.6	91.0	94.5	108.2	86.0	93.3	87.9	83.3	84.8
0950, 0980	Ceilings & Acoustic Treatment	85.6	127.3	113.7	98.8	78.5	85.2	86.4	83.9	84.7	98.0	91.0	93.3	97.0	86.0	89.6	87.4	83.3	84.6
0960	Flooring	73.8	147.3	94.2	94.5	104.6	97.3	89.6	104.6	93.7	98.8	82.4	94.2	89.8	95.6	91.4	80.4	94.4	84.3
0970, 0990	Wall Finishes & Painting/Coating	90.8	145.9	122.9	96.7	111.0	105.1	96.3	91.8	93.7	96.7	111.0	105.1	90.6	88.6	89.4	91.2	108.8	101.5
09	FINISHES	80.4	132.4	108.7	99.6	87.5	93.0	89.8	90.0	89.9	100.3	92.5	96.0	92.0	88.3	90.0	84.8	90.0	87.6
COVERS	DIVS. 10 - 14, 25, 28, 41, 43, 44, 46	100.0	113.0	102.9	100.0	96.9	99.3	100.0	99.5	99.9	100.0	100.4	100.1	100.0	97.7	99.5	100.0	96.8	99.3
21, 22, 23	FIRE SUPPRESSION, PLUMBING & HVAC	96.7	127.8	110.0	96.9	83.0	91.0	99.7	100.5	100.1	96.9	90.9	94.3	100.2	92.0	96.7	97.2	95.7	96.5
26, 27, 3370	ELECTRICAL, COMMUNICATIONS & UTIL.	92.3	127.0	110.3	96.2	110.6	103.7	91.1	92.4	91.8	96.2	110.6	103.7	95.9	88.7	92.2	93.6	91.7	92.6
MF2016	WEIGHTED AVERAGE	94.0	123.5	106.9	99.1	93.5	96.7	93.1	96.8	94.7	98.6	99.9	99.2	97.3	95.2	96.3	94.4	96.8	95.4

For customer support on your Electrical Costs with RSMeans data, call 800.448.8182.

541

PENNSYLVANIA

DIVISION		INDIANA 157			JOHNSTOWN 159			KITTANNING 162			LANCASTER 175 - 176			LEHIGH VALLEY 180			MONTROSE 188		
		MAT.	INST.	TOTAL	MAT.	INST.	TOTAL	MAT.	INST.	TOTAL	MAT.	INST.	TOTAL	MAT.	INST.	TOTAL	MAT.	INST.	TOTAL
015433	CONTRACTOR EQUIPMENT		114.1	114.1		114.1	114.1		115.8	115.8		115.0	115.0		115.8	115.8		115.8	115.8
0241, 31 - 34	SITE & INFRASTRUCTURE, DEMOLITION	99.9	100.9	100.6	106.4	101.1	102.7	90.2	102.4	98.8	83.1	100.4	95.2	90.8	103.0	99.4	89.5	102.2	98.4
0310	Concrete Forming & Accessories	83.2	89.9	89.0	82.1	83.1	83.0	85.6	93.9	92.7	89.0	87.7	87.9	92.5	127.5	122.7	81.8	90.6	89.4
0320	Concrete Reinforcing	95.0	120.1	107.7	96.4	119.9	108.2	99.0	120.2	109.7	98.6	114.6	106.7	98.4	111.3	104.9	103.1	119.9	111.6
0330	Cast-in-Place Concrete	101.8	97.7	100.2	112.9	91.2	104.6	90.5	98.0	93.3	77.6	100.9	86.5	90.5	109.4	97.7	88.7	97.8	92.2
03	CONCRETE	95.7	99.1	97.3	104.3	93.7	99.5	81.4	101.1	90.4	83.1	98.4	90.1	89.9	118.8	103.1	88.6	99.6	93.6
04	MASONRY	102.2	92.4	96.1	103.1	87.3	93.3	102.6	92.3	96.2	106.1	88.9	95.4	94.3	109.9	104.0	94.3	95.5	95.0
05	METALS	97.2	121.2	104.6	97.2	119.8	104.1	91.8	122.4	101.2	95.5	121.4	103.4	97.6	120.9	104.8	95.9	123.0	104.2
06	WOOD, PLASTICS & COMPOSITES	84.7	88.8	87.0	83.1	81.2	82.0	81.9	93.6	88.4	90.9	86.4	88.4	93.2	130.3	113.8	82.0	87.3	85.0
07	THERMAL & MOISTURE PROTECTION	99.4	95.4	97.7	99.6	91.4	96.2	100.3	96.0	98.5	95.2	104.4	99.1	101.4	113.8	106.7	101.1	90.6	96.6
08	OPENINGS	96.3	93.5	95.7	96.3	89.3	94.7	87.9	100.5	90.8	91.4	89.3	90.9	94.7	118.9	100.3	91.4	94.4	92.1
0920	Plaster & Gypsum Board	100.8	88.4	92.5	99.3	80.7	86.8	88.5	93.4	91.8	106.2	86.0	92.6	90.5	131.0	117.7	88.3	86.9	87.4
0950, 0980	Ceilings & Acoustic Treatment	98.8	88.4	91.8	98.0	80.7	86.3	89.9	93.4	92.2	88.5	86.0	86.8	87.4	131.0	116.8	89.1	86.9	87.6
0960	Flooring	95.5	104.6	98.0	94.5	82.4	91.1	83.7	104.6	89.5	88.3	101.1	91.8	85.8	99.5	89.6	81.1	104.6	87.6
0970, 0990	Wall Finishes & Painting/Coating	96.7	111.0	105.1	96.7	111.0	105.1	86.8	111.0	100.9	88.6	88.6	88.6	91.2	62.5	74.4	91.2	108.8	101.5
09	FINISHES	99.2	94.3	96.5	99.0	85.5	91.6	86.6	97.2	92.4	86.5	89.4	88.1	87.0	116.8	103.2	85.6	94.2	90.3
COVERS	DIVS. 10 - 14, 25, 28, 41, 43, 44, 46	100.0	99.5	99.9	100.0	97.9	99.5	100.0	100.2	100.0	100.0	97.9	99.5	100.0	102.9	100.7	100.0	98.4	99.7
21, 22, 23	FIRE SUPPRESSION, PLUMBING & HVAC	96.9	83.3	91.1	96.9	82.1	90.6	96.7	93.3	95.2	97.2	92.3	95.1	97.2	113.0	103.9	97.2	98.6	97.8
26, 27, 3370	ELECTRICAL, COMMUNICATIONS & UTIL.	96.2	110.6	103.7	96.2	110.6	103.7	89.5	110.5	100.5	90.4	94.7	92.6	93.6	136.6	116.0	92.8	96.2	94.6
MF2016	WEIGHTED AVERAGE	97.4	97.8	97.6	98.7	94.7	96.9	91.9	101.2	95.9	93.2	96.4	94.6	94.9	117.4	104.7	93.8	99.5	96.3

PENNSYLVANIA

DIVISION		NEW CASTLE 161			NORRISTOWN 194			OIL CITY 163			PHILADELPHIA 190 - 191			PITTSBURGH 150 - 152			POTTSVILLE 179		
		MAT.	INST.	TOTAL	MAT.	INST.	TOTAL	MAT.	INST.	TOTAL	MAT.	INST.	TOTAL	MAT.	INST.	TOTAL	MAT.	INST.	TOTAL
015433	CONTRACTOR EQUIPMENT		115.8	115.8		100.1	100.1		115.8	115.8		100.3	100.3		102.6	102.6		115.0	115.0
0241, 31 - 34	SITE & INFRASTRUCTURE, DEMOLITION	88.0	102.7	98.3	97.0	100.2	96.8	86.6	101.2	96.8	98.8	100.6	100.1	106.0	100.4	102.1	86.1	100.6	96.3
0310	Concrete Forming & Accessories	85.6	94.6	93.4	84.1	126.7	120.8	85.6	93.4	92.3	101.5	141.9	136.4	99.0	98.9	98.9	80.6	88.9	87.8
0320	Concrete Reinforcing	97.8	101.9	99.8	99.5	126.6	113.2	99.0	97.7	98.3	103.3	126.7	115.1	96.9	120.6	108.8	97.8	114.5	106.2
0330	Cast-in-Place Concrete	87.9	99.2	92.2	83.4	134.1	102.6	85.3	97.6	90.0	91.2	137.7	108.9	111.3	99.9	107.0	82.7	101.3	89.8
03	CONCRETE	79.3	98.7	88.2	85.9	128.5	105.4	77.8	96.9	86.5	98.9	136.6	116.2	104.8	103.0	104.0	86.5	99.1	92.2
04	MASONRY	99.6	96.6	97.7	110.2	127.3	120.8	98.9	88.4	92.4	95.7	130.8	117.5	99.0	101.0	100.2	99.4	91.3	94.4
05	METALS	91.8	116.3	99.3	99.2	114.8	104.0	91.8	114.7	98.8	101.9	115.1	106.0	98.6	107.9	101.5	95.7	121.4	103.6
06	WOOD, PLASTICS & COMPOSITES	81.9	94.0	88.6	82.0	126.6	106.8	81.9	93.6	88.4	96.3	144.5	123.1	102.6	99.0	99.6	81.0	85.6	83.6
07	THERMAL & MOISTURE PROTECTION	100.2	95.7	98.3	104.6	128.2	114.6	100.1	93.2	97.2	101.5	136.9	116.5	99.8	99.1	99.5	95.3	103.1	98.6
08	OPENINGS	87.9	93.0	89.1	86.9	129.8	96.7	87.9	95.8	89.7	97.1	140.3	107.0	98.7	103.5	99.8	91.5	95.0	92.3
0920	Plaster & Gypsum Board	88.5	93.7	92.0	86.5	127.3	113.9	88.5	93.4	91.8	98.0	145.5	129.9	98.0	98.8	98.5	101.0	85.2	90.4
0950, 0980	Ceilings & Acoustic Treatment	89.9	93.7	92.5	91.0	127.3	115.4	89.9	93.4	92.2	101.3	145.5	131.1	92.6	98.8	96.8	88.5	85.2	86.3
0960	Flooring	83.7	111.3	91.4	81.8	147.3	99.9	83.7	106.1	89.9	96.7	147.3	110.8	104.0	107.7	105.0	84.3	101.1	89.0
0970, 0990	Wall Finishes & Painting/Coating	86.8	111.0	100.9	87.9	145.9	121.7	86.8	111.0	100.9	94.6	150.2	127.0	99.8	111.0	106.3	88.6	108.8	100.4
09	FINISHES	86.5	99.1	93.3	84.4	132.5	110.6	86.4	97.4	92.4	100.4	144.5	124.4	101.1	101.5	101.3	84.8	92.3	88.9
COVERS	DIVS. 10 - 14, 25, 28, 41, 43, 44, 46	100.0	100.8	100.2	100.0	113.2	102.9	100.0	100.1	100.0	100.0	120.0	104.4	100.0	101.8	100.4	100.0	95.2	98.9
21, 22, 23	FIRE SUPPRESSION, PLUMBING & HVAC	96.7	100.9	98.5	96.9	126.8	109.7	96.7	96.4	96.6	100.1	136.7	115.7	99.9	98.9	99.5	97.2	97.6	97.4
26, 27, 3370	ELECTRICAL, COMMUNICATIONS & UTIL.	90.1	100.7	95.6	93.7	150.0	123.0	91.7	109.1	100.8	100.1	155.2	128.7	98.5	111.6	105.3	88.8	91.0	89.9
MF2016	WEIGHTED AVERAGE	91.5	100.9	95.6	94.5	127.6	109.0	91.3	99.6	95.0	99.7	134.6	115.0	100.3	103.0	101.5	93.0	97.9	95.1

PENNSYLVANIA

DIVISION		READING 195 - 196			SCRANTON 184 - 185			STATE COLLEGE 168			STROUDSBURG 183			SUNBURY 178			UNIONTOWN 154		
		MAT.	INST.	TOTAL	MAT.	INST.	TOTAL	MAT.	INST.	TOTAL	MAT.	INST.	TOTAL	MAT.	INST.	TOTAL	MAT.	INST.	TOTAL
015433	CONTRACTOR EQUIPMENT		122.9	122.9		115.8	115.8		115.0	115.0		115.8	115.8		115.8	115.8		114.1	114.1
0241, 31 - 34	SITE & INFRASTRUCTURE, DEMOLITION	101.2	113.7	110.0	94.0	102.2	99.8	84.4	100.6	95.8	88.5	103.0	98.7	98.7	101.6	100.7	100.4	101.7	101.3
0310	Concrete Forming & Accessories	100.1	90.2	91.6	99.2	90.8	92.0	84.8	83.4	83.6	87.2	95.5	94.3	92.2	86.1	86.9	76.5	92.6	90.4
0320	Concrete Reinforcing	100.8	143.2	122.2	101.4	120.1	110.8	99.6	104.2	102.0	101.7	120.2	111.0	100.6	114.4	107.6	95.7	120.1	108.0
0330	Cast-in-Place Concrete	74.7	102.2	85.1	92.4	98.1	94.5	89.1	91.4	90.0	87.0	104.8	93.8	90.4	99.7	94.0	101.8	98.9	100.7
03	CONCRETE	84.8	105.0	94.0	92.6	99.8	95.9	92.8	91.3	92.1	87.4	104.2	95.1	89.9	97.2	93.3	95.3	100.7	97.8
04	MASONRY	99.2	95.4	96.8	94.6	97.9	96.6	100.3	87.5	92.3	92.4	109.7	103.1	99.9	84.4	90.3	117.8	92.8	102.3
05	METALS	99.5	132.9	109.7	100.0	123.3	107.1	95.6	115.9	101.8	97.7	123.6	105.7	95.4	121.1	103.3	96.9	121.2	104.4
06	WOOD, PLASTICS & COMPOSITES	99.4	86.4	92.2	101.3	87.3	93.5	88.7	81.3	84.5	87.7	87.3	87.5	88.9	85.6	87.1	77.6	91.2	85.2
07	THERMAL & MOISTURE PROTECTION	104.9	107.9	106.2	101.6	95.1	98.8	100.8	101.5	101.1	101.2	99.0	100.3	96.2	96.3	96.2	99.3	95.8	97.8
08	OPENINGS	91.2	102.2	93.8	94.2	94.4	94.2	91.2	85.6	89.9	94.7	97.3	95.3	91.6	92.2	91.7	96.3	98.7	96.8
0920	Plaster & Gypsum Board	96.3	86.0	89.4	98.5	86.9	90.7	90.4	80.7	83.9	89.2	86.9	87.6	100.2	85.2	90.1	97.7	91.0	93.2
0950, 0980	Ceilings & Acoustic Treatment	84.0	86.0	85.3	94.1	86.9	89.2	86.5	80.7	82.6	85.6	86.9	86.5	85.1	85.2	85.1	98.0	91.0	93.3
0960	Flooring	86.5	103.5	91.2	88.9	98.2	91.5	86.3	103.5	91.1	83.8	99.5	88.1	85.1	104.6	90.5	91.3	82.4	88.8
0970, 0990	Wall Finishes & Painting/Coating	86.7	106.5	98.2	91.2	113.2	104.0	91.2	111.0	102.8	91.2	108.8	101.5	88.6	108.8	100.4	96.7	108.9	103.8
09	FINISHES	86.1	92.9	89.8	90.5	93.8	92.3	86.2	89.1	87.8	85.7	96.5	91.6	85.7	91.2	88.7	97.3	92.2	94.6
COVERS	DIVS. 10 - 14, 25, 28, 41, 43, 44, 46	100.0	99.8	99.9	100.0	98.6	99.7	100.0	97.1	99.3	100.0	102.6	100.6	100.0	95.2	98.9	100.0	100.4	100.1
21, 22, 23	FIRE SUPPRESSION, PLUMBING & HVAC	100.2	111.1	104.8	100.2	98.8	99.6	97.2	91.6	94.8	97.2	104.8	100.4	97.2	89.6	93.9	96.9	85.6	92.1
26, 27, 3370	ELECTRICAL, COMMUNICATIONS & UTIL.	99.8	94.7	97.1	98.7	96.2	97.4	92.0	110.6	101.7	93.6	142.9	119.2	89.1	93.2	91.2	93.3	110.6	102.3
MF2016	WEIGHTED AVERAGE	96.1	105.2	100.1	97.2	100.0	98.4	94.4	96.5	95.3	94.3	110.4	101.3	93.9	95.1	94.4	97.6	98.6	97.8

City Cost Indexes

PENNSYLVANIA

DIVISION		WASHINGTON 153 MAT.	INST.	TOTAL	WELLSBORO 169 MAT.	INST.	TOTAL	WESTCHESTER 193 MAT.	INST.	TOTAL	WILKES-BARRE 186-187 MAT.	INST.	TOTAL	WILLIAMSPORT 177 MAT.	INST.	TOTAL	YORK 173-174 MAT.	INST.	TOTAL
015433	CONTRACTOR EQUIPMENT		114.1	114.1		115.8	115.8		100.1	100.1		115.8	115.8		115.8	115.8		115.0	115.0
0241, 31 - 34	SITE & INFRASTRUCTURE, DEMOLITION	100.5	102.0	101.6	96.0	101.6	99.9	102.9	101.1	101.6	86.5	102.2	97.5	89.3	101.7	98.0	86.9	100.4	96.3
0310	Concrete Forming & Accessories	83.3	94.6	93.1	85.8	85.9	85.9	90.8	127.3	122.3	89.8	89.4	89.4	88.8	86.8	87.1	84.0	87.7	87.1
0320	Concrete Reinforcing	95.7	120.4	108.2	99.6	119.9	109.8	98.5	143.9	121.4	100.4	120.1	110.3	99.8	114.5	107.2	100.6	114.6	107.6
0330	Cast-in-Place Concrete	101.8	99.3	100.8	93.3	92.7	93.1	92.3	135.2	108.6	83.6	97.9	89.0	76.5	93.6	83.0	83.1	100.9	89.9
03	CONCRETE	95.8	101.9	98.6	95.0	95.7	95.3	93.4	132.2	111.1	84.5	99.0	91.1	78.5	95.5	86.2	87.6	98.4	92.5
04	MASONRY	102.2	97.4	99.2	101.4	84.9	91.2	104.9	129.4	120.1	107.7	97.5	101.4	91.4	88.0	89.3	101.0	88.9	93.5
05	METALS	96.9	122.2	104.7	95.7	122.2	103.8	99.2	121.5	106.1	95.8	123.2	104.2	95.5	120.8	103.2	97.1	121.3	104.6
06	WOOD, PLASTICS & COMPOSITES	84.8	93.6	89.7	86.3	85.6	85.9	88.6	126.6	109.7	90.0	85.6	87.6	85.5	86.4	86.0	84.3	86.4	85.5
07	THERMAL & MOISTURE PROTECTION	99.4	97.7	98.6	101.8	88.8	96.3	105.0	124.7	113.3	101.0	94.7	98.4	95.6	90.8	93.6	95.3	104.4	99.2
08	OPENINGS	96.2	100.5	97.2	94.1	93.1	93.9	86.9	134.5	97.8	91.4	93.5	91.9	91.6	92.7	91.8	91.4	89.3	90.9
0920	Plaster & Gypsum Board	100.6	93.4	95.7	89.0	85.2	86.4	87.1	127.3	114.1	89.9	85.2	86.7	101.0	86.0	90.9	101.3	86.0	91.0
0950, 0980	Ceilings & Acoustic Treatment	98.0	93.4	94.9	86.5	85.2	85.6	91.0	127.3	115.4	89.1	85.2	86.4	88.5	86.0	86.8	87.6	86.0	86.5
0960	Flooring	95.6	82.4	91.9	82.8	101.1	87.9	84.6	147.3	102.0	84.6	98.2	88.4	84.0	91.7	86.2	85.6	101.1	89.9
0970, 0990	Wall Finishes & Painting/Coating	96.7	108.9	103.8	91.2	108.8	101.5	87.9	142.1	119.5	91.2	106.5	100.1	88.6	106.5	99.0	88.6	88.6	88.6
09	FINISHES	99.1	93.6	96.1	86.2	91.0	88.8	85.9	132.7	111.3	86.7	92.0	89.5	85.4	88.9	87.3	85.1	89.4	87.5
COVERS	DIVS. 10 - 14, 25, 28, 41, 43, 44, 46	100.0	100.7	100.2	100.0	92.9	98.4	100.0	111.2	102.5	100.0	98.2	99.6	100.0	97.5	99.5	100.0	97.8	99.5
21, 22, 23	FIRE SUPPRESSION, PLUMBING & HVAC	96.9	107.0	97.0	97.2	89.0	93.7	96.9	128.1	110.2	97.2	98.6	97.8	97.2	91.5	94.7	100.2	92.3	97.0
26, 27, 3370	ELECTRICAL, COMMUNICATIONS & UTIL.	95.6	110.6	103.4	92.8	93.6	93.2	93.6	109.4	101.9	93.6	91.7	92.6	89.6	75.9	82.4	90.4	81.2	85.6
MF2016	WEIGHTED AVERAGE	97.3	102.2	99.4	95.5	94.7	95.1	95.5	123.7	107.8	94.1	98.9	96.2	91.9	92.8	92.3	94.4	94.5	94.4

PUERTO RICO / RHODE ISLAND / SOUTH CAROLINA

DIVISION		SAN JUAN 009 MAT.	INST.	TOTAL	NEWPORT 028 MAT.	INST.	TOTAL	PROVIDENCE 029 MAT.	INST.	TOTAL	AIKEN 298 MAT.	INST.	TOTAL	BEAUFORT 299 MAT.	INST.	TOTAL	CHARLESTON 294 MAT.	INST.	TOTAL
015433	CONTRACTOR EQUIPMENT		87.1	87.1		101.1	101.1		101.1	101.1		106.8	106.8		106.8	106.8		106.8	106.8
0241, 31 - 34	SITE & INFRASTRUCTURE, DEMOLITION	131.9	89.9	102.5	88.7	102.8	98.6	90.6	102.8	99.1	128.2	89.6	101.2	123.3	89.0	99.3	108.2	89.3	95.6
0310	Concrete Forming & Accessories	91.0	21.7	31.3	98.6	119.9	117.0	99.6	119.9	117.1	95.7	65.0	69.3	94.7	39.8	47.4	93.9	65.0	69.0
0320	Concrete Reinforcing	189.9	18.4	103.4	102.0	114.6	108.4	104.8	114.6	109.7	95.8	62.6	79.0	94.8	26.6	60.4	94.6	66.1	80.2
0330	Cast-in-Place Concrete	95.0	32.8	71.4	72.7	118.7	90.2	93.2	118.7	102.9	90.9	67.6	82.1	90.9	67.2	81.9	106.7	67.6	91.8
03	CONCRETE	96.0	26.0	64.0	84.3	118.1	99.8	90.6	118.1	103.2	104.7	67.2	87.6	102.0	49.6	78.0	97.6	67.8	84.0
04	MASONRY	88.8	20.8	46.6	97.3	123.6	113.6	99.3	123.6	114.4	78.0	66.7	71.0	92.2	66.7	76.4	93.6	66.7	76.9
05	METALS	116.6	38.7	92.7	97.0	111.8	101.5	101.2	111.8	104.4	97.0	91.0	95.1	97.0	81.0	92.1	98.9	92.1	96.8
06	WOOD, PLASTICS & COMPOSITES	102.4	21.0	57.1	101.3	119.0	111.2	101.1	119.0	111.1	95.0	66.8	79.3	93.7	33.9	60.4	92.6	66.8	78.2
07	THERMAL & MOISTURE PROTECTION	130.6	25.9	86.3	99.6	117.7	107.2	103.1	117.7	109.3	99.1	66.3	85.2	98.8	58.1	81.5	97.9	65.9	84.3
08	OPENINGS	154.0	19.4	123.1	98.8	118.8	103.4	98.7	118.8	103.3	95.3	63.9	88.1	95.4	39.6	82.6	99.2	64.8	91.3
0920	Plaster & Gypsum Board	168.8	18.9	68.2	91.4	119.2	110.0	103.3	119.2	113.9	99.3	65.8	76.8	102.6	32.0	55.2	104.1	65.8	78.4
0950, 0980	Ceilings & Acoustic Treatment	250.5	18.9	94.4	96.7	119.2	111.8	101.6	119.2	113.4	86.4	65.8	72.5	89.0	32.0	50.5	89.0	65.8	73.3
0960	Flooring	203.1	21.3	152.6	89.9	127.6	100.3	87.3	127.6	98.5	95.9	94.9	95.6	97.3	80.1	92.6	97.0	81.6	92.7
0970, 0990	Wall Finishes & Painting/Coating	188.0	22.6	91.5	90.5	116.3	105.6	87.0	116.3	104.1	94.9	60.0	79.8	94.9	59.5	74.2	94.9	69.0	79.3
09	FINISHES	196.9	21.8	101.7	90.1	121.5	107.2	91.0	121.5	107.6	91.7	71.1	80.5	92.6	48.1	68.4	90.8	68.9	78.8
COVERS	DIVS. 10 - 14, 25, 28, 41, 43, 44, 46	100.0	19.8	82.1	100.0	107.5	101.7	100.0	107.5	101.7	100.0	72.1	93.8	100.0	76.9	94.9	100.0	72.0	93.8
21, 22, 23	FIRE SUPPRESSION, PLUMBING & HVAC	103.9	18.2	67.2	100.3	108.7	103.9	100.1	108.7	103.8	97.4	57.1	80.2	97.4	57.0	80.1	100.5	59.4	82.9
26, 27, 3370	ELECTRICAL, COMMUNICATIONS & UTIL.	124.8	17.5	69.0	100.9	96.6	98.6	101.1	96.6	98.7	100.9	65.3	82.4	104.6	32.6	67.2	103.0	60.1	80.7
MF2016	WEIGHTED AVERAGE	120.7	27.7	79.9	96.3	112.0	103.2	98.1	112.0	104.2	98.0	69.2	85.4	98.7	56.8	80.3	98.9	68.8	85.7

SOUTH CAROLINA / SOUTH DAKOTA

DIVISION		COLUMBIA 290 - 292 MAT.	INST.	TOTAL	FLORENCE 295 MAT.	INST.	TOTAL	GREENVILLE 296 MAT.	INST.	TOTAL	ROCK HILL 297 MAT.	INST.	TOTAL	SPARTANBURG 293 MAT.	INST.	TOTAL	ABERDEEN 574 MAT.	INST.	TOTAL
015433	CONTRACTOR EQUIPMENT		106.8	106.8		106.8	106.8		106.8	106.8		106.8	106.8		106.8	106.8		100.3	100.3
0241, 31 - 34	SITE & INFRASTRUCTURE, DEMOLITION	106.9	89.3	94.6	117.4	89.1	97.6	112.4	89.4	96.3	110.2	88.7	95.1	112.2	89.4	96.3	102.0	98.2	99.3
0310	Concrete Forming & Accessories	95.3	65.0	69.2	82.1	64.9	67.3	93.4	65.0	68.9	91.5	64.7	68.4	96.2	65.0	69.3	94.1	46.1	52.7
0320	Concrete Reinforcing	99.9	66.1	82.8	94.2	66.1	80.0	94.1	66.1	80.0	95.0	66.1	80.4	94.1	66.1	80.0	96.7	72.5	84.5
0330	Cast-in-Place Concrete	107.9	67.6	92.6	90.9	67.4	82.0	90.8	67.6	82.0	90.8	67.5	82.0	90.8	67.6	82.0	112.5	78.9	99.8
03	CONCRETE	95.8	67.8	83.0	96.5	67.6	83.3	95.1	67.8	82.6	93.0	67.6	81.4	95.3	67.8	82.7	99.3	63.6	83.0
04	MASONRY	87.3	66.7	74.5	78.0	66.7	71.0	75.6	66.7	70.1	99.6	66.7	79.2	78.0	66.7	71.0	113.6	73.4	88.7
05	METALS	96.1	92.1	94.9	97.8	91.5	95.9	97.8	92.1	96.0	97.0	91.8	95.4	97.8	92.1	96.0	100.9	83.0	95.4
06	WOOD, PLASTICS & COMPOSITES	93.1	66.8	78.5	79.9	66.8	72.7	92.2	66.8	78.1	90.9	66.8	77.5	96.1	66.8	79.8	92.4	34.7	60.3
07	THERMAL & MOISTURE PROTECTION	94.2	65.8	82.2	98.2	65.9	84.5	98.1	65.9	84.5	98.0	61.4	82.5	98.1	65.9	84.5	99.8	74.9	89.3
08	OPENINGS	95.4	64.8	88.4	95.4	64.8	88.4	95.3	64.8	88.3	95.4	64.8	88.3	95.4	64.8	88.3	95.7	41.4	83.2
0920	Plaster & Gypsum Board	94.8	65.8	75.3	93.6	65.8	74.9	98.4	65.8	76.5	98.4	65.8	76.4	101.0	65.8	77.4	112.0	33.3	59.1
0950, 0980	Ceilings & Acoustic Treatment	89.7	65.8	73.6	87.3	65.8	72.8	86.4	65.8	72.5	86.4	65.8	72.5	86.4	65.8	72.5	94.2	33.3	53.1
0960	Flooring	87.1	80.2	85.2	88.4	80.1	86.1	94.7	80.1	90.6	93.6	80.1	89.9	96.1	80.1	91.6	92.3	46.3	79.6
0970, 0990	Wall Finishes & Painting/Coating	89.0	69.0	77.3	94.9	69.0	79.8	94.9	69.0	79.8	94.9	69.0	79.8	94.9	69.0	79.8	90.4	34.9	58.0
09	FINISHES	86.8	68.6	76.9	87.6	68.6	77.3	89.6	68.6	78.2	88.9	68.6	77.9	90.3	68.6	78.5	91.4	43.0	65.1
COVERS	DIVS. 10 - 14, 25, 28, 41, 43, 44, 46	100.0	72.1	93.8	100.0	72.1	93.8	100.0	72.1	93.8	100.0	72.0	93.8	100.0	72.1	93.8	100.0	79.6	95.5
21, 22, 23	FIRE SUPPRESSION, PLUMBING & HVAC	100.1	58.7	82.4	100.5	58.7	82.7	100.5	58.7	82.7	97.4	57.1	80.2	100.5	58.7	82.7	100.2	54.8	80.8
26, 27, 3370	ELECTRICAL, COMMUNICATIONS & UTIL.	103.1	63.0	82.3	100.8	63.0	81.1	103.0	61.8	81.6	103.0	61.8	81.6	103.0	61.8	81.6	101.4	65.9	82.9
MF2016	WEIGHTED AVERAGE	96.9	69.1	84.7	97.0	69.0	84.8	97.1	68.9	84.8	97.1	68.3	84.5	97.3	68.9	84.9	99.7	64.5	83.1

For customer support on your Electrical Costs with RSMeans data, call 800.448.8182.

543

City Cost Indexes

SOUTH DAKOTA

DIVISION		MITCHELL 573			MOBRIDGE 576			PIERRE 575			RAPID CITY 577			SIOUX FALLS 570 - 571			WATERTOWN 572		
		MAT.	INST.	TOTAL	MAT.	INST.	TOTAL	MAT.	INST.	TOTAL	MAT.	INST.	TOTAL	MAT.	INST.	TOTAL	MAT.	INST.	TOTAL
015433	CONTRACTOR EQUIPMENT		100.3	100.3		100.3	100.3		100.3	100.3		100.3	100.3		101.3	101.3		100.3	100.3
0241, 31 - 34	SITE & INFRASTRUCTURE, DEMOLITION	98.8	98.1	98.3	98.8	98.2	98.3	100.4	97.6	98.4	100.3	97.6	98.4	93.4	100.0	98.0	98.7	98.1	98.3
0310	Concrete Forming & Accessories	93.3	46.2	52.7	84.4	46.6	51.8	96.5	47.5	54.2	101.5	59.1	64.9	100.7	64.5	69.5	81.2	44.9	49.9
0320	Concrete Reinforcing	96.2	72.7	84.3	98.6	72.7	85.6	98.6	97.0	97.8	90.4	97.2	93.8	96.8	97.3	97.0	93.5	41.1	67.1
0330	Cast-in-Place Concrete	109.4	54.8	88.6	109.4	79.0	97.8	104.2	79.8	94.9	108.5	80.1	97.7	92.5	80.8	88.1	109.4	57.9	89.8
03	CONCRETE	97.0	55.3	77.9	96.7	63.9	81.7	93.4	68.6	82.0	96.3	74.0	86.1	87.8	76.8	82.8	95.9	50.4	75.1
04	MASONRY	101.4	77.9	86.8	110.1	75.9	88.9	112.1	75.7	89.6	110.2	77.2	89.7	92.5	77.9	83.4	137.5	76.9	99.9
05	METALS	99.9	83.0	94.7	100.0	83.5	94.9	103.2	90.0	99.1	102.7	90.8	99.1	102.9	92.9	99.8	99.9	72.0	91.4
06	WOOD, PLASTICS & COMPOSITES	91.3	35.2	60.1	81.0	34.9	55.4	99.4	35.5	63.8	96.0	50.2	70.5	98.2	57.2	75.4	77.5	34.7	53.7
07	THERMAL & MOISTURE PROTECTION	99.6	76.2	89.7	99.7	78.8	90.8	101.9	75.7	90.8	100.3	80.4	91.9	98.4	81.9	91.4	99.4	75.7	89.4
08	OPENINGS	94.8	41.6	82.6	97.3	41.0	84.4	100.5	59.3	91.0	99.4	67.5	92.1	102.0	71.3	95.0	94.8	34.0	80.8
0920	Plaster & Gypsum Board	110.4	33.8	58.9	105.1	33.5	57.0	102.2	34.0	56.4	111.4	49.2	69.6	104.1	56.4	72.1	102.8	33.3	56.1
0950, 0980	Ceilings & Acoustic Treatment	89.9	33.8	52.1	94.2	33.5	53.3	94.2	34.0	53.6	95.9	49.2	64.4	89.7	56.4	67.2	89.9	33.3	51.7
0960	Flooring	91.9	46.3	79.3	87.5	49.1	76.9	93.3	33.0	76.6	91.7	76.8	87.6	95.8	77.8	90.8	86.1	46.3	75.1
0970, 0990	Wall Finishes & Painting/Coating	90.4	38.2	60.0	90.4	39.2	60.5	95.1	98.0	96.8	90.4	98.0	94.8	97.4	98.0	97.7	90.4	34.9	58.0
09	FINISHES	90.0	43.1	64.5	88.8	43.6	64.2	94.6	47.1	68.8	91.4	64.7	76.9	95.0	69.0	80.9	87.2	42.5	62.9
COVERS	DIVS. 10 - 14, 25, 28, 41, 43, 44, 46	100.0	79.6	95.5	100.0	79.5	95.4	100.0	81.4	95.9	100.0	83.0	96.2	100.0	83.8	96.4	100.0	44.8	87.7
21, 22, 23	FIRE SUPPRESSION, PLUMBING & HVAC	97.1	51.0	77.4	97.1	70.8	85.8	100.1	79.9	91.5	100.2	80.0	91.5	100.1	71.1	87.7	97.1	53.8	78.6
26, 27, 3370	ELECTRICAL, COMMUNICATIONS & UTIL.	99.7	65.9	82.1	101.4	42.0	70.5	104.1	50.4	76.1	97.9	50.4	73.2	102.5	65.9	83.5	98.9	46.1	71.4
MF2016	WEIGHTED AVERAGE	97.5	63.1	82.4	98.2	65.0	83.7	100.4	70.5	87.3	99.5	74.4	88.5	98.5	76.4	88.8	98.7	57.6	80.7

TENNESSEE

DIVISION		CHATTANOOGA 373 - 374			COLUMBIA 384			COOKEVILLE 385			JACKSON 383			JOHNSON CITY 376			KNOXVILLE 377 - 379		
		MAT.	INST.	TOTAL	MAT.	INST.	TOTAL	MAT.	INST.	TOTAL	MAT.	INST.	TOTAL	MAT.	INST.	TOTAL	MAT.	INST.	TOTAL
015433	CONTRACTOR EQUIPMENT		107.3	107.3		101.9	101.9		101.9	101.9		108.2	108.2		101.0	101.0		101.0	101.0
0241, 31 - 34	SITE & INFRASTRUCTURE, DEMOLITION	102.4	98.4	99.6	88.0	88.5	88.4	93.5	85.3	87.7	96.6	98.7	98.1	108.8	84.7	91.9	88.9	87.6	88.0
0310	Concrete Forming & Accessories	96.0	59.9	64.9	81.9	64.7	67.0	82.0	32.9	39.7	88.8	42.6	49.0	82.5	60.0	63.1	94.2	64.5	68.6
0320	Concrete Reinforcing	96.9	67.1	81.9	88.5	64.7	76.5	88.5	64.2	76.2	88.4	66.0	77.1	97.5	63.6	80.4	96.9	63.7	80.2
0330	Cast-in-Place Concrete	94.9	64.4	83.4	91.1	62.3	80.2	103.3	40.9	79.6	100.8	69.2	88.8	76.4	59.6	70.0	89.2	65.9	80.4
03	CONCRETE	93.2	64.5	80.1	92.1	65.6	80.0	102.2	43.7	75.4	93.0	58.2	77.1	103.2	62.2	84.5	90.8	66.5	79.7
04	MASONRY	99.5	58.7	74.2	116.4	56.4	79.2	111.3	39.8	67.0	116.7	42.7	70.8	113.4	42.1	69.2	77.9	52.8	62.4
05	METALS	96.9	88.9	94.5	90.2	89.6	90.0	90.3	87.6	89.5	92.6	89.6	91.7	94.1	86.7	91.9	97.4	87.2	94.3
06	WOOD, PLASTICS & COMPOSITES	104.1	58.9	78.9	72.3	66.4	69.0	72.5	30.8	49.3	88.5	41.5	62.3	77.4	65.6	70.8	90.4	65.6	76.6
07	THERMAL & MOISTURE PROTECTION	99.4	64.1	84.4	93.8	62.5	80.5	94.3	50.8	75.9	96.2	56.2	79.3	95.1	54.8	78.1	92.8	63.2	80.2
08	OPENINGS	101.2	60.4	91.8	92.4	55.5	83.9	92.4	36.7	79.6	99.6	45.8	87.2	97.5	62.6	89.5	94.7	58.0	86.3
0920	Plaster & Gypsum Board	78.9	58.3	65.1	85.4	66.0	72.3	85.4	29.4	47.8	87.3	40.4	55.8	99.1	65.2	76.3	105.9	65.2	78.5
0950, 0980	Ceilings & Acoustic Treatment	108.2	58.3	74.6	79.6	66.0	70.4	79.6	29.4	45.7	89.0	40.4	56.2	104.5	65.2	78.0	105.4	65.2	78.3
0960	Flooring	100.0	58.4	88.5	79.9	57.0	73.6	80.0	52.9	72.5	79.5	57.3	73.3	94.0	46.4	80.8	99.5	52.5	86.5
0970, 0990	Wall Finishes & Painting/Coating	96.9	62.2	76.7	83.0	59.0	69.0	83.0	59.0	69.0	84.6	59.0	69.7	93.7	61.4	74.9	93.7	60.2	74.2
09	FINISHES	96.6	59.4	76.3	84.6	62.7	72.7	85.0	37.9	59.4	84.9	45.5	63.5	100.0	58.0	77.2	93.6	61.8	76.3
COVERS	DIVS. 10 - 14, 25, 28, 41, 43, 44, 46	100.0	70.5	93.4	100.0	69.5	93.2	100.0	37.2	86.0	100.0	63.7	91.9	100.0	78.8	95.3	100.0	82.6	96.1
21, 22, 23	FIRE SUPPRESSION, PLUMBING & HVAC	100.2	59.7	82.9	98.0	75.9	88.6	98.0	69.8	86.0	100.1	61.1	83.4	99.9	57.3	81.7	99.9	63.0	84.1
26, 27, 3370	ELECTRICAL, COMMUNICATIONS & UTIL.	100.1	87.5	93.6	92.9	53.4	72.3	94.4	63.2	78.2	99.0	54.4	75.8	91.1	43.1	66.1	96.3	58.0	76.4
MF2016	WEIGHTED AVERAGE	98.6	70.3	86.2	94.2	68.4	82.9	95.6	57.7	79.0	97.1	60.6	81.1	98.8	60.3	82.0	95.3	66.2	82.5

TENNESSEE / TEXAS

DIVISION		MCKENZIE 382			MEMPHIS 375, 380 - 381			NASHVILLE 370 - 372			ABILENE 795 - 796			AMARILLO 790 - 791			AUSTIN 786 - 787		
		MAT.	INST.	TOTAL	MAT.	INST.	TOTAL	MAT.	INST.	TOTAL	MAT.	INST.	TOTAL	MAT.	INST.	TOTAL	MAT.	INST.	TOTAL
015433	CONTRACTOR EQUIPMENT		101.9	101.9		100.8	100.8		105.2	105.2		93.5	93.5		93.5	93.5		92.5	92.5
0241, 31 - 34	SITE & INFRASTRUCTURE, DEMOLITION	93.2	85.6	87.8	89.6	94.3	92.9	96.7	97.6	97.3	98.2	92.2	94.0	98.4	91.4	93.5	96.7	90.8	92.6
0310	Concrete Forming & Accessories	89.7	35.4	42.9	96.0	64.8	69.1	94.5	65.9	69.8	98.9	61.9	67.0	97.6	54.0	60.0	94.4	53.6	59.2
0320	Concrete Reinforcing	88.6	66.1	77.2	104.1	68.1	85.9	103.4	64.9	84.0	90.4	51.2	70.6	96.8	51.1	73.8	101.1	48.7	74.6
0330	Cast-in-Place Concrete	101.1	54.1	83.2	92.7	77.6	86.9	87.3	65.7	79.1	87.4	68.2	80.1	80.8	68.1	76.0	93.6	68.2	84.0
03	CONCRETE	100.7	49.7	77.4	95.6	70.9	84.3	94.0	67.0	81.6	82.8	63.0	73.7	82.3	59.4	71.8	88.6	58.7	74.9
04	MASONRY	115.4	47.0	73.0	98.0	57.8	73.1	84.6	57.8	68.0	95.1	63.2	75.3	94.5	62.7	74.8	97.3	63.2	76.1
05	METALS	90.3	88.9	89.9	99.8	84.4	95.1	96.9	85.9	93.5	107.8	70.7	96.4	102.7	70.5	92.8	97.2	69.0	88.5
06	WOOD, PLASTICS & COMPOSITES	80.9	33.1	54.3	98.0	65.1	79.7	91.2	66.7	77.6	104.6	63.7	81.8	103.2	53.3	75.4	93.1	52.4	70.5
07	THERMAL & MOISTURE PROTECTION	94.2	53.6	77.0	94.3	67.2	82.8	95.6	64.3	82.4	99.6	66.2	85.5	97.8	64.4	83.7	96.6	66.0	83.7
08	OPENINGS	92.4	38.7	80.1	97.7	61.8	89.4	96.0	63.4	88.5	102.2	59.1	92.3	103.4	53.3	91.9	102.8	50.6	90.8
0920	Plaster & Gypsum Board	88.3	31.8	50.4	90.4	64.4	72.9	91.6	66.0	74.4	84.9	63.1	70.2	94.1	52.4	66.1	88.6	51.6	63.8
0950, 0980	Ceilings & Acoustic Treatment	79.6	31.8	47.4	98.1	64.4	75.4	102.6	66.0	77.9	91.7	63.1	72.4	95.1	52.4	66.3	94.3	51.6	65.5
0960	Flooring	82.8	57.0	75.7	102.7	58.4	90.4	95.1	58.4	84.9	92.9	71.5	87.0	93.9	67.4	86.6	87.7	67.4	82.1
0970, 0990	Wall Finishes & Painting/Coating	83.0	46.2	61.5	91.3	59.6	72.8	100.7	70.3	83.0	91.0	54.0	69.4	86.6	54.0	67.6	92.4	46.2	65.5
09	FINISHES	86.2	39.3	60.7	98.1	62.5	78.8	98.0	64.6	79.8	83.7	63.1	72.5	90.6	56.1	71.8	89.6	54.7	70.6
COVERS	DIVS. 10 - 14, 25, 28, 41, 43, 44, 46	100.0	24.7	83.3	100.0	82.8	96.2	100.0	83.2	96.3	100.0	80.6	95.7	100.0	75.4	94.5	100.0	79.1	95.3
21, 22, 23	FIRE SUPPRESSION, PLUMBING & HVAC	98.0	61.0	82.2	100.0	71.3	87.7	100.0	76.6	90.0	100.3	49.3	78.5	99.9	53.2	79.9	100.0	59.1	82.5
26, 27, 3370	ELECTRICAL, COMMUNICATIONS & UTIL.	94.2	58.5	75.6	99.1	65.7	81.7	96.1	63.2	79.0	98.5	56.1	76.5	101.5	62.2	81.1	95.2	59.4	76.6
MF2016	WEIGHTED AVERAGE	95.8	56.8	78.7	98.4	70.7	86.3	96.7	71.7	85.8	97.7	62.6	82.3	97.6	62.2	82.1	96.7	62.7	81.8

For customer support on your Electrical Costs with RSMeans data, call 800.448.8182.

DIVISION		TEXAS																
		BEAUMONT 776 - 777			BROWNWOOD 768			BRYAN 778			CHILDRESS 792			CORPUS CHRISTI 783 - 784			DALLAS 752 - 753		
		MAT.	INST.	TOTAL	MAT.	INST.	TOTAL	MAT.	INST.	TOTAL	MAT.	INST.	TOTAL	MAT.	INST.	TOTAL	MAT.	INST.	TOTAL
015433	CONTRACTOR EQUIPMENT		97.5	97.5		93.5	93.5		97.5	97.5		93.5	93.5		103.8	103.8		107.7	107.7
0241, 31 - 34	SITE & INFRASTRUCTURE, DEMOLITION	87.6	96.3	93.7	106.2	92.0	96.3	79.0	96.5	91.2	109.4	90.4	96.1	142.0	87.1	103.6	108.7	97.1	100.6
0310	Concrete Forming & Accessories	103.2	57.2	63.5	98.2	61.5	66.5	82.1	60.9	63.8	97.3	61.4	66.3	98.6	54.5	60.6	94.7	62.6	67.0
0320	Concrete Reinforcing	96.9	64.0	80.3	81.3	51.1	66.1	99.3	52.8	75.9	90.6	51.1	70.7	89.2	48.6	68.7	84.7	51.3	67.8
0330	Cast-in-Place Concrete	88.0	68.3	80.5	107.7	61.5	90.1	70.5	62.3	67.4	89.7	61.5	79.0	114.2	70.4	97.6	108.7	70.4	94.1
03	CONCRETE	90.9	63.2	78.3	94.5	60.5	78.9	76.0	60.9	69.1	90.5	60.4	76.7	96.5	61.0	80.3	99.7	64.9	83.8
04	MASONRY	102.5	64.3	78.8	128.3	63.2	87.9	142.8	63.2	93.4	99.2	62.7	76.6	85.5	63.3	71.7	100.0	61.8	76.3
05	METALS	93.5	77.2	88.5	95.5	70.3	87.8	93.3	72.9	87.0	105.2	70.2	94.5	93.5	84.7	90.8	97.6	83.5	93.3
06	WOOD, PLASTICS & COMPOSITES	114.0	56.6	82.0	104.7	63.7	81.8	79.1	62.1	69.7	104.0	63.7	81.6	122.5	53.9	84.3	95.3	64.0	77.9
07	THERMAL & MOISTURE PROTECTION	102.0	66.8	87.1	94.7	64.7	82.0	93.9	66.3	82.2	100.3	63.1	84.6	102.4	65.2	86.6	92.6	67.6	82.0
08	OPENINGS	94.9	57.8	86.4	98.0	59.1	89.1	95.3	58.2	86.8	97.4	59.1	88.6	103.9	51.6	91.9	101.2	59.3	91.6
0920	Plaster & Gypsum Board	97.1	55.9	69.4	85.3	63.1	70.4	85.0	61.5	69.3	84.5	63.1	70.1	96.1	52.9	67.1	93.8	63.1	73.2
0950, 0980	Ceilings & Acoustic Treatment	98.5	55.9	69.8	78.6	63.1	68.1	90.5	61.5	71.0	90.0	63.1	71.8	95.6	52.9	66.8	97.3	63.1	74.2
0960	Flooring	121.1	78.9	109.4	77.9	57.7	72.3	90.5	70.8	85.0	91.4	57.7	82.0	104.4	71.6	95.3	97.3	69.1	89.5
0970, 0990	Wall Finishes & Painting/Coating	88.7	56.3	69.8	89.7	54.0	68.9	86.8	60.0	71.2	91.0	54.0	69.4	104.9	46.2	70.7	102.2	54.0	74.1
09	FINISHES	94.7	60.6	76.2	76.5	60.3	67.7	81.8	62.3	71.2	84.1	60.3	71.1	98.2	56.5	75.5	97.4	62.8	78.6
COVERS	DIVS. 10 - 14, 25, 28, 41, 43, 44, 46	100.0	80.9	95.8	100.0	76.4	94.8	100.0	80.1	95.6	100.0	76.5	94.8	100.0	81.0	95.8	100.0	81.5	95.9
21, 22, 23	FIRE SUPPRESSION, PLUMBING & HVAC	100.2	64.3	84.8	97.1	47.3	75.8	97.1	63.8	82.8	97.2	53.2	78.4	100.2	50.7	79.0	99.9	60.6	83.1
26, 27, 3370	ELECTRICAL, COMMUNICATIONS & UTIL.	97.0	66.9	81.4	96.5	48.7	71.6	95.2	66.9	80.5	98.5	58.9	77.9	92.8	57.0	74.2	97.8	61.9	79.2
MF2016	WEIGHTED AVERAGE	96.5	68.0	84.0	96.9	60.2	80.8	94.0	67.3	82.3	97.6	62.7	82.3	98.7	62.4	82.8	99.2	67.6	85.4

DIVISION		TEXAS																
		DEL RIO 788			DENTON 762			EASTLAND 764			EL PASO 798 - 799, 885			FORT WORTH 760 - 761			GALVESTON 775		
		MAT.	INST.	TOTAL	MAT.	INST.	TOTAL	MAT.	INST.	TOTAL	MAT.	INST.	TOTAL	MAT.	INST.	TOTAL	MAT.	INST.	TOTAL
015433	CONTRACTOR EQUIPMENT		92.5	92.5		103.4	103.4		93.5	93.5		93.6	93.6		93.5	93.5		110.9	110.9
0241, 31 - 34	SITE & INFRASTRUCTURE, DEMOLITION	122.7	90.1	99.9	105.2	83.9	90.3	109.2	90.1	95.8	95.7	92.6	93.5	101.6	92.9	95.5	103.0	95.1	97.4
0310	Concrete Forming & Accessories	95.3	53.5	59.2	105.4	61.6	67.6	98.9	61.3	66.4	97.9	58.2	63.7	95.2	62.1	66.7	91.7	58.4	63.0
0320	Concrete Reinforcing	89.8	48.5	69.0	82.6	51.1	66.7	81.5	51.1	66.2	96.7	51.3	73.8	87.5	51.2	69.2	98.8	61.6	80.0
0330	Cast-in-Place Concrete	123.0	61.3	99.6	83.3	62.6	75.4	113.8	61.4	93.9	81.6	68.0	76.4	99.1	68.3	87.4	93.6	63.4	82.1
03	CONCRETE	116.3	56.3	88.8	74.2	62.0	68.6	99.1	60.3	81.4	82.7	61.2	72.9	89.6	63.1	77.5	92.1	62.8	78.7
04	MASONRY	100.5	63.1	77.3	137.5	61.8	90.6	96.9	63.2	76.0	85.9	64.4	72.6	95.9	61.8	74.7	98.6	63.3	76.7
05	METALS	93.3	68.2	85.6	95.2	86.0	92.3	95.3	70.1	87.6	102.7	70.1	92.7	97.1	70.9	89.1	94.9	92.9	94.3
06	WOOD, PLASTICS & COMPOSITES	100.9	52.8	74.1	116.1	63.8	87.0	111.8	63.7	85.0	88.7	58.5	71.9	95.1	63.7	77.6	96.5	58.5	75.3
07	THERMAL & MOISTURE PROTECTION	98.9	65.1	84.6	92.7	65.2	81.1	95.1	64.7	82.2	95.6	65.6	82.9	91.2	66.6	80.8	93.1	66.9	82.0
08	OPENINGS	99.2	50.1	87.9	115.6	58.7	102.5	70.6	59.1	67.9	98.6	54.0	88.3	99.7	59.1	90.4	99.4	58.3	90.0
0920	Plaster & Gypsum Board	92.4	51.9	65.2	89.9	63.1	71.9	85.3	63.1	70.4	92.3	57.7	69.1	95.5	63.1	73.7	91.6	57.7	68.8
0950, 0980	Ceilings & Acoustic Treatment	92.0	51.9	65.0	84.6	63.1	70.1	78.6	63.1	68.1	93.4	57.7	69.4	87.8	63.1	71.1	94.8	57.7	69.8
0960	Flooring	89.9	57.7	81.0	73.7	57.7	69.3	98.6	57.7	87.3	91.1	73.7	86.3	94.8	68.1	87.4	106.4	70.8	96.5
0970, 0990	Wall Finishes & Painting/Coating	92.7	44.5	64.6	99.7	51.1	71.4	91.1	54.0	69.5	95.5	49.9	68.9	92.0	54.1	69.9	97.5	56.1	73.4
09	FINISHES	90.8	52.8	70.1	75.7	60.1	67.2	83.3	60.3	70.8	89.3	60.0	73.4	89.0	62.4	74.6	91.6	59.7	74.3
COVERS	DIVS. 10 - 14, 25, 28, 41, 43, 44, 46	100.0	76.7	94.8	100.0	75.9	94.6	100.0	74.7	94.4	100.0	75.9	94.6	100.0	80.6	95.7	100.0	81.4	95.9
21, 22, 23	FIRE SUPPRESSION, PLUMBING & HVAC	97.0	61.2	81.7	97.1	58.1	80.4	97.1	47.2	75.8	99.9	66.3	85.5	100.0	57.7	81.9	100.0	81.4	95.9
26, 27, 3370	ELECTRICAL, COMMUNICATIONS & UTIL.	94.8	63.2	78.3	98.8	61.9	79.6	96.4	61.9	78.4	97.5	52.3	74.0	97.8	61.9	79.1	96.7	67.0	81.3
MF2016	WEIGHTED AVERAGE	99.5	62.9	83.4	96.8	65.3	83.0	93.6	61.8	79.7	96.0	64.7	82.3	96.6	65.1	82.8	96.2	69.0	84.3

DIVISION		TEXAS																
		GIDDINGS 789			GREENVILLE 754			HOUSTON 770 - 772			HUNTSVILLE 773			LAREDO 780			LONGVIEW 756		
		MAT.	INST.	TOTAL	MAT.	INST.	TOTAL	MAT.	INST.	TOTAL	MAT.	INST.	TOTAL	MAT.	INST.	TOTAL	MAT.	INST.	TOTAL
015433	CONTRACTOR EQUIPMENT		92.5	92.5		104.3	104.3		102.5	102.5		97.5	97.5		92.5	92.5		95.1	95.1
0241, 31 - 34	SITE & INFRASTRUCTURE, DEMOLITION	107.8	90.3	95.6	99.6	87.1	90.9	101.9	94.7	96.8	94.2	96.2	95.6	101.3	90.8	93.9	99.4	93.8	95.5
0310	Concrete Forming & Accessories	92.7	53.5	58.9	86.6	59.8	63.5	97.1	58.6	63.9	89.3	56.6	61.1	95.3	54.6	60.2	83.1	61.5	64.5
0320	Concrete Reinforcing	90.3	48.6	69.3	85.2	51.1	68.0	98.6	52.9	75.5	99.6	52.7	75.9	89.8	48.6	69.0	84.1	51.1	67.4
0330	Cast-in-Place Concrete	104.2	61.5	88.0	108.1	62.7	90.9	86.9	64.5	78.4	96.9	62.2	83.7	87.9	68.2	80.4	125.8	61.5	101.4
03	CONCRETE	93.6	56.4	76.6	93.4	61.1	78.6	91.3	61.0	77.4	98.4	58.9	80.3	88.0	59.2	74.8	110.5	60.4	87.6
04	MASONRY	107.9	63.2	80.2	155.5	61.8	97.4	97.7	63.3	76.4	141.0	63.2	92.8	93.8	63.2	74.8	151.7	61.7	95.9
05	METALS	92.7	68.7	85.4	95.1	84.6	91.9	97.6	78.4	91.7	93.2	72.6	86.9	95.8	69.1	87.6	88.3	69.3	82.5
06	WOOD, PLASTICS & COMPOSITES	100.0	52.8	73.7	85.5	61.3	72.0	101.9	58.7	77.9	87.9	56.6	70.5	100.9	53.8	74.7	80.3	63.8	71.1
07	THERMAL & MOISTURE PROTECTION	99.4	65.2	84.9	92.6	64.3	80.6	95.1	68.1	83.7	95.1	65.7	82.6	98.2	66.1	84.6	94.2	63.7	81.3
08	OPENINGS	98.3	50.8	87.4	97.8	57.8	88.6	102.5	56.3	91.9	95.3	55.1	86.1	99.2	51.4	88.2	88.4	58.7	81.6
0920	Plaster & Gypsum Board	91.4	51.9	64.9	85.2	60.3	68.5	90.8	57.7	68.6	89.0	55.9	66.7	93.2	52.9	66.2	84.2	63.1	70.0
0950, 0980	Ceilings & Acoustic Treatment	92.0	51.9	65.0	92.8	60.3	70.9	91.7	57.7	68.8	90.5	55.9	67.2	95.4	52.9	66.8	88.5	63.1	71.4
0960	Flooring	89.2	57.7	80.4	91.1	57.7	81.8	107.5	71.6	97.5	95.0	57.7	84.7	89.8	67.4	83.6	96.4	57.7	85.7
0970, 0990	Wall Finishes & Painting/Coating	92.7	46.2	65.6	96.3	54.0	71.6	95.4	58.3	73.8	86.8	58.3	70.1	92.7	46.2	65.6	87.0	51.1	66.0
09	FINISHES	89.4	53.0	69.6	91.9	58.9	74.0	97.3	60.6	77.3	84.6	56.6	69.4	89.9	55.5	71.2	97.3	60.0	77.0
COVERS	DIVS. 10 - 14, 25, 28, 41, 43, 44, 46	100.0	73.2	94.0	100.0	76.7	94.8	100.0	81.9	96.0	100.0	72.6	93.9	100.0	78.9	95.3	100.0	76.7	94.8
21, 22, 23	FIRE SUPPRESSION, PLUMBING & HVAC	97.1	62.6	82.3	96.9	58.9	80.7	100.1	65.2	85.2	97.1	63.7	82.8	100.1	61.3	83.5	96.9	62.6	82.2
26, 27, 3370	ELECTRICAL, COMMUNICATIONS & UTIL.	92.0	58.7	74.7	94.7	61.9	77.7	98.7	68.5	83.0	95.2	62.4	78.2	94.9	60.3	76.9	95.1	52.8	73.1
MF2016	WEIGHTED AVERAGE	96.1	62.6	81.4	98.6	65.2	84.0	98.3	68.0	85.0	97.4	65.1	83.3	96.1	63.5	81.8	98.9	62.8	83.1

For customer support on your Electrical Costs with RSMeans data, call 800.448.8182.

545

TEXAS

DIVISION		LUBBOCK 793 - 794			LUFKIN 759			MCALLEN 785			MCKINNEY 750			MIDLAND 797			ODESSA 797		
		MAT.	INST.	TOTAL	MAT.	INST.	TOTAL	MAT.	INST.	TOTAL	MAT.	INST.	TOTAL	MAT.	INST.	TOTAL	MAT.	INST.	TOTAL
015433	CONTRACTOR EQUIPMENT		106.2	106.2		95.1	95.1		103.9	103.9		104.3	104.3		106.2	106.2		93.5	93.5
0241, 31 - 34	SITE & INFRASTRUCTURE, DEMOLITION	123.4	91.3	100.9	94.0	95.4	95.0	146.7	87.2	105.0	95.9	87.1	89.8	126.6	90.4	101.2	98.5	92.9	94.6
0310	Concrete Forming & Accessories	97.7	54.4	60.3	86.0	56.7	60.7	99.5	53.6	59.9	85.8	59.8	63.3	101.5	61.9	67.4	98.9	62.0	67.1
0320	Concrete Reinforcing	91.7	51.2	71.3	85.6	61.5	73.4	89.3	48.6	68.8	85.2	51.1	68.0	92.8	51.2	71.8	90.4	51.2	70.6
0330	Cast-in-Place Concrete	87.6	69.3	80.6	112.5	61.4	93.1	124.1	62.7	100.8	101.4	62.7	86.7	93.1	69.3	84.0	87.4	68.2	80.1
03	CONCRETE	81.7	61.1	72.2	101.8	59.9	82.7	104.2	57.9	83.0	88.4	61.1	75.9	85.7	64.4	76.0	82.8	63.1	73.8
04	MASONRY	94.5	62.8	74.8	116.7	63.1	83.4	100.1	63.3	77.3	166.9	61.8	101.7	111.1	62.8	81.1	95.1	62.7	75.0
05	METALS	111.7	86.5	104.0	95.2	72.4	88.2	93.2	84.5	90.5	95.0	84.6	91.8	109.9	86.4	102.7	107.2	70.8	96.0
06	WOOD, PLASTICS & COMPOSITES	103.8	53.4	75.8	87.6	56.8	70.4	121.2	52.9	83.2	84.6	61.3	71.6	108.5	63.8	83.6	104.6	63.7	81.8
07	THERMAL & MOISTURE PROTECTION	88.9	65.3	78.9	94.0	62.8	80.8	102.1	66.0	86.8	92.4	64.3	80.6	89.2	66.4	79.6	99.6	65.5	85.2
08	OPENINGS	109.8	53.4	96.9	67.2	57.7	65.0	103.1	50.1	90.9	97.8	57.7	88.6	111.3	59.2	99.3	102.2	59.1	92.3
0920	Plaster & Gypsum Board	85.1	52.4	63.1	83.2	55.9	64.8	96.8	51.9	66.7	85.2	60.3	68.5	87.0	63.1	70.9	84.9	63.1	70.2
0950, 0980	Ceilings & Acoustic Treatment	92.6	52.4	65.5	84.2	55.9	65.1	95.4	51.9	66.1	92.8	60.3	70.9	90.9	63.1	72.1	91.7	63.1	72.4
0960	Flooring	86.9	70.7	82.4	129.3	57.7	109.4	103.9	76.9	96.4	90.7	57.7	81.5	88.1	67.4	82.4	92.9	67.4	85.9
0970, 0990	Wall Finishes & Painting/Coating	101.4	54.0	73.8	87.0	54.0	67.7	104.9	44.5	69.7	96.3	54.0	71.6	101.4	54.0	73.8	91.0	54.0	69.4
09	FINISHES	86.1	56.9	70.2	106.1	56.6	79.2	98.5	56.8	75.8	91.6	58.9	73.8	86.8	62.3	73.5	83.7	62.2	72.0
COVERS	DIVS. 10 - 14, 25, 28, 41, 43, 44, 46	100.0	79.8	95.5	100.0	73.0	94.0	100.0	79.0	95.3	100.0	76.7	94.8	100.0	76.8	94.8	100.0	76.5	94.8
21, 22, 23	FIRE SUPPRESSION, PLUMBING & HVAC	99.7	49.3	78.7	96.9	60.4	81.3	97.1	50.7	77.2	96.9	58.9	80.7	96.6	49.4	76.4	100.3	55.3	81.0
26, 27, 3370	ELECTRICAL, COMMUNICATIONS & UTIL.	97.2	62.2	79.0	96.3	64.8	79.9	92.6	33.4	61.8	94.8	61.9	77.7	97.2	62.2	79.0	98.6	62.2	79.6
MF2016	WEIGHTED AVERAGE	99.3	63.4	83.6	95.7	64.9	82.2	99.6	58.6	81.7	98.4	65.2	83.9	100.0	64.8	84.6	97.6	64.5	83.1

TEXAS

DIVISION		PALESTINE 758			SAN ANGELO 769			SAN ANTONIO 781 - 782			TEMPLE 765			TEXARKANA 755			TYLER 757		
		MAT.	INST.	TOTAL	MAT.	INST.	TOTAL	MAT.	INST.	TOTAL	MAT.	INST.	TOTAL	MAT.	INST.	TOTAL	MAT.	INST.	TOTAL
015433	CONTRACTOR EQUIPMENT		95.1	95.1		93.5	93.5		93.3	93.3		93.5	93.5		95.1	95.1		95.1	95.1
0241, 31 - 34	SITE & INFRASTRUCTURE, DEMOLITION	99.5	94.1	95.4	102.4	92.5	95.4	101.1	93.5	95.8	90.3	92.3	91.7	88.3	96.6	94.1	98.6	93.9	95.3
0310	Concrete Forming & Accessories	77.6	61.6	63.8	98.5	53.8	59.9	93.6	54.2	59.6	101.9	53.5	60.2	92.7	62.2	66.4	87.6	61.7	65.3
0320	Concrete Reinforcing	83.5	51.1	67.1	81.2	51.2	66.1	96.0	48.6	72.1	81.4	48.6	64.8	83.3	51.2	67.1	84.1	51.1	67.5
0330	Cast-in-Place Concrete	102.8	61.5	87.1	101.6	68.2	88.9	86.2	69.6	79.9	83.2	61.4	74.9	103.6	68.3	90.2	123.5	61.6	100.0
03	CONCRETE	102.6	60.5	83.4	90.1	59.3	76.0	90.4	59.3	76.2	76.6	56.4	67.4	94.1	63.1	79.9	109.8	60.5	87.3
04	MASONRY	110.8	61.7	80.4	124.7	63.2	86.6	93.5	63.3	74.8	136.5	63.2	91.0	171.1	61.7	103.3	161.3	61.9	99.6
05	METALS	95.0	61.4	87.1	95.7	70.6	88.0	97.6	66.7	88.1	95.4	66.7	87.2	88.2	70.0	82.6	94.9	69.5	87.1
06	WOOD, PLASTICS & COMPOSITES	79.3	63.8	70.7	104.9	52.8	75.9	97.6	53.0	72.8	114.5	52.8	80.2	90.6	63.8	75.7	89.0	63.8	75.0
07	THERMAL & MOISTURE PROTECTION	94.6	64.8	82.0	94.5	65.7	82.3	93.0	67.5	82.2	94.1	64.7	81.7	93.7	66.1	82.0	94.4	64.8	81.9
08	OPENINGS	67.2	59.1	65.3	98.0	53.1	87.7	101.6	50.9	90.0	67.2	52.5	63.8	88.3	59.1	81.6	67.1	59.1	65.3
0920	Plaster & Gypsum Board	80.9	63.1	68.9	85.3	51.9	62.9	91.3	51.9	64.9	85.3	51.9	62.9	87.8	63.1	71.2	83.2	63.1	69.7
0950, 0980	Ceilings & Acoustic Treatment	84.2	63.1	70.0	78.6	51.9	60.6	94.5	51.9	65.8	78.6	51.9	60.6	88.5	63.1	71.4	84.2	63.1	70.0
0960	Flooring	120.2	57.7	102.8	78.0	57.7	72.3	101.6	71.6	93.3	100.2	57.7	88.4	104.8	67.4	94.4	131.5	57.7	111.0
0970, 0990	Wall Finishes & Painting/Coating	87.0	54.0	67.7	89.7	54.0	68.9	91.5	46.2	65.1	91.1	46.2	64.9	87.0	54.0	67.7	87.0	54.0	67.7
09	FINISHES	103.6	60.3	80.1	76.3	53.9	64.1	99.3	55.9	75.7	82.5	53.0	66.5	99.7	62.3	79.4	107.1	60.3	81.7
COVERS	DIVS. 10 - 14, 25, 28, 41, 43, 44, 46	100.0	76.7	94.8	100.0	79.2	95.4	100.0	79.2	95.4	100.0	75.3	94.5	100.0	80.9	95.7	100.0	80.8	95.7
21, 22, 23	FIRE SUPPRESSION, PLUMBING & HVAC	96.9	57.6	80.1	97.1	49.3	76.7	100.1	64.0	84.6	97.1	53.6	78.5	96.9	56.7	79.7	96.9	60.5	81.3
26, 27, 3370	ELECTRICAL, COMMUNICATIONS & UTIL.	92.6	51.1	71.0	100.4	57.6	78.2	97.3	62.0	79.0	97.6	54.9	75.4	96.2	57.6	76.1	95.1	56.7	75.1
MF2016	WEIGHTED AVERAGE	95.0	62.7	80.9	96.4	60.7	80.8	97.8	64.4	83.2	91.9	60.4	78.1	97.9	64.4	83.2	98.9	64.4	83.7

TEXAS / UTAH

DIVISION		VICTORIA 779			WACO 766 - 767			WAXAHACHIE 751			WHARTON 774			WICHITA FALLS 763			LOGAN 843		
		MAT.	INST.	TOTAL	MAT.	INST.	TOTAL	MAT.	INST.	TOTAL	MAT.	INST.	TOTAL	MAT.	INST.	TOTAL	MAT.	INST.	TOTAL
015433	CONTRACTOR EQUIPMENT		109.5	109.5		93.5	93.5		104.3	104.3		110.9	110.9		93.5	93.5		96.6	96.6
0241, 31 - 34	SITE & INFRASTRUCTURE, DEMOLITION	107.8	92.2	96.9	99.1	92.7	94.6	97.7	87.4	90.5	112.8	94.5	100.0	99.9	92.9	95.0	101.4	93.9	96.1
0310	Concrete Forming & Accessories	91.0	53.6	58.8	100.4	62.0	67.3	85.8	61.6	65.0	86.3	55.5	59.7	100.4	62.0	67.3	102.2	67.1	71.9
0320	Concrete Reinforcing	94.9	48.6	71.5	81.1	48.7	64.7	85.2	51.1	68.0	98.7	52.2	75.2	81.1	51.2	66.0	102.1	86.6	94.3
0330	Cast-in-Place Concrete	105.1	63.3	89.2	90.1	68.2	81.8	107.0	62.9	90.2	107.9	62.3	90.6	96.2	68.3	85.6	86.3	75.2	82.1
03	CONCRETE	99.4	58.3	80.6	83.0	62.6	73.7	92.4	62.0	78.5	103.5	59.4	83.3	85.9	63.1	75.5	103.3	73.8	89.8
04	MASONRY	116.5	63.3	83.5	94.9	63.2	75.2	156.0	61.9	97.6	99.8	63.2	77.1	95.4	62.7	75.1	108.4	67.9	83.3
05	METALS	93.4	87.8	91.7	97.7	69.4	89.0	95.1	84.9	92.0	94.9	88.8	93.0	97.6	70.8	89.4	107.9	84.0	100.6
06	WOOD, PLASTICS & COMPOSITES	99.9	52.9	73.7	112.8	63.7	85.4	84.6	63.9	73.1	90.0	55.2	70.6	112.8	63.7	85.4	84.2	67.1	74.6
07	THERMAL & MOISTURE PROTECTION	96.9	64.1	83.0	95.0	66.9	83.1	92.5	64.5	80.6	93.5	66.2	81.9	95.0	65.5	82.5	96.5	71.6	85.9
08	OPENINGS	99.3	52.1	88.4	78.7	58.5	74.1	97.8	59.2	89.0	99.4	54.2	89.0	78.7	59.1	74.2	92.4	66.9	86.5
0920	Plaster & Gypsum Board	88.5	51.9	63.9	85.9	63.1	70.5	85.5	63.1	70.4	87.3	54.3	65.1	85.9	63.1	70.5	77.8	66.2	70.0
0950, 0980	Ceilings & Acoustic Treatment	95.6	51.9	66.2	81.2	63.1	69.0	94.5	63.1	73.3	94.8	54.3	67.5	81.2	63.1	69.0	103.2	66.2	78.2
0960	Flooring	104.9	57.7	91.8	99.5	67.4	90.6	90.7	57.7	81.5	103.4	70.4	94.2	100.3	76.0	93.6	98.5	59.3	87.6
0970, 0990	Wall Finishes & Painting/Coating	97.7	58.3	74.7	91.1	54.0	69.5	96.3	54.0	71.6	97.5	58.3	74.6	93.5	54.0	70.4	93.6	56.7	72.1
09	FINISHES	89.3	54.4	70.3	83.3	62.2	71.9	92.2	60.5	74.9	90.8	57.9	72.9	83.8	64.0	73.0	93.4	64.0	77.4
COVERS	DIVS. 10 - 14, 25, 28, 41, 43, 44, 46	100.0	72.5	93.9	100.0	80.6	95.7	100.0	77.0	94.9	100.0	72.8	93.9	100.0	76.5	94.8	100.0	85.0	96.7
21, 22, 23	FIRE SUPPRESSION, PLUMBING & HVAC	97.1	63.6	82.8	100.2	59.1	82.6	96.9	52.5	77.9	97.0	61.0	81.6	100.2	52.7	79.9	99.9	70.0	87.1
26, 27, 3370	ELECTRICAL, COMMUNICATIONS & UTIL.	101.4	57.0	78.3	100.5	55.6	77.2	94.8	61.9	77.7	100.0	62.5	80.7	102.3	56.1	78.3	96.4	69.2	82.3
MF2016	WEIGHTED AVERAGE	98.3	64.9	83.7	93.6	64.4	80.8	98.5	64.3	83.5	98.2	66.2	84.2	94.2	63.3	80.7	100.2	72.9	88.3

For customer support on your Electrical Costs with RSMeans data, call 800.448.8182.

City Cost Indexes

UTAH / VERMONT

DIVISION		OGDEN 842, 844 MAT.	INST.	TOTAL	PRICE 845 MAT.	INST.	TOTAL	PROVO 846-847 MAT.	INST.	TOTAL	SALT LAKE CITY 840-841 MAT.	INST.	TOTAL	BELLOWS FALLS 051 MAT.	INST.	TOTAL	BENNINGTON 052 MAT.	INST.	TOTAL
015433	CONTRACTOR EQUIPMENT		96.6	96.6		95.6	95.6		95.6	95.6		96.6	96.6		98.8	98.8		98.8	98.8
0241, 31 - 34	SITE & INFRASTRUCTURE, DEMOLITION	88.8	93.9	92.4	98.8	92.1	94.1	97.5	92.2	93.8	88.5	93.8	92.2	92.2	101.5	98.7	91.5	101.5	98.5
0310	Concrete Forming & Accessories	102.2	67.1	71.9	104.6	66.9	72.1	103.9	67.1	72.1	104.5	67.1	72.2	93.8	106.3	104.6	91.6	106.2	104.1
0320	Concrete Reinforcing	101.7	86.6	94.1	109.4	86.6	97.9	110.4	86.6	98.4	103.9	86.6	95.2	80.9	81.4	81.1	80.9	81.4	81.1
0330	Cast-in-Place Concrete	87.6	75.2	82.9	86.4	75.2	82.1	86.4	75.2	82.2	95.7	75.2	88.0	90.7	115.5	100.1	90.7	115.5	100.1
03	CONCRETE	92.9	73.8	84.2	104.4	73.7	90.4	102.9	73.8	89.6	112.0	73.8	94.5	85.2	104.7	94.2	85.1	104.6	94.0
04	MASONRY	102.9	67.9	81.2	114.4	67.9	85.6	114.5	67.9	85.6	116.9	67.9	86.5	104.2	89.8	95.3	115.7	89.8	99.7
05	METALS	108.4	84.0	100.9	105.0	83.8	98.5	106.0	84.0	99.3	112.1	84.0	103.4	93.4	89.7	92.3	93.4	89.6	92.2
06	WOOD, PLASTICS & COMPOSITES	84.2	67.1	74.6	87.2	67.1	76.0	85.5	67.1	75.2	85.9	67.1	75.4	108.3	111.9	110.3	105.6	111.9	109.1
07	THERMAL & MOISTURE PROTECTION	95.3	71.6	85.3	98.3	71.6	87.0	98.4	71.6	87.0	102.8	71.6	89.6	98.0	92.6	95.7	98.0	92.6	95.7
08	OPENINGS	92.4	66.9	86.5	96.3	63.1	88.7	96.4	66.9	89.6	94.3	66.9	88.0	100.4	100.8	100.5	100.4	100.8	100.5
0920	Plaster & Gypsum Board	77.8	66.2	70.0	80.5	66.2	70.9	78.1	66.2	70.1	90.8	66.2	74.3	105.6	111.9	109.8	103.9	111.9	109.3
0950, 0980	Ceilings & Acoustic Treatment	103.2	66.2	78.2	103.2	66.2	78.2	103.2	66.2	78.2	95.5	66.2	75.8	93.8	111.9	106.0	93.8	111.9	106.0
0960	Flooring	96.4	59.3	86.1	99.7	59.3	88.5	99.4	59.3	88.2	100.4	59.3	89.0	91.1	104.5	94.8	90.4	104.5	94.3
0970, 0990	Wall Finishes & Painting/Coating	93.6	56.7	72.1	93.6	56.7	72.1	93.6	56.7	72.1	96.7	56.7	73.3	86.2	106.1	97.8	86.2	106.1	97.8
09	FINISHES	91.5	64.0	76.6	94.6	64.0	78.0	94.0	64.0	77.7	93.7	64.0	77.6	88.1	107.0	98.4	87.6	107.0	98.2
COVERS	DIVS. 10 - 14, 25, 28, 41, 43, 44, 46	100.0	85.0	96.7	100.0	85.0	96.7	100.0	85.0	96.7	100.0	85.0	96.7	100.0	95.7	99.0	100.0	95.6	99.0
21, 22, 23	FIRE SUPPRESSION, PLUMBING & HVAC	99.9	70.0	87.1	98.3	65.7	84.4	99.9	70.0	87.1	100.1	70.0	87.2	97.2	90.8	94.5	97.2	90.8	94.5
26, 27, 3370	ELECTRICAL, COMMUNICATIONS & UTIL.	96.7	69.2	82.4	101.6	69.2	84.7	97.0	69.2	82.5	99.3	69.2	83.6	105.2	84.0	94.1	105.1	57.5	80.4
MF2016	WEIGHTED AVERAGE	98.2	72.9	87.2	100.8	71.7	88.1	100.6	72.8	88.5	102.8	72.9	89.7	96.0	95.3	95.7	96.5	91.6	94.3

VERMONT

DIVISION		BRATTLEBORO 053 MAT.	INST.	TOTAL	BURLINGTON 054 MAT.	INST.	TOTAL	GUILDHALL 059 MAT.	INST.	TOTAL	MONTPELIER 056 MAT.	INST.	TOTAL	RUTLAND 057 MAT.	INST.	TOTAL	ST. JOHNSBURY 058 MAT.	INST.	TOTAL
015433	CONTRACTOR EQUIPMENT		98.8	98.8		98.8	98.8		98.8	98.8		98.8	98.8		98.8	98.8		98.8	98.8
0241, 31 - 34	SITE & INFRASTRUCTURE, DEMOLITION	93.1	101.5	99.0	96.9	101.4	100.0	91.1	96.9	95.2	95.4	101.4	99.6	95.6	101.4	99.7	91.2	100.5	97.7
0310	Concrete Forming & Accessories	94.0	106.3	104.6	97.3	82.6	84.6	92.0	99.4	98.4	97.3	105.3	104.2	94.3	82.6	84.2	90.7	99.8	98.6
0320	Concrete Reinforcing	80.0	81.4	80.7	99.4	81.2	90.2	81.6	81.2	81.4	99.4	81.3	90.2	101.1	81.2	91.1	80.0	81.3	80.7
0330	Cast-in-Place Concrete	93.5	115.5	101.9	108.2	114.7	110.7	87.9	106.3	94.9	108.1	114.7	110.6	88.8	114.7	98.6	87.9	106.5	94.9
03	CONCRETE	87.2	104.7	95.2	96.9	93.7	95.4	82.9	98.4	90.0	96.8	104.0	100.1	87.7	93.7	90.4	82.6	98.7	90.0
04	MASONRY	115.2	89.8	99.5	113.8	89.0	98.4	115.6	74.5	90.1	110.1	89.0	97.0	94.8	89.0	91.2	144.6	74.5	101.2
05	METALS	93.4	89.8	92.3	101.1	88.8	97.4	93.4	88.6	91.9	99.1	89.0	96.0	99.2	88.8	96.0	93.4	89.0	92.1
06	WOOD, PLASTICS & COMPOSITES	108.7	111.9	110.5	100.6	81.5	89.9	105.1	111.9	108.9	97.3	111.9	105.5	108.8	81.5	93.6	100.5	111.9	106.9
07	THERMAL & MOISTURE PROTECTION	98.1	92.6	95.8	105.1	88.9	98.2	97.9	85.6	92.7	105.0	92.2	99.6	98.2	88.9	94.3	97.8	85.6	92.6
08	OPENINGS	100.4	101.0	100.5	102.9	80.2	97.7	100.4	97.1	99.6	103.7	97.1	102.2	103.6	80.2	98.2	100.4	97.1	99.6
0920	Plaster & Gypsum Board	105.6	111.9	109.8	105.9	80.6	88.9	111.5	111.9	111.8	105.5	111.9	109.8	106.0	80.6	88.9	112.8	111.9	112.2
0950, 0980	Ceilings & Acoustic Treatment	93.8	111.9	106.0	100.8	80.6	87.1	93.8	111.9	106.0	101.0	111.9	108.3	98.4	80.6	86.4	93.8	111.9	106.0
0960	Flooring	91.2	104.5	94.9	97.3	104.5	99.3	94.0	104.5	96.4	98.4	104.5	100.1	91.1	104.5	94.8	97.1	104.5	99.2
0970, 0990	Wall Finishes & Painting/Coating	86.2	106.1	97.8	95.8	92.4	93.8	86.2	92.4	89.8	92.2	92.4	92.3	86.2	92.4	89.8	86.2	92.4	89.8
09	FINISHES	88.2	107.0	98.4	95.8	87.2	91.1	89.7	101.6	96.2	95.7	105.3	100.9	89.2	87.2	88.1	90.8	101.6	96.7
COVERS	DIVS. 10 - 14, 25, 28, 41, 43, 44, 46	100.0	95.7	99.0	100.0	92.0	98.2	100.0	90.4	97.9	100.0	95.3	99.0	100.0	92.0	98.2	100.0	90.4	97.9
21, 22, 23	FIRE SUPPRESSION, PLUMBING & HVAC	97.2	90.8	94.5	100.0	68.2	86.4	97.2	60.7	81.6	96.9	68.2	84.7	100.3	68.2	86.6	97.2	60.7	81.6
26, 27, 3370	ELECTRICAL, COMMUNICATIONS & UTIL.	105.1	84.0	94.1	105.0	55.6	79.3	105.2	57.5	80.4	104.3	55.6	79.0	105.2	55.6	79.4	105.2	57.5	80.4
MF2016	WEIGHTED AVERAGE	96.9	95.3	96.2	101.0	80.9	92.2	96.4	81.1	89.7	99.8	85.9	93.7	98.1	80.9	90.6	97.8	81.5	90.7

VERMONT / VIRGINIA

DIVISION		WHITE RIVER JCT. 050 MAT.	INST.	TOTAL	ALEXANDRIA 223 MAT.	INST.	TOTAL	ARLINGTON 222 MAT.	INST.	TOTAL	BRISTOL 242 MAT.	INST.	TOTAL	CHARLOTTESVILLE 229 MAT.	INST.	TOTAL	CULPEPER 227 MAT.	INST.	TOTAL
015433	CONTRACTOR EQUIPMENT		98.8	98.8		108.6	108.6		107.3	107.3		107.3	107.3		111.5	111.5		107.3	107.3
0241, 31 - 34	SITE & INFRASTRUCTURE, DEMOLITION	95.7	100.6	99.1	115.4	92.9	99.6	125.6	90.6	101.1	109.8	89.9	95.9	114.2	92.4	99.0	113.1	90.5	97.3
0310	Concrete Forming & Accessories	88.9	100.3	98.8	90.8	71.5	74.1	89.8	71.4	73.9	86.0	66.9	69.5	84.5	48.5	53.5	81.7	71.0	72.5
0320	Concrete Reinforcing	80.9	81.3	81.1	85.2	82.8	84.0	96.2	80.6	88.3	90.0	83.0	86.3	95.5	70.3	82.8	96.2	80.5	88.3
0330	Cast-in-Place Concrete	93.5	107.3	98.8	107.7	78.3	96.5	104.8	78.3	94.7	104.3	46.4	82.3	108.5	78.8	97.2	107.3	78.1	96.2
03	CONCRETE	88.7	99.2	93.5	100.0	77.3	89.6	104.5	76.9	91.9	100.8	62.1	83.1	101.5	65.0	84.8	98.5	76.6	88.5
04	MASONRY	129.7	76.0	96.4	91.2	73.7	80.3	106.3	73.7	86.1	95.3	47.3	65.5	120.2	56.9	80.9	107.2	73.7	86.4
05	METALS	93.4	89.0	92.1	103.4	100.5	102.5	102.0	99.6	101.3	100.8	93.8	98.7	101.1	95.8	99.5	101.2	98.8	100.4
06	WOOD, PLASTICS & COMPOSITES	102.6	111.9	107.8	94.6	69.7	80.7	91.3	69.7	79.2	84.5	73.2	78.2	83.1	41.6	60.0	81.9	69.7	75.1
07	THERMAL & MOISTURE PROTECTION	98.2	86.2	93.1	102.6	81.4	93.6	104.7	80.7	94.5	104.0	60.7	85.7	103.6	69.4	89.1	103.8	80.7	94.0
08	OPENINGS	100.4	97.1	99.6	98.5	74.1	92.9	96.6	73.6	91.3	99.6	67.1	92.2	97.8	55.8	88.2	98.1	73.6	92.5
0920	Plaster & Gypsum Board	102.6	111.9	108.8	103.9	68.7	80.3	100.6	68.7	79.2	97.1	72.3	80.4	97.1	39.1	58.2	97.3	68.7	78.1
0950, 0980	Ceilings & Acoustic Treatment	93.8	111.9	106.0	94.5	68.7	77.1	92.8	68.7	76.5	91.9	72.3	78.7	91.9	39.1	56.3	92.8	68.7	76.5
0960	Flooring	89.3	104.5	93.5	96.3	78.7	91.4	94.7	78.7	90.3	91.2	58.9	82.3	89.7	58.9	81.2	89.7	78.7	86.7
0970, 0990	Wall Finishes & Painting/Coating	86.2	92.4	89.8	114.7	75.0	91.5	114.7	75.0	91.5	101.5	56.6	75.3	101.5	60.0	77.3	114.7	60.0	82.8
09	FINISHES	87.5	102.0	95.4	94.4	72.4	82.4	94.3	72.4	82.4	91.0	65.3	77.1	90.6	49.5	68.3	91.3	70.7	80.1
COVERS	DIVS. 10 - 14, 25, 28, 41, 43, 44, 46	100.0	90.9	98.0	100.0	89.2	97.6	100.0	86.7	97.0	100.0	77.6	95.0	100.0	82.1	96.0	100.0	86.7	97.0
21, 22, 23	FIRE SUPPRESSION, PLUMBING & HVAC	97.2	61.4	81.9	100.4	85.3	93.9	100.4	85.2	93.9	97.3	47.7	76.1	97.3	70.4	85.8	97.3	70.6	85.9
26, 27, 3370	ELECTRICAL, COMMUNICATIONS & UTIL.	105.2	55.6	79.4	98.0	99.6	98.8	95.7	101.2	98.6	97.6	38.4	66.8	97.6	72.2	84.4	97.3	70.6	85.9
MF2016	WEIGHTED AVERAGE	97.7	81.6	90.7	99.8	84.7	93.2	100.7	84.5	93.7	98.6	60.5	82.0	99.9	69.3	86.5	99.1	80.5	91.0

For customer support on your Electrical Costs with RSMeans data, call 800.448.8182.

547

City Cost Indexes

DIVISION		FAIRFAX 220-221 MAT.	INST.	TOTAL	FARMVILLE 239 MAT.	INST.	TOTAL	FREDERICKSBURG 224-225 MAT.	INST.	TOTAL	GRUNDY 246 MAT.	INST.	TOTAL	HARRISONBURG 228 MAT.	INST.	TOTAL	LYNCHBURG 245 MAT.	INST.	TOTAL
015433	CONTRACTOR EQUIPMENT		107.3	107.3		111.5	111.5		107.3	107.3		107.3	107.3		107.3	107.3		107.3	107.3
0241, 31-34	SITE & INFRASTRUCTURE, DEMOLITION	124.3	90.6	100.7	108.4	91.4	96.5	112.7	90.4	97.1	107.6	89.2	94.7	121.3	89.0	98.6	108.5	90.6	96.0
0310	Concrete Forming & Accessories	84.5	71.3	73.1	95.5	45.6	52.5	84.5	68.5	70.7	89.1	38.7	45.6	80.7	39.4	45.1	86.1	73.0	74.8
0320	Concrete Reinforcing	96.2	80.6	88.3	94.9	69.9	82.3	96.9	82.7	89.8	94.9	45.4	69.9	96.2	59.1	77.5	95.5	70.8	83.1
0330	Cast-in-Place Concrete	104.8	78.2	94.7	106.5	88.0	99.5	106.4	78.1	95.6	104.3	52.0	84.5	104.7	55.0	85.8	104.3	77.6	94.2
03	CONCRETE	104.2	76.8	91.7	97.2	66.6	83.2	98.3	75.9	88.0	99.4	46.6	75.3	102.0	50.6	78.5	99.3	75.7	88.5
04	MASONRY	106.2	73.7	86.0	104.3	52.1	71.9	106.3	69.6	83.6	96.1	55.7	71.1	104.2	60.3	77.0	111.7	65.8	83.2
05	METALS	101.3	99.5	100.7	98.0	91.8	96.1	101.2	100.3	100.9	100.8	76.1	93.2	101.1	88.8	97.3	101.0	95.9	99.5
06	WOOD, PLASTICS & COMPOSITES	84.5	69.7	76.2	93.8	40.2	64.0	84.5	67.7	75.1	87.1	31.9	56.4	81.0	35.5	55.7	84.5	74.6	79.0
07	THERMAL & MOISTURE PROTECTION	104.4	76.4	92.6	104.2	56.0	83.8	103.8	78.7	93.1	103.9	48.3	80.4	104.2	64.2	87.3	103.8	72.6	90.6
08	OPENINGS	96.6	73.6	91.3	97.6	45.3	85.6	97.8	72.0	91.9	99.6	31.6	84.0	98.1	45.3	86.0	98.1	67.9	91.2
0920	Plaster & Gypsum Board	97.3	68.7	78.1	106.4	37.7	60.3	97.3	66.7	76.7	97.1	29.9	52.0	97.1	33.6	54.5	97.1	73.7	81.4
0950, 0980	Ceilings & Acoustic Treatment	92.8	68.7	76.5	87.6	37.7	53.9	92.8	66.7	75.2	91.9	29.9	50.1	91.9	33.6	52.6	91.9	73.7	79.7
0960	Flooring	91.5	78.7	88.0	92.0	53.4	81.3	91.5	76.5	87.4	92.5	29.8	75.1	89.5	78.7	86.5	91.2	69.1	85.1
0970, 0990	Wall Finishes & Painting/Coating	114.7	75.0	91.5	99.7	59.8	76.4	114.7	59.6	82.6	101.5	32.7	61.3	114.7	50.5	77.3	101.5	56.6	75.3
09	FINISHES	92.9	72.4	81.7	90.5	47.3	67.0	91.8	68.3	79.0	91.2	35.0	60.7	91.7	46.2	67.0	90.8	70.6	79.9
COVERS	DIVS. 10-14, 25, 28, 41, 43, 44, 46	100.0	86.7	97.0	100.0	77.7	95.0	100.0	81.5	95.9	100.0	75.8	94.6	100.0	62.5	91.6	100.0	81.2	95.8
21, 22, 23	FIRE SUPPRESSION, PLUMBING & HVAC	97.3	85.2	92.1	97.4	46.9	75.8	97.3	83.2	91.3	97.3	67.4	84.5	97.3	68.1	84.8	97.3	70.7	85.9
26, 27, 3370	ELECTRICAL, COMMUNICATIONS & UTIL.	98.7	96.9	97.8	92.2	45.1	67.7	95.9	96.9	96.4	97.6	41.7	68.5	97.8	91.2	94.4	98.6	71.6	84.6
MF2016	WEIGHTED AVERAGE	100.0	83.8	92.9	97.4	58.6	80.4	98.7	82.2	91.5	98.5	56.0	79.9	99.5	67.3	85.4	99.2	75.2	88.7

VIRGINIA

DIVISION		NEWPORT NEWS 236 MAT.	INST.	TOTAL	NORFOLK 233-235 MAT.	INST.	TOTAL	PETERSBURG 238 MAT.	INST.	TOTAL	PORTSMOUTH 237 MAT.	INST.	TOTAL	PULASKI 243 MAT.	INST.	TOTAL	RICHMOND 230-232 MAT.	INST.	TOTAL
015433	CONTRACTOR EQUIPMENT		111.6	111.6		112.2	112.2		111.5	111.5		111.5	111.5		107.3	107.3		111.5	111.5
0241, 31-34	SITE & INFRASTRUCTURE, DEMOLITION	107.3	92.4	96.9	105.7	93.5	97.1	111.0	92.4	97.9	105.9	91.7	95.9	106.9	89.7	94.8	102.2	92.4	95.3
0310	Concrete Forming & Accessories	94.8	62.5	67.0	100.4	62.6	67.8	88.8	63.7	67.2	85.0	49.0	53.9	89.1	41.4	48.0	95.2	82.3	84.1
0320	Concrete Reinforcing	94.7	67.5	81.0	102.5	67.5	84.8	94.3	70.9	82.5	94.3	66.8	80.5	94.9	58.2	76.4	105.2	70.9	87.9
0330	Cast-in-Place Concrete	103.5	78.0	93.8	107.5	79.1	96.7	109.9	79.3	98.3	102.5	62.6	87.3	104.3	86.4	97.5	98.4	79.3	91.1
03	CONCRETE	94.5	70.5	83.6	96.5	71.0	84.8	99.7	72.1	87.1	93.4	59.0	77.7	99.4	62.2	82.4	92.2	80.6	86.9
04	MASONRY	98.6	64.5	77.4	96.6	64.5	76.7	112.9	64.5	82.9	104.6	49.4	70.4	90.8	56.6	69.6	97.3	64.5	77.0
05	METALS	100.3	94.1	98.7	102.0	95.2	99.9	98.0	96.8	97.7	99.3	93.6	97.5	100.9	87.4	96.7	102.0	97.0	100.5
06	WOOD, PLASTICS & COMPOSITES	92.8	61.4	75.3	96.1	61.4	76.8	85.2	62.4	72.5	82.0	46.9	62.5	87.1	33.5	57.3	95.9	87.4	91.2
07	THERMAL & MOISTURE PROTECTION	104.1	71.6	90.4	101.4	71.8	88.9	104.1	73.6	91.2	104.2	62.1	86.4	103.9	59.0	84.9	102.5	76.4	91.4
08	OPENINGS	98.0	60.3	89.3	96.2	61.6	88.2	97.3	67.3	90.4	98.0	49.9	87.0	99.7	38.3	85.6	100.2	81.1	95.8
0920	Plaster & Gypsum Board	107.4	59.5	75.2	100.3	59.5	72.9	100.7	60.5	73.7	101.2	44.6	63.2	97.1	31.5	53.1	101.4	86.2	91.2
0950, 0980	Ceilings & Acoustic Treatment	91.8	59.5	70.0	95.4	59.5	71.2	88.4	60.5	69.6	91.8	44.6	60.0	91.9	31.5	51.2	91.4	86.2	87.9
0960	Flooring	92.0	69.1	85.7	87.7	69.1	82.5	88.0	74.3	84.2	85.0	58.9	77.8	92.5	58.9	83.2	87.3	74.3	83.7
0970, 0990	Wall Finishes & Painting/Coating	99.7	60.0	76.5	97.4	60.0	75.6	99.7	60.0	76.5	99.7	60.0	76.5	101.5	50.5	71.8	92.7	60.0	73.6
09	FINISHES	91.3	63.0	75.9	90.7	63.0	75.6	88.9	64.6	75.7	88.4	50.5	67.8	91.2	42.9	64.9	91.0	79.4	84.7
COVERS	DIVS. 10-14, 25, 28, 41, 43, 44, 46	100.0	80.5	95.7	100.0	80.5	95.7	100.0	83.9	96.4	100.0	66.5	92.6	100.0	76.0	94.7	100.0	86.7	97.0
21, 22, 23	FIRE SUPPRESSION, PLUMBING & HVAC	100.5	64.0	84.9	100.1	67.3	86.1	97.4	69.8	85.6	100.5	64.3	85.1	97.3	66.2	84.0	100.1	69.8	87.1
26, 27, 3370	ELECTRICAL, COMMUNICATIONS & UTIL.	94.7	63.7	78.6	97.9	63.8	80.2	94.8	72.2	83.1	93.1	63.8	77.9	97.6	56.7	76.3	98.3	72.2	84.7
MF2016	WEIGHTED AVERAGE	98.2	70.4	86.0	98.5	71.3	86.6	98.3	73.9	87.6	97.7	64.3	83.1	98.2	62.7	82.7	98.5	77.9	89.5

DIVISION		VIRGINIA ROANOKE 240-241 MAT.	INST.	TOTAL	STAUNTON 244 MAT.	INST.	TOTAL	WINCHESTER 226 MAT.	INST.	TOTAL	WASHINGTON CLARKSTON 994 MAT.	INST.	TOTAL	EVERETT 982 MAT.	INST.	TOTAL	OLYMPIA 985 MAT.	INST.	TOTAL
015433	CONTRACTOR EQUIPMENT		107.2	107.2		111.5	111.5		107.3	107.3		91.5	91.5		102.2	102.2		102.2	102.2
0241, 31-34	SITE & INFRASTRUCTURE, DEMOLITION	107.5	90.6	95.7	110.7	91.3	97.1	119.9	90.5	99.3	107.5	89.6	94.9	92.9	110.9	105.5	93.0	110.9	105.5
0310	Concrete Forming & Accessories	95.2	73.1	76.1	88.8	50.2	55.5	82.9	69.5	71.4	111.4	67.0	73.1	116.2	102.0	104.5	113.8	111.8	112.8
0320	Concrete Reinforcing	95.9	70.8	83.2	95.5	70.0	82.7	95.6	82.8	89.1	112.0	98.3	105.1	108.9	111.8	110.4	90.8	109.0	97.7
0330	Cast-in-Place Concrete	118.5	87.8	106.8	108.5	88.6	100.9	104.7	67.1	90.5	87.8	84.1	86.4	98.6	109.1	102.6	86.8	105.8	95.5
03	CONCRETE	103.9	79.3	92.6	100.8	68.9	86.2	101.5	72.6	88.3	95.4	78.7	87.8	90.1	105.9	97.3	86.8	105.8	95.5
04	MASONRY	97.4	65.8	77.8	107.0	55.6	75.1	101.6	72.7	83.7	100.8	84.8	90.9	112.7	103.6	107.1	107.1	103.6	104.9
05	METALS	103.2	96.0	101.0	101.1	91.6	98.2	101.2	100.1	100.9	91.3	87.6	90.1	103.3	96.7	101.3	103.0	96.3	101.0
06	WOOD, PLASTICS & COMPOSITES	95.4	74.6	83.8	87.1	45.5	64.0	83.1	67.9	74.6	107.4	61.2	81.7	112.0	101.6	106.2	97.0	101.6	99.6
07	THERMAL & MOISTURE PROTECTION	103.7	76.1	92.0	103.5	57.5	84.1	104.3	78.3	93.3	158.3	81.7	125.8	111.9	105.8	109.3	111.6	101.9	107.5
08	OPENINGS	98.5	67.9	91.5	98.1	48.1	86.6	99.7	71.8	93.3	118.1	68.3	106.7	106.1	104.0	105.6	110.5	104.3	109.0
0920	Plaster & Gypsum Board	103.9	73.7	83.6	97.1	43.2	60.9	97.3	66.9	76.9	149.2	60.1	89.3	112.7	101.8	105.4	107.0	101.8	103.5
0950, 0980	Ceilings & Acoustic Treatment	94.5	73.7	80.5	91.9	43.2	59.1	92.8	66.9	75.3	108.8	60.1	75.9	105.7	101.8	103.1	105.7	101.8	103.1
0960	Flooring	96.3	69.1	88.8	92.1	35.0	76.2	91.0	78.7	87.6	86.7	81.9	85.3	105.2	101.5	104.2	93.1	101.5	95.4
0970, 0990	Wall Finishes & Painting/Coating	101.5	56.6	75.3	101.5	32.1	61.0	114.7	81.1	95.1	78.6	75.4	76.7	91.5	95.7	93.9	87.1	95.7	92.1
09	FINISHES	93.4	70.6	81.0	91.1	44.5	65.8	92.3	71.8	81.1	105.8	69.0	85.8	103.3	101.3	102.2	94.0	101.3	98.0
COVERS	DIVS. 10-14, 25, 28, 41, 43, 44, 46	100.0	81.1	95.8	100.0	78.8	95.3	100.0	86.1	96.9	100.0	92.9	98.4	100.0	100.0	100.0	100.0	102.5	100.6
21, 22, 23	FIRE SUPPRESSION, PLUMBING & HVAC	100.4	66.4	85.9	97.3	60.1	81.4	97.3	84.7	91.9	97.7	83.9	91.8	100.2	101.8	100.9	100.1	101.8	100.8
26, 27, 3370	ELECTRICAL, COMMUNICATIONS & UTIL.	97.6	57.8	76.9	96.6	72.2	83.9	96.3	96.9	96.6	86.6	98.0	92.6	108.2	98.4	103.1	106.5	98.3	102.2
MF2016	WEIGHTED AVERAGE	100.3	73.0	88.3	99.0	65.8	84.5	99.3	82.9	92.2	100.6	83.4	93.1	102.0	102.4	102.2	100.7	102.3	101.4

City Cost Indexes

WASHINGTON

DIVISION		RICHLAND 993			SEATTLE 980-981, 987			SPOKANE 990-992			TACOMA 983-984			VANCOUVER 986			WENATCHEE 988		
		MAT.	INST.	TOTAL	MAT.	INST.	TOTAL	MAT.	INST.	TOTAL	MAT.	INST.	TOTAL	MAT.	INST.	TOTAL	MAT.	INST.	TOTAL
015433	CONTRACTOR EQUIPMENT		91.5	91.5		103.7	103.7		91.5	91.5		102.2	102.2		98.4	98.4		102.2	102.2
0241, 31 - 34	SITE & INFRASTRUCTURE, DEMOLITION	109.4	89.7	95.6	99.3	110.1	106.9	108.8	89.7	95.4	96.1	110.9	106.5	106.0	97.8	100.3	105.3	108.3	107.4
0310	Concrete Forming & Accessories	111.5	80.0	84.4	112.3	102.9	104.2	116.8	79.6	84.7	106.5	102.5	103.0	107.1	94.7	96.4	108.4	78.9	82.9
0320	Concrete Reinforcing	107.4	98.4	102.9	114.0	111.9	112.9	108.1	98.3	103.2	107.4	111.8	109.6	108.5	110.9	109.7	108.5	98.4	103.4
0330	Cast-in-Place Concrete	88.0	85.5	87.0	102.5	108.8	104.9	91.4	85.3	89.1	101.3	109.0	104.3	113.2	100.7	108.4	103.5	93.8	99.8
03	CONCRETE	95.0	85.1	90.5	98.2	106.2	101.9	97.1	84.8	91.5	91.7	105.8	98.2	101.1	99.4	100.3	98.8	87.4	93.6
04	MASONRY	101.5	86.8	92.4	119.1	103.6	109.5	102.0	86.8	92.6	112.2	103.6	106.9	111.7	87.8	96.9	115.6	84.4	96.3
05	METALS	91.7	88.2	90.6	103.7	99.9	102.5	94.0	87.8	92.1	105.1	96.3	102.4	102.7	96.4	100.8	102.5	87.9	98.0
06	WOOD, PLASTICS & COMPOSITES	107.6	77.0	90.6	107.3	101.6	104.1	116.2	77.0	94.4	101.5	101.6	101.6	93.0	93.9	93.5	103.2	76.8	88.5
07	THERMAL & MOISTURE PROTECTION	159.4	85.7	128.2	107.1	105.2	106.3	155.6	87.1	126.6	111.6	102.0	107.6	112.0	97.3	105.8	111.1	87.9	101.3
08	OPENINGS	116.1	75.9	106.9	106.5	104.9	106.1	116.7	79.5	108.2	106.9	104.3	106.3	103.2	98.7	102.2	106.4	76.7	99.5
0920	Plaster & Gypsum Board	149.2	76.3	100.2	108.6	101.8	104.0	139.6	76.3	97.1	108.8	101.8	104.1	106.9	94.1	98.3	111.7	76.3	87.9
0950, 0980	Ceilings & Acoustic Treatment	115.5	76.3	89.1	112.0	101.8	105.1	110.7	76.3	87.5	109.1	101.8	104.2	104.2	94.1	97.4	101.3	76.3	84.4
0960	Flooring	87.0	81.9	85.6	103.6	101.5	103.0	86.0	81.9	84.9	97.9	101.5	98.9	103.6	83.6	98.1	100.9	81.9	95.6
0970, 0990	Wall Finishes & Painting/Coating	78.6	79.9	79.3	105.4	95.7	99.7	78.7	79.9	79.4	91.5	95.7	93.9	94.0	75.2	83.0	91.5	74.1	81.4
09	FINISHES	107.4	79.3	92.1	106.9	101.3	103.9	105.0	79.3	91.0	101.4	101.3	101.3	99.2	90.1	94.2	102.1	77.9	88.9
COVERS	DIVS. 10 - 14, 25, 28, 41, 43, 44, 46	100.0	95.3	99.0	100.0	102.6	100.6	100.0	95.2	98.9	100.0	102.5	100.6	100.0	98.0	99.6	100.0	94.0	98.7
21, 22, 23	FIRE SUPPRESSION, PLUMBING & HVAC	100.8	112.6	105.9	100.0	113.5	105.8	100.0	86.6	94.7	100.2	101.8	100.9	100.3	101.5	100.8	97.1	94.2	95.8
26, 27, 3370	ELECTRICAL, COMMUNICATIONS & UTIL.	84.6	98.0	91.6	107.4	113.1	110.4	83.0	81.4	82.1	108.0	98.3	103.0	113.8	101.6	107.5	108.8	93.3	100.7
MF2016	WEIGHTED AVERAGE	101.1	92.7	97.4	103.6	107.3	105.2	101.4	85.0	94.2	102.4	102.3	102.3	103.3	97.2	100.6	102.6	89.5	96.9

WASHINGTON / WEST VIRGINIA

DIVISION		YAKIMA 989			BECKLEY 258-259			BLUEFIELD 247-248			BUCKHANNON 262			CHARLESTON 250-253			CLARKSBURG 263-264		
		MAT.	INST.	TOTAL	MAT.	INST.	TOTAL	MAT.	INST.	TOTAL	MAT.	INST.	TOTAL	MAT.	INST.	TOTAL	MAT.	INST.	TOTAL
015433	CONTRACTOR EQUIPMENT		102.2	102.2		107.3	107.3		107.3	107.3		107.3	107.3		107.3	107.3		107.3	107.3
0241, 31 - 34	SITE & INFRASTRUCTURE, DEMOLITION	98.6	109.4	106.2	101.5	91.8	94.7	101.5	91.8	94.7	107.7	91.7	96.5	100.7	92.8	95.2	108.5	91.7	96.7
0310	Concrete Forming & Accessories	107.0	98.0	99.2	86.9	90.2	89.7	86.0	89.9	89.3	85.5	86.4	86.3	97.7	91.3	92.2	83.0	86.4	85.9
0320	Concrete Reinforcing	108.0	98.5	103.2	91.8	89.4	90.6	94.5	70.8	82.6	95.1	94.3	94.7	99.3	89.7	94.4	95.1	97.3	96.2
0330	Cast-in-Place Concrete	108.2	84.6	99.3	104.4	93.1	100.1	102.0	92.9	98.6	101.7	93.4	98.5	101.0	94.5	98.5	111.5	91.6	103.9
03	CONCRETE	96.2	92.9	94.7	95.9	92.0	94.1	95.4	88.7	92.3	98.3	91.3	95.1	94.5	93.1	93.8	102.4	91.2	97.3
04	MASONRY	105.1	84.5	92.3	93.0	88.8	90.4	92.5	88.8	90.2	104.5	86.5	93.3	88.7	94.2	92.1	108.4	86.5	94.8
05	METALS	103.2	88.7	98.7	102.8	104.5	103.3	101.1	97.9	100.1	101.3	106.1	102.8	101.4	104.9	102.5	101.3	107.1	103.1
06	WOOD, PLASTICS & COMPOSITES	101.9	101.6	101.8	85.1	90.5	88.1	86.3	90.5	88.6	85.6	85.9	85.8	92.3	90.5	91.3	82.3	85.9	84.3
07	THERMAL & MOISTURE PROTECTION	111.7	88.9	102.1	106.5	88.6	98.9	103.7	88.6	97.3	104.0	87.6	97.1	103.1	90.6	97.8	103.9	87.3	96.9
08	OPENINGS	106.3	101.3	105.2	96.3	85.6	93.8	100.1	81.3	95.7	100.1	84.1	96.4	94.4	85.6	92.4	100.1	84.8	96.6
0920	Plaster & Gypsum Board	108.4	101.8	103.9	98.9	90.1	93.0	96.4	90.1	92.1	96.8	85.4	89.1	99.9	90.1	93.3	94.1	85.4	88.3
0950, 0980	Ceilings & Acoustic Treatment	103.2	101.8	102.2	84.1	90.1	88.1	90.2	90.1	90.1	91.9	85.4	87.5	95.2	90.1	91.7	91.9	85.4	87.5
0960	Flooring	98.8	81.9	94.1	91.6	100.6	94.1	88.9	100.6	92.2	88.7	96.5	90.8	94.0	100.6	95.8	87.6	96.5	90.1
0970, 0990	Wall Finishes & Painting/Coating	91.5	79.9	84.7	92.6	91.7	92.1	101.5	91.7	95.8	101.5	87.9	93.6	88.2	91.7	90.3	101.5	87.9	93.6
09	FINISHES	100.6	93.2	96.6	88.2	92.6	90.6	89.3	92.6	91.1	90.1	88.3	89.1	93.4	93.2	93.3	89.4	88.3	88.8
COVERS	DIVS. 10 - 14, 25, 28, 41, 43, 44, 46	100.0	99.3	99.8	100.0	93.8	98.6	100.0	93.8	98.6	100.0	92.4	98.3	100.0	94.1	98.7	100.0	92.4	98.3
21, 22, 23	FIRE SUPPRESSION, PLUMBING & HVAC	100.2	111.8	105.1	97.5	91.4	94.9	97.3	87.1	92.9	97.3	91.2	94.7	100.1	92.8	97.0	97.3	91.1	94.7
26, 27, 3370	ELECTRICAL, COMMUNICATIONS & UTIL.	111.0	98.1	104.3	94.6	87.0	90.7	96.6	87.0	91.6	98.0	91.9	94.8	99.7	88.8	94.0	98.0	91.9	94.8
MF2016	WEIGHTED AVERAGE	102.5	98.2	100.6	97.3	91.8	94.8	97.5	89.6	94.0	98.8	91.5	95.6	97.9	93.3	95.9	99.4	91.6	96.0

WEST VIRGINIA

DIVISION		GASSAWAY 266			HUNTINGTON 255-257			LEWISBURG 249			MARTINSBURG 254			MORGANTOWN 265			PARKERSBURG 261		
		MAT.	INST.	TOTAL	MAT.	INST.	TOTAL	MAT.	INST.	TOTAL	MAT.	INST.	TOTAL	MAT.	INST.	TOTAL	MAT.	INST.	TOTAL
015433	CONTRACTOR EQUIPMENT		107.3	107.3		107.3	107.3		107.3	107.3		107.3	107.3		107.3	107.3		107.3	107.3
0241, 31 - 34	SITE & INFRASTRUCTURE, DEMOLITION	105.2	91.7	95.7	106.2	92.8	96.8	117.2	91.8	99.4	105.5	92.5	96.4	102.5	92.6	95.5	110.7	92.7	98.1
0310	Concrete Forming & Accessories	84.9	86.3	86.1	98.6	91.0	92.1	83.3	89.5	88.6	87.0	77.7	79.0	83.3	86.5	86.1	87.6	86.8	86.9
0320	Concrete Reinforcing	95.1	94.1	94.6	93.1	93.6	93.3	95.1	70.6	82.7	91.8	94.2	93.0	95.1	97.3	96.2	94.5	94.3	94.4
0330	Cast-in-Place Concrete	106.5	92.4	101.1	113.9	98.4	108.0	102.1	92.8	98.6	109.4	86.9	100.8	101.7	93.4	98.5	104.0	93.5	100.0
03	CONCRETE	98.7	90.9	95.1	101.2	95.0	98.3	105.0	88.4	97.4	99.6	84.8	92.8	95.1	91.9	93.6	100.4	91.5	96.3
04	MASONRY	109.4	84.8	94.1	91.9	96.7	94.9	95.5	85.9	89.6	94.6	87.6	90.2	126.6	86.5	101.7	82.0	88.8	86.2
05	METALS	101.2	105.8	102.6	105.3	106.6	105.7	101.2	97.3	100.0	103.2	99.5	102.1	101.3	107.2	103.1	102.0	106.1	103.2
06	WOOD, PLASTICS & COMPOSITES	84.9	85.9	85.4	95.8	90.0	92.6	82.6	90.5	87.0	85.1	75.3	79.6	82.6	85.9	84.4	85.7	85.4	85.5
07	THERMAL & MOISTURE PROTECTION	103.7	87.0	96.6	106.6	91.7	100.3	104.6	87.8	97.5	106.7	83.0	96.7	103.8	87.6	96.9	103.8	88.4	97.3
08	OPENINGS	98.2	84.1	95.0	95.6	86.2	93.4	100.1	81.3	95.8	98.2	68.9	91.4	101.4	84.8	97.6	99.0	83.9	95.5
0920	Plaster & Gypsum Board	96.1	85.4	88.9	106.2	89.6	95.1	94.1	90.1	91.4	99.3	74.4	82.6	94.1	85.4	88.3	97.1	84.9	88.9
0950, 0980	Ceilings & Acoustic Treatment	91.9	85.4	87.5	85.9	89.6	88.4	91.9	90.1	90.7	85.9	74.4	78.2	91.9	85.4	87.5	91.9	84.9	87.2
0960	Flooring	88.4	100.6	91.8	99.0	109.2	101.8	87.7	100.6	91.3	91.6	96.6	93.0	87.7	96.5	90.2	91.6	97.2	93.2
0970, 0990	Wall Finishes & Painting/Coating	101.5	91.7	95.7	92.6	91.7	92.1	101.5	67.4	81.6	92.6	87.5	89.7	101.5	87.9	93.6	101.5	87.5	93.3
09	FINISHES	89.7	89.6	89.6	91.8	94.6	93.3	90.4	89.9	90.2	88.8	81.6	84.9	89.0	88.3	88.6	91.1	88.6	89.8
COVERS	DIVS. 10 - 14, 25, 28, 41, 43, 44, 46	100.0	92.4	98.3	100.0	93.9	98.6	100.0	67.7	92.8	100.0	84.0	96.4	100.0	86.7	97.0	100.0	93.0	98.4
21, 22, 23	FIRE SUPPRESSION, PLUMBING & HVAC	97.3	86.5	92.7	100.6	89.8	96.0	97.3	91.2	94.7	97.5	84.0	91.7	97.3	91.2	94.7	100.3	89.2	95.6
26, 27, 3370	ELECTRICAL, COMMUNICATIONS & UTIL.	98.0	88.8	93.2	98.2	91.0	94.4	94.2	87.0	90.5	100.2	76.0	87.6	98.1	91.9	94.9	98.1	78.3	87.8
MF2016	WEIGHTED AVERAGE	98.8	89.9	94.9	99.7	93.8	97.1	99.1	88.9	94.6	98.8	84.4	92.5	99.4	91.6	96.0	98.8	89.5	94.8

For customer support on your Electrical Costs with RSMeans data, call 800.448.8182.

549

City Cost Indexes

WEST VIRGINIA / WISCONSIN

DIVISION		PETERSBURG 268			ROMNEY 267			WHEELING 260			BELOIT 535			EAU CLAIRE 547			GREEN BAY 541-543	
	MAT.	INST.	TOTAL	MAT.	INST.	TOTAL	MAT.	INST.	TOTAL	MAT.	INST.	TOTAL	MAT.	INST.	TOTAL	MAT.	INST.	TOTAL
015433 CONTRACTOR EQUIPMENT		107.3	107.3		107.3	107.3		107.3	107.3		101.9	101.9		102.7	102.7		100.3	100.3
0241, 31-34 SITE & INFRASTRUCTURE, DEMOLITION	101.7	92.5	95.3	104.6	92.6	96.2	111.4	92.6	98.2	95.8	106.6	103.4	97.0	104.3	102.1	100.3	99.8	99.9
0310 Concrete Forming & Accessories	86.4	77.8	79.0	82.5	78.2	78.8	89.1	86.8	87.1	96.4	92.3	92.9	95.0	105.8	104.4	103.6	105.6	105.3
0320 Concrete Reinforcing	94.5	94.1	94.3	95.1	94.3	94.7	93.9	97.3	95.6	95.5	131.6	113.7	91.4	110.6	101.1	89.4	106.3	97.9
0330 Cast-in-Place Concrete	101.7	87.0	96.1	106.5	87.1	99.1	104.0	96.6	101.2	106.4	99.1	103.6	100.2	99.8	100.0	103.6	103.8	103.7
03 CONCRETE	95.2	85.2	90.6	98.6	85.4	92.6	100.4	93.1	97.1	96.1	101.7	98.7	91.8	104.5	97.6	94.7	105.1	99.5
04 MASONRY	99.5	86.5	91.4	95.8	87.6	90.7	107.6	87.0	94.8	100.3	102.3	101.6	92.0	98.7	96.1	123.3	99.7	108.6
05 METALS	101.4	105.4	102.6	101.4	105.9	102.8	102.1	107.3	103.7	96.5	110.6	100.8	96.3	104.5	98.8	98.9	103.9	100.4
06 WOOD, PLASTICS & COMPOSITES	86.5	75.3	80.3	81.8	75.3	78.2	87.1	85.9	86.4	93.5	88.8	90.9	98.3	107.8	103.6	103.0	107.8	105.7
07 THERMAL & MOISTURE PROTECTION	103.8	82.9	94.9	103.9	83.9	95.4	104.2	88.8	97.7	100.7	94.0	97.9	103.5	98.9	101.6	105.8	100.8	103.7
08 OPENINGS	101.4	78.2	96.1	101.3	78.2	96.0	99.8	84.8	96.4	96.3	106.1	98.5	101.8	107.2	103.1	98.2	107.0	100.2
0920 Plaster & Gypsum Board	96.8	74.4	81.8	93.8	74.4	80.8	97.1	85.4	89.2	97.3	88.9	91.7	110.1	108.4	109.0	106.9	108.4	107.9
0950, 0980 Ceilings & Acoustic Treatment	91.9	74.4	80.1	91.9	74.4	80.1	91.9	85.4	87.5	86.3	88.9	88.1	90.3	108.4	102.5	84.6	108.4	100.7
0960 Flooring	89.5	96.5	91.4	87.5	98.5	90.6	92.5	96.5	93.6	91.3	125.2	100.7	80.8	112.6	89.6	96.5	121.4	103.4
0970, 0990 Wall Finishes & Painting/Coating	101.5	87.9	93.6	101.5	87.9	93.6	101.5	87.9	93.6	96.8	104.5	101.3	84.9	79.9	82.0	94.0	83.6	88.0
09 FINISHES	89.8	82.0	85.6	89.1	82.4	85.5	91.4	88.5	89.8	92.0	99.0	95.8	87.5	105.0	97.0	91.9	107.0	100.1
COVERS DIVS. 10-14, 25, 28, 41, 43, 44, 46	100.0	55.1	90.0	100.0	91.2	98.0	100.0	87.8	97.3	100.0	94.1	98.7	100.0	96.4	99.2	100.0	96.3	99.2
21, 22, 23 FIRE SUPPRESSION, PLUMBING & HVAC	97.3	86.7	92.8	97.3	86.6	92.7	100.4	91.4	96.6	100.1	95.8	98.3	100.1	87.3	94.7	100.4	85.6	94.1
26, 27, 3370 ELECTRICAL, COMMUNICATIONS & UTIL.	101.1	76.0	88.0	100.4	76.0	87.7	95.4	91.9	93.6	99.3	85.2	91.9	104.0	84.8	94.0	98.6	81.3	89.6
MF2016 WEIGHTED AVERAGE	98.5	85.0	92.6	98.6	86.4	93.3	100.0	91.9	96.5	97.7	98.6	98.1	97.6	97.3	97.5	99.6	96.5	98.2

WISCONSIN

DIVISION		KENOSHA 531			LA CROSSE 546			LANCASTER 538			MADISON 537			MILWAUKEE 530, 532			NEW RICHMOND 540	
	MAT.	INST.	TOTAL	MAT.	INST.	TOTAL	MAT.	INST.	TOTAL	MAT.	INST.	TOTAL	MAT.	INST.	TOTAL	MAT.	INST.	TOTAL
015433 CONTRACTOR EQUIPMENT		99.8	99.8		102.7	102.7		101.9	101.9		101.9	101.9		89.4	89.4		103.0	103.0
0241, 31-34 SITE & INFRASTRUCTURE, DEMOLITION	101.3	104.1	103.3	90.7	104.3	100.2	94.9	106.4	103.0	93.9	107.3	103.3	93.2	95.3	94.6	95.9	104.4	101.9
0310 Concrete Forming & Accessories	104.3	111.2	110.3	82.6	105.7	102.5	95.8	96.3	96.2	100.5	95.7	96.3	100.8	112.4	110.8	90.5	93.1	92.8
0320 Concrete Reinforcing	95.3	113.9	104.7	91.1	103.0	97.1	97.0	102.8	99.9	95.7	103.2	99.5	102.9	114.2	108.6	88.4	110.2	99.4
0330 Cast-in-Place Concrete	115.7	108.3	112.9	90.1	101.6	94.5	105.8	98.7	103.1	102.4	104.8	103.3	93.9	110.9	100.4	104.3	79.8	95.0
03 CONCRETE	101.0	110.4	105.3	83.4	103.8	92.7	95.9	98.4	97.0	93.7	100.3	96.7	95.7	111.2	102.8	89.8	91.9	90.7
04 MASONRY	97.8	110.7	105.8	91.1	98.7	95.8	100.3	95.9	97.6	95.6	99.6	98.1	105.8	114.7	111.3	118.1	95.7	104.2
05 METALS	97.4	105.5	99.9	96.2	102.5	98.1	93.9	99.3	95.5	98.3	100.6	99.0	97.9	97.4	97.8	96.6	103.5	98.7
06 WOOD, PLASTICS & COMPOSITES	97.3	110.8	104.8	84.3	107.8	97.4	92.8	95.7	94.4	93.9	93.4	93.6	98.8	111.3	105.7	87.8	91.7	90.0
07 THERMAL & MOISTURE PROTECTION	100.9	107.2	103.6	103.0	98.7	101.1	100.5	93.2	97.4	98.3	101.2	99.5	102.8	109.9	105.8	104.6	88.3	97.7
08 OPENINGS	90.9	110.8	95.5	101.8	100.2	101.4	92.3	93.5	92.5	102.3	98.3	101.3	101.0	111.1	103.3	87.9	91.8	88.8
0920 Plaster & Gypsum Board	87.3	111.6	103.6	105.0	108.4	107.3	96.5	96.1	96.2	104.3	93.6	97.1	100.3	111.6	107.9	95.6	92.0	93.2
0950, 0980 Ceilings & Acoustic Treatment	86.3	111.6	103.4	89.5	108.4	102.2	82.9	96.1	91.8	89.6	93.6	92.3	96.5	111.6	106.7	57.0	92.0	80.5
0960 Flooring	108.3	121.4	112.0	74.8	118.0	86.8	90.8	105.3	94.9	89.5	113.7	96.2	96.6	117.4	102.3	89.6	109.8	95.2
0970, 0990 Wall Finishes & Painting/Coating	108.0	118.8	114.3	84.9	81.2	82.8	96.8	102.1	99.9	90.1	104.5	98.5	106.8	123.4	116.5	96.2	85.4	89.9
09 FINISHES	96.8	114.3	106.3	84.3	106.3	96.3	91.1	99.0	95.4	91.7	99.6	96.0	100.5	114.7	108.2	82.4	95.7	89.7
COVERS DIVS. 10-14, 25, 28, 41, 43, 44, 46	100.0	101.5	100.3	100.0	96.4	99.2	100.0	86.7	97.0	100.0	99.7	99.9	100.0	102.9	100.6	100.0	88.9	97.5
21, 22, 23 FIRE SUPPRESSION, PLUMBING & HVAC	100.3	97.0	98.9	100.1	87.3	94.6	97.0	87.1	92.8	100.0	96.7	98.6	99.9	105.5	102.3	96.5	87.4	92.6
26, 27, 3370 ELECTRICAL, COMMUNICATIONS & UTIL.	99.6	97.3	98.4	104.3	84.8	94.2	99.0	84.8	91.6	100.0	94.1	96.9	99.1	101.5	100.4	102.1	84.8	93.1
MF2016 WEIGHTED AVERAGE	98.4	105.0	101.3	95.9	96.9	96.4	96.0	93.9	95.1	98.0	98.9	98.4	99.3	106.6	102.5	95.6	92.7	94.3

WISCONSIN

DIVISION		OSHKOSH 549			PORTAGE 539			RACINE 534			RHINELANDER 545			SUPERIOR 548			WAUSAU 544	
	MAT.	INST.	TOTAL	MAT.	INST.	TOTAL	MAT.	INST.	TOTAL	MAT.	INST.	TOTAL	MAT.	INST.	TOTAL	MAT.	INST.	TOTAL
015433 CONTRACTOR EQUIPMENT		100.3	100.3		101.9	101.9		101.9	101.9		100.3	100.3		103.0	103.0		100.3	100.3
0241, 31-34 SITE & INFRASTRUCTURE, DEMOLITION	92.1	99.4	97.2	85.8	106.2	100.1	95.6	108.1	104.3	104.1	99.5	100.9	92.6	104.1	100.7	87.9	100.7	96.9
0310 Concrete Forming & Accessories	87.1	89.6	89.3	88.5	91.5	91.1	96.8	111.3	109.3	84.9	90.3	89.6	88.6	89.6	89.5	86.6	105.0	102.4
0320 Concrete Reinforcing	89.6	98.5	94.1	97.1	102.9	100.0	95.5	114.0	104.8	89.7	96.8	93.3	88.4	106.9	97.8	89.7	102.8	96.3
0330 Cast-in-Place Concrete	96.1	94.2	95.4	90.8	99.3	94.0	104.3	108.1	105.8	108.9	96.9	104.4	98.3	101.9	99.6	89.6	95.8	91.9
03 CONCRETE	85.1	93.2	88.8	83.9	96.4	89.6	95.2	110.4	102.1	96.6	94.2	95.5	84.6	97.3	90.4	80.3	101.4	90.0
04 MASONRY	105.5	97.3	100.4	99.1	97.0	97.8	100.2	110.7	106.7	123.9	94.7	105.8	117.4	100.8	107.1	105.0	94.8	98.7
05 METALS	96.8	99.4	97.6	94.6	99.7	96.2	98.2	105.5	100.4	96.6	99.1	97.4	97.6	103.8	99.5	96.5	101.8	98.1
06 WOOD, PLASTICS & COMPOSITES	83.9	88.9	86.7	83.7	88.8	86.5	93.7	110.8	103.2	81.6	88.9	85.7	86.3	87.5	87.0	83.4	107.8	97.0
07 THERMAL & MOISTURE PROTECTION	104.8	82.2	95.2	99.9	88.9	95.3	100.8	106.8	103.3	105.7	82.7	96.0	104.2	89.4	98.0	104.6	84.2	96.2
08 OPENINGS	94.6	89.4	93.4	92.4	95.4	93.1	96.3	110.8	99.6	94.6	88.2	93.2	87.4	93.4	88.8	94.8	100.6	96.2
0920 Plaster & Gypsum Board	94.4	88.9	90.7	89.6	88.9	89.1	97.3	111.6	106.9	94.4	88.9	90.7	95.4	87.7	90.2	94.4	108.4	103.8
0950, 0980 Ceilings & Acoustic Treatment	84.6	88.9	87.5	85.4	88.9	87.8	86.3	111.6	103.4	84.6	88.9	87.5	57.8	87.7	78.0	84.6	108.4	100.7
0960 Flooring	87.6	121.4	97.0	87.2	114.8	94.9	91.3	121.4	99.6	87.0	111.7	93.9	90.7	127.2	100.8	87.5	111.7	94.2
0970, 0990 Wall Finishes & Painting/Coating	90.9	83.6	86.6	96.8	102.1	99.9	96.8	121.2	111.0	90.9	83.6	86.6	85.3	110.6	100.1	90.9	83.6	86.6
09 FINISHES	86.7	95.2	91.3	89.0	96.6	93.1	92.0	114.6	104.3	87.5	92.7	90.3	81.8	98.6	90.9	86.4	105.1	96.5
COVERS DIVS. 10-14, 25, 28, 41, 43, 44, 46	100.0	86.4	97.0	100.0	87.4	97.2	100.0	101.5	100.3	100.0	87.6	97.2	100.0	87.2	97.2	100.0	96.3	99.2
21, 22, 23 FIRE SUPPRESSION, PLUMBING & HVAC	97.3	80.9	90.3	97.0	96.5	96.8	100.1	97.1	98.8	97.3	86.4	92.6	96.5	91.5	94.4	97.3	86.8	92.8
26, 27, 3370 ELECTRICAL, COMMUNICATIONS & UTIL.	102.6	76.3	88.9	102.7	94.1	98.2	99.1	103.5	101.4	102.0	77.6	89.3	106.9	98.7	102.6	103.8	80.2	91.5
MF2016 WEIGHTED AVERAGE	95.7	89.2	92.9	94.4	96.6	95.4	97.8	106.2	101.5	98.3	90.1	94.7	95.4	97.1	96.1	95.0	94.5	94.8

For customer support on your Electrical Costs with RSMeans data, call 800.448.8182.

City Cost Indexes

WYOMING

DIVISION		CASPER 826 MAT.	INST.	TOTAL	CHEYENNE 820 MAT.	INST.	TOTAL	NEWCASTLE 827 MAT.	INST.	TOTAL	RAWLINS 823 MAT.	INST.	TOTAL	RIVERTON 825 MAT.	INST.	TOTAL	ROCK SPRINGS 829 - 831 MAT.	INST.	TOTAL
015433	CONTRACTOR EQUIPMENT		97.4	97.4		97.4	97.4		97.4	97.4		97.4	97.4		97.4	97.4		97.4	97.4
0241, 31 - 34	SITE & INFRASTRUCTURE, DEMOLITION	100.6	94.7	96.5	96.1	94.7	95.1	88.0	94.2	92.4	102.6	94.2	96.7	95.8	94.1	94.6	92.2	94.2	93.6
0310	Concrete Forming & Accessories	100.5	55.2	61.4	101.0	68.4	72.9	92.2	73.8	76.4	96.3	74.1	77.1	91.3	62.9	66.8	98.3	63.9	68.7
0320	Concrete Reinforcing	111.9	85.2	98.4	107.4	85.3	96.2	115.3	85.0	100.0	115.0	85.0	99.9	116.1	85.0	100.4	116.1	83.2	99.5
0330	Cast-in-Place Concrete	104.4	77.5	94.2	98.6	77.6	90.6	99.6	76.5	90.8	99.6	76.7	90.9	99.6	76.5	90.8	99.6	76.5	90.8
03	CONCRETE	100.5	68.9	86.1	98.0	74.9	87.5	98.3	77.0	88.6	111.3	77.1	95.7	106.2	72.0	90.6	98.8	72.2	86.6
04	MASONRY	103.9	66.1	80.5	107.3	67.3	82.5	103.9	61.8	77.8	103.9	61.8	77.8	103.9	61.8	77.8	168.0	57.5	99.5
05	METALS	103.6	81.5	96.8	106.1	81.8	98.6	102.3	81.2	95.8	102.3	81.4	95.9	102.4	81.1	95.9	103.2	80.3	96.1
06	WOOD, PLASTICS & COMPOSITES	94.7	49.8	69.7	91.0	67.7	78.0	82.2	76.2	78.8	85.8	76.2	80.4	81.2	61.5	70.3	90.5	62.8	75.1
07	THERMAL & MOISTURE PROTECTION	102.5	67.1	87.5	97.1	69.4	85.4	98.7	66.3	85.0	100.3	77.9	90.8	99.7	64.8	84.9	98.8	67.9	85.7
08	OPENINGS	103.1	61.3	93.5	104.0	71.2	96.5	108.1	75.9	100.7	107.7	75.9	100.4	107.9	67.8	98.7	108.4	68.1	99.2
0920	Plaster & Gypsum Board	103.4	48.5	66.5	91.7	66.8	75.0	88.5	75.5	79.8	88.9	75.5	79.9	88.5	60.5	69.7	108.4	68.1	99.2
0950, 0980	Ceilings & Acoustic Treatment	103.1	48.5	66.3	98.7	66.8	77.2	100.6	75.5	83.7	100.6	75.5	83.7	100.6	60.5	73.6	100.6	61.8	74.5
0960	Flooring	105.8	73.6	96.9	105.0	73.6	96.3	98.7	47.1	84.4	101.6	47.1	86.5	98.2	63.7	88.7	103.9	47.1	88.1
0970, 0990	Wall Finishes & Painting/Coating	91.5	61.0	73.7	97.4	61.0	76.2	94.1	61.0	74.8	94.1	61.0	74.8	94.1	61.0	74.8	94.1	61.0	74.8
09	FINISHES	99.0	57.8	76.6	95.8	68.3	80.8	91.0	67.6	78.3	93.3	67.6	79.3	91.7	62.3	75.7	94.2	59.7	75.4
COVERS	DIVS. 10 - 14, 25, 28, 41, 43, 44, 46	100.0	84.5	96.6	100.0	86.5	97.0	100.0	94.5	98.8	100.0	94.5	98.8	100.0	80.5	95.7	100.0	84.7	96.6
21, 22, 23	FIRE SUPPRESSION, PLUMBING & HVAC	100.1	72.6	88.3	100.0	72.6	88.3	98.3	71.8	87.0	98.3	71.8	87.0	98.3	71.7	87.0	100.0	80.9	90.5
26, 27, 3370	ELECTRICAL, COMMUNICATIONS & UTIL.	101.9	63.7	82.0	104.0	64.6	83.5	102.7	63.3	82.2	102.7	63.3	82.2	102.7	62.4	81.8	99.9	71.7	87.9
MF2016	WEIGHTED AVERAGE	101.4	70.3	87.8	101.3	73.5	89.2	99.9	73.1	88.2	102.1	73.5	89.6	101.2	70.7	87.8	103.9	72.7	90.2

WYOMING / CANADA

DIVISION		SHERIDAN 828 MAT.	INST.	TOTAL	WHEATLAND 822 MAT.	INST.	TOTAL	WORLAND 824 MAT.	INST.	TOTAL	YELLOWSTONE NAT'L PA 821 MAT.	INST.	TOTAL	BARRIE, ONTARIO MAT.	INST.	TOTAL	BATHURST, NEW BRUNSWICK MAT.	INST.	TOTAL
015433	CONTRACTOR EQUIPMENT		97.4	97.4		97.4	97.4		97.4	97.4		97.4	97.4		103.0	103.0		102.8	102.8
0241, 31 - 34	SITE & INFRASTRUCTURE, DEMOLITION	96.2	94.7	95.1	93.0	94.2	93.8	90.2	94.2	93.0	90.3	94.2	93.0	119.2	99.9	105.7	107.0	95.7	99.1
0310	Concrete Forming & Accessories	99.0	63.7	68.6	94.1	50.6	56.6	94.1	63.9	68.0	94.2	63.9	68.1	124.2	86.3	91.5	103.6	60.6	66.5
0320	Concrete Reinforcing	116.1	85.0	100.4	115.3	84.4	99.8	116.1	85.0	100.4	118.1	85.0	101.4	175.3	89.7	132.1	137.7	59.4	98.2
0330	Cast-in-Place Concrete	103.0	77.5	93.3	103.9	76.4	93.5	99.6	76.5	90.8	99.6	76.5	90.8	158.9	86.5	131.4	117.7	59.1	95.4
03	CONCRETE	106.3	72.7	91.0	103.2	66.3	86.3	98.5	72.4	86.6	98.8	72.5	86.8	139.8	87.3	115.8	113.6	60.8	89.5
04	MASONRY	104.2	63.3	78.9	104.3	53.4	72.7	103.9	61.8	77.8	104.0	61.8	77.8	167.9	93.1	121.5	162.5	59.9	98.9
05	METALS	105.9	81.3	98.3	102.2	80.3	95.5	102.4	81.1	95.9	103.0	80.6	96.1	110.3	93.5	105.2	113.7	75.0	101.8
06	WOOD, PLASTICS & COMPOSITES	92.2	63.7	75.1	83.9	45.0	62.2	83.9	62.8	72.2	83.9	62.8	72.2	118.0	84.7	99.5	98.4	60.6	77.4
07	THERMAL & MOISTURE PROTECTION	100.0	67.6	86.3	99.0	60.3	82.6	98.8	66.3	85.0	98.2	66.3	84.7	113.8	90.3	103.9	109.4	61.3	89.0
08	OPENINGS	108.6	67.8	99.2	106.6	58.6	95.6	108.3	68.5	99.1	101.3	68.1	93.7	92.4	85.0	90.7	86.4	53.8	78.9
0920	Plaster & Gypsum Board	112.5	60.5	77.6	88.5	43.5	58.3	88.5	61.8	70.6	88.7	61.8	70.7	151.5	84.2	106.3	126.6	59.5	81.5
0950, 0980	Ceilings & Acoustic Treatment	103.5	60.5	74.5	100.6	43.5	62.1	100.6	61.8	74.5	101.5	61.8	74.7	95.3	84.2	87.9	112.5	59.5	76.7
0960	Flooring	102.5	63.7	91.7	100.3	46.3	85.3	100.3	47.1	85.5	100.3	47.1	85.5	118.1	92.3	110.9	98.6	43.8	83.4
0970, 0990	Wall Finishes & Painting/Coating	96.3	61.0	75.7	94.1	61.0	74.8	94.1	61.0	74.8	94.1	61.0	74.8	104.4	87.6	94.6	109.9	50.1	75.0
09	FINISHES	98.6	63.0	79.2	91.8	49.0	68.5	91.5	59.7	74.2	91.7	59.8	74.4	109.8	87.7	97.8	104.7	56.5	78.5
COVERS	DIVS. 10 - 14, 25, 28, 41, 43, 44, 46	100.0	85.8	96.8	100.0	83.5	96.3	100.0	79.3	95.4	100.0	79.4	95.4	109.8	87.7	97.8	104.7	56.5	78.5
21, 22, 23	FIRE SUPPRESSION, PLUMBING & HVAC	98.3	72.6	87.3	98.3	71.7	86.9	98.3	71.7	87.0	98.3	71.7	87.0	139.2	67.6	123.3	131.1	60.2	115.4
26, 27, 3370	ELECTRICAL, COMMUNICATIONS & UTIL.	105.3	62.4	83.0	102.7	80.9	91.4	102.7	80.9	91.4	101.6	90.9	96.1	116.1	87.6	101.3	104.6	67.5	88.7
MF2016	WEIGHTED AVERAGE	102.8	71.5	89.1	100.6	69.2	86.9	100.1	73.0	88.2	99.4	74.4	88.4	117.0	81.0	105.6	111.3	65.0	91.0

CANADA

DIVISION		BRANDON, MANITOBA MAT.	INST.	TOTAL	BRANTFORD, ONTARIO MAT.	INST.	TOTAL	BRIDGEWATER, NOVA SCOTIA MAT.	INST.	TOTAL	CALGARY, ALBERTA MAT.	INST.	TOTAL	CAP-DE-LA-MADELEINE, QUEBEC MAT.	INST.	TOTAL	CHARLESBOURG, QUEBEC MAT.	INST.	TOTAL
015433	CONTRACTOR EQUIPMENT		105.8	105.8		103.1	103.1		102.6	102.6		128.2	128.2		103.3	103.3		103.3	103.3
0241, 31 - 34	SITE & INFRASTRUCTURE, DEMOLITION	134.2	98.8	109.4	119.6	100.3	106.1	103.7	97.5	99.3	123.1	120.8	121.5	99.2	98.9	99.0	99.2	98.9	99.0
0310	Concrete Forming & Accessories	143.5	69.4	79.6	123.5	93.5	97.6	96.8	71.2	74.7	123.2	99.2	102.5	129.2	83.0	89.4	129.2	83.0	89.4
0320	Concrete Reinforcing	192.0	64.9	123.5	169.0	88.3	128.3	144.0	49.2	96.2	135.2	84.5	109.6	144.0	74.8	109.1	144.0	74.8	109.1
0330	Cast-in-Place Concrete	118.3	74.1	101.5	135.8	107.5	125.0	139.5	70.4	113.3	143.2	108.3	129.9	109.6	92.2	103.0	109.6	92.2	103.0
03	CONCRETE	127.0	69.7	100.8	127.0	97.6	113.6	125.3	68.0	99.1	129.2	100.5	116.1	111.7	85.2	99.6	111.7	85.2	99.6
04	MASONRY	221.4	63.3	123.3	168.8	97.4	124.5	164.2	68.4	104.8	200.4	90.6	132.3	164.9	79.6	112.0	164.9	79.6	112.0
05	METALS	133.5	80.4	117.2	112.4	94.3	106.8	111.7	77.6	101.2	129.5	104.8	121.9	110.8	86.9	103.5	110.8	86.9	103.5
06	WOOD, PLASTICS & COMPOSITES	150.9	70.0	105.9	119.8	92.0	104.3	90.7	70.6	79.5	96.4	98.8	97.8	130.5	82.8	103.9	130.5	82.8	103.9
07	THERMAL & MOISTURE PROTECTION	131.5	72.1	106.4	118.6	96.0	109.0	113.2	70.9	95.3	129.9	97.8	116.3	111.9	88.3	101.9	111.9	88.3	101.9
08	OPENINGS	102.5	63.0	93.2	90.3	91.1	90.5	84.5	64.4	79.9	83.5	87.8	84.4	91.6	76.0	88.0	91.6	76.0	88.0
0920	Plaster & Gypsum Board	113.1	68.8	83.3	116.6	91.7	99.9	122.4	69.8	87.1	121.5	97.9	105.6	144.8	82.2	102.8	144.8	82.2	102.8
0950, 0980	Ceilings & Acoustic Treatment	119.2	68.8	85.2	103.2	91.7	95.5	103.2	69.8	80.6	142.1	97.9	112.3	103.2	82.2	89.0	103.2	82.2	89.0
0960	Flooring	130.3	65.4	112.3	112.1	92.3	106.6	94.7	62.3	85.7	117.9	88.6	109.8	112.1	90.9	106.2	112.1	90.9	106.2
0970, 0990	Wall Finishes & Painting/Coating	117.0	56.5	81.8	109.0	96.2	101.5	109.0	62.0	81.6	117.1	111.2	113.7	109.0	87.4	96.4	109.0	87.4	96.4
09	FINISHES	118.4	67.6	90.8	106.3	93.8	99.5	101.1	68.8	83.6	122.2	99.1	109.6	109.4	85.2	96.3	109.4	85.2	96.3
COVERS	DIVS. 10 - 14, 25, 28, 41, 43, 44, 46	131.1	62.7	115.9	131.1	66.9	117.5	131.1	63.3	116.0	131.1	95.5	123.2	131.1	78.8	119.5	131.1	78.8	119.5
21, 22, 23	FIRE SUPPRESSION, PLUMBING & HVAC	104.7	82.2	95.1	104.6	100.4	102.9	104.6	82.8	95.3	105.5	93.3	100.3	105.1	87.7	97.6	105.1	87.7	97.6
26, 27, 3370	ELECTRICAL, COMMUNICATIONS & UTIL.	111.1	66.6	88.0	111.4	87.1	98.7	116.0	62.0	87.9	107.3	99.9	103.1	110.4	68.9	88.8	110.4	68.9	88.8
MF2016	WEIGHTED AVERAGE	123.6	73.8	101.8	114.3	94.9	105.8	112.3	73.4	95.3	119.5	98.8	110.4	111.7	83.6	99.4	111.7	83.6	99.4

For customer support on your Electrical Costs with RSMeans data, call 800.448.8182.

551

CANADA

DIVISION		CHARLOTTETOWN, PRINCE EDWARD ISLAND			CHICOUTIMI, QUEBEC			CORNER BROOK, NEWFOUNDLAND			CORNWALL, ONTARIO			DALHOUSIE, NEW BRUNSWICK			DARTMOUTH, NOVA SCOTIA		
		MAT.	INST.	TOTAL	MAT.	INST.	TOTAL	MAT.	INST.	TOTAL	MAT.	INST.	TOTAL	MAT.	INST.	TOTAL	MAT.	INST.	TOTAL
015433	CONTRACTOR EQUIPMENT		120.7	120.7		103.4	103.4		104.0	104.0		103.1	103.1		102.6	102.6		102.9	102.9
0241, 31 - 34	SITE & INFRASTRUCTURE, DEMOLITION	133.6	108.5	116.0	103.9	98.6	100.2	138.7	96.1	108.8	117.7	99.7	105.1	102.8	95.5	97.7	110.5	71.2	76.6
0310	Concrete Forming & Accessories	118.4	55.4	64.1	131.9	91.0	96.7	120.7	79.3	85.0	121.3	86.5	91.3	102.4	60.8	66.5	184.1	49.2	116.0
0320	Concrete Reinforcing	152.2	48.0	99.6	103.9	94.2	99.0	176.3	50.4	112.8	169.0	88.0	128.2	147.3	59.4	102.9	115.9	59.1	94.3
0330	Cast-in-Place Concrete	149.0	59.1	114.9	110.8	97.0	105.6	137.7	67.0	110.8	122.1	97.9	112.9	118.1	60.9	91.9	141.6	68.1	108.0
03	CONCRETE	134.7	57.3	99.3	104.6	93.8	99.7	159.1	70.7	118.7	120.4	91.1	107.0	164.5	59.9	99.6	232.1	68.4	130.6
04	MASONRY	177.7	57.2	103.0	164.8	90.3	118.6	217.1	77.5	130.5	167.6	89.3	119.0	106.9	75.1	97.2	134.6	77.8	117.1
05	METALS	137.2	81.7	120.2	113.4	92.4	107.0	133.9	77.1	116.5	112.2	93.0	106.3	106.9	75.1	97.2	117.5	70.6	91.4
06	WOOD, PLASTICS & COMPOSITES	99.9	54.8	74.8	131.2	91.4	109.0	129.3	85.4	104.9	118.1	85.5	100.0	97.5	60.6	77.0	133.2	71.0	106.9
07	THERMAL & MOISTURE PROTECTION	128.5	59.1	99.1	109.8	98.0	104.8	136.2	69.2	107.9	118.4	90.5	106.6	116.2	61.3	92.9	133.2	71.0	106.9
08	OPENINGS	85.1	48.0	76.6	90.5	77.8	87.5	108.7	70.0	99.8	91.6	84.7	90.0	87.3	53.8	79.6	92.6	64.4	86.1
0920	Plaster & Gypsum Board	119.5	53.0	74.9	144.8	91.0	108.7	147.0	84.8	105.3	173.2	85.1	114.1	130.5	59.5	82.8	142.6	69.8	93.7
0950, 0980	Ceilings & Acoustic Treatment	128.0	53.0	77.4	111.7	91.0	97.8	120.0	84.8	96.3	105.7	85.1	91.8	106.5	59.5	74.8	126.9	69.8	88.4
0960	Flooring	106.3	58.4	93.0	114.3	90.9	107.8	112.4	52.4	95.7	112.1	90.9	106.2	100.6	67.1	91.3	107.3	62.3	94.8
0970, 0990	Wall Finishes & Painting/Coating	117.5	41.7	73.3	109.9	105.2	107.1	116.9	59.0	83.1	109.0	89.6	97.7	112.5	50.1	76.1	116.9	62.0	84.9
09	FINISHES	116.7	54.8	83.0	111.5	93.4	101.7	118.1	73.7	94.0	114.4	87.8	99.9	105.9	61.1	81.6	116.0	68.8	90.3
COVERS	DIVS. 10 - 14, 25, 28, 41, 43, 44, 46	131.1	60.4	115.4	131.1	82.5	120.3	131.1	63.1	116.0	131.1	67.3	116.9	131.1	60.1	115.3	131.1	63.3	116.1
21, 22, 23	FIRE SUPPRESSION, PLUMBING & HVAC	104.9	61.4	86.3	104.6	81.5	94.7	104.7	68.7	89.3	105.1	98.3	102.2	104.7	67.5	88.8	104.7	82.8	95.4
26, 27, 3370	ELECTRICAL, COMMUNICATIONS & UTIL.	114.3	50.1	80.9	108.7	85.6	96.7	108.8	54.4	80.5	111.7	88.0	99.4	113.9	55.8	83.6	113.0	62.0	86.5
MF2016	WEIGHTED AVERAGE	120.9	62.8	95.5	111.1	88.9	101.4	128.0	71.5	103.3	114.2	91.3	104.2	111.3	66.5	91.1	124.6	73.5	102.2

CANADA

DIVISION		EDMONTON, ALBERTA			FORT MCMURRAY, ALBERTA			FREDERICTON, NEW BRUNSWICK			GATINEAU, QUEBEC			GRANBY, QUEBEC			HALIFAX, NOVA SCOTIA		
		MAT.	INST.	TOTAL	MAT.	INST.	TOTAL	MAT.	INST.	TOTAL	MAT.	INST.	TOTAL	MAT.	INST.	TOTAL	MAT.	INST.	TOTAL
015433	CONTRACTOR EQUIPMENT		128.5	128.5		105.5	105.5		120.7	120.7		103.3	103.3		103.3	103.3	118.3	118.3	118.3
0241, 31 - 34	SITE & INFRASTRUCTURE, DEMOLITION	129.3	121.0	123.5	124.8	101.0	108.1	113.6	110.3	111.3	99.0	98.9	99.0	99.5	98.9	99.1	117.4	85.0	89.4
0310	Concrete Forming & Accessories	119.7	99.2	102.0	121.4	91.7	95.7	120.2	61.4	69.5	129.2	82.9	89.3	129.2	82.9	89.3	152.2	72.3	111.9
0320	Concrete Reinforcing	135.7	84.5	109.9	156.5	84.3	120.1	151.1	59.6	105.0	152.2	74.8	113.1	152.2	74.7	113.1	152.2	72.3	111.9
0330	Cast-in-Place Concrete	147.7	108.4	132.7	180.3	103.5	151.1	116.8	59.5	95.0	108.1	92.2	102.1	111.7	92.2	104.3	106.9	86.0	98.9
03	CONCRETE	131.2	100.6	117.2	146.3	94.7	122.7	119.4	62.1	93.2	112.1	85.2	99.8	113.8	85.1	100.7	114.0	84.3	100.4
04	MASONRY	196.0	90.6	130.7	209.5	87.3	133.7	182.5	61.2	107.3	164.8	79.6	112.0	165.1	79.6	112.1	138.4	98.1	126.0
05	METALS	136.2	105.0	126.6	139.9	92.8	125.4	134.2	87.1	119.7	110.8	86.8	103.4	111.0	86.7	103.6	100.5	84.0	91.3
06	WOOD, PLASTICS & COMPOSITES	96.8	98.8	97.9	114.2	90.9	101.2	109.4	60.9	82.4	130.5	82.8	103.9	130.5	82.8	103.9	129.0	86.1	110.8
07	THERMAL & MOISTURE PROTECTION	129.7	97.8	116.2	127.1	93.9	113.0	123.7	61.3	97.3	111.9	88.3	101.9	111.9	86.7	101.2	129.0	86.1	110.8
08	OPENINGS	81.8	87.8	83.2	91.6	83.4	89.7	86.1	52.8	78.5	91.6	71.4	87.0	91.6	71.4	87.0	88.6	76.8	85.9
0920	Plaster & Gypsum Board	116.7	97.9	104.1	117.4	90.3	99.2	123.5	59.5	80.5	114.5	82.2	92.8	114.5	82.2	92.8	118.1	83.3	94.7
0950, 0980	Ceilings & Acoustic Treatment	137.2	97.9	110.7	110.8	90.3	97.0	133.9	59.5	83.7	103.2	82.2	89.0	103.2	82.2	89.0	128.5	83.3	98.0
0960	Flooring	114.0	88.6	107.0	112.1	88.6	105.6	109.5	70.0	98.6	112.1	90.9	106.2	112.1	90.9	106.2	99.6	83.8	95.2
0970, 0990	Wall Finishes & Painting/Coating	108.3	111.2	110.0	109.1	92.3	99.3	114.8	63.8	85.1	109.0	87.4	96.4	109.0	87.4	96.4	112.8	91.7	100.5
09	FINISHES	115.6	99.1	106.6	109.2	91.4	99.5	118.5	63.4	88.5	105.3	85.2	94.4	105.3	85.2	94.4	110.4	85.9	97.1
COVERS	DIVS. 10 - 14, 25, 28, 41, 43, 44, 46	131.1	95.5	123.2	131.1	92.3	122.5	131.1	60.7	115.5	131.1	78.8	119.5	131.1	78.8	119.5	131.1	68.7	117.2
21, 22, 23	FIRE SUPPRESSION, PLUMBING & HVAC	105.5	93.3	100.3	105.1	97.1	101.7	105.0	76.7	92.9	105.1	87.7	97.6	104.6	87.7	97.4	105.3	82.4	95.5
26, 27, 3370	ELECTRICAL, COMMUNICATIONS & UTIL.	110.3	99.1	104.5	105.6	81.8	93.3	111.9	73.6	92.0	110.4	68.9	88.8	111.0	68.9	89.1	109.4	87.0	97.7
MF2016	WEIGHTED AVERAGE	120.4	98.8	110.9	123.4	91.9	109.7	118.2	72.5	98.2	111.4	83.4	99.1	111.6	83.3	99.2	117.4	87.2	104.2

CANADA

DIVISION		HAMILTON, ONTARIO			HULL, QUEBEC			JOLIETTE, QUEBEC			KAMLOOPS, BRITISH COLUMBIA			KINGSTON, ONTARIO			KITCHENER, ONTARIO		
		MAT.	INST.	TOTAL	MAT.	INST.	TOTAL	MAT.	INST.	TOTAL	MAT.	INST.	TOTAL	MAT.	INST.	TOTAL	MAT.	INST.	TOTAL
015433	CONTRACTOR EQUIPMENT		118.4	118.4		103.3	103.3		103.3	103.3		106.7	106.7		105.4	105.4		105.3	105.3
0241, 31 - 34	SITE & INFRASTRUCTURE, DEMOLITION	112.1	113.0	112.7	99.0	98.9	99.0	99.7	98.9	99.2	122.5	102.7	108.6	117.7	103.7	107.9	113.1	88.6	92.0
0310	Concrete Forming & Accessories	122.8	95.7	99.5	129.2	82.9	89.3	129.2	83.0	89.4	121.4	86.3	91.1	121.4	86.5	91.3	104.2	98.7	101.4
0320	Concrete Reinforcing	150.1	98.8	124.2	152.2	74.8	113.1	144.0	74.8	109.1	112.8	78.3	95.4	169.0	88.0	128.2	113.4	99.5	108.1
0330	Cast-in-Place Concrete	118.5	103.5	112.8	108.1	92.2	102.1	112.6	92.0	104.9	97.5	96.8	97.2	122.1	97.9	112.9	100.5	94.4	97.7
03	CONCRETE	116.3	99.6	108.7	112.1	85.2	99.8	113.2	85.2	100.4	121.2	88.9	106.4	122.3	91.1	108.0	100.5	94.4	97.7
04	MASONRY	170.9	101.7	128.0	164.8	79.6	112.0	165.2	79.6	112.1	172.2	87.6	119.7	174.7	89.3	121.8	145.6	99.0	113.8
05	METALS	131.7	106.4	123.9	111.0	86.8	103.6	111.0	86.9	103.6	112.8	89.2	105.6	113.5	92.9	107.2	121.0	97.5	113.8
06	WOOD, PLASTICS & COMPOSITES	99.9	94.9	97.0	130.5	82.8	103.9	130.5	82.8	103.9	102.5	84.5	92.5	118.1	85.6	100.0	107.6	86.4	95.8
07	THERMAL & MOISTURE PROTECTION	122.0	101.6	113.4	111.9	88.3	101.9	111.9	88.3	101.9	128.2	85.8	110.3	118.4	91.6	107.1	115.0	98.8	108.2
08	OPENINGS	85.6	93.8	87.5	91.6	71.4	87.0	91.6	76.0	88.0	88.4	82.5	87.1	91.6	84.4	90.0	83.1	87.4	84.1
0920	Plaster & Gypsum Board	121.1	94.1	103.0	114.5	82.2	92.8	144.8	82.2	102.8	102.2	83.7	89.8	176.2	85.2	115.1	105.6	86.0	92.4
0950, 0980	Ceilings & Acoustic Treatment	130.0	94.1	105.8	103.2	82.2	89.0	103.2	82.2	89.0	103.2	83.7	90.0	118.5	85.2	96.1	105.1	86.0	92.2
0960	Flooring	109.7	97.8	106.4	112.1	90.9	106.2	112.1	90.9	106.2	111.0	52.0	94.6	112.1	90.9	106.2	100.1	97.8	99.4
0970, 0990	Wall Finishes & Painting/Coating	102.9	105.4	104.3	109.0	87.4	96.4	109.0	87.4	96.4	109.0	79.9	92.0	109.0	82.8	93.7	103.1	95.1	98.4
09	FINISHES	113.5	97.1	104.6	105.3	85.2	94.4	109.4	85.2	96.3	105.8	79.6	91.5	117.3	87.1	100.9	99.1	90.5	94.4
COVERS	DIVS. 10 - 14, 25, 28, 41, 43, 44, 46	131.1	93.3	122.7	131.1	78.8	119.5	131.1	78.8	119.5	131.1	87.9	121.5	131.1	67.3	116.9	131.1	90.7	122.2
21, 22, 23	FIRE SUPPRESSION, PLUMBING & HVAC	105.5	92.2	99.8	104.6	87.7	97.4	104.6	87.7	97.4	104.6	91.2	98.9	105.1	98.4	102.2	104.8	90.9	98.8
26, 27, 3370	ELECTRICAL, COMMUNICATIONS & UTIL.	108.4	102.7	105.4	112.3	68.9	89.7	111.0	68.9	89.1	114.3	78.3	95.6	111.7	86.8	98.7	110.8	100.1	105.2
MF2016	WEIGHTED AVERAGE	116.0	99.6	108.8	111.5	83.4	99.2	111.9	83.6	99.5	114.1	87.2	102.3	115.3	91.4	104.8	109.1	95.1	103.0

552

City Cost Indexes

CANADA

DIVISION		LAVAL, QUEBEC			LETHBRIDGE, ALBERTA			LLOYDMINSTER, ALBERTA			LONDON, ONTARIO			MEDICINE HAT, ALBERTA			MONCTON, NEW BRUNSWICK		
		MAT.	INST.	TOTAL	MAT.	INST.	TOTAL	MAT.	INST.	TOTAL	MAT.	INST.	TOTAL	MAT.	INST.	TOTAL	MAT.	INST.	TOTAL
015433	CONTRACTOR EQUIPMENT		103.3	103.3		105.5	105.5		105.5	105.5		119.8	119.8		105.5	105.5		102.8	102.8
0241, 31 - 34	SITE & INFRASTRUCTURE, DEMOLITION	99.5	98.9	99.1	117.2	101.6	106.3	117.1	101.0	105.8	104.6	112.2	109.9	115.8	101.0	105.5	106.4	96.0	99.1
0310	Concrete Forming & Accessories	129.5	82.9	89.3	122.8	91.7	96.0	120.9	82.0	87.3	121.9	90.0	94.4	122.7	81.9	87.5	103.6	63.2	68.8
0320	Concrete Reinforcing	152.2	74.8	113.1	156.5	84.3	120.1	156.5	84.3	120.1	150.1	97.6	123.6	156.5	84.3	120.1	137.7	71.0	104.1
0330	Cast-in-Place Concrete	111.7	92.2	104.3	135.2	103.5	123.2	125.5	99.8	115.7	117.1	101.6	111.2	125.5	99.8	115.7	113.3	69.7	96.7
03	CONCRETE	113.8	85.2	100.7	125.1	94.8	111.2	120.3	89.1	106.0	115.6	96.2	106.7	120.4	89.0	106.1	111.5	68.0	91.6
04	MASONRY	165.1	79.6	112.1	183.2	87.3	123.7	164.1	80.9	112.5	179.6	99.9	130.2	164.1	80.9	112.5	162.1	61.4	99.7
05	METALS	110.9	86.8	103.5	134.6	92.8	121.8	112.8	92.7	106.6	131.7	107.7	124.3	113.0	92.6	106.7	113.7	85.9	105.2
06	WOOD, PLASTICS & COMPOSITES	130.6	82.8	104.0	117.6	90.9	102.7	114.2	81.2	95.8	106.2	87.7	95.9	117.6	81.2	97.3	98.4	62.6	78.4
07	THERMAL & MOISTURE PROTECTION	112.4	88.3	102.2	124.2	93.9	111.4	120.7	89.3	107.4	124.0	99.5	113.6	127.4	89.3	111.3	113.9	65.5	93.4
08	OPENINGS	91.6	71.4	87.0	91.6	83.4	89.7	91.6	78.0	88.5	80.4	88.8	82.4	91.6	78.0	88.5	86.4	61.5	80.7
0920	Plaster & Gypsum Board	114.8	82.2	92.9	106.7	90.3	95.7	102.7	80.3	87.6	123.0	86.9	98.8	105.0	80.3	88.4	126.6	61.5	82.8
0950, 0980	Ceilings & Acoustic Treatment	103.2	82.2	89.0	110.8	90.3	97.0	103.2	80.3	87.8	138.5	86.9	103.7	103.2	80.3	87.8	112.5	61.5	78.1
0960	Flooring	112.1	90.9	106.2	112.1	88.6	105.6	112.1	88.6	105.6	112.1	97.8	103.9	112.1	88.6	105.6	98.6	68.7	90.3
0970, 0990	Wall Finishes & Painting/Coating	109.0	87.4	96.4	108.9	100.8	104.2	109.1	78.5	91.3	109.3	102.0	105.0	108.9	78.5	91.2	109.9	51.4	75.8
09	FINISHES	105.3	85.2	94.4	106.8	92.4	99.0	104.9	82.6	92.7	116.2	92.3	103.2	105.0	82.6	92.8	104.7	62.9	82.0
COVERS	DIVS. 10 - 14, 25, 28, 41, 43, 44, 46	131.1	78.8	119.5	131.1	92.3	122.5	131.1	89.0	121.8	131.1	92.1	122.5	131.1	89.0	121.8	131.1	61.8	115.7
21, 22, 23	FIRE SUPPRESSION, PLUMBING & HVAC	104.8	87.7	97.5	104.9	93.8	100.2	105.1	93.7	100.2	105.5	89.4	98.6	104.6	90.5	98.6	104.6	69.9	89.8
26, 27, 3370	ELECTRICAL, COMMUNICATIONS & UTIL.	109.3	68.9	88.3	107.1	81.8	94.0	104.8	81.8	92.9	105.3	100.1	102.6	104.8	81.8	92.9	115.8	60.8	87.2
MF2016	WEIGHTED AVERAGE	111.5	83.4	99.2	118.3	91.4	106.5	112.7	88.1	101.9	115.6	97.0	107.5	112.9	87.4	101.7	111.5	69.3	93.0

CANADA

DIVISION		MONTREAL, QUEBEC			MOOSE JAW, SASKATCHEWAN			NEW GLASGOW, NOVA SCOTIA			NEWCASTLE, NEW BRUNSWICK			NORTH BAY, ONTARIO			OSHAWA, ONTARIO		
		MAT.	INST.	TOTAL	MAT.	INST.	TOTAL	MAT.	INST.	TOTAL	MAT.	INST.	TOTAL	MAT.	INST.	TOTAL	MAT.	INST.	TOTAL
015433	CONTRACTOR EQUIPMENT		123.5	123.5		101.7	101.7		102.9	102.9		102.8	102.8		103.5	103.5		105.3	105.3
0241, 31 - 34	SITE & INFRASTRUCTURE, DEMOLITION	114.0	113.5	113.6	117.1	95.1	101.7	119.1	97.7	104.1	107.0	95.7	99.1	134.5	99.6	110.0	108.3	103.9	105.2
0310	Concrete Forming & Accessories	124.4	91.6	96.1	106.3	57.2	64.0	110.5	71.2	76.6	103.6	66.0	66.7	144.3	84.0	92.3	118.5	88.7	92.8
0320	Concrete Reinforcing	126.8	94.3	110.4	110.4	63.4	86.7	176.3	49.2	112.2	137.7	59.4	98.2	207.8	87.6	147.1	165.2	93.6	129.1
0330	Cast-in-Place Concrete	130.9	99.1	118.8	121.7	67.0	100.9	132.9	70.4	109.2	117.7	59.1	95.4	128.0	84.7	111.5	131.1	87.2	114.4
03	CONCRETE	119.3	95.5	108.4	105.8	62.7	86.1	140.6	68.1	107.4	113.6	61.0	89.5	142.4	85.3	116.3	122.7	89.4	107.5
04	MASONRY	176.1	90.3	122.9	162.8	58.8	98.3	216.7	68.4	124.7	162.5	59.9	98.9	223.0	85.4	137.7	148.6	91.9	113.4
05	METALS	139.5	104.4	128.7	109.8	76.7	99.6	131.8	77.8	115.2	113.7	75.2	101.9	132.7	92.8	120.4	111.5	96.2	106.8
06	WOOD, PLASTICS & COMPOSITES	105.6	91.9	98.0	100.1	55.7	75.4	117.5	70.6	91.4	98.4	60.6	77.4	153.9	84.2	115.1	114.1	87.4	99.3
07	THERMAL & MOISTURE PROTECTION	119.2	98.2	110.3	111.3	63.0	90.9	133.2	71.0	106.9	113.9	61.3	91.6	139.8	86.8	117.4	116.0	91.1	105.4
08	OPENINGS	84.8	79.9	83.7	87.6	54.0	79.9	92.6	64.4	86.1	86.4	53.8	78.9	100.6	82.3	96.4	88.5	88.9	88.6
0920	Plaster & Gypsum Board	125.4	91.0	102.3	99.6	54.5	69.3	140.9	69.8	93.1	126.6	59.5	81.5	134.7	83.7	100.4	108.8	87.1	94.2
0950, 0980	Ceilings & Acoustic Treatment	140.4	91.0	107.1	103.2	54.5	70.3	119.2	69.8	85.9	112.5	59.5	76.7	119.2	83.7	95.3	101.7	87.1	91.8
0960	Flooring	110.0	93.1	105.3	102.4	56.7	89.7	107.3	62.3	94.8	98.6	67.1	89.9	130.3	90.9	119.4	102.9	100.3	102.2
0970, 0990	Wall Finishes & Painting/Coating	104.5	105.2	104.9	109.0	63.8	82.6	116.9	62.0	84.9	109.9	50.1	75.0	116.9	88.9	100.6	103.1	109.5	106.9
09	FINISHES	116.1	94.2	104.2	101.3	57.5	77.5	114.2	68.8	89.5	104.7	61.1	81.0	121.1	86.1	102.1	100.3	92.9	96.3
COVERS	DIVS. 10 - 14, 25, 28, 41, 43, 44, 46	131.1	83.7	120.6	131.1	59.8	115.3	131.1	63.3	116.1	131.1	60.2	115.4	131.1	66.1	116.7	131.1	90.4	122.1
21, 22, 23	FIRE SUPPRESSION, PLUMBING & HVAC	105.6	81.6	95.4	105.1	73.3	91.5	104.7	82.8	95.4	104.6	67.5	88.7	104.7	96.3	101.1	104.8	101.9	103.5
26, 27, 3370	ELECTRICAL, COMMUNICATIONS & UTIL.	108.4	85.7	96.6	113.3	59.5	85.3	109.1	62.0	84.6	111.4	59.0	84.1	109.4	87.9	98.2	111.9	90.3	100.7
MF2016	WEIGHTED AVERAGE	118.0	91.7	106.5	110.1	66.7	91.1	122.6	73.5	101.1	111.3	65.7	91.3	125.6	89.2	109.7	111.6	94.7	104.2

CANADA

DIVISION		OTTAWA, ONTARIO			OWEN SOUND, ONTARIO			PETERBOROUGH, ONTARIO			PORTAGE LA PRAIRIE, MANITOBA			PRINCE ALBERT, SASKATCHEWAN			PRINCE GEORGE, BRITISH COLUMBIA		
		MAT.	INST.	TOTAL	MAT.	INST.	TOTAL	MAT.	INST.	TOTAL	MAT.	INST.	TOTAL	MAT.	INST.	TOTAL	MAT.	INST.	TOTAL
015433	CONTRACTOR EQUIPMENT		121.6	121.6		103.0	103.0		103.1	103.1		105.4	105.4		101.7	101.7		106.7	106.7
0241, 31 - 34	SITE & INFRASTRUCTURE, DEMOLITION	108.4	114.4	112.6	119.2	99.7	105.6	119.6	99.7	105.7	118.3	98.5	104.5	112.6	95.3	100.5	126.0	102.7	109.7
0310	Concrete Forming & Accessories	122.2	90.0	94.4	124.2	82.5	88.2	123.5	85.0	90.3	122.9	68.9	76.3	106.3	57.0	63.8	111.6	81.3	85.4
0320	Concrete Reinforcing	130.3	97.6	113.8	175.3	89.7	132.1	169.0	88.1	128.2	156.5	56.3	105.9	115.3	63.3	89.1	112.8	78.3	95.4
0330	Cast-in-Place Concrete	126.9	104.4	118.4	158.9	80.6	129.2	135.8	86.2	116.9	125.5	73.4	105.7	110.3	66.9	93.8	122.2	96.8	112.5
03	CONCRETE	117.8	97.3	108.4	139.8	83.6	114.1	127.0	86.4	108.4	114.2	69.3	93.7	101.0	62.6	83.4	132.3	86.7	111.4
04	MASONRY	172.3	102.5	129.0	167.9	90.8	120.1	168.8	91.8	121.0	167.3	62.3	102.2	161.9	58.8	98.0	174.2	87.6	120.5
05	METALS	132.4	109.0	125.2	110.3	93.3	105.1	112.4	93.1	106.4	113.0	80.2	102.9	109.8	76.5	99.6	112.8	89.2	105.6
06	WOOD, PLASTICS & COMPOSITES	106.4	86.9	95.4	118.0	80.9	97.4	119.8	82.8	99.2	117.5	70.0	91.1	100.1	55.7	75.4	102.5	77.7	88.7
07	THERMAL & MOISTURE PROTECTION	130.7	100.6	118.0	113.8	87.6	102.7	118.6	92.4	107.5	111.8	71.6	94.8	111.2	61.9	90.3	122.0	85.0	106.4
08	OPENINGS	89.1	88.1	88.9	92.4	81.6	89.9	90.3	84.0	88.9	91.6	63.0	85.1	86.5	54.0	79.1	88.4	78.7	86.2
0920	Plaster & Gypsum Board	129.7	85.7	100.2	151.5	80.4	103.8	116.6	82.4	93.6	104.6	68.8	80.6	99.6	54.5	69.3	102.2	76.6	85.0
0950, 0980	Ceilings & Acoustic Treatment	134.2	85.7	101.5	95.3	80.4	85.3	103.2	82.4	89.1	103.2	68.8	80.0	103.2	54.5	70.3	103.2	76.6	85.3
0960	Flooring	102.3	93.2	99.8	118.1	92.3	110.9	112.1	90.9	106.2	112.1	65.4	99.2	102.4	56.7	89.7	107.7	71.2	97.6
0970, 0990	Wall Finishes & Painting/Coating	110.0	96.7	102.2	104.4	87.6	94.6	109.0	91.1	98.6	109.1	56.5	78.4	109.0	54.4	77.1	109.0	79.9	92.0
09	FINISHES	116.1	90.8	102.4	118.9	84.9	96.3	106.3	86.7	95.6	105.0	67.3	84.5	101.3	56.5	76.9	104.8	78.8	90.7
COVERS	DIVS. 10 - 14, 25, 28, 41, 43, 44, 46	131.1	90.7	122.1	139.2	66.4	123.0	131.1	67.5	117.0	131.1	62.3	115.8	131.1	59.8	115.3	131.1	87.2	121.4
21, 22, 23	FIRE SUPPRESSION, PLUMBING & HVAC	105.6	91.1	99.4	104.5	96.2	101.0	104.6	99.8	102.6	104.6	81.7	94.8	105.1	66.0	88.4	104.6	91.2	98.9
26, 27, 3370	ELECTRICAL, COMMUNICATIONS & UTIL.	108.5	100.9	104.5	117.6	86.7	101.5	111.4	87.5	99.0	113.0	57.5	84.1	113.3	59.5	85.3	111.4	78.3	94.2
MF2016	WEIGHTED AVERAGE	117.2	98.0	108.8	117.1	89.2	104.9	114.3	91.1	104.1	112.7	72.2	94.9	109.2	65.0	89.8	115.1	86.6	102.6

For customer support on your Electrical Costs with RSMeans data, call 800.448.8182.

553

City Cost Indexes

CANADA

DIVISION		QUEBEC CITY, QUEBEC			RED DEER, ALBERTA			REGINA, SASKATCHEWAN			RIMOUSKI, QUEBEC			ROUYN-NORANDA, QUEBEC			SAINT HYACINTHE, QUEBEC		
		MAT.	INST.	TOTAL	MAT.	INST.	TOTAL	MAT.	INST.	TOTAL	MAT.	INST.	TOTAL	MAT.	INST.	TOTAL	MAT.	INST.	TOTAL
015433	CONTRACTOR EQUIPMENT		124.2	124.2		105.5	105.5		124.2	124.2		103.3	103.3		103.3	103.3		103.3	103.3
0241, 31 - 34	SITE & INFRASTRUCTURE, DEMOLITION	115.4	113.3	113.9	115.8	101.0	105.5	127.4	115.0	118.7	99.3	98.5	98.8	99.0	98.9	99.0	99.5	98.9	99.1
0310	Concrete Forming & Accessories	122.5	91.7	95.9	136.9	81.9	89.5	124.8	95.6	99.6	129.2	91.0	96.3	129.2	82.9	89.3	129.2	82.9	89.3
0320	Concrete Reinforcing	129.3	94.3	111.7	156.5	84.3	120.1	147.1	91.7	119.1	108.6	94.2	101.3	152.2	74.8	113.1	152.2	74.8	113.1
0330	Cast-in-Place Concrete	132.7	98.9	119.9	125.5	99.8	115.7	144.3	100.7	127.7	113.7	97.0	107.4	108.1	92.2	102.1	111.7	92.2	104.3
03	CONCRETE	120.4	95.6	109.0	121.3	89.0	106.6	129.7	97.4	114.9	109.2	93.7	102.1	112.1	85.2	99.8	113.8	85.2	100.7
04	MASONRY	170.9	90.3	120.9	164.1	80.9	112.5	194.2	91.1	130.3	164.5	90.3	118.5	164.8	79.6	112.0	165.0	79.6	112.1
05	METALS	139.5	106.1	129.3	113.0	92.6	106.7	143.2	103.9	131.1	110.5	92.3	105.0	111.0	86.8	103.6	111.0	86.8	103.6
06	WOOD, PLASTICS & COMPOSITES	111.7	91.8	100.6	117.6	81.2	97.3	109.2	96.2	102.0	130.5	91.4	108.7	130.5	82.8	103.9	130.5	82.8	103.9
07	THERMAL & MOISTURE PROTECTION	115.8	98.4	108.4	137.7	89.3	117.2	132.7	90.0	114.7	111.9	98.0	106.0	111.9	88.3	101.9	112.2	88.3	102.1
08	OPENINGS	88.5	87.4	88.2	91.6	78.0	88.5	86.9	83.4	86.1	91.2	77.8	88.1	91.6	71.4	87.0	91.6	71.4	87.0
0920	Plaster & Gypsum Board	128.6	91.0	103.4	105.0	80.3	88.4	134.8	95.6	108.5	144.6	91.0	108.6	114.3	82.2	92.8	114.3	82.2	92.8
0950, 0980	Ceilings & Acoustic Treatment	137.2	91.0	106.1	103.2	80.3	87.8	141.5	95.6	110.6	102.3	91.0	94.7	102.3	82.2	88.8	102.3	82.2	88.8
0960	Flooring	110.8	93.1	105.9	114.5	88.6	107.3	113.4	97.8	109.1	113.4	90.9	107.2	112.1	90.9	106.2	112.1	90.9	106.2
0970, 0990	Wall Finishes & Painting/Coating	122.1	105.2	112.2	108.9	78.5	91.2	115.1	93.1	102.2	112.1	105.2	108.1	109.0	87.4	96.4	109.0	87.4	96.4
09	FINISHES	117.5	94.1	104.8	105.8	82.6	93.2	125.2	96.7	109.7	109.9	93.4	100.9	105.1	85.2	94.3	105.1	85.2	94.3
COVERS	DIVS. 10 - 14, 25, 28, 41, 43, 44, 46	131.1	83.4	120.5	131.1	89.0	121.8	131.1	71.8	117.9	131.1	82.5	120.3	131.1	78.8	119.5	131.1	78.8	119.5
21, 22, 23	FIRE SUPPRESSION, PLUMBING & HVAC	105.5	81.6	95.3	104.6	90.5	98.6	105.2	90.5	98.9	104.6	81.5	94.7	104.6	87.7	97.4	100.8	87.7	95.2
26, 27, 3370	ELECTRICAL, COMMUNICATIONS & UTIL.	109.6	85.7	97.2	104.8	81.8	92.9	111.4	96.6	103.7	111.0	85.6	97.8	111.0	68.9	89.1	111.6	68.9	89.4
MF2016	WEIGHTED AVERAGE	118.5	92.2	107.0	113.3	87.4	102.0	122.8	95.4	110.8	111.3	88.9	101.5	111.4	83.4	99.1	110.8	83.4	98.8

CANADA

DIVISION		SAINT JOHN, NEW BRUNSWICK			SARNIA, ONTARIO			SASKATOON, SASKATCHEWAN			SAULT STE MARIE, ONTARIO			SHERBROOKE, QUEBEC			SOREL, QUEBEC		
		MAT.	INST.	TOTAL	MAT.	INST.	TOTAL	MAT.	INST.	TOTAL	MAT.	INST.	TOTAL	MAT.	INST.	TOTAL	MAT.	INST.	TOTAL
015433	CONTRACTOR EQUIPMENT		102.8	102.8		103.1	103.1		101.5	101.5		103.1	103.1		103.3	103.3		103.3	103.3
0241, 31 - 34	SITE & INFRASTRUCTURE, DEMOLITION	107.1	97.4	100.3	118.2	99.8	105.3	114.1	97.9	102.7	107.9	99.4	101.9	99.5	98.9	99.1	99.7	98.9	99.2
0310	Concrete Forming & Accessories	123.5	66.3	74.1	122.2	91.9	96.1	106.0	95.1	96.6	111.8	87.9	91.2	129.2	82.9	89.3	129.2	83.0	89.4
0320	Concrete Reinforcing	137.7	71.6	104.4	120.7	89.5	104.9	117.9	91.6	104.6	108.4	88.2	98.2	152.2	74.8	113.1	144.0	74.8	109.1
0330	Cast-in-Place Concrete	115.9	71.4	99.0	125.3	99.4	115.5	119.1	98.1	111.1	112.6	84.8	102.0	111.7	92.2	104.3	112.6	92.2	104.9
03	CONCRETE	114.1	70.1	94.0	115.8	94.3	105.9	106.2	95.5	101.3	101.5	87.3	95.0	113.8	85.2	100.7	113.2	85.2	100.4
04	MASONRY	183.9	71.5	114.2	180.5	94.4	127.1	173.5	91.1	122.4	165.5	93.9	121.1	165.1	79.6	112.1	165.2	79.6	112.1
05	METALS	113.6	87.3	105.5	112.3	93.5	106.5	107.8	90.6	102.5	111.5	95.2	106.5	110.8	86.8	103.4	111.0	86.9	103.6
06	WOOD, PLASTICS & COMPOSITES	120.4	64.3	89.2	118.7	90.9	103.3	97.6	95.7	96.6	106.6	88.7	96.7	130.5	82.8	103.9	130.5	82.8	103.9
07	THERMAL & MOISTURE PROTECTION	114.2	72.4	96.5	118.7	96.0	109.1	112.3	89.2	102.5	117.4	91.2	106.3	111.9	88.3	101.9	111.9	88.3	101.9
08	OPENINGS	86.3	61.3	80.6	92.9	87.8	91.7	87.1	83.2	86.2	84.7	87.1	85.2	91.6	71.4	87.0	91.6	76.0	88.0
0920	Plaster & Gypsum Board	139.3	63.3	88.2	139.9	90.7	106.8	115.6	95.6	102.2	107.8	88.4	94.8	114.3	82.2	92.8	144.6	82.2	102.7
0950, 0980	Ceilings & Acoustic Treatment	117.6	63.3	81.0	107.4	90.7	96.1	123.5	95.6	104.7	103.2	88.4	93.2	102.3	82.2	88.8	102.3	82.2	88.8
0960	Flooring	110.1	68.7	98.6	112.1	99.5	108.6	105.0	97.8	103.0	105.2	96.4	102.7	112.1	90.9	106.2	112.1	90.9	106.2
0970, 0990	Wall Finishes & Painting/Coating	109.9	84.2	94.9	109.0	103.0	105.5	112.5	93.1	101.2	109.0	95.4	101.1	109.0	87.4	96.4	109.0	87.4	96.4
09	FINISHES	110.1	68.1	87.7	110.3	94.8	101.9	109.3	96.3	102.3	102.3	90.5	95.9	105.1	85.2	94.3	109.2	85.2	96.2
COVERS	DIVS. 10 - 14, 25, 28, 41, 43, 44, 46	131.1	63.0	116.0	131.1	68.8	117.3	131.1	71.0	117.8	131.1	89.5	121.9	131.1	78.8	119.5	131.1	78.8	119.5
21, 22, 23	FIRE SUPPRESSION, PLUMBING & HVAC	104.6	77.9	93.2	104.6	105.8	105.1	105.0	90.3	98.7	104.6	94.0	100.1	105.1	87.7	97.6	104.6	87.7	97.4
26, 27, 3370	ELECTRICAL, COMMUNICATIONS & UTIL.	118.5	86.3	101.7	114.3	89.7	101.5	112.6	96.6	104.3	113.0	87.9	100.0	111.0	68.9	89.1	111.0	68.9	89.1
MF2016	WEIGHTED AVERAGE	113.8	77.0	97.7	114.3	95.5	106.0	110.8	92.5	102.7	109.6	91.7	101.8	111.7	83.4	99.3	111.9	83.6	99.5

CANADA

DIVISION		ST CATHARINES, ONTARIO			ST JEROME, QUEBEC			ST JOHNS, NEWFOUNDLAND			SUDBURY, ONTARIO			SUMMERSIDE, PRINCE EDWARD ISLAND			SYDNEY, NOVA SCOTIA		
		MAT.	INST.	TOTAL	MAT.	INST.	TOTAL	MAT.	INST.	TOTAL	MAT.	INST.	TOTAL	MAT.	INST.	TOTAL	MAT.	INST.	TOTAL
015433	CONTRACTOR EQUIPMENT		103.1	103.1		103.3	103.3		121.7	121.7		103.1	103.1		102.9	102.9		102.9	102.9
0241, 31 - 34	SITE & INFRASTRUCTURE, DEMOLITION	97.4	101.4	100.2	99.0	98.9	99.0	121.6	111.7	114.7	97.5	101.0	99.9	129.3	95.0	105.3	114.8	97.7	102.9
0310	Concrete Forming & Accessories	111.1	94.9	97.1	129.2	82.9	89.3	122.8	88.0	92.8	107.1	89.9	92.3	110.8	55.1	62.7	110.5	71.2	76.6
0320	Concrete Reinforcing	105.1	98.7	101.9	152.2	74.8	113.1	161.2	85.0	122.8	106.0	102.5	104.3	173.8	48.0	110.3	176.3	49.2	112.2
0330	Cast-in-Place Concrete	108.3	101.4	105.6	108.1	92.2	102.1	140.6	101.1	125.6	109.1	98.0	104.9	126.2	57.7	100.1	102.5	70.4	90.4
03	CONCRETE	98.1	97.9	98.0	112.1	85.2	99.8	132.7	93.0	114.5	98.4	95.0	96.8	148.5	55.9	106.2	126.2	68.1	99.6
04	MASONRY	145.2	101.3	118.0	164.8	79.6	112.0	184.6	93.7	128.2	145.3	97.6	115.7	215.9	57.2	117.5	214.2	68.4	123.8
05	METALS	111.4	97.4	107.1	111.0	86.8	103.6	135.7	101.0	125.1	110.8	97.0	106.6	131.8	71.0	113.1	131.8	77.8	115.2
06	WOOD, PLASTICS & COMPOSITES	105.3	94.2	99.1	130.5	82.8	103.9	102.1	85.8	93.0	101.5	88.7	94.4	117.9	54.3	82.5	117.5	70.6	91.4
07	THERMAL & MOISTURE PROTECTION	115.0	103.0	109.9	111.9	88.3	101.9	139.2	90.0	118.4	114.5	97.5	107.3	132.6	60.1	101.9	133.2	71.0	106.9
08	OPENINGS	82.6	92.1	84.8	91.6	71.4	87.0	85.7	77.0	83.7	83.3	87.6	84.3	104.6	47.8	91.5	92.6	64.4	86.1
0920	Plaster & Gypsum Board	97.0	94.0	95.0	114.3	82.2	92.8	139.6	84.8	102.8	98.8	88.4	91.8	141.5	53.0	82.1	140.9	69.8	93.1
0950, 0980	Ceilings & Acoustic Treatment	101.7	94.0	96.5	102.3	82.2	88.8	129.6	84.8	99.4	96.6	88.4	91.1	119.2	53.0	74.6	119.2	69.8	85.9
0960	Flooring	98.9	94.5	97.7	112.1	90.9	106.2	111.1	54.2	95.3	97.1	96.4	96.9	107.4	58.4	93.8	107.3	62.3	94.8
0970, 0990	Wall Finishes & Painting/Coating	103.1	105.4	104.4	109.0	87.4	96.4	112.5	103.1	107.0	103.1	96.1	99.0	116.9	41.7	73.0	116.9	62.0	84.9
09	FINISHES	96.9	96.0	96.4	105.1	85.2	94.3	119.6	83.3	99.8	95.6	91.5	93.4	115.3	54.5	82.2	114.2	68.8	89.5
COVERS	DIVS. 10 - 14, 25, 28, 41, 43, 44, 46	131.1	70.5	117.7	131.1	78.8	119.5	131.1	69.7	117.5	131.1	90.8	122.2	131.1	59.4	115.2	131.1	63.3	116.1
21, 22, 23	FIRE SUPPRESSION, PLUMBING & HVAC	104.8	90.8	98.8	104.6	87.7	97.4	105.3	86.3	97.2	104.5	90.2	98.4	104.7	61.3	86.2	104.7	82.8	95.4
26, 27, 3370	ELECTRICAL, COMMUNICATIONS & UTIL.	112.4	100.9	106.4	111.7	68.9	89.4	109.3	85.9	97.1	112.5	101.3	106.7	108.2	50.1	78.0	109.1	62.0	84.6
MF2016	WEIGHTED AVERAGE	107.1	96.1	102.3	111.4	83.4	99.1	120.7	90.3	107.4	106.9	94.9	101.6	125.1	60.5	96.8	120.6	73.5	100.0

City Cost Indexes

		CANADA																
DIVISION		THUNDER BAY, ONTARIO			TIMMINS, ONTARIO			TORONTO, ONTARIO			TROIS RIVIERES, QUEBEC			TRURO, NOVA SCOTIA			VANCOUVER, BRITISH COLUMBIA		
		MAT.	INST.	TOTAL	MAT.	INST.	TOTAL	MAT.	INST.	TOTAL	MAT.	INST.	TOTAL	MAT.	INST.	TOTAL	MAT.	INST.	TOTAL
015433	CONTRACTOR EQUIPMENT		103.1	103.1		103.1	103.1		121.8	121.8		103.7	103.7		102.6	102.6		137.0	137.0
0241, 31 - 34	SITE & INFRASTRUCTURE, DEMOLITION	102.2	101.3	101.6	119.6	99.3	105.4	119.5	115.3	116.6	115.2	99.2	104.0	103.9	97.5	99.4	114.7	124.5	121.6
0310	Concrete Forming & Accessories	118.4	93.2	96.6	123.5	84.0	89.4	122.8	101.0	104.0	151.9	83.1	92.5	96.8	71.2	74.7	122.6	88.8	93.5
0320	Concrete Reinforcing	94.0	98.2	96.1	169.0	87.6	127.9	150.1	100.6	125.1	176.3	74.8	125.1	144.0	49.2	96.2	123.7	93.7	108.6
0330	Cast-in-Place Concrete	119.2	100.6	112.1	135.8	84.6	116.3	116.9	113.6	115.6	106.2	92.3	100.9	141.0	70.4	114.2	125.3	93.5	113.2
03	CONCRETE	105.4	96.7	101.4	127.0	85.3	107.9	115.6	105.9	111.1	128.5	85.3	108.7	126.0	68.0	99.5	118.7	92.7	106.8
04	MASONRY	145.8	101.5	118.3	168.8	85.4	117.1	170.9	108.4	132.1	218.4	79.6	132.3	164.3	68.4	104.8	170.9	88.0	119.5
05	METALS	111.4	96.4	106.8	112.4	92.7	106.3	132.3	109.7	125.3	131.0	87.1	117.5	111.7	77.6	101.2	139.8	111.4	131.1
06	WOOD, PLASTICS & COMPOSITES	114.1	91.5	101.6	119.8	84.1	99.9	105.5	98.5	101.6	168.1	82.8	120.7	90.7	71.6	79.5	100.9	86.8	93.1
07	THERMAL & MOISTURE PROTECTION	115.3	100.2	108.9	118.6	86.8	105.1	133.0	108.0	122.4	131.7	88.3	113.3	113.2	70.9	95.3	133.0	88.3	114.1
08	OPENINGS	81.9	90.0	83.8	90.3	82.3	88.5	84.1	98.0	87.3	102.2	76.0	96.2	84.5	64.4	79.9	83.0	87.2	84.0
0920	Plaster & Gypsum Board	124.9	91.3	102.3	116.6	83.7	94.5	119.4	98.1	105.1	169.3	82.2	110.8	122.4	69.8	87.1	114.2	85.4	94.8
0950, 0980	Ceilings & Acoustic Treatment	96.6	91.3	93.0	103.2	83.7	90.0	128.1	98.1	107.9	118.3	82.2	94.0	103.2	69.8	80.6	135.5	85.4	101.7
0960	Flooring	102.9	101.2	102.4	112.1	90.9	106.2	104.5	103.7	104.3	130.3	90.9	119.4	94.7	62.3	85.7	119.7	91.9	112.0
0970, 0990	Wall Finishes & Painting/Coating	103.1	97.1	99.6	109.0	88.9	97.3	100.8	109.5	105.9	116.9	87.4	99.7	109.0	62.0	81.6	110.8	89.2	98.2
09	FINISHES	101.2	95.0	97.9	106.3	86.1	95.3	110.6	102.3	106.1	124.8	85.2	103.3	101.1	66.8	83.6	119.1	89.3	102.9
COVERS	DIVS. 10 - 14, 25, 28, 41, 43, 44, 46	131.1	70.7	117.7	131.1	66.0	116.7	131.1	95.5	123.2	131.1	78.8	119.5	131.1	63.3	116.0	131.1	92.0	122.4
21, 22, 23	FIRE SUPPRESSION, PLUMBING & HVAC	104.8	91.0	98.9	104.6	96.3	101.1	105.4	99.0	102.7	104.7	87.7	97.5	104.6	82.8	95.3	105.5	78.6	94.0
26, 27, 3370	ELECTRICAL, COMMUNICATIONS & UTIL.	110.8	99.5	104.9	113.0	87.9	100.0	106.2	103.1	104.6	109.4	68.9	88.3	110.7	62.0	85.4	107.1	83.0	94.6
MF2016	WEIGHTED AVERAGE	108.3	95.4	102.7	114.4	89.2	103.4	115.9	104.2	110.8	123.3	83.6	105.9	111.9	73.4	95.1	118.0	91.2	106.3

		CANADA																
DIVISION		VICTORIA, BRITISH COLUMBIA			WHITEHORSE, YUKON			WINDSOR, ONTARIO			WINNIPEG, MANITOBA			YARMOUTH, NOVA SCOTIA			YELLOWKNIFE, NWT		
		MAT.	INST.	TOTAL	MAT.	INST.	TOTAL	MAT.	INST.	TOTAL	MAT.	INST.	TOTAL	MAT.	INST.	TOTAL	MAT.	INST.	TOTAL
015433	CONTRACTOR EQUIPMENT		109.7	109.7		130.0	130.0		103.1	103.1		123.9	123.9		102.9	102.9		129.9	129.9
0241, 31 - 34	SITE & INFRASTRUCTURE, DEMOLITION	125.5	106.8	112.4	146.2	116.0	125.0	93.8	101.3	99.0	117.4	111.2	113.1	118.9	97.7	104.1	154.5	122.1	131.8
0310	Concrete Forming & Accessories	110.8	87.6	90.8	133.6	59.2	69.5	118.4	91.4	95.1	125.4	66.7	74.8	110.5	71.2	76.6	135.9	79.3	87.1
0320	Concrete Reinforcing	115.3	93.5	104.3	159.8	65.3	112.1	103.0	97.5	100.2	142.1	61.3	101.3	176.3	49.2	112.2	156.6	67.9	111.9
0330	Cast-in-Place Concrete	123.4	90.8	111.0	179.5	73.1	139.1	110.9	102.1	107.6	144.8	73.4	117.6	131.5	70.4	108.3	177.1	90.8	144.3
03	CONCRETE	134.7	90.0	114.2	155.6	67.0	115.1	99.5	96.4	98.1	129.3	69.6	102.0	139.9	68.1	107.1	154.2	82.6	121.5
04	MASONRY	176.5	88.0	121.6	250.8	60.4	132.8	145.4	100.5	117.5	185.3	66.2	111.4	216.6	68.4	124.7	244.4	71.8	137.4
05	METALS	109.0	92.3	103.9	148.5	92.5	131.3	111.4	96.9	106.9	143.2	90.0	126.8	131.8	77.8	115.2	147.4	96.0	131.6
06	WOOD, PLASTICS & COMPOSITES	101.3	86.2	93.2	123.4	57.7	86.8	114.1	89.8	100.6	107.8	67.3	85.2	111.5	70.6	91.4	129.7	80.6	102.4
07	THERMAL & MOISTURE PROTECTION	122.2	89.3	108.3	142.8	64.9	109.9	115.1	99.6	108.5	125.3	71.0	102.3	133.2	71.0	106.6	144.1	81.4	117.5
08	OPENINGS	88.6	82.6	87.2	104.4	55.6	93.2	81.7	89.5	83.5	83.9	61.0	78.7	92.6	64.4	86.1	96.5	68.8	90.1
0920	Plaster & Gypsum Board	108.8	85.4	93.1	167.4	55.6	92.3	109.8	89.5	96.2	121.9	65.4	84.0	140.9	69.8	93.1	172.4	79.2	109.8
0950, 0980	Ceilings & Acoustic Treatment	105.6	85.4	92.0	161.4	55.6	90.1	96.6	89.5	91.8	143.8	65.4	91.0	119.2	69.8	85.9	155.1	79.2	103.9
0960	Flooring	110.4	71.2	99.5	125.4	58.3	106.8	102.9	98.5	101.7	109.0	70.4	98.3	107.3	62.3	94.8	123.4	87.6	113.5
0970, 0990	Wall Finishes & Painting/Coating	112.5	89.2	98.9	119.7	55.6	82.3	103.1	98.1	100.2	107.7	54.4	76.6	116.9	62.0	84.9	118.9	79.3	95.8
09	FINISHES	108.0	85.3	95.6	142.3	58.4	96.6	98.8	93.4	95.9	119.8	66.4	90.8	114.2	68.8	89.5	138.6	80.3	106.9
COVERS	DIVS. 10 - 14, 25, 28, 41, 43, 44, 46	131.1	67.4	116.9	131.1	62.6	115.9	131.1	70.1	117.6	131.1	64.7	116.4	131.1	63.3	116.1	131.1	66.1	116.7
21, 22, 23	FIRE SUPPRESSION, PLUMBING & HVAC	104.7	76.6	92.7	105.3	74.6	92.1	104.8	90.9	98.8	105.5	65.5	88.4	104.7	82.8	95.4	105.4	92.1	99.7
26, 27, 3370	ELECTRICAL, COMMUNICATIONS & UTIL.	112.8	82.4	97.0	132.0	60.3	94.7	115.6	101.0	108.0	112.8	65.1	88.0	109.1	62.0	84.6	121.5	81.8	100.8
MF2016	WEIGHTED AVERAGE	115.3	85.7	102.3	135.9	71.3	107.6	107.6	95.2	102.2	121.3	72.0	99.7	122.5	73.5	101.1	133.4	86.2	112.7

Costs shown in RSMeans cost data publications are based on national averages for materials and installation. To adjust these costs to a specific location, simply multiply the base cost by the factor and divide by 100 for that city. The data is arranged alphabetically by state and postal zip code numbers. For a city not listed, use the factor for a nearby city with similar economic characteristics.

STATE/ZIP	CITY	MAT.	INST.	TOTAL
ALABAMA				
350-352	Birmingham	95.9	71.4	85.2
354	Tuscaloosa	96.0	72.3	85.6
355	Jasper	96.5	71.2	85.5
356	Decatur	96.0	71.3	85.2
357-358	Huntsville	96.0	70.7	84.9
359	Gadsden	96.1	71.0	85.1
360-361	Montgomery	94.8	72.3	85.0
362	Anniston	94.5	68.5	83.1
363	Dothan	94.9	72.9	85.3
364	Evergreen	94.5	72.6	84.9
365-366	Mobile	95.4	70.2	84.4
367	Selma	94.6	73.6	85.4
368	Phenix City	95.4	72.7	85.4
369	Butler	94.8	72.8	85.2
ALASKA				
995-996	Anchorage	118.9	114.8	117.1
997	Fairbanks	120.1	115.3	118.0
998	Juneau	118.8	114.8	117.0
999	Ketchikan	130.8	114.8	123.8
ARIZONA				
850,853	Phoenix	99.5	74.1	88.4
851,852	Mesa/Tempe	98.2	72.5	87.0
855	Globe	99.2	72.4	87.4
856-857	Tucson	97.1	71.8	86.0
859	Show Low	99.4	72.5	87.6
860	Flagstaff	101.7	72.1	88.8
863	Prescott	99.3	72.2	87.5
864	Kingman	97.9	72.5	86.8
865	Chambers	97.9	75.0	87.9
ARKANSAS				
716	Pine Bluff	96.9	65.0	82.9
717	Camden	95.0	60.5	79.9
718	Texarkana	95.7	60.7	80.3
719	Hot Springs	94.3	62.0	80.1
720-722	Little Rock	95.3	66.6	82.7
723	West Memphis	94.2	66.6	82.1
724	Jonesboro	94.5	63.7	81.0
725	Batesville	92.6	60.8	78.7
726	Harrison	93.9	58.9	78.6
727	Fayetteville	91.5	61.6	78.5
728	Russellville	92.7	59.4	78.1
729	Fort Smith	95.0	63.1	81.0
CALIFORNIA				
900-902	Los Angeles	99.9	129.3	112.8
903-905	Inglewood	95.5	125.5	108.6
906-908	Long Beach	97.1	126.1	109.8
910-912	Pasadena	96.0	125.3	108.8
913-916	Van Nuys	99.1	125.3	110.6
917-918	Alhambra	98.0	125.9	110.2
919-921	San Diego	101.0	119.5	109.1
922	Palm Springs	97.2	123.9	108.9
923-924	San Bernardino	94.8	125.7	108.4
925	Riverside	99.0	125.6	110.7
926-927	Santa Ana	96.8	122.9	108.3
928	Anaheim	99.1	125.9	110.8
930	Oxnard	97.6	126.5	110.2
931	Santa Barbara	96.9	126.3	109.8
932-933	Bakersfield	98.3	121.3	108.3
934	San Luis Obispo	98.1	124.4	109.6
935	Mojave	95.2	121.4	106.7
936-938	Fresno	98.2	128.2	111.3
939	Salinas	98.8	134.8	114.6
940-941	San Francisco	106.3	158.5	129.1
942,956-958	Sacramento	100.0	131.6	113.8
943	Palo Alto	98.5	148.0	120.2
944	San Mateo	100.8	147.7	121.3
945	Vallejo	99.5	139.6	117.0
946	Oakland	102.5	148.9	122.8
947	Berkeley	102.5	148.9	122.8
948	Richmond	101.7	143.0	119.8
949	San Rafael	103.8	148.4	123.3
950	Santa Cruz	104.3	135.1	117.8

STATE/ZIP	CITY	MAT.	INST.	TOTAL
CALIFORNIA (CONT'D)				
951	San Jose	102.4	150.4	123.4
952	Stockton	100.5	127.7	112.4
953	Modesto	100.4	126.9	112.0
954	Santa Rosa	101.1	147.3	121.3
955	Eureka	102.4	130.0	114.4
959	Marysville	101.6	128.8	113.5
960	Redding	109.1	128.8	117.7
961	Susanville	108.8	127.0	116.8
COLORADO				
800-802	Denver	102.8	74.6	90.5
803	Boulder	98.8	75.9	88.8
804	Golden	101.0	72.6	88.6
805	Fort Collins	102.1	74.3	89.9
806	Greeley	99.3	75.1	88.7
807	Fort Morgan	99.4	71.2	87.0
808-809	Colorado Springs	101.0	73.2	88.8
810	Pueblo	101.3	68.8	87.1
811	Alamosa	103.3	69.0	88.3
812	Salida	103.0	65.8	86.7
813	Durango	103.7	65.5	87.0
814	Montrose	102.4	67.8	87.3
815	Grand Junction	105.5	73.0	91.3
816	Glenwood Springs	103.5	65.9	87.0
CONNECTICUT				
060	New Britain	97.4	117.9	106.3
061	Hartford	98.4	118.6	107.3
062	Willimantic	97.9	118.5	106.9
063	New London	94.7	118.5	105.1
064	Meriden	96.6	117.8	105.9
065	New Haven	98.9	118.9	107.6
066	Bridgeport	98.5	118.4	107.2
067	Waterbury	98.1	118.7	107.1
068	Norwalk	98.0	118.6	107.0
069	Stamford	98.1	125.5	110.1
D.C.				
200-205	Washington	101.1	87.6	95.2
DELAWARE				
197	Newark	97.9	112.4	104.3
198	Wilmington	97.6	112.5	104.1
199	Dover	98.0	112.4	104.3
FLORIDA				
320,322	Jacksonville	94.7	65.3	81.8
321	Daytona Beach	94.9	71.0	84.4
323	Tallahassee	95.6	66.2	82.7
324	Panama City	96.2	65.1	82.5
325	Pensacola	98.8	67.1	84.9
326,344	Gainesville	96.4	65.0	82.7
327-328,347	Orlando	96.8	67.0	83.8
329	Melbourne	97.6	71.2	86.0
330-332,340	Miami	95.6	64.6	82.0
333	Fort Lauderdale	95.1	66.0	82.4
334,349	West Palm Beach	94.2	64.0	81.0
335-336,346	Tampa	96.6	67.6	83.9
337	St. Petersburg	99.0	65.9	84.5
338	Lakeland	96.2	67.8	83.7
339,341	Fort Myers	95.4	65.4	82.3
342	Sarasota	98.1	67.9	84.9
GEORGIA				
300-303,399	Atlanta	98.7	75.0	88.3
304	Statesboro	98.7	66.3	84.5
305	Gainesville	97.2	67.6	84.2
306	Athens	96.6	69.1	84.5
307	Dalton	98.3	72.2	86.9
308-309	Augusta	96.8	74.2	86.9
310-312	Macon	93.8	75.1	85.6
313-314	Savannah	95.4	74.4	86.2
315	Waycross	95.3	69.5	84.0
316	Valdosta	95.1	67.3	82.9
317,398	Albany	95.0	73.4	85.5
318-319	Columbus	94.9	73.8	85.7

Location Factors - Commercial

STATE/ZIP	CITY	MAT.	INST.	TOTAL
HAWAII				
967	Hilo	116.9	116.0	116.5
968	Honolulu	121.1	116.0	118.9
STATES & POSS.				
969	Guam	139.1	53.4	101.6
IDAHO				
832	Pocatello	101.5	79.5	91.9
833	Twin Falls	102.8	77.8	91.8
834	Idaho Falls	100.0	78.5	90.6
835	Lewiston	108.2	86.3	98.6
836-837	Boise	100.5	79.8	91.4
838	Coeur d'Alene	108.1	84.7	97.8
ILLINOIS				
600-603	North Suburban	98.9	140.6	117.1
604	Joliet	98.8	142.0	117.7
605	South Suburban	98.9	140.6	117.1
606-608	Chicago	100.8	145.8	120.5
609	Kankakee	95.6	133.5	112.2
610-611	Rockford	98.2	128.3	111.4
612	Rock Island	96.4	101.9	98.8
613	La Salle	97.6	126.5	110.3
614	Galesburg	97.4	108.9	102.5
615-616	Peoria	99.3	112.2	104.9
617	Bloomington	96.8	113.0	103.9
618-619	Champaign	100.3	110.6	104.8
620-622	East St. Louis	94.4	109.4	100.9
623	Quincy	96.7	103.9	99.8
624	Effingham	95.8	108.7	101.5
625	Decatur	97.3	109.0	102.4
626-627	Springfield	97.8	108.8	102.6
628	Centralia	93.4	108.8	100.1
629	Carbondale	93.2	108.8	100.0
INDIANA				
460	Anderson	95.1	83.3	89.9
461-462	Indianapolis	98.3	83.2	91.7
463-464	Gary	96.5	110.4	102.6
465-466	South Bend	96.3	85.0	91.4
467-468	Fort Wayne	95.7	77.7	87.8
469	Kokomo	93.5	81.7	88.3
470	Lawrenceburg	92.8	79.6	87.0
471	New Albany	94.0	78.0	87.0
472	Columbus	96.3	80.8	89.5
473	Muncie	96.3	81.7	89.9
474	Bloomington	97.9	81.3	90.6
475	Washington	94.8	86.5	91.1
476-477	Evansville	95.5	85.5	91.1
478	Terre Haute	96.3	85.4	91.5
479	Lafayette	96.0	81.6	89.7
IOWA				
500-503,509	Des Moines	97.2	88.6	93.5
504	Mason City	95.8	76.0	87.1
505	Fort Dodge	95.9	73.0	85.9
506-507	Waterloo	97.2	80.2	89.7
508	Creston	96.2	82.1	90.0
510-511	Sioux City	97.9	79.2	89.7
512	Sibley	97.1	61.0	81.3
513	Spencer	98.7	62.1	82.7
514	Carroll	95.8	79.9	88.8
515	Council Bluffs	99.0	81.1	91.2
516	Shenandoah	96.4	76.5	87.7
520	Dubuque	97.3	82.8	91.0
521	Decorah	96.8	74.0	86.8
522-524	Cedar Rapids	98.2	86.5	93.1
525	Ottumwa	96.7	76.2	87.7
526	Burlington	96.0	82.3	90.0
527-528	Davenport	97.3	98.6	97.9
KANSAS				
660-662	Kansas City	98.4	99.3	98.8
664-666	Topeka	99.1	78.1	89.9
667	Fort Scott	97.3	79.1	89.3
668	Emporia	97.4	75.8	87.9
669	Belleville	99.1	69.9	86.3
670-672	Wichita	99.2	72.1	87.3
673	Independence	99.6	75.6	89.1
674	Salina	99.6	70.8	87.0
675	Hutchinson	95.0	70.5	84.3
676	Hays	99.1	71.3	86.9
677	Colby	99.8	73.5	88.3

STATE/ZIP	CITY	MAT.	INST.	TOTAL
KANSAS (CONT'D)				
678	Dodge City	100.9	72.9	88.7
679	Liberal	99.0	70.6	86.6
KENTUCKY				
400-402	Louisville	95.1	79.1	88.1
403-405	Lexington	94.8	80.3	88.5
406	Frankfort	96.0	78.5	88.3
407-409	Corbin	92.7	77.7	86.2
410	Covington	94.7	80.5	88.5
411-412	Ashland	93.5	93.4	93.5
413-414	Campton	94.7	80.2	88.3
415-416	Pikeville	95.9	81.4	89.6
417-418	Hazard	94.1	78.7	87.4
420	Paducah	92.6	84.5	89.1
421-422	Bowling Green	94.6	79.9	88.2
423	Owensboro	94.7	84.3	90.2
424	Henderson	92.4	83.0	88.3
425-426	Somerset	91.9	79.2	86.4
427	Elizabethtown	91.5	77.2	85.3
LOUISIANA				
700-701	New Orleans	98.7	68.4	85.5
703	Thibodaux	95.6	67.4	83.3
704	Hammond	93.2	65.1	80.9
705	Lafayette	94.9	70.1	84.0
706	Lake Charles	95.1	70.9	84.5
707-708	Baton Rouge	96.3	70.1	84.8
710-711	Shreveport	97.9	66.7	84.2
712	Monroe	96.7	66.9	83.7
713-714	Alexandria	96.9	68.2	84.3
MAINE				
039	Kittery	93.0	84.9	89.5
040-041	Portland	98.7	85.6	93.0
042	Lewiston	96.3	85.6	91.6
043	Augusta	98.4	79.8	90.3
044	Bangor	95.8	85.0	91.1
045	Bath	94.8	82.2	89.3
046	Machias	94.2	79.2	87.7
047	Houlton	94.4	79.2	87.8
048	Rockland	93.5	79.2	87.2
049	Waterville	94.7	79.8	88.2
MARYLAND				
206	Waldorf	97.9	85.2	92.3
207-208	College Park	97.9	86.8	93.0
209	Silver Spring	97.2	85.8	92.2
210-212	Baltimore	101.4	84.7	94.1
214	Annapolis	98.9	84.2	92.4
215	Cumberland	96.9	86.3	92.3
216	Easton	98.6	76.6	88.9
217	Hagerstown	97.2	87.7	93.0
218	Salisbury	99.0	69.8	86.2
219	Elkton	95.8	87.4	92.1
MASSACHUSETTS				
010-011	Springfield	97.8	109.0	102.7
012	Pittsfield	97.3	106.5	101.3
013	Greenfield	95.7	109.0	101.5
014	Fitchburg	94.4	120.8	105.9
015-016	Worcester	97.7	120.8	107.8
017	Framingham	93.7	127.9	108.7
018	Lowell	97.1	127.1	110.3
019	Lawrence	98.0	128.5	111.4
020-022, 024	Boston	101.2	133.1	115.2
023	Brockton	97.9	121.8	108.4
025	Buzzards Bay	92.8	119.5	104.5
026	Hyannis	95.1	119.5	105.8
027	New Bedford	97.0	119.8	107.0
MICHIGAN				
480,483	Royal Oak	94.3	100.5	97.0
481	Ann Arbor	96.2	102.4	98.9
482	Detroit	99.2	103.0	100.9
484-485	Flint	95.9	92.3	94.4
486	Saginaw	95.6	88.7	92.6
487	Bay City	95.7	88.3	92.5
488-489	Lansing	97.0	90.1	94.0
490	Battle Creek	94.8	83.7	89.9
491	Kalamazoo	95.1	82.1	89.4
492	Jackson	93.4	91.8	92.7
493,495	Grand Rapids	96.4	82.9	90.5
494	Muskegon	93.6	81.2	88.2

For customer support on your Electrical Costs with RSMeans data, call 800.448.8182.

557

Location Factors - Commercial

STATE/ZIP	CITY	MAT.	INST.	TOTAL
MICHIGAN (CONT'D)				
496	Traverse City	92.9	78.4	86.6
497	Gaylord	94.1	80.7	88.2
498-499	Iron Mountain	96.0	82.4	90.0
MINNESOTA				
550-551	Saint Paul	98.9	114.1	105.6
553-555	Minneapolis	100.4	113.0	105.9
556-558	Duluth	99.8	102.7	101.1
559	Rochester	98.8	100.9	99.7
560	Mankato	96.0	98.0	96.9
561	Windom	94.6	92.3	93.6
562	Willmar	94.3	97.4	95.6
563	St. Cloud	95.2	111.3	102.3
564	Brainerd	95.9	97.5	96.6
565	Detroit Lakes	97.8	92.0	95.3
566	Bemidji	97.0	96.0	96.6
567	Thief River Falls	96.6	89.2	93.4
MISSISSIPPI				
386	Clarksdale	94.7	55.2	77.4
387	Greenville	98.3	65.6	83.9
388	Tupelo	96.2	55.9	78.6
389	Greenwood	96.0	53.2	77.3
390-392	Jackson	97.3	65.3	83.3
393	Meridian	95.1	65.9	82.3
394	Laurel	96.7	56.3	79.0
395	Biloxi	96.7	65.5	83.1
396	McComb	95.0	53.7	76.9
397	Columbus	96.6	55.8	78.7
MISSOURI				
630-631	St. Louis	100.5	106.2	103.0
633	Bowling Green	98.7	95.0	97.1
634	Hannibal	97.6	93.6	95.9
635	Kirksville	101.1	88.5	95.6
636	Flat River	99.6	95.7	97.9
637	Cape Girardeau	99.1	91.5	95.8
638	Sikeston	98.0	89.1	94.1
639	Poplar Bluff	97.5	89.7	94.1
640-641	Kansas City	100.3	102.7	101.4
644-645	St. Joseph	98.9	92.8	96.2
646	Chillicothe	96.5	95.1	95.9
647	Harrisonville	95.9	102.8	98.9
648	Joplin	97.8	79.6	89.8
650-651	Jefferson City	97.8	92.5	95.5
652	Columbia	98.1	92.5	95.7
653	Sedalia	98.6	93.2	96.2
654-655	Rolla	96.2	96.9	96.5
656-658	Springfield	99.6	81.7	91.7
MONTANA				
590-591	Billings	100.0	76.4	89.6
592	Wolf Point	99.9	76.5	89.7
593	Miles City	97.8	76.7	88.6
594	Great Falls	101.3	75.3	89.9
595	Havre	98.9	75.2	88.5
596	Helena	99.6	75.3	88.9
597	Butte	99.8	75.3	89.1
598	Missoula	97.4	74.5	87.4
599	Kalispell	96.8	75.0	87.2
NEBRASKA				
680-681	Omaha	98.4	80.8	90.7
683-685	Lincoln	98.3	80.2	90.4
686	Columbus	97.3	81.4	90.3
687	Norfolk	98.8	78.6	90.0
688	Grand Island	98.5	78.4	89.7
689	Hastings	98.5	79.0	89.9
690	McCook	98.0	71.5	86.4
691	North Platte	97.8	75.2	87.9
692	Valentine	100.5	70.5	87.4
693	Alliance	100.4	71.9	87.9
NEVADA				
889-891	Las Vegas	103.1	105.2	104.0
893	Ely	101.9	94.1	98.5
894-895	Reno	101.8	85.3	94.6
897	Carson City	101.0	84.9	93.9
898	Elko	100.6	81.2	92.1
NEW HAMPSHIRE				
030	Nashua	97.7	92.6	95.5
031	Manchester	98.3	92.7	95.9

STATE/ZIP	CITY	MAT.	INST.	TOTAL
NEW HAMPSHIRE (CONT'D)				
032-033	Concord	97.9	92.4	95.5
034	Keene	94.8	79.6	88.1
035	Littleton	94.9	79.8	88.3
036	Charleston	94.3	76.3	86.4
037	Claremont	93.5	76.3	86.0
038	Portsmouth	95.1	92.4	93.9
NEW JERSEY				
070-071	Newark	99.1	137.9	116.1
072	Elizabeth	96.8	138.2	114.9
073	Jersey City	95.9	137.6	114.1
074-075	Paterson	97.4	138.2	115.3
076	Hackensack	95.5	137.9	114.0
077	Long Branch	95.0	136.2	113.1
078	Dover	95.7	138.2	114.3
079	Summit	95.8	138.2	114.3
080,083	Vineland	95.0	134.4	112.2
081	Camden	97.0	133.3	112.9
082,084	Atlantic City	95.7	134.7	112.8
085-086	Trenton	98.5	133.9	114.0
087	Point Pleasant	96.8	135.8	113.9
088-089	New Brunswick	97.4	137.8	115.1
NEW MEXICO				
870-872	Albuquerque	98.5	75.1	88.3
873	Gallup	98.9	75.1	88.5
874	Farmington	99.1	75.1	88.6
875	Santa Fe	99.1	75.1	88.6
877	Las Vegas	97.2	75.1	87.5
878	Socorro	96.8	75.1	87.3
879	Truth/Consequences	96.6	71.8	85.7
880	Las Cruces	96.4	71.7	85.6
881	Clovis	98.9	75.0	88.4
882	Roswell	100.3	75.1	89.3
883	Carrizozo	101.1	75.1	89.7
884	Tucumcari	99.5	75.0	88.8
NEW YORK				
100-102	New York	100.6	178.2	134.6
103	Staten Island	96.4	172.2	129.6
104	Bronx	94.7	173.4	129.2
105	Mount Vernon	95.1	145.6	117.2
106	White Plains	94.6	145.6	116.9
107	Yonkers	98.9	150.8	121.6
108	New Rochelle	95.5	145.5	117.4
109	Suffern	95.1	136.0	113.0
110	Queens	99.6	174.8	132.6
111	Long Island City	101.2	174.8	133.5
112	Brooklyn	101.6	172.5	132.7
113	Flushing	101.9	174.8	133.8
114	Jamaica	100.2	174.8	132.9
115,117,118	Hicksville	99.6	158.9	125.6
116	Far Rockaway	102.0	174.8	133.9
119	Riverhead	100.4	159.0	126.1
120-122	Albany	95.8	111.2	102.5
123	Schenectady	96.1	111.2	102.7
124	Kingston	100.0	131.6	113.8
125-126	Poughkeepsie	99.2	142.4	118.1
127	Monticello	98.5	130.8	112.7
128	Glens Falls	91.4	106.4	98.0
129	Plattsburgh	96.6	100.9	98.5
130-132	Syracuse	97.6	102.0	99.5
133-135	Utica	95.7	100.8	97.9
136	Watertown	97.3	100.7	98.8
137-139	Binghamton	97.2	101.7	99.2
140-142	Buffalo	101.2	111.6	105.8
143	Niagara Falls	97.6	111.9	103.9
144-146	Rochester	99.5	102.7	100.9
147	Jamestown	96.6	93.9	95.4
148-149	Elmira	96.5	103.0	99.3
NORTH CAROLINA				
270,272-274	Greensboro	98.6	68.9	85.6
271	Winston-Salem	98.4	68.7	85.4
275-276	Raleigh	96.9	68.3	84.4
277	Durham	100.2	68.9	86.5
278	Rocky Mount	96.3	68.6	84.2
279	Elizabeth City	97.2	70.4	85.5
280	Gastonia	98.4	68.7	85.4
281-282	Charlotte	98.1	68.3	85.0
283	Fayetteville	101.3	68.4	86.9
284	Wilmington	97.2	67.5	84.2
285	Kinston	95.8	68.2	83.7

For customer support on your Electrical Costs with RSMeans data, call 800.448.8182.

Location Factors - Commercial

STATE/ZIP	CITY	MAT.	INST.	TOTAL
NORTH CAROLINA (CONT'D)				
286	Hickory	96.1	69.3	84.4
287-288	Asheville	97.6	68.3	84.8
289	Murphy	96.9	67.5	84.0
NORTH DAKOTA				
580-581	Fargo	100.1	81.8	92.1
582	Grand Forks	100.1	79.0	90.8
583	Devils Lake	100.1	78.2	90.5
584	Jamestown	100.2	77.1	90.1
585	Bismarck	100.3	81.4	92.0
586	Dickinson	100.8	80.3	91.8
587	Minot	100.0	81.0	91.7
588	Williston	99.3	80.2	90.9
OHIO				
430-432	Columbus	98.1	83.3	91.6
433	Marion	94.4	81.6	88.8
434-436	Toledo	97.5	93.1	95.6
437-438	Zanesville	95.0	82.5	89.5
439	Steubenville	96.2	91.8	94.3
440	Lorain	98.6	86.9	93.5
441	Cleveland	98.9	93.4	96.5
442-443	Akron	99.7	88.2	94.7
444-445	Youngstown	98.9	83.3	92.1
446-447	Canton	99.1	81.0	91.1
448-449	Mansfield	96.8	82.8	90.7
450	Hamilton	96.6	79.4	89.1
451-452	Cincinnati	97.7	80.9	90.3
453-454	Dayton	96.7	77.9	88.5
455	Springfield	96.7	80.2	89.5
456	Chillicothe	96.2	86.1	91.7
457	Athens	99.1	83.7	92.4
458	Lima	99.4	81.1	91.4
OKLAHOMA				
730-731	Oklahoma City	97.7	67.2	84.4
734	Ardmore	96.7	66.7	83.6
735	Lawton	98.7	66.9	84.8
736	Clinton	97.9	66.6	84.2
737	Enid	98.3	67.0	84.6
738	Woodward	96.7	63.8	82.3
739	Guymon	97.8	63.8	82.9
740-741	Tulsa	96.0	66.4	83.0
743	Miami	92.9	65.8	81.0
744	Muskogee	95.1	65.3	82.1
745	McAlester	92.6	61.1	78.8
746	Ponca City	93.3	65.1	81.0
747	Durant	93.2	65.6	81.1
748	Shawnee	94.7	66.3	82.3
749	Poteau	92.5	65.4	80.6
OREGON				
970-972	Portland	99.4	100.8	100.0
973	Salem	100.9	98.7	100.0
974	Eugene	99.1	98.8	99.0
975	Medford	100.6	96.1	98.6
976	Klamath Falls	101.0	96.9	99.2
977	Bend	100.0	98.7	99.4
978	Pendleton	96.0	99.7	97.6
979	Vale	93.8	87.7	91.1
PENNSYLVANIA				
150-152	Pittsburgh	100.3	103.0	101.5
153	Washington	97.3	102.2	99.4
154	Uniontown	97.6	98.6	98.1
155	Bedford	98.7	92.2	95.8
156	Greensburg	98.6	99.9	99.2
157	Indiana	97.4	97.8	97.6
158	Dubois	99.1	93.5	96.7
159	Johnstown	98.7	94.7	96.9
160	Butler	91.4	101.5	95.8
161	New Castle	91.5	100.9	95.6
162	Kittanning	91.9	101.2	95.9
163	Oil City	91.3	99.6	95.0
164-165	Erie	93.1	96.8	94.7
166	Altoona	93.2	95.1	94.0
167	Bradford	94.8	100.2	97.2
168	State College	94.4	96.5	95.3
169	Wellsboro	95.5	94.7	95.1
170-171	Harrisburg	97.3	95.2	96.3
172	Chambersburg	94.4	90.8	92.8
173-174	York	94.4	94.5	94.4
175-176	Lancaster	93.2	96.4	94.6

STATE/ZIP	CITY	MAT.	INST.	TOTAL
PENNSYLVANIA (CONT'D)				
177	Williamsport	91.9	92.8	92.3
178	Sunbury	93.9	95.1	94.4
179	Pottsville	93.0	97.9	95.1
180	Lehigh Valley	94.9	117.4	104.7
181	Allentown	96.5	107.7	101.4
182	Hazleton	94.4	96.8	95.4
183	Stroudsburg	94.3	110.4	101.3
184-185	Scranton	97.2	100.0	98.4
186-187	Wilkes-Barre	94.1	98.9	96.2
188	Montrose	93.8	99.5	96.3
189	Doylestown	94.0	123.5	106.9
190-191	Philadelphia	99.7	134.6	115.0
193	Westchester	95.5	123.7	107.8
194	Norristown	94.5	127.6	109.0
195-196	Reading	96.1	105.2	100.1
PUERTO RICO				
009	San Juan	120.7	27.7	79.9
RHODE ISLAND				
028	Newport	96.3	112.0	103.2
029	Providence	98.1	112.0	104.2
SOUTH CAROLINA				
290-292	Columbia	96.9	69.1	84.7
293	Spartanburg	97.3	68.9	84.9
294	Charleston	98.9	68.8	85.7
295	Florence	97.0	69.0	84.8
296	Greenville	97.1	68.9	84.8
297	Rock Hill	97.1	68.3	84.5
298	Aiken	98.0	69.2	85.4
299	Beaufort	98.7	56.8	80.3
SOUTH DAKOTA				
570-571	Sioux Falls	98.5	76.4	88.8
572	Watertown	98.7	57.6	80.7
573	Mitchell	97.5	63.1	82.4
574	Aberdeen	99.7	64.5	84.3
575	Pierre	100.4	70.5	87.3
576	Mobridge	98.2	65.0	83.7
577	Rapid City	99.5	74.4	88.5
TENNESSEE				
370-372	Nashville	96.7	71.7	85.8
373-374	Chattanooga	98.6	70.3	86.2
375,380-381	Memphis	98.4	70.7	86.3
376	Johnson City	98.8	60.3	82.0
377-379	Knoxville	95.3	66.2	82.5
382	McKenzie	95.8	56.8	78.7
383	Jackson	97.1	60.6	81.1
384	Columbia	94.2	68.4	82.9
385	Cookeville	95.6	57.7	79.0
TEXAS				
750	McKinney	98.4	65.2	83.9
751	Waxahackie	98.5	64.3	83.5
752-753	Dallas	99.2	67.6	85.4
754	Greenville	98.6	65.2	84.0
755	Texarkana	97.9	64.4	83.2
756	Longview	98.9	62.8	83.1
757	Tyler	98.9	64.2	83.7
758	Palestine	95.0	62.7	80.9
759	Lufkin	95.7	64.9	82.2
760-761	Fort Worth	96.6	65.1	82.8
762	Denton	96.8	65.3	83.0
763	Wichita Falls	94.2	63.3	80.7
764	Eastland	93.6	61.8	79.7
765	Temple	91.9	60.4	78.1
766-767	Waco	93.6	64.4	80.8
768	Brownwood	96.9	60.2	80.8
769	San Angelo	96.4	60.7	80.8
770-772	Houston	98.3	68.0	85.0
773	Huntsville	97.4	65.1	83.3
774	Wharton	98.2	66.2	84.2
775	Galveston	96.2	69.0	84.3
776-777	Beaumont	96.5	68.0	84.0
778	Bryan	94.0	67.3	82.3
779	Victoria	98.3	64.9	83.7
780	Laredo	96.1	63.5	81.8
781-782	San Antonio	97.8	64.4	83.2
783-784	Corpus Christi	98.7	62.4	82.8
785	McAllen	99.6	58.6	81.7
786-787	Austin	96.7	62.7	81.8

For customer support on your Electrical Costs with RSMeans data, call 800.448.8182.

559

Location Factors - Commercial

STATE/ZIP	CITY	MAT.	INST.	TOTAL
TEXAS (CONT'D)				
788	Del Rio	99.5	62.9	83.4
789	Giddings	96.1	62.6	81.4
790-791	Amarillo	97.6	62.2	82.1
792	Childress	97.6	62.7	82.3
793-794	Lubbock	99.3	63.4	83.6
795-796	Abilene	97.7	62.6	82.3
797	Midland	100.0	64.8	84.6
798-799,885	El Paso	96.0	64.7	82.3
UTAH				
840-841	Salt Lake City	102.8	72.9	89.7
842,844	Ogden	98.2	72.9	87.2
843	Logan	100.2	72.9	88.3
845	Price	100.8	71.7	88.1
846-847	Provo	100.6	72.8	88.5
VERMONT				
050	White River Jct.	97.7	81.6	90.7
051	Bellows Falls	96.0	95.3	95.7
052	Bennington	96.5	91.6	94.3
053	Brattleboro	96.9	95.3	96.2
054	Burlington	101.0	80.9	92.2
056	Montpelier	99.8	85.9	93.7
057	Rutland	98.1	80.9	90.6
058	St. Johnsbury	97.8	81.5	90.7
059	Guildhall	96.4	81.1	89.7
VIRGINIA				
220-221	Fairfax	100.0	83.8	92.9
222	Arlington	100.7	84.5	93.7
223	Alexandria	99.8	84.7	93.2
224-225	Fredericksburg	98.7	82.2	91.5
226	Winchester	99.3	82.9	92.2
227	Culpeper	99.1	80.5	91.0
228	Harrisonburg	99.5	67.3	85.4
229	Charlottesville	99.8	69.3	86.5
230-232	Richmond	98.5	77.9	89.5
233-235	Norfolk	98.5	71.3	86.6
236	Newport News	98.2	70.4	86.0
237	Portsmouth	97.7	64.3	83.1
238	Petersburg	98.3	73.9	87.6
239	Farmville	97.4	58.6	80.4
240-241	Roanoke	100.3	73.0	88.3
242	Bristol	98.6	60.5	82.0
243	Pulaski	98.2	62.7	82.7
244	Staunton	99.0	65.8	84.5
245	Lynchburg	99.2	75.2	88.7
246	Grundy	98.5	56.0	79.9
WASHINGTON				
980-981,987	Seattle	103.6	107.3	105.2
982	Everett	102.0	102.4	102.2
983-984	Tacoma	102.4	102.3	102.3
985	Olympia	100.7	102.3	101.4
986	Vancouver	103.3	97.2	100.6
988	Wenatchee	102.6	89.5	96.9
989	Yakima	102.5	98.2	100.6
990-992	Spokane	101.4	85.0	94.2
993	Richland	101.1	92.7	97.4
994	Clarkston	100.6	83.4	93.1
WEST VIRGINIA				
247-248	Bluefield	97.5	89.6	94.0
249	Lewisburg	99.1	88.9	94.6
250-253	Charleston	97.9	93.3	95.9
254	Martinsburg	98.8	84.4	92.5
255-257	Huntington	99.7	93.8	97.1
258-259	Beckley	97.3	91.8	94.8
260	Wheeling	100.0	91.9	96.5
261	Parkersburg	98.8	89.5	94.8
262	Buckhannon	98.8	91.5	95.6
263-264	Clarksburg	99.4	91.6	96.0
265	Morgantown	99.4	91.6	96.0
266	Gassaway	98.8	89.9	94.9
267	Romney	98.6	86.4	93.3
268	Petersburg	98.5	85.0	92.6
WISCONSIN				
530,532	Milwaukee	99.3	106.6	102.5
531	Kenosha	98.4	105.0	101.3
534	Racine	97.8	106.2	101.5
535	Beloit	97.7	98.6	98.1
537	Madison	98.0	98.9	98.4

STATE/ZIP	CITY	MAT.	INST.	TOTAL
WISCONSIN (CONT'D)				
538	Lancaster	96.0	93.9	95.1
539	Portage	94.4	96.6	95.4
540	New Richmond	95.6	92.7	94.3
541-543	Green Bay	99.6	96.5	98.2
544	Wausau	95.0	94.5	94.8
545	Rhinelander	98.3	90.1	94.7
546	La Crosse	95.9	96.9	96.4
547	Eau Claire	97.6	97.3	97.5
548	Superior	95.4	97.1	96.1
549	Oshkosh	95.7	89.2	92.9
WYOMING				
820	Cheyenne	101.3	73.5	89.2
821	Yellowstone Nat'l Park	99.4	74.4	88.4
822	Wheatland	100.6	69.2	86.9
823	Rawlins	102.1	73.5	89.6
824	Worland	100.1	73.0	88.2
825	Riverton	101.2	70.7	87.8
826	Casper	101.4	70.3	87.8
827	Newcastle	99.9	73.1	88.2
828	Sheridan	102.8	71.5	89.1
829-831	Rock Springs	103.9	72.7	90.2
CANADIAN FACTORS (reflect Canadian currency)				
ALBERTA				
	Calgary	119.5	98.8	110.4
	Edmonton	120.4	98.8	110.9
	Fort McMurray	123.4	91.9	109.7
	Lethbridge	118.3	91.4	106.5
	Lloydminster	112.7	88.1	101.9
	Medicine Hat	112.9	87.4	101.7
	Red Deer	113.3	87.4	102.0
BRITISH COLUMBIA				
	Kamloops	114.1	87.2	102.3
	Prince George	115.1	86.6	102.6
	Vancouver	118.0	91.2	106.3
	Victoria	115.3	85.7	102.3
MANITOBA				
	Brandon	123.6	73.8	101.8
	Portage la Prairie	112.7	72.2	94.9
	Winnipeg	121.3	72.0	99.7
NEW BRUNSWICK				
	Bathurst	111.3	65.0	91.0
	Dalhousie	111.3	65.2	91.1
	Fredericton	118.2	72.5	98.2
	Moncton	111.5	69.3	93.0
	Newcastle	111.3	65.7	91.3
	St. John	113.8	77.0	97.7
NEWFOUNDLAND				
	Corner Brook	128.0	71.5	103.3
	St. Johns	120.7	90.3	107.4
NORTHWEST TERRITORIES				
	Yellowknife	133.4	86.2	112.7
NOVA SCOTIA				
	Bridgewater	112.3	73.4	95.3
	Dartmouth	124.6	73.5	102.2
	Halifax	117.4	87.2	104.2
	New Glasgow	122.6	73.5	101.1
	Sydney	120.6	73.5	100.0
	Truro	111.9	73.4	95.1
	Yarmouth	122.5	73.5	101.1
ONTARIO				
	Barrie	117.0	91.0	105.6
	Brantford	114.3	94.9	105.8
	Cornwall	114.2	91.3	104.2
	Hamilton	116.0	99.6	108.8
	Kingston	115.3	91.4	104.8
	Kitchener	109.1	95.1	103.0
	London	115.6	97.0	107.5
	North Bay	125.6	89.2	109.7
	Oshawa	111.6	94.7	104.2
	Ottawa	117.2	98.0	108.8
	Owen Sound	117.1	89.2	104.9
	Peterborough	114.3	91.1	104.1
	Sarnia	114.3	95.5	106.0
ONTARIO (CONT'D)				

STATE/ZIP	CITY	MAT.	INST.	TOTAL
	Sault Ste. Marie	109.6	91.7	101.8
	St. Catharines	107.1	96.1	102.3
	Sudbury	106.9	94.9	101.6
	Thunder Bay	108.3	95.4	102.7
	Timmins	114.4	89.2	103.4
	Toronto	115.9	104.2	110.8
	Windsor	107.6	95.2	102.2
PRINCE EDWARD ISLAND				
	Charlottetown	120.9	62.8	95.5
	Summerside	125.1	60.5	96.8
QUEBEC				
	Cap-de-la-Madeleine	111.7	83.6	99.4
	Charlesbourg	111.7	83.6	99.4
	Chicoutimi	111.1	88.9	101.4
	Gatineau	111.4	83.4	99.1
	Granby	111.6	83.3	99.2
	Hull	111.5	83.4	99.2
	Joliette	111.9	83.6	99.5
	Laval	111.5	83.4	99.2
	Montreal	118.0	91.7	106.5
	Quebec City	118.5	92.2	107.0
	Rimouski	111.3	88.9	101.5
	Rouyn-Noranda	111.4	83.4	99.1
	Saint-Hyacinthe	110.8	83.4	98.8
	Sherbrooke	111.7	83.4	99.3
	Sorel	111.9	83.6	99.5
	Saint-Jerome	111.4	83.4	99.1
	Trois-Rivieres	123.3	83.6	105.9
SASKATCHEWAN				
	Moose Jaw	110.1	66.7	91.1
	Prince Albert	109.2	65.0	89.8
	Regina	122.8	95.4	110.8
	Saskatoon	110.8	92.5	102.7
YUKON				
	Whitehorse	135.9	71.3	107.6

R011105-05   Tips for Accurate Estimating

1. Use pre-printed or columnar forms for orderly sequence of dimensions and locations and for recording telephone quotations.
2. Use only the front side of each paper or form except for certain pre-printed summary forms.
3. Be consistent in listing dimensions: For example, length x width x height. This helps in rechecking to ensure that the total length of partitions is appropriate for the building area.
4. Use printed (rather than measured) dimensions where given.
5. Add up multiple printed dimensions for a single entry where possible.
6. Measure all other dimensions carefully.
7. Use each set of dimensions to calculate multiple related quantities.
8. Convert foot and inch measurements to decimal feet when listing. Memorize decimal equivalents to .01 parts of a foot (1/8" equals approximately .01').
9. Do not "round off" quantities until the final summary.
10. Mark drawings with different colors as items are taken off.
11. Keep similar items together, different items separate.
12. Identify location and drawing numbers to aid in future checking for completeness.
13. Measure or list everything on the drawings or mentioned in the specifications.
14. It may be necessary to list items not called for to make the job complete.
15. Be alert for: Notes on plans such as N.T.S. (not to scale); changes in scale throughout the drawings; reduced size drawings; discrepancies between the specifications and the drawings.
16. Develop a consistent pattern of performing an estimate. For example:
    a. Start the quantity takeoff at the lower floor and move to the next higher floor.
    b. Proceed from the main section of the building to the wings.
    c. Proceed from south to north or vice versa, clockwise or counterclockwise.
    d. Take off floor plan quantities first, elevations next, then detail drawings.
17. List all gross dimensions that can be either used again for different quantities, or used as a rough check of other quantities for verification (exterior perimeter, gross floor area, individual floor areas, etc.).
18. Utilize design symmetry or repetition (repetitive floors, repetitive wings, symmetrical design around a center line, similar room layouts, etc.). Note: Extreme caution is needed here so as not to omit or duplicate an area.
19. Do not convert units until the final total is obtained. For instance, when estimating concrete work, keep all units to the nearest cubic foot, then summarize and convert to cubic yards.
20. When figuring alternatives, it is best to total all items involved in the basic system, then total all items involved in the alternates. Therefore, you work with positive numbers in all cases. When adds and deducts are used, it is often confusing whether to add or subtract a portion of an item; especially on a complicated or involved alternate.

R011105-10   Unit Gross Area Requirements

The figures in the table below indicate typical ranges in square feet as a function of the "occupant" unit. This table is best used in the preliminary design stages to help determine the probable size requirement for the total project.

Building Type	Unit	Gross Area in S.F. 1/4	Median	3/4
Apartments	Unit	660	860	1,100
Auditorium & Play Theaters	Seat	18	25	38
Bowling Alleys	Lane		940	
Churches & Synagogues	Seat	20	28	39
Dormitories	Bed	200	230	275
Fraternity & Sorority Houses	Bed	220	315	370
Garages, Parking	Car	325	355	385
Hospitals	Bed	685	850	1,075
Hotels	Rental Unit	475	600	710
Housing for the elderly	Unit	515	635	755
Housing, Public	Unit	700	875	1,030
Ice Skating Rinks	Total	27,000	30,000	36,000
Motels	Rental Unit	360	465	620
Nursing Homes	Bed	290	350	450
Restaurants	Seat	23	29	39
Schools, Elementary	Pupil	65	77	90
Junior High & Middle		85	110	129
Senior High		102	130	145
Vocational		110	135	195
Shooting Ranges	Point		450	
Theaters & Movies	Seat		15	

R011105-50   Metric Conversion Factors

Description: This table is primarily for converting customary U.S. units in the left hand column to SI metric units in the right hand column. In addition, conversion factors for some commonly encountered Canadian and non-SI metric units are included.

	If You Know		Multiply By		To Find
Length	Inches	x	25.4[a]	=	Millimeters
	Feet	x	0.3048[a]	=	Meters
	Yards	x	0.9144[a]	=	Meters
	Miles (statute)	x	1.609	=	Kilometers
Area	Square inches	x	645.2	=	Square millimeters
	Square feet	x	0.0929	=	Square meters
	Square yards	x	0.8361	=	Square meters
Volume (Capacity)	Cubic inches	x	16,387	=	Cubic millimeters
	Cubic feet	x	0.02832	=	Cubic meters
	Cubic yards	x	0.7646	=	Cubic meters
	Gallons (U.S. liquids)[b]	x	0.003785	=	Cubic meters[c]
	Gallons (Canadian liquid)[b]	x	0.004546	=	Cubic meters[c]
	Ounces (U.S. liquid)[b]	x	29.57	=	Milliliters[c, d]
	Quarts (U.S. liquid)[b]	x	0.9464	=	Liters[c, d]
	Gallons (U.S. liquid)[b]	x	3.785	=	Liters[c, d]
Force	Kilograms force[d]	x	9.807	=	Newtons
	Pounds force	x	4.448	=	Newtons
	Pounds force	x	0.4536	=	Kilograms force[d]
	Kips	x	4448	=	Newtons
	Kips	x	453.6	=	Kilograms force[d]
Pressure, Stress, Strength (Force per unit area)	Kilograms force per square centimeter[d]	x	0.09807	=	Megapascals
	Pounds force per square inch (psi)	x	0.006895	=	Megapascals
	Kips per square inch	x	6.895	=	Megapascals
	Pounds force per square inch (psi)	x	0.07031	=	Kilograms force per square centimeter[d]
	Pounds force per square foot	x	47.88	=	Pascals
	Pounds force per square foot	x	4.882	=	Kilograms force per square meter[d]
Flow	Cubic feet per minute	x	0.4719	=	Liters per second
	Gallons per minute	x	0.0631	=	Liters per second
	Gallons per hour	x	1.05	=	Milliliters per second
Bending Moment Or Torque	Inch-pounds force	x	0.01152	=	Meter-kilograms force[d]
	Inch-pounds force	x	0.1130	=	Newton-meters
	Foot-pounds force	x	0.1383	=	Meter-kilograms force[d]
	Foot-pounds force	x	1.356	=	Newton-meters
	Meter-kilograms force[d]	x	9.807	=	Newton-meters
Mass	Ounces (avoirdupois)	x	28.35	=	Grams
	Pounds (avoirdupois)	x	0.4536	=	Kilograms
	Tons (metric)	x	1000	=	Kilograms
	Tons, short (2000 pounds)	x	907.2	=	Kilograms
	Tons, short (2000 pounds)	x	0.9072	=	Megagrams[e]
Mass per Unit Volume	Pounds mass per cubic foot	x	16.02	=	Kilograms per cubic meter
	Pounds mass per cubic yard	x	0.5933	=	Kilograms per cubic meter
	Pounds mass per gallon (U.S. liquid)[b]	x	119.8	=	Kilograms per cubic meter
	Pounds mass per gallon (Canadian liquid)[b]	x	99.78	=	Kilograms per cubic meter
Temperature	Degrees Fahrenheit	(F-32)/1.8		=	Degrees Celsius
	Degrees Fahrenheit	(F+459.67)/1.8		=	Degrees Kelvin
	Degrees Celsius	C+273.15		=	Degrees Kelvin

[a]The factor given is exact
[b]One U.S. gallon = 0.8327 Canadian gallon
[c]1 liter = 1000 milliliters = 1000 cubic centimeters
1 cubic decimeter = 0.001 cubic meter

[d]Metric but not SI unit
[e]Called "tonne" in England and "metric ton" in other metric countries

R011105-60   Weights and Measures

Measures of Length
1 Mile = 1760 Yards = 5280 Feet
1 Yard = 3 Feet = 36 inches
1 Foot = 12 Inches
1 Mil = 0.001 Inch
1 Fathom = 2 Yards = 6 Feet
1 Rod = 5.5 Yards = 16.5 Feet
1 Hand = 4 Inches
1 Span = 9 Inches
1 Micro-inch = One Millionth Inch or 0.000001 Inch
1 Micron = One Millionth Meter + 0.00003937 Inch

Surveyor's Measure
1 Mile = 8 Furlongs = 80 Chains
1 Furlong = 10 Chains = 220 Yards
1 Chain = 4 Rods = 22 Yards = 66 Feet = 100 Links
1 Link = 7.92 Inches

Square Measure
1 Square Mile = 640 Acres = 6400 Square Chains
1 Acre = 10 Square Chains = 4840 Square Yards = 43,560 Sq. Ft.
1 Square Chain = 16 Square Rods = 484 Square Yards = 4356 Sq. Ft.
1 Square Rod = 30.25 Square Yards = 272.25 Square Feet = 625 Square Lines
1 Square Yard = 9 Square Feet
1 Square Foot = 144 Square Inches
An Acre equals a Square 208.7 Feet per Side

Cubic Measure
1 Cubic Yard = 27 Cubic Feet
1 Cubic Foot = 1728 Cubic Inches
1 Cord of Wood = 4 x 4 x 8 Feet = 128 Cubic Feet
1 Perch of Masonry = 16½ x 1½ x 1 Foot = 24.75 Cubic Feet

Avoirdupois or Commercial Weight
1 Gross or Long Ton = 2240 Pounds
1 Net or Short Ton = 2000 Pounds
1 Pound = 16 Ounces = 7000 Grains
1 Ounce = 16 Drachms = 437.5 Grains
1 Stone = 14 Pounds

Power
1 British Thermal Unit per Hour = 0.2931 Watts
1 Ton (Refrigeration) = 3.517 Kilowatts
1 Horsepower (Boiler) = 9.81 Kilowatts
1 Horsepower (550 ft-lb/s) = 0.746 Kilowatts

Shipping Measure
For Measuring Internal Capacity of a Vessel:
  1 Register Ton = 100 Cubic Feet

For Measurement of Cargo:
  Approximately 40 Cubic Feet of Merchandise is considered a Shipping Ton, unless that bulk would weigh more than 2000 Pounds, in which case Freight Charge may be based upon weight.

40 Cubic Feet = 32.143 U.S. Bushels = 31.16 Imp. Bushels

Liquid Measure
1 Imperial Gallon = 1.2009 U.S. Gallon = 277.42 Cu. In.
1 Cubic Foot = 7.48 U.S. Gallons

R011110-30 Engineering Fees

Typical **Structural Engineering Fees** based on type of construction and total project size. These fees are included in Architectural Fees.

Type of Construction	Total Project Size (in thousands of dollars)			
	$500	$500-$1,000	$1,000-$5,000	Over $5000
Industrial buildings, factories & warehouses	Technical payroll times 2.0 to 2.5	1.60%	1.25%	1.00%
Hotels, apartments, offices, dormitories, hospitals, public buildings, food stores		2.00%	1.70%	1.20%
Museums, banks, churches and cathedrals		2.00%	1.75%	1.25%
Thin shells, prestressed concrete, earthquake resistive		2.00%	1.75%	1.50%
Parking ramps, auditoriums, stadiums, convention halls, hangars & boiler houses		2.50%	2.00%	1.75%
Special buildings, major alterations, underpinning & future expansion		Add to above 0.5%	Add to above 0.5%	Add to above 0.5%

For complex reinforced concrete or unusually complicated structures, add 20% to 50%.

Typical **Mechanical and Electrical Engineering Fees** are based on the size of the subcontract. The fee structure for both is shown below. These fees are included in Architectural Fees.

Type of Construction	Subcontract Size							
	$25,000	$50,000	$100,000	$225,000	$350,000	$500,000	$750,000	$1,000,000
Simple structures	6.4%	5.7%	4.8%	4.5%	4.4%	4.3%	4.2%	4.1%
Intermediate structures	8.0	7.3	6.5	5.6	5.1	5.0	4.9	4.8
Complex structures	10.1	9.0	9.0	8.0	7.5	7.5	7.0	7.0

For renovations, add 15% to 25% to applicable fee.

R012153-10 Repair and Remodeling

Cost figures are based on new construction utilizing the most cost-effective combination of labor, equipment and material with the work scheduled in proper sequence to allow the various trades to accomplish their work in an efficient manner.

The costs for repair and remodeling work must be modified due to the following factors that may be present in any given repair and remodeling project.

1. Equipment usage curtailment due to the physical limitations of the project, with only hand-operated equipment being used.

2. Increased requirement for shoring and bracing to hold up the building while structural changes are being made and to allow for temporary storage of construction materials on above-grade floors.

3. Material handling becomes more costly due to having to move within the confines of an enclosed building. For multi-story construction, low capacity elevators and stairwells may be the only access to the upper floors.

4. Large amounts of cutting and patching and attempting to match the existing construction is required. It is often more economical to remove entire walls rather than create many new door and window openings. This sort of trade-off has to be carefully analyzed.

5. Cost of protection of completed work is increased since the usual sequence of construction usually cannot be accomplished.

6. Economies of scale usually associated with new construction may not be present. If small quantities of components must be custom fabricated due to job requirements, unit costs will naturally increase. Also, if only small work areas are available at a given time, job scheduling between trades becomes difficult and subcontractor quotations may reflect the excessive start-up and shut-down phases of the job.

7. Work may have to be done on other than normal shifts and may have to be done around an existing production facility which has to stay in production during the course of the repair and remodeling.

8. Dust and noise protection of adjoining non-construction areas can involve substantial special protection and alter usual construction methods.

9. Job may be delayed due to unexpected conditions discovered during demolition or removal. These delays ultimately increase construction costs.

10. Piping and ductwork runs may not be as simple as for new construction. Wiring may have to be snaked through walls and floors.

11. Matching "existing construction" may be impossible because materials may no longer be manufactured. Substitutions may be expensive.

12. Weather protection of existing structure requires additional temporary structures to protect building at openings.

13. On small projects, because of local conditions, it may be necessary to pay a tradesman for a minimum of four hours for a task that is completed in one hour.

All of the above areas can contribute to increased costs for a repair and remodeling project. Each of the above factors should be considered in the planning, bidding and construction stage in order to minimize the increased costs associated with repair and remodeling jobs.

565

For customer support on your Electrical Costs with RSMeans data, call 800.448.8182.

R012153-60   Security Factors

Contractors entering, working in, and exiting secure facilities often lose productive time during a normal workday. The recommended allowances in this section are intended to provide for the loss of productivity by increasing labor costs. Note that different costs are associated with searches upon entry only and searches upon entry and exit. Time spent in a queue is unpredictable and not part of these allowances. Contractors should plan ahead for this situation.

Security checkpoints are designed to reflect the level of security required to gain access or egress. An extreme example is when contractors, along with any materials, tools, equipment, and vehicles, must be physically searched and have all materials, tools, equipment, and vehicles inventoried and documented prior to both entry and exit.

Physical searches without going through the documentation process represent the next level and take up less time.

Electronic searches—passing through a detector or x-ray machine with no documentation of materials, tools, equipment, and vehicles—take less time than physical searches.

Visual searches of materials, tools, equipment, and vehicles represent the next level of security.

Finally, access by means of an ID card or displayed sticker takes the least amount of time.

Another consideration is if the searches described above are performed each and every day, or if they are performed only on the first day with access granted by ID card or displayed sticker for the remainder of the project. The figures for this situation have been calculated to represent the initial check-in as described and subsequent entry by ID card or displayed sticker for up to 20 days on site. For the situation described above, where the time period is beyond 20 days, the impact on labor cost is negligible.

There are situations where tradespeople must be accompanied by an escort and observed during the work day. The loss of freedom of movement will slow down productivity for the tradesperson. Costs for the observer have not been included. Those costs are normally born by the owner.

General Requirements   R0129 Payment Procedures

R012909-80   Sales Tax by State

State sales tax on materials is tabulated below (5 states have no sales tax). Many states allow local jurisdictions, such as a county or city, to levy additional sales tax.

Some projects may be sales tax exempt, particularly those constructed with public funds.

State	Tax (%)	State	Tax (%)	State	Tax (%)	State	Tax (%)
Alabama	4	Illinois	6.25	Montana	0	Rhode Island	7
Alaska	0	Indiana	7	Nebraska	5.5	South Carolina	6
Arizona	5.6	Iowa	6	Nevada	6.85	South Dakota	5
Arkansas	6.5	Kansas	6.5	New Hampshire	0	Tennessee	7
California	7.25	Kentucky	6	New Jersey	7	Texas	6.25
Colorado	2.9	Louisiana	4	New Mexico	5.125	Utah	5.95
Connecticut	6.35	Maine	5.5	New York	4	Vermont	6
Delaware	0	Maryland	6	North Carolina	4.75	Virginia	5.3
District of Columbia	5.75	Massachusetts	6.25	North Dakota	5	Washington	6.5
Florida	6	Michigan	6	Ohio	5.75	West Virginia	6
Georgia	4	Minnesota	6.875	Oklahoma	4.5	Wisconsin	5
Hawaii	4	Mississippi	7	Oregon	0	Wyoming	4
Idaho	6	Missouri	4.225	Pennsylvania	6	Average	5.11 %

Sales Tax by Province (Canada)

GST - a value-added tax, which the government imposes on most goods and services provided in or imported into Canada. PST - a retail sales tax, which five of the provinces impose on the prices of most goods and some

services. QST - a value-added tax, similar to the federal GST, which Quebec imposes. HST - Three provinces have combined their retail sales taxes with the federal GST into one harmonized tax.

Province	PST (%)	QST (%)	GST (%)	HST (%)
Alberta	0	0	5	0
British Columbia	7	0	5	0
Manitoba	8	0	5	0
New Brunswick	0	0	0	15
Newfoundland	0	0	0	15
Northwest Territories	0	0	5	0
Nova Scotia	0	0	0	15
Ontario	0	0	0	13
Prince Edward Island	0	0	0	15
Quebec	0	9.975	5	0
Saskatchewan	6	0	5	0
Yukon	0	0	5	0

R012909-85 Unemployment Taxes and Social Security Taxes

State unemployment tax rates vary not only from state to state, but also with the experience rating of the contractor. The federal unemployment tax rate is 6.0% of the first $7,000 of wages. This is reduced by a credit of up to 5.4% for timely payment to the state. The minimum federal unemployment tax is 0.6% after all credits.

Social security (FICA) for 2018 is estimated at time of publication to be 7.65% of wages up to $127,200.

R012909-86 Unemployment Tax by State

Information is from the U.S. Department of Labor, state unemployment tax rates.

State	Tax (%)	State	Tax (%)	State	Tax (%)	State	Tax (%)
Alabama	6.74	Illinois	7.75	Montana	6.12	Rhode Island	9.79
Alaska	5.4	Indiana	7.474	Nebraska	5.4	South Carolina	5.46
Arizona	8.91	Iowa	8	Nevada	5.4	South Dakota	9.5
Arkansas	6.0	Kansas	7.6	New Hampshire	7.5	Tennessee	10.0
California	6.2	Kentucky	10.0	New Jersey	5.8	Texas	7.5
Colorado	8.9	Louisiana	6.2	New Mexico	5.4	Utah	7.2
Connecticut	6.8	Maine	5.4	New York	8.5	Vermont	8.4
Delaware	8.0	Maryland	7.50	North Carolina	5.76	Virginia	6.27
District of Columbia	7	Massachusetts	11.13	North Dakota	10.72	Washington	5.7
Florida	5.4	Michigan	10.3	Ohio	8.7	West Virginia	7.5
Georgia	5.4	Minnesota	9.0	Oklahoma	5.5	Wisconsin	12.0
Hawaii	5.6	Mississippi	5.4	Oregon	5.4	Wyoming	8.8
Idaho	5.4	Missouri	9.75	Pennsylvania	10.89	Median	7.47 %

R012909-90 Overtime

One way to improve the completion date of a project or eliminate negative float from a schedule is to compress activity duration times. This can be achieved by increasing the crew size or working overtime with the proposed crew.

To determine the costs of working overtime to compress activity duration times, consider the following examples. Below is an overtime efficiency and cost chart based on a five, six, or seven day week with an eight through twelve hour day. Payroll percentage increases for time and one half and double times are shown for the various working days.

Days per Week	Hours per Day	Production Efficiency					Payroll Cost Factors	
		1st Week	2nd Week	3rd Week	4th Week	Average 4 Weeks	@ 1-1/2 Times	@ 2 Times
5	8	100%	100%	100%	100%	100%	1.000	1.000
	9	100	100	95	90	96	1.056	1.111
	10	100	95	90	85	93	1.100	1.200
	11	95	90	75	65	81	1.136	1.273
	12	90	85	70	60	76	1.167	1.333
6	8	100	100	95	90	96	1.083	1.167
	9	100	95	90	85	93	1.130	1.259
	10	95	90	85	80	88	1.167	1.333
	11	95	85	70	65	79	1.197	1.394
	12	90	80	65	60	74	1.222	1.444
7	8	100	95	85	75	89	1.143	1.286
	9	95	90	80	70	84	1.183	1.365
	10	90	85	75	65	79	1.214	1.429
	11	85	80	65	60	73	1.240	1.481
	12	85	75	60	55	69	1.262	1.524

R013113-40   Builder's Risk Insurance

Builder's risk insurance is insurance on a building during construction. Premiums are paid by the owner or the contractor. Blasting, collapse and underground insurance would raise total insurance costs.

R013113-50   General Contractor's Overhead

There are two distinct types of overhead on a construction project: Project overhead and main office overhead. Project overhead includes those costs at a construction site not directly associated with the installation of construction materials. Examples of project overhead costs include the following:

1. Superintendent
2. Construction office and storage trailers
3. Temporary sanitary facilities
4. Temporary utilities
5. Security fencing
6. Photographs
7. Cleanup
8. Performance and payment bonds

The above project overhead items are also referred to as general requirements and therefore are estimated in Division 1. Division 1 is the first division listed in the CSI MasterFormat but it is usually the last division estimated. The sum of the costs in Divisions 1 through 49 is referred to as the sum of the direct costs.

All construction projects also include indirect costs. The primary components of indirect costs are the contractor's main office overhead and profit. The amount of the main office overhead expense varies depending on the following:

1. Owner's compensation
2. Project managers' and estimators' wages
3. Clerical support wages
4. Office rent and utilities
5. Corporate legal and accounting costs
6. Advertising
7. Automobile expenses
8. Association dues
9. Travel and entertainment expenses

These costs are usually calculated as a percentage of annual sales volume. This percentage can range from 35% for a small contractor doing less than $500,000 to 5% for a large contractor with sales in excess of $100 million.

R013113-55   Installing Contractor's Overhead

Installing contractors (subcontractors) also incur costs for general requirements and main office overhead.

Included within the total incl. overhead and profit costs is a percent mark-up for overhead that includes:

1. Compensation and benefits for office staff and project managers
2. Office rent, utilities, business equipment, and maintenance
3. Corporate legal and accounting costs
4. Advertising
5. Vehicle expenses (for office staff and project managers)
6. Association dues
7. Travel, entertainment
8. Insurance
9. Small tools and equipment

R013113-60   Workers' Compensation Insurance Rates by Trade

The table below tabulates the national averages for workers' compensation insurance rates by trade and type of building. The average "Insurance Rate" is multiplied by the "% of Building Cost" for each trade. This produces the "Workers' Compensation" cost by % of total labor cost, to be added for each trade by building type to determine the weighted average workers' compensation rate for the building types analyzed.

Trade	Insurance Rate (% Labor Cost) Range		Average	% of Building Cost Office Bldgs.	Schools & Apts.	Mfg.	Workers' Compensation Office Bldgs.	Schools & Apts.	Mfg.
Excavation, Grading, etc.	2.7 % to	20.1%	8.5%	4.8%	4.9%	4.5%	0.41%	0.42%	0.38%
Piles & Foundations	5.3 to	29.8	13.4	7.1	5.2	8.7	0.95	0.70	1.17
Concrete	4.1 to	28.0	11.8	5.0	14.8	3.7	0.59	1.75	0.44
Masonry	3.9 to	49.3	13.8	6.9	7.5	1.9	0.95	1.04	0.26
Structural Steel	5.3 to	59.1	21.2	10.7	3.9	17.6	2.27	0.83	3.73
Miscellaneous & Ornamental Metals	3.3 to	24.4	10.6	2.8	4.0	3.6	0.30	0.42	0.38
Carpentry & Millwork	4.4 to	32.4	13.0	3.7	4.0	0.5	0.48	0.52	0.07
Metal or Composition Siding	5.5 to	107.2	19.0	2.3	0.3	4.3	0.44	0.06	0.82
Roofing	5.5 to	120.3	29.0	2.3	2.6	3.1	0.67	0.75	0.90
Doors & Hardware	3.2 to	32.4	11.0	0.9	1.4	0.4	0.10	0.15	0.04
Sash & Glazing	4.7 to	25.5	12.1	3.5	4.0	1.0	0.42	0.48	0.12
Lath & Plaster	3.0 to	31.6	10.7	3.3	6.9	0.8	0.35	0.74	0.09
Tile, Marble & Floors	2.7 to	18.3	8.7	2.6	3.0	0.5	0.23	0.26	0.04
Acoustical Ceilings	2.4 to	46.3	8.5	2.4	0.2	0.3	0.20	0.02	0.03
Painting	3.3 to	38.8	11.2	1.5	1.6	1.6	0.17	0.18	0.18
Interior Partitions	4.4 to	32.4	13.0	3.9	4.3	4.4	0.51	0.56	0.57
Miscellaneous Items	2.3 to	97.7	11.2	5.2	3.7	9.7	0.58	0.42	1.09
Elevators	1.3 to	13.7	4.7	2.1	1.1	2.2	0.10	0.05	0.10
Sprinklers	2.0 to	15.5	6.7	0.5	—	2.0	0.03	—	0.13
Plumbing	1.7 to	14.0	6.3	4.9	7.2	5.2	0.31	0.45	0.33
Heat., Vent., Air Conditioning	3.3 to	17.8	8.3	13.5	11.0	12.9	1.12	0.91	1.07
Electrical	1.9 to	11.6	5.2	10.1	8.4	11.1	0.53	0.44	0.58
Total	1.3 % to	120.3%	—	100.0%	100.0%	100.0%	11.71%	11.15%	12.52%

Overall Weighted Average   11.79%

Workers' Compensation Insurance Rates by States

The table below lists the weighted average Workers' Compensation base rate for each state with a factor comparing this with the national average of 11.8%.

State	Weighted Average	Factor	State	Weighted Average	Factor	State	Weighted Average	Factor
Alabama	15.0%	127	Kentucky	10.4%	88	North Dakota	6.2%	53
Alaska	10.4	88	Louisiana	18.7	158	Ohio	7.2	61
Arizona	9.6	81	Maine	10.4	88	Oklahoma	8.9	75
Arkansas	7.0	59	Maryland	11.3	96	Oregon	9.3	79
California	22.2	188	Massachusetts	11.2	95	Pennsylvania	21.1	179
Colorado	7.5	64	Michigan	8.2	69	Rhode Island	13.7	116
Connecticut	17.5	148	Minnesota	16.9	143	South Carolina	16.5	140
Delaware	13.9	118	Mississippi	11.8	100	South Dakota	11.8	100
District of Columbia	9.1	77	Missouri	12.4	105	Tennessee	8.6	73
Florida	11.1	94	Montana	8.8	75	Texas	6.6	56
Georgia	31.9	270	Nebraska	13.5	114	Utah	7.4	63
Hawaii	8.5	72	Nevada	7.5	64	Vermont	10.9	92
Idaho	9.4	80	New Hampshire	12.0	102	Virginia	6.9	58
Illinois	21.1	179	New Jersey	14.8	125	Washington	9.1	77
Indiana	4.1	35	New Mexico	13.3	113	West Virginia	4.5	38
Iowa	13.7	116	New York	19.2	163	Wisconsin	12.2	103
Kansas	6.5	55	North Carolina	15.8	134	Wyoming	5.6	47

Weighted Average for U.S. is   11.8% of payroll = 100%

The weighted average skilled worker rate for 35 trades is 11.8%. For bidding purposes, apply the full value of Workers' Compensation directly to total labor costs, or if labor is 38%, materials 42% and overhead and profit 20% of total cost, carry 38/80 x 11.8% = 6.0% of cost (before overhead and profit) into overhead. Rates vary not only from state to state but also with the experience rating of the contractor.

Rates are the most current available at the time of publication.

R013113-80   Performance Bond

This table shows the cost of a Performance Bond for a construction job scheduled to be completed in 12 months. Add 1% of the premium cost per month for jobs requiring more than 12 months to complete. The rates are "standard" rates offered to contractors that the bonding company considers financially sound and capable of doing the work. Preferred rates are offered by some bonding companies based upon financial strength of the contractor. Actual rates vary from contractor to contractor and from bonding company to bonding company. Contractors should prequalify through a bonding agency before submitting a bid on a contract that requires a bond.

Contract Amount		Building Construction Class B Projects			Highways & Bridges					
					Class A New Construction			Class A-1 Highway Resurfacing		
		$25.00 per M			$15.00 per M			$9.40 per M		
First $ 100,000 bid										
Next 400,000 bid	$ 2,500	plus	$15.00	per M	$ 1,500	plus	$10.00	per M	$ 940	plus	$7.20 per M
Next 2,000,000 bid	8,500	plus	10.00	per M	5,500	plus	7.00	per M	3,820	plus	5.00 per M
Next 2,500,000 bid	28,500	plus	7.50	per M	19,500	plus	5.50	per M	15,820	plus	4.50 per M
Next 2,500,000 bid	47,250	plus	7.00	per M	33,250	plus	5.00	per M	28,320	plus	4.50 per M
Over 7,500,000 bid	64,750	plus	6.00	per M	45,750	plus	4.50	per M	39,570	plus	4.00 per M

R015113-65   Temporary Power Equipment

Cost data for the temporary equipment was developed utilizing the following information.

1) Re-usable material-services, transformers, equipment and cords are based on new purchase and prorated to three projects.
2) PVC feeder includes trench and backfill.
3) Connections include disconnects and fuses.
4) Labor units include an allowance for removal.
5) No utility company charges or fees are included.
6) Concrete pads or vaults are not included.
7) Utility company conduits not included.

R015423-10   Steel Tubular Scaffolding

On new construction, tubular scaffolding is efficient up to 60' high or five stories. Above this it is usually better to use hung scaffolding if construction permits. Swing scaffolding operations may interfere with tenants. In this case, the tubular is more practical at all heights.

In repairing or cleaning the front of an existing building the cost of tubular scaffolding per S.F. of building front goes up as the height increases above the first tier. The first tier cost is relatively high due to leveling and alignment.

The minimum efficient crew for erecting and dismantling is three workers. They can set up and remove 18 frame sections per day up to 5 stories high. For 6 to 12 stories high, a crew of four is most efficient. Use two or more on top and two on the bottom for handing up or hoisting. They can

also set up and remove 18 frame sections per day. At 7' horizontal spacing, this will run about 800 S.F. per day of erecting and dismantling. Time for placing and removing planks must be added to the above. A crew of three can place and remove 72 planks per day up to 5 stories. For over 5 stories, a crew of four can place and remove 80 planks per day.

The table below shows the number of pieces required to erect tubular steel scaffolding for 1000 S.F. of building frontage. This area is made up of a scaffolding system that is 12 frames (11 bays) long by 2 frames high.

For jobs under twenty-five frames, add 50% to the rental cost. Rental rates will be lower for jobs over three months duration. Large quantities for long periods can reduce rental rates by 20%.

Description of Component	Number of Pieces for 1000 S.F. of Building Front	Unit
5' Wide Standard Frame, 6'-4" High	24	Ea.
Leveling Jack & Plate	24	
Cross Brace	44	
Side Arm Bracket, 21"	12	
Guardrail Post	12	
Guardrail, 7' section	22	
Stairway Section	2	
Stairway Starter Bar	1	
Stairway Inside Handrail	2	
Stairway Outside Handrail	2	
Walk-Through Frame Guardrail	2	

Scaffolding is often used as falsework over 15' high during construction of cast-in-place concrete beams and slabs. Two foot wide scaffolding is generally used for heavy beam construction. The span between frames depends upon the load to be carried with a maximum span of 5'.

Heavy duty shoring frames with a capacity of 10,000#/leg can be spaced up to 10' O.C. depending upon form support design and loading.

Scaffolding used as horizontal shoring requires less than half the material required with conventional shoring.

On new construction, erection is done by carpenters.

Rolling towers supporting horizontal shores can reduce labor and speed the job. For maintenance work, catwalks with spans up to 70' can be supported by the rolling towers.

R015433-10   Contractor Equipment

Rental Rates shown elsewhere in the data set pertain to late model high quality machines in excellent working condition, rented from equipment dealers. Rental rates from contractors may be substantially lower than the rental rates from equipment dealers depending upon economic conditions; for older, less productive machines, reduce rates by a maximum of 15%. Any overtime must be added to the base rates. For shift work, rates are lower. Usual rule of thumb is 150% of one shift rate for two shifts; 200% for three shifts.

For periods of less than one week, operated equipment is usually more economical to rent than renting bare equipment and hiring an operator.

Costs to move equipment to a job site (mobilization) or from a job site (demobilization) are not included in rental rates, nor in any Equipment costs on any Unit Price line items or crew listings. These costs can be found elsewhere. If a piece of equipment is already at a job site, it is not appropriate to utilize mob/demob costs in an estimate again.

Rental rates vary throughout the country with larger cities generally having lower rates. Lease plans for new equipment are available for periods in excess of six months with a percentage of payments applying toward purchase.

Rental rates can also be treated as reimbursement costs for contractor-owned equipment. Owned equipment costs include depreciation, loan payments, interest, taxes, insurance, storage, and major repairs.

Monthly rental rates vary from 2% to 5% of the cost of the equipment depending on the anticipated life of the equipment and its wearing parts. Weekly rates are about 1/3 the monthly rates and daily rental rates are about 1/3 the weekly rates.

The hourly operating costs for each piece of equipment include costs to the user such as fuel, oil, lubrication, normal expendables for the equipment, and a percentage of the mechanic's wages chargeable to maintenance. The hourly operating costs listed do not include the operator's wages.

The daily cost for equipment used in the standard crews is figured by dividing the weekly rate by five, then adding eight times the hourly operating cost to give the total daily equipment cost, not including the operator. This figure is in the right hand column of the Equipment listings under Equipment Cost/Day.

Pile Driving rates shown for the pile hammer and extractor do not include leads, cranes, boilers or compressors. Vibratory pile driving requires an added field specialist during set-up and pile driving operation for the electric model. The hydraulic model requires a field specialist for set-up only. Up to 125 reuses of sheet piling are possible using vibratory drivers. For normal conditions, crane capacity for hammer type and size is as follows.

Crane Capacity	Hammer Type and Size		
	Air or Steam	Diesel	Vibratory
25 ton	to 8,750 ft.-lb.		70 H.P.
40 ton	15,000 ft.-lb.	to 32,000 ft.-lb.	170 H.P.
60 ton	25,000 ft.-lb.		300 H.P.
100 ton		112,000 ft.-lb.	

Cranes should be specified for the job by size, building and site characteristics, availability, performance characteristics, and duration of time required.

Backhoes & Shovels rent for about the same as equivalent size cranes but maintenance and operating expenses are higher. The crane operator's rate must be adjusted for high boom heights. Average adjustments: for 150' boom add 2% per hour; over 185', add 4% per hour; over 210', add 6% per hour; over 250', add 8% per hour and over 295', add 12% per hour.

Tower Cranes of the climbing or static type have jibs from 50' to 200' and capacities at maximum reach range from 4,000 to 14,000 pounds. Lifting capacities increase up to maximum load as the hook radius decreases.

Typical rental rates, based on purchase price, are about 2% to 3% per month.

Erection and dismantling run between 500 and 2000 labor hours. Climbing operation takes 10 labor hours per 20' climb. Crane dead time is about 5 hours per 40' climb. If crane is bolted to side of the building add cost of ties and extra mast sections. Climbing cranes have from 80' to 180' of mast while static cranes have 80' to 800' of mast.

Truck Cranes can be converted to tower cranes by using tower attachments. Mast heights over 400' have been used.

A single 100' high material **Hoist and Tower** can be erected and dismantled in about 400 labor hours; a double 100' high hoist and tower in about 600 labor hours. Erection times for additional heights are 3 and 4 labor hours

per vertical foot respectively up to 150', and 4 to 5 labor hours per vertical foot over 150' high. A 40' high portable Buck hoist takes about 160 labor hours to erect and dismantle. Additional heights take 2 labor hours per vertical foot to 80' and 3 labor hours per vertical foot for the next 100'. Most material hoists do not meet local code requirements for carrying personnel.

A 150' high **Personnel Hoist** requires about 500 to 800 labor hours to erect and dismantle. Budget erection time at 5 labor hours per vertical foot for all trades. Local code requirements or labor scarcity requiring overtime can add up to 50% to any of the above erection costs.

Earthmoving Equipment: The selection of earthmoving equipment depends upon the type and quantity of material, moisture content, haul distance, haul road, time available, and equipment available. Short haul cut and fill operations may require dozers only, while another operation may require excavators, a fleet of trucks, and spreading and compaction equipment. Stockpiled material and granular material are easily excavated with front end loaders. Scrapers are most economically used with hauls between 300' and 1-1/2 miles if adequate haul roads can be maintained. Shovels are often used for blasted rock and any material where a vertical face of 8' or more can be excavated. Special conditions may dictate the use of draglines, clamshells, or backhoes. Spreading and compaction equipment must be matched to the soil characteristics, the compaction required and the rate the fill is being supplied.

R015433-15   Heavy Lifting

Hydraulic Climbing Jacks

The use of hydraulic heavy lift systems is an alternative to conventional type crane equipment. The lifting, lowering, pushing, or pulling mechanism is a hydraulic climbing jack moving on a square steel jackrod from 1-5/8" to 4" square, or a steel cable. The jackrod or cable can be vertical or horizontal, stationary or movable, depending on the individual application. When the jackrod is stationary, the climbing jack will climb the rod and push or pull the load along with itself. When the climbing jack is stationary, the jackrod is movable with the load attached to the end and the climbing jack will lift or lower the jackrod with the attached load. The heavy lift system is normally operated by a single control lever located at the hydraulic pump.

The system is flexible in that one or more climbing jacks can be applied wherever a load support point is required, and the rate of lift synchronized.

Economic benefits have been demonstrated on projects such as: erection of ground assembled roofs and floors, complete bridge spans, girders and trusses, towers, chimney liners and steel vessels, storage tanks, and heavy machinery. Other uses are raising and lowering offshore work platforms, caissons, tunnel sections and pipelines.

R015436-50   Mobilization

Costs to move rented construction equipment to a job site from an equipment dealer's or contractor's yard (mobilization) or off the job site (demobilization) are not included in the rental or operating rates, nor in the equipment cost on a unit price line or in a crew listing. These costs can be found consolidated in the Mobilization section of the data and elsewhere in particular site work sections. If a piece of equipment is already on the job site, it is not appropriate to include mob/demob costs in a new estimate that requires use of that equipment. The following table identifies approximate sizes of rented construction equipment that would be hauled on a towed trailer. Because this listing is not all-encompassing, the user can infer as to what size trailer might be required for a piece of equipment not listed.

3-ton Trailer	20-ton Trailer	40-ton Trailer	50-ton Trailer
20 H.P. Excavator	110 H.P. Excavator	200 H.P. Excavator	270 H.P. Excavator
50 H.P. Skid Steer	165 H.P. Dozer	300 H.P. Dozer	Small Crawler Crane
35 H.P. Roller	150 H.P. Roller	400 H.P. Scraper	500 H.P. Scraper
40 H.P. Trencher	Backhoe	450 H.P. Art. Dump Truck	500 H.P. Art. Dump Truck

R024119-10   Demolition Defined

Whole Building Demolition - Demolition of the whole building with no concern for any particular building element, component, or material type being demolished. This type of demolition is accomplished with large pieces of construction equipment that break up the structure, load it into trucks and haul it to a disposal site, but disposal or dump fees are not included. Demolition of below-grade foundation elements, such as footings, foundation walls, grade beams, slabs on grade, etc., is not included. Certain mechanical equipment containing flammable liquids or ozone-depleting refrigerants, electric lighting elements, communication equipment components, and other building elements may contain hazardous waste, and must be removed, either selectively or carefully, as hazardous waste before the building can be demolished.

Foundation Demolition - Demolition of below-grade foundation footings, foundation walls, grade beams, and slabs on grade. This type of demolition is accomplished by hand or pneumatic hand tools, and does not include saw cutting, or handling, loading, hauling, or disposal of the debris.

Gutting - Removal of building interior finishes and electrical/mechanical systems down to the load-bearing and sub-floor elements of the rough building frame, with no concern for any particular building element, component, or material type being demolished. This type of demolition is accomplished by hand or pneumatic hand tools, and includes loading into trucks, but not hauling, disposal or dump fees, scaffolding, or shoring. Certain mechanical equipment containing flammable liquids or ozone-depleting refrigerants, electric lighting elements, communication equipment components, and other building elements may contain hazardous waste, and must be removed, either selectively or carefully, as hazardous waste, before the building is gutted.

Selective Demolition - Demolition of a selected building element, component, or finish, with some concern for surrounding or adjacent elements, components, or finishes (see the first Subdivision (s) at the beginning of appropriate Divisions). This type of demolition is accomplished by hand or pneumatic hand tools, and does not include handling, loading,

storing, hauling, or disposal of the debris, scaffolding, or shoring. "Gutting" methods may be used in order to save time, but damage that is caused to surrounding or adjacent elements, components, or finishes may have to be repaired at a later time.

Careful Removal - Removal of a piece of service equipment, building element or component, or material type, with great concern for both the removed item and surrounding or adjacent elements, components or finishes. The purpose of careful removal may be to protect the removed item for later re-use, preserve a higher salvage value of the removed item, or replace an item while taking care to protect surrounding or adjacent elements, components, connections, or finishes from cosmetic and/or structural damage. An approximation of the time required to perform this type of removal is 1/3 to 1/2 the time it would take to install a new item of like kind (see Reference Number R260105-30). This type of removal is accomplished by hand or pneumatic hand tools, and does not include loading, hauling, or storing the removed item, scaffolding, shoring, or lifting equipment.

Cutout Demolition - Demolition of a small quantity of floor, wall, roof, or other assembly, with concern for the appearance and structural integrity of the surrounding materials. This type of demolition is accomplished by hand or pneumatic hand tools, and does not include saw cutting, handling, loading, hauling, or disposal of debris, scaffolding, or shoring.

Rubbish Handling - Work activities that involve handling, loading or hauling of debris. Generally, the cost of rubbish handling must be added to the cost of all types of demolition, with the exception of whole building demolition.

Minor Site Demolition - Demolition of site elements outside the footprint of a building. This type of demolition is accomplished by hand or pneumatic hand tools, or with larger pieces of construction equipment, and may include loading a removed item onto a truck (check the Crew for equipment used). It does not include saw cutting, hauling or disposal of debris, and, sometimes, handling or loading.

R024119-20   Dumpsters

Dumpster rental costs on construction sites are presented in two ways.

The cost per week rental includes the delivery of the dumpster; its pulling or emptying once per week, and its final removal. The assumption is made that the dumpster contractor could choose to empty a dumpster by simply bringing in an empty unit and removing the full one. These costs also include the disposal of the materials in the dumpster.

The Alternate Pricing can be used when actual planned conditions are not approximated by the weekly numbers. For example, these lines can be used when a dumpster is needed for 4 weeks and will need to be emptied 2 or 3 times per week. Conversely the Alternate Pricing lines can be used when a dumpster will be rented for several weeks or months but needs to be emptied only a few times over this period.

R053100-10   Decking Descriptions

General - All Deck Products

A steel deck is made by cold forming structural grade sheet steel into a repeating pattern of parallel ribs. The strength and stiffness of the panels are the result of the ribs and the material properties of the steel. Deck lengths can be varied to suit job conditions, but because of shipping considerations, are usually less than 40 feet. Standard deck width varies with the product used but full sheets are usually 12", 18", 24", 30", or 36". The deck is typically furnished in a standard width with the ends cut square. Any cutting for width, such as at openings or for angular fit, is done at the job site.

The deck is typically attached to the building frame with arc puddle welds, self-drilling screws, or powder or pneumatically driven pins. Sheet to sheet fastening is done with screws, button punching (crimping), or welds.

Composite Floor Deck

After installation and adequate fastening, a floor deck serves several purposes. It (a) acts as a working platform, (b) stabilizes the frame, (c) serves as a concrete form for the slab, and (d) reinforces the slab to carry the design loads applied during the life of the building. Composite decks are distinguished by the presence of shear connector devices as part of the deck. These devices are designed to mechanically lock the concrete and deck together so that the concrete and the deck work together to carry subsequent floor loads. These shear connector devices can be rolled-in embossments, lugs, holes, or wires welded to the panels. The deck profile can also be used to interlock concrete and steel.

Composite deck finishes are either galvanized (zinc coated) or phosphatized/painted. Galvanized deck has a zinc coating on both the top and bottom surfaces. The phosphatized/painted deck has a bare (phosphatized) top surface that will come into contact with the concrete. This bare top surface can be expected to develop rust before the concrete is placed. The bottom side of the deck has a primer coat of paint.

A composite floor deck is normally installed so the panel ends do not overlap on the supporting beams. Shear lugs or panel profile shapes often prevent a tight metal to metal fit if the panel ends overlap; the air gap caused by overlapping will prevent proper fusion with the structural steel supports when the panel end laps are shear stud welded.

Adequate end bearing of the deck must be obtained as shown on the drawings. If bearing is actually less in the field than shown on the drawings, further investigation is required.

Roof Deck

A roof deck is not designed to act compositely with other materials. A roof deck acts alone in transferring horizontal and vertical loads into the building frame. Roof deck rib openings are usually narrower than floor deck rib openings. This provides adequate support of the rigid thermal insulation board.

A roof deck is typically installed to endlap approximately 2" over supports. However, it can be butted (or lapped more than 2") to solve field fit problems. Since designers frequently use the installed deck system as part of the horizontal bracing system (the deck as a diaphragm), any fastening substitution or change should be approved by the designer. Continuous perimeter support of the deck is necessary to limit edge deflection in the finished roof and may be required for diaphragm shear transfer.

Standard roof deck finishes are galvanized or primer painted. The standard factory applied paint for roof decks is a primer paint and is not intended to weather for extended periods of time. Field painting or touching up of abrasions and deterioration of the primer coat or other protective finishes is the responsibility of the contractor.

Cellular Deck

A cellular deck is made by attaching a bottom steel sheet to a roof deck or composite floor deck panel. A cellular deck can be used in the same manner as a floor deck. Electrical, telephone, and data wires are easily run through the chase created between the deck panel and the bottom sheet.

When used as part of the electrical distribution system, the cellular deck must be installed so that the ribs line up and create a smooth cell transition at abutting ends. The joint that occurs at butting cell ends must be taped or otherwise sealed to prevent wet concrete from seeping into the cell. Cell interiors must be free of welding burrs, or other sharp intrusions, to prevent damage to wires.

When used as a roof deck, the bottom flat plate is usually left exposed to view. Care must be maintained during erection to keep good alignment and prevent damage.

A cellular deck is sometimes used with the flat plate on the top side to provide a flat working surface. Installation of the deck for this purpose requires special methods for attachment to the frame because the flat plate, now on the top, can prevent direct access to the deck material that is bearing on the structural steel. It may be advisable to treat the flat top surface to prevent slipping.

A cellular deck is always furnished galvanized or painted over galvanized.

Form Deck

A form deck can be any floor or roof deck product used as a concrete form. Connections to the frame are by the same methods used to anchor floor and roof decks. Welding washers are recommended when welding a deck that is less than 20 gauge thickness.

A form deck is furnished galvanized, prime painted, or uncoated. A galvanized deck must be used for those roof deck systems where a form deck is used to carry a lightweight insulating concrete fill.

R078413-30   Firestopping

Firestopping is the sealing of structural, mechanical, electrical and other penetrations through fire-rated assemblies. The basic components of firestop systems are safing insulation and firestop sealant on both sides of wall penetrations and the top side of floor penetrations.

Pipe penetrations are assumed to be through concrete, grout, or joint compound and can be sleeved or unsleeved. Costs for the penetrations and sleeves are not included. An annular space of 1" is assumed. Escutcheons are not included.

A metallic pipe is assumed to be copper, aluminum, cast iron or similar metallic material. An insulated metallic pipe is assumed to be covered with a thermal insulating jacket of varying thickness and materials.

A non-metallic pipe is assumed to be PVC, CPVC, FR Polypropylene or similar plastic piping material. Intumescent firestop sealant or wrap strips are included. Collars on both sides of wall penetrations and a sheet metal plate on the underside of floor penetrations are included.

Ductwork is assumed to be sheet metal, stainless steel or similar metallic material. Duct penetrations are assumed to be through concrete, grout or joint compound. Costs for penetrations and sleeves are not included. An annular space of 1/2" is assumed.

Multi-trade openings include costs for sheet metal forms, firestop mortar, wrap strips, collars and sealants as necessary.

Structural penetrations joints are assumed to be 1/2" or less. CMU walls are assumed to be within 1-1/2" of the metal deck. Drywall walls are assumed to be tight to the underside of metal decking.

Metal panel, glass or curtain wall systems include a spandrel area of 5' filled with mineral wool foil-faced insulation. Fasteners and stiffeners are included.

R224000-10   Water Consumption Rates

Fixture Type	Water Supply Fixture Unit Value		
	Hot Water	Cold Water	Combined
Bathtub	1.0	1.0	1.4
Clothes Washer	1.0	1.0	1.4
Dishwasher	1.4	0.0	1.4
Kitchen Sink	1.0	1.0	1.4
Laundry Tub	1.0	1.0	1.4
Lavatory	0.5	0.5	0.7
Shower Stall	1.0	1.0	1.4
Water Closet (tank type)	0.0	2.2	2.2
Full Bath Group			
w/bathtub or shower stall	1.5	2.7	3.6
Half Bath Group			
w/W.C. and Lavatory	0.5	2.5	2.6
Kitchen Group			
w/Dishwasher and Sink	1.9	1.0	2.5
Laundry Group			
w/Clothes Washer and			
Laundry Tub	1.8	1.8	2.5
Hose bibb (sillcock)	0.0	2.5	2.5

Notes:
Typically, WSFU = 1GPM

Supply loads in the building water-distribution system shall be determined by total load on the pipe being sized, in terms of water supply fixture units (WSFU) and gallons per minute (GPM) flow rates. For fixtures not listed, choose a WSFU value of a fixture with similar flow characteristics. Water Fixture Supply Units determined the required water supply to fixtures and their service systems. Fixture units are equal to one (1) cubic foot of water drained in a 1-1/4″ pipe per minute. It is not a flow rate unit but a design factor.

Excerpted from the 2012 *International Plumbing Code*, Copyright 2011.
Washington, D.C.: International Code Council. Reproduced with permission.
All rights reserved. www.ICCSAFE.org

R224000-20 Fixture Demands in Gallons per Fixture per Hour

The table below is based on a 140°F final temperature except for dishwashers in public places where 180°F water is mandatory.

Supply Systems for Flush Tanks			Supply Systems for Flushometer Valves		
Load	Demand		Load	Demand	
WSFU	GPM	CU. FT.	WSFU	GPM	CU. FT.
1.0	3.0	0.041040	–	–	–
2.0	5.0	0.068400	–	–	–
3.0	6.5	0.868920	–	–	–
4.0	8.0	1.069440	–	–	–
5.0	9.4	1.256592	5.0	15	2.0052
6.0	10.7	1.430376	6.0	17.4	2.326032
7.0	11.8	1.577424	7.0	19.8	2.646364
8.0	12.8	1.711104	8.0	22.2	2.967696
9.0	13.7	1.831416	9.0	24.6	3.288528
10.0	14.6	1.951728	10.0	27	3.60936
11.0	15.4	2.058672	11.0	27.8	3.716304
12.0	16.0	2.138880	12.0	28.6	3.823248
13.0	16.5	2.205720	13.0	29.4	3.930192
14.0	17.0	2.272560	14.0	30.2	4.037136
15.0	17.5	2.339400	15.0	31	4.14408
16.0	18.0	2.906240	16.0	31.8	4.241024
17.0	18.4	2.459712	17.0	32.6	4.357968
18.0	18.8	2.513184	18.0	33.4	4.464912
19.0	19.2	2.566656	19.0	34.2	4.571856
20.0	19.6	2.620218	20.0	35.0	4.678800
25.0	21.5	2.874120	25.0	38.0	5.079840
30.0	23.3	3.114744	30.0	42.0	5.611356
35.0	24.9	3.328632	35.0	44.0	5.881920
40.0	26.3	3.515784	40.0	46.0	6.149280
45.0	27.7	3.702936	45.0	48.0	6.416640
50.0	29.1	3.890088	50.0	50.0	6.684000

Notes:
When designing a plumbing system that utilizes fixtures other than, or in addition to, water closets, use the data provided in the Supply Systems for Flush Tanks section of the above table.

To obtain the probable maximum demand, multiply the total demands for the fixtures (gal./fixture/hour) by the demand factor. The heater should have a heating capacity in gallons per hour equal to this maximum. The storage tank should have a capacity in gallons equal to the probable maximum demand multiplied by the storage capacity factor.

Excerpted from the 2012 *International Residential Code*, Copyright 2011. Washington, D.C.: International Code Council. Reproduced with permission. All rights reserved. www.ICCSAFE.org

Fig. R238313-11

R238313-10   Heat Trace Systems

Before you can determine the cost of a HEAT TRACE installation, the method of attachment must be established. There are four (4) common methods:

1. Cable is simply attached to the pipe with polyester tape every 12'.
2. Cable is attached with a continuous cover of 2" wide aluminum tape.
3. Cable is attached with factory extruded heat transfer cement and covered with metallic raceway with clips every 10'.
4. Cable is attached between layers of pipe insulation using either clips or polyester tape.

Example: Components for method 3 must include:
 A. Heat trace cable by voltage and watts per linear foot.
 B. Heat transfer cement, 1 gallon per 60 linear feet of cover.
 C. Metallic raceway by size and type.
 D. Raceway clips by size of pipe.

When taking off linear foot lengths of cable add the following for each valve in the system. (E)

In all of the above methods each component of the system must be priced individually.

SCREWED OR WELDED VALVE:			FLANGED VALVE:			BUTTERFLY VALVES:		
1/2″	=	6″	1/2″	=	1' -0″	1/2″	=	0'
3/4″	=	9″	3/4″	=	1' -6″	3/4″	=	0'
1″	=	1' -0″	1″	=	2' -0″	1″	=	1' -0″
1-1/2″	=	1' -6″	1-1/2″	=	2' -6″	1-1/2″	=	1' -6″
2″	=	2'	2″	=	2' -6″	2″	=	2' -0″
2-1/2″	=	2' -6″	2-1/2″	=	3' -0″	2-1/2″	=	2' -6″
3″	=	2' -6″	3″	=	3' -6″	3″	=	2' -6″
4″	=	4' -0″	4″	=	4' -0″	4″	=	3' -0″
6″	=	7' -0″	6″	=	8' -0″	6″	=	3' -6″
8″	=	9' -6″	8″	=	11' -0″	8″	=	4' -0″
10″	=	12' -6″	10″	=	14' -0″	10″	=	4' -0″
12″	=	15' -0″	12″	=	16' -6″	12″	=	5' -0″
14″	=	18' -0″	14″	=	19' -6″	14″	=	5' -6″
16″	=	21' -6″	16″	=	23' -0″	16″	=	6' -0″
18″	=	25' -6″	18″	=	27' -0″	18″	=	6' -6″
20″	=	28' -6″	20″	=	30' -0″	20″	=	7' -0″
24″	=	34' -0″	24″	=	36' -0″	24″	=	8' -0″
30″	=	40' -0″	30″	=	42' -0″	30″	=	10' -0″

R238313-10 Heat Trace Systems (cont.)

Add the following quantities of heat transfer cement to linear foot totals for each valve:

Nominal Valve Size	Gallons of Cement per Valve
1/2"	0.14
3/4"	0.21
1"	0.29
1-1/2"	0.36
2"	0.43
2-1/2"	0.70
3"	0.71
4"	1.00
6"	1.43
8"	1.48
10"	1.50
12"	1.60
14"	1.75
16"	2.00
18"	2.25
20"	2.50
24"	3.00
30"	3.75

The following must be added to the list of components to accurately price HEAT TRACE systems:

1. Expediter fitting and clamp fasteners (F)
2. Junction box and nipple connected to expediter fitting (G)
3. Field installed terminal blocks within junction box
4. Ground lugs
5. Piping from power source to expediter fitting
6. Controls
7. Thermostats
8. Branch wiring
9. Cable splices
10. End of cable terminations
11. Branch piping fittings and boxes

Deduct the following percentages from labor if cable lengths in the same area exceed:

150' to 250'	10%	351' to 500'	20%
251' to 350'	15%	Over 500'	25%

Add the following percentages to labor for elevated installations:

15' to 20' high	10%	31' to 35' high	40%
21' to 25' high	20%	36' to 40' high	50%
26' to 30' high	30%	Over 40' high	60%

R238313-20 Spiral-Wrapped Heat Trace Cable (Pitch Table)

In order to increase the amount of heat, occasionally heat trace cable is wrapped in a spiral fashion around a pipe, increasing the number of feet of heater cable per linear foot of pipe.

Engineers first determine the heat loss per foot of pipe (based on the insulating material, its thickness, and the temperature differential across it). A ratio is then calculated by the formula:

$$\text{Feet of Heat Trace per Foot of Pipe} = \frac{\text{Watts/Foot of Heat Loss}}{\text{Watts/Foot of the Cable}}$$

The linear distance between wraps (pitch) is then taken from a chart or table. Generally, the pitch is listed on a drawing leaving the estimator to calculate the total length of heat tape required. An approximation may be taken from this table.

Feet of Heat Trace Per Foot of Pipe																
Nominal Pipe Size in Inches																
Pitch In Inches	1	1¼	1½	2	2½	3	4	6	8	10	12	14	16	18	20	24
3.5	1.80															
4	1.65															
5	1.46	1.60	1.80													
6	1.34	1.45	1.55	1.75												
7	1.25	1.35	1.43	1.57	1.75											
8	1.20	1.28	1.34	1.45	1.60	1.80										
9	1.16	1.23	1.28	1.37	1.51	1.68										
10	1.13	1.19	1.24	1.32	1.44	1.57	1.82									
15	1.06	1.08	1.10	1.15	1.21	1.29	1.42	1.78								
20	1.04	1.05	1.06	1.08	1.13	1.17	1.25	1.49	1.73							
25		1.04	1.04	1.06	1.08	1.11	1.17	1.33	1.51	1.72						
30				1.04	1.05	1.07	1.12	1.24	1.37	1.54	1.70					
35					1.06	1.06	1.09	1.17	1.28	1.42	1.54	1.80				
40						1.05	1.07	1.14	1.22	1.33	1.44	1.52	1.78			
50							1.05	1.09	1.15	1.22	1.29	1.35	1.44	1.75		
60								1.06	1.11	1.16	1.21	1.25	1.31	1.53	1.64	1.83
70								1.05	1.08	1.12	1.17	1.19	1.24	1.39	1.46	1.62
80									1.06	1.09	1.13	1.15	1.19	1.30	1.35	1.47
90									1.04	1.06	1.10	1.13	1.16	1.24	1.28	1.38
100										1.05	1.08	1.10	1.13	1.19	1.23	1.32

Note: Common practice would normally limit the lower end of the table to 5% of additional heat. Above 80% an engineer would likely opt for two (2) parallel cables.

579

R260102-20   Labor Adjustment Factors

Labor Adjustment Factors are provided for Divisions 26, 27, and 28 to assist the mechanical estimator account for the various complexities and special conditions of any particular project. While a single percentage has been entered on each line of Division 26 01 02.20, it should be understood that these are just suggested midpoints of ranges of values commonly used by mechanical estimators. They may be increased or decreased depending on the severity of the special conditions.

The group for "existing occupied buildings" has been the subject of requests for explanation. Actually there are two stages to this group: buildings that are existing and "finished" but unoccupied, and those that also are occupied. Buildings that are "finished" may result in higher labor costs due to the

workers having to be more careful not to damage finished walls, ceilings, floors, etc. and may necessitate special protective coverings and barriers. Also corridor bends and doorways may not accommodate long pieces of pipe or larger pieces of equipment. Work above an already hung ceiling can be very time consuming. The addition of occupants may force the work to be done on premium time (nights and/or weekends), eliminate the possible use of some preferred tools such as pneumatic drivers, powder charged drivers, etc. The estimator should evaluate the access to the work area and just how the work is going to be accomplished to arrive at an increase in labor costs over "normal" new construction productivity.

R260105-30   Electrical Demolition (Removal for Replacement)

The purpose of this reference number is to provide a guide to users for electrical "removal for replacement" by applying the rule of thumb: 1/3 of new installation time (typical range from 20% to 50%) for removal. Remember to use reasonable judgment when applying the suggested percentage factor. For example:

Contractors have been requested to remove an existing fluorescent lighting fixture and replace with a new fixture utilizing energy saver lamps and electronic ballast:

In order to fully understand the extent of the project, contractors should visit the job site and estimate the time to perform the renovation work in accordance with applicable national, state and local regulation and codes.

The contractor may need to add extra labor hours to his estimate if he discovers unknown concealed conditions such as: contaminated asbestos ceiling, broken acoustical ceiling tile and need to repair, patch and touch-up paint all the damaged or disturbed areas, tasks normally assigned to general contractors. In addition, the owner could request that the contractors salvage the materials removed and turn over the materials to the owner or dispose of the materials to a reclamation station. The normal removal item is 0.5 labor-hour for a lighting fixture and 1.5 labor-hours for new installation time. Revise the estimate times from 2 labor-hours work up to a minimum 4 labor-hours work for just a fluorescent lighting fixture.

For removal of large concentrations of lighting fixtures in the same area, apply an "economy of scale" to reduce estimating labor hours.

Figure R260110-10   Typical Overhead Service Entrance

Figure R260110-11   Typical Commercial Electric System

A = 1" Conduit w/1-#1/0 Wire XHHW
B = 2-3" Conduits w/4-350kcmil XHHW in Each
C = 1¼" Conduit w/4 #3 THHN
D = 1¼" Conduit w/3 #1 XHHW
E = 1" Conduit w/4 #6 THHN
F = ½" Conduit w/2 #12 THHN

For customer support on your Electrical Costs with RSMeans data, call 800.448.8182.

581

Figure R260110-12 Preliminary Procedure

1. Determine building size and use.
2. Develop total load in watts.
   a. Lighting
   b. Receptacle
   c. Air Conditioning
   d. Elevator
   e. Other power requirements
3. Determine best voltage available from utility company.
4. Determine cost from tables for loads (a) through (e) above.
5. Determine size of service from formulas (R260110-17).
6. Determine costs for service, panels, and feeders from tables.

Figure R260110-13 Office Building 90' x 210', 3 story, w/garage

Garage Area = 18,900 S.F.
Office Area = 56,700 S.F.
Elevator = 2 @ 125 FPM

Tables	Power Required	Watts
D5010-1151	Garage Lighting .5 Watt/S.F.	9,450
	Office Lighting 3 Watts/S.F.	170,100
D5010-1151	Office Receptacles 2 Watts/S.F.	113,400
D5010-1151, R262213-27	Low Rise Office A.C. 4.3 Watts/S.F.	243,810
D5010-1152, 1153	Elevators - 2 @ 20 HP = 2 @ 17,404 Watts/Ea.	34,808
D5010-1151	Misc. Motors + Power 1.2 Watts/S.F.	68,040
	Total	639,608 Watts

Voltage Available
277/480V, 3 Phase, 4 Wire

Formula

$$\text{D5010-116} \qquad \text{Amperes} = \frac{\text{Watts}}{\text{Volts x Power Factor x 1.73}} = \frac{639,608}{480V \times .8 \times 1.73} = 963 \text{ Amps}$$

Use 1200 Amp Service

System	Description	Unit Cost	Unit	Total
D5020 210 0200	Garage Lighting (Interpolated)	$2.72	S.F.	$ 51,408
D5020 210 0540	Office Lighting (Using 32 W lamps)	8.24	S.F.	467,208
D5020 115 0880	Receptacle-Undercarpet	3.92	S.F.	222,264
D5020 140 0280	Air Conditioning	0.65	S.F.	36,855
D5020 135 0320	Misc. Pwr.	0.35	S.F.	19,845
D5020 145 2120	Elevators - 2 @ 20HP	2,775.00	Ea.	5,550
D5010 130 1250	Underground Electric Service - 1200 Amp	60,100.00	Ea.	60,100
D5010 230 0480	Feeder - Assume 200 feet	313.00	L.F.	62,600
D5010 240 0580	Panels - 277/480V, 1200 Amp	35,275.00	Ea.	35,275
D5030 910 0400	Fire Detection	35,400.00	Ea.	35,400
			Total	$996,505
			or	$996,500

Table R260110-14 Nominal Watts Per S.F. for Electric Systems for Various Building Types

Type Construction	1. Lighting	2. Devices	3. HVAC	4. Misc.	5. Elevator	Total Watts
Apartment, luxury high rise	2	2.2	3	1		
Apartment, low rise	2	2	3	1		
Auditorium	2.5	1	3.3	.8		
Bank, branch office	3	2.1	5.7	1.4		
Bank, main office	2.5	1.5	5.7	1.4		
Church	1.8	.8	3.3	.8		
College, science building	3	3	5.3	1.3		
College, library	2.5	.8	5.7	1.4		
College, physical education center	2	1	4.5	1.1		
Department store	2.5	.9	4	1		
Dormitory, college	1.5	1.2	4	1		
Drive-in doughnut shop	3	4	6.8	1.7		
Garage, commercial	.5	.5	0	.5		
Hospital, general	2	4.5	5	1.3		
Hospital, pediatric	3	3.8	5	1.3		
Hotel, airport	2	1	5	1.3		
Housing for the elderly	2	1.2	4	1		
Manufacturing, food processing	3	1	4.5	1.1		
Manufacturing, apparel	2	1	4.5	1.1		
Manufacturing, tools	4	1	4.5	1.1		
Medical clinic	2.5	1.5	3.2	1		
Nursing home	2	1.6	4	1		
Office building, high rise	3	2	4.7	1.2		
Office building, low rise	3	2	4.3	1.2		
Radio-TV studio	3.8	2.2	7.6	1.9		
Restaurant	2.5	2	6.8	1.7		
Retail store	2.5	.9	5.5	1.4		
School, elementary	3	1.9	5.3	1.3		
School, junior high	3	1.5	5.3	1.3		
School, senior high	2.3	1.7	5.3	1.3		
Supermarket	3	1	4	1		
Telephone exchange	1	.6	4.5	1.1		
Theater	2.5	1	3.3	.8		
Town Hall	2	1.9	5.3	1.3		
U.S. Post Office	3	2	5	1.3		
Warehouse, grocery	1	.6	0	.5		

Rule of Thumb: 1 KVA = 1 HP (Single Phase)

Three Phase:

Watts = 1.73 x Volts x Current x Power Factor x Efficiency

$$\text{Horsepower} = \frac{\text{Volts x Current x 1.73 x Power Factor}}{746 \text{ Watts}}$$

See R260120 for an additional reference.

Table R260110-15 Horsepower Requirements for Elevators with 3 Phase Motors

Type	Maximum Travel Height in Ft.	Travel Speeds in FPM	Capacity of Cars in Lbs.		
			1200	1500	1800
Hydraulic	70	70	10	15	15
		85	15	15	15
		100	15	15	20
		110	20	20	20
		125	20	20	20
		150	25	25	25
		175	25	30	30
		200	30	30	40
Geared Traction	300	200			
		350			
			2000	2500	3000
Hydraulic	70	70	15	20	20
		85	20	20	25
		100	20	25	30
		110	20	25	30
		125	25	30	40
		150	30	40	50
		175	40	50	50
		200	40	50	60
Geared Traction	300	200	10	10	15
		350	15	15	23
			3500	4000	4500
Hydraulic	70	70	20	25	30
		85	25	30	30
		100	30	40	40
		110	40	40	50
		125	40	50	50
		150	50	50	60
		175	60		
		200	60		
Geared Traction	300	200	15		23
		350	23		35

The power factor of electric motors varies from 80% to 90% in larger size motors. The efficiency likewise varies from 80% on a small motor to 90% on a large motor.

Table R260110-16 Watts per Motor

90% Power Factor & Efficiency @ 200 or 460V			
HP	Watts	HP	Watts
10	9024	30	25784
15	13537	40	33519
20	17404	50	41899
25	21916	60	49634

For customer support on your Electrical Costs with RSMeans data, call 800.448.8182.

583

Table R260110-17 Electrical Formulas

Ohm's Law

Ohm's Law is a method of explaining the relation between voltage, current, and resistance in an electrical circuit. It is practically the basis of all electrical calculations. The term "electromotive force" is often used to designate pressure in volts. This formula can be expressed in various forms.

To find the current in amperes:

$$\text{Current} = \frac{\text{Voltage}}{\text{Resistance}} \quad \text{or} \quad \text{Amperes} = \frac{\text{Volts}}{\text{Ohms}} \quad \text{or} \quad I = \frac{E}{R}$$

The flow of current in amperes through any circuit is equal to the voltage or electromotive force divided by the resistance of that circuit.

To find the pressure or voltage:

$$\text{Voltage} = \text{Current x Resistance} \quad \text{or} \quad \text{Volts} = \text{Amperes x Ohms}$$
$$\text{or} \quad E = I \times R$$

The voltage required to force a current through a circuit is equal to the resistance of the circuit multiplied by the current.

To find the resistance:

$$\text{Resistance} = \frac{\text{Voltage}}{\text{Current}} \quad \text{or} \quad \text{Ohms} = \frac{\text{Volts}}{\text{Amperes}} \quad \text{or} \quad R = \frac{E}{I}$$

The resistance of a circuit is equal to the voltage divided by the current flowing through that circuit.

Power Formulas

$$\text{One horsepower} = 746 \text{ watts} \qquad \text{One kilowatt} = 1000 \text{ watts}$$

The power factor of electric motors varies from 80% to 90% in the larger size motors.

Single-Phase Alternating Current Circuits

Power in Watts = Volts x Amperes x Power Factor

To find current in amperes:

$$\text{Current} = \frac{\text{Watts}}{\text{Volts x Power Factor}} \quad \text{or}$$

$$\text{Amperes} = \frac{\text{Watts}}{\text{Volts x Power Factor}} \quad \text{or} \quad I = \frac{W}{E \times PF}$$

To find current of a motor, single phase:

$$\text{Current} = \frac{\text{Horsepower x 746}}{\text{Volts x Power Factor x Efficiency}} \quad \text{or}$$

$$I = \frac{HP \times 746}{E \times PF \times Eff.}$$

To find horsepower of a motor, single phase:

$$\text{Horsepower} = \frac{\text{Volts x Current x Power Factor x Efficiency}}{746 \text{ Watts}}$$

$$HP = \frac{E \times I \times PF \times Eff.}{746}$$

To find power in watts of a motor, single phase:

Watts = Volts x Current x Power Factor x Efficiency   or
Watts = E x I x PF x Eff.

To find single phase kVA:

$$1 \text{ Phase kVA} = \frac{\text{Volts x Amps}}{1000}$$

Three-Phase Alternating Current Circuits

Power in Watts = Volts x Amperes x Power Factor x 1.73

To find current in amperes in each wire:

$$\text{Current} = \frac{\text{Watts}}{\text{Voltage x Power Factor x 1.73}} \quad \text{or}$$

$$\text{Amperes} = \frac{\text{Watts}}{\text{Volts x Power Factor x 1.73}} \quad \text{or} \quad I = \frac{W}{E \times PF \times 1.73}$$

To find current of a motor, 3 phase:

$$\text{Current} = \frac{\text{Horsepower x 746}}{\text{Volts x Power Factor x Efficiency x 1.73}} \quad \text{or}$$

$$I = \frac{HP \times 746}{E \times PF \times Eff. \times 1.73}$$

To find horsepower of a motor, 3 phase:

$$\text{Horsepower} = \frac{\text{Volts x Current x 1.73 x Power Factor}}{746 \text{ Watts}}$$

$$HP = \frac{E \times I \times 1.73 \times PF}{746}$$

To find power in watts of a motor, 3 phase:

Watts = Volts x Current x 1.73 x Power Factor x Efficiency   or
Watts = E x I x 1.73 x PF x Eff.

To find 3 phase kVA:

$$3 \text{ phase kVA} = \frac{\text{Volts x Amps x 1.73}}{1000} \quad \text{or}$$

$$kVA = \frac{V \times A \times 1.73}{1000}$$

Power Factor (PF) is the percentage ratio of the measured watts (effective power) to the volt-amperes (apparent watts)

$$\text{Power Factor} = \frac{\text{Watts}}{\text{Volts x Amperes}} \times 100\%$$

Table R260120-10 ASHRAE Climate Zone Map

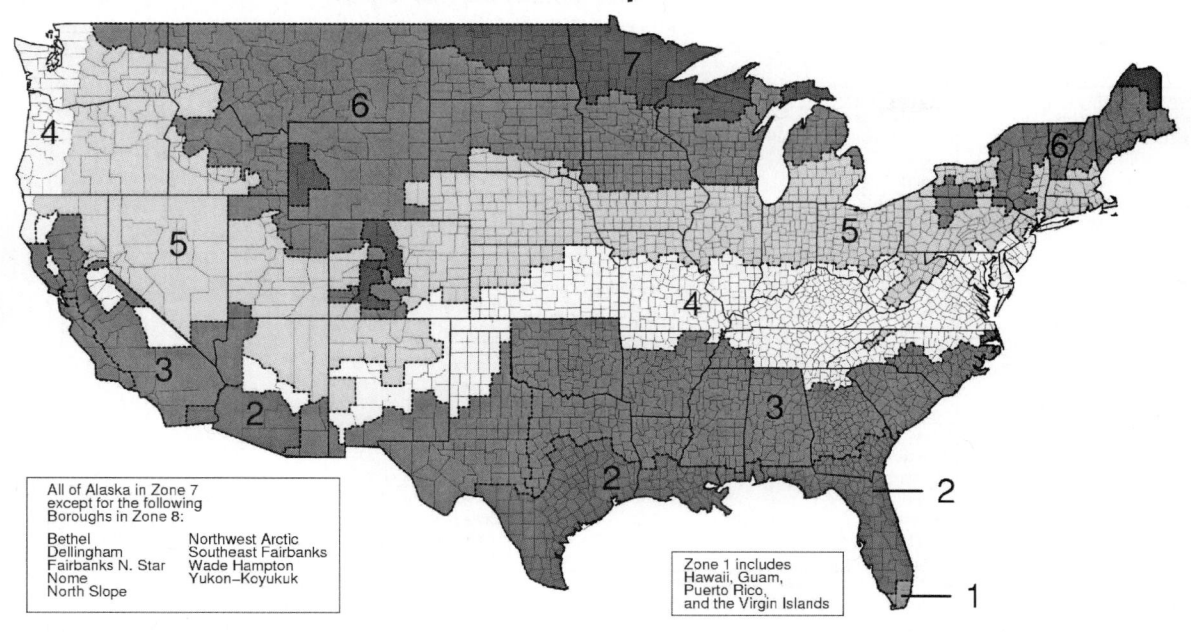

All of Alaska in Zone 7
except for the following
Boroughs in Zone 8:

Bethel	Northwest Arctic
Dellingham	Southeast Fairbanks
Fairbanks N. Star	Wade Hampton
Nome	Yukon–Koyukuk
North Slope	

Zone 1 includes
Hawaii, Guam,
Puerto Rico,
and the Virgin Islands

Table R260120-20 Watts per Square Foot for Lighting and Misc. Power by Building Type & Space

Building / Space	Percent	Lighting	Misc	Building / Space	Percent	Lighting	Misc	Building / Space	Percent	Lighting	Misc
Apartment, Low: Area = 20,000 SF; Floors = 2; Heating = DX				**Auditorium: Area = 70,000 SF; Floors = 1; Heating = Hot Water**				**Day Care: Area = 2,500 SF; Floors = 1; Heating = Furnace;**			
Coils; Cooling = DX Coils				**Coils; Cooling = Chilled Water Coils**				**Cooling = DX Coils**			
Room	85%	0.50	0.10	Theater	70%	0.90		Classroom	45%	1.45	
Storage	8%	1.19	0.00	Lobby	15%	1.50		Corridor	15%	0.57	
Laundry	7%	1.28	0.15	Office	5%	1.30		Library	15%	1.67	
Weighted Average		**0.61**	**0.10**	Corridor	5%	0.60		Food Prep	10%	1.19	
Apartment, Mid: Area = 40,000 SF; Floors = 5; Heating = DX				MEP	5%	0.70		Restrooms	7%	0.77	
Coils; Cooling = DX Coils				**Weighted Average**		**0.99**		Office	5%	1.49	
Room	71%	0.50	0.30	**Bank: Area = 25,000 SF; Floors = 2; Heating = Furnace; Cooling =**				Storage	3%	1.19	
Corridor	16%	0.57	0.00	**DX Coils**				**Weighted Average**		**1.27**	
Storage	7%	1.19	0.00	Bank	60%	2.33	0.45	**Fitness Center: Area = 20,000 SF; Floors = 1; Heating = Furnace;**			
Laundry	6%	1.28	0.15	Corridor	10%	0.57	0.00	**Cooling = DX Coils**			
Average		**0.61**	**0.22**	Office	10%	1.49	0.75	Exercise	80%	1.13	0.10
Apartment, High: Area = 72,000 SF; Floors = 8; Heating = DX				Lobby	5%	1.52	0.25	Corridor	5%	0.57	0.00
Coils; Cooling = DX Coils				Restrooms	5%	0.77	0.10	MEP	5%	0.81	0.10
Room	71%	0.50	0.30	Conference	4%	0.92	0.10	Office	5%	1.24	0.75
Corridor	16%	0.57	0.00	MEP	4%	0.81	0.10	Restrooms	7%	0.77	0.10
Storage	7%	1.19	0.00	Production	2%	1.64	0.70	**Weighted Average**		**1.07**	**0.13**
Laundry	6%	1.28	0.15	**Weighted Average**		**1.82**	**0.38**	**Hospital: Area = 250,000 SF; Floors = 4; Heating = Hot Water**			
Weighted Average		**0.61**	**0.22**	**Community Center: Area = 35,000 SF; Floors = 1; Heating =**				**Coils; Cooling = Chilled Water Coils**			
Assisted Living: Area = 10,000 SF; Floors = 1; Heating = Furnace;				**Furnace; Cooling = DX Coils**				Medical	60%	1.34	1.00
Cooling = DX Coils				Convention	50%	1.27		Laboratory	15%	3.19	1.00
Room	70%	1.46	0.25	Dining	15%	1.85		Corridor	10%	0.57	0.00
Dining	15%	1.85	0.10	Food Prep	15%	1.19		Laundry	5%	1.28	0.15
Food Prep	10%	1.19	1.00	Storage	10%	1.19		MEP	5%	0.81	0.10
Care	5%	1.34	1.00	Corridor	5%	0.57		Restrooms	5%	0.77	0.10
Weighted Average		**1.49**	**0.34**	Restrooms	5%	0.77		**Weighted Average**		**1.48**	**0.77**
				Weighted Average		**1.28**		**Hotel: Area = 180,000 SF; Floors = 10; Heating = Furnace;**			
				Convention Center: Area = 150,000 SF; Floors = 2; Heating = Hot				**Cooling = DX Coils**			
				Water Coils; Cooling = Chilled Water Coils				Room	75%	1.46	
				Convention	65%	1.27	0.25	Meeting Center	15%	1.27	
				Conference	20%	0.92	0.10	Corridor	8%	0.57	
				Dining	5%	1.85	0.10	MEP	2%	0.81	
				Corridor	4%	0.57	0.00	**Weighted Average**		**1.35**	
				Food Prep	3%	1.19	1.00				
				Restrooms	3%	0.77	0.10				
				Weighted Average		**1.18**	**0.22**				

Steps:
1. Determine climate zone using R260120-10
2. Use R260120-20 to determine lighting and miscellaneous power for appropriate building type
3. Use corresponding building type and climate zone in R260120-30 to determine other systems

*Values for R260120-20 & -30 were derived by performing an energy analysis using the eQUEST® software from James J. Hirsch & Associates. eQUEST is a trademark of James J. Hirsch
**Additional systems may be necessary

585

For customer support on your Electrical Costs with RSMeans data, call 800.448.8182.

Table R260120-20 Watts per Square Foot for Lighting and Misc. Power by Building Type & Space (cont)

Manufacturing, General: Area = 100,000 SF; Floors = 10; Heating = Furnace; Cooling = DX Coils

Building / Space	Percent	Lighting	Misc
Work	60%	2.66	1.00
Storage	25%	1.19	0.00
Office	10%	1.24	0.75
Restrooms	3%	0.77	0.10
MEP	2%	0.81	0.10
Weighted Average		**2.06**	**0.68**

Manufacturing, High Tech: Area = 200,000 SF; Floors = 1; Heating = Furnace; Cooling = DX Coils

Building / Space	Percent	Lighting	Misc
Work	45%	3.19	1.00
Office	27%	1.24	0.75
Corridor	18%	0.57	0.00
Dining	3%	1.85	0.10
Conference	2%	0.92	0.10
Server	2%	1.24	5.00
MEP	2%	0.81	0.10
Food Prep	1%	1.19	1.00
Weighted Average		**2.00**	**0.77**

Medical Clinic: Area = 20,000 SF; Floors = 1; Heating = Furnace; Cooling = DX Coils

Building / Space	Percent	Lighting	Misc
Clinic	50%	1.40	
General Office	15%	1.30	
Private Office	10%	1.30	
Laboratory	10%	1.30	
Storage	5%	0.60	
Restrooms	5%	0.60	
MEP	5%	0.70	
Weighted Average		**1.25**	

Motel: Area = 10,000 SF; Floors = 1; Heating = DX Coils; Cooling = DX Coils

Building / Space	Percent	Lighting	Misc
Room	95%	1.46	0.25
Office	3%	1.24	0.75
MEP	2%	0.81	0.10
Weighted Average		**1.44**	**0.26**

Museum: Area = 20,000 SF; Floors = 1; Heating = Hot Water Coils; Cooling = Chilled Water Coils

Building / Space	Percent	Lighting	Misc
Exhibit Area	65%	1.03	0.25
Theater	15%	1.02	0.25
Lobby	5%	1.77	0.25
MEP	5%	0.81	0.10
Office	5%	1.49	0.75
Restrooms	5%	0.77	0.10
Weighted Average		**1.06**	**0.26**

Office, Low: Area = 25,000 SF; Floors = 2; Heating = Furnace; Cooling = DX Coils

Building / Space	Percent	Lighting	Misc
Office	70%	1.30	0.75
Corridor	10%	0.60	0.00
Lobby	5%	1.10	0.25
Restrooms	5%	0.60	0.10
Conference	4%	1.60	0.10
MEP	4%	0.70	0.10
Production	2%	1.50	0.70
Weighted Average		**1.18**	**0.56**

Office, Mid: Area = 125,000 SF; Floors = 4; Heating = Hot Water Coils; Cooling = Chilled Water Coils

Building / Space	Percent	Lighting	Misc
Office Open	40%	1.24	0.75
Office Private	30%	1.49	0.75
Corridor	10%	0.57	0.00
Lobby	5%	1.52	0.25
Restrooms	5%	0.77	0.10
Conference	4%	0.92	0.10
MEP	4%	0.81	0.10
Production	2%	1.64	0.70
Weighted Average		**1.22**	**0.56**

Office, High: Area = 250,000 SF; Floors = 8; Heating = Hot Water Coils; Cooling = Chilled Water Coils

Building / Space	Percent	Lighting	Misc
Office Open	45%	1.24	0.75
Office Private	25%	1.49	0.75
Corridor	10%	0.57	0.00
Lobby	5%	1.52	0.25
Conference	4%	0.92	0.10
Production	2%	1.64	0.70
Restrooms	5%	0.77	0.10
MEP	4%	0.81	0.10
Weighted Average		**1.20**	**0.56**

Restaurant, Bar/Lounge: Area = 4,000 SF; Floors = 1; Heating = Furnace; Cooling = DX Coils

Building / Space	Percent	Lighting	Misc
Lounge	35%	1.85	0.10
Dining	30%	1.85	0.10
Food Prep	25%	1.19	1.00
MEP	5%	0.81	0.10
Restrooms	5%	0.77	0.10
Weighted Average		**1.58**	**0.33**

Restaurant, Full: Area = 5,000 SF; Floors = 1; Heating = Furnace; Cooling = DX Coils

Building / Space	Percent	Lighting	Misc
Dining	60%	1.85	0.10
Food Prep	30%	1.19	1.00
Restrooms	5%	0.77	0.10
MEP	5%	0.81	0.10
Weighted Average		**1.55**	**0.37**

Restaurant, Quick: Area = 2,500 SF; Floors = 1; Heating = Furnace; Cooling = DX Coils

Building / Space	Percent	Lighting	Misc
Dining	50%	1.85	0.10
Food Prep	35%	1.19	1.00
Restrooms	10%	0.77	0.10
MEP	5%	0.81	0.10
Weighted Average		**1.46**	**0.42**

Store, Convenience: Area = 3,000 SF; Floors = 1; Heating = Furnace; Cooling = DX Coils

Building / Space	Percent	Lighting	Misc
Retail	80%	2.38	0.25
Office	10%	1.49	0.75
Restrooms	10%	0.77	0.10
Weighted Average		**2.13**	**0.29**

Store, Department: Area = 120,000 SF; Floors = 2; Heating = Hot Water Coils; Cooling = Chilled Water Coils

Building / Space	Percent	Lighting	Misc
Retail	55%	2.38	0.25
Exhibit Area	15%	1.03	0.25
Storage	16%	1.19	0.00
Office	5%	1.24	0.75
Smoking Lounge	2%	1.85	0.10
Restrooms	2%	0.77	0.10
MEP	5%	0.81	0.10
Weighted Average		**1.81**	**0.22**

Store, Large: Area = 70,000 SF; Floors = 1; Heating = Furnace; Cooling = DX Coils

Building / Space	Percent	Lighting	Misc
Retail	65%	2.38	0.25
Storage	20%	1.19	0.00
Office	5%	1.24	0.75
Dining	3%	1.85	0.10
Food Prep	1%	1.19	1.00
Lounge	2%	1.85	0.10
Restrooms	2%	0.77	0.10
MEP	2%	0.81	0.10
Weighted Average		**1.98**	**0.22**

Store, Small: Area = 5,000 SF; Floors = 1; Heating = Furnace; Cooling = DX Coils

Building / Space	Percent	Lighting	Misc
Retail	58%	2.38	0.25
Exhibit Area	15%	1.03	0.25
Storage	15%	1.19	0.00
Office	5%	1.24	0.75
Restrooms	5%	0.77	0.10
MEP	2%	0.81	0.10
Weighted Average		**1.83**	**0.23**

Store, Strip Mall: Area = 15,000 SF; Floors = 1; Heating = Furnace; Cooling = DX Coils

Building / Space	Percent	Lighting	Misc
Retail	38%	2.38	0.25
Atrium	20%	1.55	0.13
Exhibit Area	15%	1.03	0.25
Storage	15%	1.19	0.00
Office	5%	1.24	0.75
Restrooms	5%	0.77	0.10
MEP	2%	0.81	0.10
Weighted Average		**1.66**	**0.20**

Store, Warehouse: Area = 150,000 SF; Floors = 1; Heating = Furnace; Cooling = DX Coils

Building / Space	Percent	Lighting	Misc
Retail	75%	2.38	0.25
Storage	20%	1.19	0.00
Office	4%	1.24	0.75
Restrooms	1%	0.77	0.10
Weighted Average		**2.08**	**0.22**

School, Elementary: Area = 35,000 SF; Floors = 1; Heating = Furnace; Cooling = DX Coils

Building / Space	Percent	Lighting	Misc
Classroom	50%	1.45	0.50
Corridor	15%	0.57	0.00
Gym	10%	1.13	0.10
Dining	5%	1.85	0.10
Food Prep	5%	1.19	1.00
Library	5%	1.78	0.25
Auditorium	5%	1.62	0.25
Restrooms	5%	0.77	0.10
Weighted Average		**1.28**	**0.35**

School, Middle: Area = 75,000 SF; Floors = 1; Heating = Furnace; Cooling = DX Coils

Building / Space	Percent	Lighting	Misc
Classroom	50%	1.45	0.50
Corridor	14%	0.57	0.00
Gym	9%	1.13	0.10
Auditorium	7%	1.62	0.25
Dining	6%	1.85	0.10
Food Prep	5%	1.19	1.00
Library	5%	1.78	0.25
Restrooms	4%	0.77	0.10
Weighted Average		**1.31**	**0.35**

School, High: Area = 150,000 SF; Floors = 2; Heating = Furnace; Cooling = DX Coils

Building / Space	Percent	Lighting	Misc
Classroom	45%	1.45	0.50
Corridor	10%	0.57	0.00
Gym	10%	1.13	0.10
Dining	10%	1.85	0.10
Food Prep	5%	1.19	1.00
Library	5%	1.78	0.25
Auditorium	10%	1.62	0.25
Restrooms	5%	0.77	0.10
Weighted Average		**1.36**	**0.34**

School, College: Area = 250,000 SF; Floors = 4; Heating = Furnace; Cooling = DX Coils

Building / Space	Percent	Lighting	Misc
Classroom	50%	1.45	0.50
Office	15%	1.49	0.75
Corridor	12%	0.57	0.00
Gym	10%	1.13	0.10
Library	5%	1.78	0.25
Dining	3%	1.85	0.10
Restrooms	3%	0.77	0.10
Food Prep	2%	1.19	1.00
Weighted Average		**1.32**	**0.41**

Service Station: Area = 2,000 SF; Floors = 1; Heating = Furnace; Cooling = DX Coils

Building / Space	Percent	Lighting	Misc
Workshop	70%	1.31	0.50
Office	10%	1.49	0.75
Restrooms	10%	0.77	0.10
Retail	10%	2.38	0.25
Weighted Average		**1.38**	**0.46**

Storage, High Bay: Area = 200,000 SF; Floors = 1; Heating = Furnace; Cooling = DX Coils

Building / Space	Percent	Lighting	Misc
Storage	85%	1.19	0.00
Office	10%	1.24	0.75
Restrooms	3%	0.77	0.10
MEP	2%	0.81	0.10
Weighted Average		**1.17**	**0.08**

Storage, Low Bay: Area = 70,000 SF; Floors = 1; Heating = Furnace; Cooling = DX Coils

Building / Space	Percent	Lighting	Misc
Storage	85%	1.19	0.00
Office	10%	1.24	0.75
Restrooms	3%	0.77	0.10
MEP	2%	0.81	0.10
Weighted Average		**1.17**	**0.08**

Table R260120-30 Watts per S.F. for Various Systems by Building Type & Climate Zone

Building / Zone	Heating	Cooling	Heat rejection	Auxiliary (pumps)	Vent fan	Supp heat pump
Apartment, Low						
Zone 1	0.42	3.65	0.00	0.00	0.12	1.47
Zone 2	0.63	3.53	0.00	0.00	0.12	4.01
Zone 3	0.64	3.45	0.00	0.00	0.13	4.63
Zone 4	0.65	3.09	0.00	0.00	0.13	7.41
Zone 5	0.67	2.88	0.00	0.00	0.12	10.30
Zone 6	0.69	2.66	0.00	0.00	0.13	10.68
Zone 7	0.85	2.11	0.00	0.00	0.12	11.76
Average	**0.65**	**3.05**	**0.00**	**0.00**	**0.13**	**7.18**
Apartment, Mid						
Zone 1	0.00	1.90	0.00	0.00	0.11	0.00
Zone 2	0.43	1.82	0.00	0.00	0.11	2.12
Zone 3	0.48	1.69	0.00	0.00	0.11	2.32
Zone 4	0.67	1.67	0.00	0.00	0.12	4.25
Zone 5	0.70	1.54	0.00	0.00	0.12	4.79
Zone 6	0.72	1.45	0.00	0.00	0.12	5.07
Zone 7	0.93	1.39	0.00	0.00	0.12	5.11
Average	**0.56**	**1.63**	**0.00**	**0.00**	**0.12**	**3.38**
Apartment, High						
Zone 1	0.00	2.11	0.00	0.00	0.14	0.00
Zone 2	0.00	2.03	0.00	0.00	0.13	0.00
Zone 3	0.48	1.88	0.00	0.00	0.14	1.75
Zone 4	0.57	1.77	0.00	0.00	0.14	4.22
Zone 5	0.66	1.64	0.00	0.00	0.14	5.40
Zone 6	0.71	1.63	0.00	0.00	0.15	5.64
Zone 7	0.91	1.18	0.00	0.00	0.14	5.75
Average	**0.48**	**1.75**	**0.00**	**0.00**	**0.14**	**3.25**
Assisted Living						
Zone 1	0.00	2.65	0.00	0.04	0.31	0.00
Zone 2	0.00	2.63	0.00	0.04	0.31	0.00
Zone 3	0.00	2.57	0.00	0.04	0.31	0.00
Zone 4	0.00	2.50	0.00	0.04	0.31	0.00
Zone 5	0.00	2.42	0.00	0.04	0.30	0.00
Zone 6	0.00	1.99	0.00	0.04	0.32	0.00
Zone 7	0.00	1.90	0.00	0.04	0.30	0.00
Average	**0.00**	**2.38**	**0.00**	**0.04**	**0.31**	**0.00**
Auditorium						
Zone 1	0.00	2.32	0.34	0.37	1.03	0.00
Zone 2	0.00	2.32	0.34	0.38	1.03	0.00
Zone 3	0.00	2.29	0.33	0.38	1.04	0.00
Zone 4	0.00	2.26	0.29	0.35	1.05	0.00
Zone 5	0.00	2.20	0.34	0.39	1.04	0.00
Zone 6	0.00	1.71	0.14	0.34	1.05	0.00
Zone 7	0.00	1.38	0.23	0.36	1.04	0.00
Average	**0.00**	**2.07**	**0.29**	**0.37**	**1.04**	**0.00**
Bank						
Zone 1	0.00	3.32	0.00	0.00	0.60	0.00
Zone 2	0.00	3.28	0.00	0.00	0.61	0.00
Zone 3	0.00	3.28	0.00	0.00	0.61	0.00
Zone 4	0.00	3.21	0.00	0.00	0.60	0.00
Zone 5	0.00	3.06	0.00	0.00	0.59	0.00
Zone 6	0.00	2.82	0.00	0.00	0.61	0.00
Zone 7	0.00	2.66	0.00	0.00	0.60	0.00
Average	**0.00**	**3.09**	**0.00**	**0.00**	**0.60**	**0.00**

Building / Zone	Heating	Cooling	Heat rejection	Auxiliary (pumps)	Vent fan	Supp heat pump
Community Center						
Zone 1	0.00	2.70	0.00	0.00	0.44	0.00
Zone 2	0.00	2.66	0.00	0.00	0.44	0.00
Zone 3	0.00	2.66	0.00	0.00	0.45	0.00
Zone 4	0.00	2.44	0.00	0.00	0.45	0.00
Zone 5	0.00	2.42	0.00	0.00	0.45	0.00
Zone 6	0.00	2.16	0.00	0.00	0.44	0.00
Zone 7	0.00	1.91	0.00	0.00	0.44	0.00
Average	**0.00**	**2.42**	**0.00**	**0.00**	**0.45**	**0.00**
Convention Center						
Zone 1	0.00	1.46	0.25	0.27	0.31	0.00
Zone 2	0.00	1.44	0.25	0.27	0.29	0.00
Zone 3	0.00	1.43	0.24	0.26	0.30	0.00
Zone 4	0.00	1.42	0.23	0.26	0.28	0.00
Zone 5	0.00	1.30	0.29	0.31	0.22	0.00
Zone 6	0.00	1.17	0.12	0.29	0.16	0.00
Zone 7	0.00	0.90	0.20	0.31	0.16	0.00
Average	**0.00**	**1.30**	**0.22**	**0.28**	**0.25**	**0.00**
Day Care						
Zone 1	0.00	3.47	0.00	0.02	0.29	0.00
Zone 2	0.00	3.42	0.00	0.02	0.28	0.00
Zone 3	0.00	3.40	0.00	0.02	0.29	0.00
Zone 4	0.00	3.29	0.00	0.02	0.29	0.00
Zone 5	0.00	3.08	0.00	0.02	0.29	0.00
Zone 6	0.00	2.29	0.00	0.02	0.29	0.00
Zone 7	0.00	2.12	0.00	0.02	0.24	0.00
Average	**0.05**	**3.01**	**0.05**	**0.07**	**0.29**	**0.05**
Fitness Center						
Zone 1	0.00	4.97	0.00	0.00	0.59	0.00
Zone 2	0.00	4.94	0.00	0.00	0.59	0.00
Zone 3	0.00	4.80	0.00	0.00	0.60	0.00
Zone 4	0.00	4.76	0.00	0.00	0.60	0.00
Zone 5	0.00	4.75	0.00	0.00	0.59	0.00
Zone 6	0.00	4.13	0.00	0.00	0.60	0.00
Zone 7	0.00	3.27	0.00	0.00	0.60	0.00
Average	**0.00**	**4.52**	**0.00**	**0.00**	**0.59**	**0.00**
Hospital						
Zone 1	0.00	1.54	0.23	0.26	0.80	0.00
Zone 2	0.00	1.47	0.22	0.25	0.80	0.00
Zone 3	0.00	1.44	0.22	0.25	0.81	0.00
Zone 4	0.00	1.42	0.22	0.25	0.81	0.00
Zone 5	0.00	1.37	0.23	0.27	0.81	0.00
Zone 6	0.00	1.26	0.15	0.25	0.82	0.00
Zone 7	0.00	1.07	0.19	0.26	0.82	0.00
Average	**0.00**	**1.36**	**0.21**	**0.26**	**0.81**	**0.00**
Hotel						
Zone 1	0.00	1.77	0.00	0.01	0.26	0.00
Zone 2	0.00	1.64	0.00	0.01	0.26	0.00
Zone 3	0.00	1.62	0.00	0.01	0.27	0.00
Zone 4	0.00	1.62	0.00	0.01	0.27	0.00
Zone 5	0.00	1.51	0.00	0.01	0.27	0.00
Zone 6	0.00	1.25	0.00	0.01	0.28	0.00
Zone 7	0.00	1.23	0.00	0.01	0.28	0.00
Average	**0.00**	**1.52**	**0.00**	**0.01**	**0.27**	**0.00**

For customer support on your Electrical Costs with RSMeans data, call 800.448.8182.

587

Table R260120-30 Watts per S.F. for Various Systems by Building Type & Climate Zone (cont.)

Building / Zone	Heating	Cooling	Heat rejection	Auxiliary (pumps)	Vent fan	Supp heat pump	Building / Zone	Heating	Cooling	Heat rejection	Auxiliary (pumps)	Vent fan	Supp heat pump
Manufacturing, General							**Office, Mid**						
Zone 1	0.00	1.46	0.00	0.00	0.13	0.00	Zone 1	0.00	1.61	0.31	0.33	0.45	0.00
Zone 2	0.00	1.45	0.00	0.00	0.13	0.00	Zone 2	0.00	1.60	0.29	0.32	0.46	0.00
Zone 3	0.00	1.43	0.00	0.00	0.13	0.00	Zone 3	0.00	1.59	0.28	0.34	0.42	0.00
Zone 4	0.00	1.39	0.00	0.00	0.13	0.00	Zone 4	0.00	1.56	0.27	0.33	0.43	0.00
Zone 5	0.00	1.39	0.00	0.00	0.13	0.00	Zone 5	0.00	1.55	0.34	0.36	0.37	0.00
Zone 6	0.00	1.21	0.00	0.00	0.13	0.00	Zone 6	0.00	1.29	0.16	0.34	0.31	0.00
Zone 7	0.00	1.00	0.00	0.00	0.13	0.00	Zone 7	0.00	1.10	0.22	0.35	0.32	0.00
Average	**0.00**	**1.33**	**0.00**	**0.00**	**0.13**	**0.00**	**Average**	**0.00**	**1.47**	**0.27**	**0.34**	**0.40**	**0.00**
Manufacturing, High Tech							**Office, High**						
Zone 1	0.00	2.15	0.00	0.00	0.17	0.00	Zone 1	0.00	1.70	0.30	0.33	0.44	0.00
Zone 2	0.00	2.12	0.00	0.00	0.17	0.00	Zone 2	0.00	1.59	0.29	0.32	0.41	0.00
Zone 3	0.00	2.04	0.00	0.00	0.17	0.00	Zone 3	0.00	1.59	0.32	0.34	0.42	0.00
Zone 4	0.00	2.01	0.00	0.00	0.17	0.00	Zone 4	0.00	1.58	0.28	0.35	0.42	0.00
Zone 5	0.00	1.88	0.00	0.00	0.17	0.00	Zone 5	0.00	1.58	0.32	0.36	0.41	0.00
Zone 6	0.00	1.63	0.00	0.00	0.17	0.00	Zone 6	0.00	1.11	0.16	0.29	0.33	0.00
Zone 7	0.00	1.50	0.00	0.00	0.17	0.00	Zone 7	0.00	0.86	0.23	0.31	0.37	0.00
Average	**0.00**	**1.90**	**0.00**	**0.00**	**0.17**	**0.00**	**Average**	**0.00**	**1.43**	**0.27**	**0.33**	**0.40**	**0.00**
Medical Clinic							**Restaurant, Bar/Lounge**						
Zone 1	0.00	3.15	0.00	0.00	0.54	0.00	Zone 1	0.00	3.07	0.00	0.06	0.24	0.00
Zone 2	0.00	3.09	0.00	0.00	0.53	0.00	Zone 2	0.00	2.87	0.00	0.06	0.25	0.00
Zone 3	0.00	3.07	0.00	0.00	0.54	0.00	Zone 3	0.00	2.85	0.00	0.06	0.24	0.00
Zone 4	0.00	3.02	0.00	0.00	0.54	0.00	Zone 4	0.00	2.83	0.00	0.06	0.24	0.00
Zone 5	0.00	2.99	0.00	0.00	0.54	0.00	Zone 5	0.00	2.80	0.00	0.06	0.24	0.00
Zone 6	0.00	2.58	0.00	0.00	0.55	0.00	Zone 6	0.00	1.88	0.00	0.06	0.24	0.00
Zone 7	0.00	2.45	0.00	0.00	0.52	0.00	Zone 7	0.00	1.87	0.00	0.06	0.23	0.00
Average	**0.00**	**2.91**	**0.00**	**0.00**	**0.54**	**0.00**	**Average**	**0.00**	**2.59**	**0.00**	**0.06**	**0.24**	**0.00**
Motel							**Restaurant, Full**						
Zone 1	0.97	2.68	0.00	0.00	0.19	2.15	Zone 1	0.00	3.37	0.00	0.05	0.38	0.00
Zone 2	1.19	2.52	0.00	0.00	0.18	4.43	Zone 2	0.00	3.30	0.00	0.05	0.39	0.00
Zone 3	1.19	2.52	0.00	0.00	0.19	4.81	Zone 3	0.00	3.25	0.00	0.05	0.39	0.00
Zone 4	1.19	2.42	0.00	0.00	0.21	7.39	Zone 4	0.00	3.20	0.00	0.05	0.39	0.00
Zone 5	1.28	2.26	0.00	0.00	0.21	7.82	Zone 5	0.00	3.03	0.00	0.05	0.38	0.00
Zone 6	1.29	2.07	0.00	0.00	0.23	8.89	Zone 6	0.00	2.35	0.00	0.05	0.39	0.00
Zone 7	1.39	1.97	0.00	0.00	0.23	9.51	Zone 7	0.00	2.33	0.00	0.05	0.38	0.00
Average	**1.21**	**2.35**	**0.00**	**0.00**	**0.21**	**6.43**	**Average**	**0.00**	**2.98**	**0.00**	**0.05**	**0.38**	**0.00**
Museum							**Restaurant, Quick**						
Zone 1	0.00	3.77	0.00	0.11	0.67	0.00	Zone 1	0.00	3.78	0.00	0.08	0.48	0.00
Zone 2	0.00	3.71	0.00	0.12	0.77	0.00	Zone 2	0.00	3.72	0.00	0.08	0.50	0.00
Zone 3	0.00	3.59	0.00	0.12	0.72	0.00	Zone 3	0.00	3.64	0.00	0.08	0.50	0.00
Zone 4	0.00	3.47	0.00	0.14	0.74	0.00	Zone 4	0.00	3.60	0.00	0.08	0.50	0.00
Zone 5	0.00	3.42	0.00	0.16	0.51	0.00	Zone 5	0.00	3.42	0.00	0.08	0.49	0.00
Zone 6	0.00	2.39	0.00	0.15	0.23	0.00	Zone 6	0.00	2.74	0.00	0.08	0.51	0.00
Zone 7	0.00	1.69	0.00	0.17	0.24	0.00	Zone 7	0.00	2.69	0.00	0.08	0.49	0.00
Average	**0.00**	**3.15**	**0.00**	**0.14**	**0.55**	**0.00**	**Average**	**0.00**	**3.37**	**0.00**	**0.08**	**0.50**	**0.00**
Office, Low							**Store, Convenience**						
Zone 1	0.00	2.90	0.00	0.02	0.29	0.00	Zone 1	0.00	4.08	0.00	0.02	0.29	0.00
Zone 2	0.00	2.90	0.00	0.02	0.30	0.00	Zone 2	0.00	3.66	0.00	0.02	0.30	0.00
Zone 3	0.00	2.81	0.00	0.02	0.30	0.00	Zone 3	0.00	3.60	0.00	0.02	0.30	0.00
Zone 4	0.00	2.70	0.00	0.02	0.29	0.00	Zone 4	0.00	3.46	0.00	0.02	0.29	0.00
Zone 5	0.00	2.66	0.00	0.02	0.29	0.00	Zone 5	0.00	3.28	0.00	0.02	0.29	0.00
Zone 6	0.00	2.04	0.00	0.02	0.30	0.00	Zone 6	0.00	2.66	0.00	0.02	0.27	0.00
Zone 7	0.00	1.82	0.00	0.02	0.29	0.00	Zone 7	0.00	2.50	0.00	0.02	0.24	0.00
Average	**0.00**	**2.55**	**0.00**	**0.02**	**0.29**	**0.00**	**Average**	**0.00**	**3.32**	**0.00**	**0.02**	**0.28**	**0.00**

Table R260120-30 Watts per S.F. for Various Systems by Building Type & Climate Zone (cont.)

Building / Zone	Heating	Cooling	Heat rejection	Auxiliary (pumps)	Vent fan	Supp heat pump	Building / Zone	Heating	Cooling	Heat rejection	Auxiliary (pumps)	Vent fan	Supp heat pump
Store, Department							**School, Middle**						
Zone 1	0.00	1.65	0.20	0.23	0.63	0.00	Zone 1	0.00	2.48	0.00	0.00	0.17	0.00
Zone 2	0.00	1.61	0.20	0.23	0.63	0.00	Zone 2	0.00	2.45	0.00	0.00	0.17	0.00
Zone 3	0.00	1.59	0.19	0.22	0.64	0.00	Zone 3	0.00	2.41	0.00	0.00	0.18	0.00
Zone 4	0.00	1.52	0.19	0.23	0.64	0.00	Zone 4	0.00	2.35	0.00	0.00	0.17	0.00
Zone 5	0.00	1.51	0.21	0.24	0.63	0.00	Zone 5	0.00	2.26	0.00	0.00	0.17	0.00
Zone 6	0.00	1.32	0.13	0.21	0.63	0.00	Zone 6	0.00	1.61	0.00	0.00	0.17	0.00
Zone 7	0.00	0.98	0.17	0.23	0.63	0.00	Zone 7	0.00	1.49	0.00	0.00	0.17	0.00
Average	**0.00**	**1.45**	**0.18**	**0.23**	**0.63**	**0.00**	**Average**	**0.00**	**2.15**	**0.00**	**0.00**	**0.17**	**0.00**
Store, Large							**School, High**						
Zone 1	0.00	2.68	0.00	0.00	0.16	0.00	Zone 1	0.00	3.24	0.00	0.01	0.22	0.00
Zone 2	0.00	2.61	0.00	0.00	0.17	0.00	Zone 2	0.00	3.08	0.00	0.01	0.22	0.00
Zone 3	0.00	2.61	0.00	0.00	0.17	0.00	Zone 3	0.00	3.05	0.00	0.01	0.23	0.00
Zone 4	0.00	2.56	0.00	0.00	0.16	0.00	Zone 4	0.00	3.00	0.00	0.01	0.23	0.00
Zone 5	0.00	2.53	0.00	0.00	0.16	0.00	Zone 5	0.00	2.94	0.00	0.01	0.23	0.00
Zone 6	0.00	1.84	0.00	0.00	0.16	0.00	Zone 6	0.00	1.98	0.00	0.01	0.23	0.00
Zone 7	0.00	1.50	0.00	0.00	0.16	0.00	Zone 7	0.00	1.81	0.00	0.01	0.23	0.00
Average	**0.00**	**2.33**	**0.00**	**0.00**	**0.16**	**0.00**	**Average**	**0.00**	**2.73**	**0.00**	**0.01**	**0.23**	**0.00**
Store, Small							**School, College**						
Zone 1	0.00	3.60	0.00	0.01	0.25	0.00	Zone 1	0.00	2.92	0.00	0.00	0.20	0.00
Zone 2	0.00	3.51	0.00	0.01	0.26	0.00	Zone 2	0.00	2.91	0.00	0.00	0.20	0.00
Zone 3	0.00	3.38	0.00	0.01	0.25	0.00	Zone 3	0.00	2.78	0.00	0.00	0.20	0.00
Zone 4	0.00	3.32	0.00	0.01	0.25	0.00	Zone 4	0.00	2.72	0.00	0.00	0.20	0.00
Zone 5	0.00	3.23	0.00	0.01	0.24	0.00	Zone 5	0.00	2.43	0.00	0.00	0.20	0.00
Zone 6	0.00	2.49	0.00	0.01	0.23	0.00	Zone 6	0.00	1.77	0.00	0.00	0.20	0.00
Zone 7	0.00	2.17	0.00	0.01	0.21	0.00	Zone 7	0.00	1.51	0.00	0.00	0.19	0.00
Average	**0.00**	**3.10**	**0.00**	**0.01**	**0.24**	**0.00**	**Average**	**0.00**	**2.43**	**0.00**	**0.00**	**0.20**	**0.00**
Store, Strip Mall							**Service Station**						
Zone 1	0.00	3.36	0.00	0.02	0.25	0.00	Zone 1	0.00	10.42	0.00	0.03	0.47	0.00
Zone 2	0.00	3.30	0.00	0.02	0.25	0.00	Zone 2	0.00	9.01	0.00	0.03	0.46	0.00
Zone 3	0.00	3.14	0.00	0.02	0.25	0.00	Zone 3	0.00	7.82	0.00	0.03	0.49	0.00
Zone 4	0.00	3.14	0.00	0.02	0.25	0.00	Zone 4	0.00	7.02	0.00	0.03	0.47	0.00
Zone 5	0.00	2.24	0.00	0.02	0.24	0.00	Zone 5	0.00	6.13	0.00	0.03	0.46	0.00
Zone 6	0.00	1.96	0.00	0.02	0.25	0.00	Zone 6	0.00	4.32	0.00	0.03	0.45	0.00
Zone 7	0.00	1.64	0.00	0.02	0.24	0.00	Zone 7	0.00	4.18	0.00	0.03	0.40	0.00
Average	**0.00**	**2.52**	**0.00**	**0.02**	**0.25**	**0.00**	**Average**	**0.00**	**6.99**	**0.00**	**0.03**	**0.46**	**0.00**
Store, Warehouse							**Storage, High Bay**						
Zone 1	0.00	2.99	0.00	0.00	0.21	0.00	Zone 1	0.00	1.98	0.00	0.00	0.17	0.00
Zone 2	0.00	2.97	0.00	0.00	0.21	0.00	Zone 2	0.00	1.97	0.00	0.00	0.18	0.00
Zone 3	0.00	2.94	0.00	0.00	0.21	0.00	Zone 3	0.00	1.90	0.00	0.00	0.18	0.00
Zone 4	0.00	2.83	0.00	0.00	0.21	0.00	Zone 4	0.00	1.89	0.00	0.00	0.18	0.00
Zone 5	0.00	2.70	0.00	0.00	0.21	0.00	Zone 5	0.00	1.85	0.00	0.00	0.17	0.00
Zone 6	0.00	2.10	0.00	0.00	0.21	0.00	Zone 6	0.00	1.29	0.00	0.00	0.17	0.00
Zone 7	0.00	1.78	0.00	0.00	0.21	0.00	Zone 7	0.00	1.10	0.00	0.00	0.17	0.00
Average	**0.00**	**2.61**	**0.00**	**0.00**	**0.21**	**0.00**	**Average**	**0.00**	**1.71**	**0.00**	**0.00**	**0.17**	**0.00**
School, Elementary							**Storage, Low Bay**						
Zone 1	0.00	2.73	0.00	0.01	0.21	0.00	Zone 1	0.00	1.87	0.00	0.00	0.16	0.00
Zone 2	0.00	2.69	0.00	0.01	0.21	0.00	Zone 2	0.00	1.85	0.00	0.00	0.16	0.00
Zone 3	0.00	2.63	0.00	0.01	0.21	0.00	Zone 3	0.00	1.82	0.00	0.00	0.16	0.00
Zone 4	0.00	2.61	0.00	0.01	0.20	0.00	Zone 4	0.00	1.77	0.00	0.00	0.16	0.00
Zone 5	0.00	2.51	0.00	0.01	0.20	0.00	Zone 5	0.00	1.76	0.00	0.00	0.16	0.00
Zone 6	0.00	1.71	0.00	0.01	0.20	0.00	Zone 6	0.00	1.20	0.00	0.00	0.16	0.00
Zone 7	0.00	1.64	0.00	0.01	0.20	0.00	Zone 7	0.00	1.00	0.00	0.00	0.16	0.00
Average	**0.00**	**2.36**	**0.00**	**0.01**	**0.20**	**0.00**	**Average**	**0.00**	**1.61**	**0.00**	**0.00**	**0.16**	**0.00**

For customer support on your Electrical Costs with RSMeans data, call 800.448.8182.

589

R260519-90   Wire

Wire quantities are taken off by either measuring each cable run or by extending the conduit and raceway quantities times the number of conductors in the raceway. Ten percent should be added for waste and tie-ins. Keep in mind that the unit of measure of wire is C.L.F. not L.F. as in raceways so the formula would read:

$$\frac{(\text{L.F. Raceway} \times \text{No. of Conductors}) \times 1.10}{100} = \text{C.L.F.}$$

Price per C.L.F. of wire includes:
1. Setting up wire coils or spools on racks
2. Attaching wire to pull in means
3. Measuring and cutting wire
4. Pulling wire into a raceway
5. Identifying and tagging

Price does not include:
1. Connections to breakers, panelboards, or equipment
2. Splices

Job Conditions: Productivity is based on new construction to a height of 10′ in an unobstructed area. Material staging is assumed to be within 100′ of work being performed.

Economy of Scale: If more than three wires at a time are being pulled, deduct the following percentages from the labor of that grouping:

4-5 wires	25%
6-10 wires	30%
11-15 wires	35%
over 15	40%

If a wire pull is less than 100′ in length and is interrupted several times by boxes, lighting outlets, etc., it may be necessary to add the following lengths to each wire being pulled:

Junction box to junction box	2 L.F.
Lighting panel to junction box	6 L.F.
Distribution panel to sub panel	8 L.F.
Switchboard to distribution panel	12 L.F.
Switchboard to motor control center	20 L.F.
Switchboard to cable tray	40 L.F.

Measure of Drops and Riser: It is important when taking off wire quantities to include the wire for drops to electrical equipment. If heights of electrical equipment are not clearly stated, use the following guide:

	Bottom A.F.F.	Top A.F.F.	Inside Cabinet
Safety switch to 100A	5′	6′	2′
Safety switch 400 to 600A	4′	6′	3′
100A panel 12 to 30 circuit	4′	6′	3′
42 circuit panel	3′	6′	4′
Switch box	3′	3′6″	1′
Switchgear	0′	8′	8′
Motor control centers	0′	8′	8′
Transformers - wall mount	4′	8′	2′
Transformers - floor mount	0′	12′	4′

R260519-92   Minimum Copper and Aluminum Wire Size Allowed for Various Types of Insulation

Minimum Wire Sizes

Amperes	Copper THW THWN or XHHW	Copper THHN XHHW*	Aluminum THW XHHW	Aluminum THHN XHHW*	Amperes	Copper THW THWN or XHHW	Copper THHN XHHW*	Aluminum THW XHHW	Aluminum THHN XHHW*
15A	#14	#14	#12	#12	195	3/0	2/0	250kcmil	4/0
20	#12	#12	#10	#10	200	3/0	3/0	250kcmil	4/0
25	#10	#10	#10	#10	205	4/0	3/0	250kcmil	4/0
30	#10	#10	#8	#8	225	4/0	3/0	300kcmil	250kcmil
40	#8	#8	#8	#8	230	4/0	4/0	300kcmil	250kcmil
45	#8	#8	#6	#8	250	250kcmil	4/0	350kcmil	300kcmil
50	#8	#8	#6	#6	255	250kcmil	4/0	400kcmil	300kcmil
55	#6	#8	#4	#6	260	300kcmil	4/0	400kcmil	350kcmil
60	#6	#6	#4	#6	270	300kcmil	250kcmil	400kcmil	350kcmil
65	#6	#6	#4	#4	280	300kcmil	250kcmil	500kcmil	350kcmil
75	#4	#6	#3	#4	285	300kcmil	250kcmil	500kcmil	400kcmil
85	#4	#4	#2	#3	290	350kcmil	250kcmil	500kcmil	400kcmil
90	#3	#4	#2	#2	305	350kcmil	300kcmil	500kcmil	400kcmil
95	#3	#4	#1	#2	310	350kcmil	300kcmil	500kcmil	500kcmil
100	#3	#3	#1	#2	320	400kcmil	300kcmil	600kcmil	500kcmil
110	#2	#3	1/0	#1	335	400kcmil	350kcmil	600kcmil	500kcmil
115	#2	#2	1/0	#1	340	500kcmil	350kcmil	600kcmil	500kcmil
120	#1	#2	1/0	1/0	350	500kcmil	350kcmil	700kcmil	500kcmil
130	#1	#2	2/0	1/0	375	500kcmil	400kcmil	700kcmil	600kcmil
135	1/0	#1	2/0	1/0	380	500kcmil	400kcmil	750kcmil	600kcmil
150	1/0	#1	3/0	2/0	385	600kcmil	500kcmil	750kcmil	600kcmil
155	2/0	1/0	3/0	3/0	420	600kcmil	500kcmil		700kcmil
170	2/0	1/0	4/0	3/0	430		500kcmil		750kcmil
175	2/0	2/0	4/0	3/0	435		600kcmil		750kcmil
180	3/0	2/0	4/0	4/0	475		600kcmil		

*Dry Locations Only

Notes:

1. Size #14 to 4/0 is in AWG units (American Wire Gauge).
2. Size 250 to 750 is in kcmil units (Thousand Circular Mils).
3. Use next higher ampere value if exact value is not listed in table.
4. For loads that operate continuously increase ampere value by 25% to obtain proper wire size.
5. Table R260519-92 has been written for estimating only. It is based on an ambient temperature of 30°C (86° F). For an ambient temperature other than 30°C (86° F), ampacity correction factors will be applied.

R260519-93   Metric Equivalent, Wire

United States		European	
Size AWG or kcmil	**Area Cir. Mils. (cmil) mm²**	**Size mm²**	**Area Cir. Mils.**
18	1620/.82	.75	1480
16	2580/1.30	1.0	1974
14	4110/2.08	1.5	2961
12	6530/3.30	2.5	4935
10	10,380/5.25	4	7896
8	16,510/8.36	6	11,844
6	26,240/13.29	10	19,740
4	41,740/21.14	16	31,584
3	52,620/26.65	25	49,350
2	66,360/33.61	–	–
1	83,690/42.39	35	69,090
1/0	105,600/53.49	50	98,700
2/0	133,100/67.42	–	–
3/0	167,800/85.00	70	138,180
4/0	211,600/107.19	95	187,530
250	250,000/126.64	120	236,880
300	300,000/151.97	150	296,100
350	350,000/177.30	–	–
400	400,000/202.63	185	365,190
500	500,000/253.29	240	473,760
600	600,000/303.95	300	592,200
700	700,000/354.60	–	–
750	750,000/379.93	–	–

U.S. vs. European Wire – Approximate Equivalents

R260519-94   Size Required and Weight (Lbs./1000 L.F.) of Aluminum and Copper THW Wire by Ampere Load

Amperes	Copper Size	Aluminum Size	Copper Weight	Aluminum Weight
15	14	12	24	11
20	12	10	33	17
30	10	8	48	39
45	8	6	77	52
65	6	4	112	72
85	4	2	167	101
100	3	1	205	136
115	2	1/0	252	162
130	1	2/0	324	194
150	1/0	3/0	397	233
175	2/0	4/0	491	282
200	3/0	250	608	347
230	4/0	300	753	403
255	250	400	899	512
285	300	500	1068	620
310	350	500	1233	620
335	400	600	1396	772
380	500	750	1732	951

R260526-80 Grounding

Grounding When taking off grounding systems, identify separately the type and size of wire.

Example:

Bare copper & size

Bare aluminum & size

Insulated copper & size

Insulated aluminum & size

Count the number of ground rods and their size.

Example:

1.   8′ grounding rod – 5/8″ dia. 20 Ea.
2. 10′ grounding rod – 5/8″ dia. 12 Ea.
3. 15′ grounding rod – 3/4″ dia. 4 Ea.

Count the number of connections; the size of the largest wire will determine the productivity.

Example:

Braze a #2 wire to a #4/0 cable

The 4/0 cable will determine the L.H. and cost to be used.

Include individual connections to:

1. Ground rods
2. Building steel
3. Equipment
4. Raceways

Price does not include:

1. Excavation
2. Backfill
3. Sleeves or raceways used to protect grounding wires
4. Wall penetrations
5. Floor cutting
6. Core drilling

Job Conditions: Productivity is based on a ground floor area, using cable reels in an unobstructed area. The material staging area is assumed to be within 100′ of work being performed.

R260533-20   Conduit To 10′ High

List conduit by quantity, size, and type. Do not deduct for lengths occupied by fittings, since this will be allowance for scrap. Example:

A. Aluminum — size
B. Rigid galvanized — size
C. Steel intermediate (IMC) — size
D. Rigid steel, plastic-coated 20 Mil. — size
E. Rigid steel, plastic-coated 40 Mil. — size
F. Electric metallic tubing (EMT) — size
G. PVC Schedule 40 — size

Types (A) through (E) listed above contain the following per 100 L.F.:

1. (11) Threaded couplings
2. (11) Beam-type hangers
3. (2) Factory sweeps
4. (2) Fiber bushings
5. (4) Locknuts
6. (2) Field threaded pipe terminations
7. (2) Removal of concentric knockouts

Type (F) contains per 100 L.F.:

1. (11) Set screw couplings
2. (11) Beam clamps
3. (2) Field bends on 1/2″ and 3/4″ diameter
4. (2) Factory sweeps for 1″ and above
5. (2) Set screw steel connectors
6. (2) Removal of concentric knockouts

Type (G) contains per 100 L.F.:

1. (11) Field cemented couplings
2. (34) Beam clamps

3. (2) Factory sweeps
4. (2) Adapters
5. (2) Locknuts
6. (2) Removal of concentric knockouts

Labor-hours for all conduit to 10′ include:

1. Unloading by hand
2. Hauling by hand to an area up to 200′ from loading dock
3. Installation of conduit and fittings as described in Conduit models (A) through (G)

Not included in the material and labor are:

1. Staging rental or purchase
2. Structural modifications
3. Wire
4. Junction boxes
5. Fittings in excess of those described in conduit models (A) through (G)
6. Painting of conduit

Fittings

Only those fittings listed above are included in the linear foot totals, although they should be listed separately from conduit lengths, without prices, to ensure proper quantities for material procurement.

If the fittings required exceed the quantities included in the model conduit runs, then material and labor costs must be added to the difference. If actual needs per 100 L.F. of conduit are: (2) sweeps, (4) LBs and (1) field bend, then, (4) LBs and (1) field bend must be priced additionally.

R260533-21 Hangers

It is sometimes desirable to substitute an alternate style of hanger if the support being used is not the type described in the conduit models.

One approach is the substitution method:

1. Find the cost of the type hanger described in the conduit model.
2. Calculate the cost of the desired type hanger (it may be necessary to calculate individual components such as drilling, expansion shields, etc.).
3. Calculate the cost difference (delta) between the two types of hangers.
4. Multiply the cost delta by the number of hangers in the model.
5. Divide the total delta cost for hangers in the model by the length of the model to find the delta cost per L.F. for that model.
6. Modify the given unit costs per L.F. for the model by the delta cost per L.F.

Another approach to hanger configurations would be to start with the conduit only and add all the supports and any other items as separate lines. This procedure is most useful if the project involves racking many runs of conduit on a single hanger; for instance, a trapeze type hanger.

Example: Five (5) 2″ RGS conduits, 50 L.F. each, are to be run on trapeze hangers from one pull box to another. The run includes one 90° bend.

1. List the hangers' components to create an assembly cost for each 2′-wide trapeze.
2. List the components for the 50′ conduit run, noting that 6 trapeze supports will be required.

Job Conditions: Productivities are based on new construction to 10′ high, using scaffolding, scissor lift, bucket lift in an unobstructed area. Material storage is assumed to be within 100′ of work being performed.

Add to labor for elevated installations:

10′ to 14.5′ High–10%	30′ to 35′ High–40%
15′ to 20′ High–20%	35′ to 40′ High–50%
20′ to 25′ High–25%	Over 40′ High–55%
25′ to 30′ High–35%	

Add these percentages to the L.F. labor cost, but not to fittings. Add these percentages only to quantities exceeding the different height levels, rather than the total conduit quantities.

Linear foot price for labor does not include penetrations in walls or floors and must be added to the estimate.

R260533-22   Conductors in Conduit

The table below lists the maximum number of conductors for various sized conduit using THW, TW or THWN insulations.

Copper Wire Size	Conduit Size																											
	1/2″			3/4″			1″			1-1/4″			1-1/2″			2″			2-1/2″			3″		3-1/2″		4″	
	TW	THW	THWN	TW	THW	THWN	TW	THW	THWN	TW	THW	THWN	TW	THW	THWN	TW	THW	THWN	TW	THW	THWN	THW	THWN	THW	THWN	THW	THWN
#14	9	6	13	15	10	24	25	16	39	44	29	69	60	40	94	99	65	154	142	93		143		192			
#12	7	4	10	12	8	18	19	13	29	35	24	51	47	32	70	78	53	114	111	76	164	117		157			
#10	5	4	6	9	6	11	15	11	18	26	19	32	36	26	44	60	43	73	85	61	104	95	160	127		163	
#8	2	1	3	4	3	5	7	5	9	12	10	16	17	13	22	28	22	36	40	32	51	49	79	66	106	85	136
#6		1	1		2	4		4	6		7	11		10	15		16	26		23	37	36	57	48	76	62	98
#4		1	1		1	2		3	4		5	7		7	9		12	16		17	22	27	35	36	47	47	60
#3		1	1		1	1		2	3		4	6		6	8		10	13		15	19	23	29	31	39	40	51
#2		1	1		1	1		2	3		4	5		5	7		9	11		13	16	20	25	27	33	34	43
#1					1	1		1	1		3	3		4	5		6	8		9	12	14	18	19	25	25	32
1/0					1	1		1	1		2	3		3	4		5	7		8	10	12	15	16	21	21	27
2/0					1	1		1	1		1	2		3	3		5	6		7	8	10	13	14	17	18	22
3/0					1	1		1	1		1	1		2	3		4	5		6	7	9	11	12	14	15	18
4/0						1		1	1		1	1		1	2		3	4		5	6	7	9	10	12	13	15
250 kcmil								1	1		1	1		1	1		2	3		4	4	6	7	8	10	10	12
300								1	1		1	1		1	1		2	3		3	4	5	6	7	8	9	11
350									1		1	1		1	1		1	2		3	3	4	5	6	7	8	9
400											1	1		1	1		1	1		2	3	4	5	5	6	7	8
500											1	1		1	1		1	1		1	2	3	4	4	5	6	7
600												1		1	1		1	1		1	1	3	3	4	4	5	5
700														1	1		1	1		1	1	2	3	3	4	4	5
750														1	1		1	1		1	1	2	2	3	3	4	4

R260533-25 Conduit Weight Comparisons (Lbs. per 100 ft.) with Maximum Cable Fill*

Type	1/2″	3/4″	1″	1-1/4″	1-1/2″	2″	2-1/2″	3″	3-1/2″	4″	5″	6″
Rigid Galvanized Steel (RGS)	104	140	235	358	455	721	1022	1451	1749	2148	3083	4343
Intermediate Steel (IMC)	84	113	186	293	379	611	883	1263	1501	1830		
Electrical Metallic Tubing (EMT)	54	116	183	296	368	445	641	930	1215	1540		

*Conduit & Heaviest Conductor Combination

R260533-30 Labor for Couplings and Fittings

The labor included in the unit price lines for couplings is for their installation separate from a fitting. However, the labor included in the unit price lines for fittings covers their complete installation, which may include the labor installation of the coupling(s). If coupling(s) are required to complete a fitting installation, only the material for the required number of couplings needs to be added to the appropriate unit price lines.

R260533-60   Wireway

When "taking off" Wireway, list by size and type.

Example:
1. Screw cover, unflanged + size
2. Screw cover, flanged + size
3. Hinged cover, flanged + size
4. Hinged cover, unflanged + size

Each 10' length on Wireway contains:
1. 10' of cover either screw or hinged type
2. (1) Coupling or flange gasket
3. (1) Wall type mount

All fittings must be priced separately.

Substitution of hanger types is done the same as described in R260533-21, "HANGERS," keeping in mind that the wireway model is based on 10' sections instead of a 100' conduit run.

Labor-hours for wireway include:
1. Unloading by hand
2. Hauling by hand up to 100' from loading dock
3. Measuring and marking
4. Mounting wall bracket using two (2) anchor type lead fasteners
5. Installing wireway on brackets, to 10' high (for higher elevations use factors in R260533-20)

Job Conditions: Productivity is based on new construction, to a height of 10' in an unobstructed area.

Material staging area is assumed to be within 100' of work being performed.

R260533-65   Outlet Boxes

Outlet boxes should be included on the same takeoff sheet as branch piping or devices to better explain what is included in each circuit.

Each unit price in this section is a stand alone item and contains no other component unless specified. For example, to estimate a duplex outlet, components that must be added are:
1. 4" square box
2. 4" plaster ring
3. Duplex receptacle
4. Device cover

The method of mounting outlet boxes is two (2) plastic shield fasteners.

Outlet boxes plastic, labor-hours include:
1. Marking box location on wood studding
2. Mounting box

Economy of Scale – For large concentrations of plastic boxes in the same area deduct the following percentages from labor-hour totals:

1	to	10	0%
11	to	25	20%
26	to	50	25%
51	to	100	30%
	over	100	35%

Note: It is important to understand that these percentages are not used on the total job quantities, but only areas where concentrations exceed the levels specified.

R260533-70   Pull Boxes and Cabinets

List cabinets and pull boxes by NEMA type and size.

Example:	TYPE	SIZE
	NEMA 1	6"W x 6"H x 4"D
	NEMA 3R	6"W x 6"H x 4"D

Labor-hours for wall mount (indoor or outdoor) installations include:
1. Unloading and uncrating
2. Handling of enclosures up to 200' from loading dock using a dolly or pipe rollers
3. Measuring and marking
4. Drilling four (4) anchor type lead fasteners using a hammer drill
5. Mounting and leveling boxes

Note: A plywood backboard is not included.

Labor-hours for ceiling mounting include:
1. Unloading and uncrating
2. Handling boxes up to 100' from loading dock

3. Measuring and marking
4. Drilling four (4) anchor type lead fasteners using a hammer drill
5. Installing and leveling boxes to a height of 10' using rolling staging

Labor-hours for free standing cabinets include:
1. Unloading and uncrating
2. Handling of cabinets up to 200' from loading dock using a dolly or pipe rollers
3. Marking of floor
4. Drilling four (4) anchor type lead fasteners using a hammer drill
5. Leveling and shimming

Labor-hours for telephone cabinets include:
1. Unloading and uncrating
2. Handling cabinets up to 200' using a dolly or pipe rollers
3. Measuring and marking
4. Mounting and leveling, using four (4) lead anchor type fasteners

Fig. R260536-11

R260536-10   Cable Tray

Cable Tray - When taking off cable tray it is important to identify separately the different types and sizes involved in the system being estimated. (Fig. R260536-11)

A. – Ladder Type, galvanized or aluminum
B. – Trough Type, galvanized or aluminum
C. – Solid Bottom, galvanized or aluminum

The unit of measure is calculated in linear feet. Do not deduct from this footage any length occupied by fittings. This will be the only allowance for scrap. Be sure to include all vertical drops to panels, switch gear, etc.

Hangers – Included in the linear footage of cable tray is

D. – 1 – Pair of connector plates per 12 L.F.
E. – 1 – Pair clamp type hangers and 4' of 3/8" threaded rod per 12 L.F.

Not included are structural supports, which must be priced in addition to the hangers.

Fittings – Identify separately the different types of fittings

A. – Ladder Type, galvanized or aluminum
B. – Trough Type, galvanized or aluminum
C. – Solid Bottom Type, galvanized or aluminum

The configuration, radius and rung spacing must also be listed. The unit of measure is "Ea."

D. – Elbow, vertical Ea.
E. – Elbow, horizontal Ea.
F. – Tee, vertical Ea.
G. – Cross, horizontal Ea.
H. – Wye, horizontal Ea.
I. – Tee, horizontal Ea.
J. – Reducing fitting Ea.

Depending on the use of the system other examples of units which must be included are:

K. – Divider strip L.F.
L. – Drop-outs Ea.

M. – End caps Ea.
N. – Panel connectors Ea.

Wire and cable are not included and should be taken off separately. See Unit Price sections.

Job Conditions – Unit prices are based on a new installation to a work plane of 10' in an unobstructed area.

Add to labor for elevated installations:

10' to 14.5' High	10%
15' to 20' High	20%
20' to 25' High	25%
25' to 30' High	35%
30' to 35' High	40%
35' to 40' High	50%
Over 40' High	55%

Add these percentages for L.F. totals but not to fittings. Add percentages to only those quantities that fall in the different elevations. In other words, if the total quantity of cable tray is 200' but only 75' is above 15' then the 10% is added to the 75' only.

Linear foot costs do not include penetrations through walls and floors which must be added to the estimate.

Cable Tray Covers

Covers – Cable tray covers are taken off in the same manner as the tray itself, making distinctions as to the type of cover. (Fig. R260536-11)

O. – Vented, galvanized or aluminum
P. – Solid, galvanized or aluminum

Cover configurations are taken off separately noting type, specific radius and widths.

Note: Care should be taken to identify from plans and specifications exactly what is being covered. In many systems only vertical fittings are covered to retain wire and cable.

R260539-30 Conduit In Concrete Slab

List conduit by quantity, size and type.

Example:

  A. Rigid galvanized steel + size
  B. P.V.C. + size

Rigid galvanized steel (A) contains per 100 L.F.:

1. (20) Ties to slab reinforcing
2. (11) Threaded steel couplings
3. (2) Factory sweeps
4. (2) Field threaded conduit terminations
5. (2) Fiber bushings + locknuts
6. (2) Removal of concentric knockouts

P.V.C. (B) contains per 100 L.F.:

1. (20) Ties to slab reinforcing
2. (11) Field cemented couplings
3. (2) Factory sweeps
4. (2) Adapters
5. (2) Removal of concentric knockouts

R260539-40 Conduit In Trench

Conduit in trench is galvanized steel and contains per 100 L.F.:

1. (11) Threaded couplings
2. (2) Factory sweeps
3. (2) Fiber bushings + (4) locknuts
4. (2) Field threaded conduit terminations
5. (2) Removal of concentric knockouts

Note:

Conduit in Unit Price sections do not include:

1. Floor cutting
2. Excavation or backfill
3. Grouting or patching

Conduit fittings in excess of those listed in the above Conduit model must be added. (Refer to R260533-20 for Procedure example.)

For customer support on your Electrical Costs with RSMeans data, call 800.448.8182.

599

Fig. R260543-51

R260543-50   Underfloor Duct

When pricing Underfloor Duct it is important to identify and list each component, since costs vary significantly from one type of fitting to another. Do not deduct boxes or fittings from linear foot totals; this will be your allowance for scrap.

The first step is to identify the system as either:

      FIG. R260543-51 Single Level

      FIG. R260543-52 Dual Level

Single Level System

Include on your "takeoff sheet" the following unit price items, making sure to distinquish between Standard and Super duct:

- A. Feeder duct (blank) in L.F.
- B. Distribution duct (Inserts 2′ on center) in L.F.
- C. Elbows (Vertical) Ea.
- D. Elbows (Horizontal) Ea.
- E. Cabinet connector Ea.
- F. Single duct junction box Ea.
- G. Double duct junction box Ea.
- H. Triple duct junction box Ea.
- I. Support, single cell Ea.
- J. Support, double cell Ea.
- K. Support, triple cell Ea.
- L. Carpet pan Ea.
- M. Terrazzo pan Ea.
- N. Insert to conduit adapter Ea.
- O. Conduit adapter Ea.
- P. Low tension outlet Ea.
- Q. High tension outlet Ea.
- R. Galvanized nipple Ea.
- S. Wire per C.L.F.
- T. Offset (Duct type) Ea.

Fig. R260543-52

R260543-50 Underfloor Duct (cont.)

Dual Level System + Labor

Dual Level

Include the following when "taking off" Dual Level systems:
Distinguish between Standard and Super duct.

A. Feeder duct (blank) in L.F.
B. Distribution duct (Inserts 2' on center) in L.F.
C. Elbows (Vertical), Ea.
D. Elbows (Horizontal), Ea.
E. Cabinet connector, Ea.
F. Single duct, 2 level, junction box, Ea.
G. Double duct, 2 level, junction box, Ea.
H. Support, single cell, Ea.
I. Support, double cell, Ea.
J. Support, triple cell, Ea.
K. Carpet pan, Ea.
L. Terrazzo pan, Ea.
M. Insert to conduit adapter, Ea.
N. Conduit adapter, Ea.
O. Low tension outlet, Ea.
P. High tension outlet, Ea.
Q. Wire per C.L.F.

Note: Make sure to include risers in linear foot totals. High
tension outlets include box, receptacle, covers and related
mounting hardware.

Labor-hours for both Single and Dual Level systems include:
1. Unloading and uncrating
2. Hauling up to 200' from loading dock
3. Measuring and marking
4. Setting raceway and fittings in slab or on grade
5. Leveling raceway and fittings

Labor-hours do not include:
1. Floor cutting
2. Excavation or backfill
3. Concrete pour
4. Grouting or patching
5. Wire or wire pulls
6. Additional outlets after concrete is poured
7. Piping to or from Underfloor Duct

Note: Installation is based on installing up to 150' of duct. If quantities
exceed this, deduct the following percentages:
1. 150' to 250' - 10%
2. 250' to 350' - 15%
3. 350' to 500' - 20%
4. over 500' - 25%

Deduct these percentages from labor only.

Deduct these percentages from straight sections only.

Do not deduct from fittings or junction boxes.

Job Conditions: Productivity is based on new construction.

Underfloor duct to be installed on first three floors.

The material staging area is within 100' of work being performed.

The area unobstructed and duct are not subject to physical damage.

R260580-75   Motor Connections

Motor connections should be listed by size and type of motor.
Included in the material and labor cost are:

1. (2) Flex connectors
2. 18″ of flexible metallic wireway
3. Wire identification and termination
4. Test for rotation
5. (2) or (3) Conductors

Price does not include:

1. Mounting of motor
2. Disconnect switch
3. Motor starter
4. Controls
5. Conduit or wire ahead of flex

Note: When "taking off" motor connections, it is advisable to list connections on the same quantity sheet as motors, motor starters and controls.

R260913-80   Switchboard Instruments

Switchboard instruments are added to the price of switchboards according to job specifications. This equipment is usually included when ordering "gear" from the manufacturer and will arrive factory installed.

Included in the labor cost are:

1. Internal wiring connections
2. Wire identification
3. Wire tagging

Transition sections include:

1. Uncrating
2. Hauling sections up to 100′ from loading dock
3. Positioning sections
4. Leveling sections

5. Bolting enclosures
6. Bolting vertical bus bars

Price does not include:

1. Equipment pads
2. Steel channels embedded or grouted in concrete
3. Special knockouts
4. Rigging

Job Conditions: Productivity is based on new construction and equipment to be installed on the first floor within 200′ of the loading dock.

R262213-10 Electric Circuit Voltages

General: The following method provides the user with a simple non-technical means of obtaining comparative costs of wiring circuits. The circuits considered serve the electrical loads of motors, electric heating, lighting and transformers, for example, that require low voltage 60 Hertz alternating current.

The method used here is suitable only for obtaining estimated costs. It is **not** intended to be used as a substitute for electrical engineering design applications.

Conduit and wire circuits can represent from twenty to thirty percent of the total building electrical cost. By following the described steps and using the tables the user can translate the various types of electric circuits into estimated costs.

Wire Size: Wire size is a function of the electric load which is usually listed in one of the following units:

1. Amperes (A)
2. Watts (W)
3. Kilowatts (kW)
4. Volt amperes (VA)
5. Kilovolt amperes (kVA)
6. Horsepower (HP)

These units of electric load must be converted to amperes in order to obtain the size of wire necessary to carry the load. To convert electric load units to amperes one must have an understanding of the voltage classification of the power source and the voltage characteristics of the electrical equipment or load to be energized. The seven A.C. circuits commonly used are illustrated in Figures R262213-11 through R262213-17 showing the tranformer load voltage and the point of use voltage at the point on the circuit where the load is connected. The difference between the source and point of use voltage is attributed to the circuit voltage drop and is considered to be approximately 4%.

Motor Voltages: Motor voltages are listed by their point of use voltage and not the power source voltage.

For example: 460 volts instead of 480 volts
200 instead of 208 volts
115 volts instead of 120 volts

Lighting and Heating Voltages: Lighting and heating equipment voltages are listed by the power source voltage and not the point of wire voltage.

For example: 480, 277, 120 volt lighting
480 volt heating or air conditioning unit
208 volt heating unit

Transformer Voltages: Transformer primary (input) and secondary (output) voltages are listed by the power source voltage.

For example: Single phase 10 kVA
Primary 240/480 volts
Secondary 120/240 volts

In this case, the primary voltage may be 240 volts with a 120 volts secondary or may be 480 volts with either a 120V or a 240V secondary.

For example: Three phase 10 kVA
Primary 480 volts
Secondary 208Y/120 volts

In this case the transformer is suitable for connection to a circuit with a 3 phase, 3 wire or a 3 phase, 4 wire circuit with a 480 voltage. This application will provide a secondary circuit of 3 phase, 4 wire with 208 volts between phase wires and 120 volts between any phase wire and the neutral (white) wire.

R262213-11

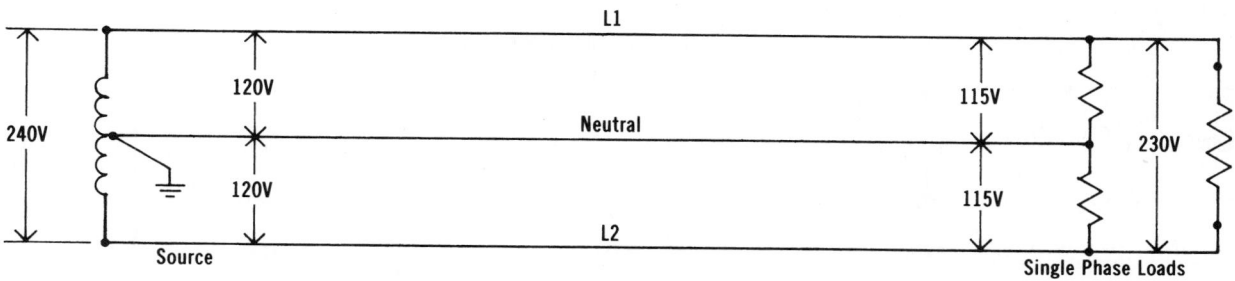

3 Wire, 1 Phase, 120/240 Volt System

R262213-12

4 Wire, 3 Phase, 208Y/120 Volt System

603

R262213-13 Electric Circuit Voltages (cont.)

3 Wire, 3 Phase 240 Volt System

R262213-14

4 Wire, 3 Phase, 240/120 Volt System

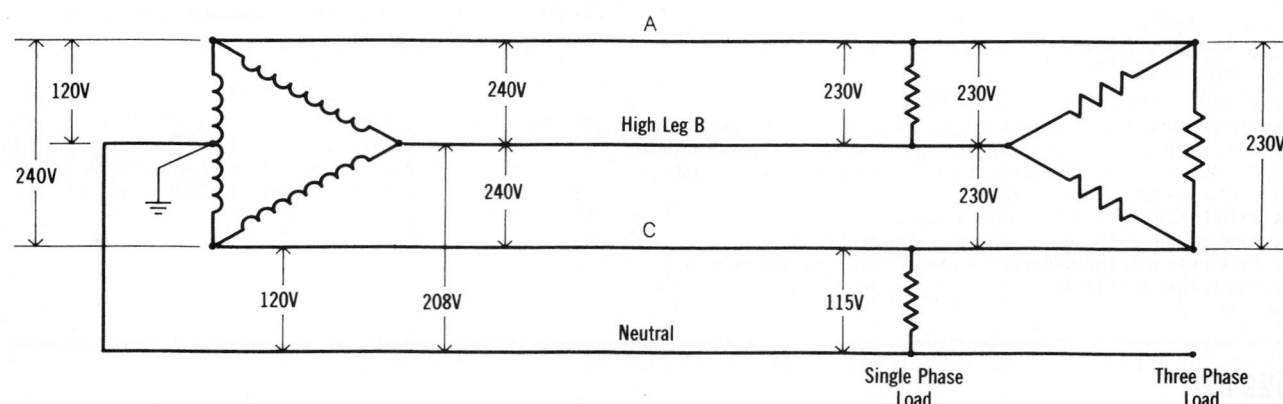

R262213-15

3 Wire, 3 Phase 480 Volt System

R262213-16

4 Wire, 3 Phase, 480Y/277 Volt System

R262213-17　Electric Circuit Voltages (cont.)

3 Wire, 3 Phase, 600 Volt System

R262213-20　kW Value/Cost Determination

General: Lighting and electric heating loads are expressed in watts and kilowatts.

Cost Determination:

The proper ampere values can be obtained as follows:

1. Convert watts to kilowatts
   (watts 1000 ÷ kilowatts)
2. Determine voltage rating of equipment.
3. Determine whether equipment is single phase or three phase.
4. Refer to Table R262213-21 to find ampere value from kW, Ton and BTU/hr. values.
5. Determine type of wire insulation – TW, THW, THWN.
6. Determine if wire is copper or aluminum.
7. Refer to Table R260519-92 to obtain copper or aluminum wire size from ampere values.
8. Next refer to Table R260533-22 for the proper conduit size to accommodate the number and size of wires in each particular case.
9. Next refer to unit cost data for the per linear foot cost of the conduit.
10. Next refer to unit cost data for the per linear foot cost of the wire. Multiply cost of wire per L.F. x number of wires in the circuits to obtain total wire cost per L.F.

11. Add values obtained in Step 9 and 10 for total cost per linear foot for conduit and wire x length of circuit = Total Cost.

Notes:

1. 1 Phase refers to single phase, 2 wire circuits.
2. 3 Phase refers to three phase, 3 wire circuits.
3. For circuits which operate continuously for 3 hours or more, multiply the ampere values by 1.25 for a given kW requirement.
4. For kW ratings not listed, add ampere values.

   For example: Find the ampere value of 9 kW at 208 volt, single phase.

$$\begin{array}{r} 4 \text{ kW} = 19.2A \\ 5 \text{ kW} = 24.0A \\ \hline 9 \text{ kW} = 43.2A \end{array}$$

5. "Length of Circuit" refers to the one way distance of the run, not to the total sum of wire lengths.

R262213-21 Ampere Values as Determined by kW Requirements, BTU/HR or Ton, Voltage and Phase Values

			Ampere Values						
			120V	208V		240V		277V	480V
kW	Ton	BTU/HR	1 Phase	1 Phase	3 Phase	1 Phase	3 Phase	1 Phase	3 Phase
0.5	.1422	1,707	4.2A	2.4A	1.4A	2.1A	1.2A	1.8A	0.6A
0.75	.2133	2,560	6.2	3.6	2.1	3.1	1.9	2.7	.9
1.0	.2844	3,413	8.3	4.9	2.8	4.2	2.4	3.6	1.2
1.25	.3555	4,266	10.4	6.0	3.5	5.2	3.0	4.5	1.5
1.5	.4266	5,120	12.5	7.2	4.2	6.3	3.1	5.4	1.8
2.0	.5688	6,826	16.6	9.7	5.6	8.3	4.8	7.2	2.4
2.5	.7110	8,533	20.8	12.0	7.0	10.4	6.1	9.1	3.1
3.0	.8532	10,239	25.0	14.4	8.4	12.5	7.2	10.8	3.6
4.0	1.1376	13,652	33.4	19.2	11.1	16.7	9.6	14.4	4.8
5.0	1.4220	17,065	41.6	24.0	13.9	20.8	12.1	18.1	6.1
7.5	2.1331	25,598	62.4	36.0	20.8	31.2	18.8	27.0	9.0
10.0	2.8441	34,130	83.2	48.0	27.7	41.6	24.0	36.5	12.0
12.5	3.5552	42,663	104.2	60.1	35.0	52.1	30.0	45.1	15.0
15.0	4.2662	51,195	124.8	72.0	41.6	62.4	37.6	54.0	18.0
20.0	5.6883	68,260	166.4	96.0	55.4	83.2	48.0	73.0	24.0
25.0	7.1104	85,325	208.4	120.2	70.0	104.2	60.0	90.2	30.0
30.0	8.5325	102,390		144.0	83.2	124.8	75.2	108.0	36.0
35.0	9.9545	119,455		168.0	97.1	145.6	87.3	126.0	42.1
40.0	11.3766	136,520		192.0	110.8	166.4	96.0	146.0	48.0
45.0	12.7987	153,585			124.8	187.5	112.8	162.0	54.0
50.0	14.2208	170,650			140.0	208.4	120.0	180.4	60.0
60.0	17.0650	204,780			166.4		150.4	216.0	72.0
70.0	19.9091	238,910			194.2		174.6		84.2
80.0	22.7533	273,040			221.6		192.0		96.0
90.0	25.5975	307,170					225.6		108.0
100.0	28.4416	341,300							120.0

R262213-25   kVA Value/Cost Determination

General: Control transformers are listed in VA. Step-down and power transformers are listed in kVA.

Cost Determination:

1. Convert VA to kVA. Volt amperes (VA) ÷ 1000 = Kilovolt amperes (kVA).
2. Determine voltage rating of equipment.
3. Determine whether equipment is single phase or three phase.
4. Refer to Table R262213-26 to find ampere value from kVA value.
5. Determine type of wire insulation – TW, THW, THWN.
6. Determine if wire is copper or aluminum.
7. Refer to Table R260519-92 to obtain copper or aluminum wire size from ampere values.

8. Next refer to Table R260533-22 for the proper conduit size to accommodate the number and size of wires in each particular case.
9. Next refer to unit price data for the per linear foot cost of the conduit.
10. Next refer to unit price data for the per linear foot cost of the wire. Multiply cost of wire per L.F. x number of wires in the circuits to obtain total wire cost.
11. Add values obtained in Step 9 and 10 for total cost per linear foot for conduit and wire x length of circuit = Total Cost.

Example: A transformer rated 10 kVA 480 volts primary, 240 volts secondary, 3 phase has the capacity to furnish the following:

1. Primary amperes = 10 kVA x 1.20 = 12 amperes
   (from Table R262213-26)
2. Secondary amperes = 10 kVA x 2.40 = 24 amperes
   (from Table R262213-26)

Note: Transformers can deliver generally 125% of their rated kVA. For instance, a 10 kVA rated transformer can safely deliver 12.5 kVA.

R262213-26   Multiplier Values for kVA to Amperes Determined by Voltage and Phase Values

Volts	Multiplier for Circuits	
	2 Wire, 1 Phase	3 Wire, 3 Phase
115	8.70	
120	8.30	
230	4.30	2.51
240	4.16	2.40
200	5.00	2.89
208	4.80	2.77
265	3.77	2.18
277	3.60	2.08
460	2.17	1.26
480	2.08	1.20
575	1.74	1.00
600	1.66	0.96

R262213-60   Oil Filled Transformers

Transformers in this section include:
1. Rigging (as required)
2. Rental of crane and operator
3. Setting of oil filled transformer
4. (4) Anchor bolts, nuts and washers in concrete pad

Price does not include:
1. Primary and secondary terminations
2. Transformer pad
3. Equipment grounding
4. Cable
5. Conduit locknuts or bushings

Transformers in Unit Price sections for dry type, back-boost and isolating transformers include:
1. Unloading and uncrating
2. Hauling transformer to within 200' of loading dock
3. Setting in place
4. Wall mounting hardware
5. Testing

Price does not include:
1. Structural supports
2. Suspension systems
3. Welding or fabrication
4. Primary & secondary terminations

Add the following percentages to the labor for ceiling mounted transformers:

10' to 15'	= + 15%
15' to 25'	= + 30%
Over 25'	= + 35%

Job Conditions: Productivities are based on new construction. Installation is assumed to be on the first floor, in an obstructed area to a height of 10'. Material staging area is within 100' of final transformer location.

R262416-50   Load Centers and Panelboards

When pricing load centers list panels by size and type. List breakers in a separate column of the "quantity sheet," and define by phase and ampere rating.

Material and labor prices include breakers; for example: a 100A, 3-Wire, 102/240V, 18 circuit panel w/ main breaker, as described in the unit cost section, contains 18 single pole 20A breakers.

If you do not choose to include a full panel of single pole breakers, use the following method to adjust material and labor costs.

Example: In an 18 circuit panel only 16 single pole breakers are desired, requiring that the cost of 2 breakers be subtracted from the panel cost.

1. Go to the appropriate unit cost section of the data set to find the unit prices of the given circuit breaker type.
2. Modify those costs as follows:
   Bare material price x 0.50
   Bare labor cost x 0.60.
3. Multiply those modified bare costs by 2 (breakers in this example).
4. Subtract the modified costs for the 2 breakers from the given cost of the panel.

Labor-hours for load center installation includes:
1. Unloading, uncrating, and handling enclosures 200' from unloading area
2. Measuring and marking
3. Drilling (4) lead anchor type fasteners using a hammer drill
4. Mounting and leveling panel to a height of 6'
5. Preparation and termination of feeder cable to lugs or main breaker
6. Branch circuit identification
7. Lacing using tie wraps
8. Testing and load balancing
9. Marking panel directory

Not included in the material and labor are:
1. Modifications to enclosure
2. Structural supports
3. Additional lugs
4. Plywood backboards
5. Painting or lettering

Note: Knockouts are included in the price of terminating pipe runs and need not be added to the load center costs.

Job Conditions: Productivity is based on new construction to a height of 6', in an unobstructed area. Material staging area is assumed to be within 100' of work being performed.

Fig. R262419-61

R262419-60   Motor Control Centers

When taking off motor control centers, list the size, type and height of structures.

Example:
1. 600A, 22,000 RMS, 72″ high
2. 600A, back to back, 72″ high

Next take off individual starters; the number of structures can also be determined by adding the height in inches of starters divided by the height of the structure, and list on the same quantity sheet as the structures. Identify starters by type, horsepower rating, size and height in inches.

Example:
A. Class 1, Type B, FVNR starter, 25 H.P., 18″ high.
    Add to the list with starters, factory installed controls.

Example:

B. Pilot lights
C. Push buttons
D. Auxiliary contacts

Identify starters and structures as either copper or aluminum and by the NEMA type of enclosure.

When pricing starters and structures, be sure to add or deduct adjustments, using lines in the unit price section.

Included in the cost of motor control structures are:
1. Uncrating
2. Hauling to location within 100′ of loading dock
3. Setting structures
4. Leveling
5. Aligning
6. Bolting together structure frames
7. Bolting horizontal bus bars

Labor-hours do not include:
1. Equipment pad
2. Steel channels embedded or grouted in concrete
3. Pull boxes
4. Special knockouts
5. Main switchboard section
6. Transition section
7. Instrumentation
8. External control wiring
9. Conduit or wire

Material for starters include:
1. Circuit breaker or fused disconnect
2. Magnetic motor starter
3. Control transformer
4. Control fuse and fuse block

Labor-hours for starters include:
1. Handling
2. Installing starter within structure
3. Internal wiring connections
4. Lacing within enclosure
5. Testing
6. Phasing

Job Conditions: Productivity is based on new construction. The motor control location is assumed to be on first floor in an unobstructed area. The material staging area is within 100′ of the final location.

Note: Additional labor-hours must be added if M.C.C. is to be installed on floors other than the first, or if rigging is required.

R262419-65   Motor Starters and Controls

Motor starters should be listed on the same "quantity sheet" as motors and motor connections. Identify each starter by:

1. Size
2. Voltage
3. Type

Example:

  A. FVNR, 480V, 2HP, size 00
  B. FVNR, 480V, 5HP, size 0, combination type
  C. FVR, 480V, size 2

The NEMA type of enclosure should also be identified.

Included in the labor-hours are:

1. Unloading, uncrating and handling of starters up to 200′ from loading area
2. Measuring and marking
3. Drilling (4) anchor type lead fasteners using a hammer drill
4. Mounting and leveling starter
5. Connecting wire or cable to line and load sides of starter (when already lugged)
6. Installation of (3) thermal type heaters
7. Testing

The following are not included unless specified in the unit price description.

1. Control transformer
2. Controls, either factory or field installed
3. Conduit and wire to or from starter
4. Plywood backboard
5. Cable terminations

The following material and labor were included for combination type starters:

1. Unloading, uncrating and handling of starter up to 200′ of loading dock
2. Measuring and marking
3. Drilling (4) anchor type lead fasteners using a hammer drill
4. Mounting and leveling
5. Connecting prepared cable conductors
6. Installation of (3) dual element cartridge type fuses
7. Installation of (3) thermal type heaters
8. Test for rotation

MOTOR CONTROLS

When pricing motor controls make sure you consider the type of control system being utilized. If the controls are factory installed and located in the enclosure itself, then you would add to your starter price the items 5200 through 5800 in the Unit Price section. If control voltage is different from line voltage, add the material and labor cost of a control transformer.

For external control of starters include the following items:

1. Raceways
2. Wire & terminations
3. Control enclosures
4. Fittings
5. Push button stations
6. Indicators

Job Conditions: Productivity is based on new construction to a height of 10′.

Material staging area is assumed to be within 100′ of work being performed.

611

Fig. R262419-81

R262419-80　Distribution Section

After "taking off" the switchboard section, include on the same "quantity sheet" the distribution section; identify by:
1. Voltage
2. Ampere rating
3. Type

Example:

(Fig. R262419-81) (B) Distribution Section

Included in the labor costs of the distribution section are:
1. Uncrating
2. Hauling to 200′ of loading dock
3. Setting of distribution panel
4. Leveling & shimming

5. Anchoring of equipment to pad or floor
6. Bolting of horizontal bus bars between
7. Testing of equipment

Not included in the Distribution section are:
1. Breakers (C)
2. Equipment pads
3. Steel channels embedded or grouted in concrete
4. Pull boxes
5. Special knockouts
6. Transition section
7. Conduit or wire

R262419-82　Feeder Section

List quantities on the same sheet as the distribution section. Identify breakers by: (Fig. R262419-81)
1. Frame type
2. Number of poles
3. Ampere rating

Installation includes:
1. Handling

2. Placing breakers in distribution panel
3. Preparing wire or cable
4. Lacing wire
5. Marking each phase with colored tape
6. Marking panel legend
7. Testing
8. Balancing

R262419-84 Switchgear

It is recommended that "switchgear" or those items contained in the following sections be quoted from equipment manufacturers as a package price.

Switchboard instruments

Switchboard distribution sections

Switchboard feeder sections

Switchboard Service Disc.

Switchboard In-plant Dist.

Included in these sections are the most common types and sizes of factory assembled equipment.

The recommended procedure for low voltage switchgear would be to price (Fig. R262419-81) (A) Main Switchboard

Identify by:
1. Voltage
2. Amperage
3. Type

Example:
1. 120/208V, 4-wire, 600A, nonfused
2. 277/480V, 4-wire, 600A, nonfused
3. 120/208V, 4-wire, 400A w/fused switch & CT compartment
4. 277/480V, 4-wire, 400A, w/fused switch & CT compartment

5. 120/208V, 4-wire, 800A, w/pressure switch & CT compartment
6. 277/480V, 4-wire, 800A, w/molded CB & CT compartment

Included in the labor costs for switchboards are:
1. Uncrating
2. Hauling to 200' of loading dock
3. Setting equipment
4. Leveling and shimming
5. Anchoring
6. Cable identification
7. Testing of equipment

Not included in the switchboard price are:
1. Rigging
2. Equipment pads
3. Steel channels embedded or grouted in concrete
4. Special knockouts
5. Transition or Auxiliary sections
6. Instrumentation
7. External control
8. Conduit and wire
9. Conductor terminations

Fig. R16450-101

R262513-10 Aluminum Bus Duct

When taking off bus duct identify the system as either:
1. Aluminum
2. Copper

List straight lengths by type and size
  A. Feeder — 800 A
  B. Plug in — 800 A

Do not measure through fittings as you would on conduit, since there is no allowance for scrap in bus duct systems.

If upon taking off linear foot quantities of bus duct you find your quantities are not divisible evenly by 10 ft., then the remainder must be priced as a special item and quoted from the manufacturer. Do not use the bare material cost per L.F. for these special items. You can, however, safely use the bare labor cost per L.F. for the entire length.

Identify fittings by type and ampere rating.

Example:

C. Switchboard stub, 800 A

D. Elbows, 800 A

E. End box, 800 A

F. Cable tap box, 800 A

G. Tee Fittings, 800 A

H. Hangers

Plug-in Units — List separately plug-in units and identify by type and ampere rating

I. Plug-in switches 600 Volt 3 phase 60 A

J. Plug-in molded case C.B. 60 A

K. Combination starter FVNR NEMA 1

L. Combination contactor & fused switch NEMA 1

M. Combination fusible switch & lighting control 60 A

Labor-hours for feeder and plug-in sections include:
1. Unloading and uncrating
2. Hauling up to 200 ft. from loading dock
3. Measuring and marking
4. Installing hangers
5. Hanging and bolting sections
6. Aligning and leveling
7. Testing

Labor-hours do not include:
1. Modifications to existing structure for hanger supports
2. Threaded rod in excess of 2 ft.
3. Welding
4. Penetrations through walls
5. Staging rental

Deduct the following percentages from labor only:

  150 ft. to 250 ft. — 10%
  251 ft. to 350 ft. — 15%
  351 ft. to 500 ft. — 20%
    Over 500 ft. — 25%

Deduct percentage only if runs are contained in the same area.

Example: If the job entails running 100 ft. in 5 different locations do not deduct 20%, but if the duct is being run in 1 area and the quantity is 500 ft. then you would deduct 20%.

Deduct only from straight lengths, not fittings or plug-in units.

R262513-10   Aluminum Bus Duct (cont.)

Add to labor for elevated installations:

10 ft. to 14.5 ft. high	10%
15 ft. to 20 ft. high	20%
21 ft. to 25 ft. high	25%
26 ft. to 30 ft. high	35%
31 ft. to 35 ft. high	40%
36 ft. to 40 ft. high	50%
Over 40 ft. high	55%

Bus Duct Fittings:

Labor-hours for fittings include:
1. Unloading and uncrating
2. Hauling up to 200 ft. from loading dock
3. Installing, fitting, and bolting all ends to in-place sections

Plug-in units include:
1. Unloading and uncrating
2. Hauling up to 200 ft. from loading dock
3. Installing plug-in into in-place duct
4. Connecting load wire to lugs
5. Marking wire
6. Checking phase rotation

Labor-hours for plug-ins do not include:
1. Conduit runs from plug-in
2. Wire from plug-in
3. Conduit termination

Economy of Scale – For large concentrations of plug-in units in the same area deduct the following percentages:

11	to	25	15%
26	to	50	20%
51	to	75	25%
76	to	100	30%
100	and	over	35%

Job Conditions: Productivities are based on new construction in an unobstructed first floor area to a height of 10′ in an unobstructed area.

Material staging area is within 100 ft. of work being performed.
Add to the duct fittings and hangers:
    Plug-in switches (fused)
    Plug-in breakers
    Combination starters
    Combination contactors
    Combination fusible switch and lighting control

Labor-hours for duct and fittings include:
1. Unloading and uncrating
2. Hauling up to 200 ft. from loading dock
3. Measuring and marking
4. Installing duct runs
5. Leveling
6. Sound testing

Labor-hours for plug-ins include:
1. Unloading and uncrating
2. Hauling up to 200 ft. from loading dock
3. Installing plug-ins
4. Preparing wire
5. Wire connections and marking

R262716-40   Standard Electrical Enclosure Types

NEMA Enclosures

Electrical enclosures serve two basic purposes: they protect people from accidental contact with enclosed electrical devices and connections, and they protect the enclosed devices and connections from specified external conditions. The National Electrical Manufacturers Association (NEMA) has established the following standards. Because these descriptions are not intended to be complete representations of NEMA listings, consultation of NEMA literature is advised for detailed information.

The following definitions and descriptions pertain to NONHAZARDOUS locations.

NEMA Type 1: General purpose enclosures intended for use indoors, primarily to prevent accidental contact of personnel with the enclosed equipment in areas that do not involve unusual conditions.

NEMA Type 2: Dripproof indoor enclosures intended to protect the enclosed equipment against dripping noncorrosive liquids and falling dirt.

NEMA Type 3: Dustproof, raintight and sleet-resistant (ice-resistant) enclosures intended for use outdoors to protect the enclosed equipment against wind-blown dust, rain, sleet, and external ice formation.

NEMA Type 3R: Rainproof and sleet-resistant (ice-resistant) enclosures intended for use outdoors to protect the enclosed equipment against rain. These enclosures are constructed so that the accumulation and melting of sleet (ice) will not damage the enclosure and its internal mechanisms.

NEMA Type 3S: Enclosures intended for outdoor use to provide limited protection against wind-blown dust, rain, and sleet (ice) and to allow operation of external mechanisms when ice-laden.

NEMA Type 4: Watertight and dust-tight enclosures intended for use indoors and out – to protect the enclosed equipment against splashing water, seepage of water, falling or hose-directed water, and severe external condensation.

NEMA Type 4X: Watertight, dust-tight, and corrosion-resistant indoor and outdoor enclosures featuring the same provisions as Type 4 enclosures, plus corrosion resistance.

NEMA Type 5: Indoor enclosures intended primarily to provide limited protection against dust and falling dirt.

NEMA Type 6: Enclosures intended for indoor and outdoor use – primarily to provide limited protection against the entry of water during occasional temporary submersion at a limited depth.

NEMA Type 6R: Enclosures intended for indoor and outdoor use – primarily to provide limited protection against the entry of water during prolonged submersion at a limited depth.

NEMA Type 11: Enclosures intended for indoor use – primarily to provide, by means of oil immersion, limited protection to enclosed equipment against the corrosive effects of liquids and gases.

NEMA Type 12: Dust-tight and driptight indoor enclosures intended for use indoors in industrial locations to protect the enclosed equipment against fibers, flyings, lint, dust, and dirt, as well as light splashing, seepage, dripping, and external condensation of noncorrosive liquids.

NEMA Type 13: Oil-tight and dust-tight indoor enclosures intended primarily to house pilot devices, such as limit switches, foot switches, push buttons, selector switches, and pilot lights, and to protect these devices against lint and dust, seepage, external condensation, and sprayed water, oil, and noncorrosive coolant.

The following definitions and descriptions pertain to HAZARDOUS, or CLASSIFIED, locations:

NEMA Type 7: Enclosures intended to use in indoor locations classified as Class 1, Groups A, B, C, or D, as defined in the National Electrical Code.

NEMA Type 9: Enclosures intended for use in indoor locations classified as Class 2, Groups E, F, or G, as defined in the National Electrical Code.

R262726-90   Wiring Devices

Wiring devices should be priced on a separate takeoff form which includes boxes, covers, conduit and wire.

Labor-hours for devices include:
1. Stripping of wire
2. Attaching wire to device using terminators on the device itself, lugs, set screws etc.
3. Mounting of device in box

Labor-hours do not include:
1. Conduit
2. Wire
3. Boxes
4. Plates

Economy of Scale – for large concentrations of devices in the same area deduct the following percentages from labor-hours:

1	to	10	0%
11	to	25	20%
26	to	50	25%
51	to	100	30%
	over	100	35%

R262726-90   Wiring Devices (cont.)

NEMA No.	15 R	20 R	30 R	50 R	60 R
1 125V 2 Pole, 2 Wire					
2 250V 2 Pole, 2 Wire					
5 125V 2 Pole, 3 Wire					
6 250V 2 Pole, 3 Wire					
7 277V, AC 2 Pole, 3 Wire					
10 125/250V 3 Pole, 3 Wire					
11 3 Phase 250V 3 Pole, 3 Wire					
14 125/250V 3 Pole, 4 Wire					
15 3 Phase 250V 3 Pole, 4 Wire					
18 3 Phase 208Y/120V 4 Pole, 4 Wire					

617

R262726-90   Wiring Devices (cont.)

NEMA No.	15 R	20 R	30 R	NEMA No.	15 R	20 R	30 R
L 1 125V 2 Pole, 2 Wire				**L 13** 3 Phase 600V 3 Pole, 3 Wire			
L 2 250V 2 Pole, 2 Wire				**L 14** 125/250V 3 Pole, 4 Wire			
L 5 125 V 2 Pole, 3 Wire				**L 15** 3 Phase 250 V 3 Pole, 4 Wire			
L 6 250 V 2 Pole, 3 Wire				**L 16** 3 Phase 480V 3 Pole, 4 Wire			
L 7 227 V, AC 2 Pole, 3 Wire				**L 17** 3 Phase 600 V 3 Pole, 4 Wire			
L 8 480 V 2 Pole, 3 Wire				**L 18** 3 Phase 208Y/120V 4 Pole, 4 Wire			
L 9 600 V 2 Pole, 3 Wire				**L 19** 3 Phase 480Y/277V 4 Pole, 4 Wire			
L 10 125/250V 3 Pole, 3 Wire				**L 20** 3 Phase 600Y/347V 4 Pole, 4 Wire			
L 11 3 Phase 250 V 3 Pole, 3 Wire				**L 21** 3 Phase 208Y/120V 4 Pole, 5 Wire			
L 12 3 Phase 480 V 3 Pole, 3 Wire				**L 22** 3 Phase 480Y/277V 4 Pole, 5 Wire			
				L 23 3 Phase 600Y/347V 4 Pole, 5 Wire			

R262816-80 Safety Switches

List each safety switch by type, ampere rating, voltage, single or three phase, fused or nonfused.

Example:

A. General duty, 240V, 3-pole, fused
B. Heavy Duty, 600V, 3-pole, nonfused
C. Heavy Duty, 240V, 2-pole, fused
D. Heavy Duty, 600V, 3-pole, fused

Also include NEMA enclosure type and identify as:

1. Indoor
2. Weatherproof
3. Explosionproof

Installation of safety switches includes:

1. Unloading and uncrating
2. Handling disconnects up to 200′ from loading dock
3. Measuring and marking location
4. Drilling (4) anchor type lead fasteners using a hammer drill
5. Mounting and leveling safety switch
6. Installing (3) fuses
7. Phasing and tagging line and load wires

Price does not include:

1. Modifications to enclosure
2. Plywood backboard
3. Conduit or wire
4. Fuses
5. Termination of wires

Job Conditions: Productivities are based on new construction to an installed height of 6′ above finished floor.

Material staging area is assumed to be within 100′ of work in progress.

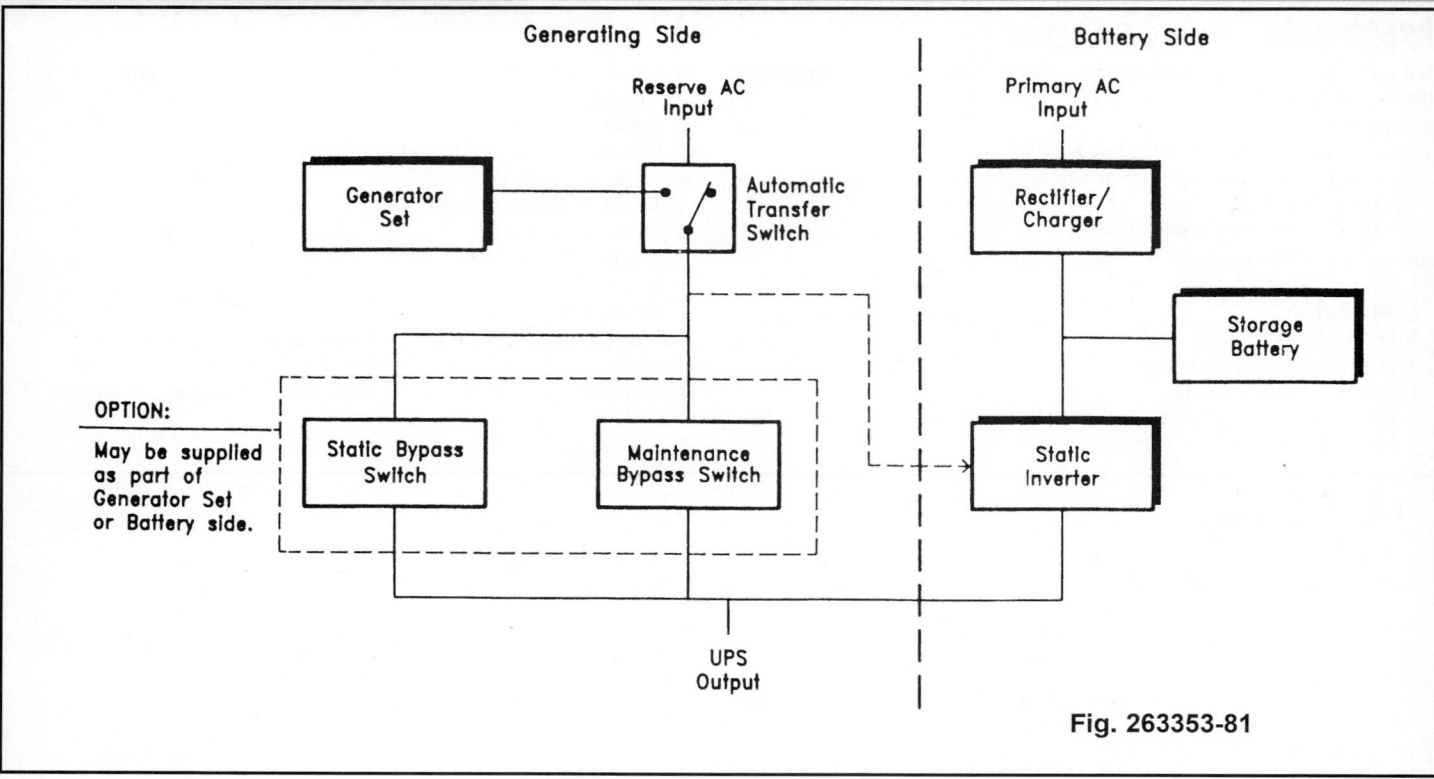

Fig. 263353-81

R263353-80 Uninterruptible Power Supply Systems

General: Uninterruptible Power Supply (UPS) Systems are used to provide power for legally required standby systems. They are also designed to protect computers and provide optional additional coverage for any or all loads in a facility. Figure R263353-81 shows a typical configuration for a UPS System with a generator set.

Cost Determination:

It is recommended that UPS System material costs be obtained from equipment manufacturers as a package. Installation costs should be obtained from a vendor or contractor.

The recommended procedure for pricing UPS Systems would be to identify:
1. Frequency – by Hertz (Hz.)
    For example: 50, 60, and 400/415 Hz.
2. Apparent power and real power
    For example: 200 kVA/170 kW
3. Input/Output voltage
    For example: 120V, 208V or 240V

4. Phase – either single or three-phase
5. Options and accessories – For extended run time more batteries would be required. Other accessories include battery cabinets, battery racks, remote control panel, power distribution unit (PDU), and warranty enhancement plans.

Note:
1. Larger systems can be configured by paralleling two or more standard small size single modules.
    For example: two 15 kVA modules combined together and configured to 30 kVA.
2. Maximum input current during battery recharge is typically 15% higher than normal current.
3. UPS Systems weights vary depending on options purchased and power rating in kVA.

620

For customer support on your Electrical Costs with RSMeans data, call 800.448.8182.

R263413-30   HP Value/Cost Determination

General: Motors can be powered by any of the seven systems shown in Figure R262213-11 through Figure R262213-17 provided the motor voltage characteristics are compatible with the power system characteristics.

Cost Determination:

Motor Amperes for the various size H.P. and voltage are listed in Table R263413-31. To find the amperes, locate the required H.P. rating and locate the amperes under the appropriate circuit characteristics.

For example:

   A. 100 H.P., 3 phase, 460 volt motor = 124 amperes (Table R263413-31)

   B. 10 H.P., 3 phase, 200 volt motor = 32.2 amperes (Table R263413-31)

Motor Wire Size: After the amperes are found in Table R263413-31 the amperes must be increased 25% to compensate for power losses. Next refer to Table R260519-92. Find the appropriate insulation column for copper or aluminum wire to determine the proper wire size.

For example:

   A. 100 H.P., 3 phase, 460 volt motor has an ampere value of 124 amperes from Table R263413-31

   B. 124A x 1.25 = 155 amperes

   C. Refer to Table R260519-92 for THW or THWN wire insulations to find the proper wire size. For a 155 ampere load using copper wire a size 2/0 wire is needed.

   D. For the 3 phase motor three wires of 2/0 size are required.

Conduit Size: To obtain the proper conduit size for the wires and type of insulation used, refer to Table R260533-22.

For example: For the 100 H.P., 460V, 3 phase motor, it was determined that three 2/0 wires are required. Assuming THWN insulated copper wire, use Table R260533-22 to determine that three 2/0 wires require 1-1/2″ conduit.

Material Cost of the conduit and wire system depends on:

1. Wire size required
2. Copper or aluminum wire
3. Wire insulation type selected
4. Steel or plastic conduit
5. Type of conduit raceway selected.

Labor Cost of the conduit and wire system depends on:

1. Type and size of conduit
2. Type and size of wires installed
3. Location and height of installation in building or depth of trench
4. Support system for conduit.

Magnetic starters, switches, and motor connection:

To complete the cost picture from H.P. to Costs additional items must be added to the cost of the conduit and wire system to arrive at a total cost.

1. Assembly Table D5020-160 Magnetic Starters Installed Cost lists the various size starters for single phase and three phase motors.
2. Assembly Table D5020-165 Heavy Duty Safety Switches Installed Cost lists safety switches required at the beginning of a motor circuit and also one required in the vicinity of the motor location.
3. Assembly Table D5020-170 Motor Connection lists the various costs for single and three phase motors.

Worksheet to obtain total motor wiring costs:

It is assumed that the motors or motor driven equipment are furnished and installed under other sections for this estimate and the following work is done under this section:

1. Conduit
2. Wire (add 10% for additional wire beyond conduit ends for connections to switches, boxes, starters, etc.)
3. Starters
4. Safety switches
5. Motor connections

621

R263413-31 Ampere Values Determined by Horsepower, Voltage and Phase Values

H.P.	Amperes							
	Single Phase			Three Phase				
	115V	208V	230V	200V	208V	230V	460V	575V
1/6	4.4A	2.4A	2.2A					
1/4	5.8	3.2	2.9					
1/3	7.2	4.0	3.6					
1/2	9.8	5.4	4.9	2.5A	2.4A	2.2A	1.1A	0.9A
3/4	13.8	7.6	6.9	3.7	3.5	3.2	1.6	1.3
1	16	8.8	8	4.8	4.6	4.2	2.1	1.7
1-1/2	20	11	10	6.9	6.6	6.0	3.0	2.4
2	24	13.2	12	7.8	7.5	6.8	3.4	2.7
3	34	18.7	17	11.0	10.6	9.6	4.8	3.9
5	56	30.8	28	17.5	16.7	15.2	7.6	6.1
7-1/2	80	44	40	25.3	24.2	22	11	9
10	100	55	50	32.2	30.8	28	14	11
15				48.3	46.2	42	21	17
20				62.1	59.4	54	27	22
25				78.2	74.8	68	34	27
30				92	88	80	40	32
40				120	114	104	52	41
50				150	143	130	65	52
60				177	169	154	77	62
75				221	211	192	96	77
100				285	273	248	124	99
125				359	343	312	156	125
150				414	396	360	180	144
200				552	528	480	240	192
250							302	242
300							361	289
350							414	336
400							477	382

For customer support on your Electrical Costs with RSMeans data, call 800.448.8182.

Figure R263413-32

Item	Type	Size	Quantity	Cost Unit	Total
Worksheet for Motor Circuits					
Wire					
Conduit			.		
Switch					
Starter					
Switch					
Motor Connection					
Other					
Total Cost					

R263413-33 Maximum Horsepower for Starter Size by Voltage

Starter Size	Maximum HP (3φ) 208V	240V	480V	600V
00	1½	1½	2	2
0	3	3	5	5
1	7½	7½	10	10
2	10	15	25	25
3	25	30	50	50
4	40	50	100	100
5		100	200	200
6		200	300	300
7		300	600	600
8		450	900	900
8L		700	1500	1500

623

For customer support on your Electrical Costs with RSMeans data, call 800.448.8182.

Electrical   R2636 Transfer Switches

R263623-60   Automatic Transfer Switches

When taking off automatic transfer switches identify by voltage, amperage and number of poles.

Example: Automatic transfer switch, 480V, 3 phase, 30A, NEMA 1 enclosure

Labor-hours for transfer switches include:
1. Unloading, uncrating and handling switches up to 200' from loading dock
2. Measuring and marking
3. Drilling (4) lead type anchors using a hammer drill
4. Mounting and leveling
5. Circuit identification
6. Testing and load balancing

Labor-hours do not include:
1. Modifications in enclosure
2. Structural supports
3. Additional lugs
4. Plywood backboard
5. Painting or lettering
6. Conduit runs to or from transfer switch
7. Wire
8. Termination of wires

Electrical   R2651 Interior Lighting

R265113-40   Interior Lighting Fixtures

When taking off interior lighting fixtures, it is advisable to set up your quantity worksheet to conform to the lighting schedule as it appears on the print. Include the alpha-numeric code plus the symbol on your worksheet.

Take off a particular section or floor of the building and count each type of fixture before going on to another type. It would also be advantageous to include on the same worksheet the pipe, wire, fittings and circuit number associated with each type of lighting fixture. This will help you identify the costs associated with any particular lighting system and in turn make material purchases more specific as to when and how much to order under the classification of lighting.

By taking off lighting first you can get a complete "walk through" of the job. This will become helpful when doing other phases of the project.

Materials for a recessed fixture include:
1. Fixture
2. Lamps
3. 6' of jack chain
4. (2) S hooks
5. (2) Wire nuts

Labor for interior recessed fixtures include:
1. Unloading by hand
2. Hauling by hand to an area up to 200' from loading dock
3. Uncrating
4. Layout
5. Installing fixture
6. Attaching jack chain & S hooks
7. Connecting circuit power
8. Reassembling fixture
9. Installing lamps
10. Testing

Material for surface mounted fixtures include:
1. Fixture
2. Lamps
3. Either (4) lead type anchors, (4) toggle bolts, or (4) ceiling grid clips
4. (2) Wire nuts

Material for pendent mounted fixtures include:
1. Fixture
2. Lamps
3. (2) Wire nuts
4. Rigid pendents as required by type of fixtures

5. Canopies as required by type of fixture

Labor hours include the following for both surface and pendent fixtures:
1. Unloading by hand
2. Hauling by hand to an area up to 200' from loading dock
3. Uncrating
4. Layout and marking
5. Drilling (4) holes for either lead anchors or toggle bolts using a hammer drill
6. Installing fixture
7. Leveling fixture
8. Connecting circuit power
9. Installing lamps
10. Testing

Labor for surface or pendent fixtures does not include:
1. Conduit
2. Boxes or covers
3. Connectors
4. Fixture whips
5. Special support
6. Switching
7. Wire

Economy of Scale: For large concentrations of lighting fixtures in the same area deduct the following percentages from labor:

25	to	50	fixtures	15%
51	to	75	fixtures	20%
76	to	100	fixtures	25%
101 and over				30%

Job Conditions: Productivity is based on new construction in an unobstructed first floor location, to 10' high.

Material staging is assumed to be within 100' of work being performed.

Add the following percentages to labor for elevated installations:

10'	to	14.5' high	10%
15'	to	20' high	20%
21'	to	25' high	25%
26'	to	30' high	35%
31'	to	35' high	40%
36'	to	40' high	50%
41' and over			55%

R265723-10 For Other than Regular Cool White (CW) Lamps

Multiply Material Costs as Follows:					
Regular	Cool white deluxe (CWX)	x 1.35	Energy	Cool white (CW/ES)	x 1.35
Lamps	Warm white deluxe (WWX)	x 1.35	Saving	Cool white deluxe (CWX/ES)	x 1.65
	Warm white (WW)	x 1.30	Lamps	Warm white (WW/ES)	x 1.55
	Natural (N)	x 2.05		Warm white deluxe (WWX/ES)	x 1.65

R271323-40   Fiber Optics

Fiber optic systems use optical fiber such as plastic, glass, or fused silica, a transparent material, to transmit radiant power (i.e., light) for control, communication, and signaling applications. The types of fiber optic cables can be nonconductive, conductive, or composite. The composite cables contain fiber optics and current-carrying electrical conductors. The configuration for one of the fiber optic systems is as follows:

The transceiver module acts as transmitting and receiving equipment in a common house, which converts electrical energy to light energy or vice versa.

Pricing the fiber optic system is not an easy task. The performance of the whole system will affect the cost significantly. New specialized tools and techniques decrease the installing cost tremendously. In the fiber optic section of Means Electrical Cost Data, a benchmark for labor-hours and material costs is set up so that users can adjust their costs according to unique project conditions.

Units for Measure: Fiber optic cable is measured in hundred linear feet (C.L.F.) or industry units of measure - meter (m) or kilometer (km). The connectors are counted as units (EA.)

Material Units: Generally, the material costs include only the cable. All the accessories shall be priced separately.

Labor Units: The following procedures are generally included for the installation of fiber optic cables:

- Receiving
- Material handling
- Setting up pulling equipment
- Measuring and cutting cable
- Pulling cable

These additional items are listed and extended: Terminations

Takeoff Procedure: Cable should be taken off by type, size, number of fibers, and number of terminations required. List the lengths of each type of cable on the takeoff sheets. Total and add 10% for waste. Transfer the figures to a cost analysis sheet and extend.

Communications | **R2715 Communications Horizontal Cabling**

R271513-75   High Performance Cable

There are several categories used to describe high performance cable. The following information includes a description of categories CAT 3, 5, 5e, 6, and 7, and details classifications of frequency and specific standards. The category standards have evolved under the sponsorship of organizations such as the Telecommunication Industry Association (TIA), the Electronic Industries Alliance (EIA), the American National Standards Institute (ANSI), the International Organization for Standardization (ISO), and the International Electrotechnical Commission (IEC), all of which have catered to the increasing complexities of modern network technology. For network cabling, users must comply with national or international standards. A breakdown of these categories is as follows:

Category 3: Designed to handle frequencies up to 16 MHz.

Category 5: (TIA/EIA 568A) Designed to handle frequencies up to 100 MHz.

Category 5e: Additional transmission performance to exceed Category 5.

Category 6: Development by TIA and other international groups to handle frequencies of 250 MHz.

Category 7: Development to handle a frequency range up to 600 MHz.

R337116-60   Average Transmission Line Material Requirements (Per Mile)

		Flat				Rolling				Mountain			
Terrain:													
Item		69kV	161kV	161kV	500kV	69kV	161kV	161kV	500kV	69kV	161kV	161kV	500kV
Pole Type	**Unit**	Wood	Wood	Steel	Steel	Wood	Wood	Steel	Steel	Wood	Wood	Steel	Steel
Conductor: 397,500 – Cir. Mil., 26/7–ACSR													
Structures	Ea.	12[1]				9[2]				7[2]			
Poles	Ea.	12				18				14			
Crossarms	Ea.	24[3]				9[4]				7[4]			
Conductor	Ft.	15,990				15,990				15,990			
Insulators[5]	Ea.	180				135				105			
Ground Wire	Ft.	5,330				10,660				10,660			
Conductor: 636,000 – Cir. Mil., 26/7–ACSR													
Structures	Ea.	13[1]	11[6]			10[2]	9[2]	6[9]		8[2]	8[2]	6[9]	
Excavation	C.Y.	–	–			–	–	120		–	–	120	
Concrete	C.Y.	–	–			–	–	10		–	–	10	
Steel Towers	Tons	–	–			–	–	32		–	–	32	
Poles	Ea.	13	11			20	18	–		16	16	–	
Crossarms	Ea.	26[3]	33[7]			10[4]	9[8]	–		8[4]	8[8]	–	
Conductor	Ft.	15,990	15,990			15,990	15,990	15,990		15,990	15,990	15,990	
Insulators[5]	Ea.	195	165			150	297	297		120	264	297	
Ground Wire	Ft.	5330	5330			10,660	10,660	10,660		10,660	10,660	10,660	
Conductor: 954,000 – Cir. Mil., 45/7–ACSR													
Structures	Ea.	14[1]	12[6]		4[11]	10[2]	9[2]	6[9]	4[11]	8[2]	8[2]	6[9]	4[13]
Excavation	C.Y.	–	–		200	–	–	125	214	–	–	125	233
Concrete	C.Y.	–	–		20	–	–	10	21	–	–	10	21
Steel Towers	Tons	–	–		57	–	–	33	57	–	–	33	63
Poles	Ea.	14	12		–	20	18	–	–	16	16	–	–
Crossarms	Ea.	28[10]	36[7]		–	10[4]	9[8]	–	–	8[4]	8[8]	–	–
Conductor	Ft.	15,990	15,990		47,970[12]	15,990	15,990	15,990	47,970[12]	15,990	15,990	15,990	47,970[12]
Insulators[5]	Ea.	210	180		288	150	297	297	288	120	264	297	576
Ground Wire	Ft.	5330	5330		10,660	10,660	10,660	10,660	10,660	10,660	10,660	10,660	10,660
Conductor: 1,351,500 – Cir. Mil., 45/7–ACSR													
Structures	Ea.		8[14]				8[14]						
Excavation	C.Y.		220				220						
Concrete	C.Y.		28				28						
Steel Towers	Ton		46				46						
Conductor	Ft.		31,680[15]				31,680[15]						
Insulators	Ea.		528				528						
Ground Wire	Ft.		10,660				10,660						

1. Single pole two-arm suspension type construction
2. Two-pole wood H-frame construction
3. 4¾" x 5¾" x 8' and 4¾" x 5¾" x 10' wood crossarms
4. 6" x 8" x 26'-0" wood crossarm
5. 5¾" x 10" disc insulator
6. Single pole construction with 3 fiberglass crossarms (5 fog-type insulators per phase)
7. 7'-0" fiberglass crossarms
8. 6" x 10" x 35'-0" wood crossarm
9. Laced steel tower, single circuit construction
10. 5" x 7" x 8' and 5" x 7" x 10' wood crossarms
11. Laced steel tower, single circuit 500-kV construction
12. Bundled conductor (3 sub-conductors per phase)
13. Laced steel tower, single circuit restrained phases (500-kV)
14. Laced steel tower, double circuit construction
15. Both sides of double circuit strung

Note: To allow for sagging, a mile (5280 Ft.) of transmission line uses 5330 Ft. of conductor per wire (called a wire mile).

R337119-30 Concrete for Conduit Encasement

The table below lists C.Y. of concrete for 100 L.F. of trench. Conduits separation center to center should meet 7.5" (N.E.C.).

Number of Conduits	1	2	3	4	6	8	9	Number of Conduits
Trench Dimension	11.5" x 11.5"	11.5" x 19"	11.5" x 27"	19" x 19"	19" x 27"	19" x 38"	27" x 27"	Trench Dimension
Conduit Diameter 2.0"	3.29	5.39	7.64	8.83	12.51	17.66	17.72	Conduit Diameter 2.0"
2.5"	3.23	5.29	7.49	8.62	12.19	17.23	17.25	2.5"
3.0"	3.15	5.13	7.24	8.29	11.71	16.59	16.52	3.0"
3.5"	3.08	4.97	7.02	7.99	11.26	15.98	15.84	3.5"
4.0"	2.99	4.80	6.76	7.65	10.74	15.30	15.07	4.0"
5.0"	2.78	4.37	6.11	6.78	9.44	13.57	13.12	5.0"
6.0"	2.52	3.84	5.33	5.74	7.87	11.48	10.77	6.0"

Reprinted with permission from NFPA 70-2014, *National Electrical Code®*, Copyright © 2013, National Fire Protection Association, Quincy, MA. This reprinted material is not the complete and official position of the NFPA on the referenced subject, which is represented solely by the standard in its entirety.

Change Orders

Change Order Considerations

A change order is a written document usually prepared by the design professional and signed by the owner, the architect/engineer, and the contractor. A change order states the agreement of the parties to: an addition, deletion, or revision in the work; an adjustment in the contract sum, if any; or an adjustment in the contract time, if any. Change orders, or "extras", in the construction process occur after execution of the construction contract and impact architects/ engineers, contractors, and owners.

Change orders that are properly recognized and managed can ensure orderly, professional, and profitable progress for everyone involved in the project. There are many causes for change orders and change order requests. In all cases, change orders or change order requests should be addressed promptly and in a precise and prescribed manner. The following paragraphs include information regarding change order pricing and procedures.

The Causes of Change Orders

Reasons for issuing change orders include:

- Unforeseen field conditions that require a change in the work
- Correction of design discrepancies, errors, or omissions in the contract documents
- Owner-requested changes, either by design criteria, scope of work, or project objectives
- Completion date changes for reasons unrelated to the construction process
- Changes in building code interpretations, or other public authority requirements that require a change in the work
- Changes in availability of existing or new materials and products

Procedures

Properly written contract documents must include the correct change order procedures for all parties—owners, design professionals, and contractors—to follow in order to avoid costly delays and litigation.

Being "in the right" is not always a sufficient or acceptable defense. The contract provisions requiring notification and documentation must be adhered to within a defined or reasonable time frame.

The appropriate method of handling change orders is by a written proposal and acceptance by all parties involved. Prior to starting work on a project, all parties should identify their authorized agents who may sign and accept change orders, as well as any limits placed on their authority.

Time may be a critical factor when the need for a change arises. For such cases, the contractor might be directed to proceed on a "time and materials" basis, rather than wait for all paperwork to be processed—a delay that could impede progress. In this situation, the contractor must still follow the prescribed change order procedures including, but not limited to, notification and documentation.

Lack of documentation can be very costly, especially if legal judgments are to be made, and if certain field personnel are no longer available. For time and material change orders, the contractor should keep accurate daily records of all labor and material allocated to the change.

Owners or awarding authorities who do considerable and continual building construction (such as the federal government) realize the inevitability of change orders for numerous reasons, both predictable and unpredictable. As a result, the federal government, the American Institute of Architects (AIA), the Engineers Joint Contract Documents Committee (EJCDC), and other contractor, legal, and technical organizations have developed standards and procedures to be followed by all parties to achieve contract continuance and timely completion, while being financially fair to all concerned.

Pricing Change Orders

When pricing change orders, regardless of their cause, the most significant factor is when the change occurs. The need for a change may be perceived in the field or requested by the architect/engineer *before* any of the actual installation has begun, or may evolve or appear *during* construction when the item of work in question is partially installed. In the latter cases, the original sequence of construction is disrupted, along with all contiguous and supporting systems. Change orders cause the greatest impact when they occur *after* the installation has been completed and must be uncovered, or even replaced. Post-completion changes may be caused by necessary design changes, product failure, or changes in the owner's requirements that are not discovered until the building or the systems begin to function.

Specified procedures of notification and record keeping must be adhered to and enforced regardless of the stage of construction: *before, during,* or *after* installation. Some bidding documents anticipate change orders by requiring that unit prices including overhead and profit percentages—for additional as well as deductible changes—be listed. Generally these unit prices do not fully take into account the ripple effect, or impact on other trades, and should be used for general guidance only.

When pricing change orders, it is important to classify the time frame in which the change occurs. There are two basic time frames for change orders: *pre-installation change orders*, which occur before the start of construction, and *post-installation change orders*, which involve reworking after the original installation. Change orders that occur between these stages may be priced according to the extent of work completed using a combination of techniques developed for pricing *pre-* and *post-installation* changes.

Factors To Consider When Pricing Change Orders

As an estimator begins to prepare a change order, the following questions should be reviewed to determine their impact on the final price.

General

- *Is the change order work* pre-installation *or* post-installation?

  Change order work costs vary according to how much of the installation has been completed. Once workers have the project scoped in their minds, even though they have not started, it can be difficult to refocus. Consequently they may spend more than the normal amount of time understanding the change. Also, modifications to work in place, such as trimming or refitting, usually take more time than was initially estimated. The greater the amount of work in place, the more reluctant workers are to change it. Psychologically they may resent the change and as a result the rework takes longer than normal. Post-installation change order estimates must include demolition of existing work as required to accomplish the change. If the work is performed at a later time, additional obstacles, such as building finishes, may be present which must be protected. Regardless of whether the change occurs

For customer support on your Electrical Costs with RSMeans data, call 800.448.8182.

629

pre-installation or post-installation, attempt to isolate the identifiable factors and price them separately. For example, add shipping costs that may be required pre-installation or any demolition required post-installation. Then analyze the potential impact on productivity of psychological and/or learning curve factors and adjust the output rates accordingly. One approach is to break down the typical workday into segments and quantify the impact on each segment.

Change Order Installation Efficiency

The labor-hours expressed (for new construction) are based on average installation time, using an efficiency level. For change order situations, adjustments to this efficiency level should reflect the daily labor-hour allocation for that particular occurrence.

- *Will the change substantially delay the original completion date?*

  A significant change in the project may cause the original completion date to be extended. The extended schedule may subject the contractor to new wage rates dictated by relevant labor contracts. Project supervision and other project overhead must also be extended beyond the original completion date. The schedule extension may also put installation into a new weather season. For example, underground piping scheduled for October installation was delayed until January. As a result, frost penetrated the trench area, thereby changing the degree of difficulty of the task. Changes and delays may have a ripple effect throughout the project. This effect must be analyzed and negotiated with the owner.

- *What is the net effect of a deduct change order?*

  In most cases, change orders resulting in a deduction or credit reflect only bare costs. The contractor may retain the overhead and profit based on the original bid.

Materials

- *Will you have to pay more or less for the new material, required by the change order, than you paid for the original purchase?*

  The same material prices or discounts will usually apply to materials purchased for change orders as new construction. In some

instances, however, the contractor may forfeit the advantages of competitive pricing for change orders. Consider the following example:

A contractor purchased over $20,000 worth of fan coil units for an installation and obtained the maximum discount. Some time later it was determined the project required an additional matching unit. The contractor has to purchase this unit from the original supplier to ensure a match. The supplier at this time may not discount the unit because of the small quantity, and he is no longer in a competitive situation. The impact of quantity on purchase can add between 0% and 25% to material prices and/or subcontractor quotes.

- *If materials have been ordered or delivered to the job site, will they be subject to a cancellation charge or restocking fee?*

  Check with the supplier to determine if ordered materials are subject to a cancellation charge. Delivered materials not used as a result of a change order may be subject to a restocking fee if returned to the supplier. Common restocking charges run between 20% and 40%. Also, delivery charges to return the goods to the supplier must be added.

Labor

- *How efficient is the existing crew at the actual installation?*

  Is the same crew that performed the initial work going to do the change order? Possibly the change consists of the installation of a unit identical to one already installed; therefore, the change should take less time. Be sure to consider this potential productivity increase and modify the productivity rates accordingly.

- *If the crew size is increased, what impact will that have on supervision requirements?*

  Under most bargaining agreements or management practices, there is a point at which a working foreman is replaced by a nonworking foreman. This replacement increases project overhead by adding a nonproductive worker. If additional workers are added to accelerate the project or to perform changes while maintaining the schedule, be sure to add additional supervision time if warranted. Calculate the

hours involved and the additional cost directly if possible.

- *What are the other impacts of increased crew size?*

  The larger the crew, the greater the potential for productivity to decrease. Some of the factors that cause this productivity loss are: overcrowding (producing restrictive conditions in the working space) and possibly a shortage of any special tools and equipment required. Such factors affect not only the crew working on the elements directly involved in the change order, but other crews whose movements may also be hampered. As the crew increases, check its basic composition for changes by the addition or deletion of apprentices or nonworking foreman, and quantify the potential effects of equipment shortages or other logistical factors.

- *As new crews, unfamiliar with the project, are brought onto the site, how long will it take them to become oriented to the project requirements?*

  The orientation time for a new crew to become 100% effective varies with the site and type of project. Orientation is easiest at a new construction site and most difficult at existing, very restrictive renovation sites. The type of work also affects orientation time. When all elements of the work are exposed, such as concrete or masonry work, orientation is decreased. When the work is concealed or less visible, such as existing electrical systems, orientation takes longer. Usually orientation can be accomplished in one day or less. Costs for added orientation should be itemized and added to the total estimated cost.

- *How much actual production can be gained by working overtime?*

  Short term overtime can be used effectively to accomplish more work in a day. However, as overtime is scheduled to run beyond several weeks, studies have shown marked decreases in output. The following chart shows the effect of long term overtime on worker efficiency. If the anticipated change requires extended overtime to keep the job on schedule, these factors can be used as a guide to predict the impact on time and cost. Add project overhead, particularly supervision, that may also be incurred.

Days per Week	Hours per Day	Production Efficiency					Payroll Cost Factors	
		1st Week	2nd Week	3rd Week	4th Week	Average 4 Weeks	@ 1-1/2 Times	@ 2 Times
5	8	100%	100%	100%	100%	100%	100%	100%
	9	100	100	95	90	96.25	105.6	111.1
	10	100	95	90	85	92.50	110.0	120.0
	11	95	90	75	65	81.25	113.6	127.3
	12	90	85	70	60	76.25	116.7	133.3
6	8	100	100	95	90	96.25	108.3	116.7
	9	100	95	90	85	92.50	113.0	125.9
	10	95	90	85	80	87.50	116.7	133.3
	11	95	85	70	65	78.75	119.7	139.4
	12	90	80	65	60	73.75	122.2	144.4
7	8	100	95	85	75	88.75	114.3	128.6
	9	95	90	80	70	83.75	118.3	136.5
	10	90	85	75	65	78.75	121.4	142.9
	11	85	80	65	60	72.50	124.0	148.1
	12	85	75	60	55	68.75	126.2	152.4

Effects of Overtime

Caution: Under many labor agreements, Sundays and holidays are paid at a higher premium than the normal overtime rate.

The use of long-term overtime is counterproductive on almost any construction job; that is, the longer the period of overtime, the lower the actual production rate. Numerous studies have been conducted, and while they have resulted in slightly different numbers, all reach the same conclusion. The figure above tabulates the effects of overtime work on efficiency.

As illustrated, there can be a difference between the *actual* payroll cost per hour and the *effective* cost per hour for overtime work. This is due to the reduced production efficiency with the increase in weekly hours beyond 40. This difference between actual and effective cost results from overtime work over a prolonged period. Short-term overtime work does not result in as great a reduction in efficiency and, in such cases, effective cost may not vary significantly from the actual payroll cost. As the total hours per week are increased on a regular basis, more time is lost due to fatigue, lowered morale, and an increased accident rate.

As an example, assume a project where workers are working 6 days a week, 10 hours per day. From the figure above (based on productivity studies), the average effective productive hours over a 4-week period are:

$$0.875 \times 60 = 52.5$$

Depending upon the locale and day of week, overtime hours may be paid at time and a half or double time. For time and a half, the overall (average) *actual* payroll cost (including regular and overtime hours) is determined as follows:

$$\frac{40 \text{ reg. hrs.} + (20 \text{ overtime hrs.} \times 1.5)}{60 \text{ hrs.}} = 1.167$$

Based on 60 hours, the payroll cost per hour will be 116.7% of the normal rate at 40 hours per week. However, because the effective production (efficiency) for 60 hours is reduced to the equivalent of 52.5 hours, the effective cost of overtime is calculated as follows:

For time and a half:

$$\frac{40 \text{ reg. hrs.} + (20 \text{ overtime hrs.} \times 1.5)}{52.5 \text{ hrs.}} = 1.33$$

The installed cost will be 133% of the normal rate (for labor).

Thus, when figuring overtime, the actual cost per unit of work will be higher than the apparent overtime payroll dollar increase, due to the reduced productivity of the longer work week. These efficiency calculations are true only for those cost factors determined by hours worked. Costs that are applied weekly or monthly, such as equipment rentals, will not be similarly affected.

Equipment

- *What equipment is required to complete the change order?*

Change orders may require extending the rental period of equipment already on the job site, or the addition of special equipment brought in to accomplish the change work. In either case, the additional rental charges and operator labor charges must be added.

Summary

The preceding considerations and others you deem appropriate should be analyzed and applied to a change order estimate. The impact of each should be quantified and listed on the estimate to form an audit trail.

Change orders that are properly identified, documented, and managed help to ensure the orderly, professional, and profitable progress of the work. They also minimize potential claims or disputes at the end of the project.

Back by customer demand!

You asked and we listened. For customer convenience and estimating ease, we have made the 2018 Project Costs available for download at **www.RSMeans.com/2018books**. You will also find sample estimates, an RSMeans data overview video and book registration form to receive quarterly data updates throughout 2018.

Estimating Tips

- The cost figures available in the download were derived from hundreds of projects contained in the RSMeans database of completed construction projects. They include the contractor's overhead and profit. The figures have been adjusted to January of the current year.

- These projects were located throughout the U.S. and reflect a tremendous variation in square foot (S.F.) costs. This is due to differences, not only in labor and material costs, but also in individual owners' requirements. For instance, a bank in a large city would have different features than one in a rural area. This is true of all the different types of buildings analyzed. Therefore, caution should be exercised when using these Project Costs. For example, for courthouses, costs in the database are local courthouse costs and will not apply to the larger, more elaborate federal courthouses.

- None of the figures "go with" any others. All individual cost items were computed and tabulated separately. Thus, the sum of the median figures for plumbing, HVAC, and electrical will not normally total up to the total mechanical and electrical costs arrived at by separate analysis and tabulation of the projects.

- Each building was analyzed as to total and component costs and percentages. The figures were arranged in ascending order with the results tabulated as shown. The 1/4 column shows that 25% of the projects had lower costs and 75% had higher. The 3/4 column shows that 75% of the projects had lower costs and 25% had higher. The median column shows that 50% of the projects had lower costs and 50% had higher.

- Project Costs are useful in the conceptual stage when no details are available. As soon as details become available in the project design, the square foot approach should be discontinued and the project priced as to its particular components. When more precision is required, or for estimating the replacement cost of specific buildings, the current edition of *RSMeans Square Foot Costs* should be used.

- In using the figures in this section, it is recommended that the median column be used for preliminary figures if no additional information is available. The median figures, when multiplied by the total city construction cost index figures (see City Cost Indexes) and then multiplied by the project size modifier at the end of this section, should present a fairly accurate base figure, which would then have to be adjusted in view of the estimator's experience, local economic conditions, code requirements, and the owner's particular requirements. There is no need to factor the percentage figures, as these should remain constant from city to city.

- The editors of this data would greatly appreciate receiving cost figures on one or more of your recent projects, which would then be included in the averages for next year. All cost figures received will be kept confidential, except that they will be averaged with other similar projects to arrive at square foot cost figures for next year.

See the website above for details and the discount available for submitting one or more of your projects.

50 17 00 | Project Costs		UNIT	UNIT COSTS			% OF TOTAL		
			1/4	MEDIAN	3/4	1/4	MEDIAN	3/4
0000	**Auto Sales with Repair**	S.F.							01
0100	Architectural		98.50	110	119	58%	64%	67%
0200	Plumbing		8.25	8.65	11.55	4.84%	5.20%	6.80%
0300	Mechanical		11.05	14.80	16.35	6.40%	8.70%	10.15%
0400	Electrical		17	21	26.50	9.05%	11.70%	15.90%
0500	Total Project Costs		165	173	177			
0000	**Banking Institutions**	S.F.							02
0100	Architectural		149	183	222	59%	65%	69%
0200	Plumbing		6	8.35	11.60	2.12%	3.39%	4.19%
0300	Mechanical		11.90	16.45	19.40	4.41%	5.10%	10.75%
0400	Electrical		29	35	54	10.45%	13.05%	15.90%
0500	Total Project Costs		247	278	340			
0000	**Court House**	S.F.							03
0100	Architectural		78.50	78.50	78.50	54.50%	54.50%	54.50%
0200	Plumbing		2.96	2.96	2.96	2.07%	2.07%	2.07%
0300	Mechanical		18.55	18.55	18.55	12.95%	12.95%	12.95%
0400	Electrical		24	24	24	16.60%	16.60%	16.60%
0500	Total Project Costs		143	143	143			
0000	**Data Centers**	S.F.							04
0100	Architectural		177	177	177	68%	68%	68%
0200	Plumbing		9.70	9.70	9.70	3.71%	3.71%	3.71%
0300	Mechanical		24.50	24.50	24.50	9.45%	9.45%	9.45%
0400	Electrical		23.50	23.50	23.50	9%	9%	9%
0500	Total Project Costs		261	261	261			
0000	**Detention Centers**	S.F.							05
0100	Architectural		164	174	184	52%	53%	60.50%
0200	Plumbing		17.35	21	25.50	5.15%	7.10%	7.25%
0300	Mechanical		22	31.50	37.50	7.55%	9.50%	13.80%
0400	Electrical		36	42.50	55.50	10.90%	14.85%	17.95%
0500	Total Project Costs		278	293	345			
0000	**Fire Stations**	S.F.							06
0100	Architectural		97	120	174	49%	55.50%	63%
0200	Plumbing		9.45	13.30	15.75	4.86%	5.80%	6.30%
0300	Mechanical		12.80	18.65	24	5.45%	8.25%	9.70%
0400	Electrical		21.50	27.50	32.50	8.35%	12.80%	14.70%
0500	Total Project Costs		195	222	294			
0000	**Gymnasium**	S.F.							07
0100	Architectural		82	108	108	57%	64.50%	64.50%
0200	Plumbing		2.01	6.60	6.60	1.58%	3.48%	3.48%
0300	Mechanical		3.09	28	28	2.42%	14.65%	14.65%
0400	Electrical		10.10	19.60	19.60	7.95%	10.35%	10.35%
0500	Total Project Costs		128	189	189			
0000	**Hospitals**	S.F.							08
0100	Architectural		99.50	164	178	43%	47.50%	48%
0200	Plumbing		7.30	13.95	30	6%	7.45%	7.65%
0300	Mechanical		48.50	54.50	71	14.20%	17.95%	23.50%
0400	Electrical		22	44	57.50	10.95%	13.75%	16.85%
0500	Total Project Costs		233	345	375			
0000	**Industrial Buildings**	S.F.							09
0100	Architectural		42	66.50	216	46%	54%	56.50%
0200	Plumbing		1.62	6.10	12.30	2%	3.06%	6.30%
0300	Mechanical		4.47	8.50	40.50	4.77%	5.55%	14.80%
0400	Electrical		6.80	7.80	65	7.85%	13.55%	16.20%
0500	Total Project Costs		74.50	97	400			
0000	**Medical Clinics & Offices**	S.F.							10
0100	Architectural		79.50	118	152	48.50%	55.50%	62.50%
0200	Plumbing		8.10	12.20	19.65	4.44%	6.40%	8.05%
0300	Mechanical		13.70	25	42.50	8.30%	11.05%	17.25%
0400	Electrical		18.60	25	36	8.55%	12.10%	14.65%
0500	Total Project Costs		156	208	272			

50 17 00 | Project Costs		UNIT	UNIT COSTS			% OF TOTAL		
			1/4	MEDIAN	3/4	1/4	MEDIAN	3/4
11	0000	**Mixed Use**	S.F.						
	0100	Architectural		85	120	182	45.50%	52.50%	70%
	0200	Plumbing		5.70	8.70	11	2.84%	3.44%	4.18%
	0300	Mechanical		14.05	36.50	65	8.05%	13.75%	20%
	0400	Electrical		14.85	36.50	49	8.30%	12.95%	15.80%
	0500	Total Project Costs		179	201	320			
12	0000	**Multi-Family Housing**	S.F.						
	0100	Architectural		70.50	106	158	56.50%	61.50%	66.50%
	0200	Plumbing		5.90	10.20	13.90	4.81%	6.30%	7.40%
	0300	Mechanical		6.60	8.80	35.50	4.92%	6.90%	11.20%
	0400	Electrical		9.20	14.35	17.40	6.20%	8.45%	10.25%
	0500	Total Project Costs		104	194	233			
13	0000	**Nursing Home & Assisted Living**	S.F.						
	0100	Architectural		66.50	87	116	51.50%	56.50%	69%
	0200	Plumbing		6.75	10.45	11.70	6.05%	7.40%	8.80%
	0300	Mechanical		5.90	7.55	17.05	4.04%	6.70%	9.55%
	0400	Electrical		9.75	12.80	21.50	7%	10.05%	12.20%
	0500	Total Project Costs		114	135	179			
14	0000	**Office Buildings**	S.F.						
	0100	Architectural		90.50	120	168	56.50%	62%	71.50%
	0200	Plumbing		4.73	7.05	11.45	2.62%	3.48%	5.85%
	0300	Mechanical		10.25	15.80	23.50	5.50%	8.20%	11.10%
	0400	Electrical		11.85	20	31.50	7.75%	10%	12.70%
	0500	Total Project Costs		150	186	262			
15	0000	**Parking Garage**	S.F.						
	0100	Architectural		29.50	36	37.50	70%	79%	88%
	0200	Plumbing		.97	1.01	1.90	2.05%	2.70%	2.83%
	0300	Mechanical		.75	1.16	4.39	2.11%	3.62%	3.81%
	0400	Electrical		2.58	2.83	5.90	5.30%	6.35%	7.95%
	0500	Total Project Costs		36	43.50	47			
16	0000	**Parking Garage/Mixed Use**	S.F.						
	0100	Architectural		95.50	104	106	61%	62%	65.50%
	0200	Plumbing		3.06	4.01	6.15	2.47%	2.72%	3.66%
	0300	Mechanical		13.10	14.75	21	7.80%	13.10%	13.60%
	0400	Electrical		13.70	19.70	20.50	8.20%	12.65%	18.15%
	0500	Total Project Costs		156	162	168			
17	0000	**Police Stations**	S.F.						
	0100	Architectural		108	121	152	49%	56.50%	61%
	0200	Plumbing		14.25	17.10	17.20	5.05%	5.55%	9.05%
	0300	Mechanical		32.50	45	46.50	13%	14.55%	16.55%
	0400	Electrical		24.50	26.50	28	9.15%	12.10%	14%
	0500	Total Project Costs		202	248	282			
18	0000	**Police/Fire**	S.F.						
	0100	Architectural		105	105	320	55.50%	66%	68%
	0200	Plumbing		8.40	8.70	32	5.45%	5.50%	5.55%
	0300	Mechanical		12.85	20.50	73.50	8.35%	12.70%	12.80%
	0400	Electrical		14.65	18.70	84	9.50%	11.75%	14.55%
	0500	Total Project Costs		154	159	580			
19	0000	**Public Assembly Buildings**	S.F.						
	0100	Architectural		119	148	200	58%	61.50%	64%
	0200	Plumbing		5.65	7.40	11.45	2.60%	3.32%	4.01%
	0300	Mechanical		14.95	23	33	6.90%	9.15%	12.75%
	0400	Electrical		19.70	24.50	38.50	8.65%	10.75%	13%
	0500	Total Project Costs		188	242	335			
20	0000	**Recreational**	S.F.						
	0100	Architectural		103	162	219	53.50%	59%	66%
	0200	Plumbing		8.40	15.30	20	3.12%	4.76%	7.40%
	0300	Mechanical		12.65	19.25	30.50	5.35%	7.90%	12.75%
	0400	Electrical		14.15	24.50	33	7.35%	8.50%	10.65%
	0500	Total Project Costs		183	272	375			

50 17 00 | Project Costs	UNIT	UNIT COSTS			% OF TOTAL			
		1/4	MEDIAN	3/4	1/4	MEDIAN	3/4	
0000 Restaurants	S.F.							21
0100 Architectural		50.50	179	231	57.50%	60.50%	78%	
0200 Plumbing		9.45	29.50	37	7.35%	7.75%	8.95%	
0300 Mechanical		7.25	18.30	44.50	6.50%	10.80%	18.30%	
0400 Electrical		11.65	21.50	48.50	6.75%	8.45%	12.45%	
0500 Total Project Costs		65	283	400				
0000 Retail	S.F.							22
0100 Architectural		52	57.50	104	60%	60%	64.50%	
0200 Plumbing		5.35	8.05	10.15	5.05%	6.70%	9%	
0300 Mechanical		4.88	7	8.60	5.05%	6.15%	6.60%	
0400 Electrical		6.85	10.95	17.60	7.90%	10.15%	12.25%	
0500 Total Project Costs		87	106	173				
0000 Schools	S.F.							23
0100 Architectural		88.50	114	152	52.50%	57%	61.50%	
0200 Plumbing		7.10	9.85	14.40	3.75%	4.82%	7%	
0300 Mechanical		16.85	23	35.50	9.50%	12.25%	14.55%	
0400 Electrical		16.65	23	29	9.45%	11.40%	13.05%	
0500 Total Project Costs		151	206	272				
0000 University, College & Private School Classroom & Admin Buildings	S.F.							24
0100 Architectural		115	142	179	50.50%	55%	59.50%	
0200 Plumbing		6.55	10.15	14.35	2.74%	4.30%	6.35%	
0300 Mechanical		24.50	35.50	43	10.10%	12.15%	14.70%	
0400 Electrical		18.55	26	31.50	7.65%	9.50%	11.55%	
0500 Total Project Costs		191	264	350				
0000 University, College & Private School Dormitories	S.F.							25
0100 Architectural		75	132	140	54.50%	65%	68.50%	
0200 Plumbing		9.95	14.05	21	6.45%	7.30%	9.15%	
0300 Mechanical		4.46	18.95	30	4.13%	9%	12.05%	
0400 Electrical		5.30	18.40	28	4.75%	7.35%	12.30%	
0500 Total Project Costs		111	211	249				
0000 University, College & Private School Science, Eng. & Lab Buildings	S.F.							26
0100 Architectural		129	137	178	50.50%	56.50%	58%	
0200 Plumbing		8.90	9.30	24.50	3.29%	3.95%	8.40%	
0300 Mechanical		40.50	64	64	13.20%	21.50%	23.50%	
0400 Electrical		26	30	35.50	8.95%	9.70%	12.85%	
0500 Total Project Costs		265	271	295				
0000 University, College & Private School Student Union Buildings	S.F.							27
0100 Architectural		69.50	102	102	51%	59.50%	59.50%	
0200 Plumbing		5	23	23	4.27%	11.45%	11.45%	
0300 Mechanical		11.25	29.50	29.50	9.60%	14.55%	14.55%	
0400 Electrical		15.40	25.50	25.50	12.80%	13.15%	13.15%	
0500 Total Project Costs		117	201	201				
0000 Warehouses	S.F.							28
0100 Architectural		44	68.50	162	61.50%	67.50%	72%	
0200 Plumbing		2.28	4.88	9.40	2.82%	3.72%	5%	
0300 Mechanical		2.70	15.40	24	4.56%	8.20%	10.70%	
0400 Electrical		4.91	18.45	30.50	7.50%	10.10%	18.30%	
0500 Total Project Costs		65.50	116	226				

For customer support on your Electrical Costs with RSMeans data, call 800.448.8182.

Square Foot Project Size Modifier

One factor that affects the S.F. cost of a particular building is the size. In general, for buildings built to the same specifications in the same locality, the larger building will have the lower S.F. cost. This is due mainly to the decreasing contribution of the exterior walls plus the economy of scale usually achievable in larger buildings. The Area Conversion Scale shown below will give a factor to convert costs for the typical size building to an adjusted cost for the particular project.

The Square Foot Base Size lists the median costs, most typical project size in our accumulated data, and the range in size of the projects.

The Size Factor for your project is determined by dividing your project area in S.F. by the typical project size for the particular Building Type. With this factor, enter the Area Conversion Scale at the appropriate Size Factor and determine the appropriate cost multiplier for your building size.

Example: Determine the cost per S.F. for a 152,600 S.F. Multi-family housing.

$$\frac{\text{Proposed building area} = 152,600 \text{ S.F.}}{\text{Typical size from below} = 76,300 \text{ S.F.}} = 2.00$$

Enter Area Conversion scale at 2.0, intersect curve, read horizontally the appropriate cost multiplier of .94. Size adjusted cost becomes .94 x $194.00 = $182.36 based on national average costs.

Note: For Size Factors less than .50, the Cost Multiplier is 1.1
For Size Factors greater than 3.5, the Cost Multiplier is .90

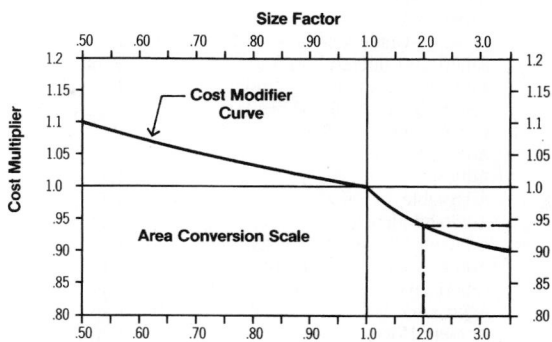

System	Median Cost (Total Project Costs)	Typical Size Gross S.F. (Median of Projects)	Typical Range (Low – High) (Projects)
Auto Sales with Repair	$173.00	25,300	8,200 – 28,700
Banking Institutions	278.00	26,300	3,300 – 28,700
Detention Centers	293.00	42,000	12,300 – 183,300
Fire Stations	222.00	14,600	6,300 – 29,600
Hospitals	345.00	137,500	54,700 – 410,300
Industrial Buildings	$97.00	16,900	5,100 – 200,600
Medical Clinics & Offices	208.00	5,500	2,600 – 327,000
Mixed Use	201.00	49,900	14,400 – 49,900
Multi-Family Housing	194.00	76,300	12,500 – 1,161,500
Nursing Home & Assisted Living	135.00	16,200	1,500 – 242,600
Office Buildings	186.00	10,000	1,100 – 930,000
Parking Garage	43.50	174,600	99,900 – 287,000
Parking Garage/Mixed Use	162.00	5,300	5,300 – 318,000
Police Stations	248.00	15,400	15,400 – 31,600
Public Assembly Buildings	242.00	30,500	2,200 – 235,300
Recreational	272.00	2,300	1,500 – 223,800
Restaurants	283.00	6,100	5,500 – 42,000
Retail	106.00	28,700	5,800 – 61,000
Schools	206.00	30,000	5,500 – 410,800
University, College & Private School Classroom & Admin Buildings	264.00	89,200	9,400 – 196,200
University, College & Private School Dormitories	211.00	50,800	1,500 – 126,900
University, College & Private School Science, Eng. & Lab Buildings	271.00	39,800	36,000 – 117,600
Warehouses	116.00	2,100	600 – 303,800

Abbreviations

A	Area Square Feet; Ampere	Brk., brk	Brick	Csc	Cosecant
AAFES	Army and Air Force Exchange	brkt	Bracket	C.S.F.	Hundred Square Feet
	Service	Brs.	Brass	CSI	Construction Specifications
ABS	Acrylonitrile Butadiene Stryrene;	Brz.	Bronze		Institute
	Asbestos Bonded Steel	Bsn.	Basin	CT	Current Transformer
A.C., AC	Alternating Current;	Btr.	Better	CTS	Copper Tube Size
	Air-Conditioning;	BTU	British Thermal Unit	Cu	Copper, Cubic
	Asbestos Cement;	BTUH	BTU per Hour	Cu. Ft.	Cubic Foot
	Plywood Grade A & C	Bu.	Bushels	cw	Continuous Wave
ACI	American Concrete Institute	BUR	Built-up Roofing	C.W.	Cool White; Cold Water
ACR	Air Conditioning Refrigeration	BX	Interlocked Armored Cable	Cwt.	100 Pounds
ADA	Americans with Disabilities Act	°C	Degree Centigrade	C.W.X.	Cool White Deluxe
AD	Plywood, Grade A & D	c	Conductivity, Copper Sweat	C.Y.	Cubic Yard (27 cubic feet)
Addit.	Additional	C	Hundred; Centigrade	C.Y./Hr.	Cubic Yard per Hour
Adh.	Adhesive	C/C	Center to Center, Cedar on Cedar	Cyl.	Cylinder
Adj.	Adjustable	C-C	Center to Center	d	Penny (nail size)
af	Audio-frequency	Cab	Cabinet	D	Deep; Depth; Discharge
AFFF	Aqueous Film Forming Foam	Cair.	Air Tool Laborer	Dis., Disch.	Discharge
AFUE	Annual Fuel Utilization Efficiency	Cal.	Caliper	Db	Decibel
AGA	American Gas Association	Calc	Calculated	Dbl.	Double
Agg.	Aggregate	Cap.	Capacity	DC	Direct Current
A.H., Ah	Ampere Hours	Carp.	Carpenter	DDC	Direct Digital Control
A hr.	Ampere-hour	C.B.	Circuit Breaker	Demob.	Demobilization
A.H.U., AHU	Air Handling Unit	C.C.A.	Chromate Copper Arsenate	d.f.t.	Dry Film Thickness
A.I.A.	American Institute of Architects	C.C.F.	Hundred Cubic Feet	d.f.u.	Drainage Fixture Units
AIC	Ampere Interrupting Capacity	cd	Candela	D.H.	Double Hung
Allow.	Allowance	cd/sf	Candela per Square Foot	DHW	Domestic Hot Water
alt., alt	Alternate	CD	Grade of Plywood Face & Back	DI	Ductile Iron
Alum.	Aluminum	CDX	Plywood, Grade C & D, exterior	Diag.	Diagonal
a.m.	Ante Meridiem		glue	Diam., Dia	Diameter
Amp.	Ampere	Cefi.	Cement Finisher	Distrib.	Distribution
Anod.	Anodized	Cem.	Cement	Div.	Division
ANSI	American National Standards	CF	Hundred Feet	Dk.	Deck
	Institute	C.F.	Cubic Feet	D.L.	Dead Load; Diesel
APA	American Plywood Association	CFM	Cubic Feet per Minute	DLH	Deep Long Span Bar Joist
Approx.	Approximate	CFRP	Carbon Fiber Reinforced Plastic	dlx	Deluxe
Apt.	Apartment	c.g.	Center of Gravity	Do.	Ditto
Asb.	Asbestos	CHW	Chilled Water;	DOP	Dioctyl Phthalate Penetration Test
A.S.B.C.	American Standard Building Code		Commercial Hot Water		(Air Filters)
Asbe.	Asbestos Worker	C.I., CI	Cast Iron	Dp., dp	Depth
ASCE	American Society of Civil Engineers	C.I.P., CIP	Cast in Place	D.P.S.T.	Double Pole, Single Throw
A.S.H.R.A.E.	American Society of Heating,	Circ.	Circuit	Dr.	Drive
	Refrig. & AC Engineers	C.L.	Carload Lot	DR	Dimension Ratio
ASME	American Society of Mechanical	CL	Chain Link	Drink.	Drinking
	Engineers	Clab.	Common Laborer	D.S.	Double Strength
ASTM	American Society for Testing and	Clam	Common Maintenance Laborer	D.S.A.	Double Strength A Grade
	Materials	C.L.F.	Hundred Linear Feet	D.S.B.	Double Strength B Grade
Attchmt.	Attachment	CLF	Current Limiting Fuse	Dty.	Duty
Avg., Ave.	Average	CLP	Cross Linked Polyethylene	DWV	Drain Waste Vent
AWG	American Wire Gauge	cm	Centimeter	DX	Deluxe White, Direct Expansion
AWWA	American Water Works Assoc.	CMP	Corr. Metal Pipe	dyn	Dyne
Bbl.	Barrel	CMU	Concrete Masonry Unit	e	Eccentricity
B&B, BB	Grade B and Better;	CN	Change Notice	E	Equipment Only; East; Emissivity
	Balled & Burlapped	Col.	Column	Ea.	Each
B&S	Bell and Spigot	CO₂	Carbon Dioxide	EB	Encased Burial
B.&W.	Black and White	Comb.	Combination	Econ.	Economy
b.c.c.	Body-centered Cubic	comm.	Commercial, Communication	E.C.Y	Embankment Cubic Yards
B.C.Y.	Bank Cubic Yards	Compr.	Compressor	EDP	Electronic Data Processing
BE	Bevel End	Conc.	Concrete	EIFS	Exterior Insulation Finish System
B.F.	Board Feet	Cont., cont	Continuous; Continued, Container	E.D.R.	Equiv. Direct Radiation
Bg. cem.	Bag of Cement	Corkbd.	Cork Board	Eq.	Equation
BHP	Boiler Horsepower;	Corr.	Corrugated	EL	Elevation
	Brake Horsepower	Cos	Cosine	Elec.	Electrician; Electrical
B.I.	Black Iron	Cot	Cotangent	Elev.	Elevator; Elevating
bidir.	bidirectional	Cov.	Cover	EMT	Electrical Metallic Conduit;
Bit., Bitum.	Bituminous	C/P	Cedar on Paneling		Thin Wall Conduit
Bit., Conc.	Bituminous Concrete	CPA	Control Point Adjustment	Eng.	Engine, Engineered
Bk.	Backed	Cplg.	Coupling	EPDM	Ethylene Propylene Diene
Bkrs.	Breakers	CPM	Critical Path Method		Monomer
Bldg., bldg	Building	CPVC	Chlorinated Polyvinyl Chloride	EPS	Expanded Polystyrene
Blk.	Block	C.Pr.	Hundred Pair	Eqhv.	Equip. Oper., Heavy
Bm.	Beam	CRC	Cold Rolled Channel	Eqlt.	Equip. Oper., Light
Boil.	Boilermaker	Creos.	Creosote	Eqmd.	Equip. Oper., Medium
bpm	Blows per Minute	Crpt.	Carpet & Linoleum Layer	Eqmm.	Equip. Oper., Master Mechanic
BR	Bedroom	CRT	Cathode-ray Tube	Eqol.	Equip. Oper., Oilers
Brg., brng.	Bearing	CS	Carbon Steel, Constant	Equip.	Equipment
Brhe.	Bricklayer Helper		Shear Bar Joist	ERW	Electric Resistance Welded
Bric.	Bricklayer				

638

E.S.	Energy Saver	H	High Henry	Lath.	Lather
Est.	Estimated	HC	High Capacity	Lav.	Lavatory
esu	Electrostatic Units	H.D., HD	Heavy Duty; High Density	lb.; #	Pound
E.W.	Each Way	H.D.O.	High Density Overlaid	L.B., LB	Load Bearing; L Conduit Body
EWT	Entering Water Temperature	HDPE	High Density Polyethylene Plastic	L. & E.	Labor & Equipment
Excav.	Excavation	Hdr.	Header	lb./hr.	Pounds per Hour
excl	Excluding	Hdwe.	Hardware	lb./L.F.	Pounds per Linear Foot
Exp., exp	Expansion, Exposure	H.I.D., HID	High Intensity Discharge	lbf/sq.in.	Pound-force per Square Inch
Ext., ext	Exterior; Extension	HEPA	High Efficiency Particulate Air	L.C.L.	Less than Carload Lot
Extru.	Extrusion		Filter	L.C.Y.	Loose Cubic Yard
f.	Fiber Stress	Hg	Mercury	Ld.	Load
F	Fahrenheit; Female; Fill	HIC	High Interrupting Capacity	LE	Lead Equivalent
Fab., fab	Fabricated; Fabric	HM	Hollow Metal	LED	Light Emitting Diode
FBGS	Fiberglass	HMWPE	High Molecular Weight	L.F.	Linear Foot
F.C.	Footcandles		Polyethylene	L.F. Hdr	Linear Feet of Header
f.c.c.	Face-centered Cubic	HO	High Output	L.F. Nose	Linear Foot of Stair Nosing
f'c.	Compressive Stress in Concrete;	Horiz.	Horizontal	L.F. Rsr	Linear Foot of Stair Riser
	Extreme Compressive Stress	H.P., HP	Horsepower; High Pressure	Lg.	Long; Length; Large
F.E.	Front End	H.P.F.	High Power Factor	L & H	Light and Heat
FEP	Fluorinated Ethylene Propylene	Hr.	Hour	LH	Long Span Bar Joist
	(Teflon)	Hrs./Day	Hours per Day	L.H.	Labor Hours
F.G.	Flat Grain	HSC	High Short Circuit	L.L., LL	Live Load
F.H.A.	Federal Housing Administration	Ht.	Height	L.L.D.	Lamp Lumen Depreciation
Fig.	Figure	Htg.	Heating	lm	Lumen
Fin.	Finished	Htrs.	Heaters	lm/sf	Lumen per Square Foot
FIPS	Female Iron Pipe Size	HVAC	Heating, Ventilation & Air-	lm/W	Lumen per Watt
Fixt.	Fixture		Conditioning	LOA	Length Over All
FJP	Finger jointed and primed	Hvy.	Heavy	log	Logarithm
Fl. Oz.	Fluid Ounces	HW	Hot Water	L-O-L	Lateralolet
Flr.	Floor	Hyd.; Hydr.	Hydraulic	long.	Longitude
Flrs.	Floors	Hz	Hertz (cycles)	L.P., LP	Liquefied Petroleum; Low Pressure
FM	Frequency Modulation;	I.	Moment of Inertia	L.P.F.	Low Power Factor
	Factory Mutual	IBC	International Building Code	LR	Long Radius
Fmg.	Framing	I.C.	Interrupting Capacity	L.S.	Lump Sum
FM/UL	Factory Mutual/Underwriters Labs	ID	Inside Diameter	Lt.	Light
Fdn.	Foundation	I.D.	Inside Dimension; Identification	Lt. Ga.	Light Gauge
FNPT	Female National Pipe Thread	I.F.	Inside Frosted	L.T.L.	Less than Truckload Lot
Fori.	Foreman, Inside	I.M.C.	Intermediate Metal Conduit	Lt. Wt.	Lightweight
Foro.	Foreman, Outside	In.	Inch	L.V.	Low Voltage
Fount.	Fountain	Incan.	Incandescent	M	Thousand; Material; Male;
fpm	Feet per Minute	Incl.	Included; Including		Light Wall Copper Tubing
FPT	Female Pipe Thread	Int.	Interior	M²CA	Meters Squared Contact Area
Fr	Frame	Inst.	Installation	m/hr.; M.H.	Man-hour
F.R.	Fire Rating	Insul., insul	Insulation/Insulated	mA	Milliampere
FRK	Foil Reinforced Kraft	I.P.	Iron Pipe	Mach.	Machine
FSK	Foil/Scrim/Kraft	I.P.S., IPS	Iron Pipe Size	Mag. Str.	Magnetic Starter
FRP	Fiberglass Reinforced Plastic	IPT	Iron Pipe Threaded	Maint.	Maintenance
FS	Forged Steel	I.W.	Indirect Waste	Marb.	Marble Setter
FSC	Cast Body; Cast Switch Box	J	Joule	Mat; Mat'l.	Material
Ft., ft	Foot; Feet	J.I.C.	Joint Industrial Council	Max.	Maximum
Ftng.	Fitting	K	Thousand; Thousand Pounds;	MBF	Thousand Board Feet
Ftg.	Footing		Heavy Wall Copper Tubing, Kelvin	MBH	Thousand BTU's per hr.
Ft lb.	Foot Pound	K.A.H.	Thousand Amp. Hours	MC	Metal Clad Cable
Furn.	Furniture	kcmil	Thousand Circular Mils	MCC	Motor Control Center
FVNR	Full Voltage Non-Reversing	KD	Knock Down	M.C.F.	Thousand Cubic Feet
FVR	Full Voltage Reversing	K.D.A.T.	Kiln Dried After Treatment	MCFM	Thousand Cubic Feet per Minute
FXM	Female by Male	kg	Kilogram	M.C.M.	Thousand Circular Mils
Fy.	Minimum Yield Stress of Steel	kG	Kilogauss	MCP	Motor Circuit Protector
g	Gram	kgf	Kilogram Force	MD	Medium Duty
G	Gauss	kHz	Kilohertz	MDF	Medium-density fibreboard
Ga.	Gauge	Kip	1000 Pounds	M.D.O.	Medium Density Overlaid
Gal., gal.	Gallon	KJ	Kilojoule	Med.	Medium
Galv., galv	Galvanized	K.L.	Effective Length Factor	MF	Thousand Feet
GC/MS	Gas Chromatograph/Mass	K.L.F.	Kips per Linear Foot	M.F.B.M.	Thousand Feet Board Measure
	Spectrometer	Km	Kilometer	Mfg.	Manufacturing
Gen.	General	KO	Knock Out	Mfrs.	Manufacturers
GFI	Ground Fault Interrupter	K.S.F.	Kips per Square Foot	mg	Milligram
GFRC	Glass Fiber Reinforced Concrete	K.S.I.	Kips per Square Inch	MGD	Million Gallons per Day
Glaz.	Glazier	kV	Kilovolt	MGPH	Million Gallons per Hour
GPD	Gallons per Day	kVA	Kilovolt Ampere	MH, M.H.	Manhole; Metal Halide; Man-Hour
gpf	Gallon per Flush	kVAR	Kilovar (Reactance)	MHz	Megahertz
GPH	Gallons per Hour	KW	Kilowatt	Mi.	Mile
gpm, GPM	Gallons per Minute	KWh	Kilowatt-hour	MI	Malleable Iron; Mineral Insulated
GR	Grade	L	Labor Only; Length; Long;	MIPS	Male Iron Pipe Size
Gran.	Granular		Medium Wall Copper Tubing	mj	Mechanical Joint
Grnd.	Ground	Lab.	Labor	m	Meter
GVW	Gross Vehicle Weight	lat	Latitude	mm	Millimeter
GWB	Gypsum Wall Board			Mill.	Millwright
				Min., min.	Minimum, Minute

Misc.	Miscellaneous	PCM	Phase Contrast Microscopy	SBS	Styrene Butadiere Styrene
ml	Milliliter, Mainline	PDCA	Painting and Decorating	SC	Screw Cover
M.L.F.	Thousand Linear Feet		Contractors of America	SCFM	Standard Cubic Feet per Minute
Mo.	Month	P.E., PE	Professional Engineer;	Scaf.	Scaffold
Mobil.	Mobilization		Porcelain Enamel;	Sch., Sched.	Schedule
Mog.	Mogul Base		Polyethylene; Plain End	S.C.R.	Modular Brick
MPH	Miles per Hour	P.E.C.I.	Porcelain Enamel on Cast Iron	S.D.	Sound Deadening
MPT	Male Pipe Thread	Perf.	Perforated	SDR	Standard Dimension Ratio
MRGWB	Moisture Resistant Gypsum	PEX	Cross Linked Polyethylene	S.E.	Surfaced Edge
	Wallboard	Ph.	Phase	Sel.	Select
MRT	Mile Round Trip	P.I.	Pressure Injected	SER, SEU	Service Entrance Cable
ms	Millisecond	Pile.	Pile Driver	S.F.	Square Foot
M.S.F.	Thousand Square Feet	Pkg.	Package	S.F.C.A.	Square Foot Contact Area
Mstz.	Mosaic & Terrazzo Worker	Pl.	Plate	S.F. Flr.	Square Foot of Floor
M.S.Y.	Thousand Square Yards	Plah.	Plasterer Helper	S.F.G.	Square Foot of Ground
Mtd., mtd., mtd	Mounted	Plas.	Plasterer	S.F. Hor.	Square Foot Horizontal
Mthe.	Mosaic & Terrazzo Helper	plf	Pounds Per Linear Foot	SFR	Square Feet of Radiation
Mtng.	Mounting	Pluh.	Plumber Helper	S.F. Shlf.	Square Foot of Shelf
Mult.	Multi; Multiply	Plum.	Plumber	S4S	Surface 4 Sides
MUTCD	Manual on Uniform Traffic Control	Ply.	Plywood	Shee.	Sheet Metal Worker
	Devices	p.m.	Post Meridiem	Sin.	Sine
M.V.A.	Million Volt Amperes	Pntd.	Painted	Skwk.	Skilled Worker
M.V.A.R.	Million Volt Amperes Reactance	Pord.	Painter, Ordinary	SL	Saran Lined
MV	Megavolt	pp	Pages	S.L.	Slimline
MW	Megawatt	PP, PPL	Polypropylene	Sldr.	Solder
MXM	Male by Male	P.P.M.	Parts per Million	SLH	Super Long Span Bar Joist
MYD	Thousand Yards	Pr.	Pair	S.N.	Solid Neutral
N	Natural; North	P.E.S.B.	Pre-engineered Steel Building	SO	Stranded with oil resistant inside
nA	Nanoampere	Prefab.	Prefabricated		insulation
NA	Not Available; Not Applicable	Prefin.	Prefinished	S-O-L	Socketolet
N.B.C.	National Building Code	Prop.	Propelled	sp	Standpipe
NC	Normally Closed	PSF, psf	Pounds per Square Foot	S.P.	Static Pressure; Single Pole; Self-
NEMA	National Electrical Manufacturers	PSI, psi	Pounds per Square Inch		Propelled
	Assoc.	PSIG	Pounds per Square Inch Gauge	Spri.	Sprinkler Installer
NEHB	Bolted Circuit Breaker to 600V.	PSP	Plastic Sewer Pipe	spwg	Static Pressure Water Gauge
NFPA	National Fire Protection Association	Pspr.	Painter, Spray	S.P.D.T.	Single Pole, Double Throw
NLB	Non-Load-Bearing	Psst.	Painter, Structural Steel	SPF	Spruce Pine Fir; Sprayed
NM	Non-Metallic Cable	P.T.	Potential Transformer		Polyurethane Foam
nm	Nanometer	P. & T.	Pressure & Temperature	S.P.S.T.	Single Pole, Single Throw
No.	Number	Ptd.	Painted	SPT	Standard Pipe Thread
NO	Normally Open	Ptns.	Partitions	Sq.	Square; 100 Square Feet
N.O.C.	Not Otherwise Classified	Pu	Ultimate Load	Sq. Hd.	Square Head
Nose.	Nosing	PVC	Polyvinyl Chloride	Sq. In.	Square Inch
NPT	National Pipe Thread	Pvmt.	Pavement	S.S.	Single Strength; Stainless Steel
NQOD	Combination Plug-on/Bolt on	PRV	Pressure Relief Valve	S.S.B.	Single Strength B Grade
	Circuit Breaker to 240V.	Pwr.	Power	sst, ss	Stainless Steel
N.R.C., NRC	Noise Reduction Coefficient/	Q	Quantity Heat Flow	Sswk.	Structural Steel Worker
	Nuclear Regulator Commission	Qt.	Quart	Sswl.	Structural Steel Welder
N.R.S.	Non Rising Stem	Quan., Qty.	Quantity	St.; Stl.	Steel
ns	Nanosecond	Q.C.	Quick Coupling	STC	Sound Transmission Coefficient
NTP	Notice to Proceed	r	Radius of Gyration	Std.	Standard
nW	Nanowatt	R	Resistance	Stg.	Staging
OB	Opposing Blade	R.C.P.	Reinforced Concrete Pipe	STK	Select Tight Knot
OC	On Center	Rect.	Rectangle	STP	Standard Temperature & Pressure
OD	Outside Diameter	recpt.	Receptacle	Stpi.	Steamfitter, Pipefitter
O.D.	Outside Dimension	Reg.	Regular	Str.	Strength; Starter; Straight
ODS	Overhead Distribution System	Reinf.	Reinforced	Strd.	Stranded
O.G.	Ogee	Req'd.	Required	Struct.	Structural
O.H.	Overhead	Res.	Resistant	Sty.	Story
O&P	Overhead and Profit	Resi.	Residential	Subj.	Subject
Oper.	Operator	RF	Radio Frequency	Subs.	Subcontractors
Opng.	Opening	RFID	Radio-frequency Identification	Surf.	Surface
Orna.	Ornamental	Rgh.	Rough	Sw.	Switch
OSB	Oriented Strand Board	RGS	Rigid Galvanized Steel	Swbd.	Switchboard
OS&Y	Outside Screw and Yoke	RHW	Rubber, Heat & Water Resistant;	S.Y.	Square Yard
OSHA	Occupational Safety and Health		Residential Hot Water	Syn.	Synthetic
	Act	rms	Root Mean Square	S.Y.P.	Southern Yellow Pine
Ovhd.	Overhead	Rnd.	Round	Sys.	System
OWG	Oil, Water or Gas	Rodm.	Rodman	t.	Thickness
Oz.	Ounce	Rofc.	Roofer, Composition	T	Temperature; Ton
P.	Pole; Applied Load; Projection	Rofp.	Roofer, Precast	Tan	Tangent
p.	Page	Rohe.	Roofer Helpers (Composition)	T.C.	Terra Cotta
Pape.	Paperhanger	Rots.	Roofer, Tile & Slate	T & C	Threaded and Coupled
P.A.P.R.	Powered Air Purifying Respirator	R.O.W.	Right of Way	T.D.	Temperature Difference
PAR	Parabolic Reflector	RPM	Revolutions per Minute	TDD	Telecommunications Device for
P.B., PB	Push Button	R.S.	Rapid Start		the Deaf
Pc., Pcs.	Piece, Pieces	Rsr	Riser	T.E.M.	Transmission Electron Microscopy
P.C.	Portland Cement; Power Connector	RT	Round Trip	temp	Temperature, Tempered, Temporary
P.C.F.	Pounds per Cubic Foot	S.	Suction; Single Entrance; South	TFFN	Nylon Jacketed Wire

Abbreviations

TFE	Tetrafluoroethylene (Teflon)	U.L., UL	Underwriters Laboratory	w/	With
T. & G.	Tongue & Groove;	Uld.	Unloading	W.C., WC	Water Column; Water Closet
	Tar & Gravel	Unfin.	Unfinished	W.F.	Wide Flange
Th., Thk.	Thick	UPS	Uninterruptible Power Supply	W.G.	Water Gauge
Thn.	Thin	URD	Underground Residential	Wldg.	Welding
Thrded	Threaded		Distribution	W. Mile	Wire Mile
Tilf.	Tile Layer, Floor	US	United States	W-O-L	Weldolet
Tilh.	Tile Layer, Helper	USGBC	U.S. Green Building Council	W.R.	Water Resistant
THHN	Nylon Jacketed Wire	USP	United States Primed	Wrck.	Wrecker
THW.	Insulated Strand Wire	UTMCD	Uniform Traffic Manual For Control	WSFU	Water Supply Fixture Unit
THWN	Nylon Jacketed Wire		Devices	W.S.P.	Water, Steam, Petroleum
T.L., TL	Truckload	UTP	Unshielded Twisted Pair	WT., Wt.	Weight
T.M.	Track Mounted	V	Volt	WWF	Welded Wire Fabric
Tot.	Total	VA	Volt Amperes	XFER	Transfer
T-O-L	Threadolet	VAT	Vinyl Asbestos Tile	XFMR	Transformer
tmpd	Tempered	V.C.T.	Vinyl Composition Tile	XHD	Extra Heavy Duty
TPO	Thermoplastic Polyolefin	VAV	Variable Air Volume	XHHW	Cross-Linked Polyethylene Wire
T.S.	Trigger Start	VC	Veneer Core	XLPE	Insulation
Tr.	Trade	VDC	Volts Direct Current	XLP	Cross-linked Polyethylene
Transf.	Transformer	Vent.	Ventilation	Xport	Transport
Trhv.	Truck Driver, Heavy	Vert.	Vertical	Y	Wye
Trlr	Trailer	V.F.	Vinyl Faced	yd	Yard
Trlt.	Truck Driver, Light	V.G.	Vertical Grain	yr	Year
TTY	Teletypewriter	VHF	Very High Frequency	Δ	Delta
TV	Television	VHO	Very High Output	%	Percent
T.W.	Thermoplastic Water Resistant	Vib.	Vibrating	~	Approximately
	Wire	VLF	Vertical Linear Foot	Ø	Phase; diameter
UCI	Uniform Construction Index	VOC	Volatile Organic Compound	@	At
UF	Underground Feeder	Vol.	Volume	#	Pound; Number
UGND	Underground Feeder	VRP	Vinyl Reinforced Polyester	<	Less Than
UHF	Ultra High Frequency	W	Wire; Watt; Wide; West	>	Greater Than
U.I.	United Inch			Z	Zone

For customer support on your Electrical Costs with RSMeans data, call 800.448.8182.

Index

Index

653

Division Notes

	CREW	DAILY OUTPUT	LABOR-HOURS	UNIT	BARE COSTS				TOTAL INCL O&P
					MAT.	LABOR	EQUIP.	TOTAL	

Division Notes

	CREW	DAILY OUTPUT	LABOR-HOURS	UNIT	BARE COSTS				TOTAL INCL O&P
					MAT.	LABOR	EQUIP.	TOTAL	

Division Notes

	CREW	DAILY OUTPUT	LABOR-HOURS	UNIT	BARE COSTS				TOTAL INCL O&P
					MAT.	LABOR	EQUIP.	TOTAL	

Division Notes

	CREW	DAILY OUTPUT	LABOR-HOURS	UNIT	BARE COSTS				TOTAL INCL O&P
					MAT.	LABOR	EQUIP.	TOTAL	

Division Notes

	CREW	DAILY OUTPUT	LABOR-HOURS	UNIT	BARE COSTS				TOTAL INCL O&P
					MAT.	LABOR	EQUIP.	TOTAL	

Division Notes

	CREW	DAILY OUTPUT	LABOR-HOURS	UNIT	BARE COSTS				TOTAL INCL O&P
					MAT.	LABOR	EQUIP.	TOTAL	

Division Notes

	CREW	DAILY OUTPUT	LABOR-HOURS	UNIT	BARE COSTS				TOTAL INCL O&P
					MAT.	LABOR	EQUIP.	TOTAL	

Table of Contents
Annual Cost Guides
Online Estimating Solution
Seminars and Professional Development

more information visit our website at www.RSMeans.com

Unit prices according to the latest MasterFormat®

Cost Data Selection Guide

The following table provides definitive information on the content of each cost data publication. The number of lines of data provided in each unit price or assemblies division, as well as the number of crews, is listed for each data set. The presence of other elements such as reference tables, square foot models, equipment rental costs, historical cost indexes, and city cost indexes, is also indicated. You can use the table to help select the RSMeans data set that has the quantity and type of information you most need in your work.

Unit Cost Divisions	Building Construction	Mechanical	Electrical	Commercial Renovation	Square Foot	Site Work Landsc.	Green Building	Interior	Concrete Masonry	Open Shop	Heavy Construction	Light Commercial	Facilities Construction	Plumbing	Residential
1	590	411	428	530	0	524	200	331	473	589	527	273	1065	421	178
2	777	280	86	733	0	991	207	399	214	776	733	481	1220	287	274
3	1688	340	230	1081	0	1480	986	354	2034	1688	1690	482	1788	316	389
4	960	21	0	920	0	725	180	615	1158	928	615	533	1175	0	447
5	1901	158	155	1093	0	852	1799	1106	729	1901	1037	979	1918	204	746
6	2453	18	18	2111	0	110	589	1528	281	2449	123	2141	2125	22	2661
7	1596	215	128	1634	0	580	763	532	523	1593	26	1329	1697	227	1049
8	2140	80	3	2733	0	255	1140	1813	105	2142	0	2328	2966	0	1552
9	2107	86	45	1931	0	309	455	2193	412	2050	15	1756	2356	54	1521
10	1090	17	10	685	0	234	32	899	136	1090	34	589	1181	237	224
11	1097	201	166	541	0	135	56	925	29	1064	0	231	1117	164	110
12	548	0	2	299	0	219	147	1551	14	515	0	273	1574	23	217
13	744	149	158	253	0	366	125	254	78	720	267	109	760	115	104
14	273	36	0	223	0	0	0	257	0	273	0	12	293	16	6
21	130	0	41	37	0	0	0	296	0	130	0	121	668	688	259
22	1165	7557	160	1226	0	1572	1063	849	20	1154	1681	875	7505	9414	719
23	1194	6995	581	938	0	157	898	787	38	1177	110	886	5235	1917	480
25	0	0	14	14	0	0	0	0	0	0	0	0	0	0	0
26	1512	491	10456	1293	0	811	644	1159	55	1438	600	1360	10237	399	636
27	94	0	447	101	0	0	0	71	0	94	39	67	388	0	56
28	143	79	223	124	0	0	28	97	0	127	0	70	209	57	41
31	1510	733	610	806	0	3265	288	7	1217	1455	3276	604	1569	660	613
32	836	49	8	905	0	4472	353	405	314	808	1889	440	1751	142	487
33	1246	1076	534	252	0	3021	38	0	237	523	3058	128	1698	2085	154
34	107	0	47	4	0	190	0	0	31	62	212	0	136	0	0
35	18	0	0	0	0	327	0	0	0	18	442	0	84	0	0
41	62	0	0	33	0	8	0	22	0	61	31	0	68	14	0
44	75	79	0	0	0	0	0	0	0	0	0	0	75	75	0
46	23	16	0	0	0	274	261	0	0	23	264	0	33	33	0
48	10	0	38	2	0	0	23	0	0	10	17	10	23	0	10
Totals	26089	19087	14588	20502	0	20877	10275	16450	8098	24858	16686	16077	50914	17570	12933

Assem Div	Building Construction	Mechanical	Electrical	Commercial Renovation	Square Foot	Site Work Landscape	Assemblies	Green Building	Interior	Concrete Masonry	Heavy Construction	Light Commercial	Facilities Construction	Plumbing	Asm Div	Residential
A		15	0	188	164	577	598	0	0	536	571	154	24	0	1	378
B		0	0	848	2554	0	5661	56	329	1976	368	2094	174	0	2	211
C		0	0	647	954	0	1334	0	1642	146		844	255	0	3	588
D		1057	941	712	1859	72	2538	330	825	0	0	1345	1105	1088	4	851
E		0	0	86	261	0	301	0	5	0	0	258	5	0	5	391
F		0	0	0	114	0	143	0	0	0	0	114	0	0	6	357
G		527	447	318	312	3377	792	0	0	534	1349	205	293	677	7	307
															8	760
															9	80
															10	0
															11	0
															12	0
Totals		1599	1388	2799	6218	4026	11367	386	2801	3192	2288	5014	1856	1765		3923

Reference Section	Building Construction Costs	Mechanical	Electrical	Commercial Renovation	Square Foot	Site Work Landscape	Assem.	Green Building	Interior	Concrete Masonry	Open Shop	Heavy Construction	Light Commercial	Facilities Construction	Plumbing	Resi.
Reference Tables	yes	yes	yes	yes	no	yes	yes	yes	yes	yes	yes	yes	yes	yes	yes	yes
Models					111			25					50			28
Crews	578	578	578	556		578		578	578	578	555	578	555	556	578	555
Equipment Rental Costs	yes	yes	yes	yes		yes	yes	yes	yes	yes	yes	yes	yes	yes	yes	yes
Historical Cost Indexes	yes	yes	yes	yes	yes	yes	yes	yes	yes	yes	yes	yes	yes	yes	yes	no
City Cost Indexes	yes	yes	yes	yes	yes	yes	yes	yes	yes	yes	yes	yes	yes	yes	yes	yes

For more information visit our website at www.RSMeans.com

Our Online Estimating Solution

Competitive Cost Estimates Made Easy

Our online estimating solution is a web-based service that provides accurate and up-to-date cost information to help you build competitive estimates or budgets in less time.

Quick, intuitive, easy to use and automatically updated, you'll gain instant access to hundreds of thousands of material, labor, and equipment costs from RSMeans' comprehensive database, delivering the information you need to build budgets and competitive estimates every time.

With our online estimating solutions, you can perform quick searches to locate specific costs and adjust costs to reflect prices in your geographic area. Tag and store your favorites for fast access to your frequently used line items and assemblies and clone estimates to save time. System notifications will alert you as updated data becomes available. This data is automatically updated throughout the year.

Our visual, interactive estimating features help you create, manage, save and share estimates with ease! You'll enjoy increased flexibility with customizable advanced reports. Easily edit custom report templates and import your company logo onto your estimates.

	Core	Advanced	Complete
Unit Prices	✓	✓	✓
Assemblies	⊘	✓	✓
Sq Foot Models	⊘	⊘	✓
Editable Sq Foot Models	⊘	⊘	✓
Editable Assembly Components	⊘	✓	✓
Custom Cost Data	⊘	✓	✓
User Defined Components	⊘	✓	✓
Advanced Reporting & Customization	⊘	✓	✓
Union Labor Type	✓	✓	✓

Continue to check our website at www.RSMeans.com for more product offerings.

Estimate with Precision
Find everything you need to develop complete, accurate estimates.
- Verified costs for construction materials
- Equipment rental costs
- Crew sizing, labor hours and labor rates
- Localized costs for U.S. and Canada

Save Time & Increase Efficiency
Make cost estimating and calculating faster and easier than ever with secure, online estimating tools.
- Quickly locate costs in the searchable database
- Create estimates in minutes with RSMeans cost lines
- Tag and store favorites for fast access to frequently used items

Improve Planning & Decision-Making
Back your estimates with complete, accurate and up-to date cost data for informed business decisions.
- Verify construction costs from third parties
- Check validity of subcontractor proposals
- Evaluate material and assembly alternatives

Increase Profits
Use our online estimating solution to estimate projects quickly and accurately, so you can gain an edge over yo competition.
- Create accurate and competitive bids
- Minimize the risk of cost overruns
- Reduce variability

Online Estimating Solution   Unit prices according to the latest MasterFormat®

Access the data online

Search for unit prices by keyword   Leverage the most up-to-date data   Build and export estimates

Try it free for 30 days   www.rsmeans.com/2018freetrial

2018 Seminar Schedule   📞 877-620-6245

Note: call for exact dates, locations, and details as some cities are subject to change.

Location	Dates	Location	Dates
Seattle, WA	January and August	San Francisco, CA	June
Dallas/Ft. Worth, TX	January	Bethesda, MD	June
Austin, TX	February	El Segundo, CA	August
Anchorage, AK	March and September	Dallas, TX	September
Las Vegas, NV	March	Raleigh, NC	October
New Orleans, LA	March	Salt Lake City, UT	October
Washington, DC	April and September	Baltimore, MD	November
Phoenix, AZ	April	Orlando, FL	November
Toronto	May	San Diego, CA	December
Denver, CO	May	San Antonio, TX	December

Gordian also offers a suite of online RSMeans data self-paced offerings. Check our website www.RSMeans.com for more information.

Self-Paced Professional Development Courses

Training on how to use RSMeans data and estimating tools, as well as Professional Development courses on industry topics are now offered in a convenient self-paced format. These courses are on-demand and allow more flexibility to learn around your busy schedule, while saving the cost of travel and time.

Current course offerings include:

Facilities Construction Estimating—our best-selling live class now available as an on-demand training course! Let the subject matter experts of construction estimating—the RSMeans Engineering Staff—walk you through the basics and much more of estimating for renovation and facilities construction.

RSMeansOnline.com Training—learn the ins and outs of the flagship delivery method of RSMeans data!

The Construction Process—how much do you and your team really know about the ins and outs of the "contract-side" of a construction project? This self-paced course will clarify best practices for items such as schedules, change orders, and project closeout.

These self-paced training courses can be completed over the course of 45 days and are comprised of multiple lessons with documentation, video presentation, software simulation, assessment quizzes and certificate of completion.

Site Work Estimating with RSMeans data

This new one-day program focuses directly on site work costs, a unique portion of most construction projects that often is the wild card in determining whether you have developed a good estimate or not. Accurately scoping, quantifying, and pricing site preparation, underground utility work, and improvements to exterior site elements are often the most difficult estimating tasks on any project. The program takes the participant from preparing a never-developed site through underground utility installation, pad preparation, paving and sidewalks, and landscaping. Attendees will use the full array of site work cost data through the RSMeans online program and participate in exercises to strengthen their estimating skills.

Some of what you'll learn:
- Evaluation of site work and understanding site scope of work.
- Site work estimating topics including: site clearing, grading, excavation, disposal and trucking of materials, erosion control devices, backfill and compaction, underground utilities, paving, sidewalks, fences & gates, and seeding & planting.
- Unit price site work estimates—Correct use of RSMeans site work cost data to develop a cost estimate.
- Using and modifying assemblies—Save valuable time when estimating site work activities using custom assemblies.

Who should attend: Engineers, contractors, estimators, project managers, owner's representatives, and others who are concerned with the proper preparation and/or evaluation of site work estimates.

Please bring a laptop with ability to access the internet.

Training for our Online Estimating Solution

Construction estimating is vital to the decision-making process at each state of every project. Our online solution works the way you do. It's systematic, flexible and intuitive. In this one-day class you will see how you can estimate any phase of any project faster and better.

Some of what you'll learn:
- Customizing our online estimating solution
- Making the most of RSMeans "Circle Reference" numbers
- How to integrate your cost data
- Generating reports, exporting estimates to MS Excel, sharing, collaborating and more

Also offered as a self-paced or on-site training program!

Facilities Construction Estimating

In this two-day course, professionals working in facilities management can get help with their daily challenges to establish budgets for all phases of a project.

Some of what you'll learn:
- Determining the full scope of a project
- Identifying the scope of risks and opportunities
- Creative solutions to estimating issues
- Organizing estimates for presentation and discussion
- Special techniques for repair/remodel and maintenance projects
- Negotiating project change orders

Who should attend: facility managers, engineers, contractors, facility tradespeople, planners, and project managers.

Construction Cost Estimating: Concepts and Practice

This one-day introductory course to improve estimating skills and effectiveness starts with the details of interpreting bid documents and ends with the summary of the estimate and bid submission.

Some of what you'll learn:
- Using the plans and specifications to create estimates
- The takeoff process—deriving all tasks with correct quantities
- Developing pricing using various sources; how subcontractor pricing fits in
- Summarizing the estimate to arrive at the final number
- Formulas for area and cubic measure, adding waste and adjusting productivity to specific projects
- Evaluating subcontractors' proposals and prices
- Adding insurance and bonds
- Understanding how labor costs are calculated
- Submitting bids and proposals

Who should attend: project managers, architects, engineers, owners' representatives, contractors, and anyone who's responsible for budgeting or estimating construction projects.

Maintenance & Repair Estimating for Facilities

This two-day course teaches attendees how to plan, budget, and estimate the cost of ongoing and preventive maintenance and repair for existing buildings and grounds.

Some of what you'll learn:
- The most financially favorable maintenance, repair, and replacement scheduling and estimating
- Auditing and value engineering facilities
- Preventive planning and facilities upgrading
- Determining both in-house and contract-out service cost
- Annual, asset-protecting M&R plan

Who should attend: facility managers, maintenance supervisors, buildings and grounds superintendents, plant managers, planners, estimators, and others involved in facilities planning and budgeting.

Practical Project Management for Construction Professionals

In this two-day course, acquire the essential knowledge and develop the skills to effectively and efficiently execute the day-to-day responsibilities of the construction project manager.

Some of what you'll learn:
- General conditions of the construction contract
- Contract modifications: change orders and construction change directives
- Negotiations with subcontractors and vendors
- Effective writing: notification and communications
- Dispute resolution: claims and liens

Who should attend: architects, engineers, owners' representatives, and project managers.

Mechanical & Electrical Estimating

This two-day course teaches attendees how to prepare more accurate and complete mechanical/electrical estimates, avoid the pitfalls of omission and double-counting, and understand the composition and rationale within the RSMeans mechanical/electrical database.

Some of what you'll learn:
- The unique way mechanical and electrical systems are interrelated
- M&E estimates—conceptual, planning, budgeting, and bidding stages
- Order of magnitude, square foot, assemblies, and unit price estimating
- Comparative cost analysis of equipment and design alternatives

Who should attend: architects, engineers, facilities managers, mechanical and electrical contractors, and others who need a highly reliable method for developing understanding, and evaluating mechanical and electrical contracts.

Professional Development

Unit Price Estimating

This interactive two-day seminar teaches attendees how to interpret project information and process it into final, detailed estimates with the greatest accuracy level.

The most important credential an estimator can take to the job is the ability to visualize construction and estimate accurately.

Some of what you'll learn:
- Interpreting the design in terms of cost
- The most detailed, time-tested methodology for accurate pricing
- Key cost drivers—material, labor, equipment, staging, and subcontracts
- Understanding direct and indirect costs for accurate job cost accounting and change order management

Who should attend: corporate and government estimators and purchasers, architects, engineers, and others who need to produce accurate project estimates.

Training for our CD Estimating Solution

This one-day course helps users become more familiar with the functionality of the CD. Each menu, icon, screen, and function found in the program is explained in depth. Time is devoted to hands-on estimating exercises.

Some of what you'll learn:
- Searching the database using all navigation methods
- Exporting RSMeans data to your preferred spreadsheet format
- Viewing crews, assembly components, and much more
- Automatically regionalizing the database

This training session requires you to bring a laptop computer to class.

When you register for this course you will receive an outline for your laptop requirements.

Also offered as a self-paced or on-site training program!

Facilities Estimating Using the CD

This two-day class combines hands-on skill-building with best estimating practices and real-life problems. You will learn key concepts, tips, pointers, and guidelines to save time and avoid cost oversights and errors.

Some of what you'll learn:
- Estimating process concepts
- Customizing and adapting RSMeans cost data
- Establishing scope of work to account for all known variables
- Budget estimating: when, why, and how
- Site visits: what to look for and what you can't afford to overlook
- How to estimate repair and remodeling variables

This training session requires you to bring a laptop computer to class.

Who should attend: facility managers, architects, engineers, contractors, facility tradespeople, planners, project managers, and anyone involved with JOC, SABRE, or IDIQ.

Life Cycle Cost Estimating for Facility Asset Managers

Life Cycle Cost Estimating will take the attendee through choosing the correct RSMeans database to use and then correctly applying RSMeans data to their specific life cycle application. Conceptual estimating through RSMeans new building models, conceptual estimating of major existing building projects through RSMeans renovation models, pricing specific renovation elements, estimating repair, replacement and preventive maintenance costs today and forward up to 30 years will be covered.

Some of what you'll learn:
- Cost implications of managing assets
- Planning projects and initial & life cycle costs
- How to use RSMeans data online

Who should attend: facilities owners and managers and anyone involved in the financial side of the decision making process in the planning, design, procurement, and operation of facility real assets.

Please bring a laptop with ability to access the internet.

Assessing Scope of Work for Facilities Construction Estimating

This two-day practical training program addresses the vital importance of understanding the scope of projects in order to produce accurate cost estimates for facility repair and remodeling.

Some of what you'll learn:
- Discussions of site visits, plans/specs, record drawings of facilities, and site-specific lists
- Review of CSI divisions, including means, methods, materials, and the challenges of scoping each topic
- Exercises in scope identification and scope writing for accurate estimating of projects
- Hands-on exercises that require scope, take-off, and pricing

Who should attend: corporate and government estimators, planners, facility managers, and others who need to produce accurate project estimates.

Building Systems and the Construction Process

This one-day course was written to assist novices and those outside the industry in obtaining a solid understanding of the construction process - from both a building systems and construction administration approach.

Some of what you'll learn:
- Various systems used and how components come together to create a building
- Start with foundation and end with the physical systems of the structure such as HVAC and Electrical
- Focus on the process from start of design through project closeout

This training session requires you to bring a laptop computer to class.

Who should attend: building professionals or novices to help make the crossover to the construction industry; suited for anyone responsible for providing high level oversight on construction projects.

Professional Development

673

For more information visit our website at www.RSMeans.com

Registration Information

Register early and save up to $100!

Register 30 days before the start date of a seminar and save $100 off your total fee. Note: This discount can be applied only once per order. It cannot be applied to team discount registrations or any other special offer.

How to register

By Phone
Register by phone at 877-620-6245

Online
Register online at
www.RSMeans.com/products/seminars.aspx

Note: Purchase Orders or Credits Cards are required to register.

Two-day seminar registration fee - $1,045.

One-Day Construction Cost Estimating or Building Systems and the Construction Process - $630.

Government pricing

All federal government employees save off the regular seminar price. Other promotional discounts cannot be combined with the government discount.

Team discount program

For over five attendee registrations. Call for pricing: 781-422-5115

Refund policy

Cancellations will be accepted up to ten business days prior to the seminar start. There are no refunds for cancellations received later than ten working days prior to the first day of the seminar. A $150 processing fee will be applied for all cancellations. Written notice of the cancellation is required. Substitutions can be made at any time before the session starts. No-shows are subject to the full seminar fee.

Note: Pricing subject to change.

AACE approved courses

Many seminars described and offered here have been approved for 14 hours (1.4 recertification credits) of credit by the AACE International Certification Board toward meeting the continuing education requirements for recertification as a Certified Cost Engineer/Certified Cost Consultant.

AIA Continuing Education

We are registered with the AIA Continuing Education System (AIA/CES) and are committed to developing quality learning activities in accordance with the CES criteria. Many seminars meet the AIA/CES criteria for Quality Level 2. AIA members may receive 14 learning units (LUs) for each two day RSMeans course.

Daily course schedule

The first day of each seminar session begins at 8:30 a.m. and ends at 4:30 p.m. The second day begins at 8:00 a.m and ends at 4:00 p.m. Participants are urged to bring a hand-held calculator since many actual problems will be worked out in each session.

Continental breakfast

Your registration includes the cost of a continental breakfast and a morning and afternoon refreshment break. These informal segments allow you to discuss topics of mutual interest with other seminar attendees. (You are free to make your own lunch and dinner arrangements.)

Hotel/transportation arrangements

We arrange to hold a block of rooms at most host hotels. To take advantage of special group rates when making your reservation, be sure to mention that you are attending the RSMeans Institute data seminar. You are, of course, free to stay at the lodging place of your choice. (Hotel reservation and transportation arrangements should be made directly by seminar attendees.)

Important

Class sizes are limited, so please register as soon as possible.

Professional Development